Essentials of Modern Communications

Essentials of Modern Communications

Djafar K. Mynbaev
New York City College of Technology of the City University of New York

Lowell L. Scheiner
Late of New York University Tandon School of Engineering

This edition first published 2020
© 2020 John Wiley & Sons, Inc.

All rights reserved. No part of this publication may be reproduced, stored in a retrieval system, or transmitted, in any form or by any means, electronic, mechanical, photocopying, recording or otherwise, except as permitted by law. Advice on how to obtain permission to reuse material from this title is available at http://www.wiley.com/go/permissions.

The right of Djafar K. Mynbaev and Lowell L. Scheiner to be identified as the authors of this work has been asserted in accordance with law.

Registered Office
John Wiley & Sons, Inc., 111 River Street, Hoboken, NJ 07030, USA

Editorial Office
111 River Street, Hoboken, NJ 07030, USA

For details of our global editorial offices, customer services, and more information about Wiley products visit us at www.wiley.com.

Wiley also publishes its books in a variety of electronic formats and by print-on-demand. Some content that appears in standard print versions of this book may not be available in other formats.

Limit of Liability/Disclaimer of Warranty
MATLAB® is a trademark of The MathWorks, Inc. and is used with permission. The MathWorks does not warrant the accuracy of the text or exercises in this book. This work's use or discussion of MATLAB® software or related products does not constitute endorsement or sponsorship by The MathWorks of a particular pedagogical approach or particular use of the MATLAB® software. While the publisher and authors have used their best efforts in preparing this work, they make no representations or warranties with respect to the accuracy or completeness of the contents of this work and specifically disclaim all warranties, including without limitation any implied warranties of merchantability or fitness for a particular purpose. No warranty may be created or extended by sales representatives, written sales materials or promotional statements for this work. The fact that an organization, website, or product is referred to in this work as a citation and/or potential source of further information does not mean that the publisher and authors endorse the information or services the organization, website, or product may provide or recommendations it may make. This work is sold with the understanding that the publisher is not engaged in rendering professional services. The advice and strategies contained herein may not be suitable for your situation. You should consult with a specialist where appropriate. Further, readers should be aware that websites listed in this work may have changed or disappeared between when this work was written and when it is read. Neither the publisher nor authors shall be liable for any loss of profit or any other commercial damages, including but not limited to special, incidental, consequential, or other damages.

Library of Congress Cataloging-in-Publication Data

Names: Mynbaev, Djafar K., author. | Scheiner, Lowell L., author.
Title: Essentials of modern communications / Djafar K. Mynbaev, New York City
   College of Technology of the City University of New York, Lowell L.
   Scheiner, Late of New York University, Tandon School of Engineering.
Description: Hoboken, NJ, USA : Wiley, 2020. | Includes bibliographical
   references and index.
Identifiers: LCCN 2019053579 (print) | LCCN 2019053580 (ebook) | ISBN
   9781119521495 (hardback) | ISBN 9781119521464 (adobe pdf) | ISBN
   9781119521457 (epub)
Subjects: LCSH: Telecommunication.
Classification: LCC TK5101 .M96 2020 (print) | LCC TK5101 (ebook) | DDC
   621.382–dc23
LC record available at https://lccn.loc.gov/2019053579
LC ebook record available at https://lccn.loc.gov/2019053580

Cover design by Wiley
Cover image: © KTSDESIGN/SCIENCE PHOTO LIBRARY/Getty Images

Set in 9.5/12.5pt STIXTwoText by SPi Global, Chennai, India

To Bronia

Contents

About the Authors xxi
Preface xxiii
Acknowledgments xxvii

1 Modern Communications: What It Is? 1
    Objectives and Outcomes of Chapter 1 1
1.1 What and Why of Modern Communications 4
    Objectives and Outcomes of Section 1.1 4
    1.1.1 What Is Modern Communications? 5
    1.1.2 General Block Diagram of a Communication System 6
    1.1.3 Operation of a Communication System 7
    1.1.4 Why Do We Need Modern Communications? 8
    1.1.5 From Today to Tomorrow – Two Examples 9
        1.1.5.1 The Internet of Things (IoT) 10
        1.1.5.2 Data Centers 12
    Questions and Problems for Section 1.1 13
1.2 Communication Technology on a Fast Track 16
    Objectives and Outcomes of Section 1.2 16
        Sidebar 1.2.S.1 Brief Notes on History of Telegraph, Telephone, Radio, and Television 22
    1.2.1 The Internet 28
        1.2.1.1 Basics of Networks 28
        1.2.1.2 The Internet: From a Point-to-Point Link to a Network of Networks 37
    1.2.2 Optical Communications 42
        1.2.2.1 Introduction to Optical Communications 43
        1.2.2.2 Developments in Optical Communications: From First Inventions to Modern Advances 46
    1.2.3 Wireless Communications 49
        1.2.3.1 Introduction to Wireless Communications 49
        1.2.3.2 Contemporary Wireless Communications Technologies 54
        1.2.3.3 Mobile Cellular Communications 57

1.2.4 Satellite Communications *59*
  1.2.4.1 Historical Notes *59*
  1.2.4.2 Principle of Operation of Satellite Communication Systems *60*
  1.2.4.3 Satellite Orbits *62*
Questions and Problems for Section 1.2 *67*

1.3 Fundamental Laws and Principles of Modern Communications *75*
  1.3.1 Fundamental Laws of Modern Communications *75*
    1.3.1.1 Hartley's Information Law *75*
    1.3.1.2 Signal Bandwidth and Transmission Bandwidth from the Transmission Standpoint *76*
    1.3.1.3 Bandwidth and Bit Rate, Nyquist's Formula, and Hartley's Capacity Law *77*
    1.3.1.4 Shannon's Law (Limit) *79*
    1.3.1.5 More Clarifications of the Shannon Law *82*
    1.3.1.6 The Shannon Law for Digital Communications *83*
  1.3.2 Fundamental Principles of Modern Communications *86*
    1.3.2.1 Channel Capacity, Bandwidth, and Carrier Frequency *86*
    1.3.2.2 Bandwidth-Length Product *90*
    1.3.2.3 Power-Bandwidth Trade-Off *91*
    1.3.2.4 Spectral Efficiency and Transmission Technology *92*
    1.3.2.5 Bit Rate vs. Bandwidth in Digital Transmission *93*
  1.3.3 Laws, Principles, and Models – Importance, Limitations, and Applications *94*
    1.3.3.1 Limitations and Applications of the Laws and Principles *94*
    1.3.3.2 Models *96*
    1.3.3.3 Modeling and Simulation *98*
Questions and Problems for Section 1.3 *99*

2 Analog Signals and Analog Transmission *103*
Objectives and Outcomes of Chapter 2 *103*

2.1 Analog Signals – Basics *104*
Objectives and Outcomes of Section 2.1 *104*
  2.1.1 Definitions *104*
    2.1.1.1 Waveforms *104*
    2.1.1.2 Analog and Digital Signals *108*
  2.1.2 Sinusoidal Signal *110*
    2.1.2.1 The Waveform of a Sinusoidal Signal *110*
    2.1.2.2 Period and Frequency *111*
    2.1.2.3 Frequency, Radian (Angular) Frequency and Angle *115*
    2.1.2.4 Phase Shift (Initial Phase) *117*
    2.1.2.5 Amplitude *121*
Questions and Problems for Section 2.1 *125*

2.2 Analog Signals – Introduction *129*
Objectives and Outcomes of Section 2.2 *129*
  2.2.1 More About a Sinusoidal Signal *130*
    2.2.1.1 Considering All Three Parameters – the Formula for a Sinusoidal Signal *130*

		2.2.1.2	The Phase of a Sinusoidal Signal: a Detailed Look *132*

              2.2.1.2   The Phase of a Sinusoidal Signal: a Detailed Look   *132*
              2.2.1.3   Cosine and Sine Signals   *138*
              Sidebar 2.2.S.1   Phasor and Sinusoidal Signal   *139*
              Sidebar 2.2.S.2   Signal and Function   *146*
      2.2.2   Frequency Domain and Bandwidth   *151*
              2.2.2.1   Frequency Domain   *151*
              2.2.2.2   Cosine and Sine Signals in Frequency Domain   *151*
              2.2.2.3   Bandwidth   *156*
              2.2.2.4   Bandwidth: a Sophisticated Entity   *159*
    Questions and Problems for Section 2.2   *162*
2.3   Analog Signals – Advanced Study   *167*
    Objectives and Outcomes of Section 2.3   *167*
      2.3.1   Revisiting the Waveforms   *168*
              2.3.1.1   More about Waveforms   *168*
              2.3.1.2   Waveform and Signal's Power   *174*
      2.3.2   Waveforms and Phasors   *178*
              2.3.2.1   Practically Realizable Waveforms   *178*
              2.3.2.2   Phasors and Phasor Diagrams   *178*
              2.3.2.3   Waveforms and Phasors for a Resistor, an Inductor, and a Capacitor   *181*
              2.3.2.4   Impedances and Phasors   *185*
    Questions and Problems for Section 2.3   *189*
      2.3.A   Mathematical Foundation of Phasor Presentation   *191*
              2.3.A.1   Phasors and Complex Numbers   *191*
              2.3.A.2   Applications of Phasor Presentation to the Analysis of Electronic Communications Circuitry   *195*
                      2.3.A.2.1   Summation of Signals   *195*
    Optional: Questions and Problems for Appendix 2.3.A   *200*

3   Digital Signals and Digital Transmission   *203*

    Objectives and Outcomes of Chapter 3   *203*
3.1   Digital Communications – Basics   *203*
    Objectives and Outcomes of Section 3.1   *203*
      3.1.1   Why Go to Digital Communications   *204*
              3.1.1.1   Main Advantage of Digital Transmission over the Analog   *204*
              3.1.1.2   Case Study 1: The Advantages of Using Digital Signals in Transmission   *207*
              3.1.1.3   Case Study 2 of Digital Communications: Transmission with Integrated-Circuit Digital Logic Families   *210*
              3.1.1.4   Why Go to Digital Communications: A Summary   *214*
      3.1.2   How to Go to Digital Communications   *215*
              3.1.2.1   From Characters to Bits   *215*
              3.1.2.2   From Bits to Electrical Pulses   *222*
              3.1.2.3   How to Go Digital Communications: A Summary   *224*
    Questions and Problems for Section 3.1   *225*
      3.1.A   Brief History of Character Codes   *229*
              3.1.A.1   International Morse Code   *229*
              3.1.A.2   Baudot Code   *230*

3.2   Digital Signals and Digital Transmission – Introduction   232
        Objectives and Outcomes of Section 3.2   232
        3.2.1   Ideal Digital Signal and Characteristics of Digital Transmission   233
                3.2.1.1   The Waveform of an Ideal Digital Signal   233
                3.2.1.2   Pulse Interval and Transmission Rate; Bit Time and Bit Rate   235
                3.2.1.3   Important Note: The Definition of Bit Time   237
                3.2.1.4   Bit Rate and Channel (Shannon's) Capacity   237
        3.2.2   Parameters of a Real Digital Signal and the Characteristics of Digital Transmission   239
                3.2.2.1   Waveform of an Actual Digital Signal   239
                3.2.2.2   Amplitude and Pulse Width   240
                3.2.2.3   Rise Time and Fall Time   241
                3.2.2.4   Rise/Fall Time and Bit Rate   244
                3.2.2.5   More on Timing Parameters of a Digital Signal: Bit Time Revisited   247
                3.2.2.6   Duty Cycle   250
        Questions and Problems for Section 3.2   253

4   Analog-to-Digital Conversion (ADC) and Digital-to-Analog Conversion (DAC)   259

        Objectives and Outcomes of Chapter 4   259
4.1   Analog-to-Digital Conversion, ADC   259
        Objectives and Outcomes of Section 4.1   259
        4.1.1   The Need for ADC and DAC   261
        4.1.2   Three Major Steps of ADC   263
        4.1.3   Sample-and-Hold (S&H) Operation   263
                4.1.3.1   Sampling (S&H) Technique and the Nyquist Theorem   263
                4.1.3.2   Aliasing   267
        4.1.4   Quantization in ADC   272
                4.1.4.1   Quantization Process   272
                4.1.4.2   Quantization Errors and Quantization Noise   284
        4.1.5   Encoding   285
        Questions and Problems for Section 4.1   291
        4.1.A   Decimal and Binary Numbering Systems   299
                4.1.A.1   Decimal Numbering System   299
                4.1.A.2   Binary Numbering System   300
                4.1.A.3   Conversion from the Decimal Number System to the Binary   301
4.2   Digital-to-Analog Conversion, DAC, Pulse-Amplitude Modulation, PAM, and Pulse-Code Modulation, PCM   303
        Objectives and Outcomes of Section 4.2   303
        4.2.1   Digital-to-Analog Conversion, DAC   304
        4.2.2   Pulse Amplitude Modulation, PAM   304
        4.2.3   Pulse Code Modulation, PCM   306
                4.2.3.1   PCM: Principle of Operation   306
                4.2.3.2   PCM: Advantages and Drawbacks   308
                4.2.3.3   PCM Applications   309
        Questions and Problems for Section 4.2   309

    4.2.A Modes of Digital Transmission *311*
        4.2.A.1 Simplex, Half Duplex and Full Duplex Transmission *311*
        4.2.A.2 Serial and Parallel Transmissions *312*
        4.2.A.3 The General Formula for Bit Rate *314*
        4.2.A.4 The Need for Synchronization in Digital Transmission *315*
            4.2.A.4.1 Digital Signals and Timing *315*
            4.2.A.4.2 Timing in Digital Transmission *316*
            4.2.A.4.3 Time Discrepancy Between Transmitter and Receiver Clocks *317*
            4.2.A.4.4 How Time Discrepancy Between Transmitter and Receiver Clocks Deteriorates the Quality of Digital Transmission *319*
            4.2.A.4.5 A Short Summary on Synchronization Issues *320*
        4.2.A.5 Asynchronous and Synchronous Transmission *320*
            4.2.A.5.1 Asynchronous Transmission *321*
            4.2.A.5.2 Synchronous Transmission *323*

5 Filters *325*
    Objectives and Outcomes of Chapter 5 *325*
  5.1 Filtering – Basics *326*
    Objectives and Outcomes of Section 5.1 *326*
    5.1.1 Filtering: What and Why *327*
    5.1.2 RC Low-Pass Filter (LPF) *330*
        5.1.2.1 Frequency Responses of a Resistor, R, and a Capacitor, C *330*
        5.1.2.2 RC Low-Pass Filter: Principle of Operation *333*
        5.1.2.3 Output Waveforms of an RC LPF *334*
        5.1.2.4 An RC LPF: Formulas for Attenuation and Phase Shift *335*
        5.1.2.5 Frequency Response of an RC LPF *339*
        5.1.2.6 Cutoff (Critical) Frequency of an RC LPF *342*
        Sidebar 5.1.S Filter's Characteristics in Absolute Values and in dB *345*
    5.1.3 Filter Operation in Time Domain and Frequency Domain *347*
        5.1.3.1 Waveform Change and Frequency Response *347*
        5.1.3.2 Bandwidth of an RC LPF *349*
        5.1.3.3 Characterization of an RC LPF *349*
        5.1.3.4 The Role of R and C Parameters in Characterization of an RC LPF *352*
    5.1.4 General Filter Specifications *354*
        5.1.4.1 Amplitude Specifications *354*
        5.1.4.2 Phase Specifications *359*
    Questions and Problems for Section 5.1 *360*
  5.2 Filtering – Introduction *365*
    Objectives and Outcomes of Section 5.2 *365*
    5.2.1 High-Pass Filter (HPF), Band-Pass Filter (BPF), and Band-Stop Filter (BSF) *366*
        5.2.1.1 High-Pass Filter (HPF) *367*
        5.2.1.2 Band-Pass Filter (BPF) *371*
        5.2.1.3 Band-Stop Filter (BSF) *378*

		5.2.1.4	Applications of RC Filters *380*

- 5.2.1.4 Applications of RC Filters *380*
- 5.2.1.5 Final Notes on RC Filters *380*
- 5.2.2 Transfer Function of a Filter *381*
  - 5.2.2.1 Input and Output of a Filter *381*
  - 5.2.2.2 Transfer Function of an RC LPF *384*
  - 5.2.2.3 Graphical Presentation of a Transfer Function: Bode Plots *387*
- Questions and Problems for Section 5.2 *394*
- 5.2.A RL Filter and Resonance Circuits as Filters *400*
  - 5.2.A.1 RL Filter *400*
  - 5.2.A.2 Resonance Circuits as Filters *402*
    - 5.2.A.2.1 Resonance Circuits: A Review *402*
    - 5.2.A.2.2 Quality Factor *405*
    - 5.2.A.2.3 Resonance Circuit as a Band-Pass Filter *406*
    - 5.2.A.2.4 Resonance Circuit as a Band-Stop Filter *407*

5.3 Active and Switched-Capacitor Filters *409*
- Objectives and Outcomes of Section 5.3 *409*
- 5.3.1 Active Filters *410*
  - 5.3.1.1 Drawbacks of Passive Filters *410*
  - 5.3.1.2 Operational Amplifier *413*
  - 5.3.1.3 Active Filters: Concept and Circuits *418*
  - 5.3.1.4 Transfer Functions of an Active Filter: General View *419*
  - 5.3.1.5 Specific Types of Active Filters *420*
  - 5.3.1.6 Concluding Remarks on Active Filters *424*
- 5.3.2 Switched-Capacitor Filters *424*
  - 5.3.2.1 Switched-Capacitor Filters: Concept and Circuits *424*
  - 5.3.2.2 Applications of Switched-Capacitor Filters *428*
- Questions and Problems for Section 5.3 *431*
- 5.3.A Active BPF and BSF *436*
  - 5.3.A.1 Active BPF *436*
  - 5.3.A.2 Active BSF *439*

5.4 Filter Prototypes and Filter Design *441*
- Objectives and Outcomes of Section 5.4 *441*
- 5.4.1 Filter Prototypes *444*
  - 5.4.1.1 The Problem in the Filter Design – The Need for the Filter Prototypes *444*
  - 5.4.1.2 Another Problem for Filter's Designer: Relationship Between Amplitude and Phase Responses *445*
  - 5.4.1.3 Main Filter Prototypes – What and Why *446*
  - 5.4.1.4 Transfer Function of the Butterworth Filter *450*
  - 5.4.1.5 Amplitude Response of the Butterworth Filter *451*
  - 5.4.1.6 Amplitude Response of the Butterworth Filter in Logarithmic Scale *453*
  - 5.4.1.7 Phase Response (Shift) and Time Group Delay of the Butterworth Filter *456*
  - 5.4.1.8 Poles of the Butterworth Filter's Transfer Function *457*

   5.4.2 Introduction to Filter Design *459*
     5.4.2.1 Two Main Steps in Filter Design *459*
     5.4.2.2 Automated Design Options *460*
     5.4.2.3 Design of a Second-order Butterworth Filter *462*
     5.4.2.4 Using the Poles of a Transfer Function *468*
   5.4.3 The Design Process: Key Questions, Answers, and Salient Points *469*
     5.4.3.1 Questions and Answers *469*
     5.4.3.2 Salient Points *470*
     5.4.3.3 Choosing Filter Technology *471*
  Questions and Problems for Section 5.4 *472*
  5.4.A Tables of the Butterworth Polynomials *478*
 5.5 Digital Filters *479*
  Objectives and Outcomes of Section 5.5 *479*
   5.5.1 What are Digital Filters? *479*
     5.5.1.1 Digital Filters – Principle of Operation *479*
     5.5.1.2 ADC and DAC Operations Revisited *481*
     5.5.1.3 Digital Filters – Difference Equation, Order, and Coefficients *484*
     5.5.1.4 Recursive (IIR) and Nonrecursive (FIR) Digital Filters and Their Difference Equations *486*
     5.5.1.5 Impulse Response of Digital Filters *487*
     5.5.1.6 Transfer Function of a Digital Filter *488*
   5.5.2 Conclusive Remarks on Digital and Analog Filters *491*
     5.5.2.1 Some Final Comments on Digital Filters *491*
     5.5.2.2 Adaptive Filters *491*
     5.5.2.3 Comparison of Analog and Digital Filters *492*
     5.5.2.4 Summary of Applications of Various Filter Technologies *492*
  Questions and Problems for Section 5.5 *494*
  What are Digital Filters? *494*

6 Spectral Analysis 1 – The Fourier Series in Modern Communications *497*
  Objectives and Outcomes of Chapter 6 *497*
 6.1 Basics of Spectral Analysis *498*
  Objective and Outcomes of Section 6.1 *498*
   6.1.1 Time Domain and Frequency Domain *498*
     6.1.1.1 Periodic and Nonperiodic Signals *498*
     6.1.1.2 Time Domain and Frequency Domain Revisited *500*
     6.1.1.3 Signal Spectrum *509*
   6.1.2 The Fourier Series *511*
     6.1.2.1 The Fourier Theorem *511*
     Sidebar 6.1.S.1 Calculating the Coefficients of a Fourier Series *515*
     6.1.2.2 Spectral Analysis – From the Whole to the Parts *519*
   6.1.3 Spectral Synthesis *520*
     6.1.3.1 Spectral Synthesis – From Parts to the Whole *520*
  Questions and Problems for Section 6.1 *528*

- 6.2 Introduction to Spectral Analysis 534
  - Objectives and Outcomes of Section 6.2 534
  - 6.2.1 More About the Fourier Series 534
    - 6.2.1.1 Coefficients of the Fourier Series 534
    - 6.2.1.2 Amplitude and Phase Spectra 537
    - Sidebar 6.2.S.1 Using the Signal's Symmetry for Finding the Fourier Series Coefficients 542
    - 6.2.1.3 Finding the Fourier Series of Various Signals 544
  - 6.2.2 Effect of Filtering on Signals 546
    - 6.2.2.1 Statement of the Problem 546
    - 6.2.2.2 Filtering a Single Harmonic 552
    - 6.2.2.3 Filtering a Periodic Signal – Time and Frequency Domains 554
    - 6.2.2.4 Filtering a Signal – The Entire Picture 560
    - 6.2.2.5 A Final Note on Effect of Filtering on Signals 566
  - 6.2.3 Harmonic Distortion 566
  - Questions and Problems for Section 6.2 572
- 6.3 Spectral Analysis of Periodic Signals: Advanced Study 578
  - Objectives and Outcomes of Section 6.3 578
  - 6.3.1 Mathematical Foundation of the Fourier Series 579
    - 6.3.1.1 The Fourier Series in Exponential and Phasor Forms 579
    - Sidebar 6.3.S.1 The Other Forms of an Exponential Fourier Series 587
    - 6.3.1.2 Two-Sided and One-Sided Spectra and Three Equivalent Forms of the Fourier Series 588
  - 6.3.2 Conditions for Application of the Fourier Series 591
    - Sidebar 6.3.S.2 Convergence of the Fourier Series 591
    - 6.3.2.1 Gibbs Phenomenon 593
  - 6.3.3 Power Spectrum of a Periodic Signal 594
    - 6.3.3.1 Power and Energy Signals 594
    - 6.3.3.2 Parseval's Theorem 595
    - 6.3.3.3 A Signal's Bandwidth and Transmission Issues Associated with a Power Spectrum 598
  - Questions and Problems for Section 6.3 609
  - 6.3.A Fourier Coefficients of a Two-sided and a One-sided Spectrum of the Periodic Pulse Train for Example 6.3.2. 613

7 Spectral Analysis 2 – The Fourier Transform in Modern Communications 615
- Objectives and Outcomes of Chapter 7 615
- 7.1 Basics of the Fourier Transform 616
  - Objectives and Outcomes of Section 7.1 616
  - 7.1.1 The Fourier Transform in Spectral Analysis 617
    - 7.1.1.1 From a Periodic to a Nonperiodic Signal 617
    - 7.1.1.2 From the Fourier Series to the Fourier Transform 628
    - 7.1.1.3 The Fourier Transform Briefly Explained 629
  - 7.1.2 First Examples of the Fourier Transform Applications 632
    - 7.1.2.1 A Rectangular Pulse 632
    - 7.1.2.2 Basics of the Spectral Analysis of a Nonperiodic Signal 635
    - 7.1.2.3 Rayleigh Energy Theorem 639
  - Summary of Section 7.1 642
  - Questions and Problems for Section 7.1 643

7.2 Continuous-Time Fourier Transform: A Deeper Look *644*
   Objectives and Outcomes of Section 7.2 *644*
   7.2.1 Definition and Existence of the Fourier Transform *645*
   7.2.2 The Concept of Function and the Transform *646*
      Sidebar 7.2.S.1 Dirac Delta Function *649*
   7.2.3 Table of the Fourier Transform *654*
   7.2.4 Properties of the Fourier Transform *656*
      7.2.4.1 Units *656*
      7.2.4.2 Linearity *657*
      7.2.4.3 Duality *657*
      7.2.4.4 Modulation *657*
      7.2.4.5 Convolution in Time and in Frequency and a Transfer Function *658*
      7.2.4.6 Time Differentiation *659*
      7.2.4.7 Other Properties of the Fourier Transform *659*
   7.2.5 Example of Using the Fourier Transform *659*
      Sidebar 7.2.S.2 The Impulse Response of an RC LPF *662*
      Sidebar 7.2.S.3 Alternative Methods of Finding a Transfer Function *667*
7.3 The Fourier Transforms and Digital Signal Processing *670*
   Objectives and Outcomes of Section 7.3 *670*
   7.3.1 Signals and the Fourier Transformations *671*
      Sidebar 7.3.S.1 A Word About DSP *677*
   7.3.2 Determining the Fourier Transform Required for DSP *681*
   7.3.3 Digital Signal Processing (DSP) and Discrete Fourier Transform (DFT) *681*
      7.3.3.1 The Problem: Choosing the Best Type of FT for DSP *681*
      7.3.3.2 How Discrete Fourier Transform (DFT) Works *682*
      7.3.3.3 Can DFT Work with Any Signal? *690*
   7.3.4 Relationship Among All Fourier Transforms *697*
   7.3.5 Fast Fourier Transform (FFT) *699*

8 Analog Transmission with Analog Modulation *707*
   Objectives and Outcomes of Chapter 8 *707*
8.1 Basics of Analog Modulation *708*
   Objectives and Outcomes of Section 8.1 *708*
   8.1.1 Why We Need Modulation: Baseband and Broadband Transmission *710*
      8.1.1.1 Baseband Transmission and Its Major Problems *710*
      8.1.1.2 Solution to the Problems of Baseband Transmission – Broadband Transmission *712*
   8.1.2 Basics of Amplitude Modulation *715*
      8.1.2.1 What Type of Analog Modulation Can We Have? *715*
      8.1.2.2 What Is Amplitude Modulation (AM) *715*
      8.1.2.3 Modulation Index *719*
      8.1.2.4 Relationship Between Frequencies of Information and Carrier Signals *722*
      8.1.2.5 The Formula for an AM Signal and It Instantaneous Value *723*
      8.1.2.6 The Spectrum of an AM Signal *725*

		8.1.2.7	Power Distribution in an AM Signal *728*

- 8.1.2.7 Power Distribution in an AM Signal *728*
- 8.1.2.8 AM Modulation and Demodulation *730*
- 8.1.2.9 The Main Drawback of Amplitude Modulation *732*

8.1.3 Basics of Frequency Modulation (FM) *733*
- 8.1.3.1 Frequency Modulation: Why and What *733*
- 8.1.3.2 The Frequency of an FM Signal *734*
- 8.1.3.3 Modulation Index of an FM Signal *738*
- 8.1.3.4 The Spectrum and Bandwidth of an FM Signal *740*
- 8.1.3.5 Relationship Between Parameters of Message and Carrier Signals in FM Transmission *746*
- 8.1.3.6 FM Modulation and Demodulation *746*

8.1.4 Basics of Phase Modulation (PM) *750*
- 8.1.4.1 How to Generate a Phase-Modulated Signal *750*
- 8.1.4.2 Instantaneous Value of a Sinusoidal PM Signal *754*

Questions and Problems for Section 8.1 *754*

8.1.A Drawbacks of Baseband Transmission *759*

8.2 Analog Modulation for Analog Transmission – An Advanced Study *762*

Objectives and Outcomes of Section 8.2 *762*

8.2.1 Classification of Modulation Revisited *763*

8.2.2 Advanced Consideration of Amplitude Modulation, AM, and Its Application in Analog Transmission *766*
- 8.2.2.1 Full (Double-Sideband Transmitted Carrier, DSB-TC) Amplitude Modulation *766*
- 8.2.2.2 Problems of Full AM Transmission *774*
- 8.2.2.3 Double-Sideband Suppressed Carrier (DSB-SC) AM *774*
- 8.2.2.4 Single-Sideband Suppressed Carrier (SSB-SC) AM *779*
- 8.2.2.5 Full AM, DSB, or SSB – Which Type to Choose? *782*
- 8.2.2.6 Applications of AM Transmission *784*

8.2.3 Advanced Consideration of Angular (Phase and Frequency) Modulation and Its Application in Analog Transmission *784*
- 8.2.3.1 Angular Modulation *784*
- 8.2.3.2 Sinusoidal (Single-Tone) Frequency Modulation (FM) *788*
- 8.2.3.3 The Spectrum of a Single-Tone FM Signal, the Main Properties of the Bessel Functions, and Narrowband and Wideband FM *790*
- 8.2.3.4 The Bandwidth of a Single-Tone FM Signal *793*
- 8.2.3.5 General Case of an FM Signal (An Arbitrary Message Signal) *799*
- 8.2.3.6 Effect of Noise on an FM Signal *807*

Questions and Problems for Section 8.2 *810*

8.2.A Finding the Spectrum of an FM Signal with MATLAB *814*

9 Digital Transmission with Binary Modulation *823*

Objectives and Outcomes of Chapter 9 *823*

9.1 Digital Transmission – Basics *824*

Objectives and Outcomes of Section 9.1 *824*

	9.1.1	Essentials of Digital Transmission Revisited *827*	
		9.1.1.1	Block Diagram of a Communication System *827*
		9.1.1.2	Characteristics of a Transmitter, Tx *828*
		9.1.1.3	Characteristics of a Receiver, Rx *829*
		9.1.1.4	Characteristics of a Transmission Channel (Link) *830*
		9.1.1.5	The Model of Noise in Shannon's Law *835*
		9.1.1.6	An Amplifier in a Transmission Channel: Internal Noise, SNR, and Noise Figure *839*
	9.1.2	Assessing the Quality of Digital Transmission: The Gaussian (Bell) Curve and the Probability Value *843*	
		9.1.2.1	Gaussian (Bell) Normal Probability Distribution *843*
		9.1.2.2	Finding the Probability Value with the Bell Curve *844*
		9.1.2.3	Standard Normal Probability Distribution *847*
		9.1.2.4	The Gaussian Curve and Q-Function *850*
	9.1.3	Assessing the Quality of Digital Transmission: Bit Error Rate and More *852*	
		9.1.3.1	Decision-Making Procedure in the Presence of Noise *852*
		9.1.3.2	The Probability of Error in Detecting the Received Signal: Bit Error Rate (Ratio) *855*
		9.1.3.3	BER: A Discussion *858*
	9.1.4	Eye Diagram *860*	
		9.1.4.1	Eye Diagram: The Concept *860*
		9.1.4.2	Estimating Transmission Quality with an Eye Diagram *865*
	Questions and Problems for Section 9.1 *869*		
9.2	Introduction to Digital Transmission – Binary Shift-Keying Modulation *878*		
	Objectives and Outcomes of Section 9.2 *878*		
	9.2.1	Digital Signal over a Sinusoidal Carrier – Binary Shift-Keying Modulation *881*	
	9.2.2	Binary Amplitude-Shift Keying (ASK) *881*	
		9.2.2.1	ASK Concept and Waveform *881*
		9.2.2.2	Mathematical Description of ASK *883*
		9.2.2.3	ASK Spectrum *884*
		9.2.2.4	ASK Bandwidth *888*
		9.2.2.5	Bandwidth and Bit Rate of ASK *893*
		9.2.2.6	Bit Error Ratio, BER, of ASK System *895*
		9.2.2.7	ASK Advantages, Drawbacks, and Applications *898*
		9.2.2.8	Detection (Demodulation) of an ASK Signal *900*
	9.2.3	Binary Frequency-Shift Keying (FSK) *901*	
		9.2.3.1	FSK Concept and Waveform *901*
		9.2.3.2	Mathematical Description of FSK *903*
		9.2.3.3	FSK Spectrum and Bandwidth with Square Wave Message *904*
		9.2.3.4	FSK Spectrum and Bandwidth with a Rectangular Pulse-Train Message *906*
		9.2.3.5	Bit Error Ratio, BER, and Remarks on our BFSK Discussion *908*

		9.2.3.6	Discontinuous-Phase FSK (DPFSK) and Continuous-Phase FSK (CPFSK) *910*

                9.2.3.6   Discontinuous-Phase FSK (DPFSK) and Continuous-Phase FSK (CPFSK) *910*
                9.2.3.7   Mathematical Description of a CPFSK Signal *911*
                9.2.3.8   Detection (Demodulation) of an FSK Signal *916*
                9.2.3.9   BFSK: Advantages, Drawbacks, and Applications *921*
          9.2.4   Binary Phase-Shift Keying (PSK) *922*
                9.2.4.1   PSK Concept and Waveform *922*
                9.2.4.2   PSK Mathematical Description; PSK Spectrum and Bandwidth with a Square Wave Message *925*
                9.2.4.3   Demodulation of a Binary PSK Signal *926*
                9.2.4.4   Bit Error Ratio, BER, of a BPSK Transmission *929*
                9.2.4.5   BPSK Advantages and Applications *932*
                9.2.4.6   Comparison of Binary ASK, FSK, and PSK *932*
      Questions and Problems for Section 9.2 *932*
          9.2.A   Jitter *940*

10 Digital Transmission with Multilevel Modulation *943*

      Objectives and Outcomes of Chapter 10 *943*
   10.1   Quadrature Modulation Systems *943*
      Objectives and Outcomes of Section 10.1 *943*
          10.1.1   Multilevel (*M*-ary) Modulation Formats – What and Why *945*
                10.1.1.1   The Concept of Multilevel Modulation *945*
                10.1.1.2   Symbols and Bits *948*
          10.1.2   Quadrature Phase-Shift Keying, QPSK *951*
                10.1.2.1   Introduction to Quadrature Phase-Shift Keying, QPSK *951*
                10.1.2.2   QPSK Signal: Waveform and Constellation Diagram *953*
                10.1.2.3   Generating (Modulating) a QPSK Signal *957*
          10.1.3   Working with QPSK Signaling *964*
                10.1.3.1   Properties of a QPSK Signal *964*
                10.1.3.2   QPSK Demodulation *965*
                10.1.3.3   Assessing the Quality of QPSK Transmission *967*
                10.1.3.4   Offset QPSK, Differential QPSK, and Minimum SK *968*
      Questions and Problems for Section 10.1 *970*
   10.2   Multilevel PSK and QAM Modulation *974*
      Objectives and Outcomes of Section 10.2 *974*
          10.2.1   Multilevel (*M*-ary) PSK *975*
                10.2.1.1   Introduction to *M*-ary PSK *975*
                10.2.1.2   BER of *M*-ary PSK *977*
          10.2.2   Multilevel Quadrature Amplitude Modulation, *M*-QAM *981*
                10.2.2.1   The Concept of Multilevel Quadrature Amplitude Modulation, *M*-QAM *981*
                10.2.2.2   BER of *M*-QAM *984*
          10.2.3   Final Thoughts *991*
                10.2.3.1   Spectral Efficiency, Signal-to-Noise Ratio, and Multilevel Modulation *991*
                10.2.3.2   Bandwidth-Power Trade-off *994*
                10.2.3.3   Applications of Multilevel Signaling *995*

Questions and Problems for Section 10.2   *995*
10.A   Multiplexing   *999*
    10.A.1   Multiplexing: Definition and Advantages   *999*
    10.A.2   Time-Based Multiplexing Principles   *1000*
        10.A.2.1   Synchronous Time-Division Multiplexing, sync-TDM   *1000*
        Sidebar 10.A.2.S   Two sync-TDM Systems: T and Synchronous Optical Network (SONET)   *1002*
        10.A.2.2   Statistical (Asynchronous) Time-Division Multiplexing, stat-TDM   *1008*
    10.A.3   Frequency-Based Multiplexing Techniques   *1010*
        10.A.3.1   Frequency-Division Multiplexing, FDM   *1010*
        10.A.3.2   Orthogonal Frequency Division Multiplexing, OFDM   *1011*
        10.A.3.3   Wavelength-Division Multiplexing, WDM   *1016*
            10.A.3.3.1   Why We Need WDM and How WDM Works   *1016*
            10.A.3.3.2   WDM Technology   *1018*
        10.A.3.4   CWDM and Other Types of Multiplexing in Optical Communications   *1020*
    10.A.4   Code-Division Multiplexing, CDM   *1023*
        10.A.4.1   CDM: The Principle of Operation   *1023*
        10.A.4.2   Spread-Spectrum Technique   *1024*
        10.A.4.3   CDM: Benefits and Applications   *1026*

Bibliography   *1029*

Specialized Bibliographies   *1037*

Index   *1043*

About the Authors

Djafar K. Mynbaev

Djafar K. Mynbaev graduated from Leningrad Electrical Engineering Institute, Russia, with MS and PhD degrees. He worked for Russian academic institutions and industrial concerns for a number of years. In the United States, he worked at Bell Communications Research (Bellcore), where he conducted research in various aspects of telecommunications. Since 1996, he has been working at New York City College of Technology (CUNY), where he was the coordinator of the telecommunications program. Currently, he is a professor (and past chairman) in the college's Electrical Engineering and Telecommunications Technologies Department. His area of research is optical communications and related fields. He has published about 150 technical and educational papers, holds 26 patents, and has delivered numerous presentations at international conferences.

Lowell L. Scheiner

Lowell L. Scheiner was a technical writer and editor for more than 20 years on engineering magazines, including *Plastics Technology, Plastics World*, and *Modern Packaging*. In addition, he has written for *Design Engineering, Semiconductor Products, Technology in Focus,* and *Solid State Design*, among other publications. He has also served in a public relations capacity for a number of major corporations, including IBM, AlliedSignal, AMF, Borg Warner, Engelhard, and American Can. For the last 37 years, he was a tenured professor at New York University's Tandon School of Engineering (formerly Polytechnic University), where he taught courses in technical writing, corporate communications, and science journalism. Over the last 17 years, he has focused on writing on subject matter related to modern communications. He holds two advanced degrees from Columbia University, including an MS degree from the Graduate School of Journalism.

Professors Mynbaev and Scheiner coauthored the textbook *Fiber-Optic Communication Technology,* published by Prentice Hall in 2001. The book was reprinted by several international publishing companies and translated into Chinese; it has been adopted by many colleges and universities throughout the globe.

Preface

Rationale (Modern Communications)

This book is about modern communications, which is the exchange of information among people that is facilitated by advanced electronic technology. Human society cannot exist without communications, and history shows that the progress of humankind is very much associated with improvement in communication technology. Writing, printing, telegraph, and especially telephone, television, computers, and the Internet – all these means of communications have greatly contributed to the development of our civilization.

Today, we continuously need to deliver information in any forms from one point to another, and this is what modern communications do. So since our society produces ever-increasing reams of information, communications is of necessity its lifeblood.

It is safe to predict that in the years to come, communications will become even more ubiquitous, and more people will have access to the modern forms of communications. This is because all aspects of our life – economic, political, and health care (to cite just a few) – depend increasingly on access to the global communications infrastructure. In short, access to modern communications allows people to improve all aspects of their lives.

Advances in communications depend largely on advances in all sectors of science and technology. On the other hand, continuously increasing demands of modern communications stimulate further progress in science and technology by setting out new challenges for these fields.

Many areas of communication technology, however, are approaching their limit. Electronic industry works today at the scale of atomic distances (about 0.3 nm), and it is hard to imagine how it can progress to smaller scales. Optical communications faces the limit dictated by information theory, which does not depend on technological improvements. Wireless communications faces the spectral restrictions, and we cannot get more spectrum than Mother Nature gave us. Communication satellite systems are limited by satellite transmitter power on the downlinks and receiver sensitivity on the uplinks, and these technological issues restrict further progress. We firmly believe, however, that solutions to all these problems will be found, because the seeds of the growth of emerging breakthroughs have already been planted in the world's R&D labs, where scientists and engineers are busily engaged in efforts that promise accomplishments in communication technologies as remarkable as what we have witnessed till date.

All in all, modern communications technology is developing so rapidly that even our youngest readers can point to significant improvements they remember. How, then, such a traditional teaching tool as a textbook can help its readers to enter this dynamic field and equip them with the knowledge that would not become obsolete in a few years? The immediate answer is almost palpable: Since modern communications technology relies on fundamental principles that change very little and – if they do – very slowly, we need to teach fundamentals. But teaching pure theory on fundamentals in an engineering course does not work either because many students may lose interest quickly. Thus, finding the right balance between teaching the theoretical fundamentals of modern communications and their practical applications is the primary objective of this book. Hopefully, our many years of industry, research, and teaching experience have helped in achieving this objective.

Our Approach

Traditionally, a textbook serves as a source of information by introducing physical laws, deriving equations, and explaining how devices and systems work. This function is still valuable, but in the Internet era, when all information is just a click away, its importance is diminishing, which calls for changing the approach to writing textbooks. Our book, while still providing necessary information, *teaches* the reader why real-life engineering problems surface and how they are solved.

To achieve this goal, we first show the readers that laws, equations, circuitry, algorithms, and virtually all engineering and scientific advancements are the results of the continuous process of exploration and discovery primarily driven by the desire to find specific solutions to real-world problems. Historical notes and short biographies of key scientists and engineers also help to show the students that the problems discussed in this text stemmed from real-life situations and that solving them required tremendous efforts by people who created the technology that we enjoy today. Hence, the readers (students) should perceive this new knowledge not just as information to be memorized but mainly as the result of hard work of our predecessors and as inspiration for what they could achieve through intellectual striving.

Second, as we review specific challenges, we explain the circumstances in which they arise, show possible approaches to addressing them, discuss available methods of resolution, and consider their realizable implementation. In other words, we do not merely present an equation, a system, or a device but encourage our readers to participate in finding and applying solutions. The objective is to teach our readers to problem-solving approaches, skills that remain with a professional throughout his or her career regardless of changes in technology.

In solving a problem or designing a device or a system, an engineer must ask: What could go wrong? What are the boundary conditions for this approach (or equation), beyond which it would not work? Is this the best solution? By asking these types of questions, the engineer will ensure that the device or system that he or she is designing will function in all possible situations. The ability to ask these questions is critical for an engineer or a technologist and so nurturing this ability is one of our objectives. We typically start every new topic by asking questions about the need for discussing that topic. Furthermore, questions appear through the course of a discussion of them; the questions are also included in each "Questions and Problems" section. We encourage readers to see what real-life problems are posted there and how to solve them.

The book fosters the approach mentioned previously by concentrating on the fundamentals of the subject and relates these fundamentals to professional responsibilities through discussions, case studies, and – mainly – examples. Such an approach is hardly innovative, but what makes our book unique is its consistent application throughout the text. In particular, our examples are not merely the exercises in plugging given numbers into equations, but the presentation of real-world problems similar to those that students will meet in their professional careers. Besides, each example is accompanied by a thorough discussion that pinpoints the advantages, drawbacks, limitations, and implications of the obtained solution. Thus, our examples serve as essential teaching tools.

As a result, the reader will not merely learn the theoretical framework for a topic in communications but will also probe into the sources of real-world engineering problems discussed in a given topic, the steps taken for a possible solution, and the limitations of solutions considered. What is more, the reader will acquire the habit of asking questions about possible reasons for problems and feasible solutions. Most important, the reader will develop a professional approach to analyzing and solving technical problems through curiosity and critical and logical thinking.

Pedagogical Features

The book follows a *system approach* in presenting the material. This is why block diagrams rather than specific circuits are used for illustrating the discussion of devices. When considering a particular device, we always highlight its function as part of a system, thus helping to place each component into the broader view of the overall system.

In this book, *mathematics* is used only as a tool enabling us to achieve specific goals, which is exactly how engineering utilizes mathematics. At the same time, using mathematics fosters the habit of thinking logically in solving a problem or finding the answer to a question. This approach necessarily constrains the complexity of mathematics used in the book, which does not, however, limit the level and depth of explanations and discussions.

As mentioned previously, the extensive *discussion of examples* helps to put the obtained solutions and answers in perspective; it forces students to always carefully evaluate the correctness, limitations, and implications of obtained results. This methodology encourages students to think outside the box.

The *shaded sidebars*, dispersed throughout the book, provide additional explanations of hard-to-tackle issues. Some of these sidebars present rigorous mathematical investigations of specific problems, whereas others introduce real-life examples of communication systems.

Several sections include *appendixes* that complement the main material of the sections.

To help the reader, we have highlighted major points by *shading* the text and using *italics*.

Objectives and outcomes precede each chapter and each section. The objectives highlight the main focus points of a unit, and outcomes present in a concise form of what you are expected to learn from this segment.

Every section is accompanied by *questions and problems* that are intended to be assigned as homework. They are based on real-life issues that students might encounter in the workplace. Also, these problems require students to comprehend an entire concept, not just solve an equation or understand how a specific device works. Such method helps students develop a professional approach to solving practical problems. The assignment segments also include questions that require essay answers, which helps students to learn how to present their results in writing, a long-lost skill that is high in demand in the industry these days. These sections additionally include questions and problems that test not just the student's ability to plug numbers into memorized formulas but also gauge his or her knowledge of theory through its applications to real-world practice. Some questions do not have a straightforward unique answer; such ambiguity is part of engineering practice, and students must be familiar with it. Of course, design-oriented and mini-project questions and problems are integral parts of these assignments because we think design-oriented and project-oriented approaches must be the hallmarks of every engineer.

To further engage our readers into the process of becoming the part of vibrant communication community, the book refers to open issues in theory and practice of modern communications that are being debated by industry and academia.

Audience (Organization of Our Textbook)

This textbook is intended for *undergraduate students*. As for specific recommendations, this textbook represents a significant departure from the traditional mode of teaching communications courses in the following aspects:

Most of the book's 10 chapters are made up of three sections. Each section deepens and broadens the topic, as the table of contents shows. This three-level structure allows the students to start at the basic level and strive for higher and higher levels.

Since almost every section is self-sufficient, an instructor can choose the sections to suit the needs of an individual class. The possible approaches to the selection are

- As the first sections of almost every chapter cover the foundations of the topic, an instructor can merge these sections to create a basic course. This course can be used as a freshman course at any college, including a community college. Similarly, sophomore, junior, and senior courses can be devised from the introductory and advanced sections.
- The scope of our book far exceeds the need for a one- and even two-semester course in communications. Fortunately, the breadth of the material allows for the use of the book in a variety of communications courses. By choosing the proper sections from Chapter 1 to Chapter 10, an instructor can prepare a course concentrated on a specific aspect of modern communications.

Adopting the three-level sections in their entirety will enable the instructor to implement the concept of *personalized and adaptive teaching*. This implementation can be done by giving an individual student the assignments according to his or her level of initial preparation and guiding the student at his or her pace to the successful competition of the course. Also, since the basics of each topic are introduced in each chapter, students can readily refresh their memory without the need for additional sources.

I trust the book will be appealing to *professionals* who want to refresh their memory on the subject matter and take a fresh look at their everyday work.

The bibliography is not only a necessary source of useful information but also an essential instrument in showing the students what kind of information sources they should look for. This book contains two parts of this section: *General bibliography and references* and *specialized bibliographies*. General bibliography shows the list of books and articles that have been used in the writing of this volume. Among these sources, there are those that have been cited in the text. This is why this list is called bibliography and references. We collected all these sources in one place to avoid repeating the same titles if a bibliography were to be put in each chapter. *Specialized bibliographies*, devoted to optical communications, wireless communications, and satellite communications, display the sources that can be helpful to those readers who want to delve into learning a specific subject.

By default, the book relies on extensive use of *modern technology* in general and *teaching technologies* in particular; we avoid, nonetheless, specific references to what technologies should be utilized and how they can be employed. Exceptions are *MATLAB* and *Multisim* that are extensively used to support our discussions and to demonstrate practical examples.

New Jersey
28 July 2019

Djafar K. Mynbaev

Acknowledgments

I am deeply obliged to many people who helped in the preparation of this book.

The first person whom I must mention in this regard is my coauthor, long-term collaborator, and friend, Professor Lowell L. Scheiner, who passed away in July 2018. He shared with me the larger part of this long, challenging, but ultimately rewarding journey. Only when I had to work alone at the end of this project, did I fully realize that his collaboration was priceless.

Without the support of President of New York City College of Technology Dr. Russel K. Hotzler and the whole administration of our college, this book wouldn't appear.

My colleagues in the Department of Electrical and Telecommunications Engineering Technology backed me throughout this venture.

Many people assisted me in writing this book in various ways, and I am indebted to them for their aid. For voluntarily giving up their time and comfort to help me overcome numerous problems, obstacles, and hurdles, I am especially grateful to Dr. Boris Amusin, Mr. Brian Chu, Mr. Sergey Genkin, Dr. Mohammed Kouar, Professor Linda Lerner, Dr. Michael Levit, Dr. Zory Marantz, Dr. Karim Mynbaev, Mr. Alex Ovrutsky, Mr. Wolf Perlov, and Dr. Michael Shur.

I am very appreciative to Mr. Boris Ratner who spent substantial time aiding me to find the correct interpretations of certain statements and verify the solutions of many problems.

After the death of my coauthor, Mr. Michael Ratner provided invaluable assistance in finishing the project. For this, I am most sincerely thankful to him.

The use of MATLAB in this book would not be possible without the work done by Ms. Ina Tsikhanava, who developed the vast majority of MATLAB and Simulink examples and prepared most of the MATLAB-based figures and tables. Dr. Zory Marantz and Mr. Vitaly Sukharenko also contributed to the use of MATLAB. I am indebted to these colleagues for their work.

I sincerely thank the reviewers who devoted their time and efforts to assess the proposal for this manuscript.

I am greatly appreciative of the assistance provided by many people, companies, and organizations in granting me the permissions to reproduce their graphical materials.

Finally, I want to thank my students whose curiosity about the subject, desire for more in-depth learning, and general reaction to my teaching inspired me to write this book.

March 2019   *Djafar K. Mynbaev*

1

Modern Communications: What It Is?

Objectives and Outcomes of Chapter 1

Objectives

- To understand the concept of modern communications and its societal implications.
- To become familiar with the Internet and the three pillars of modern communications – optical, wireless, and satellite communications.
- To learn the fundamental laws and principles of modern communications and get to know their applications and limitations.

Outcomes

- Learn that communications is the exchange of information, where information is knowledge that had not been known before. Realize that modern communications, also called telecommunications, employs electronic, photonic, and other technologies to deliver information over a significant distance.
- Understand that a communication system includes a source of information, a transmitter, a transmission link, a receiver, and a destination point (end user). Comprehend that the source, transmitter, link, receiver, and destination point are not merely the members of the set of components, but they make up a communication *system*, which means they must work in concert.
- Remember that an end user deals only with a received signal and he or she does not know how close the delivered message is to original information generated by the source because noise and other hurdles can alter the transmitted signal beyond recognition.
- Study societal implications of modern communications and realize its importance for economic development and national defense. Get to know that modern communications not only consumes production of other industries but also drives innovations and new developments in many segments of national and global economy. Realize that since our society produces, shares, and utilizes ever-increasing reams of information, modern communications is of necessity its lifeblood.
- Become familiar with main milestones of communications history and understand importance of globalization of modern communications from historical perspectives.
- Learn the basics of networking; specifically, realize that a network is a set of nodes connected by links and that a communication network is a symbiosis of a tangible, physical side (hardware) and intelligent, logical side (software). Get to know that hardware part of a network includes user devices (computers, tablets, smartphones, etc.), transmission links, switches and routers, amplifiers and repeaters, servers, and many more modules and components.

Essentials of Modern Communications, First Edition. Djafar K. Mynbaev and Lowell L. Scheiner.
© 2020 John Wiley & Sons, Inc. Published 2020 by John Wiley & Sons, Inc.

- Understand that the intelligent side of a network is based on the rules (protocols) governing the traffic transmission; the network software contains these protocols along with instructions on their executions. Know that an increasing presence of artificial intelligence enables the network software to execute the protocols flexibly and efficiently. Learn that the main task of an intelligent side of a network is switching or routing the communication traffic through the network.
- Comprehend that switching, besides being a general term about relaying messages through a network, also describes the use of switches to change the circuit-to-circuit connections. Realize that routing is also a general term about redirecting the stream of information; however, routing means choosing a path for forwarding information through a network and from one network to the other, provided that all connections are established.
- Get to know that there are two types of networks, depending on the form of directing information flow in modern communications: circuit switching and packet switching (routing). Be aware that the Internet is the packet-switching network.
- Study that any network includes control plane and data plane, and that a control plane is a set of software and/or hardware that executes control and management functions and a data plane is a set of hardware and software that actually carriers the user traffic. Understand that we need two planes in order to separate control and transport functions of a network, thus making the data plane protocol-independent.
- Get to know that the Internet is a communication network, specifically, the worldwide system of interconnected communication networks enabling any user in the world to reach any other user. Be aware that, thanks to the scale and nature of its operation, the Internet is called the network of networks.
- Learn that the World Wide Web, WWW, is an information space, where all knowledge is codified and organized to make any particular entry readily accessible. Get to know that today the Web and the Internet are associated so closely that we often confuse one with another; thus, let it be clear that the Internet is the global communication network and the World Wide Web is an information space.
- Learn that the set of protocols governing the Internet transmission is called Transmission Control Protocol and Internet Protocol, TCP/IP. Know that all data transmission in the Internet is governed by the TCP suite and that all addressing issues are controlled by the IP suite.
- Be aware that the Internet transmission relies on two principles: the "black boxes," which implies that the Internet simply transfers packets to their destination points without knowing the packets' contents, and "the best efforts," which means that if a packet failed to arrive to the destination point, it must be retransmitted.
- Remember that the Internet is simply a name for the assemblage of all connected networks, which means that without its parts – the worldwide optical network, the innumerable set of wireless transmission systems, and the global satellite communication complex – the whole entity, the Internet, would not exist.
- Learn that optical (fiber-optic) communications is the backbone of modern communications infrastructure and that vast majority of global communication traffic is delivered by optical networks.
- Understand that the optical communications industry emerged as a new main form of global communications because optical communication networks provides the highest possible transmission rate (the number of bits per second) because the optical signal has the highest frequency that can be utilized in practice.

- Remember that today optical fiber spans the entire globe, and optical communications, combining the signal propagation speed close to the speed of light with the enormous transmission rate of 10^{15} b/s, capable of delivering tremendous amount of information at the most remote corner of the world almost instantly.
- Realize that the wireless communications – the second of the three pillars of modern communication industry – is growing at the fastest pace fueled by the advent of ever-increasing number of mobile applications and new technologies like the Internet of Things (IoT).
- Get to know that the wireless communication systems deliver information by transmitting electromagnetic (EM) waves in the radiofrequency (RF) spectrum from 3 kHz to 300 GHz through air. Understand that information transmission to and from wireless communication base stations is provided by optical communication networks.
- Learn that the main problem of modern wireless communications is the scarcity of spectrum because the range of frequencies that can be employed in practice (the usable spectrum) is determined by the combination of technological, economic, and legal factors. Know that the latter refers to the government regulations that are necessary to control and maximize the use of the spectrum for a whole nation. Be aware that the most promising solution to this problem is employing the bands of frequencies higher than 3 GHz.
- Become familiar with technologies of wireless communications, including 5 G systems, Wi-Fi, Bluetooth, ZigBee, optical wireless, and Li-Fi.
- Comprehend that wireless communications can be fixed and mobile.
- Grasp the concept of mobile communications, particularly the terrestrial cellular networks. Understand that the entire wireless communication network blanketing a geographic region is divided into a number of cells. Know that a cell tower that transmits and receives the electromagnetic signals to and from mobile devices within its cell is located in the center of each cell. Comprehend that the main advantage of a cellular architecture is the frequency reuse achieved by assigning to every cell its own frequency band and ensuring that none of the adjacent cells have the same band.
- Get to know that, thanks to proximity of mobile devices to a nearby cell tower, all cellular operations are done in a low-power mode. Realize that due to this transmission mode (i) low-power cell tower's signals diminish when they reach the cell borders and do not interfere with frequencies of the adjacent cells and (ii) only small transmission power from mobile devices is required enabling their batteries to be compact.
- Become familiar with brief history and the principle of operation of the third pillar of modern communications – satellite communications.
- Learn that communication satellites are the spacecrafts orbiting the earth and serving as intermediate nodes (active mirrors) for redirecting the EM signals from one point on the earth surface to the other. Be aware that the satellites carry an appropriate equipment powered by solar panels.
- Be aware that the frequencies used in satellite communications are concentrated in the microwave region of the EM spectrum. Know that particular bands of this spectrum is employed in their specific applications, such as support of global positioning system (GPS), weather forecast, TV broadcasting, mobile communications, and aircraft and satellites communications.
- Study that satellites can orbit the earth in circular and elliptic orbits and that three main types of geocentric (around the earth) orbits – high, medium, and low – differ in their distances (that is, altitudes) from the earth center. Learn that each type of orbits has its advantage and drawbacks, which determines its specific applications.

- Become familiar with two main problems of satellite communication signals – power loss and propagation delay – and available solutions to these problems.
- Remember that the satellite communications enables us to reach any distant corner of the globe without wireline connections and that this unique property of satellite communications makes it the major player in international mobile communications.
- Comprehend that all three major sectors of modern communication industry – optical, wireless, and satellite – work together and make up the global communication infrastructure whose existence continues to improve the quality of our life.

1.1 What and Why of Modern Communications

Objectives and Outcomes of Section 1.1

Objectives

- To understand what modern communications is and why we need to study this subject.
- To comprehend the operation of a modern communication system.
- To become familiar, through examples, with modern communications and its societal implications.

Outcomes

- Understand that communications is the exchange of information, where information is knowledge that had not been known before.
- Learn that to exchange information we always need three basic components: a transmitter, a transmission link, and a receiver.
- Realize that there is direct communications among human beings that does not involve any technology, and electronic communications where electronics is an integral part of the process of exchanging information. Comprehend that the latter is necessary to deliver information over a significant distance; it is also called telecommunications.
- Get to know that a general block diagram of a communication system includes a source of information, a transmitter, a transmission link, a receiver, and a destination point (end user).
- Remember that an end user deals only with a received signal and he or she does not know how close the delivered message is to original information generated by the source.
- Comprehend that the source, transmitter, link, receiver, and destination point are not merely the members of the set of components, but they make up a communication *system*, which means they must work in concert. Understand that, for example, an excellent transmitter does not make a perfect communication system if the link and the receiver cannot support the excellent operation of the transmitter.
- Realize that modern communications is a central part of our personal and professional life.
- Review the examples of the role that modern communications plays in our everyday life and find similar examples from your personal experience.
- Study societal implications of technological achievements of modern communications and realize the importance of modern communications for economic development and national defense.
- Learn that modern communications is a valuable sector of American economy, which not only consumes production of other industries but also drives innovations and new developments in many segments of national and global economy.

- Realize that since our society produces, shares, and utilizes ever-increasing reams of information, modern communications is of necessity its lifeblood.
- Familiarize yourself with the concepts that underlie the operation of the IoT and of data centers, as examples of cutting-edge developments in modern communications.
- Get to know that security of delivering information remains one of the important issues in modern communications.

1.1.1 What Is Modern Communications?

What is communications? Exchange of information. But what do we mean by *information*? We will put off the scientific definition of this term until later chapters, but for now, we can say that *information* is basically *something new, knowledge that had not been known before*. News, new facts, new data – these are typical examples of information. It is we human beings who generate information, who *inform* each other via various technical means about the new things.

To exchange, that is deliver information, we always need three basic components: transmitter, Tx, receiver, Rx, and transmission link. Figure 1.1.1 shows a basic block diagram of a communication system. This block diagram emphasizes the concept of communications; it is true regardless of the communication technology. (Consider your oral conversation and identify a transmitter, a receiver, and a link.)

Historically, we could communicate only orally; next development was the invention of writing. Telegraph, devised by Samuel Morse and used to transmit the first message over distance by electrical signal in 1844, heralded the advent of a new era – that of telecommunications or electronic communications. Telephone, conceived by Alexander Graham Bell in 1876, followed, as well as radio, through which Marconi was able to send the first signal over Atlantic Ocean in 1901.

This new era heralded the beginning of the *exchange of information over distance by using electronic devices*. To describe this new form of communications, we interchangeably use terms like *electronic communications, telecommunications,* or simply *modern communications*.

The term *communications* in its strict sense implies the use of any of our senses – sight, sound, touch, taste, and smell (yes, even the last two) – in presenting or acquiring information. Let us refer to this type of communications as *direct communications*, which is simply exchange of information among human beings directly, without involving any technology. The simplest example of *direct communications* is a personal conversation between two people. *Electronic communications* refers to electronics as an integral part of the process of exchanging information because we cannot deliver information over a significant distance without electronic devices. The simplest example of electronic communications is a conversation between two people via the smartphones.

In this regard, the term *telecommunications* emphasizes the difference being discussed very clearly. This term consists of two words: the Greek word *tele*, which means far, apart, over a distance, and communications, which means the exchange of information. Carrying this concept a step further, we define telephone as sound over a distance and television as seeing over a distance. Thus, telecommunications, or modern communications, in its basic sense is tele + communications; that is, sharing information over a distance. Two important consequences follow from this definition. First, sharing means delivering. Second, in order to deliver information

Figure 1.1.1 Basic block diagram of a communication system.

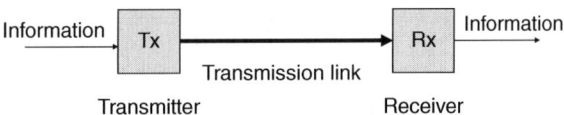

over a distance, we have to employ some technical device, because, since information in its original form is generated by a human being, it can be seen, heard, or felt only within a very limited range. The history of communications, in fact, is about efforts to extend the reach of the information we humans generate. Typical examples are human's first attempts to communicate by optical means, such as fire or sunlight, by sound, with instruments such as African drums, or by manually operated mechanical semaphores (optical telegraph towers with movable arms for messaging). All these rudimentary means of communications over distance are followed, as mentioned above, by telegraph, telephone, radio, television, and the wealth of modern telecommunication technology we are so familiar with today: optical (fiber-optic), wireless and the satellite, through which we disseminate information across the globe.

1.1.2 General Block Diagram of a Communication System

If you carefully review the examples of communication systems shown in the preceding subsection, you will notice that they all have the same basic components: the transmitter, the unit that sends information; the transmission medium, the link through which transmission occurs; and the receiver, the unit that ends up with the information. However, the general block diagram of a communication system contains several more elements, as presented in Figure 1.1.2.

A communication system starts with a source of information that produces a message. As stressed early in this section, it is we, human beings who originally generate information. However, the secondary source can be, for example, your mobile device or your computer. We produce logical (intelligent) information, and an electronic device converts this information into an electrical signal. Today, however, we can task our smartphone or computer to generate information (find, for example, the location of our traveling destination point), which will be done in the ready-for-transmission electronic format. Rapid development of *artificial intelligence, AI*, capable of performing really sophisticated intelligent operations, further blurs the distinction between human and artificial sources of information.

In any event, it is the *transmitter* that eventually prepares the message signal for transmission. The transmitter also does necessary manipulations – such as amplification, modulation, conversion into an optical beam or electromagnetic waves – to prepare the information signal for transmission. It is the transmitter that sends the prepared signal through a transmission link.

The *transmission link*, also called a channel, connects the transmitter and the receiver and delivers the information signal (message) to the destination point. A transmission link, however, causes some *degradations of a transmitted signal*. First, as Figure 1.1.2 shows, noise harmfully affects an information signal, reducing the probability of extracting the correct information from the received signal. Second, the signal loses its power and arrives at the receiver front end much weaker than

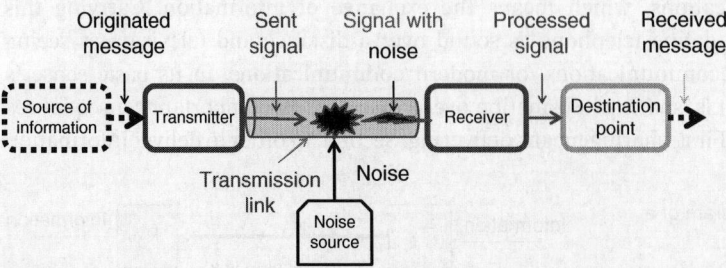

Figure 1.1.2 General block diagram of a communication system.

when it left the transmitter. Third, the signal might lose its ability to support the required bit rate. There are other aspects of signal's degradation during transmission, which we will discuss in later chapters.

A *receiver* processes the arrived signal to extract the information sent by the transmitter. This processing includes amplification, filtering (removing as much noise as possible), changing the signal's format into one suitable to the end user, etc. Finally, the extracted information is presented to the destination point (end user).

We must remember a very fundamental point:

> The end user deals only with the received signal and he or she doesn't know how close the delivered message is to original information generated by the source.

1.1.3 Operation of a Communication System

The brief description of the function of each component of a communications system shown in Figure 1.1.2 naturally raises the question: How do these components work together? To answer this question, let us consider the analogy between a communication system and a water supply system. In this analogy, a source of information plays the same role as a natural water source, such as a spring, river, or reservoir in the water supply system. A transmitter is analogous to a water pump system that purifies the natural water, collects it in a cistern, and pumps it into a pipe under a certain pressure. Accordingly, a transmission link is similar to a pipe connecting the pump and the point of consumption. Finally, a receiver is like the destination point that also processes water to prepare it for final consumption.

The main property of the transmitter is its ability to send the required volume of information per second; similarly, the main property of the system's pump is its ability to send the required volume of water per second through a pipe. The main feature of a transmission link is its capacity to deliver the required volume of information per second. Similarly, the pipe in a water supply system must be able to deliver the adequate volume of water to a destination. If transmitting capacity of the link is low, it will take a long time to deliver the whole volume of information to a receiver. In our analogy, if the pipe is too small, it will not be able to timely deliver the required volume of water even though a pump will supply enough. Finally, a receiver must be able to process the required volume of information per second, just as equipment at the consumption point of a water system must be able to process all the received water in timely manner. Bear this useful analogy in mind when reading the subsequent material.

Though the analogy between a communications system and a water supply system is helpful in understanding the operation of the former, you have to always remember an important point: *Analogy is a partial similarity*, where the word *partial* is the key.

What we should learn from this description is that all components of a communication system, as well as components of a water supply system and any other system for that matter, must have the proper characteristics in order to make the system operate successfully. Indeed, the communication system could have an excellent transmitter capable of supplying information at any required rate, but if a transmission link cannot support this rate, the receiver would not receive the needed volume of information in time. Always bear this point in mind when considering the operation of a communication system.

Solid understanding of the structure and operation of basic communication system is really a necessary condition for successful study of the subject. Our strong recommendation is to always refer (mentally or physically) to the block diagram shown in Figure 1.1.2 when considering any communications issue. In particular, constantly refer to this figure and accompanying discussion when studying Sections 1.2 and 1.3.

1.1.4 Why Do We Need Modern Communications?

Today, we cannot imagine our life without fresh air, heat, running water, and electricity. Can we imagine our life without modern communications? No. The permanent access to the global communications infrastructure from any location has becomes a central part of our personal and professional life.

Connecting to the Internet – which is the network of networks; that is, a complex of globally interconnected communication networks – provides us with a link to any online point in this world. Access to the WWW brings us to the information space from which we can retrieve any needed document. Besides these two major instruments, we have many other, also very important applications, such as e-mail, GPS, Skype, Facebook, Instagram, WhatsApp, and YouTube, to name a few. We enjoy a new way of living thanks to modern communications.

Think about how we would live without reliable and fast networks that deliver worldwide zettabits of information every second: The individual computers would be standalone units and we would need to physically carry such storage devices as CDs or USBs from one computer to another to share information. Satellites would be isolated space objects; to obtain information from them, we would need to physically access them either by landing them or by sending astronauts to their orbits. Of course, there would be no navigation guidance in real time and around the globe. Our smart phones would be simply the toys enabling us to deal with information stored on their memory cards. There would be no access to any information online because no Internet would be in existence. We choose only these few examples to stress the point: Our life would not be as comfortable and full as it is today without modern communications. Look back (or ask your parents to do so) ten and twenty years and compare life then to the life today. We encourage you to analyze how you use modern communications in your everyday life in both professional and personal areas.

Another important point is our tools for accessing communication networks. Twenty years ago, they were only desktop or laptop computers. Today, in addition to these much-improved devices, we have plenty of mobile gadgets (e.g. tablets and smart phones) performing the same and many other tasks. Consider your smart cellular phone: It has turned today into a multitasking instrument enabling us to not only use voice or text messaging for interpersonal communication but also performed numerous other tasks, such as e-mailing, web search, and using GPS. Other mobile platforms (tablets) perform the same and many additional tasks; in fact, the border between smartphones and mobile platforms diminishes with every new generation of these devices.

Implications of all these technological achievements are numerous. A person buried under building ruins after an earthquake was saved thanks to the signal from his wearable device – an example of life-saving function of modern communications. A street vendor in a rural area selling his/her small personal produce scanned the payment from a cell phone of a buyer – an example of a convenience provided by modern communications. These simple examples should inspire you to find more cases of necessity and usefulness of modern communications. Analyzing the effect of modern communications on our life, we see that it not only makes our everyday life more convenient but changes the way how our society is organized. For the first time in human history, an ordinary person can make his or her voice to be heard by millions of others by tweeting, blogging, and broadcasting their views, creative works, and personal news. As all humans get connected to one another in real time throughout the world, new way of existing through cooperation will emerged.

We should not underestimate importance of communications to the military area. In fact, communications is considered as important as armament and logistics. The center of war operations is shifting from direct combat to the war in communications space because interruptions in communications make all armament useless. Therefore, the defense of our country depends on modern communications.

The economic development and prosperity of a district, city, region, and a whole country depends today on the level of their connectivity with the global communication infrastructure. Having high-speed connections to the Internet is a necessary condition for the success of a company, an industry, and a nation's economy.

Delivering information today requires the use of the newest technological developments in computer, transmission, and networking technologies, advances that have been brought about by the semiconductor, electronic, and photonic industries. The installation and maintenance of communications networks involve the construction and maintenance sectors of the industry. To make technology work, money must flow in and out of the industry. The point here is that modern communications is a significant part of global economy. Producing, delivering, and processing massive amounts of information rapidly and accurately – this is what drives the American economy and the economies of other developed countries, and will continue to do so in the years ahead. In this regard, it is important to understand that the communications sector not only consumes production of other industries but also drives innovations and new developments in many segments of the national and global economy.

In short, modern communications underlies operations of all private and public businesses, including government. Communication industry – a valuable sector of the US economy – is continuing to grow, making a significant contribution to the expansion of the nation's economy.

(**Question**: *Can you name the countries with the highest density of installed fiber-optic communications networks, bearing in mind that these networks provide the highest communication speed? To check your answer, go online and find the appropriate maps.*)

So, the answer to the question, "Why do we need modern communications?" is simple: *Since our society produces, shares, and utilizes ever-increasing reams of information, modern communications is of necessity its lifeblood.*

1.1.5 From Today to Tomorrow – Two Examples

Obviously, nobody can predict the future. Nevertheless, in technology it is safe to say that the seeds of the future are growing in today's research labs. An example: In 1995, one of the executives of Bellcore, the company that was a spin-off of the Bell Labs[1] and served as an R&D arm of regional telephone companies, sent an e-mail claiming that the distance will soon cease to be a factor in telecommunications in general and in telephony in particular. These were the days when customers had to pay for telephone connections depending on distance, and an international call to a remote country, such as China, could cost up to $10 per minute. Merely 25 years later, we not only could place a telephone call but have a video conversation with anyone located at every point on the globe

---

1 Alexander Bell, the inventor of the telephone, founded a company best known as Bell Systems that dominated the telephone industry for the most of twentieth century. This company, once being the biggest corporation in the world, subsidized a huge R&D institution commonly known as Bell Labs. The level of research there was incomparable. Suffice to say that arguably the greatest invention of the twentieth century – the transistor – came to life there, as did the laser, the concept of information theory, and the Unix operating system, the C programming language, and much more. The Labs produced 13 Nobel laureates, a feat no other corporation in the world has come close to matching. In 1984, both the Bell System and its Bell Labs were split. Today, the remaining core of the institution continuous its research and innovations under the name Nokia Bell Labs.

via Skype, and we could do it for free. (Nowadays, there are several other applications that provide similar types of free connections.)

Today, the commercialization of research results occurs much faster; in fact, it becomes a dynamic process. While a first prototype of an innovation goes to manufacturing, research and development continue to improve the product, and its commercialization accelerates and expands. What is more, commercial success brings new demands to properties and quality of the product, which stimulates further progress in invention, development, and production. Here a product means an outcome of both hardware and software areas of production.

Let us consider two developments that are available today and promise even more ubiquitous and advanced presence tomorrow.

1.1.5.1 The Internet of Things (IoT)

The IoT is a first example of such a development. The IoT is the communication paradigm that, on the one hand, has been partly developed and commercialized and, on the other hand, is more promising than the existing technology. We, naturally, focus on a communication aspect of the IoT.

So what is the IoT? Let us start with "things." A *thing is any natural or man-made object, fixed or mobile, that is able to transmit and receive data through the Internet.* Examples include transceivers (transmitter and receiver combined in one unit) installed on trucks of a national transportation company, smart collars worn by animals in a livestock farm, security cameras set around a building, and traffic monitoring cameras mounted around a city.

At this point, you might say, wait a minute. We do have such systems based on a radiofrequency identification (RFID) setup that wirelessly transmits information about every object to which an RFID tag is attached. Yes, you are right, at the first glance RFID and the IoT looks similar. The difference – and the big one – is that the IoT elevates the concept of operating with individual objects to a much higher level.

An RFID system is local by its very nature. For example, consider an RFID system operating at an assembly line: RFID tags are attached to every part to be supplied to the assembly process. Information from these RFID tags about the locations and volumes of all these parts comes to the operator's computer to inform the operator of possible shortages and delays. The operator uses this information to prevent any interruptions to the assembly process. So the RFID system heavily relies on human intervention. Now, imagine that all the tags attached to the items to be assembled are connected directly to the Internet. Then, information about all these objects becomes accessible not only to computers at the factory floor but also to computers of factory management, the company headquarters, to the factories around the globe, where these items are manufactured, to the mills where steel, plastic, and other materials for these parts are produced, and so on and so forth. All the computers involved in this process communicate with each other of course, and keep the manufacturing process running without interruption. Such an arrangement minimizes human intervention and greatly reduces waste, loss, and cost. (In fact, we are considering an example of the Industrial Internet of Things, the IIoT.)

So the IoT is a system or a complex of various things combined by their locations, functions, users, owners, or any other features; the main property of these things is that all of them are connected to the Internet.

Imagine you live in a smart – means based on the IoT – home. You can see and talk to someone who rings your doorbell even when you are far away from your home; your home's door gets unlocked when you approach it; the light, heat, air conditioning, and all other home systems are controlled remotely; your refrigerator e-mails you a list of items that need to be replenished, and so on and so forth. A *smart building*, *smart district*, and a *smart city* – all these entities have already been attempted, and are in the process of rapid development. Even though the goal to build an IoT-based smart city would seem more an academic curiosity than a real-life need, the accelerating global urbanization process calls for innovative approaches to building and managing the cities whose growing population sets unimaginable challenges to sustainable development of the entire world. In this regard, the IoT promises to find optimal solutions to all aspects of urban life.

The IoT finds its applications not only in urban environment but also in all other areas of society, such as healthcare, education, finance, agriculture, food industry, transportation, and utilities.

It is clear that the IoT is a sophisticated *communication system*. Even though intelligent cloud storage, intelligent processing of accumulated data, and intelligent remote control of processes are important qualities of the IoT, intelligent *communications* is the key feature of this system. In particular, in the IoT, machine-to-machine (M2M) communications supported by artificial intelligence (AI) is the main form of sharing information.

The type of communication technology used by the IoT for connection to the Internet depends on its range of operation. Based on this range, the IoT networks are grouped into four categories: a wireless personal area network (PAN), operating within several meters; wireless and/or wired local area network (LAN), operating within a hundred meters; wireless neighborhood area network (NAN), operating within 20 km; wide area network (WAN), ranging globally and relying on a complex mix of wireless and wired communication systems. Figure 1.1.3 shows how a typical IoT system is connected to the Internet; the figure lists technologies employed at every stage of connections. (Do not be confused if you are not familiar with these technologies: We will discuss many of them in the subsequent sections.)

The IoT is an important and rapidly developing area of modern technology; a lot of research and, consequently, numerous publications are devoted to this new field. This short review of the IoT gives an idea of how modern communications supports and contributes to the latest technological advances.

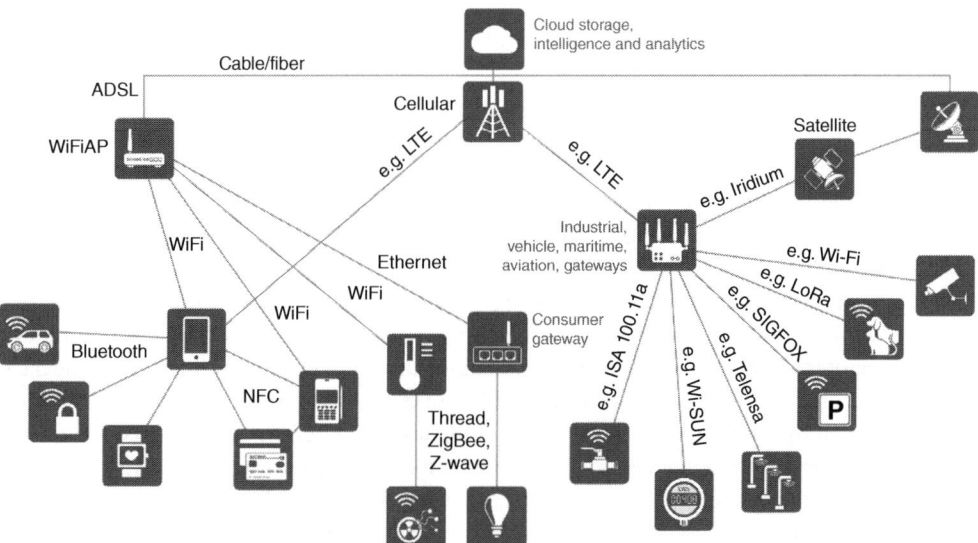

Figure 1.1.3 The IoT connections to the Internet (cloud). Source: © Keysight Technologies, Inc. (2016) *The Internet of Things: Enabling Technologies and Solutions for Design and Test*. Application Note. Reproduced with permission, Courtesy of Keysight Technologies.

1.1.5.2 Data Centers

Let us now discuss another example of the continuing development in communications: the **data centers**. When we mention *cloud* in reference to the Internet, we actually mean an information space where all globally generated data is collected, saved, and processed. We can save our files to the cloud – the Internet, that is – and we can retrieve them anywhere by using any computing device. We can also share information with any device regardless of its location. The cloud takes care of directing our messages to intended recipients or making them publicly available.

The word "cloud" evokes immaterial white puffs. The data, however, travels in the form of electrical or optical pulses and therefore this cloud must be the tangle of hardware that receives, saves, and processes these pulses. This hardware is millions of specialized computers (servers) installed in the data centers.

The data centers are warehouses containing rows and rows of servers. The scale of these warehouses varies from a "small" building of the size of a regular supermarket to a building occupying the area of a football field. They consume tremendous amounts of electrical power to supply it to thousands of servers and, mainly, to cool those servers, as they radiate an enormous amount of heat. By some estimates, the American data centers are responsible for more than 2% of country's total electrical power consumption. The electrical-power expenses are so serious that big companies such as Facebook, Google, Microsoft, and Amazon move their data centers to the states with minimal cost of electricity. What is more, some of these companies build their data centers outside the USA, in polar regions to make cooling them easier.

Figure 1.1.4 pictures an inside view of a data center and a detail of a server backplane.

Beside energy consumption, the data centers experience constant pressure to improve their computational operations. This pressure is caused by exponentially increasing global volume of information, to which the IoT makes a significant contribution. One of the main requirements for this improvement is reduction of *propagation latency*, the time delay in transporting pulses from one point to another. For transmission outside and within a data center, the transportation latency is caused by the finite velocity at which pulses travel through a transmission medium. Though today, a typical delay is about several milliseconds, this number will become unacceptable with the growth of total communication traffic in general and of the IoT in particular. (You can skip to Section 10.1, where latency is discussed.)

To reduce the propagation delays in outside communications, the data centers must be moved closer to the areas being served; this solution, however, contradicts the requirement to reduce electrical power expenses. The propagation latencies within data centers are small compare to the outside delays due to shorter distances of transmission. Still, since the traffic volume in data centers is doubling every 12 months (by some estimates), seriousness of the latency problem will increase.

The two current developments – the IoT and data centers – are among the key factors that will determine the future of communication industry. Three out of four levels of the IoT networks rely on the wireless access to the Internet, which emphasizes importance of the future progress in rapidly developing wireless communications. Long-distance connections of the IoTs to the data centers depend on fiber-optic networks and satellite links. These established industry segments undergo further transformations to meet future demands. Together, three segments – wireless, optical, and satellite – work in concert to make up the modern communication industry.

Analyzing the operation of modern communications, we always need to think of how secure is the delivery of information. Our personal data, corporate expertise, and military secrets must be protected from possible theft. Latest advances in computer science, artificial intelligence (AI), and communication technology, including quantum communications, enable the industry to protect transmitted information better and better. Unhappily, criminal hackers also use these advances to penetrate into most vulnerable points of communication infrastructure to steal or modify

(a)

(b)

Figure 1.1.4 Data centers: (a) A view of the interior of construction of a data center; (b) A detail of a backplane fiber-optic connections in a server. Source: (a) Reprinted with permission https://betterbuildingsinitiative.energy.gov/challenge/sector/data-centers. (b) Reprinted with permission https://www.dreamstime.com/stock-photo-fiber-optic-servers-technology-data-center-image54827788.

information for their malicious purposes. The security issue is far from being finally resolved; this is another dynamic process in modern communications.

This brief examination of basics of modern communications, accompanied by the consideration of its societal implications and the examples of future developments, gives a glimpse into the role of modern communications and into why it is important to study this fascinating discipline.

Questions and Problems for Section 1.1

- Questions marked with an asterisk require a systematic approach to finding the solution.
- Many questions and problems, including those marked with an asterisk, imply that you, in addition to reading the textbook, will do your research to find the answers. Consider such questions as mini-projects.

What and Why of Modern Communications

1 Consider the term *communications*:

a) What is communications?
b) What is information?
c) What is a communication system?

2 Sketch the basic block diagram of a communications system and explain the function of each component.

3 Why is the set of three components – transmitter, receiver, and a link – called a *system*?

4 Consider your smart phone as an example of a communications system and identify three basic system components.

5 What is the difference between communications, electronic communications, and telecommunications?

6 Figures 1.1.1 and 1.1.2 show the block diagram of a communication system. What is the difference between these figures?

7 Figure 1.1.2 shows that a communication system includes a source of information and a transmitter of information: Why does not the system transmit information directly from the source? In other words, why do we need a separate unit for sending a message?

8 The text says that a transmission link shown in Figure 1.1.2 causes degradation of the transmitted signal. Why do not we eliminate this component of a communication system?

9 Figure 1.1.2 shows that noise affects the signal in the transmission link. Can noise affect the signal within a transmitter and a receiver? Explain.

10 Refer to Figure 1.1.2: Why do we need a separate unit – a receiver – to present a received message to a destination point? In other words, can we connect directly a transmission link and a destination point? Explain. Give examples.

11 Consider the operation of a communication system: The text says that an end user deals only with a received message and does not know what message has been sent. How then can the user be sure that the received signal brings the original message correctly? (*Hint*: Think over the entire operation of a communication system.)

12 The text introduces the analogy between a communication system and a water pumping system and warns that the analogy is only a partial similarity:
a) What features of both systems prevent the analogy from being the complete similarity? (*Hint*: Compare every pair of components (transmitter vs. pump, link vs. pipe, and receiver vs. consumption equipment) and the operations of the systems as a whole.
b) Another popular analogy to a communication system is a road system of a developed country. Describe this analogy on component-to-component basis.

13 Suppose you are watching a streaming video on your laptop:
a) In general terms, describe what communication system you would have to establish for this process.

b) If you show experiences interruptions, what could be – in qualitative terms – the reasons for these faults?

14 Explain the difference between the Internet and the World Wide Web, WWW.

15 Discussing the need for modern communications, the text says: "The permanent access to the global communications infrastructure from any location becomes a central part of our personal and professional life." Demonstrate the truth of this statement based on your personal experience. Give examples of the role of modern communications in your life.

16 Imagine that we have all the latest communication devices and gadgets but without connections to the Internet: How would our life change? Give two or three examples. (*Hint*: consider yourself in an open sea or on the trail in a remote unpopulated a geographic area.)

17 Explain why modern communications is so important for military operations and therefore for the country's defense.

18 General media and professional publications give numerous examples of how economic development of a district, a city, a region, or a whole country have improved after they established high-speed (called broadband) connections to the Internet. Why is this so? Go online and find examples of such changes.

19 The text posed a question "Why do we need modern communications?" Write down the list of reasons answering this question.

20 Do your research to write an essay on history of Bell Systems and another one on history of Bell Labs.

21 Consider the Internet of Things, IoT:
a) What is it?
b) What are those "things?"
c) Why does this name refer to the Internet?

22 The Internet of Things enables us to create an environment called a "smart home." What is this?

23 Imagine that you live in a settlement called *smart city*:
a) List the main features of this city.
b) Find the examples of implementation of this concept.

24 Examine Figure 1.1.3 and highlight wireless and wireline connections separately. (*Hint*: You need to do the search online and you may want to jump to Section 1.2.)

25 Figure 1.1.3 symbolically shows the Internet in the form of a cloud. Why so?

26 Today, we use terms cloud computing and cloud storage. What do they mean? Where are the underlying operations performed physically?

27 Consider the data centers:
  a) What are the data centers?
  b) Why do we need them?
  c) What problems do the data centers encounter? What solutions to these problems are found?

28 The text refers to importance of the issue of security in modern communications. Consider your everyday communications and analyze what security measures are taken to protect your information.

1.2 Communication Technology on a Fast Track

This book is devoted to the essentials of modern communications. We focus on the fundamental principles, laws, and rules that govern modern communications regardless of the specifics of technology being used. In other words, the textbook does not concentrate on particular communication technologies. This approach is justified by the stability of communications fundamentals compared with ever rapidly changing technology. Nevertheless, the brief introduction to the main types of modern communication technology, which is the objective of this section, is necessary to demonstrate where and how the fundamentals are applied and used.

In this section, we concisely discuss the networks, optical communications, wireless communications, and satellite communications. They are the main sectors of modern communications and important fields in their own right. Their research, industrial, and business activities are reflected in numerous books, journal articles, technical documents, and – most importantly – in working devices and systems, thanks to which we have such a technological marvel as modern communications. We also present a brief overview of history of each sector of communications technology, its growth in scale, its ever-increasing importance, and its impact on the globalization of modern communications.

This laconic introduction to the main technologies of modern communications should help in understanding the content of this book; if you become interested in deeper study of any of these topics, turn to specialized literature.

Objectives and Outcomes of Section 1.2

Objectives

- To become familiar with modern communication technology from historical perspectives.
- To comprehend the operational principles and applications of the Internet and the pillars of modern communication technology – optical communications, wireless communications, and satellite communications.
- To understand how all types of modern communications work together to make the Internet and the World Wide Web the central part of our professional and personal lives.

Outcomes

- Become familiar with main milestones of communications history, such as telegraph, telephone, radio, and television. Understand importance of globalization of modern communications from historical perspectives.

- Learn the basics of networking, starting with the difference between a point-to-point (PPT) link and a network.
- Understand that a network is a set of nodes connected by links.
- Realize that a communication network is a symbiosis of a tangible, physical side (hardware) and intelligent, logical part side (software). Comprehend that in this regard a communication network is analogous to a computer, which is also a symbiosis of hardware and software.
- Get to know that hardware part of a network includes its transmission links that can be cables (wires) or air, where the cables can be copper-based (twisted pair or coaxial) or fiber-optic. Comprehend that we can have either wireline (through the cables) or wireless (through air) transmissions. Be aware that network hardware also includes user devices (computers, tablets, smartphones, etc.), switches and routers, amplifiers and repeaters, servers, and many more modules and components.
- Understand that the intelligent side of a network is based on the rules (protocols) governing the traffic transmission; the network software contains these protocols along with instructions on their executions. Know that an increasing presence of artificial intelligence enables the network software execute the protocols flexibly and efficiently.
- Learn that the main task of an intelligent side of a network is switching or routing the communication traffic through the network.
- Comprehend that switching, besides being a general term about relaying messages through a network, also describes the use of switches to change the circuit-to-circuit connections. Realize that routing is also a general term about redirecting the stream of information; however, routing means choosing a path for forwarding information through a network and from one network to the other, provided that all connections are established.
- Understand that the physical layout of a network can be built in five main topologies: bus, ring, star, tree, and mesh (see Figure 1.2.3). Be aware that the flow of information, which is a logical topology, can be different from physical topology of a network.
- Realize that there are three main levels of network hierarchy: access (local) networks, providing access to the metro networks for its end users; metro (regional) networks, accepting traffic from the access network, providing communications within themselves, and connecting metro traffic to the next level, long-distance networks; the long-distance (core, trunk, or backbone) networks linking all networks globally. Understand that access networks cover small territory and serve to local community; the metro networks cover a substantial territory, such as a big city, and serve to millions of users; the long-distance networks cover countries, continents, and the whole globe and serve to billions of users. Comprehend that the level of network hierarchy determines the number of users and therefore the volume of traffic and the quality of network equipment.
- Get to know that there are two types of networks, depending on the form of directing information flow in modern communications: circuit switching and packet switching (routing). Understand that in a circuit-switching network, the switches at every node connect the input and output links in such a way that two end users become directly connected for the duration of their communications, and that after the communication session ends, the connections at every node are disjoined. Realize that in a packet-switching network all network nodes are permanently connected, a transmitted message is disassembled into packets, these packets are directed through the network by routers through the various paths, and, once arrived at a destination point, the packets are reassembled back into the original message.
- Be aware that any network includes control plane and data plane, and that a control plane is a set of software and/or hardware that executes control and management functions and a data plane is a set of hardware and software that actually carriers the user traffic. Understand that we need

two planes in order to separate control and transport functions of a network, thus making the data plane protocol-independent.
- Learn that the current architecture of a packet-switched Internet network relies on routers that perform both intelligent (choosing the path for a packet) and physical (sending the packet) operations, and that the advent of routers made the Internet a reality. Get to know that the ever-increasing growth of traffic volume calls for new network architecture called software-defined networking (SDN) which further decouples the control and data planes and relegates the functions of the control plane (network intelligence) to an application called a controller.
- Be aware of importance of network security and become familiar with some methods to provide the network security, such as firewall, encryption, and application of quantum communications.
- Get to know that **the Internet**, in essence, is a communication network. Realize that the Internet, however, is not just a network, but the worldwide system of interconnected communication networks enabling any user in the world to reach any other user. Be aware that, thanks to the scale and nature of IT operation, the Internet is called the network of networks.
- Learn that the World Wide Web (WWW) is an information space, where all knowledge is codified and organized to make any particular entry readily accessible. Be aware that today the Web and the Internet are associated so closely that we often confuse one with another; thus, let it be clear that the Internet is the global communication network and the World Wide Web is an information space.
- Understand that the Internet's hardware is all links, routers, switches, servers, and other tangible modules, units, and components of the existing access, metro, and long-distance national and international networks. Remember that the Internet is simply a name for the assemblage of all connected networks, which means that without its parts – the worldwide optical network, the innumerable set of wireless transmission systems, and the global satellite communication complex – the whole entity, the Internet, would not exist.
- Learn that the set of protocols governing the Internet transmission is called TCP/IP. Get to know that all data transmission in the Internet is governed by the TCP suite and the all-addressing issues are controlled by the IP suite. Become familiar with samples of IPv4 and IPv6 addresses.
- Study the basics of the Internet operation concerning with data transmission through the Internet. Learn that a user device (a transmitter) sends the request for e-mail, web search, or chat communication to the Internet Service Provider (ISP) that directs the request to a Domain Name Server (DNS). Know that for each request, the DNS finds a corresponding IP address and forwards the packets to a proper domain, where core routers continue direct them to the destination point (see Figure 1.2.7).
- Be aware that the Internet transmission relies on two principles: the "black boxes," which implies that the Internet simply transfers packets to their destination points without looking into the packets' contents, and "the best efforts," which means that if a packet failed to arrive to the destination point, it must be retransmitted.
- Learn that **optical (fiber-optic) communications** is the backbone of modern communications infrastructure, and that vast majority of global communication traffic is delivered by optical networks.
- Get to know that the basic block diagram of a fiber-optic communication system includes a transmitter that converts electrical information signals into optical pulses, an optical fiber through which these pulses travel, and a receiver that converts light into electrical current containing a received message. Be aware that this optical PPT link is a building block of optical communication network.
- Understand that the optical communications industry emerged as a new main form of global communications because optical communication networks provide the highest possible

transmission rate (the number of bits per second). Know that this main advantage of optical networks exists thanks to (i) the fundamental principle stating that the transmission rate of any communication system is proportional to the frequency of an information-carrying signal, and (ii) to the fact that the optical signal has the highest frequency that can be utilized in practice. Be aware that modern optical communication networks are capable of delivering information at mind-boggling rate 10^{15} b/s (petabits per second, Pb/s) and still have a potential for the further raise.
- Remember that today optical fiber spans the entire globe, and optical signals travel inside an optical fiber at the velocity close to the speed of light in vacuum, which – combining this propagation speed with the enormous transmission rate – makes optical communications capable of delivering tremendous amount of information at the most remote corner of the world almost instantly.
- Get to know that the scale of operation of optical communications spans from intra-chip connections measured in nanometers to intercontinental submarine networks measured in thousands of kilometers.
- Learn that the quality metrics of an optical fiber are its attenuation and bandwidth. Grasp that attenuation is loss in decibels per unit length in kilometers, and that only after optical fiber attenuation had been reduced to 0.2 dB/km, the commercialization of optical communications started in serious. Be taught that a single-mode fiber (SMF) – the type served for long-distance transmission – has the widest (best) bandwidth (about 8 THz) among the two other wired transmission media (copper wires and a coaxial cable).
- Master the point that optical communications not only grasps all the latest developments in communications technology, such as SDN and programmable optical switches, but also stimulates many of new advances by demanding from photonics, electronics, computers, material science, and other related industries better hardware, software, and firmware products.
- Understand the relationship between optical communications and wireless communications – currently the fastest growing sector of communications industry – in which wireless communications provides mobile and fixed connections within short distances, whereas optical communications covers all the remaining distances for the remote destination points.
- Be familiar with brief history of two major developments in optical communications: manufacturing a low-attenuation optical fiber and creating a miniature, low-power-consuming laser diodes operating at room temperature. Know that both developments won Nobel prizes. Also, be aware that many other advances in optical communication technology – such as optical amplifiers and wavelength-division multiplexing (WDM) – have turned the optical communications from an exotic form of traffic transmission to the backbone of modern communications infrastructure.
- Get to know that transmission rate supported by optical communication networks has increased ten million times over last 25 years but the demand for increasing this rate continues. Be aware that the industry is now approaching to the fundamental limits in information-carrying capacity of optical fiber, so the further advances would be more difficult to achieve.
- Realize that the **wireless communications** is the second of the three pillars of modern communication industry and that this sector is growing at the fastest pace fueled by the advent of ever-increasing number of mobile applications and new technologies like the IoT.
- Get to know that the wireless communication systems transmit information by radiating electromagnetic (EM) waves in the radiofrequency (RF) spectrum from 3 kHz to 300 GHz (see Figure 1.2.11). Comprehend the specifics of wireless transmission: An information signal, delivered by a digital (or analog) carrier, arrives at a transmitting point, from which it is radiated by an antenna; at the receiving point, these waves are accepted by the antenna, processed, amplified,

converted to an optical signal, and sent to a communication network through an optical fiber. Understand that both end points can perform both transmitting and receiving tasks.
- Understand that the RF waves travel through air, regardless of the end points locations. Realize that using air as a transmission medium is, on the one hand, a great advantage because such a link is always available everywhere and, thanks to this feature, wireless systems can support mobile communications; on the other hand, this medium creates additional problem because we cannot control the air conditions. Remember that in wireline communication systems, we can improve a coaxial cable or an optical fiber and can protect them from external noise, but in wireless system, we can do nothing with air.
- Learn that the RF waves can travel through air in three main propagation modes: ground wave, sky wave, and light-of-sight (LoS), as shown in Figure 1.2.13.
- Be aware that there is a fundamental reason for power loss in wireless communications, regardless of the propagation mode and air condition, which is the spread of RF signal as it travels down the transmission distance. Know that this loss of signal power is called free space loss (FSL) and is described by (1.2.5). Understand that, in addition to FSL, there are a number of other mechanisms causing the loss of the signal power.
- Learn that the main problem of modern wireless communications is the scarcity of spectrum because the range of frequencies that can be employed in practice (the usable spectrum) is determined by the combination of factors, among which are air attenuation, a transparency of obstacles, the size of an antenna, a transmission distance, an adequate bandwidth to support the required transmission rate, availability of electronics to build cost-effective hardware, and – atop of all that – the government regulations that are necessary to control and maximize the use of the spectrum for a whole nation.
- Be aware that the control and management of the spectrum in the USA are executed by the Federal Communications Commission (FCC) and the National Telecommunications and Information Administration (NTIA) and that the licensing of the spectrum bands for commercial wireless communications is provided by the FCC. Know that at the international level, recommendations on the use of the EM spectrum are issued by International Telecommunication Union – Radiocommunications (ITU-R) sector.
- Learn that the further development of wireless communications requires more spectrum, but the spectrum is finite and its best bands are almost totally occupied (licensed), which threats to limit the future advances in wireless communications. Know that the most promising solution to this dilemma is employing the bands of frequencies higher than 3 GHz.
- Comprehend that the use of new ranges of spectrum will open the possibility for development and deployment of new types of wireless communications, the first among which is the fifth generation of traditional, RF-based wireless communications called *5G*. Become familiar with advantages of 5G systems, such as high bit rate up to 20 Gb/s and many advances in other transmission aspects.
- Learn the principle of operations and the range of applications such contemporary wireless communication systems as Wi-Fi, Bluetooth, ZigBee, optical wireless, and Li-Fi.
- Comprehend that wireless communications can be fixed and mobile.
- Grasp the concept of mobile communications, particularly with the terrestrial cellular networks. Understand that the entire wireless communication network blanketing a geographic region is divided into a number of cells. Know that a cell tower, called *base transceiver station (BTS)*, that transmits and receives the electromagnetic signals to and from mobile devices within its cell, is located in the center of each cell. Comprehend that the main advantage of a cellular architecture is the frequency reuse achieved by assigning to every cell its own frequency band and ensuring

that none of the adjacent cells have the same band. Examine Figure 1.2.16. Master the point: Without a cellular architecture, the mobile communications would require assigning an individual frequency to each customer to avoid signal interferences, which would severely restrict the number of mobile phones because the number of frequencies (spectrum) available for wireless communications is strictly limited.
- Become familiar with another advantage of a cellular architecture: low-power operation thanks to proximity of mobile devices to a nearby BTS. Realize that this advantage is twofold because (i) low-power BTS signals diminish when they reach the cell borders and do not interfere with frequencies of the adjacent cells and (ii) a cell's compact area and low-power operation are fulfilled by small transmission power from mobile devices enabling their batteries to be compact.
- Become familiar with brief history and the principle of operation of the third pillar of modern communications – **satellite communications**.
- Learn that communication satellites are the spacecrafts orbiting the earth and serving as intermediate nodes (active mirrors) for redirecting the EM signals from one point on the earth surface to the other. Realize that to perform this task, communication satellites receive signals from the ground, amplify them, change their frequencies, and transmit the signals back to the intended area on the earth surface. Be aware that the satellites carry an appropriate equipment powered by solar panels (see Figure 1.2.18).
- Be aware that the frequencies used in satellite communications are concentrated in the microwave region of the EM spectrum and that three bands – L-band, C-band, and Ku-band – are mostly employed (see Figure 1.2.11). Know that each band has its specific applications in performing certain communication tasks, such as support of GPS, weather forecast, TV broadcasting, mobile communications, and aircraft and satellites communications.
- Study that satellites can orbit the earth in circular and elliptic orbits and that tree main types of geocentric (around the earth) orbits – high, medium, and low – differ in their distances (that is, altitudes) from the earth center.
- Learn that the most-used high-altitude orbit is a geostationary earth orbit (GSO) whose distance from the earth surface is approximately 35 700 km. Get to know that a satellite placed on this orbit travels the orbit for 24 hours and therefore is stationary with respect to a point on the earth surface, which easies their communications. Know that the example of a *medium earth orbit (MEO)* is a *semi-synchronous orbit* at which the satellites travel the full orbit for 12 hours at about 20 200 km above the earth surface. Be aware that the low-earth orbits, LEOs, range in altitude from 160 to 2000 km and that the LEO satellites, thanks to their proximity to the earth surface, perform a variety of functions that require immediate responses. Realize that since these satellites rotate around the earth at short intervals (from 88 to 127 minutes), they pass over the same ground point many times a day and therefore a constellation of satellites must be on the orbit to support constant satellite communications. Understand that the LEO satellite systems today support all kinds of mobile connections.
- Become familiar with two main problems of satellite communication signals: power loss and propagation delay. Know that the power loss can be calculated by using (1.2.4) and (1.2.5) and the propagation delay is estimated by simply dividing a distance by the speed of light.
- Remember that the satellite communications enables us to reach any distant corner of the globe without cables and that this unique property of satellite communication makes it the major player in international mobile communications.
- Comprehend that all three major sectors of modern communication industry – optical, wireless, and satellite – work together and make up the global communication infrastructure whose existence continues to improve the quality of our life.

1 Modern Communications: What It Is?

> **Sidebar 1.2.S.1 Brief Notes on History of Telegraph, Telephone, Radio, and Television**
>
> *1.2.S.1.1 The Historical View on Globalization of Communications*
> Delivering information must be global by its very nature. Indeed, people at any point on the globe need to be informed as quickly as possible. We take for granted the ability of modern telecommunications to perform this task, but this not always was the case. The speed and the reach of delivering information change with the times and these changes constitute important turning points in the history of civilization.
>
> Because telecommunications is about delivering information over a distance, people have tried to extend the reach of transmission links as far as possible ever since the advent of this technology.
>
> The scale of telecommunications networks has grown from local to regional to national to international. These developments occurred in sequence and in parallel; that is, some intercontinental connections were achieved even before the continental networks had been completed.
>
> The point is this: As soon as new telecommunications technology appears, this technology is immediately used in an attempt to extend the global reach of the telecommunications network.
>
> *Globalization* means extending the reach of telecommunications beyond the national and continental scale. The meaning of this term has changed over the years. For example, building the telegraph link between England and France in the nineteenth century was certainly an example of globalization, while building a fiber-optic link between these two countries today could be considered a regional development.
>
> The need for quick communications between the continents can best be comprehended by considering an event that occurred shortly before the invention of a practical telegraph: In the war of 1812, General Andrew Jackson continued to fight French troops in New Orleans even after the peace treaty had been signed in Paris, France. This was because it took more than a week for a ship to deliver the news from Europe to America.
>
> This historical anecdote stresses one important point: We almost forget that before the advent of modern telecommunications, all information was stored on paper and disseminated in the form of letters, newspapers, magazines, and books. This material had to be delivered by a human being. It took a messenger, a horse carriage, a ship, a train, or a plane to deliver it. Modern communications has changed all that.
>
> *1.2.S.1.2 The Telegraph*
> Telecommunications in the modern sense came into existence with the invention of the telegraph – the first electrical system that delivered information. Even though the first electrical telegraph was installed in England in 1837, the era of *practical telegraphy* began in 1844, when *Samuel Morse* transmitted his first message from Baltimore, MD, to Washington, DC (see Figure 1.2.S.1.1).
>
> Morse's telegraph operates in a discrete mode: When the key is pressed, two end terminals of the circuit get connected and the pulse of electrical current flows through the circuit, delivering a character. The character is presented by a dot or a dash. (Figure 1.2.S.1.2). The shorter pulse delivers a dot; the longer pulse delivers a dash. (**Exercise:** Sketch the schematic of a telegraph circuit.) Samuel Morse,

Figure 1.2.S.1.1 Samuel Morse. Source: Mathew Brady Studio, c. 1860. Salted paper print. National Portrait Gallery, Smithsonian Institution. Open access.

Figure 1.2.S.1.2 Morse telegraph key. It is also called Morse–Vail telegraph key to reflect the decisive innovation made by Morse collaborator named Alfred Vail. Interestingly, Alfred Vail's first cousin, Theodor Vail, became the President of American Telephone and Telegraph Company and transformed the company into Bell System, the corporation that formed the modern telephone industry. Source: Reprinted with permission Division of Work and Industry, National Museum of American History, Smithsonian Institution.

(Continued)

Sidebar 1.2.S.1 (Continued)

who was trained as a painter and knew very little about electricity, learned about this circuit from Joseph Henry,[2] then a professor at Princeton University.

In 1851, the first telegraph cable linked England and France – albeit a small first step, but telecommunications had begun its global expansion. Six years later, after several unsuccessful attempts, the first transatlantic telegraph cable was installed by an English company with the support of the British and American navies. This event heralded the advent of the true globalization of telecommunications. Unfortunately, though, this cable had an inherent design flaw and it was damaged almost immediately by the high-voltage signal that the chief engineer forced the operator to apply; within three weeks the cable ceased to work. Three years later, in 1861, the first American transcontinental telegraph cable was laid. Eventually, by the end of nineteenth century, the reliable telegraph communications between the United States and Europe was established.

To understand what impact this new type of communications made, suffice to say that messages *seemed* to arrive hours before they were sent because of time-zone differences. People did not realize this difference in time prior to the advent of the telegraph because all communications up to that era had been delivered by ships crossing the Atlantic. Clearly, everyone became excited over this new means of communication. On the down side, however, was the fact that this new service was extremely expensive: The initial rate was $1 a character, payable in gold. At that time, the monthly wage for a laborer was about $20.

The point is that with these first transatlantic cables a new era in global telecommunications had started: In 1865, the telegraph linked India and Europe; in 1871, Australia and Europe and in 1902, New Zealand and Canada. By 1888, there were 107 000 mi of undersea telegraph cable linking all parts of the world. The globalization of telecommunications was now well under way.

Interestingly, the telegraph was still important means of communications up to second half of the twentieth century even though fax and teletype were already in use. The pay rate was per word; naturally, senders tried to deliver maximum information by using the minimum words. This is how a telegraphic style of writing was born.

1.2.S.1.3 The Telephone

Alexander Graham Bell's invention of the telephone in 1876 had a far greater impact on modern telecommunications, obviously, than Morse's telegraph. And today, while the telegraph has been largely consigned to a museum shelf, breakthrough technological (and business) developments have brought to life mobile communications that stems from the telephone industry.

Almost immediately following his invention – in 1876 – Bell[3] tried to use a transatlantic telegraph cable to conduct a telephone conversation. That attempt – and many others – failed.

---

2 Joseph Henry (1791–1878), self-educated American physicist, started his professional career as a schoolteacher in his native Albany. Through creative work and continuing study, he became a leading scientist in the then new field of electromagnetism. He ended up as a secretary of the Smithsonian Institution. "Joseph Henry became a friend of Abraham Lincoln and died in 1878, revered as America's greatest scientist" (Bodanis, p. 52). He gained international recognition long after his death: Today, we measure inductance in the unit (*henry*) named after him.

3 Alexander Graham Bell (1847–1922) was born in Scotland and moved to the United States in 1871, where he became a professor of vocal physiology in Boston. He experimented with various electro-acoustical devices to help deaf people. He sent the first telephone message to his assistant from the second floor to the basement of his house on June 10, 1875. He patented a telephone in 1876, and this eventually gave birth to a new era of telecommunications. The Bell System – the main provider and developer of telephone systems throughout the last century and at one time the largest corporation in the world – proudly carried his name.

However, in 1877, the first long-distance phone conversation was held between two parties separated by 60 mi in California. The era of the telephone literally exploded onto the scene.

By 1886, there were 250 000 telephones in service worldwide; however, almost all of them were connected locally and complete long-distance service was still some years away. By 1924, the Bell System – the company that actually monopolized telephone service in the United States – alone had 15 million phones in operation. Today, the number of wired telephones in the United Sates declined, after reaching its peak at about 300 million telephones by the end of twentieth century, due to emergence of mobile smartphones. And the Bell System no longer exists.

Figure 1.2.S.1.3 shows the portrait of Alexander Graham Bell and his early telephone equipment. If you compare these historic pieces of Bell's original device with modern smartphones, you can surely appreciate the progress telecommunications technology has made.

(a) (b)

Figure 1.2.S.1.3 Alexander Graham Bell (a) and its early telephone equipment (b). Sources: (a) Bell's portrait: By Moffett Studio – Library and Archives Canada/C-017335, Public domain. (b) Smithsonian Libraries.

The first telephone cable between America and England was not laid until 1956. That was when the first wireline transatlantic telephone conversation occurred.

Note that with telegraphy, it took only 14 years (from 1844 to 1858) to move from the first practical transmission to transatlantic transmission. However, it took 80 years (from 1876 to 1956) to make the leap from local telephone conversation to transatlantic cable connection.

Why? Telegraph transmission, in Morse code, is, by its very nature, digital. The receiver must only discern the presence and duration of a signal. In other words, the telegraph's simple on/off mode of operation made long-distance transmission possible even with rudimentary technology. On the other hand, telephone transmission, until the 1960s, was strictly analog. The signal delivered information by virtue of variations in its amplitude. Because of power loss and noise, the signal became distorted after only a short transmission distance. To deliver information, an analog electrical signal had to be boosted and, at that time, there were no means to do it.

(Continued)

Sidebar 1.2.S.1 (Continued)

This is why Alexander Graham Bell was unable to transmit voice over a transatlantic telegraph cable.

In fact, the first transatlantic telephone cable (called TAT-1) contained 102 unidirectional electronic repeaters built on vacuum tubes; they repeatedly boosted the weak electrical signal traveling over the transatlantic span. It is interesting to note that in 1966, after 10 years of service, all 1608 of the original tubes in the repeaters were still working. Very impressive, indeed, especially if we take into account that each vacuum tube repeater contained 5000 parts and cost about $100 000. The total cost of the TAT-1 system was $42 million. Two cables were laid; they contained 36 two-way circuits. The first day in operation the system carried 588 calls, 75% more than a radio telephone system handled in an average day at the time.

It should be noted that over the first half of the twentieth century, the telephone system extended globally with the help of radio transmission. The first commercial transatlantic radio telephone service began in 1927; a three-minute call was $75. By 1937, US customers could call 68 countries via high-frequency radio. Until the advent of optical fiber in the mid-1980s, terrestrial long-distance telephone transmission relied on microwave links. Today, all continents and countries are connected with optical fiber and satellite systems, which deliver all forms of information in digital format anywhere in the world.

1.2.S.1.4 Radio

Let us consider the other great invention that dramatically changed the telecommunications industry. In 1895, Guglielmo Marconi, an Italian-born physicist, inventor, and industrialist, transmitted a radio signal over more than a mile, and in 1901, he achieved the first transatlantic radio transmission (Figure 1.2.S.1.4). Independently, in 1896, Russian physicist Alexander Popov wirelessly transmitted his first message over the distance of 250 m. In 1899, Popov established reliable wireless communication between a stranded ship and a shore separated by 25 mi.

Figure 1.2.S.1.4 Guglielmo Marconi and his radio equipment. Source: Reprinted with permission of Smithsonian Libraries.

Nevertheless, the main glory for invention of radio justifiably went to Marconi who was just 23 years old when he started the first company to commercialize his invention. Radio quickly became the major means of communication among ships and between ships and shore.

A later development – high-frequency (HF) radio, also called short-wavelength radio, operating at tens of megahertz – enables the transmission of radio signals on a global scale by bouncing the signals off the ionosphere (see Figure 1.2.13). This was the other major step in globalization of telecommunications.

During radios "golden age," from the mid-1930s to the mid-1950s, it was the main means of communications worldwide. Radio is still widely used and the recent birth of commercial satellite radio brings a "new wine in old bottles."

Even though everyone knows about radio (or at least uses it), not everyone is familiar with the scientific developments that eventually brought this technology to life.

Morse's telegraph and Bell's telephone used wires to deliver a signal from one point to another. Very few people at the time of those inventions really understood the physics behind the means of transmission, but everyone saw a tangible means of delivering a signal. Not so with radio! The signal was delivered by mysterious waves that nobody could see. It was Michael Faraday[4] who proved experimentally that electromagnetic forces existed and could do some type of work at a distance. (Joseph Henry did experiments in electromagnetism, too, but his work was less known in Europe.) James Clark Maxwell[5], basing his work on Faraday's experiments, derived the set of equations that predict the existence of electromagnetic waves, but not every physicist at that time accepted Maxwell's prediction. It was Heinrich Hertz[6] who experimentally demonstrated these waves. Marconi, in his lecture on receiving the Nobel Prize in 1909, said that his goal was to determine "whether it would be possible by means of Hertzian waves to transmit to a distance without the aid of connecting wires." Today, we are so much accustomed to radio transmission that we cannot imagine that anyone could doubt in such a possibility.

1.2.S.1.5 Television

It was not many years later after the widespread of the radio that a new communications technology – television – came on the scene, delivering not simply voice but video as well. Commercial black and white TV broadcasting started in the United States in 1941. In 1954, the

(Continued)

---

4 Michael Faraday (1791–1867), a self-educated English physicist who in his famous series *Experimental Researches in Electricity* reported a wide range of discoveries, most notably the relationship between electrical and magnetic fields. He discovered the law (named after him) of electromagnetic induction. All of his work was done at the Royal Institution in London, where he also delivered educational speeches and demonstrations. Today, electrical engineers remember Faraday when they measure a capacitance, since the unit (*farad*) is named after him.

5 James Clerk Maxwell (1831–1879), the first professor of experimental physics at Cambridge University and a founder of Cavendish Laboratory, studied at Edinburgh and Cambridge. In his *A Treatise on Electricity and Magnetism* (1873), Maxwell predicted theoretically the existence of electromagnetic waves. His results laid a solid theoretical foundation for further work in understanding nature, specifically for Albert Einstein's theory of relativity. On the practical note, his work paved the way for modern telecommunications technology.

6 Heinrich Rudolf Hertz (1857–1894), a German physicist, studied under Kirchhoff and Helmholtz in Berlin. In 1887, he devised and performed the experiments that proved the existence of electromagnetic waves, as Maxwell had predicted. He transmitted electrical signals over a distance without wires, which was the required proof. His experiments gained international recognition immediately and eight years later, Marconi put these results into practice.

> **Sidebar 1.2.S.1 (Continued)**
>
> first color TV set went on sale. Price: $1000. The Japanese demonstrated analog high-definition TV in 1987; a consortium of US companies in 1990 also announced high-definition digital TV.
>
> Invention of TV is a long and dramatic story filled with fierce competitions, patent wars, and technological and commercial revolutions. In contrast to telegraph, telephone, and radio, it is impossible to name even several major players who formed the basis of modern TV. (You probably have heard the name *Vladimir Zworykin*, a Russian-born American engineer, in reference to invention of television. He made a major contribution into development of commercial TV, specifically by inventing the electronic iconoscope capable to receive and display moving images and building a completely electronic TV system in 1939. Still, he was not alone; in his work he relied on many other inventions and developments.)
>
> Until advent of optical fiber in the mid-1980s, TV could only broadcast its signals by using electromagnetic waves at a higher frequency than radio. Since it is impossible to utilize the same bouncing principle that has been used for many years for HF radio, TV was mainly local. Also, it was impossible to use the intercontinental telephone cables to transmit video directly because TV transmission requires a much broader bandwidth than voice transmission. It was not until the advent of the satellite era that TV went global. But the globalization of TV transmission truly came into its own with the introduction of fiber-optic communications technology in the mid-1980s. That was when an epoch of cable TV started.
>
> Initially, TV was completely analog. With the advances in electronics, computers, and digital signal processing, TV became digital. However, transition from *analog to digital TV* had been a process rather than an act. The United States mainly completed its transition in 2009; the international agreement set 2015 as a mandate date for shutting off analog signals. Nevertheless, this process still continues in many countries.
>
> Today, the advances in TV technology and combining the TV and the Internet bring new era of smart TV. Also, traditional broadcasting and even fiber-optic transmissions give the way to streaming TV when the content is delivering from the Internet in the process of watching. With all these new trends, it is most likely that traditional broadcasting and even cable TV will be moved to a quiet corner of our information space, similar to the place where today's radio is positioned.

1.2.1 The Internet

1.2.1.1 Basics of Networks

The block diagram of a communication system shown in Figure 1.1.1 demonstrates the simplest type of connections between a transmitter (Tx) and a receiver (Rx): a **PPT link**. In a PPT link, *information flows directly from Tx to Rx*. Since both ends of the link can serve as Tx and Rx, the more general diagram of this simple connection is presented in Figure 1.2.1a. A PPT link is a building block of any sophisticated communication system. However, such a block is not sufficient to build the modern communication infrastructure.

In general, a communication system provides exchanging information between many users. Think about the Internet: While you are sending a message to its recipient, millions of other people also use the Internet to send and receive messages, or visit websites, or download files, or watch movies. The Internet supports all these communication activities regardless of the locations of all the parties and the timings of their connections. How is it possible? This is because the Internet is a ***network***, not just a PPT link. What is a network?

A network is a set of nodes connected by links.

A node is any lumped element, such as computer, switch, router, telephone, printer, amplifier, repeater. Links are the means of contact between nodes; they are made of various media, such as copper wire, coaxial cable, air for electromagnetic waves, and optical fiber.

This very general definition stresses the main feature of a network: It consists of nodes and links. Figure 1.2.1b illustrates this definition. (Review Figure 1.1.3, which is an example of a communication network.)

The above definition of a network emphasizes the physical setup through which some messages can be transmitted. This definition can be applied to any type of interconnected system, such as electrical circuit, electrical power grid, transport network, or logistic network. However, in communications, we need to deliver information; therefore, a communication network must be able to transmit information signals. To do so, two necessary conditions must be met: First, the network must "know" where and when a signal must be sent; second, the network must arrange for the signal to be delivered to its destination.

The first condition implies that a network must be able to read and understand the destination address, find the best route for delivering a message, and verify that the message is securely delivered. But the words *read, understand, find, and verify* imply involvement of intelligence. And yes, communication networks possess the intelligent (logical) properties enabling them to perform all of the above and many more functions.

A simple example: Suppose Figure 1.2.1b shows the local network of our organization. A worker at Node 3 wants to communicate with a colleague at Node 4. The message from Node 3 must be sent to Node 1, from which it can travel directly to Node 4 or – if that line is busy – through Node 5, or – if necessary – through Nodes 2 and 5. How will the message be sent in reality? An intelligent (logical) agent of Node 1 identifies the destination point and decides which first route to choose, depending on the availability of routes. If other nodes will be involved, each of them must make a similar decision. So a network must have a measure of intelligence; its intelligent agents are located in the network's nodes.

But what if we need to communicate with someone outside of our organization? This is where the Internet comes to the rescue. Local, regional, and other types of networks are interlinked through a cloud (the Internet), as Figure 1.2.2 shows. This is why the Internet is called the network of networks.

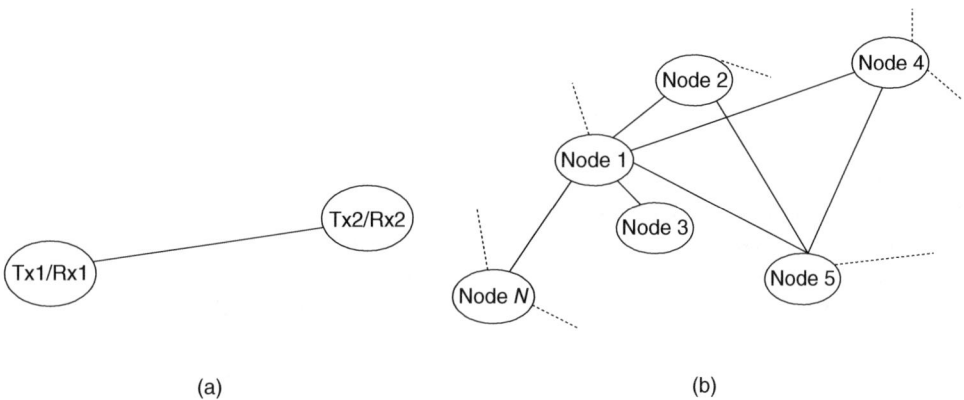

Figure 1.2.1 Basic layouts of communication systems: (a) point-to-point link; (b) a network.

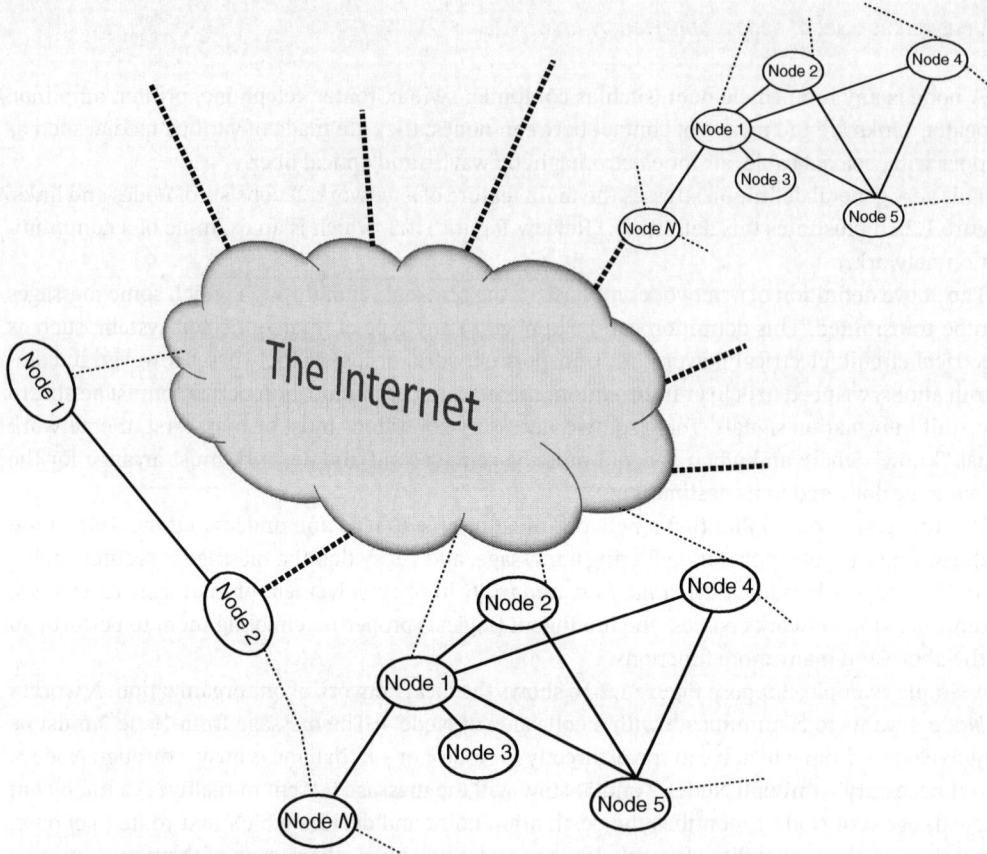

Figure 1.2.2 Local and regional networks and the Internet (cloud).

Examining Figure 1.2.2, bear in mind that the set of nodes connected by solid lines constitute one network, whereas the dashed lines show connections between different networks. Thus, we can see, for example, several Nodes 2 because each of them belongs to a different network.

Interestingly, our connection to the Internet (cloud) is determined not by our geographical location but by our IP address, where IP stands for Internet Protocol. For example, if you communicate with a visitor from a foreign country through his/her laptop, your and the visitor's messages go through the cloud even though the visitor might be sitting next to you.

We know that today's communications is mainly *digital*, which means that information is presented as a stream of 1s and 0s and these bits are physically carried by electrical or optical pulses. Then, the analysis of our example of communications among nodes should raise an important question: How can a node understand a message if this message is simply a sequence of 1s and 0s? The answer lies in establishing a rule which all the nodes must follow. For example, the rule might say that the message must be sliced in packets with 720 bits each, out of which the first 24 bits must contain the address and the remaining 696 bits must carry the data. (In digital communication parlance, the first 24 bits are called *header* and the 696 bits are called *payload*.) Then all nodes reading the message will understand where to direct the message by deciphering its first twenty-four 1s and 0s. Though the numbers in this example are fictional, the concept is real: Without such a rule, modern communications would be impossible. The rules governing transmission in communication networks are called *protocols*. The Internet is governed by the *TCP/IP*. This is a suite (set) of communication protocols used to interconnect network devices and support their communications on the Internet.

In spite of oversimplification, the above examples emphasize the two main features of a telecommunications network: First, the network must provide **physical** connections among all communicating parties; second, communication is possible if and only if the communicating parties follow certain rules (protocols). But connections are something physical, tangible, whereas protocols are something logical, **intelligent**. Therefore,

> *a modern communication network has both physical and logical (intelligent) features.*

Clearly, physical features consist of the network's *hardware*, such as transmitters, receivers, switches, links. What devices support logical (intelligent) operations? The general answer is *computers in their various forms*. Specifically, we refer to electronic machines that are able to perform arithmetic and logic operations; today, they are *servers, processors, switches, routers, amplifiers,* etc. (To better understand the nature of a modern communication network as a combination of hardware and software, consider a computer: Without software, which is a collection of computer programs [instructions, that is], a computer is just a set of parts but without this set there is no place to execute those instructions.)

The intelligent side of network operation is based on the protocols. At its most basic level, the network compares the current transmission situation with the protocols and decides when and where to direct the communication traffic. But real-life situations are more complicated than the presented simplistic picture; therefore, network software must be able to resolve any ambiguous situation. Fortunately, progress in development of artificial intelligence and its application to network logical operations has helped to increase network flexibility and efficiency.

After a node understands the address to which a message needs be delivered, it must direct the message. Here comes another operation without which a communication network cannot exist – **switching (routing)**. In a PPT link, two nodes are connected directly, and information flows from one point to the other without any ambiguity; thus, no switching is needed. In a network, information cannot be delivered to all the nodes without switching. Therefore,

switching (routing) ability is the main logical (intelligent) feature of a communication network.

A word about terminology: *Switching* is a general term that refers to the operation of relaying messages through a network. However, today the term *switching* is also used to describe the *use of switches to change the circuit-to-circuit connections* within a network *for transmitting the signals*.

Routing is also a general term that refers to redirecting the stream of information; however, *routing means choosing a path for forwarding information* through a network and from one network to the other, provided that all connections are established. Thus, to execute routing, we need not switches but *routers*. The routers perform many functions: They choose the best route for information to travel, analyze the messages being transmitted, change their packaging, send messages to another network, and execute the security and control operations.

We will touch on these terms again shortly.

(**Question**: *Why do switching and routing require intelligent (logical) ability from a communication network?*)

Next topic in networking is **network topologies**, which refers to a network's physical layout and the routes of the flow of information. Figure 1.2.3 demonstrates five main network topologies.

In a *bus topology*, information flowing along the main line is accessible to all connected nodes. It is effectively broadcasting messages. In a *ring layout*, information flows sequentially from one

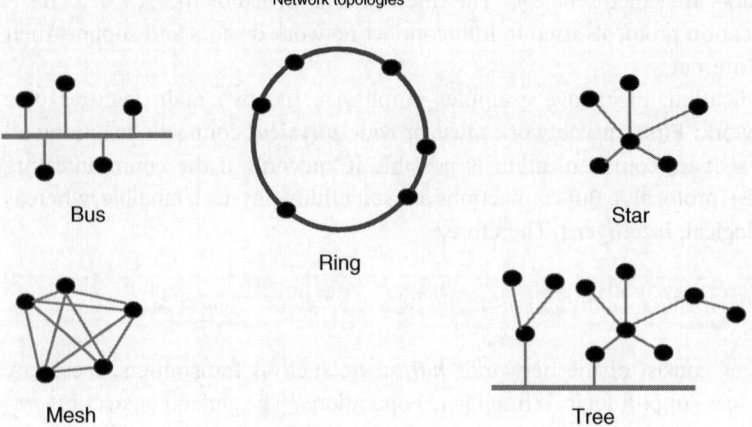

Figure 1.2.3 Network topologies.

node to the adjacent one; thus, for example, the message from the first node addressed to the fourth node must travel through the second and third nodes.

(**Question**: *How would the second and third nodes know that this message must be transferred through them? How would the content of this message be prevented from being read at the second and third nodes?*)

Star configuration implies that all communications goes through a central hub (also called switch or concentrator). *Tree topology* is a combination of bus and star topologies, the combination that allows for network expansion without modification of the network core. (**Exercise**: Describe how information flows in a tree network.) Finally, in a *mesh network topology*, all nodes are interconnected so that each node can communicate directly to every other node.

Even though this discussion of network topologies combines the physical layout with the flow of information, there is, in fact, a difference between *physical and logical topologies*. For example, a star hub can broadcast information to all its nodes; in this case, the network with a star physical layout will operate in a bus logical topology. You are encouraged to figure out more examples of networks whose physical and logical topologies might be different.

Depending on the range of *transmission distance*, networks are loosely divided into three categories. This division is often called **network hierarchy**. The *access networks* serve the shortest range; they connect subscribers (users) to their local service provider. This is the users' first pathway to the global communication infrastructure. Think about your home or office computer, tablet, or smartphone: They access the Internet through the network of your organization or neighborhood. The service range of access networks is up to a few tens of kilometers.

Next level of the hierarchy is *metro networks* that covers the range from tens to a hundred kilometers, serving a metropolitan area. Think about the network of a big corporation or a university that is spread across a big city, such as New York City, Beijing, Paris, Cairo, or Sydney. This type of network accommodates all incoming traffic from its access networks, circulates the traffic within, and directs it to the outside web.

Finally, all metro networks are connected to the *long-distance (long-range, core, backbone, or trunk) networks* that make up the global communication infrastructure (the Internet, that is). The range of long-distance operations extends to thousands of kilometers; the networks can be terrestrial or submarine (undersea). Examples of long-distance networks are transatlantic fiber-optic

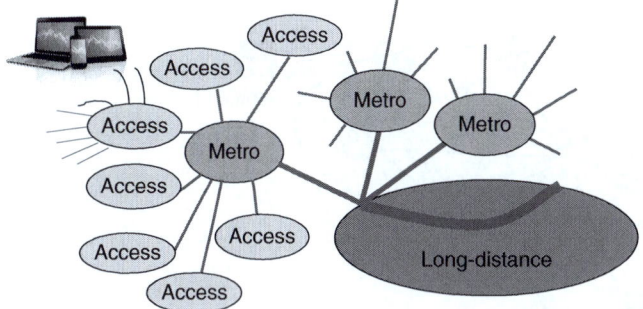

Figure 1.2.4 The concept of network hierarchy.

cables connecting America to Europe and Africa, transpacific cables linking America to Asia and Australia and New Zeeland, and fiber-optic networks binding the East and West coasts of the USA.

Figure 1.2.4 visualizes the concept of network hierarchy.

It is easy to understand that the volume of traffic carried by networks at every level is determined by their range because the range implies the geographical area, which in turn determines the number of customers (*users* or *clients*). The access networks have the lightest traffic because they serve thousands of customers, the traffic in metro networks are heavier because they are used by millions, and the long-distance networks experience the heaviest traffic as they connect billions of clients. The number of customers indicates the level of revenue, and the volume of traffic puts demands on the quality of networks and their components and modules. Thus, the long-distance networks need the best (read the most expensive) components, modules, and the entire grids, whereas the access networks can afford only inexpensive hardware and software because their revenue is minimal.

Figure 1.2.4 shows the lines connecting different network levels. Thickness of these lines depicts the traffic volume (and consequently, transmission speed); the thickest line delivers the highest traffic volume and has the fastest transmission rate.

Bear this hierarchical relationship in mind for all our succeeding discussions.

(**Question**: *Why does an access network have the minimum revenue, whereas a long-distance network have the maximum revenue?*)

Let's return to **switching or routing**, the distinct network property that separates a network from a PPT link. How are these operations actually performed in networks? There are two modes of directing information flow in modern communications: **circuit switching** and **packet switching (routing)**. Examine Figure 1.2.5, where these types of switching are illustrated.

In a *circuit-switching network* shown in Figure 1.2.5a, the first operation is to establish a virtual dedicated line between end users **M** and **N**. Switches at every node connect the input and output links in such a way that users **M** and **N** become directly connected for the duration of their communication. Once the line within a network is set up, information between points **M** and **N** flows only through this line, even though there are many other possible routes for transmission. When the communication session ends, the connections at every node are disjoined, and the network becomes ready for the next operation. A classic example of a circuit-switched network is our wireline telephone system, called *public switched telephone network (PSTN)*. The advantage of a circuit-switching network is high quality of transmission due to a dedicated link between the users. Disadvantage – and the big one – is extremely inefficient use of the network capacity. For example, while we are occupying a network by our telephone conversation, which occurs at rate of 64 kilobits per second, the network, capable of delivering at terabit-per-second transmission rate,

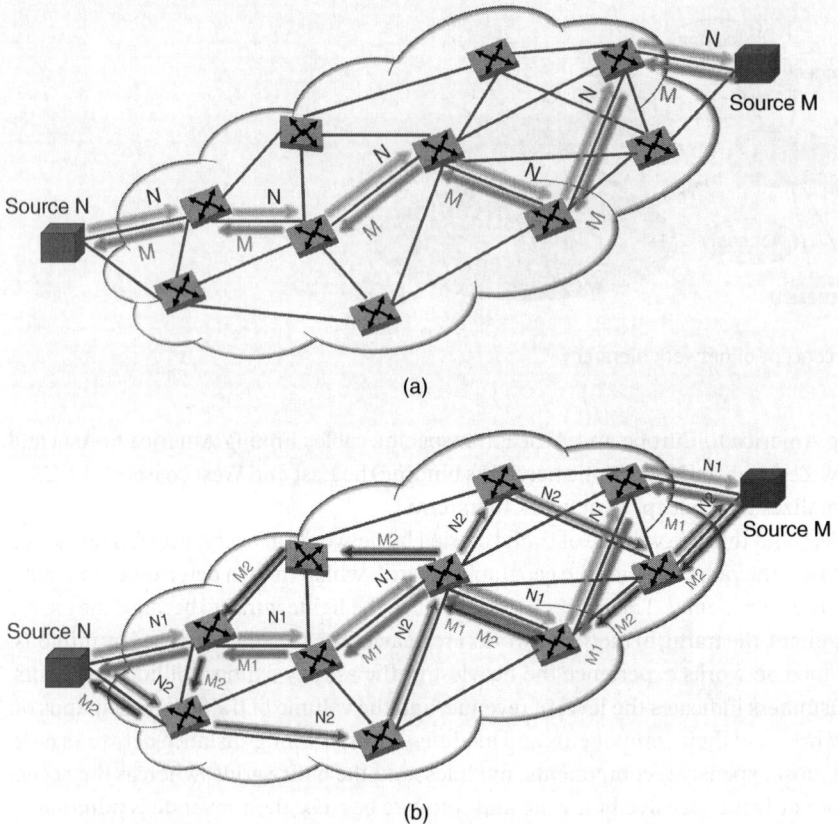

Figure 1.2.5 Two types of switching in networks: (a) circuit-switching network; (b) packet-switching (-routing) network.

actually stands idle. The other drawback of a circuit-switching network is that its *survivability* is low: If a significant part of the network is destroyed by natural catastrophe or military attack, it would not be able to support reliable communications.

A *packet-switching network* operates very differently: First, all network nodes are permanently connected and therefore no dedicated line can be established for a communication between users M and N specifically. Second, a message is broken down to pieces of data called *packets*. These packets travel through the network by various routes. Once arrived at a destination point, they are assembled back into the original message. When a packet reaches a router, the router, having read the overhead and learned the packet destination, chooses the best route (in terms of routes availability, time to travel, security to deliver, etc.), and send the packet further. This packet's examination and directing are done by every router on its path. See an example shown in Figure 1.2.5b. Here messages M and N broke into two packets and each are transmitted through the packet-switching network.

The operation of a packet-routing network looks sophisticated and cumbersome, but modern digital electronics make all these operations easy to perform and fast to complete. Nevertheless, the complexity and inherent lack of reliability delivering information are among the drawbacks of packet-switching networking. But this type of networking has a gigantic advantage over a circuit-switching type: *the most efficient use of network's transmitting capacity*. This efficiency stems from the nature of packet switching: Each router utilizes every possible path to convey information. In addition, a router looks for the shortest available route for each packet enabling its fastest delivery.

There is another huge advantage to a packet-switching network – *resilience*. Even if a significant part of it gets destroyed, the network will stay operational. *The Internet operates in a packet-switching mode*, and now you know why.

(**Question**: Consider Figure 1.2.5: Can a third end user be connected to conversation between end users **M** and **N** in a circuit-switching network? In a packet-switching network? Explain.)

The last short note on networking: **Control plane and data plane** whose concept is demonstrated in Figure 1.2.6.

Any network includes two planes – the control plane and the data plane. A control plane is a set of software and/or hardware that executes control and management functions. Implementation of control plane depends on protocols. Examples of control plane protocols include *open shortest path first (OSPF)* – the routing protocol used in IP networks – and *generalized multiprotocol label switching (GMPLS)* protocol, the transport protocol. Examples of hardware are routers, such as a label edge router (LER) or a label switched router (LSR) for GMPLS control plane.

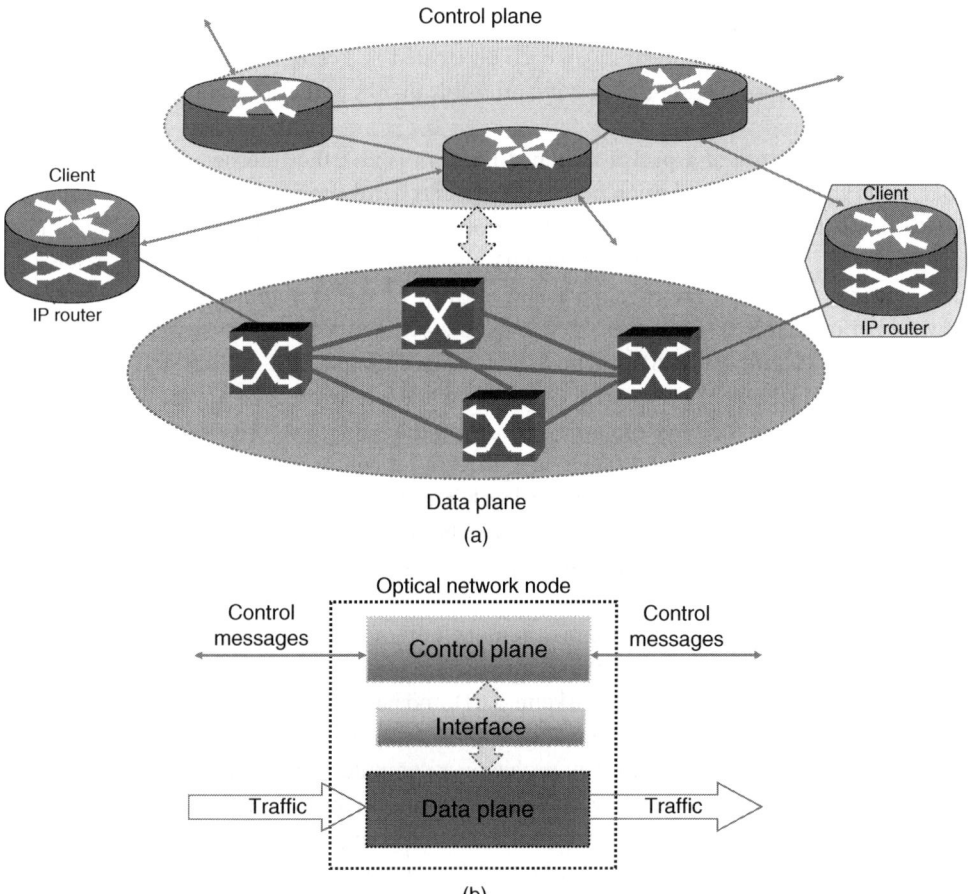

Figure 1.2.6 Control plane and data plane of a communication network: (a) general view; (b) hierarchical view.

Data (or forwarding) plane is a set of hardware and software that actually carries the user traffic. An example of hardware is a switch, such as an optical cross-connect (OXC) and an example of a protocols implemented in software is the TCP suite.

Hardware for both control and data planes resides in the network nodes. To distinguish between these planes, it is necessary to look at the nature of their packets: *Control plane packets are originated with a router and heading to a router, whereas data plane packets originate with an end user and travel through a router to a final destination point.*

How do the control and data plane interact? The control plane generates routing and label tables and shares this information with the data plane to enable transportation. In other words, *control plane protocols enable IP routers to forward traffic correctly*.

How are these two planes implemented? Are they two separate physical networks? The answer is simple: in a variety of forms. There are networks whose control and data planes are two physical parallel networks interconnected at each node. This is how the legacy PSTN operates. There are networks where one module performs both functions; this is how Internet routers work. And there are networks that use a mix of both approaches.

Why do we need two planes? To separate control and transport functions of a network. Specifically, separation of the control and data planes makes the data plane protocol-independent. We remember that data is a stream of 1s and 0s and to make sense of such a stream we need to know a protocol, the rule defining the meaning of each bit located in a certain position. The data plane transmits all data packets regardless of specific protocols thanks to the control plane, which deals with this issue.

The current architecture of a packet-switched network (think the Internet) relies on routers. The *routers* are the devices combining both software and hardware in one unit called *firmware*. Thanks to this combination, the routers can perform both intelligent (e.g. choosing the route for a packet) and physical (sending the packet) operations. The advent of routers made the Internet a reality. However, the ever-increasing growth of traffic volume puts new demands on network performance. To meet these demands, the communication industry has introduced new network architecture called *SDN*. The SDN further decouples the control and data planes and relegates the functions of the control plane (network intelligence) to an application called a *controller*. By doing so, the SDN replaces the routers that supports both network control and data traffic forwarding with the centralized software-based control unit. In this programmable network, the task of forwarding data traffic is relegated to simple switches. By adopting cloud and SDN, modern communication networks substantially boost the efficiency of their operations.

(**Question**: *Consider a packet-switching network (the Internet): Where are the control plane and the data plane in this network?*)

As brief as this review is, we must still mention one other vitally important aspect of networking – **security**. During the time of standalone computers, any kind of malware could be brought in only in a physical storage device, such as diskette, a CD, and later a USB drive. Today's proliferation of wired and wireless communication networks, while – along with bringing economic and social benefits to billions of people – causes new problem: A computer virus inserted in one machine can spread almost instantly around the world. Nowadays, we encounter the situations when global corporations and even entire industries were blocked by shutting down their computer networks, which actually stops regular business operations and even people's normal life for days, weeks, and even months. We can easily imagine all the difficulties we would face if our communication networks were blocked: There would be no cell phone connections, no credit cards operations, no GPS, no e-mail, and so on. And what would happen if our "cloud," that is the data centers, suffered from such a vicious attack? Not only processing of communication traffic would be stopped, but we would lose all our data stored on the data centers' servers.

The first level of defense is, obviously, prevention from physical intrusion into our homes, offices, and data centers. The other level is precluding logical intrusion into the networks. Everybody knows the term *firewall,* which is the first logical "fence" around our local networks. Still another level of providing security in communications is *encryption* of all information. Encryption is encoding information in such a way that only those who possess a code (key) can decode the information. For example, communicating in foreign language is encryption (encoding) of information. (During World War II, Japanese army considered their native language as a safe code; in their radio and vocal communications they spoke Japanese without additional security measures. American army soon learned how to use this simplistic approach to its advantage.) This way of encoding has an obvious weak point: protection of the code. Since the code must be also transmitted through a communication network, it can be hacked too. Today, there are systems that use *quantum communications,* which promises theoretically unbreakable security, to deliver encryption codes. In the future, quantum communications might enable us to build absolutely secure networks.

Since today more and more people increasingly rely on cloud storage, it is easy to imagine how important it is to provide the required security level in that "cloud," that is, in the data centers.

Communication security is a crucial aspect of communication process; as such, it is a big area of research, development, and implementation. Unhappily, we have to leave further consideration of this captivating topic to specialized literature.

Today's networks mainly operate merely as transport arteries, similar to highways and local roads, simply transferring the communication traffic. The trend in their development is to make them smarter and capable of dynamically reacting to changes. This trend is implemented in SDN.

In the future, a communication network will not only provide connectivity but also implement contextual processing and intelligible interfaces. Nevertheless, most of the fundamental laws and principles will still be in use because after all data still has to physically travel between the nodes of networks.

1.2.1.2 The Internet: From a Point-to-Point Link to a Network of Networks

What is the Internet? The Internet is a communication network. It is not simply a network, however. *The Internet is a worldwide system of interconnected communication networks*; it interlinks any types of local, regional, national, or international networks enabling any user in the world to reach any other user. This is why the *Internet is called a network of networks.*

The History of the Internet The Internet, as we know it today, was created by many prominent scientists and engineers supported by public organizations and private businesses. The history of the Internet is a fascinating story of collaboration and competition, generosity in sharing knowledge and bitter struggles for monopolizing concepts and technology. In the absence of serious scientific research, it is impossible to list the names of even the most outstanding contributors without missing someone or wrongly attributing a certain achievement. In short, history of the Internet is a big subject in its own right and we can afford only a brief overview focused on the main Internet features. (To start, see the article written by the group of founders of the Internet in 1997.[7] The list of (arguably) the main events of Internet history in chronological order, which includes names of principal contributors, can be found on the website.[8]).

---

7 https://www.internetsociety.org/resources/doc/2017/brief-history-internet/.
8 https://www.explainthatstuff.com/internet.html.

Development of the Internet started in 1950s and early 1960s, when the advent of electronic computers immediately caused the need to share information produced and stored on one computer with another computer. Originally, information could only be transferred by carrying the physical memory media such as magnetic tapes or even punch cards. To make this transfer automatic, the computers had to be physically connected. But even installing an electronic or an optical link could not make data transmission possible. Why? Computers had to "understand" each other; in other words, they had to obey the same communication rules, that is, protocols. So these first attempts of creating a communication network immediately demonstrated that the network had to have both physical and intelligent (logical) levels.

After both physical and logical connections were developed, in 1969, the first long-haul transmission link between computer laboratories of the University of California, Los Angeles (UCLA), and Stanford University was established. This event is considered as the date of birth of the Internet.

Extensive research, many failed attempts and breakthrough discoveries preceded and followed this epochal event. One of the decisive moments was, of course, the development of theory and practice of a *packet-switching* approach, and the construction of a *packet-switching* network. (American scientist, Dr Leonard Kleinrock is credited with developing the mathematical foundation of the concept of packet-switching.) The need for packet-switching became apparent after the first attempts of using telephone lines for data transmission demonstrated the inadequacy of this approach. What is more, the development of a packet-switching network had been urged by military demand caused by the network's ability to support communications even if a significant part of the network were destroyed – a vitally important trait for defense applications. (Since these first developments were supported by DARPA, the research agency of the US Department of Defense, when the Cold War [confrontation between the USSR and the USA] was at its peak, many considered the development of the Internet a product of military demand for a highly resilient communication network.)

As more and more individual computers and local computer networks with a variety of software joined the main network, the need for the universal protocol became urgent. Two American scientists, Robert Kahn and Vinton Cerf developed the basis of the Internet protocol suite known as *TCP/IP*. This protocol suite was adapted in 1983; from that date, assembling the global communication network, which is now the Internet, proceeded at an accelerated pace.

The next milestone happened in 1990 when British scientist *Tim Berners-Lee*, then at the European Organization for Nuclear Research (CERN), developed the *World Wide Web, WWW*. The Web is an information space where all knowledge is codified and organized so as to make any particular entry readily accessible. In WWW, all resources, organized into web pages, are identified by Uniform Resource Locators, or URLs. (https://www.history.com/news/who-invented-the-internet is an example of a URL.) The search for information is done by using a *browser*; almost everyone is familiar with such browsers as Google Chrome, Internet Explorer, and Firefox.

One of the most important features of WWW is *hypertext*, the text enabling access to another text or web page connected by a *hyperlink*. The hypertext appears either underlined or in a different color. A click on a hypertext results in an instant transfer to the web page described by the hypertext thanks to *hypertext transfer protocol (HTTP)*. In WWW many pages are written in *Hypertext Markup Language (HTML)*. It is WWW that makes the Internet such a useful tool for accessing and exchanging any kind of information.

> *Today, the Web and the Internet are associated so closely that we often confuse one with the other. Let it be clear that the* Internet *is the global communication* network *and the* World Wide Web *is an* information space.

Much more new developments have occurred over the past years; these developments have made the Internet and the Web more convenient and efficient tools. Nevertheless, the most fundamental principles laid out by the Internet pioneers still work well, regardless of all the advances in technology. From this standpoint, it is interesting to revisit the ground rules set by one of the developers of the TCP/IP suite, Dr Robert Kahn:

- "Each distinct network would have to stand on its own and no internal changes could be required to any such network to connect it to the Internet.
- Communications would be on a best effort basis. If a packet didn't make it to the final destination, it would shortly be retransmitted from the source.
- Black boxes would be used to connect the networks; these would later be called *gateways* and *routers*. There would be no information retained by the gateways about the individual flows of packets passing through them, thereby keeping them simple and avoiding complicated adaptation and recovery from various failure modes.
- There would be no global control at the operations level."

It is amazing that all these principles still govern the Internet.

The Internet Operation How does the Internet work? The basic parts that make up any communication network in general and the Internet in particular are its tangible components, which are designated as *hardware*. The list of the Internet hardware starts with our computers, tablets, smartphones, and other *user devices* that convert logical (intelligent) information into electrical or optical signals. To transmit these signals, we need *transmission links*. As mentioned in the *Networks* subsection, they are implemented in copper or fiber-optic cables or can be simply air for wireless transmission. In addition to providing access to the Internet, the transmission links also connect all *nodes* of this network. The Internet nodes are *routers* that direct and redirect the packets carrying messages and *servers* (specialized computers) that store and process the information.

Remember, the Internet is simply a name for the assemblage of all connected networks. This means that the Internet relies on existing transmission links and other hardware of those participating networks. Without its parts – the worldwide optical network, the innumerable set of wireless transmission systems, and the global satellite communication complex – the whole entity (the Internet, that is) would not exist. But it is important to note that certain hardware, such as routers, were built specifically to support the Internet operation.

As discussed in the *Networks* subsection, the network is a symbiosis of two sides – tangible and intelligent (logical). The intelligent side is especially important for the Internet because it must make the variety of all worldwide networks work together and this task seems to be impossible to achieve. The Internet, however, does operate successfully; it does so thanks to the *TCP/IP protocol suite*, as mentioned above.

Why do we need the suite of protocols for the Internet? Imagine you work in the United States and your friend or business partner works in another part of the globe, for instance, in Germany. Your computer and the targeted computer are built with different technologies, they have different (possibly proprietary) software, and you are typing your messages in different languages. To transfer your messages through the Internet, numerous operations must be performed, and each of them must obey a certain rule (protocol, that is). The combination of all these rules combined constitute the *TCP/IP suite*. This is why we refer to the *suite of protocols,* as both TCP and IP include many individual protocols, such as *hypertext transfer protocol* (http) or *simple mail transfer protocol* (smtp).

The main principle underpinning Internet transmission is mentioned above among the TCP/IP ground rules as the "black boxes" rule. At its most basic level, the rule says that the Internet simply transfers packets to their destination points without looking into the packets' contents; that is, regardless of whether these packets contain e-mails, or Web pages, or Skype images, or anything else. (In this regard, the Internet operates similarly to the mail system that delivers letters and parcels without knowing what is inside those envelopes or boxes.) All data transmission in the Internet is governed by the suite of *Transmission Control Protocol (TCP)*.

The example of how TCP administers data transmission is the *flow control* that prevents a receiver from being overwhelmed with traffic. The task is to make a sender transmit data at such a rate that a receiver can reliably accept and process incoming packets. The data (bit) rate is the number of bits per second; implementing the flow control is done by changing the number of bits being sent at once. This number is called *flow control window size*. In reality, this control is a process of exchanging information between a transmitter and a receiver: The transmitter sends a packet – whose size is equal to a flow control window – and a receiver sends back acknowledgment and information of how many bits it can accept at the next step. Thus, the duration of every cycle of this communication is equal to the round-trip time between two end devices. The number of bits transmitted for that time is equal to window size plus acknowledgement, whose size is negligible. Therefore, the bit rate is equal to

$$\text{Bit rate (b/s)} = \text{Window size (b)}/\text{Round-trip time (s)} \tag{1.2.1}$$

By changing the window size, end devices ensure that communication occurs at a desired rate.

What is the role of *the Internet Protocol, IP*? It is an addressing system. *Every user device connected to the Internet is identified by its IP address*, as mentioned above. You certainly hear about IPv4 and IPv6, which refers to the IP Version 4 and IP Version 6 addressing systems. The Internet founders developed IPv4 based on a 32-bit address; this system can accommodate 2^{32} or approximately 4.3 billion of addresses. (Here is an example of *IPv4 address* in decimal format: 59.48.136.279. You might want to jump to Appendix 3.A.1 to learn about binary and decimal number systems.) When the Internet started to expand at an accelerating rate, it became apparent that this system would soon be exhausted. In 1991, the Internet Engineering Task Force (IETF) – the open international community of network designers, operators, vendors, and researchers dedicated to developing open standards – decided to develop a new system, IPv6, that uses a 128-bit long address. This address is arranged in eight 16-bit groups. Each group is presented as four *hexadecimal* digits, and the groups are separated by colons. (An example of a *IPv6 address* in decimal format is[9]: **FE80:BD00:0000:0CDE:1257:0000:211E:729C**.)

With IPv6, we will have 2^{128} or approximately 10^{38} the Internet addresses. That should be enough for the foreseeable future, even after adopting such systems as the IoT (see Section 1.1 on the IoT). In addition, IPv6 simplifies the use of the addresses and has a number of other advantages over IPv4. In 2017, IPv6 became an Internet standard; it does not mean, however, that this system becomes mandatory. There is no deadline for transition from IPv4 to IPv6; this is a long process because it requires significant changes in both hardware and software of the Internet network.

Finally, let us briefly consider how *data is transmitted through the Internet*. This is a big topic in its own right; its thorough discussion lies outside the scope of this textbook. For this reason, we will skip the discussion of logical side of data transmission related to the four-layer TCP/IP model, detailed discussion of the IP addressing system, thorough examination of the structure of Internet packets, etc. We will simply review this process in general terms.

---

9 https://internetofthingsagenda.techtarget.com/definition/IPv6-address.

As introduced above, the transmission process starts at end devices, including computers, tablets, smartphones, various sensors, cameras, and similar gadgets. From the logical view, these devices generate messages (files); from the hardware view, these messages are carried by electrical or optical signals.

As mentioned several times previously, these files (messages) are disassembled into individual packets for transmission through the Internet. In the IPv6 system, each packet can carry a 64-kbyte payload and a 40-byte header. (To recall, 1 byte contains 8 bits.) This string of bits starts with a *header*, the packet's section describing its source and destination addresses. The header also contains other information regarding the transmission conditions, such as traffic class (transmission priority), payload length (the size of actual data), hop limit (the maximum number of network hops), and, of course, the next header (ID of the header following the given packet). It is TCP that controls both the disassembly of a message into packets at the transmitter end and the reassembly of the packets into the original message at the destination point.

Table 1.2.1 briefly describes the concept of TCP/IP application for packet transmission through a router.

Now, consider Figure 1.2.7 that conceptually depicts data transmission through the Internet. A transmitter sends the request for e-mail, web search, or chat communication wirelessly or through the wireline to the *ISP*. (Term *ISP* denotes both the organization or company providing access to the Internet and other associated services; it also designates the equipment used to technically implement this access and services.) The ISP router sends the request to a *DNS*. Why? When we want to send an e-mail or reach a certain website, we request access by typing in the address, for example, as http://www.something.com or some.body@domain-name.net. We have learned, however, that all Internet end users are identified by their IP addresses. How, then, can our web search or e-mail reach a destination point? With the help of the DNS. This server translates our request into IP addresses; more accurately, for each written request, it finds a corresponding IP address. After finding the required IP address, the DNS forwards our packets to a proper domain, where *core routers* continue to direct them to the destination point. The packets, as we know, travel by different

Table 1.2.1 Conceptual description of routing operation.

Event	Physical operation	Logical operation	Basis
A packet arrives	Accepting the packet	Reading a header	TCP
		Finding the forwarding direction (required domain) by comparing the destination address with the routing table	TCP
		Finding the best route to forward the packet by reviewing the information about immediate neighboring routes and then throughout the network	IP routing protocols, such as OSPF (for IPv4 only) and EIGRP
The packet forwarded to the next router through the best route	Sending a packet		

OSFP, Open Shortest Path First and EIGRP, Enhanced Interior Gateway Routing Protocol.

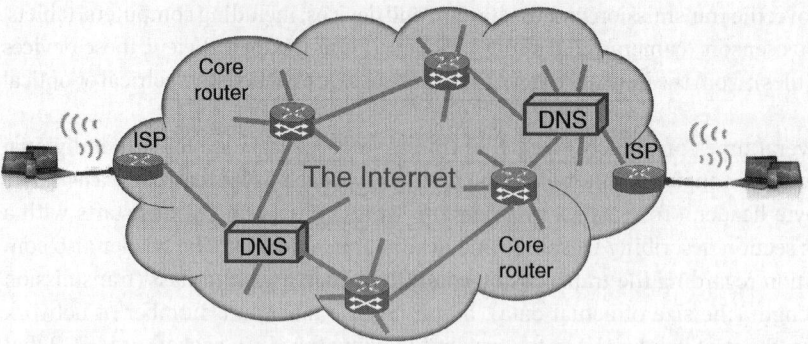

Figure 1.2.7 Conceptual view at data transmission through the Internet.

paths and might reach the destination at different times. The difference, of course, is measured by no more than milliseconds for e-mails and seconds for huge files.

What if a packet or several packets belonging to one file get lost or dropped or corrupted? Since information about the order of packets within a specific file and the total number of packets is carried by each packet, the destination device can check the status of the file and request retransmission of the needed packets. This approach is the implementation of the other ground rule: "Communications would be on a best effort basis. If a packet didn't make it to the final destination, it would shortly be retransmitted from the source." In fact, this rule is another basis for making the Internet a very effective communication network: A retransmission of a couple of packets is much easier and faster than the retransmission of an entire file.

(**Question**: *Waiting for acknowledgments and resending dropped packets take time and delay the throughput message transmission. Do we see this delay in our everyday work with the Internet?*)

Regarding routers and servers, which are considered previously, it is worth mentioning that there are several levels of these devices. For example, there are ISP routers and core routers, as Figure 1.2.7 shows. Also, it should be clear that routers logically interconnect different networks, which enables them at every instant to choose the best path for the transmission of a current packet. This arrangement allows for the fastest transmission of an entire file because its pieces travel through the Internet by different paths simultaneously.

It is worth mentioning that, thanks to the intelligibility of the Internet transmission, we can add new applications without restructuring the Internet. If we wanted to develop, for example, YouTube, we would need to convert the images and sounds into digital signals, break these streams of bits into packets, transmit these packets through the Internet, reassemble them, and convert the received streams of bits back into images and sounds. This is it. No changes in the network hardware or software are needed. This is why new applications appear on the Internet almost every other day.

To sum up, this brief overview should give you an idea about how the Internet operates; more importantly, it should inspire you to continue to learn about this technological marvel, which decisively shapes our life today and will continue to do so for the foreseeable future.

1.2.2 Optical Communications

Optical (fiber-optic) communications is the backbone of modern communications infrastructure. Vast majority of global communication traffic is delivered by optical networks. All the features of modern communications that transform our personal, social, and professional life have been brought in by optical communications. It is astonishing that the vital role of optical communications in our life is mainly unknown to the general public and very few people know what it is.

This short review of optical communications can provide only a cursory survey of the field and its role in modern communications; you are urged to learn more about this compelling field from specialized literature. (See subsection *Optical Communications* in the *Bibliography* section of this textbook.) We start with an introduction to the principles of operation of an optical communication system and the short historical notes.

1.2.2.1 Introduction to Optical Communications

The basic block diagram of a fiber-optic communication system is shown in Figure 1.2.8. A message in electronic format enters a transmitter whose electronics process the signal and convert it into a serial stream of pulses. This stream drives a laser diode (LD) that emits optical pulses carrying the submitted message. The train of optical pulses travels through an optical fiber, which is a conduit of light signal. At the receiver end, the optical signal falls on a photodiode (PD) that converts light into electrical current containing the received message. The receiver's electronics process the electrical signal and present the result to a user. This optical PPT link is a building block of an optical communication network. (Additional discussion of optical communications is in Section 10.1.)

Why do we need to convert electrical signals to optical? Why has it been worth to build the whole new sector of communication industry that has replaced the traditional copper-based transmission systems? It is because optical communication networks can provide the highest possible transmission rate, that is the number of bits transmitted per one second. The physics underlying this statement is that the transmission rate of any communication system is directly proportional to the frequency of an information-carrying wave, and the optical wave has the highest frequency that can be utilized in practice. Modern optical communication networks are capable of delivering information at a mind-boggling rate of 10^{15} b/s (petabits per second, Pb/s) and still have potential for the further increases. To put this rate in perspective, realize that at 1 Pb/s transmission rate it is possible to send 5000 two-hour long HDTV videos in a single second. This huge transmission rate is a key enabler of hundreds of new online applications. (**Exercise:** List all online applications you use and know. For example, streaming video, Google search, Facebook, Instagram, and WhatsApp.)

And how does optical communications eliminate the problem of distance in modern communications? Optical fiber spans the entire globe, and optical signals travel inside an optical fiber at the velocity close to the speed of light in vacuum. Combining this propagation speed with the enormous transmission rate, optical communications is capable of delivering tremendous amount of information to the most remote corner of the world almost instantly.

*(**Question**: The text says that optical communications eliminates the problem of distance in modern communications. How important is this fact for the functioning of the Internet operation? Explain.)*

Figure 1.2.8 Basic block diagram of a fiber-optic communication system.

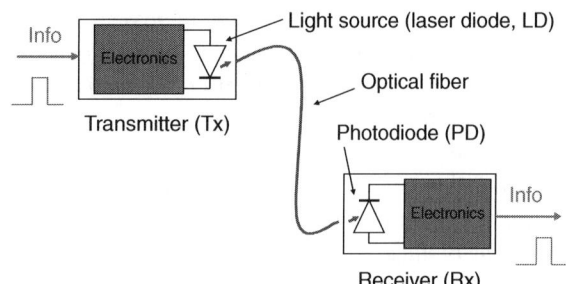

But optical communications is responsible not only for long-distance transmissions. In fact, the scale of operation of optical communication systems range from micro- and even nanometers in optical interconnects to several kilometers in access networks and data centers to thousands of kilometers in transcontinental links (see, for example, Mynbaev 2010, pp. 23–42.). Figure 1.2.9 shows that all continents are interconnected by the massive global network of undersea (submarine) cables. You can easily imagine (or better, find it online) the map of terrestrial cables covering the territory of the United States or any other developed country.

What parameter determines the transmission distance in any type of communications? First and foremost, the *loss* introduced by transmission medium. The loss in decibels (dB) of an optical fiber is given by

$$\text{Loss}_{OF} \text{ (dB)} = 10 \log_{10} \left(\frac{P_{\text{out (W)}}}{P_{\text{in (W)}}} \right) \tag{1.2.2}$$

where P_{in} (W) and P_{out} (W) are the powers of signals launched into and emerged out of the optical fiber, respectively. This formula, however, does not include the length of the optical fiber, whereas we know that the greater the length, the greater the loss. To resolve this issue, communication industry assesses the quality of a transmission medium not by its losses but by *attenuation*, which is the loss per unit of length. Thus, attenuation of an optical fiber, A_{OF} (dB/km), can be calculated as

$$A_{OF} \text{ (dB/km)} = \frac{-\text{Loss(dB)}}{\text{Length (km)}} \tag{1.2.3}$$

For example, if the input and output powers are $P_{\text{in}} = 1\,\text{mW}$ and $P_{\text{out}} = 1\,\mu\text{W}$, and there is Length = 100 km, then $A_{OF} = 0.3\,\text{dB/km}$. These numbers are typical for modern optical communications. They show that the present-day optical fiber operates with very-low-power signals and it can transmit these signals over 100 km without amplification. (Of course, for delivering optical signals around the globe, amplification is used.) Attenuation is a universal measure of loss-related property of a transmission medium; it allows us to make comparisons not only between the different optical fibers but also between the transmission media of different nature. (You can jump to Figure 1.3.2 to see such a comparison.) You should be aware that communication industry uses terms *loss* and *attenuation* interchangeably despite their different meanings; thus, bear that in mind when reading the industry documents.

The second important characteristic of a transmission medium is its *bandwidth*, BW (Hz). This is the band of frequencies within which transmission is restricted due to various factors, the main of which is usually the acceptable value of attenuation. The optical fiber designed for long-distance transmission (*SMF*) has the widest bandwidth (about 8 THz) among the two other types of wired transmission media: twisted pair of copper wires and coaxial cable. The optical fiber used for short distances (*multimode fiber*) has a smaller bandwidth, which limits its transmission distance. Nonetheless, since this fiber is now widely employed in data centers, its importance in modern optical communications has greatly increased. This stimulates intensive and productive industry efforts to improve the overall quality of multimode fiber, bandwidth including. We will discuss these topics – losses and bandwidth of transmission media – in succeeding chapters; specifically, Section 10.1 might be especially helpful.

(**Question**: *An SMF, a singlemode optical fiber used for long-distance transmissions, has lower attenuation and higher bandwidth than a multimode optical fiber (MMF) employed in short-distance networks. It seems natural to use the better optical fiber in all applications, but optical communication industry still uses MMF. Why? Hint: Refer to the discussion of network hierarchy in the preceding subsection.*)

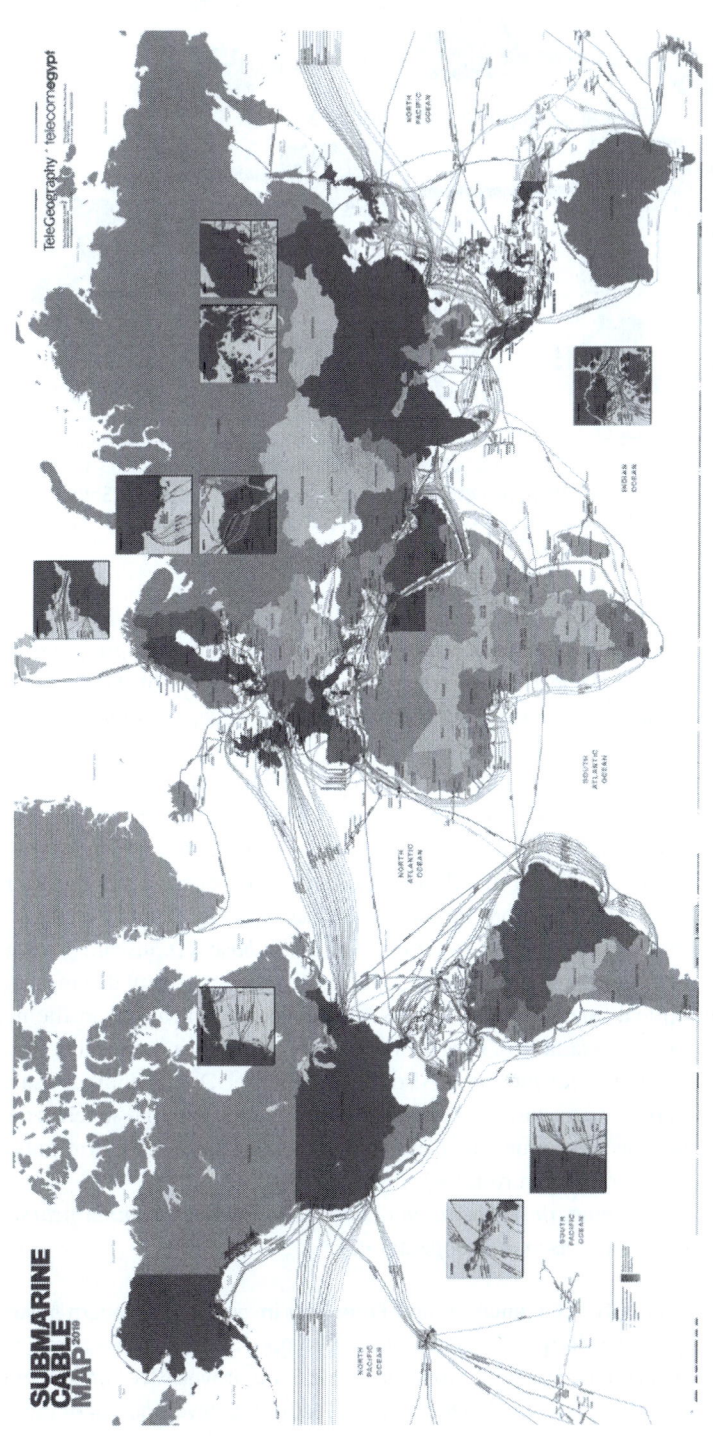

Figure 1.2.9 Map of undersea (submarine) fiber-optic cables. Source: Reproduced with permission of ©TeleGeography. www.telegeography.com.

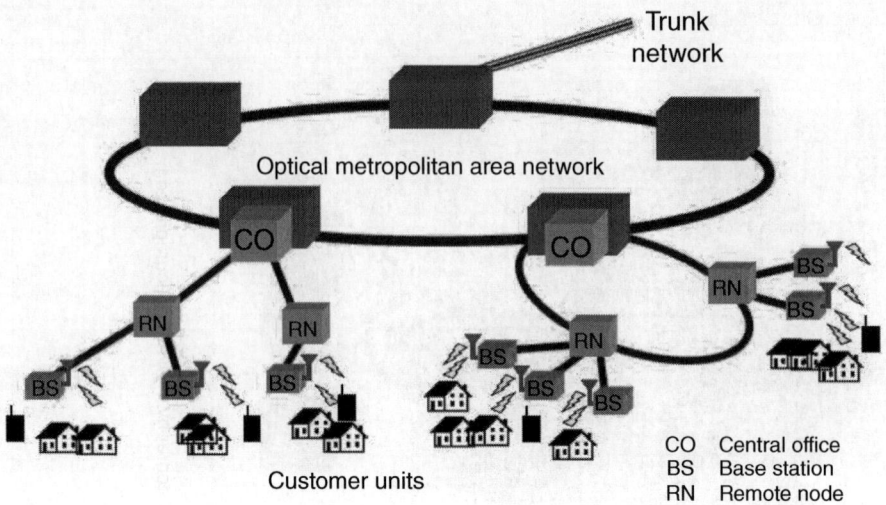

Figure 1.2.10 Example of interconnections between wireless and optical networks. Source: Reprinted with permission of Lim et al. (2010) "Fiber-Wireless Networks and Subsystem Technologies." *Journal of Lightwave Technology*. Vol. 28, No. 4. February 15, 2010. Pages 390–405. © 2010 IEEE.

Returning to general discussion, optical communications has adapted the *latest developments in theory and practice of communication networking*, such as SDN architecture, programmable optical switches, etc. What is more, optical communications stimulates many of the latest advances by demanding from photonics, electronics, computers, material science, and other related industries better hardware, software, and firmware products.

Wait a minute, you might say, reading this text, I have an impression that optical communications is the only type of communications we have, but when we use our smartphones to view the news online, there is no optical fiber in sight. Yes, are right. Wireless communications become more and more important as the demand for mobile and fixed short-range connections grow. But bear in mind that terrestrial wireless connections for mobile devices typically extend over short distances; then, the radio-frequency signals reach the antennas at nearest *base stations* and are transferred to optical networks. In other words, information destined to travel a significant distance is eventually transmitted through optical fiber. (We will discuss wireless communications in the next subsection.) Figure 1.2.10 shows an example of interconnections of wireless and optical networks, where fiber-optic cables are shown as *n* solid lines. Also, review Figure 1.1.3 as another example of the linkage of wireless and fiber-optic communications. Both figures show that wireless communications provide short-range access connections and optical networks transmit information over the distances. (It helps to re-examine Figure 1.2.4.)

(***Question***: *Figure 1.2.10 shows the cooperation of optical and wireless communications. What is the role of satellite communications in this cooperation?*)

1.2.2.2 Developments in Optical Communications: From First Inventions to Modern Advances

In 1955, Narinder Kapany, working in England, developed optical fiber with a core and cladding. Dr Charles Kao, also working in England, proposed in 1966 the provocative concept of using optical fiber for communications. (For his pioneering work leading to development of modern optical communications, Charles Kao was awarded Nobel Prize in Physics in 2009.) Taking up the challenge, Corning Corporation developed the process of manufacturing optical fiber in 1970 with attenuation below 20 dB/km. The era of optical communications had arrived. (Interestingly, today's commercial optical fiber has attenuation slightly less than 0.2 dB/km, but this number has not practically changed over last 30 years.)

Development of a low-attenuation optical fiber was, of course, the main trigger for explosive growth of fiber-optic communications, but it became immediately clear that an optical communication system is a complex structure whose all parts must work in concert.

First component needed for building this system was a proper light source. Initially, it was a light emitting diode (LED); its main advantages were small size and relatively low consumption of electric power. But the only suitable type of optical fiber for long-distance transmission was (and still is) a SMF whose inner (core) diameter was smaller than 10 μm. No LED could generate a light beam with such a directivity; in other words, it was possible to launch only a small portion of an LED entire beam. Given that the LED's entire output power was inherently low, the launched light beam diminished over a short transmission distance. In addition to power problem, an LED could not support a transmission rate even in the order of hundreds of megabits per second, let alone much anticipated gigabits per second. And yet another LED problem was that its beam contained many wavelengths, which put further limits on transmission rate and distance. Thus, using an LED as a light source, the optical communication systems could not realize its main advantage – high-bit-rate transmission over long distance.

Reading these explanations, you may naturally want to ask, why not use a laser? Among all existing types of laser, only laser diodes – thanks to their small sizes and potentially low-power consumption – could be suitable light source for optical communications. Unhappily, the first generation of laser diodes could operate only under cryogenic (very low) temperature and consumed tremendous amount of electric power. So the task was to create a small diode capable of operating at room temperature, consume little electric power, and generate a laser beam with all its advantages, such as big optical power, high directivity, and few wavelengths. Fortunately, such a laser diode had been created in research laboratories by the time when the need arose. The needs of optical communications industry accelerated the process of commercialization of these laser diodes. Today, various types of laser diodes are the only light sources for optical communication systems; they support transmission at all the required distances from meters in data centers to thousands of kilometers in intercontinental networks. In 2000, two scientists, Zhores I. Alferov from Russia and Herbert Kroemer from the United States were awarded Nobel Prize for their contribution into developing modern laser diodes.

Besides optical fiber and laser diode, optical communication networks include many other components, whose invention and development can readily be a topic of another historical review. We will just touch two of them: optical amplifier and WDM.

The modern era of worldwide telecommunications can be traced back to the installation of optical submarine networks. The first such cable was laid between England and Belgium in 1986. The initial transatlantic fiber-optic cable, called TAT-8, was laid two years later. To boost optical signals, this cable employed electronic repeaters installed every 50 km; these repeaters performed optical-to-electrical and electrical-to-optical (OEO) conversions – a costly operation that slows the speed of transmission. What is more, these electronic repeaters consumed tremendous amount of electric power. On the positive side, this cable can support 40 000 voice conversation, 10 times more than the preceding transatlantic coaxial cable. TAT-8 was taken out of service in 2002 because the many new optical cables installed proved far more efficient and less costly to operate.

Why new cables were much better than TAT-8? Mainly thanks to replacement of electronic repeaters by *optical amplifiers* that amplified optical signals without OEO conversions. The advent of optical amplifiers not only dramatically increased efficiency and reduced cost of optical networks, but it also enabled deployment of new technology – WDM.

In communications, the multiplexing means transmitting several signals through one link. To avoid interference among these signals, they must differ in one of their properties, such as transmission time slot, frequency, or space (see Appendix 10.A). The WDM is, in essence, a frequency-division multiplexing, but since this multiplexing is done in optical domain, the industry prefers to use *wavelength*.

Be aware that wavelength, λ (m), and frequency, f (Hz), are related through the famous equation,

$$\lambda \cdot f = c \tag{1.2.4}$$

where $c = 3 \times 10^8$ m/s is the speed of light in vacuum. If one wavelength can transmit at 100 gigabits per second, Gb/s, then multiplexing 80 wavelengths, we transmit 8000 Gbs/s = 8 Tb/s, where Tb/s stands for terabits per second. The WDM has been a main driving force in incredible progress of optical communications for the last 20 years, the progress measured by an increase in transmission rates by 10^7 (10 million!) times.

There have been many other big and small advances in optical communication technology, along with laser diodes, optical amplifiers, and WDM. The result is that today the global optical network carries most of the world's traffic. The network is interconnected with the world-wide wireless and satellite communications systems providing access to the global network, national, regional, and local networks to deliver information to our fixed and mobile devices wherever we need it. In short, the optical network is the backbone of the Internet and the linchpin of the global communication infrastructure.

What does **the future hold for optical communications**? As indicated earlier, the industry has increased its data transmission rate by 10 million times within the last 25 years. It is easy to understand how many breakthrough improvements had to be made at all levels of fiber-optic communication technology from a smallest component to the entire global network in order to achieve such a result. However, the volume of global communication traffic continues to increase at an exponential rate. (It is interesting to note that this accelerating growth is stimulated by the availability of a tremendous transmission rate over any imaginable distance provided by optical communications. Thus, this relationship has created a feedback loop: The more transmission capacity is available, the more demand for new capacity appears.) To meet the future demands, optical communication industry must continue to progress in all aspects of its technology, starting with transmission rate. But the industry is now approaching fundamental limits in information-carrying capacity of optical fiber, so further advances would be more difficult to achieve (see, for example, Mynbaev 2016, pp. 1640010-1–1640010-21).

Optical communications is a huge industry; numerous books, innumerable journal and magazine articles, industrial documents, and online resources are devoted to this subject. (See the bibliography section devoted to optical communications.) This concise overview can give you only a glimpse on this mammoth field; the objective of this book and the restrictions on its volume prevent us from more rigorous presentation of this fascinating discipline. We will, however, refer to optical communications through the course of this textbook to support and exemplify our discussions of specific topics.

1.2.3 Wireless Communications

1.2.3.1 Introduction to Wireless Communications

The wireless communications is the second of the three pillars of modern communication industry; today, this sector is growing at the fastest pace. This growth is caused by the advent of ever-increasing new mobile applications and new technologies like the IoT. As a result of these new developments, the number of mobile devices will exponentially increase and reach billions in the very near future.

Wireless communications use radio-frequency (RF) waves. These waves occupy the wide range of the electromagnetic (EM) spectrum from 3 kHz to 300 GHz; they are further classified into individual categories, depending on their frequencies. Figure 1.2.11 demonstrates this spectrum and shows the standard designations for each band.

Figure 1.2.11 also shows the wavelengths corresponding to the given frequencies. Conversion between them can be done according to (1.2.4).

The wireless communication systems, as the name suggests, transmit information without wired connections. It performs this transmission by radiating electromagnetic (EM) waves from one point to another. Figure 1.2.12 shows the principle of operation of a wireless communication system.

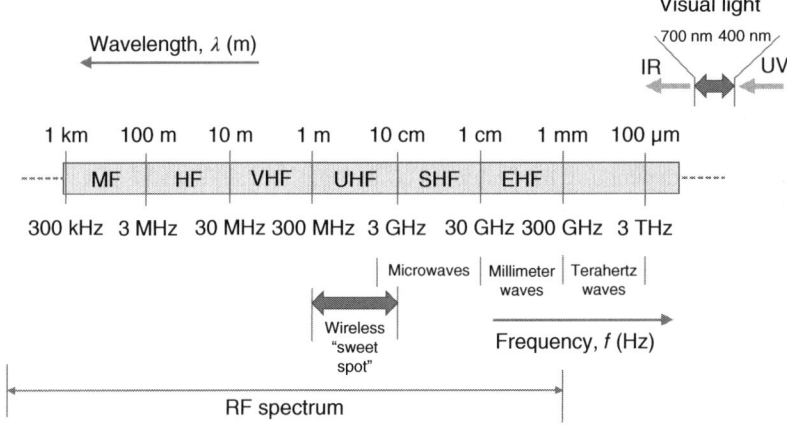

Figure 1.2.11 Electromagnetic (EM) spectrum. MF, medium frequency; HF, high frequency; VHF, very high frequency; UHF, ultra high frequency; SHF, super high frequency; EHF, extremely high frequency; IR, infrared; UV, ultraviolet. Not shown: ELF, extremely low frequency (3 Hz $< f <$ 30 Hz); SLF, super low frequency (30 Hz $< f <$ 300 Hz); ULF, ultra low frequency (300 Hz $< f <$ 3000 Hz); VLF, very low frequency (3 kHz $< f <$ 30 kHz), and LF, low frequency (30 kHz $< f <$ 300 kHz).

Figure 1.2.12 Principle of operation of a wireless communication system.

Information carried by a digital (or analog) signal arrives at the transmitting tower; here the signal is processed, amplified, and radiated by the antenna in form of RF waves. These waves reach the receiving antenna; at its tower these waves are processed, amplified, converted into an electrical or optical signal, and sent to a communication network through an optical fiber. Obviously, these end points can perform both transmitting and receiving tasks.

Figure 1.2.12 shows antennas installed on towers, which is typical for outdoor systems such as cellular phone networks; nonetheless, wireless communication systems successfully operate inside buildings, subway stations, and all other indoor locations. In these installations, antennas and processing electronics are parts of mobile or fixed devices.

The main point to draw from Figure 1.2.12 is that RF signal travels through air, regardless of the Tx and Rx locations. Utilizing **air as a transmission medium** is, on the one hand, a great advantage because such a link is always available everywhere and thanks to this feature wireless systems can provide mobile communications; on the other hand, this medium creates a problem because we cannot control air conditions. Indeed, in wireline communication systems, we can improve a coaxial cable or an optical fiber and can protect them from external noise, but in a wireless system, we can do nothing with air.

Depending on the antenna type and RF frequency, radiation patterns emitting from an antenna can vary from omnidirectional (all around, no directivity at all) to a light beam (totally directed radiation). Directivity is typically described by a *beam width*, which is the area where most of the radiated power is concentrated. *Half power beam width is the angle* at which relative beam's power is at more than 50% of its peak power. Beam width is a measure of radiation directivity; this is the characteristic of an *antenna*.

Wireless antennas exist in a variety of construction principles. For our purpose, it is important to know that the size of an antenna depends on the frequency (wavelength) of a radiated EM wave. To estimate the order of magnitude of an antenna's size, set it to half the wavelength and use (1.2.1) to compute the frequency. For example, for $f = 300\,\text{MHz}$, the *antenna size* is 0.5 m because λ is 1 m, as Equation (1.2.4) and Figure 1.2.11 show.

The RF waves can travel through air in three main *propagation modes* shown in Figure 1.2.13.

Ground Wave Propagation This propagation mode uses RF waves whose frequencies are up to 2 MHz. In their propagation, these waves follow the contour (curvature) of the Earth far beyond the horizon, as shown in Figure 1.2.13a. They travel this way because they induce current on the Earth's surface and experience diffraction effect due to the size of their wavelengths. (**Exercise**: Compute the wavelength of an RF wave whose frequency is 2 MHz.) Both phenomena tilt the wavefront of traveling RF waves toward the direction of propagation because the wave part that touches the earth surface travels slower than the part that travels through air. The best-known application of ground-wave communications is AM radio broadcasting; other applications include amateur radio, RFID, submarine communication, and long-range navigation. This mode of propagation can cover distances up to 100 km; the limit, of course, depends on atmospheric conditions.

Sky Wave Propagation In this mode, as Figure 1.2.13b shows, the RF waves bounce back and forth between ionosphere and the earth surface due to reflection. (The ionosphere is the layer of the Earth's atmosphere containing the charged particles, as its designation suggests.) This propagation type exists because the RF waves, whose frequencies range from 2 to 30 MHz, become more directional. Typical applications include amateur radio, citizens band (CB) or short-distance radio, international (long-range) broadcasting, military communications, and long-range aircrafts and

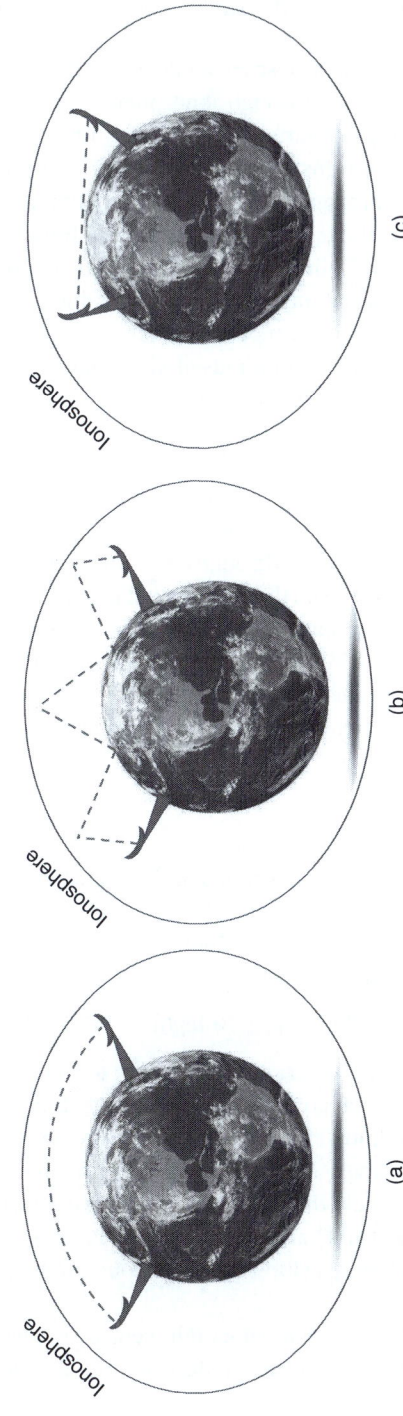

Figure 1.2.13 Propagation modes of EM waves for wireless communications. (a) Ground wave propagation, (b) sky wave propagation, and (c) line-of sight propagation.[10]

---

10 After Olenewa and Ciampa (2006), p. 144.

ships communications. Sky wave propagation can cover distances in the range of thousands of kilometers.

(**Question**: *International radio broadcasting is often called short-wavelength broadcasting. Why?*)

Line-of-Sight (LoS) Propagation The RF waves whose frequencies are higher than 30 MHz do not follow the earth's curvature and do penetrate through ionosphere; that is, they cannot support either ground wave or sky wave types of propagation. These waves travel in a line-of-sight propagation mode, as shown in Figure 1.2.13c. A transmitter and a receiver operating in this mode must "see" each other, that is the term for this propagation. Applications of the LoS propagation include cellular communications (starting with cellular phones, of course), FM broadcasting, TV broadcasting, personal communication services (PCS), fixed-wireless links, and all types of wireless optical communications. For these terrestrial applications, the LoS transmission can cover distances from meters to tens of kilometers. In addition, this is the mode of operation for satellite communications, where transmission distances vary from tens to hundreds of kilometers. Propagation distance for terrestrial outdoor LoS operation can be approximately estimated as

$$D\,(\mathrm{km}) \approx 3.57\sqrt{h}\,(\mathrm{m}) \tag{1.2.5}$$

where h (m) is antenna height. For example, if $h = 25$ m for both the Tx and Rx the antennas, then $D \approx 18$ km. For indoor installations, LoS operates at distances from meters to hundreds of meters.

Since wireless communications transmit signals through air, it would seem that the loss of signal power, and therefore the decrease in transmission distance, would first and foremost depend on the air quality. Yes, air condition is a big factor, particularly in outdoor transmission; however, there is a fundamental reason for power loss in wireless communications, regardless of the propagation mode and air quality. This reason is the *spread of an RF signal as it travels down the transmission distance*. Figure 1.2.12 conceptually depicts this spread; corresponding loss of signal power is called a *FSL*. FSL is given by

$$\mathrm{FSL} = \frac{P_t\,(\mathrm{W})}{P_r\,(\mathrm{W})} = \frac{(4\pi f d)^2}{c^2} = \frac{(4\pi d)^2}{\lambda^2} \tag{1.2.6}$$

Here P_t (W) and P_r (W) are the powers of the transmitted and received signal respectively, f (Hz) and λ (m) are the frequency and wavelength of the signal, d (m) is the transmission distance, and c (m/s) $= 3 \times 10^8$ m/s is the speed of light in vacuum. Communication industry measures loss in decibels, dB; thus, in reference to frequency we find

$$\mathrm{FSL\,(dB)} = P_t\,(\mathrm{dBm}) - P_r\,(\mathrm{dBm}) = 20\,\log(f) + 20\,\log(d) - 147.56\,\mathrm{dB} \tag{1.2.7}$$

Note that (1.2.6) calculates loss as the ratio of transmitted power to received power, whereas the definition of loss – the measure of how much power is lost due to travelling along the transmission path – requires the reverse order. Get to know that both definitions are acceptable. The loss definition given in (1.2.7) produces loss in decibels as a positive number; the inverse formula gives FSL (dB) as a negative number. Carefully review the loss definition in every case to avoid confusion. (**Exercise**: Take reasonable values of P_t (mW) and P_r (mW), select corresponding frequency and transmission distance for terrestrial LoS propagation type, and compute FSL in absolute numbers and in decibels. See Problem 48.)

Equation (1.2.6) shows that the FSL is determined simultaneously by two factors – signal frequency and transmission distance. This fact always underlies more detailed analysis of loss in wireless communications.

We need to firmly realize that (1.2.6) determines the loss of signal power only due to one yet fundamental reason – the spread of an EM beam. In addition to this, there are a number of other mechanisms causing signal loss. In particular, the same message can be simultaneously delivered by the RF waves traveling through various directions, which is called *multipath propagation*. The reasons for splitting one radiated signal into several transmitted signals include reflection of RF waves from such obstacles as buildings and mountains and their redirection, reflection, and refraction due to various atmospheric turbulences. (There are no waveguides in wireless communications, remember?) The losses and other detrimental effects caused by multipath propagation must be calculated by equations other than (1.2.6).

What problems do modern wireless communications meet? The main one is the **scarcity of spectrum**. Examining Figure 1.2.11 suggests that wireless communications can use any frequency from 300 kHz to 300 GHz, but such a superficial conclusion would be very far from reality. The fact is that the range of frequencies that can be employed (the usable spectrum, that is) is determined by a combination of factors, among which are air attenuation, transparency of obstacles, the size of an antenna, transmission distance, adequate bandwidth to support the required transmission rate, and – a very practical – availability of electronics to build cost-effective hardware. On top of these technical requirements, there are government regulations that are necessary to make use of the spectrum for the whole nation.

In general, spectrum is considered a national treasure just as land, water, and other natural resources. To prevent interference between users and to avoid chaotic use of the spectrum, every country controls its spectrum. At the global level, the International Telecommunications Union (ITU) supports worldwide regulatory activities through its sectors: Radiocommunications (ITU-R), Telecommunication Standardization (ITU-T), and Telecommunication Development (ITU-D). The ITU is the agency of the United Nations.

Control and management of the spectrum in the USA are executed by the *FCC* and the *National Telecommunications and Information Administration (NTIA)*. The FCC administers spectrum for non-Federal use (i.e. state, local government, commercial, private internal business, and personal use), whereas the NTIA administers spectrum for Federal use (e.g. the use by the military, the FAA, and the FBI, and other branches of Federal Government). For our discussion, we need to know that the licensing of spectrum bands for commercial wireless communications is provided by the FCC. (For complete information regarding spectrum allocation and licensed bands visit the FCC website at www.fcc.gov.)

At present time, wireless communications mostly use the spectrum from 300 MHz to 3 GHz due to a combination of all technological factors. This band is termed *sweet spot* or *beachfront spectrum* (see Figure 1.2.11). Within this band, some ranges are occupied more densely, others less densely, and some bands are reserved for future needs. The problem is that rapid growth of wireless communications requires more spectrum because of (i) the sheer increase in the number of users and expansion of the types of cell-phone services and (ii) the improvement in transmission quality. Today, it is difficult to find a person without a cellular phone in almost any country in the world. What is more, cellular phones become smarter and demand a variety of new services (think about an increase in the use of video), for which more spectrum (bandwidth) is required. Regarding the transmission quality, it suffices to refer to an increase in transmission rate (bit rate) in mobile communications that 5G and other new technologies promise, the increase which will also demand more bandwidth (spectrum).

Thus, on the one hand, the development of wireless communications requires more spectrum and, on the other hand, the spectrum is finite and its best bands are almost totally occupied (licensed). This contradiction threatens to limit further development of wireless communication. The industry is searching for new solutions to resolve this dilemma; the most promising is to use the ranges of frequencies higher than 3 GHz. However, two next bands that range from 3 to

300 GHz – microwaves and millimeter waves shown in Figure 1.2.11 – include regions where EM radiation experiences high losses. These regions are approximately determined by frequency bands of 21–25 GHz, 57–64 GHz, and 164–200 GHz. Also, air attenuation (loss in decibels per kilometer) tends to increase as frequency gets higher. (You can skip to Figure 1.3.2, where the spectral attenuation for all transmission media is shown.) Nevertheless, the use of this open spectrum is the only real opportunity for wireless communications to meet the future demand.

The use of new ranges of spectrum will open the possibility for development and deployment of **new types of wireless communications**. The one of them is, of course, the fifth generation of traditional, RF-based wireless communications called *5G*. It is loosely defined as the next generation of wireless communication technology with a high bit rate (up to 20 Gb/s) combined with enhancements to other transmission aspects; the term *5G* has been introduced by ITU. It is projected that 5G systems will use the bands from 600 MHz to 6 GHz and from 24 to 84 GHz. All major companies of the wireless communication industry are involved in developing this new technology, which creates diversity in the technical approaches and the competition in their commercialization. ITU plans for 2020 to be the launch year for commercial deployment of 5G. Since 5G dramatically increases the data rate in mobile communications, exchange of information between all the users will occur much faster than it does today. All 5G-enabled devices will communicate and react faster than a human being could, and therefore every task will appear to be performed instantly. The IoT, autonomous vehicles, health care system, education, sports, entertainment – all aspects of our life will dramatically change with the advent of new generation of wireless communications. 5G arguably is the most active research and development area in modern communications. Comprehensive information on the state of this supremely important work can be found, for example, on the website of the Institute of Electrical and Electronics Engineers (IEEE).

1.2.3.2 Contemporary Wireless Communications Technologies

As revolutionary as the 5G and other new wireless technologies are, they all stem from current wireless techniques. Let us consider the most popular modern wireless communications technologies.

The first technology to examine is, of course, **Wi-Fi** *(or WiFi)* whose principle of operation is shown in Figure 1.2.14. Signals from the Internet are delivered to a *Wi-Fi access point (AP)* through

Figure 1.2.14 Wi-Fi principle of operation.

a fiber-optic or coaxial cable. The AP can be a *router* or other similar device that distributes the accepted signals wirelessly to all the users. Figure 1.2.14 depicts a home Wi-Fi network; it shows that signals are transmitted from the Wi-Fi's AP to a computer, tablet, smartphone, a printer, a TV set, a video game console, and even to a 3D printer.

Wi-Fi networks mainly have star topology in which the AP is a hub; the end devices can, however, communicate with each other directly, provided that they are equipped with Wi-Fi adaptors. (Direct communications between devices are called *device-to-device, D2D, communications.*) Wi-Fi networks support two-way communications; they operate in LoS mode, and cover up to 100 m. To provide Internet connections in large buildings, airports, subways, and outdoors (think about smart cities), many APs (also called hot spots) are deployed.

Wi-Fi employs 2.4 and 5.8-GHz bands; the first band can only support data rate up to 10 Mb/s, and the second can provide up to 300 Mb/s but at shorter distances. The latest advances, such as *multiple-input multiple output (MIMO)* technique, increases the bit rate up to 3 Gb/s; the next step – moving the operation into 60-GHz band – promises to achieve 7-Gb/s transmission rate.

The big advantage of Wi-Fi is that it is inherently compatible with the TCP/IP suite of protocols, that is, with the Internet. Nevertheless, the software required for processing this form of communications is relatively big in size, complicated, and power-hungry. Power consumption is a problem for the mobile devices that run on batteries. The tremendous progress of modern electronics, however, alleviates all these issues, paving the path to further advances in Wi-Fi applications.

A word about the term: It is a common misconception to interpret Wi-Fi as a short name for *wireless fidelity*. In fact, Wi-Fi is simply a trademark of the Wi-Fi Alliance that manages the IEEE 802.11 standard.[11]

Wi-Fi is the most ubiquitous technology providing wireless connections to the Internet. It is so widespread that we often refer to Wi-Fi meaning the Internet connection. The exponentially increasing demand for high-speed wireless Internet connectivity from mobile and fixed devices (see discussion of the IoT, smart homes, and smart cities in Section 1.1) assures further advances in the Wi-Fi technology.

Today, Wi-Fi and some other wireless technologies (Wi-Max and LTE) use new approach called *MIMO*. A MIMO system includes several antennas at both transmitter and receiver sides. This architecture enables the system transmit one high-rate signal as a set of independent low-rate signals (data streams). Each receiving antenna accepts all low-rate signals but from different directions and at slightly different times because an individual data stream (a low-rate signal) travels slightly different path to the destination. Employing the capability of modern electronics to intelligently process received signals, a receiver recovers the original message from many data streams obtained. If M is the number of independent data streams, than the total channel capacity, C (b/s) is multiplied by M; that is, Shannon's law for MIMO is read as C (b/s) = $M \cdot BW \cdot \log_2(1 + SNR)$. There are many important details of this technology to be considered but we have to leave them for your independent study.

The next ubiquitous current wireless communication technology is **Bluetooth**, which provides connections between our personal devices. For example, most modern cars are equipped with Bluetooth technology enabling us to talk on the phones even when we are driving because the phone conversations become hands-free. Wireless headsets are the most common Bluetooth application. The new additions to our personal devices, such as wearable gadgets, increase the use and importance of Bluetooth in our everyday life.

---

11 https://www.webopedia.com/DidYouKnow/Computer_Science/wifi_explained.asp.

Bluetooth standards are controlled by the Bluetooth Special Interest Group (SIG); this industrial consortium also oversees the licensing of Bluetooth word mark and trademark.

Bluetooth operates in the 2.4-GHz range; it supports up to 3 Mb/s bit rate at a distance of 10 m. Bluetooth provides mainly peer-to-peer communications; that is, it works as a PPT link. It obviously transmits in LoS propagation mode and it can simultaneously support up to 10 devices in a star topology.

Rapid development of the IoT opens up a new area for Bluetooth applications, which, naturally, implies likely further advances in the Bluetooth technology.

Another wireless short-distance technology is called **ZigBee**, and its main application is communications in mesh networks connecting thousands of small devices. This technology predominantly operates in the 2.4-GHz spectrum band; it requires low power and supports low (up to 250 kb/s) bit rates. ZigBee devices can work on coin cell batteries for years; the latest versions of them are even provided with *energy harvesting* ability. It should be clear from this description that ZigBee is the best candidate for linking the IoT numerous sensors, but this network needs an external device working as the Internet gateway. The operational standard is maintained by ZigBee Alliance, whose website contains the detailed information about this technology.

There are a number of other important and interesting wireless technologies, which we leave for your self-study if your professional career will so require.

The next type of wireless technologies is called **optical wireless communications** because it uses the light spectrum for wireless communications. (Do not confuse the optical [fiber-optic] communications, whose signals are transmitted through an optical fiber [a light conduit], with optical wireless communications, which transmits its signals through air without cables. Also, be aware that term *optical* refers to the spectrum from infrared [µm] to ultraviolet [nm] wavelengths – see Figure 1.2.11.) Within optical wireless communications, we distinguish three technologies: *Free space optical (FSO) wireless communications, light fidelity (Li-Fi) communications*, and *visual light communications (VLC)*.

The **FSO** is the outdoor method of optical communications; its principle of operation is easy to understand by reviewing Figure 1.2.8, where an optical fiber must be substituted by air. Replacing the antennas by a laser Tx and a photodiode Rx on the top of the towers in Figure 1.2.12 gives the block diagram of an FSO communication system. A laser radiates in the infrared (IR) region to minimize the light beam absorption. (**Exercise**: Sketch the block diagram of a FSO system.)

The most popular FSO *terrestrial application is communications* between high buildings in big cities. For a corporation occupying two or more skyscrapers in downtown of New York City, for example, it is much easier and less expensive to connect their buildings by an FSO system than dig the city's streets and install fiber-optic cables. From the capacity standpoint, an FSO system can deliver information at the same bit rate as a regular fiber-optic system. In this application, however, an FSO suffers from two major drawbacks: First, weather condition can drastically affect the quality of transmission; rain and especially snow could even completely interrupt the transmission. Second, high buildings are in constant motion, and FSO's laser Tx and Rx must be dynamically aligned to stay in the line-of-sight.

The other important FSO application is *space communications* between satellites. Here, the transmission conditions are almost ideal and alignment between satellites Tx and Rx is not a problem.

Li-Fi *communications* utilizes the entire light spectrum from infrared (IR) to ultraviolet, UV (see Figure 1.2.11). The principle of Li-Fi operation is shown in Figure 1.2.15. Data stream modulates the driver of a LED or a laser diode (LD), which results in modulation of intensity of emitting light. This modulated light falls on a receiver's PD that converts it back to electrical signal. Thus, the original message reaches our laptops, tablets, or smartphones.

Figure 1.2.15 Principle of operation of a Li-Fi system.

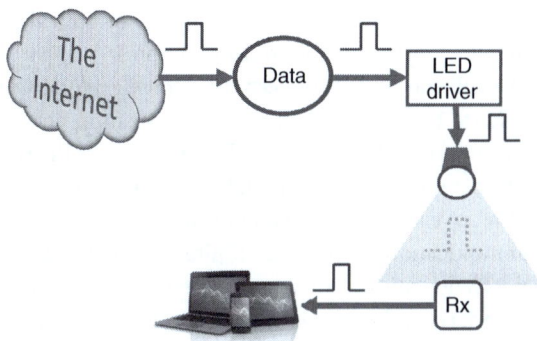

This communications system is called Li-Fi thanks to its similarity to Wi-Fi. The difference between these two lies in the nature of a signal carrier: Li-Fi utilizes a high-frequency optical carrier, whereas Wi-Fi relies on a RF carrier. This difference determines the advantages and disadvantages of both systems. Li-Fi supports much higher data rates; its transmission is less susceptible to interference and is much more secure. On the other hand, Li-Fi, as Figure 1.2.15 shows, provides one-way transmission, works only in the LoS mode, and covers much smaller area due to generating a very focused beam. Its transmission distance is limited by 10 m, and it is mainly deployed indoors.

The third optical wireless technology is **VLC**. This is basically a variant of Li-Fi whose carrier waves are restricted by visual-light frequencies. Its principle of operation is presented in Figure 1.2.15; the minor variation is that the LED is used simultaneously for both illumination and data transmission. In other words, the VLC takes advantage of trend to replace incandescent lamps by LEDs in factories, offices, and homes. Modulation of the LED light by high-frequency communication signal does not affect its illumination function because a human eye cannot follow the high-frequency variations in light intensity. (Refer to our personal experience: An incandescent lamp becomes totally dark 120 times during each second and we do not notice these changes.) VLC has the same advantages and disadvantages as Li-Fi, and the former is even more restricted in carrier frequencies and Tx's locations than the latter.

Optical wireless communications is an important segment of the wireless communications industry.

In our preceding discussions, we considered *PPT links*; today, with an exponentially increasing number of end users (devices), **wireless communications networks** start to play the main role in wireless communications. The IoT shown in Figure 1.1.2 is an example of such a network. When we use a router at our home to provide access to the Internet for many fixed and mobile devices, we, in fact, employ a wireless access network (see Figure 1.2.12). You can readily find many other examples that confirm the ubiquity of wireless networks around us. The format of this book prevents us from delving deeper into this topic but the overviews of networks and wireless communications given in this section should prepare you for further study of wireless networks if the need arises.

1.2.3.3 Mobile Cellular Communications

Discussion of wireless communications cannot be complete without a short review of *mobile cellular communications*. Today, billions of people carry their phones and other mobile devices in order to access the Internet while on the road, in train, in an aircraft, or aboard a ship. It is a common means of communication for local or cross-country communities, but they are also used for global connections by taking advantage of optical and satellite communications. (See preceding and following subsections of this section.)

We will concentrate on *terrestrial mobile cellular communications*. A *cellular network* is depicted in Figure 1.2.16A. The entire communication network blanketing a geographic region is divided into a number of cells; each cell covers part of the whole region. A cell tower called *BTS* is located in the center of each cell. (A BTS is sometimes called *base station, BS*.) It is the BTS that transmits and receives the electromagnetic signals to and from mobile devices within its cell. This wireless communications uses a very limited band of radio frequencies. Figure 1.2.16B shows an individual BTS operating at the band f_1 (left-hand side) and also the frequency assignments in a cellular network (right-hand side). The main feature of this arrangement, which enables *frequency reuse*,

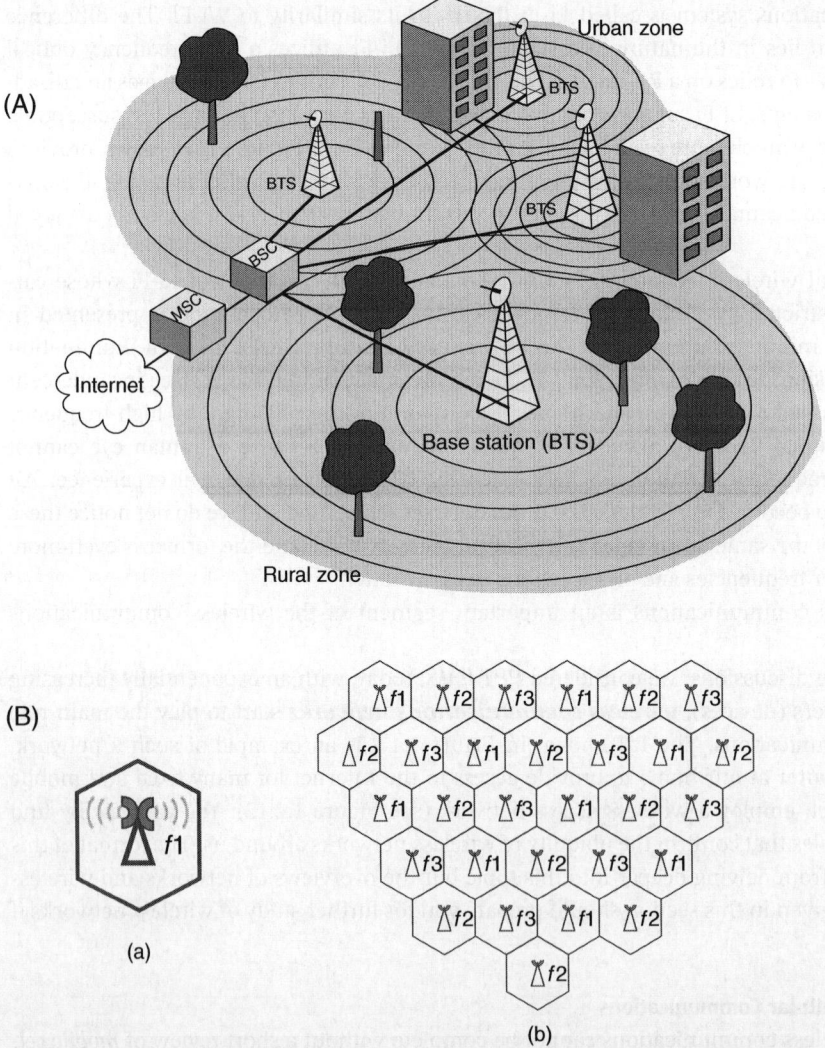

Figure 1.2.16 Cellular mobile networks: (A) General architecture; (B) an individual cell (a) and frequency assignment in a cell array (b). BTS, base transceiver station; BSC, base station controller; and MSC, mobile switching station. Source: (A) Reprinted with permission of http://www.thelifenetwork.org/about.html.

is that every cell has its own frequency band and none of the adjacent cells have the same band. Examine Figure 1.2.16B closely: It takes only three frequency bands, f_1, f_2, and f_3 to cover the array of 29 cells. This example shows that frequency reuse enables a finite number of frequencies to cover an almost infinite number of customers. Why is this a main feature of today's mobile communications? Without a cellular architecture, mobile communications would require assigning an individual frequency to each customer to avoid signal interferences. But then the number of frequencies (spectrum) available for wireless communications is strictly limited; therefore, the number of mobile phones would be limited too. Before the advent of cellular networks, mobile communications was called radio-communications and, indeed, every customer had been assigned an individual frequency. As you can guess, there were very few such persons in each area and, typically, they were either highly ranked public officials or very wealthy individuals. What a contrast to today's situation when the majority of world's population carry mobile phones.

Because of their proximity to a nearby antenna, mobile devices do not require high-power signals from a BTS. This is also very important feature because low-power BTS signals (on the order of 10 W) diminish when they reach the cell borders and, thus, do not interfere with frequencies of the adjacent cells. There are, of course, signals overlapping at the borders, as Figure 1.2.16A shows, but our smart phones "know" how to choose the right frequency. Another advantage of a cell's compact area and low-power operation is that it requires very little power from a mobile device for transmission, thus allowing the cell-phone battery to be so compact.

Today, to improve mobile connectivity, especially indoors, the hierarchical cell architecture takes further steps placing *microcells, picocells,* and *femtocells* inside a single cell in the indicated consecutive order. These mini-cellular base stations are designed to cover small areas and serve a restricted number of customers. Specifically, a smallest femtocell radiates about 0.1 W of power, has coverage radius of 60 ft, and serves less than ten customers. A picocell radiates 1 W, covers 750 ft, and serves 60–70 customers. A microcell, which can also operate outdoors, radiates 5 W, covers 1000 ft, and serves up to 200 users. These small cells not only improve connectivity but also add additional capacity in densely populated coverage areas.

Returning to Figure 1.2.16A, we notice that all BTSs are connected to a *base station controller (BSC)* which mediates the communications between BTSs and the *mobile switching station, MSC*. The MSC is a gateway of this cell array: It prepares the signals received from the cell array for transmission and sends them to the Internet; it also accepts information from the Internet, processes it, and transmits to the cells through a BSC.

All connections in the cell networks from BTSs to BSC to MSC and to the Internet utilize fiber-optic cables, shown in Figure 1.2.16A in solid lines.

Mobile technology rapidly evolves and new generations of smart phone appear on the market regularly, but this cell network architecture is still in use even though it was developed many years ago. This network architecture's stability is due to its optimal structure and good functionality.

1.2.4 Satellite Communications

1.2.4.1 Historical Notes

In 1957, the USSR launched the first satellite, Sputnik. In April 1961, Soviet cosmonaut Yuri Gagarin became the first man to orbit the Earth. The United States took up the challenge and started to develop its own space program. The first communication satellite was American Echo 1 launched into an orbit in 1960. It was a metallized balloon that worked as a mirror, passively reflecting signals from the earth back to another point on the earth's surface. Then, in 1962, the Telstar satellite relayed the first transatlantic television picture. The globalization of television had

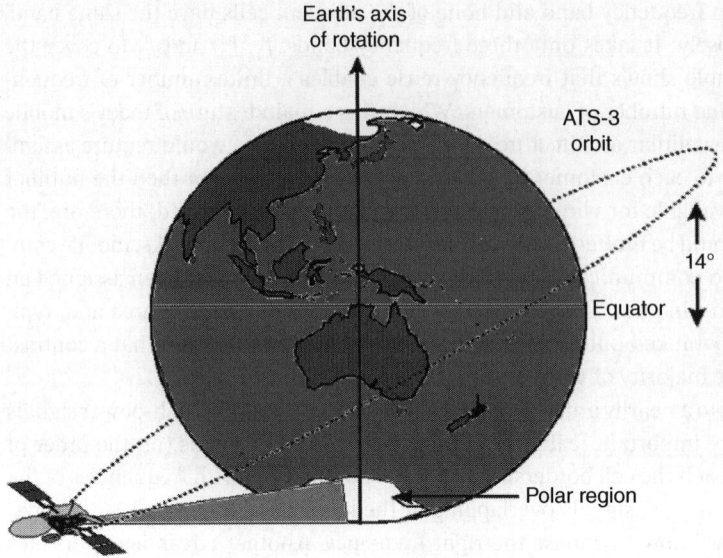

Figure 1.2.17 ATS-3 communication satellite orbiting Earth. Source: http://ctd.grc.nasa.gov.

begun. The Applications Technology Satellite (ATS-3) was launched five years later, providing live TV coverage of the 1968 Mexico City Olympics and obtaining the first color pictures of the entire disk of the earth. Figure 1.2.17 depicts the ATS-3 communications satellite orbiting the earth.

A number of interesting American projects, such as Iridium, Globalstar, Teledesic, and GTS, had launched over these years, making the satellite communications a vitally important component of our life. Many other countries joined the space race launching their satellites onto various orbits. Today, even private companies can afford to deliver satellites on geocentric orbits. In 2018, NASA counted 20 000 satellites orbiting our planet. Unfortunately, in addition to regular satellites purposely launched onto orbits, there are hundreds of thousands of small objects (debris, in fact) in space.

Remember, satellite communications enables us to reach any distant corner of the globe without installing a cable there. This unique property of satellite communication makes it the major player in international mobile communications. Today, hundreds of communication satellites are in orbit, providing truly global connectivity for citizens of the Earth.

1.2.4.2 Principle of Operation of Satellite Communication Systems

Communication satellites are spacecraft orbiting the earth. They serve as intermediate nodes (active mirrors) for redirecting the EM signals from one point on the earth's surface to another. The first communication satellite worked as a mirror, merely reflecting signals back to earth. Today, all communication satellites are the active spacecraft that receive signals from the ground, amplify them, change their frequencies, and transmit the signals back to the predetermined area on the earth's surface. To perform these tasks, a satellite carries appropriate equipment, including receiving and transmitting antennas. This equipment is powered by solar panels. Figure 1.2.18 presents the concept of satellite communications.

Transmission from an earth ground station to a satellite is called *uplink*; transmission from a satellite to another ground station is called *downlink*. Satellite's equipment that processes the received signal and transmits it back to earth is called *transponder*; this term combines words

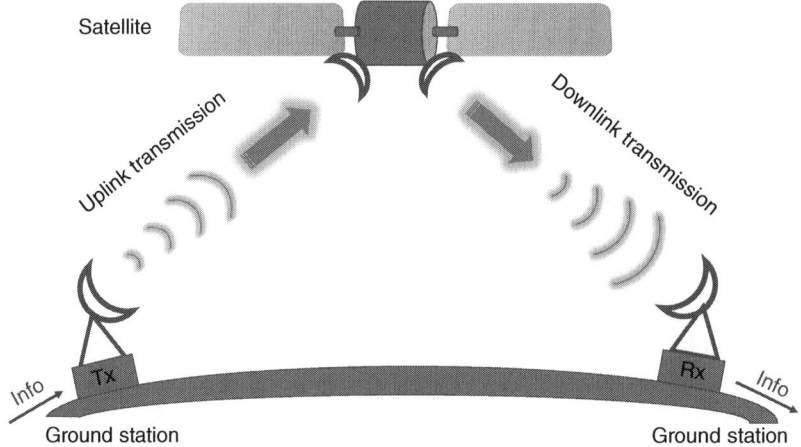

Figure 1.2.18 Satellite communication system.

transmitter and *responder* – two main blocks of the unit. A transponder connects satellite's receiving and transmitting antennas. It is a transponder that accepts the received signal, processes and amplifies it, changes the carrier frequency to avoid interference with internal and external signals, and transmits the signal back to another ground station. A transponder is a sophisticated module that must not only filter, clean of noise, and decipher a weak received signal but also produce a powerful (from tens to a hundred watts) output signal. Given that the distance between earth stations and satellites can reach 36 000 km and more, it is understandable that a big power output is needed to assure that the surface station receives a recognizable message. This is why a transponder is a power-hungry device; it is fed by energy constantly generated by solar panels, which are the necessary attributes of any communication satellite. And, again, the transponder has to preserve the received information while performing all of its operations.

Regarding the ***frequencies used in satellite communications***, they are concentrated in the microwave region of the EM spectrum (see Figure 1.2.11). Three frequency bands – L-band, C-band, and Ku-band – are mostly employed.

The *L-band* lies between 1 and 2 GHz. Since its carrier frequencies are low, L-band has restricted transmission bandwidth; on the other hand, its low-frequency range enables easier implementation of both hardware and software modules. Applications of L-band are numerous; they start with *GPS*, without which we cannot imagine today's traveling. This band is also used for *mobile communications*. For example, have you heard of LTE, long-term evolution, which is the highest-speed version of the *fourth-generation (4G) mobile communications*? This standard utilizes L-band frequencies. The *Iridium* satellite mobile phones that communicate directly with an Iridium satellite (to be discussed shortly) operate in L-band too. In addition, L-band is used in aircraft communications to ground stations and between them, in amateur radio, in digital audio broadcasting, and in astronomy.

The *C-band* ranges from 4 to 8 GHz and covers large areas of earth. Since its beam is widely spread, the ground stations receive low-power signals, which requires large ground antennas and other equipment to achieve high-quality reception. Typically, the C-band transmits uplink at 5.925–6.425 GHz and downlink transmission occurs at 3.7–4.2 GHz frequencies; there are, however, other versions of uplink and downlink frequencies within the given margins. C-band – the first frequency range used for satellite communications – has been in service for more than 40 years. Today, according to ITU-D, about 180 satellites in orbit provide more than 2000 transponders utilizing C-band frequencies; this band plays a key role in the global communications infrastructure.

Applications of C-band are too many to be discussed in detail; we just list the main of them: provide weather forecasts, support humanitarian activities and e-learning and e-healthcare systems, enable businesses (maritime operations, oil and gas remote platforms, ATM machines, and mobile-phones communications), and broadcast TV. The popularity of C-band stems from its ability to cover large areas by a single beam and withstand the rain attenuation. Due to these and other technological advantages, the C-band is actively used not only for satellite but also for terrestrial wireless communications. The band becomes overcrowded, and this has become an issue at national and international levels.

The *Ku-band* operates on frequencies from 12 to 18 GHz. Its beam is more focused and covers small ground areas. Thanks to the high directivity of its beam, Ku-band delivers more power to surface stations, which relaxes the requirements for ground-station equipment, enabling the use of small-size antennas. This is why the Ku-band is widely employed for direct-to-home service, the application that most of us are familiar with because dish antennas as small as 60 cm in diameter are seen on many houses and buildings. For this service, called *direct broadcast satellite (DBS)*, the uplink transmission occurs at 17.3–17.8 GHz and downlink is at 12.2–12.7 GHz.

In addition to a very focused beam delivering enough power for individual dish antennas, Ku-band withstands the interference with terrestrial wireless transmissions better than other bands. The signals in Ku-band range, however, are susceptible to *rain fading*; that is, the variations in signal attenuation due to rain. Also, the Ku-band focused beam might be a disadvantage when the coverage of a large geographical area is needed. In addition to DBS service, applications of Ku-band include fixed satellite services, FSSs, and space communications with International Space Station.

There are several other frequency bands used in satellite communications, which we leave to specialized sources.

(**Exercise**: Summarize all applications of the satellite communications discussed above.)

(**Question**: *You probably notice that uplink transmission in any band always occurs at the higher frequency than the downlink. Why is this so?*)

1.2.4.3 Satellite Orbits

How, you may reasonably ask, can **satellite orbit the earth** and what **orbits are best for communications**? To answer these questions, let us consider Figure 1.2.19.[12]

A satellite is a projectile; that is, a body on which a single force – gravity – is exerted. If we dropped a satellite well above earth's surface and pushed it at the horizontal velocity of 8000 m/s, it would seem that the satellite should move to a new position shown as a shaded body in Figure 1.2.19A. Gravity, however, makes it fall toward the earth center. The result of these two motions (shown in dashed arrows) is that the satellite's trajectory will be a curve. If over one second the satellite travels horizontally 8000 m and falls to the earth center by 5 m, it will keep a constant distance from the earth because the earth curves at the same rate. This is how a satellite moves in a circular orbit. If satellite's velocity exceeds 8000 m/s, its orbit will be elliptic and the earth will be one of two foci. Circular and elliptic orbits are shown in Figure 1.2.19B. If satellite's velocity is smaller than 8000 m/s, the satellite's spiral-like trajectory will end on the earth surface.

---

12 https://www.physicsclassroom.com/class/circles/Lesson-4/Circular-Motion-Principles-for-Satellites.

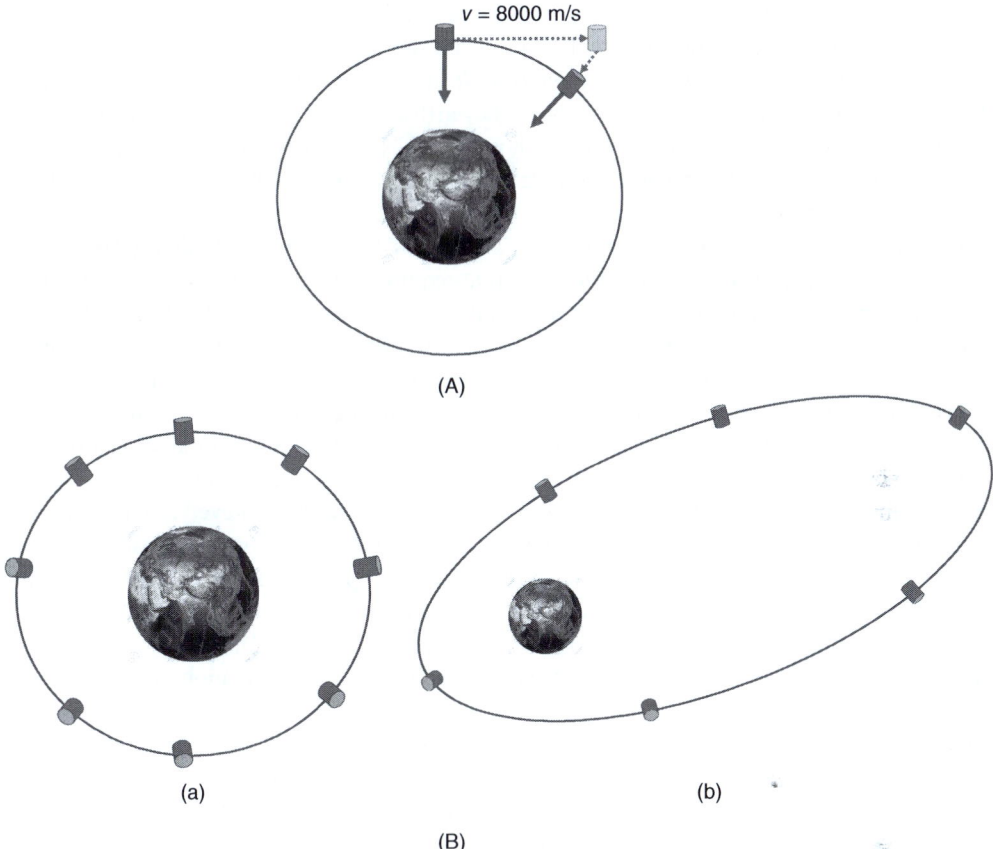

Figure 1.2.19 Satellite for communications: (A) Satellite motion (not to scale); (B) circular (a) and elliptic (b) satellite orbits.

Both elliptic and circular orbits used for satellite communications are represented by the term *geocentric orbits*, which means "orbits around the earth." Three main types of *geocentric orbits* – high, medium, and low – differ in their distances (that is, altitudes) from the earth center.

High-altitude orbits start with *geostationary earth orbit, GSO*, whose distance from the Earth's surface is approximately 35 700 km. This circular orbit lies on the equatorial plane. A satellite placed into this orbit is stationary with respect to a point on the earth's surface. Thus, this surface point and the satellite are constantly in the line-of-sight, which eases their communications. The concept of such satellite communications was devised by prominent futurist writer and the originator of many brilliant ideas Artur C. Clark in 1945. He calculated that the satellite placed at such an orbit will rotate at the same angular velocity as the earth and will therefore remain fixed with respect to a ground point. (In other words, the given earth's surface and the GSO satellite cover the same angle over the same time interval; put in another way, both the earth point and the GSO satellite complete their full rotational cycle in sync for 24 hours.) This orbit is the most used in modern satellite communications. The GSO satellites deliver the majority of weather monitoring and forecasting information; the weather-related images we see on TV or online are provided by these satellites.

The GSO satellites operate mostly in Ku frequency band, though C-band and L-band are also employed at this orbit.

There is a version of a GSO orbit called *geosynchronous earth orbit (GEO)*. This orbit has some inclination and may be eccentric; nevertheless, this orbit serves the same purpose as the GSO.

An example of a *MEO*, is a *semi-synchronous orbit* at which satellites travel the full orbit in 12 hours. Such orbits are about 20200 km above the earth's surface. The GPS satellites rotate on this orbit. There are a number of other MEOs whose altitudes vary from 2000 to 35 700 km.

The *low earth orbits (LEOs)* range in altitude from 160 to 2000 km. (There is no standard for the range of the LEO altitudes; you may find many versions of these distances. Nevertheless, the given numbers define the order of magnitudes.) The LEO satellites, thanks to their proximity to the earth's surface, perform a variety of functions that require immediate responses. They include weather information, mobile communications, and Internet connectivity. The International Space Station is on this orbit. Since these satellites rotate around the earth at short intervals (from 88 to 127 minutes), they pass over the same ground point many times a day; thus, to support constant satellite communications, a constellation of satellites must be on the orbit. Compare: Three GSO satellites can cover the whole Earth, but it takes 15 LEO satellites to do the same job. (The best-known example of the LEO satellite system is the Iridium Communications' cluster whose 66 active satellites cover the whole globe, providing connections to and from every point on Earth. This program was established to launch 77 low-orbit satellites. Seventy-seven is the atomic number of the element *iridium*, hence, the name of the project. Originally, devised as a mobile phone network, this system today provides all kind of mobile connections.) The LEO orbits exist with all inclinations from zero (in equator plane) to 90° (polar orbits).

(**Exercise**: Sketch a figure showing the earth and three LEO where inclinations are equal to 0°, 45° and 90°.)

All three orbit types employ all available frequency bands in different proportions. There is no rule for choosing a specific band for a specific orbit; it is determined by the task and technological and economic variables. For example, Ku-band is mostly used by GSO satellites partly because of its high directivity; nevertheless, this band is also in use for LEO satellites communications, though in a smaller volume.

The approximate distribution of satellites among the orbits in 2018 have been as follows[13]:

2% are in elliptical orbits, 29% are in geostationary earth orbits (GSO), 6% are in MEOs, and 63% are in LEO.

Why do we need the orbits of various altitudes and shapes? Consider, for example, GSO satellites: From the preceding discussion, we can deduce that they are needed when the satellites and their ground stations must be in permanent contact. In addition, the beam of a GSO satellite covers large ground area; that is, one satellite can serve many customers. Does it mean that GSO satellites are the best, and we should employ only them? What drawbacks do these satellites have?

It is intuitively clear that the higher the satellite orbit, the greater the **signal power loss**. Indeed, Equations (1.2.6) and (1.2.7) – applicable for any wireless transmission, including the satellite – show that the FSL (dB) directly depends on distance. To apply this rule to all three orbit types, let us take $f = 12$ GHz and distances $d_{LEO} = 2000$ km, $d_{MEO} = 20\,200$ km, and $d_{GSO} = 35\,700$ and compute FSL (dB) by using (1.2.6) as follows:

$$FSL_{LEO}(dB) = 20\ \log(f) - 147.56 + 20\ \log(d_{LEO}) = 180.02\ dB$$

$$FSL_{MEO}(dB) = 20\ \log(f) - 147.56 + 20\ \log(d_{MEO}) = 200.13\ dB$$

$$FSL_{GSO}(dB) = 20\ \log(f) - 147.56 + 20\ \log(d_{GSO}) = 205.07\ dB$$

---

13 https://www.pixalytics.com/sats-orbiting-the-earth-2018/.

The obtained, well-predictable results show that the longer the distance, the greater the loss. This relationship implies that a ground station and a satellite transponder must generate the greatest signal power for GSO communications, the lesser for the MEO, and the least for the LEO.

In this example, we keep frequency constant to emphasize the dependence of FSL on distance only. Also, be aware that the wireless communications industry, as (1.2.6) attests, considers loss as a ratio of transmitted (input) power to the received (output) power. This is why in our example the FSLs are positive numbers. Optical communications industry, as mentioned previously, practices the inverse approach, Loss = $\frac{P_{out}(W)}{P_{in}(W)}$; this produces Loss (dB) as a negative number.

Bear in mind that Equations (1.2.6) and (1.2.7) describe only the power loss caused by one fundamental reason – the spread of the EM wave due to propagation over distance. Even though the calculated FSLs are close in the orders of magnitude to the actual values, the realistic analysis of the satellite link budget required a much more sophisticated approach. Here are some numbers from an example of such calculations[14]: The downlink transmission of an LEO satellite whose altitude is 800 km and carrier frequency is $f = 437$ MHz experiences FSL = 151.4 dB, generates a transmitter power $P_t = 27$ dBm, and delivers received power $P_r = -116.74$ dBm. The detailed calculations for power budgets of uplink and downlink transmissions considering many transmission impedances can be found in specialized literature. (See section *Satellite Communications* in the textbook's bibliography).

Another consequence of placing satellites on various orbits is the different **propagation delay,** t_d, of a signal. It is clear that an increase in transmission distance increases the time interval needed for a signal to reach its destination point. Satellite communications operate at thousands of kilometers; therefore, we can expect significant propagation delays. To calculate t_d for a specific ground station, it is necessary to know the exact distance between the satellite's transmitting antenna and the ground receiving antenna. There are formulas enabling these calculations that are derived from the straightforward geometrical considerations. We can, however, estimate the delay's minimum value by simply dividing an orbit's distance by the signal's velocity, which can be approximated by the speed of light in vacuum, $c = 3 \times 10^8$ m/s. That is,

$$t_d(s) = \frac{d_{LEO}(m)}{c(m/s)} \tag{1.2.8}$$

Thus, for distances considered in the preceding example, $d_{LEO} = 2000$ km, $d_{MEO} = 20\,200$ km, and $d_{GSO} = 35\,700$ km, we find

$$t_d(LEO) = \frac{d_{LEO}(m)}{c(m/s)} = 0.0066 \text{ s.}$$

$$t_d(MEO) = \frac{d_{MEO}(m)}{c(m/s)} = 0.0673 \text{ s.}$$

$$t_d(GSO) = \frac{d_{GSO}(m)}{c(m/s)} = 0.119 \text{ s.}$$

Given that the delay is doubled due to round-trip transmission, the wait of 0.238 s is significant even for voice communications, let alone more urgent messages, such as navigational information for autonomous vehicles. Nothing can be done to decrease a given propagation delay because we cannot exceed the speed of light, $c = 3 \times 10^8$ m/s. Thus, to minimize the delay, we must shorten the distance; that is, lower the satellite orbit. This is another reason for having a variety of orbit altitudes.

---

14 https://www.itu.int/en/ITU-R/space/workshops/2016-small-sat/Documents/Link_budget_uvigo.pdf.

(**Question**: *The decrease in propagation delay is the obvious reason to keep the communication satellite at a low orbit. Do you know the reason for keeping the satellites at the medium and high orbits? Explain.*)

How do **satellite communications** relate to the other two sectors of communications industry – **optical and wireless**? Optical communications can deliver signals only to stationary points located on the earth's surface. When a satellite signal reaches a ground antenna, the signal is converted into optical and gets transmitted over an optical fiber to any other destination point connected to the optical network. As for the relationship between the satellite and wireless communications, the situation is slightly ambiguous. On the one hand, both segments transmit their signal wirelessly; from this standpoint, satellite communications can be considered a segment of the wireless sector. (Many textbooks devoted to wireless communications indeed include satellite communications in their content.) On the other hand, the realm of satellite communications is outer space, whereas wireless communications employ the thin lowest layer of the earth's atmosphere. Thus, we can say that satellite communications are the space wireless communications and wireless communications is terrestrial communications.

It is worth mentioning that it is ITU that coordinates the placement of communication satellites and the frequencies they use for communications.

The last and may be the most significant point regarding satellite communications is its importance for **military operations**. All stationary military installations (bases, camps, stations, ports, airfields, command centers, etc.) are connected by closed optical communication networks. These installations, of course, have access to the Internet through these networks. All mobile military objects (vehicles, ships, aircrafts, and spacecrafts) communicate wirelessly, where terrestrial and satellite communications are equally important. Satellite communications connect military personal engaged in field, sea, and air operations with their command centers, provide them with navigational information, and coordinate the logistical actions; it also supports the distant control of weaponry and autonomous vehicles, such as drones. Of course, we cannot overestimate such an important function of military satellites as reconnaissance. To put it plainly, satellite communications are vitally important to military operations; the interruptions of these connections undermine the military's ability to perform its tasks. As you can guess, this importance of the satellite communications for defense makes the developed countries take all possible measures to protect their satellites and ground equipment in order to sustain reliable communications.

Satellite communications are a well-developed branch of modern communications industry based on fundamental scientific achievements and the state-of-the-art hardware and software technologies. It is impossible to fully describe this sector in such a brief overview. Nevertheless, it is our hope that the glimpse of the modern satellite communications presented here will give you an idea about this subject and inspire you to a deeper study of this captivating topic.

Figure 1.2.20 summarizes our discussion of operation of the three main branches of communications industry – optical, wireless, and satellite – by demonstrating an example of interconnections of all three types. It shows that in the center of the entire communication infrastructure lies a core fiber-optic network that links all other types of communications. Communications between satellites occurs by optical wireless system and between satellites and earth it is done by RF waves. All terrestrial stations are linked to the core network by fiber-optic cables. Then the end users are reached via optical wireless communications (e.g. between tall buildings), via Wi-Fi links (e.g. between a router and individual devices) and even via Li-Fi (e.g. between lighting LED and personal devices). Of course, this figure does not show the entire modern communication infrastructure; nevertheless, it demonstrates how various types of modern communication technologies work together.

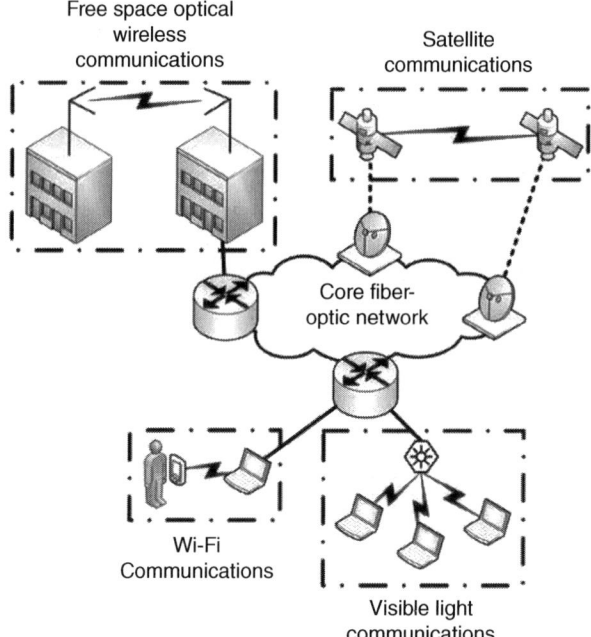

Figure 1.2.20 Interconnections of optical, wireless, and satellite communication system. Dashed lines – wireless RF communications, zigzag lines – optical wireless communications, solid lines – fiber-optic communications. Source: Borah et al. (2012). A review of communication-oriented optical wireless systems. *EURASIP Journal on Wireless Communications and Networking* 2012:91 http://jwcn.eurasipjournals.com/content/2012/1/91 doi: 10.1186/1687-1499-2012-91.

Questions and Problems for Section 1.2

- Questions marked with an asterisk require a systematic approach to finding the solution.
- Many questions and problems, including those marked with an asterisk, imply that you, in addition to reading the textbook, will do your research to find the answers. Consider such questions as mini-projects.

The Internet

1 Name three pillars of modern communications and explain the role of each of them.

2 Compare a point-to-point (PPT) link and a network:
   a) Sketch their basic layouts.
   b) Explain their principle of operations.

3 In a network's layout, two adjacent nodes and their link make up a PPT; thus, every network is built from the PPTs. Why, then, does a network constitute a different from PPT entity?
   *According to the text, "*a network is a set of nodes connected by links:*"
   a) What are the nodes? Give the examples.
   b) What are the links? Give the examples.

c) Consider a communication network in your home or in your academic laboratory: Sketch its layout and identify nodes and links.

4   What is the difference between a network and a communication network?

5   *The text explains that a communication network, in addition to physical layout, must have an intelligent (logical) ability:
a) Why does a communication network need intelligence?
b) The network's intelligent agents are located in its nodes. Why? Where else they might be located?

6   Examine Figure 1.2.2: You find three Nodes. How can you distinguish among them?

7   *Discuss a communication protocol:
a) What is it?
b) Why do we need protocols?
c) What protocol do you follow sending an e-mail?

8   Explain the difference between switching and routing.

9   Consider five main network topologies:
a) Sketch their physical layout.
b) Explain intended logical topology (the flow of information) of each network.
c) Explain how can a network with mesh physical topology execute star logical topology.
d) *Sketch point-to-multipoint topology. This is a version of which of the man topologies?

10   Examine Figure 1.2.4 showing the network hierarchy:
a) Why is an access network also called local network?
b) Is the network of your organization of the access or metro level? Explain.
c) *Find the example of a long-distance network. (*Hint*: Go online and find the example of a submarine (undersea) network.)
d) *The text says that the level of a network in the hierarchy determines the quality of the network's components. Why?

11   A circuit-switching network is shown in Figure 1.2.5a:
a) How does a circuit-switching network operate? How does information flow in this network?
b) PSTN is an example of a circuit-switching network. Explain why.
c) What are the main advantage and main drawbacks of a circuit-switching network?
d) *It is known that the first Internet transmissions were done through the existing telephone network and show inadequacy of PSTN for digital transmission. Why?

12   Consider a packet-switching (routing) network depicted in Figure 1.2.5b:
a) How does a packet-switching network operate? How does information flow in this network?
b) The Internet is an example of a circuit-switching network. Explain why.
c) What are the main advantages and main drawbacks of a packet-switching network?

13 The concept of control plane and data plane of a network can be discussed based on Figure 1.2.6:
   a) What is a control plane? What is its function?
   b) What is a data plane? What is its function?
   c) Why do we need two planes?
   d) Are the control plane and data plane two physically separate networks or one network? Explain.
   e) *The text says that the control plane makes a data plane protocol-independent. How do you understand this statement? Explain, considering transmission through the Internet.

14 *Discuss the network security:
   a) Why is the network security important? Give an example.
   b) Does the advent of the Internet make the networks more or less secure? Explain.
   c) List and discuss all the measures providing network security.
   d) What is the weakest point in encryption? What measure can be taken to improve the situation?

15 What is the Internet?

16 Consider the Internet and World Wide Web, the Web;
   a) What is the difference between them?
   b) Why do people often confuse one with another?
   c) Give a detailed explanation.

17 *It says that the Internet is the network of networks. Does it mean that the special network was built for the Internet?

18 The text says that the Internet successfully operates thanks to TCP/IP protocol suite:
   a) What is TCP/IP stands for?
   b) Why is it called *suite*?
   c) *List all the functions of TCP suite do you know.
   d) *List all the functions of IP suite do you know.

19 Consider the IP addressing system:
   a) What is an IP address?
   b) What are IPv4 and IPv6?
   c) Why do we need two addressing systems?

20 We know that a message (file) to be transmitted through the Internet is disassembled into packets:
   a) Describe the general structure of the packet.
   b) What is the function of a header?
   c) Why do we need a payload section of the packet?
   d) In percentage, what is the ratio between the header and the payload in a single IPv6 packet?

21 *Follow the path of a message through the Internet based on Figure 1.2.7:
   a) In what form – as a file or packet – the message travel from your computer (or smartphone) to the ISP router? To DNS? To a core router? To the receiving end user?

b) What is an ISP? Why do we need it?
c) What is DNS? What is its role in the Internet transmission?
d) What is the function of a core router?

22 *Consider packet transmission through Internet:
a) When a router inspects a packet, what information does it retrieve?
b) How does a destination point know that the received message is complete?
c) If a packet gets lost or corrupted, what measure does the network take?

23 *Consider the flow control in the Internet transmission: If the original TCP window size is equal to 65 535 bytes and an entire round-trip time given as 31 ms, what is data rate for this communications? (*Hints*: 1. One byte is 8 bits. 2. Refer to Eq. (1.2.1). Note: Actually, the window of 65 535 bytes is transmitted from a client to a server in small portions (flow control window sizes) and the round-trip time for each cycle is much shorter than 31 ms. Thus, the obtained bit rate is the throughput number.)

Optical Communications

24 Starting from the mid-1980s, optical communications quickly, for less than 15 years, replaced electronic – copper-based wired and microwave wireless – communications at every distance from several hundreds of meters to thousands of kilometers and became the linchpin of the global communications infrastructure. Why?

25 *Consider the basic block diagram of an optical communication system shown in Figure 1.2.8:
a) Suppose that there is no light at the frontend of a photodiode. What are the possible faults?
b) If a bit stream is presented to a transmitter and the laser diode emits unmodulated light, what are the possible faults?
c) If the photodiode can detect light but cannot decipher the message, what are the possible faults?

26 *Consider Figure 1.2.9 showing the world-wide map of submarine cables. If one cable can carry 10 terabits per second (10×10^{12} Tb/s), why do we need so many cables in general and transatlantic cables in particular?

27 Suppose that power of light launched into an optical fiber is $P_{in} = 5$ mW and light power emerged from the optical fiber is $P_{out} = 5$ µW. What is the loss introduced by this optical fiber in absolute numbers and in decibels?

28 If one manufacturer offers you an optical fiber with loss of -30 dB and the other offers the optical fiber with loss of -50 dB, which optical fiber will you choose?

29 Consider the power losses in an optical fiber:
a) If $P_{in} = 5$ mW, $P_{out} = 5$ µW, and the fiber length is $L = 120$ km, what is the optical fiber's attenuation?
b) If one manufacturer offers you an optical fiber with $A_1 = 0.3$ dB/km and the other offers the optical fiber with $A_2 = 0.5$ dB/km, which optical fiber will you choose?

30 What is the maximum transmission distance the optical fiber whose $A = 0.25$ dB/km can support if $P_{in} = 5$ mW and $P_{out} = 3$ µW?

31 *Compare the bandwidth of an optical fiber with two other wireline media – twisted pair and coaxial cable – based on the data shown in Figure 1.3.3a reproduced here:

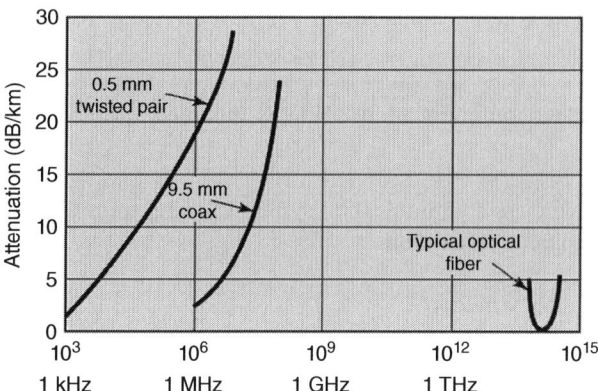

Figure 1.2.P31 (1.3.3aR) Bandwidths of copper wire (twisted pair), coaxial cable, and optical fiber.

a) Figure 1.3.3a shows that the twisted pair introduces attenuation of about 30 dB/km at 10 MHz. We know that the optical fiber has minimal attenuation 0.2 dB/km. Calculations show that a signal becomes a thousand times weaker after transmission over $L_{TP} = 1$ km for the twisted pair and $L_{OF} = 150$ km for optical fiber. Perform these calculations.

b) Examination of Figure 1.3.3a makes an impression that a twisted pair has bandwidth from 1 kHz to 10 MHz, a coaxial cable's bandwidth ranges from 1 to 100 MHz, and the bandwidth of an optical fiber varies approximately from 100 to 500 THz. Is this correct estimation of the bandwidths?

32 If optical communications is everywhere, why are we using wireless communications at an increasing pace? (Optical communication is fixed and wireless communications is mobile.)

33 Describe the flow of information in the network shown in Figure 1.2.10. Identify all shown optical networks in terms of their place in the network hierarchy.

34 *Consider the history of development of optical communications:
a) Why the development of a low-attenuation optical fiber was the decisive point in this history?
b) There are a number of laser types (solid-state laser, gas laser, laser diode, fiber laser, to name a few) invented over the last 50 years. Why only development of laser diode was recognized by the Nobel Prize?

35 The optical long-distance transmission occurs at the set of wavelengths centered around 1550 nm. What is the frequency of this wavelength?

36 Using wavelength-division multiplexing (WDM) we can transmit more bits per second over the same optical fiber. If a single wavelength carries 120 Gb/s, what will be the total bit rate after multiplexing 120 wavelengths?

37 *What obstacles might restrict further progress in development optical communications?

Wireless Communications

38 The rapid growth of wireless communications is caused by the advent of ever-increasing new mobile applications. List and explain new mobile applications you have started to use over the last five years.

39 What is the main advantage of wireless communications? Why is this filed advancing so rapidly?

40 Sketch the basic block diagram of a wireless communication system showing transmission from one smartphone to the other. Explain its operation.

41 Why is this type of communications called wireless?

42 The transmission medium for wireless communications is air. Is this an advantage or a drawback? Explain.

43 A radiation pattern emitting from an antenna can vary from omnidirectional to a light beam:
   a) Sketch the diagrams showing these two patterns and add two other intermediate radiation patterns.
   b) Show the beam width for each pattern.

44 Estimate the size of an antenna radiating at $f = 600$ MHz.

45 Figure 1.2.13 shows three main propagation modes of EM waves used in wireless communication:
   a) What factor does determine which mode will actually occur? Explain.
   b) Why do EM waves propagate differently depending on radiated frequencies?

46 If an antenna height is $h = 10$ m, what is the propagation distance for this antenna?

47 What is the mechanism of free space loss (FSL)? Sketch the diagram supporting your explanation.

48 *The minimum received power needed to support streaming video is $P_r = -67$ dBm. If operating frequency is 600 MHz and the transmission distance is 800 m, what must be the transmitted power, P_t in dBm and in absolute numbers? What is FSL (dB)? (*Hint*: P (dBm) $= 10 \log \left(\frac{P(mW)}{1(mW)} \right)$.)

49 *What is the maximum transmission distance covered by a base transceiver station (BTS) whose signal power is $P_t = 18$ dBm, the required received power is $P_r = -67$ dBm, and operating frequency is $f = 600$ MHz?

50 The segment of the EM spectrum suitable for wireless transmission is severely restricted and must be regulated:
   a) What agency controls this segment in the USA?
   b) What agency regulates the use of the EM spectrum at the international level?
   c) In what forms is this control executed? Explain.

51 Discuss scarcity of spectrum for wireless communications:
   a) Why do developments of wireless communications require more spectrum?
   b) If the available spectrum is occupied, how can the further development of wireless communications progress?

52 Much expected new type of wireless communications is 5G – the fifth generation of traditional, RF-based wireless communications. Describe main advantageous features of 5G.

53 Explain the principle of operations and applications of the following contemporary wireless communications technologies:
   a) Wi-Fi.
   b) Bluetooth.
   c) ZigBee
   d) Free space optical communications (FSO), light fidelity communications (Li-Fi), and visual light communications (VLC).

54 Explain why has Li-Fi become popular technology today?

55 Examine Figure 1.2.20 and discuss how contemporary wireless communications technology work in conjunction with other modern communications technologies.

56 Explain the concept of a cellular wireless network with reference to Figure 1.2.16A. What is the main advantage of this architecture?

57 The mobile communications, that we have today, has become a reality thanks to the concept of frequency reuse. What is this concept?

58 Mobile cellular communications occurs in a low-power mode. Why so?

59 Consider government regulations of communications industry:
   a) Why did the federal government have to intervene with regulations in the affairs of the communications industry?
   b) What federal agency is responsible for implementing government policy in modern communications? Explain.

Satellite Communications

60 *Consider the operation of satellite communications:
   a) What is the difference between satellites and communication satellites?
   b) What is the function of a communication satellite?
   c) Why does a satellite's transponder change the frequency of a downlink transmission compared to a carrier frequency of an uplink transmission?
   d) What other operations does a satellite's transponder perform?
   e) A satellite's transponder is an active device. Where is its power supply located?

61 Review the frequencies used in satellite communications:
   a) What are three main frequency bands? List the range of frequencies for each of them.

1 Modern Communications: What It Is?

b) Group the applications of each band. Explain why is each application used with a given frequency band.
c) Build a table to summarize your responses.

62 Carrier frequency of an uplink transmission is always higher than that in downlink transmission, which holds true for any frequency band. Explain why?

63 Discuss celestial mechanics of satellite communications:
a) A satellite is a heavy object. How, then, does it not fall on the earth but keeps orbiting it?
b) What orbits are called geocentric? Are circular and elliptic satellite orbits geocentric? Explain.
c) How can we make a satellite to move on a circular or an elliptic orbit?
d) *It says that the velocity of 8 km/s is the minimal that keeps a satellite on a geocentric orbit. What happens if the velocity will be less than 8 km/s? What if it will be greater than 8 km/s; e.g. $v = 11$ km/s? If much greater; e.g. $v = 16$ km/s?
e) What is the inclination of an orbit? Sketch a figure showing the earth and three circular orbits with inclinations equal to 0°, 45°, and 90°.

64 There are three main types of *geocentric orbits* – high, medium, and low:
a) How can we distinguish among them?
b) Explain the difference between geostationary earth orbit (GSO) and geosynchronous earth orbit (GEO). Explain their applications.
c) Give an example of an MEO orbit and its applications.
d) Explain the use of LEO orbits.
e) Why do we need the orbits of various altitudes and shapes?
f) Add to the table built for Problem 60 three main orbits and their applications. The objective of building this table is to summarize all our findings regarding the satellite communications.

65 Compare the free space loss, FSL (dB), for high, medium, and low orbits:
a) Compute FSL_{LEO} (dB), FSL_{MEO} (dB), and FSL_{GSO} (dB) if $f = 10$ GHz and $d_{LEO} = 2000$ km, $d_{MEO} = 20\,000$ km, and $d_{GSO} = 36\,000$ km.
b) If the minimum received power is $P_r = -91.24$ dBm for the LEO satellite whose $f = 10$ GHz and $d_{LEO} = 2000$ km, what is its transmitted power, P_t?

66 Consider propagation delay in satellite communications:
a) What is this?
b) Calculate round-trip propagation delay for $d_{LEO} = 800$ km, $d_{MEO} = 20\,000$ km, and $d_{GSO} = 36\,000$ km.
c) We calculate one-way propagation delay by using (1.2.8), $t_d = \frac{d(m)}{c(m/s)}$, where d (m) is the satellite's altitude. Is this minimum or maximum delay? Explain.
d) *Equation (1.2.8) approximates the velocity of EM waves in air by the speed of light in vacuum. How accurate this approximation?

67 *Figure 1.2.20 shows, in particular, the satellite communications. Why is the communications between the satellites shown as lighting, but between the satellites and ground as dashed line?

68 Discuss Figure 1.2.20, explaining every block, every component, unit, and module, and relationship among all them. The result of this discussion should be a picture of the global communications infrastructure.

1.3 Fundamental Laws and Principles of Modern Communications

Several fundamental laws and principles govern modern communications regardless of the status of technological development at any given time. In this section, we will briefly discuss those of them that are most relevant to the objective of this book. In addition, we will consider such important parameters as signal-to-noise ratio (SNR) and spectral efficiency. We also discuss the limitations in applications of these laws and principles and briefly examine the relationship between real objects and their models.

In the subsequent chapters, all these principles and their applications will be considered in greater detail.

It is worth noting that the fundamentals of modern communications were laid out in the twentieth century by many scientists and engineers; among them the three towering figures – Ralph Hartley, Harry Nyquist, and Claude Shannon – played the pivotal role in this development.

1.3.1 Fundamental Laws of Modern Communications

1.3.1.1 Hartley's Information Law

In communications, we are concerned with the relationship among the amount of information to be transmitted, the speed of its transmission, and the time required for its transmission. This relationship is intuitively clear:

> *The greater the amount of information to be transmitted during a given time interval, the higher the transmission rate (speed) must be. Thus,*
>
> $$H(b) = BR(b/s) \times T(s) \tag{1.3.1}$$
>
> *where H (b) is the amount of information, BR (b/s) is the bit rate (transmission speed), and T (s) is the transmission time. (Notation (b) stands for bit.)*

Pay attention to the units: The volume (amount) of information is measured in *bits*, which stands for **binary units**; the transmission speed – formally called *bit rate* or *transmission rate* or *data rate* – is measured in *bits per second, b/s*; and time is measured *in second, s*. Notice that the communication industry uses notations *bit/s, b/s, or bps* for bit rate. We will mainly use *b/s*, but sometimes refer to the other designations.

This relationship is known as *Hartley's information law*, after R.V.L. Hartley,[15] a prominent American scientist who was among the pioneers of information theory. Of particular note is that he introduced *bit* as a unit of information volume in 1928.

---

15 R.V.L. (Ralph) Hartley (1888–1970) worked for Western Electric and Bell Labs. He made seminal contributions into both electronics and theory of information. He invented an electronic oscillator bearing his name and he developed the formal approach to quantification of information.

Example 1.3.1 Application of Hartley's Information Law

Problem

Assume that a professor delivered lectures using slides prepared in Microsoft PowerPoint format. The size of the entire file was 15 Mbytes. (You will recall that 1 byte is 8 bits.) He placed this file on the department's website. How long will it take to download this file to a student's platform in the following three cases:

1. The platform is connected to the Internet with an obsolete dial-up modem whose transmission speed is 56 kb/s? (Nobody uses this modem anymore, but we refer to this museum artifact for an illustrative purpose.)
2. The platform uses another old-fashion digital subscriber line (DSL) transmitting at 1.5 Mbit/s? (A digital subscriber line (DSL) is the technique that has been used to transmit a computer signal over a telephone line at high speed.)
3. The platform employs modern passive optical network (PON) line with bit rate 10 Gb/s? (PON is the optical network on which modern access lines are based.)

Solution
Rearrange (1.3.1) as

$$T(s) = H(\text{bit})/C(\text{bit/s}).$$

Thus,
With a dial-up modem (do not forget to convert bytes to bits),

$$T = [8 \times 15 \times 10^6 \text{ bit}]/[56 \times 10^3 \text{ bit/s}] = 2142.9 \text{ s} = 35.7 \text{ minutes}$$

With a DSL modem,

$$T = 80 \text{ s} = 1.3 \text{ minutes}$$

With PON,

$$T = 12 \times 10^{-3} \text{ s} = 12 \text{ ms}$$

Discussion

These simple computations show the paramount importance of transmission speed for timely delivering information.

Hartley's information law emphasizes the main point: *Communications networks must provide high-speed transmission*. If you look at the history of electronic communications – from the Morse telegraph to modern optical networks – you will see one major force driving the development of this technology: the quest for increasing the transmission speed. But on what parameters the transmission speed (bit rate, remember?) depends. To answer this question, we need to consider the relationship between bandwidth and bit rate.

1.3.1.2 Signal Bandwidth and Transmission Bandwidth from the Transmission Standpoint

First, we have to distinguish between bandwidth and bit rate. We recall that bandwidth is a frequency range measured in hertz. The terms *frequency* and *hertz* are applied only to the analog format. Thus, *bandwidth is a characteristic of both analog signal and analog transmission*.

Table 1.3.1 Analog and digital transmission characteristics.

	Analog	Digital
Signal	Signal bandwidth (Hz)	Bit rate (bit/s)
Link	Channel bandwidth (Hz)	Channel capacity (bit/s)

Note that signal bandwidth is also called signal spectrum.

Bit rate, on the other hand, is the number of bits per second. The terms *bits* and *bit-per-second* are applied only to the digital format. Thus, *bit rate is a characteristic of both digital signal and digital transmission*.

Second, we have to distinguish between signal bandwidth and transmission (channel) bandwidth on the one hand, and between bit rate and channel capacity on the other hand.

We use the term *bandwidth* in two meanings: as the range of frequencies that a given signal occupies (signal bandwidth) and as a measure of the transmitting capacity of a link (channel bandwidth). Thus, signal bandwidth is the characteristic of an analog information signal presented for transmission; the similar characteristic of a digital signal is bit rate. On the other hand, transmission bandwidth is a characteristic of a transmission link (channel) designed for transmitting analog signals; it is a channel's ability to accommodate a certain range of frequencies. The similar characteristic of a digital transmission is channel capacity. Table 1.3.1 clarifies this point:

Unhappily, in the communications industry, the use of the term *bandwidth* in the sense of channel capacity is applied today to both analog and digital transmissions. But since digital transmission is the predominant form of today's transmission, using the term *bandwidth* usually does not cause any ambiguity. For example, it is commonplace today to use the term *bandwidth* when talking about broadband transmission. However, broadband transmission means using a high-speed connection to the Internet, which, obviously, can be done only in digital format (the computer is a digital machine, remember?). Thus, the term *bandwidth*, as applied to broadband transmission, means the channel capacity of the link that provides access of home or office computers to the Internet.

1.3.1.3 Bandwidth and Bit Rate, Nyquist's Formula, and Hartley's Capacity Law

Is there any relationship between signal bandwidth and bit rate? In general, it depends on the transmission code. Let us consider, for example, the simplest code, NRZ. (NRZ stands for nonreturn-to-zero. This code will be discussed in Chapter 3.) Suppose we transmit a pulse train where 1s and 0s appear one after the other. We can inscribe a sinusoidal signal in the way shown in Figure 1.3.1. It is clearly seen that for one period of an analog sinusoidal signal, a digital, the NRZ signal delivers two bits. Therefore, the bit rate (BR) in this case is twice as much as the bandwidth, BW; thus, we can write (Mynbaev and Scheiner 2001, pp. 69–71)

$$BW\ (Hz) = BR\ (b/s)/2 \tag{1.3.2a}$$

This intuitive approach leads to the answer to the more fundamental question: What bit rate can a link with given bandwidth support? It follows from (1.3.2a) that the first answer is

$$BR\ (b/s) = 2BW\ (Hz) \tag{1.3.2b}$$

This formula states that a transmission channel must have bandwidth 2BW to transmit bit rate BR. For example, if the bandwidth of a Wi-Fi channel is 40 MHz, then, according to (1.3.2b), it can support 80 Mb/s bit rate.

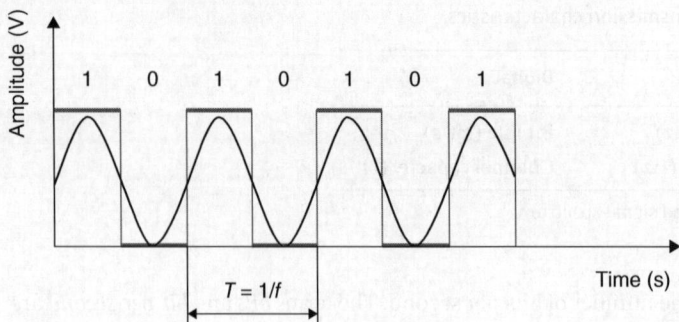

Figure 1.3.1 Bandwidth and bit rate: Two bits are delivered per cycle of a sinusoidal signal with NRZ code.

However, a more thorough look reveals a more complicated relationship between bandwidth and transmitting capacity. We owe this development to Harry Nyquist,[16] who in 1927 came up with his renowned formula,

$$\text{BR (b/s)} \leq 2\text{BW (Hz)} \tag{1.3.3}$$

which states that the *bit rate is not equal but is restricted by the transmission link bandwidth*. Soon after, in 1928, Ralph Hartley further developed this idea, by expanding the Nyquist formula to the multilevel binary coding in his celebrated law,

$$\text{BR (b/s)} = 2\text{BW (Hz)} \log_2 M \tag{1.3.4}$$

where M is the number of discrete levels carrying binary information. If we substitute *maximum* channel capacity, C (b/s), for bit rate, we obtain *Hartley's capacity law* in its original form:

$$C \text{ (b/s)} = 2\text{BW (Hz)} \log_2 M \tag{1.3.5}$$

(Note that in Eq. (1.3.1), we introduced the other Hartley law, the one concerned with the volume of information, transmitting capacity, and transmission time. That law is sometimes called the *Hartley Information Law*. Both of these laws are rightly attributed to Hartley.)

For simple binary coding, where one level carries 1 bit, we have $M = 2$, which brings us back to (1.3.2b):

$$\text{BR (b/s)} = 2\text{BW (Hz)} \log_2 M = 2\text{BW (Hz)} \log_2(2) = 2\text{BW (Hz)}$$

A word about multilevel coding: The idea is to make one symbol (level, or signal) carry more than 1 bit, or – put it another way – we can encode many bits in one symbol. If we managed, for example, to encode 4 bits in one level, then we will need $M = 16$ because $2^4 = 16$. We use base 2 because we are coding binary information. We tackle the role and meaning of M in Sections 3.2 and 10.1.

---

16 Harry Nyquist (1889–1976) was an American scientist who worked for Bell Laboratories from 1917 to 1954. He was born in Sweden and immigrated to the United States in 1908. He earned his PhD in physics from Yale University. He is famous for his foundational contribution to communication and control theory and practice. He obtained many significant results, but his truly fundamental contributions were the sampling theorem, the stability criterion, and the theory of thermal (Nyquist–Johnson) noise. He also developed a filter prototype that bears his name. In addition to his theoretical work, he obtained 138 patents relating to communications.

Equation (1.3.5) determines the transmitting capacity of a channel, depending on its bandwidth. However, this formula does not take into account the influence of noise; in other words, it holds true only for a *noiseless channel*. It was Claude Shannon, who further developed this idea and came up with his famous formula, which is discussed next.

1.3.1.4 Shannon's Law (Limit)

In July and October 1948 issues of the *Bell System Technical Journal*, which published the results of research conducted at Bell Laboratories, a seminal article, "A Mathematical Theory of Communication," appeared. In 1998, the global telecommunications community celebrated 50 years of the tremendous accomplishments of information theory that have been the outgrowth of that article by Claude Shannon [17]. (Interestingly that Shannon entitled the second edition of his work published in 1949 *The Mathematical Theory of Communication* to reflect the grown confidence in his theory which was generally accepted by research and industrial communities after year of intense discussion.)

The content of Shannon's work is fundamental. We cannot afford to discuss all its aspects in this section; however, we return to this law in later chapters (particularly in Sections 3.2 and 9.1), where both the theory and applications of Shannon's work will be considered. For now, we will briefly review Shannon's law:

$$C\,(\text{b/s}) = \text{BW}\,(\text{Hz})\,\log_2\,(1+\text{SNR}) \qquad (1.3.6)$$

where C (b/s) is the *maximum transmission rate* (speed) that a communication channel can reliably provide; BW (Hz) is the *available channel bandwidth*, which can be effectively used for transmission; and SNR (or S/N) is the *signal-to-noise ratio*. But (1.3.6) not simply add noise into another formula. In effect,

> Shannon's law establishes the upper limit of the transmission rate of a communication system to reliably deliver information when both bandwidth of a transmission channel and SNR are given. This equation based on information theory holds true regardless of the technology used.

Understandably, Shannon's law is often referred to as Shannon's limit because it says that an actual transmission (bit) rate, BR (b/s), should not exceed channel capacity, C (b/s), to support *error free* transmission. This means that BR (b/s) $\leq C$ (b/s). This condition justifies the term *limit* and can be traced to Nyquist criterion (1.3.3).

We need to know, however, that C (b/s) in (1.3.6) is not an *absolute limit*. We can transmit at the bit rate higher than C (b/s) but with the increasing probability of errors. In other words, Shannon's law delimits the bit rate supporting a reliable transmission whose *probability of error tends to zero*. This is why (1.3.6) is often called Shannon's law for *error-free transmission*.

Clearly, the practical error-free bit rate, BR (b/s), is always smaller than C (b/s) given by (1.3.6); that is,

$$\text{BR}\,(\text{b/s}) < C\,(\text{b/s}) \qquad (1.3.7)$$

---

17 Claude E. Shannon (1916–2001) earned his degrees in electrical engineering and mathematics from MIT. While with Bell Laboratories (1941–1972), he did his milestone work on information theory. After leaving Bell Labs, he held a professorial position at MIT. His work laid a strong foundation for the development of both communications and computer theory. He was primarily interested in solving problems rather than seeking new applications. A shy, reserved scientist, his death in February 2001 was noted only by the academic scientific community.

It is important to know that (1.3.6) is derived under the condition that the noise is simply added to a signal and its energy is evenly spread along the whole spectrum (spectrally flat); such noise model is called *additive white Gaussian noise (AWGN)*.

Bear in mind that, taking into account the history of its development presented above, Eq. (1.3.6) is often referred to as the *Shannon–Hartley theorem*.

Again, in-depth consideration of this formula is the subject of many academic treatises; we concentrate here on its pragmatic engineering applications.

(Revisit the part of Section 1.2 where the block diagram of a communication system and its operation are discussed. It helps to fully understand the following considerations.)

Note that the channel capacity depends on the *SNR* defined as the ratio of a signal's power to noise's power; that is,

$$\text{SNR} = \frac{P_{\text{signal}}(W)}{P_{\text{noise}}(W)} \tag{1.3.8}$$

The Shannon law shows that the greater this ratio, the higher the communications system's capacity – that is, the transmission speed.

Example 1.3.2   The Role of *SNR* in Shannon's Law

Problem

1. For more than a hundred years, voice (the telephone signal) has been transmitted over a copper telephone line (a twisted pair), in which the available bandwidth was equal to 4 kHz. What was the capacity of this channel if SNR was equal to 1000? To 10 000?
2. Modern wireless technology called Wi-Fi provides high-speed mobile access to the Internet for personal devices as cell phones, tablets, and laptops (see Section 1.2). The Wi-Fi bit rate could be as high as 0.5 Gb/s. We all enjoy this technology, but everyone is familiar with its main problem: The signal is nonreliable. This is because noise, which interferes the signal, is typically high for a radio-wave transmission. Consider the typical values of the signal power at 1 μW and the channel bandwidth at 40 MHz, compute the maximum bit rate for Wi-Fi transmission if the noise power is 0.03 μW.

Solution
1. Formula 1.3.6 shows

$$C\,(b/s) = BW\,(Hz)\,\log_2(1 + SNR)$$

(For convenience, we convert the base-2 logarithm into base-10 one as follows: $\log_2 M = \log_{10} M / \log_{10} 2 = 3.32 \log_{10} M$.)

Thus, we compute

$$C = 4 \times 10^3 \times 3.32 \log_{10}(1001) = 39.8 \text{ kb/s}$$

With a better SNR, we can achieve

$$C = 4 \times 10^3 \times 3.32 \log_{10}(10001) \approx 53.1 \text{ kb/s}$$

2. To apply (1.3.6), we first need to compute SNR by using (1.3.8). We obtain

$$\text{SNR} = \frac{P_S(W)}{P_N(W)} = \frac{1\,\mu W}{0.03\,\mu W} = 33.3.$$

Plugging the SNR and BW values into (1.3.6), we compute

$$C\,(b/s) = BW\,(Hz)\,\log_2(1 + SNR) = 203.88 \text{ Mb/s}$$

The problem is solved.

Discussion

- This example demonstrates the role of SNR: The transmission speed increases with the increase of SNR, but logarithmically, not linearly. In this example, C (b/s) increases only 1.3 times when SNR increases 10 times.
- For Wi-Fi transmission, the computed channel capacity (the maximum bit rate) is far less than the declared 500Mb/s speed. What can be done to increase the BR (b/s)? Hypothetically, the bandwidth should be increased, but in practice the bandwidth in any wireless transmission, including Wi-Fi, is strongly regulated and cannot be changed. Therefore, we need to increase SNR. What SNR is needed in order to achieve $C = 400$ Mb/s? Rearranging (1.3.6), we compute

$$\text{SNR} = 2^{(C/BW)} - 1 = 5792.$$

Thus, to increase the bit rate 2.5 times (from about 200 to 500 Mb/s), we need to increase SNR more than 176 times, from 33.3 to 5792. This is another example of the difficulty in achieving a high-speed transmission.

- How can SNR be increased? Since we do not control the noise power, the only solution is increasing the signal power. This solution, however, meets its practical limitations because generating more signal power requires bigger power source, which is not always desired approach. For example, for wireless communications systems, such as Wi-Fi, it is simply impossible to significantly increase the size and weight of our mobile devises by including a heavier battery. Nevertheless, today's communications technology manages to achieve this and even higher goals in increasing the transmission rates of various communications systems. Note that wired transmission systems (e.g. telephone lines or optical fiber lines) have much better SNR than the wireless systems – just compare the SNR values given in this example. This difference is determined by the nature of a transmission link.
- Last, but not least point: What if we have a noiseless transmission channel? In such a hypothetical case, when noise power is zero, SNR goes to infinity and a channel capacity, according to (1.3.6), should go to infinity too, even at a very small bandwidth. More rigorous analysis shows, however, that Shannon's law works only for a noisy channel; for noiseless channel, the Nyquist formula, C (b/s) = 2BW (Hz) given in (1.3.3), would be applied. In reality, obviously, noise always exist.

Another important consequence of the Shannon formula is that channel capacity is proportional to channel bandwidth. Note that the channel bandwidth here is the range of frequencies within which the channel can transmit a signal without significant distortion (error-free channel, remember?). Thus, *the amount of information that can be transmitted over a channel per second (bit rate, that is) depends on the channel's bandwidth available for transmission.*

Example 1.3.3   The Role of Channel Bandwidth in Shannon's Law
Problem

1. Example 1.3.2 shows that the maximum transmission speed over a copper telephone line whose SNR = 10 000 is only 53.1 kb/s; however, Example 1.3.1 refers to the transmission speed of 1.5 Mb/s that a DSL modem can support over the same old-fashioned telephone line. The contradiction is obvious. How can this be explained?
2. Today, vast majority of all telecommunications traffic is delivered by fiber-optic communications (optical communications) in which the available bandwidth is equal to 8 THz. What is the transmitting capacity of an optical communications system if SNR is equal to 100?

Solution

1. Analyzing (1.3.6), we realize that increasing C (bit/s) can be achieved by either increasing BW (Hz) or SNR. In this example, SNR is given. Thus, for the transmission of 1.5 Mbit/s, the bandwidth must be properly increased. Let us compute the required bandwidth:
Referring to (1.3.6), C (b/s) = BW (Hz) $3.32 \log_{10}(1 + \text{SNR})$, and plugging in the given numbers, we find

$$\text{BW (Hz)} = \frac{1.5 \times 10^6 (\text{bit/s})}{3.32 \log(100\,01)} = 0.11 \times 10^6 \text{ Hz} = 0.11 \text{ MHz}$$

Indeed, a real DSL modem occupied bandwidth from 0.1 to 1.1 MHz.

2. Applying (1.3.6) again, we compute:

$$C = 53.2 \times 10^{12} \text{ b/s} = 53.2 \text{ Tb/s}.$$

Discussion

- This and the previous examples clearly show that the bandwidth plays the major role in Shannon's law. Indeed, when channel's bandwidth equals 4 kHz, the transmitting capacity reaches only 53.1 kb/s and when $BW_{channel} = 0.11$ MHz, $C = 1.5$ Mb/s – in both cases at SNR = 10 000. At the same time, when $BW_{channel} = 8$ THz, $C = 53.2$ Tb/s in spite of SNR = 100. This result clearly follows from the structure of (1.3.6): *System's transmitting capacity is directly proportional to channel's bandwidth and depends on SNR only logarithmically.*
- It seems that with increasing bandwidth to infinity, the channel capacity should go to infinity too. However, this is not so because an increase in bandwidth results in the increase in noise power simply because the channel with greater bandwidth will collect more noise. (You can skip to Section 9.1 to learn more about noise.) Therefore, there is trade-off between BW and SNR, which is not shown explicitly in the Shannon formula.
- How can a DSL system utilize 1.1-MHz bandwidth over a telephone line if we continue to stress that this line has 4-kHz bandwidth? The answer is that 4-kHz restriction is caused not by a twisted pair (traditional telephone line) itself but the whole transmission system. The twisted-pair cable allows for transmission of the higher frequencies. Using multilevel modulation (discussed in Section 3.2 and especially in Section 10.1) also helps to significantly increase the bit rate.

The question that crops up now is what limits the bandwidth? We postpone answering to this question to the later subsection; for now, let us consider two important points regarding Shannon's law.

1.3.1.5 More Clarifications of the Shannon Law

Look closely at the Shannon law, (1.3.6): You will surely notice that channel capacity, C (b/s), is measured in bit-per-second, b/s, whereas the bandwidth, BW (Hz), is measured in hertz, Hz. What may puzzle you is how different dimensions can be equal to one another given that logarithm is the dimensionless quantity. Formally, this is possible because b/s is the same as 1/s, since bit stands for the *number of bits* and a number carries no dimension. At the same time, Hz is also 1/s, since hertz is the *number of cycles* per second; thus, both sides have the same dimensions: 1/s.

Another point to be clarified regarding (1.3.6) is that channel capacity, C, is measured in bits per second, b/s, for digital transmission; when communications is based on analog signals, however, the measure of channel capacity is bandwidth (BW) measured in hertz. Unhappily, industry uses both terms – *bandwidth and channel capacity* – interchangeably, which can be confusing. This confusion stems from the fact that the term *bandwidth* carries many meanings. For example, bandwidth can be the range of frequencies that a given signal occupies; thus, here BW is a signal's bandwidth. Bandwidth, however, can be the range of frequencies needed for transmission of a given signal and this is the required bandwidth of a transmission channel. The practical example is the bandwidth of a current optical fiber. This medium can transmit signals within the entire optical band; that is, from 400 to 1600 nm, which ranges from 750 to 187.5 THz. (Remember, λ (m)·f (Hz) = c (m/s) = 3×10^8 (m/s) as given in (1.2.4).) However, only "small" 8-THz band located around 200 THz region is available for communication transmission because only within those 8 THz an optical fiber has the minimum attenuation. Obviously, the communication industry has chosen 1550 nm (193.5 THz) as a center wavelength (frequency) and has used 8-THz band to add more wavelengths (frequencies) for transmission. So what is the bandwidth of an optical fiber? To be able to transmit an optical signal, the optical fiber must have 208-THz bandwidth; however, to deliver information, the available bandwidth of an optical fiber must be 8 THz. It is BW = 8 THz that we have to plug into Shannon's law to calculate the fiber-optic channel capacity.

You have to carefully analyze the content and formulas of the text you are reading to understand what meaning of bandwidth is involved.

To further delve into content of Shannon's law, we urge you to refer to *Questions and Problems* for this section.

1.3.1.6 The Shannon Law for Digital Communications

The Shannon law can be presented in the form specific for digital communications, which is important because modern communications is mostly digital.

Let us introduce E_b (J) – energy in joules needed to deliver a single bit. This is energy per bit that a transmitter provides on its side. If we remind that power of a bit generated by the transmitter is P_b (W) and introduce the bit rate, BR (b/s), of the transmitter, then

$$E_b \text{ (J)} = \frac{P_s \text{ (W)}}{BR \text{ (b/s)}} = \frac{P_s}{BR} \text{ (W s)} \tag{1.3.9}$$

Why? You are reminded that power is the rate at which energy is transmitted; that is, for a single bit we can write

$$Pb \text{ (W)} = \frac{E_b \text{ (J)}}{T_b \text{ (s)}} \tag{1.3.10}$$

But bit rate, BR (s), is inversely proportional to the bit duration, T_b (s), because the shorter the T_b (s), the greater the number of bits transmitted for one second; that is,

$$T_b \text{ (s)} = \frac{1}{BR \text{ (b/s)}} \tag{1.3.11}$$

Thus, (1.3.9) follows from (1.3.10) and (1.3.11). In words, the energy of a single bit, E_b (J), is equal to the power of the bit, P_b (W), divided by the number of bits generated by a transmitter per second, BR (b/s).

Let us stress again that BR (b/s) is the transmitter bit rate, whereas C (b/s) is the communication's channel capacity. A transmitter could generate 100 Gb/s, but the system could transmit only 40 Gb/s, which means that a receiver would "see" only transmission speed equals 40 Gb/s.

Since bit is a unit of information transmitted digitally, any equation involving E_b (J) can be applied to digital transmission only.

Another quantity we need to introduce is the *noise spectral density*, N_0 (W s), as

$$N_0 \text{ (W s)} = \frac{P_N \text{ (W)}}{\text{BW (Hz)}} \quad (1.3.12)$$

You reminded that P_N (W) is power of noise and BW (Hz) is the available (used) bandwidth. Thus, the noise spectral density is the noise power measured within 1 Hz of bandwidth. We need to realize that BW (Hz) here is the bandwidth occupied by the noisy signal being transmitted through a channel. (Notice that hypothetically a channel could have BW = 100 MHz, but the transmitted signal occupies on 60 MHz. In this case, noise obviously can interfere only with 60 MHz signal.)

Refer to SNR given in (1.3.8) and take into consideration (1.3.10)–(1.3.12) to obtain:

$$\text{SNR} = \frac{P_b \text{ (W)}}{P_N \text{ (W)}} = \frac{E_b \text{ (J)} \cdot \text{BR (b/s)}}{N_0 \text{(W s)} \cdot \text{BW (Hz)}} \quad (1.3.13)$$

Thus, we derived that for digital communications, the SNR can be presented through the ratio $\frac{E_b}{N_0}$. Equation (1.3.13) shows that $\frac{E_b}{N_0}$ is the dimensionless quantity, as an SNR should be.

The ratio E_b/N_0 is called *digital, or normalized SNR* and is considered a figure of merit for digital communications.

Now, we can derive the Shannon law for digital communications from (1.3.6) and (1.3.13) as

$$C \text{ (b/s)} = \text{BW (Hz)} \log_2 (1 + \text{SNR}) = \text{BW (Hz)} \log_2 \left(1 + \frac{E_b \text{ (J)} \cdot \text{BR (b/s)}}{N_0 \text{(W s)} \cdot \text{BW(Hz)}}\right) \quad (1.3.14a)$$

This form of Shannon's law brings us to understanding that this law sets another limit – this time the limit on $\frac{E_b (J)}{N_0 \text{ (W s)}}$. Indeed, assume we manage to transmit binary digital signal at the maximum bit rate; that is, BR (b/s) = C (b/s). Then (1.3.14a) can be rewritten as

$$C \text{ (b/s)} = \text{BW (Hz)} \log_2 \left(1 + \frac{E_b \text{ (J)} \cdot C \text{ (b/s)}}{N_0 \text{(W s)} \cdot \text{BW (Hz)}}\right)$$

This equation can be rearranged as

$$\frac{C}{\text{BW}} = \log_2 \left(1 + \frac{E_b C}{N_0 \text{BW}}\right) \quad (1.3.14b)$$

(We temporarily omit the units to simplify presentation of the formulas.) From here, the digital SNR can be obtained in the form

$$\frac{E_b}{N_0} = \frac{\text{BW}}{C} \left(2^{\frac{C}{\text{BW}}} - 1\right) \quad (1.3.14c)$$

Equation (1.3.14c) shows relationship between digital SNR, E_b/N_0 (J/(W s)) and the ratio C (b/s)/BW (Hz) called *spectral efficiency*, SE ((b/s)/Hz). (SE will be discussed shortly.) Figure 1.3.2 graphically demonstrates this relationship.

Analysis of (1.3.14c) reveals that digital SNR has its minimal value, $\left(\frac{E_b}{N_0}\right)_{\text{limit}} = -1.59$ dB, at which SE approaches zero; that is, Shannon's law cannot be applied beyond this limit. In other words, when $\left(\frac{E_b}{N_0}\right)$ approaches −1.59 dB, "no error-free communications is possible, regardless of the channel capacity.[18]" This digital SNR value is sometimes called *Shannon's limit*. In fact, this is one of the limits hidden in Shannon's law.

Interestingly that when $\left(\frac{E_b}{N_0}\right)$ goes to infinity, SE theoretically goes to infinity too.

Let us consider an example regarding Eq. (1.3.14a).

---

18 http://www.dip.ee.uct.ac.za/~nicolls/lectures/eee482f/04_chancap_2up.pdf.

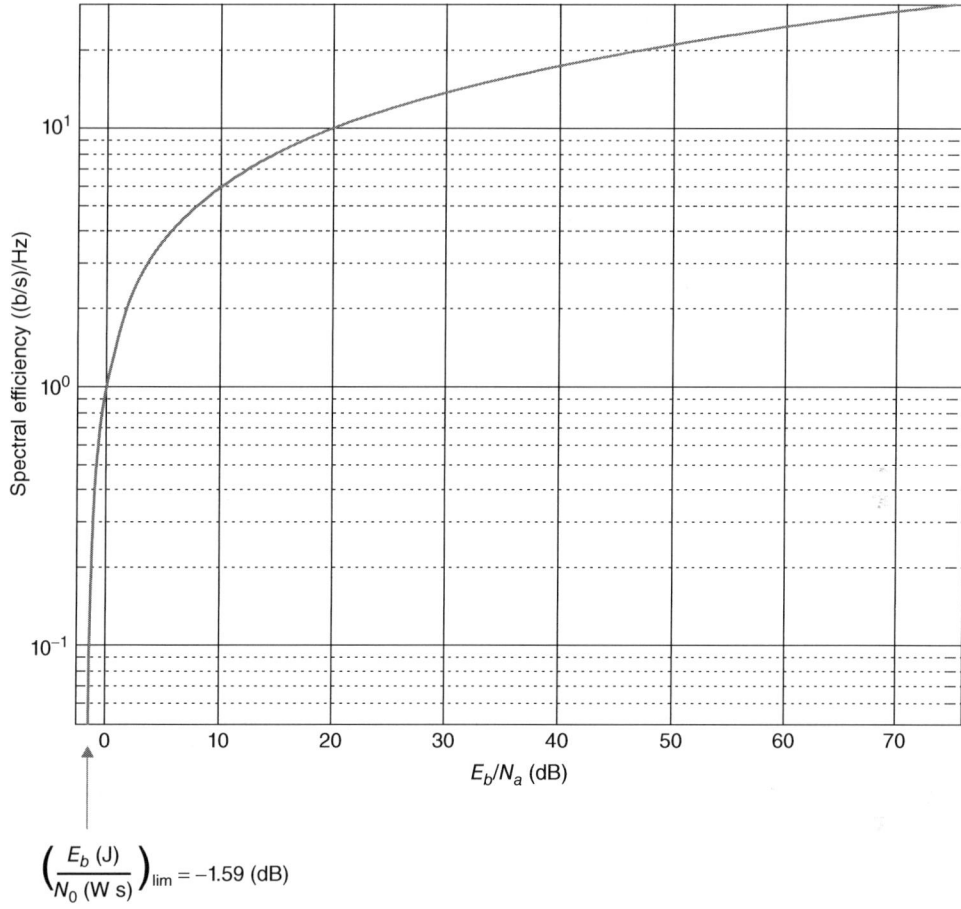

Figure 1.3.2 Spectral efficiency vs. digital SNR.

Example 1.3.4 Shannon's Limit in Digital Transmission

Problem

Compute the Shannon limit for digital communications if $E_b/N_0 = 100$, BR = 100 Gb/s, and BW = 100 GHz.

Solution

Solution is straightforward: Just plug the numbers into (1.3.14a) and compute:

$$C\,(\text{b/s}) = \text{BW (Hz)} \log_2 \left(1 + \frac{E_b\,(\text{J}) \cdot \text{BR (b/s)}}{N_0(\text{W s}) \cdot \text{BW(Hz)}}\right) = 665.4\,\text{Gb/s}$$

Discussion

- This example again demonstrates the main point that follows from the Shannon law: Capacity of a digital channel is determined by its bandwidth and logarithmically depends on digital SNR. (**Question:** $C = 655.4\,\text{Gb/s}$ seems to be a very good capacity for 100-GHz bandwidth and SNR = 100. Have you expected such a good result? What number have you expected?)

- There is another interesting fact regarding energy of a single bit (Mynbaev 2016, p. 1640010-1). Information is physical because it is stored in physical, electronic, and optical systems and therefore, it must obey the laws of physics, specifically, thermodynamics. Thermodynamic entropy, S, has been related to information entropy, H, by Claude Shannon in 1948 as

$$S = k_B H \qquad (1.3.15)$$

where $k_B = 1.38 \times 10^{-23}$ J/K is Boltzmann's constant. Rolf Landauer, then at IBM, derived from (1.3.15) the formula that determines minimum amount of energy needed to erase one bit of information; it is given by Landauer (1966, pp. 183–191)

$$E_{\min} \text{ (J)} = k_B \cdot T \cdot \ln(2) \approx 3 \times 10^{-21} \text{ J} \qquad (1.3.16)$$

where T is room absolute temperature in K. When we reach this limit, we would not be able to process a digital signal with modern CMOS-based electronic devices. To take this number in perspective, the energy of a photon in optical transmission frequency is about 1.3×10^{-19} J, and thus, optical communications always sustains bit energy above the Landauer limit.

The importance of the Landauer limit is that information, regardless of the form in which it is transmitted, must be eventually presented in the form in which a human being can perceive it; that is, the end equipment must be physical, which today means electronic.

At the time of publishing the Shannon's paper in 1948, the limit he imposed on the transmission rate looked more like the point of academic curiosity rather than a practical guideline. Today, just 70 years later, optical communications – the industry that delivered most of the Internet traffic world wide – approaches to Shannon limit very closely. To meet the exponentially increasing traffic demand, the industry must continue to raise the transmission rate of the global optical communication network by the orders of magnitude. To achieve this, there are only two ways, as (1.3.6) shows: either increase the bandwidth of a transmission channel or decrease SNR. The channel in this case is an optical fiber and increasing its bandwidth faces severe technological restrictions that stemmed from the optical fiber's nature. In any event, the bandwidth could be increase at most three times, not an order of magnitude. Practically achievable decrease in SNR would affect C (b/s) even in smaller scale.

Today, the optical communication industry is seeking the solutions to this problem. One of the solutions is to actively employ multilevel modulation, the method to increase the channel capacity without increasing its bandwidth (see Section 10.2). Besides that, the industry is looking for the answers outside the traditional, based on information theory approach. One of the possible directions, where the Shannon limit is not applied, is to involve the other dimension, such as space, to build new types of communication system. Even more promising direction, that could drastically change the communication landscape, is developing quantum communications governing by absolutely different laws and principles.

1.3.2 Fundamental Principles of Modern Communications

1.3.2.1 Channel Capacity, Bandwidth, and Carrier Frequency

Let us review: Channel capacity (maximum transmission rate) is proportional to the bandwidth of a transmission channel. But what does channel bandwidth depend on, we asked in the preceding subsection? Now, it is time to answer this question.

First, let us clarify an important point: A transmission channel is made up of a transmission medium. This is the physical medium through which the signal is transmitted, as was discussed in Section 1.2. To remind, these media are copper wire (twisted pair), coaxial cable, air, and optical fiber. Due to the properties of the material from which a medium is composed, each of these media has its specific bandwidth. This bandwidth is determined by the medium's attenuation, which is the loss of signal power per kilometer of transmission. For example, a copper wire cannot transmit the signals even in a megahertz range over a significant distance because losses become intolerable. A coaxial cable can transmit megahertz signals over a longer distance than a copper wire can, but still has a severe restricted bandwidth. We can transmit low-gigahertz-range signal through air over thousands of kilometers, as satellite communication systems do. Air can even transmit terahertz-range optical signal but over a short distance to provide a reliable communication. An optical fiber is capable to transmit signals in 200-THz frequency range around the globe, but it can do this only within a relatively narrow 8-THz band. Why bandwidths of these media are restricted? Because, again, signals rapidly lose their power outside of their bandwidths. Figure 1.3.3 shows how much signals attenuate traveling through different transmission media. (We will elaborate on this topic in Section 10.1.)

So the channel bandwidths are predetermined by the properties of the channel materials, specifically, by their spectral attenuations.

However, to understand a transmission operation in its entirety, we need to consider the spectrum of a signal being transmitted. We have to know that the signal consists of a message signal containing the information to be delivered and a carrier that carries this message signal. A message signal, by its very nature, is always a low-frequency (baseband) signal. In contrast, a carrier signal must be a high-frequency signal because the higher the *carrier frequency*, f_C (Hz), the more information this signal can carry. (You can skip to Sections 9.1 and 10.1 to read more on modulation principle.)

How does the bandwidth of a transmission channel relate to a carrier frequency? Refer to an optical fiber whose transmission bandwidth is about 8 THz concentrated around 200-THz central wavelength. This means that the frequency of a carrier to be used in fiber-optic communications must be about 200 THz. Indeed, we cannot send a carrier signal in radio-frequency range (about 50 MHz) through an optical fiber because attenuation will be so high that this signal will completely diminish after traveling a very short distance (see Figure 1.3.3a). Similarly, we cannot transmit optical carrier with frequency about 200 THz through a copper wire whose bandwidth is in a low-megahertz range.

Even more straightforward example: If we set the maximum allowed attenuation for a communication system as 20 dB/km, then using a twisted pair, we can transmit signals from 1 kHz to 1 MHz, as Figure 1.3.3 shows. Thus, its available transmission bandwidth is BW_{tp} (MHz) = 10 − 0.001 = 9.999 MHz. But at the same attenuation of 20 dB/km, a coaxial cable can transmit signal from 1 MHz to approximately 100 MHz and therefore its transmission bandwidth is given by BW_{cc} (MHz) = 99 MHz. Important point to realize is that the twisted pair *can* transmit signals at the higher than 1-MHz frequencies; similarly, the coaxial cable *can* transmit higher than 100-MHz signals, but – since the channel must meet a 20 dB/km requirement – their bandwidths that can be employed for the transmission are restricted by 9.999 and 99 MHz, respectively.

The above discussion brings us to the important deduction:

The greater is the needed transmission bandwidth, the higher must be the frequency of a carrier wave.

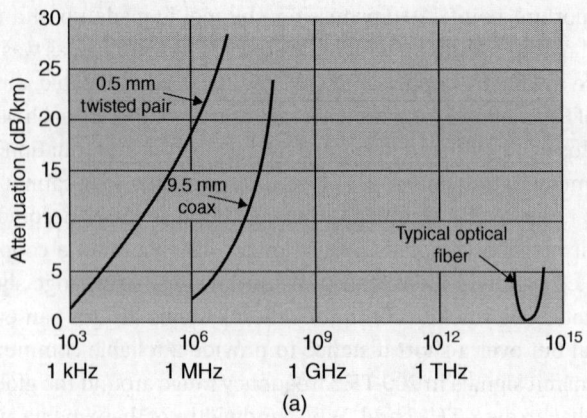
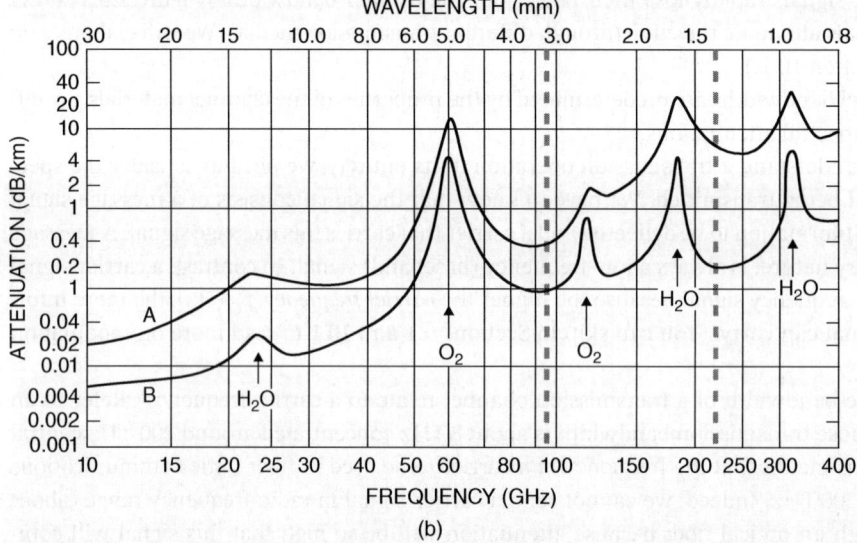

Figure 1.3.3 Bandwidths of four transmission media: (a) Copper wire (twisted pair), coaxial cable, and optical fiber; (b) air. Source: Stallings (2014), William, Data and Computer Communications, 10th Ed. Reprinted by permission of Pearson Education, Inc., New York. (b) U.S. Department of Transportation - Federal Highway Administration (2007).

Therefore, the bit rate – the key parameter of a communication system in this consideration – is proportional to the carrier frequency; that is,

$$\text{BR (b/s)} \sim f_C \text{ (Hz)} \tag{1.3.17}$$

In practice, (1.317) is often replaced by the following rule of thumb: The available transmission bandwidth is approximately equal to one-tenth of a carrier's frequency; that is,

$$\text{BW (Hz)} \sim 0.1 f_C \text{ (Hz)}. \tag{1.3.18}$$

We can summarize this fundamental relationship as follows:

$$C \text{ (bit/s)} \sim \text{BW (Hz)} \sim f_C \text{ (Hz)} \tag{1.3.19}$$

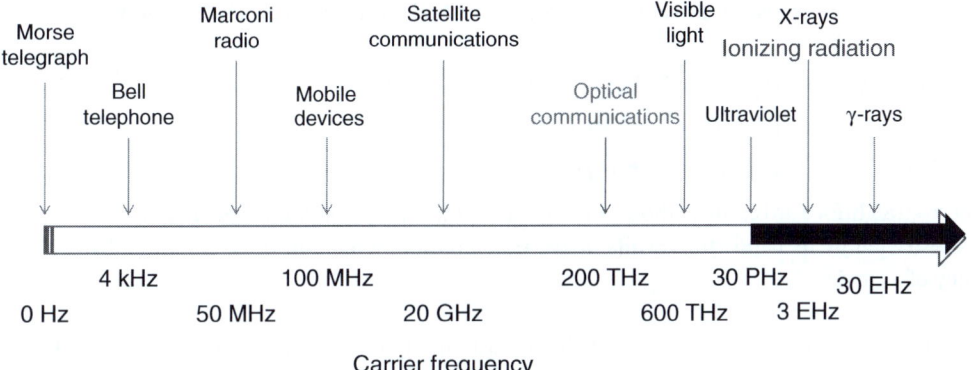

Figure 1.3.4 The progress in modern communications and a historical increase in the carrier frequency.

The principle (1.3.9) demonstrates why communication industry kept increasing the carrier frequency through the history of its development: Using the radio frequencies ranged from 300 kHz to 300 MHz, we can convey voice and music; using frequencies from 300 MHz to 3 GHz, we can transmit a television signal that delivers video information; using optical frequencies in 200 THz range, we can provide streaming services, which transmit online everything we want to watch in real time.

Figure 1.3.4 shows that the communication industry consistently increases the frequency of the carrier, f_C (Hz), to achieve the greater transmission capacity. Consider the EM spectrum from the carrier-frequency perspective: The Morse telegraph uses DC current, the Bell telephone utilizes the spectrum up to 4 kHz, the Marconi radio employs the spectrum between hundreds of kilohertz to hundreds of megahertz, mobile devices use the upper range of radio frequencies, satellite communications utilizes tens of gigahertz, and modern optical communications operates at about 200 THz. (Figure 1.3.4 shows just the range of frequencies, not precise values.)

(**Exercise**: Build the tables showing all major types of modern communications and the range of their carrier frequencies. Referring to Section 1.2, show how the advances in communication technologies relate to the carrier frequencies. (*Hint*: Consider, for example, 5G mobile communications.)

Now we comprehend why, for the last three decades, optical communications has become the major transmission technology: It is simply because light (optical signal) has the highest frequency among all possible carriers.

Example 1.3.5   A Carrier Frequency and Channel Capacity

Problem

A SNR is equal to 10 000. What capacity, C (b/s), can be achieved if a carrier's frequency equals 200 kHz? 200 MHz? 200 THz?

Solution
Plug the given frequency figures in (1.3.18) and find that the transmission bandwidths that can be employed with these carrier frequencies are approximately equal to 20 kHz, 20 MHz, and

20 THz. Now we must use (1.3.6) and compute. (See calculations in Examples 1.3.2 and 1.3.3.) The results are

For BW = 20 kHz, $C = BW \log_2(1 + SNR) = 265.6$ kb/s

For BW = 20 MHz, $C = 265.6$ Mb/s

For BW = 20 THz, $C = 265.6$ Tb/s

These straightforward computations highlight the point we discussed earlier: The channel capacity is restricted by its bandwidth available for transmission, which, in turn, is limited by the frequency of a carrier.

To sum up: Equation (1.3.19) describes the first fundamental principle of communications: *The achievable bit rate is restricted by the carrier frequency.*

(**Question:** *Can we use a carrier whose frequency would be higher than the optical? Justify your answer.*)

1.3.2.2 Bandwidth-Length Product

This fundamental principle of communications states that for any given transmission medium, bandwidth-length product is constant; that is,

$$BW (Hz) \times L (km) = \text{constant} \qquad (1.3.20)$$

where BW (Hz) is the bandwidth as a measure of capacity (transmission bandwidth) and L (km) is the length of the transmission medium. This principle states that there is a trade-off between bandwidth and length for any given medium.

An example of the implementation of this fundamental principle can be found in the computer-hardware area. One company had developed a new chip enabled data transmission within a computer circuit board at the bit rate as high as a trillion bits per second. A key to this success was putting the transmitter and receiver chips in direct contact, thus eliminating the need for connecting wires. In other words, this design shortens the transmission distance from millimeters to micrometers enabling the company to increase the transmission speed substantially. This is how the bandwidth-length-product principle works in practice. (We will return to justification and discussion of this principle in the subsequent chapters.)

Example 1.3.6 Application of the Bandwidth-Length-Product Principle

Problem

A manufacturer specifies the bandwidth-length characteristic of a multimode optical fiber as 300 MHz km. What is the transmission bandwidth of this fiber if the distance is equal to 2 km?

Solution

The answer is found by applying (1.3.20): The bandwidth in question times the given length must be equal to 300 MHz-km; that is,

$$BW (MHz) \times 2 \text{ km} = 300 \text{ MHz} \times \text{km}$$

Thus,

$$BW (MHz) = 300 \text{ MHz} \times \text{km} / 2 \text{ km} = 150 \text{ MHz}$$

The problem is solved.

Discussion

Can we use this optical fiber to transmit at 600 MHz? (Remember, the bandwidth here is the measure of capacity of this optical fiber.) Yes, we can but only if the length of the optical fiber would be equal to

$$L \text{ (km)} = 300 \text{ MHz} \times \text{km}/600 \text{ MHz} = 0.5 \text{ km}$$

This is because, again,

$$600 \text{ MHz} \times L \text{ (km)} = 300 \text{ MHz} \times \text{km}$$

This is a practical example of implementation of the bandwidth-length-product principle.

Consider another example of the use of this principle: Standards regulating transmission in LANs, which connects several computers within a short distance, always specify bandwidth and transmission distances for the medium being used. The popular medium in LAN connections, Category 5 cable (which is the unshielded twisted pair that is a copper wire) can be used for transmission of 100 MHz up to 800 m (2624 ft). If we need to transmit 200 MHz, we must either halve the distance or, to save the distance, use a better medium, such as an optical fiber.

Still another example: A cooper wire (twisted pair), the link traditionally associated with low-bit-rate transmission, is still in use in the modern data centers transmitting data stream at 25 Gb/s. The secret? It does so only over a few meters.

Thus, when someone asks what is the bandwidth of this or that medium, our response must be, "For what distance?"

Why does the product of so different entities as bandwidth and length impose a limit on the capacity of a transmission channel? As discussed above, the channel's bandwidth is the measure of the channel ability to transmit data flow at a certain bit rate. If we consider a transmission channel (link) as a pipe, then the bandwidth corresponds to the diameter of this pipe. Then, the greater the diameter (bandwidth, that is), the faster the data flow this pipe (link) can afford. In this analogy, the bandwidth-length product is the volume of our pipe; this volume determines how many bits in total this pipe (link) can accommodate. This number – the maximum amount of bits a given channel (link) can accommodate – is constant and determined by the physical properties of a medium from which the link is made up.

In computer networks, the important parameter is a message latency (delay) rather than the channel length. Nevertheless, since the latency is mainly determined by the link length, the principle BW (Hz) × L (km) = constant can be also applied in this case.

1.3.2.3 Power-Bandwidth Trade-Off

Referring to definition of SNR given in (1.3.8), we can rewrite the Shannon law given in (1.3.6) as

$$C \text{ (b/s)} = \text{BW (Hz)} \log_2 \left(1 + \frac{P_S(W)}{P_N(W)}\right) \tag{1.3.21}$$

Since C (b/s) is the limit (constant), (1.3.21) determines the power-bandwidth trade-off. Indeed, we cannot control the noise power; hence, we consider P_N (W) as a given number. Thus, increasing the signal power, P_S (W), must cause reducing the bandwidth, BW (Hz) in order to satisfy (1.3.21).

The well-known application of this principle is the *spread-spectrum transmission technology*. Once classified and only military-applicable technique, the spread-spectrum communications today is the basis for GPS and for many mobile transmission technologies, including 4G and Bluetooth, all of which are available for civilian service.

The technology spreads a transmitted signal over the bandwidth much larger than the signal's original bandwidth. As a result, the power of individual signal's component gets smaller than the power of noise so that SNR becomes less than 1. This exchange power for bandwidth enables the technology provide secure and reliable communications, as we – the users of GPS, cell phones, and Bluetooth – can attest.

Under condition SNR = $\frac{P_S(\text{W})}{P_N(\text{W})} \ll 1$, Eq. (1.3.21) can be reduced to[19]

$$\text{BW (Hz)} \cdot P_S (\text{W}) \approx C \text{ (b/s)} \cdot P_N (\text{W}) \tag{1.3.22}$$

Since C (b/s) is the limit and P_N (W) is fixed for a given communications, (1.3.22) shows that the bandwidth-power product in a spread-spectrum technology is constant.

Another important example of employing this principle is *multilevel modulation* (signaling). This communications technique packs several bits in one symbol for transmission at a given bit rate but within a smaller bandwidth. Saving the transmission bandwidth is achieved by injecting more power in a transmitted signal. Section 10.1 provides a detailed discussion of multilevel signaling; one of its subsections is devoted to the use of bandwidth-power trade-off in this communication technology. Chapters 7 and 9 further clarify this trade-off by considering it from the spectral-analysis standpoint.

1.3.2.4 Spectral Efficiency and Transmission Technology

Spectral efficiency is not a principle, but the measure – the figure of merit, if you wish – of modern electronic communications. First, let us introduce the concept of spectral efficiency, SE: This measure determines how efficiently a given bandwidth is used for transmission of a required data stream. The more bits per second we can transmit per 1 Hz of available bandwidth, the more efficiently we use this bandwidth. Thus, spectral efficiency is defined as follows:

$$\text{SE ((b/s)/Hz)} = \frac{\text{Actual bit rate (b/s)}}{\text{Available bandwidth (Hz)}} = \frac{\text{BR (b/s)}}{\text{BW (Hz)}} \tag{1.3.23}$$

To fully understand any term, we should analyze the dimension of this term: Here, the unit (b/s) is the measure of an actual transmission (bit) rate and (Hz) is the measure of the bandwidth available for transmission of this data stream. A thorough look at the SE dimension reveals that this quantity is, in fact, dimensionless because b/s = 1/s and Hz = 1/s. The industry, however, always keeps the dimension ((b/s)/Hz) to emphasize the meaning of spectral efficiency.

The importance of SE cannot be overestimated: Bandwidth is a commodity; all frequency spectra usable for electronic communication in the United States is controlled by the FCC and the spectra are mostly assigned or licensed (refer to Section 1.2). A provider of mobile phone service, for example, needs to purchase a license for certain portion of the spectrum, which then will be locked in with a given bandwidth. But the company needs to constantly increase the data rate because the volume of telecommunications traffic continuously grows. Ultimately, the faster a provider can transmit data, the greater will be its revenue. One way to increase bit rate is to use more bandwidth, as Shannon's law shows, but this approach does not work since the bandwidth is fixed. Therefore, in reality, the only way to increase the data rate is increasing spectral efficiency, as (1.3.23) indicates.

The principle concerning spectral efficiency can be formulated as follows:

The greater the spectral efficiency of a transmission system, the better the system.

The term *better* can be quantified by various measures but all of them will stem from the key point: the spectral efficiency shows how efficiently we use an available transmission bandwidth, which is commodity.

---

19 https://www.maximintegrated.com/en/app-notes/index.mvp/id/1890.

This parameter and its applications will be discussed in later chapters, especially in Chapters 9 and 10; here we need to make one point:

How can we improve (increase) the spectral efficiency? Equation (1.3.23) shows the possible ways: Either increase the bit rate over the fixed bandwidth or reduce the bandwidth with the given bit rate or (ideally) do both.

Increasing the bit rate (BR) implies squeezing more bits per second. We can increase the BR by reducing bit time, T_b, the time occupied by each bit. However, this straightforward approach confronts a serious practical hurdle: At 100-Gb/s bit rate, the bit time becomes equal 0.01 ns, or 10 ps, and modern electronic technology experiences difficulty in directly generating such a bit stream without significant errors. To resolve this problem, the industry employs various methods to directly raise bit rate; the most popular among them is multiplexing – transmitting several (many) channels over the same transmission line.

Reducing the required bandwidth can be achieved by using multilevel modulation techniques and by employing filters, the circuitry that filters out unnecessary (unwanted or negligible) parts of the spectrum.

In any event, increasing (improving) the spectral efficiency relies on using better (that is, more sophisticated and, therefore, more expensive) equipment; hence, implementation of this principle depends on technology. Thus, we see the consequence of this principle:

The better the spectral efficiency, the more sophisticated (complex) the equipment that must be used in a transmission system.

This statement shows that there is a trade-off between the spectral efficiency of transmission and the complexity of the technology used for this transmission.

Spectral efficiency is one of the most important characteristics of communication systems, but it is not the only one. For example, an updated DSL transmission mentioned above could have an SE up to 14 (b/s)/Hz, but nobody wants to replace the modern optical access networks with a DSL line just to achieve higher spectral efficiency. This is because the optical network yields a much higher transmission rate, which is the decisive factor in choosing the link type to provide the access from an individual customer to the global communications infrastructure. Nevertheless, increasing the spectral efficiency is one of the major objectives of a designer of modern communications networks.

Finally, it must be noted that Shannon's law (1.3.6) imposes the limit on the spectral efficiency. Indeed, dividing channel capacity, C (b/s) by channel bandwidth, BW (Hz), in (1.3.6) yields

$$\frac{C\,(\text{b/s})}{\text{BW (Hz)}} = \text{SE}_{max} = \log_2(1 + \text{SNR}) \tag{1.3.24}$$

It follows from (1.3.24) that SE_{lim} is the measure of capacity per unit bandwidth and it is limited by the channel's SNR. (Refer to discussion of Figure 1.3.2.) Do not forget that C (b/s) determines the *maximum bit rate* for error-free transmission.

We will discuss spectral efficiency in later chapters.

1.3.2.5 Bit Rate vs. Bandwidth in Digital Transmission

To summarize our discussion of the fundamental principles of modern communications, we need to discuss why *signal's bit rate is proportional to channel's bandwidth*, as predicted by the main communication laws (1.3.3), (1.3.4), and (1.3.6).

As we will learn in Chapter 7, the pulse bandwidth, BW (Hz), is inversely proportional to the *pulse duration*, T_b (s); that is, the narrower the pulse, the larger the signal's bandwidth and vice versa. This is a well-known phenomenon called *reciprocal spreading*. The physics behind this phenomenon is that *the faster the time variations of a signal the greater the number of frequencies that its spectrum contains and vice versa*. A pulse perfectly exemplifies this point: Pulse's width, T_b (s), shows how fast the pulse changing in time. Specifically, the smaller the T_b (s), the faster the changes; hence, a narrower pulse has a greater number of frequencies in its spectrum than a wider one. Therefore, T_b (s) is a measure of the broadness of a pulse spectrum; that is, its bandwidth.

On the other hand, pulse's width, T_b (s), is inversely proportional to bit rate, BR (b/s); that is, BR (b/s) = $1/T_b$ (s). Now, it would seem that we arrive to a familiar relationship: Both bandwidth and bit rate are inversely proportional to T_b (s), and therefore, one is proportional to the other. But we need fully understand that it is a *digital signal's bandwidth, BW_{sig} (Hz), that is inversely proportional to pulse width, T_b (s)*, whereas Shannon's and other fundamental laws refer to a channel's bandwidth, $BW_{channel}$ (Hz). This logical gap is closed by the following reasoning: *An increase in BR (b/s) requires the decrease in T_b (s), which causes the increase in BW_{sig} (Hz), which – in turn – leads to the increase in $BW_{channel}$ (Hz).*

1.3.3 Laws, Principles, and Models – Importance, Limitations, and Applications

1.3.3.1 Limitations and Applications of the Laws and Principles

Every principle, law, and equation are based on assumptions and have their limitations, that is, the boundary of their applicability.

Let us start with **Ohm's law**, the first law that electrical engineering students meet in their specialty classes. The law states

$$V(V) = R(\Omega) \cdot I(A) \tag{1.3.25}$$

But even for DC circuit, this law holds true only if the ambient temperature is constant. This is the first condition (or limitation, if you wish) to observe when performing measurements. Why temperature is a factor in Ohm's law? Refer to basic physics underlining this law: Current is a flow of electrons; the electrons, on their way from one circuit terminal to the other, bouncing among heavy ions that make up the structure of a conductor. These ions constantly vibrate. When the electrons collide with the ions, they are reflected or deflected; in either event, the electron flow gets impeded. The level of this impedance is the measure of a conductor resistance. When temperature rises, amplitudes of ion vibrations increases; consequently, impedance to electron flow raises, which is measured as an increase in conductor's resistance.

In discussion of the meaning and limitations of any equation, it is always useful to consider the extreme conditions. Let us return to Ohm's law: How (1.3.25) will work if ambient temperature approaches the absolute zero? (Reminder: Absolute zero means 0 K.) What if ambient temperature is equal to 5778 K? (Get to know that 5778 K is the temperature at the sun's surface.) Answering these questions (qualitatively, of course) help you to fully understand limitations of Ohm's law in regard to temperature. (Be aware that as conductor's temperature approaching 0 K, its resistance drops to zero, the condition called superconductivity. This phenomenon enables many modern technological marvels, the chief among them being Large Hadron Collider [LHC]. This is the world's largest and most powerful particle accelerator and is, in fact, the largest machine in the world. For its operation, LHC needs electric current up to 12 000 amperes (!), for which the superconducting cables have to be used. Ohm's law given in (1.3.25) certainly cannot be applied here. Go online to learn more about this unique apparatus operating at the very boundary of scientific knowledge.)

Another limitation in Ohm's law concerns with signal's frequency. Electrical engineering students usually study Ohm's law in analyses of DC and AC circuits, where AC typically refers to the signals in a low-frequency range. But will Ohm's work for high-frequency signals? The answer to this question depends on our interpretation of Ohm's law. If we define Ohm's law as Eq. (1.3.25), the answer is *no* because this equation is fully true only for DC circuits. It can be used for low-frequency AC circuits as long as we are satisfied with this approximate approach. Also, this consideration implies that the law is applied to a circuit with lumped elements. However, if we consider Ohm's law as the rule of linearity between current-related and voltage-related quantities, then the answer is *yes* because (1.3.25) can been generalized as

$$E \text{ (V/m)} = \rho \text{ } (\Omega \text{ m}) \cdot J \text{ (A/m}^2) \tag{1.3.26}$$

Here E (V/m) is the electric field strength, ρ (Ω m) is the resistivity, and J (A/m^2) is the current density. Equation (1.3.26), however, describes not circuits but electromagnetic (EM) field, where high-frequency signals belong. Bear in mind, too, that in EM field there is no lumped elements and its property (resistivity in this example) is distributed in space.

Is there a strict frequency value that would border the low-frequency and high-frequency ranges in terms of applications of (1.3.25) and (1.3.26)? No. Engineers have to decide which model serve better for every specific problem they encounter.

(Question: Can Ohm's law (1.3.25) be applied to an electrical circuitry whose operational voltage is 100 kV? How about 100 A? Explain.)

Let us now consider **Shannon's law** – the fundamental relationship governing modern communications regardless of its technology. Equation (1.3.6) states the maximum channel capacity, C (b/s), is determined by the channel's bandwidth, BW (Hz), and depends on SNR.

$$C \text{ (b/s)} = \text{BW (Hz)} \log_2(1 + \text{SNR}) \tag{1.3.6R}$$

> We studied that Shannon's law puts, in fact three limits:
>
> 1. A communication system's bit rate, BR (b/s), cannot exceed the channel capacity, C (b/s), to support error-free transmission. See Eq. (1.3.6).
> 2. A digital SNR has the minimum value, $\left(\frac{E_b}{N_0}\right)_{\text{limit}} = -1.59 \text{ dB}$, delimiting the possibility of error-free transmission. See Figure 1.3.2.
> 3. Spectral efficiency, SE ((b/s)/Hz), is limited by the channel's SNR; that is, $\frac{C \text{ (b/s)}}{\text{BW (Hz)}} = \text{SE}_{\text{lim}} = \log_2(1 + \text{SNR})$. See Eq. (1.3.24).

Besides these limitations, it appears that application of Shannon's laws has its boundaries too: Transmission channels of communication technology of 1940s, when Shannon presented his law, were copper wires and cables. These communication links have been adequately described by linear models; that is, their properties (resistance, capacitance, and inductance) do not depend on the strength of transmitted signals.

Situation has changed completely with the advent of optical fiber. Today, vast majority of the global communication traffic is delivered optically, whereas copper-based transmission links play minor role. An optical fiber is a nonlinear transmission channel because its refractive index depends on the strength of an optical signal being transmitted. (See, for example, Mynbaev and Scheiner 2001, pp. 195–204.) When the power of an input signal is small, Shannon's law works well; when, however, the input power is big, Shannon's equation does not hold true. To explain, we refer to spectral efficiency given in (1.3.24) that states that SE should grow indefinitely with an increase in a signal power because this increase raises SNR. Including nonlinear effect in analysis of transmission capacity of an optical fiber (Mitra and Stark 2001, pp. 1027–1030) shows that the

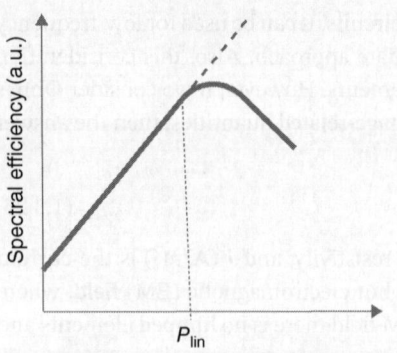

Figure 1.3.5 Spectral efficiency vs. signal power in optical fiber. (a.u. stands for arbitrary units.)

capacity of nonlinear channel has its limit. The graphs depicting a linear course (dotted line) and nonlinear Shannon's law (solid line) are in Figure 1.3.5. It shows that when signal power exceeds a linear value, P_{lin}, the spectral efficiency not only stops growing but starts declining. The further increase of the signal power causes bigger decrease in spectral efficiency. Therefore, Shannon's law in form (1.3.24) does not hold true.

This example demonstrates that even such fundamental law as Shannon's limit must be applied only after verifications of all conditions and assumptions under which the law holds true.

1.3.3.2 Models

We constantly refer to *model* of a component, system, or other entity but have never defined what it is. In engineering,

a model *is a mathematical description of a real system or a real process.*

Here is the simple example: The model of a circular wire resistance, which every electrical engineering student knows very well. Referring, for certainty, to a popular textbook on circuit analysis (Boylestad 2016, pp. 63–75.), we find the following model that emerged from brief explanations:

$$R\,(\Omega) = \rho\,(\Omega\,\text{cm}) \frac{l\,(\text{cm})}{A\,(\text{cm}^2)} \tag{1.3.27}$$

Here, $R\,(\Omega)$ is the resistance, $\rho\,(\Omega\,\text{cm})$ is the material resistivity, l and A are the length and the cross area of the wire. This model gives the correct resistance value if the ambient temperature is constant. To consider the temperature effect, more sophisticated model given as

$$R\,(\Omega) = \rho\,(\Omega\,\text{cm}) \frac{l\,(\text{cm})}{A\,(\text{cm}^2)} (1 + \alpha \Delta T) \tag{1.3.28}$$

is needed. Here α (1/°C) is the temperature coefficient of resistance and $\Delta T\,(°C) = T - 20\,°C$ with $T\,(°C)$ being an actual ambient temperature. Using (1.3.28), our calculations of the resistance produce more accurate result than model given in (1.3.27). But we must be aware that both of the above models give us an average resistance, whereas in reality its value changes every instant. (See subsection "The model of noise in Shannon's law" in Section 9.1.) What if we need to know an instantaneous resistance value? We must further develop model determined by (1.3.25) to reflect this phenomenon. If we need to take into account the effect of the material (contents of alloy), we must add new equation to (1.3.25). So, to increase the accuracy of our model, new equations or new members of the existing equations must be added. Eventually, the model of resistance becomes very accurate but too complicated for the practical use. So at what stage the further increase in a model accuracy should be stopped? Obviously, it depends on the application: For rough estimation of resistance at an entry-level student laboratory (1.3.27) would work perfectly; for precise, nano-scale measurements the most accurate model would be needed.

The conclusion we should draw from this discussion is that choosing the good model is more art than science.

So should we avoid modeling altogether? No, we must create models because the real-life situations are too sophisticated, and they are affected by so many factors that we simply cannot describe these situations by reasonably sized mathematics. What is the solution to this problem? Create a model that, from the one hand, provides the best description of a real system and, from the other hand, be computational and capable to deliver the results at the reasonable expenditure in both time and cost. How? Again, there is no equation or scientific method to follow; this is not a science but art based on the knowledge, experience, intuition, deep understanding of the situation, and eventually the talent of an engineer and scientist.

Let us consider the classical example from the history of physics related to the fundamental laws of nature: As we know, special theory of relativity (1905) and general theory of relativity (1915), the mathematical theories of processes in our nature, were discovered by Albert Einstein. Naturally, their creation was a difficult journey; in developing these theories, Einstein tried various approaches, abandoned some of them, tried new ones, and repeated his attempts again and again. Fortunately, he succeeded on both efforts; the results were the equations describing the subjects of his research. While working at special theory of relativity, Einstein learned that the famous French mathematician named Henri Poincare was also developing equations describing the same problem. Similar situation appeared again in 1915, when Einstein got to know that another prominent mathematician, David Gilbert from Germany, was working at the developing field equations, the core results of general theory of relativity. But today, we rightly consider Albert Einstein as an author of both theories, even though the mathematical results of Henri Poincare and David Gilbert are very close to the equations discovered by Einstein. This is because Einstein's results are closer to reality; the most predictions of his theories are confirmed today by the experiments. Why Einstein was more successful in these cases than his rivals who seemed to be more mathematically qualified for the research? In developing his theories (models, in essence), he mainly relied on his intuition and deep understanding of physical processes. This approach enabled him to correctly decide what features of the processes must be included into his mathematical models and what might be neglected for simplifying the mathematics without increasing the gap between the models and the real processes. This approach allowed him to develop the correct mathematical description of the concepts of physics. The point we can draw from these historical episodes is that successful describing real-life processes and situations by mathematical equations requires the reasonable approach, but the term *reasonable* cannot be described by a mathematical equation.

We cannot underestimate, however, the role of mathematics in revealing the real processes of our world. Here there are two examples, again regarding the fundamental laws of the universe. Every engineer and physicist know *positron*, an elementary particle that has the same properties as electron except of the positive electric charge. The existence of positron was predicted in 1928 by Paul Dirac, prominent British scientist and one of the fathers of quantum mechanics. This prediction appeared because Dirac's equation required the existence of a particle which would mirror an electron but must be positively charged. It was even initially called anti-electron. Again, positron appeared under the Dirac's pen. In fact, by his equation, Dirac predicted the existence of anti-matter, the result that changes our view of the universe. Positron was discovered experimentally in 1932; the existence of anti-matter has also been confirmed in numerous experiments since then. Anti-matter and its interaction with matter play important role in understanding the natural processes in our universe; as for positron, it finds its applications in medicine and science. And all these vitally important results emerge from one mathematical equation.

The second example: In 1964, British physicist Peter Higgs published a paper in which he mathematically predicted existence of what is now called Higgs field and Higgs boson (elementary particle). The Higgs's theory fundamentally changes our understanding of how the universe "works." This is why it must be proved or disapproved, for which a particle accelerator must be created. It took 13 years (from 1995 to 2008), about $10 billion, and collaboration of the world-wide scientific and engineering community to build and put in operation Large Hadron Collider (LHC). This is, as mentioned previously, the greatest scientific instrument ever built. In 2012, after series of experiments and long period of calculations and verifications of the results, it was announced that Higgs boson had been discovered. Incidentally, calculations were done simultaneously in 170 computer centers located in 36 countries through the grid-based computer network. To constantly analyze all collision events produced by 2012, 25 petabytes (25×10^{15} bytes) collision data had to be transmitted annually. And all these expenditures, intellectual and material global efforts were put in action based only on the mathematical prediction!

These two modern examples show that mathematical models can initiate the practical steps that change our understanding of the universe and result in creating new practical applications. As a scientist puts it, regarding LHC story, "Before the elusive Higgs boson could be discovered – a smashing success – it had to be imagined.[20]"

1.3.3.3 Modeling and Simulation

Nowadays, the role of mathematical models in engineering practice keeps increasing. This is because an enormous progress in computer science in both software and hardware areas enables the industry replace the tangible prototyping by accurate computer simulation. Traditionally, engineers first built a physical prototype of a designed circuitry, machine, or system. Then they tested it, changed the pieces or even whole design, tested again, and repeated this process many times before starting the mass production. It required huge amount of efforts, resources, and – most importantly – time to complete the design process from an idea to the final product. Fifty years ago, this cycle typically took from 5 to10 years. Fortunately, these days long gone. Today, for a typical midsize product, such a cycle can take from hours to days to months. Of course, designing a new spacecraft will take much longer period of time, but, still, it will be very short interval comparing with the design of such an object 50 years ago. The advantage of application of digital technology in the real manufacturing processes enabling industrial companies not only optimize these processes but also hone the usage of their products. For example, General Electric (GE) is creating virtual replica (called digital twin) for each its aircraft engine being in use. This allows GE to follow the aging of each individual engine and prevent their possible malfunctions.

All this progress in designing and manufacturing processes has been achieved thanks – let us repeat – precise computer simulation of a real systems and processes. But – this is the main point – computer simulation is based on the mathematical models of a real product, and the *simulation accuracy cannot be better than the accuracy of the model*. Returning to the story about the Higgs boson and the LHC operation, to verify the discovery, all obtained experimental data were compared with simulation data and only those experimental data that coincided with the simulated ones were selected for the further processing. This fact highlights the role of a mathematical model in such an epochal experiment. The success of this experiment is the triumph of power of a correct mathematical model.

---

20 Greene (2013).

We must remember, nevertheless, that there is always a gap between the model and its real-life object and even smallest deviations the model from the real object could result in erroneous outcomes.

The conclusion we should draw from this subsection is that the laws and principles used as guiding instruments in analysis of real-life engineering tasks are not more than models having their limitations and constrains; therefore, we have to use them mindfully.

Questions and Problems for Section 1.3

- Questions marked with an asterisk require a systematic approach to finding the solution.
- Many questions and problems, including those marked with an asterisk, imply that you, in addition to reading the textbook, will do your research to find the answers. Consider such questions as mini-projects.

Fundamentals Laws of Modern Communications

1. Average size of an HD (high-definition) movie is approximately 7.5 gigabytes. How long will it take to download this movie if the maximum transmission rate of an access link is 500 Mb/s? (Use (1.3.1), $T = H/BR = 8 \times 7500/500 = 120$ s.)

2. What bit rate is required to deliver 32-megabyte file for 4 ms? (BR $= 32 \times 8/4 = 64$ Gb/s)

3. Typical real transmission speed provided by PON (passive optical network) is 10 Gb/s. What volume of information can be downloaded with this technique for 1 ms? 10 ms?

4. In Section 1.2, we compare a communication point-to-point link with a water supply system. Based on this analogy, derive the equation governing the relationship among water volume, the velocity of water stream, and time of supply process.

5. The bandwidth of a Bluetooth channel is 20 kHz. What bit rate it can support?

6. It would seem that Eqs. (1.3.2b) and (1.3.3) state basically the same: The bit rate that a transmission link can support is either equal or less than the link's bandwidth. Why then, Nyquist formula, (1.3.3), is considered as a fundamental statement?

7. Compare (1.3.3) and (1.3.5): Why does the text say that Hartley's capacity law, (1.3.5), followed from Nyquist formula, (1.3.3) given that the former is an equation and the latter contain an inequality sign?

8. *Consider Hartley's capacity law, (1.3.5): As we will learn in Section 10.1, the number of bits per level (symbol), N_b, relates to the number of levels (symbols), M, as $M = 2^{N_b}$, or $N_b = \log_2 M$ (see Eqs. (10.1.2a) and (10.1.2b).)

9. If we use multilevel modulation in which one level carries two bits, what will be channel capacity, C (b/s), with respect to channel bandwidth, BW (Hz)?

10 We know that a channel bandwidth, BW (Hz) is restricted by the properties of a transmission medium and by regulations, but it seems that (1.3.5) states that we can increase the channel capacity indefinitely by increasing M. Are there limitations on M in reality? Give the answer on the intuitive level, providing your reasoning. Then, for extra credit, jump to Section 10.1 to study the topic.

11 Equation (1.3.5) works only for a noiseless channel, the abstraction that cannot exist in reality. What is, then, the value of this equation?

12 Discuss Shannon's law (limit). Explain the meaning of each member of this formula.

13 *Consider Shannon's law (1.3.6):
   a) The bandwidth of a twisted copper pair is 10 MHz and the signal-to-noise ratio is 1000. What is the channel capacity of this medium?
   b) Assume the capacity of a fiber-optic link is 100 Tbit/s. What bandwidth this link provides if its signal-to-noise ratio is 6200?
   c) What signal-to-noise ratio is required to provide channel capacity equals bandwidth of the link? Do you want this equality? Explain.
   d) What will be the ratio of channel capacity to bandwidth for signal-to-noise ratio equals 100? 1000? 10 000? Comment your results.

14 The most popular wavelength used in fiber-optic communications technology is 1550 nm, which is approximately 193.5 THz. What bandwidth do you expect of an optical fiber operating at this wavelength?

15 *The strict formulation of Shannon's law states: If bit (data) rate, BR (b/s), does not exceed the maximum channel capacity, C (b/s), then the transmission can be provided with arbitrarily low error probability by using intelligent coding techniques. This statement is usually simplified as follows: If BR (b/s) $\leq C$ (b/s), then transmission is error free. What if BR (b/s) > C (b/s)?

16 *Shannon's law, (1.3.6), states that the channel capacity, C (b/s), is proportional to the channel bandwidth, BW (Hz). Why? What physics does underlie this phenomenon?

17 If BW = 8 THz and SNR = 14, then C = 31.24 Tb/s. If SNR increases to 28 and BW decreases to 6.43 THz, then C will stay the same. Why?

18 *Equation (1.3.14c) shows relationship between digital SNR, E_b (J)/N_0 (W s), and spectral efficiency, SE = C (b/s)/BW (Hz), as $\frac{E_b}{N_0} = \frac{BW}{C}\left(2^{\frac{C}{BW}} - 1\right)$:
   a) Prove that digital SNR has its minimal value. Find this value.
   b) What will be the value of SE when digital SNR reaches its minimum?
   c) According to the text, Shannon's law cannot be applied beyond the minimal value of SE. Why?

19 Compute the Landauer limits at the following values of ambient temperature:
   a) Room temperature of $T = 27\,°C$.

b) Cryogenic temperature of –271.3 °C, at which the main magnets of the Large Hadron Collider (LHC) operate.
c) Sun' surface of 6000 °C.
d) Comment on your results. (*Hint*: $T°K = T°C + 273$.)

20 To increase achievable channel capacity, C (b/s), we need, according to Shannon's law, increase the channel bandwidth. The text explains that increasing optical fiber's bandwidth faces severe technological restrictions that stemmed from the optical fiber's nature. What restrictions does the text refer to?

Fundamental Principles of Modern Communications

21 Examine Figure 1.3.3a:
   a) Can a 0.5-mm twisted pair transmit at 1 GHz?
   b) Is it possible to transmit a signal at 1 GHz through a 9.5-mm coaxial cable?
   c) Can a typical optical fiber deliver a signal at 1 GHz?
   d) Explain your answers.

22 Figure 1.3.3a shows spectral attenuation of a typical optical fiber: At $f_C = 100$ THz, $A_1 \approx 5$ dB/km and when $f_C = 200$ THz, $A_2 \approx 0.2$ dB/km. What transmission distances can be covered with A_1 and A_2 if losses in both cases are equal to -30 dB? (*Hint*: $A\text{ (dB/km)} = \frac{-\text{Loss (dB)}}{\text{Distance (km)}}$.)

23 Review Figure 1.3.3a: In the order of magnitudes, what channel capacities can a twisted pair, a coaxial cable, and optical fiber support provided that each medium utilizes its minimal attenuation? (*Hint*: Refer to (1.3.19).)

24 Wireless transmission, as explained in Section 1.2, needs to expand its spectrum to high-frequency range. What frequency windows on the graphs shown in Figure 1.3.3b would you choose for this expansion and why?

25 History of communication technology shows that to increase the maximum transmission rate, C (b/s), we need to increase a carrier frequency, f_C. Why, then, modern communications does not pursue this trend and stop at optical frequencies?

26 A manufacturer specifies the bandwidth-length product of a multimode optical fiber as 600 MHz-km. Over what distance can this fiber transmit 1200 MHz? 300 MHz?

27 Category 6 cable, which is unshielded twisted-pair (UTP) cable, can transmit 250 MHz over 800 m (2624 ft). What distance you can reach if you need to transmit 500 MHz? 125 MHz?

28 Usually standards for local area networks (LAN) specify transmission speed for these networks. For instance, a specific type of LAN called Ethernet was specified as 10-Mbit/s and 100-Mbit/s Ethernet. However, latest standards developed by IEEE (Institute of Electrical and Electronics Engineers) for LANs specify the length of a cable rather than transmission speed. For example, Gigabit Ethernet is specified as 2 km transmission length with multimode optical fiber and 10-Gigabit Ethernet allows for 300-m distance. Comment on these specifications.

29 Equation (1.3.22) shows that, under condition SNR = $\frac{P_S \text{ (W)}}{P_N \text{ (W)}} \ll 1$, the bandwidth-power product is constant. How this constancy is used in a spread-spectrum technology?

30 A multilevel modulation packs several bits in one symbol for transmission at a given bit rate but within a smaller bandwidth. How bandwidth-power trade-off is used in this technique?

31 Consider spectral efficiency, SE ((b/s)/Hz), as defined in (1.3.23):
   a) What is the meaning of this parameter?
   b) Do you want big or small SE? Explain.
   c) List all the directions for increasing SE ((b/s)/Hz) and show their limitations.
   d) Is there a limit for increasing spectral efficiency? (See (1.3.24).)

32 Calculate spectral efficiency of various communication systems:
   a) A Wi-Fi transmission uses 22-MHz channel bandwidth to transmit at BR = 54 Mbps. What is SE of this transmission?
   b) A fiber-optic communication system support BR = 10 Tb/s over 4.45-THz bandwidth. Find the SE.
   c) Which system does utilize its bandwidth more efficiently? Explain.

33 Consider dimension of spectral efficiency, (b/s)/Hz:
   a) The text says that the spectral efficiency is a dimensionless quantity. Why?
   b) If SE ((b/s)/Hz) is the dimensionless, why does communications industry continue to show this dimension?

34 *Compare Eqs. (1.3.14c) and (1.3.24): Do they describe the same phenomenon? Explain.

35 We distinguish between bandwidths of a signal and a channel. What other bandwidths we need to take into account considering a communication system?

Laws, Principles, and Models – Importance, Limitations, and Applications

35 Consider limitation in application of Shannon's law shown in Figure 1.3.5:
   a) Equation (1.3.24) states that $\frac{C \text{ (b/s)}}{\text{BW (Hz)}} = \text{SE}_{\text{lim}} = \log_2(1 + \text{SNR})$. Where is signal's power, P_{signal}, in this equation?
   b) How will this equation change after P_{signal} becomes greater than P_{lin}?

36 Equations (1.3.3)–(1.3.6) are the mathematical models of a communication system. Which of these models better describe a real-life communication system? Why?

37 Describe the relationship between modeling and simulation of real systems and processes. Give examples.

2

Analog Signals and Analog Transmission

Objectives and Outcomes of Chapter 2

Objectives

- To understand the mathematical description of an analog signal and to know its main characteristics in time domain, in frequency domain, and in phasor domain.
- To learn how to apply the phasor presentation of an analog signal to the analysis of electronic communication circuits.
- To become familiar with how the characteristics of an analog signal affect the capacity and quality of analog transmission.

Outcomes

- Gain knowledge of the nature of an analog signal and its use in analog transmission.
- Become familiar with the characteristics of analog signals and the characteristics of analog transmission.
- Learn the mathematical description of an analog signal.
- Understand the concept of presenting an analog signal in time domain, frequency domain, and phasor domain and study the relationship among its three main forms.
- Study in depth a sinusoidal signal, its waveform, period and frequency, phase shift, and amplitude. Understand that in analog communications the message is delivered by the waveform of an analog signal. Realize that this is done by superimposing the message waveform onto a carrier (this process is called modulation) and become familiar with such an example of analog transmission as amplitude modulation.
- Acquire in-depth knowledge of the phasor presentation of an analog signal as a rotating phasor form and a polar and understand the relationship between a phasor and the waveform of a sinusoidal signal.
- Learn how to apply the phasor presentation to the analysis of electronic communication circuitry.
- Be aware that analog signals can be classified as power and energy signals.
- Learn how the frequency-domain presentation of an analog signal leads to the concepts of spectrum and bandwidth.

Get to know the difference between a signal bandwidth and a transmission bandwidth. Understand that the transmission bandwidth determines the transmission rate of a communication system.

2.1 Analog Signals – Basics

Objectives and Outcomes of Section 2.1

Objectives

- To become familiar with an analog signal, its main characteristics and the role of these characteristics in analog transmission.
- To study in depth a sinusoidal signal, its waveform, period and frequency, phase shift, and amplitude.

Outcomes

- Learn about a signal's waveform, its mathematical and graphical presentations and its importance for signal characterization.
- Master the difference between analog and digital signals.
- Study in depth a sinusoidal signal, its waveform, and its mathematical description.
- Gain an understanding of the relationship (i) between a signal's period and its frequency, (ii) among linear frequency, radian frequency, and a signal's angle, (iii) between the phase shift (initial phase) and a signal's phase.
- Distinguish between a signal's magnitude and amplitude, learn about peak-to-peak value and its significance in defining the signal's amplitude and offset, and study the complexity in defining a signal's amplitude.

2.1.1 Definitions

2.1.1.1 Waveforms

All electrical signals used in telecommunications are divided into two classes: analog and digital. How do we distinguish between them? Simply by analyzing their waveforms. A waveform is a graph that shows how a signal changes with respect to time. Figure 2.1.1 shows examples of waveforms of analog and digital signals.

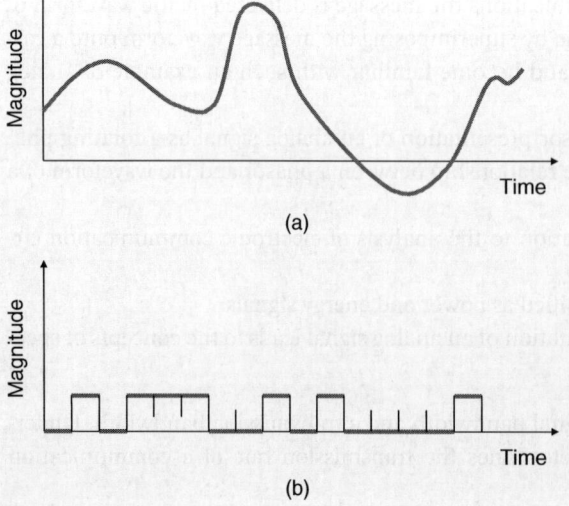

Figure 2.1.1 Waveforms: (a) Analog signal; (b) digital signal.

It's useful to know that there is the Institute of Electrical and Electronics Engineers, IEEE, the main professional organization in this field. Among its many activities, the institute develops a dictionary to standardize the terms used in electricity and electronics. This is how the *IEEE Standard Dictionary* defines the term *waveform*:

"A representation of a *signal* (for example, a graph, plot, oscilloscope presentation, discrete time series, equations, or table of values). This term refers to a measured or otherwise-defined estimate of the physical phenomenon or *signal*."

Keep in mind this definition for the future discussion of this important entity.

Formally speaking,

> *A waveform is the mathematical model of a signal; it represents an instantaneous signal's value, which is the signal's magnitude, as a function of time. Typically, the waveform is presented as a formula (e.g. v(t) (V) = A cos (ωt + θ)) or as a graph (e.g. see Figure 2.1.1).*

"Instantaneous" means the value at every instant of time. This concept is visualized in Figure 2.1.2. For example, as we change our voice from a low to a high level, we essentially change its magnitude from a small to a large value. Figure 2.1.2 exemplifies this by showing that at instant t_1 the magnitude of the signal is $M_1 = 9\,V$ and at instant t_2 the magnitude is $M_2 = 7\,V$. Note that the magnitude can be a negative number. (See, for example, the value $M_4 = -4\,V$ in Figure 2.1.2.)

Amplitude, A (V), is the maximum value of a magnitude

For example, the amplitude of the analog signal shown in Figure 2.1.2 is $M_3 = 29\,V$. Incidentally, what is the amplitude of the digital signal shown in Figure 2.1.2b? Obviously, in this case, the magnitude and amplitude coincide. (Bear in mind, however, that this is not true for *any* digital signal, as we will learn in Chapter 3.)

Now let us consider the origin of a waveform: Imagine that you are speaking into a microphone and the microphone's output is connected to an oscilloscope. A microphone converts sound waves

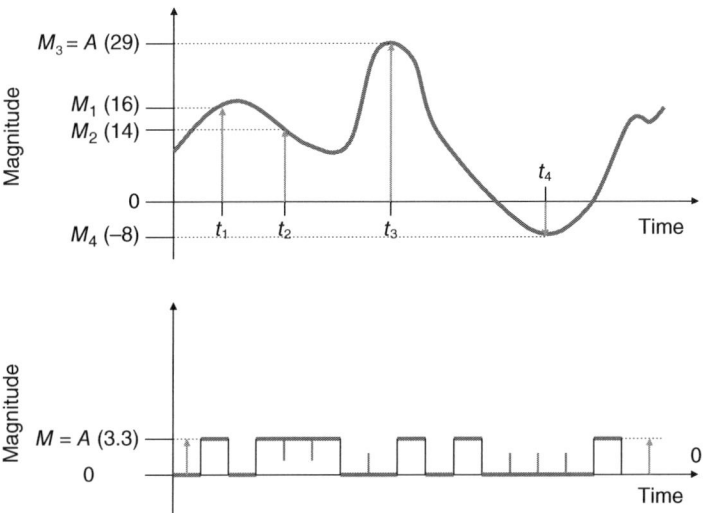

Figure 2.1.2 Waveforms – magnitude and amplitude: (a) Magnitudes and amplitudes of an analog signal; (b) amplitude of a digital signal.

2 Analog Signals and Analog Transmission

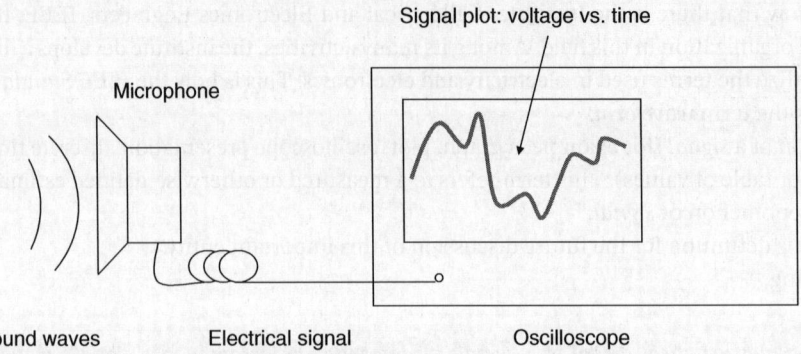

Figure 2.1.3 Building the waveform of a voice signal.

to an electrical signal; an oscilloscope shows the plot of this signal vs. time, which is a waveform. This setup is shown in Figure 2.1.3.

Since a waveform is the plot of a signal's magnitude vs. time, we can build a waveform by finding the value of the signal's magnitude for each instant, that is, via a point-by-point method. Connecting the values of these magnitudes with a solid line, we draw a plot, which is the signal's *waveform*. Consider this plot as a graphical representation of data arranged in a table. Here is an example:

Example 2.1.1 Building the Waveform of a Signal

Problem

Given the following table of recorded values of a signal's magnitude as a function of time, build the plot of the signal:

Time (s)	Magnitude (V)
1	3
2	4
3	6
4	8
5	4
6	2
7	0
8	3
9	6
10	9
11	11
12	12
13	8
14	6
15	4
16	1
17	3

Time (s)	Magnitude (V)
18	4
19	7
20	7

Solution

Use any computerized tool (MS Excel, MATLAB, Mathematica, etc.) to build the plot. The graph shown in Figure 2.1.4a was built with MS Excel.

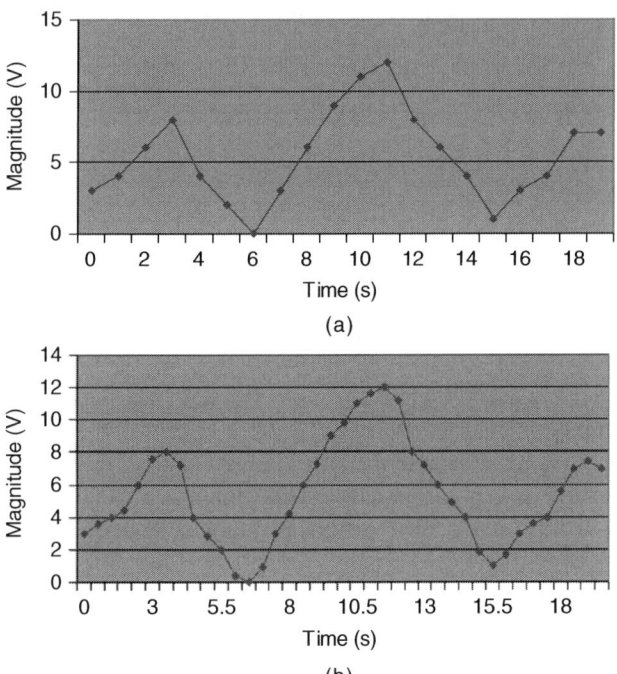

Figure 2.1.4 Building the waveforms by a point-by-point method: (a) Coarse plot; (b) a smoother plot.

Discussion

Note that the table gives the magnitude value for every second and these values are transferred to the chart. When these discrete points are connected, we obtain a continuous plot. However, this plot doesn't look as smooth as the graphs presented in Figures 2.1.1a and 2.1.3. To achieve a better graph, we need a more detailed description of the signal. Let us now record the magnitude of the signal every 0.5 second in contrast to every second, as we did previously. The table with the calculation samples is presented below.

Time (s)	Magnitude (V)
1	3
1.5	3.6

Time (s)	Magnitude (V)
2	4
⋮	⋮
19	7
19.5	7.4
20	7

Now let us build a graph. The result is shown in Figure 2.1.4b. Obviously, this graph looks much smoother than the one we sketched in Figure 2.1.4a. The point is simply this: The more dots involved in building the plot, the more uniform the graph. Eventually, when all the dots are adjacent to one another, the graph will be a smooth curve, as shown in Figures 2.1.1a and 2.1.3. An oscilloscope builds a plot exactly in this manner.

Generally, we denote a waveform as $v(t)$ (V). The most common example of a waveform is an electrical sinusoidal signal with the following formula:

$$v(t) \text{ (V)} = A\, \cos(\omega t + \theta) \qquad (2.1.1)$$

where cos stands for a cosine function with the following parameters: A (V), the signal's amplitude; ω (rad/s), the radian frequency; and θ (rad), the initial phase. A sinusoidal signal will be discussed shortly.

Waveforms are very important characteristics of a telecommunications signal. In fact, we always start to analyze a signal through an examination of its waveform.

We can't overestimate the importance of the waveforms in modern communications. Here is an example: One of the sections of a copy of IEEE Communications Magazine[1] is entitled "New Waveforms for 5G Networks," where "5G Networks" stands for the 5th generation of modern mobile communications technology.

2.1.1.2 Analog and Digital Signals

Study Figure 2.1.1 more closely: An analog signal (Figure 2.1.1a) develops continuously over time, whereas a digital signal (Figure 2.1.1b) changes its magnitude abruptly. This is the main difference between them. Thus, we arrive at the following definitions:

*A signal is called **analog** if its magnitude changes continuously with respect to time; in other words, the waveform of an analog signal is a continuous graph.*

*A signal is called **digital** if its magnitude changes in discrete values as time changes; in other words, the waveform of a digital signal is a discontinuous, step-like graph.*

(The *continuity of signals along the time axis* is described by the terms *continuous* and *discrete*, whereas the *continuity of signals along the magnitude axis* is referred to as *analog* and *digital*. Here, however, we will use the terms *analog* and *digital* informally, leaving more formal definitions to future study.)

Figure 2.1.1a shows an arbitrary analog signal; the digital signal shown in Figure 2.1.1b is a typical on/off signal. In fact, the purpose of this waveform of a digital signal is just to show the signal's levels (HIGH and LOW) at every instant of time. In addition to these examples, Figure 2.1.5 depicts other samples of analog and digital signals.

---

[1] *IEEE Communications Magazine*, vol. 54, no. 11, 2016, pp. 64–113.

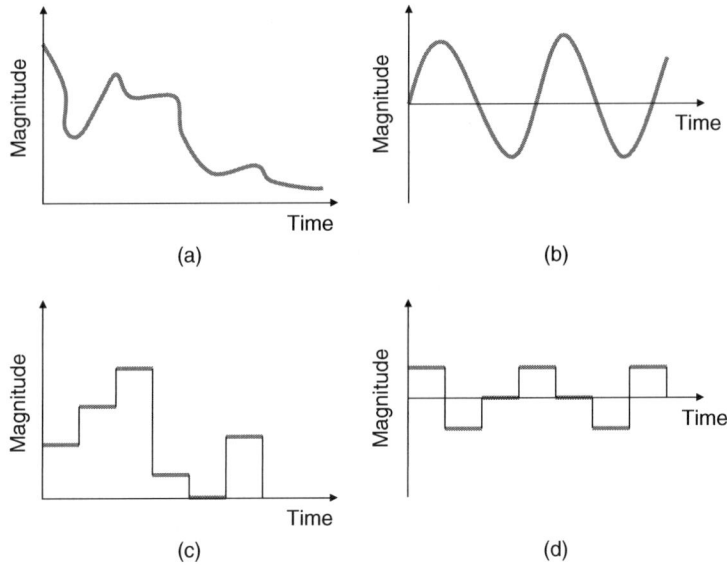

Figure 2.1.5 Examples of analog and digital signals: (a) An arbitrary analog signal; (b) a sinusoidal analog signal; (c) a M-ary (multilevel) digital signal; (d) a bipolar three-level digital signal.

The analog signal shown in Figure 2.1.5a could be a record of temperature change over time, a record of the change in an aircraft's altitude, or a record of any other analog value. The analog signal shown in Figure 2.1.5b is a special one: It is a *sinusoidal signal* widely used in communications. In fact, this signal is so important to communication technology that we will devote Section 2.1.2 to its consideration. The digital signal in Figure 2.1.5c is a multilevel signal; observe that it complies with the definition of a digital signal because its waveform changes discretely. Another example of a digital signal is shown in Figure 2.1.5d, where the signal's polarity changes. This type of signal is called a *bipolar* signal. One example of such a signal is an alternate mark inversion (AMI), shown in Figure 2.1.5d; it is used in copper-wire telephone circuits for the transmission of a digitized voice signal.

Interestingly, the waveform of an electrical analog signal is often the same as the waveform of the original non-electric signal (the outside temperature, an aircraft's altitude, etc.) so that the electrical signal is analogous to the original continuous signal. Typical examples are the air pressure produced by a voice, which is an original signal, and the electrical signal that is generated by a microphone in response to the voice. (See Figure 2.1.3.) This is why this type of signal is called analog.

It follows from this consideration that every single value of an analog signal's magnitude is important for delivering information. As an example, consider an electrical output signal, $v(t)$ (V), of an analog thermostat. Assume that the thermostat is calibrated in the following way: The magnitude value of this signal, $V_1 = 60$ V, corresponds to 60 °F, the magnitude, $V_1 = 70$ V, corresponds to 70 °F, and so on. Thus, we estimate the value of an original signal (the temperature, in our example) based on the value of the magnitude of a received analog electrical signal.

The main problem in delivering information by analog signal is that its magnitude can be changed (and it usually is changed) by unwanted external signals, such as noise. This distortion results in the changing (that is, the deterioration) of the analog signal and, consequently, in the delivery of wrong information. In the above example, suppose the actual temperature is 60 °F but we observe the thermostat's output of 61 V because the received signal was changed by noise.

Then, we mistakenly assume that the temperature is 61 °F. Therefore, in receiving 61 V instead of the 60 V that was sent, we obtain the wrong information.

In contrast, a digital signal carries information only by virtue of its level. In the simplest case of a two-level digital signal, the only thing we need to know is whether it is a high-state (HIGH) or a low-state (LOW) signal. But we can easily understand whether the level of a digital signal is high or low even when its received amplitude will be more or less than the sent value. For example, suppose we know that all amplitudes from 2.8 to 5 V represent a high-state signal. Therefore, receiving 3.2 V means HIGH, regardless of what actual amplitude within the margin was sent, meaning that we have received correct information. This is a main advantage of a digital signal over an analog one. We discuss this point in greater detail in Chapter 3.

2.1.2 Sinusoidal Signal

2.1.2.1 The Waveform of a Sinusoidal Signal

The sinusoidal wave is an extremely important analog signal: Most analog transmission in modern communications is based on using this type of signal. We, of course, are familiar with sinusoidal signals from our high school days. There are two signals called sinusoidal: cosine and sine. The basic waveforms of a sinusoidal signal are described by the following formulas:

$$v(t)\,(V) = A\,\cos(\omega t) = A\,\cos(2\pi f t) \qquad (2.1.2a)$$

and

$$v(t)\,(V) = A\,\sin(\omega t) = A\,\sin(2\pi f t) \qquad (2.1.2b)$$

Here cos stands for cosine and sin stands for sine; amplitude, A (V), and radian frequency, ω (rad/s), are introduced in Eq. (2.1.1). Also, here is the linear frequency, f (Hz), related to the radian frequency as ω (rad/s) $= 2\pi f$ (Hz). The waveforms of the signals defined in (2.1.2a) and (2.1.2b) are shown in Figure 2.1.6. Cosine and sine signals differ only in one feature: A cosine starts at its maximum value and descending to its zero value, whereas a sine starts at its zero value and ascending to its maximum. Example 2.1.2 stresses the relationship between the formula of a sinusoidal signal and its graph.

Example 2.1.2 Building the Waveform of a Sinusoidal Signal

Problem

Given $A = 3$ V and $\omega = 12$ rad/s, build the waveforms of sine and cosine signals with these amplitudes and radian frequencies.

Solution

Using Eqs. (2.1.2a) and (2.1.2b) and following the procedure developed in Example 2.1.1, we can build the table with the signals' magnitudes vs. time and draw the graphs. We use MS Excel as our tool and obtain the results as follows: Table 2.1.1 shows how the magnitudes, $v_1(t) = A\,\sin(\omega t)$ and $v_2(t) = A\,\cos(\omega t)$, change at every instant, $t(s)$. (To save space, we show only the three first and the three last rows of data.) Figure 2.1.6 demonstrates the graphs of both cosine and sine signals with the given amplitudes and frequencies.

The problem is solved.

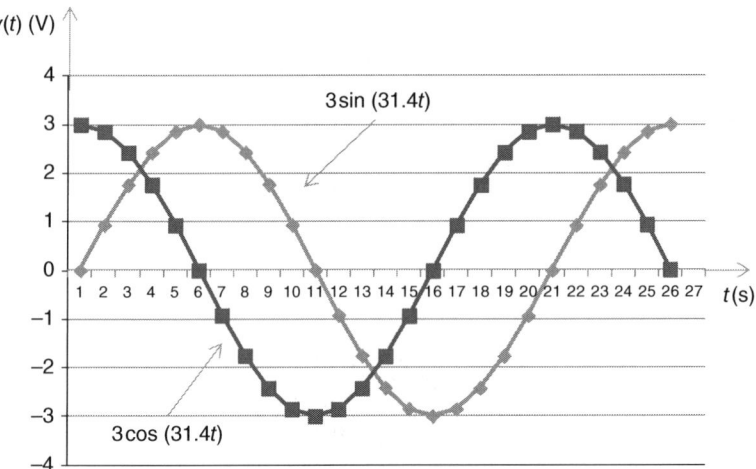

Figure 2.1.6 The waveforms of the cosine and sine signals for Example 2.1.2.

Table 2.1.1 Selected magnitudes vs. time for cosine and sine signals for Example 2.1.2.

t (s)	A (V)	ω (rad/s)	$v_1(t) = A\sin(\omega t)$	$v_2(t) = A\cos(\omega t)$
0	3	31.4	0	3
0.01	3	31.4	0.926 596 56	2.853 317 16
0.02	3	31.4	1.762 582 58	2.427 612 543
⋮	⋮	⋮	⋮	⋮
0.23	3	31.4	2.420 575 36	1.772 234 445
0.24	3	31.4	2.849 605 18	0.937 950 05
0.25	3	31.4	2.999 976 22	0.011 944 87

Discussion

It is crucial to fully understand the meaning of the formula for a waveform, $v(t)$ (V): By plugging into this formula specific values of time (see the leftmost column in Table 2.1.1) and executing the given mathematical operations, we obtain a signal's magnitude for every instant. (See the dots on the graphs in Figure 2.1.6.) Connecting these magnitudes by a solid line, we can draw the graphical presentation of the waveform given by the formula, $v(t)$. Refer to our discussion of Figures 2.1.2 and 2.1.4 and Example 2.1.1.

2.1.2.2 Period and Frequency

First, we'll concentrate on the development of a signal along the time axis.

Observe Figure 2.1.7 and note that the sinusoidal signal repeats itself after a certain time interval. This interval is called a *period*.

We usually denote a period, T, the way Figure 2.1.7 indicates. A period is measured in seconds (s) or fractions of a second, such as milliseconds (ms = 10^{-3} s), microseconds (µs = 10^{-6} s), or nanoseconds (ns = 10^{-9} s). Examining Figure 2.1.7 that shows a sine signal with two periods, we can clearly discern how a signal's period shapes its waveform.

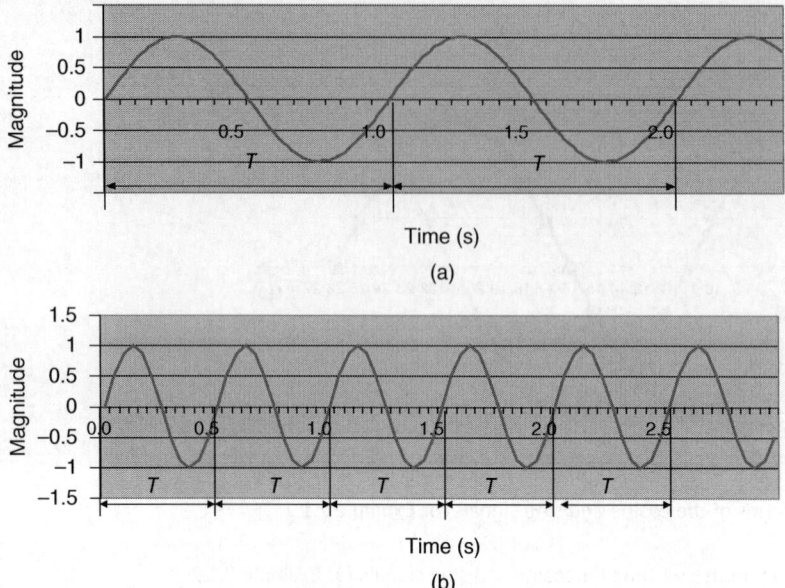

Figure 2.1.7 Waveforms of sinusoidal signals: (a) Sine signal with period $T_1 = 1$ s; (b) sine signal with period $T_2 = 0.5$ s.

A signal that repeats itself after a period is called a *periodic* signal; the sinusoid is one of the most important signals in this general category.

Note that a sinusoidal signal intersects the time axis twice during one period; however, after the first intersection, the signal doesn't repeat itself. As can be seen from Figure 2.1.7, the first time the signal crosses the time axis the sine wave decreases; however, after the second intersection, the sine starts to copy itself.

We need to identify fractions of a period. Refer to Figure 2.1.7a: The half period, which is $T_1/2 = 0.5$ s, is where the graph intersects the time axis; in Figure 2.1.7b, the half period is $T_2/2 = 0.25$ s. Accordingly, in Figure 2.1.7a, the quarter of a period, $T_1/4 = 0.25$ s, is where the magnitude of the sine signal reaches its maximum value; in Figure 2.1.7b, $T_2/4 = 0.125$ s.

We can say that a sinusoidal signal consists of cycles; *each cycle lasts one period*. The graph in Figure 2.1.7a shows two full *cycles* and the beginning of a third. The graph in Figure 2.1.7b shows five cycles and the beginning of a sixth.

It is important to remember that the number of cycles per second is inversely proportional to the value of the period. Indeed, for $T_1 = 1$ s a sinusoid has one cycle per second and for $T_2 = 0.5$ s the sinusoidal signal has two cycles per second, as Figure 2.1.7 shows. This observation brings us to an understanding of one of the fundamental characteristics of a sinusoidal wave: *frequency*.

> *Frequency is the number of cycles per second.*

Since a number doesn't have any units, frequency is expressed as *1/s*; this symbol is denoted as *hertz (Hz)*.[2] Hence, a typical notation of a frequency is *f* (Hz). Since the frequency of a sinusoidal

---

2 Heinrich Hertz (1857–1894) was a German physicist who proved the existence of electromagnetic (EM) waves, predicted by James Maxwell's theory. He was the first to broadcast and receive EM waves experimentally, which

signal is the number of cycles per second and this number is inversely proportional to the signal's period, we can readily derive the following key formula:

$$f \text{ (Hz)} = 1/T \text{ (s)} \tag{2.1.3}$$

Equation (2.1.3) is a fundamental expression that every electrical engineer has to know. This equation also allows us to compute a period if the frequency is given. Indeed, rearranging (2.1.3) in this way,

$$T \text{ (s)} = 1/f \text{ (Hz)}, \tag{2.1.4}$$

we can easily calculate the period.

Example 2.1.3 Period and Frequency of a Sinusoidal Signal

Problem

The period of a sine signal is 0.1 s. What is the signal's frequency?
The frequency of a sine signal is 60 Hz. What is its period?

Solution

(a) Use (2.1.3) and find f (Hz) $= 1/0.1$ s $= 10$ Hz.
(b) Apply (2.1.4) and compute T (s) $= 1/f$ (Hz) $= 0.0166$ s.

Discussion

From all the examples we have considered so far, it's easy to conclude that the smaller the period, the greater the frequency and vice versa. This statement usually confuses students; that is why we are presenting this idea as a set of numbers organized in the following table:

Period (s)	1	0.1	0.01	0.001
Frequency (Hz)	1	10	100	1000

Figure 2.1.8 displays two sinusoidal waveforms: $T = 1$ s and $T = 0.1$ s. Note the difference in the waveforms: The signal shown in Figure 2.1.8a has a smaller frequency than the signal shown in Figure 2.1.8b. To prove this, just count the number of cycles per second and recall the definition of "frequency." You should develop the ability to see the difference in frequencies of sine signals by simply looking at their waveforms.

A period can be shown between any two identical points within a waveform, as illustrated in Figure 2.1.8. Just remember the definition of a period: It is the time interval after which a sinusoidal signal starts to copy itself.

Note, too, that a waveform allows us to show only a period; frequency must be calculated. Therefore, when you are asked to sketch a sine signal and show its main characteristics, the only figure you can show along the time axis is a period.

Let us calculate one half and also a quarter of the given period. For $T_2 = 0.1$ s, it's easy to compute that $T_2/2 = 0.05$ s and that $T_2/4 = 0.025$ s.

We believe everyone knows this frequency value: 60 Hz is the frequency at which our home and office power networks operate. Thus, our lamps are supplied with electrical power that, in essence,

---

meant that he created the prototypes of the transmitter and the receiver for EM radiation. Fittingly, the first words transmitted by Guglielmo Marconi over the newly invented radio were "Heinrich Hertz."

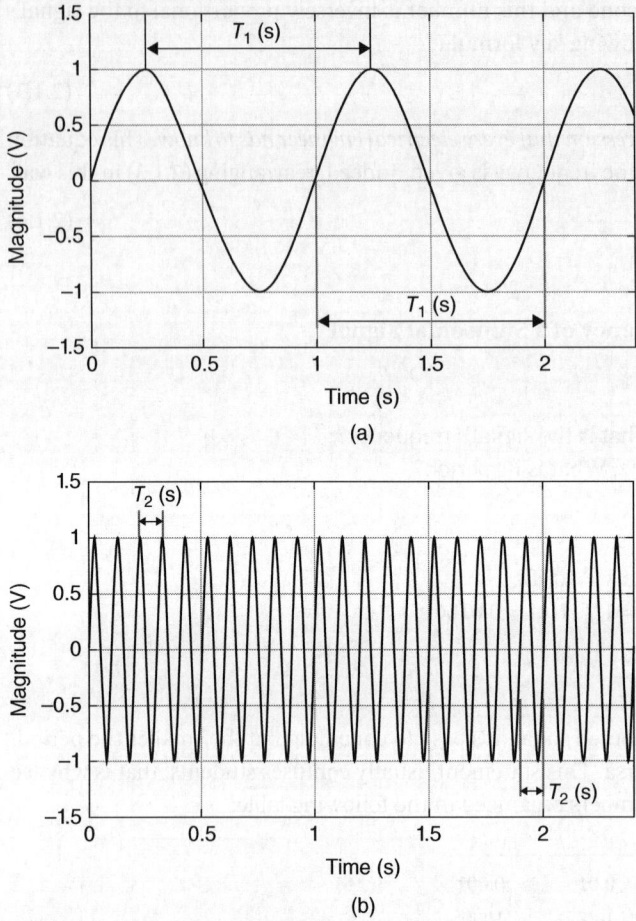

Figure 2.1.8 Sinusoidal signals with various periods (frequencies): (a) $T_1 = 1\,\text{s} \Rightarrow f_1 = 1\,\text{Hz}$; (b) $T_2 = 0.1\,\text{s} \Rightarrow f_2 = 10\,\text{Hz}$.

is a 60-Hz sinusoidal signal. Since we are now familiar with a sinusoidal waveform, we might wonder why we don't notice any change in the intensity of the light in our lamps; after all, Figures 2.1.7 and 2.1.8 show that their currents change magnitude and even reach a zero value every half period. The answer lies in the numbers: Since the period of an electrical-power signal is 0.016 66 s, this signal reaches a null value every 0.008 33 s; our eyes simply can't catch these fast changes. In other words, we see the average of light intensity.

To become familiar with the values of frequencies and periods of the signals used in modern communications, compute the periods of the following five signals:

Voice $\Rightarrow f_{\max} = 4\,\text{kHz}$
AM radio $\Rightarrow f_{\max} = 3\,\text{MHz}$
Cellular phone $\Rightarrow f_{\max} = 3\,\text{GHz}$
Satellite communications $\Rightarrow f_{\max} = 12\,\text{GHz}$
Optical communications $\Rightarrow f_{\max} = 193\,\text{THz}$

Note that while the orders of magnitude (kHz, MHz, GHz, and THz) present the range of frequencies actually used in the appropriate areas, the values of the maximum frequencies (4 kHz, 3 MHz, etc.) are shown for illustrative purposes only.

The solution is obvious: We need to apply (2.1.4) and find:

$4\,\text{kHz} = 4 \times 10^3\,\text{Hz} \Rightarrow T\,(\text{s}) = 1/f\,(\text{Hz}) = 0.25 \times 10^{-3}\,\text{s} = 0.25\,\text{ms}$
$3\,\text{MHz} = 3 \times 10^6\,\text{Hz} \Rightarrow T\,(\text{s}) = 0.33 \times 10^{-6}\,\text{s} = 0.33\,\mu\text{s}$
$3\,\text{GHz} = 3 \times 10^9\,\text{Hz} \Rightarrow T\,(\text{s}) = 0.33 \times 10^{-9}\,\text{s} = 0.33\,\text{ns}$
$12\,\text{GHz} = 12 \times 10^9\,\text{Hz} \Rightarrow T\,(\text{s}) = 0.08 \times 10^{-9}\,\text{s} = 80\,\text{ps}$
$193\,\text{THz} = 193 \times 10^{12}\,\text{Hz} \Rightarrow T\,(\text{s}) = 5.18 \times 10^{-15}\,\text{s} = 5.18\,\text{fs}$

Remember the notations for frequency units used in communications: kHz = 10^3, MHz = 10^6, GHz = 10^9, and THz = 10^{12}. Also, familiarize yourself with fractions of seconds, such as millisecond (ms = 10^{-3} s), microsecond (μs = 10^{-6} s), nanosecond (ns = 10^{-9} s), picosecond (ps = 10^{-12} s), and femtosecond (fs = 10^{-15} s).

These calculations stress again that the greater the frequency, the smaller the period and vice versa. To be sure, light – the signal with the highest frequency, 193×10^{12} Hz, used in communications – has the smallest period, $T = 5.18 \times 10^{-15}$ s = 5.18 fs.

2.1.2.3 Frequency, Radian (Angular) Frequency and Angle

Frequency, f (Hz $\equiv$ 1/s), is also called *linear, spectral, or cyclical* frequency. When somebody says "frequency," he or she usually means frequency, f (Hz). But we also have already introduced *radian frequency*, which is denoted with the Greek letter *omega*, ω (rad/s), and which is measured in radians per second.

Frequency is the number of cycles per second; radian frequency is the number of radians per second.

These measures stem historically from angular rotation: Radian frequency, ω (rad/s), is simply the measure of angular velocity; this is why it is also called *angular frequency*. (See Sidebar 2.2.S.1, "Phasor and Sinusoidal Signal.") Since each rotation cycle covers a 2π angle, measured in radians, we can easily derive the relationship between frequency, f, and radian frequency, ω:

$$\omega\,(\text{rad/s}) = 2\pi f\,(\text{Hz}) \qquad (2.1.5)$$

In fact, on the one hand, the angle achieved after one cycle (period) is equal to $\omega \times T$; on the other hand, this angle is exactly equal to 2π. Therefore,

$$\omega\,(\text{rad/s}) \times T\,(\text{s}) = 2\pi\,(\text{rad})$$

and since

$$T\,(\text{s}) = 1/f\,(\text{Hz}),$$

we obtain

$$\omega\,(\text{rad/s})/f\,(\text{Hz}) = 2\pi\,(\text{rad}) \qquad (2.1.6)$$

as in (2.1.5) above.

Given a radian frequency, we can compute a period using the following formula:

$$T(\text{s}) = 2\pi/\omega\,(\text{rad/s}) \qquad (2.1.7)$$

As an exercise, derive (2.1.7).

2 Analog Signals and Analog Transmission

We can relate the radian frequency, ω (rad/s), to angular measurements as follows: The product ω (rad/s) $\times$ t (s) gives us radians. Thus, $\omega t(\frac{\text{rad}}{\text{s}} \cdot \text{s})$ is an angle measured in radians:

$$\alpha \text{ (rad)} = \omega \text{ (rad/s)} \times t \text{ (s)} \tag{2.1.8}$$

Conversion between degrees and radians is obtained by the following well-known formulas:

$$\alpha \text{ (°)} = (180°/\pi) \times \alpha \text{ (rad)} \tag{2.1.9a}$$

and

$$\alpha \text{ (rad)} = (\pi/180°) \times \alpha \text{ (°)} \tag{2.1.9b}$$

We can memorize these formulas or, better yet, use the following simple reasoning: Since 2π is equivalent to 360°, we can build the following proportion:

2π (rad) $\leftrightarrow$ 360°

α (rad) $\leftrightarrow$ α(°)

Therefore,

$$\alpha \text{ (rad)} \times 360° = 2\pi \times \alpha \text{ (°)} \tag{2.1.10}$$

and so (2.1.9a) and (2.1.9b) can be easily derived from this formula.

The following example will help in understanding of the calculations of frequency, radian frequency, and angle.

Example 2.1.4 Frequency, Radian Frequency, and Angle

Problem

If the frequency is 1 kHz, what is the radian frequency? What angle will be attained with this frequency at $t = 1$ ms?

Solution

First, convert kHz to the basic unit, that is, Hz: 1 kHz = 1000 Hz = 10^3 Hz.

Remember, hertz is essentially 1/s. Then, apply (2.1.5) and compute:

$$\omega = 2\pi f = 6280 \text{ rad/s} = 6.28 \times 10^3 \text{ rad/s}$$

In these computations, we rounded π to 3.14.

To calculate the angle, convert ms to the basic unit: 1 ms = 0.001 s = 10^{-3} s.

Next, apply (2.1.8) and compute

$$\alpha \text{ (rad)} = \omega \text{ (rad/s)} \times t \text{ (s)} = 6.28 \text{ rad}.$$

Now, convert radians to degrees by using (2.1.9a):

$$\alpha \text{ (°)} = (180°/\pi) \times \alpha \text{ (rad)} = 359.84°.$$

The problem is solved.

Discussion

- Remember that π, as a measure of the angle, carries a dimension: radians. Thus, the unit of ω (rad/s) appears logically.

- When doing computations, it is extremely important to retain the units involved. We cannot, for instance, mix hertz and kilohertz or seconds and microseconds. To avoid errors, convert all units to the basic ones. Thus, work with hertz, seconds, radians, degrees, etc. This is what we did in the above calculations.
- The conversion of radians to degrees can result in a large angle value, as computed above. This is because one radian is equivalent to 57.3°. It is important to note that we can compute the obtained angle in degrees by simply dividing angle ωt by 2π and multiplying by 360°. In our example, we calculate $\alpha = \omega t = 6.28$ rad. Applying (2.1.9a), we obtain

$$(6.28/2\pi) \times 3600 \approx 1 \times 360° = 360°$$

The difference between this value and the previous result ($\alpha = 359.84°$) is explained by the approximation we made in the value of π.
- Clearly, we can convert radian frequency to frequency by using the following variation of (2.1.5):

$$f\,(\text{Hz}) = \omega\,(\text{rad/s})/2\pi\,(\text{rad}) \tag{2.1.11}$$

Consider this simple example: If $\omega = 6280$ rad/s, then $f = \omega/2\pi = 6280\,\text{rad/s}/6.28\,\text{rad} = 1000$ Hz $= 1$ kHz.

2.1.2.4 Phase Shift (Initial Phase)

We have considered the sinusoidal signals that have zero magnitude at zero time, as Figures 2.1.6–2.1.8 show. This is not always the case. In reality, the sinusoidal signal can cross the *zero point of a time axis* having any magnitude value or, putting it in another way, it can reach the first zero magnitude at any point in time. Figure 2.1.9 illustrates this concept.

To explain this property of a sinusoidal signal, we now introduce another main parameter, which is referred to as *phase shift* or *initial phase*. The importance of a phase shift can be seen from observing top and bottom graphs in Figure 2.1.9: The top figure plots a *sine* signal; the bottom, a *cosine* signal. In other words, a positive 90° phase shift changes the signal from sine to cosine.

> *A phase shift, Θ, is an angle by which a sinusoidal signal shifts along the time axis with respect to the time zero point. This is why the phase shift is also called an initial phase.*

Remember this: We usually denote a phase shift (initial phase) with the Greek letter *theta*, θ.

Figure 2.1.10 visualizes the concept of initial phase, or phase shift. Observe that a phase shift doesn't change the waveform of a sinusoidal signal; it simply moves the whole waveform along the time axis. Figure 2.1.10 shows that with a positive phase a sine waveform shifts to the left with respect to the zero point along the time axis (Figure 2.1.10a), whereas with a negative phase, the sine plot shifts to the right (Figure 2.1.10b). The role of a phase shift's value can be clearly seen by comparing Figure 2.1.10b,c. Also, refer to Figure 2.1.9.

(**Question:** *Why does a sinusoidal signal whose initial phase is negative shift to the right with respect to zero-point time?*)

It is essential to understand that a *phase shift (initial shift) is a constant angle* for a given sinusoidal signal. This angle is, in essence, the initial condition of a sinusoidal signal because it determines the state of the signal at $t = 0$ s.

Since the phase shift is an angle, we measure it in degrees or radians. On the other hand, you will recall that the entity ωt is an angle by its very nature, which is reflected in its dimension, also

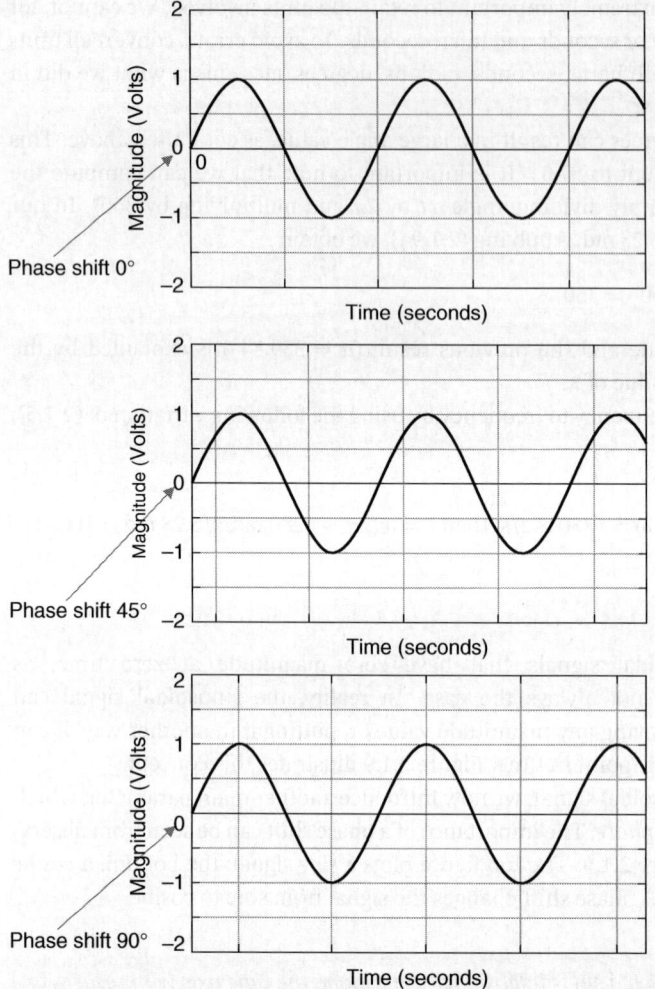

Figure 2.1.9 A sine signal with various phase shifts: Θ = 0° (top), Θ = 45° (middle), and Θ = 90° (bottom).

measured in degrees or radians. Thus, we can compare these two entities, which allows us to relate an instant of time to a phase shift:

$$\omega t_\theta \text{ (rad)} = \theta \text{ (rad)} \tag{2.1.12a}$$

which yields the simple formula

$$t_\theta \text{ (s)} = |\theta|/\omega \text{ (s)} \tag{2.1.12b}$$

Let us take, for example, a 90°, or $\pi/2$, phase shift. Substitute this value in the above equation and obtain

$$t_\theta = |\theta|/\omega = (\pi/2)/(2\pi f) = T/(4)$$

Thus, a 90°, or $\pi/2$, phase shift is equivalent to a quarter of a period (see Figures 2.1.9c and 2.1.10c). You can therefore compute that a 180° shift is equivalent to half a period and so on.

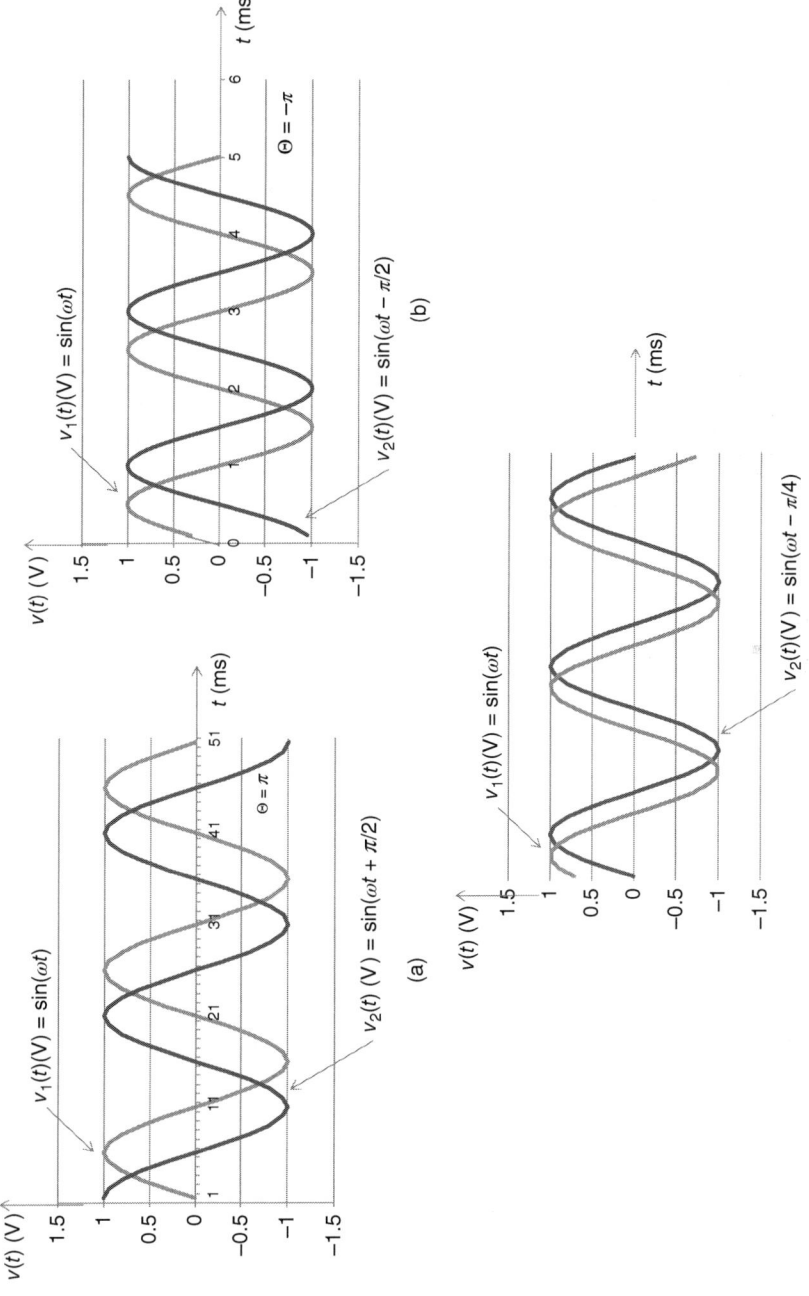

Figure 2.1.10 The impact of phase shift (initial phase) on a waveform's position: (a) $\Theta = \pi/2$; (b) $\Theta = -\pi/2$; (c) $\Theta = -\pi/4$.

Example 2.1.5 Phase Shift of a Sinusoidal Signal

Problem

Sketch a sine signal with $\theta = -90°$. How is this shift related to a signal period?

Solution

We used MS Excel to sketch the required signal in Figure 2.1.11.

Observe that at a $-90°$ phase, a sine signal reaches its minimum, that is, -1 V in our example. Apply (2.1.12b) and compute the instant, t_θ, at which this minimum is reached:

$$t_\theta = |\theta|/\omega = (-\pi/2)/(2\pi f) = T/(4)$$

The problem is solved.

Discussion

- Since the phase shift is negative in this example, we expect that the sine signal shifts to the right with respect to the zero phase-shift position. The graph in Figure 2.1.11 confirms our expectation.
- It should be clear from our discussion that, for a sinusoidal signal, any constant time shift results in a phase shift, as (2.1.12a) and (2.1.12b) show. For example, if a sinusoidal signal with a radian frequency of 1000 rad/s is delayed by 1 ms, it will result in an additional phase shift computed as follows:

$$\theta = \omega t_\theta = 1000 \text{ rad/s} \times 0.001 \text{ s} = 1 \text{ rad}$$

- There is another key point to emphasize: When we said previously that a phase shift is an angle by which a sinusoidal signal shifts along the time axis, we seemed to contradict ourselves because an angle cannot be measured in time units. Indeed, on the one hand, we refer here to a waveform, which is a signal's plot vs. time; on the other hand, we also refer to a phase shift, which, as an

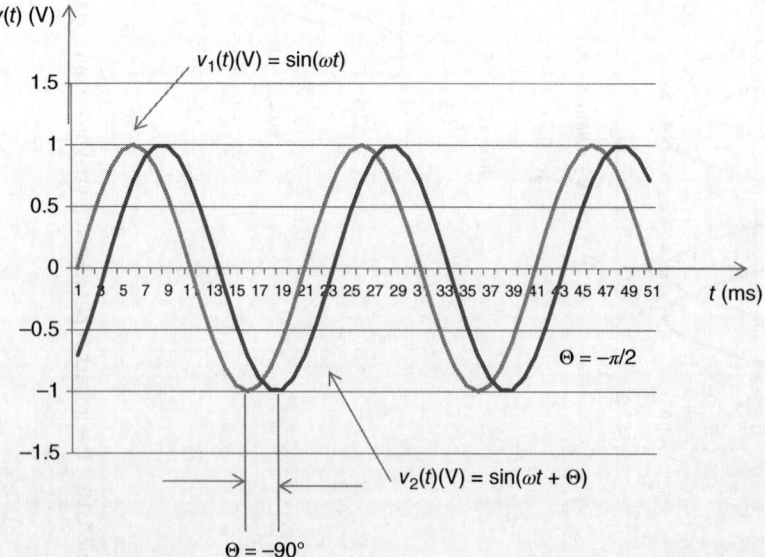

Figure 2.1.11 A sine signal with a negative 90°-phase shift for Example 2.1.5.

angle, must be measured in radians or degrees and must be shown in its proper axes. The solution to this enigma lies in (2.1.8):

$$\alpha \text{ (rad)} = \omega \text{ (rad/s)} \times t \text{ (s)}$$

In general, as you know from your math courses, the formula for a sinusoidal signal is given by

$$v = A \sin \alpha \qquad (2.1.13)$$

where α is an angle. If you compare (2.1.13) and (2.1.1), you immediately come to the conclusion that in this case

$$\alpha \text{ (rad)} = \omega t + \theta \qquad (2.1.14)$$

Since ω and θ are constants for any given signal, the only variable in (2.1.14) is time, t. This means that the only reason why angle α can change is if there is a change in t. Therefore, the waveform, v, shown in (2.1.13) as a function of α is, in fact, a function of time. Some books show this waveform as a function of ωt, which differs from our presentation in multiplying by constant ω (rad/s). All in all, we need to remember that the time axis in our waveform graphs represents, in essence, an angle; we can do this formal conversion by multiplying t and ω.

- Another very significant point is this: Eq. (2.1.14) shows that the phase angle α (rad) consists of two parts: a variable angle, ωt (rad), and a constant angle, θ (rad). Our preceding discussion has been devoted to the constant angle θ (rad), which we call a phase shift or an initial phase. It should be noted that in physics, particularly in optics, this parameter, $\omega t + \theta$, is usually referred to as a phase, which underscores the common nature of both members – a variable and a constant – of this angle. To avoid any ambiguity, we must clearly indicate which parameter – general phase α (rad) = $\omega t + \theta$, a variable angle ωt (rad), or a phase shift θ (rad) – we are referring to in any given situation.

2.1.2.5 Amplitude

The second main parameter of a sinusoidal signal is its amplitude. As a sinusoidal waveform shows, a magnitude changes from zero to the maximum value and back to zero, goes to negative values, reaches the minimum (negative maximum) value, and finally returns to the zero value again during a single period. Thus, *magnitude* is a variable, as we explained at the beginning of this subsection. But we need a constant number to describe how big a sinusoidal signal is. To characterize a sinusoidal signal from this standpoint, we introduce *amplitude*. Figures 2.1.2 and 2.1.12 illustrate the definition of amplitude: They show that

> *the amplitude is the value of a signal's magnitude from zero to its maximum (or minimum) value.*

Thus, magnitude is a variable (instantaneous) value of a signal and amplitude is a constant, extreme value of a signal. Amplitude is the main measure of a signal's strength.

Amplitude is measured either from the zero level to the maximum value in a positive direction or from zero to the minimum value in a negative direction along the axis of the magnitude. Figure 2.1.12 demonstrates that defining the amplitude of a given sinusoidal signal seems to be a simple task: Just measure the signal's maximum value either in a positive direction or in a negative direction.

Along with amplitude, we use another measure: peak-to-peak value, App (V).

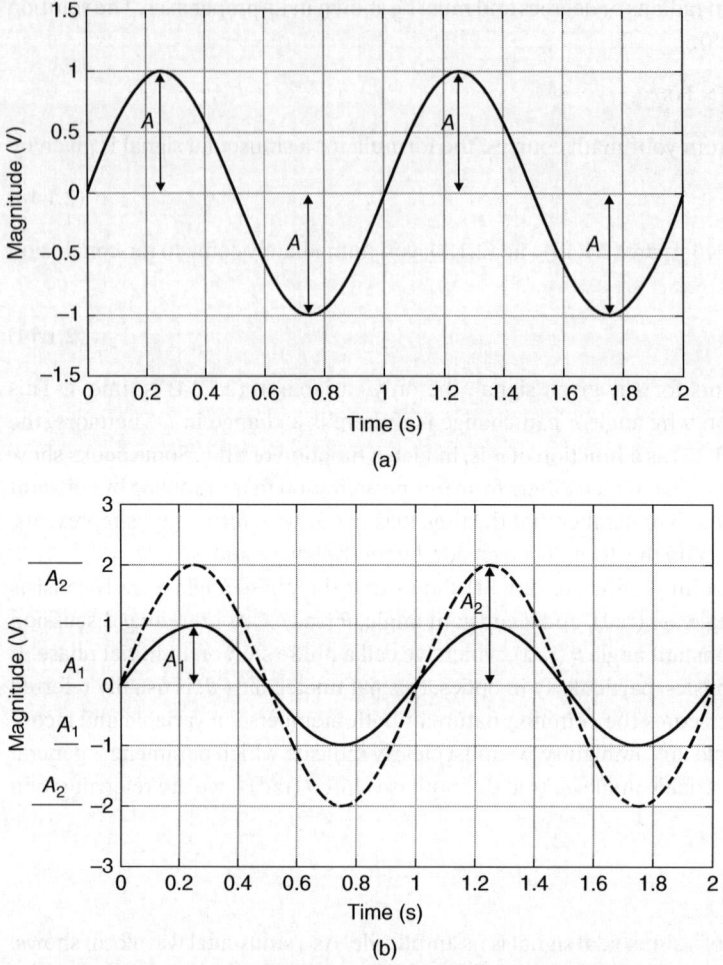

Figure 2.1.12 Amplitude of a sine signal: (a) Definition; (b) $A_1 = 1$ V; $A_2 = 2$ V.

Peak-to-peak amplitude is a signal's value measured from its minimum to its maximum values; that is,

$$A_{pp} (V) = A_{max} - A_{min} \tag{2.1.15}$$

To become familiar with the use of (2.1.15), consider, for example, signals A_1 and A_2 in Figure 2.1.12b: $A_{1max} = 1$ V and $A_{1min} = -1$ V; hence $A_{1pp} = A_{1max} - A_{1min} = 1$ V $- (-1$ V$) = 2$ V. Similarly, $A_{2pp} = A_{2max} - A_{2min} = 2$ V $- (-2$ V$) = 4$ V.

Examine Figure 2.1.13, where the definition of A_{pp} (V) is visualized. What's the purpose of introducing such a simple measure if amplitude comprehensively describes the strength of a sinusoidal signal? In fact, from reviewing the sinusoidal waveforms given above in Figures 2.1.5 to 2.1.8, we might reasonably conclude that the peak-to-peak value, A_{pp}, is twice a signal's amplitude; that is,

$$A_{pp} (V) = 2A (V), \tag{2.1.16}$$

and introducing this value doesn't produce new information. True enough, but A_{pp} (V) is an important and valuable measure that helps us to characterize a sinusoidal signal in a general sense.

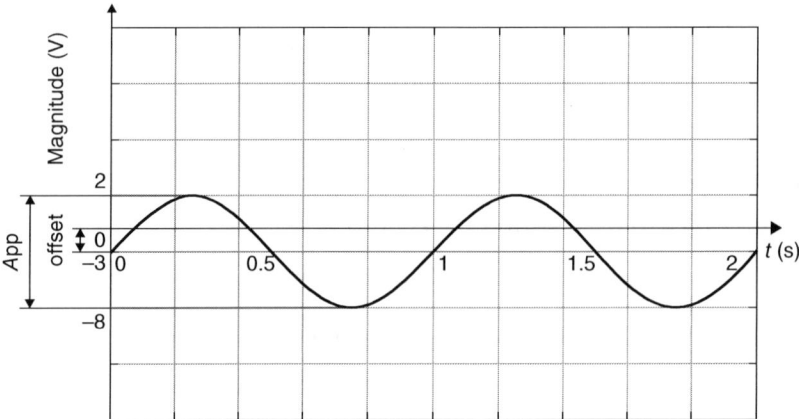

Figure 2.1.13 Peak-to-peak value: It helps to characterize a sinusoidal signal and leads to the definition of amplitude.

Figure 2.1.13 shows such a situation: If a signal's waveform is asymmetrical with respect to the horizontal axis (in other words, $A_{max} \neq A_{min}$), then the peak-to-peak amplitude helps us to characterize the signal's amplitude correctly as

$$A(V) = \frac{A_{pp}(V)}{2} \tag{2.1.17}$$

Example 2.1.6 clarifies this point.

Example 2.1.6 Amplitude of a Sinusoidal Signal

Problem

1. What is the amplitude of the sine signal depicted in Figure 2.1.13 if its A_{pp} is 10 V?
2. What is the minimum value of this signal if its maximum value is 2 V?

Solution

1. Apply (2.1.17) and compute

$$A = A_{pp}/2 = \frac{10\,V}{2} = 5\,V$$

2. Rearrange (2.1.15) and find

$$A_{min} = A_{max} - A_{pp} = 2\,V - 10\,V = -8\,V$$

Discussion

- Imagine that we are examining a sine signal using an oscilloscope and obtain a graph like the one shown in Figure 2.1.13. Clearly, we need to determine the signal's amplitude. If we didn't know anything about peak-to-peak value, we would have to try to determine the line of symmetry and measure the amplitude. Imagine how difficult that job would be and how inaccurate would be the result. The peak-to-peak value solves such a problem easily and effectively. Fortunately, today's oscilloscopes can measure peak-to-peak values, calculate amplitudes, and display the required results.

- The maximum and minimum values of the signal enable us to introduce a measure of the signal's shift along the magnitude axis, in other words, a measure of the signal's asymmetry in this direction. This measure is called the signal's offset, or dc shift, and it is defined as

$$\text{Offset (V)} = (A_{max} + A_{min})/2 \tag{2.1.18}$$

In this example, the signal's offset is

$$\text{Offset} = (A_{max} + A_{min})/2 = (2\,V + (-8\,V))/2 = -3\,V$$

Therefore, this signal is shifted by 3 V in the negative direction along the magnitude axis.

The peak-to-peak concept allows us to present the formal definition of the amplitude of a sinusoidal signal:

> *The amplitude of a sinusoidal signal is half the difference between the maximum and minimum values of a waveform; that is, $A\,(V) = A_{pp}/2$, as Eq. (2.1.17) states.*

If, for example, the maximum value is 3 V and the minimum value is -7 V, the amplitude is equal to $(3-(-7))\,V/2 = 5\,V$. Incidentally, the offset of this is equal to $(3-7)/2\,V = -2\,V$. Refer to Figure 2.1.13 to better understand this example. Another example: $A_{max} = 12\,V$ and $A_{min} = 2\,V$; thus, $A_{pp} = 10\,V$, $A = 5\,V$, and Offset $= 7\,V$. Now, sketch the waveform of this signal to better understand all the important values introduced in this subsection.

We can now appreciate the true usefulness of the peak-to-peak value: This concept gives us a general view of the nature of the amplitude of a sinusoidal signal and it is a general measure of a signal's strength regardless of the signal's shift along the magnitude axis.

Final note: Still find the concept of amplitude confusing? This is because, after comparing two examples of the amplitudes shown in Figures 2.1.2a and 2.1.13, we find that they seem to be contradictory. (We reproduce these figures in Figure 2.1.14 with small alterations.) If we count the amplitudes in Figure 2.1.14a from the zero level, we find $+A = 18\,V$ and $-A = -4\,V$. But if we follow the amplitude's definition, (2.1.17), we compute $A_{avg}\,(V) = \frac{A_{pp}(V)}{2} = \pm 11\,V$. These are the amplitudes of the arbitrary signal given in Figure 2.1.14a when we count it from the average level of this signal. What's more, Figure 2.1.14b shows a similar conflict. Are the amplitudes of this sinusoidal signal $+A = +2\,V$ and $-A = -8\,V$ or $A_{avg} = \pm 5\,V$? Which case is correct?

The answers to these questions lie in the definitions of amplitude given in this section. Careful examination of these definitions shows the following:

The value of a signal's amplitude depends on its definition.

For an arbitrary signal, the amplitude is the maximum value of the magnitude. This definition implies that the amplitude value counts from the zero level. In Figure 2.1.14a, this definition gives $A = 18\,V$ and $A = -4\,V$.

For a sinusoidal signal, the amplitude has been defined as half of a peak-to-peak value. This definition relies on the symmetrical property of a sinusoidal signal. In this case, the amplitude value counts from an average level of a sinusoidal signal. Thus, in Figure 2.1.14b the amplitudes are $A_{avg} = \pm 5\,V$.

Other definitions of amplitude exist, and we must clearly understand what type of amplitude we have to deal with in every specific case.

Our discussion is focused on electrical signals only. There are numerous approaches to defining the amplitude in various branches of science, such as physics, astronomy, mechanical engineering, etc.

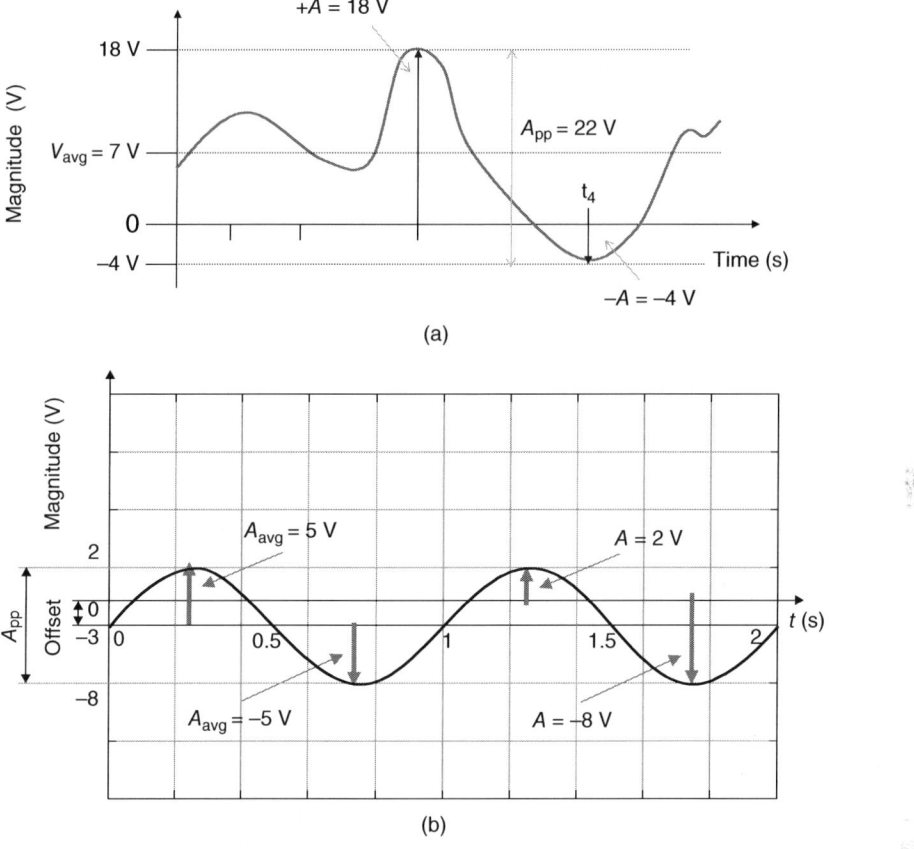

Figure 2.1.14 Definition of amplitude: (a) For an arbitrary analog signal – see Figure 2.1.2a; (b) for a sinusoidal signal – see Figure 2.1.13.

This concludes the consideration of a sinusoidal signal in this section. We will take a deeper look at this subject in Section 2.2, where Sidebar 2.2.S.1 will help you to fully understand the concept of a sinusoidal signal.

Questions and Problems for Section 2.1

- Questions marked with an asterisk require a systematic approach to finding the solution.
- Many questions and problems, including those marked with an asterisk, imply that you, in addition to reading the textbook, will do your research to find the answers. Consider such questions as mini-projects.

Definitions

1 Define and explain a waveform. Give examples.

2 What is the difference between the magnitude and the amplitude of a signal? Explain using a signal's waveform. Sketch the graph.

3 *Sketch an example of an analog non-sinusoidal signal's waveform.

4 *Explain how we can generate an analog electrical signal.

5 What measuring instrument can display a signal's waveform? Explain.

6 Given the following table of recorded values of a signal's magnitude:
   a) Build the plot of the signal.
   b) Determine the amplitude of this signal.

Time (s)	Magnitude (V)
1	2
2	3
3	5
4	7
5	3
6	2
7	1
8	4
9	5
10	8
11	10
12	12
13	9
14	7
15	5
16	2
17	4
18	5
19	7
20	8

7 Sketch the waveform of a digital signal. Explain how analog and digital signals differ.

8 Define and explain what an analog signal is. Give examples.

9 Give examples of several analog signals you have encountered in your everyday experience.

10 *What is the main advantage of using the analog signals employed in communications? What's the main drawback?

11 Define and explain what a digital signal is. Give examples.

12 Give examples of digital signals you have encountered in your everyday experience.

13 *What is the main advantage of using the digital signals employed in communications? What's the main drawback?

Sinusoidal Signal

14 Plotting a sinusoidal signal:
   a) Sketch a cosine signal whose amplitude is 5 V and whose period is 0.01 s.
   b) Sketch a cosine signal whose amplitude is 5 V and whose frequency is 100 Hz. Compare this signal with that sketched in the preceding question.

15 Given $A = 5$ V and $\omega = 36$ rad/s:
   a) Build the waveforms of sine and cosine signals using the point-by-point method.
   b) Determine the period of these sinusoidal signals.

16 Period and frequency:
   a) The period of a sinusoidal signal is 1 ms. What is its frequency?
   b) The frequency of a sinusoid is equal to 1000 Hz. What is its period?
   c) *Examine Figure 2.1.14: Can you identify a period and find its value of the signal shown in Figure 2.1.14a? Figure 2.1.14b? Explain.
   d) Can the terms "period" and "frequency" be applied to only sinusoidal signal or to any periodic signal? Explain. Give examples.

17 Calculate the frequencies of sinusoidal signals whose periods are 1 μs and 1 ns. To present the results, use the scientific notation, that is, powers of 10.

18 Calculate the periods of sinusoidal signals whose frequencies are equal to 5 kHz, 5 MHz, and 5 GHz. Use the scientific notation to present the results.

19 Explain why in Eq. (2.1.3), f (Hz) $= 1/T$ (s), the units on the left-hand side are equal to the units on the right-hand side.

20 *A cosine signal has $A = 12$ V and $\omega = 24$ rad/s:
   a) Plot this signal.
   b) On this waveform, show its period, T (s), $T/2$ (s), and $T/4$ (s).
   c) How many cycles per second does this signal have?

21 The angular frequencies of the signals used in modern communications have the following typical values:
Voice $\Rightarrow f_{max} = 4$ kHz
AM radio $\Rightarrow f_{max} = 3$ MHz
Cellular phone $\Rightarrow f_{max} = 3$ GHz
Satellite communications $\Rightarrow f_{max} = 12$ GHz
Optical communications $\Rightarrow f_{max} = 193$ THz
Compute their radian frequencies

22 Explain why in Eq. (2.1.5), ω (rad/s) $= 2\pi f$ (Hz), the units on the left-hand side are equal to the units on the right-hand side.

23 *A cosine signal has $A = 4$ V and $f = 3$ kHz:
   a) What is its radian frequency?
   b) What angle in radians will it attain at $t = 2$ ms? Compute this angle in degrees.
   c) What will be the magnitude of this signal at $t = 2$ ms?
   d) Show all the calculations.

24 *Consider a cosine signal with $A = 3$ V, $f = 5$ kHz, and $\theta = -60°$:
   a) Build its waveform.
   b) How is the phase shift related to a signal's period?
   c) What time delay is caused by this phase shift?

25 *Define "amplitude." Show the amplitudes of the sine signals in Figures 2.1.4b–2.1.13.

26 Two sine signals have identical parameters except for their amplitudes: One signal has $A = 3$ V and the other has $A = 6$ V. Sketch the signals on the same axes.

27 Define peak-to-peak value. Show the peak-to-peak values of the signals shown in Figure 2.1.4.

28 What is the purpose of introducing the peak-to-peak value of a sinusoidal signal? Explain. Sketch a waveform.

29 Calculate the peak-to-peak value if the amplitude is 5 V.

30 A sinusoidal signal has $A_{pp} = 24$ V. What is its amplitude?

31 *The peak-to-peak value of a sinusoidal signal is 12 V and its minimum value is −6 V. What is the maximum value? What is the offset of the signal? The amplitude? Sketch the signal's waveform.

32 *Define peak-to-peak value and answer these questions:
   a) Can it be applied only to a sinusoidal signal or to any analog signal?
   b) We apply peak-to-peak value to an analog signal. Can we apply it to a digital signal? Explain.

33 A sinusoidal signal has $A_{max} = 4$ V and $A_{min} = -8$ V. What is the signal's amplitude? Offset? A_{pp}?

34 *An analog signal has $A_{max} = -4$ V and $A_{min} = -8$ V:
   a) Plot the signal.
   b) What is the signal's A_{pp}? Amplitude?

35 Can a signal's amplitude be a negative number? Explain.

36 What instrument can display a signal's waveform and measure its parameters, such as amplitude, peak-to-peak value, and offset? Explain.

37 Consider Figure 2.1.14Rb, reproduced here: How many types of amplitude does this signal have? Explain.

38 Can the amplitude be universally defined for any analog signal or must it be defined for each specific signal? Explain.

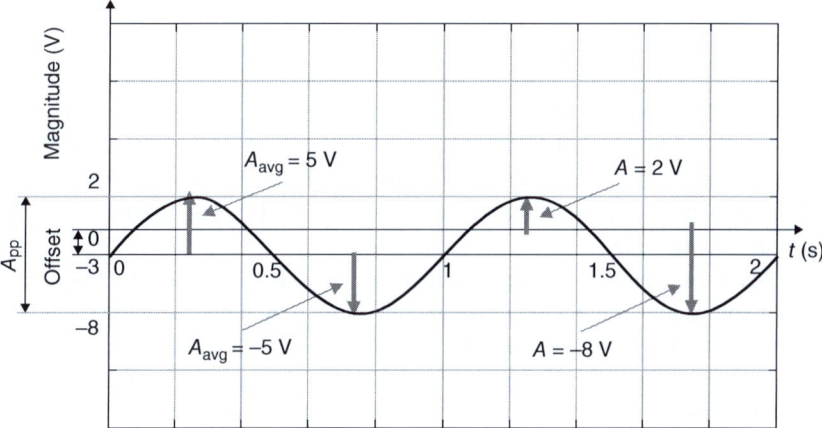

Figure 2.1.14R Definition of amplitude: (b) for a sinusoidal signal – see Figure 2.1.13.

2.2 Analog Signals – Introduction

Objectives and Outcomes of Section 2.2

Objectives

- Deepen and broaden your knowledge of the characteristics of a sinusoidal signal.
- Acquire a solid mathematical interpretation of an analog signal.
- Become familiar with frequency domain and bandwidth.

Outcomes

- Learn that a sinusoidal signal is determined by its three parameters: amplitude, frequency, and phase shift. Understand that using these parameters through the equation $v(t)$ (V) $= A \sin(\omega t + \theta) = A \sin(2\pi f t + \theta)$, the waveform of a sinusoidal signal can be built in a point-by-point manner.
- Realize that the term "phase" refers to a whole angle of a sinusoidal signal, the angle consisting from a variable part, ωt (rad), and a constant angle, θ (rad) and that the latter is called a phase shift or initial phase.
- Learn that a phase shift (initial phase) determines the initial value of a sinusoidal signal and study how to plot a sinusoidal signal with different phase shifts.
- Get to know that a constant time delay in a sinusoidal signal is equivalent to an additional phase shift.
- Learn that the difference between two types of a sinusoidal signal – cosine and sine – is that a cosine is considered a reference signal and a sine is shifted by −90° angle with respect to the cosine. Get to know that this relationship between cosine and sine can be clearly shown in phasor domain.

- Understand that an electrical signal is an implementation of a mathematical entity called function.
- Master the concept of frequency domain as an amplitude vs. frequency, A–f, space. Gather a strong understanding that in time domain a signal is presented by its waveform, $v(t)$ (V) $= A \cos(2ft + \Theta)$, and in frequency domain this signal is depicted by a single line whose height is the signal's amplitude, A (V), and whose frequency is f (Hz) $= 1/T$ (s). Gain knowledge that it is the relationship between period and frequency, T (s) $= 1/f$ (Hz), that enables us to link time domain and frequency domain.
- Learn that a cosine signal is shown in frequency domain by only an amplitude- vs. -frequency (A–f) line, whereas a sine signal is depicted in frequency domain by both A–f and Θ–f lines, where $\Theta = -90°$. Realize that the complete description of a signal in frequency domain requires introduction of two plots: amplitude vs. frequency and phase vs. frequency. Understand that therefore a frequency-domain signal's presentation needs to refer to an amplitude *spectrum* and a *phase spectrum*.
- Become thoroughly comfortable with understanding the concept of bandwidth as a range of frequencies and the bandwidth's visualization in frequency domain as a difference between two lines showing f_{max} (Hz) and f_{min} (Hz); that is, BW (Hz) $= f_{max}$ (Hz) $- f_{min}$ (Hz).
- Learn that there is a signal bandwidth (the range of frequencies occupied by the signal) and transmission bandwidth (the range of frequencies available in the transmission link).
  Understand that the transmission bandwidth must be able to accommodate the signal's bandwidth, but it also depends on the specifics of the transmission.
- Understand that the transmission bandwidth determines the transmission speed of a communications system.
- Be aware that there are six technical (engineering) definitions of bandwidth plus a legal definition as well, established by the Federal Communications Commission.
- Realize that bandwidth is a range of frequencies and it is always measured in hertz and therefore, strictly speaking, the term "bandwidth" should be applied only to an analog signal and to analog transmission. Know that communication industry uses the term in a much broader sense, such as a measure of the transmission capacity of any type of link – analog or digital.

2.2.1 More About a Sinusoidal Signal

2.2.1.1 Considering All Three Parameters – the Formula for a Sinusoidal Signal

At this point, we are ready for a more formal discussion of a sinusoidal signal. In Section 2.1, we thoroughly discussed a sinusoidal signal and its three main parameters: *period (frequency), phase shift, and amplitude*. Now we can examine in unison all three parameters. To remind you of the basics, the signal's development along the time axis is described by its *period* (which can be easily related to frequency) and *phase*; the signal's strength is described by the *amplitude* and the *peak-to-peak value*, both measured along the magnitude axis.

The formula that completely describes a sinusoidal signal is

$$v(t) \text{ (V)} = A \sin(\omega t + \theta) = A \sin(2\pi f t + \theta) \quad (2.1.1R)$$

where $v(t)$ (V) is the instantaneous value of a sine signal measured in volts, A (V) is the sine amplitude, ω (rad/s) is the radian frequency, f (Hz) is the linear (spectral, or cyclical) frequency, and θ (rad) is the phase shift; the symbol sin stands for the sine function. Clearly, similar formula works for a cosine function.

Formula (2.1.1R) allows us to build a given sine signal point by point, just as we built an analog signal in Examples 2.1.1 and 2.1.2. To do this, we need to know the basic parameters of the sine signal: frequency, phase shift, and amplitude. Next, we assign specific values to time, t; then, plugging the computed figures into Formula (2.1.1R), we obtain specific values of $v(t)$. Example 2.2.1 will help us do this in a practical manner.

Example 2.2.1 Building a Sinusoidal Waveform from Three Signal Parameters

Problem

Given the following parameters of the signal $A = 3\,\text{V}, f = 40\,\text{kHz}$, and $\theta = 30°$, build the waveform of the sine signal.

Solution

To start, we need to calculate the radian frequency:

$$\omega = 2\pi f = 2\pi \times 40 \times 10^3 \approx 250 \times 10^3 \text{ rad/s}$$

Note how scientific notations help us do the calculations and present their results.

To build the graph of the signal in question, we have to take the set of time instants and compute the corresponding signal's magnitudes. But first we need to have both members of the signal's argument, $(\omega t + \theta)$, in the same unit. Thus, we must convert θ to radians or, alternatively, convert ωt to degrees. Let us convert θ to radians by using (2.1.9b):

$$\theta = (\pi/180°) \times \theta\,(°) = 3.14/180° \times 300 = 0.523 \text{ rad}$$

Therefore, the phase of the sine signal is given by

$$\omega t + \theta = 250 \times 10^3 \text{ rad/s} \times t \times 10^{-6}\text{ s} + 0.523 \text{ rad} = 0.25 \times t\,(\mu s) + 0.523 \text{ rad}$$

where time is given in microseconds for the convenience of the computations.

Now we can compute the instantaneous values of a sine signal as the following examples demonstrate:

at $t = 0\,\mu s$, $v(t) = A\sin(\omega t + \theta) = 3\sin(0.523) = 1.498\,\text{V}$;
at $t = 4\,\mu s$, $v(t) = A\sin(\omega t + \theta) = 3\sin(0.25\cdot 4 + 0.523) = 2.997\,\text{V}$, and so on.
at $t = 17\,\mu s$, the magnitude is $-2.994\,\text{V}$, and so forth (see Table 2.2.1).

Table 2.2.1 Samples of data for building the graph of a sinusoidal signal in Example 2.2.1.

Time (μs)	ωt (rad)	$\omega t + \theta$ (rad)	$\sin(\omega t + \theta)$	$3\sin(\omega t + \theta)$ (V)
0	0	0.523	0.499	1.498
1	0.25	0.773	0.698	2.095
2	0.5	1.023	0.854	2.561
⋮	⋮	⋮	⋮	⋮
57	14.25	14.773	0.805	2.414
58	14.5	15.023	0.633	1.898
59	14.75	15.273	0.421	1.264

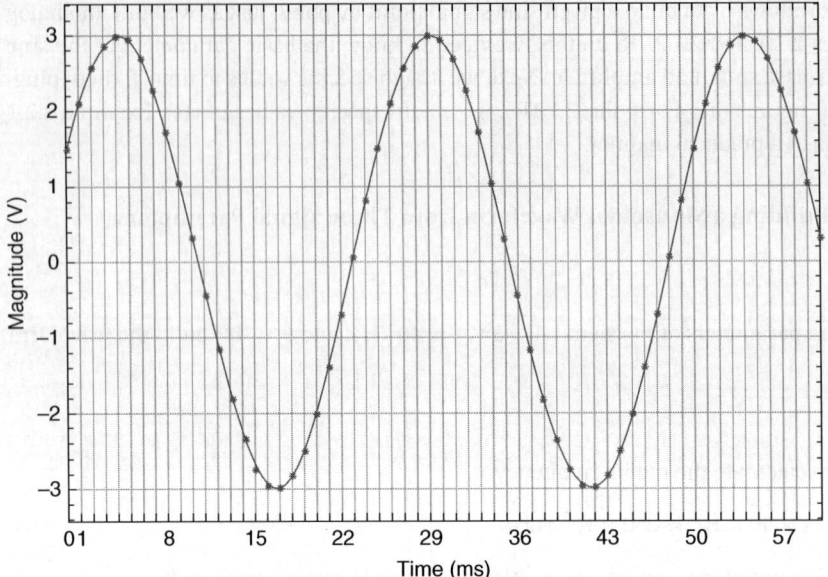

Figure 2.2.1 A sine signal for Example 2.2.1.

Of course, we need to automate these computations by using a software tool such as MS Excel, MATLAB, or Mathematica. Here, we have used MS Excel to compute all the numbers (samples of which are shown in the table below) and MATLAB to build the required graph, presented in Figure 2.2.1.

Discussion

- This exercise gives you a deeper understanding of the concept of function in general and a sinusoidal signal in particular. In addition, you gain an insight of how any computerized tool automatically build such graphs.
- This graph and table allow us to find the magnitude of a sine signal at any given time. This is what is called an instantaneous value of a sine signal. Refer to Sidebar 2.2.S.2 in this section to review the concept of a function: The dependent variable follows a change in the independent variable according to the given function. As this example shows, a sine is a specific function; its waveform is the graphical representation of this function. In mathematics, we know the formula and so we need only plug in the numbers to obtain a table and a graph (as explained in Sidebar 2.2.S.1 and worked out in this example); in engineering, however, this is not always the case. Very often, we build a signal plot experimentally and, to obtain the analytical expression of the function, we have to do a lot more work.

2.2.1.2 The Phase of a Sinusoidal Signal: a Detailed Look

Now we are better equipped to discuss the phase of a sine signal. Let us recall Formulas (2.1.1), (2.1.13), and (2.1.14):

$$v(t)\,(\text{V}) = A\,\sin(\omega t + \theta) \tag{2.1.1R}$$

$$v = A\,\sin\alpha \tag{2.1.13R}$$

and

$$\alpha \text{ (rad)} = \omega t + \theta \tag{2.1.14R}$$

As discussed in Section 2.1, these formulas tell us that

1. The waveform v given in (2.1.13) as a function of α is, in fact, a function of time, as (2.1.1) shows, because ω (rad/s) and θ (rad) are constants.
2. The phase angle, α (rad), consists of two parts: a variable angle, ωt (rad), and a constant angle, θ (rad).

Our discussion in Section 2.1. is devoted to the constant angle, that is, to the phase shift, θ (rad); here, we want to discuss the sinusoidal phase in its entirety.

As for the phase shift itself, it is, on the one hand, the angle that determines the initial value of a sine signal and, on the other hand, the angle by which a sine signal is shifted along the time axis with respect to the zero point.

Let us consider the initial value first. This is the value of a sine signal when $t = 0$ s. Plugging this time value into Formula (2.1.1), we obtain

$$v(t) = A\sin(\omega t + \theta) = A\sin(2\pi f t + \theta)$$
$$v(0)(V) = A\sin(0 + \theta) = A\sin\theta \tag{2.2.1}$$

Formula (2.2.1) says that the magnitude of a sine signal is equal to $A\sin\theta$ at $t = 0$. This statement is illustrated in Example 2.2.2:

Example 2.2.2 Initial Value of a Sinusoidal Signal

Problem

Given the sine signal $v(t) = 3\sin(250 \times 10^3 t + \theta)$, find the initial values of this signal at the following phase shifts: $\theta = 0$; $\theta = 45°$; $\theta = 90°$; $\theta = 180°$; and $\theta = 270°$. Sketch the signal at all initial values.

Solution

The calculations are easy: Just plug the phase-shift figures at $t = 0$ into the formula for the signal and obtain the initial values:

$\theta = 0 \Rightarrow v(0) = 3\sin(250 \times 10^3 \times 0 + 0) = 0\,\text{V}$
$\theta = 45° \Rightarrow v(0) = 3\sin(45°) = 3\sin(0.785\,\text{rad}) = 3 \times 0.707 = 2.121\,\text{V}$
$\theta = 90° \Rightarrow v(0) = 3\sin(90°) = 3\sin(1.57\,\text{rad}) = 3 \times 1 = 3\,\text{V}$
$\theta = 180° \Rightarrow v(0) = 3\sin(180°) = 3\sin(\pi) = 3 \times 0 = 0\,\text{V}$
$\theta = 270° \Rightarrow v(0) = 3\sin(270°) = 3\sin(4.712\,\text{rad}) = 3 \times (-1) = -3\,\text{V}.$

The solution shows that for computations of the initial values of the sine signal, we did not need to convert the unit for phase shift from degrees to radians. However, in general, we have to present the sine phase in the same unit. This is why we have shown these conversions in the above computations.

Using the same method for building the sine signal that we presented in Example 2.2.1, we can easily sketch all the graphs. They are shown in Figure 2.2.2.

Discussion

- Observe the initial values shown in Figure 2.2.2 and spend some time analyzing this figure to make sure that you understand the concept of phase shift completely. If necessary, return to the previous subsection, where the concept of phase shift is discussed.
- An initial value is a good tool to verify your calculations of a sinusoidal signal's parameters. Suppose, for example, we work with the signal

$$v(t) = 5 \sin(33 \times 10^6 \times t + 90°)$$

and we are not sure whether this sine signal shifts left or right with respect to a magnitude axis. Just substitute $t = 0$ and compute $v(0) = 5 A \sin(90°)$. Since $\sin 90° = 1$, this signal shifts left, as Figure 2.2.2c shows. What if a signal is given by $v(t) = 5 \sin(33 \times 10^6 \times t - 90°)$? Again, at $t = 0$,

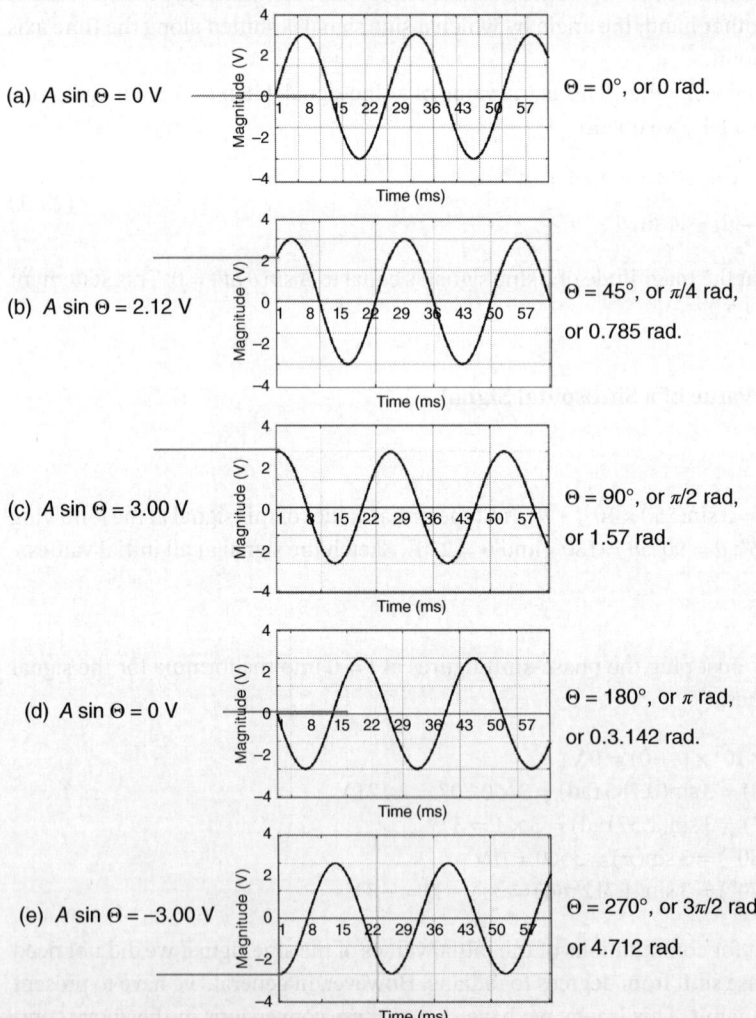

Figure 2.2.2 Initial values of a sine signal with various phase shifts: (a) $A \sin \theta = 0$ V at $\theta = 0°$; (b) $A \sin \theta = 2.1$ V at $\theta = 45°$; (c) $A \sin \theta = 3.0$ V at $\theta = 90°$; (d) $A \sin \theta = 0$ V at $\theta = 180°$; (e) $A \sin \theta = -3$ V at $\theta = 270°$.

simply compute $v(0) = 5 \sin(-90°) = -5 \sin(90°) = -5$ V. Therefore, this signal shifts right so that its negative peak is at the zero point of the time axis (see Figures 2.2.2e and 2.1.11). Incidentally, refer again to Figure 2.2.2e. Does this figure depict what we have just said about an initial value? Yes, it does. Simply compare Figures 2.1.11 and 2.2.2e to see this point for yourself.

Now let us turn to the other side of a phase shift. By definition, phase shift is the angle itself. Where is this angle in Figure 2.2.2? The answer lies in understanding the nature of the argument of a sine signal. So consider again the general formula for a sine signal,

$$v(t) = A \sin(\omega t + \theta) = A \sin(2\pi f t + \theta) \tag{2.1.1R}$$

Time changes independently, and this change causes a change in the entire signal. However, the sine is the function of the whole independent variable, $\alpha(\text{rad}) = (\omega t + \theta) = (2\pi f t + \theta)$, not just time alone. In other words, if we change ω (or f) and θ, we will obtain a different sine signal. This means, strictly speaking, that we should sketch the graph of a sine signal as a function of the entire phase; that is, α (rad) $= \omega t + \theta = 2\pi f t + \theta$. Quite clearly, if we know f and θ, computations in time domain and in angle domain will be equivalent. To elaborate on this discussion, consider the following example:

Example 2.2.3 Plotting Sine Signals with Given Phase Shifts

Problem

Sketch the sine signal described in Example 2.2.2 and show the phase shifts on the graphs.

Solution

We will draw the graph of the sine signal so that the *x*-axis represents an angle. Then we can easily identify the phase shift along the *x*-axis. The result of these operations is shown in Figure 2.2.3.

Discussion

- We built the graphs in Figure 2.2.3 to emphasize the following point: A phase shift is the angle by which a sine wave is shifted along the angle's axis, a point we have stressed several times already. Look closely at Figure 2.2.3 and compare it with Figures 2.1.9–2.1.11, which are the same figures drawn in time domain. We should now understand completely the relationship between time domain and angle domain in plotting a sinusoidal signal; therefore, we would not need to return to this topic again. (Refer to Sidebar 2.2.S.1 and the problems at the end of this section for more exercises on this material.)
- What if the phase shift is 360°, or 2π? The sine signal returns to the starting point of the next period, where it begins to repeat itself. You cannot distinguish between phase shifts of, say, 90° and 450° because 450° = 90° + 360°.
- What if a phase shift will not stay constant but will vary? Well, in this case, the frequency of the sinusoidal signal will change. A variation in phase shift is equivalent to a change in frequency. Equation (2.1.1) states this explicitly: $v(t) = A \sin(\omega t + \theta) = A \sin(2\pi f t + \theta)$. If θ changes discretely, the sine signal will experience a discrete phase shift, as Figures 2.2.2 and 2.2.3 show. If θ changes constantly, the phase shift becomes time dependent, and we would not be able to distinguish between a variable phase shift, $\theta(t)$, and a frequency member, $\omega t = 2\pi f t$. Thus, the whole phase, $\omega t + \theta = 2\pi f t + \theta$, of a sinusoidal signal becomes variable. This last statement can be proved as follows:

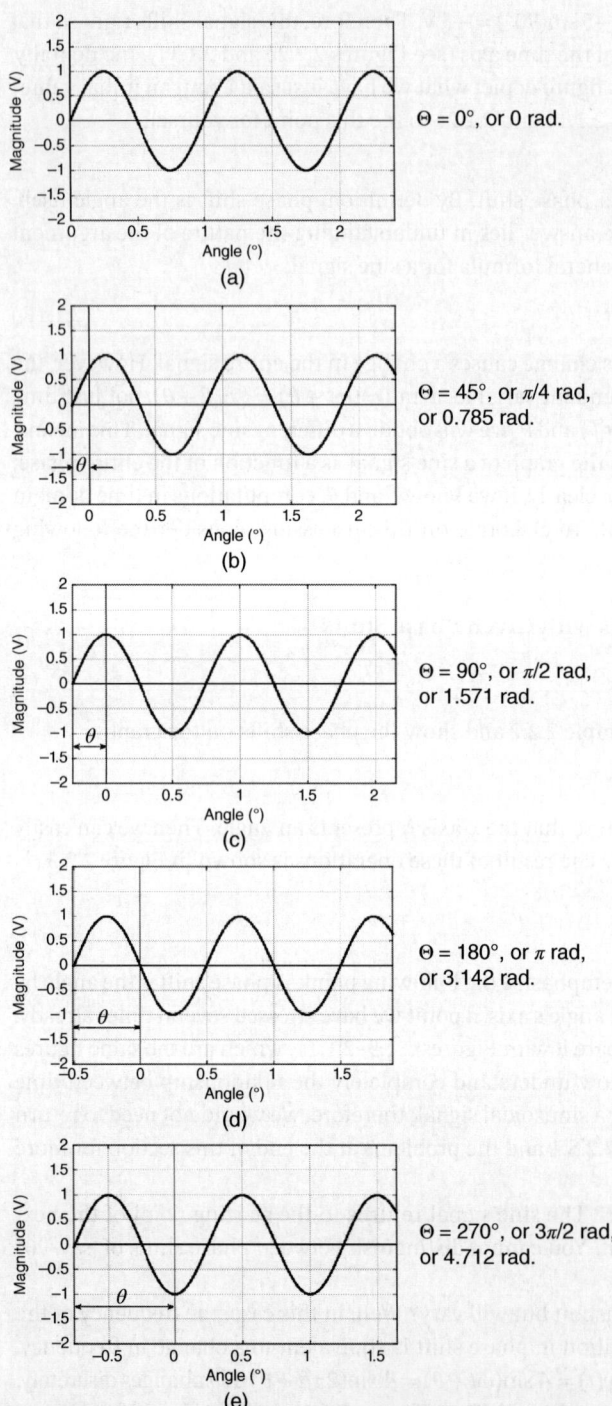

Figure 2.2.3 Various initial phases (phase shifts) of a sine signal: (a) $\theta = 0°$; (b) $\theta = 45°$; (c) $\theta = 90°$; (d) $\theta = 180°$; (e) $\theta = 270°$.

By definition, a signal's radian frequency is the derivative of its phase; i.e. ω (rad/s) = $\frac{d(\alpha)}{dt}$, where the phase angle, α (rad), is given by $\alpha = \omega t + \theta$ for a sinusoidal signal. If θ = constant, then the signal's radian frequency is given as ω (rad/s) = $\frac{d(\alpha)}{dt} = \frac{d(\omega t + \Theta)}{dt} = \omega$ (rad/s). If, however, the phase shift changes with time, $\theta = \theta(t)$, then the radian frequency become ω (rad/s) = $\frac{d(\alpha)}{dt} = \frac{d(\omega t + \Theta(t))}{dt} = \left[\omega + \frac{d\Theta(t)}{dt}\right]$ (rad/s). In this case, a sinusoidal signal has a new value of its radian frequency. For example, assume that $\theta(t) = kt$ (rad), where k (rad/s) is a constant. Then $\left[\frac{d\Theta(t)}{dt}\right] = k$ (rad/s) and the radian frequency becomes $[\omega + k]$ (rad/s) instead of just ω (rad/s).

What if a radian frequency, ω (rad/s), is not constant? Well, in this case, the waveform of the signal would not be a sinusoidal because the formula for a sinusoidal signal will be changed. For example, if $\omega(t) = mt$, where m (rad/s²) is a constant, then $\cos(\omega(t)t) = \cos(mt^2)$, which is not a cosine signal. (Construct this waveform to verify this statement. Also, refer to Sidebar 2.2.S.1 for a deeper understanding of this point.)

- A final note: Suppose a sinusoidal signal passes through a communication system and experiences a time delay. This time delay results in an additional phase shift of the output signal, as mentioned previously in this subsection. Putting this statement into formulas, we can present the input and output signals of a communication system as

$$v_{in}(t) = A \sin(2\pi f_{in} t + \theta_{in})$$

and

$$v_{out}(t) = A \sin(2\pi f_{in}(t + t_d) + \theta_{in})$$

where t_d is a constant time delay. Therefore, the output signal takes the form

$$v_{out}(t) = A \sin(2\pi f_{in}(t + t_d) + \theta_{in}) = A \sin(2\pi f_{in} t + (2\pi f_{in} t_d) + \theta_{in})$$

Since $(2\pi f_{in} t_d)$ is a constant and its dimension is radian, we can denote it as θ_d; that is,

$$(2\pi f_{in} t_d) = (\omega_{in} t_d) = \theta_d \tag{2.2.2a}$$

Hence, the time-delayed signal has the waveform

$$v_{out}(t) = A \sin(2\pi f_{in} t + \theta_{in} + \theta_d) \tag{2.2.2b}$$

Therefore, we have proved our statement: For a sinusoidal signal, a constant time delay is equivalent to an additional phase shift. The mathematical expression of this fact is given by

$$A \sin(2\pi f_{in}(t + t_d) + \theta_{in}) = A \sin(2\pi f_{in} t + \theta_{in} + \theta_d) \tag{2.2.2c}$$

A telecommunication engineer should have a deep understanding of phase shift. Any variations in phase during the transmission of a telecommunications signal will distort the signal and may lead to errors in the transmitted information.

We now need to introduce one more important concept concerning phase shift. Consider the two sinusoidal signals shown in Figures 2.2.2a,b. Let us define a starting point as the point where a signal begins changing its values in a positive direction for the first time. Then, observing these signals, we can see that the signal in Figure 2.2.2b starts earlier than the signal in Figure 2.2.2a because its starting point occurs first. Based on this observation, we can say that

> *the signal $v_b(t)$ leads the signal $v_a(t)$ because its starting point occurs first; or, conversely, the signal $v_a(t)$ lags the signal $v_b(t)$ because its starting point occurs second.*

Taking a formal approach, we need to say these signals have the same frequency, but the signal in Figure 2.2.2b is shifted with respect to the signal in Figure 2.2.2a by the phase angle, θ. Converting this statement to formulas, we can write

$$v_a(t) = A \sin(\omega t) = A \sin(2\pi ft)$$

and

$$v_b(t) = A \sin(\omega t + \theta) = A \sin(2\pi ft + \theta).$$

Now we can say that the signal $v_b(t)$ is leading the signal $v_a(t)$ by the angle θ; on the other hand, the signal $v_a(t)$ is lagging the signal $v_b(t)$ by the same angle. We can check this relationship by setting t (s) = 0: We immediately find that $v_a(0) = 0$, but $v_b(0) = A \sin(\theta)$, which means that the signal's starting point comes before t (s) = 0. We will discuss these terms – *leading* and *lagging* – again when we consider phasors.

2.2.1.3 Cosine and Sine Signals

We already touched on this topic in Section 2.1 (see Example 2.1.2). As you no doubt have gathered from our discussion, there are two similar periodic sinusoidal signals – sine and cosine. They are often called harmonic signals; we have referred to both of them as sinusoidal signals. They have identical shapes and, in fact, their other features are similar as well: They repeat themselves after one period (cycle) and are characterized by their amplitude and period (frequency). Cosine and sine signals differ only in phase shift. The question is which signal should we take as a reference in terms of phase shift? If we take the sine, then the cosine signal will have a 90° phase shift; if we take the cosine, then the sine will be shifted by a −90° angle with respect to the reference signal (see Figures 2.2.2, 2.2.3, and 2.1.6). Putting this in mathematical format, we obtain

$$\cos \alpha = \sin(\alpha + 90°) \quad (2.2.3a)$$

and

$$\sin \alpha = \cos(\alpha - 90°) \quad (2.2.3b)$$

From this consideration, there are no evident preferences in choosing any signal as a reference point. However, the engineering and scientific community has selected the cosine as the phase reference signal. (We explain the justification for this choice in later chapters. For now, we can only say that when a sinusoidal signal is presented as a complex number, the cosine is obtained as a real part of this number.)

Therefore, *starting from this point, we will use*

$$v(t) = A \cos(\omega t + \theta) = A \cos(2\pi ft + \theta)$$

as the reference to a sinusoidal (harmonic) signal.

Again, do not be confused by the term "sinusoidal": This term refers to both signals – sine and cosine.

To fully understand a sinusoidal signal in general and relationship between sine and cosine in particular, it is worth knowing how it can be related to a rotating vector (phasor). Sidebar 2.2.S.1 gives this explanation.

Sidebar 2.2.S.1 Phasor and Sinusoidal Signal

A sinusoidal signal can be generated by a rotating vector; this procedure is the topic of this sidebar. The understanding of this operation reveals important details about the nature of a sinusoidal signal.

Consider the vector shown in Figure 2.2.S.1.1. Here is the set of conditions determining its behavior:

- A vector of constant length A rotates in one plane around a fixed zero point.
- The initial position of this vector is arbitrary. (In Figure 2.2.S.1.1, where a cosine signal is considered, we set it in the positive vertical position, implying that the phase shift is zero; i.e. θ (rad) = 0.)
- The vector rotates in a counterclockwise (CCW) direction with a constant angular velocity, ω (rad/s).
- A current vector's position is determined by angle, α (rad), defined below. This angle counts from the initial position; it is considered positive if the angle counts in a CCW direction. Clearly, angle counting in a clockwise (CW) direction is considered negative.
- This vector is called a *phasor*.

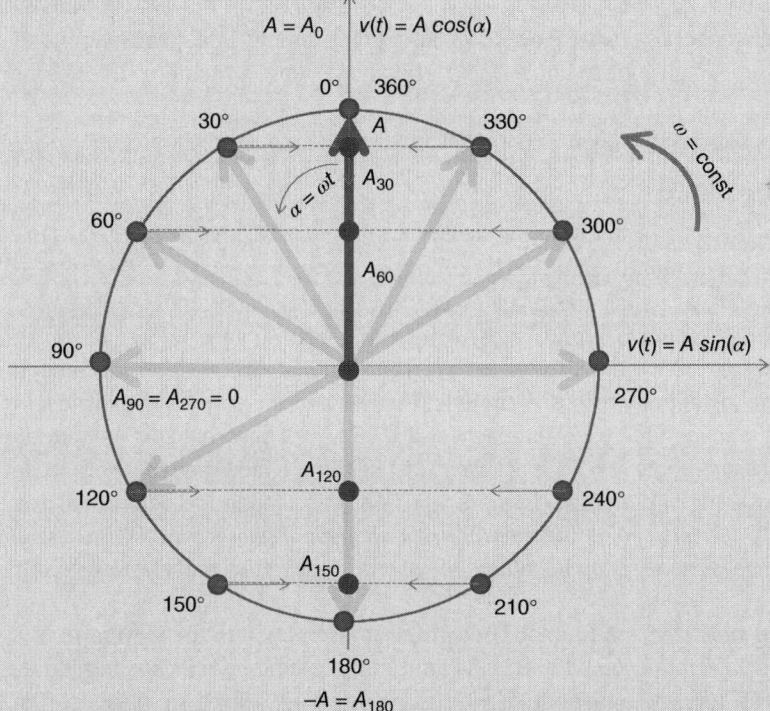

Figure 2.2.S.1.1 Phasor: a graphical presentation.

We need to realize that the phasor's position is a function of angle α (rad), but this angle depends on the phasor's angular velocity, ω (rad/s), and time, t (s), as

$$\alpha \text{ (rad)} = \omega \text{ (rad/s)} \cdot t \text{ (s)} \qquad (2.2.S.1.1)$$

(Continued)

Sidebar 2.2.S.1 (Continued)

Since ω (rad/s) is constant, the phasor's position is eventually determined by time. Let us discuss the main phasor's features, as shown in Figure 2.2.S.1.1.

The phasor's projection on a vertical axis, $v(t_\alpha)$, at an angle α is given by

$$v(t_\alpha) = A\,\cos(\alpha) = A\,\cos(\omega t) = A_\alpha \qquad (2.2.S.1.2)$$

which follows from the trigonometric definitions of a cosine. This projection is a *magnitude*, A_α, of the cosine signal. Since we assume the phasor's initial position to be vertical, we find that $v(0) = A_0 = A$. When the phasor turns 30° in the CCW direction, its magnitude (projection at the vertical axis, that is) becomes $v(t_{30}) = A\,\cos(30°) = A_{30} = 0.87A$. Similar calculations give $A_{60} = 0.5A$, $A_{90} = 0$, $A_{120} = -0.5A$, and so on. Figure 2.2.S.1.1, which shows the cosine's instantaneous values (its *magnitude*, A_α) for a set of angles, helps you follow this procedure. Since the phasor's vertical projection is a cosine function, as (2.2.S.1.2) states, we have developed a method to find the instantaneous value (magnitude) of a cosine signal, $v(t_\alpha) = A\,\cos(\alpha) = A\,\cos(\omega t) = A_\alpha$.

A close review of Figure 2.2.S.1.1 reveals, however, what seems to be a confusing point: The magnitude at angle *30°* is equal to the magnitude at angle *330°*; that is, $A_{30} = A_{330} = 0.87A$. Similarly, $A_{60} = A_{300} = 0.5A$, $A_{90} = A_{270} = 0$, $A_{120} = A_{240} = -0.5A$ and $A_{150} = A_{210} = -0.87A$. How can we distinguish between these pairs? How do we know, for example, that given magnitude 0.87A belongs to angle 30° but not to angle 330°? The answer lies in timing of the phasor rotation.

The phasor reaches any given angle at the instant

$$t_\alpha\,(\text{s}) = \alpha\,(\text{rad})/\omega\,(\text{rad/s}) \qquad (2.2.S.1.3)$$

which is derived from (2.2.S.1.1). If, for example, $\omega = 10$ rad/s, then the angle $\alpha = 30° = \pi/6$ rad is reached for $t_{30}\,(\text{s}) = \pi/6$ rad/10 rad/s $= 0.05$ s. Consequently, it takes $t_{30}\,(\text{s}) = 0.05$ s to attain the angle $\alpha = 30°$, at which the phasor's vertical projection becomes $A_{30} = 0.87A$. Similarly, it will take $t_{60}\,(\text{s}) = \pi/3$ rad/10 rad/s $= 0.1$ s to reach angle $\alpha = 60°$, at which $v(t_{60}) = A\,\cos(60°) = A_{60} = 0.5A$. In this manner, we can establish the relationship among the phasor's angle, the instant at which this angle is reached, and the cosine's magnitude. In our example, when the magnitude 0.87A is reached at 0.05 s, we know that this magnitude corresponds to the phasor's angle $\alpha = 30°$ because $t_{30}\,(\text{s}) = 0.05$ s. If, however, the magnitude 0.87A is reached at $t_{330}\,(\text{s}) = 11\pi/6$ rad/10 rad/s $= 0.56$ s, this magnitude belongs to the phasor's angle of 330°. *Understanding the interconnection among a phasor's angle, its timing, and a cosine's magnitude is the key point in relating the phasor presentation to the waveform of a sinusoidal signal.*

The next – and the *main step* – is to relate this phasor presentation to the waveform of a sinusoidal signal. The waveform, you will recall, is built in the v–t plane, which is formed by a time (horizontal) axis, t(s), and a magnitude (vertical) axis, $v(t)$ (V). Since we establish the relationship between the phasor's magnitude and its instant in the phasor presentation, we can place the magnitudes of the cosine signal against the proper instants in the v–t plane and construct the waveform of a cosine signal. Figure 2.2.S.1.2a shows how the cosine signal is built from the phasor presented in Figure 2.2.S.1.1. To directly relate the phasor presentation to the corresponding waveform, we combine both presentations shown in Figures 2.2.S.1.1 and 2.2.S.1.2a in Figure 2.2.S.1.2b.

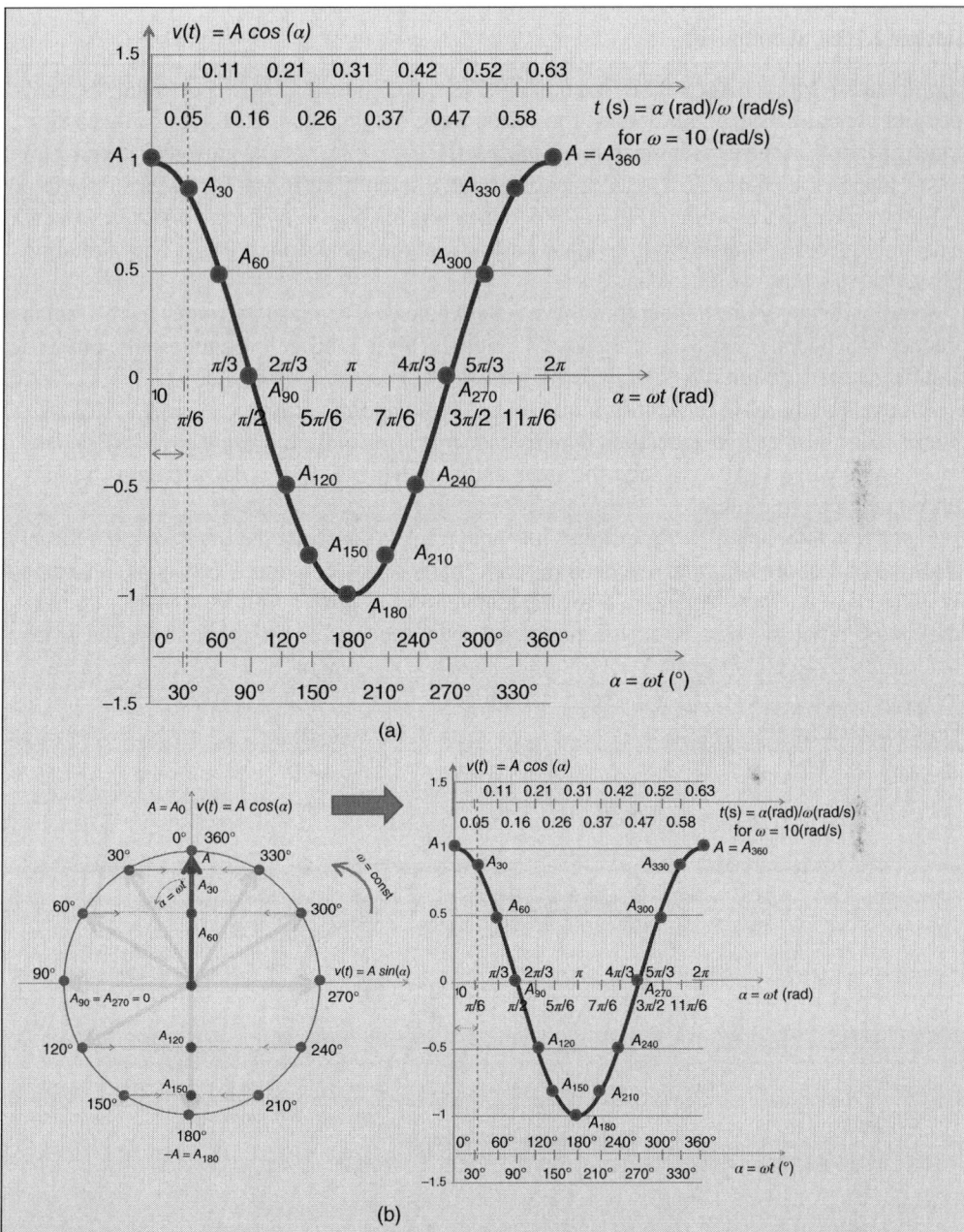

Figure 2.2.S.1.2 Constructing a cosine signal from a phasor presentation: (a) Building the cosine's waveform by transferring magnitudes and their corresponding instants from the phasor presentation; (b) the phasor presentation and the signal waveform in one figure.

Specifically, constructing a cosine signal from a phasor is performed in Figure 2.2.S.1.2a as follows: When $t = 0$ s, the phasor's magnitude A_0 (which is equal to the amplitude, A, here) is located at the $v(t)$ axis at $t = 0$ s. When the phasor reaches $\alpha = 30° = \pi/6$ rad at the instant t_{30} (s) $= 0.05$ s, its magnitude becomes $v(t_{30}) = A_{30} = 0.87A$. Placing the dot with these

(Continued)

Sidebar 2.2.S.1 (Continued)

coordinates – $v(t_{30}) = A_{30} = 0.87A$ V and $t = 0.05$ s – in Figure 2.2.S.1.2a, we build the next point of the cosine signal. All dots on the cosine waveform in Figure 2.2.S.1.2a correspond to the proper dots in the phasor diagram presented in Figure 2.2.S.1.1. Obviously, we can take as many dots on the phasor's circle as needed to build a smooth, clear graph of a cosine signal.

To summarize this procedure, Figure 2.2.S.1.2b shows the phasor presentation (the left-hand side) and the cosine waveform (the right-hand side) of the same signal. This is how a rotating phasor can generate a sinusoidal signal.

Focus particularly on the fact that the angular velocity of a phasor, ω (rad/s), *is* the radian frequency, ω (rad/s) = $2\pi f$ (Hz), of a sinusoidal signal. This is another link between the phasor's and the signal's presentations.

Observe that there are three horizontal axes on the waveform's graphs in Figures 2.2.S.1.2a,b. These axes present three equivalent dimensions: time in seconds, the angle $\alpha = \omega \cdot t$ in radians for the assumed $\omega = 10$ rad/s, and the same angle in degrees. This is done to help you fully understand this procedure.

In addition to building a cosine signal, it is important to construct the graph of a *sine signal* from the phasor presentation. We need to recall that a cosine is a reference signal and a sine is shifted by −90° with respect to a cosine, as given in (2.2.3b). Therefore, in phasor domain, the sine's initial position is on the horizontal axis. (Do you understand why?) The sine's magnitude is calculated as $v(t_\alpha) = A\sin(\alpha) = A\sin(\omega t) = A_\alpha^s$, where the angles are the same as in Figures 2.2.S.1.1 and 2.2.S.1.2. These magnitudes for various phasor positions are shown in Figure 2.2.S.1.3.

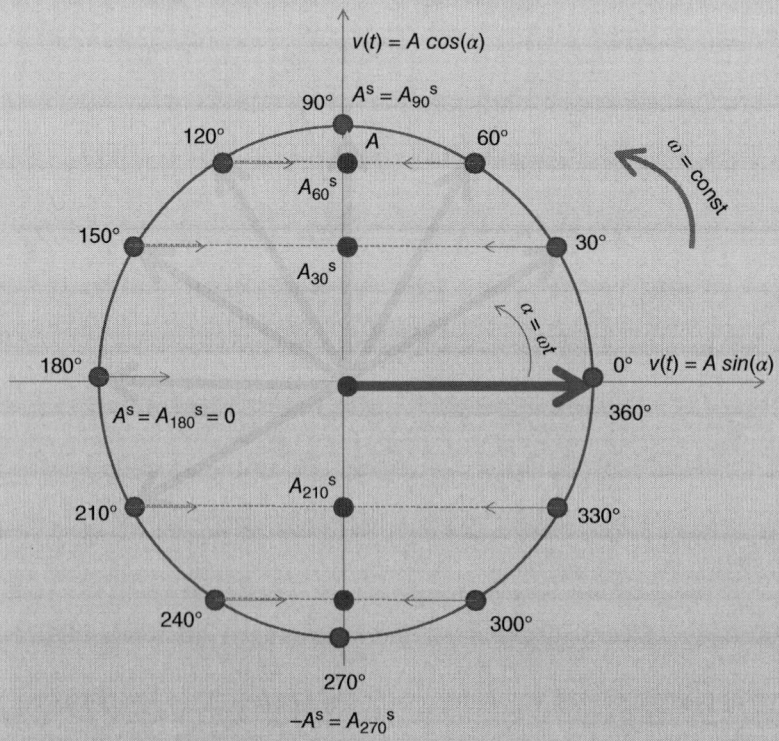

Figure 2.2.S.1.3 Phasor and the magnitudes of a sine signal.

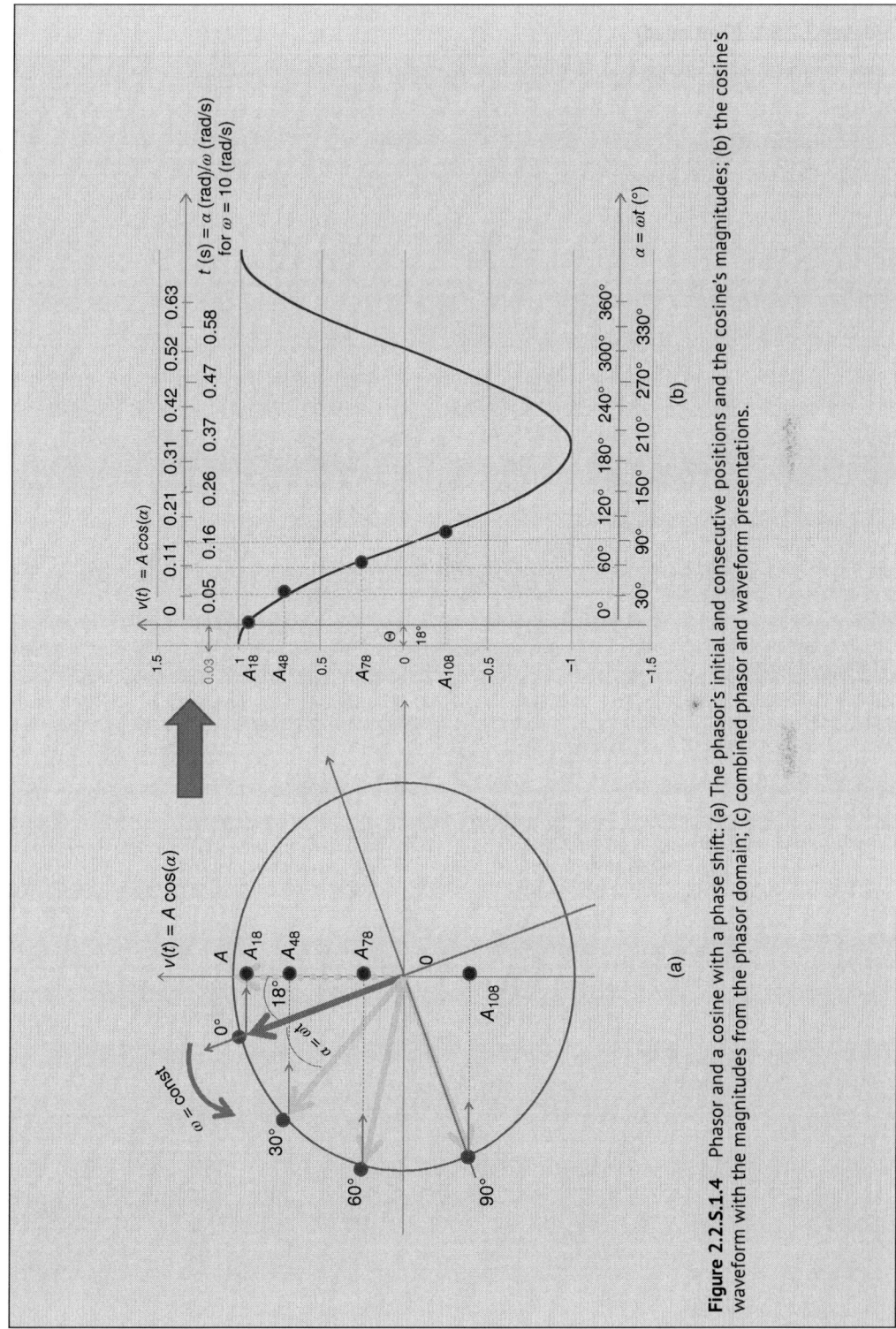

Figure 2.2.S.1.4 Phasor and a cosine with a phase shift: (a) The phasor's initial and consecutive positions and the cosine's magnitudes; (b) the cosine's waveform with the magnitudes from the phasor domain; (c) combined phasor and waveform presentations.

(Continued)

Sidebar 2.2.S.1 (Continued)

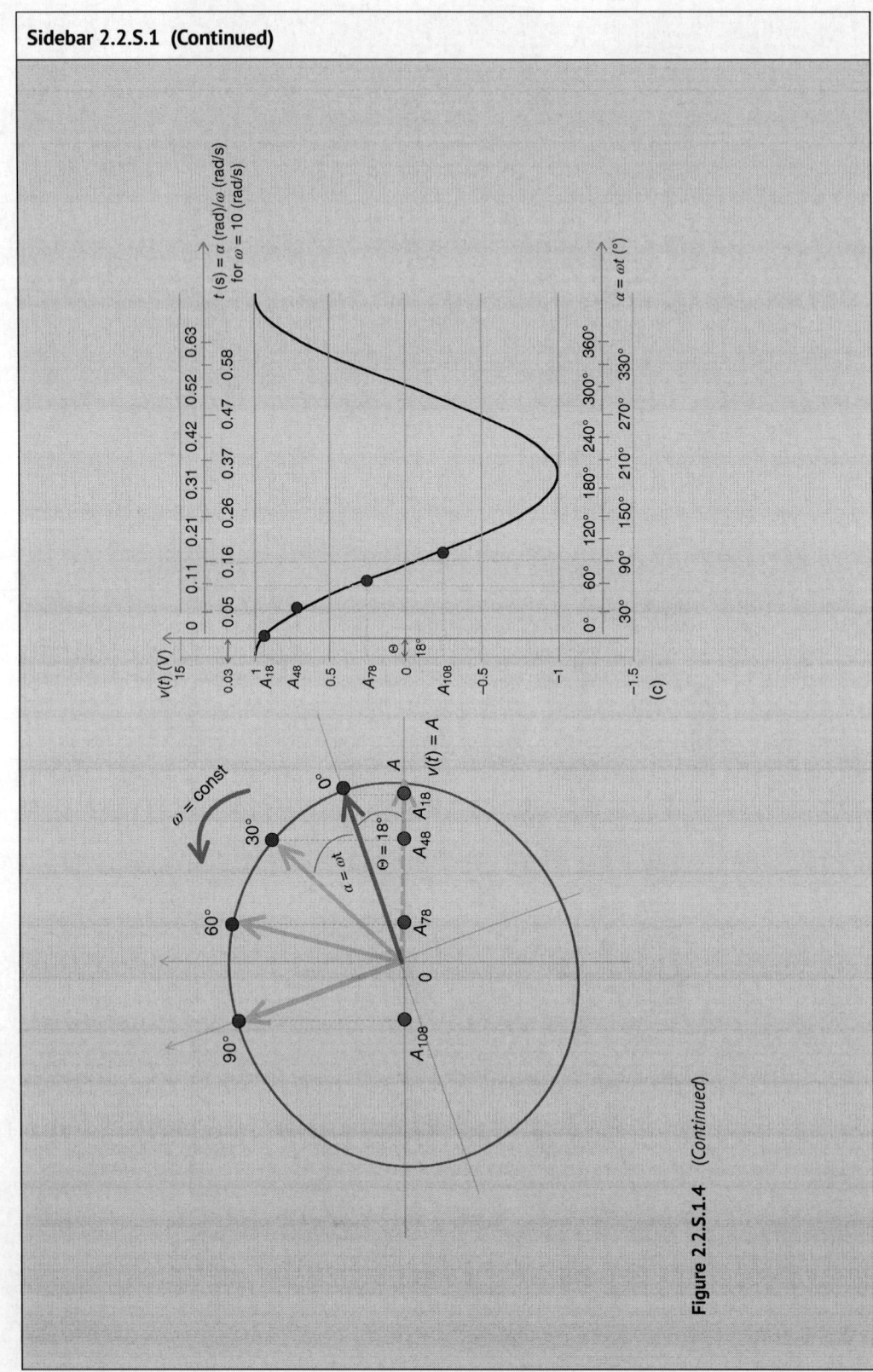

Figure 2.2.S.1.4 (Continued)

We observe in Figure 2.2.S.1.3 that at $t_0 = 0$ s, the sine's magnitude – that is, its projection on the vertical axis – is zero; at $t_{30} = 0.05$ ms, the sine's magnitude is $v(t_{30}) = A \sin(30°) = A_{30}^s = 0.5A$ V; at $t_{60} = 0.11$ ms the magnitude is $A_{60}^s = 0.87A$ V; and so on. Following this procedure, we can build the waveform of a sine signal. (**Exercise**: Build the waveform of a sine signal generated by the rotating phasor shown in Figure 2.2.S.1.3.)

Several implications of this approach to constructing a sinusoidal signal are worth emphasizing:

- *Initial position*: We said previously that a phasor's initial position can be one of many. Consider, for example, the phasor turned by the angle $\theta = 18°$ in the CCW direction with respect to a vertical axis, as shown in Figure 2.2.S.1.4a. Now, this phasor's new position *is* its initial position. Therefore, the phasor's vertical projection (that is, the cosine magnitude) at $t_{18} = 0$ s is equal to $A \cos(18°) = A_{18} = 0.95A$, as shown in Figure 2.2.S.1.4a. This figure also shows the phasor's vertical projections for its consecutive turns at angles 30°, 60°, and 90° with respect to the current initial position; these positions correspond to angles 48°, 78°, and 108° with respect to the vertical axis. Thus, the cosine's magnitude for a 30° turn is equal to $A \cos(18° + 30°) = A_{48} = 0.67A$. Similarly, $A_{78} = 0.21A$ and $A_{108} = -0.31A$.

Obviously, the phasor covers a 30°-angle over the same time interval as before, t_{30} (s) $= \pi/6$ rad/10 rad/s $= 0.05$ s, provided that the phasor's angular velocity remains the same, $\omega = 10$ rad/s, as in the preceding discussion. This information allows us to transfer the cosine's magnitudes and their corresponding instants (angles) to the v–t plane for building the cosine's waveform. The construction of the cosine signal for the four phasor angles (instants) is shown in Figure 2.2.S.1.4b.

The correspondence between a phasor's position, its timing, and the cosine's magnitude can be observed in Figure 2.2.S.1.4c, which combines both preceding figures into one. Therefore, here we follow the same procedure for constructing the cosine waveform generated by a phasor's rotation as we did in Figure 2.2.S.1.2. The only difference is in the initial position of the phasor.

(**Exercise:** Based on Figure 2.2.S.1.4, finish constructing the cosine's waveform for the whole period.)

- *A phasor's phase shift and a signal's initial position*: Examine Figure 2.2.S.1.4c again: The phasor's phase shift (the left-hand side graph) is the angle θ, at which the phasor turns from its vertical position. But this is the same angle at which the cosine's waveform is shifted along the time axis (the right-hand side graph). Since we now know how the phasor's position relates to the cosine's magnitude, we can fully understand the relationship between the phase shift and the initial value of a cosine signal, discussed in this section in Examples 2.2.2 and 2.2.3. It would be well worth your returning to these examples to strengthen your knowledge of this topic.

Note that the initial phase angle θ in the left-hand side graph in Figure 2.2.S.1.4c is positive because it turns in the CCW direction. Now, observe the corresponding cosine's waveform in the right-hand side graph in Figure 2.2.S.1.4c: Do you have a visual impression that the waveform is shifted to the left with respect to the vertical axis, $v(t)$? If you do, you are correct. *This observation explains the rule introduced in Section 2.1 and confirmed in this section: If the phase shift θ of a sinusoidal signal is positive, then its waveform moves in the negative direction*

(Continued)

Sidebar 2.2.S.1 (Continued)

in the v–t domain. (In fact, a phasor's positive phase shift means that the starting point, $t = 0$ s, of the time axis is shifted to the right with respect to the previous zero point. Therefore, the word "impression" precisely reflects the situation with all these movements.)

- *Period and cycle*: The relationship between a phasor and a sinusoidal signal, as illustrated in Figures 2.2.S.1.2 and 2.2.S.1.4, highlights the meaning of the terms *cycle* and *period*, introduced in Section 2.1 and described in this section as well. Examining the phasor's graph in Figure 2.2.S.1.1, we see that after the phasor covers 360°, it returns to the original position. In other words, rotating at 360°, the phasor completes a *cycle* of its motion. But a sinusoidal signal, as shown in Figures 2.2.S.1.2 and 2.2.S.1.4, after covering 360°, starts repeating itself; that is, it completes its *period*. This is why, *for a periodic signal, the number of cycles is always the same as the number of periods*, as was pointed out in Section 2.1.
- *Constancy of a phase shift and an angular velocity*: We have to fully understand that all the considerations of an analog signal given in Section 2.1 and in this section hold true if and only if both the phase shift and the angular velocity are constants. Look at Figure 2.2.S.1.4 again: If the phase shift, θ, varies with time, the phasor will cover different angles over the same time intervals. The reason for this change is that the phasor's angular velocity also varies with a variable phase shift, as explained in this section. It was shown that if we assume $\theta(t) = kt$ (rad), where k (rad/s) is a constant, then $\left[\frac{d\Theta(t)}{dt}\right] = k$ (rad/s) and the phasor's angular velocity becomes $[\omega + k]$ (rad/s) instead of $[\omega]$ (rad/s). The result is a change in the covered angle for the same time interval. For example, assuming $\omega = 10$ rad/s and $k = 5$ rad/s, we compute the phasor's angle for $t_{30} = 0.05$ s as $\alpha_k = (\omega + k)t = 15$ rad/s·0.05 s = 0.75 rad instead of $\alpha = \pi/6 = 0.52$ rad, as found for the constant phase shift. Clearly, such variations in a phasor's angles result in changing the corresponding waveform of a sinusoidal signal.

(Question: What if the angular velocity, ω (rad/s), becomes variable? Hint: Assume, for example, $\omega = mt$ (rad/s); take a specific number for m and compute the angles that the phasor will cover over equal time intervals. Generalize your result.)

This sidebar presents a sinusoidal signal from a new standpoint, which should aid you in gaining further insight into this topic.

Sinusoidal signals are the example of a general group of signals – analog signals. These signals play an important role in electrical engineering in general and in modern communications in particular. From the mathematical point of view, however, these signals are the subset of a general set of entities called *function*. It is essential to become familiar with this mathematical interpretation, which is presented in Sidebar 2.2.S.2.

Sidebar 2.2.S.2 Signal and Function

In electrical engineering, we are interested in *signals*, but a signal, in fact, is a specific case of a very general entity called function. We have mentioned the term "function" many times; specifically, we said that *a signal is a function of time*. In this sidebar, we discuss *function* from a more rigorous, mathematical point of view.

Function is a mathematical operation that produces a relevant number every time we submit another number to the function.

For example, function $y = 3x$ gives $y = 6$ when we submit $x = 2$, and function $y = \cos x$ produces $y = 1$ when we enter $x = 0$.

The *number submitted to the function* is called an *independent variable*; the *number produced by the function* in response to the presented independent variable is called a *dependent variable*. There are other terms used for these variables; for example an independent variable can be called an input or a cause, whereas a dependent variable can be called an output or a response.

A function can be presented as a formula, a table, and a graph.

To see how a function can be defined by a *formula*, consider a sinusoidal signal commonly used in electrical circuits. This signal is described as

$$v(t) \, (V) = A \, \cos(\omega t).$$

Here, time, t (s), is the *independent variable* and $v(t)$ (V) is the *dependent variable*, called the magnitude of a signal; $v(t)$ varies according to changes in t. A cosine, *cos*, is a specific *kind of function* whose properties are known. The amplitude, A (V), and radian frequency, ω (rad/s) $= 2\pi f$,[3] are constants called *parameters*. The structure of this formula is shown in Figure 2.2.S.2.1a. The numerical example of this function is

$$v(t) \, (V) = 3 \, \cos(10t)$$

This function produces the dependent variable $v(t)$, which is equal to 3 V when the independent variable, t, takes the value zero; when $t = 0.03$ s, then $v(t)$ equals 2.9 V; when $t = 0.06$ s, $v(t) = 2.5$ V; and so on.

A function can be presented as *a table*, in which one column (usually the left) contains the numbers presented to the function and the other column (typically the right) displays the corresponding numbers produced by the function. The first 17 values of t and the corresponding values of $v(t) = 3\cos(10t)$ are shown in Figure 2.2.S.2.1b. This figure also shows that the *set of independent variables constitutes the function's domain* and the *set of dependent variables makes up the function's range*.

In Figure 2.2.S.2.1c, the segment of the function $v(t) = 3\cos(10t)$ is displayed as a *graph*, which, again, is another way to show a function. Figure 2.2.S.2.1c shows that for every given value of *time in seconds*, we find the appropriate value of a *signal's magnitude, v(t), in volts*. Such a graph is a visual device for delivering a function's shape. From this graph, we can find each value of $v(t)$ corresponding to the value of t; however, a table does this job much better. At any rate, it is clear now that a *waveform is simply a graphical presentation of a signal*.

The concept of function discussed above and illustrated in Figure 2.2.S.2.1 can be formulated mathematically as follows:

A *function is a rule* that assigns to each element of an input set of numbers a unique element in the output set of numbers. The input set is called the *function domain* and the output set is called the *range of the function*.

(Continued)

---

[3] Reminder: The units of radian frequency, rad/s, are drawn from its definition: ω (rad/s) $= 2\pi$(rad)$\cdot f$(1/s) because $1/s \equiv Hz$.

Sidebar 2.2.S.2 (Continued)

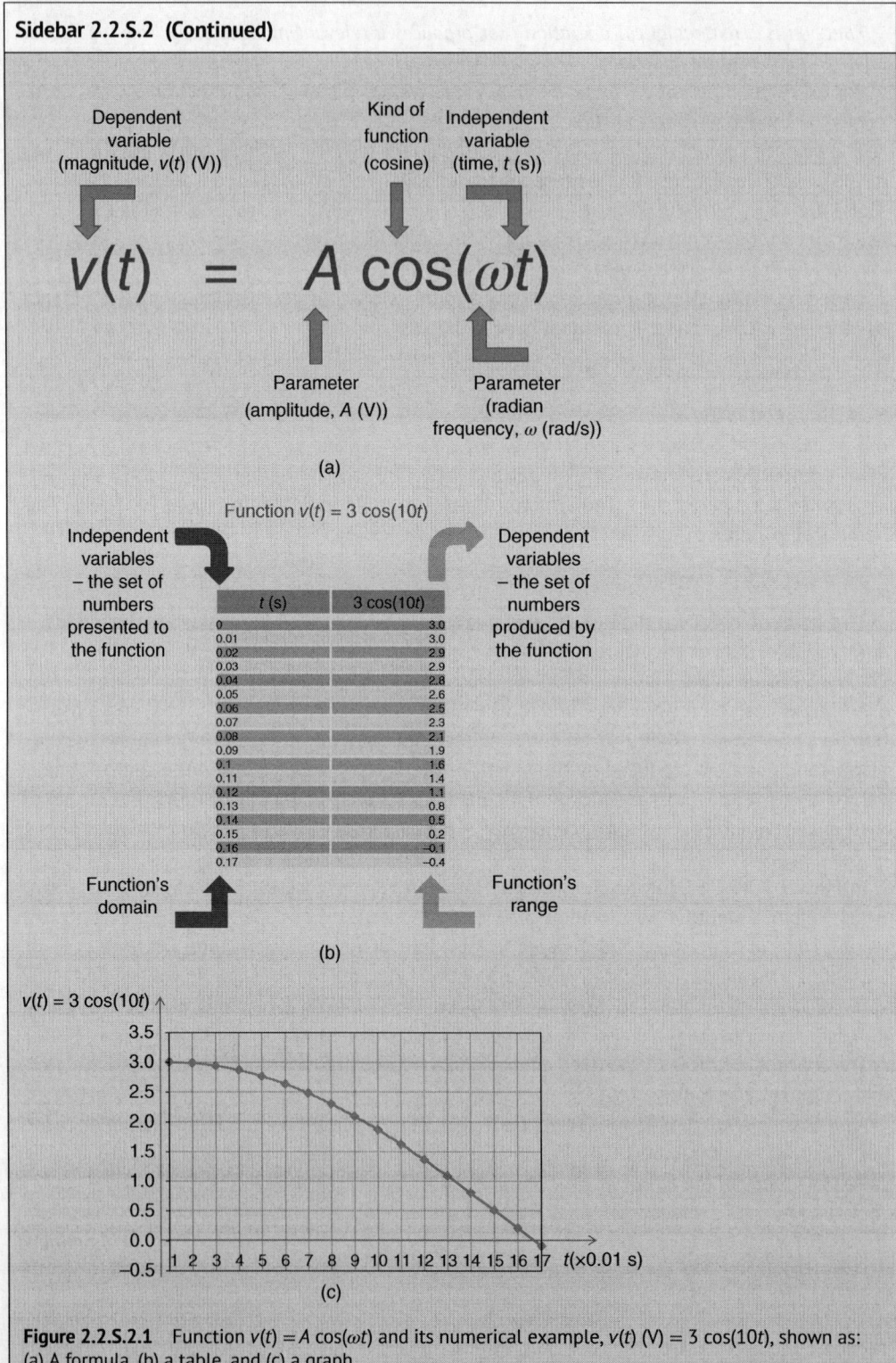

Figure 2.2.S.2.1 Function $v(t) = A\cos(\omega t)$ and its numerical example, $v(t)$ (V) $= 3\cos(10t)$, shown as: (a) A formula, (b) a table, and (c) a graph.

This mathematical definition is shown in drawings in Figure 2.2.S.2.2. The meaning is that if x is an arbitrary element of the input set and y is its unique assigned element in the output set, then we say that *y is the function of x* and denote it as

$$y = f(x)$$

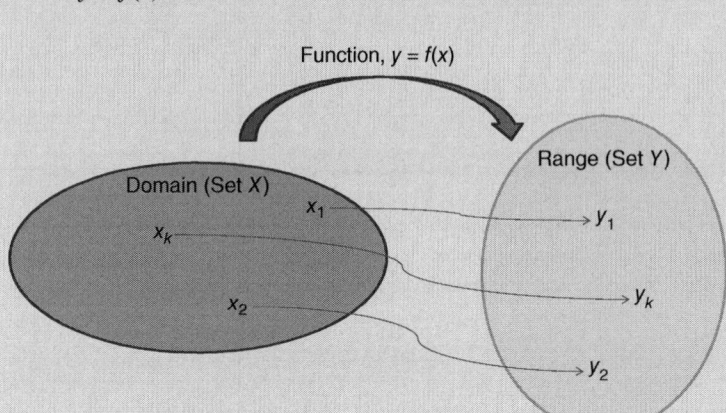

Figure 2.2.S.2.2 Function is a rule assigning to each element from a domain set a unique element from a range set.

Again, element x varies independently over the input set (the function's domain) and it is called an *independent variable;* element y varies over the output set (the function's range) according to the rule f (the function, that is) and it is called a *dependent variable* because it depends on x and f.

We must know that if y is the function of x, $y = f(x)$, then x *may or may not* be an *inverse function* of y denoted as $x = f^{-1}(y)$. The simplest example is the function $y = x^2$. Its inverse function, $x = f^{-1}(y)$, should be $x = \sqrt{y}$; however, $y = x^2$ corresponds to both $x = +\sqrt{y}$ and $x = -\sqrt{y}$. But our definition of function requires that the function must assign to an element in its domain a *unique* element in its range. Thus, the inverse expression $x = \sqrt{y}$ does not meet the definition of a function unless we introduce this restriction: For the function $y = x^2$, the inverse function $f^{-1} = +\sqrt{y}$. Today, when saying $x = \sqrt{y}$, we actually mean $x = +\sqrt{y}$.

It is necessary to mention one more characteristic of a function: *continuity*, an important property of electrical signals. We distinguish between continuous and discrete signals (functions, that is) because each of these signal types determines a different technology used in practical circuits, devices, and systems. An informal but quite clear definition of a continuous signal is:

A signal is continuous if we can plot its waveform without lifting pen from paper.

This intuitively plausible definition is, in fact, based on a strict mathematical consideration: *A function is continuous if two dependent variables become indistinguishable when two corresponding independent variables tend to coincide with each other.* Consider a function $y = f(x)$: Let us take fixed point a at the x axis and arbitrary point x near a. Then we have two dependent values of the function, $f(a)$ and $f(x)$, corresponding to a and x. As we move x to a, the dependent point $f(x)$ will move to $f(a)$. In the limit, when x tends to coincide with a, $f(x)$ goes to $f(a)$. Formally, these words can be presented as

$$y = f(x) \text{ is continuous if } \lim_{x \to a}(f(x)) \to f(a) \qquad (2.2.S.2.1)$$

(Continued)

Sidebar 2.2.S.2 (Continued)

To repeat: Eq. (2.2.S.2.1) says that the function $y = f(x)$ is continuous if y becomes $f(a)$ when the independent variable x coincides with the specific point a.

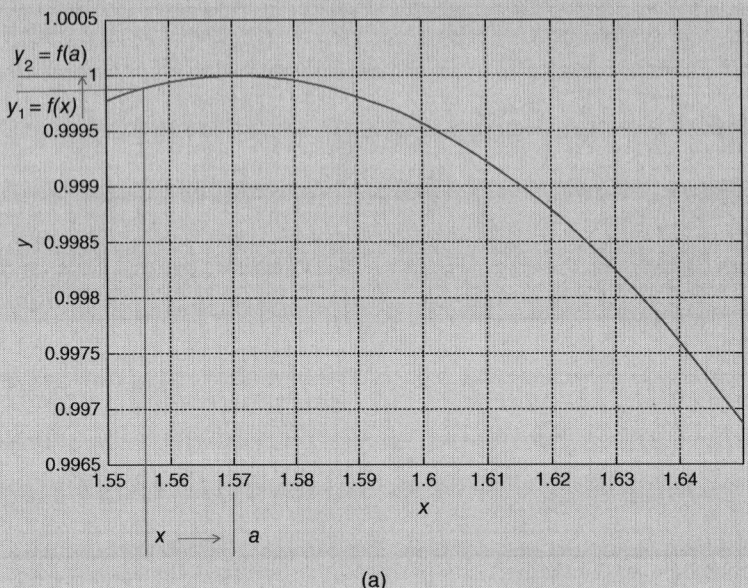

(a)

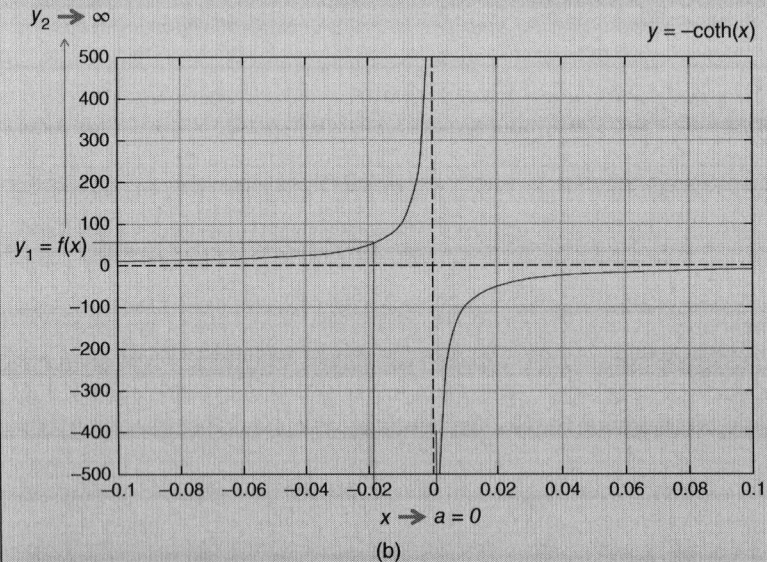

(b)

Figure 2.2.S.2.3 Continuity of a function: (a) A continuous function, $y = \sin(x)$, and (b) a discontinuous function, $y = -\cotan(x)$.

Figure 2.2.S.2.3 is the graphical interpretation of this formula exemplified by two functions, $y = \sin(x)$ and $y = -\cotan(x)$. The sine function in Figure 2.2.S.2.3a is *continuous* because it satisfies (2.2.S.2.1) as follows: *When an arbitrary independent variable, $x = 1.556$, approaches* *point a = 1.57, the dependent variable, y_1 = sin(1.556) = 0.99, tends to coincide with its corresponding dependent variable, y_2 = sin(1.57) = 1.* In contrast, the negative cotangent function in Figure 2.2.S.2.3b is discontinuous (discrete) because when *x* tends to *a*, the dependent variable, $y_1 = f(x)$, tends to coincide with its corresponding dependent variable, y_2, but y_2 goes to infinity and therefore condition (2.2.S.2.1) will never be met.

There are many other important details pertaining to the definition and interpretation of a function's continuity, but we leave them to specialized mathematical textbooks.

This brief excursion into the mathematical realm will enhance your knowledge of signals, their waveforms, and their properties. We will return to this subject – function (signal) and its properties – many times in this text.

2.2.2 Frequency Domain and Bandwidth

2.2.2.1 Frequency Domain

Let us consider two sinusoidal signals like those shown in Figure 2.1.6. We can easily see that such signals may differ only in their frequencies and amplitudes. But, in this case, why should we sketch the waveforms of two almost identical signals? Would it not be meaningless to draw the same waveforms with slightly different parameters? We can develop a better presentation of such similar signals by introducing an important concept: *frequency domain*.

Let the *x*-axis show how frequency changes and the *y*-axis show how amplitude changes. This new *A–f* space is called *frequency domain*. Recall that previously in this subsection we used *time domain*, that is *v–t* space, where we showed the signal waveforms. *To present a signal in frequency domain, we need to show its two main parameters: amplitude and frequency*. Thus, we sketch a line along the *y*-axis with a height equal to the signal's amplitude; along the *x*-axis this line is located at the signal's frequency. Therefore,

in time domain, a sinusoidal signal is presented by its waveform, whereas in frequency domain this signal is depicted as a line whose height is the signal's amplitude, A (V), and whose frequency is f (Hz) = 1/T (s).

Consider, for example, a signal whose amplitude equals 3 V and whose frequency equals 40 kHz: that is,

$$v(t) \text{ (V)} = 3 \sin((2\pi \times 40 \times 10^3)t)$$

In time domain, this signal is displayed by its waveform; in frequency domain, it is depicted as a line whose height is 3 V and whose frequency is 40 kHz. Both graphs are shown in Figure 2.2.4.

Figure 2.2.5 shows several sinusoidal signals in time domain and frequency domain. Observe that the amplitudes of these signals remain the same in both domains, whereas periods and frequencies follow the rule given in (2.1.1R) and (2.1.4): f (Hz) = $1/T$ (s) and T (s) = $1/f$ (Hz).

Time domain and frequency domain presentations are equally important in modern communications, and a telecommunication engineer must be comfortable in developing each of these presentations and in switching from one domain to the other.

2.2.2.2 Cosine and Sine Signals in Frequency Domain

The preceding discussion shows that a sinusoidal signal can be depicted in frequency domain by a line determined by the amplitude and frequency of the signal. But a sinusoidal signal means either

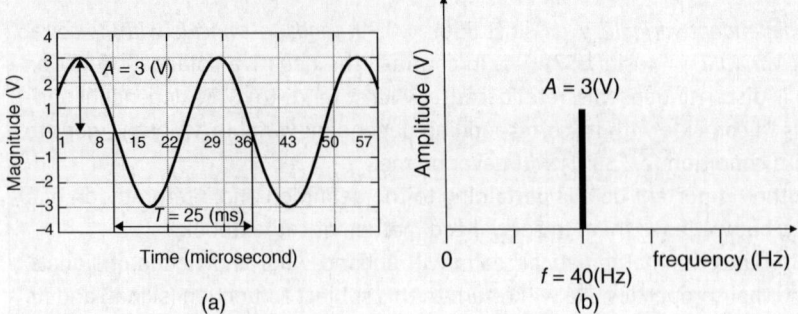

Figure 2.2.4 Sinusoidal signal, $v(t)$ (V) = $3 \sin((2\pi \times 40 \times 10^3)t)$, in (a) time domain and (b) frequency domain.

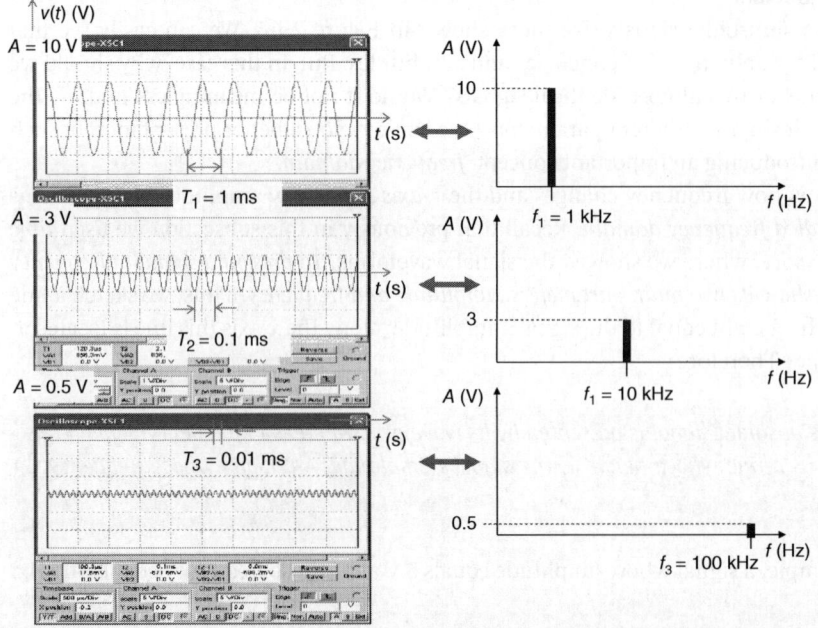

Figure 2.2.5 Time domain (left) and frequency domain (right) of sinusoidal signals.

cosine or sine. How can we distinguish between them in frequency domain? You will recall that sine and cosine functions differ only in their 90° phase shift (see (2.2.3a) and (2.2.3b) and the subsection "Sine and Cosine Signals" in this section). However, the A–f space, as shown in Figure 2.2.4, does not include the phase; therefore, we need to enhance our frequency-domain concept. We can do this by introducing two plots in frequency domain: amplitude vs. frequency (A–f space) and phase vs. frequency (θ–f space). With this modification, we are able to depict the spectra of cosine and sine signals, as Figure 2.2.6 shows.

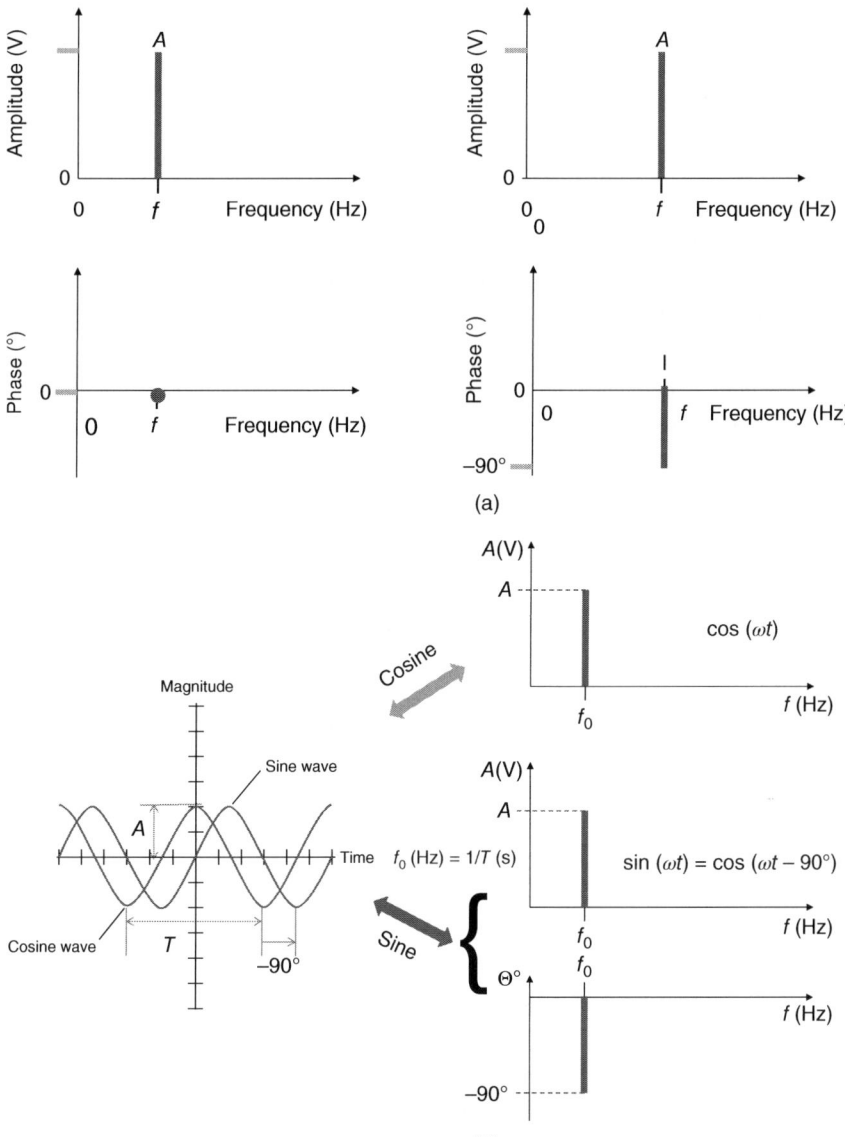

Figure 2.2.6 Cosine and sine signals in time domain and frequency domain: (a) Amplitude and phase spectra of cosine and sine signals; (b) waveforms and spectra of cosine and sine signals.

A cosine wave is a reference signal and, *by convention, its phase shift is zero*. Therefore, it is depicted in frequency domain only by an amplitude-vs.-frequency (*A–f*) line.

A sine wave, by definition, *has a −90°-phase shift with respect to a cosine wave,* as (2.2.3b) shows. Therefore, it is depicted in frequency domain by both *A–f* and Θ–*f* lines.

In fact, Figure 2.2.6a is a graphical representation in frequency domain of the following cosine and sine signals:

$$v_1(t) = A\cos(\omega t) = A\cos(2\pi f t)$$

and

$$v_2(t) = A\sin(\omega t) = A\sin(2\pi f t) = A\cos(2\pi f t - 90°)$$

Since the *cosine* is considered a phase-reference signal, Figure 2.2.6a shows a *zero phase shift* for this signal. The *sine* signal is shown in Figure 2.2.6a with amplitude A_2 and with a *negative 90° phase shift*. Thus, the rule to remember:

> *A cosine signal is presented in frequency domain by a single line in an A–f coordinate system, whereas a sine-signal presentation additionally includes the indication of a −90°-phase shift in a θ–f coordinate system.*

Pay particular attention to Figure 2.2.6b: It shows cosine and sine signals in both time domain and frequency domain. This illustration should help you to fully understand the relationship between a signal's presentation in time domain (that is, its *waveform*) and in frequency domain (that is its *spectrum*). In general, Figure 2.2.6 shows that *in frequency domain we basically depict a cosine line, adding a 90°-phase shift for a sine signal.*

Therefore, for the complete description of a signal in frequency domain, we need to introduce two plots: amplitude vs. frequency and phase vs. frequency. This is why, when considering a frequency-domain signal's presentation, we refer to an amplitude *spectrum* and a *phase spectrum*. The following example should help in the understanding of this concept.

Example 2.2.4 Waveform and Spectrum of an Analog Signal

Problem

Sketch the waveform of the following signal and present its line spectrum:

$$v(t) = 8 - 12\cos((2\pi \times 30)t - 30°) + 5\sin((2\pi \times 90)t + 40°)$$

Solution

The waveform of this signal is depicted in Figure 2.2.7a. It was built using MATLAB. The signal's spectrum is drawn in Figure 2.2.7b.

Discussion

- Observe the waveform of this signal and understand that it is a graphical presentation of the signal's formula. (It would be beneficial for you to return to our waveform discussion in Section 2.1. and in this section.)
  (**Question**: Is the signal shown in Figure 2.2.7a a periodic? Explain.)
- Focus on the *spectrum of this signal*, shown in Figure 2.2.7b. How was this spectrum drawn? Answer: by depicting the spectrum of each member of the waveform's formula. The first spectral component is the constant, or DC, member at zero frequency. Its amplitude is 8 V, as the formula

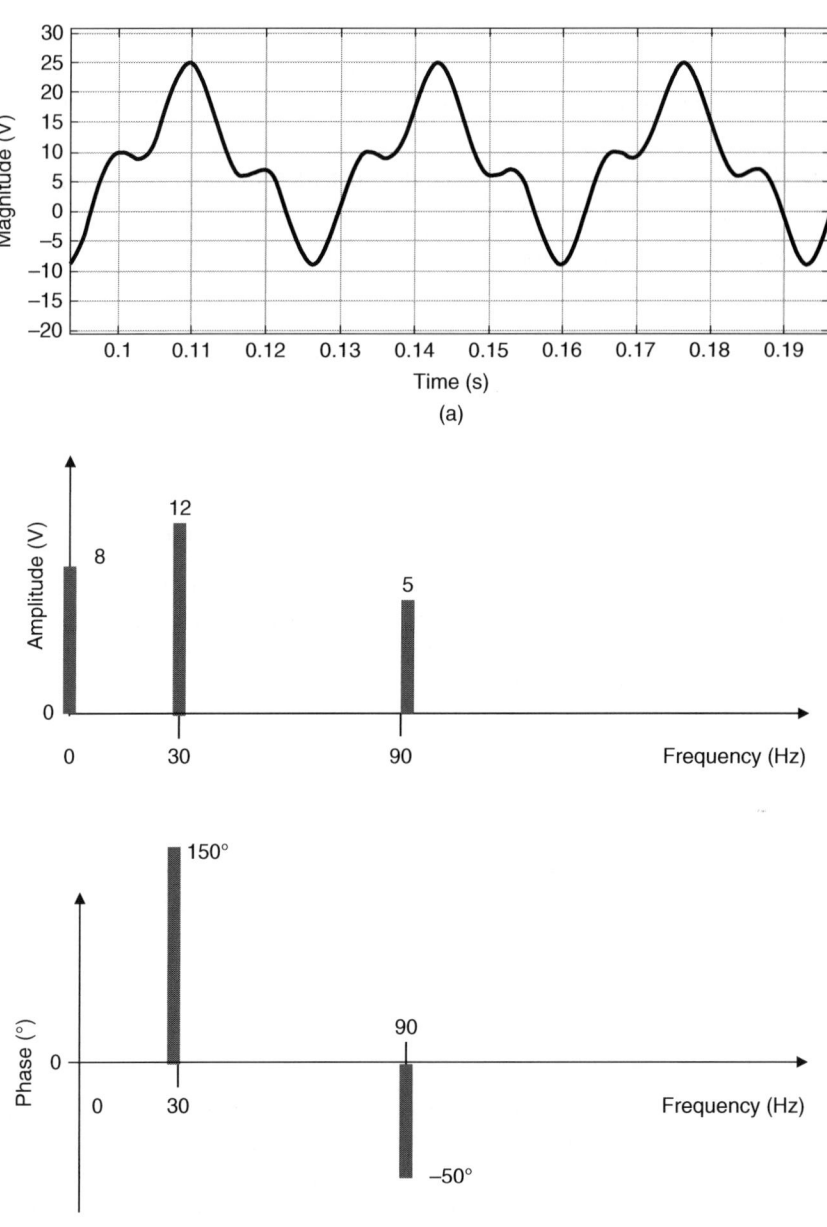

Figure 2.2.7 Composite signal $v(t) = 8 - 12\cos((2\pi \times 30)t - 30°) + 5\sin((2\pi \times 90)t + 40°)$: (a) Waveform; (b) spectrum.

for the signal states. Clearly, the phase of this component is zero. The second spectral component of this composite signal has its amplitude, frequency, and phase equal to 12 V, 30 Hz, and 150°, respectively. Based on its frequency (see our previous discussion, specifically Example 2.2.2), you can conclude that this member should be a component presented in time domain as $-12\cos$

$((2\pi \times 20)t - 30°)$. However, instead of the negative 12 V specified by its waveform, we depict a positive 12-V amplitude and, instead of $-30°$, we show $150°$. To solve this enigma, we need to know that our presentation shown in Figure 2.2.7b is the result of the conventional method:

> When depicting a signal in frequency domain, its amplitude is regarded as a positive quantity and the negative sign is absorbed by the phase as follows:
>
> $$-A \cos(\omega t + \theta) = A \cos(\omega t + \theta \pm 180°) \qquad (2.2.4)$$

Specifically, this is how we use (2.2.4) to convert a member with a negative amplitude into a positive-amplitude component: $-12\cos[(2\pi \times 30)t - 30°)] = 12\cos[(2\pi \times 20)t + 150°)]$.

The third spectral component in Figure 2.2.7b has an amplitude of 5 V, a frequency of 90 Hz, and a phase of $-50°$. We can easily identify it as a component rated at $5\sin((2\pi \times 90)t + 40°)$. The phase shift of $-50°$ provides the representation of a sine signal through a cosine as follows:

$$5 \sin((2\pi \times 90)t + 40°) = 5 \cos((2\pi \times 90)t + 40° - 90°) = 5 \cos((2\pi \times 90)t - 50°)$$

For a better understanding of the manipulations with the phase angles, refer to Sidebar 2.2.S.1

- Final point: We use the term *line spectrum* when a spectrum is presented as a set of lines, as is the case in Figure 2.2.7 and in the preceding figure. This form of a spectrum stems from the fact that each sinusoidal signal is presented in frequency domain by a line. Subsequent chapters will show that nonsinusoidal signals might have *continuous spectra*.

2.2.2.3 Bandwidth

Having introduced frequency domain, we can now focus on another important concept: *bandwidth*. (Refer to Section 1.3.1.2, where introductory remarks on this topic were made.) Suppose we need to transmit a signal over a transmission link and this signal includes two sinusoids,

$$v_1(t) = A_1 \sin(2\pi f_1 t)$$

and

$$v_2(t) = A_2 \sin(2\pi f_2 t)$$

We need to understand what frequency range this signal occupies and what frequency band the link must have for the successful transmission of this signal. To solve such a problem, the frequency-domain presentation is invaluable.

To begin, let us present our signal in frequency domain by showing two lines with amplitudes A_1 and A_2 at the frequencies f_1 and f_2 – see Figure 2.2.8. In sketching this figure, we assume that $f_2 > f_1$ and $A_1 > A_2$. Figure 2.2.8 depicts the *frequency spectrum* of a signal, that is, the set of frequencies contained in the signal.

Let us now establish what frequency range the signal shown in Figure 2.2.8 occupies. From Figure 2.2.8, we can readily see that this range extends from f_1 to f_2. This consideration brings us to the concept of *bandwidth*.

In general, *bandwidth is a range of frequencies*. As to what frequencies, that depends on the bandwidth we need to work with. In our example, the answer looks quite simple: The bandwidth, BW (Hz), occupied by this signal (consisting of two sinusoids, remember) is

$$\text{BW (Hz)} = f_2 - f_1 \qquad (2.2.5)$$

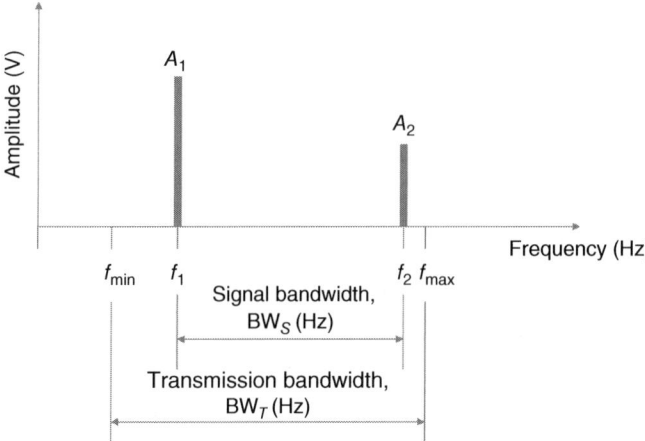

Figure 2.2.8 Frequency domain and bandwidth.

However, in general, bandwidth is defined as the difference between the maximum and the minimum frequencies; that is,

$$\text{BW (Hz)} = f_{max} - f_{min} \tag{2.2.6}$$

Referring again to Figure 2.2.8, we see that the maximum frequency is clearly f_2, but what is the minimum? Is it f_1 or 0 or any other frequency? Depending on the answer, we will have two different bandwidths, as Figure 2.2.8 shows. It is quite clear that the *signal's bandwidth* is

$$\text{BW}_S \text{ (Hz)} = f_2 - f_1 \tag{2.2.7}$$

However, to transmit this signal over a link, the link must have a bandwidth that at least covers all the signal's frequencies and, in general, a broader range to include service signals. Hence, *transmission bandwidth* – the frequency band that a transmission channel must have for successful transmission of a given signal – is

$$\text{BW}_T \text{ (Hz)} = f_{max} - f_{min} \tag{2.2.8}$$

Both bandwidths are shown in Figure 2.2.8. Let us consider the following example to help you become thoroughly knowledgeable when working with these new concepts.

Example 2.2.5 Frequency and Bandwidth of an Analog Signal

Problem

The signal that consists of three sinusoids having the following amplitudes and frequencies: $A_1 = 15$ V and $f_1 = 9 \times 10^6$ Hz, $A_2 = 50$ V and $f_2 = 9.5 \times 10^6$ Hz, and $A_3 = 30$ V and $f_3 = 11 \times 10^6$ Hz:

1. Depict its spectrum.
2. What is the bandwidth of this signal?
3. What transmission bandwidth do we need to deliver this signal?

Solution

1. The frequency-domain presentation of the signal (the spectrum, that is) in question is shown in Figure 2.2.9.

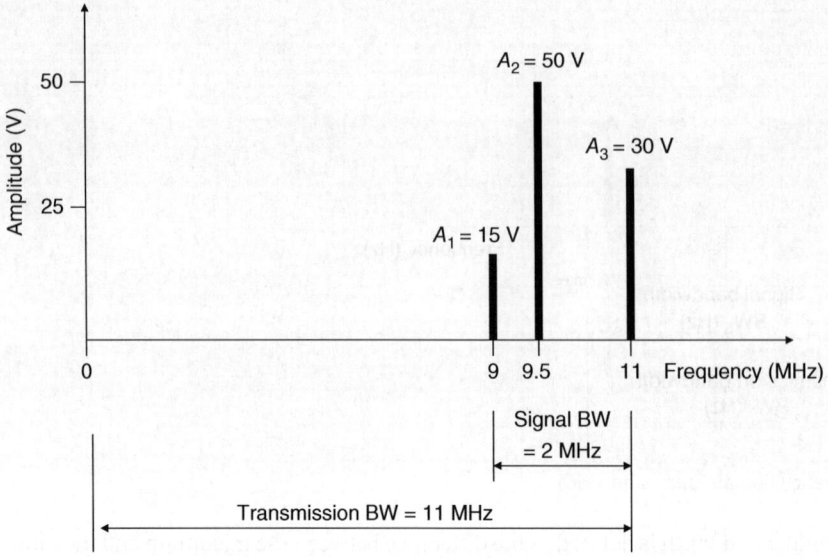

Figure 2.2.9 Signal bandwidth and transmission bandwidth for Example 2.2.5.

2. The bandwidth of this signal (the group of sinusoids) is the range of frequencies it occupies. Therefore, this bandwidth is given by Formula (2.2.5):

$$BW_S = f_3 - f_1 = 11\,\text{MHz} - 9\,\text{MHz} = 2\,\text{MHz}$$

3. Since we don't know anything about the transmission conditions of this group, we should consider the worst-case scenario, which means the need for the broadest bandwidth. Thus, we assume $f_{min} = 0$ MHz and use (2.2.6) to arrive at the following transmission-bandwidth value:

$$BW_T = f_{max} - f_{min} = 11\,\text{MHz} - 0\,\text{MHz} = 11\,\text{MHz}$$

Both bandwidths are shown in Figure 2.2.9.

Discussion

Three important points to be highlighted here:

- First, suppose we are given Figure 2.2.9 without any further explanation. Can we determine what kinds of signals this figure displays? In other words, how do we know that these lines show some real signals, whose waveforms we can actually see on an oscilloscope screen? The answer is that the engineering and scientific community throughout the world has agreed to display cosine and sine signals in frequency domain, as shown in Figures 2.2.4–2.2.9. Thus, we have to realize that *the straight lines in the A–f space represent cosine or sine waveforms only*. In the real world, we will encounter many waveforms other than sinusoidal signals; how to find and show the spectra of these nonsinusoidal signals will be discussed in later chapters.
- Second, the term *bandwidth* has been used throughout this and the previous sections in connection with different meanings. Even in Figures 2.2.8 and 2.2.9, we define bandwidth in two ways: as a band of frequencies between two individual sinusoids and as the entire range from the minimal to the highest frequency being taken into consideration. Which definition is correct? <u>Both</u>.

This is why we have introduced two types of bandwidth: signal bandwidth and transmission bandwidth. In fact, bandwidth is a sophisticated concept, as next subsection partly reveals, and this book will consider its complex aspects.
- Finally, observe the three lines in Figure 2.2.9. They present a signal containing three sinusoids in frequency domain. This presentation, again, is a visual format describing the *spectrum* of a signal, that is, the number and the value of frequencies that a given signal contains. Formally speaking, a spectrum is a set of frequencies related to one another, that is, frequencies that have something in common. In Figure 2.2.9, these frequencies belong to one signal; this is why they constitute the *spectrum* of the signal. The term "spectrum" is used throughout the book and we will certainly need to become thoroughly familiar with it.

2.2.2.4 Bandwidth: a Sophisticated Entity

As defined above, bandwidth is a range of frequencies. We have distinguished between signal bandwidth and transmission bandwidth. The composite signal $v(t) = 15\cos(2\pi \times 9 \times 10^6) + 50\sin(2\pi \times 9.5 \times 10^6) + 30\sin(2\pi \times 11 \times 10^6)$, discussed in Example 2.2.5, has a signal bandwidth of 2 MHz, but might need a transmission bandwidth of 11 MHz.

This is another example of the difference between signal bandwidth and transmission bandwidth. Suppose you want to send three groups of signals, the first group spanning the frequency range from 0 Hz to 1 MHz, the second group occupying the band from 5 to 6 MHz, and the third group covering the range from 9 to 10 MHz. Thus, the bandwidth of each group is 1 MHz. The question now boils down to which transmission bandwidth we need to transmit each of these groups. Clearly, to transmit the signals between 0 and 1 MHz, we need a link capable of transmitting at least 1 MHz; transmitting the signals occupying the 9–10 MHz range also requires a transmission bandwidth of 1 MHz However, transmitting all three groups needs a bandwidth of 10 MHz. (Can you sketch the spectra and show the values of the signal bandwidths and the transmission bandwidth in this example? Also, can you give the values of the signal bandwidths and the transmission bandwidths for the signals shown in Figures 2.2.6–2.2.9?)

In conjunction with a bandwidth's concept, the important question is, what is the bandwidth of a single sinusoidal signal, $v(t) = A\cos(\omega t + \theta) = A\cos(2\pi f t + \theta)$? Reviewing Figures 2.2.8 and 2.2.9, we can easily conclude that a sinusoidal signal occupies a single frequency, not a range. Therefore, *the bandwidth of a single harmonic signal is zero. This property explains the wide use of a sinusoidal signal as a carrier of information in communications.*

(**Question**: *Why does zero bandwidth of a sinusoidal signal enable this signal to be a main carrier in modern communication?*)

A deeper look reveals that the term *bandwidth* requires additional clarification, which is further explained through the following example.

Example 2.2.6 The Bandwidths in Wi-Fi Transmission

Problem

Wi-Fi, as explained in Section 1.2, is wireless communication system used by smartphones, tablets, laptops, sensors, and many other fixed and mobile devices. Wi-Fi employs several RF bands. One of the major Wi-Fi bands ranges from 2400 to 2500 MHz. It is called 2.4-GHz band, and it is subdivided into 14 channels, out of which only first 11 channels are allowed in North America. Table 2.2.2 shows the frequencies of all these channels. Analysis of this table reveals that the frequencies of all adjacent channels overlap, which results in severe interference among their signals. To sustain

Table 2.2.2 Frequencies of 2.4-GHz Wi-Fi band (https://www.electronics-notes.com/articles/connectivity/wifi-ieee-802-11/channels-frequencies-bands-bandwidth.php).

Channel number	Lower frequency (MHz)	Center frequency (MHz)	Upper frequency (MHz)
1	2401	2412	2423
2	2406	2417	2428
3	2411	2422	2433
4	2416	2427	2438
5	2421	2432	2443
6	2426	2437	2448
7	2431	2442	2453
8	2436	2447	2458
9	2441	2452	2463
10	2446	2457	2468
11	2451	2462	2473
12	2456	2467	2478
13	2461	2472	2483
14	2473	2484	2495

an acceptable transmission quality, a router must employ only nonoverlapping (noninterfering) channels. Questions:

1. What is the bandwidth of 2.4 GHz band?
2. What is the bandwidth of each channel?
3. How much do the frequencies of the adjacent channels overlap?
4. Does 2.4 GHz band contain noninterfering channels?
5. If the answer to the preceding question is yes, how many sets (groups) of non-interfering channels can you find?

Solution

1. The bandwidth of 2.4 GHz band is 2500 MHz − 2400 MHZ = 100 MHz.
2. From Table 2.2.2, we find for Channel 1: 2401 MHz − 2423 MHz = 22 MHz. Similar calculations show that all channels have the same bandwidth.
3. To compute overlapping, subtract the lower frequency of channel $N+1$, $f_{lf}(N+1)$, from the upper frequency of channel N, $f_{uf}(N)$. For example, $f_{uf}(1) - f_{lf}(2) = 2423$ MHz − 2406 MHz = 17 MHz. The overlapping interval of 17 MHz holds true for all channels except of Channels 13 and 14, for which this interval is 10 MHz. (Incidentally, by what interval the center frequencies of adjacent channels are separated? Show your calculations.)
4. Yes. The criterion is $f_{lf}(M) - f_{uf}(N) > 0$ for $M > N$. For example, $f_{lf}(6) - f_{uf}(1) = 2426 - 2423 = 2$ MHz > 0. (Explain this criterion.)
5. From analysis of Table 2.2.2, we find the following five combinations of non-interfering (non-overlapping) Wi-Fi channels:
   1, 6, 11, or 2, 7, 12, or 3, 8, 13 or 4, 9, 14 or 5, 10, 14.

   These combinations can be generally used together as sets. However, since only 11 channels are allowed for the use in North America, there is a single set – 1, 6, and 11 channels – that can be employed in the United States.

(**Exercise**: Develop a graphical presentation for Table 2.2.2. Show first four channels beneath each other shifted according the values of their frequencies. Shade the overlapping segments. Indicate the shifts among the center frequencies of the channels.)

Discussion

- Based on this example, we understand that transmission bandwidth must include the highest and the lowest frequencies of a transmitted signal. Indeed, since the signal bandwidth determines the required transmission capacity of a link, it is necessary for the link to be capable of transmitting all signal's frequencies.
- It's important to understand that the requirement for a transmission bandwidth to include a signal bandwidth is merely the necessary condition. This requirement can be rephrased by saying that the transmission bandwidth must *at least* cover the entire signal bandwidth. The specifics of a real transmission, however, can impose additional, sufficient requirements on the transmission bandwidth.
- The demonstration of how a transmission bandwidth depends on transmission specifics can be derived from this example: Here, if all signals can be sent simultaneously (in parallel), the transmission bandwidth should cover the sum of all bandwidths of the signals and be equal to 22 MHz·14 (channels) = 308 MHz. In reality, however, the given transmission bandwidth is only 100 MHz; this is why this Wi-Fi system has overlapping problem.
- Yet another important factor in determining a transmission bandwidth is the properties of the transmission medium. This point is discussed in Sections 1.2 and 1.3 (see Figure 1.3.3). In this example, we need to know the air attenuation at 2.4 GHz band.

Summarizing our discussion regarding transmission bandwidth, we should realize that, in fact, BW_T (MHz) is limited in both – minimum and maximum – values. Its minimum value is determined by signal's bandwidth and its maximum value is restricted by technological or regulatory factors. Thus,

$$BW_S \text{ (MHz)} \leq BW_T \text{ (MHz)} \leq BW_{max} \text{ (MHz)} \tag{2.2.9}$$

The requirement for minimum BW_T (MHz) is clear; let us consider BW_{max} (MHz). Example 2.2.6 demonstrates that 14 channels each having 22 MHz bandwidth would require 308 MHz transmission bandwidth to be transmitted in parallel. But what if we would obtain 616-MHz transmission bandwidth at our disposal? Do we need it? How can we use this excessive bandwidth? The answer can be found in Nyquist formula,

$$BR \text{ (b/s)} \leq 2BW_T \text{ (Hz)} \tag{1.3.3R}$$

and in Hartley information law,

$$H \text{ (b)} = BR \text{ (b/s)} \times T \text{ (s)} \tag{1.3.1R}$$

Here H (b) is the amount of information, BR (b/s) is the bit rate, BW_T (Hz) is the bandwidth of a transmission channel, and T (s) is the transmission time. Combining these formulas, we derive for the maximum bit rate

$$H \text{ (b)} = 2BW_T \text{ (Hz)} \times T \text{ (s)} \tag{2.2.10}$$

This equation states that increasing the transmission bandwidth we can either deliver more information for the same time interval or shorten the transmission time to convey the same volume of information.

Return to our example: if BR (b/s) = 2BW$_T$ (Hz), then for one second, the system can transmit $H = 616$ bits of information. If the bandwidth doubles, then H doubles too and becomes $H = 1232$ bits. Or the Wi-Fi system can deliver the initial 616 bits for 0.5 second. Therefore, we demonstrate that doubling the transmission bandwidth will result in shortening the transmission time by a factor of two. Generalizing this statement, we can conclude that *the greater the transmission bandwidth, the faster the transmission*. This is a very important and far-reaching result, which we will discuss throughout the book.

To sum up: We use the term "bandwidth" to refer to the frequency band occupied by a signal, that is, a signal's bandwidth; we also use this term in the sense of a link's capacity to accommodate the signals being transmitted, that is, *the transmission bandwidth*. Bear in mind that (i) transmission bandwidth must *at least* cover the bandwidth of the signals being transmitted and (ii) the greater the transmission bandwidth, the greater the amount of information that can be transmitted, or the shorter the transmission time.

To add another note to complexity of understanding the bandwidth, we need to know that *there are six technical (engineering) definitions of bandwidth*; what is more, there is a *legal definition* as well, established by the Federal Communications Commission. What this means is that we will have to return to this topic many times in this book.

Note well: Bandwidth is a range of frequencies and it is always measured in hertz. Therefore, strictly speaking,

the term bandwidth can be applied only to an analog signal and to analog transmission.

We also must keep in mind that today communication specialists use the term in a much broader sense, such as, for example, a measure of the transmitting capacity of any type of link – analog or digital. This informal usage has become industry-wide practice. Be well aware of this, but do not forget the formal definition.

It may be well worthwhile for you to reread Sections 1.1 and 1.2. With this new view of bandwidth, you will further reinforce the fundamental principles of modern communications you learned in those sections.

Questions and Problems for Section 2.2

- Questions marked with an asterisk require a systematic approach to finding the solution.
- Many questions and problems, including those marked with an asterisk, imply that you, in addition to reading the textbook, will do your research to find the answers. Consider such questions as miniprojects.

More About a Sinusoidal Signal

1. Consider the following sinusoidal signal: $v(t)$ (V) $= 12\cos(628t + 30°)$:
   a) In a point-by-point manner, build its waveform for the three periods. Show all your work for the first 11 μs, including the table with parameters calculated at every given instant and the graph showing every calculated point.
   b) Build the whole signal for the three periods using MATLAB.
   c) Compare both graphs.

2. Consider the signal $v(t)$ (V) $= 12\cos(628 \times 10^3 t + 60°)$. What is its magnitude at $t = 2$ ms?

3. Is there any difference between a magnitude and an instantaneous value of a sinusoidal signal? Explain.

Questions and Problems for Section 2.2

4 Consider the signal $v(t)$ (V) $= 12\cos(628 \times 10^3 t + 60°)$:
  a) What is its initial value?
  b) Depict the signal. Does this cosine signal shift left or right with respect to a magnitude's axis? Explain.

5 *Consider the signal $v(t)$ (V) $= 12\cos(628 \times 10^3 t + 60°)$:
  a) This signal has a phase shift of 60°. How can you show this angle by depicting the signal's waveform?
  b) Depict the signal's waveform if its phase shift is $-60°$.
  c) Compare the signals shown in points a and b of this problem and comment on their difference.

6 If the phase shift has changed by $+2\pi$, what changes will you see in the signal's waveform? If the change is -2π?

7 Consider the signal $v(t)$ (V) $= 12\cos(628 \times 10^3 t + 60°)$:
  a) If a constant time delay, $t_d = 0.2\,\mu s$, is introduced, explain how it will change the signal.
  b) Give the numerical answer to the preceding question.

8 Consider two signals, $v_1(t)$ (V) $= 12\cos(628t)$ and $v_2(t)$ (V) $= 12\cos(628t + 30°)$: Which signal leads the other? Which one lags?

9 Depict two sinusoidal signals, $v_1(t)$ (V) $= 12\cos(628t + 30°)$ and $v_2(t)$ (V) $= 12\sin(628t + 30°)$, and comment on their difference.

Phasor and Sinusoidal Signal (See Sidebar 2.2.S.1)

10 Consider a phasor's graphical presentation, as in Figure 2.2.S.1.1:
  a) Show a phasor when $\alpha = \omega t = -30°$.
  b) If $\omega = 20$ rad/s, how long will it take to reach this position, $\alpha = -30°$, provided that initially the phasor is in the vertical up position?

11 Consider a phasor rotating at $\omega = 20$ rad/s:
  a) Show the phasor in a current position of $\alpha = 0.7\pi$.
  b) Show the phasor when $t_\alpha = 0.11$ s.
  c) Show its vertical projection at $t_\alpha = 0.11$ s.
  d) Show its horizontal projection at $t_\alpha = 0.11$ s.

12 Consider a phasor like the one shown in Figure 2.2.S.1.1. Knowing the phasor operation and recalling the definitions of period, T (s), and frequency, f (Hz), prove the fundamental equation T (s) $= 1/f$ (Hz).

13 Explain the relationship between an angular velocity of a phasor and a radian frequency of a sinusoidal signal.

14 Prove that the units on the right-hand side and the left-hand side of equation ω (rad/s) $= 2\pi f$ (Hz) are the same.

15 Given $\omega = 20$ rad/s, t = 0.05 s, and $A = 2$ V, find the instantaneous value of the horizontal projection of this phasor. The magnitude of what signal does this projection present?

16 *Consider the phasor shown in Figure 2.2.S.1.1 and give it the phase shift $\theta = -90°$:
    a) Show this phasor.
    b) What type of sinusoidal signal – cosine or sine – can you construct using this phasor? Give your reasoning.

17 *Given the phasor with $\omega = 20$ rad/s and $A = 2$ V, build a sine signal corresponding to this phasor. (*Hint*: Refer to Figure 2.2.S.1.2.)

18 *Analyzing Figure 2.2.S.1.2, we, of course, note that there are two equal magnitudes for every current phasor's position (except for the vertical positions). For example, the phasor's tip at $\alpha = 30°$ has the same vertical projection as the phasor's tip at $\alpha = 330°$. How do we know which of these two dots belongs to the descending part of the cosine signal and which one to the ascending part?

19 *Consider a sinusoidal signal corresponding to the phasor with the following parameters: $\omega = 20$ rad/s, $A = 2$ V and $\theta = -30°$.
    a) In which direction – right or left – is this signal shifted from a zero-phase-shift position?
    b) Build the signal's waveform to prove your answer.

20 *Suppose the phase shift, θ, changes with time as θ (rad) $= 3t^2$:
    a) How will this condition change the phasor's behavior?
    b) Given $\omega_0 = 20$ rad/s, show the phasor's position at $t = 0.026$ s, $t = 0.052$ s, and $t = 0.078$ s.
    c) Build the first three points of the sine's waveform corresponding to this phasor.

21 *Suppose the phase shift is constant, $\theta = 0$, but the angular velocity changes as ω (rad/s) $= 6t$:
    a) How will this condition change the phasor's behavior?
    b) Given $\omega_0 = 20$ rad/s, show the phasor position at $t = 0.026$ s, $t = 0.052$ s, and $t = 0.078$ s.
    c) Build the first three points of the sine's waveform corresponding to this phasor.

22 What feature of the phasor-sinusoid presentation given in Figure 2.2.S.1.2 did you find the most helpful for understanding a sinusoidal signal?

Signal and Function (See Sidebar 2.2.S.2)

23 What is the difference between a signal and a function? Explain.

24 Define and give examples of "function."

25 In what forms can a function be presented? What form do you prefer and why?

26 Consider the function $y = e^{2x}$.
    a) What is the independent variable? What is the dependent variable?
    b) What is the domain of this function? What is its range?

27 One of the definitions of function starts, "A function is a rule…" What is the "rule" stated in this definition?

28 Consider the function $y = e^{2x}$: Does this function have an inverse function? If the answer is yes, show this inverse function.

29 *Sidebar 2.2.S.2 gives two definitions of a continuous function: informal ("a function is continuous if we can plot its waveform without lifting pen from paper") and formal ("the function $y = f(x)$ is continuous if $\lim_{x \to a}(f(x)) \to f(a)$").
a) Prove that both definitions are equivalent.
b) Is a sinusoidal signal continuous? Why?
c) *Is a digital signal continuous? Why?

Frequency Domain and Bandwidth

30 What is the difference between time domain and frequency domain? Explain.

31 A signal's waveform is shown in what domain? Explain. Give an example.

32 Show signal $v(t)$ (V) $= 5\cos((2\pi \times 2 \times 10^3)t)$ in frequency domain.

33 Sketch a sinusoidal signal with an amplitude of 3.3 V and a frequency of 16 kHz in both time and frequency domains.

34 *The frequency-domain presentation of a signal is shown in Figure 2.2.P65. Write the formula for its waveform and draw this waveform.

35 What is the spectrum of a signal?

36 *Consider the following composite signal: $v(t)$ (V) $= 2.8\cos(2\pi \times 8)t + 3.2\sin(2\pi \times 16)t$:
a) Draw the signal's waveform.
b) Show the amplitude and the phase spectra of this signal.
c) Why is the spectrum of this signal called "line?"
d) Determine the bandwidth of this signal.

37 *Visible light is usually referred to as electromagnetic radiation whose wavelengths range from 400 to 700 nm. (1 nanometer [nm] is 10^{-9} meters [m]). Given that c (m/s) $= \lambda$ (m) $\times f$ (Hz),

Figure 2.2.P65 Signal for Problem 2.2.65.

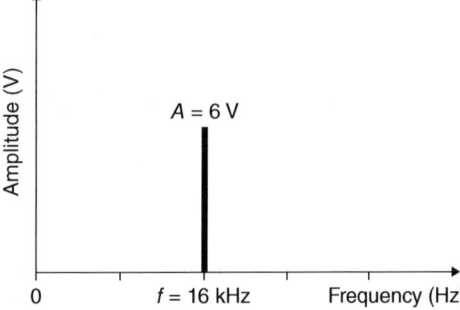

where $c = 3 \times 10^8$ m/s is the speed of light in a vacuum, λ (m) is a wavelength, and f (Hz) is the corresponding frequency, find the frequency spectrum of visible light.

38 *Is a signal's spectrum always a set of discrete lines or can it be a continuum of frequencies between f_{max} and f_{min}? Give your reasoning.

39 Distinguish between signal bandwidth and transmission bandwidth. Amplify your explanation by referring to Figures 2.2.8 and 2.2.9.

40 Draw the following signals in time domain and in frequency domain: $v_1 = 25\cos(16.48 \times 10^6)t$, $v_2 = 28\sin(19.24 \times 10^6)t$, and $v_3 = 12\cos(32.64 \times 10^6)t$. What bandwidth do these signals occupy and what bandwidth is needed to transmit them?

41 *Suppose you need to transmit a group of cosine signals whose frequencies range from 12.61 to 45.23 kHz. What is the signal bandwidth and the transmission bandwidth for this group?

42 *We define bandwidth as the range of frequencies that a given signal occupies. On the other hand, we constantly refer to bandwidth as a measure of transmission capacity. Which approach is correct? Explain.

43 What is the bandwidth of a signal given as $v(t)$ (V) $= 5\cos((2\pi \times 2 \times 10^3)t)$? Explain. Depict it.

44 The textbook states that "the bandwidth of a sinusoidal (single harmonic) signal is zero."
   a) Prove this statement.
   b) Is this an advantage or a disadvantage of a sinusoidal signal? Explain.

45 *The highest frequency of a signal presented for transmission is 6 MHz. What transmission bandwidth do we need to transport this signal? Explain.

46 Does the term "bandwidth" describe analog or digital transmission? Explain.

47 *A composite signal contains four sinusoids with frequencies of 2 kHz, 6 kHz, 8 kHz, and 12 kHz; the amplitudes of these components are 8 V, 6 V, 5 V, and 3 V, respectively:
   a) What is the spectrum of this signal? Explain. Depict it.
   b) What is the bandwidth needed to transmit all these signals simultaneously?

48 *In Figure 2.2.9, a transmission bandwidth is determined as the range of frequencies from zero to the signal's maximum frequency. Is this always the case? Explain.

49 *The transmission bandwidth of an optical fiber – the transmission medium that carries a vast majority of global telecommunications traffic – is bounded approximately between 190 and 200 THz. Frequencies of signals in satellite transmission, on the other hand, range from 12 to 14 GHz. Can we transmit the satellite signals through the optical fiber? Give your reasoning.

50 The Internet of Things (IoT) uses a variety of wireless communications technologies to deliver its signals. Among the frequency bands employed in these technologies are

433 MHz, 868 MHz, 15 MHz, and 2.4 GHz. Explain what these numbers tell you about transmission bands used in the IoT.

51 *The bandwidth of a typical signal carried by a single wavelength in optical communications is 400 GHz. The transmission bandwidth available in an optical communications system is 4 THz. How can we use this excess of transmission bandwidth to improve the signal's transmission? Accompany your explanations with proper calculations.

52 Do you want big or small transmission bandwidth from your transmission medium? What would be an ideal transmission bandwidth? Explain your answers.

2.3 Analog Signals – Advanced Study

Objectives and Outcomes of Section 2.3

Objectives

- To enhance your knowledge of a signal's waveform and its role in analog transmission.
- To deepen your understanding of the phasor presentation of a sinusoidal signal and application of this presentation in theory of modern communications.
- To show the relationship among three main forms – time-domain, phasor-domain, and frequency-domain – of analog signals.

Outcomes

- Understand that in analog communications the message is delivered by the waveform of an analog signal.
- Realize that this is done by superimposing the message waveform onto a carrier (this process is called modulation) and become familiar with such an example of analog transmission as amplitude modulation.
- Get knowledge of waveforms' classification that includes time-varying and spatial waveforms, deterministic and random waveforms, continuous and discrete waveforms, and periodic and nonperiodic waveforms. Study their man features and mathematical descriptions.
- Learn how to calculate a signal power by using waveform's amplitudes, get to know how to calculate an average power of a sinusoidal signal, and become familiar with the concept of average normalized power.
- Understand that an average power of a sinusoidal signal is its DC value.
- Become familiar with the concept of power and energy waveforms (signals).
- Learn that nor every waveform can be realized in practice and get knowledge of conditions that practically realizable waveforms must meet.
- Revisit the concept of a phasor as a means to present a sinusoidal signal in phasor domain and learn that this vector can be displayed in a phasor (time-varying) form or in a polar (steady-state) form.
- Revisit the relationship between a time-varying phasor and a sinusoidal signal (time-domain waveform).
- Learn that a phasor's polar form is mathematically based on theory of complex numbers and revisit the basics of this theory in the appendix.

2 Analog Signals and Analog Transmission

- Study the waveforms and phasors of voltages and currents in three main passive components of electronic circuits – resistor, R, capacitor, C, and inductor, L – and, in particular, analyze the frequency responses of these components.
- Get an understanding how to present the R, C, and L impedances in phasor domain and how to use this presentation for finding the sum of impedances.
- Comprehend how to use the phasor's presentation of impedances in analysis of electronic communications circuitry by considering example of *RLC* circuit.
- Study the mathematical foundation of phasor presentation in appendix to this section.

2.3.1 Revisiting the Waveforms

To study analog and digital communications, we need to have a deep understanding of analog signals in all forms of their presentations. This section considers three main presentations – time-domain, phasor-domain, and frequency-domain – of analog signals. The material discussed in this section will be actively used in all the subsequent chapters; it forms the necessary background for Chapters 4–6.

2.3.1.1 More about Waveforms

In analog transmission, we need to know the characteristics of a signal's waveform because this transmission relies on the waveform. This is why we need to review the topic of waveforms, discussed in Sections 2.1 and 2.2, from the transmission perspective. Such a review will deepen your understanding of waveforms. Here we introduce important new aspects of this topic.

In general, a communication process works as follows: The message, M_{in}, is presented to a transmitter, which generates a waveform, $v_M(t)$, on which the message is superimposed by changing the parameters of this waveform. The waveform is transmitted over a communications link; a receiver presents, at its output, the message M_{out}, which is an approximation of the message that was sent. This process is depicted in Figure 2.3.1. (This figure is an application of the general block diagram of a communications system, shown in Figure 1.1.2.)

To better understand the statement that "some parameters of this waveform depend on M_{in}," consider a sinusoidal signal as an example. We can modify its amplitude, frequency, or phase to superimpose message M on this waveform. Suppose, for instance, that you speak into a microphone. In this case, your voice submits a message, M_{in}, and the microphone generates an electrical signal to deliver this message. Suppose the microphone is connected to a generator that produces a powerful high-frequency carrier sinusoidal signal. If we decide that the message signal should modify the amplitude of this carrier, then the transmitter, which is made up in this case of the microphone and the generator, sends a carrier sinusoidal signal whose amplitude is modified (we say *modulated*) by message M_{in}. This is the waveform, $v_M(t)$. You receive such a waveform when you tune your radio set to the AM band. Here, AM stands for *amplitude modulation*, the operation briefly described above. Figure 2.3.2 illustrates the concept of amplitude modulation.

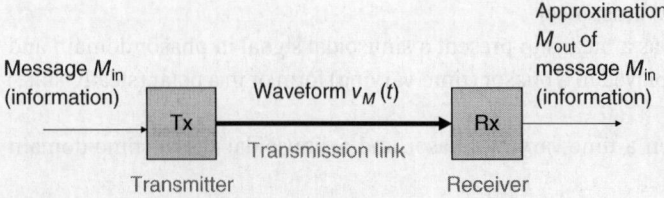

Figure 2.3.1 Transmitting a message with a waveform.

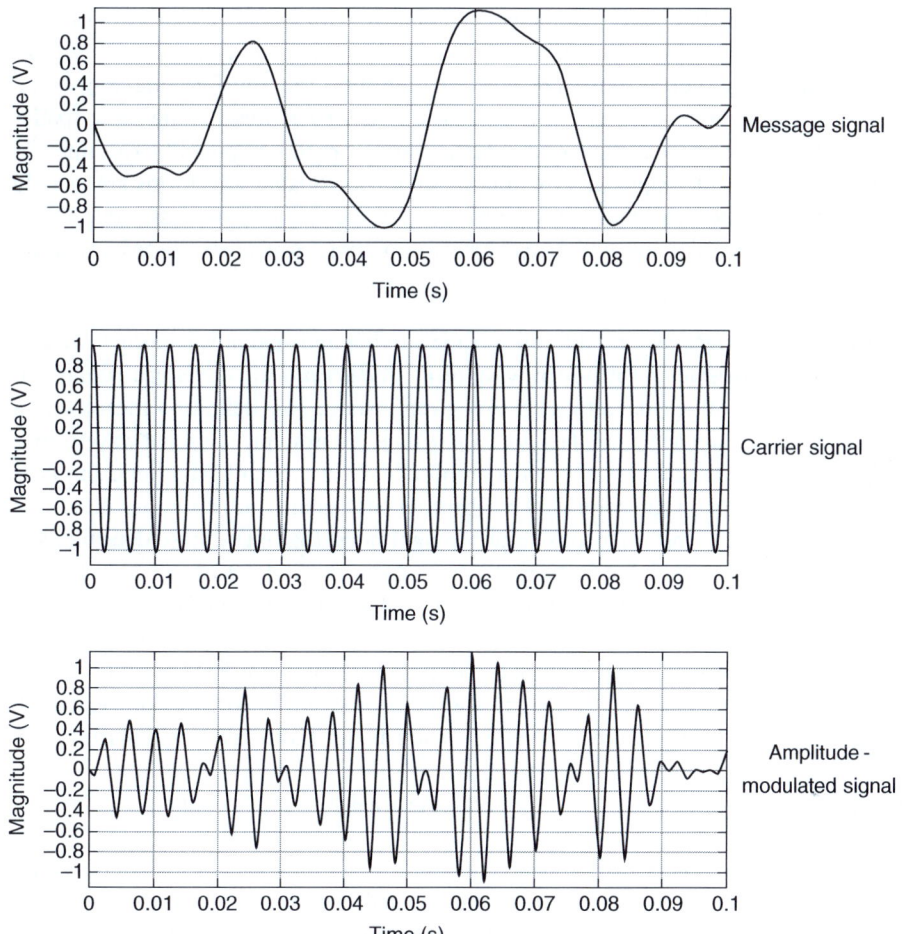

Figure 2.3.2 Amplitude-modulated sinusoidal signal.[4]

Amplitude modulation is communications by transmitting the waveform of an analog signal and therefore is an example of *analog communications*. In *digital communications*, we transmit a set of discrete signals associated with their levels, whereas in analog communication, we transmit a continuous signal associated with its waveform. However, the term "waveform" is applied to both analog and digital signals, as we noted in Section 2.1.

You will recall that the waveform is a mathematical model of a real signal (see Section 1.3, where the concept of model is considered). Let us at this point discuss the classifications of a waveform, that is, its physical characteristics and its main parameters.

Time-Varying and Spatial Waveforms Let us start with a discussion of **time-varying** and **spatial** waveforms. When we refer to a waveform, we usually imply a *time-varying waveform, v(t)*. Indeed, our discussion and examples presented up to this point were largely based on time-dependent waveforms, that is, the waveform whose magnitude changes with time. However, a waveform can be a

---

4 Professor Ram M. Narayanan, The Pennsylvania State University. Reprinted with permission.

function of spatial variables; that is, its parameters can change with a change in a signal's position in two-dimensional (2D) or three-dimensional (3D) coordinate space. Such a *spatial waveform* is described by formula v(x,y,z) or formula v(***r***), where v represents the waveform and x, y, and z are the positions (displacements) along the x, y, and z axes. Vector ***r*** represents the same spatial position of a given waveform because the coordinates x, y, and z are its components. Examples of spatial waveforms include photographs and the images in movies. Spatial waveforms are usually converted into time-varying waveforms for transmission, as in the case of broadcast, cable TV, or streaming video. In our discussion, we will focus *exclusively on time-varying waveforms* unless we state otherwise.

Deterministic and Random Waveforms In addition to time-varying and spatial waveforms, we distinguish between **deterministic** and **random** waveforms. A *deterministic* waveform is one that allows us to know its exact value (magnitude) at any given instant. This definition means that a *deterministic waveform is a completely specified function of time*. An example of such a waveform is a sinusoidal signal, $v(t) = A \cos(\omega t + \Theta)$, where the amplitude A is a constant. Another example is $v(t) = Ae^{(-\alpha t)}t$, where α and A are constants. If we substitute one instant t after another into these formulas, we can compute the instantaneous values (magnitudes, that is) of these deterministic waveforms at these instants, as we did in Sections 2.1 and 2.2 (see Examples 2.1.1, 2.1.2, and 2.2.1). The definition of a deterministic waveform also tells us that if we know the formula and parameters of a waveform, $v(t)$, we can *predict* its value at any chosen instant. Consider a sinusoidal signal again. If we repeatedly measure the magnitude of this deterministic waveform at the same instant, say at $t = T/2$, where T is the signal's period, we will always theoretically obtain the same value. (In reality, however, because of the finite precision of our instruments, these values will vary slightly from one measurement to another.)

The digital signal shown in Figure 2.1.1 is also an example of a deterministic (but not continuous) waveform.

A waveform is called random if we can determine its value (magnitude) at any instant only with some probability. Putting it another way, the waveform is called "random" if we cannot determine the exact value of the waveform's magnitude at any instant because it takes on a random (probabilistic) value. This definition of a random waveform also means that we cannot predict the waveform's value at any chosen instant; that is, we can predict its value only with some probability. If we measure the random waveform's magnitude 10 times at the same instant, even using an ideal measuring instrument, we will obtain 10 different yet close values. We can therefore say that the waveform's magnitude is a random variable. (And what do we mean by a random variable? A random variable is a variable that has a whole set of values and it could take on any of those values, randomly.[5] We will tackle this topic Section 10.1.)

Figure 2.3.3 illustrates these explanations. The two top waveforms shown in Figure 2.3.3 are deterministic ones. We can certainly determine the patterns of their development and can predict what values these waveforms will take at any instant. In contrast, the bottom waveform does not exhibit any pattern of its development and no definite predications can be made regarding its development at any instant.

In this section, we will be concerned only with deterministic waveforms.

Continuous and Discrete Waveforms Time-varying waveforms can be **continuous** and **discrete**. We have discussed them in Sections 2.1 and 2.2, particularly in Sidebar 2.2.S.2. Here we would like to present a more general and formal view of this topic.

---

[5] https://www.mathsisfun.com/data/random-variables.html.

2.3 Analog Signals – Advanced Study | 171

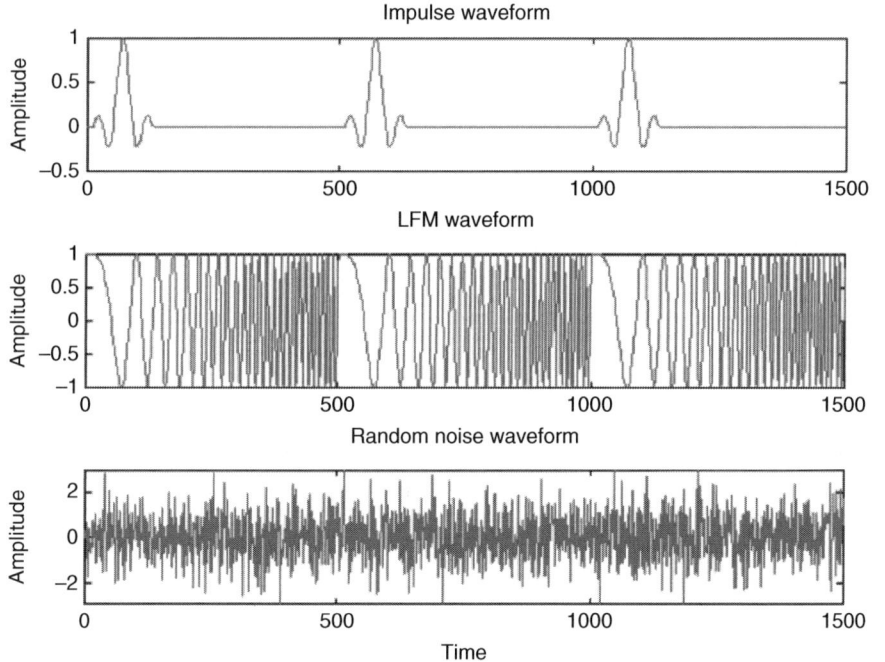

Figure 2.3.3 Deterministic and random waveforms. (Source: Professor Ram M. Narayanan, The Pennsylvania State University. Reprinted with permission.)

A waveform $v(t)$ is called *continuous* if the difference $|v(t) - v(t')|$ can be made as small as we please when an instant t approaches a given instant t'; that is, formally speaking,

$$v(t) \to v(t') \text{ as } t \to t' \tag{2.3.1}$$

This definition means that *we can draw the graph of a continuous waveform over any interval of its domain with one continuous motion of the pen.* We can also say that the *graph of a continuous waveform is an unbroken curve.*

A sinusoidal signal is, of course, an example of a continuous waveform. Indeed, let us compute the instantaneous value of a sine signal with $A = 3$ V, $f = 40 \times 10^3$ Hz, and $\theta = 0.523$ rad at the instant $t' = 4.000$ μs as

$$v(t) = A \sin(2\pi f t + \theta) = 3 \sin(2\pi \times 40 \times 10^3 \times 4 \times 10^{-6} + 0.523) = 2.997\,29 \text{ V}$$

Let us now find the magnitude of this waveform at the instant $t'' = 4.001$ μs as

$$v(t'') = V_m \sin(2\pi f\, t' + \theta) = 3 \sin(2\pi \times 40 \times 10^3 \times 4.001 \times 10^{-6} + 0.523) = 2.997\,32 \text{ V}.$$

Thus, when the difference $|t' - t|$ is equal to $|4.000 \text{ μs} - 4.001 \text{ μs}| = 0.001$ μs, the difference, $|v(t') - v(t'')|$, becomes $|2.997\,29 \text{ V} - 2.997\,32 \text{ V}| = 0.000\,03$ V. You can verify that this difference will reduce to approximately 0.000 003 V when the difference, $|t' - t''|$, will be reduced to 0.0001 μs. As you can see, in our case of a sine signal, the value $v(t'')$ can be made as close to value $v(t')$ as we want by making t'' closer to the instant t'. Therefore, this waveform, $v(t) = A \sin(2\pi f t + \theta)$, satisfies the definition (2.3.1) and is continuous.

Figure 2.3.4a depicts an example of a continuous waveform; you can readily verify that this graph can be drawn *with one continuous motion of the pencil.*

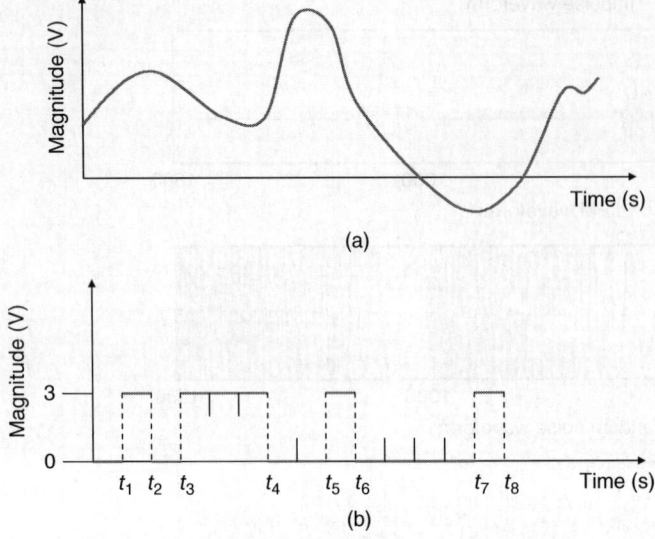

Figure 2.3.4 Continuous (a) and discrete (b) waveforms.

In contrast to a continuous signal, a *discrete* waveform belongs to the category known as *discontinuous functions*. Such a function exhibits at least one point of discontinuity where the waveform has no value at a given instant, t'. As the term suggests, *we cannot draw the graph of a discrete waveform with a continuous line because, at a discontinuity point, we must leap from one segment of the graph to another*.

The digital signal pictured in Figure 2.3.4b is a perfect example of a discrete (discontinuous) waveform. Assuredly, at intervals $0 < t < t_1$, $t_1 < t < t_2$, $t_2 < t < t_3$, and so on, the waveform's values are either equal to 0 V or to 3 V. However, there are no values of this waveform at instants $t = t_1$, $t = t_2$, and so on. Thus, there are points of discontinuity at switching times $t_1, t_2, \ldots t_8$. We can prove this statement by trying to apply the definition of a continuous waveform to this waveform: When an arbitrary instant t approaches instant t_1, the difference, $|v(t) - v(t_1)|$, cannot be made as small as we wish; on the contrary, this difference goes to infinity. This is clear from the following formal expression describing this waveform:

$$v(t) = \begin{cases} 0\,\text{V} & \text{when } 0 < t < t_1 \\ 3\,\text{V} & \text{when } t_1 < t < t_2 \\ 0\,\text{V} & \text{when } t_6 < t < t_7 \\ 3\,\text{V} & \text{when } t_7 < t < t_8 \end{cases} \tag{2.3.2}$$

Formula (2.3.2) shows that this discrete waveform is determined as a signal whose amplitude equals 0 V at the time interval between 0 and t_1 (the instant t_1 is not included in this interval), but it has no value at $t = t_1$. Similarly, the discrete signal's amplitude is equal to 3 V at the time interval between t_1 and t_2 (the instants t_1 and t_2 are not included in this interval), but it has no value at $t = t_1$ and $t = t_2$. This is true for all other intervals and instants $t_3, t_4, \ldots t_8$. From this example, we can derive another important view of discrete waveforms: We can say that *a discrete waveform assumes its values from a discrete set*. As Figure 2.3.4b and (2.3.2) show, this discrete waveform assumes its values from the set {0 V, 3 V}. It is necessary to stress that we have considered here the *mathematical model of a discrete signal*; in reality, the discrete waveform is presented by a continuous graph that sometimes changes its values abruptly. We discuss this point in greater detail in Chapter 3.

Periodic and Nonperiodic Waveforms Another important classification of waveforms is their division into **periodic and nonperiodic (aperiodic) waveforms.** A signal is called *periodic* if its waveform satisfies the following condition:

$$v(t \pm kT) = v(t) \quad \text{for all } t \tag{2.3.3}$$

where k is an integer and T is a period. This condition can be written for $k = 1$ as

$$v(t \pm T) = v(t) \quad \text{for all } t \tag{2.3.4}$$

In this case, T is the *smallest* positive number that satisfies this condition. Obviously, the signal is called *nonperiodic (aperiodic)* if it does not satisfy (2.3.3). We know that a sinusoidal signal is a periodic signal; let us confirm this by applying Eq. (2.3.4) to a sine signal. Taking any sine signal,

$$v(t) = A \sin(\omega t + \theta)$$

we want to prove that

$$A \sin(\omega t + \theta) = A \sin(\omega(t + T) + \theta) \tag{2.3.5}$$

where T is the period. Recalling from Section 2.1 that the sine's period is

$$T = 1/f = 2\pi/\omega \tag{2.3.6}$$

we substitute (2.3.6) into (2.3.5) and obtain

$$A \sin(\omega(t + T) + \theta) = A \sin(\omega(t + 2\pi/\omega) + \theta) = A \sin(\omega t + 2\pi + \theta)$$
$$= A \sin((\omega t + \theta) + 2\pi)) \tag{2.3.7}$$

Let us apply the following trigonometric identity,

$$\sin(\alpha + \beta) = \sin \alpha \cos \beta + \cos \alpha \sin \beta, \tag{2.3.8}$$

to (2.3.7); we obtain

$$A \sin((\omega t + \theta) + 2\pi) = A \sin(\omega t + \theta) \cos 2\pi + A \cos(\omega t + \theta) \sin 2\pi) \tag{2.3.9}$$

Since $\cos 2\pi = 1$ and $\sin 2\pi = 0$, we arrive at

$$A \sin(\omega(t + T) + \theta) = A \sin(\omega t + \theta) \tag{2.3.10}$$

which means that the sine signal satisfies (2.3.4). Therefore, the sine signal is a periodic signal.

The continuous waveform $v(t) = Ae^{(-\alpha t)}t$, where α and A are constants, is an example of a *nonperiodic* signal. If we apply Eq. (2.3.4) to this waveform, we can readily see that this condition cannot be satisfied, which means that this waveform is nonperiodic. By reviewing the graph $v(t) = Ae^{(-\alpha t)}t$, we can readily see that the signal does not repeat itself after any time interval. (**Exercise**: Apply the concept shown in Eqs. (2.3.4)–(2.3.10) to prove that the function $v(t) = Ae^{(-\alpha t)}t$ is aperiodic. Build the function's graph to support your proof.)

(**Question**: Refer to Figure 2.3.3: What graphs do represent the waveforms of periodic functions? Can a random waveform be periodic?)

2.3.1.2 Waveform and Signal's Power

In communications, a signal's power plays an important role in determining the quality and reliability of transmission. Specifically, the *ratio of signal power to noise power (SNR) determines the quality of the communication*, as we learned in Section 1.3. (See Shannon's law presented in (1.3.6) and its discussion.) Therefore, the **average power** that both a communication signal and a noise signal deliver to a receiver is an important parameter. To calculate such power, we need to recall that *instantaneous power*, $p(t)$ (W), in an electrical circuit is given by

$$p(t) \text{ (W)} = v(t) \text{ (V)} \cdot i(t) \text{ (A)} \tag{2.3.11}$$

You will recall that *power in watts is equal to energy (work) in joules divided by time in seconds*. Formula (2.3.11) is derived from the basic definitions you learned in your circuit analysis course:

Since $v(t)$ = work/charge and $i(t)$ = charge/time, then $p(t)$ (W) = $v(t)$ (V)·$i(t)$ (A) = work/time, which satisfies the definition of power.

We now need to know that the *average power*, P_{avg} (W), of a periodic signal can be computed by averaging (integrating) its instantaneous power over a period as follows:

$$P_{avg} \text{ (W)} = 1/T \int_{-T/2}^{T/2} p(t)\, dt = 1/T \int_{-T/2}^{T/2} (v(t)\, i(t))\, dt \tag{2.3.12}$$

Let us consider a simple example.

Example 2.3.1 Instantaneous and Average Power of a Sinusoidal Signal

Problem

A sinusoidal signal $v(t)$ (V) $= V_m \cos(\omega t)$ is delivered to a receiver, which is a pure resistive circuit:

a) Determine the formula for instantaneous power of this signal.
b) Find the signal's average power.

Solution

(a) We understand that the voltage and current at a receiver are in phase since this is a pure resistive circuit; that is $i(t) = I_m \cos(\omega t)$. With this in mind, we can apply (2.3.11) as

$$p(t) \text{ (W)} = (v(t)\, i(t)) = (V_m \cos(\omega t)\, I_m \cos(\omega t)) = V_m I_m \cos^2(\omega t)$$

Using a well-known trigonometric identity, $\cos^2(\omega t) = \tfrac{1}{2}(1 + \cos(2\omega t))$, the solution to the first part of the problem is found as

$$p(t) \text{ (W)} = (V_m I_m \cos^2(\omega t))dt = 1/2\, V_m I_m (1 + \cos(2\omega t)) \tag{2.3.13}$$

(b) Applying (2.3.12) and (2.3.13), the average power of this signal is computed as

$$P_{avg} \text{ (W)} = 1/T \int_{-T/2}^{T/2} (v(t) \cdot i(t))dt = 1/2T \int_{-T/2}^{T/2} V_m \cdot I_m (1 + \cos(2\omega t))dt = V_m \cdot I_m/2$$

because $\int_{-T/2}^{T/2} \cos(2\omega t) dt = 0$.

Therefore, the average power delivered to a resistive receiver circuit by a sinusoidal signal is equal to $V_m \cdot I_m / 2$, which is well known from a basic circuit analysis course.

If we recall that the *root-mean-square (rms)* value of a sinusoidal voltage and current is given by

$$V_{rms} = V_m/\sqrt{2} \text{ and } I_{rms} = I_m/\sqrt{2}$$

then we can readily see that the average power of such a signal can be computed as

$$P_{avg} (W) = V_m \cdot I_m/2 = V_{rms} \cdot I_{rms} \tag{2.3.14}$$

To clarify this example for yourself, compute the value of instantaneous power at any reasonable instant, t, and average power of this signal by assuming the values of V_m, I_m, and ω.

Discussion

- Can we compute the average power if we know only the voltage or only the current? Yes, of course. In our example, by applying Ohm's law, we can easily express the current through the voltage and vice versa as $i(t) = v(t)/R$, where R (Ω) is the load resistance. Now (2.3.12) takes the form

$$P_{avg} (W) = \int (v(t) \cdot i(t)) dt = 1/R \int (v^2(t)) dt = R \int (i^2(t)) dt \tag{2.3.15a}$$

It seems that we cannot find the average power by using this formula because we do not know the resistance. However, in communications (and in signal analysis in general), we use *normalized power*, the power dissipated over $R = 1\Omega$. We can also say that *normalized power is power per ohm*. Using normalized power is an important expedient that allows us to simplify computations without compromising the results. Therefore, *average normalized power* can be found by the following formula:

$$P_{avg}^{norm} (W) = \int (v^2(t) \, dt = \int (i^2(t)) \, dt \tag{2.3.15b}$$

In our example of a cosine signal, we compute

$$P_{avg}^{norm}(W) = \int (v^2(t)) dt = \int (i^2(t)) \, dt = \int_{-T/2}^{T/2} (V_m \cos(\omega t))^2 dt$$

$$= \int_{-T/2}^{T/2} (I_m \cos(\omega t))^2 dt = V_m^2/2 = I_m^2/2$$

Since $V_m = I_m$ for $R = 1\,\Omega$, our result for average normalized power correctly corresponds to the previous calculations, in which we computed the general average power as $P_{avg} = V_m I_m/2$.

The concept of normalized power also allows us to compute an instantaneous power even if we are given only the signal's voltage as $v(t) = V_m \cos(\omega t)$. Putting $R = 1\,\Omega$ and plugging $I_m = V_m/R$ into (2.3.13), we find the formula for the instantaneous power as

$$p(t) (W) = 1/2\, V_m I_m (1 + \cos(2\omega t)) = \tfrac{1}{2} V_m^2 (1 + \cos(2\omega t)) \tag{2.3.15c}$$

To compute the real average power from its normalized value, we need to properly insert the resistor's value into our final result. The concept of normalized power is important in communications because we need to consider *the signal-to-noise power ratio*. By calculating normalized power for both signal and noise, we will obtain the correct result for this ratio regardless of the resistor's value. This is because we are taking the same resistance for both signal and noise and these resistances will be canceled as soon as we compute the ratio.

- Let us consider another aspect of our calculations: By averaging, we can find another parameter of a *symmetrical waveform* – its dc value. In fact, it is intuitively clear that the average of a time-varying signal is its dc value. To prove this statement, let us first find the average value of a sinusoidal signal $v(t) = V_m \cos(\omega t)$:

$$v_{avg} = 1/T \int_{-T/2}^{T/2} (v(t)) dt = 1/T \int_{-T/2}^{T/2} (V_m \cos(\omega t)) dt = 0 \tag{2.3.16a}$$

This result – the average value of a sinusoidal signal over a period is zero – is quite clear from the signal's nature. Indeed, if you look again at the graph of a cosine (or sine) signal and consider the average of this signal, you can immediately see that its positive half wave is equal to a negative half wave and therefore its average over a period is equal to zero. To visualize this explanation, see Figure 2.3.5a.

Second, consider a cosine signal with offset B; that is,

$$v(t) = B + V_m \cos(\omega t)$$

where B is constant. Without doubt, we know that the average (dc) value of this signal is equal to B (V). Now, by substituting this expression into (2.3.16a), we compute

$$v_{avg} = 1/T \int_{-T/2}^{T/2} (B + V_m \cos(\omega t)) dt = B \qquad (2.3.16b)$$

which proves the above statement (see Figure 2.3.5b).

Finally, Figure 2.3.5c illustrates the concepts of the instantaneous and average power for $v(t)$ (V) = 0.15cos $(2\pi 1000t)$.

Power and Energy Waveforms (Signals) Let us introduce another waveform classification: power and energy waveforms. A waveform, $v(t)$, is called a *power waveform* if its average normalized power, P_{avg}^{norm} (W), is finite; that is,

$$0 < P_{avg}^{norm} (W) < \infty \qquad (2.3.17)$$

Bear in mind that the formal definition of average normalized power – which is stricter than that given in (2.3.17) – is

$$P_{avg}^{norm} (W) = \lim_{T \to \infty} 1/T \int_{-T/2}^{T/2} |v(t)|^2 dt \qquad (2.3.18)$$

Accordingly, a waveform, $v(t)$, is called an *energy waveform* if its average normalized energy, E_{avg}^{norm} (J), is finite; that is,

$$0 < E_{avg}^{norm} (J) < \infty \qquad (2.3.19)$$

Here, average normalized energy is defined as

$$E_{avg}^{norm} (J) = \int_{-\infty}^{\infty} |v(t)|^2 dt \qquad (2.3.20)$$

The use of limits in (2.3.18) and (2.3.20) enables us to clarify the definitions: If limit in (2.3.18) is finite, it means that the signal's power converges to a finite value therefore, this signal is a power one. If we plug this power signal into an energy integral, (2.3.20) it will go to infinity.

There are waveforms, however, that cannot be classified as either energy or power entities.

We need to remember that these definitions consider the mathematical models of waveforms; in reality, our instruments always average over a finite interval of time and therefore condition $t \to \infty$ is not satisfied.

(**Question**: *What do you think – is cosine a power or energy signal? Hint: Apply (2.3.18) and (2.3.20) to verify your answer.*)

Finally, remember that a waveform, $v(t)$, is simply a physical form of a signal; therefore, we can rightly substitute the term "signal" in all considerations of waveforms in this section.

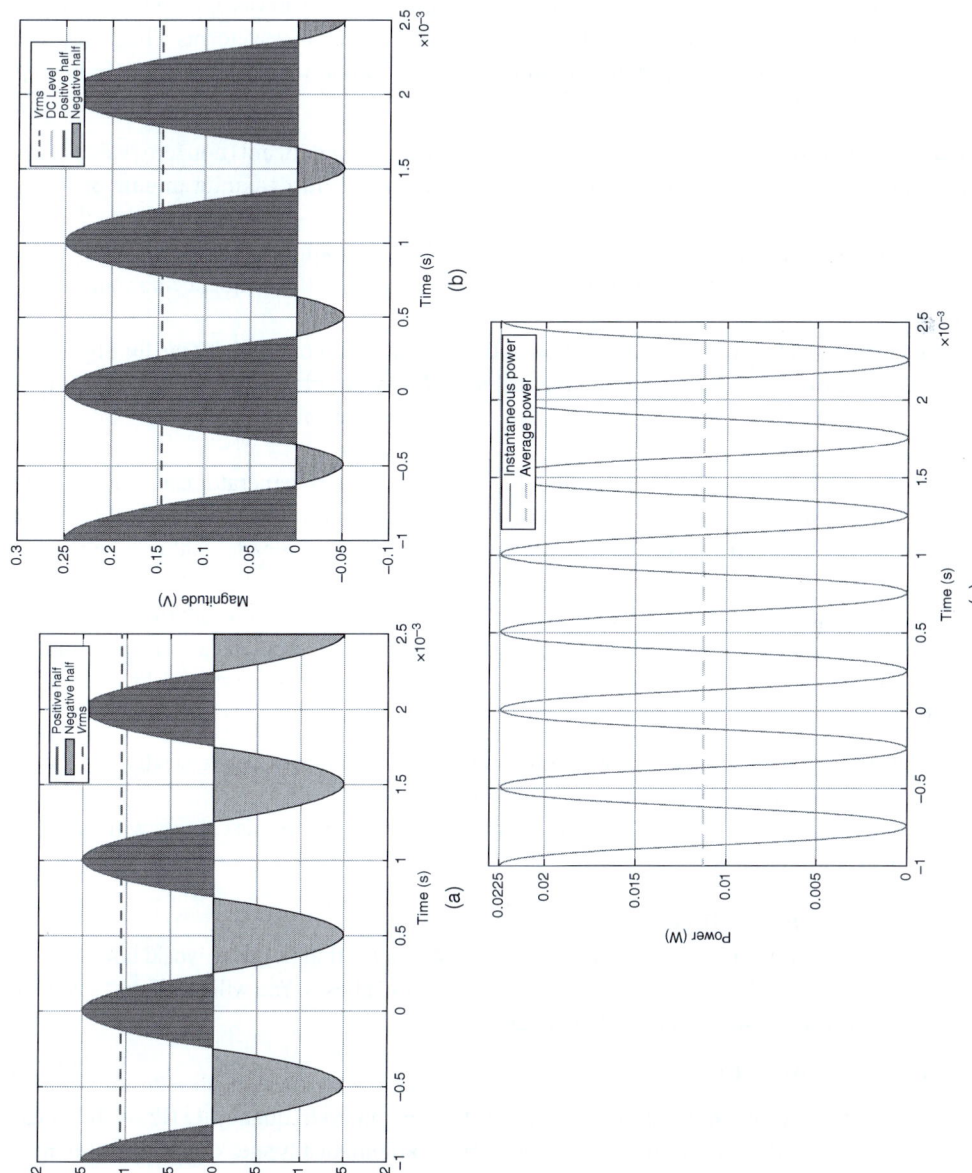

Figure 2.3.5 The average value and instantaneous power of a sinusoidal signal: (a) The average and rms values of a cosine signal $v(t)$ (V) = $0.15\cos(2\pi 1000t)$ with symmetrical positive and negative halves; (b) the average value of $v(t)$ (V) = $0.7 + 0.15\cos(2\pi 1000t)$; (c) instantaneous power of $v(t)$ (V) = $0.15\cos(2\pi 1000t)$.

2.3.2 Waveforms and Phasors

2.3.2.1 Practically Realizable Waveforms

We need to observe waveforms and measure their parameters in our laboratories and in the field; in other words, we want to deal only with practically realizable waveforms (signals). This is an important aspect in the investigation of waveforms by practicing engineers as opposed to the needs of a researcher, who might be interested only in mathematical models of waveforms. The practically realizable waveforms must satisfy certain conditions, the most important of which are the following (Couch 2012, pp. 34–40):

- The waveform must exist over a finite amount of time and have significant (nonzero) values over this interval. It follows from this condition that the waveform produces a finite amount of power or energy.
- The spectrum of a waveform must have significant values over a finite interval of frequencies. This condition treats the physical realization of a waveform from the frequency-domain standpoint.
- The waveform must be a continuous function of time. Again, this condition is true for physically realizable waveforms, whereas the mathematical models of waveforms can have discontinuities. We will discuss this condition in greater detail in Chapter 3, when we consider digital signals and digital transmission.
- The waveform must have a finite peak value. In truth, any physical generator can produce only a finite amount of energy; on the other hand, if we generate a waveform with "infinite" peak value, we can destroy our measuring instruments. Here, "infinite" means a value much above the measuring and safety capabilities of a device.
- The waveform must have only real values. This condition again stresses the fact that, in contrast to a mathematical model of a waveform, a measurable waveform cannot be a complex number. However, the *properties* of a waveform, such as its spectrum, can be described by a complex number.

These conditions establish formal restrictions on waveforms that can be physically realizable, thereby distinguishing them from their mathematical models.

Let us now briefly review the methods of analysis of telecommunications circuits with applications of the waveform models.

2.3.2.2 Phasors and Phasor Diagrams

If we look again at a sinusoid that describes ac signals, we realize that in the ac world both voltage and current are described by their amplitudes, frequencies, and phases. You will recall that a voltage sinusoidal waveform is defined in time domain as

$$v(t) \, (V) = A \, \cos(\omega t + \theta) \tag{2.1.1R}$$

where A (V) is the voltage amplitude, ω (rad/s) $= 2\pi f$ is the radian frequency, f (Hz) is the linear (cyclical) frequency, and θ is the initial phase. Therefore, mathematically speaking, a sinusoid must be represented as a two-dimensional entity showing an amplitude, measured in volts, and a phase angle, $(\omega t + \Theta)$, measured in radians. In circuit analysis, we use *phasors* for this purpose. In Sidebar 2.2.S.1, we introduced a phasor as follows:

A phasor is a constant-length vector with one end attached to the origin of a plane coordinate system. The vector rotates around this fixed zero point with constant angular velocity, ω (rad/s).

A phasor's length is equal to the amplitude of a sinusoidal signal, and its initial position is determined by the signal's initial phase, Θ. (You will recall that in Sidebar 2.2.S.1, an inherent connection

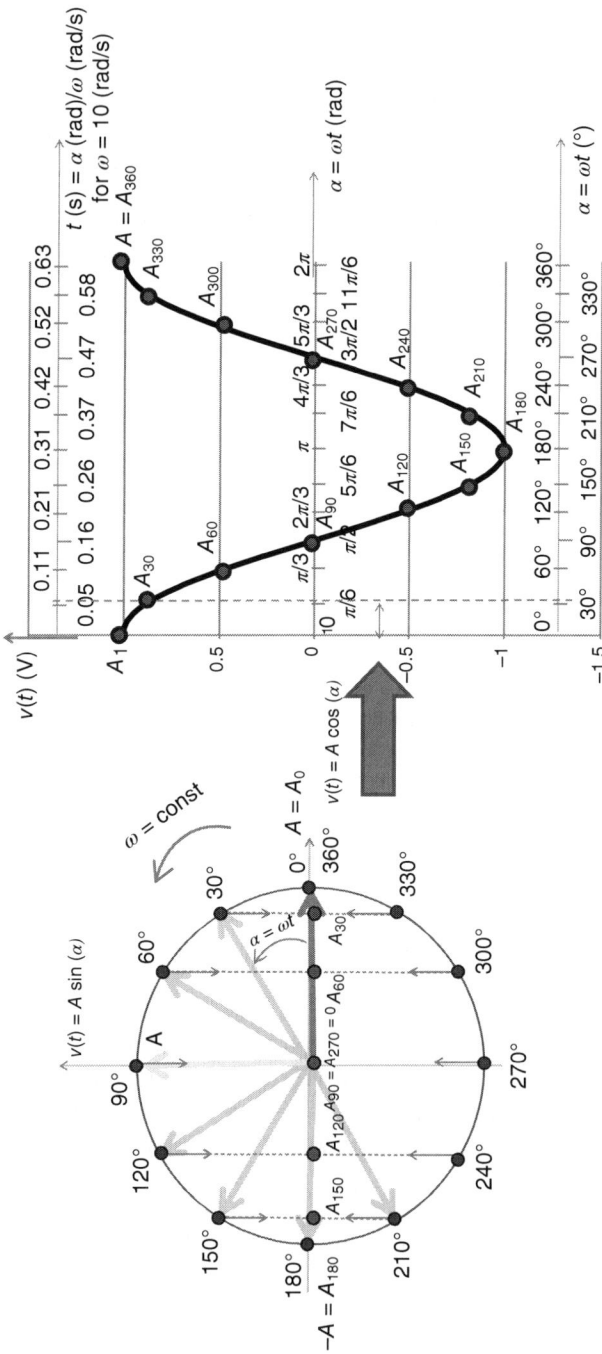

Figure 2.2.S.1.2R Constructing a cosine signal from a phasor presentation.

of a phasor and a sinusoid was thoroughly discussed. Figure 2.2.S.1.2, reproduced here, shows how a sinusoidal signal can be constructed by projecting a phasor's instantaneous positions into a v–t plane.)

Now, we have two options:

1. We can make this phasor rotate with an angular velocity, ω (rad/s), equals to the radian frequency, ω (rad/s) = $2\pi f$, of a sinusoidal signal, which gives us a time-varying presentation.
2. We can assume the rotation of our phasor – that is, the existence of this frequency, ω – by default but show the phasor in a steady position (actually, at $t = 0$) with respect to a coordinate plane.

Both of these presentations are valid and widely used, though in circuit analysis, it is customary to choose the second option. There is a problem with terminology in both academia and industry, as usual. Some sources refer to a rotating vector as a *rotating phasor*; others prefer to simply say *phasor*. As for a vector without rotation, some (a majority) sources call it a *phasor* (to confuse students even more), whereas others refer to it as a *stationary phasor* or a *steady-state phasor* or a *polar*. In essence, the meaning of this discussion boils down to the following definition:

The same sinusoidal signal,

$$v(t)\,(V) = A\,\cos(\omega t + \theta)$$

can be presented in a phasor form,

$$v(t)\,(V) = Ae^{j\omega t}e^{j\theta} \tag{2.3.21a}$$

or in a polar form,

$$v(t)\,(V) = Ae^{j\theta} = A\angle\theta \tag{2.3.21b}$$

As for us, we will use the term "phasor form" when referring to a rotating vector and the term "polar form" to denote the same vector without rotation. A *phasor form* is characterized by three parameters – A (V), ω (rad/s), and θ (rad) – and a *polar form* is described by two parameters – A (V) and θ (rad). It's customary to write a polar form as

$$\mathbf{A} = A\angle\theta \tag{2.3.22}$$

where $\mathbf{A}$ is the phasor (we use boldface font to stress that the phasor is a vector, i.e. a two-dimensional quantity), A is the phasor's amplitude, and Θ is its phase; the symbol "$\angle$" stands for angle.

How can we distinguish between cosine and sine in phasor domain? As explained in Sidebar 2.2.S.1, if the initial phase Θ (rad) is zero, then the phasor corresponding to the cosine signal is positioned along a vertical axis in the positive (up) direction. Consequently, at $\Theta = 0$ rad, the sine's phasor starts from the positive (right) horizontal axis because $\sin(\omega t) = \cos(\omega t - 90°)$.

Figure 2.3.6 visualizes these explanations by showing the waveform of a sinusoidal signal,

$$v(t) = A\,\sin(2\pi f t + \Theta) = 0.15\,\sin(2\pi 1000 t + 30°)$$

To consider its instantaneous value, we arbitrarily choose the 165° phase,

$$(2\pi 1000 t + 30°) = 165°$$

which corresponds to the instant (see (2.2.S.1.3)),

$t_{165} = ((165° - 30°)/360°) \times 10^{-3} = 0.375$ ms

Computation of the instantaneous value results in

$v(t) = 0.375$ ms $= 0.15 \sin(360 \cdot 0.375 + 30°) = 0.15 \cdot 0.26 = 0.039$ V

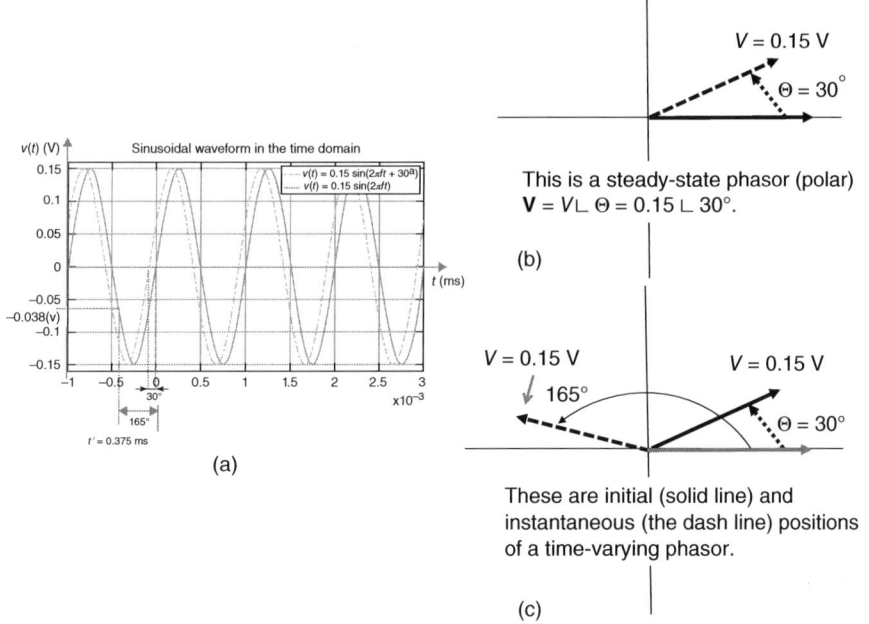

Figure 2.3.6 A sinusoidal signal presented by (a) its waveform, (b) its phasor, and (c) its polar form.

Since the sine is shifted to the left with respect to its zero-shift position, its actual instantaneous value is negative; that is, $v(t_{165}) = -0.039$ V. All these values – the 165° phase shift, instant $t = 0.375$ ms, and instantaneous value $v(t_{165}) = -0.039$ V – are shown in Figure 2.3.6a.

Figure 2.3.6b shows the phasor form of this signal (a time-varying phasor, that is), which rotates around the origin of the coordinate system at angular velocity $2\pi f = \omega = 6283$ rad/s. Its starting position is determined by the initial phase, $\Theta = 30°$, and its instantaneous position is determined by the entire phase, $2\pi ft + \Theta$, where f and Θ are constants and t is a variable. The polar form (or a steady-state phasor) is shown in Figure 2.3.6c.

To sum up, *a phasor is another form of representing a waveform (a signal in time domain).* In the example discussed, a sine signal, $v(t) = 0.15 \sin(2\pi 1000 t + 30°)$, can be presented as the phasor, $\mathbf{A} = 0.15 \angle 30°$, in *polar form*, whereas the rotation of this phasor with the angular velocity $\omega = 2\pi 1000$ must be included to show this sine in a *phasor form*.

Refer to Appendix 2.3.A for the mathematical foundation of phasor operations.

2.3.2.3 Waveforms and Phasors for a Resistor, an Inductor, and a Capacitor

In this subsection, we will apply the phasors for analyzing the frequency responses of basic passive components – resistors, capacitors, and inductors – in time, phasor, and frequency domains. This analysis will prepare us to understand the next level of our analysis of the circuitry in communications equipment.

Resistor, R: The waveform of a voltage drop across, as well as a phasor diagram for, a resistor, $R(\Omega)$, is shown in Figure 2.3.7. We recall that the current, i_R (A), flowing through a resistor is given by

$$i_R (A) = I_R \sin \omega t \qquad (2.3.23)$$

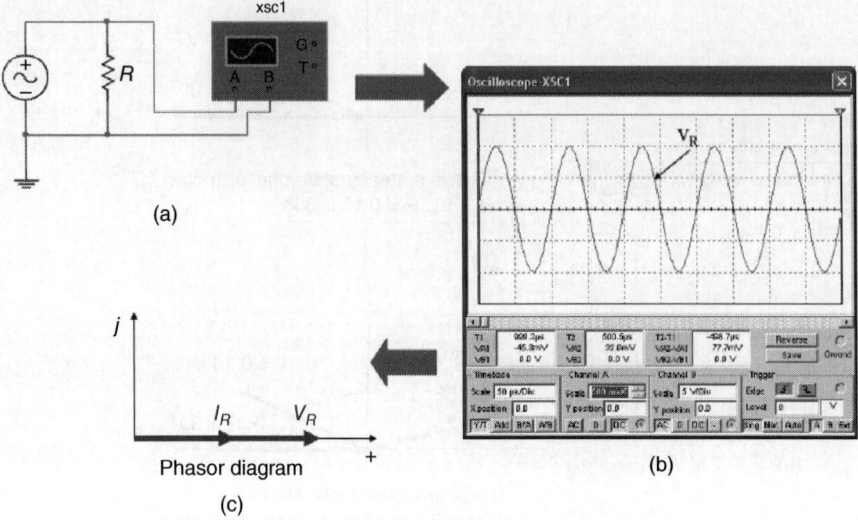

(a)

(b)

(c) Phasor diagram

Figure 2.3.7 Waveform and phasor diagram for a resistor: (a) The circuit; (b) the result of simulation, as shown on an oscilloscope; (c) the phasor diagram.

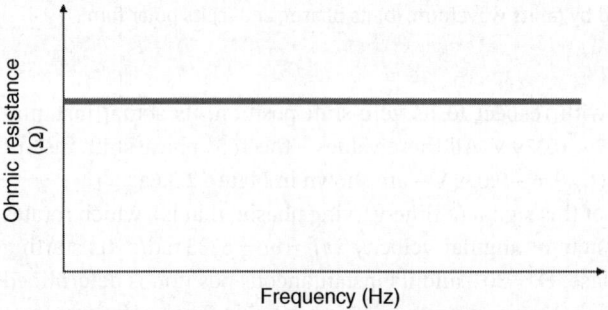

Figure 2.3.8 Frequency response of ohmic resistance.

where I_R (A) is the current amplitude, ω (rad/s) $= 2\pi f$ is the radian frequency, and f (Hz) is the linear (cyclical) frequency. We can compute the voltage drop across the resistor as

$$v_R \text{ (V)} = i_R \times R = RI \sin \omega t = V_R \sin \omega t \tag{2.3.24}$$

We can verify (2.3.23) and (2.3.24) by an experiment. The schematic of the experimental setup, the measured voltage across the resistor, and the phasor diagram are demonstrated in Figure 2.3.7. This result – where a resistor's current and voltage are in phase – can be presented in a phasor diagram, as illustrated in Figure 2.3.7c.

As an exercise, sketch the i_R *waveform* on the graph in Figure 2.3.7. Show the amplitudes I_R and V_R, the periods, and the phase shift (if any) between $i_R(t)$ and $v_R(t)$ on this graph.

(**Question**: *What domain – time or phasor – are you working in Figure* 2.3.7b?)

Of course, we remember that the resistance, R (Ω), does not depend on the frequency. Figure 2.3.8 depicts this fact.

Inductor, L: The experimental waveform of a voltage drop across an inductor – as well as a phasor diagram for an *inductor* $L(H)$ – is shown in Figure 2.3.9. The current, i_L, flowing through an

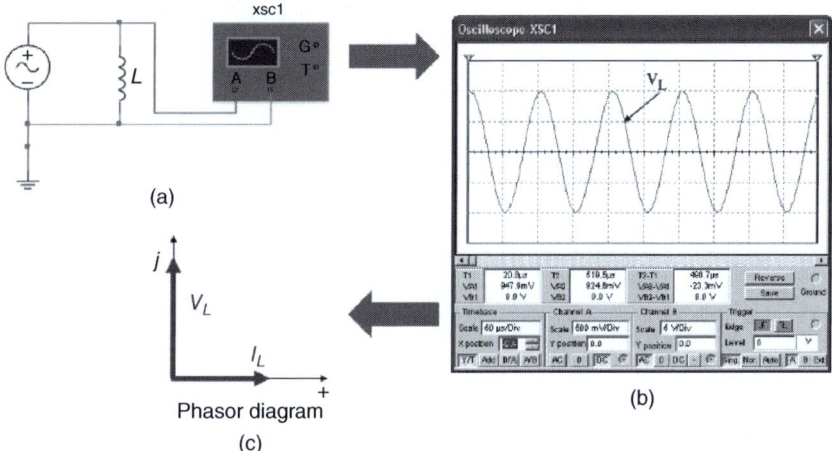

Figure 2.3.9 Waveform and phasor diagram for an inductor: (a) The circuit; (b) the result of simulation, as shown on an oscilloscope; (c) the phasor diagram.

inductor is given by

$$i_L(t) \, (A) = I_L \sin \omega t \qquad (2.3.25)$$

where I_L (A) is the current amplitude and ω (rad/s) = $2\pi f$ is the radian frequency. The voltage drop across the inductor is given by

$$v_L(t) \, (V) = V_L \sin(\omega t + 90°) \qquad (2.3.26)$$

where V_L (V) = X_L (Ω) × I_L (A) is the voltage amplitude and X_L (Ω) is the inductive reactance. We remember that *in a pure inductor, the voltage leads the current by 90°*; this fact is stressed in (2.3.26). The experimental waveform shown in Figure 2.3.9b also confirms this fact. The figure shows that the voltage drop across the inductor is a cosine, whereas the current flowing through the inductor is a sine function. The phase relationship between i_L and v_L can be written mathematically as

$$v_L(t) \, (V) = jX_L i_L(t) \qquad (2.3.27)$$

where j is an imaginary unit. Figure 2.3.9c, which is a phasor diagram, is a graphical representation of (2.3.27). It shows that the *value* of voltage v_L is equal to the product of *values* X_L and i_L; to reflect the 90° positive phase shift, we must multiply this product by j, as the phasor diagram shows.

Again, as an exercise, sketch the i_L waveform on the graph shown in Figure 2.3.9b. Show magnitudes I_L and V_L, the periods, and the phase shift between $v_L(t)$ and $i_L(t)$ on this graph. What domain – time or phasor – are you working in?

Compare the two waveforms – v_R and v_L – given in Figures 2.3.7b and 2.3.9b. You can immediately see that v_L has a 90° phase shift with respect to v_R. Analyze this fact: Is this the same 90° phase shift that v_L has with respect to i_L? If you answer yes, explain why.

As for frequency response, we need to recall that an inductive reactance is directly proportional to the frequency: X_L (Ω) = $\omega L = 2\pi f L$, where L (H) is the inductance. Figure 2.3.10 presents this relationship graphically.

Capacitor, C: The waveform for a voltage drop across a capacitor and a phasor diagram for a capacitor C (F) are shown in Figure 2.3.11.

Bear in mind that the current, $i_C(t)$ (A), flowing through a capacitor is given by

$$i_C(t) \, (A) = I_C \sin \omega t \qquad (2.3.28)$$

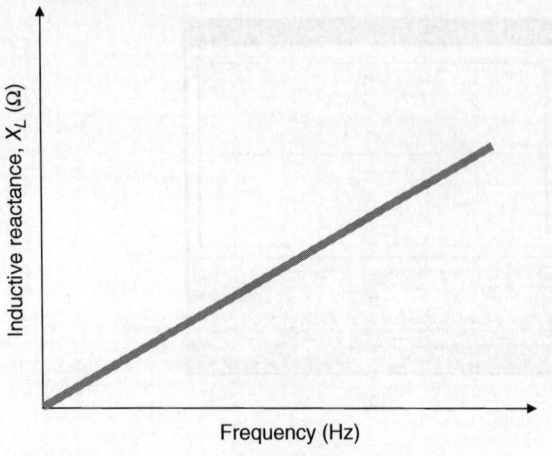

Figure 2.3.10 Frequency response of inductive reactance.

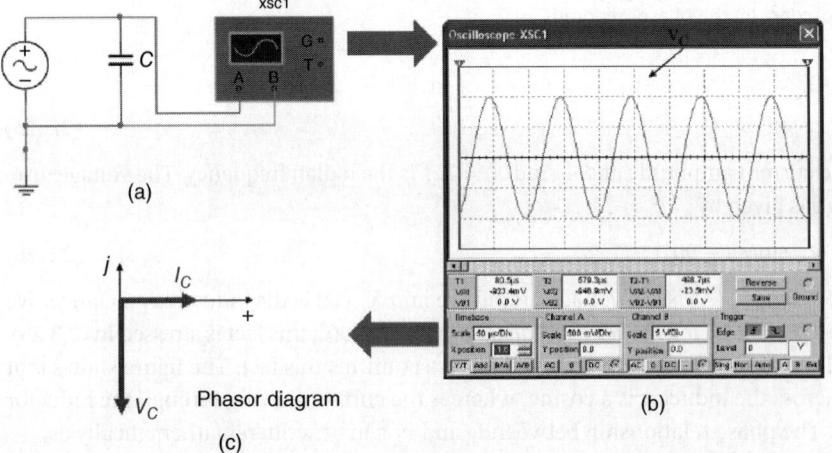

Figure 2.3.11 Waveform and phasor diagram for a capacitor: (a) The circuit; (b) the result of simulation, as shown on an oscilloscope; (c) the phasor diagram.

where I_C (A) is the current's amplitude and ω (rad/s) $= 2\pi f$ is the radian frequency. The voltage drop across the capacitor is given by

$$v_C(t)\,(V) = V_C\,\sin(\omega t - 90°) \tag{2.3.29}$$

where V_C (V) $= I_C \times X_C$ is the voltage amplitude and X_C (Ω) is the capacitive reactance. We remember, surely, that *in a pure capacitor, the current leads the voltage by 90°* or, conversely, the voltage lags the current by 90°; both conditions are reflected in (2.3.29). This phase relationship between $i_C(t)$ and $v_C(t)$ can be written mathematically as

$$v_C(t)(V) = -jX_C i_C(t) \tag{2.3.30}$$

where j is an imaginary unit and C (F) is a capacitance. Formula (2.3.30) shows that the *value* of the voltage, $v_C(t)$, is equal to the product of *values* X_C and i_C, whereas a negative imaginary unit, $-j$, reflects a 90° negative phase shift between v_C and i_C. Figure 2.3.11c, which is a phasor diagram, is a graphical representation of (2.3.30).

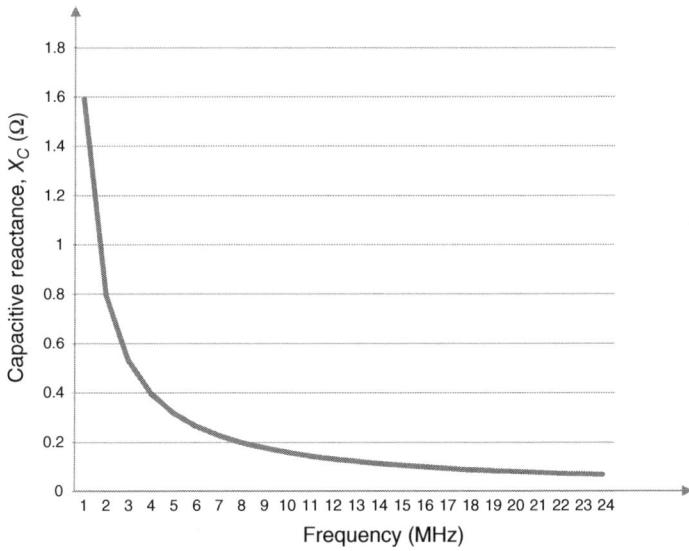

Figure 2.3.12 Frequency response of capacitive reactance.

It's important at this point to repeat that a capacitive reactance is inversely proportional to the frequency: X_C (Ω) $= 1/2\pi f C$. This relationship is shown in Figure 2.3.12 for $C = 1\,\mu\text{F}$.

Now do the same exercise you carried out for a resistor and an inductor: Sketch the i_C *waveform* on the graph shown in Figure 2.3.11b. Show magnitudes I_C and V_C, the periods, and – the main point – the phase shift between $v_C(t)$ and $i_C(t)$ on this graph. Question: What domain – time or phasor – are you working in?

Compare the three waveforms – v_R, v_L, and v_C – given in Figures 2.3.7b, 2.3.9b, and 2.3.11b. You should be able to readily explain why these three voltage signals have different initial phase shifts. Remember that the same current, $i(t)$ (A) $= I \sin(\omega t)$, was presented to all three – R, L, and C – passive components. Hint: Refer again to the phasor diagrams shown in Figures 2.3.7c, 2.3.9c, and 2.3.11c.

The main point to comprehend here is that in the ac world, the *voltage drops* across a resistor, an inductor, and a capacitor ($v_R(t)$, $v_L(t)$, and $v_C(t)$, respectively, *can't be added algebraically as mere numbers*; in other words, we have to take into account their phase relationships and add them as vectors. Phasor diagrams help us do this. Obviously, the same relationships are true for currents.

2.3.2.4 Impedances and Phasors

Impedances From a basic course in circuit analysis, we certainly remember that the *impedance, Z, of a circuit element is determined as its reactance and associated phase shift*. That is, the impedances of ohmic resistance, capacitive reactance, and inductive reactance are given by

$$\mathbf{Z}_R = R\angle 0°$$
$$\mathbf{Z}_L = X_L \angle 90°$$
$$\mathbf{Z}_C = X_C \angle -90° \tag{2.3.31}$$

Using phasor or complex-number graphical presentations, we can build the impedance diagram shown in Figure 2.3.13. In this diagram, two applications of the impedance concept are demonstrated. First, it allows us to compute the circuit's total impedance, $\mathbf{Z}_T$. Figure 2.3.13 shows how

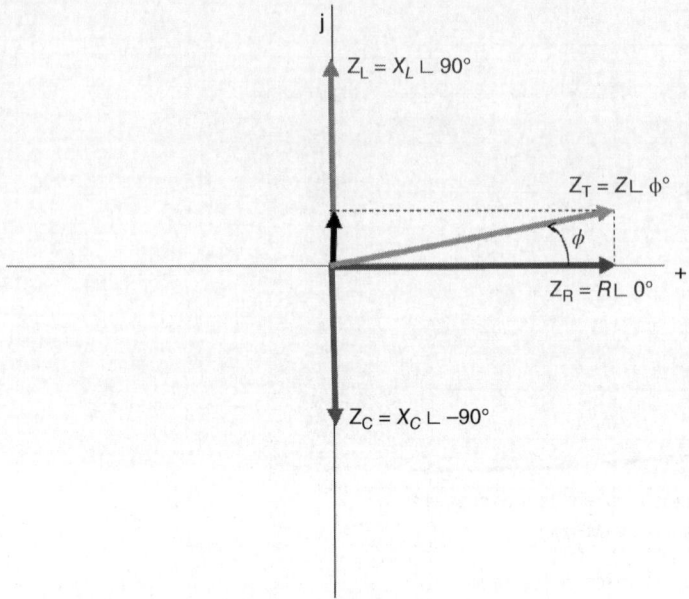

Figure 2.3.13 RLC impedance diagram.

this total impedance obtained graphically. Note that for any ac circuit, *the angle, Φ, associated with total impedance, Z_T, is the angle by which the voltage source leads the current source* (Boylestad 2016, p. 589).

Second, using impedance, we can apply Ohm's law to an ac circuit; in fact, for pure resistive, inductive, and capacitive circuits, we can write

$$\mathbf{I}_R = \mathbf{V}_R/\mathbf{Z}_R = V\angle 0°/R\angle 0° = (V/R)\angle 0°$$
$$\mathbf{I}_L = \mathbf{V}_L/\mathbf{Z}_L = V\angle 0°/X_L\angle 90° = (V/X_L)\angle -90°$$
$$\mathbf{I}_C = \mathbf{V}_C/\mathbf{Z}_C = V\angle 0°/(X_C\angle -90°) = (V/X_C)\angle 90° \quad (2.3.32)$$

Formula (2.3.32) helps us better understand the phasor diagrams pictured in Figures 2.3.7c, 2.3.9c, and 2.3.11c. These formulas and figures show that in a pure resistive circuit, the current and the voltage are in phase; in a pure inductive circuit, the voltage leads the current by 90° (or, conversely, the current lags the voltage by 90°); and in a pure capacitive circuit, the current leads the voltage by 90°. The impedances of these components must have the phase shift shown in (2.3.32).

Note that the phasors representing impedance **Z**, voltage **V**, and current **I** are given here in polar forms, which means they are not associated with any rotation. In general, the impedance, **Z**, cannot be a rotating vector by its very nature, but phasors **V** or **I** could be. Therefore, be careful when applying the phasor-based technique to the analysis of electrical circuits.

We summarize our discussion of phasors and impedances in Example 2.3.2.

Example 2.3.2 Analysis of an RLC Circuit.

Problem

Given an RLC series circuit with $R = 6\,\Omega$, $X_L = 8\,\Omega$, and $X_C = 4\,\Omega$ and the voltage source $v(t)$ (V) $= 10\sin(\omega t)$, find the current flowing through – and the voltage drop across – each component in time domain.

Solution

The circuit diagram is shown in Figure 2.3.14.

The total impedance, $\mathbf{Z}_T$, of this circuit is given by

$$\mathbf{Z}_T = \mathbf{Z}_R + \mathbf{Z}_L + \mathbf{Z}_C = R\angle 0° + X_L\angle 90° + X_C\angle -90°$$
$$= 10\angle 0° + 6\angle 90° + 8\angle -90°$$
$$= 10 + j6 - j8 = 10 - j2 = 10.2\angle -11.3°$$

The impedance diagram of this circuit is shown in Figure 2.3.15.
Current flowing through this circuit is given by

$$\mathbf{I} = \mathbf{E}/\mathbf{Z}_T = (7.07\angle 0°)/(10.2\angle -11.3°) = 0.69\angle 11.3°$$

Thus, voltages are given by

$$\mathbf{V}_R = \mathbf{I} \times \mathbf{Z}_R = (0.69\angle 11.3°) \times (10\angle 0°) = 6.9\angle 11.3°$$
$$\mathbf{V}_L = \mathbf{I} \times \mathbf{Z}_L = (0.69\angle 11.3°) \times (6\angle 90°) = 4.1\angle 101.3°$$
$$\mathbf{V}_C = \mathbf{I} \times \mathbf{Z}_C = (0.69\angle 11.3°) \times (8\angle -90°) = 5.5\angle -78.7°$$

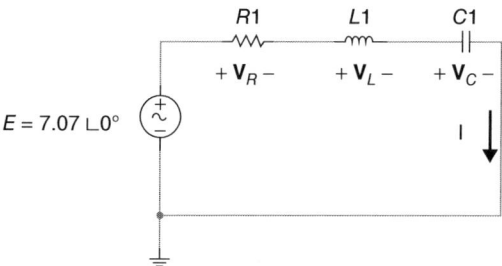

Figure 2.3.14 Series RLC circuit for Example 2.3.2.

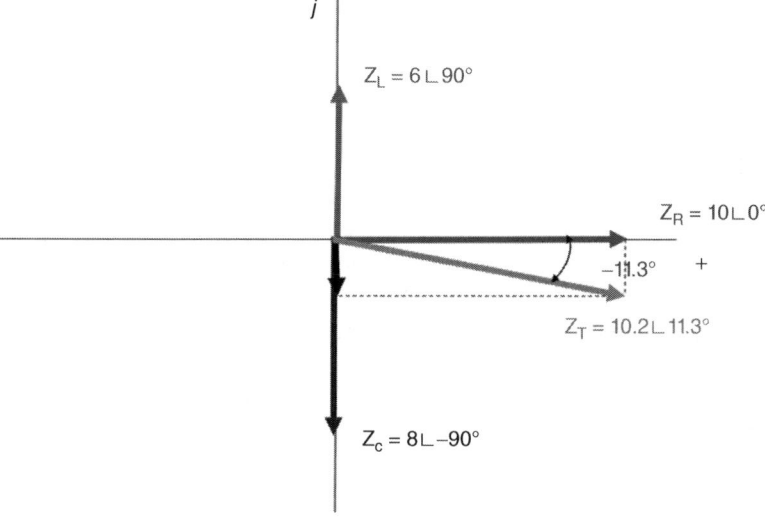

Figure 2.3.15 Impedance diagram for Example 2.3.2.

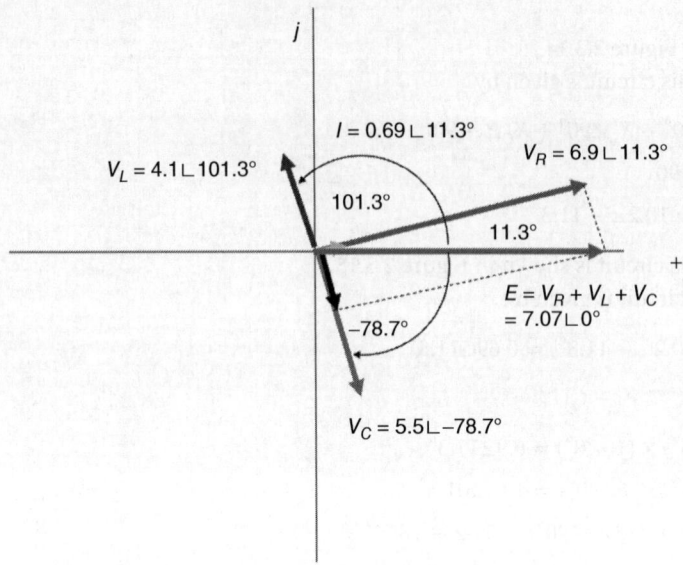

Figure 2.3.16 Phasor diagram for Example 2.3.2.

The phasor diagram for this circuit is shown in Figure 2.3.16.
To verify that our solution is correct, we can apply Kirchhoff's voltage law in phasor domain:

$$\mathbf{E} = \mathbf{V}_R + \mathbf{V}_L + \mathbf{V}_C$$

From the phasor diagram, we can easily compute two orthogonal components of the phasor $\mathbf{E}$: They are $\mathbf{V}_R = 6.9 \angle 11.3°$ and $(\mathbf{V}_C - \mathbf{V}_L) = 1.4 \angle -78.7°$; therefore, the amplitude of phasor $\mathbf{E}$ is equal to

$$E = \sqrt{(6.9^2 + 1.4^2)} = 7.04 \text{ V}$$

which is close to the assigned value: 7.07 V. The phase angle of phasor $\mathbf{E}$ with respect to phasor $\mathbf{V}_R$ is equal to

$$\angle[\tan^{-1}(-1.4/6.9)] = \angle -11.3°$$

Thus, phasor $\mathbf{E}$, with respect to the coordinate system, is given by

$$\mathbf{E} \approx 7.04 \angle 0°$$

As Figure 2.3.16 shows, phasor $\mathbf{E}$ lies along the positive real axis and its computed value is approximately equal to the assigned value. This verifies our solution.

All the above computations and graphical presentations show that the direction of current $\mathbf{I}$ coincides with that of $\mathbf{V}_R$, as theory predicts.

To complete this example, we need to find time-varying voltages. This can be done by converting corresponding phasors into time-domain quantities. We have:

$\mathbf{V}_R = 6.9 \angle 11.3° \rightarrow v_R(t) = 6.9 \sin(\omega t + 11.3°)$
$\mathbf{V}_L = 4.1 \angle 101.3° \rightarrow v_L(t) = 4.1 \sin(\omega t + 101.3°)$
$\mathbf{V}_C = 5.5 \angle -78.7° \rightarrow v_R(t) = 5.5 \sin(\omega t - 78.7°)$
These computations provide the answer to our problem.

Discussion

To verify your solution to problems like this one, compare your solution to the problem with the theory considered above. Such a comparison will give you a clear understanding of the material presented in this subsection.

Questions and Problems for Section 2.3

- Questions marked with an asterisk require a systematic approach to finding the solution.
- Many questions and problems, including those marked with an asterisk, imply that you, in addition to reading the textbook, will do your research to find the answers. Consider such questions as miniprojects.

Revisiting the Waveforms

1. Define a waveform. Give an example that satisfies your definition.

2. Consider sinusoidal signal $v(t) = A \sin(\omega t + \Theta)$:
   a) What three quantities do completely describe this signal?
   b) Are these quantities constants or variables? Explain.

3. What is the role of a waveform in analog transmission?

4. In analog transmission, the text states, a message is delivered by changing a parameter of an analog carrier signal. Taking a sinusoidal signal as the example of a carrier, explain how this concept can be implemented.

5. The text states that "… the waveform is a mathematical model of a real signal." Is a waveform only a model, what is a real signal? How can we describe it?

6. Consider the classification of waveforms:
   a) What independent variables can a waveform be a function of? Give examples.
   b) Define the main difference between deterministic and random waveforms. Give examples.
   c) Mathematically define a continuous waveform and a discrete waveform. Give examples.
   d) Can a discrete waveform be a periodic one? Prove your answer.

7. It seems that improving the quality of an analog transmission can be achieved by simply increasing a signal's power, but this is not the case if we take into account the effect of noise. What parameter related to power truly determines the quality of transmission.

8. Given $v(t)$ (V) $= 12 \cos(100t)$:
   a) Compute its average normalized power.
   b) Why is it called "normalized power"?

9. Given $v(t)$ (V) $= 10 + 12 \cos(100t)$:
   a) Without calculations, tell what is the average value of this signal?
   b) Prove your answer by sketching the signal's graph.
      i) Is the signal $v(t)$ (V) $= 12 \cos(100t)$ power or energy signal? Prove your answer.

c) *Determine whether the signal $v(t) = e^{-t} \cdot u(t)$ is power or energy.[6] (Here $u(t)$ is a Heaviside (step) function equals to zero for $t < 0$ and to 1 for $t > 0$.)

Waveforms and Phasors

10 Can any waveform be realized in practice? Explain.

11 List and comment on the conditions to build practically realizable waveform.

12 Consider the conditions of practically realizable waveforms:
 a) Can a nonperiodic waveform be practically realized? Explain.
 b) A discrete signal, as the text explains, must have at least one discontinuity point. How does this condition affect the ability of a waveform to be practically realizable?
 c) Can $v(t) = Ae^{j\Theta}$ be realized in practice? Explain.

13 Consider phasor:
 a) What is a phasor? Explain. Give an example.
 b) What phasor forms do you know? Explain.

14 What is the relationship between the waveform and a phasor of a sinusoidal signal? Explain, using an example.

15 Present the signal $v(t)$ (V) $= 12 \sin(628t + 60°)$ in phasor domain.

16 Given that a sinusoidal signal has constant amplitude, constant frequency, and a constant phase shift, what options for presenting the signal in phasor domain do we have? Sketch graphs to clearly explain your answer.

17 Given sinusoidal signal $v(t) = 3.3 \sin(2\pi 5000t + 60°)$:
 a) Illustrate the relationship between the given signal and its rotating phasor.
 b) Show this signal in a polar form and in a phasor form at $t = 0.02$ ms.

18 Given a phasor **A** $= 5.0 \angle 45°$ rotating at angular velocity $\omega = 6280$ rad/s:
 a) Write the formula for the corresponding sinusoidal signal.
 b) What is the angular frequency of this signal?

Applications of Phasor Presentation to Analysis of Electronic Communications Circuitry

19 Given the voltage drop across a resistor is $v_R = 12 \sin(120t)$, sketch its and the current's waveforms; also, depict the phasor diagram for V_R and I_R.

20 Given the voltage drop across an inductor is $v_L = 12 \sin(120t)$, sketch its and the current's waveforms; also, depict the phasor diagram for V_L and I_L.

21 Given the voltage drop across a capacitor is $v_C = 12 \sin(120t)$, sketch its and the current waveforms; also, depict the phasor diagram for V_C and I_C.

---

6 https://matel.p.lodz.pl/wee/i12zet/Signal%20energy%20and%20power.pdf.

22 In one set of axes, depict the graphs of resistance, inductive, and capacitive reactances vs. frequency. Write down the formulas and sketch the graphs for $R(f)$, $X_C(f)$, and $X_L(f)$. Explain the course of the graph of each reactance vs. frequency.

23 Consider a series RLC circuit. Compare the graphs of the waveforms of v_R, v_L, and v_C. Write the formulas for every waveform. Explain why v_L, and v_C have different initial phase shifts with respect to v_R.

24 Given voltage drops across a resistor, an inductor, and a capacitor are $v_R = 12\sin(120t)$, $v_L = 12\sin(120t)$, and $v_C = 12\sin(120t)$, find their sum. Depict all these quantities in a phasor diagram. (*Hint*: See Appendix 2.3.A.)

25 Given an RLC series circuit with voltage source $v(t) = 8\sin\omega t$ and $R = 12\,\Omega$, $X_L = 16\,\Omega$ and $X_C = 8\,\Omega$:
   a) Find the current flowing through, and the voltage drop across, each component in time domain. Sketch the phasor diagram for the circuit.
   b) Find the total impedance of the circuit. Draw the circuit and impedance diagrams.
   c) How can you prove that your solution to this problem is correct? Explain your reasoning and show your calculations.

2.3.A Mathematical Foundation of Phasor Presentation

2.3.A.1 Phasors and Complex Numbers

The mathematical foundation of a phasor's presentation lies in what we call *complex-number analysis*. Even though today students commonly use scientific calculators to perform computations with complex numbers, solid mathematical foundation underpinning these manipulations should be in the background of every electrical engineer. This appendix provides a brief introduction to the mathematics of complex numbers.

You will recall that a complex number, z, consists of a real part, x, and an imaginary part, jy, as (2.3.A.1) shows:

$$z = x \pm jy \tag{2.3.A.1}$$

where $j = \sqrt{-1}$ is an imaginary unit. Formula (2.3.A.1) represents a complex number in *rectangular* form; another popular form is called *polar* (sound familiar?) form, which is shown in (2.3.A.2):

$$z = c\angle\Phi \tag{2.3.A.2}$$

where

$$c = \sqrt{(x^2 + y^2)} \tag{2.3.A.3a}$$

and

$$\Phi = \tan^{-1}(y/x) \tag{2.3.A.3b}$$

This presentation is shown in Figure 2.3.A.1, which is self-explanatory.
We should also recall Euler's identity:

$$e^{\pm j\Phi} = \cos\Phi \pm j\sin\Phi \tag{2.3.A.4}$$

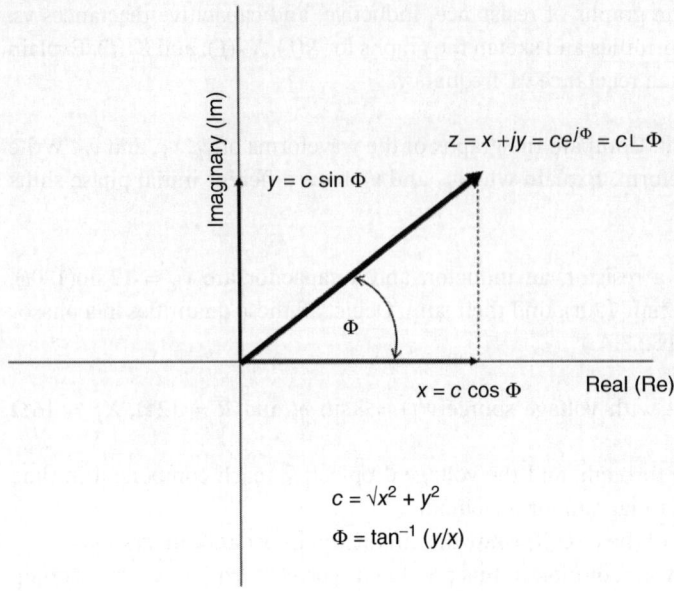

Figure 2.3.A.1 A complex number and its rectangular and polar forms.

If we convert Euler's identity from its rectangular form, given in (2.3.A.1), into the polar form, given in (2.3.A.2), we obtain

$$e^{\pm j\Phi} = \cos\Phi \pm j\sin\Phi = \sqrt{((\cos\Phi)^2 + (\sin\Phi)^2)}\,(\pm\angle\Phi) = \pm\angle\Phi$$

Therefore, an important corollary of Euler's identity is

$$e^{\pm j\Phi} = \pm\angle\Phi \qquad (2.3.A.5)$$

which allows us to write a phasor's polar form in two identical ways:

$$z = c\angle\Phi \qquad (2.3.A.6a)$$

and

$$z = ce^{j\Phi} \qquad (2.3.A.6b)$$

Equations (2.3.A.6a) and (2.3.A.6b) emphasize, in effect, the mathematical interpretations of a phasor as a polar having an amplitude, c, and a phase shift, Φ, and as a complex exponential function with a given amplitude, c, and a function, $e^{j\Phi}$. We refer you again to Figure 2.3.A.1 for an illustration of this formula.

An important consequence of Euler's identity is the possibility of presenting a sinusoid by means of an exponential function. Indeed, it is easy to derive the following presentations from (2.3.A.4):

$$\cos\Phi = \tfrac{1}{2}(e^{j\Phi} + e^{-j\Phi})$$

and

$$\sin\Phi = \tfrac{1}{2}(e^{j\Phi} - e^{-j\Phi}) \qquad (2.3.A.7)$$

The significance of (2.3.A.7) becomes apparent when differentiation and integration are involved: We know that the exponential function does not change under these operations.

We should recall, also, the basic operations with complex numbers, such as addition and subtraction; for example,

$$z_1 \pm z_2 = (x_1 \pm x_2) + j(y_1 \pm y_2) \tag{2.3.A.8}$$

To add phasors in polar form, we need to convert them into rectangular form, then add and convert them back into polar form. Multiplication and division can be done more easily in polar form:

$$z_1 z_2 = (c_1(e^{j\Phi_1}))(c_2(e^{j\Phi_2})) = c_1 c_2 (e^{j(\Phi_1 + \Phi_2)}) \tag{2.3.A.9}$$

and

$$z_1/z_2 = (c_1(e^{j\Phi_1}))/(c_2(e^{j\Phi_2})) = c_1/c_2 (e^{j(\Phi_1 - \Phi_2)}) \tag{2.3.A.10}$$

In fact, (2.3.A.9) and (2.3.A.10) are well-known laws of exponential functions. In future sections, we will be referring to other laws; the two most often used are

$$(e^a)^b = e^{ab} \tag{2.3.A.11}$$

and

$$1/(e^a) = e^{-a} \tag{2.3.A.12}$$

Of course, multiplication and division can also be done in *rectangular form*, but then we need to apply the *conjugate theorem*,

$$z \cdot z^* = (x + jy)(x - jy) = x^2 + y^2 \tag{2.3.A.13}$$

where $z^* = (x - jy)$ is a *complex conjugate* (or, simply, a *conjugate*) to the complex number $z = (x + jy)$.
Having (2.3.A.13), the multiplication becomes straightforward,

$$z_1 z_2 = (x_1 + jy_1)(x_2 + jy_2) = (x_1 x_2 - y_1 y_2) + j(x_1 y_2 + x_2 y_1) \tag{2.3.A.14}$$

whereas division requires converting a denominator into a real number by multiplying both the numerator and the denominator by a complex conjugate to the denominator:

$$\begin{aligned}
z_1/z_2 &= (x_1 + jy_1)/(x_2 + jy_2) \\
&= (x_1 + jy_1)(x_2 - jy_2)/(x_2 + jy_2)(x_2 - jy_2) \\
&= ((x_1 x_2 + y_1 y_2) + j(x_2 y_1 - x_1 jy_2))/(x_2^2 + y_2^2) \\
&= (x_1 x_2 + y_1 y_2)/(x_2^2 + y_2^2) + j(x_2 y_1 - x_1 jy_2)/(x_2^2 + y_2^2)
\end{aligned}$$

Thus, the result of such a division is another complex number,

$$z_1/z_2 = (x_1 + jy_1)/(x_2 + jy_2) = a + jb \tag{2.3.A.15}$$

where $a = (x_1 x_2 + y_1 y_2)/(x_2^2 + y_2^2)$ and $b = (x_2 y_1 - x_1 y_2)/(x_2^2 + y_2^2)$.

Example 2.3.A.1 Polar Forms and Complex Numbers

Problem

You are given two complex numbers: $z_1 = 3 - j4$ and $z_2 = 6 + j8$.

a) Find their polar forms and depict them in a complex plane.
b) Determine their sum, difference, product, and quotient.
c) Find the value of $(z_1)^3$ and its reciprocal.

Solution

a) The relationship between the rectangular and polar forms is given by (2.3.A.2). Applying this formula, we compute $c_1 = \sqrt{(3^2 + 4^2)} = 5$ and $\Phi_1 = \tan^{-1}(4/3) = -0.93$ rad = $-53.1°$. However, since our angle is in the fourth quadrant, we have to compute $\Phi_1 = 360° - 53.1° = 306.9°$. Let us consider this matter in our discussion of this example. Hence,

$$z_1 = c_1 \mathbf{L} \Phi_1 = 5 \mathbf{L} 306.9°$$

Repeating these calculations for z_2, we obtain

$$z_2 = c_2 \mathbf{L} \Phi_2 = 10 \mathbf{L} 53.1°$$

Also, referring to (2.3.A.6b), we can present these polar forms as follows:

$$z_1 = c_1 e^{j\Phi_1} = 5e^{j306.90} \text{ and } z_2 = c_2 e^{j\Phi_2} = 10e^{j53.10}$$

Figure 2.3.A.2 shows these two complex numbers.

b) The sum of these two complex numbers can be found by applying (2.3.A.8), which, in rectangular form, yields

$$z_1 + z_2 = (3 - j4) + (6 + j8) = (3 + 6) + (-j4 + j8) = (9 + j4)$$

and, in polar form, yields

$$z_1 + z_2 = 9.9 e^{j240} = 9.9 \mathbf{L} 24°$$

Similarly, applying (2.3.A.8), we find the difference as follows:

$$z_1 - z_2 = (3 - j4) - (6 + j8) = (3 - 6) + (-j4 - j8) = (-3 - j12)$$

and

$$z_1 - z_2 = 12.4 e^{j2560} = 12.4 \mathbf{L} 256°$$

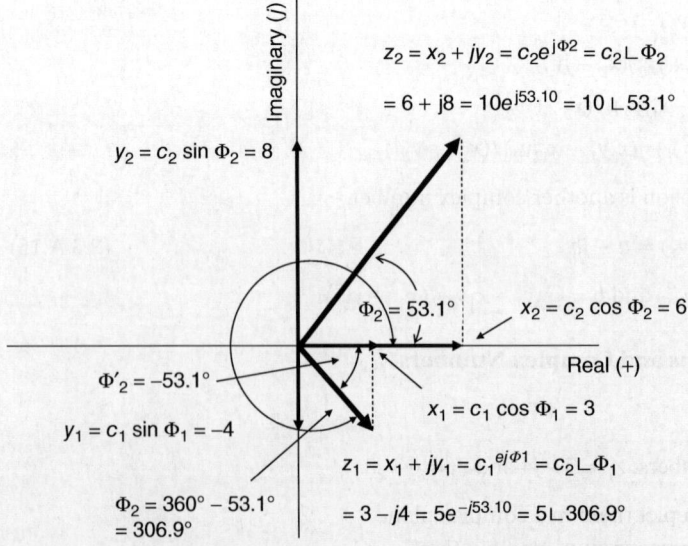

Figure 2.3.A.2 Two complex numbers examined in Example 2.3.A.1.

The product of these complex numbers can be found by applying either (2.3.A.9) or (2.3.A.14). Of course, finding this product in a polar form is much easier:

$$z_1 z_2 = c_1 c_2 e^{j(\Phi_1 + \Phi_2)} = (5 \times 10) e^{j(306.9.10 + 53.10)} = 50 e^{j0} = 50$$

Observe that the result of multiplying *these* two complex numbers is a real number even though they are not a complex conjugate. Can you explain why?

The quotient, again, can be readily obtained by applying (2.3.A.10), which yields the following:

$$z_1 / z_2 = c_1 / c_2 e^{j(\Phi_1 - \Phi_2)} = (5/10) e^{j(306.9 \times 10 - 53 \times 10)} = 0.5^{j253.80}$$

To find the power of 3 of z_1, we had better use a polar form, which allows us to apply (2.3.A.11). Doing so, we get

$$(z_1)^3 = (c_1 e^{j\Phi_1})^3 = (c_1)^3 e^{j3\Phi_1} = 125 e^{-j159.30}$$

Its reciprocal, $1/(z_1)^3$, is determined by (2.3.A.10) and (2.3.A.12):

$$1/(z_1)^3 = 1/(c_1^3 e^{j3\Phi_1}) = (1/125) e^{-j159.30} = 0.008 e^{-j159.30}$$

All these operations are straightforward, but the calculation of a correct phase angle requires a brief discussion.

Discussion

- First, as mentioned in our explanation of (2.3.30), we can obtain the polar form of the sum of and the difference between two complex numbers only by converting their rectangular forms. This is exactly what we did in this example.
- Second, we need to understand that by applying (2.3.A.3b), $\Phi = \tan^{-1}(y/x)$, we obtain an *acute angle adjacent to the x axis*. In mathematics, this is called a "reference angle;" however, the phase angle, Φ, of a complex number must be counted from the positive real axis; therefore, if our complex number is in the second, third, or fourth quadrant, we have to follow the rules presented in Table 2.3.A.1 to obtain the correct answer. The reason our calculator gives us an ambiguous answer is that the tangent is a periodic function. This explanation can be readily understood if we recall that, by definition, $\tan \Phi = y/x = c \sin \Phi / c \cos \Phi$; therefore, we achieve the correct answer from direct computation only when both x and y are positive. However, when x or y is negative, we need to examine the result of our computation to obtain the correct answer. Figure 2.3.A.3 and Table 2.3.A.1 present explanations, general rules, and an example of such calculations.

2.3.A.2 Applications of Phasor Presentation to the Analysis of Electronic Communications Circuitry

2.3.A.2.1 Summation of Signals

We need, at this point, to apply the mathematical foundation of the phasor representation discussed above to the analysis of circuits used in telecommunications. Let us use Euler's formula (see Eq. (2.3.A.4)) for a sinusoidal signal:

$$A e^{j(\omega t + \Theta)} = A \cos(\omega t + \Theta) + jA \sin(\omega t + \Theta) \quad (2.3.A.16)$$

This formula shows that the cosine member represents a real part of the complex entity $A e^{j(\omega t + \Theta)}$, and the sine member represents an imaginary part of this entity – that is,

$$A \cos(\omega t + \Theta) = \text{Re}(A e^{j(\omega t + \Theta)})$$

Table 2.3.A.1 Phase angles and complex numbers.

Quadrant	Signs of x and y	Angle $\Phi = \tan^{-1}(y/x)$	Example
I	+x	$0° \leq \Phi \leq 90°$	$x = 3, y = 12 \Rightarrow \tan \Phi = y/x = 4$
	+y	$\Phi = \tan^{-1}(y/x)$	$\Rightarrow \Phi = 76°$
II	+x	$90° \leq \Phi \leq 180°$	$x = -3, y = 12 \Rightarrow \tan \Phi = y/x = -4$
	-y	$\Phi = 180° - \tan^{-1}(y/x)$	$\Rightarrow \Phi = 180° - 76° = 104°$
	-x	$180° \leq \Phi \leq 270°$	$x = -3, y = -12 \Rightarrow \tan \Phi = y/x = 4$
	-y	$\Phi = 180° + \tan^{-1}(y/x)$	$\Rightarrow \Phi = 180° + 76° = 256°$
	+x	$270° \leq \Phi \leq 360°$	$x = 3, y = -12 \Rightarrow \tan \Phi = y/x = -4$
	-y	$\Phi = 360° - \tan^{-1}(y/x)$	$\Rightarrow \Phi = 360° - 76° = 284°$

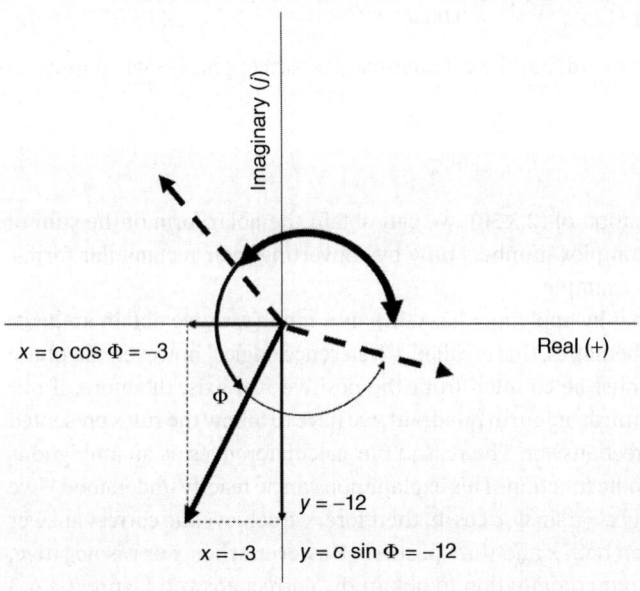

Figure 2.3.A.3 The sum of and the difference between two complex numbers in Example 2.3.A.1.

and

$$A \sin(\omega t + \Theta) = I_m(A e^{j(\omega t + \Theta)}) \qquad (2.3.A.17)$$

Next, let us consider a sinusoidal signal in exponential form. Using (2.3.A.9), we readily obtain

$$A e^{j(\omega t + \Theta)} = A e^{j\Theta} e^{j\omega t} \qquad (2.3.A.18)$$

Formula (2.3.A.18) clearly shows that a sinusoidal signal can be presented in a complex plane in two ways:

(1) as a steady-state phasor with amplitude A and initial phase Θ (the polar form) of the signal,

$$\mathbf{A} = A e^{j\Theta} \qquad (2.3.A.19)$$

and

(2) as a unit vector rotating with angular velocity, ω,

$$e^{j\omega t}$$

Formula (2.3.A.18) explains why we said in Section 2.3.2.2 that we have two options – time-varying, $e^{j\omega t}$, and steady-state, $Ae^{j\Theta}$ – in building a phasor presentation. Since in circuit analysis we use a steady-state presentation (that is, a polar form), we omit the frequency component and the phasor takes the form

$$\mathbf{A} = Ae^{j\Theta} \equiv A\angle\Theta \qquad (2.3.A.20)$$

In ac circuits, we use the rms amplitude, V_{rms}, more often than the peak value, V_m. (We remind you that $V_{rms} = V_m/\sqrt{2}$ or, conversely, $V_m = V_{rms} \times \sqrt{2}$.) Thus, the amplitude, A, in (2.3.A.20) can be A_{rms} by default. We must be sure to clarify what value – peak or rms – is used in a specific formula.

The main purpose of introducing phasors is to simplify the mathematical and graphical analysis of sinusoidal waveforms. If, for example, we need to add two sinusoids with the same frequency,

$$v_1(t) = A_1 \sin(\omega t + \Theta_1)$$

and

$$v_2(t) = A_2 \sin(\omega t + \Theta_2)$$

to obtain

$$v(t) = v_1(t) + v_2(t)$$

we can represent them in phasor domain as

$$\mathbf{A_1} = A_1 e^{j\Theta_1} \quad \text{and} \quad \mathbf{A_2} = A_2 e^{j\Theta_2}$$

add them as

$$\mathbf{A} = \mathbf{A_1} + \mathbf{A_2} = A_1 e^{j\Theta_1} + A_2 e^{j\Theta_2}$$

and convert the result back into time domain. However, to obtain this sum in explicit form, we need to take several intermediate steps. Example 2.3.A.2 shows these steps and the entire procedure.

Example 2.3.A.2 Summation of Two Sinusoidal Signals

Problem

Two sinusoidal signals,

$$v_1(t) = 12 \sin(200t + 60°) \text{ and } v_2(t) = 7 \cos(200t - 120°)$$

are presented to a circuit for summation. Find the output voltage.

Solution

Clearly, the output voltage is

$$v(t) = v_1(t) + v_2(t)$$

As a preliminary step, we need to convert cosine to sine (or vice versa). Since

$$\cos(x) = \sin(x + 90°)$$

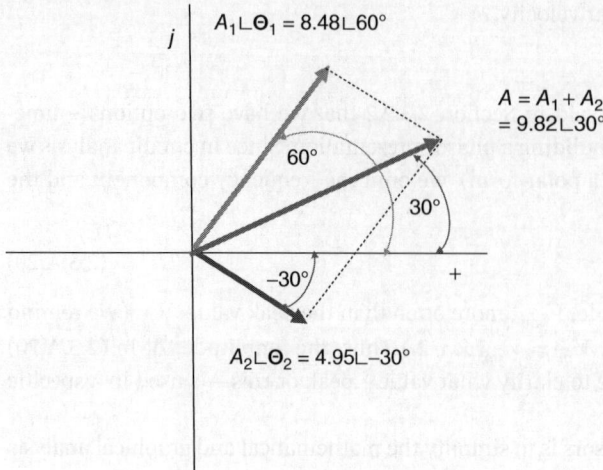

Figure 2.3.A.4 Summation of two phasors in Example 2.3.A.2.

we obtain

$$v_2(t) = 7 \cos(200t - 120°) = 7 \sin(200t - 30°)$$

The main part of the summation of two sinusoidal signals is based on converting the waveforms to phasors:

$$v(t) = A_m \sin(\omega t + \Theta) \rightarrow \mathbf{V} = A_{rms} \angle \Theta \qquad (2.3.A.21)$$

In our example, we get

$$v_1(t) = 12 \sin(200t + 60°) \Rightarrow \mathbf{A_1} = 8.48 \angle 60°$$

and

$$v_2(t) = 7 \sin(200t - 30°) \Rightarrow \mathbf{A_2} = 4.95 \angle -30°$$

Note that we have chosen to use the *rms* values of the amplitudes; that is, $A_{rms} = A/\sqrt{2} \equiv 0.707 \times V$. This is why the phasors' amplitudes are equal to $0.707 \times 12\,V = 8.48\,V$ and $0.707 \times 7\,V = 4.95\,V$, respectively.

Now we need to add these phasors. A graphical summation of these phasors is shown in Figure 2.3.A.4.

The only way to attain this sum analytically is to present each phasor as a sum of real and imaginary parts, add these parts, and turn the obtained complex number back into phasor form. This procedure can be undertaken through the following steps:

1. To represent each of these phasors as the sum of the real and imaginary parts, we need to use Euler's identity (see Eq. (2.3.A.7)). We obtain

$$\mathbf{A_1} = A_1 \angle \Theta_1 \equiv A_1 e^{j\Theta_1} = A_1 \cos \Theta_1 + jA_1 \sin \Theta_1$$

and

$$\mathbf{A_2} = A_2 e^{j\Theta_2} = A_2 \cos \Theta_2 + jA_2 \sin \Theta_2$$

2. To determine the phasor's sum in rectangular form, we add the real parts (cosines) and the imaginary parts (sines) as shown:

$$\mathbf{A} = \mathbf{A_1} + \mathbf{A_2} = A_1 \cos \Theta_1 + A_2 \cos \Theta_2 + j(A_1 \sin \Theta_1 + A_2 \sin \Theta_2)$$

2.3.A Mathematical Foundation of Phasor Presentation

3. To obtain the phasor form of the above sum, we convert the rectangular form into the phasor by applying (2.3.A.2):

$$\mathbf{A} = A_1 \cos \Theta_1 + A_2 \cos \Theta_2 + j(A_1 \sin \Theta_1 + A_2 \sin \Theta_2) \Rightarrow A \llcorner \Theta$$

4. To get the resultant sinusoidal signal, we need to convert the obtained phasor into a waveform:

$$\mathbf{A} = A \llcorner \Theta \Rightarrow v(t) = A \sin(\omega t + \Theta)$$

Let us implement these steps:

1. We first plug the numbers used in this example into the above formulas and find

$$\mathbf{A_1} = 8.48 \llcorner 60° = 8.48 \cos 60° + j8.48 \sin 60°$$

and

$$\mathbf{A_2} = 4.95 \llcorner -30° = 4.95 \cos 30° - j4.95 \sin 30°$$

2. To obtain the real and imaginary parts in numerical format, we compute the values of the cosines and sines:

$$\mathbf{A_1} = 8.48 \cos 60° + j8.48 \sin 60° = 4.24 + j7.34$$

and

$$\mathbf{A_2} = 4.95 \cos 30° - j4.95 \sin 30° = 4.29 - j2.48$$

3. Now we add the real and imaginary parts of two vectors:

$$\mathbf{A} = \mathbf{A_1} + \mathbf{A_2} = (4.24 + j7.34) + (4.29 - j2.48) = 8.53 + j4.86$$

This is the sum we are looking for of the two sinusoidal signals in phasor rectangular form.

4. We are ready to convert the rectangular form into the polar form by using (2.3.A.3a) and (2.3.A.3b):

$$\mathbf{A} = \mathbf{A_1} + \mathbf{A_2} = 8.53 + j4.86 = \sqrt{((8.53)^2 + (4.86)^2)} \llcorner (\tan^{-1}(4.86/8.53)) = 9.82 \llcorner 30°$$

This phasor is shown in Figure 2.3.A.4 as the sum of two constituents.

5. The last step in finding the sum of the two sinusoidal signals is to transfer the result obtained from phasor domain to time domain. To convert the phasor $\mathbf{A}$ into a sinusoidal form, we can employ one of the following methods:

- We can use direct conversion. In our example, we can write

$$\mathbf{A} = \mathbf{A_1} + \mathbf{A_2} = 9.82 \angle 30°$$
$$\rightarrow v(t) = v_1(t) + v_2(t) = \sqrt{2} \times 9.82 \sin(200t + 30°) = 13.89 \sin(200t + 30°)$$

- We can take the real part of a phasor for the cosine and the imaginary part for the sine. Here, we need the sine and therefore

$$\text{Im}(9.82 e^{j30}) = 9.82 \sin(30°)$$

Returning to our problem, we reinstate the frequency and peak value of the amplitude and obtain

$$\sqrt{2} \times 9.82 \sin(200t + 30°)$$

which gives us the answer:

$$v(t) = v_1(t) + v_2(t) = 13.89 \sin(200t + 30°)$$

Either way, we will find the signal in time domain.

Discussion

- The set of operations needed to add or subtract two sinusoids seems like a drawn-out, time-consuming process, but as soon as we get used to it, we will do it very quickly. Most up-to-date scientific calculators can do much of this work for us in a couple of simple steps. As for the need to convert the polar form to the rectangular form for doing the addition or subtraction, bear in mind that in the precomputer era, we did not have any other means to add the sinusoidal signals in time domain. We could only make a graphical addition, which was (and still is) much more time-consuming and cumbersome than operating in phasor domain. Of course, today, we can assign this task to our computer and add as many sinusoids as we need by using such tools as MS Excel, MATLAB, or Mathematica.
- To prove that the addition of two sinusoids with the same frequency can be done easily by using phasors, consider one more exercise in determining the sum of two signals:

$$v_1(t) = 10 \sin(50t + 60°) \quad \text{and} \quad v_2(t) = 4 \sin(50t + 30°)$$

Following the procedure discussed in detail above, in this exercise we calculate

$$\mathbf{A} = \mathbf{A}_1 + \mathbf{A}_2 = 10\angle 60° + 4\angle 30°$$
$$= 5 + j0.87 + 0.87 + j2 = 5.87 + j2.87 = 6.53\angle 26°$$

After converting this phasor into a sinusoidal signal, we get the sum as

$$v(t) = v_1(t) + v_2(t) = 6.53 \sin(50t + 26°)$$

Note that we use peak values of amplitudes. This exercise should convince us that the computations needed for the summation of two sinusoids of the same frequency are really easy and simple with the use of the polar form.

- Obviously, everything said about the relationship between the time and phasor domains for the voltage holds true as well for the current.

To bone up more thoroughly on phasors and phasor diagrams, refer to a textbook on circuit analysis.

Optional: Questions and Problems for Appendix 2.3.A

1. Given complex number, $z = 3 + j5$, find this number in polar form and depict it in a complex plane.

2. Given complex number, $z = 4\angle 270°$, depict it in a complex plane. Find it in a rectangular form.

3. Given two complex numbers, $z_1 = 3 + j4$ and $z_2 = 6 - j8$, find their sum. Depict the components and the sum in a complex plane.

4. Given two complex numbers, $z_1 = 3 + j4$ and $z_2 = 6 - j8$, find their difference. Depict the components and the difference in a complex plane.

5. Given two complex numbers, $z_1 = 3 + j4$ and $z_2 = 6 - j8$, find their product. Depict the components and the product in a complex plane.

6. Given two complex numbers, $z_1 = 3 + j4$ and $z_2 = 6 - j8$, find their quotient. Depict the components and the quotient in a complex plane.

7 Given complex number, $z = 4 + j3$, find the reciprocal to $(z)^2$.

8 Given complex number, $z = 4 + j3$, find its polar and exponential forms.

9 Given complex number, $z = 4 + j3$, find its cosine and sine forms.

10 Write down the Euler identity. What do we need this formula for? Explain.

11 Find the output voltage of a circuit that performs the summation, $v(t) = v_1(t) + v_2(t)$, and depict this sum and its constituents in the phasor domain for the following signals:
   a) $v_1(t) = 6\sin(300t - 60°)$ and $v_2(t) = 8\cos(300t + 150°)$.
   b) $v_1(t) = 9\sin(60t - 60°)$ and $v_2(t) = 3\sin(60t - 30°)$.

3

Digital Signals and Digital Transmission

Objectives and Outcomes of Chapter 3

Objectives

- To learn how to present and transmit analog information in digital format.
- To gain an understanding of the main parameters of a digital signal.
- To get to know the main parameters of a digital signal and their relationship to the characteristics of digital transmission.

Outcomes

- Know the advantages of digital signals over the analog signals in transmitting information.
- Describe the parameters of digital signals and the characteristics of digital transmission.
- Demonstrate the relationship between the digital signal's parameters on the one hand and the characteristics of digital transmission on the other.

3.1   Digital Communications – Basics

Objectives and Outcomes of Section 3.1

Objectives

- To learn how to present and transmit analog information in digital format.
- To get to know the main parameters of a digital signal and their relationship to the characteristics of digital transmission.

Outcomes

- Recall that the waveform of an analog signal is a continuous plot, whereas the waveform of a digital signal is a set of discrete dashes representing discrete values of a signal. Remember that an analog signal carries information by changing one of its parameters and, since information changes every instant, the parameters of an analog signal are in constant state of change. Realized that a digital signal has discrete values and changes from one value to another discretely.
- Comprehend that in analog transmission, any change in signal parameters causes errors, whereas digital transmission allows for signal distortion within specific margins while still preserving the correct meaning of the information being delivered. Get to know that this advantage of digital transmission justifies the need to convert analog signals into digital ones.

- Understand that the aforementioned advantage of digital signals is implemented through voltage specifications in modern digital logic families that reserve the specific margins of digital-signal distortions, within which digital signals deliver the correct information despite these distortions.
- Be aware that in transmission, there is subtle difference in terms *binary* and *digital* because binary information can be delivered by an analog signal though it is delivered mostly by a digital signal. Also, be familiar that the terms *digital* and *data communications* are almost identical – but *not* completely because there are cases where data is delivered by analog signals and cases where a digital signal delivers analog information.
- Remember that we live in an analog world, but, to provide reliable, high-quality communications, we need to deliver analog information digitally. Learn that to do so, we need to (i) convert analog characters into bits by applying a character code and (ii) convert bits into digital signals by using transmission codes.
- Study that conversion analog characters into bits are performed by all computing machines through using the two most popular character codes – American Standard Code for Information Interchange (ASCII) and Extended Binary Coded Decimal Interchange Code (EBCDIC). Know that modern software provides the compatibility of these codes and that the character codes are simply a universally accepted formats for coding.
- Comprehend that the character codes present all the information in 1s and 0s, but to convert these logical 1s and 0s into electrical signals for delivery, we must use transmission codes. Realize that these codes are also universally accepted formats. Study that such examples of transmission codes as nonreturn-to-zero (NRZ), return-to-zero (RZ), alternate mark inversion (AMI), and the Manchester code demonstrate how the presentation of bits by electrical pulses can be implemented in practice.

3.1.1 Why Go to Digital Communications

3.1.1.1 Main Advantage of Digital Transmission over the Analog

Look at Figure 2.1.1R, reproduced here again: You will recall the difference between two main classes of signals – analog and digital. The waveform of an analog signal is a continuous plot, whereas the waveform of a digital signal is a set of discrete dashes representing discrete values of a signal. An analog signal carries information by changing one of its parameters. In the case of

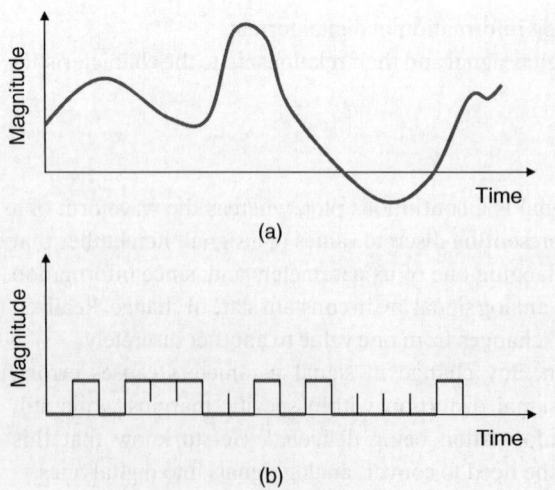

Figure 2.1.1R Waveforms: (a) analog signal; (b) digital signal.

broadband transmission, these parameters are either amplitude or frequency or phase of a carrier sinusoidal signal. In general, since information changes every instant of time, the parameters of an analog signal are in permanent change.

Remember, a digital signal has discrete values and changes from one value to the other discretely.

Let us take the most common case, one when a digital signal carries binary information; that is, this signal delivers logic 1 and logic 0. In this case, a digital signal can deliver logic 1 by a pulse with amplitude A and logic 0 by a pulse with amplitude zero. Mark these signals for yourself in Figure 2.1.1R.

Now it is time to ask a pertinent question: Why do we need to use a digital signal? Why bother to convert natural analog signals to the artificial digital? In other words, what advantage does a digital signal have over an analog? The answer lies in an analysis of analog and digital transmission. These types of transmission are shown in Figure 3.1.1.

The top portion of Figure 3.1.1 shows analog transmission. An external source generates electromagnetic waves that interfere with an analog signal being transmitted; the result is that an analog signal arriving at the receiver end is distorted. We discussed this main problem of an analog baseband transmission in Section 2.2.

The bottom portion of Figure 3.1.1 shows digital transmission. Again here, an external source generates electromagnetic waves that interfere with a digital signal being transmitted; the result is the same: A digital signal arriving at the receiver end is distorted.

If both signals are distorted in a similar way, what is the advantage of digital transmission, you may ask?

The answer: A digital signal is being distorted during the transmission in the same way as an analog but it still delivers correct information. How?

You will recall that the distortion of an analog signal results in changing one of its parameters, as the top portion of Figure 3.1.1 shows. Since in analog transmission the information is delivered

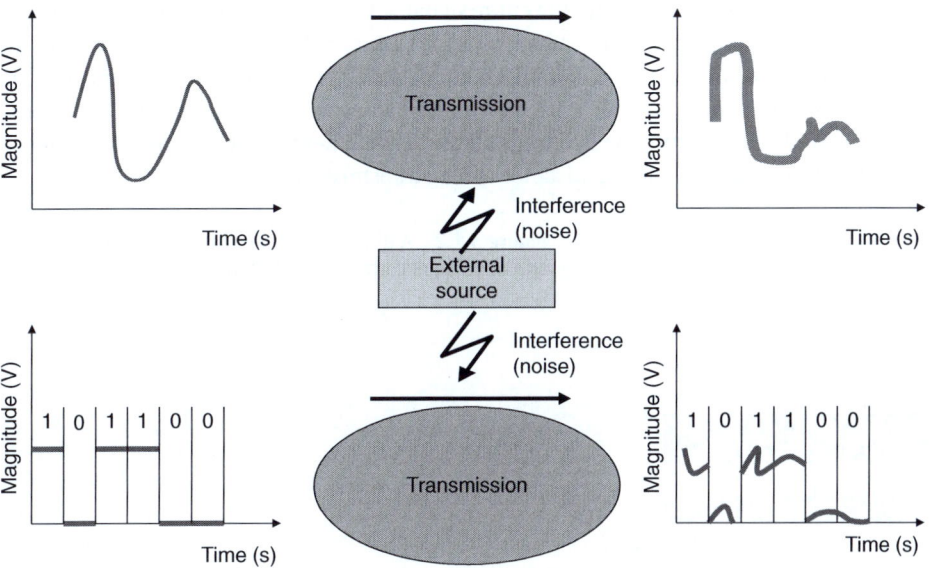

Figure 3.1.1 Digital and analog transmission.

by the exact value of a parameter, changing this value because of signal distortion results in the delivery of incorrect information. For example, a voice signal is carried by virtue of amplitude of an analog electrical signal, as Figure 2.3.2 shows. Distortion of this amplitude results in delivering inaccurate information, as we discussed in Section 2.2.

To sum up, *in analog transmission, any change in signal parameters causes errors.* In contrast to analog transmission, *digital transmission allows for signal distortion within specific margins while still preserving the correct meaning of the information being delivered.*

This concept is visualized in Figure 3.1.2.

Figure 3.1.2 shows the received signal with solid lines. In fact, this is the bottom right portion of Figure 3.1.1; the only difference is that here we show the margins within which the digital signal is allowed to be distorted. Also, with the dotted lines, we depict the sent pulses so you can easily compare the pulses at the transmitter and the receiver ends. Observe the first pulse: You can clearly see how much this pulse is deformed after transmission. However, since this signal is distorted within the given margins, it still delivers logic 1. Now observe the second pulse: The amplitude of this signal is far from 0, as it was at the sending end. However, again, since the distortion does not exceed the allowed margins, this pulse still delivers bit 0.

Herein lies the main advantage of digital transmission over analog:

Distortion of a transmitted digital signal doesn't necessarily result in errors.

Let us refer to one historical example that confirms the main advantage of digital transmission over analog. You will recall that the first transatlantic telegraph cable was installed in 1858, 14 years after the development of the system, and the first transatlantic telephone cable was laid in 1956, 80 years after Bell's invention of telephone (see Section 1.2). Why such a disparity between telegraph and telephone service in the realization of long-distance transmission?

Telegraph transmission is digital by its very nature; it works in a discrete mode of operation. To obtain information transmitted in Morse code, a telegraph receiver needs to detect only the presence and duration of an electrical signal. (To bone up on the Morse code, see the sidebar *Character codes* in this section.)

In contrast to the telegraph, original telephone transmission was analog. To garner information, a telephone receiver (speaker) must obtain the entire signal with minimum distortion. (See Section 1.2 for the basics on the original telephone transmission.) Therefore, the loss of power of a telephone signal over the distance results in the loss of information as well. This is why the transmission of analog telephone signals was impossible until electronics had been developed to the point where reliable repeaters could be built into the first transatlantic cable. These repeaters boosted the signal and provided satisfactory quality of long-distance analog transmission.

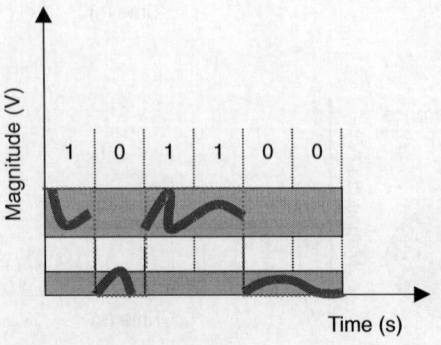

Figure 3.1.2 A digital signal after transmission: This signal delivers correct information despite its distortion.

Figure 3.1.3 Digital signal after transmission: This signal is distorted beyond its margins and delivers erroneous information.

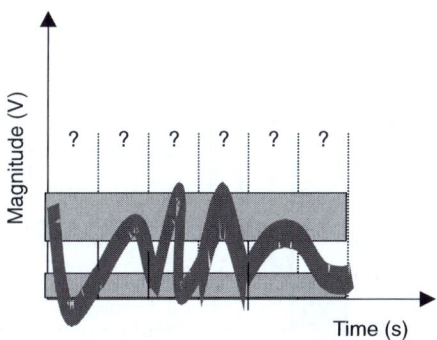

Now to return to the main point of our discussion: Can distortion of a digital signal result in errors? Yes. When a digital signal is deformed beyond the allowed margins, a receiver will not be able to recognize what information – logic 1 or logic 0 – has been transmitted. In other words, such a greatly distorted signal will deliver erroneous information. The explanation is shown in Figure 3.1.3.

Consider the pulses in Figure 3.1.3: They are deformed beyond the allowed margins. When these pulses arrive at the destination point, a receiver will not be able to recognize what bit – 1 or 0 – each pulse is supposed to deliver. Therefore, such a signal delivers erroneous information. Fortunately, this situation is relatively rare in digital transmission; thus, thanks to its tolerance of distortion of a transmitted signal, digital transmission is much more reliable than analog.

The limit within which a digital signal is allowed to be distorted during transmission is called the *noise margin*.

Let us turn to examples of digital transmission.

3.1.1.2 Case Study 1: The Advantages of Using Digital Signals in Transmission

We would like to demonstrate the advantages of employing digital signals in transmission by using, as an example, one of the first digital transmission systems, RS-232 standard. This standard, though outdated, presents a clear conceptual view at the subject. The standard was developed in 1969 by the Electronic Industries Association (EIA) for serial transmission. It was intended for one-way data transmission from a computer to a printer and other peripheral devices. These transmission systems still can be found in some of today's PCs, digital equipment, scientific instruments, and networking devices. There are updated versions of this standard that operate at a low voltage level, such as ±3.3 V. We, however, want to concentrate on its specifications rather than on the technical detail of its operation.

At the transmitter end, the RS-232 specifies an electrical pulse with any amplitude from −5 to −15 V as logic **1** and a pulse with any amplitude from +15 to +5 V as logic **0**. All this means that the RS-232 allows a transmitter to generate pulses with big changes in amplitude; however, the minimum voltages are specified as −5 and 5 V. At the receiver end, the minimum voltages of this digital signal are specified as −3 and 3 V. Therefore, the noise margin for RS-232 is equal to 2 V. The original RS-232 transmission code is shown in Figure 3.1.4.

This code demonstrates a very important concept in digital transmission: *There are two types of variations that digital signal can withstand*. First, we allow for a great *variation in a signal's amplitude* when generating the signal at the transmitter end. This is what the ranges from 15 to 5 V and from −5 to −15 V are all about. Thus, in RS-232 the first type of tolerances is 10 V in the amplitude

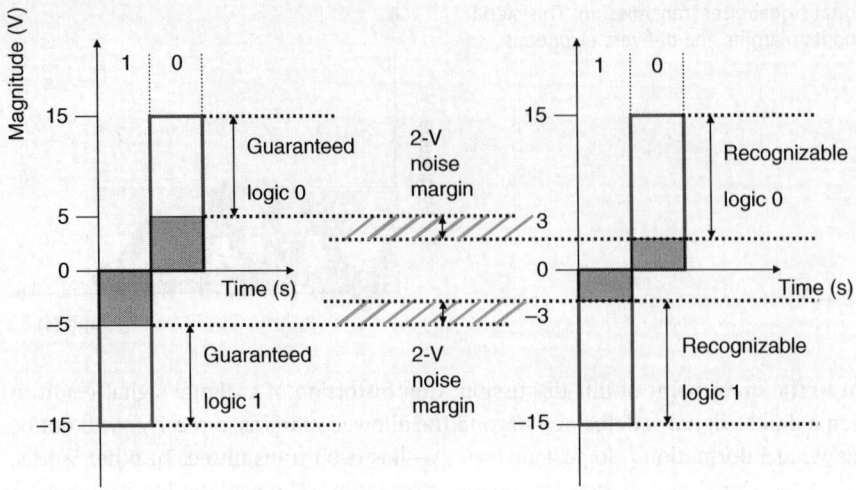

Figure 3.1.4 RS-232 transmission code.

variation of an output signal. Figure 3.1.4 shows this tolerance as voltage ranges that guarantee presentation of logic 0 and logic 1.

The second type of tolerance is the *noise margin*, the allowed change in a signal's amplitude caused by electromagnetic interference during signal transmission. In RS-232, this tolerance is presented as changing the minimum voltage from ±5 V at the sending end to ±3 V at the receiver end. Thus, the noise margin that is equal to 2 V. All this means the following: If the input pulse has any amplitude from 15 to 3 V, a receiver can recognize that this pulse delivers logic 0; if the input pulse has any amplitude from −3 to −15 V, a receiver can recognize that this pulse delivers logic 1. Figure 3.1.4 shows these tolerances as voltage ranges that are recognizable presentation of a logic 0 and logic 1.

The implementation of the RS-232 transmission code is shown in Figure 3.1.5.

Figure 3.1.5a left shows the RS-232 signal generated at the sending point. Observe that the amplitudes of the pulses are quite different; however, since they vary within the transmitter's specified borders from 15 to 5 V and from −5 to −15 V, all of these pulses present correct information. Figure 3.1.5a right shows the same signal at the receiver end. Observe that the amplitudes of the pulses changed during transmission. The majority of the pulses are within the receiver's specified borders from 15 to 3 V and from −3 to −15 V; therefore, the receiver is enabled to recognize what logic – 1 or 0 – is delivered by these pulses. However, the amplitude of the fifth pulse is approximately −2 V and the receiver is unable to recognize whether this pulse delivers logic 1 or logic 0. In other words, if pulse amplitude fails in the region between +3 and −3 V, the receiver can produce an unpredictable result. This area is called the *uncertain region*. Figure 3.1.5b shows how the industry transmits today with the RS-232 standard. Notice that the amplitude voltage reduced to ±5 V. Observe the use of several service bits (start bit, parity bit, two stop bits) needed for the real transmission. (The acronym ASCII stands for **American Standard Code for Information Interchange**; it will be discussed shortly.)

Figure 3.1.6 summarizes the traditional voltage specifications for RS-232 standard. Again, observe the two types of tolerance considerations allowed in digital transmission: the variations in the signal's amplitude at the transmitter end within which the meaning of presented logic – 1 or 0 – is guaranteed and the noise margin that determines the allowed distortion of the signal's

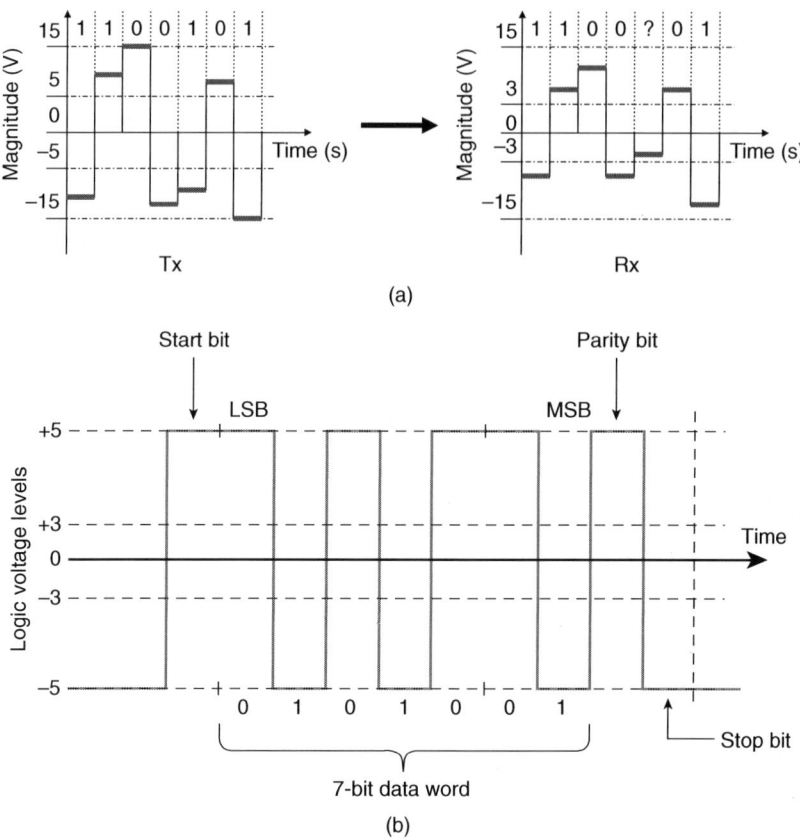

Figure 3.1.5 The RS-232 signal delivering letters in ASCII code: (a) transmitted signal and received signals delivering letter "M"; (b) example of the modern industrial RS-232 transmission. http://electronicdesign.com/what-s-difference-between/what-s-difference-between-rs-232-and-rs-485-serial-interfaces.

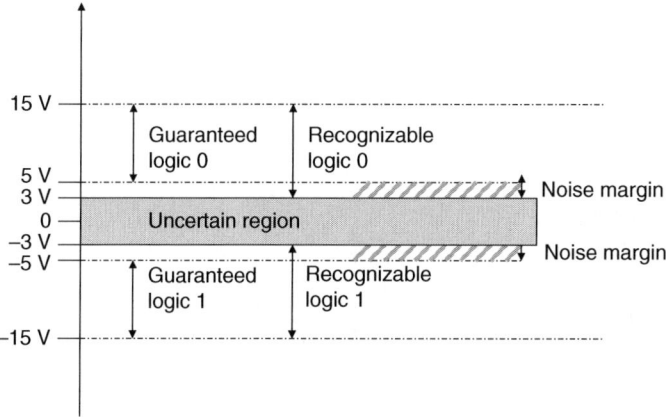

Figure 3.1.6 Summary of RS-232 voltage specifications.

amplitude during the transmission. Pay attention to the uncertain region: If the signal's amplitude falls into this voltage range, you will have an unpredictable result.

3.1.1.3 Case Study 2 of Digital Communications: Transmission with Integrated-Circuit Digital Logic Families

Let us consider the second example: digital transmission with integrated circuits (ICs), known as chips. The ICs are the main active components of any computer and other electronics. Any device surrounding you today is replete with such chips. Thanks to progress in semiconductor electronics, these ICs have become so small that they are today the most prolific components in technology; what's more, they are so powerful that the computing power of a modern smartphone is greater than the power of the supercomputers of the 1980s.

Digital ICs work, obviously, in binary format; they send and receive pulses encoding bits 1 and 0. Transmission among these digital ICs covers very short distances, such as links within a computer unit or even links among ICs placed on the same board. Electrical signals used for transmission among these ICs are standardized and we will discuss these specifications. But before we turn to the voltage specifications of these ICs, we need to introduce the terminology used in the transmission among digital logic families.

Every pulse having certain amplitude, typically from 2 to 5 V, represents logic **1** or the HIGH state of a signal; every pulse with amplitude near zero represents logic **0** or the LOW state of the signal. The voltage of a high-level pulse sent *from* an IC is denoted as V_{OH}, where "O" stands for *output* and "H" stands for *HIGH*. The voltage of a high-level pulse *entering* an IC is denoted as V_{IH}, where "I" stands for *input* and "H" stands for *HIGH*. Similar designations are used for low-level signals: V_{OL} and V_{IL}. When these pulses are transmitted, noise changes the value of these voltages; a receiver can recognize bits correctly if these changes are within a noise margin. Figure 3.1.7 shows all these voltages.

You will recall that digital-signal specifications refer to two types of tolerance: The range of voltages for sending signals and the range of voltages for receiving signals, as exemplified in

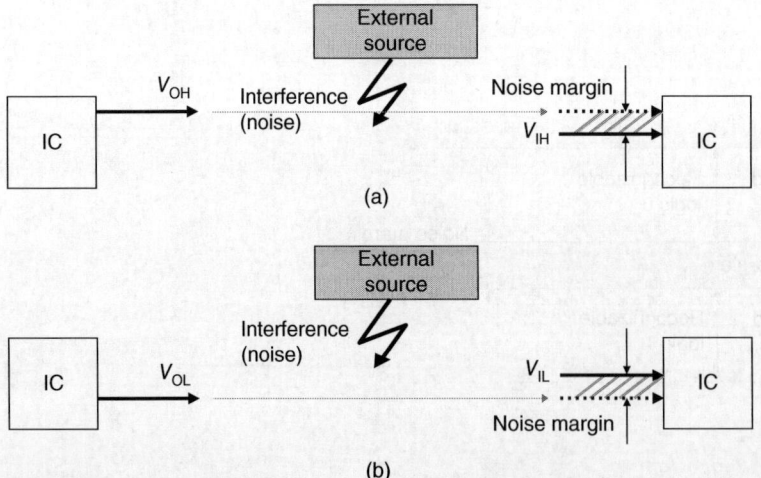

Figure 3.1.7 Digital ICs' electrical signals: (a) high-level output (sent), V_{OH}, and input (received), V_{IH}, voltages; (b) low-level output (sent), V_{OL}, and input (received), V_{IL}, voltages.

Figure 3.1.5. This means that all the above voltages, V_{OH}, V_{IH}, V_{OL}, and V_{IL}, show the limits of the values of the output and input signals. V_{OH} is the *minimum* voltage required at the output of an IC to guarantee representation of logic 1, or HIGH; V_{IH} is the *minimum* voltage required at the input of an IC to ensure the recognizable representation of logic 1, or HIGH. This statement is also true for low-level signal: V_{OL} is the *maximum* guaranteed voltage to represent logic 0, or LOW; V_{IL} is the *maximum* voltage required to represent logic 0, or LOW. Note that for a high-level signal, specified voltages show the minimum values, while for a low-level signal, specified voltages show the maximum values. This is because a high-level signal is encoded by an electrical pulse with amplitude ranges from, say, 5 to 1 V, whereas a low-level signal must have near-zero amplitude. Figure 3.1.8 graphically presents all these explanations.

Before turning to an example, it is necessary to mention that there are a number of logic families in digital electronics today. Three main families are transistor–transistor logic (TTL), complementary metal-oxide semiconductor (CMOS), and emitter-coupled logic (ECL). CMOS, TTL, and ECL refer to specific semiconductor technologies. The term *family* means that they include series of ICs that differ in their functions and characteristics. The CMOS family is the most popular today, the TTL is widely used, and ECL is the least common. The following example of TTL specs should make the above consideration more specific. Table 3.1.1 presents the examples of the binary logic levels.

Example 3.1.1 provides the detailed discussion of specifications of a TTL logic family.

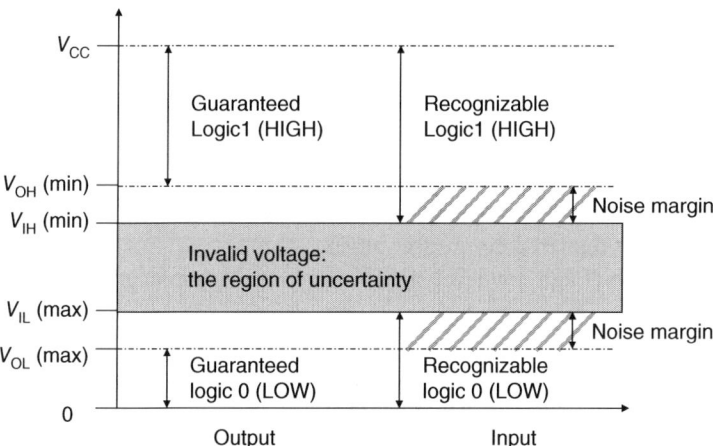

Figure 3.1.8 Digital ICs' voltage specifications.

Table 3.1.1 Examples of binary logic levels.

Technology	L voltage	H voltage	Notes
CMOS	0 V to $V_{DD}/2$	$V_{DD}/2$ to V_{DD}	V_{DD} = supply voltage
TTL	0–0.8 V	2 V to V_{CC}	V_{CC} is 4.75–5.25 V
ECL	−1.175 V to V_{EE}	0.75 to 0 V	V_{EE} is about −5.2 V. V_{CC} = Ground

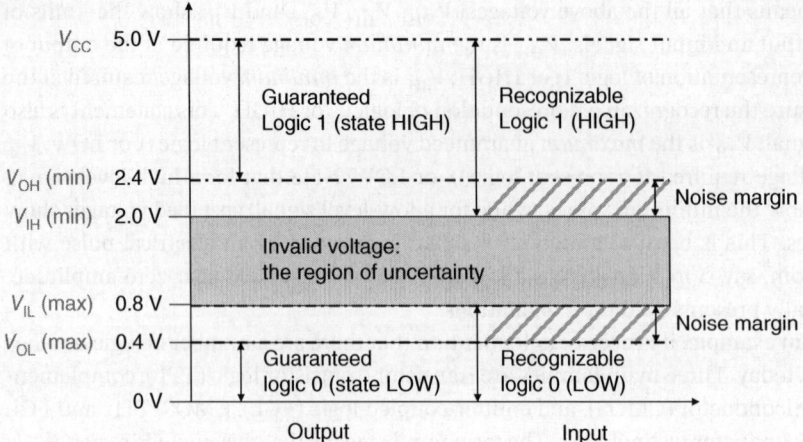

Figure 3.1.9 TTL voltage specifications.

Example 3.1.1 Voltage Specifications of the TTL Digital Logic Family

Problem

The TTL family specifies its voltages as

$V_{CC} = 5.0\,\text{V}$
$V_{OH} = 2.4\,\text{V}$
$V_{IH} = 2.0\,\text{V}$
$V_{IL} = 0.8\,\text{V}$
$V_{OL} = 0.4\,\text{V}$

Note: You might meet slightly different TTL voltage specifications, but we are interested in the principle of these specifications rather than in their precise values.

Compute noise margins for high-level and low-level signals and present graphically all these voltages.

Solution

By definition, the noise margin for a high-level signal is equal to

$$V_{OH} - V_{IH} = 2.4 - 2.0 = 0.4\,\text{V}$$

Similar computations yield the noise margin for low-level signal

$$V_{IL} - V_{OL} = 0.8 - 0.4 = 0.4\,\text{V}$$

It is industry practice to specify noise margin for ICs in millivolts rather than in volts. Hence, the answer is that the noise margin for both signals is 400 mV.

Figure 3.1.9 illustrates TTL voltage specifications.

Discussion

- Again, bear in mind that all the voltage values are true for this example only. Here, all the pulses having amplitudes from 5 to 2.4 V represent logic 1, or state HIGH. The minimum voltage at the output of an IC that still guarantees representation of logic 1 is $V_{OH} = 2.4\,\text{V}$. Similarly, all the pulses having amplitudes from 0 to 0.4 V represent logic 0, or state LOW. The maximum input voltage that still guarantees representation of logic 0 is $V_{IL} = 0.8$ V. These voltages, V_{OH} and V_{IL}, show the first type of tolerances that digital signals feature: To encode logic 1 and logic 0 at the sending end, we can have pulses with a range of amplitudes.

- What happens if the output voltage of a high-level signal will be 2.2 V? This voltage lies in the region of uncertainty, and we do not know what logic such a pulse represents. Similarly, if the output voltage for a low-level signal exceeds 0.4 V and reaches, say, 0.7 V, such a pulse falls into the region of uncertainty and we do not know what logic this pulse encodes. These voltages, $V_{OH} = 2.2$ V and $V_{IL} = 0.7$ V, exemplify the requirements for sending signals. The range from 0.8 to 2.0 V for the TTL logic family is invalid; a signal falling into this range produces uncertain results: A circuit may respond to these signals either as logic 1 or as logic 0. We have to conclude that digital transmission allows for variations in the amplitudes of sending signals, but these variations must be within a certain range.
- The second type of tolerance allowed by the digital-transmission technique is the signal's distortion during transmission. Here we send a pulse with amplitude at least of 2.4 V and we can afford receiving 2.0-V amplitude for logic 1. In other words, noise can reduce the pulse amplitude up to 2.0 V, but a receiver still will be able to recognize this pulse as logic 1. Similarly, noise can increase the amplitude of the transmitted signal up to 0.8 V, but a receiver still will be able to recognize this pulse as logic 0.
- This example clearly demonstrates the main advantage of digital transmission over analog: A receiver recognizes any pulse with amplitude ranges from 5.0 to 2.0 V as the pulse delivering logic 1 and any pulse with amplitudes ranges from 0 to 0.8 V as the pulse delivering logic 0.

  In contrast, in analog transmission, you will recall, the information is transported by the precise value of signal's parameter (e.g. either amplitude, or frequency or phase for a sinusoidal signal); any variations in the value of a parameter results in delivering erroneous information.
- Final thoughts: First, you may wonder why a high-level signal requires a minimum voltage of V_{OH} and V_{IH}, whereas a low-level signal requires the maximum voltage of V_{IL} and V_{OL}. The answer lies in the nature of these signals: Ideally, we want 5.0 V for V_{OH} but can afford the drop in this voltage up to 2.0 V. Similarly, we want 0 V for the pulse-encoding logic but can afford the increase in this voltage up to 0.8 V.

  Second, you certainly noticed that an input (received) high-level signal allows for subtracting from its minimum output (sent) voltage, whereas a received low-level signal allows for adding to its maximum transmitted voltage. This is because noise can add or subtract voltage from a signal during transmission. If noise adds some voltage to the minimum-output voltage of a high-level signal, this addition will only improve the quality of our transmission. Look at Figure 3.1.9 and mentally add any voltage up to 5 V to V_{OH}: This pulse will guarantee logic 1. But if noise subtracts voltage from V_{OH} during the transmission, then we may receive an unrecognizable pulse. This is what noise margin is all about: We allow noise to subtract only 0.4 V from V_{OH}.

  You can develop similar reasoning to explain the situation with a low-level signal.

It will be a good exercise for you to conduct an analysis similar to that given in Example 3.1.1 for TTL and CMOS logic families. Bear in mind that there are two standards for the TTL and CMOS families: One is for traditional TTL and CMOS families that feature a 5-V power supply and the second one is for new, low-voltage TTL and CMOS called LVTTL and LVCMOS, where the power supply is reduced to 3.3 V. All these specifications are shown in Table 3.1.2.

Analyze Table 3.1.2, compare it with Table 3.1.1, and do the exercises given in the Questions and Problems section. It will help you to better grasp the concept of digital transmission.

Table 3.1.2 Voltage specifications for TTL and CMOS logic families (www.tektronix.com/signal_generators).

Parameter	TTL	CMOS	LVTTL	LVCMOS
V_{CC} (V)	5	5	3.3	3.3
V_{OH} (V)	2.4	4.5	2.4	3.2
V_{IH} (V)	2.0	3.5	2.0	2.3
V_{IL} (V)	0.4	1.5	0.8	0.66
V_{OL} (V)	0.8	0.5	0.4	0.1
Noise margin (mV)	400	1000	400	900 for HIGH
				560 for LOW

3.1.1.4 Why Go to Digital Communications: A Summary

> *Digital signals deliver correct information in spite of their distortions as long as these distortions are within the specified margins.*

No wonder that digital transmission has replaced the analog everywhere from computers to telecommunications to consumer electronics. Incidentally, in the 1960s, when analog electronics reached its peak of development and analog ICs became powerful and reliable components, there was an effort to develop computing devices based on analog electronics. Since you cannot find any analog computer in use today, you can clearly understand that analog devices and analog transmission do not work reliably and fast enough to do the jobs perfectly done by modern computers. Indeed, modern personal computers do billions of operations per second, including generating billions of bits per second and moving these bits within a computer from one unit to another. Without the advantages of digital transmission described in this section, such operations could not be possible.

Nevertheless, bear in mind that we must always exercise an intelligent approach to any engineering problem. Remember, our world is analog and sometimes conversion an analog entity into digital is not justified.

Additional note regarding the terms "binary and digital," "digital and data": From the above discussion you may gather impression that binary and digital are synonyms. They are not. True, bits are the symbols of the modern industry called digital; this industry includes computers, telecommunications, and consumer electronics. True also that the computational operations are done in binary format, and this information is delivered with digital signals. However, bits can be delivered with analog signals, too.

What you have to realize is that *binary information is delivered mostly by a digital signal, but not always – because it can also be delivered by an analog signal*. (We learn more on this topic in Chapter 9.)

In this regard, we need to clarify other terms: *data and digital communications*. It is very common for these terms to be considered synonyms, but they are not. Indeed, *data is a general term that denotes any factual information*. However, when we refer to data as a subject for transmission, we imply that it is information produced by a computer. This understanding is very true because *computer is nothing more than a data processing machine*. Hence, data communications is communications among computers; this type of communications includes operations in computer networks. This is the most prolific type of telecommunications today, but it is not the only one. We can encode analog information generated by different sources with a digital signal.

Let us consider, for example, our regular telephone conversation: Our voice message, after being converted into an analog electrical signal by a microphone, is then encoded as a digital signal and transmitted to a destination point in digital form. There, this signal is converted back to analog and delivered to your calling party.

What this means is that digital and data communications are almost identical – but not completely: There are cases where data is delivered by analog signals and cases where a digital signal delivers analog information.

3.1.2 How to Go to Digital Communications

We live in an analog world. Time changes continuously; the earth revolves around the sun continuously; we speak and see continuously. Everything around us undergoes smooth, gradual change. This is exactly how we describe an analog signal in Section 2.1: The plot of an analog signal is a continuous graph. This graph shows how a value we want to observe (for example, the position of the sun above the horizon) changes with respect to time. This value is information. Thus, information originates in analog format. This is why the first and the most natural way to deliver the information that mankind has developed is by analog communication. However, analog communications, as discussed in Chapter 2 and previously in this section, suffers from a number of serious drawbacks. With technological advances, the quality of analog communication has improved, but the drawbacks inherent in this kind of transmission cannot be totally overcome.

The fundamental solution to this problem, then, is to bypass analog communications completely and turn to *digital communications*. To do so, we need to convert analog information (which, remember, exists naturally in our real world) to digital format and develop the technology for this type of communication. Since the 1960s, this is precisely what the industry has been doing. Today, our computers are digital, our watches are digital, our phones are digital, our CD players are digital – and the list goes on and on. Look around and see for yourself all the digital devices and technologies that affect your life every day. The last bastion of analog communications, cable TV, turned to digital transmission in 2006 by order of the Federal Communications Commission. What digital communications is and why it is replacing the analog format are what we are going to start discussing in this chapter. This discussion continues throughout the book, culminating in Chapter 10, which is completely devoted to digital transmission.

> *Two major steps are necessary to implement digital communications: The first one is the logical aim to present analog information in digital format mathematically; the second is the technological ability to transmit digital information physically.*

Thus, the first step is concerned with preparing digital information and the second is devoted to transmitting digital signals. In this chapter, we discuss both steps, whereas modulation issues of digital transmission will be considered in Chapter 10.

3.1.2.1 From Characters to Bits

Binary World To make digital transmission a reality, we need to develop a special means to present information. Specifically, all information in digital format is delivered by using only two symbols:

logic **1** and logic **0**. Each symbol is called a *bit*, which is an acronym for *bi*nary digi*t*.[1] Each symbol represents the state of a signal: logic **1** is usually referred to as the HIGH state and logic **0** as the LOW state. Since we use only two symbols, this way of presenting information is called *binary*. Using these two symbols, we can express all the information we need to convey, such as our words, mathematical operations, and data. What this means is that we can present in binary code any letter, numeral, punctuation mark, and control character. In other words, any characters we use in our daily communication can be given in binary format.

Bits **1** and **0** are building blocks – elementary units of information – in the binary world. Eight bits combined in one unit is called a *byte*. You are certainly familiar with this term, since it is used when referring to a computer's memory capacity, which today is measured in megabytes (10^6 bytes = 1 million bytes) and gigabytes (10^9 bytes = 1 trillion bytes). Let us work out a problem to make you more comfortable using these entities.

Example 3.1.2   Bits and Bytes

Problem

a) How many bits are in the binary word: **1 1 0 1 0 0 1**?
b) How many bytes are in the binary word: **1 1 1 1 0 0 1 0 1 1 1 1 0 0 1 1**?

Solution

(a) We need simply count the number of binary digits; thus we have 7 bits.
(b) Since 8 bits are equal to 1 byte, the answer is 2.

Discussion

Observe that both bits, **1** and **0**, must be counted. Observe, too, that a combination of bits is called a binary word. Usually, the term *binary word* is applied to the combination of a specific number of bits, such as 8, 16, 32, or 64.

To summarize: In the binary world, we operate with only two symbols: bits **1** and **0**. Using only these two symbols, we can present all the information that we use in our daily communications.

The Computer Is a Binary Machine   Now let us make a very important point: *Our computers are binary machines*. What this means is that computer electronics (hardware) operates with only two-level (HIGH and LOW) electrical signals, and computer programs (software), which tell the electronics what to do, eventually operate with bits (logic **1**s and **0**s).

The fact that the computer is a binary machine explains immediately why going binary is so important. Can you imagine life today without the computer? Of course not. It is everywhere: in our homes, in our schools, in the workplace. In fact, does a day go by that you do not connect to the Internet? The point is, whether you realize it or not, your communication through the Internet

---

1 It is interesting to note that it was Dr. John Tukey, a statistician and National Medal of Science winner, who coined the term *bit*. Another popular term – *software* – is also attributed to him. Dr. Tukey had a very productive career mostly at Princeton University and died in 2000 at the age of 85.

is done in binary format. And since the computer world is a binary world, we need to learn about this world.

> Let us begin by posing a problem: Since our real world is analog and the computer world is binary, how can we present and transmit information in binary format? The answer is as follows:
> 1. We need to present the traditional readable characters in binary format. This step is done with character codes.
> 2. We can transmit only electrical or optical signals; therefore, we need to encode bits as physical signals. This step is done with transmission codes.

Let us consider these steps more closely.

Character Codes

> To present conventional characters in binary format, we have to devise a means for the translation of characters to bits. Let us, for example, present the letter "A" by the following sequence of **1**s and **0**s: **10000001**. This is nothing more than a suggestion. If everyone on earth were to accept this suggestion, we would have a working convention that would be used for such a translation. In general, such a convention is called a code.

Obviously, we need to develop a code that translates every character (a letter, a numeral, a punctuation mark, a control character) to a specific combination of **1**s and **0**s. Such a code is called a *character,* or *alphanumerical, code.*

Since there is no law in nature for this coding, everyone was free to suggest his or her own codes; this is why several character codes were developed. Two have survived and today we use either the ASCII or EBCIDIC character codes.

ASCII and EBCIDIC Codes ASCII (pronounced as-key) is the acronym for *American Standard Code for Information Interchange*. This, the most popular character code in the world, was developed by the American National Standards Institute (ANSI). It works as follows: When you hit a button "*a*" on your computer keyboard, the computer translates this letter into the following binary format: **1100001**. For capital "*A,*" the computer sees **1000001**; the letter "*b*" it encodes as **1100010**; capital "*B*" is encoded as **1000010**, and so on. Electronic circuitry transmits these **1**s and **0**s from your keyboard to the computer; the computer converts these **1**s and **0**s into a graphical image, and you see the letters "*a,*" "*A,*" "*B,*" or "*b*" on the computer screen. (Specifically, the computer pulls out the image of a specific letter from the font file in response to receiving a certain combination of **1**s and **0**s.)

The concept behind the ASCII underlies EBCIDIC (pronounced eb-si-dik or eb-see-dick) code, which stands for **E**xtended **B**inary **C**oded **D**ecimal Interchange **C**ode. This code was developed by IBM, and it is still used with IBM computers except for IBM PC machines and workstations, for which an extended version of the ASCII code is in use. The charts of the ASCII and EBCDIC codes are shown in Tables 3.1.3 and 3.1.4.

Table 3.1.3 ASCII code.

Character	Bits 7654321	Character	Bits 7654321	Character	Bits 7654321	Character	Bits 7654321
NUL	0000000	SP	0100000	@	1000000	`	1100000
SOH	0000001	!	0100001	A	1000001	a	1100001
STX	0000010	"	0100010	B	1000010	b	1100010
ETX	0000011	#	0100011	C	1000011	c	1100011
EOT	0000100	$	0100100	D	1000100	d	1100100
ENO	0000101	%	0100101	E	1000101	e	1100101
ACK	0000110	&	0100110	F	1000110	f	1100110
BEL	0000111	'	0100111	G	1000111	g	1100111
BS	0001000	(	0101000	H	1001000	h	1101000
HT	0001001	)	0101001	I	1001001	i	1101001
LF	0001010	*	0101010	J	1001010	j	1101010
VT	0001011	+	0101011	K	1001011	k	1101011
FF	0001100	,	0101100	L	1001100	l	1101100
CR	0001101	-	0101101	M	1001101	m	1101101
SO	0001110	.	0101110	N	1001110	n	1101110
SI	0001111	/	0101111	O	1001111	o	1101111
DLE	0010000	0	0110000	P	1010000	p	1110000
DC1	0010001	1	0110001	Q	1010001	q	1110001
DC2	0010010	2	0110010	R	1010010	r	1110010
DC3	0010011	3	0110011	S	1010011	s	1110011
DC4	0010100	4	0110100	T	1010100	t	1110100
NAK	0010101	5	0110101	U	1010101	u	1110101
SYN	0010110	6	0110110	V	1010110	v	1110110
ETB	0010111	7	0110111	W	1010111	w	1110111
CAN	0011000	8	0111000	X	1011000	x	1111000
EM	0011001	9	0111001	Y	1011001	y	1111001
SUB	0011010	:	0111010	Z	1011010	z	1111010
ESC	0011011	;	0111011	[	1011011	{	1111011
ES	0011100	<	0111100	\	1011100	|	1111100
GS	0011101	=	0111101	]	1011101	}	1111101
RS	0011110	>	0111110	^	1011110	~	1111110
US	0011111	?	0111111	_	1011111	DEL	1111111

Note: Remember that *ASCII starts its transmission with the <u>most</u> significant bit*, which is bit #7, as the table shows. Thus, you have to read bits in the order "7, 6, 5, 4, 3, 2, 1," as shown in the table. For example, the letter "B" is read as 1000010 and the character "$" is read as 0100100.

Table 3.1.4 EBCIDIC code.

	4	0	0	0	0	0	0	0	0	1	1	1	1	1	1	1	1
	5	0	0	0	0	1	1	1	1	0	0	0	0	1	1	1	1
	6	0	0	1	1	0	0	1	1	0	0	1	1	0	0	1	1
	7	0	1	0	1	0	1	0	1	0	1	0	1	0	1	0	1
Bits	0123																
	0000	NUL	SON	STX	ETX	PF	HT	LC	DEL				VT	FF	CR	SO	SI
	0001	DLE	DCI	DC2	DC3	RES	NL	BS	IL	CAN	EM			IPS	IGS	IRS	IUS
	0010			FS		BYP	LF	EOB	PRE			SM			ENQ	ACK	BEL
	0011			SYN		PN	RS	UC	EOT					DC4	NAK		SUB
	0100	SP										¢		<	(	+	|
	0101	&										!	$	*	)	;	¬
	0110	-	/									|	,	%	_	>	?
	0111											\	:	#	@		=	"
	1000		a	b	c	d	e	f	g	h	i						
	1001		j	k	l	m	n	o	p	q	r						
	1010		~	s	t	u	v	w	x	y	z						
	1011																
	1100	{	A	B	C	D	E	F	G	H	I						
	1101	}	J	K	L	M	N	O	P	Q	R						
	1110			S	T	U	V	W	P	Y	Z						
	1111	0	1	2	3	4	5	6	7	8	9						?

Note: Remember, *EBCIDIC code starts its transmission with the* least *significant bit*, which is bit #0, as shown in the table. To read this table, find the character in the table, look left on the row, read the bits in the order "0, 1, 2, 3," and then look at the top of the column and read the bits in the order "4, 5, 6, 7." For example, the letter "B" is read as 11000010 and the character "$" is read as 01011011. Compare this reading with the ASCII code: You can clearly see that the EBCIDIC code reads the bits from left to right and the ASCII code reads from right to left.
Source: http://www.egrannie.com/cheatsheets/asciiebcdic.html.

The following table presents a set of ASCII's special characters:

ACK	Acknowledge	FF	Form Feed
BEL	Bell	FS	File Separator
BS	Backspace	GS	Group Separator
CAN	Cancel	HT	Horizontal Tabulation
CR	Carriage Return	LF	Line Feed
DC1	Device Control 1	NAK	Negative Acknowledgment
DC2	Device Control 2	NUL	Null
DC3	Device Control 3	RS	Record Separator
DC4	Device Control 4	SI	Shift In
DEL	Delete	SO	Shift Out
DLE	Data Link Escape	SOH	Start of Heading
EM	End of Medium	SP	Space

ENQ	Enquiry		STX		Start of Text
EOT	End of Transmission		SUB		Substitute
ESC	Escape		SYN		Synchronous Idle
ETB	End of Transmission Block		US		Unit Separator
ETX	End of Text		VT		Vertical Tabulation

When you start to work with character codes on the job, you will appreciate the importance of this chart. For now, it gives you the sense of each abbreviation used in the ASCII code.

EBCDIC code

The following table presents the set of EBCDIC special characters.

ACK	Acknowledgment	EOT	End of Transmission	PF	Punch Off
BEL	Bell	ETX	End of Text	PN	Punch On
BS	Backspace	FE	Form Feed	PRE	Prefix
BYP	Bypass	FS	File Separator	RES	Restore
CAN	Cancel	HT	Horizontal Tab	RS	Reader Stop
CR	Carriage Return	IFS	Information File Separator	SI	Shift In
DCI	Device Control I	IGS	Information Group Separator	SM	Start Message
DC2	Device Control 2	IL	Idle	SO	Shift Out
DC3	Device Control 3	IRS	Information Record Separator	SOH	Start of Heading
DC4	Device Control 4	IUS	Information Unit Separator	SP	Space
DEL	Delete	LC	Lower Case	STX	Start of Text
DLE	Data Link Escape	LF	Line Feed	SUB	Substitute
EM	End of Medium	NAK	Negative Acknowledgment	SYN	Synchronous Idle
ENQ	Enquiry	NL	New Line	UC	Upper Case
EOB	End of Block	NUL	Null	VT	Vertical Tab

Source: http://www.egrannie.com/cheatsheets/asciiebcdic.html.

You noticed, of course, that EBCIDIC code uses eight bits and ASCII code uses seven bits in their original versions presented here.

You have certainly noticed that the ASCII and EBCDIC tables are presented in different formats. This is because you may encounter either presentation in the technical literature and we would like to make you familiar with both formats. The following example should do so:

Example 3.1.3   ASCII and EBCIDIC Codes

Problem

Show the following characters in ASCII and EBCIDIC codes: F, G, f, s, Z, #, @, and $.

Solution
Using Tables 3.1.3 and 3.1.4, we obtain:

	ASCII	EBCIDIC
F	1000110	11000110
G	1000111	11000111
f	1100110	10000110
s	1110011	10100010
Z	1011010	11101001
#	0100011	01111011
@	1000000	01111100
$	0100100	01011011

Discussion

- There are several important observations you should make when analyzing these results. First, note again that ASCII is a seven-bit code, while EBCIDIC is an eight-bit code. This difference results in a different number of characters that each code is able to encode. Indeed, since $2^7 = 128$ and $2^8 = 256$, we readily see that the ASCII code enables you to encrypt 128 characters, whereas EBCDIC can encode 256. In these computations, we use base two because we operate with only two symbols: 1 and 0. Number 7 (used as a power of two in the ASCII code) and number 8 (used in EBCDIC) show the number of bits used for encoding. Compare the charts of special characters for ASCII and EBCDIC: You can see that EBCDIC code is able to encode almost twice the number of special characters that ASCII can. Now you know why. In all, ASCII is able to encode 128 characters; EBCDIC can encrypt 256 characters. Since the number of regular characters, such as letters and numerals, is the same for both codes, the number of used bits translates into the ability to encode a larger or smaller number of special characters.
- Second, these codes look like an arbitrary combination of 1s and 0s, but they are not. If you look closely at, say, capital "*F*" and the following capital, "*G*," in ASCII, you can see that they differ in only 1 bit:

F	1000110
G	1000111

The binary code for "F" corresponds to the decimal number 70 and the binary code for "G" is the decimal number 71. ("Decimal" refers to our common numbering system, founded on the base of 10, which we use in everyday life.) In fact, all capital letters are encoded in decimal numbers that range from 65 for "A" to 90 for "Z," and all lower-case letters are encoded from 97 for "a" to 122 for "z." For example, the letters "A," "B," "C," and "D" correspond to decimal numbers 65, 66, 67, and 68. Hence, the ASCII code is simply the presentation of these numbers in binary format. (You can readily find the tables with a detailed presentation of the ASCII code in decimal, octal, and hexadecimal systems; see, for example, Searle (2002).)

- Third, as we have already mentioned, the ASCII and EBCIDIC codes present their bits in different orders: ASCII starts with the most significant bit (MSB) and finishes with the least significant bit (LSB), whereas EBCIDIC does this in reverse order; that is, EBCIDIC starts with LSB and finishes with MSB. This is important to know when you convert a binary number into its decimal counterpart.

3.1.2.2 From Bits to Electrical Pulses

Transmission (Line) Codes: What They Are With character codes, we have made the first step in moving from the analog world to the digital. As a result, we have all the information presented in 1s and 0s. But we cannot transmit 1s and 0s; we can transmit only electrical or optical signals. Therefore, to complete the work, we need to present these bits as tangible signals. Let us make an agreement to present bit **1** as an electrical pulse with the amplitude 3.3 V and present bit **0** as an electrical pulse with the amplitude 0 V (that is, no pulse). This is shown graphically in Figure 3.1.10a. You are quite familiar with this presentation because we have used it in our previous examples.

> *As is true of character codes, there is no natural or scientific basis for such a presentation; this is simply a universally accepted format for coding. The industry refers to all such accords on presenting bits as electrical signals as transmission, or line, codes.*

Several most popular transmission codes are shown in Figure 3.1.10.

Thus, the conversion of character codes into electrical signals prior to their transmission completes the work of translating information from its original analog format to digital signals ready for transmission.

Transmission Codes We Can Arrange Presenting 1 and 0 as the "on" and "off" modes of an electrical signal, as we have done in Figure 3.1.10a, seems to be a natural way to encode bits. This code is called *NRZ*. However, this is not the only transmission code used in practice. *The "off" mode is*

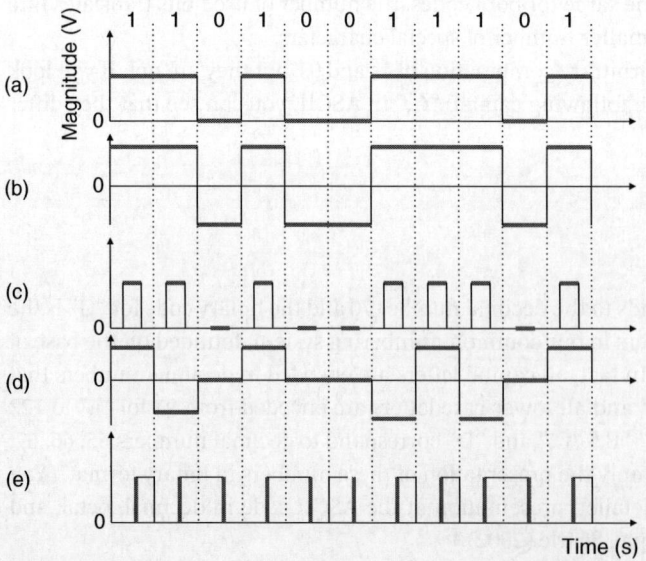

Figure 3.1.10 Transmission codes: (a) Unipolar nonreturn-to-zero (NRZ); (b) bipolar NRZ; (c) return-to-zero (RZ); (d) alternate mark inversion (AMI); (e) Manchester code.

always uncertain: It may be caused by many reasons other than the sending of 0. For example, interruptions in sending a signal, substantial signal distortion during transmission, and errors in signal reception may be erroneously seen as an "off" mode by a receiver. (Note that we have stressed that all components of a transmission system – the transmitter, the transmission link, and the receiver – can contribute to these errors.) To avoid the "off" position as a transmission code, we can use a positive pulse to transmit 1 and a negative pulse to transmit 0. Such a *bipolar* (that is, a double-polarity) code is very popular in practice. In addition to eliminating any ambiguity with respect to the "off" situation, it has a number of other benefits, such as lowering the average power of the signal being transmitted. This type of code is shown in Figure 3.1.10b.

Lowering electrical power for transmission is a significant requirement. Indeed, we need to deliver information, not power, since delivering electrical power is the realm of another industry. For the telecommunications engineer, the ideal version of a transmission system is one that delivers information and does not consume any power because power costs extra money. (But, clearly, this ideal transmission is simply hypothetical because we need to transmit electrical or optical signals, which cannot be done without consuming power.)

Let us consider an NRZ code from this standpoint. Transmitting a pulse for a single bit takes a certain amount of time (called *bit time*) and requires some electrical power. We do not need to transmit a pulse for a long time if we just want to inform a receiver about the state of a digital signal. Indeed, to deliver information as to whether a signal is in a HIGH or LOW state, we can transmit a short pulse, thus saving electrical power. Such a transmission code is called *RZ,* since this signal returns to the 0-V position during a bit-time interval. This code is shown in Figure 3.1.10c. Obviously, an RZ code can be implemented in a bipolar version.

Even the bipolar RZ code, where we use positive and negative pulses for every transmitted bit, still requires substantial electrical power for transmission. We can save even more power if we remain in the "off" position when transmitting bit 0 and invert pulse polarity with every bit **1**. This code is called *AMI*, and it is used in telephony for digital voice transmission. The AMI code is shown in Figure 3.1.10d. (Note that the AMI code uses the "off" position despite all the well-known drawbacks of this mode. However, AMI code employs a special technique called bipolar [binary] eight zero substitution, or B8ZS, to cope with all the shortcomings of the "off" transmission mode.)

All the codes discussed above present the state of a digital signal by virtue of the pulse amplitude. But we already know, from our discussion of analog transmission, that amplitude is the most vulnerable characteristic of a transmitting signal because the amplitude suffers the most from electromagnetic interference. *We can inform our receiving party about the state of a digital signal* not by the level of the signal but *by the transition of the signal from one level to another*. Let us agree that when a signal goes from the lower level (near 0 V) to the upper (near 3.3 V), it delivers bit **1**; when the signal transits from the upper level to the lower, it delivers logic **0**. With this agreement – this code, that is – we can avoid the problems associated with amplitude distortion. Such an electrical signal delivering binary information is called a *Manchester* code; it is shown in Figure 3.1.10e.

As you can see from the above discussion, we have a number of reasons to develop different types of transmission codes; in fact, different applications require different types of codes. We will return to the subject of transmission codes in Chapter 10 after we gain a deeper knowledge of transmission techniques.

Important notes

- Transmitting information with a set of transmission codes is an example of *baseband transmission*. Actually, the purest example of a baseband transmission is a verbal conversation where information is delivered in the band of the original frequencies generated by a human voice cord. Nevertheless, this term is applied to any type of transmission in which the signal is carried in its own frequency band. The transmission type is opposite to a baseband is called *band-broad* (or *broadband*). In a broadband transmission, the signal is delivered by a carrier at a frequency higher than the signal frequency; this is achieved by *modulation*. We will discuss these topics throughout the book.

- Be aware that, besides character coding and transmission coding, there are also *source coding* and *channel coding*. The source coding eliminates the redundant bits to compress the volume of a file while preserving its content; the channel coding adds redundant bits to a file in order to conserve transmitted information despite errors caused by noise, losses, and other detrimental effects. These codings are an extremely important and dynamic area of modern communications, but they are outside the scope of this book.

3.1.2.3 How to Go Digital Communications: A Summary

Review of the Basic Steps Now we can summarize the above discussion: On the one hand, we write and communicate in analog, that is, with readable characters consisting of letters, numerals, punctuation marks, and control characters. On the other hand, computers are binary machines that operate with logic 1s and 0s – bits, that is. To encode characters into bits, we use character codes, such as ASCII and EBCIDIC. To deliver these 1s and 0s from one point to another (from a keyboard to a computer screen, for example), we need to encode these bits to electrical signals. We do this by using such transmission codes as NRZ, RZ, AMI, or the Manchester code, all of which are examples of a digital signal.

Therefore, the total change from characters to bits to digital signals consists, first, of applying a character code and, secondly, of using transmission codes.

Two Additional Points: Encoding and Types of Digital Signals Note that we use the word *encode* for two distinctly different operations: for logical transformation and for technical procedure. This is because in both cases we apply a kind of agreement – a code, that is – to perform these operations.

Why do we emphasize time and again that a code is nothing more than a convention? In engineering, we use formulas and equations derived from our experiments, that is, from our observations of natural phenomena or from our theoretical assumptions. Thus, engineering formulas have a scientific foundation but codes do not. In contrast, codes are pure inventions. To make codes practical, everyone involved in their application must accept them and consider them standards. Being accepted as de facto standards, practical codes have usually passed through national and international standard-developing organizations and thus become de jure standards.

And here's another important point to note: NRZ, RZ, AMI, and Manchester codes, which encode the string of bits, are simply examples of digital signals. There are many other digital signals that change in ways other than by merely jumping between "on" and "off" positions. This is because we use a digital signal to encode not only binary words – even though a binary format is the most common way to present information today – but many other formats as well, such as the decimal format. See Figure 2.1.5c, where an example of a multilevel digital signal is shown. We always need to remember that the main feature of a digital signal is that it changes discretely.

As an example, let us agree on the following: Every decimal numeral will be encoded with an electrical pulse of appropriate amplitude. Thus, number 0 will be encoded with a 0-V pulse, number 1 will be encoded with a pulse of 1-V amplitude, and so on. This code is shown in Figure 3.1.11.

Obviously, if you need to encode any decimal number, the code shown in Figure 3.1.11 will not be accepted in practice. Why not? Because nobody wants to work with amplitudes that range in voltage from 0 to infinity. However, if you need to encode a set of numbers whose maximum value does not

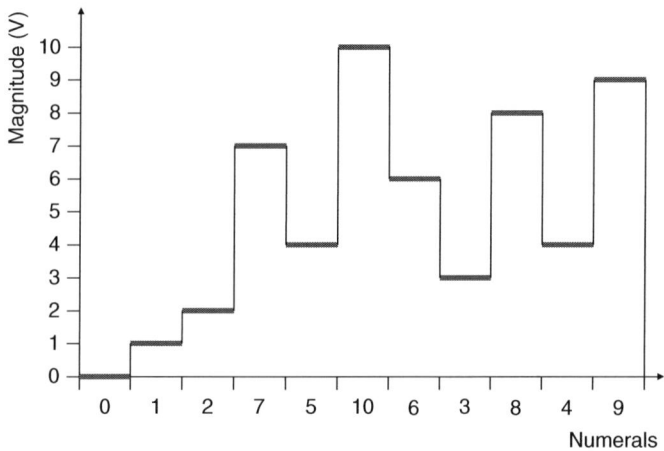

Figure 3.1.11 Example of a digital signal: encoding decimal numerals.

exceed, say, 20, this code will work well. In fact, this type of code is employed in the technique used in converting an analog signal to a digital one. These two pointers will help you to avoid confusion when referring to the words "code" and "encode."

Questions and Problems for Section 3.1

- Questions marked with an asterisk require a systematic approach to finding the solution.
- Many questions and problems, including those marked with an asterisk, imply that you, in addition to reading the textbook, will do your research to find the answers. Consider such questions as mini-projects.

Four parameters define the logic levels for digital logic families: V_{IL}, V_{IH}, V_{OL}, and V_{OH}.

- V_{IL} defines the maximum voltage level that will be interpreted as a "0" by a digital input.
- V_{IH} defines the minimum voltage level that will be interpreted as a "1" by a digital input.
- V_{OL} defines the guaranteed maximum voltage level that will appear on a digital output set to "0."
- V_{OH} defines the guaranteed minimum voltage level that will appear on a digital output set to "1."

Why Go Digital

1. What is the difference between analog and digital transmission? Explain. Sketch graphs showing this difference.

2. Does a digital signal get distorted during its transmission? Explain.

3. Why is analog transmission so sensitive to signal distortion?

4. What is the main advantage that digital transmission has over analog?

5. How can a digital signal deliver correct information despite its distortion?

6 Can a digital signal deliver erroneous information? In what case? Explain. Sketch a figure showing this.

7 What two types of variation in a digital signal can digital transmission tolerate?

8 How does the RS-232 standard encode 1s and 0s?

9 Take the letter K in ASCII and present it in RS-232 encoding. Show this in a sketch.

10 Consider RS-232 traditional format with ±15 V levels:
   a) An RS-232 transmitter sends a pulse whose amplitude is 7 V? What bit does this pulse encode? What if the amplitude is −12 V? 3.6 V? Explain.
   b) An RS-232 receiver gets an input pulse whose amplitude is −12 V. What bit has been delivered? What if the amplitude is −2 V? Explain.
   c) In the RS-232 code, what logic is presented by a pulse with amplitude of 1.5 V? −2 V? Explain.

11 For the RS-232 standard, sketch a diagram specifying the range of voltages that guarantee delivering logic 1 and logic 0, the noise margin, and the uncertain region.

12 Consider the RS-232 transmission code depicted in Figure 3.1.4: When a receiver gets a signal whose amplitude ranges from 5 to 15 V, this guarantees receiving bit **0**; but when a receiver obtains a signal between 3 and 15 V, it is considered a recognizable bit **0**. What is the difference between these two terms? What causes this difference?

13 What does the acronym IC stand for? What nickname for IC do you know?

14 What applications of ICs are you aware of?

15 What logical format – binary, decimal, or any other – do digital ICs operate in?

16 Define the following voltages: V_{OH}, V_{IH}, V_{IL}, and V_{OL}. Sketch a diagram to visualize your definitions.

17 Is V_{OH} the minimum or maximum allowed voltage? Is this an input or an output voltage? What about V_{IL}? Explain.

18 Voltage specifications for the new digital logic family called enhanced transceiver logic (ETL) are given as $V_{OH} = 2.4$ V, $V_{IH} = 1.6$ V, $V_{OL} = 0.4$ V, and $V_{IL} = 1.4$ V. Determine the guaranteed voltages, recognizable voltages, and noise margins for high-level and low-level signals. Sketch a diagram. Show the uncertain region.

19 Compare specifications for the TTL, LVTTL, CMOS, LVCMOS, and ECL logic families given in Tables 3.1.1 and 3.1.2. What are the advantages and drawbacks of each family?

20 Why are V_{OH} and V_{IH} given as minimum voltages, while V_{IL} and V_{OL} are given as maximum voltages?

21 The noise margin of a high-level signal is the result of subtracting the input voltage from the output ($V_{OH} - V_{IH}$), whereas the noise margin for a low-level signal is the result of subtracting the output voltage from the input ($V_{IL} - V_{OL}$). Why?

22 Consider the TTL logic family: What logic is delivered with an input pulse of 2.2 V amplitude? What if an input pulse's amplitude is 1.2 V?

23 What range of voltages is called invalid voltage? Why is it called invalid?

24 Determine and compare invalid-voltage ranges for the TTL, CMOS, LVTTL, LVCMOS, and ECL logic families. (Refer to Tables 3.1.1 and 3.1.2.) From this standpoint, which family is the best? Why?

25 Does noise add voltage to or subtract voltage from a digital signal? Explain.

26 Noise can add voltage to and subtract voltage from a digital signal during transmission. What effect – adding or subtracting – will be harmful?

27 Consider the LVCMOS logic family. Determine two types of tolerance states that digital transmission allows for its reliable operation. Give the numbers. Sketch a diagram visualizing your answer.

28 The noise margin for CMOS circuitry is given as 1000 mV. If the signal has been sent with an amplitude of 3.4 V and received with an amplitude of 2.6 V, what bit – **1** or **0** – does this signal deliver? Explain.

29 What is the difference between binary and digital signals?

30 Can we deliver binary information with an analog signal? Explain.

31 What is the difference between data and digital communications?

How to Go to Digital Communications

32 How does an analog signal deliver information?

33 Explain the main advantage that digital transmission has over its analog counterpart.

34 If digital communications has a big advantage over analog, why do we still need to use analog communications?

35 What are two major steps necessary to implement digital communications? Explain.

36 What does the acronym *bit* stand for?

37 Why is bit **1** referred to as HIGH and bit **0** as LOW? Explain.

38 What is the difference between bit and byte? Between binary word and byte? Explain.

39 Is the following combination of bits – **1 1 1 0 0 10 1** – a byte or a binary word?

40 Give examples of using the binary format for computations and communications.

41 What is the meaning of the term *code* in communications?

42 Why do we need to introduce character codes?

43 What is the difference between character codes and transmission codes?

44 What do the acronyms ASCII and EBCIDIC stand for?

45 What are the similarity and difference between ASCII and EBCIDIC codes? Give examples.

46 Show the following characters in ASCII and EBCIDIC codes: C, D, c, m, Y, !, &, and *.

47 Why do the letters "b" and "c" differ in 1 bit only in ASCII code?

48 Present the word "Nice" into the ASCII and EBCIDIC codes.

49 A manufacturer stores parts in a warehouse by using five-character labels like AHp-6. The computers, obviously, will read this label in binary format. Write such a label in ASCII and EBCDIC codes.

50 Why do we need transmission codes?

51 Are there any natural or scientific bases for introducing transmission codes?

52 Presenting 1 and 0 as the "on" and "off" modes of an electrical signal seems to be a natural way to encode bits. Why is this transmission code called nonreturn-to-zero (NRZ)?

53 What drawbacks of an NRZ code does a bipolar NRZ overcome? Explain.

54 What problem does the return-to-zero (RZ) code solve? Explain.

55 What is the main advantage of an alternate mark inversion (AMI)? Explain.

56 What is the distinguishing feature of a Manchester code? Explain. Sketch a figure showing this.

57 *Present the word "Code" in NRZ, RZ, AMI, and Manchester transmission codes using the ASCII coding.

58 *What is the main difference in the nature of a scientific law, such as Ohm's law, and a code, such as ASCII code? Explain.

3.1.A Brief History of Character Codes

Character codes, often called alphanumerical codes, have a storied history. Long before the advent of computers, people found the need to develop character codes to present comprehensible information through the use of symbols. The first time this need arose was when the telegraph was invented. Initially, the telegraph was a primitive electric transmission system that could make electrical current flow through a circuit or not. In other words, a receiver could only detect the presence or absence of a signal. But the absence of a signal did not represent any kind of transmission, as it does in today's codes. At the time, it simply meant that no information was being delivered because the system was either down or broken. Thus, the inventors of the telegraph found themselves with only one transmission option: sending a signal.

3.1.A.1 International Morse Code

The first practical telegraph was devised by the American inventor Samuel Morse in 1837. His was a brilliant idea: Use only two electrical signals – short and long and eliminate the absence of a signal. This two-level (binary, in modern terms) system proved to be practical because of its reliability. (One has to bear in mind, of course, the poor quality of electrical systems at that time.) It is interesting to note that traditionally in telegraphy one signal is called a *mark* and the complementary signal is called a *space*. You can see the tracks of this terminology in modern transmission systems, such as telephony.

After developing the electrical parts for his telegraph, Morse immediately realized that he needed to encode letters in a two-level format to be able to transmit messages with his system. Then, in 1838, he developed his famous code, which was later adopted throughout the world. In 1844, Morse sent the first telegraph message ("What hath God wrought!") over the line installed between Baltimore, MD, and Washington, DC.

In the interest of historical accuracy, we must point out that the first working telegraph was built in 1837 by the Englishmen Sir Charles Wheatstone and Sir William Cooke. Unfortunately, this electrical system was very complicated and unreliable. Morse, who was an accomplished painter, not an engineer, studied in London prior to the Wheatstone–Cooke invention, brushing up on the latest developments in electromagnetism by British scientists.[2]

Everyone has at least heard of the Morse code, even though it is pretty much out of vogue today. Every conventional character is represented in the code by a combination of dots and dashes. In the early days of telegraphy and radio, this was the only format for transmitting information. Morse conceived a simple but ingenious idea: Letters that are frequently used should have fewer symbols than those used less frequently. Thus, the commonly used letters "E" and "T" are encoded as "•" and "—," respectively. The modern international Morse code is shown in Table 3.1.A.1. Analyze this code to see how Morse encoded English letters.

Think about the makeup of this code. Why, for example, is the letter "*A*" presented as "• —" and the letter "*K*" coded as "• — •"? There is not any scientific or commonsense reason for this, only the arbitrary decision by Samuel Morse in 1838 to structure it this way. His code was later adopted globally and has survived all the technological revolutions that have come ever since.

---

2 Searle (2002).

Table 3.1.A.1 International Morse code (Footnote 5).

A	•—	N	—•	0	—————
B	—•••	O	———	1	•————
C	—•—•	P	•——•	2	••———
D	—••	Q	——•—	3	•••——
E	•	R	•—•	4	••••—
F	••—•	S	•••	5	•••••
G	——•	T	—	6	—••••
H	••••	U	••—	7	——•••
I	••	V	•••—	8	———••
J	•———	W	•——	9	————•
K	—•—	X	—••—	, (comma)	——••——
L	•—••	Y	—•——	. (period)	•—•—•—
M	——	Z	——••	?	••——••

3.1.A.2 Baudot Code

The next notable achievement in the development of character codes was the introduction of a 5-bit code in 1874 by the French engineer Jean-Maurice-Émile Baudot. His was the first binary code. With 5 bits, this code can encode 32 characters, since $2^5 = 32$. The code is shown in Table 3.1.A.2. To enable encoding Arabic numerals and punctuation marks, the code uses a "shift" operation to switch between two planes (columns here) of characters.

As you can see from Tables 3.1.A.1 and 3.1.A.2, the Baudot code is much more advanced than the Morse code. This code not only includes numerals that were added to the Morse code years after its inception, but it also enables the use of punctuation marks and control characters. The Baudot code was standardized by the Consultative Committee for International Telephone and Telegraph (CCITT) in 1932 as a 5-bit code for teleprinters.

It should be noted, too, that the Baudot electrical transmission system is more advanced than the telegraphic system. In particular, the Baudot system uses "on" and "off" electrical current states to transmit the binary code. The price for using the "off" state is the need to send a start and stop bit before and after transmitting each character.

A highly effective binary code for use with punched cards and tabulating machines was developed by the American Herman Hollerith and was used in computing the US census in 1890. In fact, this code and the accompanying machines were so effective that all computations associated with that census were completed less than six weeks after the raw data were gathered. Contrast this with the seven years the US Census Bureau spent processing data gathered during the previous (1880) census. The Hollerith code and his machines eventually led to the formation of the International Business Machines Corporation and the development of the EBCIDIC code.

The evolution from the Hollerith code to EBCIDIC included two intermediate steps: First, IBM developed a 4-bit BCD code and later the company developed a 6-bit BCDIC (binary coded decimal interchange code). Clearly, with the advent of new computers, neither the 4-bit nor 6-bit codes could satisfy the requirements needed to encode the number of characters necessary for the successful operation of modern computers. Thus, in 1964 IBM announced the new computer series

Table 3.1.A.2 Baudot code (Footnote 5).

Binary code	Letters	Figures
00011	A	-
11001	B	?
01110	C	:
01001	D	$
00001	E	3
01101	F	!
11010	G	&
10100	H	STOP
00110	I	8
01011	J	'
01111	K	(
10010	L	)
11100	M	.
01100	N	,
11000	O	9
10110	P	0
10111	Q	1
01010	R	4
00101	S	BELL
10000	T	5
00111	U	7
11110	V	;
100011	W	2
11101	X	/
10101	Y	6
10001	Z	"
00000	n/a	n/a
01000	CR	CR
00010	LF	LF
00100	SP	SP
11111	LTRS	LTRS
11011	FIGS	FIGS

n/a, CR, LF, SP, LTRS, and FIGS are the control characters.

System/360, where EBCIDIC was used for the first time. Since then, IBM has used this 8-bit code for its mainframes and – later – its personal computers.

IBM has developed 57 national EBCIDIC character codes and the international reference version. The latter is compatible, from a practical standpoint, with ASCII; however, the translation between EBCIDIC and ASCII has never been officially defined. This is because some ASCII characters do not exist in EBCIDIC (for example, square brackets) and vice versa (for example, the cent sign). In general, since ASCII is a universally accepted code, the problem of EBCIDIC–ASCII compatibility is essentially the point that IBM has taken care of. As of today, ASCII–EBCDIC compatibility exists factually and users do not need to know what character code their computers are using.

ASCII was initially developed by the American Standards Association in 1963. The ASCII standard was revised in 1968 and again in 1986 by the same organization, which by that time had been renamed the ANSI. To accommodate other languages, the international version of ASCII was developed and adopted in 1967 by the International Organization for Standardization (ISO). The need to accommodate more characters led to the development of an extended version of ASCII that uses 8 bits and that can encode 256 ($2^8 = 256$) characters. In 2016, by the IEEE initiative, the Milestone plague was installed in AT&T Lab building to commemorate ASCII development. The plague reads: "ASCII, a character-encoding scheme originally based on the Latin alphabet, became the most common character encoding on the World Wide Web through 2007. ASCII is the basis of most modern character-encoding schemes."[3]

The fact that today we have two character codes – ASCII and EBCDIC – in the computer world is not a great advantage; quite the contrary. This situation, in fact, has caused some problems associated with the compatibility of computer equipment manufactured by different companies. Fortunately, modern software makes the conversion from one code to the other automatically, so you are not even aware as to which code your computer is using.

3.2 Digital Signals and Digital Transmission – Introduction

Objectives and Outcomes of Section 3.2

Objectives

- To gain an understanding of the main parameters of a digital signal.
- To learn the relationship between the parameters of a digital signal and the characteristics of digital transmission.

Outcomes

- Deepen your knowledge on the waveform of a digital signal, which determines the main parameters of a digital signal. Gather that the magnitude-related parameters include the amplitude and mean levels of HIGH and LOW states. Understand that the timing parameters include bit time, bit width, symbol (pulse) time, rise time, fall time, period, and duty cycle.
- Learn that there are single-level and multilevel modulations. Study that when single pulse (symbol) carries 1 bit, the modulation is called a single-level modulation; when one pulse (symbol) carries several bits, we have a multilevel modulation. Know that in single-level modulation, bit rate is inversely proportional to duration of a pulse (symbol) or a bit; in multilevel modulation, transmission speed is determined by symbol (baud) rate, which is inversely proportional to a symbol duration. Be aware that a multilevel modulation enables us to increase bit rate or to save the bandwidth required for transmission of a given message.
- Understand that in digital transmission, we distinguish between the bit rate, BR (b/s), which is the characteristic of a digital signal, and the channel capacity, C (b/s), which is the characteristic of a transmission link. Recall that similarly, we distinguish between a signal bandwidth and a channel bandwidth in analog transmission. Revisit Shannon's law that states that the channel

---

3 IEEE Photonic Society Newsletter, August 2016, p. 24.

capacity – the maximum bit rate supporting error-free transmission – is proportional to channel bandwidth and logarithmically depends on signal-to-noise ratio (SNR).
- Realize that the waveform of a real digital signal differs from an ideal waveform due to various detrimental transmission effects. Comprehend that it is difficult to define the basic signal parameters of a real waveform and that this issue is resolved by using a simplified industrial model of the real waveform. Learn how this model enables us to define a signal's amplitude, pulse width, and rise/fall times.
- Grasp the fact that in reality, the parameters of real digital signals are random variables because these signals are distorted by noise and the instabilities of a transmission system. Understand that, consequently, the HIGH and LOW levels of a real digital signal are defined as the *mean values* of hundreds and thousands of measurements and that the pulse amplitude, A (V), is determined as a difference between these mean values.
- Know that the digital signal amplitude serves as a reference point in defining such timing parameters as the pulse width, signal period, and the rise and fall times. Be aware that the rise (or fall) time, t_r (s), determines the bit rate, BR (b/s), as BR (b/s) = α/t_r (s), where α is a constant whose value depends on a transmission code. Derive that the rise time is equal to bit time multiplied by α.
- Gain understanding that though transmitted bit stream is a random combination of 1s and 0s, we can introduce pulse repetition interval (PRI), also called signal period. Learn that the period is defined as the time interval between two consecutive rising (falling) edges measured at 50% amplitude of a "010101" bit stream, and that, in practice, a period is determined by averaging thousands of individual measurements.
- Ascertain that the signal's period plays a key role in defining another important timing parameter: duty cycle, which is given as the ratio of a pulse width to a signal period. Know that a duty cycle determines how long a bit's amplitude is not equal to 0 with respect to the signal's period, which enables us to evaluate power consumed by a specific signaling code for transmission. Be aware that, for example, RZ signaling has a much smaller duty cycle than NRZ and therefore consumes less power in transmitting the same volume of information.
- Find out that duty cycle is not the only parameter to be taken into account when selecting a signaling code for a specific transmission. Familiarize yourself with a comparison of the NRZ and RZ transmission codes, for instance, that reveals the following factors to be considered in choosing the right signaling for a digital transmission system:
    ◦ Due to its simplicity, NRZ signaling was more popular than RZ when the transmission rate was low, less than 1 Gb/s.
    ◦ RZ signaling requires more bandwidth – typically, twice that of NRZ.
    ◦ However, due to its smaller pulse width, RZ signaling is more immune to intersymbol interference (ISI) than NRZ, which makes RZ more reliable – and therefore more preferable – coding for high-speed digital transmission.
    ◦ The other significant advantage of RZ signaling over NRZ is that RZ requires less power for transmission thanks to its smaller duty cycle.
  Get to know that the designer of a transmission system must take into account all the above considerations.

3.2.1 Ideal Digital Signal and Characteristics of Digital Transmission

3.2.1.1 The Waveform of an Ideal Digital Signal
As with any signal, we have to start the study of a digital signal by examining its waveform. The waveform of a digital signal features its discrete changes from one magnitude to the other, as

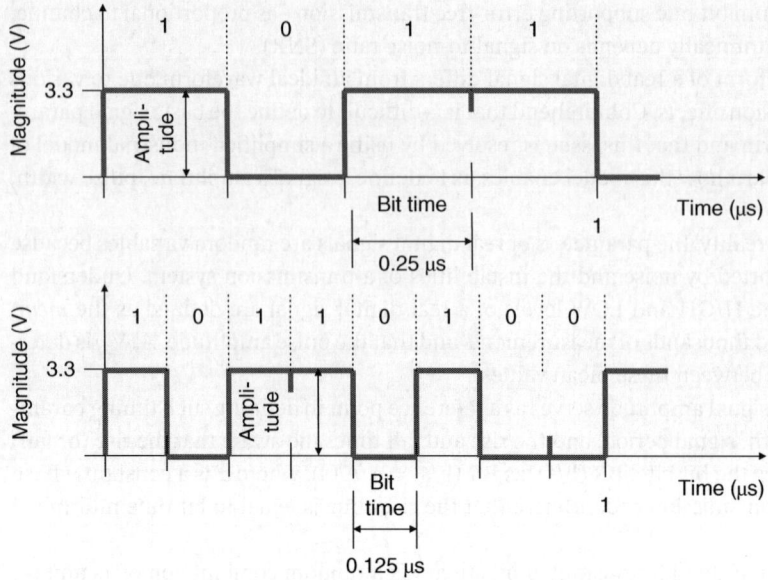

Figure 3.2.1 The amplitude and bit (pulse) time of an ideal digital signal.

Figures 3.1.1 and 3.1.2 demonstrate. These figures also show that there are three types of digital signals: (i) multilevel digital signals, (ii) three-level, or bipolar, signals, and (iii) bi-level, or two-state, or "On/Off," digital signals. It becomes clear from examining these waveforms that we need to know two types of parameters: those associated with magnitude (vertical measurements) and those associated with time (horizontal measurements).

Let us consider the digital signal shown in Figure 3.2.1. In terms of vertical measurements, the first parameter we need to know is its *amplitude*. Since this signal represents HIGH and LOW logic states, we define the digital signal's amplitude as follows:

The amplitude, A (V), is the maximum value of a pulse magnitude carrying either bit 1 or bit 0.

Figure 3.2.1 exemplifies this statement by showing the pulse amplitude is 3.3 V for bit 1 and 0 V for bit 0.

In terms of horizontal measurements, the first parameter we need to know is the *duration of an individual bit*. Since signals carrying HIGH and LOW logic states have the same duration, we define bit time as follows:

Bit time, T_b (s), is the duration of an individual bit.

The bit time of the top signal in Figure 3.2.1 is 0.25 μs and that of the bottom is 0.125 μs. Note that bit time is sometimes called *bit interval*.

Figure 3.2.1 shows an example of an *ideal digital signal* (pulses) carrying binary logic – bits 1 and 0 – information. Here, idealization implies that the amplitudes of these pulses are constant during the bit time, and these intervals are constant, too.

It is important to understand that Figure 3.2.1 shows a special case of a digital signal, a case in which *one pulse carries one bit*. What is more, here the pulse interval, T_p (s), coincides with its bit time, T_b (s). In general, however, T_p *(s) and* T_b *(s) are different timing parameters*, and we always have to bear this point in mind. We encounter this difference, for example, in studying duty cycle in this section.

3.2.1.2 Pulse Interval and Transmission Rate; Bit Time and Bit Rate

As the study of multilevel modulation shows, we *can consider a pulse as a symbol able to carry one bit or many bits.* Thus, pulse interval, T_p (s), is a symbol time, T_S (s). *Pulse (symbol) interval, T_S (s), determines how many pulses (symbols) per second a communications system can transmit.* Examine Figure 3.2.1 and count the number of symbols carried by each signal in this figure: The top signal carries 4 symbols/µs; thus, this signal transmits 4 symbols/1 µs = 4 symbols/$(1 \times 10^{-6}$ s$)$ = 4×10^6 symbols/s = 4 Msymbols/s. From another viewpoint, this signal naturally has a pulse (symbol) interval of T_S = 0.25 µs to accommodate 4 symbols in 1 ms. This results in symbol rate SR (symbol/s) = $1/T_S$ (s) = 1/(0.25 µs) = 1/$(0.25 \times 10^{-6}$ s$)$ = 4×10^6 symbols/s = 4 Msymbols/s. Similarly, the bottom signal in Figure 3.2.1 delivers 8 symbols/µs and has a symbol time of T_S = 0.125 µs; thus, its symbol rate is as follows: 1/(0.125 µs) = 1/$(125 \times 10^{-6}$ s$)$ = 8×10^6 symbols/s = 8 Msymbols/s.

The *number of symbols/s* is also referred to as *baud rate* (after Émile Baudot, inventor of the character code discussed in Section 3.1). The unit of a baud rate is *baud*, **Bd**. Thus,

$$\text{SR (Bd)} = \frac{\text{number of symbols}}{\text{s}} \tag{3.2.1a}$$

Therefore, the *symbol rate*, SR (Bd), which is also often called the *transmission rate* or *signaling rate*, is the inverse of a symbol (pulse) interval, T_S (s); that is,

$$\text{SR (Bd)} = 1/T_S \text{ (s)} \tag{3.2.1b}$$

With the same approach, we can introduce a crucial parameter of a digital communication: *bit rate*. The *bit rate determines how many bits, which is the actual information, is transmitted per second. We measure the bit rate in number of bits per second, b/s, also denoted as bps, or bit/s.*

Bit rate, BR (b/s), is the number of bits transmitted per second.

or

$$\text{BR (b/s)} = \frac{\text{number of bits}}{\text{s}} \tag{3.2.1c}$$

It is clear from the above discussion that

$$\text{BR (b/s)} = 1/T_b \text{ (s)} \tag{3.2.1d}$$

Again, in Figure 3.2.1, the bit time and symbol time coincide, and therefore the bit rate is the same as the baud rate.

To learn more about baud rate, see Appendix 3.1.A. Baud rate is what a link has to support to deliver the number of bits per second in multilevel modulation. Let us return to our example with a dial-up modem operating over a copper telephone line: Suppose this line carries 8000 pulses/s, that is, transmitting at the baud rate of 8 kBd. If we manage to code 5 bits into one pulse (symbol), then the bit rate, as we calculated previously, is 40 kb/s. Hence, the line will transmit 8000 pulses/s, but actually will deliver 40 000 bits/s. This is how we can achieve higher transmission capacity over a channel with given bandwidth.

How can we code many bits into one symbol? We will treat this topic in Chapter 10.

The following example should help you better understand the relationship between bit rate and bit time, the relationship on which we want to concentrate due to its importance.

Example 3.2.1 Bit Time and Bit Rate

Problem

(a) Bit time is 0.1 ms. What is the bit rate?
(b) Bit rate is 2.5 Gb/s. What is the bit time of an individual bit?

Solution

(a) We need to apply Formula (3.2.1d) and compute

$$\text{BR (b/s)} = 1/T_b \text{ (s)} = 1/(0.1 \text{ ms}) = 10 \text{ kb/s}$$

(b) We have to rearrange Formula (3.2.1c) in the following way:

$$T_b \text{ (s)} = 1/\text{BR (b/s)}$$

Then, we plug the numbers in and obtain

$$T_b \text{ (s)} = 1/(2.5 \text{ Gbit/s}) = 0.4 \text{ ns}$$

The problem is solved.

Discussion

- Pay attention again to the unit of bit rate: It is given in b/s. Since bit rate, by definition, is the *number* of bits per second and a number does not have any dimension, the absolute unit of bit rate is 1/s. This is why the units in Formula (3.2.1c) are correct: BR (1/s) = $(1/T_B)$ (1/s).
- If you want a high bit rate, you must arrange a short bit time, which enables us to place the greater number of bits within a second. Since the number of bits per second is a measure of the amount of information delivered per second, we certainly want to attain as high a bit rate as possible. (See Section 1.3 and, particularly, the discussion of Hartley's law to refresh your memory as to why we need a high bit rate.) But we do not have any other means to achieve a high bit rate except for making the bit time of the pulses as short as possible. Today's optical communications technology achieves bit rate as high as tens of terabits per second, Tbit/s, where *tera-* stands for 10^{12}. By (3.2.1c), such a bit rate would require the bit time of an individual pulse to be on the order of picoseconds, that is, 10^{-12} s. As one may guess, this is not an easy task. A more detailed investigation reveals, however, that the relationship between bit time and bit rate is not always as straightforward as we discuss here. We leave this in-depth consideration to the proper sections of this text. For now, it is enough to stress that modern technology enables us to relax the direct relationship between bit rate and bit time.
- The relationship between bit time and bit rate is quite similar to that between period and frequency: In the digital world, the shorter the bit time, the higher the bit rate; in the analog world, the shorter the period, the higher the frequency. See the discussion of Formulas (2.1.3) and (2.1.4) to brush up on this aspect of analog signals.
- Though in this example we consider bit time and bit rate, we can generalize this consideration to the case of symbol (baud) rate and symbol (pulse) interval.

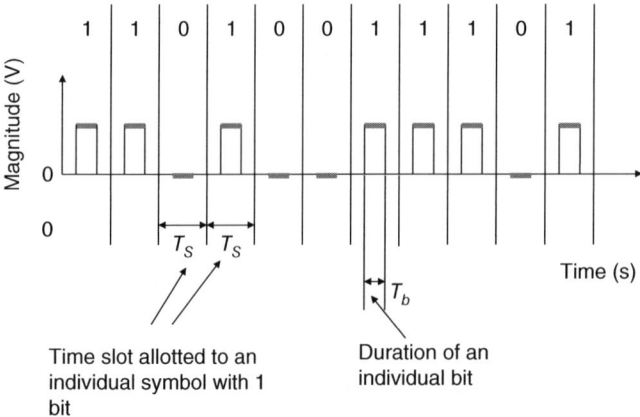

Figure 3.2.2 Bit time and bit rate for RZ signaling.

3.2.1.3 Important Note: The Definition of Bit Time

We have defined bit time as the duration of an individual bit; Figure 3.2.1 illustrates this definition. However, we know that there are a number of different transmission codes, such as NRZ, RZ, AMI, and many others. These are discussed in Section 3.1 and shown in Figure 3.1.10. Let us consider, for example, the RZ code reproduced in Figure 3.2.2. It seems that to define bit time in this case, we need to distinguish between the duration of a symbol (pulse), T_S (s), and the time slot allotted to an individual bit, T_b (s). These two durations are shown in Figure 3.2.2. Figure 3.2.2 creates the impression that the BR (b/s) of this transmission should be different from its symbol (signaling) rate, SR (Bd), because T_S (s) and T_B (s) are different.

In reality, however, a *receiver counts the number of bits received every second* and, in this case, this number is equal to $1/T_S$ (s) because each symbol (pulse) interval contains only 1 bit. Hence, in Figure 3.2.2, we have BR (b/s) = $1/T_S$ (s), which seems to contradict the definitions introduced earlier. This contradiction, however, is false. Figure 3.2.2 shows that *when one pulse carries only 1 bit, the actual duration of an individual bit, T_B (s), plays no role in the transmission speed*. Indeed, from the transmission standpoint, here BR (b/s) = SR (Bd).

Be cognizant of these details because you will encounter transmission systems in which one pulse will contain many bits. In such cases, the definitions discussed here play a crucial role in the correct calculation of bit rate.

3.2.1.4 Bit Rate and Channel (Shannon's) Capacity

At this point, you may be confused with the difference between bit rate and channel capacity. In fact, both entities are measured in bits per second and both of them show the amount of information delivered per second. We briefly mentioned their difference in Section 1.3. (See Eq. (1.3.7), BR (b/s) < C (b/s).) Now, it is time to elaborate this important issue. The difference is as follows:

> Channel (Shannon's) capacity, C (b/s) is the maximum transmission rate that a transmission link is capable of supporting, whereas bit rate, BR (b/s), is the actual number of bits per second carried by a signal being transmitted.

The channel capacity of a link is calculated by the Shannon formula when a channel's bandwidth and SNR are given:

$$C \text{ (b/s)} = \text{BW (Hz)} \log_2(1 + \text{SNR}) \tag{1.3.6R}$$

Again, channel capacity is the *maximum transmission speed* of the channel; it is called the *Shannon limit*, as explained in Section 1.3.

On the other hand, bit rate is the number of bits per second contained in the signal we want to transmit. There is no general formula for computing bit rate; we just need to count the number of bits per second in a given signal. We can compute the *required BR (b/s)* by using Hartley's information law,

$$\text{BR (b/s)} = H \text{ (b)}/T \text{ (s)} \tag{1.3.1R}$$

when the volume of information, H (b), and transmission time, T (s), are known. Thus, C *(b/s) is the characteristic of a transmission channel, whereas BR (b/s) is a parameter of a signal.*

Transmitting Many Individual Signals Through One Link (Multiplexing) Suppose we have a link with given transmission capacity, C (b/s), and we need to transmit a number of signals from different sources with each stream having the same bit rate, BR_i (b/s). (This assumption is made for the sake of simplicity.) How many streams can we transmit? The answer is obvious: We need to divide the total transmission capacity by the individual bit rate. If we denote the number of streams as N, we arrive at the following equation:

$$N = C \text{ (b/s)}/\text{BR}_i \text{ (b/s)} \tag{3.2.2}$$

where $i = 1, 2, 3, \ldots, N$.

The technique of transmitting many signals over one channel is called *multiplexing*.

Let us consider an example of multiplexing in traditional telephone transmission.

Example 3.2.2   Bit Rate and Channel Capacity

Problem

An individual voice signal, after converting into the digital format, has a bit rate of 64 kb/s. Suppose we have a channel capable of transmitting at a speed of 1.544 Mb/s.[4] How many voice signals can we multiplex over this channel?

Solution

Clearly, this channel has a capacity to convey more than one voice signal. To find the precise number, we need simply divide the channel capacity by the bit rate of an individual channel, as Formula (3.2.2) states. Thus, we compute

$$N = C \text{ (b/s)}/R_i \text{ (b/s)}$$
$$= 1.544 \text{ (Mb/s)}/(64 \text{ kb/s}) = 1.544 \times 10^6 \text{ (b/s)}/64 \times 10^3 \text{ (b/s)} = 24 \text{ channels}$$

Note that the calculations give us 24.125, but, clearly, we have to present the answer in integers.

Discussion

- Now we understand that when we control bit time, we control bit rate; this operation has nothing to do with the channel capacity. When we control the capacity of a link, this operation has nothing to do with the bit rate of the incoming information stream. Obviously, we want to use our link efficiently by matching the bit rate to the channel capacity.

---

4 The industrial applications of these numbers are shown in Appendix 10.A.

- Suppose we have a message signal whose bit rate is 600 kb/s. How many signals with this bit rate can be transmitted over a given link whose capacity is 1.544 Mb/s ? The answer is two because 1544×10^3 b/s $/600 \times 10^3$ b/s $= 2.57$, which gives two channels. (Again, do not forget to answer in integers.)
- *Important note*: Can you transmit three channels at 0.6 Mb/s each over a 1.5-Mb/s link? At first glance, the answer is no because we need a link with 3×0.6 Mb/s $= 1.8$ Mb/s capacity. However, a closer look reveals that the answer is yes, but at a certain condition. Obviously, we cannot transmit three streams at 0.6 Mb/s each over a 1.5 Mb/s link *simultaneously*. Nevertheless, a 1.8 Mb/s bit rate means that we *need* to transmit 1.8 Mbit of the total volume of information per second, whereas we *can* transmit only 1.5 Mbit/s. Hartley's information law, H (b) $= C$ (b/s) $\times T$ (s), says that there is a trade-off between transmission speed, C (b/s), and transmission time, T (s), when information volume, H (bit), is given. Therefore, we can transmit 1.5 Mbit out of 1.8 Mbit over a 1.5-Mb/s link for the first second and transmit the remaining 0.3 Mbit for a fraction of the following second. In total, we can transmit 1.8 Mbit for

$$T \text{ (s)} = H \text{ (b)}/C \text{ (b/s)} = 1.8 \text{ Mb}/1.5 \text{ Mb/s} = 1.2 \text{ s}$$

Therefore, though we can't exceed a link's transmission speed, we can, however, transmit the required volume of information by increasing the transmission time.

Back to the analogy of a communication system to a water supply, discussed in Section 1.1, since this "pipe" is too narrow for the required stream, it takes longer to deliver the entire volume. We have to always remember this principle hidden in Hartley's information law.

Technologically, the transmission of three channels at 0.6 Mb/s over a 1.5-Mb/s link will require the additional rearrangement of these signals for transmission. This means that we need additional circuitry at both the transmitting and receiving ends. This is why engineers always want to avoid such a problem by properly designing the network.

3.2.2 Parameters of a Real Digital Signal and the Characteristics of Digital Transmission

3.2.2.1 Waveform of an Actual Digital Signal

As pointed out previously, the *waveform* of a signal is the plot of the signal's magnitude vs. time. Figure 3.2.3a repeats the waveform of an ideal digital signal already shown in Figure 3.2.1. From this ideal waveform, we can clearly identify the main signal parameters, such as its amplitude, A (V), and the bit time, T_b (s). However, in reality, the waveform of a digital signal looks quite different when compared to this idealization. Figure 3.2.3b shows the real waveform with typical flaws that are accumulated along the entire transmission path starting from the transmitter, going through a link, and finishing at the receiver.

Observe that with a real waveform, we have difficulty determining the basic parameters. For example, what is the amplitude of this pulse? Is this the height between the maximum and minimum marks or is it the height between some mean (average) high and low levels? We show one of the possible ways to define the amplitude in Figure 3.2.3b; since this is only one of the possibilities, we place a question mark here. The same uncertainty appears in determining the bit time. Should we define bit time as the duration between the middle points of the pulse, as shown in Figure 3.2.3b? Or should we measure bit time between the first crossings of the low-level line?

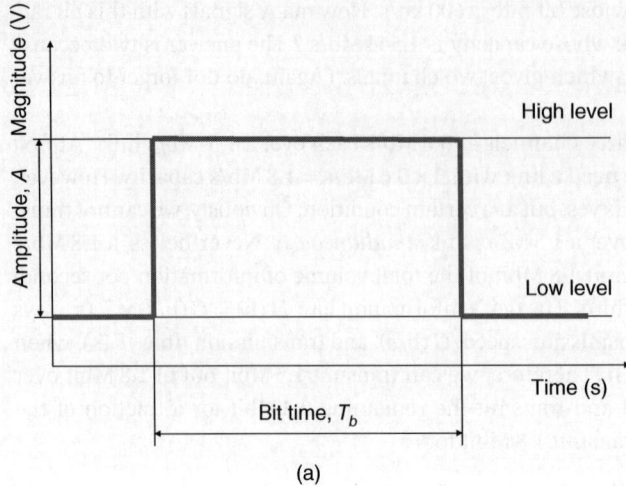

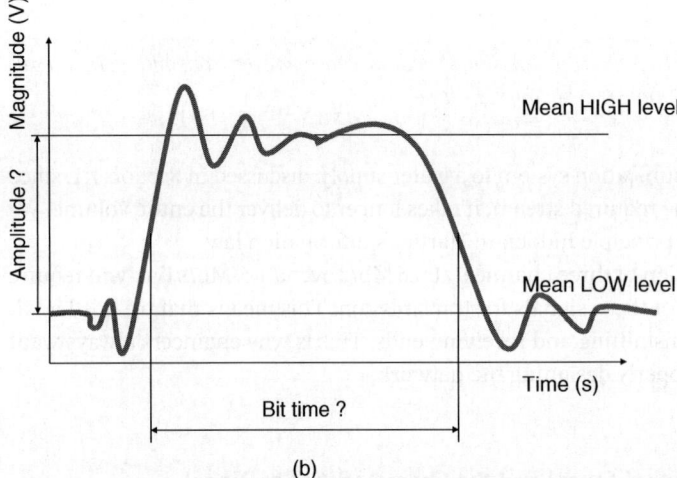

Figure 3.2.3 Waveforms of a digital signal: (a) ideal waveform; (b) real waveform.

Or by some other way? The point is this: The waveform of a real digital signal requires special definitions of the parameters of this signal.

Complete answers to these questions will be postponed until Chapters 9 and 10, where we delve deeper into digital transmission. For our current purpose, we consider the model of a digital pulse that preserves the main features of a real waveform but omits some details. This waveform is shown in Figure 3.2.4.

3.2.2.2 Amplitude and Pulse Width

Let us consider vertical (amplitude) parameters first. In the model shown in Figure 3.2.3a, amplitude, A (V), is defined as the difference between the mean values of the HIGH and LOW levels; that is,

$$A \text{ (V)} = \text{mean HIGH level} - \text{mean LOW level} \tag{3.2.3}$$

where mean HIGH and mean LOW levels are determined as the mean (average) voltages of magnitude fluctuations around the high and low levels of a specific pulse.

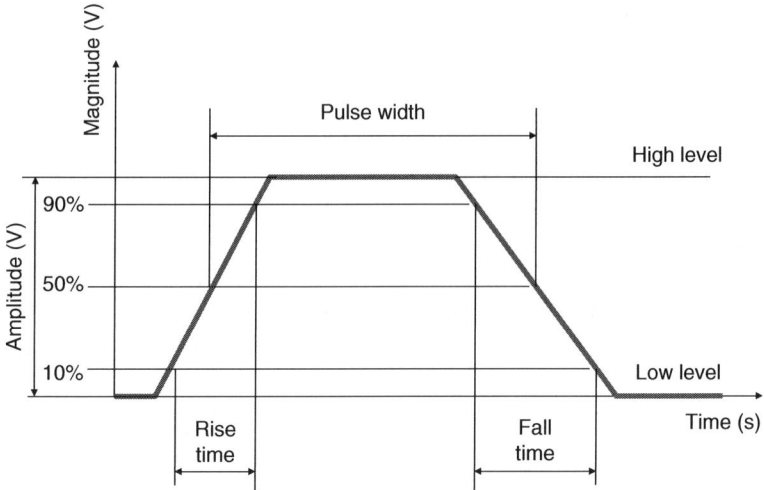

Figure 3.2.4 Basic parameters of a real digital pulse.

As for horizontal (timing) measurements, the first parameter we want to know is something similar to bit time. Looking at Figures 3.2.3 and 3.2.4, it seems natural to consider the *width of a pulse* as analog to bit time. To introduce this parameter, let us take the point of the leading edge at 50% of the pulse amplitude and a similar point of the trailing edge. The time interval between these two points is the pulse width. Refer to Figure 3.2.4.

Pulse width, T_W (s), is the time elapsed between the points of a leading pulse edge and a trailing pulse edge at 50% of the pulse maximum.

This definition of pulse width is called *full width half maximum (FWHM)*. Since pulse width is time, the unit of a pulse width is seconds. Why is pulse width important? As you have no doubt noted, we have not defined bit time. This is because the definition of bit time requires a deeper look into the properties of a digital pulse that we will discuss later in this section. By far, pulse width gives us a sense of how long it takes to transmit an individual pulse; this is a characteristic somewhat similar to bit time.

An important question to consider regarding the pulse-width definition: Why do we use the terms *leading* and *trailing* edges instead of rising and falling edges? The answer is that these definitions enable us to avoid any ambiguity. Indeed, since we can have a pulse with positive and negative voltages, we refer to this feature. Observe a bipolar pulse, such as those shown in Figure 3.1.5a, b. Thus, a leading edge can be either positive or negative; the same is true for a trailing edge. Figure 3.2.4 shows a positive pulse; therefore, its leading edge is the rising edge.

What will be the pulse width of an *ideal pulse*? Review again Figures 3.2.3 and 3.2.4 and answer this question.

3.2.2.3 Rise Time and Fall Time

The pulse width is an important timing parameter, but industry considers other quantities – *rise time* and *fall time* – as the principal timing parameters in digital transmission.

It seems that we should define the rise time as *the time it takes the rising pulse edge to transit from a LOW to a HIGH state*. However, a pulse at both levels experiences fluctuations, as Figure 3.2.3b shows, and we do not know what voltages uniquely correspond to LOW and HIGH states. Thus, we need to determine these levels first, which gives us the *signal amplitude*. Considering the *pulse amplitude as a reference parameter*, we can define the low-level reference point of a rise time as 10% of the pulse amplitude and the high-level reference point at 90% of the pulse amplitude. This approach enables us to define rise time as follows:

Rise time, t_r, is the time it takes a rising edge of a pulse to make the transition from 10% to 90% of the pulse amplitude.

Reading this definition, you may ask, why 10% and 90%? What is the reason for choosing these numbers? To answer these questions, we have to ask ourselves why we need to introduce rise time at all. Consider the ideal digital pulse in Figure 3.2.3a: How long does it take for this pulse to transit from the LOW to the HIGH level? No time at all. An ideal signal transits from one state to the other instantaneously. Hence, rise time shows how long it takes for a real pulse to transit from the LOW state to the HIGH. To avoid any uncertainties, we decided to measure this transition time by the 10–90% criterion. Thus, our definition says that

rise time is an interval needed for a pulse to change its value from 10% to 90% of its amplitude, or its final value.

The next question in this regard should be, why do not we measure rise/fall time from the mean LOW to the mean HIGH values. The answer is shown in Figure 3.2.3b. The waveform of a real pulse suffers from irregularities around these levels, which makes it difficult to measure these values. In contrast, the values of the waveform at 10% and 90% of amplitude can be measured at the higher level of certainty.

As for the numbers 10% and 90%, this is simply a convention among all engineers. We often have to rely on certain understood agreements when no natural solution exists. There are two alternative techniques to determine rise time: Most often used one is based on the different percentages of pulse amplitude (typically, 20% and 80%) and the other, less used, is based on two fixed magnitude levels (for example, from 0.2 to 2.7 V). Thus, when looking at technical documentation, pay attention to the definition of rise time for a specific measuring instrument or a circuit; most likely, it is given as 10–90% or 20–80%.

Fall time is defined similarly; for example, for 90% to 10% measured points, the definition is as follows:

Fall time, t_f, is the time it takes the falling edge of a pulse to make the transition from 90% to 10% of the pulse amplitude.

In general, rise and fall times can be of different values, as Figure 3.2.4 illustrates. Obviously, this difference is a measure of the degree of the pulse asymmetry.

The following example presents this theory in practical figures, which should help you better understand this material.

Example 3.2.3   Basic Parameters of a Real Pulse

Problem

Determine the amplitude, pulse width, and rise and fall times of the signal shown in Figure 3.2.6.

Solution

We need to start with the amplitude because all other parameters refer to this quantity. To find the amplitude, we need to determine the mean value of the high-level and low-level voltages. (See Section "Discussion" in this example to understand how to calculate the mean values.) In our example, we determine, approximately, from Figure 3.2.5, the mean low-level voltage as 0.2 V and the mean high-level voltage as 3.2 V. Thus, applying (3.2.3), we compute

$$A (V) = \text{mean HIGH level} - \text{mean LOW level} = 3.2\,V - 0.2\,V = 3.0\,V$$

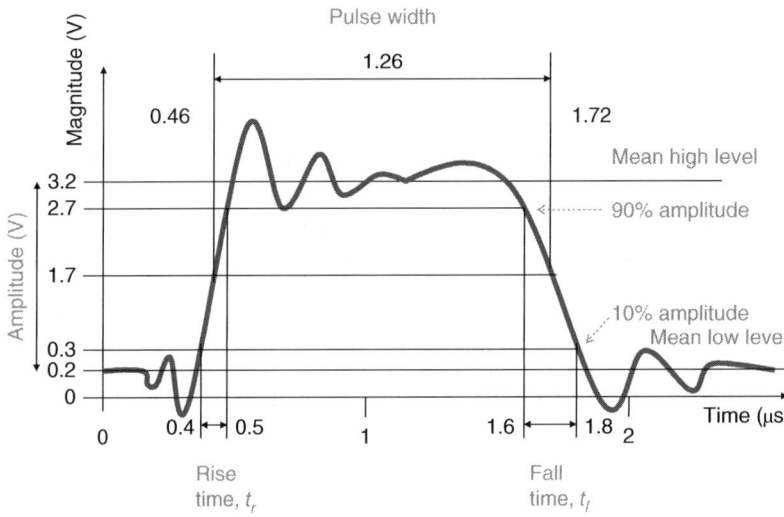

Figure 3.2.5 Basic parameters of the digital pulse in 3.2.3.

Next, we calculate the pulse width. Taking 50% of the pulse amplitude, we obtain

50% of the amplitude $= 0.5 \times 3\,\text{V} = 1.5\,\text{V}$

Since the mean low level is 0.2 V, the half maximum of this pulse is

50% of the amplitude + low-level voltage $= 1.5\,\text{V} + 0.2\,\text{V} = 1.7\,\text{V}$

This voltage is shown in Figure 3.2.5. Crossing the pulse with a horizontal line at 50% of the amplitude, we can find the pulse width. We refer to these crossing points at the time axis and find that the first crossing occurs at 0.46 μs and the second crossing at 1.72 μs. (These points are shown above the signal plot.) Thus, the pulse width is

$1.72\,\mu s - 0.46\,\mu s = 1.26\,\mu s$

Next, we calculate the 10% and 90% reference voltages to determine the rise and fall times. These voltages are as follows:

10% reference level $= 0.1 \times \text{Amplitude (V)} = 0.1 \times 3.0\,\text{V} = 0.3\,\text{V}$

90% reference level $= 2.7\,\text{V}$

Taking the 10% and 90% reference voltages at the rising edge, we determine the points where the horizontal lines cross the rising edge of the signal's waveform. Referring to these crossing points at the time axis, we find the rise time. In Figure 3.2.5, the 10% reference voltage crosses the rising edge at 0.4 μs and the 90% reference voltage crosses the rising edge at 0.5 μs. Thus, the rise time, t_r, is

$t_r = 0.1\,\mu s$

In a similar way, we find the fall time: The first crossing at 90% of the pulse amplitude occurs at 1.6 μs and the second crossing, at 10% of the pulse amplitude, occurs at 1.8 μs; therefore, the fall time, t_f, is

$t_f = 0.2\,\mu s$

Observe that the rise and fall times here have different values.

Discussion

- We have not presented the formal method to determine mean low and high levels because of its complexity. In mathematics, there is what is called the *Mean Value theorem*, which gives us the technique for calculating a mean value in a general case. In practice, there are several methods for this procedure; these methods have been developed by the companies manufacturing the measuring instruments, such as Keysight Technologies, Tektronix, Fluke, and others. You can find descriptions of these methods at the web sites of these companies.
- Review carefully all computations presented in this example. They look very simple (and, indeed, they are simple), but we must fully understand how to do these calculations. In particular, observe how we have computed the amplitude and all the other reference voltages based on the amplitude. Focus on the units of all the quantities involved in these calculations. As an additional exercise, calculate the rise and the fall times for the signal given in Figure 3.2.5, using the 20% and 80% reference voltages.
- Last, but probably the most important point: Modern oscilloscopes and other measuring instruments perform all these measurements automatically; in practice, you do not need to do these individual measurements and perform these tedious calculations. We demonstrate all these operations only to familiarize you with the meaning of these measurable parameters and operations performed by the oscilloscopes.

3.2.2.4 Rise/Fall Time and Bit Rate

Rise and fall times are very important parameters. They determine the actual bit rate that can be achieved with a given digital signal. Suppose, for example, that we have three digital pulses with different rise/fall times. (We assume, for the sake of simplicity, that the rise and fall times of each pulse are equal to each other.) How will this difference affect the bit rates that can be achieved with these signals? Look at Figure 3.2.6, where three signals are shown. These signals have the same amplitudes and the same duration of the high-level voltages; they differ only in rise time.

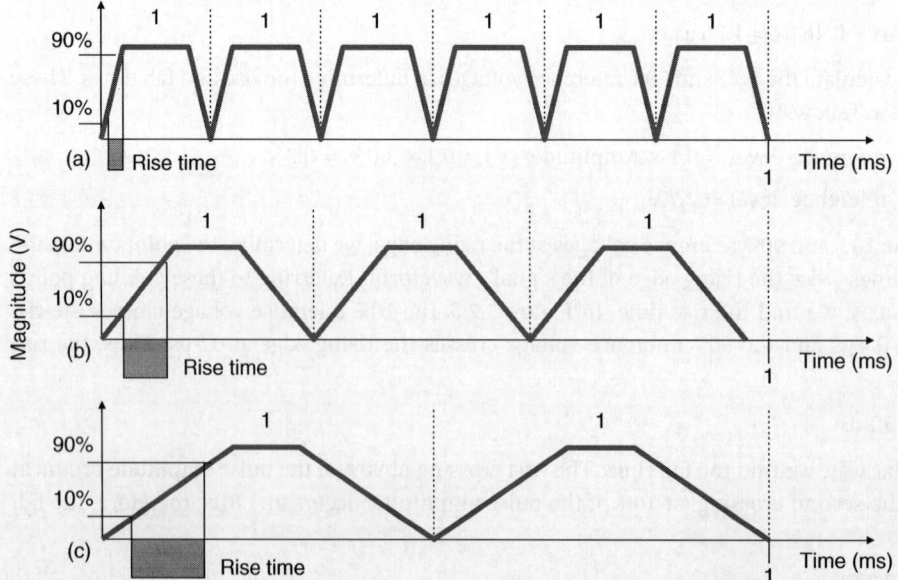

Figure 3.2.6 Rise time and bit rate: (a) short rise time; (b) intermediate rise time; (c) long rise time.

3.2 Digital Signals and Digital Transmission – Introduction

Observing the signals shown in Figure 3.2.6, we can readily say that bit rate is inversely proportional to rise time. Indeed, the top signal shown in Figure 3.2.6a allows for 6 bits/ms; that is, its bit rate is 6×10^3 b/s. Why? Because this signal has short rise time. The bottom signal shown in Figure 3.2.6c allows for only 2 bits/ms; hence, its bit rate is 2×10^3 b/s. This is because the rise time of this signal is much greater than that of the top signal. Figure 3.2.6b shows the intermediate rise time; we can readily determine that the bit rate in this case is equal to 3 kb/s.

In general, rise time and bit rate are related through the following formula:

$$\text{Bit rate} = \alpha/\text{rise time} \quad (3.2.4a)$$

or

$$\text{BR (b/s)} = \alpha/t_r \text{ (s)} \quad (3.2.4b)$$

where α is a constant. What is the value of this constant? It depends on the transmission code and coding technique we use. Typically, this value ranges from 0.35 to 2. Let us consider the following example.

Example 3.2.4 Rise Time and Bit Rate

Problem

Compute the rise time of the signals shown in Figure 3.2.6 if $\alpha = 0.5$.

Solution

We have to rearrange (3.2.4b) in this way:

$$t_r \text{ (s)} = \alpha/\text{BR (b/s)} \quad (3.2.4c)$$

Then, we compute rise time for all the cases shown in Figure 3.2.6:

(a) Rise time, $t_r^a = \alpha/\text{bit rate} = 0.5/(6 \times 10^3 \text{ b/s}) = 0.083$ ms.
(b) $t_r^b = 0.5/(3 \times 10^3 \text{ b/s}) = 0.167 \times 10^{-3}$ s $= 0.167$ ms.
(c) $t_r^c = 0.5/(2 \times 10^3 \text{ b/s}) = 0.250 \times 10^{-3}$ s $= 0.250$ ms.

Discussion

- Observe that the bit rate of the signal shown in Figure 3.2.6b is two times less than that of the signal shown in Figure 3.2.6a. The rise times of these signals are in inverse proportion to the same coefficient: 0.167 ms $= 2 \times 0.083$ ms. (Some small numerical discrepancies in this equality come from rounding the irrational number 0.083 333….) Hence, we can compute all the rise times by simply multiplying 0.083 by 2 in the case shown in Figure 3.2.6b and by 3 in case of Figure 3.2.6c. This statement bears the following mathematical confirmation: Let us denote the rise time and bit rate as $t_r^a(s)$ and BR_a (b/s) for the case shown in Figure 3.2.6a and as $t_r^b(s)$ and BR_b (b/s) for the case (3.2.6b). Applying (3.2.4c) for each case, we can write:

$$t_r^a(s) = \alpha/\text{BR}_a(b/s)$$

and

$$t_r^b(s) = \alpha/\text{BR}_b(b/s)$$

Now let us divide the top equation by the bottom. We obtain

$$t_r^a(s)/t_r^b(s) = \text{BR}_b(b/s)/\text{BR}_a(b/s)$$

3 Digital Signals and Digital Transmission

This is exactly what our observation shows:

$$0.083 \text{ ms}/0.167 \text{ ms} = 3 \text{ kb/s}/6 \text{ kb/s}$$

Therefore, we can compute $t_r^b(s)$ and $t_r^c(s)$ by simply multiplying the $t_r^a(s)$ by 2 and by 3, respectively. That is,

$$t_r^b(\text{ms}) = t_r^a(\text{ms}) \times 2 = 0.0833 \times 2 = 0.167 \text{ ms}$$

and

$$t_r^c(\text{ms}) = t_r^a(\text{ms}) \times 3 = 0.0833 \times 3 = 0.250 \text{ ms}$$

These are the numbers that we computed for all the rise times by applying (3.2.4b).
- Examining Figure 3.2.6 shows that the value of a pulse width follows a change in the rise time. If we apply the definition of pulse width illustrated in Figure 3.2.4 to the pulses shown in Figure 3.2.6, we can see that the greater the rise time, the wider the pulse. We can express this relationship through the following equation:

$$\text{Pulse width(s)} = \beta \times \text{rise time} \qquad (3.2.5)$$

where β is a constant. The value of β depends on the transmission and coding techniques; usually, this figure is more than 1. As a result, if we know the rise time and the constant β, we can compute the pulse width.
- We have defined bit rate in two ways: (3.2.1d) relates bit rate to bit time and (3.2.4b) relates bit rate to rise time:

$$\text{BR (b/s)} = 1/T_b \text{ (s)} \qquad (3.2.1\text{dR})$$

and

$$\text{BR (b/s)} = \alpha/t_r \text{ (s)}. \qquad (3.2.4\text{bR})$$

Comparing these formulas, we derive

$$T_b(s) = t_r(s)/\alpha \qquad (3.2.6\text{a})$$

or

$$\alpha \times T_b(s) = t_r(s) \qquad (3.2.6\text{b})$$

Therefore, we need to know either the rise time or the bit time and α to determine the bit rate. In other words, *either rise time or bit time can be considered as the fundamental timing parameter of a* digital signal. The definition of bit time, as we will see shortly in this section, depends on a specific transmission code, whereas the definition of rise time is the same for any transmission code. More important, it is difficult to measure and determine the bit time of a real pulse, as Figure 3.2.3 shows. Thus, the rise time is considered a primary timing parameter. Though we need to know the constant α to determine the bit rate based on rise time, in practice engineers prefer to deal with the rise/fall time.

The Use of Rise/Fall Time Why should we be concerned with rise/fall time when we have other timing characteristics, such as bit time or pulse width? Because *by measuring the pulse's rise time at the receiver end we can determine how much the pulse degrades during the transmission.* A real transmitter generates a real pulse with rise and fall times conditioned by the presence of reactive (capacitor and inductor) components in the transmitter's circuitry. The inductive and capacitive properties of this circuitry are responsible for an exponential waveform describing a pulse's transition from the initial to the final state. (Remind yourself, for example, about the charging and discharging processes of a capacitor to which a DC signal is applied.) It is only due to these properties that a real pulse does not jump step-like from the initial to the final value, as an ideal pulse does, but exhibits the waveform with a slope. However, the rise/fall time of a pulse generated by a transmitter would stay constant if and only if the transmission line would be ideal.

During transmission, the rise/fall time of the pulse will change because of the properties of the transmission link. As a transmission medium, a copper wire can be modeled by an RLC circuit; this is where reactive properties come from. While the pulse travels along such a transmission line, the rise and fall times increase because the effect of the reactive components accumulates. Thus, the rise and fall times at the receiver end will be greater than that emerging from a transmitter. This phenomenon is responsible for such a fundamental principle of communications as bandwidth-length (or bit-rate-length) trade-off. (Refer to Section 1.3.)

These general considerations are true for any type of modern communications, including wireless and optical. The underlying physical mechanisms might be different, but the results are the same: The rise and fall times exist in communication systems and their values increase as a signal travels down the transmission path.

3.2.2.5 More on Timing Parameters of a Digital Signal: Bit Time Revisited

Let us take a closer look at bit time. When we introduced the parameters of a digital signal earlier in this section, we said that the amplitude and the bit time are two fundamental parameters that we need to know about a digital pulse. (See the discussion on Figure 3.2.3.) However, a more careful look at the digital signal presented in Figures 3.2.5 and 3.2.6 reveals that it is not easy to define bit time for a real digital signal. In addition, we found that such timing parameters as the rise and fall times play the same role in determining bit rate as does bit time; what is more, the bit time gives us the most straightforward method of determining the bit rate, as (3.2.1d) states. Hence, we need to determine the bit time for a real digital signal.

Let us consider how we can define bit time for NRZ and RZ signals. These signals were defined in Figure 3.1.10. Here, Figure 3.2.7 shows models of real waveforms of these signals; it also shows the stream of bits rather than individual pulses.

Bit Time of an NRZ Signal Now let us introduce the following definition of the bit time of a NRZ signal:

NRZ bit time, $T_b^{NRZ}(s)$, *is the time interval between two points at the consecutive rising and falling edges measured at half the amplitude.*

NRZ bit time is shown in Figure 3.2.7a. Though this definition of bit time looks like that of pulse width, these two parameters are quite different, as discussed previously in this section.

Note that we can measure $T_b^{NRZ}(s)$ either between the rise and fall edges or between the fall and rise edges; in other words, the sequence of the measurements is not important.

Strictly speaking, bit time is determined by the measurement of this timing parameter over a long stream of NRZ pulses when a random sequence of 1s and 0s is being transmitted. The bit

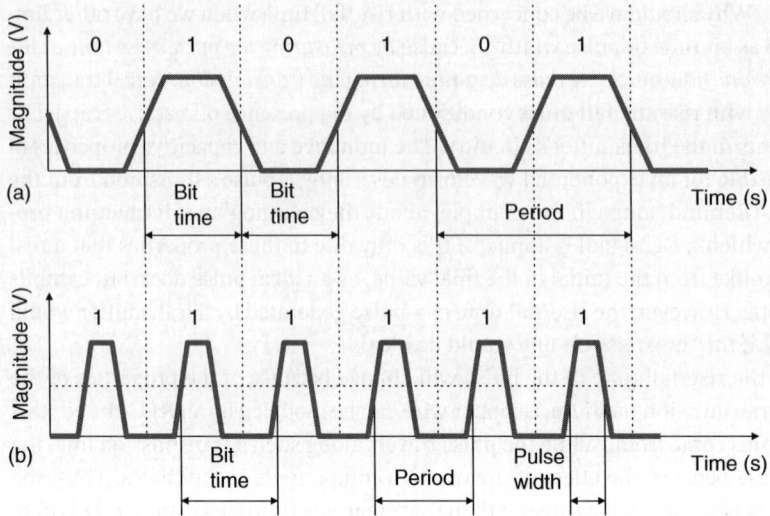

Figure 3.2.7 Timing parameters of NRZ and RZ signals: (a) NRZ signal; (b) RZ signal.

time assigned to this stream is the mean value of hundreds or even thousands of individual measurements. This is why it is important to measure bit time at half amplitude because the rise/fall edges of every individual pulse can vary and so measuring this time at the same level of amplitude provides the consistency necessary for getting reliable results.

Pulse Repetition Interval (PRI) of an NRZ Signal Though the real bit stream is a series of a random sequence of 1s and 0s, we can introduce such a timing parameter of an NRZ signal as a *PRI* also called a *period*. To start, you need to recall that a period is the time interval between two points, after which a signal starts to repeat itself. Looking at the NRZ signal, you can see that

The PRI, (period) of an NRZ signal, $T_{PRI\text{-}NRZ}(s)$, is the time interval between two consecutive rising (falling) edges measured at 50% amplitude of a "010101..." bit stream.

PRI can be unambiguously defined when the pattern "010101..." is transmitted, as illustrated by Figure 3.2.7a. The PRI, or period, of a real NRZ signal is the result of averaging over hundreds or thousands of individual measurements.

The PRI is useful for two reasons. First, it helps to determine the bit time. It is clear from Figure 3.2.7a that the *bit time, $T_b^{NRZ}(s)$, is equal to half of the PRI*; that is,

$$T_b^{NRZ}(s) = \tfrac{1}{2} T_{PRI\text{-}NRZ}(s) \tag{3.2.7}$$

For an NRZ signal, this result seems obvious, but we will see shortly that it is not so for an RZ signal. Second, the period enables us to introduce a pulse repetition frequency, $f_{PRF}(Hz) = 1/T_{PRI\text{-}NRZ}(s)$, which is necessary for finding the signal bandwidth. Then, the bandwidth of an NRZ signal can be approximated as

$$BW_{NRZ}(Hz) \approx 2 f_{PRF}(Hz) \tag{3.2.8}$$

The origin and meaning of this relationship will be treated in the chapters devoted to spectral analysis.

Bit Time and the Pulse Repetition Interval of an RZ Signal Let us now consider the RZ signal shown in Figure 3.2.7b. You will recall that the main feature of an RZ signal is that its bit 1 occupies only a part of the entire time slot allotted to a pulse; thus, its amplitude returns to zero value during this time slot. Bit 0 here has zero amplitude. (There is what is called a *bipolar RZ* signaling where bit 0 is represented by a negative pulse.)

The pulse repetition interval, PRI_{RZ}, of the RZ signal, $T_{PRI\text{-}RZ}$ (s), is the time interval between two consecutive rising (falling) edges measured at 50% amplitude of a "010101..." bit stream.

Now,

The bit time of an RZ signal, T_{BRZ}(s), is equal to half of the PRI_{RZ} of "010101..." bit stream; that is,

$$T_b^{NRZ}(s) = \tfrac{1}{2}\, T_{PRI\text{-}RZ}(s) \tag{3.2.9}$$

Both definitions are depicted in Figure 3.2.7b. As with NRZ, an RZ bit time is determined as a mean value of hundreds or thousands of individual measurements.

Though definitions of bit time for both NRZ and RZ signals look the same, their meaning is quite different. In an NRZ signal, the bit time is the same as the pulse width and therefore T_{NRZ} (s) can be determined without reference to the signal period. In an RZ signal, the bit time is NOT equal to the pulse width; for the RZ signal shown in Figure 3.2.7b, the bit time can be determined only with reference to the period.

Let us recall that, in reality, digital signals are the streams of millions and billions of pulses whose transmission patterns are random by their very nature. Of course, no pattern like *"010101..."* can be found in practice. Hence, the term *period cannot be applied to digital signals in general; this is why we use the term PRI*. This parameter helps to define and measure many of the characteristics of digital signals, as we will see in the following chapters.

Comparison of the Timing Parameters of NRZ and RZ Signals Which signal – NRZ or RZ – is better for transmission? There is no definitive answer to this question because the answer depends on a specific task and the transmission system. Here, we compare these signals by just two criteria – *signal bandwidth* and *ISI*.

The concept of signal bandwidth has been discussed in this section, and we noted that small signal bandwidth is good for a transmission system. At this point, we need to remember that the narrower the pulse, the greater the bandwidth of the digital signal. (A detailed discussion of this topic is given in Chapter 7.) Therefore, an RZ signal requires much higher bandwidth than an NRZ signal. This, in turn, means that an RZ transmission system must have better (read: more expensive) components. Thus, it seems that NRZ signal is the preferable one. Indeed, when the transmission rate of digital systems was small (<1 Gb/s), NRZ was the signal of choice in digital transmission. However, with the increase in bit rate, other factors start playing a significant role in the design of transmission systems.

The second criterion, *ISI*, is one of the most important factors that determine the reliability of transmission. Refer to (3.2.1c), which states that bit time, T_b (s), is inversely proportional to bit rate, BR (b/s):

$$T_b(s) = 1/BR\ (b/s) \tag{3.2.1cR}$$

Bit rate 10 Tb/s, which is common for modern optical communications systems, requires a bit time of 0.1×10^{-12} s. This means that bits follow one another every 0.1 ps. Given the detrimental effect of noise and the timing instability of the components of a transmission system, the probability that a preceding bit will interfere with the following one becomes significant. This phenomenon *when two or more bits overlap and become indistinguishable is called* ISI. Such an interference makes it impossible for a receiver to correctly decipher the information.

Now, return to Figure 3.2.7: It is clear that the timing interval between consecutive 1s in RZ signaling is much greater than that in the NRZ signaling. Therefore, the probability of interference among bits in RZ signaling is lower that it is in NRZ. This big advantage makes RZ signaling more and more popular in modern digital transmission systems.

3.2.2.6 Duty Cycle

An RZ signal is an example of encoding, where a bit occupies only a part of the whole-time interval dedicated to a pulse. There are many such signals in modern communications. To describe their timing characteristics, we need to introduce one of the most important timing parameters: *duty cycle*. In general, the duty cycle shows how long a system (signal) stays in OFF mode during the whole operating cycle. Here, the duty cycle is equal to the ratio of the pulse width, T_W (s), to the signal period, T (s):

Duty cycle = Pulse width (s)/Period (s)

or

$$\text{Duty cycle} = \frac{T_W(s)}{T(s)} \tag{3.2.10}$$

It is a measure of how long a bit's amplitude is not equal to zero with respect to the signal's period. Compare Figures 3.2.7a,b: For NRZ signal, the width of a pulse with nonzero amplitude (bit 1) is exactly equal to half a period; therefore, the NRZ duty cycle is 50%. In contrast, RZ pulse width is approximately a quarter of the period; thus, the RZ duty cycle is about 25%. Note that the duty cycle can be presented as a ratio or in percentage.

The importance of this parameter becomes clear if we realize that the main purpose of transmitting a signal in telecommunications is delivering information, not power. If we were able to deliver information without transmitting any power, it would be the ideal situation. In reality, without power we do not have electrical or optical signals. Following this reasoning, we immediately understand that the real goal in telecommunications is transmitting information with a minimum expenditure of power. From this standpoint, *RZ signaling has the other great advantage over the NRZ*. Duty cycle is a quantitative measure of this advantage because the smaller the duty cycle, the less power we transmit.

Let us summarize the discussion of signal timing parameters by consulting the following example.

Example 3.2.5   Determine the Timing Parameters of a Digital Signal

Problem

Determine the bit time, period, and duty cycle for the signals shown in Figure 3.2.8.

Solution

First, let us find the period. By definition, period is the time interval between two consecutive rising (falling) edges measured at 50% of the signal amplitude. Since we have three periods within a 6 ns interval for all three signals, the period is 2 ns. Thus,

$$T_{\text{NRZ}} = T_{\text{RZ}} = 2 \text{ ns}$$

Second, we need to determine bit time. Figure 3.2.8a shows that there are six NRZ bits fitting the 6-ns interval. Thus, each NRZ bit occupies 1 ns. Let us take another approach and follow the bit-time definition: NRZ bit time is equal to half of its period. From (3.2.7) we compute

$$T_b^{\text{NRZ}} = \tfrac{1}{2} T_{\text{NRZ}} = 1 \text{ ns}$$

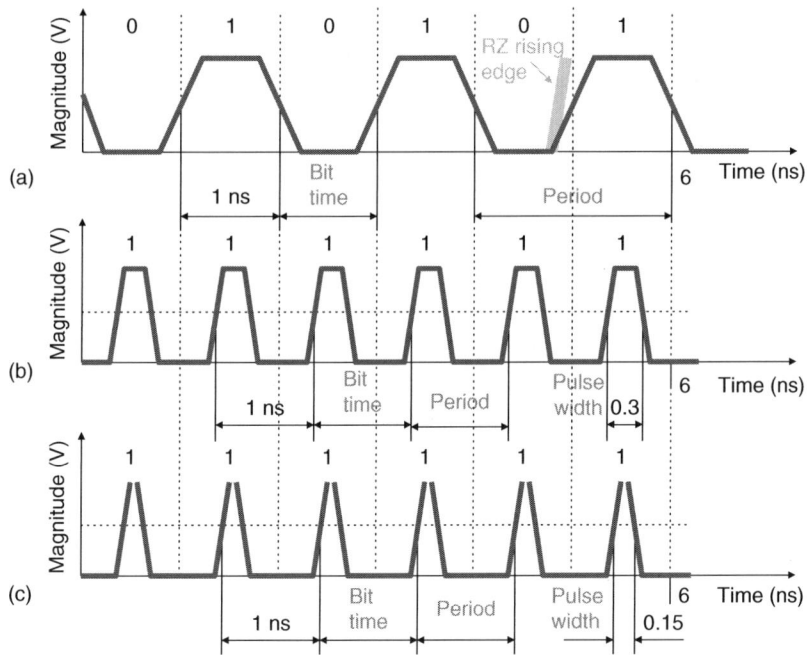

Figure 3.2.8 Timing parameters of a digital signal in 3.2.5: (a) NRZ signal; (b) RZ signal; (c) RZ signal with a smaller pulse width.

Figure 3.2.8b,c shows that there are also six RZ bits fitting the same 6-ns interval. Therefore, the bit time of the both RZ streams is 1 ns. We obtain the same number, $T_{BRZ} = 1$ ns, if we follow the definition of RZ bit time given in (3.2.9).

Third, we calculate the duty cycle as the ratio of a pulse width to the signal period. Taking into account that $T_W^{NRZ}(s) = T_b^{NRZ}(s)$ and applying (3.2.10), we compute for the NRZ signal

$$\text{Duty cycle} = \frac{\text{Pulse width}}{\text{Signal period}} = \frac{T_{WNRZ}(s)}{T_{NRZ}(s)} = \frac{1 \text{ ns}}{2 \text{ ns}} = 0.5 \text{ or } 50\%$$

For an RZ signal, the duty cycle depends on the pulse width. Figure 3.2.8b shows an RZ signal whose pulse width equals 0.5 ns and Figure 3.2.8c shows a similar signal with a pulse width of 0.25 ns. Therefore, the duty cycle of the RZ signal shown in Figure 3.2.8b is equal to

$$\text{Duty cycle} = \frac{T_{WRZ}(s)}{T_{RZ}(s)} = \frac{0.5 \text{ ns}}{2 \text{ ns}} = 0.25 \text{ or } 25\%$$

For the RZ signal in Figure 3.2.8c we compute

$$\text{Duty cycle} = \frac{T_{WRZ}(s)}{T_{RZ}(s)} = \frac{0.25 \text{ ns}}{2 \text{ ns}} = 0.125 \text{ or } 12.5\%$$

Discussion

- *Measuring duty cycle*: Modern oscilloscopes measure duty cycles and immediately display the result. Therefore, there is no need for all presented calculations; in practice, they are done automatically. But studying this material, we become familiar with what and why an oscilloscope is doing before showing the numbers on its display.

- *Bit rate*: As we have pointed out several times already, the first thing you want to know about a transmitting signal is its *bit rate*. So what are the bit rates of the signals shown in Figure 3.2.8? Formula (3.2.1d) says explicitly that

$$\text{BR (b/s)} = 1/T_b \text{ (s)}$$

Since the bit times of all three signals in Figure 3.2.8 are equal to one another, these signals have the same bit rate. Specifically,

$$\text{BR (b/s)} = 1/T_b \text{ (s)} = 1 \text{ Gb/s}$$

Visually, it seems that the RZ signals shown in Figure 3.2.8b,c have a higher bit rate than the NRZ signal shown in Figure 3.2.8a. This is so because the RZ signals clearly have the smaller rise time than the NRZ signal. To emphasize this point more, we put the RZ rising edge in Figure 3.2.8a so that you can see the difference between the NRZ and RZ rise times. Our estimation shows that rise times of the signals shown in Figure 3.2.8 are 0.4 ns for Figure 3.2.8a, 0.2 ns for Figure 3.2.8b, and 0.1 ns for Figure 3.2.8c. Thus, these signals have the same bit rate but different rise times. But we must remember that bit rate and rise time are inversely proportional, as (3.2.4b) shows. So what is the catch? Look at (3.2.4b) again:

$$\text{BR (b/s)} = \alpha/t_r \text{ (s)}$$

Now we can see that it is the constant α that makes all the difference. Indeed, for the NRZ signal (Figure 3.2.8a) we have

$$1 \text{ Gb/s} = \alpha/0.4 \text{ ns} \rightarrow \alpha = 0.4$$

For the first RZ signal (Figure 3.2.8b) we have

$$1 \text{ Gb/s} = \alpha/0.2 \text{ ns} \rightarrow \alpha = 0.2$$

For the second RZ signal (Figure 3.2.8c) we have

$$1 \text{ Gb/s} = \alpha/0.1 \text{ ns} \rightarrow \alpha = 0.1$$

Now we can fully appreciate the usefulness of bit time: To calculate bit rate based on rise time, we need to know the value of the rise time and constant α; to calculate bit rate based on bit time, we need to know the value of bit time only. However, do not take this statement at face value. Rise time is the most valuable and important parameter of a digital signal; what is more, in many circumstances, the rise time is often the only parameter that allows for finding bit rate.
- *Duty cycle, bit rate, and transmitting power*: When we look at Figure 3.2.8, we may gather the impression that the signal shown in Figure 3.2.8c carries the higher bit rate. Not so. All three signals have the same bit rate, as our calculations show. Why do we get such a wrong visual impression? Because these signals have different *duty cycles*. The importance of the duty cycle can be truly recognized if we recall that the duty cycle is a measure of how much power we use to transmit information. Clearly, with such transmission codes as NRZ and RZ we need power to transmit 1s only, whereas transmitting 0s does not require any power.

Examine Figure 3.2.9: This is, in essence, the reproduction of Figure 3.2.8 with visual highlights pertaining to the duty-cycle concept.

Be aware that power carried by an individual pulse is proportional to the area covered by the pulse. In Figure 3.2.9, we fill in all pulses with color to highlight the area that every pulse covers. We can readily see that the NRZ pulses with duty cycle 1 have the most area covered, whereas the RZ stream with duty cycle 0.25 covers the least area. Therefore, the RZ stream shown in

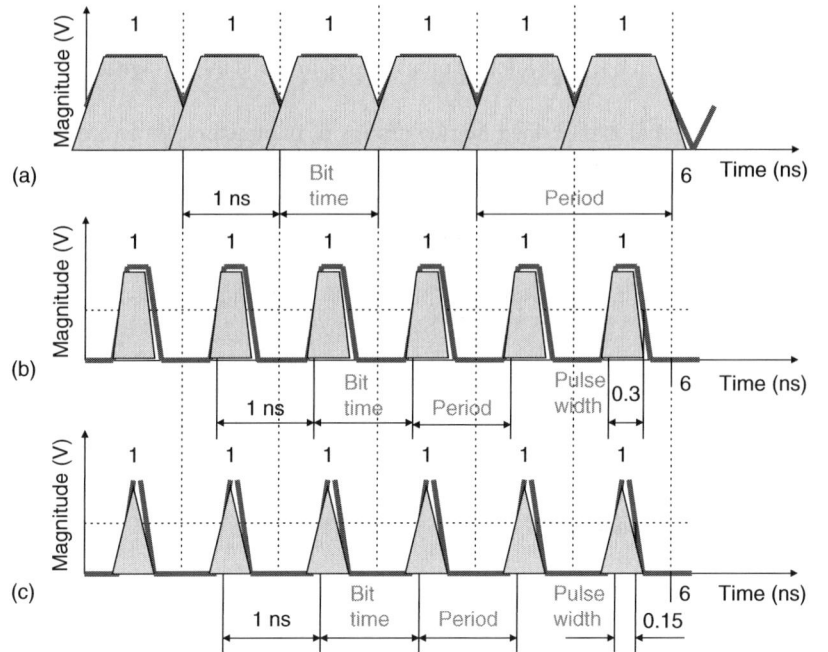

Figure 3.2.9 Duty cycles of NRZ and RZ signals: (a) an NRZ signal with a duty cycle of 1; (b) an RZ signal with a 0.5 duty cycle; (c) an RZ signal with a 0.25 duty cycle.

Figure 3.2.9c carries the same information as all the other signals in Figure 3.2.9 but requires the least power for this transmission. This characteristic of transmission is given by the duty cycle. It is not difficult to come to the conclusion that we want to minimize a duty cycle. The question is what is the limit for such minimization? The answer lies in the transmission quality. If we make the width of an RZ pulse very small (the pulse width determines the duty cycle, remember?), we will make this pulse very vulnerable; it may be severely distorted during transmission and information may be lost. Thus, some compromise in the value of pulse width must be found. Since duty cycle is the ratio of two timing quantities, it is a dimensionless parameter. Observe that duty cycle has no unit, as our calculations show.

The above consideration of the timing parameters of NRZ and RZ are just examples. As we can see from this discussion, we need to determine the timing parameters for every transmission code. Some of these parameters, such as bit time and rise/fall time, are universal and used for every code; some, such as period, are defined only for specific codes. This is also true for amplitude parameters.

There are a number of other magnitude and timing parameters of a digital signal; these parameters require a deeper analysis, which we will do in the appropriate chapters.

Questions and Problems for Section 3.2

- Questions marked with an asterisk require a systematic approach to finding the solution.
- Many questions and problems, including those marked with an asterisk, imply that you, in addition to reading the textbook, will do your research to find the answers. Consider such questions as mini-projects.

Ideal Digital Signal and Characteristics of Digital Transmission

1. What is a signal waveform? Sketch the waveforms of an analog and of a digital signal.

2. The waveform of a digital signal is described by two groups of parameters. What are those groups and what parameters does each group include?

3. What is the difference between a digital signal's amplitude and its magnitude?

4. Sketch the waveforms of NRZ digital signals with amplitude $A = 2.8$ V and bit times $T_{B1} = 1$ ms and $T_{B2} = 2$ ms.

5. Sketch the waveforms of NRZ, unipolar RZ, and AMI digital signals with the same amplitude, $A = 1.2$ V, and bit time, $T_B = 0.5$ ps. (Refer to Section 3.1 to recall transmission codes.)

6. There is an analogy between period and frequency on the one hand and bit time and bit rate on the other hand. Explain this analogy.

7. *Is there any difference between pulse interval, T_p (s), and symbol time, T_s (s)? Between pulse interval, T_p (s) and bit time, T_b (s)? Explain and sketch a figure supporting your explanations.

8. Define bit rate and baud rate.

9. Explain the difference between bit rate, BR (b/s), and baud rate, SR (Bd = symbol/s) and sketch a figure to show this difference.

10. Explain how a bit rate relates to bit time and a baud rate relates to pulse interval.

11. Bit time and symbol interval are given by $T_b = 0.1$ μs and $T_s = 0.2$ μs, and 2 bits were placed in one symbol. Compute the bit rate and the baud rate.

12. Consider the case where one pulse carries one bit, whereas $T_b = 0.2$ ms and $T_s = 0.4$ ms. (Refer to Figure 3.2.2.) What are the bit rate and the baud rate in this case?

13. Both channel capacity, C (b/s), and bit rate, BR (b/s), have the same units. Why do we need two different measures with the same units?

14. Is it possible to transmit a signal with BR = 100 Gb/s over a transmission channel with $C = 50$ Gb/s? Explain.

15. A transmission link has $C = 1$ T (b/s). How many individual bit streams with BR = 100 Gb/s each can this link multiplex? Show and explain your calculations.

16. *Plain old telephone service, POTS, transmits a 1.544 Mb/s bit stream that multiplexes many individual voice channels in digital format. How many channels does this stream carry?

17. Is it possible to transmit 200 Gb/s over a transmission channel with $C = 150$ Gb/s? If your answer is yes, how long does it take?

18 Why do we distinguish between a channel bandwidth and a signal bandwidth? What is the difference between bit rate and channel capacity?

19 According to Shannon's law, (1.3.6), the transmission capacity, C (b/s), is equal to the channel's bandwidth, BW (Hz), times the dimensionless logarithm. How can two quantities with different units be proportional?

20 Given BW = 100 GHz and SNR = 100:
   a) Compute the transmission capacity, C (b/s).
   b) Can you compute the bit rate, BR (b/s)?
   c) Can you make any conclusion about the BR (b/s) of this link?

21 A transmission link has $C = 4$ Tb/s. How many individual bit streams with BR = 100 Gb/s each can this link multiplex?

22 Plain old telephone service, POTS, transmits a 1.544 Mb/s bit stream that multiplexes many individual voice channels in digital format. How many voice channels does this stream carry?

23 Is there any direct relationship between the bandwidth and the bit rate of a signal? Explain.

24 *The Nyquist formula states that bit rate is restricted by the link bandwidth as BR (b/s) < 2 BW$_T$ (Hz). Why?

25 *We learned that one digitized voice signal has BR = 64 kb/s. How many symbols, M, do we need to place to transmit this signal through a copper-wire telephone line with bandwidth of 8 kHz? (Hint: Refer to Hartley's law, BR(b/s) = 2BW(Hz) $\log_2 M$.)

26 What is the principal difference between Hartley's law, C (b/s) = 2BW(Hz) $\log_2 M$, and Shannon's law, C (b/s) = BW(Hz) $\log_2(1 + \text{SNR})$? Explain.

Parameters of a Real Digital Signal and Characteristics of Digital Transmission

27 List the differences between the parameters of an ideal digital signal and the parameters of a real digital signal. Comment on each parameter in your list.

28 Name four basic parameters of a real digital pulse and explain their meaning.

29 Consider Figure 3.2.3b and define the amplitude of a real digital signal. Explain each term being used.

30 Consider definition of a pulse width:
   a) Why is it referred to as *full width half maximum*?
   b) Explain the importance of this parameter from the transmission standpoint.

31 What is the meaning of the terms *leading edge* and *trailing edge* of a pulse? Sketch the figure of a bipolar signal and show these edges.

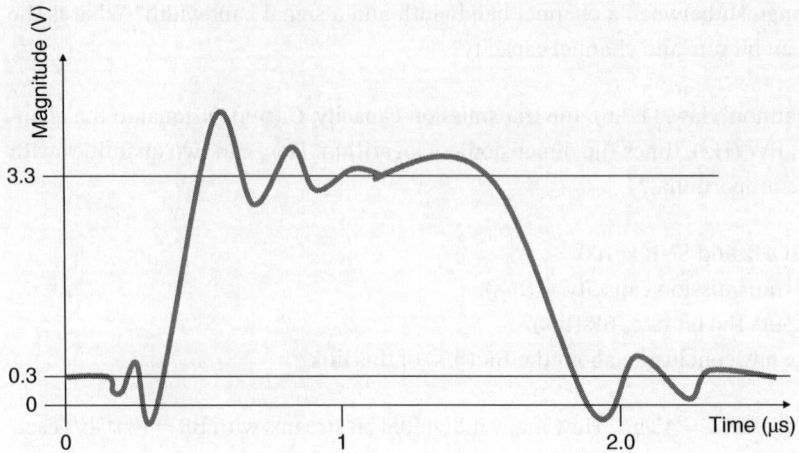

Figure 3.2.P35 The waveform of a digital pulse.

32. Consider the definition of rise time:
    a) Why is this timing parameter defined with references to 10% and 90% of a signal amplitude?
    b) Why do we need to know rise and fall times?

33. *If rise times of the same pulse are determined at 10–90% and 20–80% of the pulse's amplitude, then they will have different values. Which value is correct? Explain.

34. Suppose you make measurements of the rise time and fall time of the same pulse at 20–80% of the pulse's amplitude. Will you obtain the same values? Explain.

35. *Consider the pulse shown in Figure 3.2.P35 with a mean HIGH level of 3.3 V and a mean LOW level of 0.3 V. Find the approximate values of the amplitude, pulse width, rise time, and fall time of this signal. Use the 20–80% calculation method.

36. What is the relationship between rise time and bit rate?

37. Compute the bit rate of a digital signal if its rise time is 0.4 ms and constant $\alpha = 0.6$.

38. The rise time of a signal is 0.8 ms and its bit rate is 1 kb/s. What is the value of the constant α? What is the value of BR_2 (b/s) of a signal with $t_{r2} = 0.4$ ms?

39. If the rise time is 0.2 ms and constant β is 1.6, what is the pulse width?

40. The rise time of a pulse is 0.18 ms and the constant $\alpha = 0.6$. Compute the bit time of this pulse.

41. Explain whether there is a relationship between the rise/fall time and the pulse width of a signal. If your answer is yes, explain the nature of this relationship.

42. *NRZ and RZ signals have the same bit time of 2.5 ms. Sketch the waveforms of these signals.

3.2 Digital Signals and Digital Transmission – Introduction

43 Bit stream in a digital transmission is a random sequence of 1s and 0s. Nevertheless, we introduce the period of NRZ and RZ signals. How? Explain.

44 *What is the bandwidth of an NRZ signal if its bit time is 0.1 ms?

45 Compare advantages and drawbacks of NRZ and RZ signals: Which signal is better for digital transmission? Explain.

46 Explain which signaling code – NRZ or RZ – is more beneficial for reducing intersymbol interference.

47 Define "duty cycle" and explain the meaning of this parameter.

48 Consider the NRZ and RZ signals shown in Figure 3.2.P48a,b. Given that periods of NRZ and RZ signals are equal, $T_{NRZ} = T_{RZ} = 1$ ms, and pulse widths are given by $T_W^{NRZ} = 0.5$ ms, and $T_W^{RZ} = 0.25$ ms, find the duty cycles of both signals.

49 The duty cycle of an RZ signal is always smaller than that of an NRZ. Is this an advantage or disadvantage? Explain.

50 Compare the power consumption and bandwidth of NRZ and RZ signals and discuss which signal is preferable for transmission.

51 Consider Figure 3.2.PR51, reproduced here: What is the bit rate, BR (b/s), of each signal presented? Show your calculations and comment on your results.

52 Duty cycle is one of the parameters of a digital signal. What important information about the transmission of the digital signal does it deliver?

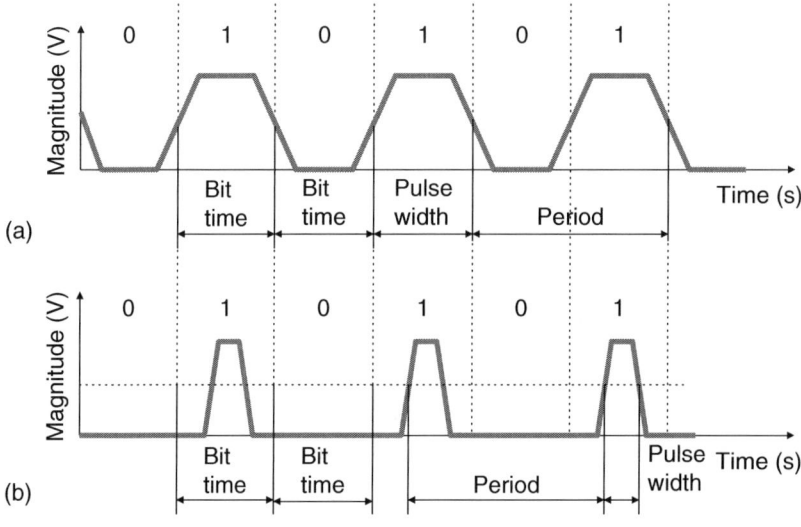

Figure 3.2.P48 Duty cycles of digital signals: (a) NRZ signal; (b) RZ signal.

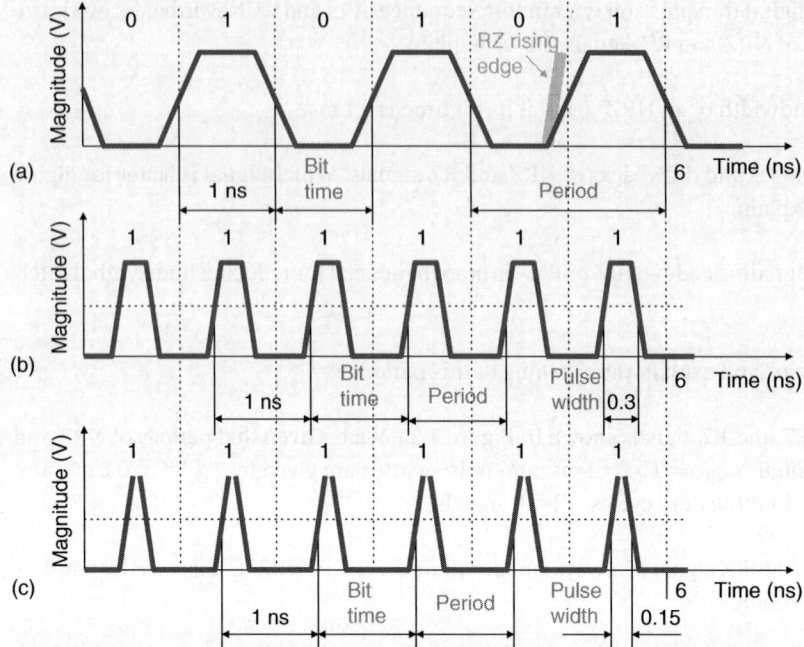

Figure 3.2.PR51 The timing parameters of a digital signal in 3.2.5: (a) NRZ signal; (b) RZ signal; (c) the RZ signal with the smaller pulse width.

53 *When presenting a digital signal for transmission, we need to take into account many of its timing parameters discussed in this section. List all of these parameters and explain the importance of each for characterizing a digital signal.

4

Analog-to-Digital Conversion (ADC) and Digital-to-Analog Conversion (DAC)

Objectives and Outcomes of Chapter 4

Objectives

- To gain an understanding of the principle of analog-to-digital conversion (ADC) and digital-to-analog conversion (DAC) and their applications in modern communications.
- To understand how other digital communications technologies – pulse-amplitude modulation (PAM) and pulse-code modulation (PCM) – are built on the principles employed in ADC and DAC.

Outcomes

- Learn that all natural and technological processes are analog in nature and information about these processes is presented as analog electrical signals. Understand that, however, the processing and transmission of these analog signals must be done in digital format to provide high-quality operation. Realize that this is the reason why modern technology in general and modern communications in particular widely use *ADC* and *DAC*.
- Be aware that ADC systems accept an analog signal as the input and convert it into a digital signal as the output.
- Know that *DAC* is the reverse operation of ADC.
- Become familiar with PAM which is another digitizing technique that produces pulses (digital signals) to represent an input analog signal.
- Learn that PCM converts an analog signal into pulses that carry the signal's information by virtue of binary codes, thus producing a truly digital signal. Grasp the idea that a PCM system is a technologically implemented version of ADC and DAC operations.

4.1 Analog-to-Digital Conversion, ADC

Objectives and Outcomes of Section 4.1

Objectives

- To learn the need for ADC, and DAC.
- To study in depth the ADC operation.

Outcomes

- Learn that information about all natural and technological processes is presented as analog electrical signals, but the processing and transmission of these analog signals must be done in digital format. Realize that this is the reason why modern communications widely use *ADC* and *DAC*.
- Be aware that ADC systems accept an analog signal as the input and convert it into a binary signal as the output.
- Study that to carry out its operation, an ADC system must perform three major steps: *sample-and-hold (S&H), quantization,* and *encoding*.
- Understand that sampling – the first step in ADC operation and, therefore, the foundation of the ADC operation – is based on Nyquist criterion that requires that a sampling frequency must be at least double of the signal maximum frequency.
- Get to know that sampling in fact includes S&H operations, which sample an input analog signal at equal time intervals and hold the measured amplitudes for these intervals. Conclude that sampling enables us to present a continuous analog signal as a set of decimal numbers – sample values.
- Learn that if an ADC system meets or exceeds the *Nyquist criterion*, the original analog signal can be, in principle, precisely restored from the collected samples.
- Realize that if an ADC system does not meet the Nyquist criterion, such an S&H operation (called *undersampling*) results in the appearance of a false signal whose frequency is equal to the difference between the frequencies of the analog and sampling signals. Be aware that this phenomenon is called *aliasing* and the false signal is called an aliased signal.
- Gain knowledge that the second step of an ADC operation called *quantization* is the process of presenting the decimal numbers, obtained after a S&H operation, in binary format.
- Learn that a quantization process starts with choosing a number of bits per sample, m (b/sample), which determines the number, $N_{ql} = 2^m$, and the size, $\Delta = \frac{A_{pp}(V)}{N_{ql}}$, of quantization steps, where A_{pp} (V) being the peak-to-peak amplitude of the sampled signal. Get to know that each quantization level corresponds to one combination of bits chosen from the m available bits.
- Comprehend that the central decimal values of the quantization levels are assigned to all samples found within each level and that the *assigned values* of samples do not coincide with the actual sampled values. Realize that the difference between assigned and actual sampled values constitutes *quantization error* and know that the stream of these errors vs. time is considered *quantization noise*.
- Be aware that the greater the number of quantization levels, N_{ql}, the more accurate the quantization of the signal because an assigned value will be closer to the actual value of the sample. Recall that since N_{ql} eventually depends on m, the key measure for increasing the accuracy of quantization – and the ADC operation as a whole – is increasing the number of bits per sample.
- Grasp the idea that the central decimal number of a quantization level is automatically coded into the proper binary codeword and that a designer can choose a system of binary codewords, such as natural binary code or two's complement code, from the systems accepted by industry.
- Conclude that an ADC operation takes an actual decimal value of a sample and generates a corresponding binary number (codeword) for each sample.
- Realize that the third step of ADC operation is *encoding* the binary codewords obtained in the quantization step into a set of electrical pulses by applying one of the *transmission codes* discussed in Section 3.1 and sending this electrical signal out from an ADC unit.

4.1.1 The Need for ADC and DAC

Note: Reviewing Appendix 4.2.A will help you better understand the material in this section.

Our world is analog, but transmission is digital, as we have stated several times previously. In Chapter 3, we discussed why it is necessary to prepare analog information for digital transmission and how to do so. In Section 3.1, we learned how to code characters (logical information) into binary words and how to encode those 1s and 0s into electrical or optical pulses. In Section 3.2, the parameters of a digital signal and their relationship with the characteristics of digital transmission were discussed. Though it seems that Chapter 3 has covered digital transmission in its entirety, one important aspect remains to be explored: Suppose we obtain information in the form of an analog signal. How is this handled electronically? Consider the work of your mobile phone, for example: Your voice is converted into an analog electrical signal by a microphone and this signal is sent out by your phone in digital format. Thus, the conversion of an analog electrical signal into a digital signal occurs within the mobile phone, and this is only one example of this kind of conversion. As we will see shortly, an ADC, is one of the most ubiquitous operations in modern electronics. This is why *the goal of this section is to consider why we need to convert analog electrical signals into digital ones and how it is done.*

Since we are talking about signals, this is entirely a technical problem, yet its importance should not be underestimated. The need for ADC and DAC in modern technology is much broader than merely supporting digital transmission. The general view of the role of ADC and DAC in today's technological world can be seen in Figure 4.1.1.

Let us recall that *all original natural* and technological processes are analog *by their very nature*. Consider the examples of natural processes: acoustic (produced by the human voice, music, thunder, etc.), thermal (brought about by ambient temperature, the temperature of objects, etc.),

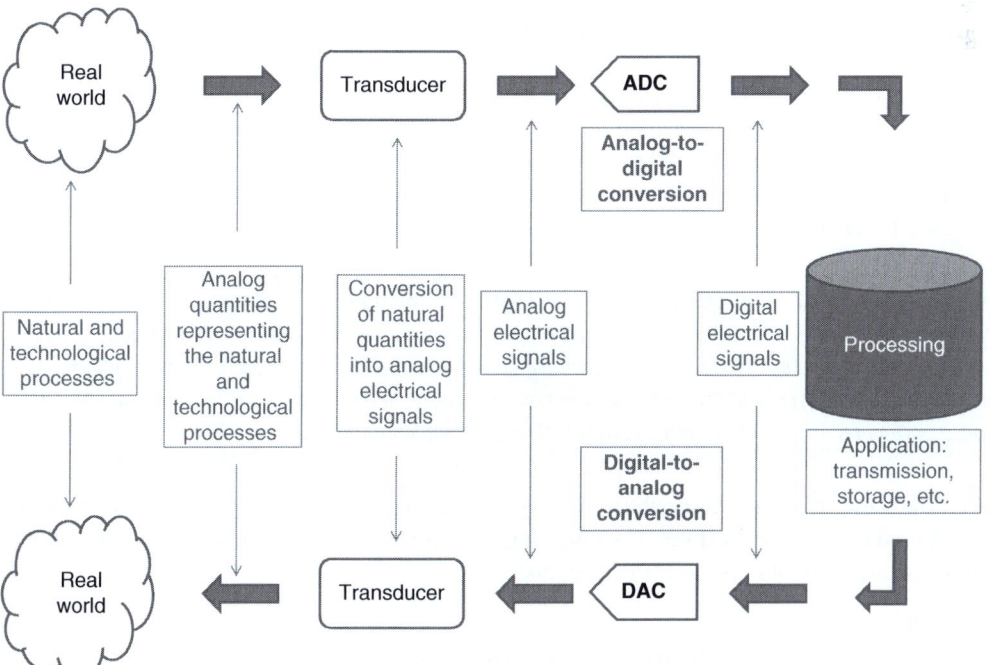

Figure 4.1.1 The need for analog-to-digital conversion (ADC) and digital-to-analog conversion (DAC).

motion (caused by the mechanical movement of bodies, waves, the fluctuations of man-made and natural objects, etc.), and radiation (generated by electromagnetic sources, including light and radio-frequency sources and emitted by atoms, including X-rays, γ and other ionized radiation). Now, the parameters of these processes are the sources of information to be processed, stored, and then delivered to destination points. For example, the parameters of the human voice (speech) are its amplitude and spectrum at every instant; the parameters of thermal processes are temperature, location, and the time of measurements; the parameters of motion are 3D coordinates and their effect on changing objects as functions of time. Bear in mind, too, that today we also have a myriad of technological processes, such as communications, manufacturing, food processing, energy conversion, and means of environmental protection – all of which are characterized by their specific parameters. All these parameters are analog quantities. (Note well: We consider here, of course, only macro-processes, that is, processes that occur and are observed in our macro world. In other words, we exclude all micro, that is, all quantum effects.)

In order to work with these parameters using modern technology, we need to obtain them in electrical form. This work is done by devices called *transducers*, which convert natural parameters into electrical signals, typically, voltage. Examples of transducers are microphones, thermometers, manometers, and various sensors measuring the parameters of different processes. Consider, for instance, a microphone that converts the acoustic waves generated by the human voice into an analog electrical signal. Another example: an electrical thermometer that converts heat generated by any thermal source into an analog electrical signal. (Here is a good exercise you can do at this point: Make a three-column table. In the first column, list all the natural and technological processes and their measurable parameters you know of and then find and write down the corresponding transducers in the second column. In the third column, briefly describe the operation of each transducer. Observe the wide variety of existing transducers.)

As follows from the above discussion, *the analog electrical signals appear in modern technology as the sources of information generated by natural and technological processes and delivered by transducers*. The storage, transmission, and processing of these analog signals with high quality require their conversion into digital signals. This is where ADC, comes into play, as Figure 4.1.1 demonstrates. Since we eventually have to present the results of digital signal processing in natural form, the need for reverse conversion, DAC, becomes clear.

To fully comprehend the role of ADC and DAC operations in modern technology, review again Figure 4.1.1, this time in conjunction with the following statement:

> *All natural and technological processes are described by analog quantities (parameters). To obtain these parameters for practical use, we must employ transducers to convert them into analog electrical signals, typically, voltage. To store, transmit, and process these signals with high fidelity, we must use digital systems and therefore convert analog signals into digital ones. To eventually present these parameters to users, we must convert them back into analog signals.*

ADC and DAC units exist everywhere in today's technology. They can be found in all devices and systems used in our everyday lives. The following are several practical examples of equipment that cannot operate without ADC and DAC systems:

- Communication
  - Any type of phone, transmitter, receiver, router, or switch.
- Consumer electronics
  - TV, audio and video systems, appliances, cars, and toys.

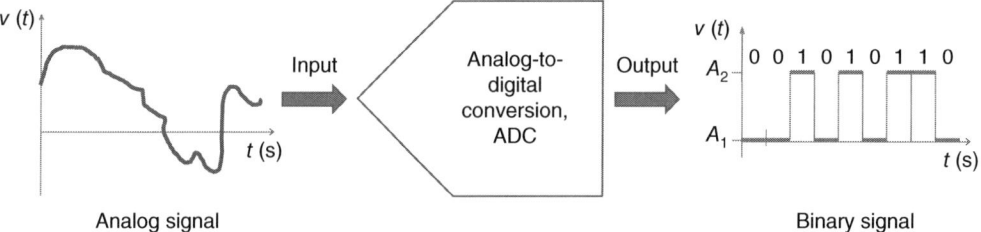

Figure 4.1.2 The concept of analog-to-digital conversion.

- Computers
  o Workstations, desktops, laptops, tablets, servers, storage, and networking electronics.
- Control systems
  o Automatic control equipment used in science and technology, in the art and entertainment world, and in home heating and cooling systems.
- Instrumentation
  o Measurement and test instruments, laboratory and medical equipment.

Your mobile phone contains tens of ADC and DAC units, making the phone a wonderful personal assistant performing hundreds of useful functions; oscilloscopes in the laboratory are replete with high-performance ADCs and DACs enabling the measurement of many signal parameters and the processing of these measurements automatically and precisely; in modern fiber-optic communications systems, the ADC units – real marvels of modern electronics – enable transmitters and receivers to perform tens of trillions (10^{12}!) of operations per second.

The above explanation states succinctly why we need to use ADC and DAC. Let us stress again that DAC is simply an inversed operation of ADC; in this section, we concentrate on ADC.

4.1.2 Three Major Steps of ADC

Figure 4.1.2 visualizes the concept of ADC: An analog signal is presented to an ADC unit; the output of this unit is a binary signal. The ADC unit is called, obviously, an *analog-to-digital converter*. There are a number of important details of the ADC process to be discussed but, for the moment, we must grasp the main idea of ADC.

First of all, how can we convert an analog signal (mathematically, a continuous function) into a set of 1s and 0s? This indeed is the main question we need to answer to design an ADC process. Taking a logical approach to solving this problem, we understand that the first and most crucial step is to present an analog signal as a set of numbers. Clearly, the use of decimal numbers will be the easiest way to perform this step, which is done in a *sampling (S&H)* operation. As soon as we have the decimal numbers, we can present them by binary numbers, that is, 1s and 0s. This goal is achieved in a *quantization* step. Finally, we need to put those 1s and 0s into electrical signals; we are already familiar with this procedure, *encoding*, which is discussed in Section 3.1. These three steps are depicted in Figure 4.1.3; let us consider it in detail.

4.1.3 Sample-and-Hold (S&H) Operation

4.1.3.1 Sampling (S&H) Technique and the Nyquist Theorem

Let us start by considering the S&H process. It would seem that presenting an analog signal as a set of decimal numbers would be an easy task: Just measure the magnitude, which is the signal value at

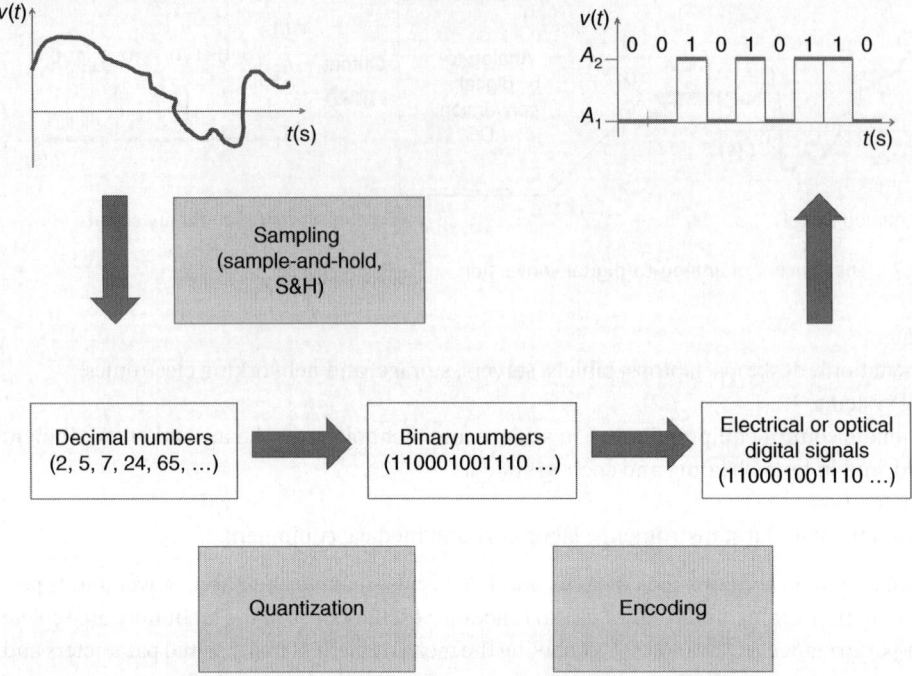

Figure 4.1.3 Three major steps in analog-to-digital conversion.

every instant. But think again: Taking the samples at every instant means that the interval between consecutive samples becomes infinitesimal, which in turn means that the results of sampling – the set of magnitudes vs. time – becomes a continuous function, that is, an analog signal. Indeed, as Figure 4.1.4a shows, as interval Δt (s) $= t_2 - t_1$ tends to zero, the difference between consecutive amplitudes, $\Delta v(t)$ (V) $= v(t_2) - v(t_1)$, goes to zero, too; thus, this signal becomes continuous (analog). (Analog signals were discussed in Chapter 2. If you need to refresh your memory on a continuous function, review the detailed discussion in Section 2.2.) Thus, this approach does not work.

Resolving this issue requires the sampling of an analog signal with fixed time intervals Δt (s). But what amplitude should we attribute to each interval – $v(t_2) = V_2$, $v(t_1) = V_1$ or something else? The industry answers this question with a simple solution: Sample an analog signal at t_1 and hold this magnitude value, $v(t_1) = V_1$, until t_2. Then sample the signal at t_2 and hold the magnitude, $v(t_2) = V_2$, until t_3, and so on. This procedure is illustrated in Figure 4.1.4b. Naturally, *this approach is called S&H*.

It is important to know that holding for an interval provides the time needed for the ADC circuitry to process the obtained value – the other important function of an S&H operation. (See the *quantization* step, which will be discussed shortly.)

The circuit performing an S&H operation is shown in Figure 4.1.4d. The logic-controlled switch instantly connects the input to the capacitor, charging the latter at a sampled voltage. The capacitor holds this voltage and feeds it to the output until the next sampled value arrives. Then the cycle repeats. Buffers, as usual, isolate their input and output circuits. Here, the input signal is sampled quickly, but at the output, each sampled value is held as long as required.

It would seem that the problem is solved, but a careful review of Figure 4.1.4b shows that this solution, unhappily, raises a new issue: We are now missing the key features of the sampled signal. For example, in the interval between t_7 and t_8, the signal's magnitude changes from a positive value

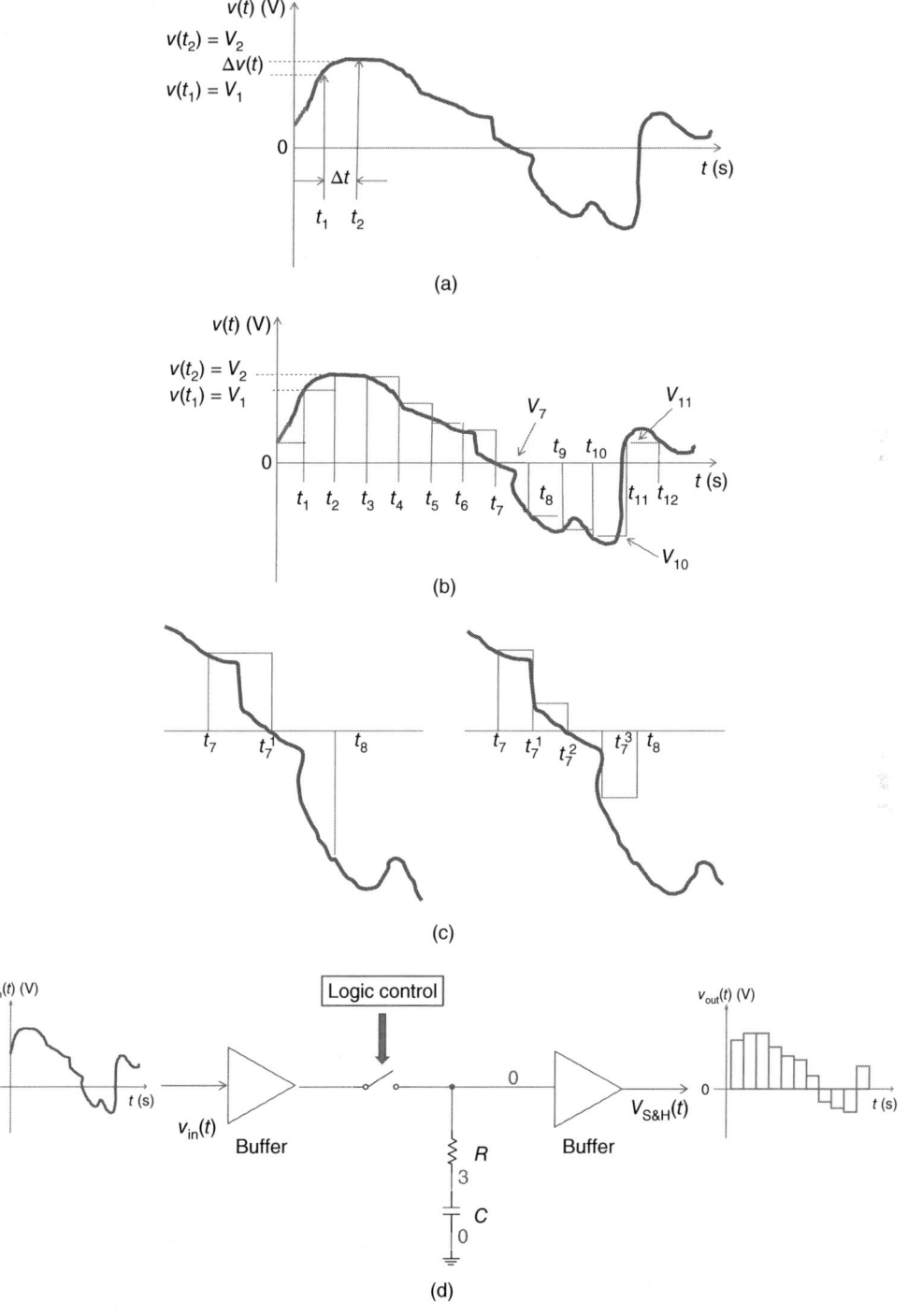

Figure 4.1.4 Sampling (sample-and-hold, S&H) technique: (a) Sampling at every instant; (b) sampling and holding; (c) sampling and holding at smaller intervals; (d) S&H circuit.

to a negative one, but the S&H rule requires us to attribute the positive value, V_7, to the whole interval. Another example is the interval between t_{10} and t_{11}, where the magnitude jumps from the negative V_{10} value to the positive V_{11}, but the rule still attributes $-V_{10}$ to the positive part of the signal. Analyze every interval in Figure 4.1.4b and observe that in practically all instances the real and attributed magnitudes differ.

The solution to this new problem would appear at the surface: Decrease the time interval. In fact, review Figure 4.1.4c, where the parts of the graph under discussion between t_7 and t_8 are shown in a larger scale. The interval between t_7 and t_8 in the graph on the left is split in two. These smaller sampling intervals produce two amplitudes instead of one and therefore provide a more accurate representation of the analog signal. The right-side graph shows that using four sampling intervals instead of two enables us to approximate the original analog signal even more accurately. Thus, it looks as though we have found the solution: *Decrease a sampling interval.* Pursuing this approach, we immediately conclude that in order to represent the analog signal in its entirety, we must take a sample (and hold it) at every instant. This means that the time interval between the adjacent samples becomes infinitesimal; that is, we have circled back to the original problem that cropped up during our discussion of Figure 4.1.4a – obtaining an analog signal instead of a digital one.

Therefore, we now face a dilemma: *Making Δt (s) too big, we cannot represent an analog information signal comprehensively; making Δt (s) too small, we come close to producing an analog signal instead of a discrete one.* The solution must lie in *choosing the proper sampling interval, Δt_s (s),* or, equivalently, the proper *sampling frequency, f_s (Hz)*. And, you should be aware, this solution was found by Harry Nyquist,[1] who in 1928 indicated that the sampling frequency, f_s (Hz), must be at least twice the maximum frequency of an analog signal, $f_{\text{signal}}^{\max}$ (Hz); that is,

$$f_s \text{ (Hz)} \geq f_{\text{signal}}^{\max} \text{ (Hz)} \qquad (4.1.1)$$

Equation (4.1.1) is known as *the Nyquist theorem, which determines the minimum value of a sampling frequency*. Since uniform sampling is a periodic signal with period T_s (s) $= \Delta t_s$ (s), we can relate the sampling frequency to the sampling interval (period) as

$$T_s \text{ (s)} = \Delta t_s \text{ (s)} = 1/f_s \text{ (Hz)} \qquad (4.1.2)$$

Equation (4.1.2) enables us to rewrite the Nyquist theorem in time domain as

$$\Delta t_s \text{ (s)} \leq 0.5\, T_{\text{signal}}^{\min} \text{ (s)} \qquad (4.1.3)$$

where $T_{\text{signal}}^{\min}$ (s) is the minimum period of the analog signal, $T_{\text{signal}}^{\min}$ (s) $= 1/f_{\text{signal}}^{\max}$ (Hz). *Equation (4.1.3) gives us the rule for choosing the right sampling interval, Δt_s (s): If we choose period Δt_s greater than $2\,T_{\text{signal}}^{\max}$ (s), we will lose the significant features (information) of the original signal; if we choose the period Δt_s smaller than $2\,T_{\text{signal}}^{\max}$ (s), we will waste additional resources by taking extra samples without acquiring additional information.* Thus, (4.1.1) and its time-domain version (4.1.3) give us the solution to this problem.

(**Exercise**: Formulate this statement in terms of sampling and maximum signal frequencies.)

The Nyquist theorem, (4.1.1), which is also called the *Nyquist sampling theorem*, was proved by Claude Shannon in 1949; this is why it is often called *the Nyquist–Shannon theorem*. Several other scientists and engineers independently arrived at similar results or contributed to the further development of the idea; thus, we should not be surprised to see other names associated with this theorem.

---

1 Short biography of Dr. Harry Nyquist is placed in Section 1.3.

4.1 Analog-to-Digital Conversion, ADC

The importance of the Nyquist theorem, (4.1.1), cannot be overestimated: It is the rule for the proper sampling of an analog signal and therefore *the key equation in ADC*. Hence, the significance of the theorem calls for its formal interpretation:

> *In order to preserve all information carried by an analog signal after conversion of the analog signal into a digital, the sampling frequency must be at least two times greater than the maximum frequency of the analog signal being converted.*

Now, let is recall that the bandwidth of a signal, BW_{signal} (Hz), is given by

$$BW_{signal}\ (Hz) = f_{signal}^{max}\ (Hz) - f_{signal}^{min}\ (Hz) \tag{4.1.4a}$$

If f_{signal}^{min} (Hz) = 0 as it often the case, then (4.1.4a) takes the form

$$BW_{signal}\ (Hz) = f_{signal}^{max}\ (Hz) \tag{4.1.4b}$$

This is why the Nyquist theorem is often presented as

$$f_s\ (Hz) \geq 2BW_{signal}\ (Hz) \tag{4.1.5}$$

This form of the theorem is also called *Nyquist–Shannon* because this clarification was made by Shannon. In fact, the industry prefers the sampling theorem in the form of (4.1.5).

We should point out that *if we follow the Nyquist rule, then the original analog signal can be, in principle, exactly restored from its S&H version*. The importance of this point will become clear shortly, when we consider a quantization step of an ADC operation.

Formal justification of the Nyquist theorem requires knowledge of spectral analysis, the topic presented in Chapter 7. Many other important and interesting details exist pertaining to the Nyquist–Shannon sampling theorem, but we leave them for your independent study.

4.1.3.2 Aliasing

Question: What if we violate the requirement of the Nyquist theorem? Obviously, we will miss some features of the original analog signal, as was explained in the previous subsection. But can we tolerate this loss? Consider Figure 4.1.5a, where a sinusoidal analog signal is sampled with different frequencies. (The purpose of using a sinusoid as the input analog signal will soon become clear.) Until the sampling frequencies, f_s (Hz), are greater or equal to the double signal maximum frequency, $2f_{max}$ (Hz), as in Cases 1–3, the original signal can be reproduced perfectly. (See the dotted-line graphs for these three cases.) However, when the sampling frequency does not meet the Nyquist limit, f_s (Hz) $\geq 2f_{signal}^{max}$ (Hz) or Δt_s (s) $\leq 0.5\ T_{signal}^{min}$ (s), as in Case 4 in Figure 4.1.5a, the reproduced analog signal strays far from the original. Looking at this situation in greater scale in Figure 4.1.5b, we observe that the sampling interval (period), Δt_s (s), becomes greater than 0.5 T_{signal}^{min} (s), which violates (4.1.3). If we connect the sample points in this case, we obtain a sinusoidal signal but with a different frequency. This new sinusoid is called an *aliased signal*, or *alias* (from *alias*, which means an assumed, false name). Our case is an example of the phenomenon called *aliasing*.

Aliasing is the effect that occurs when the signal recreated from the samples differs from the original analog signal. And, certainly, the aliased signal in Figure 4.1.5b differs substantially from the original analog one.

We should bear in mind, too, that the frequency of the aliased signal, f_{as} (Hz), is equal to the difference between the input and the sampling frequencies, f_{as} (Hz) = f_{signal} (Hz) $- f_s$ (Hz). In general, the sinusoidal signal with the frequency of such a difference appears from the beats between two

268 | 4 Analog-to-Digital Conversion (ADC) and Digital-to-Analog Conversion (DAC)

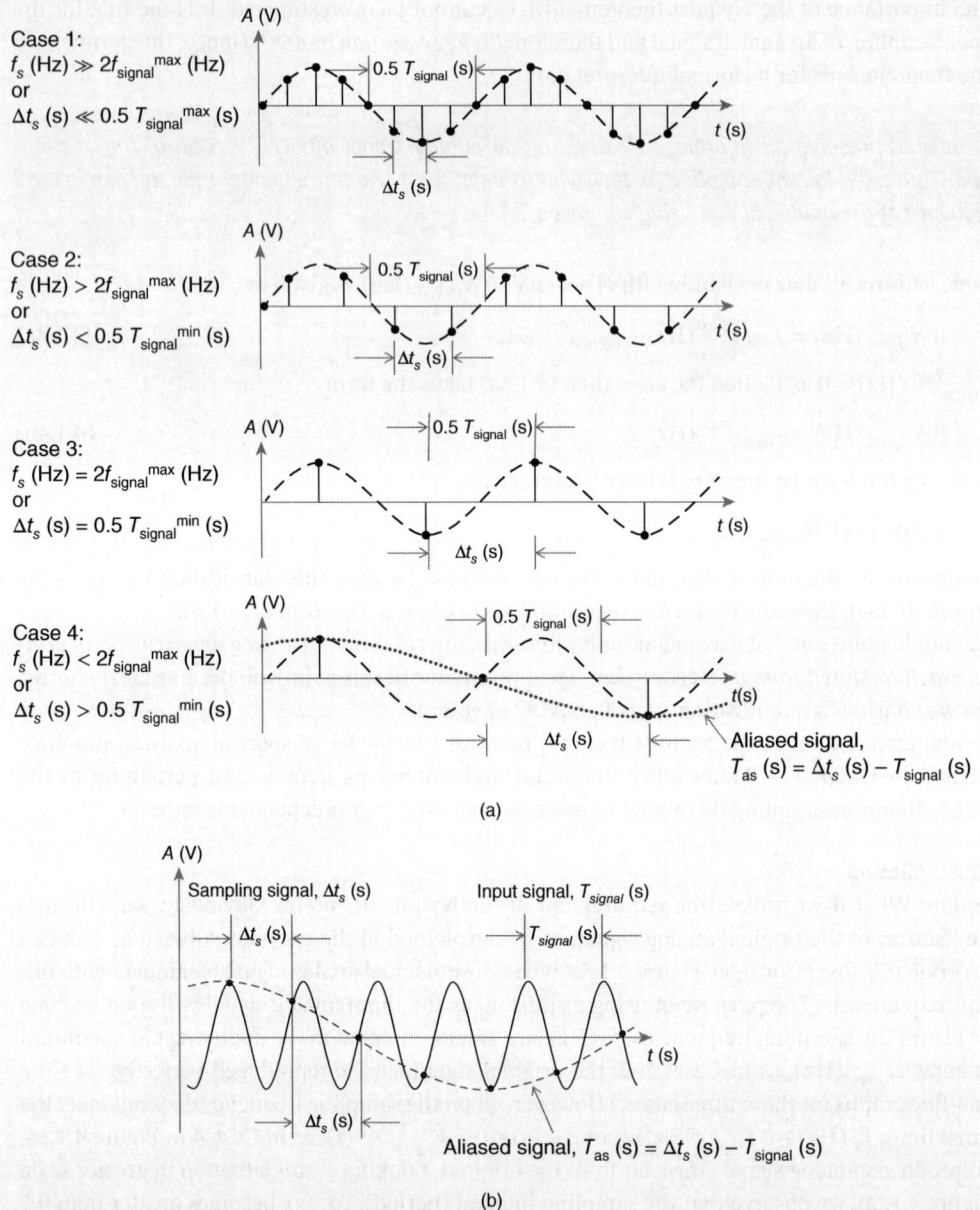

Figure 4.1.5 Aliasing: (a) The violation of the Nyquist criterion produces the wrong result; (b) aliasing results in the appearance of a false sinusoid; (c) aliasing results in receiving a distorted image; (d) the aliasing effect in frequency domain. Source: (c) Reprinted with permission of Professor Fredo Durand, MIT.

Figure 4.1.5 (Continued)

sinusoids with different frequencies. This new false signal appears as a result of the *undersampling* of an input sinusoid. Keep in mind that an alias, or aliased signal, is not always a sinusoid since the input analog signal is practically never a sinusoidal waveform. In Figure 4.1.5a,b, however, we depict sinusoids to better visualize the aliasing effect. Figure 4.1.5c demonstrates how aliasing can degrade quality of transmission.

Figure 4.1.5d depicts aliasing in frequency domain. It shows that the sampling frequency, f_s, is less than twice the signal's frequency, f_{signal}, which results in the appearance of the alias signal at the frequency of $f_s - f_{signal}$.

It follows from this discussion that aliasing is the phenomenon to be avoided in ADC because it produces false results. Fortunately, we know that it is sufficient to meet the Nyquist criterion, f_s (Hz) $\geq 2 f_{signal}^{max}$ (Hz) or Δt_s (s) $\leq 0.5\, T_{signal}^{min}$ (s), to completely avoid aliasing. Let us now consider an example of an S&H operation.

Example 4.1.1 S&H technique, the Nyquist theorem, and aliasing; oversampling and undersampling

Problem

We humans can hear a sound wave with a frequency up to 20 kHz. What are the (1) minimum, (2) maximum, and (3) unacceptable sampling frequencies, f_s (Hz), and what are the proper sampling intervals, Δt_s (s)?

Solution

(1) The minimum f_s^{min} (Hz) is obviously given by the Nyquist theorem in form (4.1.1). Simple calculations produce

$$f_s^{min} \text{ (Hz)} = 2f_{signal}^{max} \text{ (Hz)} = 2 \cdot 20 \text{ kHz} = 40 \text{ kHz}$$

Of course, all manufacturers of audio equipment know this number and often specify it on the equipment label.

To compute the proper Δt_s (s), we have to realize that it is the maximum sampling interval, Δt_s^{max} (s), that corresponds to the minimum sampling frequency, f_s^{min} (Hz). To find Δt_s^{max} (s), we need to know the period of a given analog signal, which, in this case, is given by

$$T_{signal} \text{ (s)} = 1/f_{signal}^{max} \text{ (Hz)} = 1/20 \text{ kHz} = 0.05 \text{ ms}$$

Following the limit set in (4.1.3), we find

$$\Delta t_s^{max} \text{ (s)} = 0.5 \, T_{signal} \text{ (s)} = 0.025 \text{ ms}$$

(2) How can we find the maximum sampling frequency? We know that the Nyquist theorem says that a sampling frequency must be equal to or more than twice the maximum signal frequency – but the theorem does not specify the upper limit of the sampling frequency. Is f_s (Hz) = $3f_{signal}^{max}$(Hz) the maximum sampling frequency? No, because f_s (Hz) could be $10f_{signal}$ (Hz) or $100f_{signal}$ (Hz). Therefore, *there is no formal maximum sampling frequency*.

In practice, however, manufacturers always design their equipment with a sampling frequency greater than, *not* equal to, the maximum signal frequency. This approach is called *oversampling*. How much to oversample is up to the individual manufacturer. Obviously, the higher the f_s (Hz), the better the digital signal obtained. The increase in f_s (Hz) will eventually result in better sound quality, a sharper image, and in fact any improvement in the output of an ADC operation. On the downside, an increase in f_s (Hz) will require more sophisticated expensive circuitry and processing software. This will result in additional power consumption. Bottom line: A designer has to find the balance between these two contradictory requirements to meet the given specifications and budget constraints.

This consideration can be easily applied to sampling intervals.

(3) Unacceptable values of sampling frequency and sampling intervals are obviously values that do not meet the Nyquist limit (4.1.1). The sampling operation with f_s (Hz) $< 2f_{signal}^{max}$ (Hz) is called *undersampling*. We learned that with undersampling *aliasing* occurs and the quality (or even the meaning) of the converted digital signal will be compromised. Hence, even f_s (Hz) = 39.99 kHz should not be acceptable in this example. In practice, however, the ADC operation is considered as *undersampling* when f_s (Hz) is noticeably lower than $2f_{signal}^{max}$ (Hz). For this exercise, since f_{signal}^{max} (Hz) = 20 kHz, the sampling at f_s (Hz) = 30 kHz is considered undersampling.

The problem is solved.

Discussion

- In general, the *signal's bandwidth*, BW_{signal} (Hz), *and the signal's maximum frequency*, f_{signal}^{max} (Hz), are different entities if f_{signal}^{min} (Hz) $\neq 0$. It follows from (4.1.4a) that BW_{signal} (Hz) = f_{signal}^{max} (Hz) $- f_{signal}^{min}$ (Hz). We can benefit from this fact by reducing the sampling frequency, which is allowed by the Nyquist–Shannon theorem, f_s (Hz) $\geq 2BW_{signal}$ (Hz). Consider, for instance, a signal whose frequencies range from f_{signal}^{max} (Hz) = 100 kHz to f_{signal}^{min} (Hz) = 60 kHz. Then, the signal's bandwidth is computed as BW_{signal} (Hz) = 100 kHz − 60 kHz = 40 kHz, and the required sampling frequency is given by f_s (Hz) = $2BW_{signal}$ (Hz) = 80 kHz. If we would meet the Nyquist limit, f_s (Hz) $\geq 2f_{max}$ (Hz) = 100 kHz, then we would need f_s (Hz) = 200 kHz. By using the f_s (Hz) $\geq 2BW_{signal}$ (Hz) criterion, we substantially reduce the sampling frequency and thus achieve significant improvements in the design of an ADC system, as the preceding discussion points out.
- It seems perfectly clear that undersampling is unacceptable, but hold on: A closer look reveals that this is not always true. In fact, *undersampling can be used to the benefit of an ADC operation*.

If we sample a sound signal with f_{signal}^{max} (Hz) = 20 kHz at f_s (Hz) = 30(kHz), this will, surely, be a case of undersampling. As a result, we obtain an *aliased signal* with frequency f_{as} (Hz) = f_s (Hz) $- f_{signal}^{max}$ (Hz) = 30 kHz − 20 kHz = 10 kHz. Using this aliased signal, we can find the maximum frequency of a given signal: f_{signal}^{max} (Hz) = f_s (Hz) $- f_{as}$ (Hz) = 30 kHz − 10 kHz = 20 kHz. This means we can recover the original signal. Thus, *using the lower sampling frequency, we can achieve the same goal as we could with the higher one while saving on expensive circuitry and the costly consumption of excessive electric power*.

- Having an unexpected ability to employ undersampling as a valid ADC operation, we need to choose which technique – oversampling or undersampling – should be used in the industrial environment. To decide, let us consider the advantages and disadvantages of both techniques.
  - Oversampling
    - *Advantages*: It delivers a better-quality sampled signal and yields the better overall ADC result. Other advantages include a better signal-to-noise ratio, SNR, and flexibility in the design of ADC circuits.
    - *Disadvantages*
      - Higher sampling frequency requires higher data rates from the processing circuitry, which in turn leads to *stringent requirements for components* and, in fact, for *the design* of the whole ADC system.
      - Higher sampling frequency results in a shorter hold time, which makes it *more difficult to capture and process data*; this again leads to the need *for better-quality circuitry*.
      - The requirement for better quality of the ADC circuitry ultimately results *in increasing the cost of the system and its operation*.
      - Higher sampling frequency requires more computations per second, resulting in *greater power consumption*.
  - Undersampling
    - *Advantages*: It is able to achieve the same result as the oversampling technique but uses the lower sampling frequency. The lower sampling frequency results in lower data rates and a greater hold time and lessens the strict requirements for quality circuitry and meticulous system design, thus decreasing the cost and power consumption. In other words, the advantages of undersampling are the disadvantages of oversampling.

- *Disadvantages*: It enables us to achieve good ADC results with a lower sampling frequency – but at a price:
  - This technique demands sophisticated signal processing.
  - The requirements for system filters are very stringent.
  - SNR and the total quality of the technique's output are still inferior to those of oversampling.

With the advent of modern electronics, the use of the undersampling technique has become easier to implement and more advantageous. Thus, in industrial practice today, undersampling often becomes the ADC technique of choice.

- Final note: We must fully understand that the *Nyquist rate is not a fundamental limit, but just a recommendation for optimal ADC conversion*. This is why we can use both undersampling and oversampling in the ADC process. It is worth mentioning now, however, that *digital transmission needs to meet the Nyquist criterion* to provide optimal conditions for operation of a whole communications system. This aspect will appear at the higher level of discussion of digital communications in Chapters 9 and 10.

Having read the fount of information in this subsection, you should now understand that the main point in converting an analog signal into a set of decimal numbers is the Nyquist theorem, f_s (Hz) $\geq 2f_{signal}^{max}$ (Hz), or f_s (Hz) $\geq 2BW_{signal}$ (Hz). Yet a puzzling question remains to be addressed: How do we know the maximum frequency or bandwidth of an input analog signal? In some cases, these values are certainly known. For example, the human voice does not contain a frequency higher than 4 kHz and humans cannot hear a frequency higher than 20 kHz. In most practical situations, unhappily, we know neither f_{signal}^{max} (Hz) nor BW_{signal} (Hz). We do know, however, either the expected range of frequencies or bandwidth of incoming signals or we know by what frequencies (bandwidth) we want to restrict the spectrum of a signal. In both cases, a filter that limits the input spectrum to the desired value of f_{signal}^{max} (Hz) solves the problem. Including such a filter at the input of an ADC unit enables us to correctly set the sampling frequency. In essence, this filter solves the aliasing problem; this is why it is often called an *anti-aliasing filter*. This filter also removes the noise components whose frequencies lie above f_{signal}^{max} (Hz). The specifications of an anti-aliasing filter must meet exacting requirements; it is no wonder that these filters require sophisticated circuitry. An elliptic filter of the eighth order and similar types (see Chapter 5) are typically used in this application.

4.1.4 Quantization in ADC

4.1.4.1 Quantization Process

A S&H operation solves the first ADC problem by presenting a continuous analog signal as a set of discrete decimal numbers. Our ultimate goal, however, is to obtain an electrical digital signal carrying complete information about the analog signal. Refer to Figure 4.1.3; it shows that the next step in an ADC operation must be to present the obtained decimal numbers in a binary format. This goal is achieved by quantization.

> *Quantization is representing the decimal numbers, received after a S&H operation, by binary numbers.*

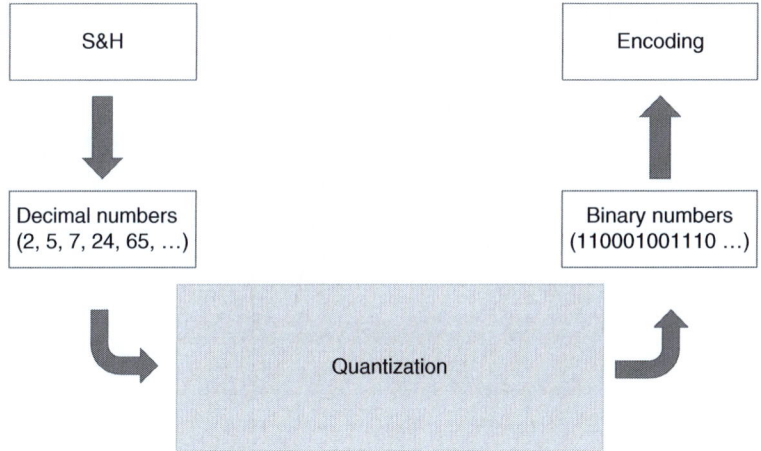

Figure 4.1.6 The quantization operation.

The concept of a quantization operation is depicted in Figure 4.1.6. Since this operation relies on the manipulation of decimal and binary numbers, a brief review of the decimal and binary number systems is given in Appendix 4.1.A. Let is now discuss the quantization process more closely.

How to approach the development of a quantization process

To remind you, the input to a quantizer – the unit performing the quantization process – is a set of decimal sample values and the output is a set of binary numbers. Presenting decimal numbers in binary format would not seem to be a problem. Just convert the sample decimal values into binary numbers, as explained in Appendix 4.1.A (see Figures 4.1.A.3 and 4.1.A.4.) It turns out, however, that this straightforward approach does not work well in an ADC operation, and so it is necessary to resolve several problems before we can achieve the goal of this process. Let us consider these problems and their solutions in a logical step-by-step manner.

Step 1: How to choose the number of bits per sample, m (b/sample)

Consider the sampled signal shown in Figure 4.1.7a: The actual sample decimal numbers, obtained after an S&H procedure, can be of any value; what is important for this operation is that each number is, in fact, the infinite. For example, suppose at first we obtain a sample with a value of 9.375 V. Yet a more accurate measurement gives us a value of the same sample as 9.375 796 V. And even more precise sampling results in 9.375 796 207 61 V. You get the point? *In principle, a sample is an infinite decimal number*. For such a sample value, we obviously would need to find the corresponding infinite binary number, a futile operation.

The infinity of a sample decimal number would not seem to be a problem: Let us truncate a measured sample value to any reasonable number, for example, instead of 9.375, take just 9. Now, simply convert this number into a binary and find the following result: 9 → **1001.** (You will recall that $9 = \mathbf{1} \times 2^3 + \mathbf{0} \times 2^2 + \mathbf{0} \times 2^1 + \mathbf{1} \times 2^0 \to \mathbf{1001}$. See the principle of converting decimal numbers into their binary counterparts in Figure 4.2.A.4. Also, be aware that this conversion is NOT how decimal numbers are represented by binary numbers in the quantization process. We need this example just to find the number of bits needed.)

This approach, however, raises a valid question: Truncating a sample value, we certainly are losing accuracy in describing an actual signal. What restrictions do we have in the number of sample

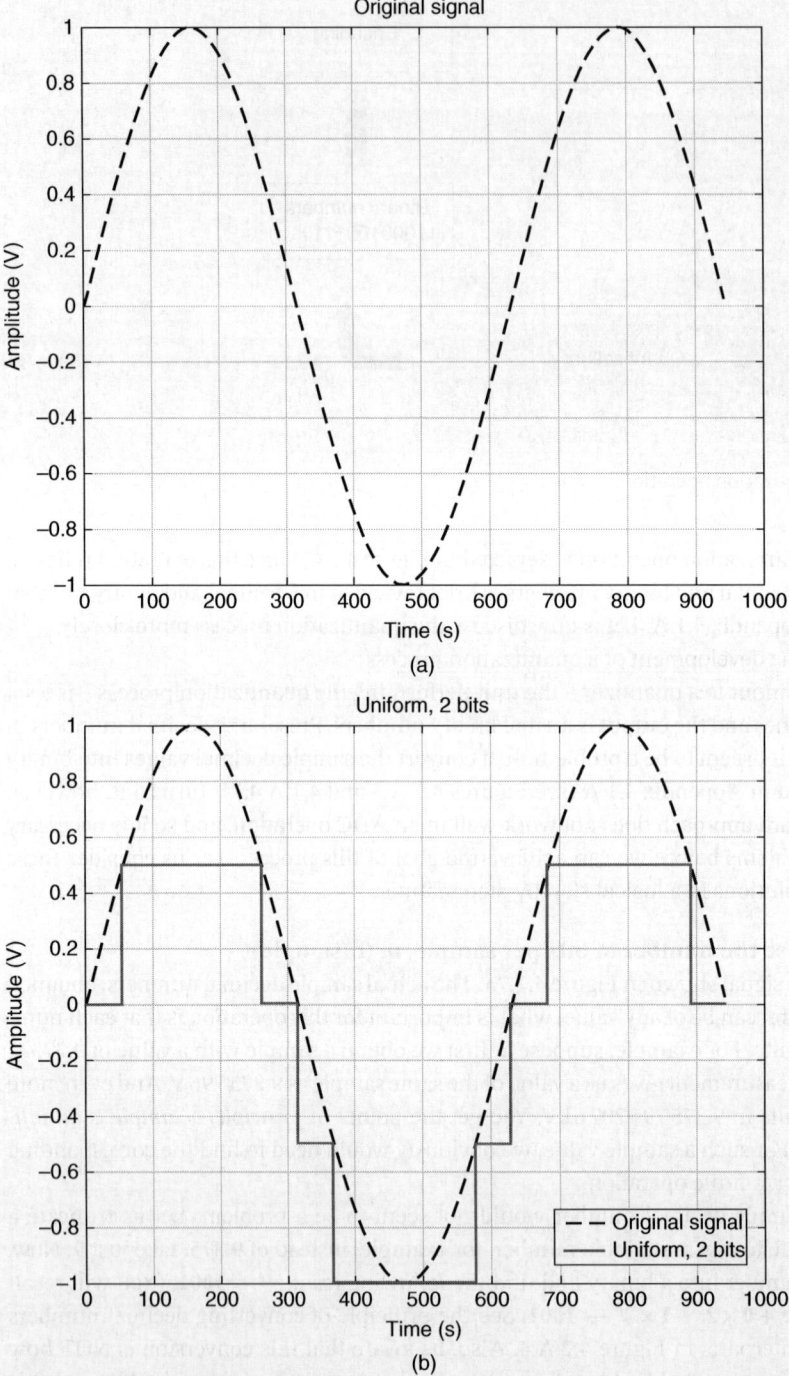

Figure 4.1.7 The quality of an ADC operation depends on the number of bits per sample: (a) An original analog signal; (b) the analog signal sampled with 2 bits per sample; (c) the same analog signal sampled with 4 bits per sample.

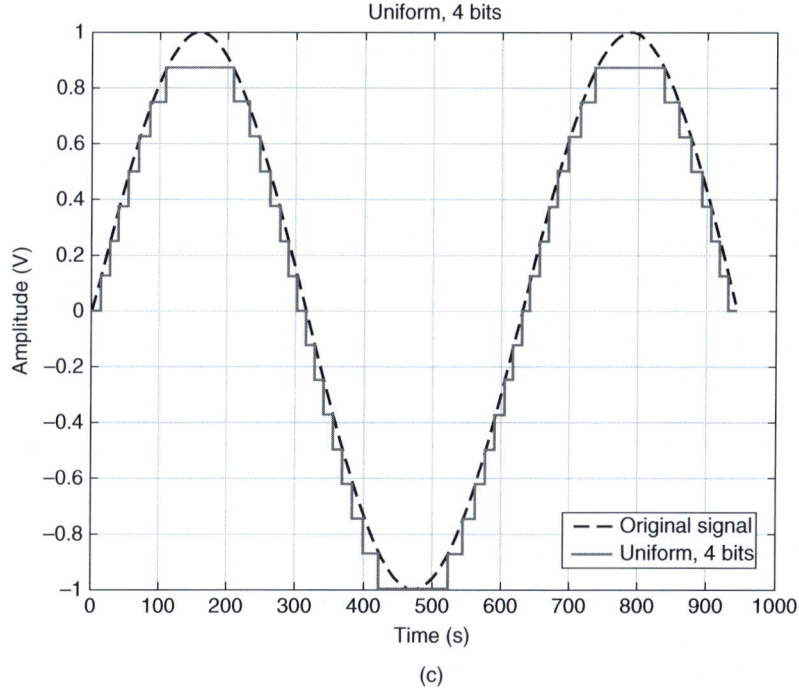

Figure 4.1.7 (*Continued*)

digits, that is, in the precision of the sample description? Let us try to keep the initial sample value, 9.375, and convert it into a binary. We obtain: 9.375 → **1001.0110.**

"Why don't we use the more accurate sampling value in binary?" you may ask. The answer lies in the required *number of bits per sample*: To digitize the sample value of 9, we need only *four bits*, but with 9.375, we need *eight bits*, as the preceding discussion shows. It is well known that the greater the number of bits, the greater the computing power required from the processing devices. Putting it another way, by increasing the number of used bits, we run into the trade-off between the computing power of a DSP system and the processing time. Bottom line: *We need to restrict the sample values to reasonable decimal numbers and, therefore, to a reasonable number of bits per sample.*

But what number should be considered reasonable? The answer: *There is a trade-off between the number of bits per sample, m (bit/sample), and ADC quality.* How does the quality of an ADC operation depend on the number of bits? A glance at Figure 4.1.7 readily shows that the greater the number of bits, the better the ADC quality. Why? This will become clear shortly, as we proceed to further consideration of the quantization process. In the meantime, to approach this problem from a different perspective, let us compare two practical examples:

Example 1: Voice transmission is digitized by using the sampling frequency of $f_s = 8000$ samples/s and the number of bits per sample, $m = 8$ b/sample, which results in a bit rate of BR_{ADC} (b/s) $= f_s m = 64\,000$ b/s. (This means that the given ADC system generates 64 kilobits every second.) Thus, for the transmission of a digitized voice, we need a channel with the bandwidth BW_T (Hz), which can accommodate a bit rate of 64 kb/s. Clearly, if we want superior voice-transmission quality, we can assign 24 bits per sample, but this will increase the bit rate to 192 kb/s.

Example 2: Blu-ray storage disc and high-definition DVD systems sample analog signals at $f_s = 192$ ksample/s and require an $m = 24$ b/sample, which results in BR_{ADC} (b/s) $= 4.608$ Mb/s.

Transmitting this bit stream will require the proper transmission bandwidth. Also, the volume of information generated by this ADC system increases to 4 608 000 bits every second.

From these examples, we can derive an important relationship: The bit rate of a digital signal generated by an ADC system, BR_{ADC} (b/s), is given by

$$BR_{ADC} \text{ (b/s)} = f_s \text{ (sample/s)} \cdot m \text{ (b/sample)} \quad (4.1.6)$$

where f_s (measured in the number of samples per second, that is, in hertz) is the sampling frequency and m (b/sample) is the number of bits per sample. Equation (4.1.6) tells us that increasing either f_s (Hz) or m (b/sample) or both will result in increasing the volume of information to be transmitted every second. This will produce a greater bit rate, which, in turn, will demand a larger transmission bandwidth. On the other hand, as we have learned in the preceding text, increasing either f_s (sample/s) or m (bit/sample) or both will result in a better-quality ADC operation.

Observe that (4.1.6) denotes a *restriction in choosing the reasonable number of bits per second: It shows that there is a trade-off between f_s (Hz) and m (b/sample) when either BR_{ADC} (b/s) or BW_T (Hz) is given.*

We can sum up the above consideration as follows:

1. *The first objectives of an ADC operation are to sample an analog signal and to represent the obtained decimal sample values in binary format.*
2. *Decimal sample values are infinite in principle; truncating them, though necessary, results in losing the accuracy of a signal's presentation in binary format.*
3. *The number of digits contained in a truncated decimal sample value determines the required number of bits per sample. The opposite is also true: By choosing the number of bits per sample, we determine the number of digits in a decimal sample value.*
4. *The key point in determining the accuracy of an ADC operation is choosing the proper number of bits per sample: The greater the m (b/sample), the more accurate the signal's presentation.*
5. *There is no natural rule or criterion for choosing the number of bits per sample, but we have to bear in mind that increasing the number of bits per sample causes the increase in the bit rate, BR_{ADC} (b/s), generated by an ADC system. The greater the BR_{ADC} (b/s), the bigger the demand for transmission bandwidth and for computing power of the DSP devices.*
6. *Thus, choosing the right number of bits per sample is the first and crucial step in a quantization operation. Choosing this number is always a designer's decision, which is based on all the abovementioned factors.*

Step 2. How to determine the number of quantization levels according to the chosen number of bits per sample and how to assign a binary codeword to each quantization level

Review Figure 4.1.8a, where sampling operation is reminded. Next, we need to realize that as soon as we have chosen the number of bits per second, we immediately determine the number of *quantization levels*. To understand what and why, let us choose, for example, 3 bits per sample. This means we have only eight combinations of these 3 bits, which can be proved as follows: Let us start from **000** and increase the value of every subsequent binary word by 1; as a result, we can obtain the following eight codewords: **000, 001, 010, 011, 100, 101, 110, 111**. (This sequence is called a *natural binary code.*) Therefore, we can choose only certain combinations, called *binary codewords*, out of the given bits. Also, having 3 bits we can distinguish only $2^3 = 8$ levels within the whole range of changing a signal's amplitude. (This range is called the *dynamic range of a signal*; in reality, it is the *peak-to-peak amplitude*, A_{pp} *[V]*. If *A* [V] is the amplitude of a symmetrical signal, the dynamic range is given as 2*A* [V]. The industry usually normalizes *A* [V] to unity.) These distinguishable levels are called *quantization levels*. Figure 4.1.8b illustrates this concept: It shows eight codewords and, corresponding to them, eight quantization levels, where the lowest level carries the minimal binary number, **000**.

It follows from the above example that with the given number of bits per sample, *m* (bit/sample), we can obtain only the following number of quantization levels, N_{ql}:

$$N_{ql} = 2^m \qquad (4.1.7)$$

What specific binary codeword to be assigned to a specific quantization level is a designer's choice. In fact, there are a number of options for this decision; Figure 4.1.8b shows a *natural binary code*. In Figure 4.1.8e, we can see another type of codewords.

Step 3. How to assign decimal numbers to quantization levels and round the actual sample decimal values to the assigned decimal numbers

In choosing the number of bits, we must meet another requirement: *All binary words used by one system must have the same number of bits*. But when we simply convert various decimal numbers into the binary system, we have a different number of bits. For example, for converting 9, we need four bits because 9 → **1001**, but for converting 17, we need five bits because **17** → **10001**. (Again, refer to Appendix 4.1.A.) An ADC processing unit, as is true of any computer, can work only with *binary words*, which are sets consisting of an equal number of bits. (Practically speaking, computers work only with binary words that are made up of $2^2 = 4$ bits, $2^3 = 8$ bits, $2^4 = 16$ bits, and so on.) Therefore, we cannot present to the same digital processing unit two binary words of different sizes – one, for example, consisting of 4 bits and another of 5 bits.

This is why the decimal numbers to be presented in binary format must contain the same number of digits. For instance, if the actual samples have the values 1.6, 1.83, 1.978, and 2.1943, the decimal values must form the set of 1.60, 1.83, 1.98, and 2.19. *This equalization can be readily achieved by rounding the actual sampled values to assigned decimal numbers containing an equal number of digits.*

In assigning the decimal numbers to the samples, we need to rely on corresponding the quantization levels because they relate the decimal numbers to the proper binary codewords. Since we do not have a natural criterion for such an assignment, it is up to the designer to choose the approach to the assignment. <u>The common industry practice is to assign the center value of a quantization level to all sample decimal numbers falling within this level</u>. For example, consider quantization level 7 in Figure 4.1.8b. Now, suppose the decimal center value of quantization level 7 is equal to 0.875 and three samples falling into this level have values of 0.986, 0.912, and 0.864. Then, by a quantization operation, the center value of 0.875 will be assigned to all three samples regardless of their actual values. Therefore, after quantization, all three samples will carry one value, 0.875, as required.

Step 4. How to represent an assigned decimal sample value by a binary code

We now need to take the final step in the quantization process: represent the assigned decimal sample values by binary numbers. It would seem that we have already taken all the necessary steps: We have chosen the number of bits per sample and we have assigned an equal number of digits to each decimal sample value. Why would we not simply convert these assigned decimals into binary

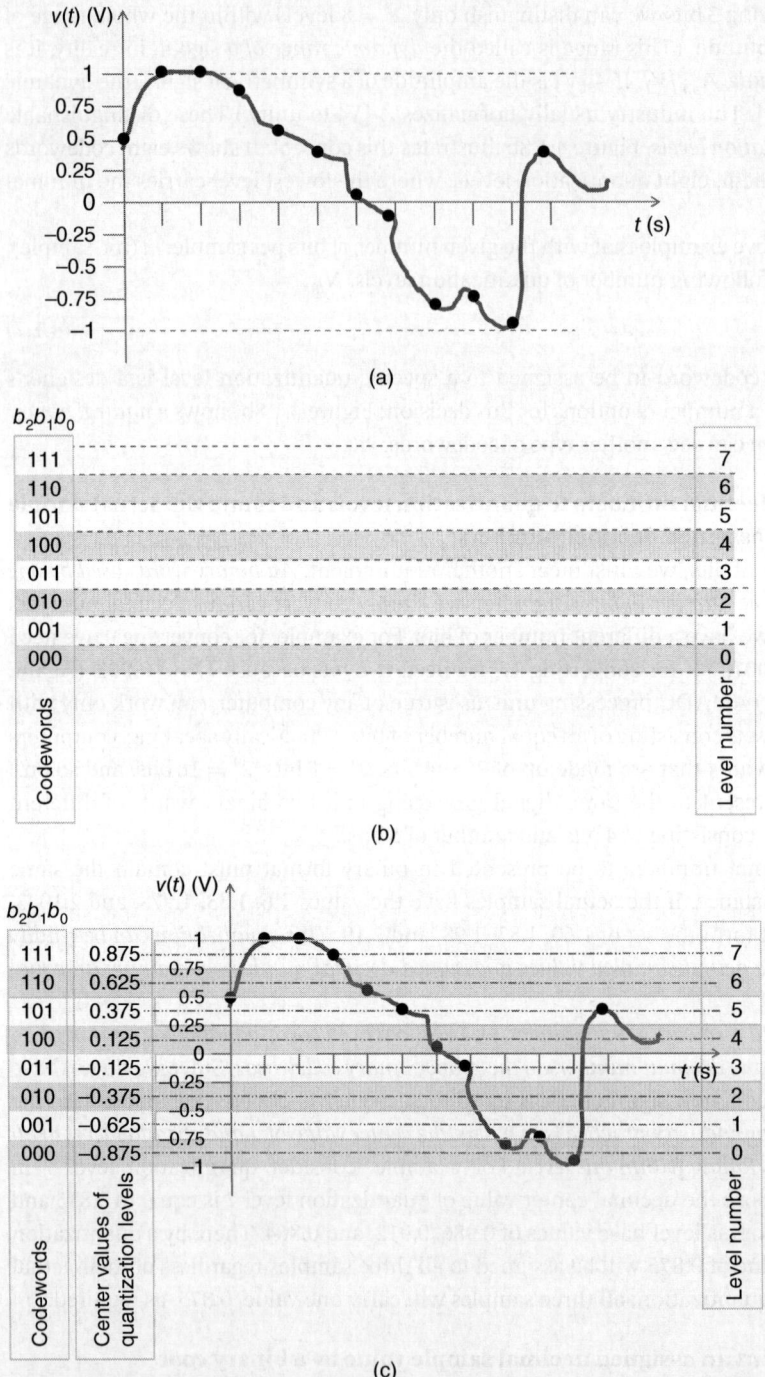

Figure 4.1.8 The principle of quantization: (a) A sampled analog signal; (b) the binary codewords and corresponding quantization levels; (c) the binary codewords and the center values of the quantization levels; (d) two binary systems for coding the decimal center values.

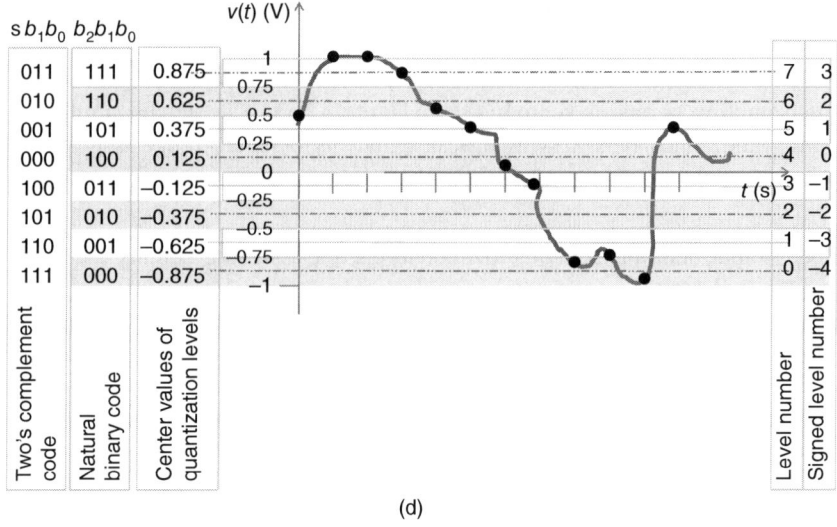

(d)

Figure 4.1.8 (*Continued*)

numbers? Let us start doing this for the positive numbers: Recalling from Appendix 4.1.A that $0.125 = \frac{0}{2^1} + \frac{0}{2^2} + \frac{1}{2^3} \rightarrow \mathbf{0.001}$ and applying this procedure to the other decimals, we find

$0.125 \rightarrow \mathbf{0.001}$
$0.375 \rightarrow \mathbf{0.011}$
$0.625 \rightarrow \mathbf{0.101}$
$0.875 \rightarrow \mathbf{0.111}$

Hence, for direct conversion of the assigned decimal sample values, we would need four bits and an agreement about the binary point; however, we have only three bits per sample in this example. What is more, to designate a negative sign for the negative decimals, we would need an additional bit and an agreement about which bit – **1** or **0** – would represent the minus sign. All this means that the direct conversion of the assigned decimal into the binary numbers does not work.

The solution to this problem is *coding*, that is, to determine what combination of binary numbers (a *binary codeword*) would represent a specific decimal number. In implementing this solution, we *superimpose the quantization levels with their binary codewords onto the sampled signal*, as shown in Figure 4.1.8c. Then, all samples in quantization level 0 will assume binary code **000**, all samples in quantization level 1 will be assigned the binary codeword **001**, and so on. This entails an agreement among all users of this approach. We need only inform these users that we are applying *natural binary coding*. With this coding, we represent the center minimal decimal number by binary codeword **000**, the next center decimal value by **001**, and so forth. Thus, in our example, the assigned decimal sample values are coded into binary words as follows:

$0.875 \rightarrow \mathbf{111}$
$0.625 \rightarrow \mathbf{110}$
$0.325 \rightarrow \mathbf{101}$
$0.125 \rightarrow \mathbf{100}$
$-0.125 \rightarrow \mathbf{011}$
$-0.375 \rightarrow \mathbf{010}$
$-0.625 \rightarrow \mathbf{001}$
$-0.875 \rightarrow \mathbf{000}$

The quantization process is complete.

Step 5. Discussion of the quantization process depicted in Figure 4.1.8

To fully understand the quantization principle, let us review this operation as it is visualized in Figure 4.1.8. We consider the signal discussed in the S&H operation and shown in Figure 4.1.4a. This signal, with its sampling points, is reproduced in Figure 4.1.8a. The signal's amplitude is normalized to unity and, since it is a symmetrical signal, its dynamic range is given by ±1 V. The remaining drawings in Figure 4.1.8 show the following steps in the quantization operation:

1. Choosing 3 bits per sample, building eight quantization levels, and assigning a binary codeword to each level – see Figure 4.1.8b.
2. Superimposing the quantization levels along with their binary codewords onto the sampled signal – see Figure 4.1.8c. This figure also shows the decimal center values of each quantization level.
3. Choosing two various methods of binary coding: two's complement code and natural binary code – see Figure 4.1.8d.

(Another good **exercise**: Write down the result of each step of the quantization operation visualized in Figure 4.1.8 and described above.)

Carefully review the above explanations of the quantization operation, by checking every point of these explanations in conjunction with Figure 4.1.8.

Important notes

- As to the question of why we cannot simply convert sample decimal values into binary numbers, the above discussion should provide the clear answer. Make sure you fully understand this point.
- Consider the relationship between the decimal numbering of quantization levels (the rightmost column) and the binary codewords (the leftmost column) in Figure 4.1.8c. Is not it the simple conversion of decimal numbers into binary? Yes. For example, number 3 is converted into **011** and number 7 into **111**. (Refer to Figure 4.1.A.4 to recall the conversion rule.). This coding system is called *natural binary code*, as mentioned previously. However, the natural binary code is not based on the numbering of quantization levels because these levels can be numbered in various ways. In fact, the *natural binary code,* as was explained previously, assigns all 0s to the minimal quantized value and consequently increases the value of the binary numbers to follow the value of the decimal numbers. Thus, the largest decimal number will correspond to the largest binary number, consisting of all 1s. In Figure 4.1.8c, the least quantized decimal number, –0.875, is encrypted into the least binary codeword, **000**, and the largest quantized decimal, 0.875, is encoded into **111**.
- Consider the two right-hand columns in Figure 4.1.8d: They show two different ways of numbering the quantization levels – as unsigned decimal numbers from 0 to 7 and as signed decimal numbers from –4 to 3. For the unsigned decimal numbering, the natural binary code is applied. To represent the signed decimal number in the binary format, there is a system called the *two's complement binary code*. This code is used today in practically all computers. The principle of two's complement binary coding is as follows: The most significant bit (MSB) (the leftmost bit) represents the sign, **0** for plus and **1** for minus. Positive decimal numbers, then, are simply converted into binary numbers and decimal zero is represented by zeros in the binary system. Thus, the increasing decimal positive numbers 0, 1, 2, and 3 (the rightmost column in Figure 4.1.8d) are converted into the two's complement binary code as **000**, **001**, **010**, and **011** (the second left-hand column in Figure 4.1.8d). For example, **011** is $1 \times 2^1 + 1 \times 2^0 = +3$. The maximum positive decimal number is given by $2^{m-1} - 1$. In our example, this number is $2^{3-1} - 1 = 3$.

Since the negative decimal numbers start with −1, the negative decimals are *coded, not* simply *converted*, into binary words. In our example, this coding is as follows:

−1 → **100**

−2 → **101**

−3 → **110**

−4 → **111**

(**Exercise**: Try to convert negative decimals into binary numbers given that the MSB bit stands for minus. What rule for the conversion of negative decimal numbers can you derive?) For negative decimal numbers, the *maximum integer* is 2^{m-1}. In our example, it is $2^{3-1} = 4$; that is, the *minimum negative number* is −4. Thus, in a negative set, we have one additional negative integer because decimal zero belongs to the positive numbers.

Step 6. A step size of a quantization level

An important parameter of quantization is the *step size of a quantization level*, Δ. Obviously, it can be found as

$$\Delta = \frac{2A\,(V)}{N_{ql}} = \frac{2A\,(V)}{2^m} \qquad (4.1.8)$$

where, as before, $2A$ (V) is the *dynamic signal range* and $N_{ql} = \frac{2A\,(V)}{2^m}$ is the number of quantization levels defined in (4.1.7). To realize the meaning of (4.1.8), consider a simple example: With $m = 3$ b/sample, we find from (4.1.7) and (4.1.8) that $N_{ql} = 2^3 = 8$ and $\Delta = \frac{2\,V}{2^3} = 0.25V$; on the other hand, it follows from (4.1.8) that $N_{ql} = \frac{2\,V}{\Delta} = \frac{2\,V}{0.25\,V} = 2^3 = 8$.

Now, we understand why

> to increase the accuracy of quantization, we need to increase the number of quantization levels, N_{ql}, or to decrease the step size of a quantization level, Δ, which will demand an increase in the number of bits per sample, m(b/sample).

(By doing so, we can achieve better accuracy of quantizing. This is because *every actual value will be represented more precisely by an assigned value since the difference between the actual and assigned values will decrease*.) Examine Figure 4.1.8 and observe that the second, third, and the fourth samples have the same *assigned value*, 0.875, whereas in reality, all of these points have different actual values. Now, imagine that you make 16 quantization levels in Figure 4.1.8. Then the second sample would have the assigned value of 0.8125 and the third sample would have 0.9375. These assigned values would represent the actual values much more accurately than they do in the case of 8 levels. (**Exercise**: Make 16 quantization levels in Figure 4.1.8, compute the border and center values, and observe how the precision of the presentation of samples will increase.) Thus, increasing the number of levels will increase the accuracy in the presentation of a real signal after quantization, thereby increasing ADC quality. But, again, the number of quantization levels depends on the number of bits per sample, as (4.1.8) states. This explains the phenomenon displayed in Figure 4.1.7: *the greater the number of bits per sample used in quantizing, the better the quality of the quantizing process*.

Step 7. Resolution

Another view of the quantization process can be presented as an input–output graph of an ADC quantization operation. This view is shown in Figure 4.1.9.

282 | 4 Analog-to-Digital Conversion (ADC) and Digital-to-Analog Conversion (DAC)

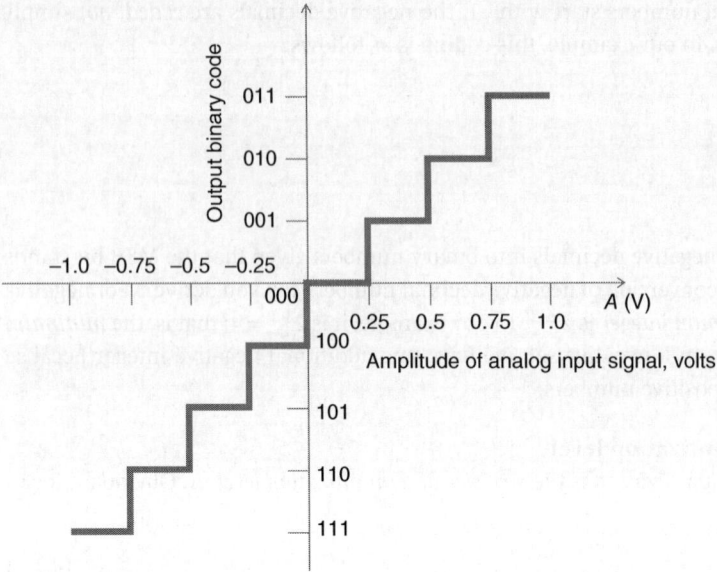

Figure 4.1.9 A staircase input–output characteristic of an ADC quantization operation with two's complement binary code.

Be aware that the general description of the relationship between the input and output of a system is called the *transfer function*. Hence, Figure 4.1.9 depicts the transfer function of an ADC system. To understand the meaning of Figure 4.1.9, refer to the example discussed previously in conjunction with Figure 4.1.8. In that example, the signal's dynamic range was ±1 V, the step size was given by $\Delta = 0.25$ V, the number of bits per sample was 3 b/sample, and the two's complement binary code was chosen. Now, Figure 4.1.9 clearly shows that all sample values that vary from −1 to −0.75 V belong to one quantization level and are coded into one binary word, **111**; all sample values that range from −075 to −0.5 V are coded into **101**, and so on. This graph illustrates that, after quantization, no voltage changes finer than 0.25 V can be resolved. From this standpoint, Figure 4.1.9 visualizes the level of resolution provided by a given ADC system. Since this resolution is determined by the number of bits per sample (do you understand why?), the industry often refers to m (b/sample) as an indicator of resolution by saying "3-bit resolution" or "8-bit resolution." (**Exercise**: Draw a figure similar to Figure 4.1.9 with 3-bit resolution but using the natural binary code.)

Step 8. Uniform and nonuniform quantization

It is worth mentioning that the quantization process presented in Figure 4.1.8 is called *uniform* because Δ holds constant at all levels. Though uniform quantization is widely used, in some applications, such as speech and music processing, a *nonuniform process*, with variable step-size quantization, is applied. The goal of nonuniform quantization is to sustain the fidelity of a signal even if its amplitude ranges widely. This is accomplished by making step sizes smaller where the signal is weak (that is, the amplitude is small) and bigger where the signal is strong (the amplitude is large and changes smoothly). This nonuniform setting makes the sampling more detailed (and therefore accurate) for weak parts of a signal and more coarse for the strong parts. Figure 4.1.10 presents examples of nonuniform quantizing. Examine and analyze the figure; observe when, how, and why the quantization steps change.

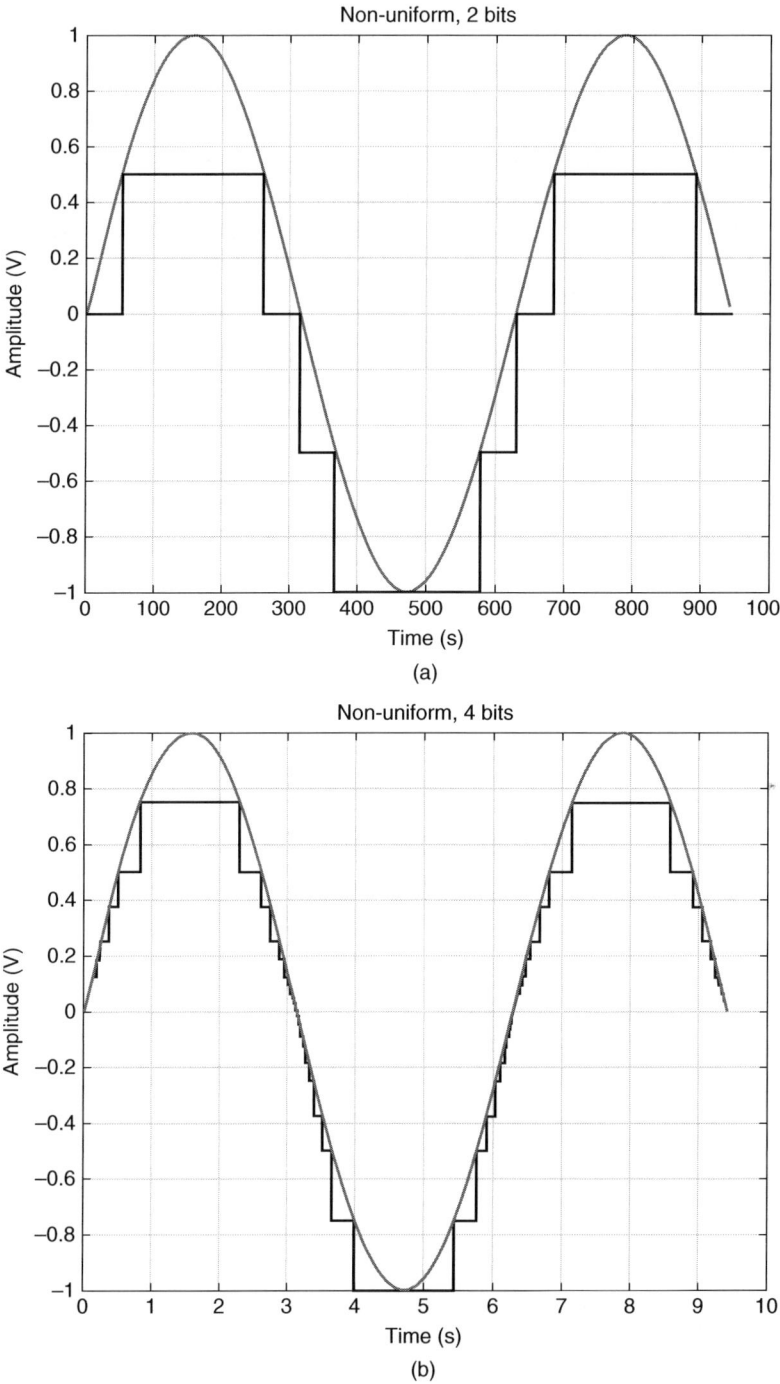

Figure 4.1.10 Nonuniform quantization: (a) Quantization with 2 bits per sample; (b) quantization with 4 bits per sample.

4 Analog-to-Digital Conversion (ADC) and Digital-to-Analog Conversion (DAC)

Figure 4.1.10 also demonstrates how the quality of the ADC process depends on the number of quantization levels. (Here is another useful **exercise** for you: Compute the number of quantization levels in each graph in Figure 4.1.10 to see the effect of N_{ql} and Δ on ADC quality.)

4.1.4.2 Quantization Errors and Quantization Noise

Let us repeat that in a quantization operation, *we compute the value at the center of each level and assign it to each sample located within this level*, as shown in Figure 4.1.8c. Thus, the sample points now assume these assigned values instead of their real values, as was indicated previously. From a careful examination of Figure 4.1.8c, we will find that most of the samples differ in value from the center value of their quantization level. Thus,

> *the quantization procedure rounds the actual values of sampling points to the assigned numbers and therefore introduces quantization errors.*

This concept is also visualized in Figure 4.1.11. The nature of quantization errors is clarified in Figure 4.1.11a. The figure demonstrates on a greater scale the difference between the assigned

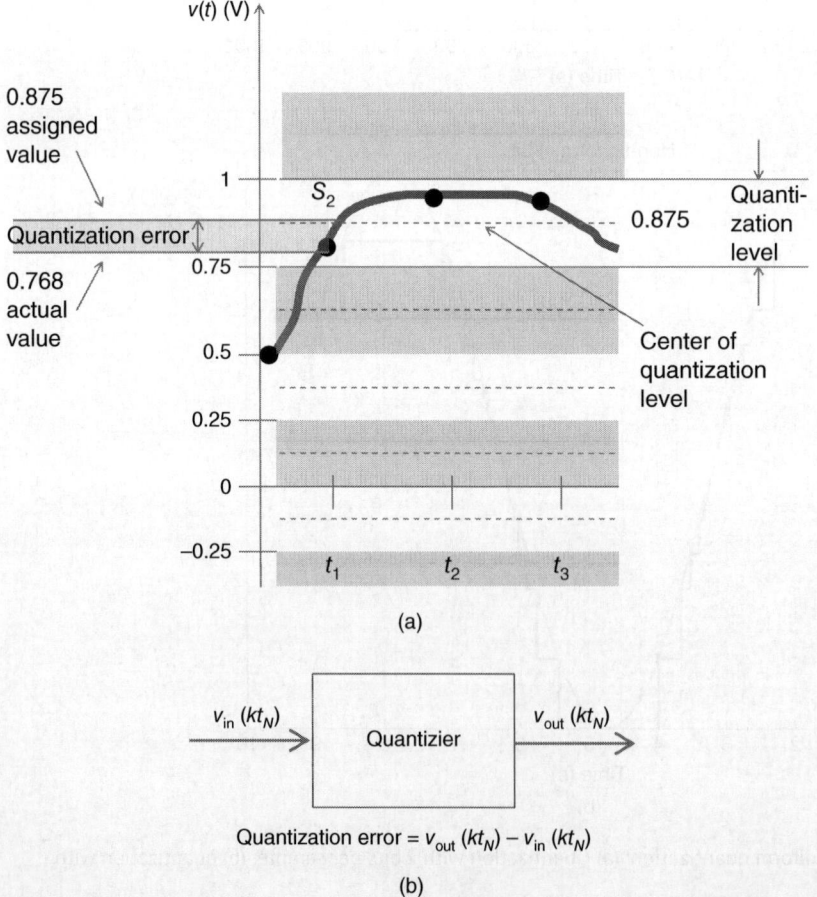

Figure 4.1.11 Quantization errors: (a) The difference between actual and assigned values of an individual sample; (b) general definition of a quantization error.

and actual values of the second sample, S_2. Here, the actual value of this sample is 0.768 V, but the assigned center value is 0.875 V. Hence, the quantization error is 0.875 V − 0.768 V = 0.107 V. In general, as Figure 4.1.11b shows, the quantization error, $\varepsilon_k(t_k)$, is the difference between the quantized output, $v_{out}(t_k)$, and the input, $v_{in}(t_k)$, signals:

$$\text{Quantization error,} \ \varepsilon_k(t_k) = v_{out}(t_k) - v_{in}(t_k) \tag{4.1.9}$$

where $k = 1,2,3, \ldots, N$ is a sample number, t_k is the sampling instant, $v_{out}(t_k)$ *is the assigned value*, and $v_{in}(t_k)$ *is the actual sampled value*. Obviously, $\varepsilon_k(t_k)$ is a random variable.

The use of the assigned, rounded sample values instead of the actual ones in the subsequent processes makes the ADC operation *irreversible* from this point. Until quantization, we can, in principle, exactly reconstruct the original analog signal provided that the Nyquist criterion is met. Therefore,

> *due to the principle of the quantization operation – the use of assigned sample values instead of the actual ones – we cannot precisely reconstruct the original signal from its ADC digital output.*

Quantization error appears at practically every sample; this stream of errors produces the signal called *quantization noise*. These errors are termed *noise* because they create an unwanted signal. If we mark all errors, similar to what is done in Figure 4.1.11a for an individual error, and sequentially connect them, we will form a graph of quantization errors vs. sampling time. Remember that quantization errors are random quantities and therefore quantization noise is a stochastic process whose description relies on the theory of probability, a topic to be taken up in Chapter 9. It is worth mentioning that there are a number of techniques that help to reduce ADC quantization noise; among them the oversampling is one of most popular (Lyons and Fugal 2014).

This ends the description of the quantization procedure.

4.1.5 Encoding

In this context, *an encoding means presenting bits for transmission as an electrical or optical signal*. This task is discussed in Section 3.1.2; you particularly need to return briefly to Figure 3.1.10 to refresh your memory on transmission codes. Retrieving this information will enable you to describe the final step in an ADC operation and will be a good exercise for you in performing an encoding operation.

Let us consider an example that summarizes the quantization procedure and shows the encoding technique.

Example 4.1.2 The quantization and encoding of a sampled analog signal

Problem

Consider the sampled signal shown in Figure 4.1.8a. Given $A = \pm 1$ V, $m = 4$ b/sample, $BW_{signal} \leq 4$ kHz, and the measured sample values (see Table 4.1.1), do the following:

1. Find the parameters listed below of the quantization process of this signal:
   a) The number of quantization levels.
   b) The assigned center decimal values for the given sampling points of the analog signal.
   c) The quantization errors for all samples.
2. Depict quantization noise.

Table 4.1.1 Actual and assigned sample values and quantization errors for Example 4.1.2.

Time instants	Actual sample values (given), S_k	Assigned sample values, S_k^A	Quantization errors $\varepsilon_k = S_k^A - S_k$, $k = 0, 1, 2, \ldots, 15$
t_0	$S_0 = 0.2741$	$S_0^A = 0.3125$	0.0384
t_1	$S_1 = 0.9268$	$S_1^A = 0.9375$	0.0107
t_2	$S_2 = 0.9984$	$S_2^A = 0.9375$	−0.0609
t_3	$S_3 = 0.9372$	$S_3^A = 0.9375$	0.0003
t_4	$S_4 = 0.7502$	$S_4^A = 0.8125$	0.0623
t_5	$S_5 = 0.5016$	$S_5^A = 0.5625$	0.0519
t_6	$S_6 = 0.3994$	$S_6^A = 0.4375$	0.0381
t_7	$S_7 = 0.3364$	$S_7^A = 0.3125$	−0.0239
t_8	$S_8 = -0.0617$	$S_8^A = -0.0625$	−0.0008
t_9	$S_9 = -0.6793$	$S_9^A = -0.6875$	0.0082
t_{10}	$S_{10} = -0.8791$	$S_{10}^A = -0.9375$	−0.0584
t_{11}	$S_{11} = -0.7496$	$S_{11}^A = -0.6875$	0.0621
t_{12}	$S_{12} = -0.9973$	$S_{12}^A = -0.9375$	0.0598
t_{13}	$S_{13} = 0.3668$	$S_{13}^A = 0.3125$	−0.0543
t_{14}	$S_{14} = 0.1893$	$S_{14}^A = 0.1875$	−0.0018
t_{15}	$S_{15} = 0.0614$	$S_{15}^A = 0.0625$	0.0011

3. Encrypt all quantized sampled values into the natural and two's complement binary code systems.
4. Sketch the sampled signal with all the obtained parameters and display the binary codes.
5. Draw the output digital signal encoded in the unipolar NRZ transmission code using the natural coding.

Solution
Though the assignment looks intimidatingly long, in fact, most of the actions take just a minute to perform:

1. The parameters of the quantization process:
   a) From (4.1.7), $N_{ql} = 2^m = 2^4 = 16$. The signal, with 16 quantization levels, is presented in Figure 4.1.12. We leave the numbering of the quantization levels as an exercise for you.
   b) From (4.1.8), $\Delta = \frac{2A \,(V)}{N_{ql}} = \frac{2\,V}{16} = 0.125\,V$. Therefore, starting from 0, the first assigned center positive decimal value, A_0, is equal to half of the Δ; that is, $A_0 = \frac{\Delta}{2} = \frac{0.125}{2} = 0.0625\,V$. The next value, A_1, is given by $A_1 = A_0 + \Delta = 0.0625 + 0.125 = 0.1875$. Thus, we compute $A_2 = A_1 + \Delta = 0.1875 + 0.125 = 0.3125$, and so on. Note that we can compute any assigned value by formula $A_n = A_0 + n \cdot \Delta$. For example, the last, greatest positive value, A_7, is computed as $A_7 = A_0 + 7\Delta = 0.0625 + 7 \cdot 0.125 = 0.9375$. To check the accuracy of this computation, observe from Figure 4.1.11a that, adding $\frac{\Delta}{2}$ to A_7, we must obtain the positive peak, $A = 1$. Indeed, $0.9375 + 0.0625 = 1$. Similarly, we determine the assigned negative decimal values. All these values – positive and negative – are shown in Figure 4.1.12.

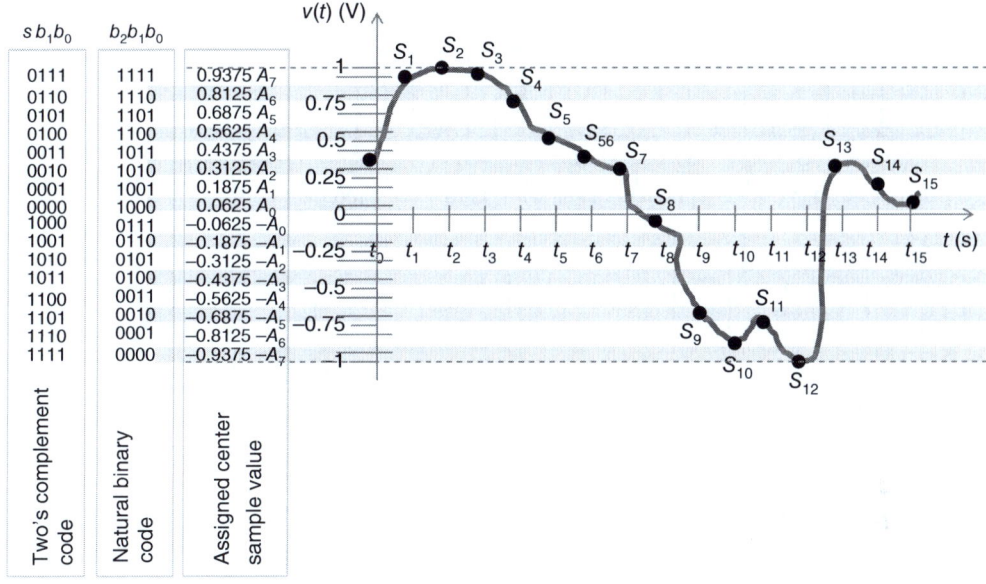

Figure 4.1.12 The signal's parameters and binary coding for Example 4.1.2.

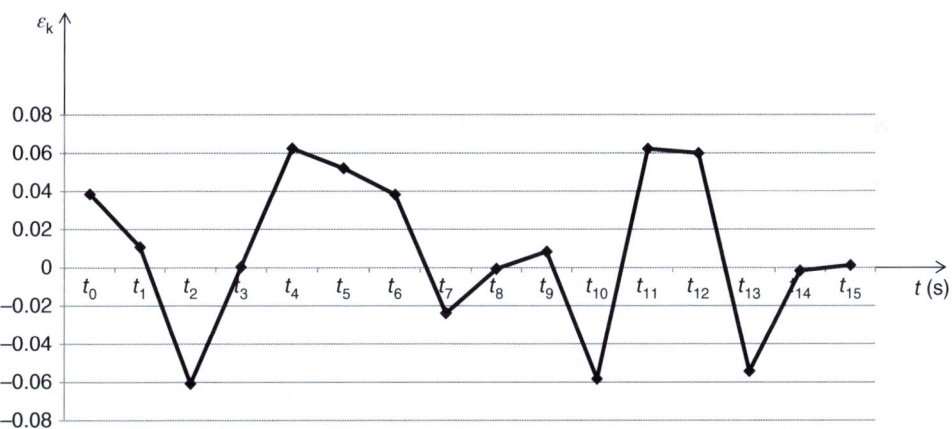

Figure 4.1.13 The error signal, $\varepsilon_k(t_k)$, is a snapshot of quantization noise.

c) The measured sample values (given) along with their assigned values are presented in Table 4.1.1, where the calculated quantization errors are also displayed. Here, sample value S_1 is taken at instant t_1, and this value holds until t_2, at which time sample S_2 is taken and holds until t_3, and so on. The quantization errors are calculated according to Equation (4.1.9): $\varepsilon_k(t_k) = v_{\text{out}}(t_k) - v_{\text{in}}(t_k)$. In Table 4.1.1, we designate the assigned sample value as S_k^A – that is, $v_{\text{out}}(t_k) = S_k^A$ – and the actual sample value as S_k, which gives $v_{\text{in}}(t_k) = S_k$.

2. The error signal, $\varepsilon_k(t_k)$, calculated in Table 4.1.1, which is the snapshot of quantization noise, is shown in Figure 4.1.13.
3. Encryption into binary codes is shown in two columns at the left-hand side in Figure 4.1.12.

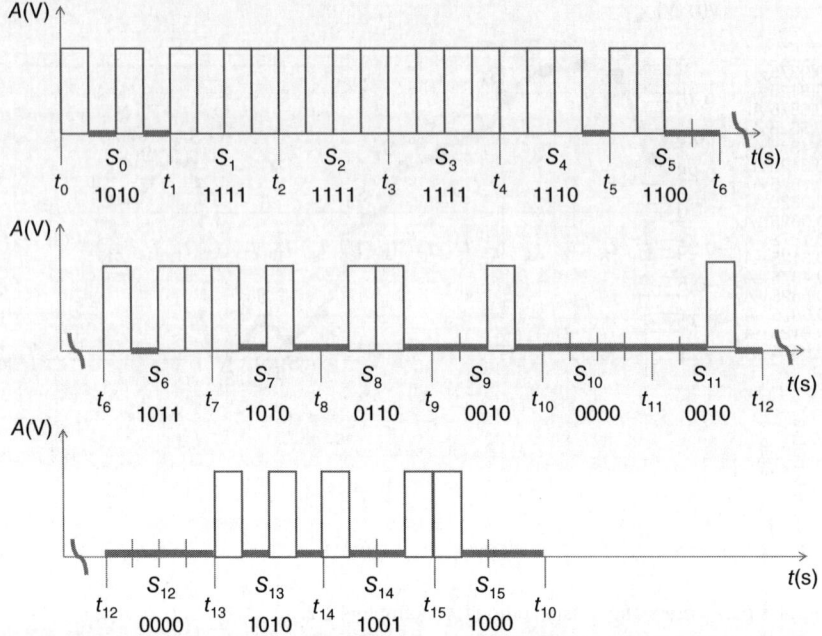

Figure 4.1.14 The digital signal encoded in an NRZ transmission code for Example 4.1.2.

A unipolar nonreturn-to-zero (NRZ) encoding assigns a pulse with amplitude A to bit 1 and a pulse with amplitude 0 to bit 0. Refer to Figure 3.1.10a. When depicting our signal, we need to bear in mind that every sample is coded in four bits. The codeword of every sample is sequentially drawn in Figure 4.1.14. To relate the NRZ signal to the ADC operation, a sample and its binary coding, the initial time instant, t_k, of each interval sample number, S_k, and the assigned codeword are shown. Remember that in ADC, the sample value holds within the time interval $t_{k+1} - t_k$ (refer to the S&H procedure) to obtain the assigned value, S_k^A, and then is encrypted into a codeword.

The problem is solved.

Discussion

- *The nature of an error signal*: If we sample the same signal 10 times, we will obtain similar but slightly different decimal sample values. The difference among the values of the same sample is an error, which, by its very nature, is a random entity. Changes in environmental conditions, variations of external electromagnetic fields and sun radiation, fluctuations in supply power, and self-heating of the electronic chips and components are among the factors that contribute to changes in the sampled values, therefore producing errors. Though we can control many of these factors to some extent and thus minimize their detrimental effects, we cannot completely eliminate them. Thus, the errors – random variables – are always present in any measurements. We need to bear this important point in mind every time we deal with measured results. Right now, we need to interpret the actual values given in Table 4.1.1 as a result of one set of measurements. This explanation also clarifies why Figure 4.1.13 refers to the graph showing error vs. time, $\varepsilon_k(t_k)$, as a *snapshot of quantization noise*.
- *Checking the accuracy of error calculations*: It follows from the above paragraph that due to the random nature of errors, we never know the exact values of the samples and, therefore, cannot check how accurately we have determined the errors. We can, however, assess the correctness of

our approach by verifying that the errors do not exceed the given margins. In the quantization process, a sampled value is in the center of a quantization level, and the actual value is within the quantization level; therefore, the error must not be more than half of a quantization level; that is,

$$\varepsilon_k \leq \frac{\Delta}{2} \qquad (4.1.10)$$

In Example 4.1.2, $\Delta = 0.125$ and therefore $\varepsilon_k \leq \frac{\Delta}{2} = 0.0625$. Indeed, Table 4.1.1 shows that the errors do not exceed this margin. Clearly, this test does not guarantee the precision of our calculations, but it at least indicates the absence of huge mistakes. As mentioned previously, error and noise are fully described only by the theory of probability; such a description is provided in Chapter 9.

- *Minimizing the quantization errors*: Quantization errors prevent us from the precise reconstruction of an input analog signal but we cannot eliminate them because of the inherent principle of an ADC operation. Can we at least minimize them? The answer is given by (4.1.10) and Figure 4.1.9: Minimize the quantization level, Δ, and therefore increase the resolution of an ADC quantization. This solution, however, comes with some unpleasant consequences. Indeed, (4.1.8), $\Delta = \frac{2A}{2^m}$, states that Δ can be minimized only by increasing the number of bits per sample, m (bit/sample), with the given dynamic range of a signal: A_{pp} (V). But increasing m will also increase the bit rate, BR_{ADC} (b/s), of the ADC signal to be delivered, as shown in (4.1.6). And it will, therefore, demand to increase the transmission channel's bandwidth, BW_T (Hz), as Shannon's theorem, BR (b/s) $\leq BW_T$ (Hz) $\log_2(1+SNR)$, requires. Thus, minimizing the quantization errors will eventually require an increase in the transmission channel's bandwidth. (This trade-off is an example of a typical dilemma confronting engineers on a daily basis.)
- *Saturation – another source of errors*: If the peak-to-peak value of an input signal exceeds the dynamic range, $2A$ (V), of an ADC unit, the device will truncate the input signal, generating another error in the quantization operation. Fortunately, we can control this issue by scaling the input signal properly.
- *Digitizing the voice signal*: In the early 1960s, when integrated circuits became available and electronics became applicable to digital signals, one of the first technical tasks was digitizing a voice signal. To appreciate the importance of this task, we need to recall that it was the era when the telephone was a major means of electronic communications and neither the Internet nor mobile communications were around. Thus, Bell Labs (the R&D arm of the old Bell System and one of the major contributors to advances in science and technology in the twentieth century, to be reminded) developed the standard for voice digitizing based on this simple reasoning: Scientists there assigned to each voice channel 4 kHz bandwidth. This value guaranteed the inclusion of practically the whole spectrum (300–3400 Hz) of the human voice. This bandwidth also provided some guard bands between the adjacent channels. Bell Labs also assigned 8 bits to each sample to ensure the high fidelity of voice transmission. (Compare Figures 4.1.8 and 4.1.11 to see the difference in ADC quality achieved with 3 and 8 bits.) Then, applying the Nyquist theorem, Bell researchers calculated the sampling frequency, f_S (Hz) = $2BW_{signal}$ (Hz) = 8 kHz, and the bit rate, BR_{voice} (b/s) = 8 kHz × 8 bit = 64 kb/s. Since that time, 64 kb/s and its multiples have been the main building blocks of the digital hierarchy used in the Public Switched Telephone Network (PSTN). Now you know the origin of the 64 kb/s bit rate initially introduced in the derivation of (4.1.6).

And so, our journey from an analog electrical signal to its presentation as a binary electrical signal is successfully completed. Figure 4.1.15 summarizes the steps in an ADC operation: It shows that an

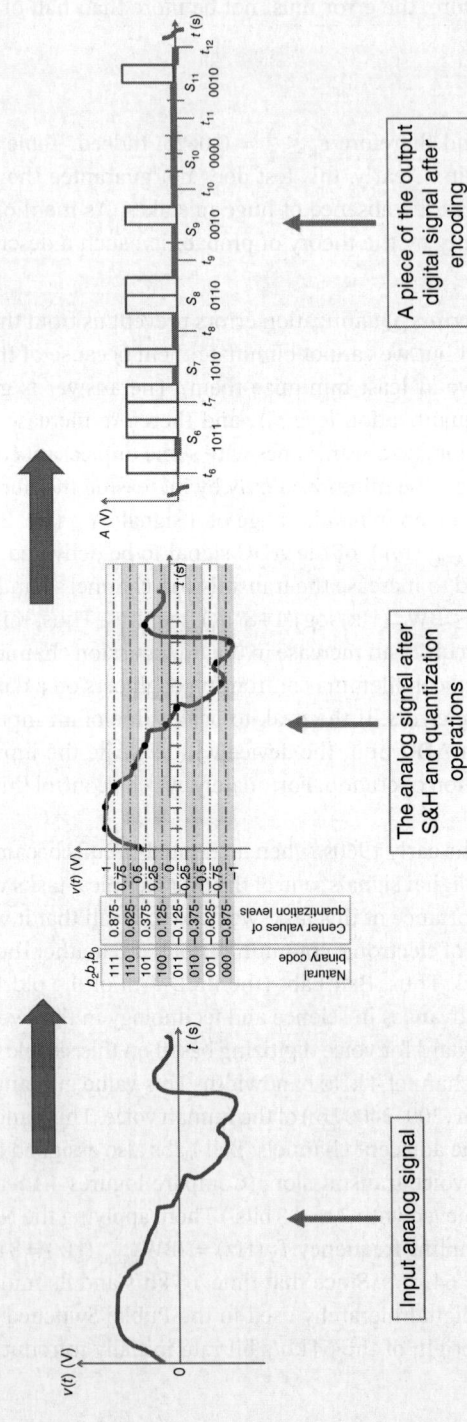

Figure 4.1.15 The ADC operation from the input analog signal to the output digital signal.

input analog signal is sampled, quantized, and coded; the result of these operations is the binary code of an analog signal. The binary code (the logical form of information) is encoded into the NRZ electrical signal (the physical carrier of information). This digital signal is transmitted to a destination point.

Obviously, all these operations are done automatically by electronic devices. An ADC operation can be performed either by a dedicated chip or a microcontroller capable of executing ADC along with many other operations. (A microcontroller is a system that integrates all essential computer components on a single chip.)

It is useful to compare Figures 4.1.3 and 4.1.15 to acquire an overall picture of an ADC operation: The conceptual view of the operation is presented in the former figure and the implementation steps are shown in the latter.

To conclude our discussion of ADC, it is noteworthy to stress again and again that ADC operations and circuits play extremely important role in modern communications.

(**Question**: *Can you find ADC application in the Internet of Things [IoT] system described in Section 1.1? Show you findings.*)

The ADC technology continues to rapidly improve; the examples of these advances can be found in the research and industry publications. (See, for instance, Buchwald 2016, Page 71 and Murmann 2016, Page 78.) The progress in ADC circuitry enables the communication industry to enhance the existing ADC applications and find the new areas for its employment.

Questions and Problems for Section 4.1

- Questions marked with an asterisk require a systematic approach to finding the solution.
- Many questions and problems, including those marked with an asterisk, imply that you, in addition to reading the textbook, will do your research to find the answers. Consider such questions as miniprojects.

The Need for ADC and DAC

1. Briefly and clearly explain what analog-to-digital conversion, ADC, and digital-to-analog conversion, DAC, do.

2. Modern electronics relies on digital technology. Why do we still need to use ADC?

3. Where do analog signals come from today?

4. Review all electronic and electrical devices in your home (laptop, cell phones, tablets, TV sets, the refrigerator, the controllers of heater, AC units, etc.) and identify where analog and where digital electrical signals are in use.

5. *Sketch a block diagram of a whole ADC + DAC system and discuss its principle of operation, showing the input and output signals and all the signals at the intermediate stages.

6. List and comment on all the applications of ADC and DAC systems of which you are aware.

7 *Give examples of ADC and DAC operations in communications, consumer electronics, computers, control systems, and instrumentation.

8 *Listening to your radio while driving a car, you certainly observe that the quality of the signal deteriorates when you go farther from the source of the radio signal. This is because modern radio broadcasting relies on analog transmission technology. Suggest a change in transmission technology to improve the quality of a radio signal. Justify your suggestion.

9 If you are asked to change an analog transmission technology into a digital one, sketch a block diagram of a digital transmission system.

Three Major Steps of ADC and Sample-and-Hold (S&H) Operation

10 Discuss the concept of analog-to-digital conversion, ADC.

11 *Consider an ADC operation: List and comment on three major steps in analog-to-digital conversion. Why, in your opinion, does ADC need these steps? Can you offer any alternative to performing an ADC operation?

12 *Consider sampling in ADC:
   a) What is the objective of this operation?
   b) What is the problem we need to solve to achieve this objective?
   c) Why do we need to sample and hold, but not simply to sample? Explain.
   d) For how long do we need to hold in an S&H operation? Why this interval? Explain.

13 Sketch a circuit performing an S&H operation and explain how it works.

14 Formulate the Nyquist (Shannon–Nyquist) sampling theorem in frequency terms and explain its meaning.

15 Discuss the presentations of the Nyquist theorem in time domain and in frequency domain. Show the relationship between these two presentations.

16 *Why is the Nyquist theorem sometimes called the Nyquist criterion? Explain.

17 If the maximum signal frequency is 48 MHz, what must be the sampling frequency? What will be the oversampling and undersampling frequencies?

18 What is aliasing in a sampling process? Explain this phenomenon. Give an example.

19 Sketch figures showing aliasing in both time domain and frequency domain. Explain the relationship between these two presentations.

20 *Consider Figure 4.1.P20, where oversampling and undersampling of an analog signal are shown in frequency domain and time domain:
   a) How many samples per period are taken in the oversampling and undersampling cases? How does this number relate to the presentation of sampling in frequency domain?

b) What is the minimum number of samples per period that must be taken to meet the Nyquist criterion?
c) How does the number of samples per period relate to the presentation of sampling in frequency domain?
d) Compare the frequency-domain presentations in Figure 4.1.P20 and explain how oversampling, undersampling, and aliasing cases are shown there.

21 Consider a signal whose bandwidth $BW_{signal} = 56\,kHz$. What are the minimum and maximum sampling frequencies? What are the minimum and maximum sampling time

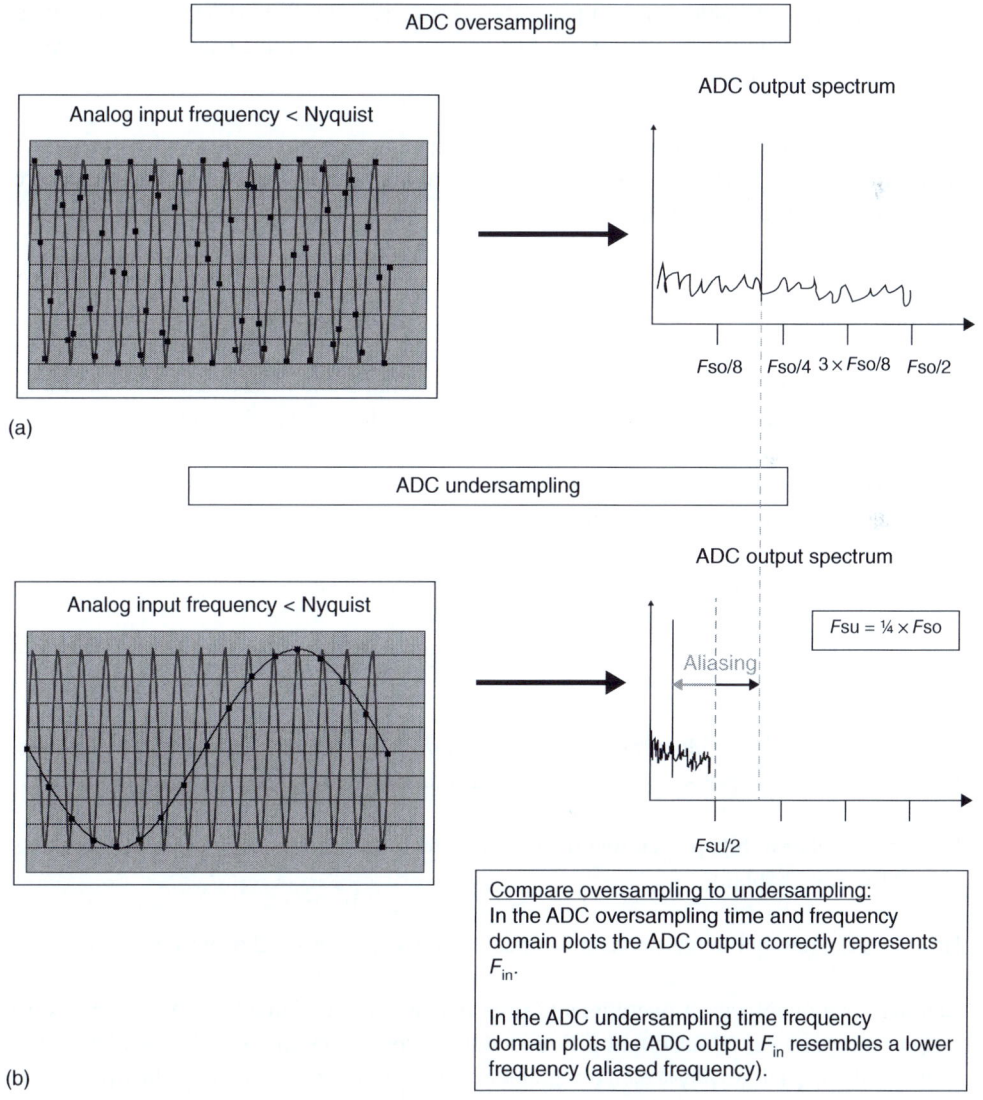

Figure 4.1.P20 Oversampling (a) and undersampling (b) in frequency domain and time domain. Legend: F_{in} (Hz), input analog frequency; F_s (Hz), sampling frequency; F_{so} (Hz) and F_{su} (Hz), oversampling and undersampling frequencies. Source: Reprinted with permission of Texas Instruments, Application Report SLAA510 – January 2011.

intervals? Explain. What happens if we use a sampling frequency lower than the minimum frequency?

22 Is there any difference between a signal's maximum frequency, f_{signal}^{max} (Hz), and the signal's bandwidth, BW_{signal} (Hz)? Explain. Give an example.

23 *Undersampling will result in aliasing and therefore looks like an unacceptable action. Nevertheless, modern communications technology manages to utilize undersampling to some benefit in an ADC operation. How is this possible? Explain and give an example.

24 *Build a two-column table that compares the advantages and disadvantages of employing oversampling and undersampling in an ADC operation. Add a third column to state your comments on each point of comparison.

25 *The Nyquist theorem requires the use of a sampling frequency greater than twice the signal's bandwidth, $2BW_{signal}$ (Hz). How do we know the value of the BW_{signal} (Hz)? What measure can we take to specify such values?

Quantization in ADC

26 Explain the principle of the quantization operation: What is the objective of this operation? What are the preceding and subsequent steps? What are the input and the output signals?

27 *Why do we need to restrict the number of digits of the input decimal values? To what number do we need to restrict a set of digits in each decimal input?

28 Do we need to have an equal number of digits in the input decimal values? Explain.

29 How does the number of bits per second, m (b/sample), relate to the number of digits in a decimal figure?

30 *The assigned number of bits per sample for a quantization process is given by $m = 8$ b/sample. How many digits of input-sampled decimal values can be used?

31 Does the number of bits per sample relate somehow to the computing-power requirements of an ADC unit? Explain.

32 How does the number of bits per sample affect the quality of an ADC operation?

33 Compare two quantization operations of the same sinusoidal signal with $m_1 = 4$ b/sample and $m_2 = 8$ b/sample: Which quantization produces a better presentation of the analog signal? Why? Explain. Qualitatively sketch two graphs showing quantization with m_1 and m_2 over one period of the sine.

34 The standard bit rate for voice transmission is equal to 64 kb/s. Explain how this number is composed.

35 What is the bit rate of voice transmission if 8 bits per sample are assigned? If 16 b/sample?

36 *Why is the bit rate of high-definition DVD (4.6 Gb/s) much greater than that of voice transmission (64 kb/s)? Explain.

37 *If $f_s = 192$ (sample/s) and $m = 24$ b/s, what is the bit rate of an ADC system? What bandwidth is required for transmission of this bit rate? Explain.

38 The bit rate of an ADC system for digitizing music is given as 840 kb/s. What is the minimum number of bits per sample, m (b/sample), provided that a human can hear a sound wave with a frequency only up to 20 kHz?

39 *We know that increasing either the sampling frequency, f_s (sample/s), or the number of bits per sample, m (b/sample), will result in improving ADC quality. However, (4.1.6) – BR_{ADC} (b/s) $= f_s$ (sample/s)·m (b/sample) – states that increasing f_s (sample/s) and/or m (b/sample) will produce the greater ADC bit rate, BR_{ADC} (b/s). Is the latter a good or bad result? Does BR_{ADC} (b/s) restrict an increase in f_s (sample/s) or m (b/sample)? Explain.

40 Is there any natural criterion for choosing the right number of bits per sample in a quantization operation? Do you have any recommendations for this selection?

41 *If you are assigned to design an ADC system, from what step would you start this design? Explain.

42 Why is it necessary to introduce quantization levels? How can the number of these levels be chosen?

43 If the number of bits per second is $m = 8$ b/sample, how many quantization levels can be obtained? Explain your calculations.

44 What is the relationship between the quantization levels and the binary codewords? Explain. Give an example.

45 How many systems of binary codes do you know? Explain their organizational principles.

46 *Consider quantization levels:
   a) How does the number of bits per sample determine the number of digits of a decimal sample value?
   b) Why do we need to assign only one decimal number to each quantization level? By what rule, if any, should this number be assigned? Explain.
   c) How do the quantization levels relate the decimal sample values to the chosen binary codewords?

47 Can we directly convert decimal sample values into binary numbers to complete an ADC process? Explain.

48 What approach in an ADC system is used to relate the binary numbers to assigned decimal numbers? Explain.

49 *How does an ADC system relate the chosen quantization levels with their binary codewords to a sampled analog signal? Explain.

50 *Explain the quantization process step by step by referring to Figure 4.1.P50 (a reproduction of 4.1.12).

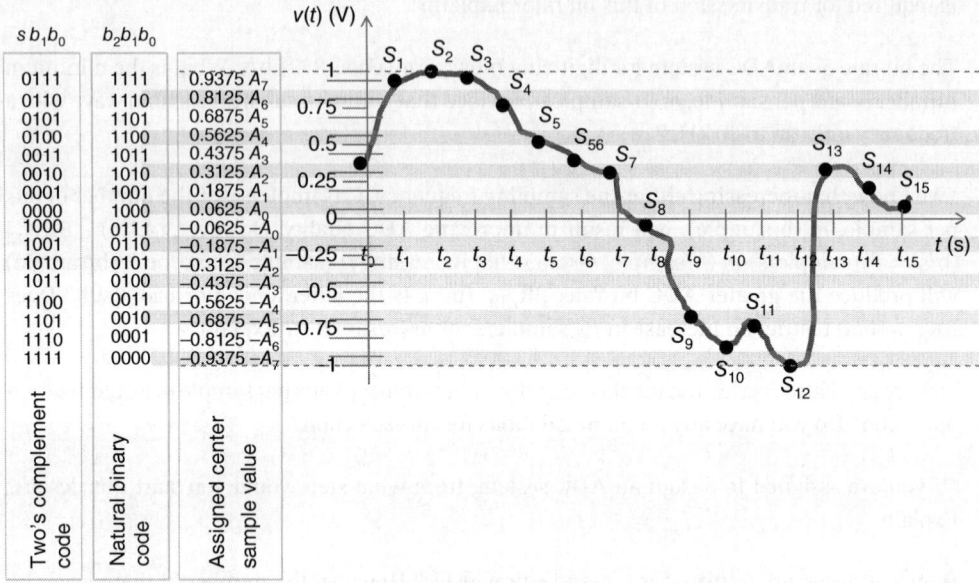

Figure 4.1.P50 (4.1.12R) The signal's parameters and binary coding for Example 4.1.2.

51 How does the step size of a quantization level relate to the number of the quantization levels? Explain.

52 If the step size is given as $\Delta = 0.125$ V, how many bits per sample, m (b/sample), must be used for the dynamic range of 2 V?

53 If the step size is 0.125 V and the center decimal number of a quantization level is 0.9375, what are the maximum and minimum values of this quantization level?

54 *Does the step size affect the accuracy of the quantization process? Explain.

55 *Figure 4.1.P54 (which is a part of Figure 4.1.7) shows that the quality of an ADC operation depends on the number of bits. Explain why.

56 *Figure 4.1.9 shows a staircase input–output characteristic of an ADC quantization operation. Why does this graph present a transfer function of an ADC unit? Explain.

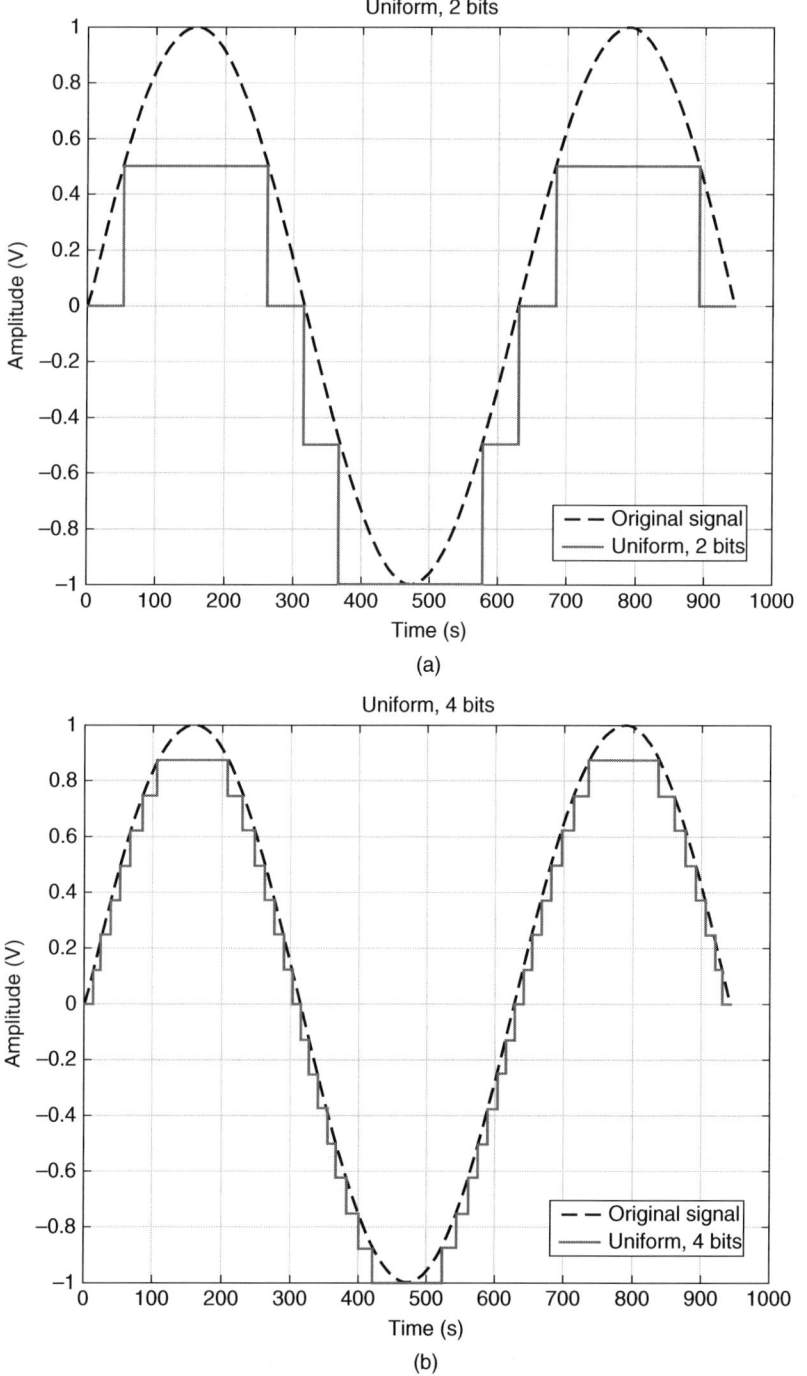

Figure 4.1.P54 The quality of an ADC operation depends on the number of bits per second.

57 Sketch a staircase input–output ADC characteristic (similar to that shown in Figure 4.1.9) with four bits per sample. Show all the quantization step sizes and binary codewords. Compare your figure with Figure 4.1.9 and comment on the quantization resolution provided by both ADC systems.

58 *How does the staircase graph of an ADC transfer function help to explain the origin of quantization errors? Give examples.

59 There are uniform and nonuniform quantization methods. Compare them and explain the advantages and disadvantages of each. Sketch the figures to visualize your explanations.

60 Define quantization error. Can it be eliminated? Explain.

61 Can we precisely predict a quantization error before the actual measurements are made? Explain.

62 *Assume that a step size is 0.125 V and the center decimal number of a quantization level is 0.9375. The measurements collect the following three sample values: 0.9486, 0.8964 and 0.9213 V.
   a) Do these values fall into the given quantization level?
   b) If the answer is yes, what are quantization errors of each of these sample values?

63 If the step size is 0.25 V, what could be the maximum quantization error? Compute and explain.

64 *We know that quantization errors cannot be completely eliminated.
   a) Can we, at least, minimize them?
   b) If the answer is yes, how?
   c) Is there any disadvantage from minimizing quantization errors?

65 Name and explain other errors in an ADC operation besides the quantization errors.

66 The result of an ADC operation is the presentation of an analog signal as a set of binary codewords. Can we precisely reconstruct the original analog signal from these codewords? Explain.

67 *When considering noise, we can always indicate the source of the noise (lightning, external electromagnetic waves, cosmic radiation, etc.). What is the source of quantization noise?

Encoding

68 The binary codewords obtained after a quantization operation must be transmitted as electrical or optical signals. Why is the conversion of binary codewords into electrical or optical pulses called encoding? Explain.

69 Dictionaries define the verbs "code" and "encode" as synonyms because they mean the same: *put in code*. What do we "put in code" in the *encode* operation? Explain.

70 *A single voice is digitized at 64 kb/s. The higher digital hierarchy line, called T1, has a bit rate of 1544 Mb/s. Given that this bit rate includes 8 service bits, the actual data rate is 1536 Mb/s. How many voice channels does T1 line carry? Explain your calculations.

71 Consider Figure 4.1.15: Explain in a step-by-step sequence how an input analog signal is eventually converted into a digital output signal.

4.1.A Decimal and Binary Numbering Systems

Since an ADC operation deals with decimal and binary numbers, it is worth reminding you about the basics of the decimal and binary number systems. After reviewing this material, you will well understand the difference between *coding decimal numbers into binary* (which is performing in ADC operation) and *converting decimal numbers into binary*.

4.1.A.1 Decimal Numbering System

Let us start with *decimal whole numbers (integers)*. We know, of course that the decimal number system uses 10 *numerals* – 0, 1, 2, 3, 4, 5, 6, 7, 8, and 9. This is a *positional-value* system because a *position of a numeral determines its value*. For example, in number 123, numeral 3 represents ones, numeral 2 represents tens, and numeral 1 represents hundreds. This concept is shown in Figure 4.1.A.1a, where the positions change in ascending order from right to left. Each position represents the power of ten; thus, the smallest position is $10^0 = 1$ and the *n*-th position is 10^n. The change from one position to the next occurs through multiplication by 10. The numeric value of a numeral depends on its position; thus, numeral 2 placed in the leftmost position in Figure 4.1.A.1a will have a value of 2 million; if this numeral were, on the other hand, in the rightmost position, its value would be just 2. (To better appreciate the importance of a numeral's position, imagine you are considering your money; thus, numeral 2 in the leftmost position in Figure 4.1.A.1a gives you $2 million, but in the rightmost, it would be just $2.) The total value of a number is the sum of all numerals weighted by their position; in the example shown in Figure 4.1.7 we have

$$2 \times 10^6 + 5 \times 10^5 + 1 \times 10^4 + 8 \times 10^3 + 3 \times 10^2 + 9 \times 10^1 + 4 \times 10^0$$
$$= 2 \times 1\,000\,000 + 5 \times 100\,000 + 1 \times 10\,000 + 8 \times 1000 + 3 \times 100 + 9 \times 10 + 4 \times 1$$
$$= 2\,158\,394$$

Since the system uses ten numerals and the positions (orders of magnitude) are determined by multiplication by 10, it is called a *base-10 system*.

Decimal fractions, which are separated from the whole part by a *decimal point*, are arranged in a similar way but with the negative power of 10. The first position following the decimal point has a weight of 10^{-1}, the weight of the second position is 10^{-2}, etc. Hence, the numeric value of the first digit after the decimal point is obtained by multiplying the first numeral by 10^{-1}, the second value is the result of multiplying the proper numeral by 10^{-2}, and so on. Figure 4.1.A.1b demonstrates this principle through the following example:

$$8 \times 10^3 + 3 \times 10^2 + 9 \times 10^1 + 4 \times 10^0 + 6 \times 10^{-1} + 2 \times 10^{-2} + 0 \times 10^{-3} + 7 \times 10^{-4}$$
$$= 8 \times 1000 + 3 \times 100 + 9 \times 10 + 4 \times 1.6 \times 0.1 + 2 \times 0.01 + 0 \times 0.001 + 7 \times 0.0001$$
$$= 8394.6207$$

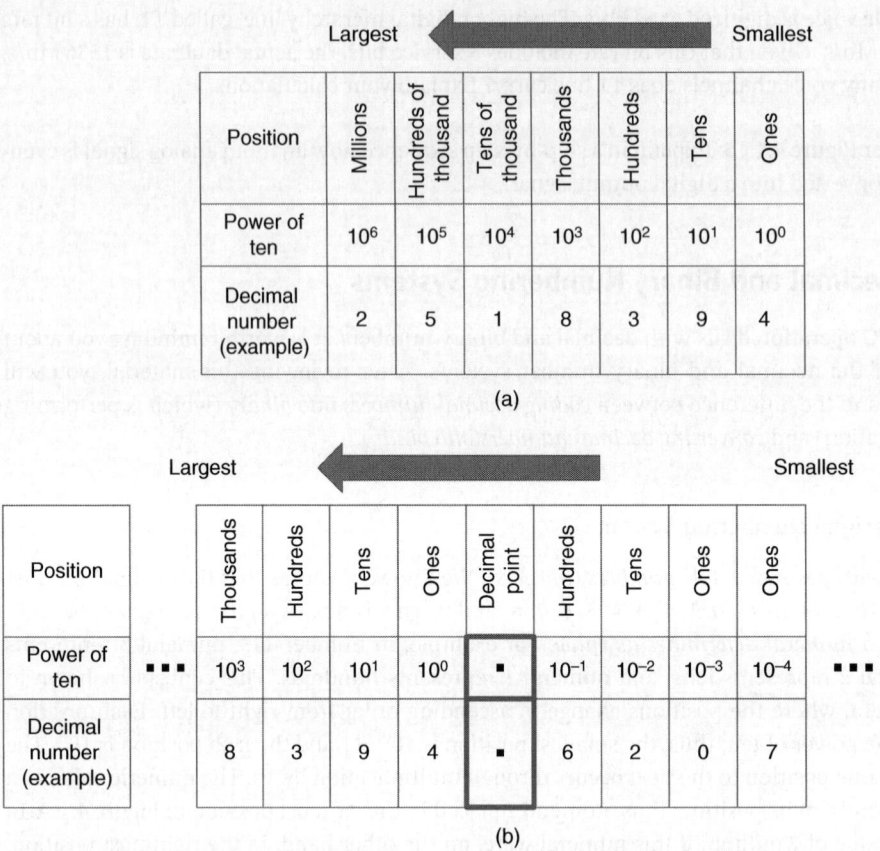

Figure 4.1.A.1 Decimal numbering system: (a) Whole numbers (integers); (b) fractional numbers.

This brief review not only refreshes your memory on decimal numbering but also ensures that you understand it clearly to follow our discussion in this section.

(Note that several terms are used to denote a *numbering* system; they are *number, numeral,* and *numeric.*)

4.1.A.2 Binary Numbering System

To figure out how to represent a decimal number in a binary numbering system, where only two numerals – 0 and 1 – are available, we need to understand the system clearly.

The concept of the binary numeric system was devised by the German scientist Gottfried Leibnitz in 1679. He did it based on the ancient Chinese approach to binary counting. Of course, the binary system that we have today is the result of further inventions and developments by many scientists and engineers in the nineteenth and twentieth centuries.

The modern binary number system is a positional system where, similar to decimal numbering, the position of each *bit* determines its numeric decimal value. (You will recall that **bit** is the term derived by combining **bi**-nary and digi-**t**.) Examine Figure 4.1.A.2a, where this concept is outlined for a whole binary number. We know that the binary system has only two numerals – 0 and 1 – and the decimal value of each position is determined by the power of two. Thus, the positions in the

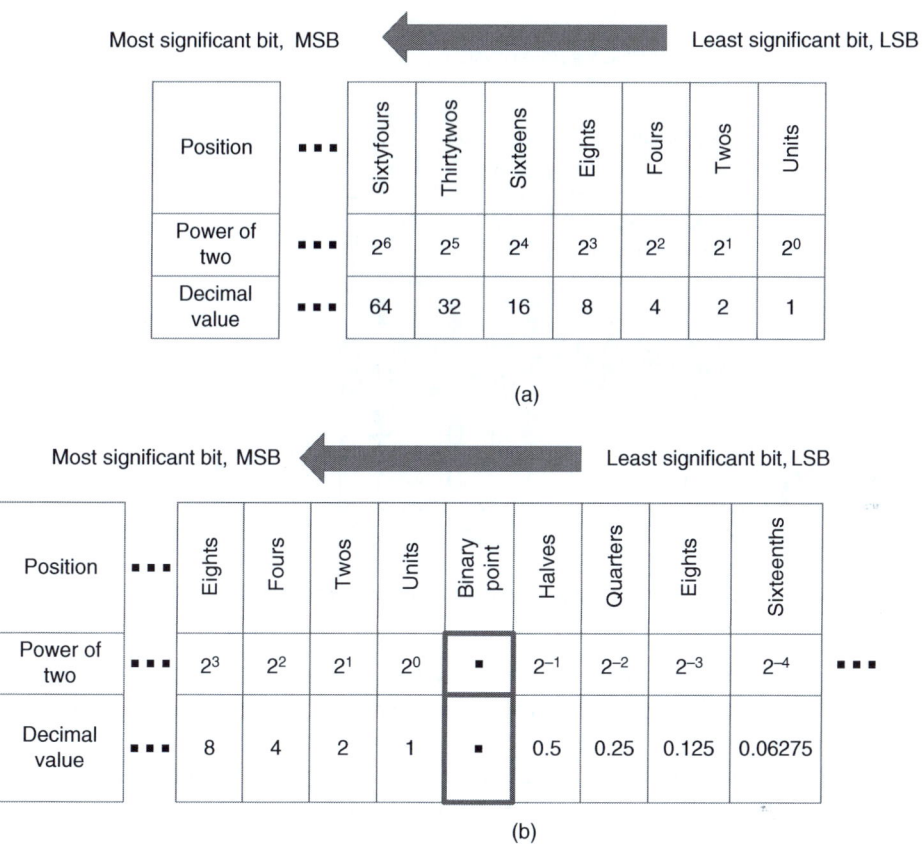

Figure 4.1.A.2 Binary numbering system: (a) The whole binary number; (b) the fractional binary number.

ascending order are given by 2^0, 2^1, 2^2, etc. This means the change from a smaller position to the next (greater) one is done by multiplying the previous one by 2. These considerations explain why the *binary system is called base-2*.

Positions are placed in ascending order from right to left:

> The bit placed in the rightmost position has the smallest decimal value and is called the least significant bit (*LSB*); the leftmost bit has the largest decimal value and is called the MSB.

Binary fractions, like the decimal, are given by their positions after the *binary point*. These positions are determined by the negative power of two, as shown in Figure 4.1.7b. With this brief introduction to the decimal and binary numbering systems, we are ready to confront the objective of this appendix – conversion of a decimal number into its binary counterpart.

4.1.A.3 Conversion from the Decimal Number System to the Binary

Let us remind ourselves that the first step in ADC is the sampling of an analog signal – the source of the original information. The result of sampling is a set of decimal numbers representing the continuous analog signal. Now, we need to convert these decimal numbers into the binary.

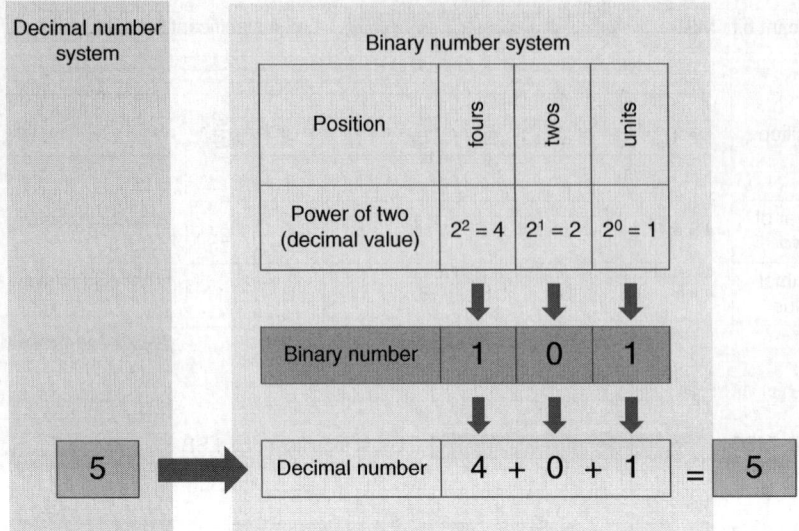

Figure 4.1.A.3 Example of converting a decimal number into its binary equivalent.

The problem we face is shown in Figure 4.1.A.3. Consider, for example, the need to convert number 5 into a set of 1s and 0s. Given our knowledge of the decimal and binary number systems, how can we solve this problem?

One approach follows from the structure of number 5: It is the product of the numeral 5 and its decimal position, $10^0 = 1$; that is, $5 = 5 \times 10^0$. In the binary system, a decimal value is given by the binary position; specifically, decimal 1 is equal to 2^0, decimal 2 is equal to 2^1, decimal 4 is equal to 2^2, etc. Hence, number 5 can be naturally obtained as $5 = 1 + 4 = 2^0 + 2^2$. But how can we arrange to skip position 2^1, having at hand only two numerals, 1 and 0? The answer is that, similar to the structure of a decimal number system, we need to assign a numeral to each binary position and this numeral must be either 1 or 0. If we agree that *assigning numeral 1 to a position means that its value will be taken into account and assigning numeral 0 to a position will nullify the digit value in this position*, we solve the problem. In our simple example, we need to do the following:

$$5 = 2^2 \cdot \mathbf{1} + 2^2 \cdot \mathbf{0} + 2^2 \cdot \mathbf{1} = 4 + 0 + 1$$

The boldfaced 1s and 0s are the binary numbers that represent a counting operation in the binary system. Thus, we convert decimal number **5** into binary number **101** because the *rightmost 1* allows decimal $2^0 = 1$ to pass, the next left **0** blocks decimal $2^1 = 2$ from passing, and the *leftmost 1* passes decimal $2^2 = 4$. This operation sets the equivalency between decimal number $5 = 4 + 0 + 1$ and binary number **1 0 1**. Figure 4.1.A.3 demonstrates this operation.

Based on the principles underlined in Figures 4.1.A.2 and 4.1.A.3, several techniques for conversions between decimal and binary systems have been developed. We will consider a simple technique called *successive division*; an example of its application is shown in Figure 4.1.A.4a. We start by dividing a given decimal number by two; the remainder of the division is a LSB. Next, we divide the first quotient by two and the remainder is the next bit in the binary number being sought. The last remainder of this division is the most significant number, MSB, of the binary number. Apply this explanation to the example given in Figure 4.1.A.4a. This example clearly demonstrates that

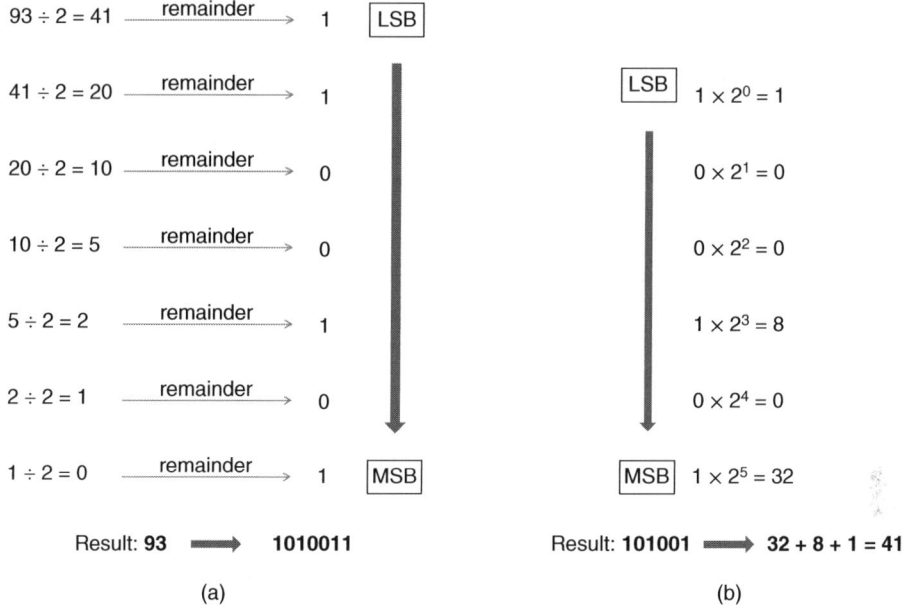

Figure 4.1.A.4 Examples of conversion: (a) Decimal to binary by the successive-division technique; (b) binary to decimal by the inverse technique.

this operation is inversed with respect to decimal-to-binary one. The conversion of fractions is done using the same technique.

This brief review should help you to well understand the material in this section.

4.2 Digital-to-Analog Conversion, DAC, Pulse-Amplitude Modulation, PAM, and Pulse-Code Modulation, PCM

Objectives and Outcomes of Section 4.2

Objectives

- To study the DAC process.
- To learn PAM and PCM.

Outcomes

- Know that *DAC*, is the reverse operation of ADC.
- Comprehend that DAC includes presenting a received digital signal to a binary decoder, which extracts the binary codewords from the input signal, converts them into decimal numbers, and sends them to an inverse S&H unit for building a S&H version of the received signal. Get to know that DAC operation is finalized by smoothing the S&H waveform with a lowpass filter, making it the restored version of the original analog signal. Realized that the reconstructed signal will inevitably differ from the original analog signal due to quantization errors.
- Become familiar with PAM which is another digitizing technique that produces pulses (digital signals) to represent an input analog signal. Study that amplitudes of PAM's pulses approximate the actual continuous waveform of the input analog signal after an S&H operation has

performed. Know that despite its simplicity, PAM not in the mainstream of modern communications because it carries information by virtue of amplitude and amplitude is the first parameter of a signal that corrupts under noise.
- Learn that PCM converts an analog signal into pulses that carry the signal's information by virtue of binary codes, thus producing a truly digital signal whose advantages were discussed in Sections 3.1 and 3.2. Grasp the idea that a PCM system is a technologically implemented version of ADC and DAC operations.
- Be aware that PCM suffers from several drawbacks such as PCM requires a big transmission bandwidth, a PCM signal is produced with errors, the main one of which is quantization noise, and PCM, like any digital transmission, suffers from intersymbol interference (ISI).
- Realize that despite its drawbacks PCM is an advantageous digital transmission system that finds today many applications in consumer electronics.

4.2.1 Digital-to-Analog Conversion, DAC

DAC is, obviously, the reverse of an analog-to-digital operation. Hence, we can easily envision a DAC operation, as shown in Figure 4.2.1.

A received digital signal is presented to a binary decoder, which gathers from it the binary codewords and converts them into decimal numbers. These numbers are the assigned sample values centered at each quantization level. Based on these sample values, an inverse S&H unit builds a S&H version of the received signal, thereby recovering the sampled version of the waveform of an analog signal. A lowpass filter smooths the waveform, making it the continuous version of the original analog signal. The result: an output version of the analog signal that has been sent. Do not forget that the reconstructed signal inevitably differs from the sent signal because the assigned sampled values do not coincide with the actual sampled values. Figure 4.2.1. qualitatively compares the two waveforms, the solid line representing the reconstructed signal and the dotted line presenting the original input signal.

(**Question:** *Modern high-performance DACs offer high speed and high resolution within a fully integrated package. Such DACs are used in research aimed at developing quantum communication systems, the systems promising to completely change our means of communication. One of these DACs has 16-bit resolution and a sample rate of 10 Gsamples/s. How do these numbers support the above statement about the high speed and high resolution of modern DACs? Prove your answer by referring to Figure 4.2.1. Indicate where these numbers must be placed in this figure.*)

Combining Figures 4.1.15 and 4.2.1 and referring to Figure 4.1.1, we obtain a graphical view of the whole process of transmitting an analog signal through a digital channel. This view summarizes our discussion of ADC and DAC operations.

(**Exercise**: Build this figure.)

4.2.2 Pulse Amplitude Modulation, PAM

PAM stands for *pulse amplitude modulation* and **PCM** stands for *pulse code modulation*.

You can expect to see a lot of technological details and a variety of devices associated with both technologies in textbooks and industrial technical literature. We will, however, restrict our discussion to their operating principles.

PAM, is a technique that produces pulses (digital signals) from the input analog signal. We can obtain a PAM signal by performing a S&H operation, as Figure 4.2.2 demonstrates. Examining this figure, we observe that PAM signal is a digital signal in which the amplitude of each pulse represents

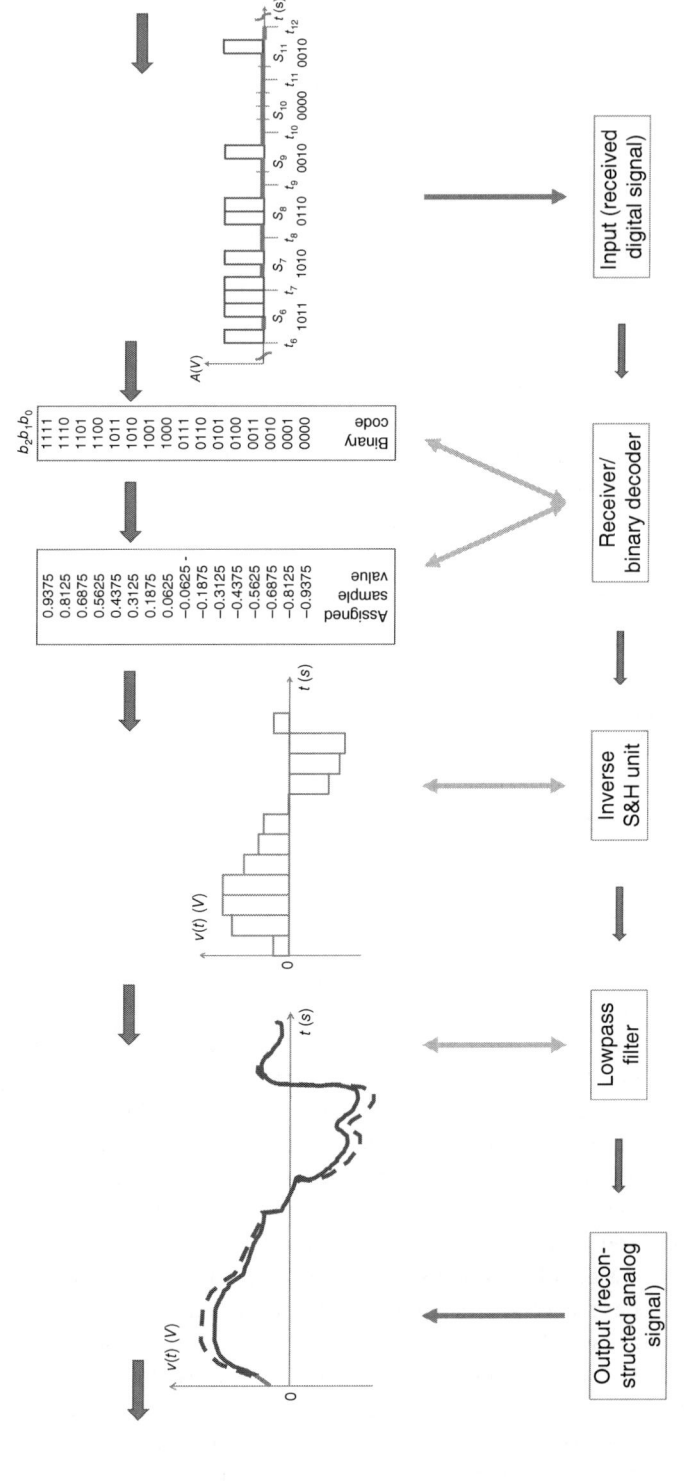

Figure 4.2.1 Digital-to-analog conversion.

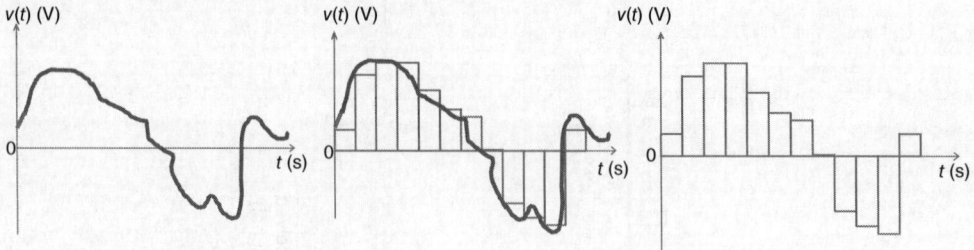

Figure 4.2.2 Pulse amplitude modulation, PAM: an input analog signal (left), the sampled analog signal (middle), and the PAM signal (right).

the amplitude of an input analog signal at the proper hold instant. This is why this transmission technique is called *PAM*. Despite its simplicity, PAM is used very seldom because it carries information by virtue of amplitude. *Amplitude is the first parameter of a signal that corrupts under noise.* This is why two other pulse analog modulation techniques – pulse width modulation (PWM) and pulse position modulation (PPM) – have been developed. In PWM, the signal's amplitude is represented by the pulse width and in PPM, as you might guess, the signal's amplitude is described by the pulse position. Today, PAM, PWM, and PPM are no longer in the mainstream of telecommunications.

4.2.3 Pulse Code Modulation, PCM

4.2.3.1 PCM: Principle of Operation

As the name suggests, this is the modulation of an analog signal into pulses that carry the signal's information by virtue of codes. What information do we want to obtain from an analog signal? For one thing, we want to know its amplitudes at given instants. This is exactly what PCM does: It delivers a code that represents the signal's amplitudes at given instants. Now, you might ask, is not it the same process that is done in an ADC operation? Yes, indeed. Review Figure 4.1.15: An analog signal – after sampling, quantizing and coding – is presented as a set of binary words that carry information about its amplitude at every instant. Hence, the coding part of a PCM operation is the same as in an ADC procedure and can be seen in Figure 4.1.15. Clearly, the decoding part of the PCM system is simply a DAC unit. There are a couple of subtle variations between PCM as a whole and ADC and DAC; these variations are shown in the block diagram of a PCM transmission system presented in Figure 4.2.3.

An input analog signal is converted into quantized form following the S&H and quantization procedures, as shown in Figure 4.2.3. (Understand that a quantized signal is not a PAM signal despite their similarity in appearance. They differ in amplitude values: The PAM signal's amplitudes are actual values measured at sampling, but the quantized signal's amplitudes are the assigned sampled values, as discussed in the "Quantization in ADC" subsection.) Binary codewords are then read and transmitted in parallel to a parallel-to-serial converter, which produces a serial version of the signal. The encoder translates the stream of binary codewords into electrical pulses, thereby generating a PCM digital signal for transmission. At the receiver end, operations are performed in reverse order: A decoder recovers the stream of binary words and transmits them in series to a serial-to-parallel converter, which produces a parallel set of binary codewords. These codewords are converted into the decimal numbers from which the S&H version of a signal is deciphered. Then, a lowpass filter constructs the output analog signal that should be, but never is, the same as the input. Note that S&H signals are shown here for illustrative purpose only; in fact, the system operates only with numbers except at the input and output points.

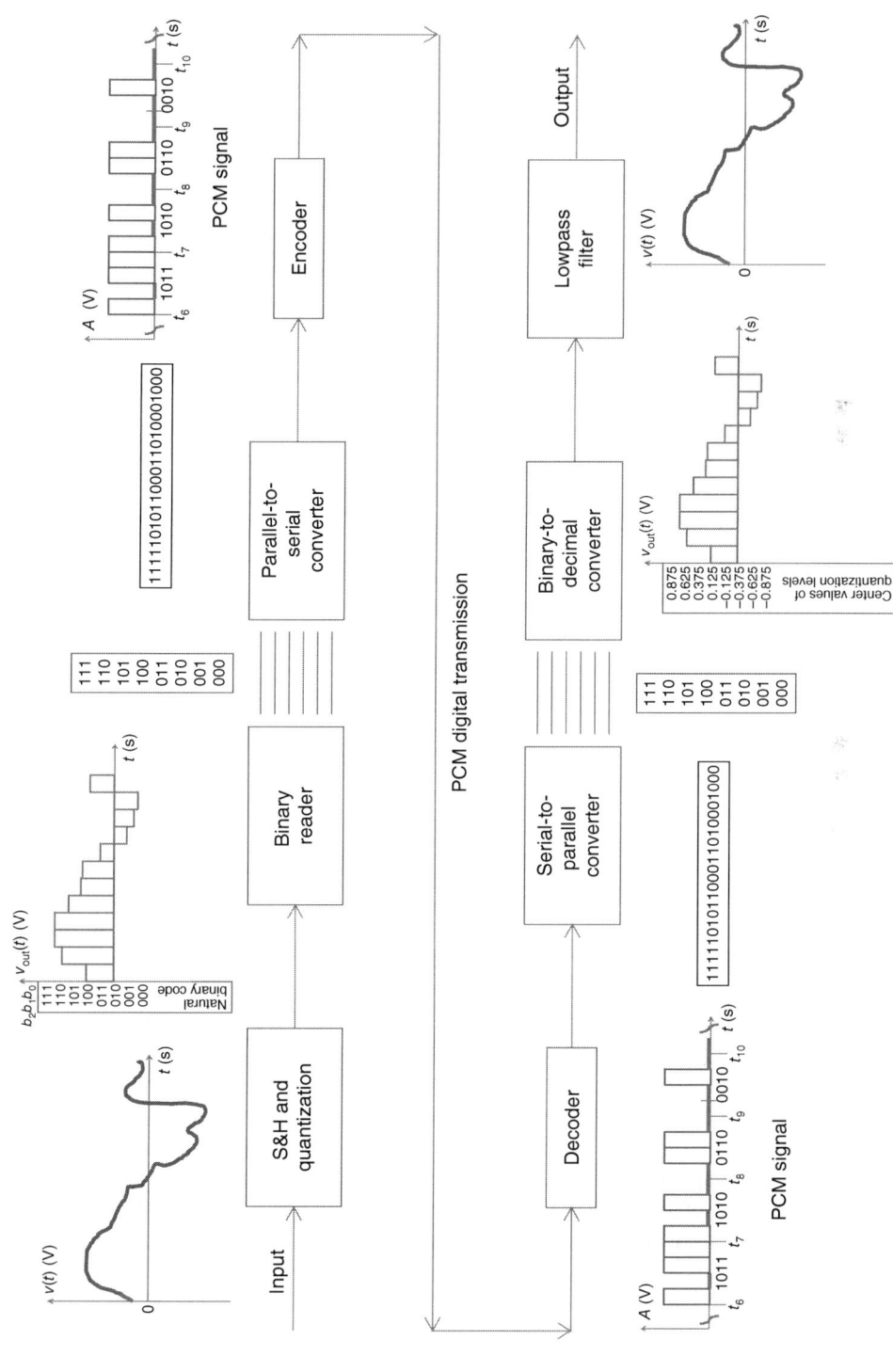

Figure 4.2.3 Block diagram of a PCM transmission system.

4.2.3.2 PCM: Advantages and Drawbacks

PCM's advantages are the same as those of any digital transmission, the main one being, of course, its immunity to corruption under noise and other detrimental factors, as considered in Sections 3.1 and 3.2. In addition, PCM signals are easy to *multiplex* (that is, to combine for transmission over a single channel) and to process because their binary form is natural for a computer.

PCM is not, however, free from drawbacks, one of them being that it requires significant transmission bandwidth, which can be calculated as

$$\text{BW}_{\text{PCM}} (\text{Hz}) = 2m\text{BW}_{\text{signal}} (\text{Hz}) \tag{4.2.1}$$

where m is the number of bits per sample, which is a dimensionless quantity. We understand from our ADC discussion that increasing m will result in improving the quality of a PCM signal. Now we can see that this improvement requires an increase in the transmission bandwidth, which in turn ends up increasing the cost of PCM transmission.

Another PCM liability is its propensity for *errors*. Obviously, there are *quantization errors* that prevent PCM from securing the precise recovery of the input analog signal.

To assess the possible detrimental role of quantization noise, let us derive the $(\text{SNR})_{\text{QN}}$ formula. You will recall that SNR is the ratio of signal power to noise power and we want it to be as big as possible. The general formula for SNR when measured parameters are in volts is given by

$$\text{SNR (dB)} = 20 \log \left(\frac{V_S}{V_N} \right) \tag{4.2.2}$$

Since a signal's dynamic range is its maximum voltage, we find that $V_S (V) = A_{\text{pp}} (V)$. The maximum noise value is given in (4.1.10) as $\varepsilon_k \leq \frac{\Delta}{2}$ (V), which means that the dynamic noise range cannot exceed a step size Δ. Consequently, the maximum noise voltage is given by $V_N (V) = \Delta$ (V). Recalling from (4.1.8) that $\Delta(V) = \frac{A_{\text{pp}}(V)}{2^m}$, we can express the noise voltage as

$$V_N (V) = \frac{A_{\text{pp}}(V)}{2^m} \tag{4.2.3}$$

Therefore,

$$(\text{SNR})_{\text{QN}} = 20 \log \left(\frac{V_S}{V_N} \right) = 20 \log(2^m) = 20 \cdot m \log(2) \approx 6.02 \, m \, \text{dB} \tag{4.2.4}$$

For example, for $m = 3$ b/sample, the $(\text{SNR})_{\text{QN}} = 18.06$ dB, and for $m = 8$, the $(\text{SNR})_{\text{QN}} = 48.16$ dB. Equation (4.2.4) shows that the *only way to increase the* $(\text{SNR})_{\text{QN}}$ *is to increase the number of bits per symbol, m (b/sample)*. This conclusion – an increase in m results in improving PCM system – is fully supported by our previous considerations of the ADC and PCM systems. Do not forget, however, that increasing m causes broadening of the required transmission bandwidth.

Another error is caused by *ISI* and it is inherent in any type of digital communications. This error stems from the fact that a pulse (symbol) is spreading in time as it travels along a transmission link. Thus, at the end of the transmission, the pulse's bit time, T_b, becomes greater than it was at the starting point. Due to this spread, pulses cross the time boundaries of adjacent pulses and exchange their energies. The energy of one pulse acts as noise for an adjacent pulse, degrading the quality of the transmission, which is why this phenomenon is called ISI. In an extreme case, pulses overlap so much that it becomes impossible to distinguish among them. This phenomenon is shown in Figure 4.2.4.

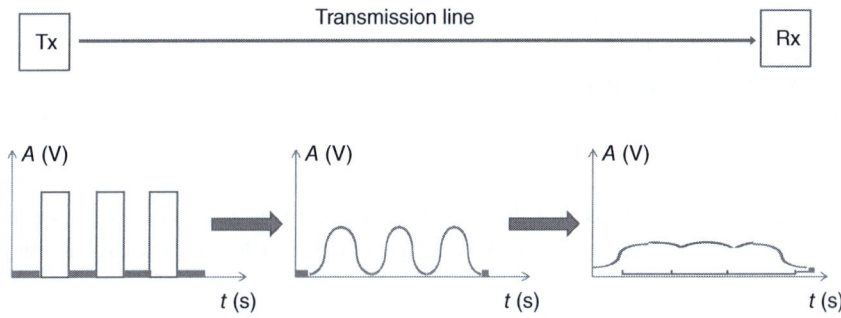

Figure 4.2.4 Intersymbol interference: Pulses are narrow and well separated at the transmitter point (left), but spread as they travel down a transmission line (middle) and completely overlap at the receiver end (right).

4.2.3.3 PCM Applications

PCM was invented by Alex Reevs in 1937 who worked at the Bell Labs office in Paris. This was a case where an idea came long before the arrival of the appropriate technology to support it; in fact, at that time electronics was based on vacuum tube elements and was not capable of performing the operations required by PCM in an acceptable time frame and at an affordable cost. Only with the advent of transistor-based integrated circuits in the early 1960s did PCM become the workhorse of digital transmission. Today, there are several versions of PCM, among them delta PCM and companding PCM. PCM and its versions are widely used in such familiar applications as MP3, Blu-ray, and DVD. Though the Internet and mobile communications employ different forms of digital transmission, the PCM concept of converting analog signals into digital still plays an important role in modern communications.

There are myriad important details about PCM transmission, including descriptions of its versions, its parameters and characteristics, and its technology. We believe that you, being equipped with the knowledge provided in this section, will be able to acquire on your own all the knowledge about ADC and DAC in general and PCM in particular that your professional responsibilities will require.

Questions and Problems for Section 4.2

- Questions marked with an asterisk require a systematic approach to finding the solution.
- Many questions and problems, including those marked with an asterisk, imply that you, in addition to reading the textbook, will do your research to find the answers. Consider such questions as miniprojects.

Digital-to-Analog Conversion, DAC

1 *Explain the concept of a DAC operation, referring to Figure 4.2.1:
   a) What is the input signal and what is the output signal of this unit?
   b) How are the bits (the logical information) extracted from the input electrical signal (the physical carrier)?

c) How does an inverse S&H unit build a sample-and-hold version of the received signal? What do we need this signal for?
d) What is the role of a lowpass filter?
e) Does the reconstructed analog output signal coincide with the original (sent) analog signal? Explain.

2. *Use Figures 4.1.15 and 4.2.1 to build a detailed block diagram of an ADC–DAC system with analog input and output signals. Compare your figure with Figure 4.1.1 and discuss any discrepancies.

PAM and PCM

3. What does the acronym PAM stand for? What are the input and output signals of a PAM unit? What operation of an ADC system is necessary to build a PAM unit?

4. What are the main advantage and the main drawback of a PAM system?

5. Do you know any methods to overcome the major drawback of a PAM system? Explain.

6. *What does the acronym PCM stand for? What are the input and output signals of a PCM unit? What operations of an ADC system are used to build a PCM unit?

7. *What are the main advantages and the main drawbacks of a PCM system?

8. *What bit rate is necessary to deliver a single digitized voice signal with PCM?

9. *What will be the signal-to-noise ratio, $(SNR)_{QN}$, of a PCM signal with 8 bits per sample? With 16 bits per sample? Compare these $(SNR)_{QN}$ values: What conclusion can you make from this comparison?

10. What does the subscript "QN" stand for in the designation $(SNR)_{QN}$? Explain.

11. What is the intersymbol interference, ISI? How does this phenomenon affect the transmission of a digital signal? Explain.

12. *Consider the signal-to-noise ratio given in (4.2.4): $(SNR)_{QN} = 6.02 \cdot m$ (dB). Industry, however, prefers to use the following formula: $(SNR)_{QN} = [6.02 \cdot m + 1.76]$ (dB). An additional 1.76 dB is caused by additional noise that includes all the detrimental phenomena causing the deterioration of a signal. In general, noise is a process developed with respect to time, as shown in Figure 4.2.P12; the level of noise is termed *noise floor*.
    a) Compute $(SNR)_{QN}$ for $m = 8$ b/sample with and without additional noise and calculate the percentage of change caused by noise.
    b) Repeat these calculations for $m = 16$ b/sample and compare the percentages. What conclusion can you derive from this comparison?

13. *Compare the applications of PAM and PCM. Which modulation system is more widely used and why? Explain.

Figure 4.2.P12 Noise floor and SNR.

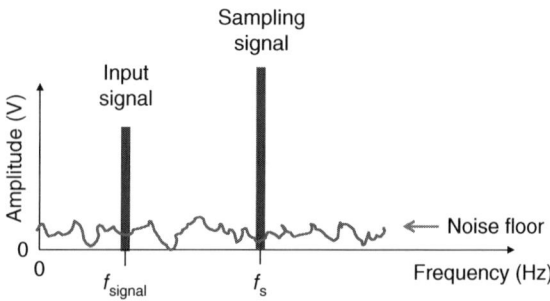

4.2.A Modes of Digital Transmission

Digital transmission can be done in different modes or techniques. It can be simplex, half duplex and full duplex, serial and parallel, synchronous and asynchronous, and it can be done in various combinations of these techniques. Additionally, these modes are interleaved with such vital techniques as framing, protocols, and error measurements. In this appendix, we introduce the basics of these techniques leaving detailed discussion to the other appropriate places in the book.

4.2.A.1 Simplex, Half Duplex and Full Duplex Transmission

These terms refer to one-way or two-way operations. In *simplex* communications, transmission is performed in one direction only. This is pure one-way communications, where one side of the link is always transmitting and the other is always receiving. The examples are security and fire-alarm systems: They send signals from sensors to the monitoring (control) station and there is no way these systems could send signals in the opposite direction. The advantage of this type of transmission is its simplicity and therefore its cost effectiveness. In fact, this is the least expensive type of transmission because the systems based on simplex transmission use a minimum number of components and rely on the simplest control system. The main disadvantage of such a system is that it enables one-way operation only. The RS-232 link, discussed in Section 3.1, is the example of simplex systems.

In *half-duplex* communications, transmission can be carried out in both directions but in only one direction at a time. Thus, when one end of the link is transmitting, the other is receiving. Examples are fax machines, credit-card verification systems, and bank automatic teller machines, ATMs. The ATMs, for instance, send a request to verify the information you entered; after receiving confirmation, they send another request, allowing for execution of the operation you required. Credit-card verification systems work the same way. Since each transmission takes time, you have to wait until your request is being processed. The half-duplex transmission systems are still economical, but more expensive than simplex systems. They need more components and more sophisticated control of their operation. For example, each side of a transmission link must contain both a transmitter and a receiver simultaneously and the control system must command which device has to work at every time instant. The downside of such a system is that its operation is relatively slow, as we all know from our experience with ATM machines. Another example is a RS-232 connection: It is a simplex system by design but can perform half-duplex operation if some circuitry is added.

Full-duplex communication allows for transmission in both directions simultaneously. The most well-known example is your telephone connection. Another example is one of the recently developed forms of communication called Li-Fi. This system uses the same LEDs for both lighting and communication, thus saving on the cost of special hardware and increasing transmission speed

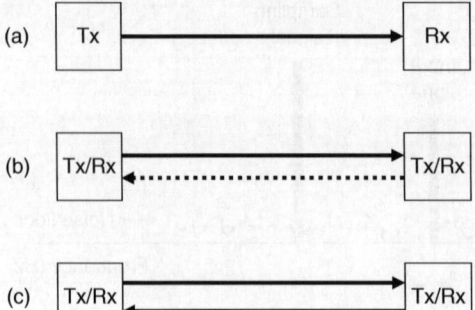

Figure 4.2.A.1 Transmission modes: (a) Simplex transmission; (b) half-duplex transmission; (c) full-duplex transmission.

up to the gigabits-per-second rate. Li-Fi can be built in half- and full-duplex mode. Today, many communication networks allow for full-duplex communications, too. Clearly, this is the best communications method; however, it requires appropriate (read: *expensive*) technology to support such a transmission.

Figure 4.2.A.1 illustrates all these transmission modes. Bear in mind that these technologies are used mostly in the United States.

In Europe and elsewhere in the world, you can encounter different usage of these terms in technical documentation. For example, in many cases, the term *simplex* is used to denote half-duplex transmission and the term *duplex* denotes full-duplex operation. This ambiguity stems from different traditions, and these different usages are supported by different standard bodies: In the United States, the terms *simplex*, *half-duplex*, and *full-duplex* are supported by the American National Standards Institute (ANSI); internationally, the terms *simplex* and *duplex* are supported by the International Telecommunications Union (ITU).

You certainly noticed from our discussion of these transmission modes that the terminology introduced above can be applied to both digital and analog transmissions.

4.2.A.2 Serial and Parallel Transmissions

The other important concept of digital and analog transmissions is serial and parallel transmissions. In serial transmission, we send signals down the link in sequence, one after the other. In parallel transmission, we send all the signals down the link simultaneously. Let us illustrate this concept by using an example of bit transmission: Suppose we need to transmit 8 bits. In serial transmission, we will transmit them one after another; in parallel transmission, we will transfer all 8 bits simultaneously. Figure 4.2.A.2 shows an example of serial and parallel digital transmission.

Observe from Figure 4.2.A.2 the difference in time required to complete transmission in the serial and parallel modes of operation. If T_b is the bit time, then in the serial mode we need $8 \times T_b$ second, while in parallel mode we need T_b second to complete the transmission of 8 bits. Figure 4.2.A.2 shows that when in serial transmission only the first bit of the stream just arrives at the destination point, all 8 bits in parallel transmission already received. The following example clarifies the matter.

Example 4.2.A.1 Transmission in serial and parallel modes

Problem

A binary word contains 16 bits. The bit time is equal to 0.01 ms. How long will it take to transmit this word in serial and parallel modes of operation?

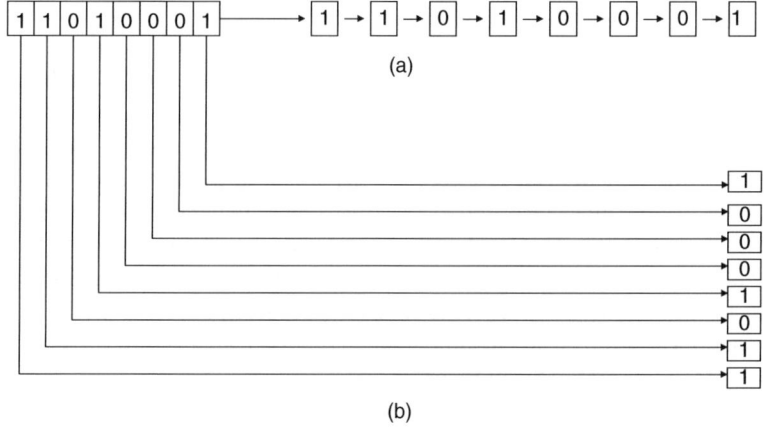

Figure 4.2.A.2 Serial and parallel transmission: (a) Serial transmission; (b) parallel transmission.

Solution

Since serial transmission occurs in sequence, the minimum transmission time, T, required to complete the operation is equal to the number of bits, N, times bit time, T_b; that is,

$$T_{\text{SER}}(s) = N \times T_b(s) \tag{4.2.A.1}$$

Thus, for our problem transmission time in serial operation is $T_{\text{SER}} = 0.16$ ms.

Since parallel transmission occurs simultaneously, transmission time is equal to bit time; that is,

$$T_{\text{PAR}}(s) = T_b(s) \tag{4.2.A.2}$$

For this problem, transmission time in parallel operation is $T_{\text{PAR}} = 0.016$ ms.

Clearly, parallel transmission is N times faster than serial.

Discussion

- From this example, the main advantage of parallel transmission over the serial becomes absolutely clear: *parallel transmission is faster*. How much faster? It depends on the length of a digital word, that is, on the number of bits to be transmitted at once. Formulas (4.2.A.1) and (4.2.A.2) show this point explicitly: The greater the number of bits in a binary word, the faster the parallel transmission in comparison with the serial. In our example, if the number of bits in the binary word were to increase to 32, the duration of the serial transmission would increase to 0.512 ms, whereas the duration of a parallel transmission would stay the same and be equal to T_b.
- The general trend in developing modern computers is to increase the length of a binary word that can be transmitted at once, that is, in parallel. Over the years this length has increased from 4 bits to 8 bits to 16 bits to 32 bits. This number is usually referred to as the main feature of a computer's processor; thus, the most popular in 1990s processors are called 32-bit processors. The rise of this number itself is one of the major reasons why modern computers operate faster and faster. Today, 64-bit processors are commercially available and these processors, due to the length of their binary word and other innovations, have elevated the computing power and speed of modern computers to the next level because all 64 bits are transmitted in parallel.
- What is the main drawback of parallel transmission? Look inside your PC. You will see unusual flat cables that combine several wires. (This type of cable is called "ribbon.") We need ribbon cables to transmit several bits in parallel. In other words, in parallel transmission, we need as

many wires for as many bits as we want to transmit. Have you ever seen such ribbon cables outside a computer unit or other electronic devices? The answer is clearly no. Nobody can afford to carry four – let alone eight or more – wires in parallel for any significant distance. Therefore, the main shortcoming of parallel transmission is the need for the link consisting of many parallel wires.

This is why we have to use one-wire link for any connections except of the ultra-short ones. (Basically, we use two wires to close an electrical loop.) But by using a single wire, we immediately are resorting to serial transmission. This is why serial transmission is the universal type of transmission except for ultra-short distances, such as within computers and similar devices.

Thus, we conclude that the major advantage of serial transmission over parallel is that serial transmission is more economical. And we already know the main disadvantage of serial transmission: It is much slower than its parallel counterpart.

Each type has found applications where its advantages are used to the maximum extent. Clearly, *most communications transmission today are done in a serial mode if any distance beyond a couple of feet is involved.*

- It is necessary to stress that any definition of serial and parallel transmissions always imply transmission of the same signal. In this example, we have discussed transmission of the same 16 bits in series and in parallel. How we can transmit different signals is a completely different story that is discussed in this chapter and in Chapters 9 and 10.

4.2.A.3 The General Formula for Bit Rate

Example 4.2.A.1 shows explicitly that

$$\text{BR (b/s)} = 1/T_b \text{ (s)} \tag{3.2.1dR}$$

is applied to *serial* transmission only. You recall that *bit rate is the number of bits transmitted per second*, which leads to (3.2.1) in the case of serial transmission. Therefore, if you transmit 16 bits serially with $T_b = 0.01$ ms each, the bit rate will be

$$\text{BR}_{\text{SERIAL}} \text{ (b/s)} = 1/T_b \text{ (s)} = 100 \text{ kb/s}$$

We can arrive at this result by the following reasoning: Since we transmit 16 bits every 0.16 ms, it is easy to compute that for one second we will transmit $(16 \text{ bits} \times 1 \text{ s})/(0.16 \times 10^{-3} \text{ s}) = 100$ kb/s.

However, if we try to apply (3.2.1) to *parallel* transmission, we obtain an incorrect result. Indeed, by direct computation, we will get 100 kb/s while, in reality, the bit rate will be 160 kb/s. We compute 160 kb/s by the same reasoning: If we transmit 16 bits per 0.01 ms, we can transmit $(16 \text{ bits} \times 1 \text{ s})/(0.01 \times 10^{-3} \text{ s}) = 160 \times 10^3$ bits per second. Thus, this bit rate is 16 times greater than the one for a serial transmission just because we transmit 16 bits per bit time, T_b. Therefore, from this example, we can draw the following conclusion: In general, *bit rate, BR, is equal to the number of bits, N, transmitted per bit time, T_b, divided by bit time,*

$$\text{BR (b/s)} = N/T_b \text{ (s)} \tag{4.2.A.3}$$

If we apply (4.2.A.3) to our example of parallel transmission, we will obtain the correct result immediately:

$$\text{BR (b/s)} = N/T_b \text{ (s)} = 16/0.01 \text{ ms} = 160 \text{ kb/s}$$

Be aware of this general formula and its applications.

4.2.A.4 The Need for Synchronization in Digital Transmission

Every term containing the word *chronous* refers to timing. The origin of this reference roots in ancient Greece mythology where Kronos was the god of time. Thus, the term *synchronous* must be applied to describe something related to time, something occurs in timing sequence, within a time frame, time-coordinated, and so on. For example, *synchronization* means aligning in time among different processes and events. Following this logic it would be reasonable to think that the term *asynchronous* must be applied to describe the processes or events without any connection to time. This reasoning is absolutely correct when you discuss some general topics in nature or society. However, being applied specifically to digital transmission, terms *synchronous* and *asynchronous* have quite different meanings. Explanation of the sense of these terms is the subject of the following two subsections.

4.2.A.4.1 Digital Signals and Timing
Let us make it crystal clear:

All processes in digital world occur under clock control.

Indeed, in digital transmission, we send electrical or optical pulses that represent bits. Each of these pulses is described by its bit time, T_b; for given transmission this bit time is the same. Therefore, each bit must be generated at the certain instance, and the time interval between these instances must be equal to T_b. This process of bit generation is controlled by the clock that "tells" the signal generator when to produce the next pulse. Figure 4.2.A.3 illustrates these explanations.

Observe in Figure 4.2.A.3 that it is the rising edge of a clock pulse that constitutes an instance for generating a data (signal) pulse. Arrows directed from clock pulses to data pulses in Figure 4.2.A.3 indicate these instances. The time interval between two adjacent rising edges of clock pulses (that is, the arrows in Figure 4.2.A.3) is equal to T_b, as it follows from definition of bit time. Observe from Figure 4.2.A.3 too that clock pulses have a finite duration equal to T_C. Clearly, T_C must be much less than T_b, as Figure 4.2.A.3b shows. To emphasize the concept of timing in digital transmission, we deliberately consider the ideal waveforms of pulses.

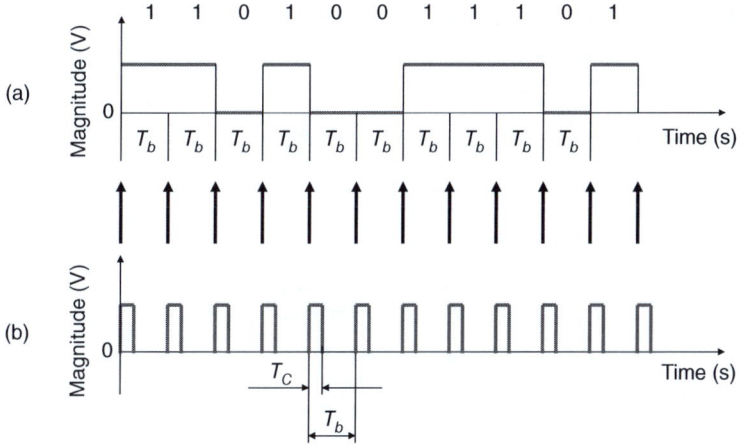

Figure 4.2.A.3 Data (signal) pulses and clock pulses: (a) Data pulses; (b) clock pulses.

Since digital signal must be related to time, as Figure 4.2.A.3 clearly shows, it seems that the digital communications must be related to time too; that is, it must be synchronous. In reality, however, digital transmission is always related to time but not always *synchronous*. Why? Let us consider it more closely.

4.2.A.4.2 Timing in Digital Transmission

When we discuss *transmission*, we need always bear in mind that transmission involves two parties: transmitter and receiver. (See Figures 1.1.1 and 1.1.2 to remind yourself that a transmission system consists of a transmitter, link, and a receiver.) Transmitter that sends a digital signal works under control of its clock and a receiver that receives this digital signal works under control of its clock. In other words, at both ends of a transmission link, the signal processes related to time – that is, synchronous – as illustrated by Figure 4.2.A.3. But it does not mean that *transmission* itself is synchronous.

> In digital transmission, synchronization means time aligning between a transmitter and a receiver of the communications link.

Why do we need such synchronization? First and foremost, a receiver must know the bit rate (that is, bit time, T_b) at which the transmitter sends data pulses; this is necessary condition to make digital transmission occur. Indeed, here is how digital transmission works: Every time interval equal to T_b a transmitter generates and sends pulses carrying bits 1 or 0 (see Figure 4.2.A.3). A receiver checks (samples) arriving pulses at the same bit rate; that is, every T_b interval. Based on this sampling, the receiver detects which bit – 1 or 0 – the incoming pulse carries. This basic concept of digital transmission is illustrated in Figure 4.2.A.4.

Clearly, if a receiver does not follow the transmitter's bit rate – or, equivalently, T_b – its readings will have nothing to do with the sent signal. This is why the transmitter and receiver clocks must work in unison, that is, synchronously.

We need to emphasize that a receiver checks (samples) the arriving pulses at their middle; that is, at the half of T_b. This technique minimizes the possible errors because (i) it gives the maximum

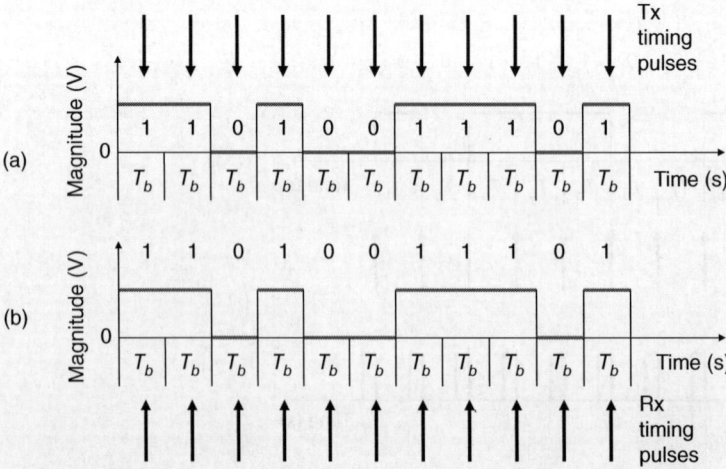

Figure 4.2.A.4 Principle of digital transmission: (a) Pulses and bits send by transmitter; (b) pulses and bits read by receiver.

tolerance to the drift of receiver's clock and (ii) there is a high probability that amplitudes of the arriving distorted pulses will reach their maximum in the middle of the bit time. This is why Figure 4.2.A.4 shows receiver clock pulses in the middle of the pulse duration. We will follow this practice throughout the book. To stress the need for time alignment between transmitter and receiver clocks, in Figure 4.2.A.4, we also sketch arrows representing transmitter clock pulses in the middle of data pulses.

Clearly, Figure 4.2.A.4 depicts the ideal situation when transmitter and receiver clocks are aligned perfectly.

What problems we encounter in timing of digital transmission? To answer, we have to understand two important points: First, all clocks in this world have the finite precision, which means that there is always the difference in time reading between any pair of clocks. Second point is that discrepancy between receiver and transmitter clocks results in obtaining erroneous information. Let us consider these points more closely.

4.2.A.4.3 Time Discrepancy Between Transmitter and Receiver Clocks

Since our clocks are implemented with some technology, they cannot be ideal; they always have a finite precision. The best clock in the United States is the national standard clock; we measure all other clocks with respect to it. This clock has been developed, is kept, and maintained by National Institute of Standards and Technology (NIST) in their Boulder, Colorado, laboratory. The latest version of this clock has been developed in 2017; it allows for one erroneous second over 15 billion years, the time span longer than the age of the universe! The other national timing laboratory is the Time Service Department of the US Naval Observatory, which is the official source of time for the Department of Defense and Global Positioning System (GPS); this laboratory also provides a standard time for the United States with the same precision as the NIST clock.

The most general time reference is called "Coordinated Universal Time" (UTC), which is the basis for the worldwide system of civil time. The International Bureau of Weights and Measures makes use of data from the national timing laboratories to provide the international standard UTC, which is accurate to approximately a nanosecond (billionth of a second) per day.[2]

This is tremendous accuracy, but we cannot use the international or national standard clocks for providing transmission between a handle of computers in our laboratory.

For digital networks, we need to use different clocks depending on the scale of the networks. Let us take for example a local network; the typical clock accuracy for this type of network is about 10^{-6}, which is much lower than the national standard. The precision equal to 10^{-6} means that after generating 10^6 seconds the clock produces one second more or one second less than the theoretical number of second must be. Let us do simple calculations: Recall that 24 hours contains $24 \times 60 \times 60 = 86400 \text{ s} = 0.0864 \times 10^6 \text{ s}$. Taking the networking clock accuracy equal to 1×10^{-6}, we will easily compute that for one day of work two communicating computers can accumulate the difference in their clocks equal to

$$0.0864 \times 10^6 \text{ s} \times 1 \times 10^{-6} = 0.0864 \text{ s}$$

This concept and computations are illustrated in Figure 4.2.A.5.

Figure 4.2.A.5 shows pulses generated every second by a transmitter and a receiver clocks. We make an obvious assumption that both clocks are synchronized at the starting instance, which we consider as zero point. Pulses # 3 of both clocks are in almost perfect alignment, as two arrows show in Figure 4.2.A.5. In other words, seconds No. 3 in both clocks are almost perfectly synchronized.

---

2 http://aa.usno.navy.mil/faq/docs/UT.html.

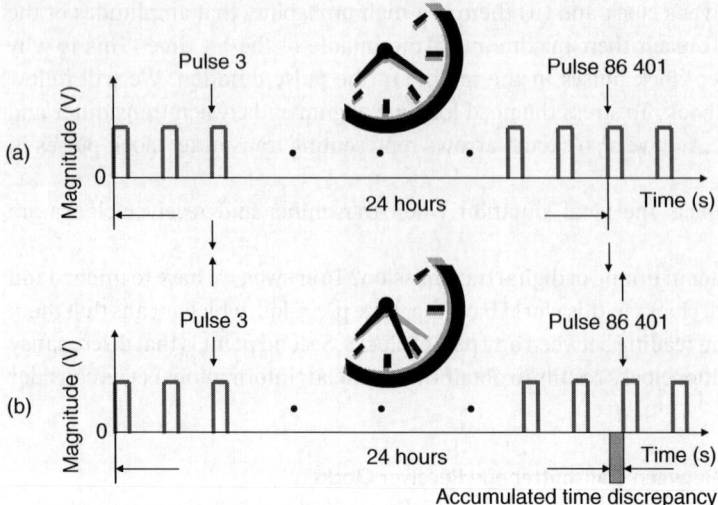

Figure 4.2.A.5 Discrepancy between transmitter and receiver clocks: (a) Clock pulses and time at transmitter end; (b) clock pulses and time at the receiver end.

After 24 hours of work, the accumulated discrepancy between transmitter and receiver clocks will be equal to 0.0864 seconds. Since 24 hours contains 86400 seconds, the next pulse after 24 hours of work will have number 86401, which is shown in Figure 4.2.A.5. Because of accumulated discrepancy between transmitter and receiver clocks, pulse # 86401 of a receiver clock will be off by 0.0864 second with respect to pulse # 86401 of a transmitter clock. Figure 4.2.A.5 shows this discrepancy as accumulated time interval at the receiver time axis and as a shift between two arrows indicating the rising edges of transmitter and receiver clock pulses. In this example, receiver clock comes ahead of the transmitter. Clearly, shown discrepancy is not to scale.

To better visualize this point, Figure 4.2.A.5 shows the transmitter and receiver clocks in cartoon drawing where everyone can clearly see that these clocks show different time. We have to bear in mind that this discrepancy is constantly accumulating over the time; it does not happen abruptly after 24 or 48 hours or any other number of working hours. In other words, even the third pulses of transmitter and receiver clocks will be off, but this discrepancy will be extremely small; specifically, 3×10^{-6} in our example. We consider the essential time interval simply to show a significant amount of timing discrepancy.

In the above explanation, we have implied that a receiver clock will drift permanently in one way with respect to a transmitter clock. In Figure 4.2.A.5, the receiver clock works faster than a transmitter clock; this is why its pulse # 86401 arrives behind of the transmitter's pulse # 86401. The permanent drift is, obviously, the worst-case scenario. In reality, time error, as any error for that matter, is a random variable; we describe its behavior by the law of its distribution. While given explanation suffice enough to stress the main point of this discussion – receiver and transmitter clocks will always have different time – the real picture is much more sophisticated.

When we say, "… receiver clocks will be … off with respect to transmitter clock," we imply that transmitter clock is a reference. It is not exactly correct; in this case, we need to consider the discrepancy between these two clocks only. Thus, equivalently, we can say that a transmitter clock is off with respect to the receiver. However, when we consider a telecommunications network, where many transmitters, receivers, and switches (we call all such network elements *nodes,* remember?) are involved, we have to have a general time reference for all the clocks working at the network nodes. In reality, every telecommunications network operates under control of its own clocks that may or may not be traced to the UTC or national standard time.

4.2.A.4.4 How Time Discrepancy Between Transmitter and Receiver Clocks Deteriorates the Quality of Digital Transmission

Now we are clear with the first point of our discussion of timing in digital transmission: We always have discrepancy between transmitter and receiver clocks. But why is this so important? In a word, because this discrepancy results in obtaining erroneous information. How?

Let us put together Figures 4.2.A.3–4.2.A.5. We need a transmitter clock to control generation of sent data pulses, as Figure 4.2.A.3 shows. We need a receiver clock to check the received data pulses, as Figure 4.2.A.4 shows. Since transmitter and receiver clocks have discrepancy, a receiver may check (sample) the wrong pulse, which results in wrong interpretation of received information; that, is error. Figure 4.2.A.6 pictures this concept; in this illustration we follow (Stallings 2014, page 236).

Figure 4.2.A.6 shows clock pulses of transmitter Tx and receiver Rx as arrows and data as NRZ pulses. You will recall that the receiver timing pulses are set in the middle of the data pulses; to stress the main point of this discussion, we put transmitter clock pulses against the receiver's ones; that is, also in the middle of data pulses. As you can see from Figure 4.2.A.6, the accumulated discrepancy between transmitter and receiver clocks results in error: *Receiver clock pulse # 86401 reads bit 0 instead of bit 1 because this receiver clock pulse is shifted with respect to appropriate (# 86401) pulse of a transmitter clock.*

Let's consider an example to comprehend the problem better.

Example 4.2.A.2 Time Difference Between Transmitter and Receiver Clocks

Problem

Calculate the time difference between transmitter and receiver clocks accumulated over an hour of transmission if the precision of both clocks is equal to 10^{-9}. Can this time difference result in error if a bit rate equals 10 Gbit/s? 40 Gbit/s?

Solution

Since one hour contains 3600 seconds, the maximum accumulated time difference is equal to $3600\,\text{s} \times 10^{-9} = 3.6\,\text{ns}$.

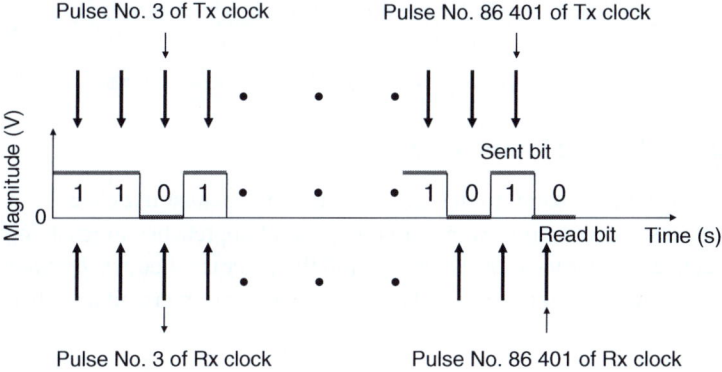

Figure 4.2.A.6 Error in received information caused by discrepancy between transmitter and receiver clocks.

To answer the question, we need to compare this shift with the bit time of a given bit stream. Applying (3.2.1c), we can compute for 10 Gb/s

$$T_b \text{ (s)} = 1/\text{BR (b/s)} = 1/10\,\text{Gb/s} = 0.1\,\text{ns}$$

Now we need to compare accumulated time difference with a bit time. From our computations we can readily see that within the accumulated time shift 36 bits can be placed. Indeed, 3.6 ns/0.1 ns = 36 bits. Therefore, the timing pulse of a receiver clock will sample a wrong bit, the bit which is 36 bits off the ideal one. Examine Figure 4.2.A.6 to visualize this result. Thus, the answer to the problem is yes, such an accumulated time difference between transmitter and receiver clocks will most likely result in error.

Clearly, this system will not work for 40-Gbit/s transmission rate.

Discussion

Thirty-six bits is a tremendous error, but this is a quite real number in many cases of digital transmission. Two points should be stressed here: The first point is that digital transmission is not error free. In a comparison of digital and analog transmission (see Section 3.2), we emphasize that digital transmission is much less error prone. This is true; however, as we see now, digital transmission has its own problems, fortunately, at much lower level of errors. Many types of errors in digital transmission relate to precision of timing support.

The second point we can draw from this example is that the higher the transmission rate, the more stringent requirements on clock precision must be put. Indeed, for 10-Gbit/s transmissions, we need clocks with at least 10^{-11} precisions. In reality, atomic clocks with such a precision are used in modern synchronous telecommunications networks.

4.2.A.4.5 A Short Summary on Synchronization Issues

Let us summarize the results of our discussion:

- *All digital operations are done under control of clock.*
- *Data (signal) pulses are generated and sent out under control of transmitter clock.*
- *Data pulses are sent from transmitter to receiver.*
- *To read the received information, receiver clock must work synchronously with a transmitter clock.*
- *Due to finite precision, transmitter and receiver clocks inevitably will have time difference.*
- *Because of this time difference, a receiver may read the wrong data pulse, which results in error of obtained information.*

4.2.A.5 Asynchronous and Synchronous Transmission

Now we clearly understand the timing problem: In digital transmission, we must use clocks, but the finite precision of these clocks causes errors. There are two general approaches to resolving this problem. One method is called *asynchronous transmission*, and the other is called *synchronous transmission*. Here, we consider how these terms are applied to transmission in circuit-switched networks.

4.2.A.5.1 Asynchronous Transmission

This type of transmission combats the errors caused by time discrepancy between transmitter and receiver by very simple means: Each character, or binary word, sent out by a transmitter is accompanied by start and stop bits. When a receiver accepts this character, it first reads the start bit. This bit commands a receiver clock to start counting time. From this bit, the receiver clock samples all the arriving bits every T_b second, as described in discussion of Figure 4.2.A.5. When a receiver reads a stop bit, its clock stops counting time and waits for the next start bit.

This explanation is supported by Figure 4.2.A.7.

Figure 4.2.A.7 shows that before and after transmission the line is in the idle state. Typically, this is so-called "mark" state that corresponds to the state of bit 1. For example, in RS-232 standard, mark state means presence of a negative voltage on the line and in TTL logic family, the mark is a positive voltage. However, the idle state can also be a state when no active signal is present, which in NRZ coding corresponds to bit 0; this state is called "space." Incidentally, these terms – mark and space – are rooted to Baudot time.

A start bit precedes a character being transmitted. If an idle state is mark (high), then the start bit is space (low) and vice versa. This transition from high to low state commands a receiver clock to start sampling arriving pulses every T_b (s). Based on this sampling, a receiver reads the arriving bits. One or two stop bits at the end of transmission command the receiver to put to an end reading the bits. After that, the line returns to the idle state.

Figure 4.2.A.7 shows the examples of transmitting characters "2" and "M" in ASCII code. Look at the direction of data flow and do not forget that ASCII transmits characters from MSB to LSB; thus, the leftmost character bit in Figure 4.2.A.7 is the MSB.

It is now clear that the change from idle state to start bit constitutes the beginning of a character. It is also clear that a receiver reads bit after bit by simply counting T_b intervals. But how a receiver knows that it has to stop reading? First, a receiver is informed what number of data pulses (bits) it supposes to receive. For example, a receiver has been "informed" that it will receive a regular ASCII characters; thus, it "knows" that it has to count seven data pulses. (In reality, transmission of ASCII character is always accompanied by one extra bit called parity bit; therefore, usually an ASCII character consists of 8 bits.) Second, receiver "looks" for the bits following the data bits:

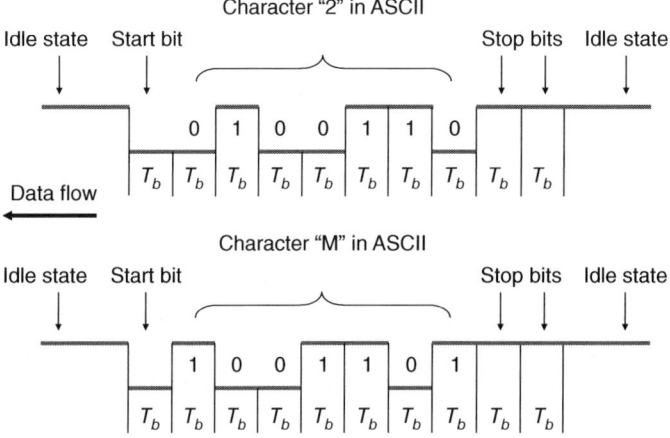

Figure 4.2.A.7 Asynchronous transmission.

They are always in the same state as the idle state of a transmission line. Based on these two features, the receiver recognizes the stop bits. Asynchronous transmission may require one or two stop bits, depending on a specific implementation. In any case, the bit or bits following the data pulses signal about finishing character transmission.

How does asynchronous transmission help to avoid errors caused by timing difference between transmitter and receiver clocks? Start bit actually brings timing information to a receiver clock: A receiver begins counting T_b from this start bit. All this means that every time when a start bit arrives, receiver clock is synchronized with a transmitter clock. In other words, a receiver clocks undergoes resynchronization with every character. Time interval needed for reading a single character is very short; with a typical clock precision at the range of 10^{-6}, accumulated time discrepancy will be very small.

Therefore, asynchronous transmission simply breaks down total transmission into short-time portions, receiver clocks resynchronizes with transmitter clocks at every character, and these clocks do not accumulate essential timing discrepancy because they work together for a very short period of time.

Example 4.2.A.3 Error Caused by Time Discrepancy Between Tx and Rx in Asynchronous Transmission

Problem

Calculate an error caused by time discrepancy between a transmitter and a receiver when one ASCII character is transmitted. Assume that the accuracy of both transmitter and receiver clocks is equal to 10^{-6}.

Solution

The plan of the solution is simple: We need to calculate time interval needed for transmission of a character and compute the maximum time shift caused by the clock's inaccuracy. Based on that, we will evaluate what error can be caused by this time difference.

Time needed to transmit a character is equal to $T_b \times$ number of bits. We assume that an ASCII character contains eight bits; thus, in our example transmission time is equal to $T_b \times 8$. Assuming $T_b = 0.0016$ ms as in Example 4.2.A.2, we compute that total time for reading a character is equal to 0.0016 ms $\times 8 = 0.0128$ ms.

Since the accuracy of the clocks is equal to 10^{-6}, the maximum time difference between transmitter and receiver is equal to $10^{-6} \times 0.0128$ ms $= 12.8 \times 10^{-12}$ s. This is a small fraction of a T_b; specifically, 12.8×10^{-12} s$/1.6 \times 10^{-6}$ s $= 8 \times 10^{-6} = 0.0008\%$. Therefore, this time difference between transmitter and receiver clocks might cause small error.

Discussion

Compare this result with that obtained in Example 4.2.A.2: In this example, we do not have any error (in spite of low precision of the clocks) because of the low speed of transmission and ultrashort transmission time. In Example 4.2.A.2, we had a significant error (despite the high precision of the clocks) because of high speed of transmission. This comparison emphasizes the main advantage of asynchronous transmission.

Why this type of transmission is called asynchronous? Because every character is transmitted at arbitrary time. An example of asynchronous transmission that everybody is familiar with is transmission from your keyboard top your computer unit. When you hit a button on your keyboard, eight bits representing a character in ASCII code are transmitted to the computer unit. Even if you are a professional typist, you do typing at a variable pace. Most likely, your goal is not typing fast but developing a meaningful text or presentation. In any case, you send all these characters at random times, not at a specific rate. Nobody, including you, can predict when the next character will be sent. In other words, there is no specific rate for transmitting these characters. This is why this transmission is called asynchronous.

The main advantage of asynchronous transmission is that this type of transmission almost eliminates errors caused by timing difference between transmitter and receiver clocks. The main drawback of asynchronous transmission is that it adds two or three service bits to every 8 data bits. Thus, we need to transport at least 25% of traffic additionally to data traffic. This is the price we have to pay for resynchronization that asynchronous transmission provides with every character. The price is clearly too high: Imagine your network operates at 100 Gb/s data rate; if you need to do it asynchronously, you need to transmit extra 25 Gb/s. Nobody can afford such a network because, obviously, every bit being transported cost money to the network operator.

Asynchronous transmission in its original form described above still finds its niche in low-speed operation like transport of the bits from a keyboard to a computer unit. You can also meet asynchronous transmission within a circuit board. However, you should be aware that this term – *asynchronous transmission* – is often used in a broad sense to denote transmission without special measures for synchronization of all clocks of the network nodes. In this meaning, asynchronous transmission is a main type of data transmission in modern communications systems and the Internet operates based on this type.

4.2.A.5.2 Synchronous Transmission

In synchronous transmission data transmitted not by characters but by big blocks of bits called *frame*. These blocks may contain thousands of bits. Each of such blocks is preceded and concluded by one or two service words. The starting service words inform the receiver that data bits begin to arrive. The concluding service words inform the receiver that transmission of a data block is completed. A word is usually an 8-bit string also called octet. Between starting and ending words an arbitrary number of data bits are transmitted. Figure 4.2.A.8 illustrates the concept of synchronous transmission.

In Figure 4.2.A.8, *flag* is a unique combination of bits that always precedes the block of characters sent in synchronous transmission. In other words, it is the flag – the first service word – that shows the start of the transmission of a data block. You can often meet flag as the following combination: 01111110; however, other combinations are used in synchronous transmission. It is a protocol that says what flag is used in any specific case. Flag is always called preamble.

Control field may contain the address field, which carries the location of a destination point, and other service information about the block being transmitted. The address is vitally important when

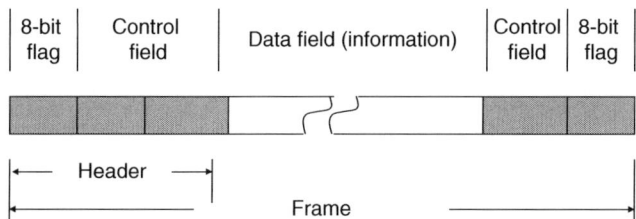

Figure 4.2.A.8 Concept of synchronous transmission.

data is synchronously transmitted through a network where the block of characters is to be directed through a number of nodes.

Flag (preamble) and control fields are usually called *header*, the term that denotes all service fields that precede a block of information bits.

The main segment of this block is, of course, *data (information) field*, the string of bits that carries actual information. The size of this field – the number of bits it contains, that is – could be variable or fixed, depending on a transmission technique. This field is also called *payload*.

The block of characters transmitted synchronously ends with another flag, the same predetermined combination of bits, as appended at the block beginning. Thus, both flags delimit the block of characters in synchronous transmission.

How synchronization occurs in this type of transmission? Well, each starting flag informs a receiver about beginning of transmission; thus, it delivers timing information. However, since block (frame) of transmitted characters can be long enough, it can take a while to complete the transmission. Therefore, transmitter and receiver clocks must keep their synchronization for a relatively long period of time, as compared to asynchronous transmission. Here lies the main problem of synchronous transmission: It puts much more stringent requirement on a network timing system. This is the price we need to pay for the main advantage of synchronous transmission: its effectiveness.

Indeed, let us compare the number of service bits needed for asynchronous and synchronous transmissions. In asynchronous transmission, we need at least two service bits – start and stop – for every eight data bits transmitted. Thus, service bits take $2/10 \times 100 = 20\%$ of the total number of 10 bits. In synchronous transmission, the number of service bit compare to the data bits is much smaller. For example, in *synchronous optical network (SONET)* system, the number of service bits is equal to 27, while the total number of transmitted bits in a block equals 810. Therefore, service bits constitute only 3.3% of the total number of transmitted bits. To appreciate this difference – 3.3% vs. 20% – even more, we need to realize that with asynchronous transmission, we could transmit only 10 bits in a second, whereas with SONET, we can transmit 10×10^9 bits (gigabits) per second.

Conclusion Let us reiterate that these terms – *asynchronous and synchronous* – are applied to digital transmission in the circuit-switching networks. As explained previously, these are the networks developed and exploited by telephone companies for connecting the traditional landline telephones. Today, a communication professional understands that the words *synchronous transmission* and *asynchronous transmission* refer to transmission through these legacy networks. Modern communications, however, rely on packet routing and this type of transmission is asynchronous by its very nature (see Section 1.2.) Therefore, the Internet transmission does not need synchronous and asynchronous classification.

5

Filters

Objectives and Outcomes of Chapter 5

Objectives

- To learn how to analyze filter's operations and present filter's characteristics in time domain and frequency domain. To grasp the concept of transfer function and its application in analysis of filters.
- To study the principles of operation and main characteristics of passive filters based on RC, RL, and RLC circuits.
- To become familiar with all major types of filters, such as low-pass filters (LPFs), high-pass filters (HPFs), band-pass filters (BPFs), and bandstop filters (BSFs).
- To gain knowledge of op-amps-based active filters, their main characteristics and applications. To comprehend the principle of operation and main applications of switched-capacitor filters.
- To understand the need for filter prototypes in filter design. To know major filter prototypes and to become familiar with the design process of filters.
- To get basics of digital filters, their principle of operations and mathematical description.

Outcomes

- Using a LPF as an example, learn the principle of operation, main characteristics, and mathematical description of passive analog filters based on RC, RL, and RLC circuits.
- Expand knowledge of an LPF to other major types of passive analog filters, such as HPF, BPF, and BSF.
- Enhance the students' knowledge of the mathematical description of passive filters by studying the filters' transfer functions and their applications.
- Learn about the drawbacks of passive filters and the need for the use of operational amplifiers in the filters' circuits.
- Understand the principles of operation, the main characteristics, the advantages, and the applications of all major types of active analog filters.
- Know the principle of operation, advantages, and main applications of active switched-capacitor filters.
- Study what filter prototypes are and why we need them for the design of filters.
- Considering the Butterworth filter prototype as an example, learn how to obtain main prototype characteristics based on its transfer function.

- Become familiar with the filter design, understand the design automated options, and learn main steps of the design process based on the design of a Butterworth filter.
- Grasp the idea how to choose filter technology for a specific application.

5.1 Filtering – Basics

Objectives and Outcomes of Section 5.1

Objectives

- To understand the concept of and the need for filtering.
- To learn the principle of operation and the basic mathematical description of an RC LPF.
- To know the presentation of a filter's operation and its characteristics in the time domain and frequency domain.
- To become familiar with filter's general specifications accepted by the industry and academia.

Outcomes

- Understand that filtering means allowing the desired signals to pass while rejecting the undesired signals, using the signals' frequencies as criteria for the selections.
- Know that there are four main filter types: a LPF, a HPF, a BPF, and a BSF. The BSF is also called a notch filter or a band-rejection filter.
- Learn that *passband* is the band of frequencies within which the signals are allowed to pass by a filter, and *stopband* is the band of frequencies within which signals are rejected by the filter.
- Understand that the strict, step-like border between passband and stopband exists only in an ideal filter; in a real filter, the transition between a passband and a stopband occurs smoothly.
- Gain an understanding in the practical implementation of an LPF by a series RC circuit, in which an output voltage is measured across the capacitor. Get to know that this circuit performs the selection or rejection of signals thanks to capacitive reactance, which becomes very high in the low-frequency range and very low in the high-frequency range.
- Be aware that a change in the output voltage of an RC LPF is caused only by a change in the input frequencies, $\omega(\text{rad/s}) = 2\pi f$.
- Understand that after passing through an RC LPF, the waveform (cosine or sine), the frequency (ω), and the initial phase (Θ_{in}) of an input sinusoidal signal, $v_{in}(t) = V_{in}\cos(\omega t + \Theta_{in})$, remain the same, but the output amplitude and the output phase shift decrease. Hence, the output signal takes the form: $v_{out}(t) = V_{out}\cos(\omega t + \Theta_{in} + \Theta_{out})$.
- Learn that the change in the output signal is assessed by the filter's attenuation, $A_v = \frac{V_{out}(V)}{V_{in}(V)}$, and the phase shift, Θ_{out} (rad). The graphs A_v and Θ_{out} (rad) vs. ω (rad/s) (or f (Hz)) are called the amplitude (magnitude) and phase frequency responses of an LPF.
- Acquire knowledge that one of the main characteristics of a filter is its cutoff frequency, f_C (Hz) (or ω_C(rad/s)); this frequency is also called the critical frequency, the corner frequency, the break frequency, or the rolloff frequency.
- Comprehend that in a real RC filter there is no natural border between passband and stopband. By convention, the cutoff frequency is the frequency at which the output amplitude decreases to 0.707 of the input amplitude; that is, at f (Hz) $= f_C$, $A_v = \frac{V_{out}(V)}{V_{in}(V)} = 0.707$ or $A_v = -3$ dB.
- Realize that the passband of an LPF is the range of frequencies from 0 Hz to f_C (Hz). This passband is the bandwidth of the LPF.

- Obtain the following formulas for the attenuation and the output phase shift in an RC LPF by applying the circuit-analysis technique: $A_v = \frac{V_{out}\,(V)}{V_{in}\,(V)} = \frac{1}{\sqrt{1+\left(\frac{f}{f_C}\right)^2}}$ and $\Theta_{out}\,(\text{rad}) = -\tan^{-1}\left(\frac{f}{f_C}\right)$. Find that the conditions $A_v = 0.707$ and $\Theta_{out} = -45°$ at $f\,(\text{Hz}) = f_C$ follow from these formulas. Also, learn that $f_C = \frac{1}{2\pi RC}$, where R and C are the values of a filter's components.
- Learn how to build graphs of an LPF's amplitude and phase responses by using the abovementioned formulas.
- Be able to explain the course of the frequency-response graphs of an LPF, specifically (i) why the amplitude response, $A_v(f)$, is equal to 1 when $f\,(\text{Hz}) = 0$ and goes to zero when $f\,(\text{Hz})$ goes to infinity and (ii) why the phase response changes from 0° to −90° when the frequency goes from zero to infinity. Also, be able to prove that at $f\,(\text{Hz}) = f_C$, $A_v(f) = 0.707$ and $\Theta_{out} = -45°$.
- Be able to relate the amplitude response, $A_v(f)$, and the phase response, $\Theta_{out}(f)$, in frequency domain to changes in the amplitude and phase shift of the waveforms of an LPF's output signal.
- Learn that for the characterization of an LPF filter we need to trace the changes in $A_v(f)$ and $\Theta_{out}(f)$ with respect to the deviation of a current frequency from a filter's cutoff frequency, specifically, as a function of $\frac{f}{f_C}$.
- Realize that the $A_v(f)$ and $\Theta_{out}(f)$ graphs in general and f_C in particular depend on the values of an LPF's circuit parameters, $R\,(\Omega)$ and $C\,(F)$, which is the key point in designing an LPF.
- Gain an understanding of the limitations in the design of an RC LPF caused by (i) uncertainties in choosing simultaneously both $R\,(\Omega)$ and $C\,(F)$ to obtain the required $f_C\,(\text{Hz})$ – the main design parameter – and (ii) the inability to improve the shape of attenuation and phase response graphs by changing the values of $R\,(\Omega)$ and $C\,(F)$.
- Learn that the filter characteristics $A_v(f)$, $\Theta(f)$, and f_C describe an idealized situation in which for every frequency value we have corresponding unique values of A_v and Θ, and f_C uniquely delimits a passband. Understand that, in contrast, a real filter (i) exhibits uncertainties in determining the values of A_v, Θ, and f_C and (ii) the shape of its frequency response is far from that of an ideal filter. Become familiar with general specifications in which the attenuation of a real filter is inscribed into ideal-like specifications producing a new approach to specify an LPF.
- Study the parameters of general filter specifications, get to know how to calculate these parameters and how to apply these parameters to specify a real filter.

5.1.1 Filtering: What and Why

In communications, we want to deliver signals carrying the intended information without errors from one point to another. In the real world, however, these desired (intended) signals interfere with many undesired signals, and the receivers will obtain a mix of both useful and harmful signals. Consider a simple example: You are trying to talk to your friend in a room filled with other people and where everyone is trying to talk to everyone else. The speech of any person other than your friend is noise to you; it interferes with your conversation and makes it difficult, if not impossible, for you to keep a normal dialogue going. In communications, we encounter a similar situation – the mixing of undesired signals with useful ones every time we deliver signals. One way to combat this undesired phenomenon is to use filtering.

Filtering, as the term suggests, *is the process of rejecting some signals while allowing others to pass through a communication channel.*

The process of filtering is performed by a device called (you guessed it) a filter. We classify filters as *passive* and *active* or as *analog* and *digital*. In this chapter, we consider all of them, but we start with passive analog filters.

We can select, or filter, signals by various criteria. Let us consider analog signals: We know that these signals have three main parameters: amplitude, frequency, and phase. Therefore, we can filter an analog signal by any of these parameters; for example, we can reject all sinusoidal signals with amplitudes of more than 10 V. However, *in modern communications, filtering means selecting signals by their frequency*.

For instance, we can reject high-frequency signals and allow low-frequency signals to pass. Or, alternatively, we can reject low-frequency signals and pass high-frequency signals. We can also either reject or allow the passage of only signals within a certain band of frequencies. The filtering concept is illustrated in Figure 5.1.1.

We need to realize that Figure 5.1.1 shows the operation of ideal filters. If a filter rejects signals with high frequencies and allows the passage of only signals with low frequencies, it is called a

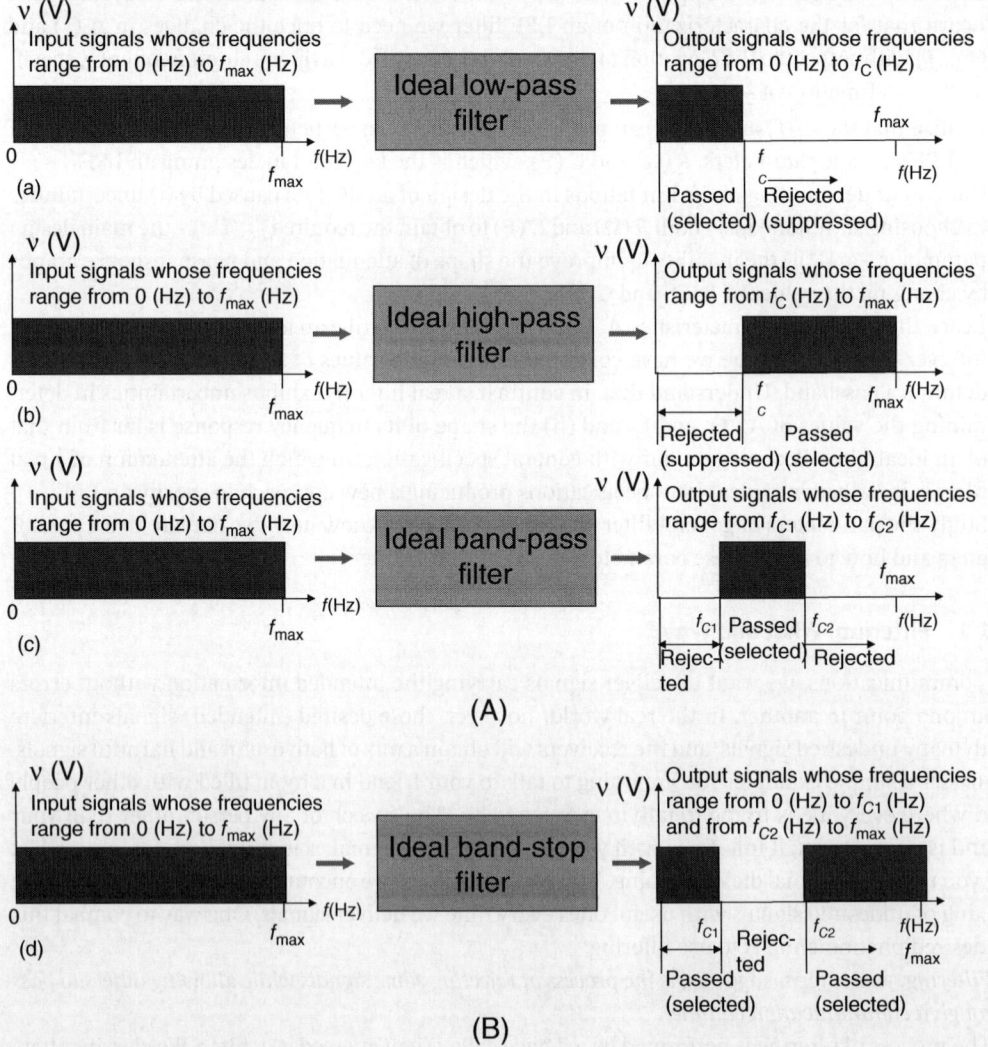

Figure 5.1.1 Concept of filtering: (a) low-pass filter; (b) high-pass filter; (c) band-pass filter; and (d) band-stop filter.

low-pass filter. Consider a filter that passes signals with frequencies ranging from 0 to 4000 Hz while rejecting signals whose frequencies are greater than 4000 Hz. This is an example of a LPF. The operation of this type of filter is shown in Figure 5.1.1a.

The cutoff frequency is the frequency at which a filter starts rejecting or passing signals. It is also called the critical frequency, or corner frequency, f_C.

(Note that the cutoff frequency is sometimes called the *rolloff* frequency.) In other words, the cutoff frequency is the border between signals being passed and rejected. In our example, 4000 Hz is the cutoff frequency of a LPF.

Clearly, a *HPF* rejects low-frequency signals and passes only signals whose frequencies are higher than the filter's critical frequency. Consider a filter that will reject all signals with frequencies from 0 to 4000 Hz and allow the passage of all signals with frequencies greater than 4000 Hz. This is an example of a HPF with a cutoff frequency of 4000 Hz. The concept of a HPF is shown in Figure 5.1.1b.

A *BPF* passes only signals whose frequencies fall into a certain band, while rejecting all signals whose frequencies are either less than a filter's low limit (f_{C1}) or greater than the filter's high limit (f_{C2}). For instance, consider a filter that rejects all signals with frequencies either lower than 4000 Hz or higher than 6000 Hz. This filter passes only signals whose frequencies fall within the band from $f_{C1} = 4000$ to $f_{C2} = 6000$ Hz; thus, this is an example of a BPF, which is shown in Figure 5.1.1c.

Finally, the fourth major type of filter rejects signals whose frequencies are restricted by a certain band and allows the passage of all other signals. This type of filter is called a *band-stop* (or band-rejection) filter; it is shown in Figure 5.1.1d. Turning to an example, consider a BSF that rejects signals with frequencies from $f_{C1} = 4000$ Hz to $f_{C2} = 6000$ Hz and selects (passes) all signals with frequencies from 0 Hz to f_{C1} and from f_{C2} to f_{max}.

It is industry practice to use the acronym LPF for a low-pass filter, HPF for a high-pass filter, BPF for a band-pass filter, and BSF for a band-stop filter.

It is necessary to stress that in reality, we present to a filter only signals whose frequencies are restricted by a certain band. Figure 5.1.1 shows, for example, input signals whose frequencies range from 0 Hz to f_{max} (Hz). Obviously, a filter can reject and allow for passage (that is, select) signals from the given frequency band only. In our examples, an LPF rejects all signals whose frequencies range from $f_C = 4000$ Hz to f_{max}, an HPF selects all signals whose frequencies range from $f_C = 4000$ Hz to f_{max}, and a BPF selects signals from $f_{C1} = 4000$ Hz to $f_{C2} = 6000$ Hz while rejecting signals from 0 to 4000 Hz and from 6000 Hz to f_{max}.

These explanations raise the question of the proper way to define low and high frequencies. Indeed, is 4000 Hz a low or a high frequency? If we compare 4000 Hz with a frequency of 60 Hz, which is the frequency of industrial ac power, then 4000 Hz looks like a high frequency. However, if we compare 4000 Hz with 4 000 000 Hz, which is in the radio-frequency range, we can naturally conclude that 4000 Hz is a low frequency. Certainly, you get the point:

There is no strict definition of low or high frequency.

Nevertheless, audio frequencies, ranging from 4 to 20 kHz, are considered low; radio frequencies, ranging from hundreds of kilohertz to hundreds of megahertz, are considered intermediate; and

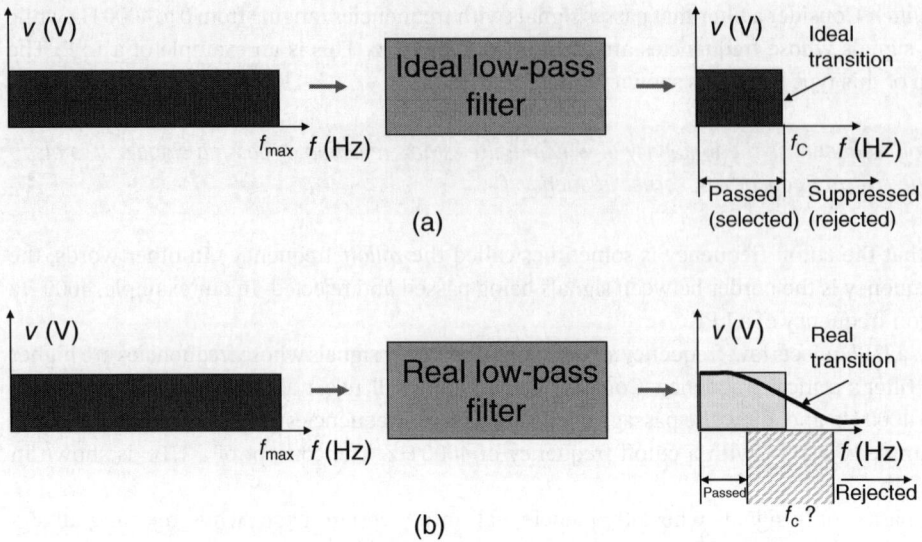

Figure 5.1.2 Example of a low-pass filter's operation: (a) ideal filter and (b) real filter.

microwave frequencies, ranging from hundreds of megahertz to tens of gigahertz, are considered high (see frequency classification in Figure 1.2.11). Clearly, optical signals that lie in the band of hundreds of terahertz are considered ultra-high. Thus, radio frequencies are high with respect to audio range but low with respect to microwave range.

As we noted above, a filter is a device that selects or rejects signals within a band of frequencies. The range of frequencies within which the filter allows the passage of signals is called the *passband*; the range of frequencies within which the filter rejects signals is called the *stopband*. Ideally, the border between the passband and the stopband should resemble a step-like (some say "wall-brick") transition, and we can easily identify the unique cutoff frequency. Such an ideal transition is shown in Figure 5.1.2a; we can now say the following:

The cutoff frequency, f_C, is the frequency separating the passband from the stopband.

In reality, however, there is always a smooth transition between selected and rejected frequency bands, as Figure 5.1.2b shows. The fact is that there is a transitional region between the passband and the stopband in a real filter; therefore, it is difficult, if not impossible, to precisely and uniquely identify the cutoff frequency.

Naturally, we want to know why we cannot achieve the ideal filter's transition. The answer lies in understanding how a filter operates. This is why we need to turn to the circuitry of a real filter.

5.1.2 RC Low-Pass Filter (LPF)

5.1.2.1 Frequency Responses of a Resistor, *R*, and a Capacitor, *C*

Let us start with a simple RC LPF. The concept of LPF is visualized in Figure 5.1.1a. To remind: The LPF selects low-frequency signals for passage and rejects high-frequency signals. The schematic of a typical RC LPF is shown in Figure 5.1.3.

To understand the principle of the filter's operation, we need to recall that a resistor, R, does not change its resistance when the frequency of a signal passing through it is changing (refer to Figure 2.3.8).

Figure 5.1.3 Schematic of a low-pass RC filter.

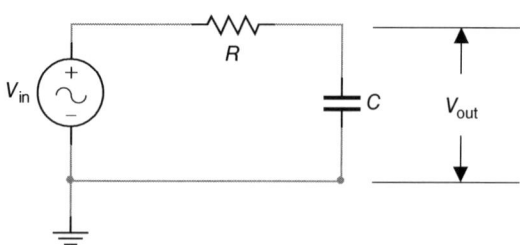

Even crucial to understanding the principle of an RC LPF operation is that the capacitor's reactance, X_C, decreases with an increase in the frequency. Surely, you remember that reactance is the measure of the opposition (resistance, if you prefer) that a capacitor exerts on a current flowing through it. The way the value of X_C changes with respect to frequency is given by

$$X_C\ (\Omega) = \frac{1}{2\pi f C} = \frac{1}{\omega C} \tag{5.1.1}$$

and it is shown in Figure 2.3.12R. In (5.1.1), f (Hz) is a signal's frequency, C (F) is the capacitance of a capacitor, $\pi \approx 3.14$ is the well-known constant, and ω (rad/s) is the signal's radian frequency given by ω (rad/s) $= 2\pi f$.

(**Reminder**: We use the capital letter R to denote both a resistor and its resistance, whereas we use C to designate a capacitor and its capacitance and X_C to denote its reactance.)

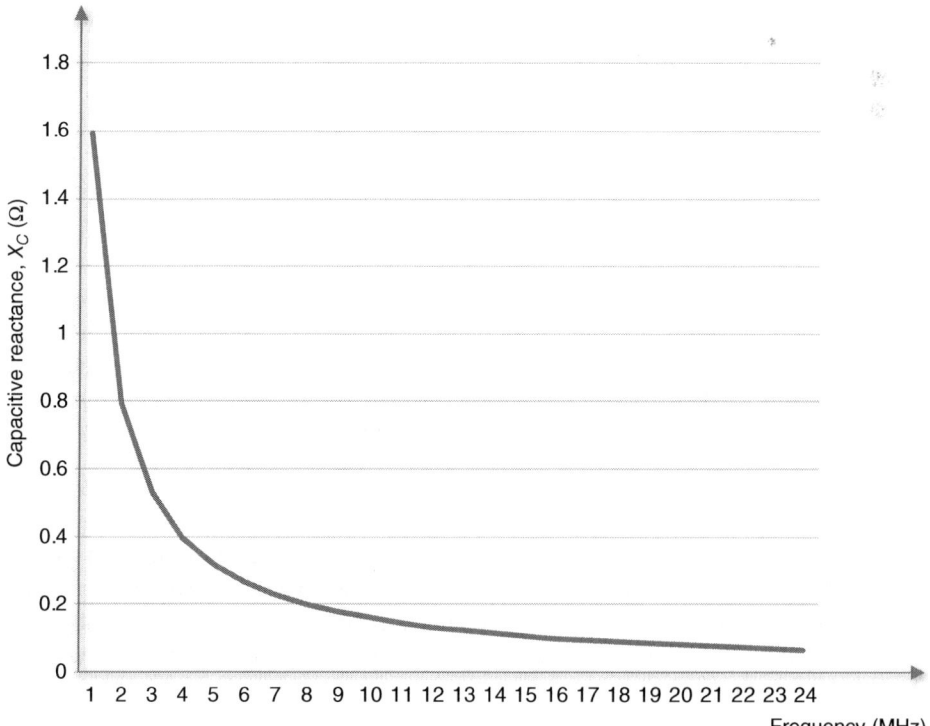

Figure 2.3.12R Frequency response of capacitive reactance.

To understand capacitor's role in the operation of an RC LPF, let us consider two extreme situations: First, let the frequency go to zero. In this case, X_C approaches infinity, according to (5.1.1):

$$f \to 0 \Rightarrow X_C (\Omega) = \frac{1}{2\pi fC} \to \infty$$

For example, a capacitor with $C = 2\,\mu F$ at a frequency of $f = 0.001$ Hz exhibits $X_C = 79.6$ MΩ. This is a huge value, and we can surely consider the capacitor as an open circuit.

In the second extreme case, let the frequency go to infinity. Then X_C approaches zero:

$$f \to \infty \Rightarrow X_C (\Omega) = 1/2\pi f_C \to 0$$

The same capacitor with $C = 2\,\mu F$ but at a frequency of $f = 10\,000\,000$ Hz will have $X_C = 0.007\,96\,\Omega$. This is a very small value, and we can surely consider this capacitor acting like a short circuit.

As (5.1.1) shows, the lower the frequency, the greater the value of X_C and vice versa.

(**Exercise**: Take the capacitance as $C = 2\,\mu F$ and choose the following set of frequencies: 0.001, 0.01, 0.1 Hz, and so on up to 10 MHz. Compute the corresponding set of reactance values and build the table and to see how X_C (Ω)-vs.-f (Hz) changes.)

(Bear in mind that the frequency responses of a resistor and a capacitor discussed earlier hold true only within the low part of RF spectrum, up to tens of megahertz. In the gigahertz range, both the resistor and capacitor would behave differently. Refer to Figure 1.2.11 for EM spectrum.)

To reinforce your understanding of the course of the R-vs.-f and X_C-vs.-f graphs, let us look at this problem from different perspectives, as given in the following example.

Example 5.1.1 Equivalent Circuits of a Resistor and a Capacitor

Problem

Build equivalent circuits for a resistor and a capacitor at low and high frequencies. (See "Discussion" section of this example for reminding the meaning of the term "equivalent circuit.")

Solution

Equivalent circuits for a resistor and a capacitor at low and high frequencies can be built based on the R and C properties discussed in this section: An ohmic resistance, R (Ω), of a resistor remains constant when the frequency of a passing signal changes, whereas the reactance value, X_C (Ω), of a capacitor decreases with an increase in frequency, as shown in Figure 2.3.12R. Therefore, an equivalent circuit for a resistor is the same at all frequencies, whereas an equivalent circuit for a capacitor is an open circuit at very low frequencies and a short circuit at very high frequencies. These equivalent circuits are shown in Figure 5.1.4. They help us to understand the filter's operation in two extreme cases: near zero frequencies and at extremely high frequencies.

Discussion

- The term "equivalent circuit" defines the circuit that behaves the same way as the real circuit, but may be built from components different from those used in an actual circuit. For example, Figure 5.1.4b shows two disconnected wires in the equivalent circuit, whereas the actual circuit is built with a capacitor. This is because these two disconnected wires operate the same way the capacitor does at zero-frequency input.

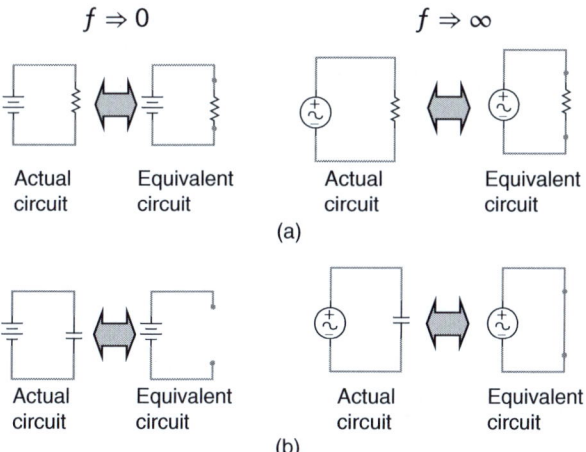

Figure 5.1.4 Equivalent circuits: (a) resistor at low frequency and at high frequency and (b) capacitor at low frequency and at high frequency.

- An equivalent circuit is a powerful tool in circuit analysis and it helps tremendously in our understanding of the operation of circuits and devices; however, it has its limitations. For instance, what could be a capacitor's equivalent circuit at an intermediate frequency? We would have to imagine a circuit consisting of components that would represent both resistive and phase-shifting properties of a capacitor; thus, we would lose the main advantage of an equivalent circuit: simplicity.
- Even though we use two disconnected wires as a capacitor's equivalent circuit at low frequencies, but, in fact, capacitive reactance does have a huge but a finite value. The same reasoning is applied to the short-circuit approximation of a capacitor at high frequencies. For example, if $C = 2\,\mu F$, then $X_C = 79.6\,M\Omega$ at $f = 0.001$ Hz, and $X_C = 0.007\,96\,\Omega$ at $f = 10$ MHz. Bear in mind this word of caution when applying the equivalent-circuit methodology to the analysis of the circuit you will encounter.

5.1.2.2 RC Low-Pass Filter: Principle of Operation

Now we can understand the principle of operation of this RC filter. Referring to Figures 5.1.3 and 5.1.4, we can build the equivalent circuit of an RC LPF in two extreme cases when f (Hz) $\Rightarrow 0$ and f (Hz) $\Rightarrow \infty$. These circuits are shown in Figure 5.1.5, and they enable us to find the output voltages in these cases: For f (Hz) $\Rightarrow 0$, the RC LPF output is an open circuit and therefore an output voltage, V_{out} (V) $\equiv V_C$, is equal to the input voltage, V_{in}. Conversely, for f (Hz) $\Rightarrow \infty$, the RC LPF output is a closed circuit and V_{out} (V) $\equiv V_C$ is zero. (Remember, an equivalent circuit is only an approximation of a real circuit; thus, we should say that "when f (Hz) goes to zero, the RC LPF circuit is approximated by an open circuit, and the output voltage approaches to V_{in} (V).")

Thus, in two extreme cases, the output voltage, V_{out} (V), takes either the value of V_{in} or approaches zero. Obviously, the value of the output voltage changes smoothly from V_{in} to 0. Thus, we can present the course of the output voltage vs. the frequency in the graph of the RC filter shown in Figure 5.1.2b. And we need to realize that *this change in V_{out} (V) is caused by changing*

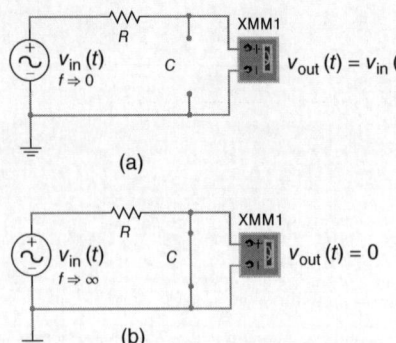

Figure 5.1.5 Equivalent circuits of an RC LPF in two extreme situations: (a) signal's frequency goes to zero and (b) signal's frequency goes to infinity.

only the *frequency* of an input signal whose amplitude and initial phase remain constant during this operation.

5.1.2.3 Output Waveforms of an RC LPF

Figures 5.1.1a and 5.1.2 show what will be the filter's output signals vs. frequency. Clearly, the graphs in these figures are built in *frequency domain*. But what will be the filter's response vs. time, that is, in *time domain*? In other words, how will the signal's waveform change after the signal passes the filter? We suggest arranging the following experiment: Present a sinusoidal signal to a low-pass RC filter, change the signal's frequency and observe on an oscilloscope screen the waveforms of the input and output signals at each frequency. The experimental setup for this observation is shown in Figure 5.1.6a and the result of the experiment is shown in Figure 5.1.6b. The circuit simulation is done with Multisim.

Figure 5.1.6b provides critical information for understanding the operation of a filter:

> *The filter simply decreases the amplitude of an output signal at the band of frequencies to be filtered.*

Why does a LPF decrease the amplitudes of the output signal with an increase in the input frequencies? This decrease occurs due to the specific property of a capacitor: The capacitive reactance, X_C, changes inversely to the change in the input frequency, as (5.1.1) and the graph in Figure 2.3.12R show. In this experiment, as the frequency (f) increases, the reactance (X_C) decreases, and thus the output-voltage drop across the capacitor decreases, too:

$$f \rightarrow \infty \Rightarrow X_C = 1/2\pi f_C \rightarrow 0 \Rightarrow V_{out} \equiv V_C = i \cdot X_C \rightarrow 0$$

All these processes occur smoothly, and there are no abrupt changes in either of these values.

The actual waveforms confirm this point very clearly: At the low frequency ($f = 100$ Hz), the output signal is indistinguishable from the input; at the intermediate frequency ($f = 1$ kHz), the output amplitude is slightly smaller than the input; and at the high frequency ($f = 10$ kHz), the output amplitude is much smaller than the input. Again, the *amplitude of the input signal* remains *the same* for all three measurements; only its frequency has been changed.

The decrease in the amplitude is not only the change in the LPF's output signal; it also experiences a *phase shift*, Θ (rad), with respect to the input signal. This effect is clearly seen in Figure 5.1.6.

(**Question**: *What is the phase shift of the output signal with respect to the input in the bottom graph (at $f = 10$ kHz) of Figure 5.1.6?*)

Figure 5.1.6 Investigation of a low-pass RC filter's operation in time domain: (a) experimental setup and (b) waveforms of the input and output signals of an RC LPF at various frequencies. Oscilloscope settings are the same for all the screens.

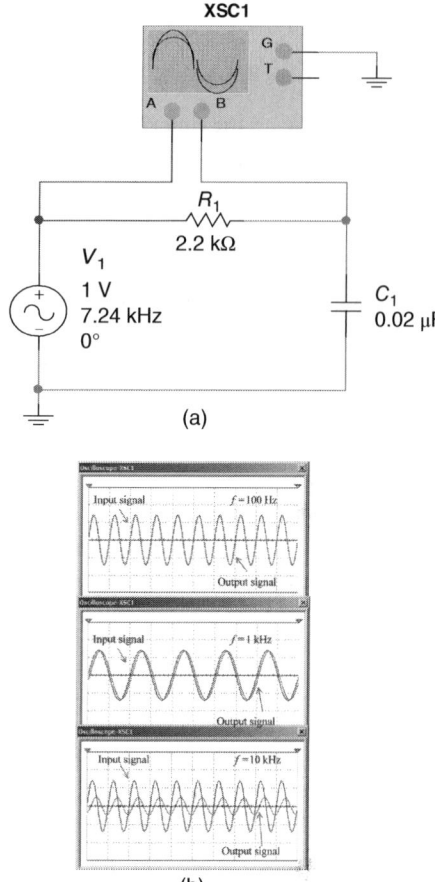

So we understand that the decrease in amplitude and the phase shift in an RC LPF are caused by properties of RC circuit. But how can we calculate these values? From what parameters do these decrease and shift depend on? To answer, we need to mathematically describe the operation of an RC LPF and derive the formulas for $A_v = V_{out}$ (V)/V_{in} (V) and Θ.

5.1.2.4 An RC LPF: Formulas for Attenuation and Phase Shift

Our goal is to obtain formulas for the output voltage, $v_{out}(t)$, and its phase shift, Θ. Such a quantitative approach is the goal of any engineering analysis.

Let us consider the RC LPF circuit shown in Figure 5.1.3. When we apply ac voltage, $v_{in}(t)$, at a certain frequency, f_1, this voltage will be distributed between a resistor and a capacitor, according to Kirchhoff's voltage law:

$$\mathbf{V_{in}} = \mathbf{V_R} + \mathbf{V_C}, \tag{5.1.2}$$

where V_R and V_C are the voltage drops across the resistor and the capacitor, respectively. In (5.1.2), the boldface type is used to show that all ac voltages are *phasors*, which means that each of them is described by both magnitude and phase. Refer to the subsection "Frequency response R, L, and C components" in Section 2.3 to brush up on this point.

We can derive the general formula for $\mathbf{V_{out}}$ by applying the voltage-divider rule to our circuit. Remember that this rule says that the voltage across impedances in a series circuit divides in direct proportion to the individual impedance. By applying this rule, we obtain

$$V_C \equiv V_{out} = V_{in} \left(\frac{Z_C}{Z_T} \right) \tag{5.1.3}$$

where $\mathbf{Z_C} = -jX_C$ is the capacitive *impedance* and $\mathbf{Z_T} = \mathbf{Z_R} + \mathbf{Z_C} = R - jX_C$ is the total impedance of the circuit. (Remember, impedance is described by its amplitude and phase, see Section 2.3.) Thus, the output voltage, in explicit form, is given by

$$V_{out} = V_{in} \left(\frac{-jX_C}{R - jX_C} \right) \tag{5.1.4}$$

We are interested in knowing, however, not only how much the output signal changes by itself but also how much this signal varies with respect to the input signal. To acquire this information, we had better consider the ratio of output voltage to input voltage rather than simply V_{out}. The ratio $\frac{V_{out}}{V_{in}}$ is a very important characteristic of any filter. Called the *filter frequency response*, it is denoted as

$$A_v = \frac{V_{out}}{V_{in}} \tag{5.1.5}$$

This characteristic includes both amplitude *attenuation*, A_v, and the *phase shift*, Θ, that result from a filter's operation. This is why we use boldface for $\mathbf{A_v}$; it is, again, the main characteristic of a filter.

From (5.1.4) and (5.1.5), we can present $\mathbf{A_v}$ in this way:

$$\mathbf{A_v} = \left(\frac{V_{out}}{V_{in}} \right) = \left(\frac{-jX_C}{R - jX_C} \right) \tag{5.1.6}$$

Incidentally, (5.1.6) can be rewritten as

$$\mathbf{A_v} = \frac{1}{1 + j2\pi RC} \tag{5.1.7}$$

because $X_C = 1/2\pi f_C$ and $j(-j) = 1$.

Let us now separate the real and the imaginary parts in (5.1.6). Multiplying both the numerator and the denominator by the denominator's complex conjugate, we obtain

$$\mathbf{A_v} = \frac{X_C^2}{R^2 + X_C^2} - \frac{-jX_C R}{R^2 + X_C^2} \tag{5.1.8}$$

Recall that these complex-numbers manipulations were discussed in Appendix 2.3.A.1. In particular, we learned that a complex number in a rectangular form,

$$z = x + jy$$

can be presented in polar form with magnitude A and phase Θ,

$$z = A \sqsubset \Theta \equiv Ae^{j\Theta}$$

where

$$A = \sqrt{(x^2 + y^2)} \text{ and } \tan \Theta = y/x$$

Applying these rules to (5.1.8), we derive the formulas for the filter's *attenuation*, A_v, and its *phase shift*, Θ, as

$$A_v = \frac{V_{out}}{V_{in}} = \frac{X_C}{\sqrt{(R^2 + X_C^2)}} = \frac{1}{\sqrt{\left(1 + \left(\frac{R}{X_C}\right)^2\right)}} \tag{5.1.9}$$

$$\Theta = -\tan^{-1} \left(\frac{R}{X_C} \right) \tag{5.1.10}$$

Now, we can present the general formula for A_v written in polar form as

$$A_v = A_v \angle \Theta \equiv A_v e^{j\Theta}$$
$$= X_C/\sqrt{(R^2 + X_C^2)} \angle [-\tan^{-1}(R/X_C)] \equiv X_C/\sqrt{(R^2 + X_C^2)}\, e^{j[-\tan^{-1}(R/X_C)]} \quad (5.1.11)$$

This formula describes an RC low-pass filter in its entirety.

> We need to stress again that the filter's attenuation, A_v, is a factor by which the filter decreases the input voltage, whereas the phase shift, Θ, is the additional phase angle that the filter introduces to the input signal.

We can derive from (5.1.9) the formula for calculating the amplitude of an output signal of an RC LPF as

$$V_{out}\,(V) = A_v \cdot V_{in}\,(V) = V_{in}\,(V) \cdot \frac{1}{\sqrt{\left(1 + \left(\frac{R}{X_C}\right)^2\right)}} \quad (5.1.12)$$

Finally, (5.1.10) and (5.1.12) enable us to mathematically describe the waveform of an output signal of an RC LPF as

$$v_{out}(t)\,(V) = V_{out}\,\cos(2\pi ft + \Theta) = V_{out}\,\cos(\omega t + \Theta) \quad (5.1.13)$$

Having (5.1.13), we achieve our goal to obtain the equations for quantitative analysis of an RC LPF. Let us apply these formulas to the practical calculations given in Example 5.1.2.

Example 5.1.2 Waveforms of an Output Signal of an RC LPF

Problem

An RC LPF whose $R = 2.2\,\text{k}\Omega$ and $C = 0.02\,\mu\text{F}$ is shown in Figure 5.1.7. The input signal is $v_{in}(t) = 14\sin(2\pi ft)$. Build the waveforms of the output signals at f equals 904, 3617, and 14 468 Hz.

Solution

Apply (5.1.12) and (5.1.13) and obtain the formula for the output waveform. Then do calculations for the given frequencies by using (5.1.1) and (5.1.10) and build the required waveforms. Thus,

$$v_{out}(t)\,(V) = V_{out}\,\cos(2\pi ft + \Theta) = V_{in}\,(V)\,\frac{1}{\sqrt{\left(1 + \left(\frac{R}{X_C}\right)^2\right)}}\,\sin(2\pi ft + \Theta)$$

$$X_C\,(\Omega) = \frac{1}{2\pi fC}$$

Figure 5.1.7 An RC LPF circuit for Example 5.1.2.

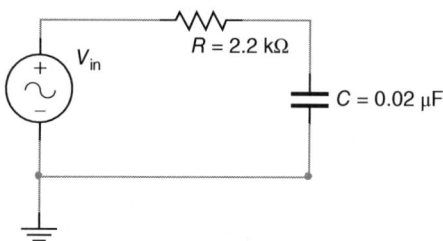

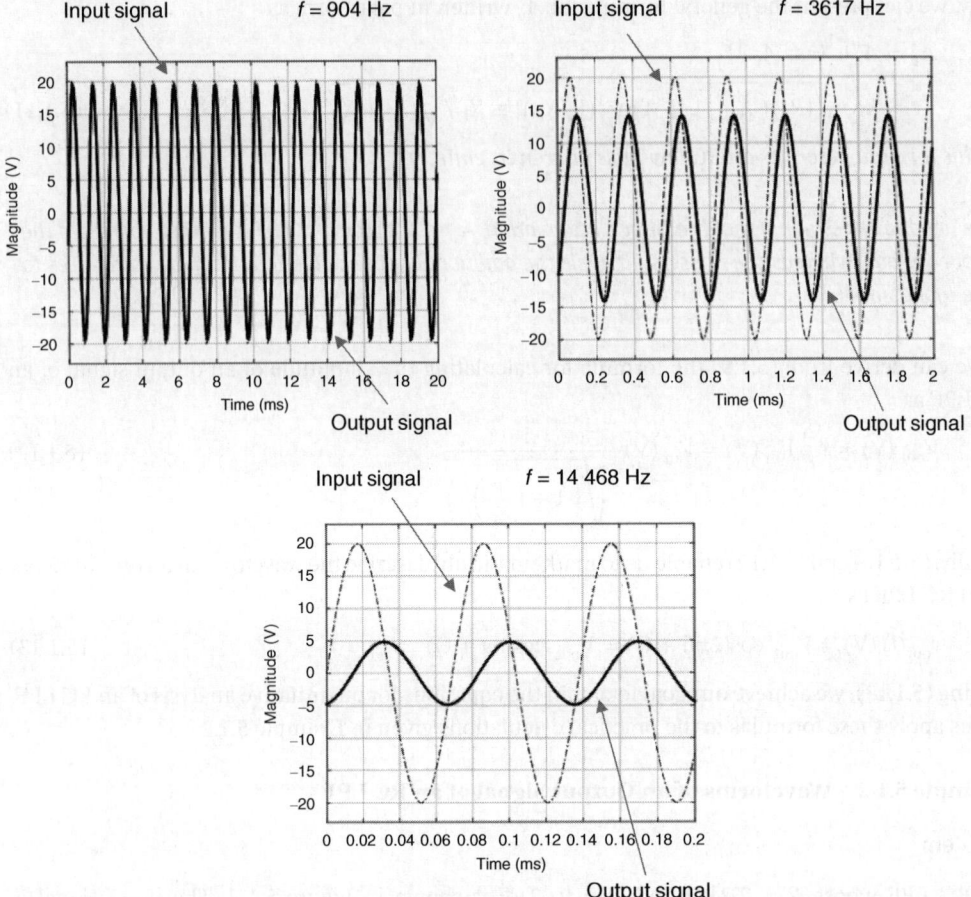

Figure 5.1.8 Three waveforms of the output signal for Example 5.1.2.

and

$$\Theta \text{ (rad)} = -\tan^{-1}\left(\frac{R}{X_C}\right)$$

Perform the calculations and obtain the required waveforms. They are shown in Figure 5.1.8, where the input signal is also shown to stress the changes in the output signals.

Discussion

- Reviewing the obtained input and output waveforms, we can readily observe that, as the input signal's frequency increases from 904 to 3617 to 14 468 Hz, the amplitude of the output signal decreases. This is exactly what a LPF must do. In effect, a *filter is a frequency-dependent voltage divider* (Horowitz and Hill 1995, p. 40). The collateral effect of the filter operation is the phase shift of the output signal, as the waveforms in Figure 5.1.8 display.
- Calculated values of V_{out}, A_v, and Θ at given frequencies are presented in Table 5.1.1. Compare these values with that shown in Figure 5.1.8 and comment on any discrepancies. Observe how the filter attenuation, A_v, follows the changes in frequency. Estimate the phase values shown in

Table 5.1.1 Calculated values of V_{out}, A_v, and Θ for Example 5.1.2.

f (Hz)	V_{out} (V) = V_{in} (V) $\frac{1}{\sqrt{1+\left(\frac{R}{X_C}\right)^2}}$	$A_v = \frac{V_{out}(V)}{V_{in}(V)} = \frac{1}{\sqrt{1+\left(\frac{R}{X_C}\right)^2}}$	Θ (rad) = $-\tan^{-1}\left(\frac{R}{X_C}\right)$	Θ (°)
904	19.40	0.97	−0.24	−14.04
3 617	14.14	0.71	−0.79	−45.00
14 468	4.85	0.24	−1.33	−75.96

Figure 5.1.8 and compare them with that presented in Table 5.1.1. How does the negative sign of Θ exhibited in Table 5.1.1 affects the waveforms shown in Figure 5.1.8?

The above discussion presents the important lesson for us: *When we present a <u>sinusoidal signal</u> to a low-pass RC filter, the filter's response will have the following important features:*

- The output will be the sinusoidal signal, too.
- The frequencies of the input and output signals will be the same.
- The amplitudes of the output signal will decrease smoothly with an increase in the input frequency, even though the input signal maintains the same amplitude.
- The output signal will experience a negative phase shift with respect to the input.

We need to realize that, in fact, we consider the input–output relationship, which is a very important approach that engineers contemplate in their everyday work.

5.1.2.5 Frequency Response of an RC LPF

The discussed waveforms given in (5.1.10) and (5.1.13) deliver the output of an RC LPF in time domain. Now, (5.1.9) and (5.1.10) enable us to characterize the RC LPF in frequency domain. We tackle this problem through Example 5.1.3.

Example 5.1.3 Frequency Response of an RC LPF

Problem

An RC LPF is built from $R = 2.2$ kΩ and $C = 0.02$ μF, as shown in Figure 5.1.8:

a) Build the graphs A_v and Θ vs. frequency and show passband and stopband on these graphs.
b) Find the output voltage at $f_1 = 3.617$ kHz if $V_{in} = 20$ V ∟ 0°.
c) Find the phase of the output signal under the same condition.

Solution
a) Using (5.1.9) and (5.1.10), we build the graphs A_v-vs.-f and Θ-vs.-f shown in Figure 5.1.9.
b) We can compute the output voltage V_{outi} at any frequency f_i by calculating X_{Ci} (Ω) = $\frac{1}{2\pi C f_i}$ at this frequency and plugging the obtained value of X_{Ci} into (5.1.12). By doing so, we compute the following value of V_{out} at $f_1 = 3.617$ kHz:

$$X_{C_1} = \frac{1}{2}\pi f_1 C = 2.2 \text{ (k}\Omega\text{)}$$

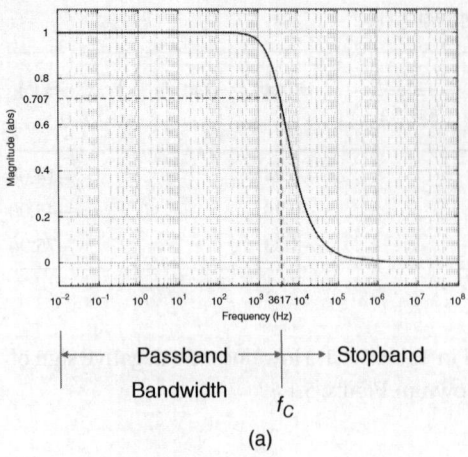

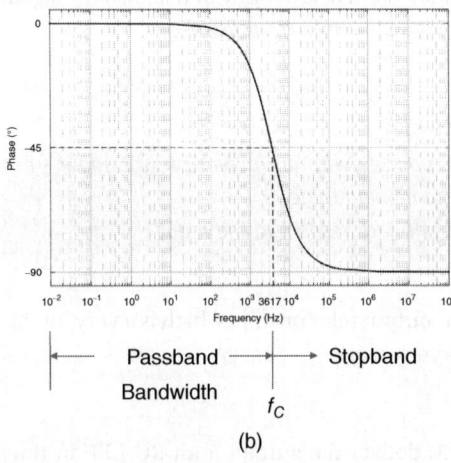

Figure 5.1.9 (a) Attenuation and (b) phase shift of the RC LPF given in Example 5.1.3.

and

$$V_{\text{out1}} = 14.14 \text{ V}$$

c) The phase of an output signal can be calculated by directly substituting the proper numbers into (5.1.10), $\Theta = -\tan^{-1}(R/X_C)$. Thus, at $f_1 = 3.617$ kHz, $R/X_C = 1.0$ and $\Theta_C = -45°$.

Discussion

Let us analyze the graphs A_v and Θ vs. frequency given in Figure 5.1.9.

First, note that they are built in semilogarithmic scale; that is, the frequency axis is shown in logarithmic scale and the amplitude (attenuation) and phase axes are built in natural scale. Bear in mind that the industry also builds such graphs in full logarithmic scale; that is, all axes – amplitude, phase, and frequency – are given in logarithmic scale.

Second, note that the course of these graphs follows exactly the course predicted by our preceding considerations (see Figure 5.1.5, for example). Specifically, the amplitude response, A_v, is equal to 1 at $f = 0$, which is equivalent to the condition $V_{\text{out}} = V_{\text{in}}$ at $f = 0$, and A_v goes to zero as f goes to infinity. The visual course of this graph (the graph's *shape*) depends on the scale of the frequency axis: If this axis is presented in semilogarithmic scale, which is common practice, then the graph exhibits a steeper transition between the low-frequency and high-frequency ranges; if this scale is natural, then the transition looks less steep. Also, the passband looks much greater when both graphs built in semilogarithmic scale. Be aware of these visual effects. Build the attenuation and phase response graphs with different frequency-axis scales and see how their shapes will change.

Third, observe that Figure 5.1.9 shows the correspondence between the given frequencies and the values of A_v and Θ calculated at these frequencies. It is important that you become comfortable with such a visual presentation. As an exercise, compute A_v and Θ at $f = 10\,\text{kHz}$. Then find these points in Figure 5.1.9 and see how close these are.

Fourth, let us now consider phase shift Θ in greater detail: This is *the phase shift that an output signal experiences with respect to an input signal*. Figure 5.1.10 shows this shift in both time domain and phasor domain. This angle changes with frequency as follows:

- When f goes to zero, the *phase shift*, Θ, goes to zero, according to the formula

$$\Theta = -\tan^{-1}(R/X_C) = -\tan^{-1}(2\pi f RC)$$

- When f goes to infinity, X_C goes to zero, the member (R/X_C) goes to infinity, and the phase shift, Θ, approaches $-90°$.

(**Questions**: Why $\Theta \Rightarrow 0$, when $f\,(\text{Hz}) \Rightarrow 0$? Why $\Theta \Rightarrow -90°$ when $f \Rightarrow \infty$? Give explicit answers.)

This course of Θ vs. f is clearly presented in Figure 5.1.10b. (For more on a phasor diagram, revisit Section 2.3.) We can also conclude that for an RC LPF, V_{out} lags V_{in} (or V_{in} leads V_{out}) when $f > 0$. This conclusion can be drawn from observation of Figure 5.1.10b.

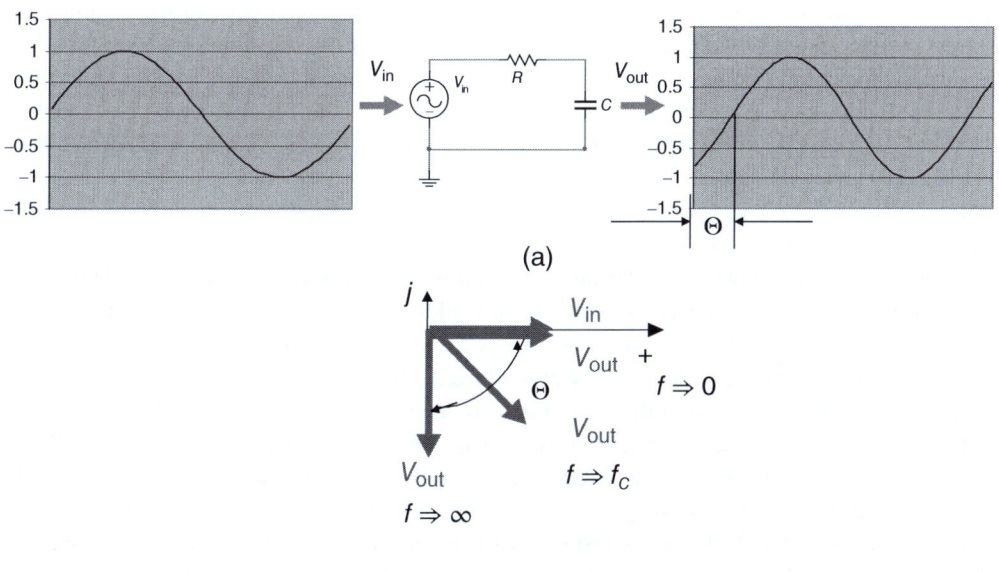

Figure 5.1.10 Phase shift in an RC low-pass filter: (a) phase shift in time domain and (b) phase shift in phasor domain.

Therefore, the real filter exhibits the smooth output-vs.-frequency response shown in Figure 5.1.2b and repeated in Figure 5.1.9, whereas the ideal filter shows the step-like output characteristic depicted in Figure 5.1.2a. Obviously, our goal is to design a filter whose output will be as close to that of the ideal filter as possible. We will return to a discussion of this challenge later in this chapter.

5.1.2.6 Cutoff (Critical) Frequency of an RC LPF

We have learned from the above considerations that A_v and Θ are two main characteristics of a LPF. Specifically, reviewing Figures 5.1.6, 5.1.8, and 5.1.9, we see from both time-domain and frequency-domain presentations that an RC LPF's output signal decreases in amplitude and attains additional phase shift with an increase in input frequency. But at what point does the LPF start to reject high-frequency signals? There is a *cutoff frequency*, f_C (Hz), which – as Figures 5.1.1 and 5.1.2 show that for *ideal filter* – determines the border between frequency bands where the signals finish to pass and start to be rejected.

A cutoff frequency is easy to determine for an ideal filter, but the real filter's A_v-vs.-f graphs presented in Figures 5.1.2b and 5.1.9 do not exhibit any visual cutoff point. This makes it difficult to detect a single frequency that would serve as a border between those signals that are allowed to pass and those that are rejected.

The solution to the problem of defining the cutoff frequency of a real filter lies in the following universally agreed-upon statement:

The frequency at which filter attenuation, A_v, becomes equal to 0.707 is called the cutoff frequency, f_C (Hz).

In other words, we agree that a real LPF starts rejecting high-frequency signals when the amplitudes of the output signals drop below 0.707 of the input-signal amplitude; that is, the condition $A_v \leq 0.707$ is equivalent to the requirement V_{out} (V) $\leq 0.707\, V_{in}$. We refer you again to Figure 5.1.6 to stress the fact that a LPF simply decreases the amplitudes of the unwanted (rejected) signals. Thus, below the cutoff frequency, the amplitude of the output signal becomes smaller than 0.707 of the input amplitude. For example, if $V_{in} = 10\,V$, then at the cutoff frequency, V_{out} is equal to 7.07 V, and just beyond f_C (Hz), V_{out} becomes equal to 7.0699 V and then continues to decrease as the frequency increases.

But why, you may ask, do we draw the line at some value of a filter's attenuation? The answer is that we do not have purely natural solutions helping us to define the *cutoff frequency of a real filter,* and so the only way we can determine the cutoff point is through an agreement. This is the approach scientists and engineers throughout the world have been taking for many years when they run up against similar situations; that is, they agree upon a definition for the troublesome term through a general consensus – an understanding, if you will.

Figure 5.1.9 visualizes the definition of a cutoff frequency. This figure shows that there is no natural border between signals that are allowed to pass, and those that are rejected by an RC LPF; this is why we need to introduce some kind of an artificial border. Consider the above example: If $V_{in} = 10\,V$, then $V_{out} = 7.07\,V$ at f_C (Hz). Is there a big or, in fact, any difference between signals with amplitudes equal to 0.707 000 1 V just before the f_C and equal to 0.706 999 9 V just beyond the f_C? Of course not. What is more, when making practical measurements, you often do not notice this difference if your instruments are not precise enough.

All these examples and considerations emphasize once more the artificial nature of our definition of the cutoff frequency for an RC LPF; however, engineers everywhere concur on this matter and therefore follows this definition. But why stick to $A_v = 0.707$? Why do not we choose any other number – for example, $A_v = 0.5$? The truth is that the number 0.707 is not chosen arbitrarily: You may remember that the rms voltage value, V_{rms}, is related to peak value, V_{pk}, as $V_{rms} = 0.707\, V_{pk}$; this relation is justified by the power relationship in ac circuits; hence, we use this value in other applications in the electrical engineering field.

How can we calculate the value of f_C? Just follow the definitions: First, attenuation, A_v, is defined as

$$A_v = \frac{1}{\sqrt{1 + \left(\frac{R}{X_C}\right)^2}} \tag{5.1.9R}$$

When $f = f_C$ (Hz), $A_v = 0.707$; that is,

$$A_v|_{f=f_C} = \frac{1}{\sqrt{1 + \left(\frac{R}{X_C}\right)^2}} = 0.707 = \frac{1}{\sqrt{2}} \tag{5.1.14}$$

Thus, $\sqrt{1 + \left(\frac{R}{X_C}\right)^2} = \sqrt{2}$ when f (Hz) $= f_C$, which means that R (Ω) must be equal to X_C (Ω), or

$$R\,(\Omega) = X_C\,(\Omega) \text{ for } f = f_C \text{ (Hz)} \tag{5.1.15}$$

Second, the definition of X_C is

$$X_C = \frac{1}{2\pi f C} \tag{5.1.1R}$$

which at f (Hz) $= f_C$ takes the form

$$R = \frac{1}{2\pi f_C C} \tag{5.1.16}$$

Therefore, the *cutoff frequency*, f_C (Hz), is determined by the following equation:

$$f_C \text{ (Hz)} = \frac{1}{2\pi RC} \tag{5.1.17}$$

Equation (5.1.17) shows that the cutoff frequency of an RC LPF is determined by its parameters, R (Ω) and C (F). We also need to remember that this formula defines the frequency at which $A_v = 0.707$.

Plugging the equality, $R = X_C$, into (5.1.10), $\Theta = -\tan^{-1}(R/X_C)$, we find that at $f = f_C$ (Hz) the phase shift induced by a filter is equal to $\Theta = -\tan^{-1}(R/X_C) = -\tan^{-1}(1) = -45°$. Therefore,

$$\Theta = -45° \quad \text{for } f = f_C \text{ (Hz)} \tag{5.1.18}$$

Conclusion
- At f (Hz) $= f_C$, $A_v = 0.707$, and $\Theta = -45°$.
- Three main parameters of a filter are its attenuation, A_v, phase shift, Θ (°), and cutoff frequency, f_C (Hz).

Example 5.1.4   Cutoff Frequency of an RC LPF

Problem

An RC LPF is built from $R = 1\,k\Omega$ and $C = 500\,pF$.

a) Determine the cutoff frequency, f_C (Hz)
b) Find the output voltage at f (Hz) $= f_C$ if $V_{in} = 20\,V$.

c) Calculate the phase shift at f (Hz) $= f_C$.
d) Illustrate the locations of A_v and Θ at f_C in Figure 5.1.9.

Solution

a) Using (5.1.17), we compute: f_C (Hz) $= \frac{1}{2\pi RC} = 318.3$ kHz.
b) Since $A_v = 0.707$ at f_C, the amplitude of an output signal is equal to

$$V_{\text{out}} = 0.707 \, V_{\text{in}} = 14.14 \text{ V}$$

c) When f (Hz) $= f_C$, $\Theta = -\tan^{-1}\left(\frac{R}{X_C}\right) = -\tan^{-1}(1) = -45°$.
d) The locations of A_v and Θ at f (Hz) $= f_C$ are shown in Figure 5.1.9.

The problem is solved.

Discussion

Simplicity of these manipulations should not divert us from the main point we can draw from this example: The cutoff frequency, f_C, of a low-pass RC filter can be determined by three interrelated conditions

- $A_v = 0.707$
- $\Theta = -45°$
- $R = X_C$.

Remember that these conditions hold true only for a series low-pass RC filter.

Now we want to show *the key role that the cutoff frequency plays in describing the operation of a low-pass RC filter*. To find A_v and Θ at different frequencies, we use (5.1.9), $A_v = \dfrac{1}{\sqrt{\left(1+\left(\frac{R}{X_C}\right)^2\right)}}$, and

(5.1.10), Θ (rad) $= -\tan^{-1}\left(\frac{R}{X_C}\right)$. However, as Examples 5.1.3 and 5.1.4 shows, calculating A_v and Θ using these formulas is a rather cumbersome process. More importantly, the role that frequency plays in changing both parameters is obscure because the frequency is not explicitly shown in these formulas. We can derive better expressions through the following manipulations:

Consider the formula for calculating a *cutoff* frequency (5.1.17),

$$f_C \text{ (Hz)} = \frac{1}{2\pi RC}$$

Let us divide both sides of this formula by a frequency, f. We obtain:

$$\frac{f_C}{f} = \frac{1}{2\pi f RC} \tag{5.1.19}$$

Recall that

$$\frac{1}{2\pi fC} = X_C$$

Therefore, (5.1.19) takes the form

$$\frac{f_C}{f} = \frac{X_C}{R}$$

Inverting this formula, we obtain

$$\frac{f}{f_C} = \frac{R}{X_C} \tag{5.1.20}$$

Plugging $\frac{f}{f_C}$ for $\frac{R}{X_C}$ into (5.1.9) and (5.1.10), we get

$$A_v = \frac{1}{\sqrt{1+\left(\frac{R}{X_C}\right)^2}} = \frac{1}{\sqrt{1+\left(\frac{f}{f_C}\right)^2}} \qquad (5.1.21)$$

and

$$\Theta \text{ (rad)} = -\tan^{-1}\left(\frac{R}{X_C}\right) = -\tan^{-1}\left(\frac{f}{f_C}\right) \qquad (5.1.22)$$

Equations (5.1.21) and (5.1.22) show explicitly the role of a current frequency, f, and the importance of the *cutoff* frequency, f_C, in shaping filter's attenuation and phase shift. These formulas, in conjunction with (5.1.17) for f_C, will be our working tools for calculating the filter's three main filter parameters.

(**Exercise**: Repeat the calculations in Examples 5.1.3 and 5.1.4 by using (5.1.21) and (5.1.22) and see how easier the computations can be done with new formulas.)

We introduced *cutoff frequency* early in this section as *the frequency separating passing and rejecting signals*. In addition, our discussion of an RC LPF, particularly Figure 5.1.9, has led us to conclude that

> the cutoff frequency is the border between the passband and the stopband of a filter.

Sidebar 5.1.S Filter's Characteristics in Absolute Values and in dB

The industry often refers to cutoff frequency at the other condition, namely $A_v = -3\,\text{dB}$. Remembering that

$$A_v \text{ (dB)} = 20 \log \frac{V_{out}(V)}{V_{in}(V)} \qquad (5.1.S.1)$$

we compute at $A_v = 0.707$:

$$A_v \text{ (dB)} = 20 \log \frac{V_{out}(V)}{V_{in}(V)} = 20 \log(0.707) = -3\,\text{dB}$$

In other words, *cutoff frequency criterion $A_v = -3\,dB$ in logarithmic scale is equivalent to $A_v = 0.707$ in natural scale*.

Today, the optical communications deliver a vast majority of telecommunication traffic. The communications where light is an information carrier measures the strength of its signals in watts rather than in volts. But what condition for the *cutoff frequency* should we use if considering the power, not the voltage, ratio of the output and input signals? It appears that *the criterion $P_{out}/P_{in} = -3\,dB$ corresponds to $P_{out}/P_{in} = 0.5$*.

Surely, the power attenuation, A_p, is given by

$$A_p \text{ (dB)} = 10 \log \frac{P_{out}(W)}{P_{in}(W)} \qquad (5.1.S.2a)$$

and, consequently,

$$A_p \text{ (dB)} = 10 \log(0.5) = -3\,\text{dB} \qquad (5.1.S.2b)$$

(Continued)

Sidebar 5.1.S (Continued)

Thus, $A_v = -3$ dB corresponds to $A_v = 0.707$ when we measure the voltages of the input and output signals, and $A_p = -3$ dB corresponds to $A_p = 0.5$ when we measure the power of these signals. The derivation of (5.1.2) is given in Example 5.1.S.1.

These formulas and explanations will help you describe a filter's amplitude response in both absolute units and in decibels.

Example 5.1.S.1   Measurement Values in Absolute Numbers and dB
Problem

a) At a certain frequency, the amplitudes of the input and output signals of an RC LPF are equal to 3.0 V and 0.3 V, respectively. What is their voltage ratio in dB? What will this ratio be in dB if $V_{in} = 3.0$ V and $V_{out} = 0.03$ V? If $V_{in} = 2$ V and $V_{out} = 0.1$ V?
b) Given signals with $V_{in} = 3.0$ V and $V_{out} = 0.3$ V, what is their power ratio in absolute numbers and in decibels?

Solution
a) Applying (5.1.S.1), we compute

$$A_v \text{ (dB)} = 20 \log \frac{V_{out}\text{ (V)}}{V_{in}\text{ (V)}} = 20 \log \frac{0.3 \text{ V}}{3.0 \text{ V}} = -20 \text{ dB}$$

For $V_{in} = 3$ V and $V_{out} = 0.03$ V, we obtain A_v (dB) $= -40$ dB.
Finally, when $V_{in} = 2$ V and $V_{out} = 0.1$ V, we calculate A_v (dB) $= -26.02$ dB.

b) By definition, the power of the input and output signals is equal to

$$P_{in}\text{ (W)} = (V_{in})^2/R_{in} \text{ and } P_{out}\text{ (W)} = (V_{out})^2/R_{out} \quad (5.1.S.3)$$

where R_{in} and R_{out} are the input and output resistances. Since in communications, we use *normalized power*, for which all resistances are set to $1\,\Omega$, we derive

$$\frac{P_{out}\text{ (W)}}{P_{in}\text{ (W)}} = \frac{(V_{out}(V/\Omega))^2}{(V_{in}(V/\Omega))^2} \quad (5.1.S.4)$$

Plugging in the given numbers, $V_{in} = 3.0$ V and $V_{out} = 0.3$ V, we calculate the power ratio in absolute numbers as

$$\frac{P_{out}\text{ (W)}}{P_{in}\text{ (W)}} = 0.01$$

Using (5.1.2), we compute the power ratio in dB:

$$A_p \text{ (dB)} = 10 \log \left(\frac{P_{out}\text{ (W)}}{P_{in}\text{ (W)}} \right) = -20 \text{ dB}$$

The problem is solved.

Discussion

- The solution to the problem regarding the ratios of the input and output signals teaches us that if the voltage ratio decreases ten times, the decibel value changes by -20 dB. Indeed, for the $\frac{V_{out}(V)}{V_{in}(V)} = 0.1$ ratio, we obtained -20 dB and for the $\frac{V_{out}(V)}{V_{in}(V)} = 0.01$ ratio, we computed

−40 dB. To fully understand this result, we recommend that you review the properties of logarithms. Also, this solution tells us that we can obtain any number in decibels for any arbitrary ratio of output and input voltages. (**Exercise**: Compute A_v (dB) if $V_{in} = 3.2$ V and $V_{out} = 0.048$ V.)

- The solution to the power-ratio problem leads us to an important conclusion:
Since the power ratio is given by (5.1.S.4), we can derive the following relationship between the power and voltage ratios in decibels:

$$10 \log \left(\frac{P_{out} (W)}{P_{in} (W)} \right) = 10 \log \frac{(V_{out} (V/\Omega))^2}{(V_{in} (V/\Omega))^2} \qquad (5.1.S.5)$$

Using the well-known property of a logarithm, $\log_m r^n = n \log_m r$, for every real number n, we can rewrite the right-hand side in this form:

$$\frac{(V_{out}(V/\Omega))^2}{(V_{in}(V/\Omega))^2} = 20 \log \frac{V_{out}(V)}{V_{in}(V)}$$

Therefore,

$$10 \log \left(\frac{P_{out} (W)}{P_{in}(W)} \right) = 20 \log \frac{V_{out} (V)}{V_{in} (V)} \qquad (5.1.S.6)$$

Formula (5.1.S.6) explains why we need to use the factor 10 when calculating the power ratio in decibels and why we need to use the factor 20 for the same operation when dealing with the voltage ratios. This equation also shows that when the power ratio is −20 dB, it means that the power decreases 100 times. Create your own example (e.g. taking $P_{in} = 3$ mW and assuming $P_{out} = 0.003$ mW) to verify this statement.

5.1.3 Filter Operation in Time Domain and Frequency Domain

5.1.3.1 Waveform Change and Frequency Response

We need to understand that in our explanation of a filter's operation, we necessarily must switch back and forth from a time-domain presentation to a frequency-domain presentation. (We refer you again to Section 2.2 to brush up on time domain and frequency domain and their interconnection.) Figures 5.1.6 and 5.1.8 demonstrate waveforms of the input and output signals for three different frequencies. Analyzing these waveforms, we can again conclude that, as the frequency increases, the output sinusoid experiences a decrease in amplitude and an additional phase shift; thus, these figures give you an idea of a LPF's operation in *time domain*. On the other hand, Figures 5.1.1, 5.1.2, 5.1.8, and 5.1.9 show, in frequency *domain*, that the filter's amplitude and phase responses decrease as the frequency increases. Clearly, all these figures depict the same process from two different perspectives. It is important that you feel secure working in both domains and are fully aware of the correspondence between these different presentations.

Figure 5.1.11 presents the operation of an LPF in both domains. This figure, in fact, is a combination of Figures 5.1.8 and 5.1.9. Waveforms – the plots of signal magnitude vs. time – show the filter operation in time domain; frequency response – amplitude (attenuation) and phase shift vs. frequency – shows the filter operation in frequency domain. Looking at the waveforms, we can again conclude that with the increase of frequency from 100 Hz to 1 kHz and 10 kHz the output signal deviates more and more significantly from the input signal. We can measure this deviation

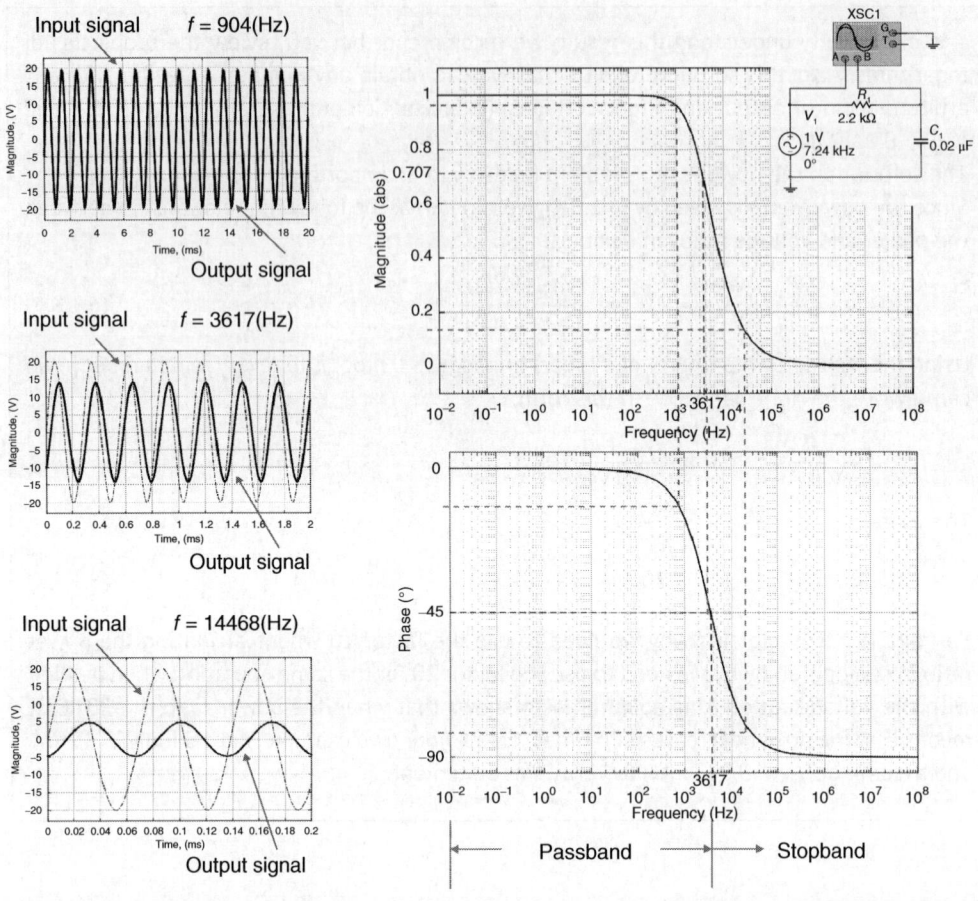

Figure 5.1.11 Waveforms (time-domain presentation) vs. attenuation and phase shift (frequency-domain presentation) of an RC LPF. (Not to scale.)

by the values of the output amplitude, V_{out}, and the phase shift, Θ, at these waveforms. If we then measure the values of attenuation, $A_v = V_{out}/V_{in}$, and phase shift, Θ, on the frequency-response graph at the same frequencies (100 Hz, 1 kHz, and 10 kHz), we will obtain the same numbers for $V_{out} = A_v V_{in}$ and Θ.

Bear in mind that the frequency-response graph is built in semilogarithmic scale. Calculated values of these parameters are given in Table 5.1.2 and the values measured in both time and frequency domains are shown in Figure 5.1.11.

Table 5.1.2 Attenuation and phase shift values for Figure 5.1.11.

$\dfrac{f(Hz)}{f_c(Hz)}$	$A_v = \dfrac{V_{out}(V)}{V_{in}(V)} = \dfrac{1}{\sqrt{1+\left(\frac{f}{f_c}\right)^2}}$	$\Theta\,(°) = -\tan^{-1}\left(\dfrac{f}{f_c}\right)$
0.1/3.62	0.99	$-1.6°$
1/3.62	0.96	$-15.4°$
10/3.62	0.34	$-70.1°$

This exercise should further enhance your understanding of the relationship between time domain and frequency domain.

5.1.3.2 Bandwidth of an RC LPF

Introducing the cutoff frequency of a low-pass RC filter allows us, at this point, to become more familiar with another important parameter of this filter – bandwidth. We discuss bandwidth in Chapters 2 and 3; you will recall that one of the bandwidth interpretations states:

Bandwidth is the range of frequencies within which a device can acceptably operate.

To apply this definition to an RC LPF, we need to remember that the function of this device is to filter certain signals; specifically, it has to allow the passage of low-frequency signals. Therefore, the *bandwidth of a LPF is the range of frequencies within which it passes signals*. But this range is determined by the cutoff frequency, f_C (Hz). Thus, we conclude

The bandwidth, BW (Hz), of a LPF is the range of frequencies from 0 Hz to f_C Hz.

This frequency band is also called the *passband*; logically, the band of frequencies beyond f_C is called the *stopband*. The bandwidth of a LPF is shown in Figure 5.1.9.

Example 5.1.5 Bandwidth of an RC LPF

Problem

Consider an RC LPF with $R = 2.2\,\text{k}\Omega$ and $C = 0.002\,\mu\text{F}$. What is its bandwidth?

Solution

Since the bandwidth of a LPF is BW (Hz) = (f_C Hz – 0 Hz), we need only compute the cutoff frequency. Applying (5.1.17), we compute:

$$f_C\,(\text{Hz}) = \frac{1}{2\pi RC} \approx 36.2\,\text{kHz}$$

Hence,

$$\text{BW}_{\text{LPF}} = 36.2\,\text{kHz}$$

Discussion

- We certainly remember that a HPF allows passage of all signals from f_C to f_{max}. (See Figure 5.1.1b.) Therefore, the bandwidth of such an HPF is BW_{HPF} (Hz) = $f_{\text{max}} - f_C$. Analyzing Figure 5.1.1c, we can easily conclude that the bandwidth of a BPF is given by BW_{BPF} (Hz) = $f_{C2} - f_{C1}$. These formulas will help to compute the bandwidth of both HPF and BPF filters. What about the bandwidth of a BSF? Now you have acquired enough knowledge to derive the formula of its bandwidth. Do so and create examples for the bandwidths of all types of filters.
- What bandwidth – big or small – do we need? Clearly, it depends on a filter application. If we want to use this RC LPF to clear a voice signal, 3.62 kHz is the adequate bandwidth value. If, however, we work with a music channel, we need a bandwidth more than 20 kHz.
- You certainly notice that a filter bandwidth simply coincides with its passband.

5.1.3.3 Characterization of an RC LPF

The cutoff frequency, conceptually, is the most important parameter of a filter because it is the only quantity we need to know to characterize a specific filter, that is, to find specific values of its amplitude and phase responses at any frequency from 0 to f_{max}. Equations (5.1.21) and (5.1.22), along with Formula (5.1.17) for the cutoff frequency, are proof of this statement. When the parameters

R and C of a low-pass RC filter are given – and they are the only parameters we need to know to build a filter in reality – we can easily calculate the main characteristics of the filter: f_C, A_v, and Θ. Example 5.1.6 illustrates this point.

Example 5.1.6 Characterization of a Low-Pass RC Filter

Problem

For the filter, whose circuit is shown in Figure 5.1.7 and whose $R = 2.2\,\text{k}\Omega$ and $C = 0.02\,\mu\text{F}$, do the following:

a) Determine the cutoff frequency.
b) Find A_v and Θ at $0.001f_C$, $0.01f_C$, $0.1f_C$, f_C, $10f_C$, and $100f_C$.
c) Build the graph A_v vs. frequency and show all the above values.
d) Build the graph Θ_{out} (°) vs. frequency and show the values of Θ at $0.001f_C$, $0.01f_C$, $0.1f_C$, f_C, $10f_C$, and $100f_C$.
e) Show the filter passband, stopband, and bandwidth on these graphs.

Solution

a) The cutoff frequency can be found by using (5.1.17), $f_C\,(\text{Hz}) = \frac{1}{2\pi RC}$, as $f_C = 3.62\,\text{kHz}$.
b) To find A_v and Θ_{out} (°) at different frequencies, we use (5.1.21) and (5.1.22), plug in the calculated f_C and the given frequencies, and compute the required quantities. The results of these computations are presented in Table 5.1.3, where calculations at $0.5f_C$, $2f_C$, $10\,000f_C$, $10^8 f_C$, $10^9 f_C$, and $10^{10} f_C$ are added to better visualize the course of the A_v and Θ graphs.
c) The graphs of A_v and Θ vs. frequency with the required values at given frequencies along with passband, stopband, and bandwidth, are shown in Figure 5.1.12.

(**Question**: If you need to instantaneously verify whether the calculations presented in Table 5.1.3 are correct, what line would you choose for this check?)

Table 5.1.3 Solution to Example 5.1.6.

f (Hz)	$\dfrac{f(\text{Hz})}{f_C(\text{Hz})}$	$A_v = \dfrac{1}{\sqrt{1+\left(\frac{f}{f_C}\right)^2}}$	$\Theta\,(°) = -\tan^{-1}\left(\dfrac{f}{f_C}\right)$
3.62	0.001	0.999 999 499	−0.057
36.2	0.01	0.999 949 925	−0.57
362	0.1	0.995 029 446	−5.7
1 810	0.5	0.894 286 609	−26.56
3 620	1	0.706 829 029	−45
7 240	2	0.446 932 63	−65.78
36 200	10	0.099 426 367	−84.3
362 000	100	0.009 991 65	−89.43
3 620 000	1 000	0.000 999 214	−89.94
36 200 000	10 000	9.99215E − 05	−89.99
3.62E + 08	10^8	9.99215E − 06	−89.999
3.62E + 09	10^9	9.99215E − 07	−89.9999
3.62E + 10	10^{10}	9.99215E − 08	−89.99999

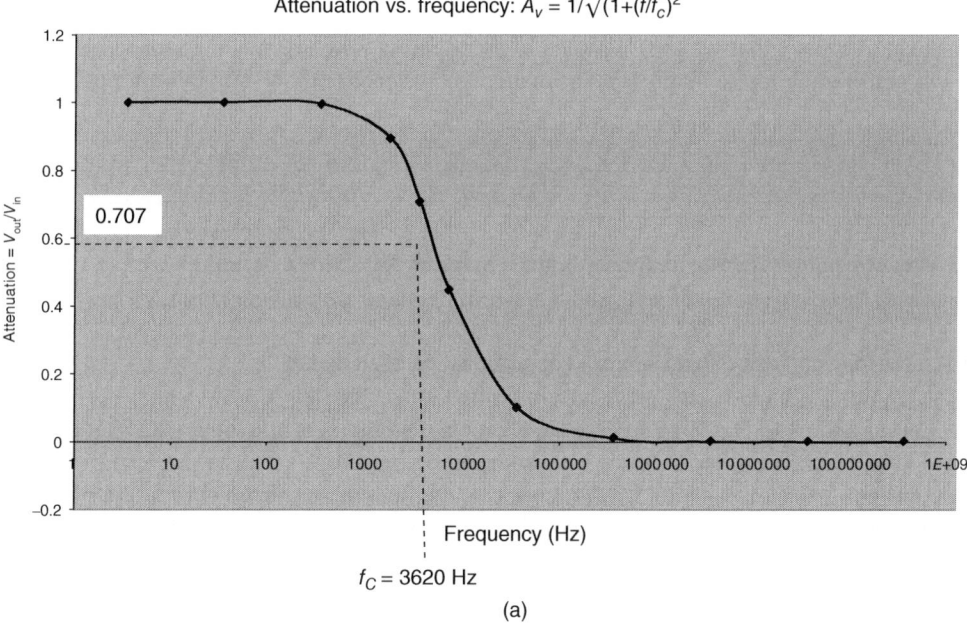

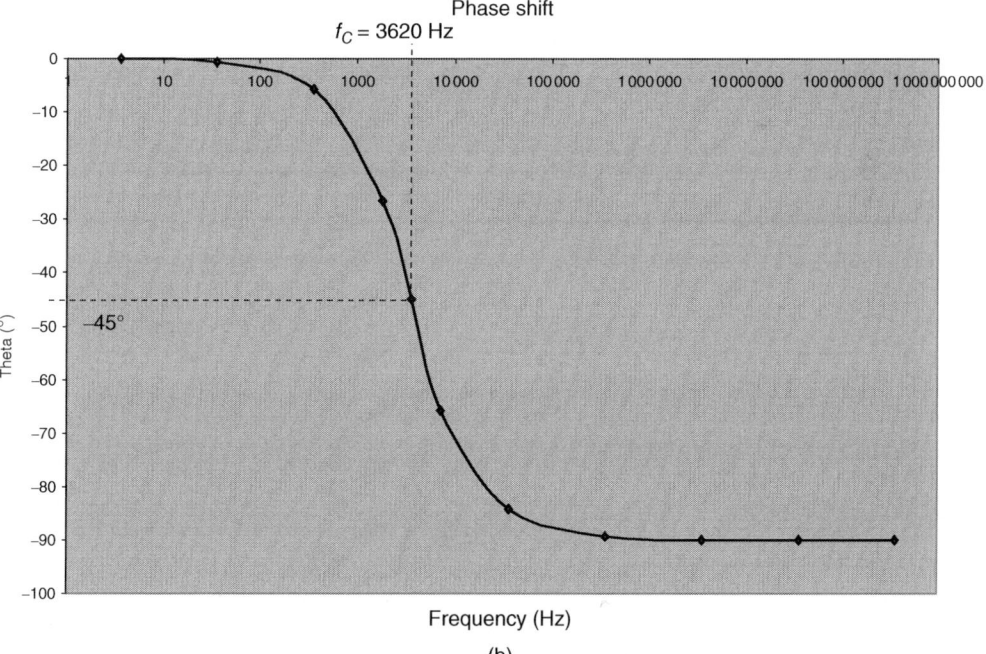

Figure 5.1.12 (a) Attenuation, A_v, and (b) phase shift, Θ, of an RC low-pass filter vs. frequency in Example 5.1.6.

Discussion

You realize, of course, that we are interested in ascertaining how two major characteristics of a filter – A_v and Θ – change vs. frequency. The main point we need to draw from this example is that

> *the evaluation of the operation of a low-pass filter should be made not with respect to absolute frequency but with respect to the deviation of the current frequency from the <u>cutoff</u> frequency.*

This is why in Table 5.1.3 the frequency is given in terms of fractions, or multiples of f_C. Become familiar with this concept; it will help you to better understand a filter's operation.

5.1.3.4 The Role of *R* and *C* Parameters in Characterization of an RC LPF

Suppose we want to build a filter with certain desired characteristics. Can we change the shape of the A_v-vs.-f and Θ-vs.-f graphs? In general, what characteristics of a low-pass RC filter can we change? Remember, this filter has only two parameters – R and C – that we can manipulate to achieve our goal. As (5.1.21) and (5.1.22) show, the attenuation and phase shift of the filter are determined by the cutoff frequency, f_C, which, in turn, is determined by R and C according to (5.1.17). Hence, by changing R and C, we will change the cutoff frequency and, as a result of this modification, we can adjust A_v and Θ. Figures 5.1.13 and 5.1.14 show how changing R and C will affect f_C and A_v. The following is a detailed discussion of these changes.

How will the filter's attenuation, A_v, change if we vary the resistance? The answer is given by (5.1.17), $f_C = 1/(2\pi RC)$, and by (5.1.21), $A_v = 1/\sqrt{(1 + (f/f_C)^2)}$: The greater the R, the smaller the f_C. For example, consider an RC LPF with $C = 0.002\,\mu\text{F}$ and let the resistance take values $R_1 = 220\,\Omega$, $R_2 = 2200\,\Omega$, and $R_3 = 22000\,\Omega$. Then its cutoff frequencies will be equal to

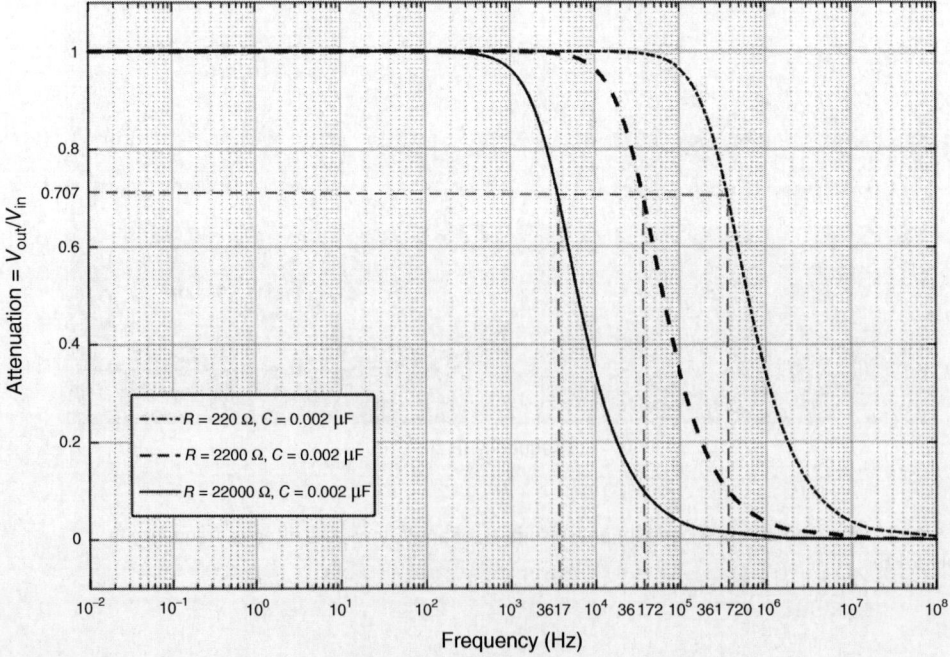

Figure 5.1.13 Changing f_C and A_v by changing R in RC LPF.

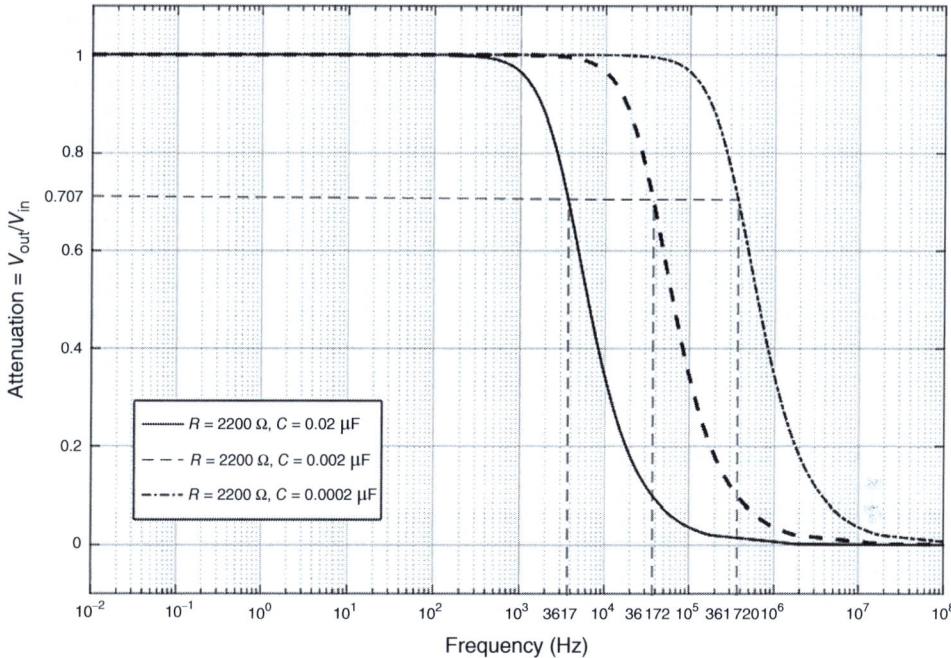

Figure 5.1.14 Changing f_C and A_v by modifying C in RC LPF.

$f_{C1} = 1/(2\pi R_1 C) = 361.7\,\text{kHz}$, $f_{C2} = 1/(2\pi R_2 C) = 36.17\,\text{kHz}$, and $f_{C3} = 1/(2\pi R_3 C) = 3.62\,\text{kHz}$, respectively. The graphs of attenuation for these values of R shown in Figure 5.1.13 confirm our reasoning. These graphs also demonstrate that the shape of attenuation changes very little within the filter passband, but the value of this band changes with a change in R and, therefore, in f_C.

How will the attenuation, A_v, change if we vary capacitance? The answer is given by the same formula: $f_C = 1/(2\pi RC)$: The smaller the C, the greater the f_C and, of course, the greater the filter's bandwidth. Specifically, if the capacitor of our LPF with $R = 2.2\,\text{k}\Omega$ takes the values $C_1 = 0.0002\,\mu\text{F}$, $C_2 = 0.002\,\mu\text{F}$, and $C_3 = 0.02\,\mu\text{F}$, then its cutoff frequencies will be equal to $f_{C1} = 1/(2\pi RC_1) = 361.7\,\text{kHz}$, $f_{C2} = 1/(2\pi RC_2) = 36.17\,\text{kHz}$, and $f_{C3} = 1/(2\pi RC_3) = 3.62\,\text{kHz}$. The graphs of attenuation for these values of C shown in Figure 5.1.14 confirm our reasoning.

(**Exercise**: For the same set of R and C values, build the graphs Θ-vs.-f of the given RC LPF. Comment on your findings.)

Figures 5.1.13 and 5.1.14, along with the calculations, help in understanding our ability to design an RC LPF with the desired characteristics. The main point we have to draw from this exercise is that *changing R or C can only change the value of the cutoff frequency; it cannot change the shape of the amplitude response in such a low-pass RC filter*. (We leave it for you to prove the similar statement concerning the phase response.) Therefore, if we need to increase the steepness of the A_v-vs.-f graph to make it more like an ideal filter characteristic, we cannot do it for this simple filter. We pursue this issue in greater detail in Sections 5.2, 5.3, and 5.4.

Two important points to emphasize in regard to the filter's design: First, it is important to realize that, in this design, we meet some uncertainties. When f_C is given (and this is the main requirement we need to meet in designing a filter), we have to compute R and C. Mathematically speaking, we have one equation, $f_C = 1/(2\pi RC)$, and two unknown variables, R and C. To solve the problem, we have to start by choosing one parameter (R, for instance) and compute the other. In addition, our task as designers is to choose reasonable values of R and C, ones that will support optimal operation of our circuit. But by what criterion should we choose this starting parameter? Well, we have to be reasonable and understand that we want to choose the smaller C at the expense of the greater R to make our design practical. This is so because a greater value of C will result in a bulky and expensive element. For example, if $f_C = 3456$ Hz, then in applying the main formula, $2\pi f_C = 1/RC$, we will find 21.71×10^3 rad $= 1/RC$. We can choose $R = 1\,\Omega$ and $C = 46.05 \times 10^{-6}$ F (46.05 μF) or $R = 1$ kΩ and $C = 46.05 \times 10^{-9}$ F (46.05 nF). Both sets give us the correct f_C, but we certainly want to choose a smaller capacitor because it will make the operation of the electrical circuit more efficient. And, of course, we have to choose *standard values* of R and C and recalculate f_C. In the above example, if we initially choose $R = 1$ kΩ and $C = 46.05$ nF, we have to change these figures to the closest standard values, which are $R = 1$ kΩ and $C = 47$ nf. Our modified cutoff frequency will be $f_C = 3386$ Hz. We have to use this f_C value in our future work with the filter we have designed.

5.1.4 General Filter Specifications

5.1.4.1 Amplitude Specifications

Our discussion of an RC LPF demonstrate that we need amplitude and phase characteristics to describe the entire filter operation. But the graphs of A_v-vs.-f and Θ-vs.-f shown in this section so far demonstrate that for every frequency value we have unique corresponding values of A_v and Θ. This is, obviously an ideal situation; in reality, every measurement will result in a close to the preceding but still new values of A_v and Θ. This uncertainty is demonstrated in industrial A_v specification shown in Figure 5.1.15.

Therefore, the characteristics of a real filter must reflect this situation. And indeed, industrial filter specifications take into account these uncertainties by giving some margins in defining amplitude and phase values of a filter. In this subsection, we consider the general filter specifications accepted by the industry.

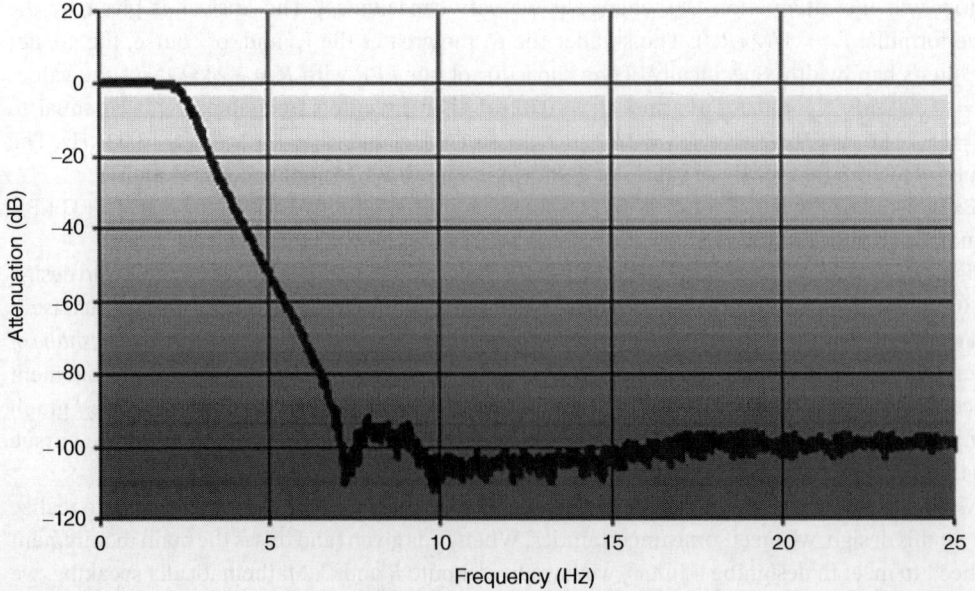

Figure 5.1.15 The example of an industrial attenuation specification of a low-pass filter. Low frequency 2.5 Hz filter response, 100 dB filter rejection. Source: Reprinted with permission of www.kemo.com.

The problem with the specifications is that the amplitude responses of real filters can take forms like that shown in Figure 5.1.16a: They might have ripples in the passband or stopband or both and transition from on band to another is very far from vertical line. How can we specify such responses? Bear in mind that we also want the characteristics of our filter to be as close as possible to the ideal version shown in Figure 5.1.16b. To resolve this dilemma, the industry develops the specifications that embrace both requirements. Such an attenuation specification of an LPF is shown in Figure 5.1.16c.

Consider Figure 5.1.16, where all the new parameters of the amplitude response are shown (Sedra and Smith 2004, p. 1086). First, Figure 5.1.16a demonstrates new features that the real filters exhibit; it is clear that we need to introduce new specifications to describe these new features. Second, we remind you that that an amplitude of an ideal filter is specified by only two parameters: The value of the amplitude, $A_v = 1$, and the cutoff (critical or corner) frequency, f_C; the former is the filter's attenuation and the latter is the border frequency that separates two frequency bands – the passband and the stopband (see Figure 5.1.16b). Third, Figure 5.1.16c illustrates how the attenuation of a real filter can be inscribed into ideal-like specifications producing a new approach to specify an LPF. Finally, Figure 5.1.16d shows the amplitude specifications applied to the amplitude response of a real LPF and accepted by the industrial and academic communities. These new specifications can be explained as follows:

- *The amplitude of the ripples*, A_P (also called passband amplitude, A_p, or A_{max}), characterizes the value of the *maximum allowed variations in the passband*. This quantity can be measured either

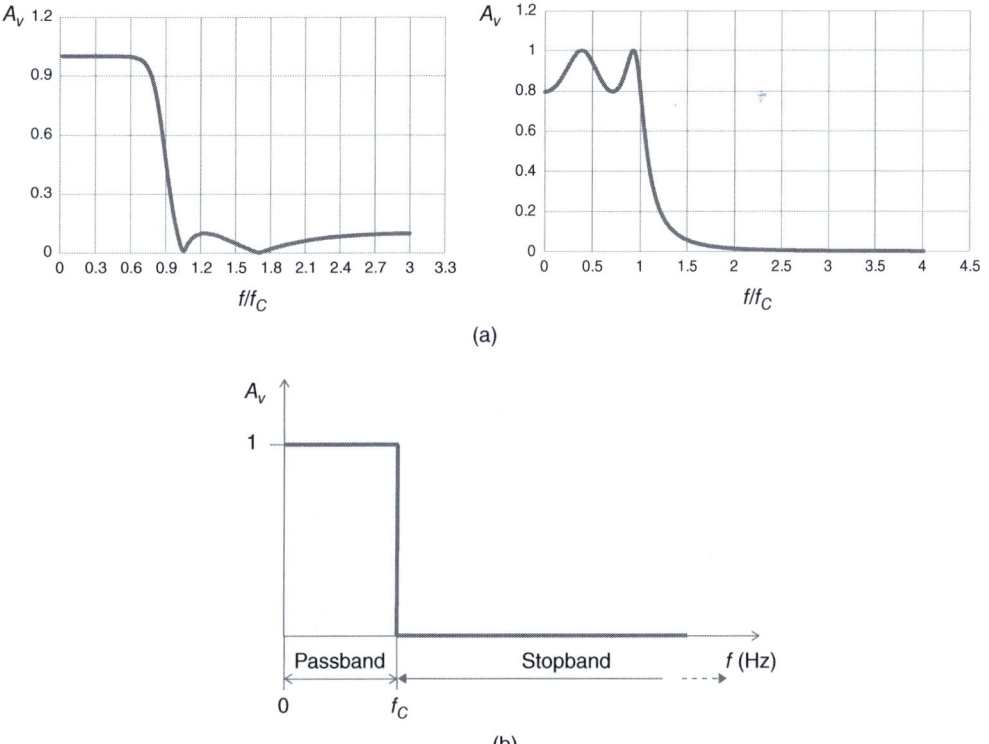

Figure 5.1.16 Specifications of the amplitude response of a LPF: (a) Examples of real amplitude responses; (b) the amplitude response of an ideal filter; (c) bands and characteristics of a real filter; and (d) specifications of the amplitude response of a real filter.

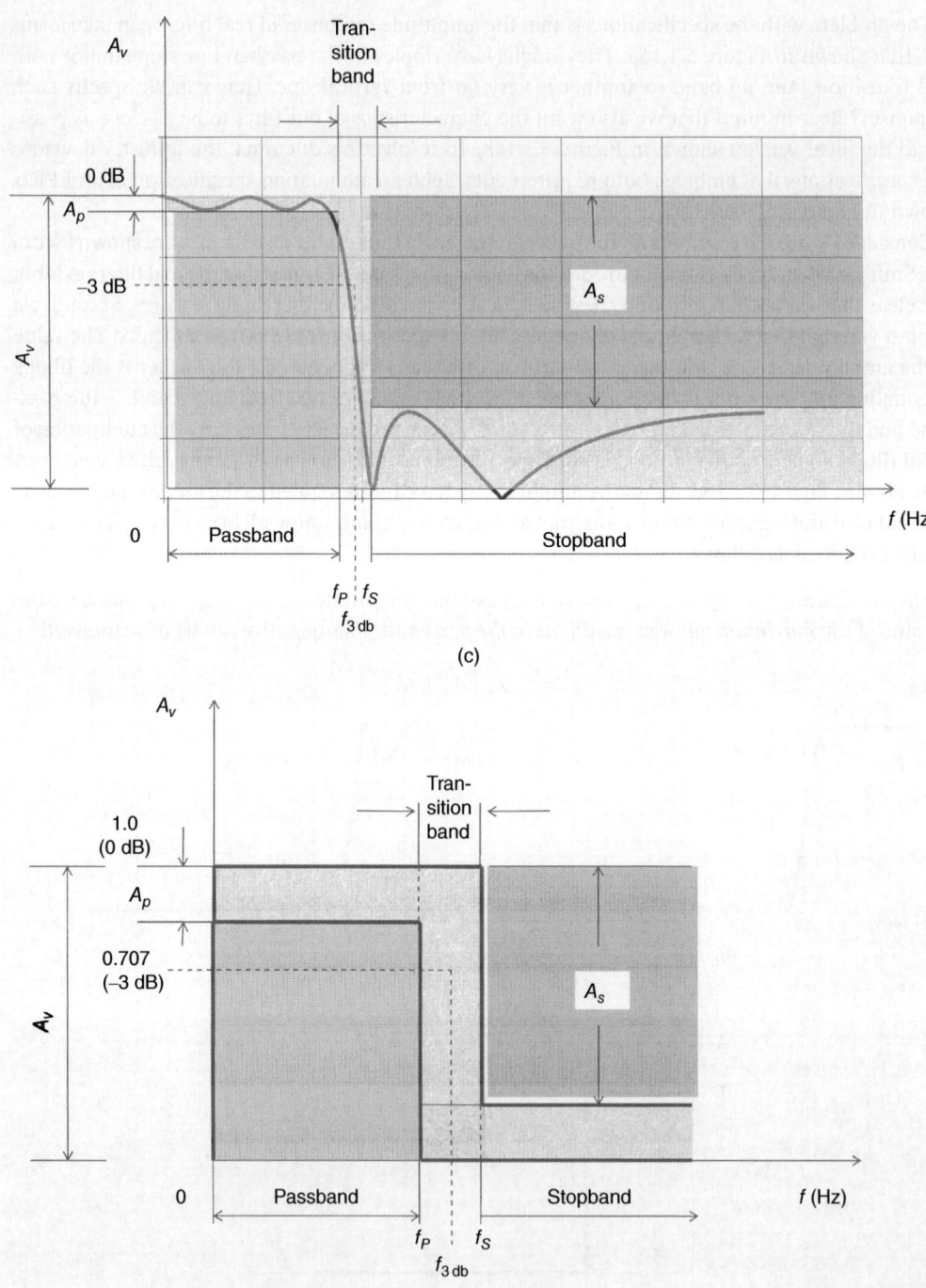

Figure 5.1.16 (*Continued*)

in absolute numbers or in decibels. It may vary from *zero*, which is an *ideal value* describing the flat amplitude response within a passband, to a significant percentage of the total amplitude, A_v, whose upper limit is equal to 1 (0 dB) for a passive filter. The importance of this measure is twofold: First, A_p delimits the passband because its intersection with the graph A_v-vs.-f determines the end frequency of a passband, f_p (Hz). Second, ripples are the price we pay for improving some parameters of the amplitude response; hence, A_p is the value of this price.
- *The stopband amplitude A_S* (also called A_{min}), specifies the value of the *minimum required attenuation in the stopband, that is, the stopband attenuation*. This is one of the most important characteristics a designer needs to know; it determines how much a filter suppresses the unwanted signals. *The stopband amplitude, A_s, is simply the filter's attenuation, A_v, measured in the stopband; i.e. A_s (dB) $= 20 \log \left(\frac{V_{out}(V)}{V_{in}(V)} \right)$ for $f \in$ stopband*. This means that ideally its value in decibels should go to minus infinity because V_{out} should be zero in the stopband. Typically, this amplitude may vary from -20 to -100 dB; it is often specified as a positive number, as shown in Figure 5.1.15.
- *The passband (edge) frequency, f_p (Hz)*, as Figure 5.1.16c,d show, is defined by the intersection of the low-limit line of the amplitude, A_p, and the graph of the amplitude response. This frequency delimits the end of a passband region for some filters, which means that f_P, not f_C, serves as the cutoff frequency for these filters. The range of frequencies from 0 to f_P is specified as *a ripple bandwidth*; this is why the f_P frequency is sometimes called *ripple frequency, f_r* (Hz).
- *The stopband (edge) frequency, f_S (Hz)*, as Figures 5.1.16c and 5.1.16d show, is defined by the intersection of the low-limit line of the stopband amplitude, A_S, and the graph of the amplitude response. This frequency marks the start of the stopband region. For some types of filters, this frequency serves as the cutoff frequency.
- *The cutoff (critical or corner) frequency, f_C (Hz)*, defined in the previous Subsections 5.1.2.6 and 5.1.3.3, does not serve here as a main parameter characterizing the frequency-selective properties of filters. Instead, we use two new characteristic frequencies: Passband edge frequency, f_P, and stopband edge frequency, f_S. The traditional cutoff frequency, f_C, defined as the frequency at which the filter attenuation, A_v, becomes equal to 0.707 (or -3 dB), is located within a transition band. The industry often denotes this frequency as the $\omega_{3\,dB}$ frequency (see Sidebar 5.1.S). Note, too, that the industry often omits the negative sign in decibel designations.
- We have several specifications by which to characterize **a transition band**:
  o One is the *width of the transition band, f_{TB} (Hz)*, which is equal to

  $$f_{TB} \text{ (Hz)} = f_s - f_P \qquad (5.1.23)$$

  Quite obviously, the narrower this band, the closer this response is to an ideal version; in other words, the smaller the transition width, the steeper the amplitude response.
  o Another measure (this time dimensionless) that characterizes the width of this band is the *selectivity factor, f_{SF}*, which is defined as

  $$f_{SF} = f_S / f_P \qquad (5.1.24)$$

  The selectivity factor is one of the major parameters that set the requirements for a filter design.
  o Still another parameter, the *shape factor, f_{ShF}*, is used to characterize the transition-band width:

  $$f_{ShF} = f_S / f_C \qquad (5.1.25)$$

Clearly, the more tightly we specify these main characteristics (that is, the smaller the A_P, the greater the A_S, and the closer f_{SF} is to unity), the closer our filter will approach the ideal state. What the price will be to realize such a close-to-ideal filter is another matter.

Finally, here is an excellent **exercise** for you: Apply all the introduced specifications to the industrial graph given in Figure 5.1.16 and find out all the required numbers. To start, determine to what specifications a manufacturer refers by saying, "Low frequency 2.5 Hz filter response" and "100 dB filter rejection." (Observe that the manufacturer uses its own terms for the specifications we introduced above. This is another indication of the unfortunate absence of standard terminology in our field.)

The specifications we have discussed above apply to a LPF. It is quite straightforward to transfer these specifications to a HPF: Just change the locations of the passband and the stopband and all associated parameters.

To specify *BPF and BSFs*, however, we need to delve deeper into consideration of these filter types, which lies outside of the scope of this book.

Now, we will consider an example that introduces the method of computations of the introduced specifications and the order of their values.

Example 5.1.7 Finding the Specifications of an LPF

Problem

Consider a LPF with the following relationship between attenuation and the characteristic frequencies:

Attenuation (dB)	Frequency (units of f_P)
−0.05	1.00
−3.00	1.05
−80.0	1.56

Assuming that $A_S = -80$ dB, find the width of the transition band, the selectivity factor, and the shape factor of this filter.

Solution

The given data should be interpreted as follows: This LPF has its ripple edge frequency, f_P, at the minimum of the amplitude of the ripples: $A_{Pmin} = 0.05$ dB. When $A_v = -3$ dB, the cutoff frequency is given by $f_C \equiv f_{3\,dB} = 1.05 f_P$. When $A_v = -80$ dB, the stopband frequency is equal to $f_S = 1.56 f_P$. Therefore, applying (5.1.23), (5.1.24), and (5.1.25), we find

The *width of the transition band*, f_{TB}:

$$f_{TB} \text{ (Hz)} = f_s - f_P = 0.56 f_P$$

The *selectivity factor*, f_{SF}:

$$f_{SF} = f_S/f_P = 1.56$$

The *shape factor*, f_{ShF}:

$$f_{ShF} = f_S/f_C = 1.56 f_P/1.05 f_P = 1.486$$

Discussion

- It is always a good idea to visualize your operations. In this case, we recommend that you qualitatively draw an amplitude response of a LPF similar to that shown in Figure 5.1.16d and mark the f_P, f_{3dB}, and f_S frequencies in accord with Figure 5.1.16c. This exercise will help you understand the calculations performed in this example.
- As for specific solutions, note carefully the following points: We want a narrow transition band, meaning that, ideally, f_{TB} (Hz), should approach zero (which means that all the frequencies, f_P, f_{3dB}, and f_S, should be equal). In our case, we have $f_{TB} = 0.56 f_P$, which is far from zero; however, to appreciate this value, we need to know that typical values of f_{TB} for other filter types are much greater. Similarly, an ideal selectivity factor should be equal to 1. Our LPF filter is 1.56; other filter types exhibit greater values of this parameter.

5.1.4.2 Phase Specifications

We have introduced the specifications of a filter's amplitude response. Now, we need to take a deeper look into the specifications of the *phase response*.

In (2.1.12) and (2.2.4) the relationship between the phase shift, Θ (rad), and the time delay, t_d (s), of a sinusoidal signal was introduced as

$$\Theta \text{ (rad)} = -2\pi f t_d = -\omega t_d \tag{2.2.4R}$$

or

$$t_d \text{ (s)} = -\frac{\theta}{\omega} \tag{5.1.26}$$

It is now necessary to clarify that the time delay defined by (5.1.26) is called the **time phase delay**, t_d **(s)**, because it relates the phase shift to the delay only for *a single sinusoid*. Thus, if we submit to an ideal filter a single sinusoidal signal, the filter introduces time delay, t_d (s), which will appear as the phase shift, Θ (rad), in an output signal.

But what if the input signal is a composition of many sinusoids? To deal with this situation, we need to introduce a more accurate measure of filter time delay, a measure called **time group delay**, T_D **(s)**:

$$T_D \text{ (s)} = -\frac{d\theta}{d\omega} \tag{5.1.27}$$

The *group delay*, T_D (s), of an ideal filter must be the same for all the sinusoids of the input signal; that is, it must be a constant. Then, it follows from (5.1.26) that the phase shift must be proportional to the frequency. To be sure, if

$$\Theta \text{ (rad)} = k\omega$$

then the differentiation of the phase shift over the frequency results in the constancy of T_D (s),

$$\frac{d\theta}{d\omega} = k = -T_D \text{ (s)}$$

as required by (5.1.26). Comparing (5.1.26) and (5.1.27), we conclude that, *for an ideal filter, phase delay, t_d (s), and group delay, T_D (s), are constant and equal*.

Note that in general, the time phase delay is the phase delay of an individual sinusoid of a composite signal and the time group delay is the delay of the envelope of this signal.

Since T_D (s) is the derivative of phase with respect to frequency, it shows how much (if any) a phase changes with a change of frequency; in this regard, group delay shows how much the phase of the output signal changes – that is, how much the output signal gets distorted due to the difference in frequencies of the input composite signal.

The industry often introduces *normalized group delay*, T_{ND} (s), which is a one second group delay at 1-Hz cutoff frequency. Knowing T_{ND} (s), we can compute the actual filter's group delay, T_{AD}, as

$$T_{AD} \text{ (s/Hz)} = \frac{T_{ND}}{\text{Actual cutoff frequency}} \tag{5.1.28}$$

Here, the actual cutoff frequency is f_P, or f_S, or f_C, depending on the frequency characterizing the type of filter in question. For example, if $T_{ND} = 1$ second and the actual cutoff frequency equals 1 kHz, then the actual group delay is equal to

$$T_{AD} = \frac{T_{ND}}{\text{Actual cutoff frequency}} = 0.001 \text{ s/Hz}$$

As you can see, the phase specifications are simpler than the amplitude specifications.

Important note on filter specifications: There is a difference between designations of the same specifications – especially amplitudes A_p and A_s – in MATLAB and in technical documentation. Be aware of that difference and always closely observe the designations used in every specific document.

Questions and Problems for Section 5.1

- Questions marked with an asterisk require a systematic approach to finding the solution.
- Many questions and problems, including those marked with an asterisk, imply that you, in addition to reading the textbook, will do your research to find the answers. Consider such questions as miniprojects.

Filtering: What and Why

1 Explain the meaning of the term "filtering." Give examples of filtering you meet in everyday life.

2 By what parameter of a signal do the filters select signals in modern communications? Explain.

3 *Sketch a figure to illustrate the concept of ideal low-pass, high-pass, band-pass, and band-stop filters.

4 Consider the cutoff frequency of a filter:
   a) Give its definition(s).
   b) Explain its importance to the filter's characterization.
   c) Give examples of cutoff frequency (or frequencies) for low-pass, high-pass, band-pass, and band-stop filters.

5 *What frequency ranges are called passband and stopband?

6 *Show the similarities and difference between the characteristics of an ideal LPF and a real LPF.

RC Low-Pass Filter (LPF)

7 *Sketch the schematic of a low-pass RC filter and explain the function of each circuit's component.

8 *Consider the frequency response of three main passive circuit components:
   a) Sketch the graph of X_C (Ω) vs. frequency and explain its course. If $C = 1$ mF, what is X_C at $f = 1$ MHz?
   b) Sketch the graph of R (Ω) vs. frequency and explain its course. If $R = 1$ kΩ, what is R at $f = 1$ MHz?
   c) Sketch the graph of X_L (Ω) vs. frequency and explain its course. If $L = 1$ mH, what is X_L at $f = 1$ MHz?

9 *Properties of reactive components:
   a) Consider two inductors, one with $L_1 = 0.01$ mH and the other with $L_2 = 0.1$ mH. Which one's reactance will be greater at frequency $f = 20$ kHz? Prove your statement.
   b) Consider two capacitors, one with $C_1 = 0.02$ μF and the other with $C_2 = 2$ nF. Which one's reactance will be greater at frequency $f = 20$ kHz? Prove your statement.

10 *Consider the principle of operation of an RC LPF based on consideration of the equivalent circuits of R and C:
   a) Build equivalent circuits of a resistor having $R = 1$ kΩ and a capacitor having $C = 1$ μF at 100 Hz and at 100 MHz. Compute resistances and capacitive reactances at these frequencies to support your reasoning.
   b) Explain the principle of operation of an RC low-pass filter (LPF) based on the equivalent-circuit model.

11 *Sketch a circuit of a low-pass RC filter:
   a) Why do we need to use a capacitor in this circuit? Explain.
   b) Is the order of the component placement important? Can we change the order?
   c) Why is the RC filter shown in Figure 5.1.3 called "low-pass?" Explain.
   d) Explain how this circuit filters (selects) the signals.

12 *Qualitatively sketch the graphs of amplitude (magnitude) and the phase frequency responses of an RC LPF and explain which parameter of an input signal causes the changes in these graphs.

13 *For a low-pass RC filter, sketch the waveforms of the output signal at the low-, intermediate-, and high-frequency input signals, provided that the amplitude of an input signal remains fixed at any frequency.

14 *Compare the output signal of an RC LPF with the input sinusoidal signal:
   a) What will change and what will be the same?
   b) How will these changes occur with the change of the input frequency – smoothly or abruptly? Explain your reasoning.

5 Filters

15 *Consider the RC LPF shown in Figure 5.1.3:
Derive the formulas for the filter's attenuation, $A_v = V_{out}/V_{in}$, and the output phase, Θ, expressed through the input signal and the circuit's parameters, R and C.
Derive the formula for a cutoff frequency, f_C (Hz).

16 *Derive the formulas $A_v = V_{out}/V_{in}$, and Θ for an RC LPF expressed through the ratio f/f_C of current frequency to cutoff frequency.

17 The text says that "a filter is a frequency-dependent voltage divider." Why "voltage divider?" Why "frequency dependent?"

18 *Consider the cutoff frequency, f_C, of an RC LPF:
  a) A cutoff frequency is the frequency at which the amplitude of an output signal decreases to 0.707 of the input sinusoid amplitude; that is, at f_C, $V_{out}/V_{in} = 0.707$. Is any natural delineator behind this definition? Explain.
  b) Discuss why the number 0.707 is chosen for the definition of a cutoff frequency.

19 The industry uses two criteria for a filter's cutoff frequency: $A_v = 0.707$ vs. $A_v = -3\,\text{dB}$.
  a) Which one is correct? Prove your answer.
  b) Is there any advantage in using one criterion over the other?

20 *Consider the cutoff frequency in an *optical* filter:
  a) Which criterion for power attenuation, $A_p = P_{out}\,(\text{W})/P_{in}\,(\text{W})$, corresponds to voltage attenuation, $A_v = V_{out}\,(\text{V})/V_{in}\,(\text{V})$, of 0.707?
  b) Which criterion for power attenuation corresponds to voltage attenuation of $-3\,\text{dB}$? (*Hint*: See Sidebar 5.1.S.)

21 The communications industry uses decibel measure more often than natural scale, therefore, we must be comfortable in working with dB:
  a) If voltage attenuation is $A_v = -3\,\text{dB}$, how much does the V_{out} (V) decrease with respect to V_{in} (V)? If $A_v = -30\,\text{dB}$?
  b) If power attenuation is $A_p = -3\,\text{dB}$, how much does the P_{out} (W) decrease with respect to P_{in} (W)? If $A_p = -30\,\text{dB}$? (*Hint*: See Sidebar 5.1.S.)

22 *For a series low-pass RC filter, what is the relationship between X_C (Ω) and R (Ω) at f_C (Hz)? Prove your answer.

23 Find the output voltage of an RC LPF at f_C with $V_{in} = 20\,\text{V}$ if:
  a) $R = 10\,\text{k}\Omega$ and $C = 0.6\,\text{nF}$.
  b) $R = 20\,\text{k}\Omega$ and $C = 0.06\,\text{nF}$.
  c) What conclusion can you make based on your results?

24 *An RC LPF is built with $R = 4.7\,\text{k}\Omega$ and $C = 0.002\,\mu\text{F}$:
  a) Sketch the schematic of this filter.
  b) Compute its cutoff frequency, f_C.
  c) Compute A_v and Θ at $f = 0.1 f_C, f = 0.5 f_C, f = f_C, f = 5 f_C$, and $f = 10 f_C$.

d) Plot the A_v-vs.-f and Θ-vs.-f graphs for this filter and show the computed values of A_v and Θ at these graphs.
e) Show the passband and the stopband regions of the filter.

25 *A low-pass RC filter has $R = 2.2\,k\Omega$, $C = 0.02\,\mu F$, and $v_{in}(t)\,V = 2\cos(4t)$:
a) Determine the V_{out} at $f = 0.1f_C$, $f = f_C$, and $f = 10f_C$.
b) Sketch the waveforms of the input and output signals at f_C.
c) How much do these waveforms change if R remains fixed but C changes to $0.002\,\mu F$?

26 *Consider an RC LPF with $f_C = 5\,kHz$:
a) Depict the graphs of attenuation and the output phase shift for this filter.
b) Show the bandwidth of this filter and indicate the range of values of A_v and Θ within the bandwidth.

27 What is the relationship between filter's bandwidth and passband?

Filter Operation in Time Domain and Frequency Domain

28 Consider an RC LPF with $f_C = 4\,kHz$:
a) Plot the attenuation graph, A_v-vs.-f, of this filter and plot the graph showing the output phase shift, Θ-vs.-f, of this filter.
b) Show the values of A_v and Θ at $f = 0.1f_C$, $f = f_C$, and $f = 10f_C$.
c) Plot three output waveforms at these values of the signal frequency, f, if $v_{in}(t) = 2\cos(8t)$.

29 Consider an RC LPF with $f_C = 4\,kHz$ and build the phasor diagram showing the phase shift values at $f = 0.1f_C$, $f = f_C$, and $f = 10f_C$.

30 What two main characteristics describing the operation of an LPF in frequency domain do you know? Give the formulas and qualitatively sketch their graphs.

31 To build the graphs showing A_v-vs.-f and Θ-vs.-f, we need to know only one number. What is it? Describe how you would plot the required graphs.

32 *Consider an RC LPF with $f_C = 4\,kHz$: What values of R (Ω) and C (F) do we need to build the filter with this cutoff frequency? Explain.

33 *Consider an RC LPF with $R = 4.7\,k\Omega$ and $C = 0.002\,\mu F$:
a) Compute f_{C1} and plot the graphs of $A_v(f)$ and $\Theta(f)$.
b) Change R from 4.7 to $47\,k\Omega$, keep $C = 0.002\,\mu F$, compute new f_{C2}, and plot the new graphs of $A_v(f)$ and $\Theta(f)$.
c) Change C from $0.002\,\mu F$ to $2\,nF$, keep $R = 4.7\,k\Omega$, compute new f_{C3}, and plot the new graphs of $A_v(f)$ and $\Theta(f)$.
d) Compare all the obtained results: What characteristics of an RC filter can you change by varying R and C?

34 Suppose you are given an RC LPF whose $f_C = 5\,kHz$, but you need to change the cutoff frequency to $10\,kHz$. How can you build a new RC LPF? Explain.

35 Figures 5.1.13 and 5.1.14 show how attenuation of an RC LPF changes by changing the values of its R and C. Sketch the graphs showing how the output phase shift of this RC LPF will change under the same changes of R and C.

36 What parameters of an RC LPF do we need to change in order to change the shape of its A_v-vs.-f graph? Explain.

37 *Figure 5.1.2 shows the attenuations for an ideal and real low-pass filter. The goal of a designer is to build a filter with characteristics as close to the ideal as possible. Can we achieve this goal with a series RC LPF?

General Filter Specifications

38 Consider the amplitude response of a low-pass filter (LPF):
   a) Sketch this response with $A_v = 1$, $A_p = 0.1$, $A_s = 0.8$, $f_P = 100$ Hz, and $f_S = 102$ Hz.
   b) Show all the proper characteristics in dB.
   c) Determine and show the passband and stopband areas of this filter.

39 Consider the following filter characteristics: A_v, A_p, A_s, ω_p, and ω_s:
   a) Explain the meaning of each, including their units.
   b) What value of each – small or big – do you want to design and why?
   c) What should be the ideal value of each of these characteristics? Can we achieve these values in practice?

40 Consider the *transition band* of an LPF's amplitude characteristic with $f_P = 100$ Hz and $f_S = 102$ Hz:
   a) What is its width, f_{TB}? Explain what an ideal width should be.
   b) Compute the selectivity factor, f_{SF}. Is this a good value? What should be its ideal value?
   c) Calculate the shape factor, f_{ShF}. Is this a good value? What should be its ideal value?

41 Consider the amplitude response of an industrial filter shown in Figure 5.1.16. Compute all its amplitude characteristics – A_v, A_p, A_s, f_p, f_S, f_{TB}, f_{SF}, and f_{ShF} – and comment on the calculated values in terms of their proximity to the ideal values.

42 The textbook says that it is straightforward to apply amplitude characteristics of a low-pass filter to a HPF. Build an attenuation graph for an HPF and show all the specifications for the HPF amplitude response.

43 You are given the following characteristics of the amplitude response of a low-pass filter:

Attenuation (dB)	Frequency (units of f_p)
−0.1	1.00
−3.00	1.1
−100.0	1.48 = f_S

Find the width of the transition band, the selectivity factor, and the shape factor of this filter. Sketch the figure that displays your results.

44 *You are given the following characteristics of the amplitude response of a HPF:

Attenuation (dB)	Frequency (units of f_p)
−0.1	1.00
−3.00	1.1
−100.0	1.48 = f_S

What is its stopband amplitude, A_S, if $f_S = 1.48 f_p$? Draw a figure to show all these characteristics.

45 *Consider the phase response of a filter: We can characterize this response by the phase shift, $\Theta(\omega)$ (rad), and the time delay, $t_d(\omega)$ (s).
a) Which characteristic is better? Why?
b) Is there any relationship between these two characteristics? If your answer is yes, show the proper formula.

46 To describe the time delay introduced by a filter, we use time phase delay, $t_d(\omega)$ (s), and time group delay, $T_D(\omega)$ (s). Why do we need two time-delay characteristics? Explain the difference between them.

47 Given that the normalized time group delay is $T_{ND}(\omega) = 2$ seconds and the actual filter's cutoff frequency is 4 kHz, calculate the actual time group delay, $T_D(\omega)$ (s).

5.2 Filtering – Introduction

Objectives and Outcomes of Section 5.2

Objectives

- To gain in-depth knowledge of the principles of operation of the main filter types (in addition to the LPF): HPF, BPF, and BSF.
- To learn how to characterize the abovementioned filters based on their mathematical description.
- To understand concept of transfer function and its application to filter analysis.
- To become familiar with the graphical presentation of a filter's transfer function based on the use of the Bode plots.

Outcomes

- Learn that an HPF, whose function is to pass high-frequency signals and reject low-frequency signals, can be built from the same two R and C components that are used in an LPF circuit by simply rearranging their circuit placement.

- Grasp the concept of an HPF operation by analyzing an HPF's equivalent circuit and the signal's output waveforms.
- Acquire knowledge of the characterization of an HPF by deriving the formulas of its amplitude and phase responses and its cutoff frequency.
- Learn that the function of a BPF is to pass the signals within a certain band of frequencies and reject the signals whose frequencies are outside of this band. Realize that a BPF can be built as a combination of LPF and HPF.
- Understand that a BPF is characterized by the following parameters: Two cutoff frequencies (a lower one, f_{C1} (Hz), set by an HPF, and a higher one, f_{C2} (Hz), set by an LPF); a central frequency, f_0 (Hz) = $\sqrt{f_{C1} \cdot f_{C2}}$; a bandwidth, BW_{BPF} (Hz) = $f_{C2} - f_{C1}$; and a quality factor, $Q_{BPF} = f_0/(f_{C2} - f_{C1}) = f_0/BW_{BPF}$.
- Become familiar with the way to conceptualize the design of a BPF based on understanding LPF and HPF operations and the BPF's parameters mentioned above.
- Learn how to analyze the frequency response of a BPF by using a Bode plotter.
- Realize that a BSF rejects the signals within a certain band of frequencies and passes the signals whose frequencies are outside of this band.
- Understand that the BSF can be built from LPF and HPF and that BSF's passband must be formed as f_{C2} (set by the HPF) $> f_{C1}$ (set by the LPF). Learn that to meet the requirement $f_{C2} > f_{C1}$, the two filters (LPF and HPF), from which a BSF is built, must be connected in parallel.
- Analyze the operation of a BSF and be able to qualitatively build the amplitude response based on this analysis.
- Get to know that the RC filters still find applications in analog communications circuits, but they are not the main type of filter used in modern electronics.
- Become familiar with how all types of filters perform their tasks in time domain by analyzing the changes a noisy sinusoidal signal experiences after passing each type of filter.
- Grasp the concept of a transfer function as the ratio of an output signal to an input signal described in frequency domain.
- Learn that a transfer function is a complex entity and can be presented as $H(f) = \frac{V_{out}(f)}{V_{in}(f)} = |H(f)|e^{-j\Theta}$, where $|H(f)|$ is the transfer function's amplitude and θ is its phase.
- Master the point that a transfer function's amplitude is the attenuation of a passive LPF and its phase is the output phase shift of the LPF. Realize that the transfer function enables us to find a filter output in frequency domain as $V_{out}(f) = V_{in}(f) \times H(f) = V_{in}(f) \times A_v e^{j\Theta}$.
- Know that after obtaining the output signal's amplitude and phase shift in frequency domain by using a transfer function, we return to time domain by deriving the equation describing the signal's waveform, which is the final result of these transformations.
- Understand that a filter's transfer function can be presented graphically as Bode plots, which are the straight-line approximations of real, smooth graphs.
- Learn in depth how to build Bode plots and what errors these graphs introduce as a price for their simplicity.

5.2.1 High-Pass Filter (HPF), Band-Pass Filter (BPF), and Band-Stop Filter (BSF)

To recap, there are four main filter types: LPF, HPF, BPF, and BSF. The BSF is also called a *band-reject* or *notch* filter. (There is also the fifth type called *all-pass* or *phase-shift* filter; at different frequencies, this filter differently shifts the phase of an output signal but does not change its amplitude.) Discussion in Section 5.1 comprehensively introduces LPF; in this section, we will consider the three remaining types of filter.

5.2.1.1 High-Pass Filter (HPF)

Concept and Circuit of an HPF We now understand that by using the frequency dependence of a capacitive reactance, we can decrease the amplitudes of a LPF's output signal at high frequencies; that is, we can reject (i.e. filter) high-frequency signals. Question: Can we use the same property of a capacitor to build an RC HPF, a filter that will allow high-frequency signals to pass and reject the low-frequency ones? Recall that to filter the low-band frequencies, we build the RC circuit shown in Figure 5.2.1a; the frequency response of this circuit is demonstrated in Figure 5.2.1b. (Remember, Figure 5.2.1b presents the frequency in *logarithmic scale,* which means that every mark shows a 10-fold increase in the frequency.) Given that we want to restrict ourselves to two components, the only series circuit we can build is schematically shown in Figure 5.2.1c. Since we want to obtain the output voltage for the high-pass RC filter shown in Figure 5.2.1d, the problem is where to place the resistor, R, and the capacitor, C, to build a circuit that will be able to filter the low-band frequencies and pass the high-frequency signals.

The solution can be found in (5.1.2):

$$V_{in} = V_R + V_C$$

Now we want the output voltage to be small at low frequencies. Remember that reactive capacitance, X_C (Ω), changes with frequency as

$$X_C (\Omega) = \frac{1}{2\pi fC} = \frac{1}{\omega C} \tag{5.1.1R}$$

whereas R (Ω) remains constant. If we place a capacitor next to the voltage source and take the voltage drop across the resistor as an output, then V_C will be high and V_R will be low at low frequencies. What is more, at high frequencies, V_C will be low and V_R will be high, precisely the result

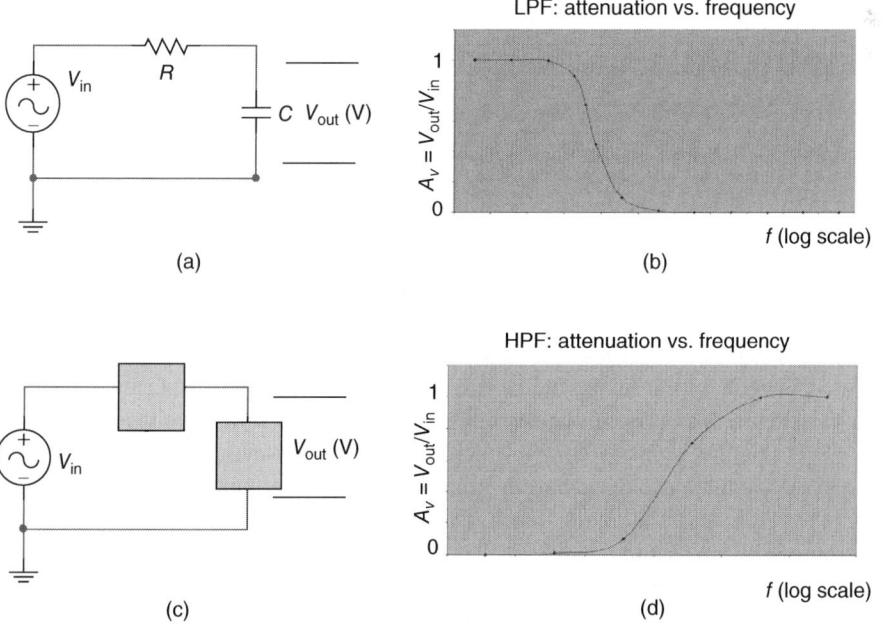

Figure 5.2.1 Concept of a low-pass RC filter and a high-pass RC filter: (a) the circuit of an RC LPF; (b) the output voltage of an RC LPF; (c) the available circuit for building a high-pass RC filter; and (d) the required output voltage of an RC HPF.

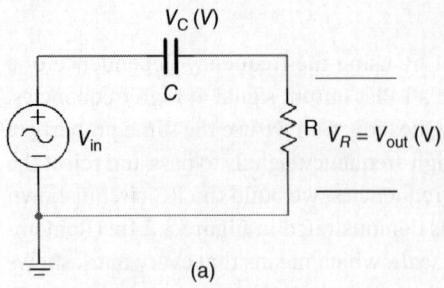

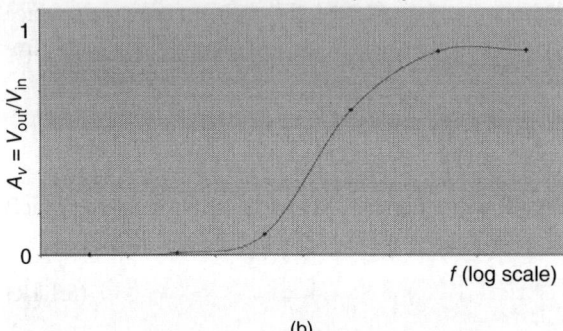

Figure 5.2.2 High-pass RC filter: (a) schematic of the filter and (b) output voltage as a function of frequency.

we want to achieve. (Do you understand why V_R changes here?) Thus, we can build a HPF with R and C components, as shown in Figure 5.2.2.

If we analyze the circuit shown in Figure 5.2.2a, it would just be a variation of the analysis we did for an RC LPF: When the frequency goes to zero, the capacitor, C, works like an open circuit, since X_C approaches infinity:

$$f \to 0 \Rightarrow X_C = 1/2\pi f_C \to \infty$$

When the frequency goes to infinity, the capacitor, C, works like a short circuit, since X_C approaches zero:

$$f \to \infty \Rightarrow X_C = 1/2\pi f_C \to 0$$

Therefore, at low frequencies, $V_R \equiv V_{out}$ approaches zero, whereas at high frequencies, $V_R \equiv V_{out}$ approaches V_{in}. These explanations are illustrated in Figure 5.2.3.

To make the operation of an RC HPF more understandable, we have to perform an experiment to observe the waveforms of output signals at low and high frequencies. These experimental waveforms are shown in Figure 5.2.4. As can be seen in Figure 5.2.4a, the output voltage at the low frequency of 50 Hz is small. When the frequency increases 10-fold, however, the amplitude of the output signal increases about fourfold and becomes almost equal to the input amplitude – the maximum value that the output amplitude could reach. This case is shown in Figure 5.2.4b. It is exactly what we want from a HPF: To respond with a small output voltage (ideally, zero) at low frequency and to show a high output voltage (ideally, equal to the input) at high frequency. Did you observe the values of the phase shifts of the output waveforms? In two extreme cases – when f_{in} (Hz) $\Rightarrow 0$ and f_{in} (Hz) $\Rightarrow \infty$ – what output phase shifts do you want?

Note that in this example we use 50 Hz as the low frequency and 500 Hz as the high frequency. Clearly, 500 Hz is a very low frequency as compared, say, with a radio frequency in the kilohertz

Figure 5.2.3 Operation of an RC HPF filter: (a) the filter's equivalent circuit at low frequency; (b) the filter's equivalent circuit at high frequency; and (c) output voltage of the filter.

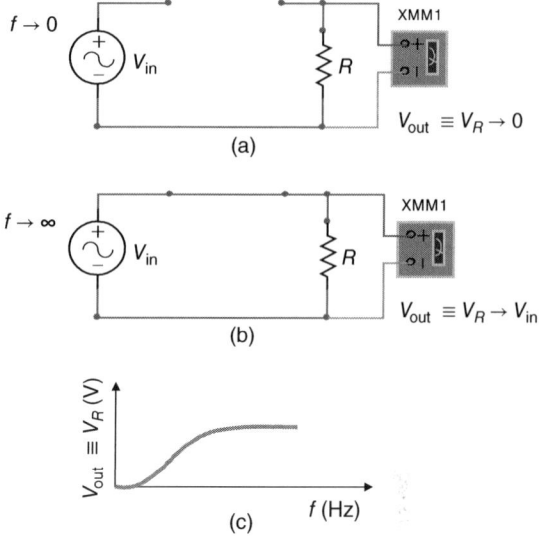

and megahertz ranges. However, compared with 50 Hz, a frequency of 500 Hz is high. This example stresses again that defining low and high frequencies, we need to have a reference point; in other words, the frequency designations "low" and "high" are relative.

Characterization of an RC HPF To derive the formulas for the HPF's amplitude and phase responses, we will follow the pattern developed for the RC LPF in Section 5.1. Consider the RC HPF circuit shown in Figure 5.2.2a. Our goal is to obtain formulas for the output voltage, V_{out}, and its phase shift, Θ_{out}. We can derive the general formula for $\boldsymbol{V}_{out}$ by applying the voltage-divider rule to our circuit as

$$V_R \equiv V_{out} = V_{in} \frac{Z_R}{Z_T} \quad (5.2.1)$$

where $Z_R = R$ is the resistive *impedance* and $Z_T = Z_R + Z_C = R - jX_C$ is the total impedance of the circuit. Bear in mind, of course, that in an RC HPF, the output voltage is the voltage across the resistor, R. Therefore,

$$A_v = \frac{V_{out}}{V_{in}} = \frac{Z_R}{Z_T} = \frac{R}{R - jX_C} \quad (5.2.2)$$

Plugging $X_C = 1/2\pi f_C$ into (5.2.2), multiplying both numerator and denominator by j, and separating the real from the imaginary parts, we obtain A_v in polar form as

$$A_v = \frac{R}{\sqrt{(R^2 + X_C^2)}} \angle \left(\tan^{-1}\left(\frac{X_C}{R}\right)\right) = \frac{R}{\sqrt{(R^2 + X_C^2)}} e^{\left(j\tan^{-1}\left(\frac{X_C}{R}\right)\right)} \quad (5.2.3)$$

Therefore, the attenuation and phase shift of an RC HPF are as follows:

$$A_v = \frac{R}{\sqrt{(R^2 + X_C^2)}} = \frac{1}{\sqrt{\left(1 + \frac{X_C}{R}\right)^2}} \quad (5.2.4a)$$

$$\Theta = \tan^{-1}\left(\frac{X_C}{R}\right) \quad (5.2.4b)$$

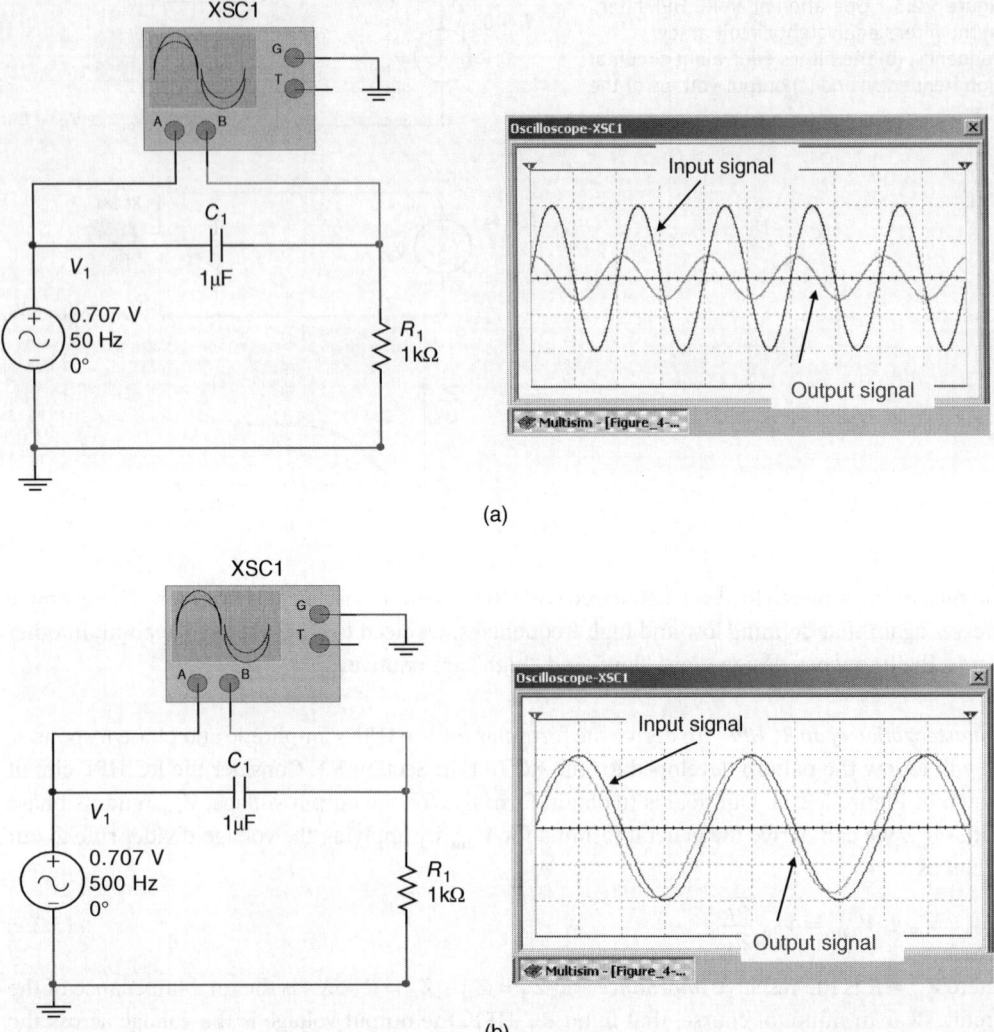

Figure 5.2.4 Experimental setup and the input and output signals of an RC HPF: (a) output voltage at $V_{in} = 1$ V and $f = 50$ Hz and (b) output voltage at $V_{in} = 1$ V and $f = 500$ Hz. (The oscilloscope amplitude settings are the same in both cases.)

Equation (5.2.3) and its corollaries (5.2.4a) and (5.2.4b) fully describe an RC HPF. Finally, let us determine the cutoff frequency of an RC HPF. Using the familiar criterion, $A_v = 0.707$, and (5.2.4a), we can immediately arrive at the condition $R = X_C$, which, in turn, results in

$$f_C = \frac{1}{2\pi RC} \qquad (5.2.5)$$

because $X_C = \frac{1}{2\pi fC}$. (See the derivation of Eq. (5.1.17).)

We need to recall from (5.1.20) that $\frac{f}{f_C} = \frac{R}{X_C}$. To keep the HPF formulas in the same format, we need to understand that the attenuation and phase shift of an HPF depend on the inverse ratio, $\frac{f_C}{f} = \frac{X_C}{R}$. Thus, using (5.2.4a) and (5.2.4b), we obtain the formulas for an HPF in a form convenient for calculations:

$$A_v = \frac{1}{\sqrt{1+\left(\frac{X_C}{R}\right)^2}} = \frac{1}{\sqrt{1+\left(\frac{f_C}{f}\right)^2}} = \frac{1}{\sqrt{1+\frac{1}{\left(\frac{f}{f_C}\right)^2}}} \quad (5.2.6)$$

and

$$\Theta = \tan^{-1}\left(\frac{X_C}{R}\right) = \tan^{-1}\left(\frac{f_C}{f}\right) = \tan^{-1}\left(\frac{1}{\frac{f}{f_C}}\right) = 90° - \tan^{-1}\left(\frac{f}{f_C}\right) \quad (5.2.7)$$

(The well-known rule, $\tan^{-1}\left(\frac{1}{x}\right) = 90° - \tan^{-1}(x)$, is used in (5.2.7).)

Compare these formulas with (5.1.21) and (5.1.22) to see the similarities and differences between the amplitude and phase responses of an RC LPF and an HPF.

Now it is time to consider an example.

Example 5.2.1   Frequency Response of an RC HPF

Problem

An RC HPF is built from $R = 2.2\,\text{k}\Omega$ and $C = 0.02\,\mu\text{F}$.

- Compute the cutoff frequency.
- Find A_v and Θ at f equals $0.001f_C$, $0.01f_C$, $0.1f_C$, f_C, $10f_C$, and $100f_C$ and build their graphs vs. frequency, showing all the computed values.
- Using Multisim, simulate the operation of this filter and show the input and output waveforms at $f = 100\,\text{Hz}$, $f = 1\,\text{kHz}$, and $f = 10\,\text{kHz}$.

Solution

We notice, of course, that this HPF is built from the same components as the LPF discussed in Examples 5.1.2 and 5.1.3.

- Applying (5.2.5), we compute the cutoff frequency as

    $$f_C = 1/2\pi RC = 3.62\,\text{kHz}$$

- The graphs of A_v and Θ based on computed values at given frequencies are shown in Figure 5.2.5.
- The circuit and the required waveforms are shown in Figure 5.2.6.

Discussion

The best way to analyze the results of this example is to review the analysis of Examples 5.1.2 and 5.1.3. Also, it is useful to compare Figures 5.1.8 and 5.1.9 with Figures 5.2.5 and 5.2.6. We leave it for you to do this.

5.2.1.2   Band-Pass Filter (BPF)
Concept and circuit of a BPF

We are now ready to move on to the next challenge: The conceptual design of a BPF. Figure 5.1.1c shows what we want from this filter: to allow the passage of signals within a certain band of frequencies and reject signals outside this band. To understand how we can build such a filter, we

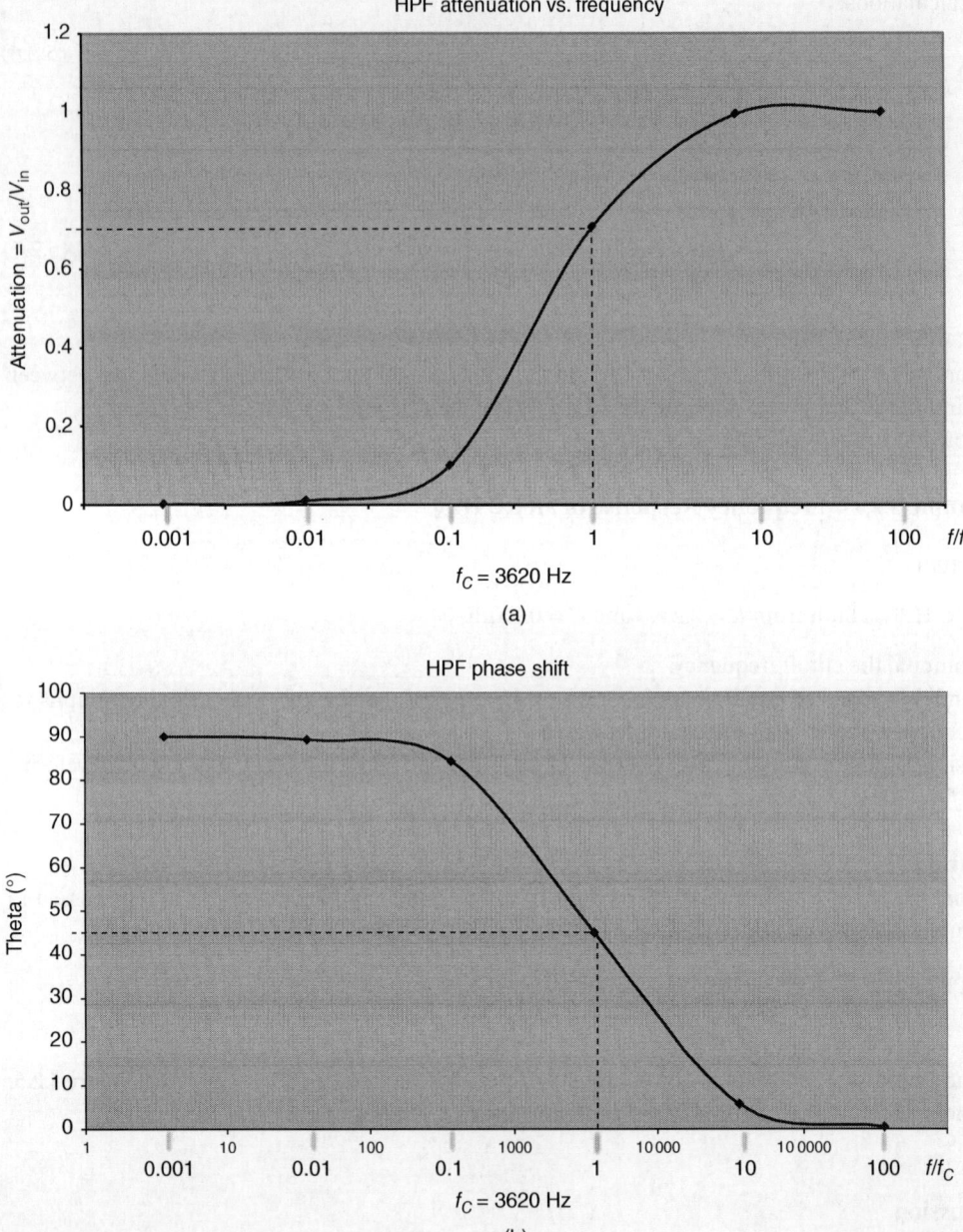

Figure 5.2.5 Graphs of (a) amplitude response and (b) phase response of the HPF in Example 5.2.1.

need to look at Figure 5.1.1a,b: Analysis of these figures clearly reveals that a BPF is a combination of a LPF and a HPF with certain cutoff frequencies. Refer to Figure 5.2.7:

An ideal LPF would reject signals whose frequencies are above its cutoff frequency, f_{C2}, whereas an ideal HPF would reject signals whose frequencies are below its cutoff frequency, f_{C1}. If we were

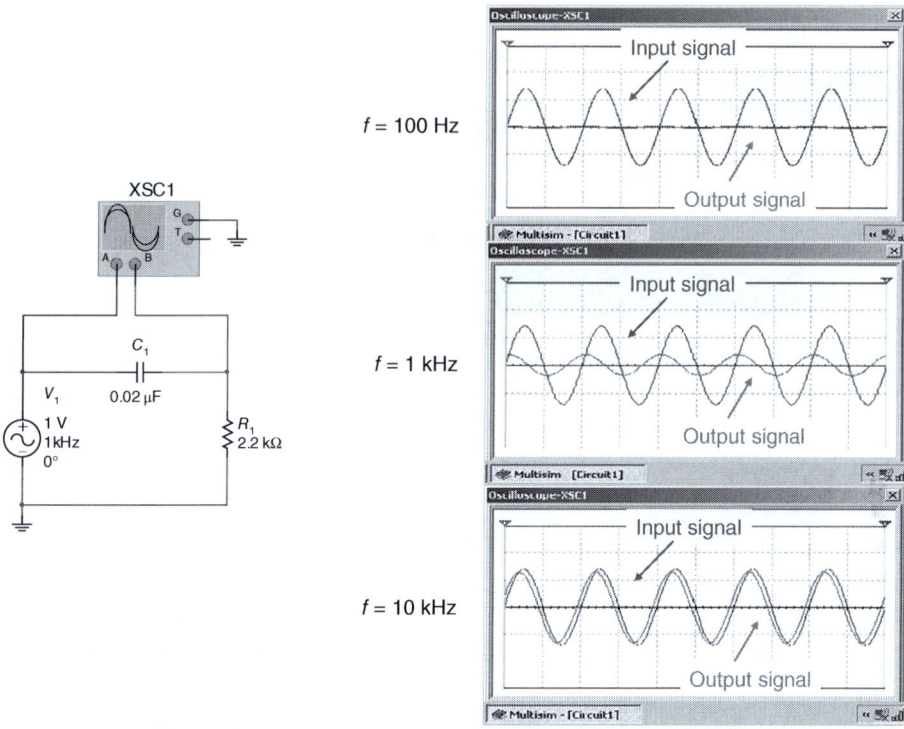

Figure 5.2.6 Experimental setup and output waveforms of the HPF at $f = 100$ Hz (top), $f = 1$ kHz (middle), and $f = 10$ kHz (bottom). Amplitude scales at the oscilloscope are the same for all three cases; time scales are different.

to combine these two filters with the aforementioned properties, we would obtain a BPF that rejects all signals whose frequencies are greater than f_{C2} and smaller than f_{C1}. Thus, this filter passes only signals whose frequencies are within the band from f_{C1} to f_{C2}. Consider, for example, an ideal LPF whose $f_{C1} = 12$ kHz and an ideal HPF whose $f_{C2} = 10$ kHz. Using these two filters, we can build an ideal BPF that will pass signals with frequencies from 10 to 12 kHz and reject all others. Applying this concept to design of a *real* BPF from a real LPF and a real HPF, we can obtain amplitude responses shown in Figure 5.2.8.

It follows from this consideration that the main requirement for building a BPF is that the cutoff frequency, f_{C1}, set by an HPF, must be below f_{C2}, set by an LPF:

$$f_{C1} \text{ (set by HPF)} < f_{C2} \text{ (set by LPF)} \tag{5.2.8}$$

(**Question**: *Why this requirement? What if we don't meet this condition?*)
If an HPF is built from R_1 and C_1 and an LPF is built from R_2 and C_2, then

$$f_{C1} \text{ (HPF)} = \frac{1}{2\pi R_1 C_1} \tag{5.2.9a}$$

$$f_{C2} \text{ (LPF)} = \frac{1}{2\pi R_2 C_2} \tag{5.2.9b}$$

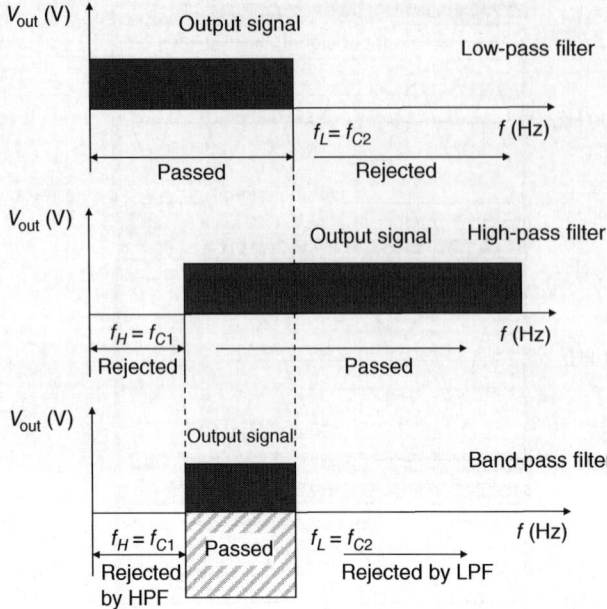

Figure 5.2.7 Concept of building an ideal band-pass filter from ideal low-pass and high-pass filters.

(Refer to Formulas 5.2.5 and 5.1.17.) It also follows from the BPF's principle of operation that in building BPF circuitry, we can place either the LPF first and the HPF second or the HPF first and the LPF second; we need only satisfy the condition set by (5.2.8).

The bandwidth, which is the bandpass of a BPF, is obviously defined by the following formula:

$$\text{BW}_{\text{BPF}} = f_{C2} - f_{C1} \tag{5.2.10}$$

It is important to introduce one more parameter specific to a BPF. It is called the *central frequency*, f_0. This is the frequency in the middle of the passband of this filter; it is the geometric mean of two cutoff frequencies and is defined as

$$f_0 = \sqrt{(f_{C1} \times f_{C2})} \tag{5.2.11}$$

The central frequency is another BPF parameter helping us to better characterize a BPF. In addition, to make the description of a BPF complete, we introduce a *quality factor*, Q, defined as

$$Q_{\text{BPF}} = \frac{f_0}{f_{C1} - f_{C2}} = \frac{f_0}{\text{BW}_{\text{BPF}}} \tag{5.2.12}$$

The quality factor describes how wide or narrow the attenuation-vs.-frequency graph of the BPF is; in other words, *the quality factor describes the sharpness of a BPF's amplitude response*. See more about the quality factor in the sidebar entitled "Resonance circuits as filters" in this section or any textbook on circuit analysis.

Designing a BPF When designing a BPF, we need to remember that the circuit of a BPF must be a combination of the circuits of a HPF and a LPF. The key factor is choosing its cutoff frequencies, f_{C1} and f_{C2}. Imagine, for instance, that we choose the LPF cutoff frequency, f_{C1}, below the HPF cutoff frequency, f_{C2}; in other words, we violate (5.2.8). Refer to Figures 5.2.7 and 5.2.8a and sketch

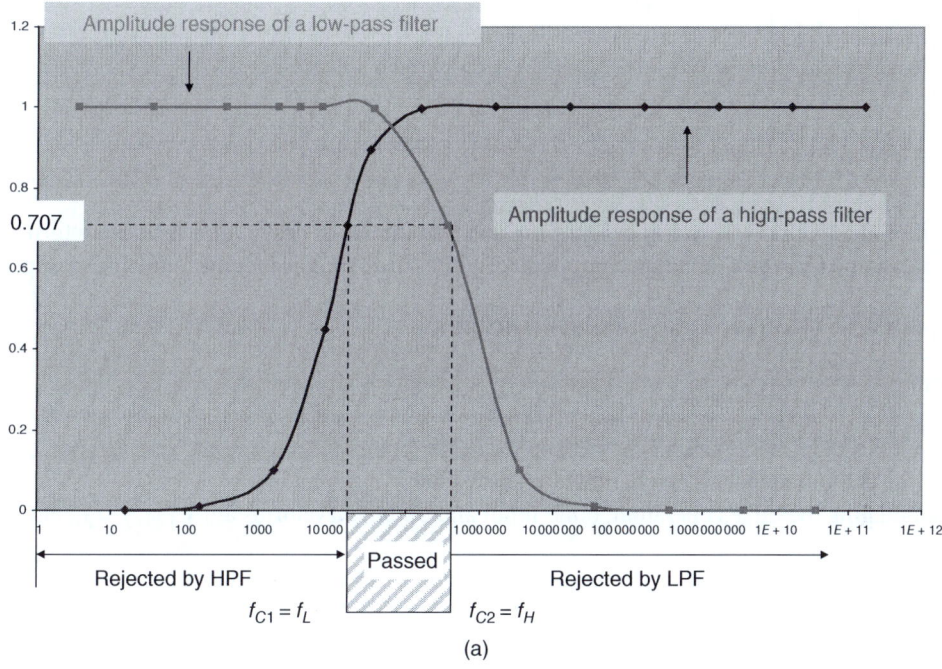

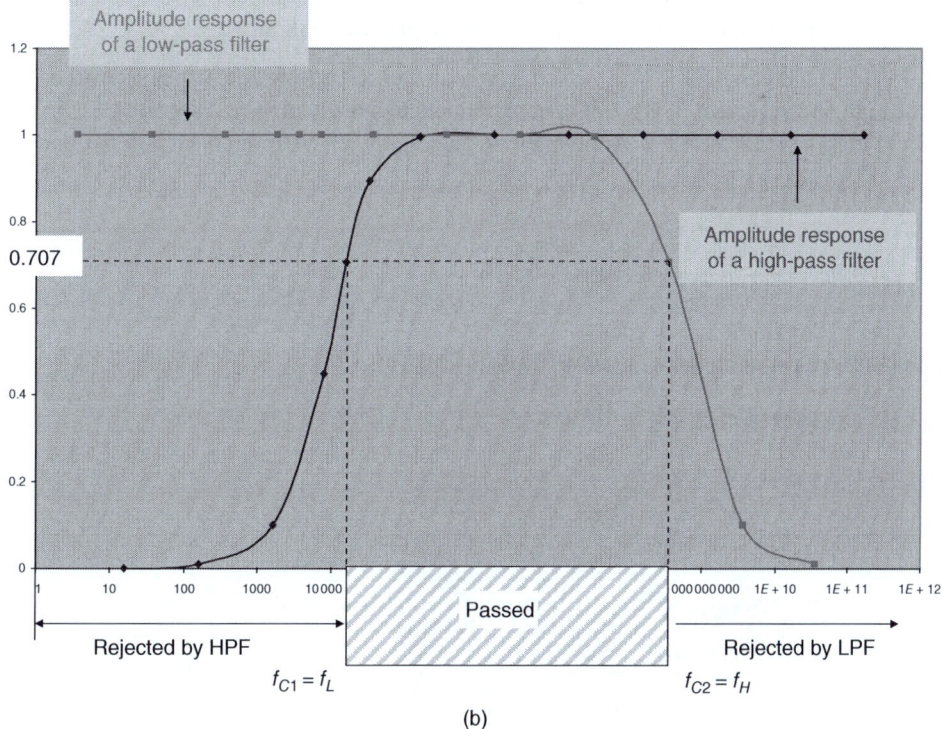

Figure 5.2.8 Real band-pass filter: (a) amplitude of the frequency response of a BPF as a combination of the amplitude responses of low-pass and high-pass filters; (b) amplitude of the frequency response of a BPF with a greater band-pass region; and (c) a BPF as a serial connection of an HPF and LPF.

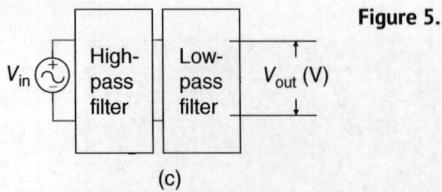

Figure 5.2.8 (Continued)

the kind of attenuation you would expect from such an improper design of a BPF. Now turn to Figure 5.2.8c and try to design the specific circuit of a BPF. To do so, consider the following example.

Example 5.2.2 Frequency Response of an RC BPF

Problem

An RC BPF is shown in Figure 5.2.9.

a) Find the cutoff frequencies and bandwidth of this filter.
b) By simulating the filter's operation, build input and output waveforms at $f < f_{C1}, f = f_0$ and at $f > f_{C2}$.
c) Build the amplitude and phase responses of this filter based on its simulation.

Solution

To solve this problem, we need to apply the knowledge we accumulated during the discussion of LPFs and HPFs and our understanding of the BPF principle of operation.

a) First, we understand that a BPF has two critical frequencies: The low one, f_{C1}, is set by the HPF and the high one, f_{C2}, is set by the LPF. These frequencies are given by (5.2.9a) and (5.2.9b) and shown in Figures 5.2.7 and 5.2.8. Applying these equations, we compute the cutoff frequencies

$$f_{C1} \text{ (HPF)} = \frac{1}{2\pi R_1 C_1} = 15.92 \text{ kHz}$$

$$f_{C2} \text{ (LPF)} = \frac{1}{2\pi R_2 C_2} = 36.17 \text{ kHz}$$

Quite clearly, Condition 5.2.8 is met.
For our BPF, we compute the bandwidth by using (5.2.10):

$$\text{BW}_{\text{BPF}} = f_{C2} - f_{C1} = 20.25 \text{ kHz}$$

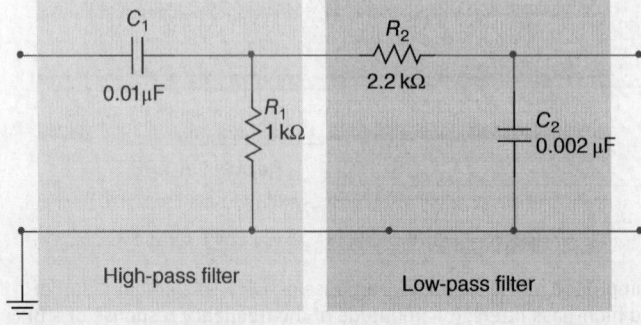

Figure 5.2.9 Schematic of a band-pass filter.

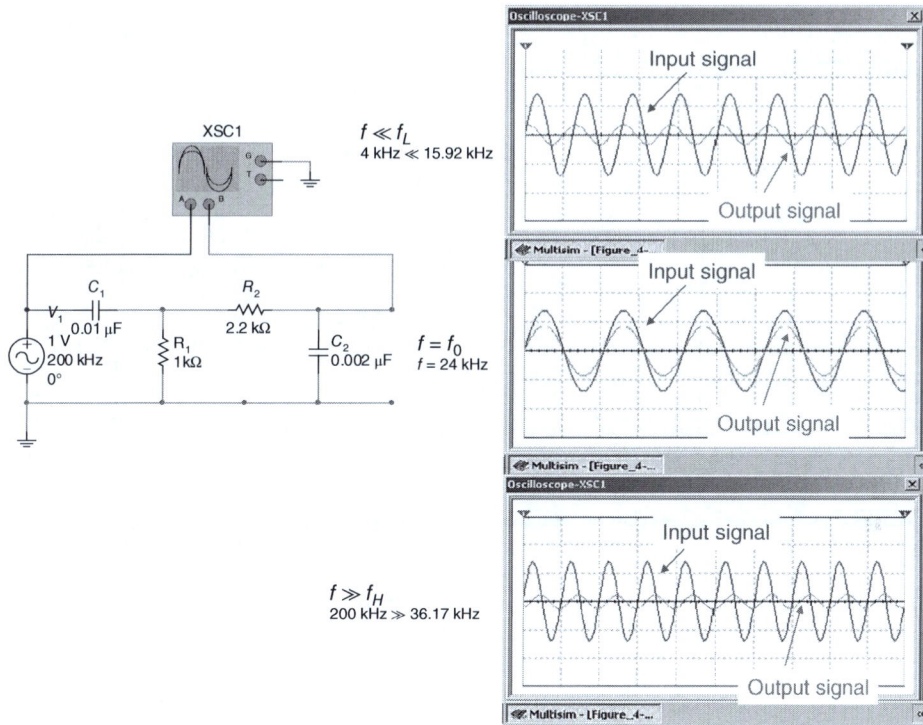

Figure 5.2.10 Experimental setup and input and output waveforms at three different frequencies of the BPF in Example 5.2.2.

For this BPF, we compute the central frequency by using (5.2.11):

$$f_0 = \sqrt{(15.92 \times 36.17 \text{ kHz})} = 24.00 \text{ kHz}$$

This number is really midway between 15.92 and 36.17 kHz.

b) The waveforms of the input and output signals at $f = 10$ kHz ($f < f_{C1}$), $f_0 = 24$ kHz, and $f = 42$ kHz ($f > f_{C2}$) obtained by the circuit simulation with Multisim are shown in Figure 5.2.10.

c) Figure 5.2.11 shows the attenuation and phase shift based on simulation of our RC BPF. Note that the attenuation graph is built in logarithmic scale and the phase graph is built in semilogarithmic scale with an instrument called a *Bode plotter*; this instrument has been connected in parallel to the oscilloscope shown in Figure 5.2.10. Settings of the Bode plotter are clearly shown in Figure 5.2.11. We will discuss a Bode plot in greater detail shortly; for now, refer to general filter specifications presented in Figure 5.1.16.

Discussion

- Examination of Figure 5.2.11: Study Figure 5.2.11 to see how the amplitude of the output signal decreases when the frequency is outside the filter's bandwidth: When we change frequency from 10 Hz to 50 MHz, the output magnitude increases from −50 to −4.297 dB and decreases back to −50 dB. Pay special attention to the fact that at $f = f_0$, the output amplitude is smaller than the input; indeed, at $f_0 = 24.048$ kHz, the magnitude is equal to −4.297 dB. (What value in dB the Bode plotter would show if the output amplitude equaled the input? Refer to Discussion of (5.1.16)) This fact follows from the conceptual view presented in Figure 5.2.8a. Specifically,

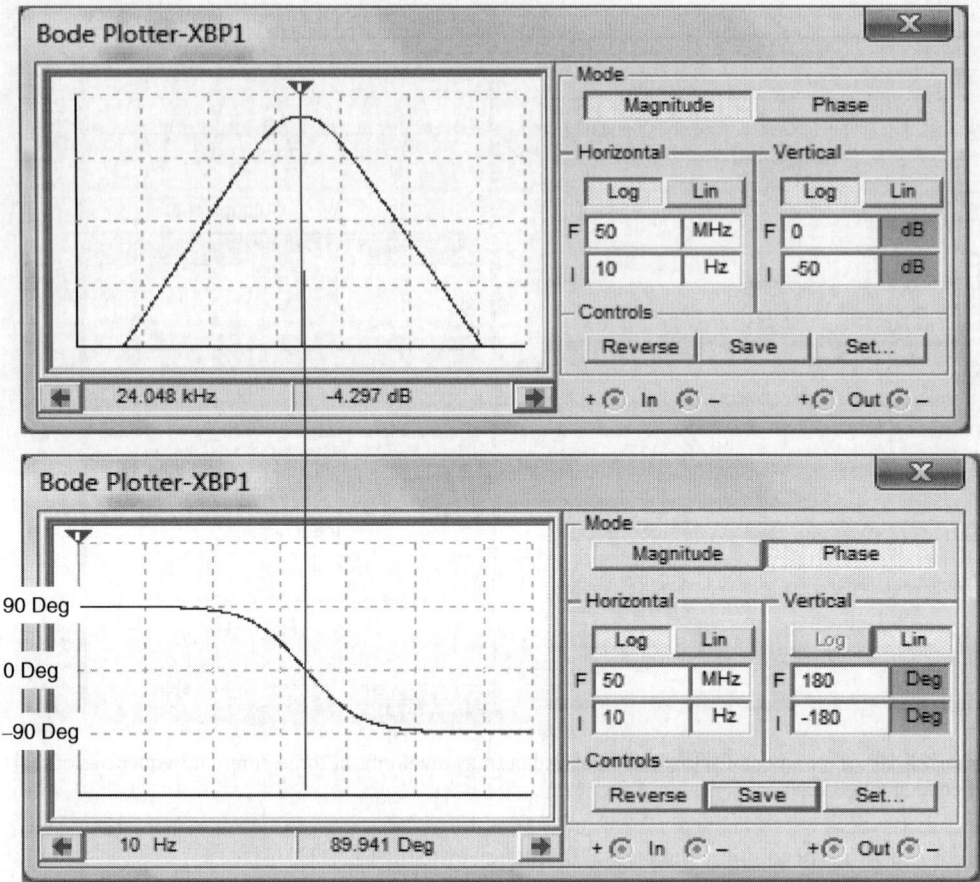

Figure 5.2.11 Bode plots of the amplitude and phase frequency responses of the BPF in Example 5.2.2.

this figure shows that the output amplitude does not reach the input because the intersection of the LPF and HPF amplitude responses does not produce $A_v = 1$. Nevertheless, the point $f = f_0$ shows not only the maximum amplitude response but also the unique phase response: At this frequency, the phase shift is zero. This situation can be clearly seen in Figure 5.2.11, where the phase response at f_0 is shown.

- Formulas for BPF attenuation and phase shift: We can obtain formulas for attenuation and phase shift of this BPF by analyzing this circuit, as described in any circuit-analysis course. We leave this derivation for you as an exercise. Then you should compute A_v and Θ at three frequencies shown in Figure 5.2.11. Obviously, the results obtained by simulation and computation should coincide.

5.2.1.3 Band-Stop Filter (BSF)

We have already considered three of main filter types: LPF, HPF, and BPF. In this subsection, we briefly introduce the last type, BSF. Based on our previous discussions, we can readily conclude that a BSF rejects signals whose frequencies fall within a certain band while allowing all signals whose frequencies lie outside this band to pass. Also, reviewing Figures 5.2.7 and 5.2.8, we can figure out the concept of building such a filter. In the same way we design a BPF, we can also design a BSF as a combination of LPF and HPFs. Again, the crucial requirement is choosing the right cutoff frequencies. Having the cutoff frequency of a LPF, f_{C1}, lower than that of a HPF, f_{C2}, would be a mistake in designing a BPF, as we discussed earlier; however, this would be the correct approach in designing a BSF.

Figure 5.2.12 explains how, conceptually, we can build a BSF; however, if you look at this figure closely, you can immediately understand that combining the LPF and the HPF *in series*, as we did for constructing a BPF, does not work. Indeed, if we connect the LPF and HPF in series and meet Condition (5.2.8), then we have built a BPF, as discussed in the previous Subsection 5.2.1.2. If we violate (5.2.8) still keeping the LPF and HPF circuits connected in series, then the HPF would reject all low-frequency signals, which would leave nothing for the LPF to select. The solution lies in placing the LPF and the HPF *in parallel* and meeting the condition

$$f_{C2} \text{ (set by the HPF)} > f_{C1} \text{ (set by the LPF)} \tag{5.2.13}$$

In such an arrangement and with this condition, both filters will do their work in parallel and one will not prevent the other from performing its specific task. Figure 5.2.13 visualizes this concept.

The input signal, containing all frequencies from 0 to f_{max}, enters both the LPF and the HPF. Each filter performs its job; that is, the LPF selects low-frequency signals and rejects high-frequency

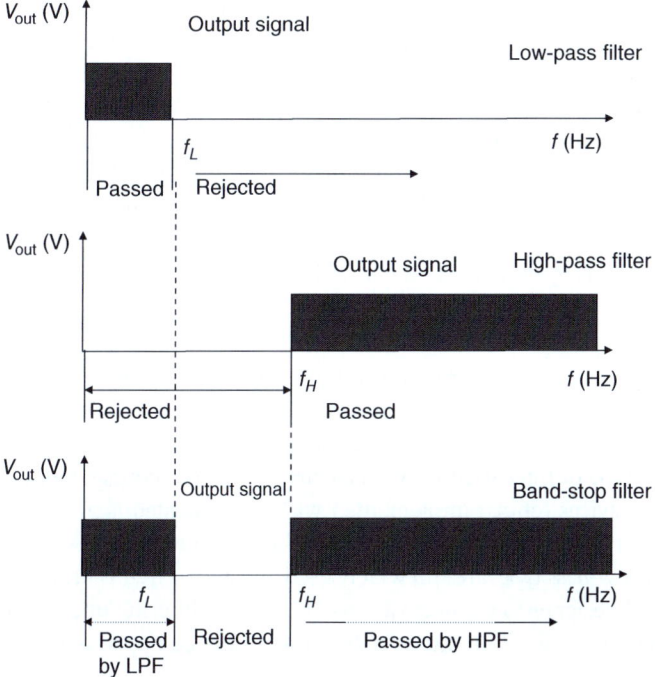

Figure 5.2.12 Building an ideal band-stop filter.

Figure 5.2.13 Block diagram of circuitry for a band-stop filter.

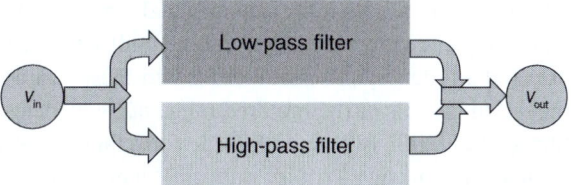

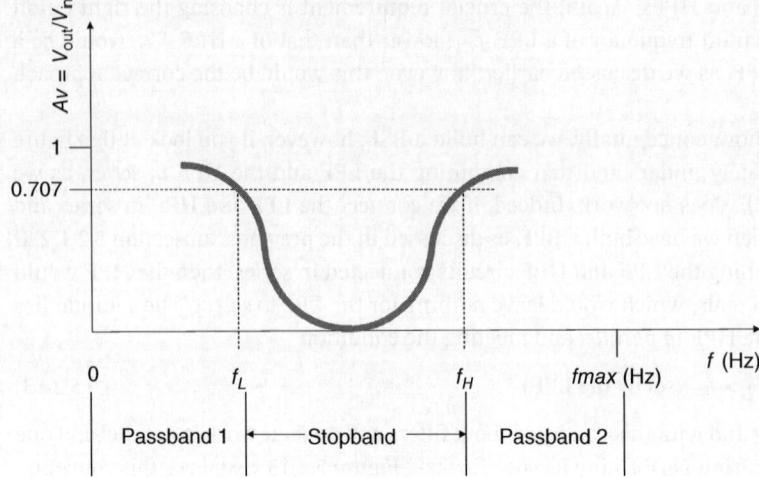

Figure 5.2.14 Attenuation of a band-stop filter.

signals, whereas the HPF does the opposite. The outputs of both filters combine and the result is that signals with frequencies between f_{C1} and f_{C2} are rejected, whereas signals whose frequencies are lower than f_{C1} and higher than f_{C2} are passed. A theoretical amplitude response of a BSF is shown in Figure 5.2.14, from which we can see that the bands of frequencies from 0 to f_{C1} and from f_{C2} to f_{max} are passbands, whereas the band of frequencies between f_{C1} and f_{C2} is a stopband. Obviously, the cutoff frequencies f_{C1} and f_{C2} are defined by Formulas 5.2.9a and 5.2.9b, the central frequency (called the *central stop frequency* or the *central frequency of rejection*) is given by (5.2.10), and the bandwidth (called the *stop bandwidth* or *bandwidth of rejection*) is determined by (5.2.11).

(**Exercise**: Prove the above propositions by deriving all the formulas mentioned.)

5.2.1.4 Applications of RC Filters

As we learned in this section, RC filters are simple and perform the required tasks. However, they are far from ideal filters and we can do very little, if anything, to improve their characteristics. Also, discrete *R* and *C* components are bulky and susceptible to noise. Because of these and other constraints, RC filters are not the main type of filter used in modern electronics. (Of course, today's filters are mainly integrated-circuit devices (chips) implemented with sophisticated analog and digital circuitry. They exhibit excellent characteristics, which are very close to the ideal ones.)

However, RC filters, especially the low-pass type, are still widely used, mostly as circuits bridging chips on a board. They allow a board designer to connect various chips while filtering undesired signals (noise, in this case) that might be induced in interconnected wires not equipped with RC filters.

5.2.1.5 Final Notes on RC Filters

Filtering is the process of selecting signals in one range of frequencies while rejecting signals in other frequency ranges. A filter can be built with such basic components as a resistor, a capacitor, and an inductor. A circuit built with these components simply decreases the amplitudes of the output signals at the required range of frequencies. Figure 5.2.15 shows how all types of the filters – LPF, HPF, BPF, and BSF – perform their tasks in time domain. We can see how the waveform of an input signal (noisy sinusoid) changes after passing each of the above filter. It would be

Figure 5.2.15 Time-domain responses (output waveforms) of LPF, HPF, BPF, and BSF to a noisy sinusoidal input signal.[1]

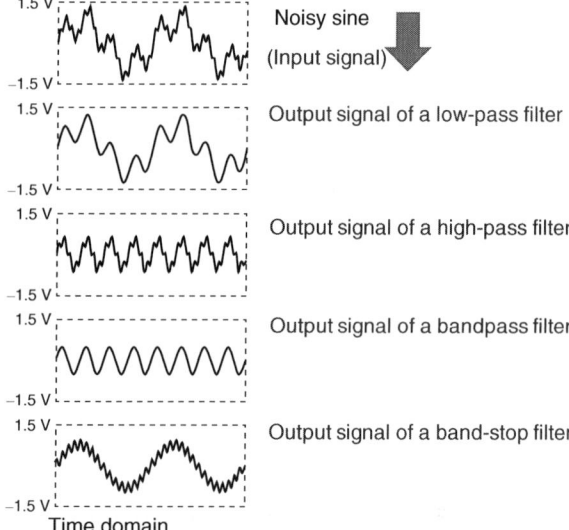

informative to compare these time-domain presentations with frequency responses of each filter type discussed in this section.

In a real filter, the decrease of unwanted components occurs smoothly, whereas, ideally, we want to achieve an abrupt, step-like change in output voltage from the desired (selected) signals to the undesired (rejected) signals. What is more, while the desired output signals should be equal to the input signals, all others should be simply reduced to zero. Unfortunately, such ideal conditions do not exist in the real world. Hence, the differences between real and ideal filtering are the key challenges confronting a filter designer. Examine again Figure 5.1.15 that exemplifies how the industrial device approaches the characteristics of ideal filters.

5.2.2 Transfer Function of a Filter

5.2.2.1 Input and Output of a Filter

Here we will discuss the filtering process from the input–output perspective. We have touched input–output relationship in RC filters; now, we will exercise more rigorous approach to this problem.

Let us summarize our findings in the preceding sections of this chapter: If we present an *input sinusoidal signal*, $v_{in}(t)$, to an RC filter, the output will be a sinusoidal signal with the same frequency but smaller amplitude and an additional phase shift. This statement is visualized in Figure 5.2.16 specifically for an RC LPF. Putting this statement into mathematical form, we can write

$$v_{out}(t) = V_{out} \sin(\omega t + \Theta_{in} + \Theta_{out}) \quad (5.2.14)$$

Equation (5.2.14) is generalization of (5.1.13). Also, from (5.1.21), (5.1.22), and (5.1.26) we find

$$V_{out} (V) = V_{in} \cdot A_v = V_{in} \cdot \frac{1}{\sqrt{\left(1 + \left(\frac{f}{f_c}\right)^2\right)}} \quad (5.2.15)$$

---

1 The waveforms from http://cas.ee.ic.ac.uk/people/dario/files/E22/L7-Active%20Filters.pdf

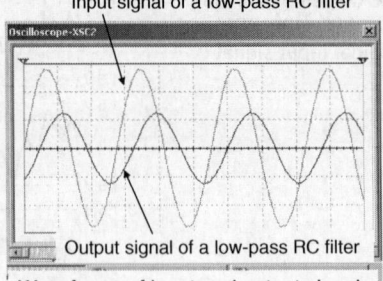

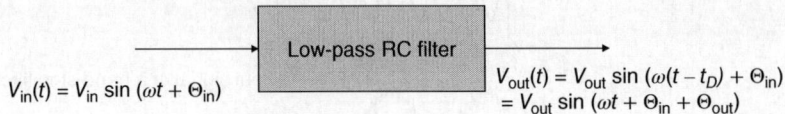

Waveform of an input signal | Waveforms of input and output signals

$V_{in}(t) = V_{in} \sin(\omega t + \Theta_{in})$ → [Low-pass RC filter] → $V_{out}(t) = V_{out} \sin(\omega(t - t_D) + \Theta_{in})$
$= V_{out} \sin(\omega t + \Theta_{in} + \Theta_{out})$

Figure 5.2.16 Input-output view of a low-pass RC filter operation.

and

$$\Theta_{out} (\text{rad}) = -2\pi f\, (\text{rad/s}) \cdot t_d\, (\text{s}) = -\tan^{-1}\left(\frac{f}{f_C}\right) \quad (5.2.16)$$

It is worth to repeat that V_{out} and Θ_{out} depend on the signal frequency.

In analysis of (5.2.16), let us focus on t_d (s), which is the phase time delay introduced by the filter circuit. (Refer to Section 5.1.4.2.) We realize, of course, that in (5.2.14) and (5.2.16), time, t, is always a variable and time delay, t_d, *is a constant at a given frequency*. However, t_d changes when input frequency, f (Hz), changes. The relationship between t_d and f can be derived as follows: We recall that

$$f_C\, (\text{Hz}) = \frac{1}{2\pi RC} \quad (5.1.17R)$$

Plugging (5.1.17R) into (5.2.16) and re-arranging the obtained formula yields

$$t_d\, (\text{s}) = -\frac{\Theta_{out}}{2\pi f} = \left(\frac{1}{2\pi f}\right) \tan^{-1}\left(\frac{f}{f_C}\right) = \left(\frac{1}{2\pi f}\right) \tan^{-1}\left(\frac{f}{2\pi RC}\right) \quad (5.2.17)$$

which shows how time delay depends on a signal's frequency, f (Hz), and the filter's parameters, R (Ω) and C (F). An example fully clarifies this point.

Example 5.2.3 Phase Shift and Time Delay of an RC LPF

Problem

An RC LPF is built from $R = 2.2\,\text{k}\Omega$ and $C = 0.02\,\mu\text{F}$ (see Example 5.1.4.) Build the graph t_d-vs.-f, and find the time delay at f (Hz) $= 0.1 f_C$, f (Hz) $= f_C$, and f (Hz) $= 5 f_C$.

Solution
Application of (5.1.17) produces

$$f_C\, (\text{Hz}) = \frac{1}{2\pi RC} = 3.62\,\text{kHz}$$

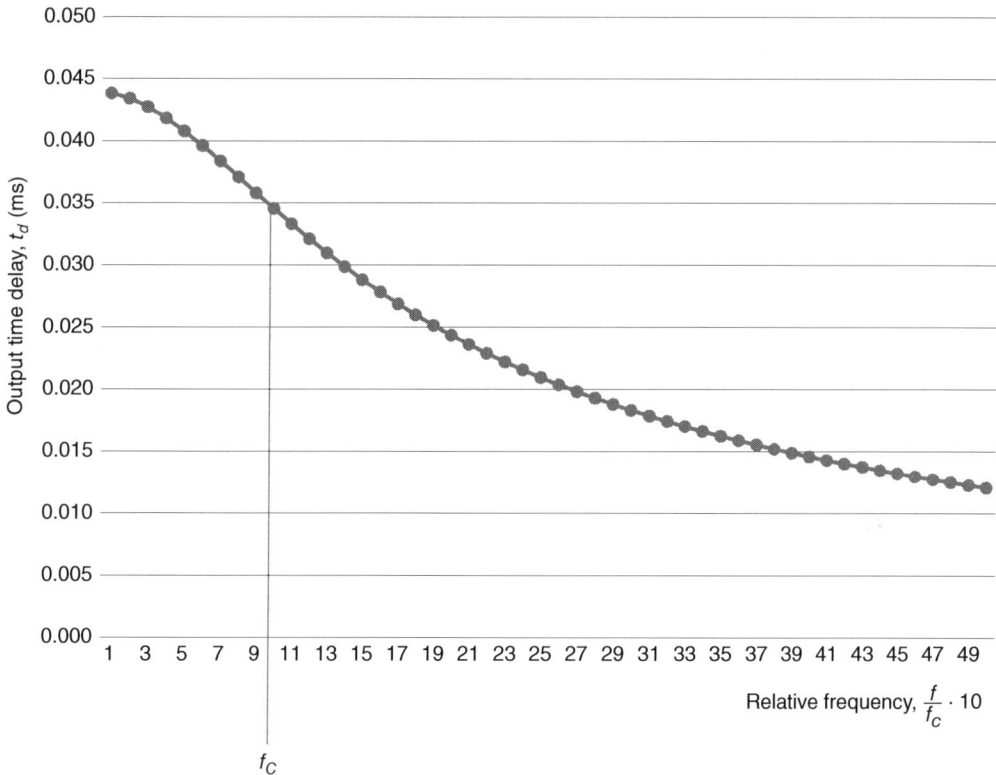

Figure 5.2.17 Time delay of an RC LPF output signal vs. frequency in Example 5.2.3.

Using (5.2.17), we build the required graph shown in Figure 5.2.17. Computed values of the time delay at the required frequencies are as follows:

$$t_{d1} = \left(\frac{1}{2\pi(0.1f_C)}\right) \tan^{-1}\left(\frac{0.1f_C}{f_C}\right) = 0.044 \text{ ms for } f \text{ (Hz)} = 0.1f_C$$

$$t_{d1} = \left(\frac{1}{2\pi f_C}\right) \tan^{-1}\left(\frac{f_C}{f_C}\right) = 0.035 \text{ ms for } f \text{ (Hz)} = f_C$$

and

$$t_{d1} = \left(\frac{1}{2\pi(5f_C)}\right) \tan^{-1}\left(\frac{5f_C}{f_C}\right) = 0.12 \text{ ms for } f \text{ (Hz)} = 5f_C$$

Discussion

Note how the time delay changes with a change in frequency: the higher the frequency, the smaller the time delay. (Can you explain why? *Hint*: Refer to the capacity properties.) This relationship between the time delay and the frequency of an input signal is not obvious because, as (5.2.17) shows, time delay is inversely proportional to frequency and, at the same time, has frequency in the formula's numerator as an argument of an arctangent. Clearly, a straight inverse frequency dependence, $\left(\frac{1}{2\pi f}\right)$, prevails over a direct frequency dependence encrypted in arctangent, $\tan^{-1}\left(\frac{f}{f_C}\right)$; the obtained results reflect this relationship.

5.2.2.2 Transfer Function of an RC LPF

The preceding discussion of the input–output filter relationship prepares us to describe this relationship mathematically by introducing the concept of *transfer function*. A transfer function represents the properties of a circuit and allows us to find the output signal when the transfer function and the input signal are given. The concept of a transfer function is visualized in Figure 5.2.18. We will introduce this concept by considering an RC LPF as an example. This approach can be easily applied to other types of filters and to other types of linear systems.

As will be explained in the following discussion, a transfer function has to be introduced in frequency domain. Thus, if $V_{in}(f)$ and $V_{out}(f)$ are input and output signals presented in *frequency domain*, then a transfer function, $H(f)$, is defined as

$$H(f) = \frac{V_{out}(f)}{V_{in}(f)} \tag{5.2.18}$$

Therefore, as stated previously,

> the output signal can be readily found if we know the input signal and the transfer function of a system:

$$V_{out}(f) = V_{in}(f) \times H(f) \tag{5.2.19}$$

If we review again the material in Sections 5.1 and 5.2, we realize that we did find the ratio of output to input signals in frequency domain – that is, its *transfer function* – for an RC LPF. In particular, this ratio is given in Eqs. (5.1.11) and (5.1.12), rewritten here by using (5.1.21) and (5.1.22) as

$$A_v(f) \equiv H(f) = \frac{V_{out}(f)}{V_{in}(f)} = \frac{1}{\sqrt{\left(1 + \left(\frac{R}{X_C}\right)^2\right)}} e^{j[-\tan^{-1}(R/X_C)]} \tag{5.2.20a}$$

We see that the transfer function of an RC LPF is a complex entity and therefore can be presented as

$$H(f) = \frac{V_{out}(f)}{V_{in}(f)} = \frac{1}{\sqrt{\left(1 + \left(\frac{f}{f_c}\right)^2\right)}} e^{-j\tan^{-1}(f/f_c)} = |H(f)|e^{j\Theta} \tag{5.2.20b}$$

Therefore, the amplitude is given by

$$|H(f)| = A_v = \frac{1}{\sqrt{\left(1 + \left(\frac{f}{f_c}\right)^2\right)}} \tag{5.2.20c}$$

and the phase of a transfer function is

$$\Theta = -\tan^{-1}\left(\frac{f}{f_c}\right) \tag{5.2.20d}$$

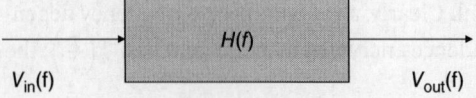

Figure 5.2.18 Concept of transfer function.

Then, the transfer function of a LPF can be presented as

$$H(f) = |H(f)|e^{-j\Theta} = A_v e^{j\Theta} \qquad (5.2.21)$$

where A_v and Θ are given above and by (5.1.21) and (5.1.22).

Now we can demonstrate the main purpose of introducing a transfer function: Combining (5.2.19) and (5.2.21), we can obtain an output signal when the input signal and the system (here, LPF) are known; that is,

$$V_{out}(f) = V_{in}(f) \times H(f) = V_{in}(f) \times A_v e^{j\Theta} \qquad (5.2.22)$$

Again and again, (5.2.22) is given in *frequency domain*.

We can present the transfer function in *phasor form* as

$$H(f) = |H(f)| \angle (H(f)) \qquad (5.2.23)$$

where the magnitude, $|H(f)| \equiv A_v$, and the phase, $\angle(H(f)) = \Theta$, of the transfer function of an RC LPF are given in (5.1.21R) and (5.1.22R).

(We refer you to Sections 2.2 and 2.3 to refresh your memory on the phasor form.)

Using the phasor form of a transfer function, we can obtain the output signal as follows:

$$V_{out}(f) = V_{in}(f) \times H(f) = V_{in}(f) \times |H(f)| \angle (H(f)) \qquad (5.2.24)$$

which, for an RC LPF, comes to

$$V_{out}(f) = V_{in}(f) \times \frac{1}{\sqrt{\left(1 + \left(\frac{f}{f_c}\right)^2\right)}} \angle - \tan^{-1}\left(\frac{f}{f_c}\right)^2 \qquad (5.2.25)$$

To find the *time-domain response* of an LPF, we need to transform the output signal, $V_{out}(f)$, found in frequency domain back into time domain. Whether you consider this transformation in traditional frequency-domain form, shown in (5.2.19), or in phasor form, shown in (5.2.23), the result will obviously be the same. We refer you to Sections 2.2 and 2.3, where all these transformations are discussed in detail.

The process of transferring from time domain to frequency domain and back is illustrated in Figure 5.2.19. Initially, we are given the time-domain input signal in the traditional format:

$$v_{in}(t) = V_{in} \cos(2\pi f t + \Theta_{in})$$

To transform the time-domain signal into frequency domain, we present this input sinusoidal waveform in frequency domain (or phasor domain) by its amplitude, V_{in}, and phase, Θ_{in}, leaving its frequency, f, as a default parameter. Since a linear system, such as RC LPF, does not change the waveform and frequency, we know that the output signal will also be a sinusoidal signal with the input frequency but with different amplitude and phase. We have found these new values of the output amplitude, V_{out}, and phase, $\Theta_{out} \equiv \Theta$, in frequency domain by using a transfer function, $H(f)$. Specifically, we calculate V_{out} and $\Theta_{out} \equiv \Theta$ by using (5.2.25). Now we can easily recover a time-domain presentation of the output signal by inserting the output amplitude and the output phase into a sinusoidal form of the input signal, which results in

$$v_{out}(t) = V_{out} \cos(2\pi f t + \Theta_{in} + \Theta_{out})$$

This formula gives us an output signal in time domain, as shown in Figure 5.2.19.

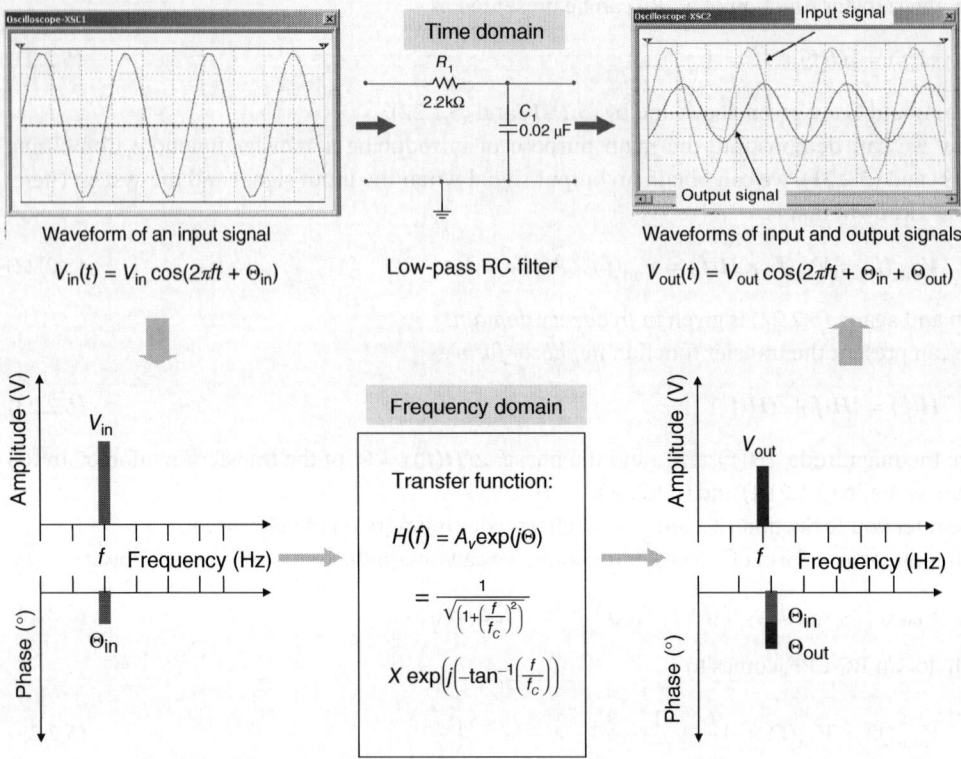

Figure 5.2.19 General diagram for finding the output signal of an RC LPF by using a transfer function.

In introducing a transfer function, we did not obtain any new formulas or facts; nevertheless, the concept of transfer function creates a new domain to operate. Specifically, it emphasizes how a system (in our case, the RC LPF) performs. We know for certain that an RC LPF reduces the amplitude of, and adds a negative phase to, an output signal. This result allows us to compute values of the output amplitudes and phases.

To sum up our discussion, consider the following example:

Example 5.2.4 The Use of the Transfer Function of an RC LPF

Problem

Consider the RC LPF discussed in Example 5.2.3 whose $R = 2.2\,\text{k}\Omega$ and $C = 0.02\,\mu\text{F}$. Present cosine signal $v(t) = 3\cos(2\pi ft)$ to this filter. Find the output signal at $f = 0.362\,\text{kHz}$, $f = 3.62\,\text{kHz}$, and $f = 36.2\,\text{kHz}$, using the filter's transfer function.

Solution

We computed the filter cutoff frequency as $f_C\,(\text{Hz}) = \frac{1}{2\pi RC} = 3.62\,\text{kHz}$. Therefore, the ratios of $\frac{f}{f_C}$ we need for these computations are 0.1, 1, and 10.

To find the output signal by using a transfer function, we need to apply (5.2.19), $\boldsymbol{V_{out}}(f) = \boldsymbol{V_{in}}(f) \times H(f)$, which has taken on a more specific form in (5.2.22): $\boldsymbol{V_{out}}(f) = \boldsymbol{V_{in}}(f) \times A_v e^{j\Theta}$. To calculate the amplitude and phase responses of an RC LPF, we apply (5.2.20c) and (5.2.20d), $A_v = \dfrac{1}{\sqrt{\left(1+\left(\frac{f}{f_C}\right)^2\right)}}$

and $\Theta = -\tan^{-1}\left(\dfrac{f}{f_C}\right)$, and compute

for f (Hz) $= 0.1 f_C$, $V_{out}(f) = 2.985$ V and $\Theta = -5.71°$

for f (Hz) $= f_C$, $V_{out}(f) = 2.121$ V and $\Theta = -45°$

and

for f (Hz) $= 10 f_C$, $V_{out}(f) = 0.292$ V and $\Theta = -84.29°$

Discussion

- Working out this problem, we must clearly understand not only the main point of introducing a transfer function – finding the signal at the output of a device with a given input signal and the device's transfer function – but also the practical way to apply this concept to specific calculations.
- Another quandary confronting us is this: What if an input signal is not a single sinusoid but any arbitrary waveform? We will postpone consideration of such a situation until we study the spectral analysis of signals in Chapters 6 and 7.

We need now to make several points to elaborate on our *discussion of the transfer function*:

- The input and output signals and the transfer functions are defined in frequency domain.
- $H(f) \equiv H(j\omega)$ because $\omega = 2\pi f$. Here, j is inserted to show that a transfer function is a complex entity. You can use just "ω" to shorten the wording and equations.
- Refer to Section 2.3, where Euler's identity, the exponential form, and the rules of manipulation of exponents are discussed.
- The concept of transfer function implies that V_{out} is measured across the *open-circuit output terminals* of the circuit or device whose transfer function is being defined.
- We must use the exponential forms of both input and output signals and the transfer function to obtain the correct results, but we can turn back to the trigonometric form of the output signal by taking <u>real</u> for cosine or <u>imaginary</u> for the sine parts of the exponential output form. Then an output signal can be written as

$$v_{out}(t) = V_{out} \cos(2\pi f t + \Theta_{in} + \Theta)$$

- We can present the input and output signals as rotating phasors; that is, $v_{in}(t) = V_{in} \times e^{j(2\pi f t + \Theta_{in})}$ and $v_{out}(t) = V_{in} A_v \, e^{j(2\pi f t + \Theta_{in} + \Theta)} = V_{out} e^{j(2\pi f t + \Theta_{in} + \Theta)}$.
- In reality, the signals we operate with are given in time domain. To obtain the output signal when the input signal and the device are known, we need to transform these signals into frequency domain and use the device's transfer function. After obtaining the output signal's amplitude and phase shift in frequency domain, we return to time domain by obtaining the equation describing the signal's waveform, which is the final result of these transformations.

5.2.2.3 Graphical Presentation of a Transfer Function: Bode Plots

We can present a transfer function in a graphical form by depicting its amplitude and phase graphs vs. frequency. In fact, we use this presentation for an RC LPF many times in the preceding sections. See, for example, Figures 5.1.9 and 5.1.10, where both the amplitude and phase-shift responses of an RC LPF are shown. These graphs are visualizations of (5.2.25) or its version (5.2.20a) and (5.2.20c).

Since these equations show how to use the transfer-function concept for finding the output signal, the A_v-vs.-f and Θ-vs.-f graphs are, in essence, the graphical forms of a filter's transfer function.

However, to visualize a transfer function, it is customary to build these graphs in logarithmic scale. Specifically, *A_v should be shown in decibels, Θ should be given in radians (or degrees), and frequency should be in logarithmic scale*. There are two main reasons for building the amplitude graph in logarithmic scale and the phase graph in semi-logarithmic scale: First, it allows us to cover, graphically, a much greater scale of the frequency change than can be done in natural scale. To be sure, showing a change in the frequency from 0.001 to 100 000 Hz takes only eight marks in logarithmic scale (see Figure 5.1.12), whereas it would be practically impossible to show this range of frequencies in one graph in natural scale. The second, and more important, reason for introducing logarithmic scale is the ease it provides for building these graphs. Such graphs enable us to build the graphical presentation of the transfer function of a complex system by combining the graphs of individual blocks of this system.

Let us consider how to build A_v-vs.-f and Θ-vs.-f graphs in logarithmic and semilogarithmic scales for an RC LPF.

We start building these graphs by recalling that the transfer function of this filter is defined by (5.1.21) and (5.1.22). By definition, the filter's amplitude response, A_v, in decibels is

$$A_v \text{ (dB)} = 20 \, \log \left(\frac{V_{\text{out}} \text{ (V)}}{V_{\text{in}} \text{ (V)}} \right) \tag{5.2.26}$$

We remember, of course, that here we employ the base-10 logarithm. Plugging (5.1.21) into (5.2.26), we find

$$A_v \text{ (dB)} = 20 \, \log \left(\frac{V_{\text{out}}(V)}{V_{\text{in}}(V)} \right) = 20 \, \log \left(\frac{1}{\sqrt{\left(1 + \left(\frac{f}{f_C}\right)^2\right)}} \right)$$

$$= 20 \, \log(1) - 20 \, \log \left(\sqrt{\left(1 + \left(\frac{f}{f_C}\right)^2\right)} \right) \tag{5.2.27}$$

To analyze this formula, let us consider two extreme cases, as we usually do in the analysis of a filter. First, when $f \ll f_C$, we get

$$A_v \text{ (dB)} = 20 \, \log(1) - 20 \, \log \sqrt{\left(1 + \left(\frac{f}{f_C}\right)^2\right)} \approx 20 \, \log(1) - 20 \, \log(1) = 0 \text{ for } f \ll f_C \tag{5.2.28}$$

Thus, in a low-frequency region, $f \ll f_C$, the A_v-vs.-f graph is a straight line parallel to the frequency axis, that is, the line with the zero slope.

The other extreme case, when $f \gg f_C$, results in

$$A_v(\text{dB}) = 20 \, \log(1) - 20 \, \log \sqrt{\left(1 + \left(\frac{f}{f_C}\right)^2\right)} \approx -20 \, \log \left(\frac{f}{f_C}\right) \text{ for } f \gg f_C \tag{5.2.29}$$

5.2 Filtering – Introduction

The term $-20\log\left(\frac{f}{f_C}\right)$ *in (5.2.29) describes a straight line for which an increase in the* $\frac{f}{f_C}$ *ratio by a factor of 10 (we call it a decade) results in a decrease of A_v by 20 dB.* To visualize this statement, consider the following sequence:

When $\frac{f}{f_C} = 1$, this member, $-20\log\left(\frac{f}{f_C}\right)$, is equal to 0 dB because $\log_{10}(1) \equiv \log_{10}(10^0) = 0$;

when $\frac{f}{f_C} = 10$, then $-20\log\left(\frac{f}{f_C}\right) = -20$ dB because $\log_{10}(10) \equiv \log_{10}(10^1) = 1$;

when $\frac{f}{f_C} = 100$, then $-20\log\left(\frac{f}{f_C}\right) = -40$ dB because $\log_{10}(100) \equiv \log_{10}(10^2) = 2$;

when $\frac{f}{f_C} = 1000$, then $-20\log\left(\frac{f}{f_C}\right) = -60$ dB because $\log_{10}(1000) \equiv \log_{10}(10^3) = 3$; and so on.

Since the cutoff frequency, f_C, is constant for a given filter, a change in the ratio $\frac{f}{f_C}$ is equivalent to a change in the frequency, f. Therefore, we can say that *the line, A_v (dB), decreases at a rate of -20 dB per decade of frequency change*.

We obtained two straight lines that graphically represent the A_v-vs.-f parts of the RC LPF's transfer function in logarithmic scale at low, $f \ll f_C$, and high, $f \gg f_C$, frequencies. These lines must intersect because this graph must cover all frequencies. And, of course, they do intersect at the point where $f = f_C$ because this is the only point where (5.2.28) and (5.2.29), describing these two lines, are equal to each other. (Prove this statement.) Let us stress one more time that we are discussing straight-line approximation of the real A_v-vs.-f graph for two extreme cases – $f \ll f_C$ and $f \gg f_C$. This graph is depicted in Figure 5.2.20a.

To build the piecewise approximation of the phase-shift graph, we need to analyze (5.1.22),

$$\Theta_{out} \equiv \Theta = -\tan^{-1}\left(\frac{f}{f_C}\right)$$

The problem with the phase graph is that it does not have natural points of intersections of its straight lines. The solution to this problem has been achieved based on the best approximation of a real smooth phase graph by a piecewise function. This solution is presented by the following rule:

1. The low-frequency line (with a 0°/decade slope) starts at f (Hz) = 0 and intersects the mid-frequency, linear decreasing line (with a $-45°$/decade slope) at $f = 0.1f_C$. In this range, $\Theta = 0°$.

2. The mid-frequency line (with a $-45°$/decade slope) starts at f (Hz) = $0.1f_C$ and intersects the high-frequency line (with a 0°/decade slope) *at f (Hz) = $10f_C$*. Therefore, in the range $0.1 \leq f/f_C \leq 10$, the phase shift changes from $\Theta = 0°$ to $-90°$.
   At $f/f_C = 1$, the output phase shift is given by $\Theta = -45°$.

3. The high-frequency line starts at f (Hz) = $10f_C$ and continues (with a 0°/decade slope) to the end of a given frequency band. Here $\Theta = -90°$.

All three lines are shown in Figure 5.2.20b.

The straight-line approximations of the real A-vs.-f and Θ-vs.-f graphs shown in Figure 5.2.20 are called *Bode plots*, after Hendrik Wade Bode, an American scientist who introduced these plots in the middle of the twentieth century. (Incidentally, Bode worked at Bell Labs, where he accomplished major developments in network analysis. The results of his pioneering work are still being applied.)

If we compare the Bode plots of the amplitude and phase of an RC LPF frequency response with

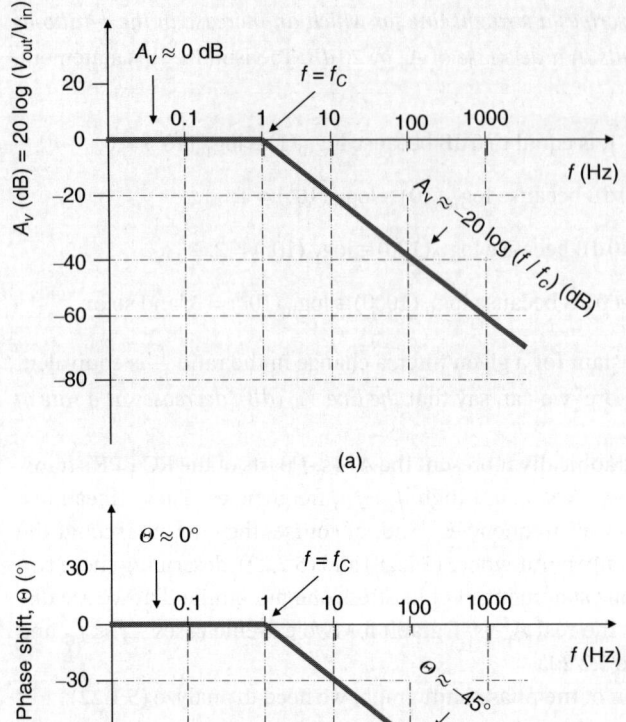

Figure 5.2.20 Straight-line approximation of (a) amplitude and (b) phase responses of an RC LPF in logarithmic scale (Bode plots).

the actual graphs – built with A_v in dB, Θ in radians (or degrees), and frequency in logarithmic scale – you can readily see how similar these graphs are. In fact, they are almost the same; the only noticeable difference is in the area where f is close to f_C and $f \gg f_C$: The real graphs are smooth curves, whereas the Bode plots are still straight lines. This can be clearly seen in Figure 5.2.21. This difference stems from the fact that the actual graphs are precise, whereas the Bode plots are approximations. In other words,

> *if we approximate the precise, smooth A_v-vs.-f and Θ-vs.-f graphs by the straight lines, we will obtain Bode plots.*

Obviously, approximation becomes particularly essential (and visible) in the regions where the slopes of the lines change.

What errors do we introduce by using Bode plots? Well, they are really quite minor. For the amplitude, the error is 0 dB when $f = 0.1 f_C$ and $f = 10 f_C$; the error is −3 dB when $f = f_C$. For phase, the

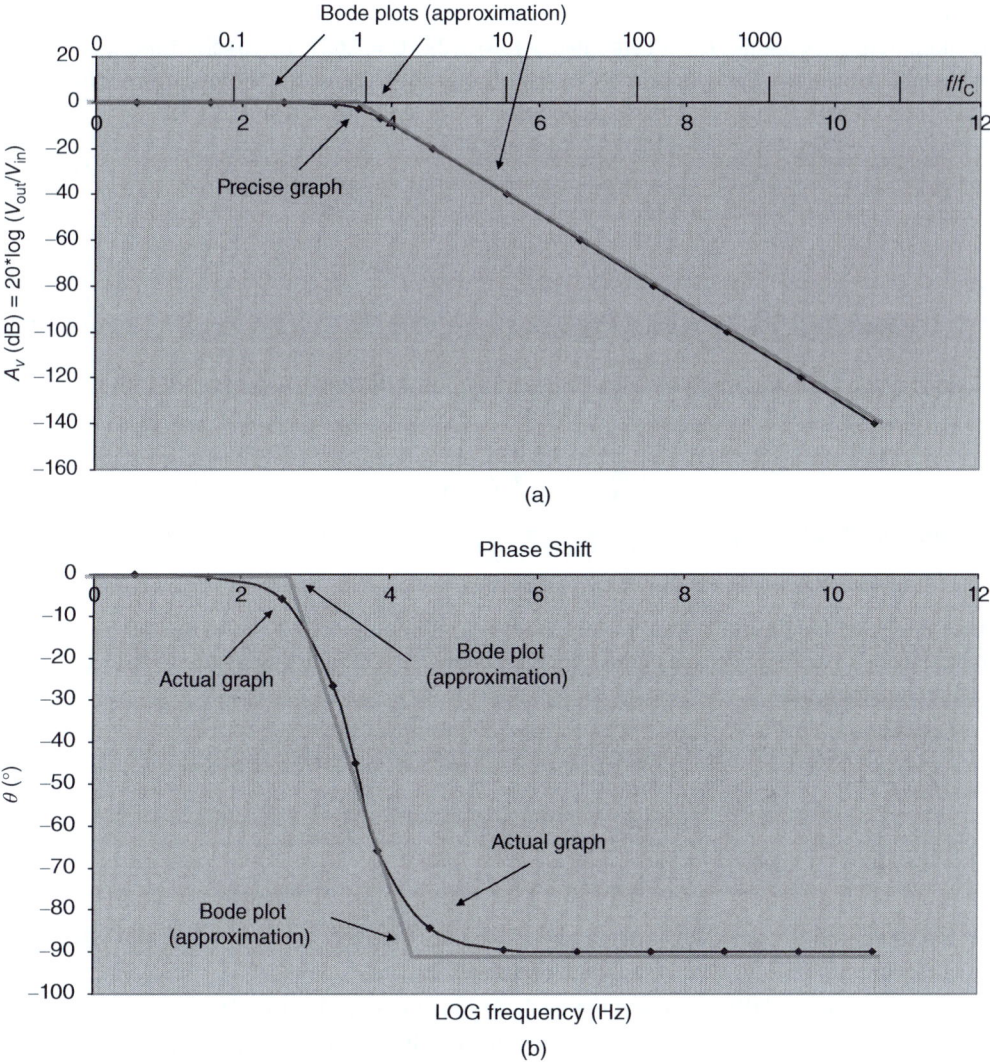

Figure 5.2.21 Actual graph and approximate Bode-plot graph of an RC LPF frequency response for (a) amplitude and (b) phase. (Not to scale.)

error is $-5.7°$ and $+5.7°$ when $f = 0.1 f_C$ and $f = 10 f_C$, respectively; the error is almost $0°$ when $f = f_C$ (Rizzoni 2009, p. 324).

To better understand the concept of Bode plots, consider the following example.

Example 5.2.5 Bode Plots of Amplitude and Phase Responses of an RC HPF

Problem

Build Bode plots for the amplitude and phase responses of the RC HPF based on the parameters $R = 2.2\,\text{k}\Omega$ and $C = 0.02\,\mu\text{F}$ (see Example 5.2.1).

Solution

The cutoff frequency of this HPF is computed as $f_C = 1/2\pi RC = 3.62$ kHz. To solve this problem, we need to refer to Section 5.2.1, where an RC HPF is discussed. The constituents of a transfer function of this filter – the amplitude and phase responses – are given by (5.2.6) and (5.2.7):

$$A_v = \frac{1}{\sqrt{1 + \left(\frac{f_C}{f}\right)^2}} \tag{5.2.6R}$$

and

$$\Theta = \tan^{-1}\left(\frac{f_C}{f}\right) = 90° - \tan^{-1}\left(\frac{f}{f_C}\right) \tag{5.2.7R}$$

Following the same pattern that we developed for building the Bode plots of an RC LPF, we derive

$$A_v \text{ (dB)} = 20 \log(1) - 20 \log\left(\sqrt{1 + \left(\frac{f_C}{f}\right)^2}\right) \tag{5.2.30}$$

Again, when $f \ll f_C$, we get

$$A_v \text{ (dB)} \approx -20 \log(f_C) + 20 \log(f)$$

(Make sure to derive this formula explicitly.) This is a straight line, as it was in the case of an LPF, but its slope is +20 dB per decade of the frequency change and its starting point is $-20 \log(f_C)$. For our filter, we compute

$$-20 \log(f_C) = -20 \log(3620) = -71.2 \text{ dB}$$

When $f \gg f_C$,

$$A_v \text{ (dB)} = 20 \log(1) - 20 \log\left(\sqrt{1 + \left(\frac{f_C}{f}\right)^2}\right) \approx 0$$

This is a straight line of zero slope. Again, these two lines intersect at $f = f_C = 3.62$ kHz. The Bode plot of the amplitude of an HPF is shown in Figure 5.2.22a.

For the phase of this HPF, we repeat the analysis done for an LPF: Since $\Theta = \tan^{-1}\left(\frac{f_C}{f}\right)$, we can approximate the actual Θ-vs.-f graph by employing the following three straight lines:

Line 1: For the low-frequency band, $f \ll f_C$, we have a straight line of zero slope. Here, the tangent goes to infinity; that is, $\Theta = 90°$. The intersection point with a mid-frequency line is at $f = 0.1 f_C$.

Line 2: For the mid-frequency band, $0.1 f_C < f < 10 f_C$, we have a straight line of a 45°/decade slope. Here, when $f = f_C$, the tangent goes to 1; that is, $\Theta = 45°$. The intersection points with two other lines are at $f = 0.1 f_C$ and at $f = 10 f_C$.

Line 3: For the high-frequency band, $f \gg f_C$, we have a straight line of zero slope. Here, the tangent goes to zero; that is, $\Theta = 0°$. The intersection point with the mid-frequency line is at $f = 10 f_C$.

The Bode plot for this phase response is shown in Figure 5.2.22b.

Discussion

Three points to discuss here:

- *First point*: Building the accurate Θ-vs.-f graph for a specific case could be a puzzle because we do not know at what points (frequency values) the graph changes its slope from zero to

Figure 5.2.22 Bode plots of (a) amplitude and (b) phase of an RC HPF in Example 5.2.5.

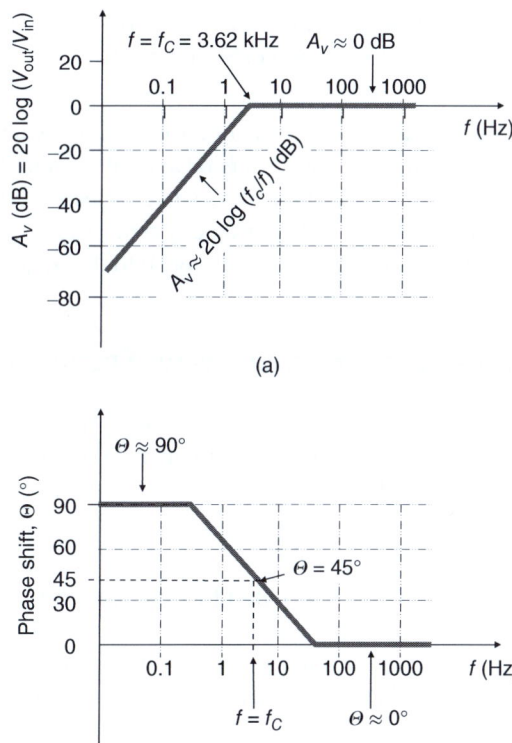

+20 dB/decade and from that slope to zero. The solution lies in the accepted rule we discussed above. To check the accuracy of building the amplitude lines, recall that they must intersect at $\frac{f_C}{f} = 1$. For the phase lines, the best check of accuracy is also the point at $\frac{f_C}{f} = 1$, where Θ must be 45°. Of course, we always must check the value of a line's slope. These checking points are sufficient to ensure the accuracy of building the entire Bode plot. See Figure 5.2.22.

- *Second point*: When building Bode plots for LPFs and HPFs, we constantly refer to conditions like $f \ll f_C$ or $f \gg f_C$. Then, we accept that $\frac{f}{f_c} = 0.1$ and $\frac{f}{f_c} = 10$ are the borders that separate low-frequency from mid-frequency ranges and mid-frequency from high-frequency ranges. However, the question remains: What ratio of f to f_C would allow us to consider that the current frequency, f, is much lower or much higher than the cutoff frequency, f_C? We need to understand that there are no unique values that define these conditions. In other words, we cannot say that $\frac{f}{f_c} = 10$ is the border at which the condition $f \gg f_C$ is satisfied; nor can we say that $\frac{f}{f_c} = 0.1$ is the threshold for the $f \ll f_C$ condition. In fact, $\frac{f}{f_c} = 9$ or $\frac{f}{f_c} = 11$ could satisfy the condition $f \gg f_C$ as well as $\frac{f}{f_c} = 10$. However, it is customary to consider that both conditions – $f \ll f_C$ and $f \gg f_C$ – can be applied when the current frequency is lower or higher than the critical frequency by an *order of magnitude*. In other words, when $\frac{f}{f_c} = 10$, we can say $f \gg f_C$, and when $\frac{f}{f_c} = 0.1$, we can say that $f \ll f_C$. Nevertheless, the conditions – much lower or much higher – are determined by the required precision of our approximation. We trust that the examples discussed in Section 5.1 and in this section provide enough information by which we can judge these conditions reasonably well.

- *Third point*: To explore this topic further, we recommend that you build Bode plots for the BPF and the BSF discussed in Section 5.2.1.

Bode plots, as we have mentioned already, are a powerful and convenient tool for analyzing and designing any systems in general and filters in particular. They were one of the major tools in the precomputer era that scientists and engineers used; they are still an important tool helping in the analysis and design of electronic systems. Figure 5.2.11 shows how the industry uses Bode plots in practice. We will delve more deeply into the subject of Bode plots in future discussions.

Questions and Problems for Section 5.2

- Questions marked with an asterisk require a systematic approach to finding the solution.
- Many questions and problems, including those marked with an asterisk, imply that you, in addition to reading the textbook, will do your research to find the answers. Consider such questions as mini-projects.

High-Pass Filter (HPF), Band-Pass Filter (BPF), and Band-Stop Filter (BSF)

1. Differentiate among low-pass, high-pass, band-pass, and band-stop filters and explain what these filters do. Qualitatively sketch attenuation vs. frequency graphs for each of these filters.

2. In reviewing the properties of filter's components, consider two inductors ($L_1 = 0.01$ mH and $L_2 = 0.1$ mH), two capacitors ($C_1 = 0.01$ μF and $C_2 = 0.1$ μF), and two resistors ($R_1 = 0.1$ kΩ and $R_2 = 0.01$ kΩ). Among all the components, which reactance or resistance will be greater at frequency $f = 20$ kHz? Prove your answers.

3. You are given only two passive components, R (Ω) and C (F). Can you build a high-pass filter with these components? Prove your answer.

4. You have a voltage source, a resistor, and a capacitor to be connected in series in any order. Based on Kirchhoff's voltage law, build an HPF from these components and explain your reasoning.

5. Show the schematic of an HPF built from a resistor and an inductor connected in series. Explain how this filter operates when f (Hz) $\to 0$ and f (Hz) $\to \infty$. Qualitatively draw its attenuation characteristic.

6. Compare two filters – one low-pass and the other high-pass; each is built with R (Ω) and C (F) components. Explain the similarities and the difference between these filters.

7. Consider an RC HPF. Derive the formulas for its attenuation, $A_v = V_{out}/V_{in}$; its phase shift, Θ; and its cutoff frequency, f_C.

8. Consider an RC HPF:
   a) The filter is built with $R = 4.7$ kΩ and $C = 0.002$ μF. Sketch the schematic of this filter. Compute its cutoff frequency, f_C, and $A_v = V_{out}/V_{in}$ and Θ at f_C.

b) Build the attenuation and phase shift characteristics of this filter by computing A_v and Θ at $f = 0.1f_C, f = 0.5f_C, f = f_C, f = 2f_C, f = 5f_C,$ and $f = 10f_C$.

c) Assuming $v_{in}(t) = 5\cos(6328t)$, qualitatively sketch the waveforms of the input and output signals on the same set of axes at $f = 0.1f_C, f = f_C, f = 2f_C,$ and $f = 10f_C$.

9 Consider an RL high-pass filter. Derive the formulas for its attenuation, A_v; its phase shift, Θ; and its cutoff frequency, f_C. (*Hint*: See Appendix 5.2.A.1.)

10 Consider an RL HPF discussed in Appendix 5.2.A.1:
   a) The filter is built with $R = 4.7\,k\Omega$ and $L = 0.2\,mH$. Sketch the schematic of this filter. Compute its critical frequency and $A_v = V_{out}/V_{in}$ and Θ at the critical frequency.
   b) Build the attenuation and phase shift characteristics of this filter by computing A_v and Θ at $f = 0.1f_C, f = 0.5f_C, f = f_C, f = 2f_C, f = 5f_C,$ and $f = 10f_C$.
   c) Assuming $v_{in}(t) = 5\cos(6328t)$, qualitatively sketch the waveforms of the input and output signals on the same set of axes at $f = 0.1f_C, f = f_C, f = 2f_C,$ and $f = 10f_C$.

11 What three main parameters do you need to know to characterize an RC HPF? Explain your reasoning.

12 What is the function of a BPF? Qualitatively sketch its attenuation characteristic to explain your answer.

13 The text says that a BPF is a combination of an LPF and an HPF. Why do we need those two types of filters to build a BPF? Draw a figure with attenuation graphs of an LPF and an HPF to explain your answer.

14 Explain why a BPF has two cutoff frequencies. Why does an LPF (or an HPF) have only one cutoff frequency? Compare a BPF with an LPF and an HPF in this regard.

15 Can you build an ideal BPF from an ideal HPF whose cutoff frequency is $f_{C1} = 14\,kHz$ and an ideal LPF whose $f_{C2} = 11\,kHz$? Explain your answer by sketching the attenuation graphs of these filters.

16 *Consider a BPF built from an LPF and an HPF:
   a) What condition governs the relationship between cutoff frequencies of the constituent filters?
   b) How does the above condition affect the values of this BPF's circuit components?

17 Consider a BPF built from an LPF and an HPF:
   a) How must these filters be connected – in parallel or in series?
   b) If the constituent filters must be connected in series, is it important as to which filter – LPF or HPF – must be placed first? Why?
   c) Sketch a block diagram to explain your answer.

18 *Consider the circuit of a BPF:
   a) Draw a schematic of this circuit.

b) Design an RC BPF whose $f_{C1} = 4\,\text{kHz}$ and whose $f_{C2} = 16\,\text{kHz}$. Show all values of the resistors and capacitors. (*Hint*: There is no unique solution to this problem. You can choose any component values to meet the requirements on cutoff frequencies.)

19 A BPF is built from an HPF whose $f_{C1} = 10\,\text{kHz}$ and an LPF whose $f_{C2} = 12\,\text{kHz}$:
  a) What is the passband of this BPF? What is its bandwidth?
  b) What can you do to increase (or decrease) the passband of a BPF? The bandwidth of a BPF?

20 *A BPF is built from an HPF whose $f_{C1} = 7\,\text{kHz}$ and an LPF whose $f_{C2} = 15\,\text{kHz}$:
  a) What is its central frequency, f_0?
  b) Why do we need a central frequency to characterize a BPF?
  c) Show all three abovementioned frequencies in an attenuation graph.

21 *Consider the quality factor, Q, of a BPF:
  a) Quality factor is a parameter usually associated with a resonance circuit. Why do we use this factor to characterize a BPF?
  b) Compute a BPF's quality factor if its cutoff frequencies are 4 and 9 kHz. (*Hint*: See Appendix 5.2.A.2.)

22 Compare two BPFs, one whose $Q_a = 20$ and the other whose $Q_b = 400$: Which BPF would you prefer to use in your application? Why?

23 *Consider the frequency response of a BPF whose $f_{C1} = 6\,\text{kHz}$ and whose $f_{C2} = 24\,\text{kHz}$:
  a) At what frequency will the amplitude response reach its maximum, V_{out}^{max}?
  b) Is V_{out}^{max} equal to V_{in}? If not, why not?
  c) What will be the phase response at the central frequency?
  d) Qualitatively sketch the amplitude and phase responses of this filter and show all the values that you can find based on the given data.

24 *If we change the values of resistors and inductors in an RL BPF, what characteristics of this filter will change? List all of them and explain their relationship to the values of R (Ω) and L (H) (see Appendix 5.2.A.1).

25 *Consider the characterization of an RC BPF:
  a) Derive the formula for the amplitude response of a BPF.
  b) Derive the formula for the phase response of a BPF.
  c) Find the values of the output amplitude and phase of a BPF at $f_{C1} = 5\,\text{kHz}, f_{C2} = 20\,\text{kHz}$, and at the central frequency.
  d) Sketch the graphs of both amplitude and phase responses.

26 *Find an example of the application of a BPF in communication devices and systems.

27 Why do we need a BSF? What possible applications of this filter can you list?

28 Explain why a BSF has two cutoff frequencies.

29 Can you build an ideal BSF from an ideal LPF whose cutoff frequency is $f_{C1} = 12\,\text{kHz}$ and an ideal HPF whose $f_{C2} = 16\,\text{kHz}$? Explain your answer by sketching the attenuation graphs of these filters.

30 Consider the concept of designing a BSF as a combination of an LPF and an HPF:
   a) What main criterion for this design would you need to set up?
   b) Sketch the amplitude responses of an ideal LPF, an ideal HPF, and an ideal BSF to explain the function of a BSF.
   c) In what topology – series or parallel – would you connect an LPF and an HPF? Explain your reasoning.

31 Sketch the block diagram of the circuitry for a BSF and explain conceptually the principle of the BSF operation.

32 Consider an RC BSF built from an LPF whose parameters are $R_1 = 4\,k\Omega$ and $C_1 = 6\,nF$ and HPF whose parameters are given by $R_2 = 2\,k\Omega$ and $C_2 = 3\,nF$:
   a) Compute the cutoff frequencies f_{C1} (HPF) and f_{C2} (LPF).
   b) Compute the central stop frequency, f_0 (Hz).
   c) Compute the stopband.

33 What can we do at the circuit level to control the stop bandwidth of a BSF?

34 *Consider the circuit of a BSF:
   a) Sketch the circuit schematic.
   b) Design an RC BSF whose LPF's $f_{C1} = 4\,kHz$ and whose HPF's $f_{C2} = 16\,kHz$. Show all the values of the resistors and the capacitors. (*Hint*: There is no unique solution to this problem. You can choose any component values to meet the requirements on cutoff frequencies.)

35 Consider a BSF built from an LPF whose $f_{C1} = 5\,kHz$ and an HPF whose $f_{C2} = 20\,kHz$: What is the stopband of this BPF? What is its stop bandwidth?

36 *A BSF is built from an LPF whose $f_{C1} = 4\,kHz$ an HPF whose $f_{C2} = 39\,kHz$:
   a) What is its central frequency, f_0?
   b) Why do we need a central frequency in the characterization of a BSF?
   c) Show all three abovementioned frequencies in an attenuation graph.

37 Consider the quality factor, Q, of a BSF:
   a) Quality factor is a parameter usually associated with a resonance circuit. Why do we use this factor to characterize a BSF?
   b) Compute a BSF's quality factor if its cutoff frequencies are 4 and 9 kHz. (*Hint*: Quality factor is discussed in Appendix 5.2.A.2.)

38 Compare two BSFs, one whose $Q_a = 20$ and the other whose $Q_b = 40$: Which BSF would you prefer to use in your application? Why?

39 *Consider the frequency response of a BSF whose $f_{C1} = 6\,kHz$ and whose $f_{C2} = 24\,kHz$:
   a) At what frequency will the amplitude response reach its minimum value, V_{out}^{min}?
   b) What would be an ideal V_{out}^{min}? Why?
   c) What will be the BSF's phase response at the central frequency? Explain your reasoning.
   d) Qualitatively sketch the amplitude and phase responses of this filter and show all the values that you can.

40 *If we change the values of resistors and inductors of an RL BSF, what characteristics of this filter will change? List all of them and explain their relationship to the values of R (Ω) and L (H) (see Appendix 5.2.A.1 for RL filters).

41 *Consider the characterization of a BSF:
a) Derive the formula for the amplitude response of the RC BSF.
b) Derive the formula for the phase response of the RC BSF.
c) Compute the values of the output amplitude and phase of this BSF at $f_{C1} = 4\,\text{kHz}$, $f_{C2} = 24\,\text{kHz}$, and at the central frequency.
d) Sketch the graphs of both the amplitude and phase responses.

42 *Find an example of using a BSF in communication devices and systems.

Transfer Function of a Filter

43 Sketch circuit diagrams of an RC LPF and an RL LPF. Show how you can present to these filters a sinusoidal signal, $v_{in}(t) = V_{in}\cos(\omega_{in}t + \theta_{in})$, and how you will measure an output signal. In subsequent problems, we will discuss this situation.

44 If we present a sinusoidal signal, $v_{in}(t) = V_{in}\cos(\omega_{in}t + \theta_{in})$, to a first-order passive filter, what happens to each parameter of the output signal? Does it increase, decrease, or not change at all? Explain why each change, if any, will occur.

45 The formula for a sinusoidal input signal is given by: $v_{in}(t) = V_{in}\cos(\omega_{in}t + \theta_{in})$. This signal passes through a first-order passive LPF. Write the formula for an output signal, compare every parameter of the input and output signals, and comment on all changes in the output signal.

46 The text states that the frequency of an input signal will not change after the signal passes the first-order passive filter. Explain why.

47 The amplitude of a signal presented to an RC LPF will decrease. Why?

48 The input amplitude of a sinusoidal signal is $V_{in} = 12$ V. This signal passes through an RC LPF with cutoff frequency $f_C = 4$ kHz. What will be the output amplitude, V_{out}, at f (Hz) $= 0.1 f_C$? At f (Hz) $= f_C$? At f (Hz) $= 10 f_C$?

49 *An RC filter introduces time delay to a signal passing the filter:
a) Why does this happen?
b) How can we measure this time delay with an oscilloscope?

50 *An RC LPF is built from the following components: $R = 4.7\,\text{k}\Omega$ and $C = 0.02\,\mu\text{F}$:
a) What will be the time delays, t_d (s), at f (Hz) $= 0.5 f_C$, at f (Hz) $= f_C$, and at f (Hz) $= 5 f_C$? Sketch the graph t_D (s) vs. f (Hz).
b) What will be the output phase shifts at each of the three above frequencies? Sketch the graph θ (rad) vs. f (Hz).
c) Comment on how the time delay and the output phase shift depend on the input signal's frequency, f (Hz).

Figure 5.2.P51 The waveforms of input and output signals of an RC LPF. Source: Reproduction of a part of Figure 5.2.16.

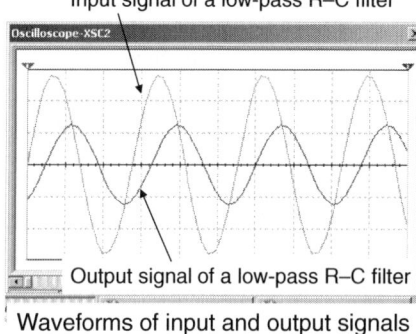

Waveforms of input and output signals

51 *Figure 5.2.P51 shows the waveforms of an input and output signal of an RC LPF whose $f_C = 8$ kHz. Determine the approximate time delay between these two waveforms if f (Hz) $= f_C$. (*Hint*: First of all, determine the output phase shift.)

52 Define the transfer function of a filter.

53 *Derive the formula of the transfer function of an RC LPF. (*Hint*: Refer to the definition of a transfer function.)

54 A sinusoidal signal, $v_{in}(t)$ (V) $= V_{in} \cos(\omega_{in} t)$, whose $V_{in} = 12$ V and whose $\omega_{in}(t) = 6280$ rad/s, is presented to an RC LPF. Using the filter's transfer function, find the formulas for the amplitude and the phase shift of the output signal and compute these values at $f/f_C = 0.5$, $f/f_C = 1$, and $f/f_C = 5$.

55 *Suppose an RC LPF delivers the output signal whose $V_{out}(f) = 6$ V (at $f = f_C$) and whose $\Theta_{out}(f) = -45°$ (at $f = f_C$). Reconstruct the waveform of the output signal.

56 *Why do we need a transfer function of a first-order LPF if we already know the formulas for its amplitude and phase responses?

57 The text says that a transfer function is defined in frequency domain, which implies that both the input and output signals are also presented in frequency domain. How then can we obtain real signals, the signals in time domain, by using the transfer function?

58 Consider the Bode plots: The text says that they are approximate presentations of the real amplitude and phase responses of a filter. Why do we need approximate presentations if we can obtain the actual responses?

59 Bode plots are graphical presentations of a filter's responses in logarithmic or semi-logarithmic scales. Why do we need logarithmic scales instead of natural scales for building these graphs?

60 *Consider an RC LPF whose $R = 4.4$ kΩ and $C = 0.04$ μF:
 a) Build a Bode plot of the filter's amplitude response in logarithmic scale for two regions: when $f \ll f_C$ and when $f \ll f_C$. Find the slopes of these two lines. Show the formulas you use.

b) Show the intersection points of these lines. Give your reasoning for this finding.
c) Build a Bode plot of its phase response.
d) Explain each step of your operations.

61 Bode plots are the approximations of the actual amplitude and phase responses. How accurate are these approximations? Give your estimations.

62 The HPF is built based on the parameters $R = 4.7\,\text{k}\Omega$ and $C = 0.03\,\mu\text{F}$:
a) Build the Bode plots of the amplitude and phase responses of this filter.
b) Estimate the errors introduced by Bode plots compared to the actual graphs at the checking points.
c) What ratios of $\frac{f}{f_C}$ are considered as borders for conditions $f \ll f_C$ and $f \gg f_C$. Give your reasoning.

63 Qualitatively build Bode plots for the amplitude and phase responses of a BPF and a BSF.

5.2.A RL Filter and Resonance Circuits as Filters

5.2.A.1 RL Filter

There is yet another passive component – the inductor, L – whose reactance depends on frequency. Inductive reactance, you recall, is given as follows:

$$X_L\,(\Omega) = \omega L = 2\pi f L \qquad (5.2.\text{A}.1.1)$$

where L is inductance in henry (H) and f is frequency in hertz (Hz).

Figure 5.2.A.1.1a presents a graph showing inductive reactance as a function of frequency. (Review Section 2.3 on the frequency response of R, L, and C components.) Obviously, we can say that an inductor works like a short circuit at low frequency and like an open circuit at high frequency. The equivalent circuits of an inductor at different frequencies are shown in Figure 5.2.A.1.1b.

Based on this property of an inductor – that inductive reactance is directly proportional to frequency – we can build a low-pass RL filter, as shown in Figure 5.2.A.1.2. Analyzing this circuit is similar to analyzing an RC LPF. First, we need to recall this fundamental fact: When we apply ac voltage, V_{in}, at a certain frequency, f_1, it will be distributed between the resistor and the inductor, according to Kirchhoff's voltage law:

$$\boldsymbol{V}_{in} = \boldsymbol{V}_L + \boldsymbol{V}_R, \qquad (5.2.\text{A}.1.2)$$

where $\boldsymbol{V}_L$ and $\boldsymbol{V}_R$ are the voltage drops across the inductor and the resistor, respectively.

When the frequency of the input signal is low, the inductor works like a short circuit and the voltage drop across the resistor approaches the input voltage because there is no other place in this circuit for the voltage drop. Thus,

$$f \to 0 \Rightarrow X_L = 2\pi f L \to 0 \Rightarrow V_{out} \equiv V_R = i \times R \to V_{in} \qquad (5.2.\text{A}.1.3\text{a})$$

When the frequency goes to infinity, the inductor works like an open circuit and the voltage drop across the resistor approaches zero because no current flows through an open circuit:

$$f \to \infty \Rightarrow X_L = 2\pi f L \to \infty \Rightarrow V_{out} \equiv V_R = i \times R \to 0 \qquad (5.2.\text{A}.1.3\text{b})$$

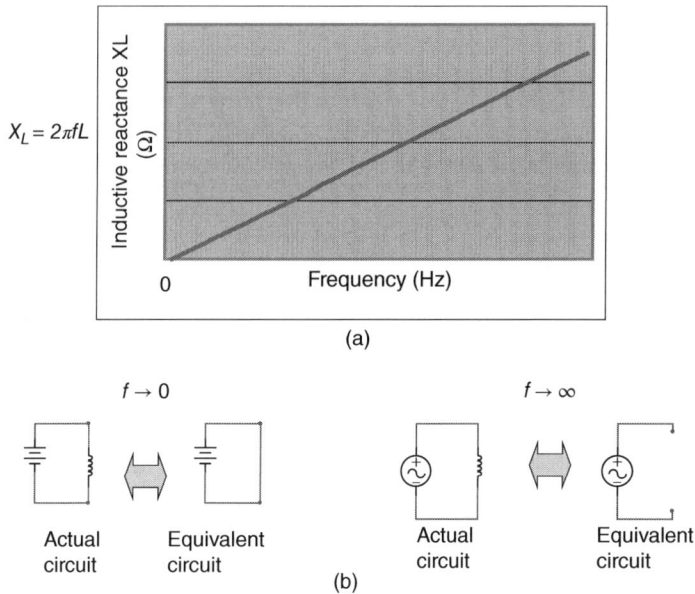

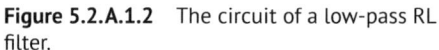

Figure 5.2.A.1.1 Inductive reactance and the inductor's equivalent circuits: (a) Inductive reactance as a function of frequency and (b) an inductor's equivalent circuits at low and high frequencies.

Figure 5.2.A.1.2 The circuit of a low-pass RL filter.

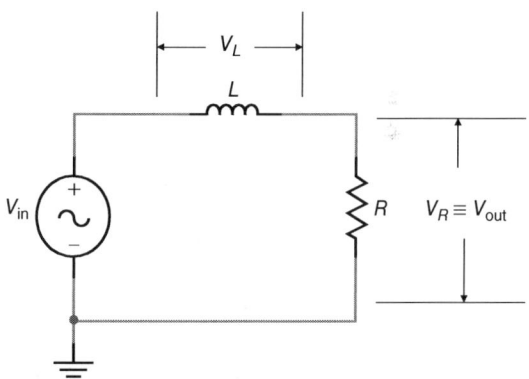

We leave it for you to work with Multisim or any other circuit-simulating software to replicate the work of a low-pass RL filter to see that the amplitudes of an output signal remain close to V_{in} at low frequencies and become small at high frequencies. We also suggest that you build HPF and BPFs based on a resistor, R, and an inductor, L.

It should be noted that RL filters are not widely used at frequencies below 100 kHz. This is mainly because, at this frequency range, the required inductors are bulkier and more expensive components than capacitors. Also, the real properties of inductors deviate from the ideal ones much more than the real properties of capacitors do. The RL filters used to be applied in the video circuits of analog television sets, where frequencies above 100 kHz are used, but those days are gone since the industry has turned to digital TV. Inductors are still in use for some exotic high-frequency circuits, but this is not in the mainstream of modern electronics.

5.2.A.2 Resonance Circuits as Filters

5.2.A.2.1 Resonance Circuits: A Review

Resonance circuits can serve as filters and they are, in fact, used for this purpose in many analog circuits. Here we will briefly review resonance conditions and show how to build filters based on resonance circuits. This review will also familiarize you with the concept and terminology related to the resonance phenomenon.

Consider the series circuit shown in Figure 5.2.A.2.1. The total impedance of this circuit is the sum of the impedances of its components; that is,

$$Z_T = Z_R + Z_L + Z_C \tag{5.2.A.2.1}$$

Placing the explicit forms for all impedances into (5.2.A.2.1), we find

$$Z_T = R + j(X_L - X_C) \tag{5.2.A.2.2}$$

Resonance in this circuit occurs when

$$X_L = X_C \tag{5.2.A.2.3a}$$

which *is* the resonance condition.

Since $X_L\,(\Omega) = 2\pi f L$ and $X_C\,(\Omega) = \frac{1}{2\pi f C}$, and at the resonance $X_L\,(\Omega) = X_C\,(\Omega)$, the resonance condition can be rewritten as

$$2\pi f_r L = \frac{1}{2\pi f_r C} \tag{5.2.A.2.3b}$$

Therefore, *resonance occurs only at a specific frequency, f_r*, which is equal to

$$f_r\,(\text{Hz}) = \frac{1}{2\pi \sqrt{(LC)}} \tag{5.2.A.2.4}$$

The concept of resonance condition can be exemplified as follows: Let $L = 1\,\mu\text{H}$ and $C = 100\,\text{pF}$. Then the resonance frequency, f_r, is given by

$$f_r = \frac{1}{2\pi \sqrt{(LC)}} = 0.5\,\text{MHz}$$

At this frequency, $X_L = X_C = 3.14\,\Omega$. This resonance is demonstrated in Figure 5.2.A.2.2.

It follows from (5.2.A.2.2) and (5.2.A.2.3a) that, at resonance, the total circuit impedance reduces to the resistance, R, which means that the impedance of the circuit is minimal and the circuit draws from the source the highest power possible.

(**Exercise**: We encourage you to do the following exercise with a RLC circuit whose $R = 5\,\Omega$: Build the phasor diagram at $f = 0.4\,\text{MHz}$, which will be a general case. Then, build the phasor diagram for this RLC circuit at resonance. (To refresh your memory about presentation of impedances in phasor domain, review Section 2.3.) You should obtain the results similar to those shown in

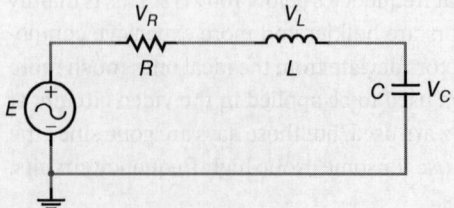

Figure 5.2.A.2.1 Series RLC resonance circuit.

Figure 5.2.A.2.2 Resonance condition.

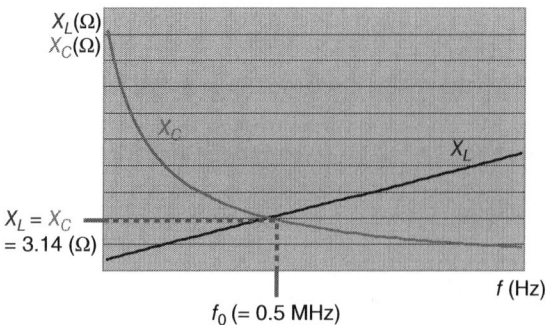

Figure 5.2.A.2.3 Phasor presentation of the resonance condition in an RLC circuit: (a) the phasor diagram in a general case and (b) the same phasors and values at resonance.

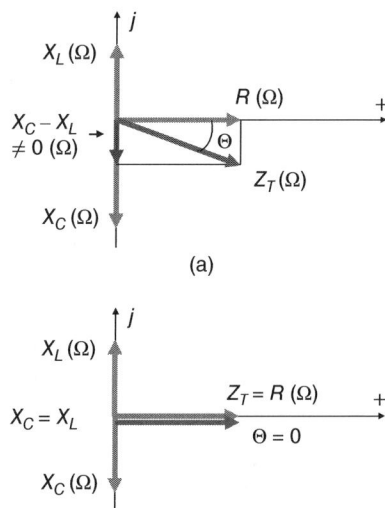

Figure 5.2.A.2.3. To complete the work, place all the necessary numbers ($X_L, X_C, R, Z_T, \theta, f$, and f_r) in the figure.)

It is important to stress that these are not the values of X_L and X_C that are critical at resonance; *the crucial point is that, at resonance, the reactive impedances, X_L and X_C, compensate for each other.* Figure 5.2.A.2.3 shows two situations for a series RLC circuit. At an arbitrary frequency (in a general case), the inductive and capacitive reactance are not equal to each other, as Figure 5.2.A.2.3a shows. This fact results in the deviation of total impedance from its minimum value; also, a phase shift exists with respect to the resistive impedance. You are reminded that, in general, the total impedance and the phase shift are given by

$$Z_T = \sqrt{(R^2 + (X_L - X_C)^2)} \tag{5.2.A.2.5a}$$

$$\Theta = \tan^{-1}\left(\frac{X_L - X_C}{R}\right) \tag{5.2.A.2.5b}$$

At resonance, when X_L and X_C compensate for each other, the total impedance reduces to pure resistance without any phase shift:

$$Z_T^r = R \tag{5.2.A.2.6a}$$

$$\Theta^r = 0 \tag{5.2.A.2.6b}$$

Quite clearly, the voltages across all components of this circuit will follow the pattern set by the impedances.

To continue this discussion in practical terms, let us consider this example:

Example 5.2.A.2.1 Resonance Condition in a Series RLC Circuit

Problem

Given the circuit in Figure 5.2.A.2.4 with $R = 790\,\Omega$, $L = 8.2\,\mu H = 8.2 \times 10^{-3}$ H, and $C = 91$ pF $= 0.91 \times 10^{-10}$ F, determine (1) the resonant frequency, f_r, (2) the total impedance, Z_T, (3) the voltage phasors of each component, V_R, V_L, V_C, and (4) the power dissipated by the circuit at resonance.

Solution

1. The resonant frequency, f_r, is equal to

$$f_r = \frac{1}{2\pi\sqrt{(LC)}} = \frac{1}{(6.283 \times \sqrt{(8.2 \times 0.91 \times 10^{-13})})} = 184.26 \text{ kHz}$$

2. At resonance, $X_L^r = 2\pi f_r L = 9490\,\Omega$.
3. At resonance, $X_C^r = \frac{1}{2\pi f_r C} = 9490\,\Omega$, as it must be. Therefore, we can write the value of X_C^r at resonance without doing the calculations since we have already computed the value of X_L^r.
4. At resonance, the total impedance $Z_T^r = R + jX_L - jX_C = 790\,\Omega$. Again, we know that, at resonance, the total impedance is equal to the circuit's resistance.
5. At resonance, power dissipation can be calculated as follows:
   a) The circuit's current is equal to $i^r = \frac{E}{Z_T^r} = 5/790 = 6.33 \times 10^{-3}$ A $= 6.33$ mA. This is the maximum current that can flow through this circuit.
   b) The voltage drops across the resistor, the inductor, and the capacitor in phasor format are equal to
   i) $V_R = i \times R = 6.33 \times 10^{-3}$ A $\times 790\,\Omega = 5\angle 0°$ V;

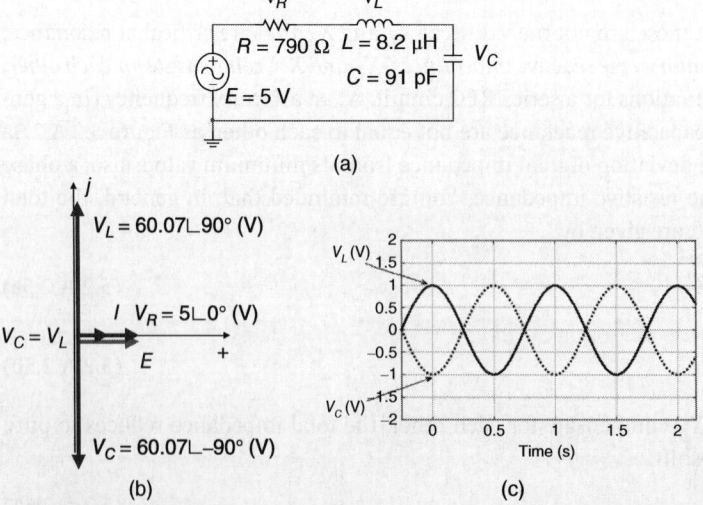

Figure 5.2.A.2.4 Resonance in a series RLC circuit in Example 5.2.A.2.1: (a) circuit diagram; (b) voltage phasors V_R, V_L, and V_C at resonance; and (c) waveforms of $v_L(t)$ and $v_C(t)$ at resonance.

ii) $V_L = i \times X_L = 6.33 \times 10^{-3}$ A $\times j9490\,\Omega = 60.07\, \angle\, 90°$ V;
iii) $V_C = i \times X_C = 6.33 \times 10^{-3}$ A $\times -j9490\,\Omega = 60.07\, \angle\, -90°$ V.

c) The power dissipated by the circuit at resonance is the power dissipated by the resistor, R:

$$P_R^r = i^2 R = ((6.33 \times 10^{-3} \text{ A})^2 \times 790\,\Omega) = 31.65 \text{ mW}$$

Discussion

You are reminded that, according to Kirchhoff's voltage law, the voltages in this circuit are given by

$$E \equiv V_{in} = V_R + V_L + V_C \quad (5.2.A.2.7)$$

This equation also holds true for a resonance condition, of course. Figure 5.2.A.2.4b depicts this equation at resonance. It clearly shows that the reactive voltages, V_L and V_C, are out of phase. Waveforms of these voltages, $v_L(t)$ and $v_C(t)$, shown in Figure 5.2.A.2.4c, demonstrate the same idea in time domain.

5.2.A.2.2 Quality Factor

Another important parameter of a resonance circuit is its *quality factor*. In general, the quality factor, Q, is given as

$$Q = \frac{\text{Reactive power}}{\text{Average power}} \quad (5.2.A.2.8)$$

(or the ratio of maximum energy stored to the energy amount lost per ac cycle). Putting this definition into a formula, we can write

$$Q = \frac{I^2 X_L}{I^2 R} = \frac{X_L}{R} \quad (5.2.A.2.9)$$

At resonance, $Q = \frac{X_L}{R} = \frac{X_C}{R}$

Referring to Example 5.2.A.2.1, we can compute the quality factor of this circuit as

$$Q = \frac{X_L}{R} = \frac{9490\,\Omega}{790\,\Omega} = 12.01$$

The quality factor shows the increase in the voltage drop across the inductor (capacitor); indeed, $V_L/E = V_C/E = 60.07 \text{ V}/5 \text{ V} = 12.01$.

Figure 5.2.A.2.5 depicts current flowing through a series RLC circuit vs. frequency. As this bell-shaped curve shows, the current clearly reaches its maximum value at resonance. *The quality factor is a measure of the sharpness of this graph.* The greater the Q, the sharper the curve; in other words, the greater the value of the quality factor, the narrower this curve will be. For a series RLC circuit, the quality factor should be on the order of tens; for a laser, the typical Q is about a million.

Another way to describe this property (the narrowness or width) of the resonance curve is to introduce the *bandwidth* of a resonance circuit.

The bandwidth of a resonance circuit is the range of frequencies of the signals that a circuit can pass at $I = 0.707\, I_{max}$.

Figure 5.2.A.2.5 shows the bandwidth as BW $= f_2 - f_1$, where f_2 and f_1 are the frequencies of the signals at which the current flowing through a circuit is equal to 0.707 of the maximum value. Note that, at this point, power consumed by the circuit is given by

$$P_{BW} \text{ (W)} = I^2 R = (0.707\, I_{max})^2 R = 0.5 P_{max}$$

This is why such a bandwidth is called a half-power (or -3 dB) bandwidth.

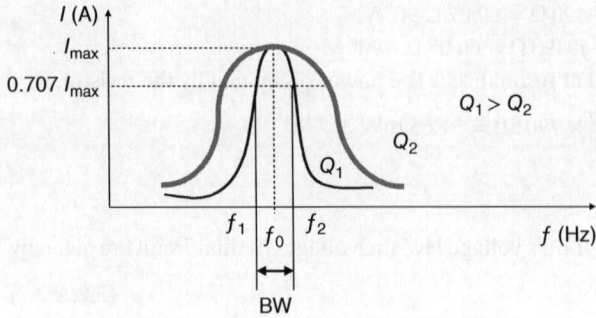

Figure 5.2.A.2.5 Quality factor of a resonance circuit.

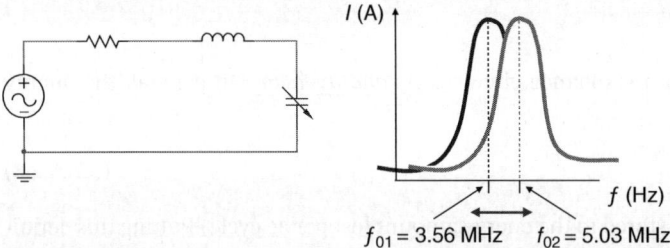

Figure 5.2.A.2.6 Tuning a resonance circuit: A tunable circuit (left) and resonance curves with different resonance frequencies (right).

Since quality factor and bandwidth describe the same quantity – the sharpness of the resonance curve – they must be closely related. Indeed,

$$Q = \frac{f_r \text{ (Hz)}}{\text{BW (Hz)}} \quad (5.2.\text{A}.2.10)$$

where f_r (Hz) is the resonance frequency at which the current flowing through a circuit reaches its maximum value.

Since the resonant frequency is given as $f_r = \frac{1}{2\pi\sqrt{(LC)}}$, changing the value of L or C will result in a change in the resonant frequency; therefore, we can _tune_ a resonance circuit. _Tuning_ is the most important property of a resonance circuit in its applications in analog electronics, and changing the value of a capacitor is the most practical method for tuning. Note that the bandwidth – the quality factor – determines the circuit's _selectivity_.

Figure 5.2.A.2.6 shows a circuit with a variable capacitor (a) and the effect of changing the value of this capacitor, that is, tuning (b). Tuning, of course, is built into analog radio and TV receivers, enabling the selection of a specific radio station or TV channel.

5.2.A.2.3 Resonance Circuit as a Band-Pass Filter

Based on understanding the principles underlying resonance circuits, we can turn to their applications in filtering. If we look at the graph of the resonance circuit shown in Figures 5.2.A.2.5 and 5.2.A.2.6, we immediately recognize in this curve the amplitude response of the BPF shown in Figure 5.2.11. Therefore, the circuit shown in Figure 5.2.A.2.7 can perform as a BPF.

Figure 5.2.A.2.7 Series RLC circuit as a band-pass filter.

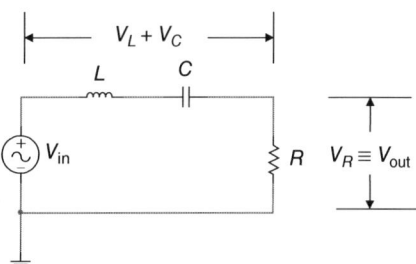

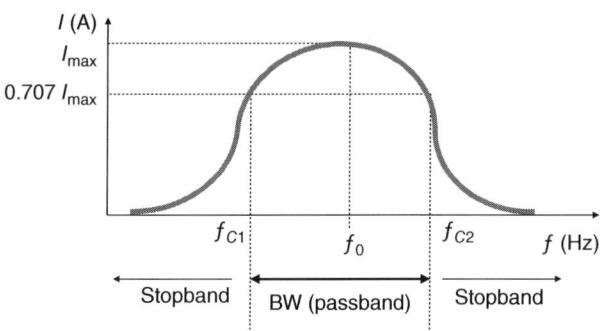

Figure 5.2.A.2.8 The amplitude response of a series resonance circuit working as a band-pass filter.

This circuit works as follows: At frequencies lower than the low critical frequency, f_{C1}, and greater than the high critical frequency, f_{C2}, the input voltage, V_{in}, is distributed among all the circuit components; the output voltage, $V_{out} \equiv V_R$, is smaller than V_{in} because part of V_{in} is distributed among L and C (see Eq. (5.2.A.2.7)). At frequencies that are equal to or close to the resonance frequency, f_r, the signals will pass through the circuit having the maximum output voltage. This is equivalent to saying that the signals whose frequencies are close to f_r will experience the minimum resistance from this circuit because its total impedance at resonance will reduce to R (see Eq. (5.2.A.2.6a) and (5.2.A.2.6b) and Figures 5.2.A.2.2–5.2.A.2.4). Therefore, this circuit passes all signals whose frequencies fall within the circuit bandwidth, BW, which lies between the cutoff frequencies, f_{C1} and f_{C2}. This bandwidth is determined by the 0.707 criterion. The circuit rejects all signals whose frequencies are smaller than f_{C1} and greater than f_{C2}. Figure 5.2.A.2.8 illustrates this point.

We ask you now to analyze this circuit and derive formulas for the amplitude and phase responses. *Hint*: The general formula for the response of this filter is

$$A_v = \frac{V_{out}}{V_{in}} = \frac{R}{R + j(X_L - X_C)} \quad (5.2.A.2.11)$$

We are sure that, based on our previous discussion, you will be quite capable of understanding the origin of (5.2.A.2.11) and how to use this formula to derive the A_v and Θ responses. For more details on this type of filter see Alexander and Sadiku (2009, pp. 362–366).

5.2.A.2.4 Resonance Circuit as a Band-Stop Filter

If we look closely at Figure 5.2.A.2.8, we realize that just by inverting the bell curve, we can obtain a BSF. To implement this idea in a series RLC circuit, the only thing we need to do is achieve the minimum output voltage at resonance. But we know that in this circuit the minimum voltage at resonance is the sum of voltages V_L and V_C; that is,

$$V_L^r + V_C^r = 0$$

Therefore, if we collect the output voltage across the LC segment of the circuit, we will attain our goal. See Figure 5.2.A.2.9a, where the required RLC circuit is shown. Review Figure 5.2.A.2.9b to see the amplitude response of this circuit.

Signals whose frequencies are close to the resonance frequency, f_r, will be suppressed because, at this frequency, the output voltage, which is the sum of the reactive voltages, will be close to zero. The farther the frequencies of the input signals will deviate from f_r, the greater will be the voltage drops across the inductor and capacitor and, consequently, the greater will be the output voltage.

Again, we leave it for you to do a detailed circuit analysis and to derive the amplitude and phase of the filter's frequency response. *Hint*: The ratio of the output and input voltages in this circuit is given by

$$A_v = \frac{V_{out}}{V_{in}} = \frac{j(X_L - X_C)}{R + j(X_L - X_C)} \quad (5.2.\text{A}.2.12)$$

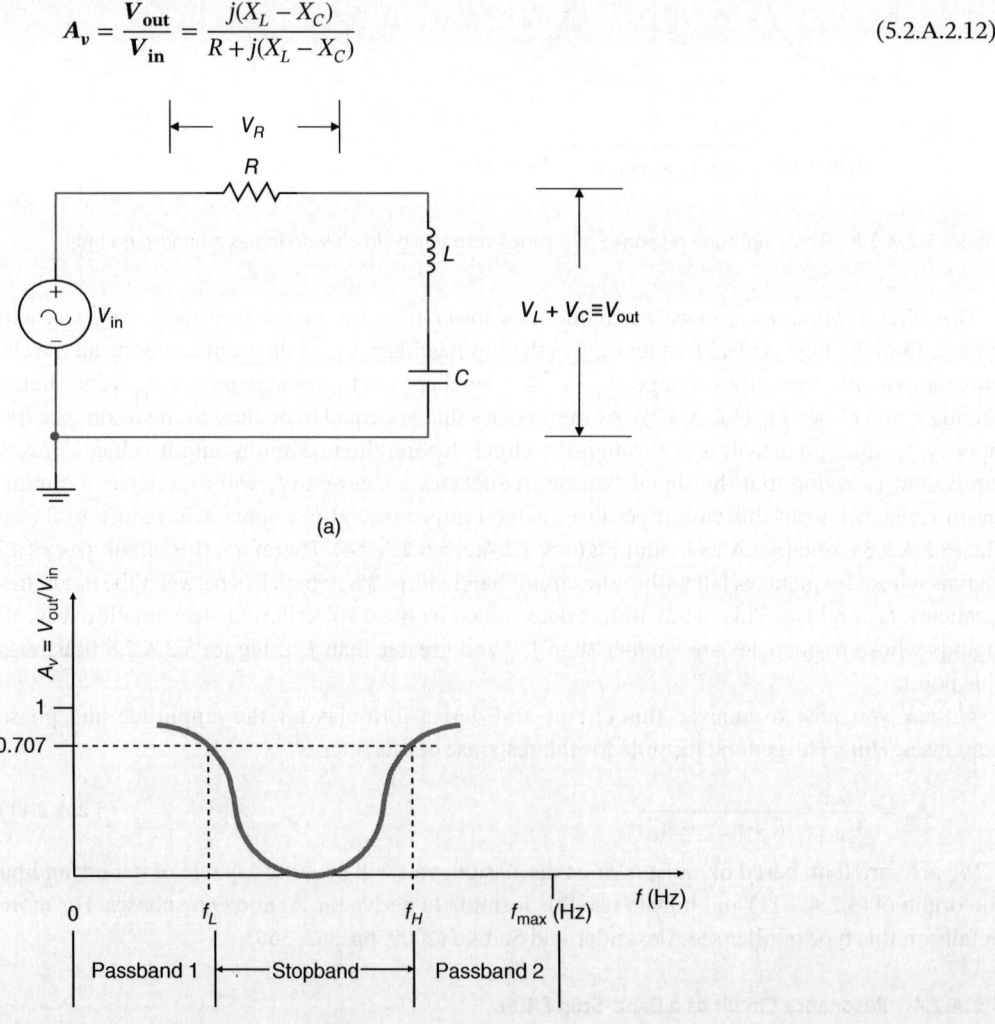

(a)

(b)

Figure 5.2.A.2.9 A series RLC resonance circuit working as a band-stop filter: (a) the circuit diagram and (b) the amplitude response.

(**Exercise**: Derive the amplitude and the phase responses from (5.2.A.2.12) in a way similar to how we did these derivations in this section.)

Our discussion in this appendix shows that it takes two reactive, or energy-storing, components (the inductor and/or the capacitor) to build both BPF and BSFs regardless of the circuit's specificity, whereas LPFs and HPFs are built with one reactive element. LPFs and HPFs are called *first-order* filters and BPFs and BSFs are called *second-order* filters, terms we will clarify in the further sections.

5.3 Active and Switched-Capacitor Filters

Objectives and Outcomes of Section 5.3

Objectives

- To gain knowledge of active analog filters, their main component – the operational amplifier – and the circuits and parameters of active LPFs, HPFs, BPFs, and BSFs.
- To become familiar with one of the most ubiquitous active filter technologies – switched-capacitor filters. To learn their principle of operation, main characteristics, advantages, and applications.

Outcomes

- Comprehend that the passive RC, RL, and RLC filters studied so far have a number of drawbacks, in particular, that they do not perform their task when a load is connected to their output terminals. Understand that this problem can be solved by the use of a special circuit to separate the filter from the load.
- Become familiar with the requirements of this special circuit: very high input impedance and extremely low output impedance; grasp the fact that the required circuit can be implemented with an operational amplifier, an op-amp.
- Revisit the op-amp's equivalent circuit and the op-amp's main parameters, such as very high input impedance, extremely low output impedance, and tremendous gain (on the order of hundreds of thousand) in an open-loop op-amp's configuration.
- Recall that huge gain of an open-loop op-amp, A_{ol}, is, in fact, a major drawback of this amplifier because it drives the circuit to a saturation regime even with very small input signal and it amplifies noise to a level much exceeding an acceptable one.
- Understand that in practice an operational amplifier is used with a negative feedback circuit made of two resistors, R_f and R_i. Learn that this configuration, called a closed-loop op-amp, eliminates the major drawback of an open-loop op-amp and, additionally, improves input and output impedances, stabilizes the gain, and increases the op-amp's bandwidth.
- Realize that the gain of a closed-loop op-amp is on the order of tens and is given by $A_{cl} = 1 + \frac{R_f}{R_i}$ for a noninverting op-amp and by $A_{cl} = -\frac{R_f}{R_i}$ for an inverting op-amp. Understand that the bandwidth-gain product of an op-amp is constant (that is, BW_{ol} (Hz)$\times A_{ol}$ = BW_{cl} (Hz)$\times A_{cl}$ = constant) and therefore a decrease in the gain of a closed-loop op-amp results in an increase in its bandwidth.
- Perceive that an output voltage cannot exceed the voltage of the power supply, $\pm V_{CC}$, because an output signal can draw power only from a power supply, whereas the input signal simply controls the amount of power to be drawn.

- Learn that by connecting in series a passive RC (or RL) frequency selective circuit (a passive filter) and a closed-loop op-amp we obtain an active filter whose cutoff frequency is determined by the passive filter.
- Appreciate that active filters eliminate all the drawbacks of passive filters and, in addition, can amplify input signals.
- Become familiar with the general forms of transfer functions of active filters; namely, $H(f) = 1 + \frac{Z_f}{Z_i}$ for a noninverting op-amp and $H(f) = -\frac{Z_f}{Z_i}$ for an inverting op-amp, where Z_f and Z_i are complex impedances of the feedback circuit.
- Learn the formulas of the transfer functions of LPF, HPF, BPF, and BSF active filters and the use of these formulas in the filters' applications.
- Understand that the most practical filters today are active filters thanks to their advantages over passive filters.
- Master a switched-capacitor filter's principle of operation and its circuit and learn how to design this type of filter.
- Study two main advantages of switched-capacitor filters: This is the only technology that allows us to build the circuit of an analog filter in completely integrated form and they have easy programmable control of the cutoff frequency.
- Become familiar with applications of switched-capacitor filters and the samples of industrial switched-capacitor filters.

5.3.1 Active Filters

5.3.1.1 Drawbacks of Passive Filters

To start, we need to define passive and active components:

An element is called <u>passive</u> if it operates with only the power of a signal, that is, without an external power supply. Naturally, a circuit built with passive components is called a passive circuit. Examples of passive components include R, L, and C elements, and examples of passive circuits are RC, RL, and RLC filters.

An element is called <u>active</u> if it needs an external power supply to operate. Examples include a transistor and an electronic amplifier. A circuit built with such components is called active; examples of an active circuit include operational amplifiers and all integrated circuits (chips).

Clearly, the RC, RL, and RLC filters we have studied so far are called *passive filters*. Though the passive filters perform their tasks, they suffer from a number of drawbacks:

✓ Their output signal is always smaller than the input signal. This is why their magnitude (amplitude) characteristic is called *attenuation*.
✓ They require discrete components, which are costly and bulky. Using discrete components is anachronistic in this era of integrated circuits; in fact, modern electronics is in dire need of ways to minimize size, reduce cost, and lessen the power consumption of its circuits.
✓ They do not perform well at frequencies below audio range, which is under 300 Hz.
✓ Passive filters work better in the higher frequency range but, for this range, the required reactive components (capacitors and inductors) become unacceptably bulky and costly.
✓ Very often, a passive filter's performance is greatly compromised by the attached *load*. Sometimes it becomes the main problem, which is why we discuss this problem in detail below.

Consider an audio speaker, whose entry circuit is a voice coil. The function of the coil is to move a diaphragm, which produces acoustic waves, that is, sound. This coil is a load for an RC LPF, as shown in Figure 5.3.1. The speaker's coil is characterized by its inductive reactance,

Figure 5.3.1 Passive low-pass RC filter with loading coil.

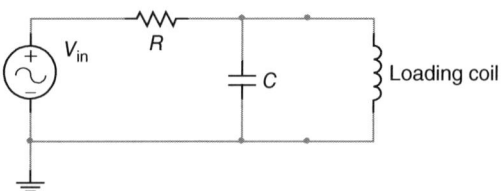

X_L. We discussed the formula $X_L\,(\Omega) = 2\pi f L$ in analytical and graphical formats in Section 5.1 and in Appendix 5.2.A.1. Now let us consider how the entire circuit – the filter plus the loading coil – operates.

At low frequencies ($f \to 0$), a capacitor works like an open circuit ($X_C\,(\Omega) \to \infty$). However, the reactance of a coil is low ($X_L\,(\Omega) \to 0$). Given that the ohmic resistance of a coil, R_L, is very low (on the order of tens of ohms), we immediately understand that X_L will shorten the output of an RC circuit, diminishing the filtering effect (see Figure 5.3.2a). At high frequencies ($f \to \infty$), a capacitor works like a short circuit ($X_C\,(\Omega) \to 0$), leaving almost no current for the loading coil (see Figure 5.3.2b). The equivalent circuits for both situations are shown in Figure 5.3.2. We leave a formal analysis of this circuit to you; here, we want to emphasize only one point:

> *A passive RC LPF does not work as intended when connected to a load, such as a voice coil. Specifically, at low frequency, the filter's output is shorted by the load and at high frequency, the filter's output is shorted by its capacitor. See Figure 5.3.2.*

Whereas *our RC LPF analysis in the preceding sections has been done on the assumption that the filter is connected to an open circuit*, in reality the filter is always connected to a load, which often is simply the next stage of a cascade-connected circuit. The load changes the filter's operation completely, as our example of a voice coil shows.

How can this problem be solved? Well, the hint to the solution can be found in the problem itself: In an ideal case, a filter's output should be an open circuit but, in reality, a load compromises this condition. Therefore, we need to insert additional circuitry between the filter and the load to provide operating conditions close to the ideal. What circuitry should it be? Clearly,

> *The additional circuitry between the filter and the load has to have high input impedance, Z_{in}, and low output impedance, Z_{out}, as shown in Figure 5.3.3.*

Figure 5.3.2 Equivalent circuits of a passive low-pass RC filter with a loading coil at (a) low frequency and (b) high frequency.

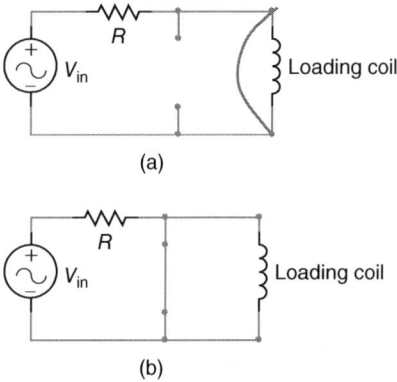

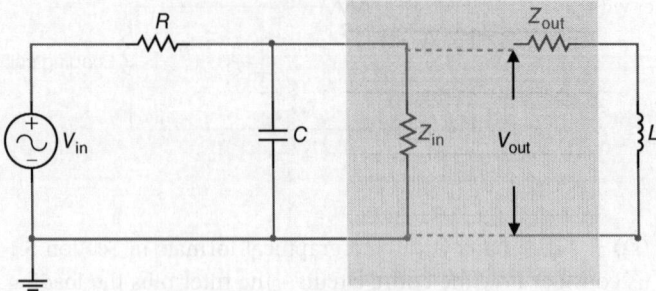

Figure 5.3.3 The loading coil must be separated from the passive RC filter by a component with high input impedance and low output impedance.

With this additional component, the filter now "sees" high Z_{in} as its load. Therefore, dropping in X_L (Ω) at low frequencies will not now affect the filter's operation. Compare Figures 5.3.2a and 5.3.3. Thus, connecting a filter's output to high impedance Z_{in} provides the filter's normal operating condition at low frequency. But this new Z_{in} would not affect the filter's output at high frequencies because a filter's capacitor still shorts the output, you may correctly argue. You are right; at high frequencies, the filter's $v_{out}(t)$ is close to zero and stays so regardless of the load. But this is exactly the output we want from a LPF. Consequently, high Z_{in} enables the filter to operate in almost ideal condition.

Why must the output impedance, Z_{out} (Ω), be very small? Because under this condition, the output voltage, V_{out} (V), will always drop mostly across the load. In the worst-case scenario, when the circuit operates at low frequencies, X_L (Ω) becomes very small and the impedance of the loading coil will be determined by its small ohmic resistance, R_L (Ω). By taking Z_{out} (Ω) much smaller than R_L (Ω), we make the most of the V_{out} (V) drops across the load. Consider, for example, a loading coil with $L = 1$ mH, $R_L = 18\,\Omega$, and $Z_{out} = 2\,\Omega$. At $f = 100$ Hz, inductive reactance $X_L = 0.62\,\Omega$ and the coil impedance will be determined by its ohmic resistance, $R_L = 18\,\Omega$. Since Z_{out} and R_L are connected in series, 90% of the output voltage will drop across the loading coil, according to the voltage-divider rule. This consideration is illustrated in Figure 5.3.4.

To sum up: A high input impedance allows for a passive RC filter to operate under an ideal condition at low frequencies; the filter will see, virtually, an open circuit, as shown in Figure 5.3.4a. At the same time, a low output impedance provides the output voltage to drop mainly across the

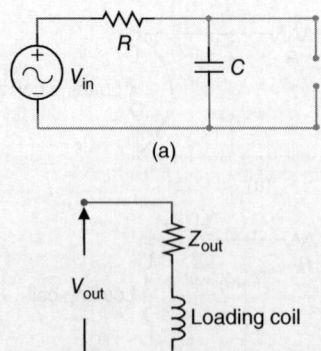

Figure 5.3.4 Operation of an RC LPF when a loading coil is separated by circuitry with high input impedance, Z_{in} (Ω), and low output impedance, Z_{out} (Ω): (a) the filter works with high Z_{in} (Ω) and (b) the load works with low Z_{out} (Ω).

load, as Figure 5.3.4b shows. Note that Z_{out} functions here essentially as an internal resistance of a source.

It has to be emphasized that a passive LPF must be separated from a load regardless of the physical nature of the load. We consider here a voice coil as a load because it clearly demonstrates the need for the separation; however, such a common load as buffer amplifier or any other buffering circuit will also change the operation of a passive filter and therefore must be separated.

Example 5.3.1   Separating the Load from a Passive RC LPF

Problem

Consider an RC LPF built with $R = 2.2\,k\Omega$ and $C = 20\,pF$. To provide the proper operating condition for this filter, it is necessary to separate the filter from a load with circuitry having a high input impedance, Z_{in} (Ω), and low output impedance, Z_{out} (Ω). Solve the following problems for this circuitry:

- The input impedance, Z_{in} (Ω), must be high regardless of the change in the input frequency so that an RC LPF's terminating impedance will be close to that of an open circuit. Assuming X_{Cmax} (Ω) $= 0.1 Z_{in}$ as the open-circuit criterion, calculate Z_{in} (Ω) if $f_{max} = 4\,kHz$.
- The output impedance, Z_{out} (Ω), must be negligible compared with the minimum inductive reactance, X_{Lmin} (Ω), of the loading coil. Calculate Z_{out} (Ω) if $L = 2\,mH$, and $R_L = 20\,\Omega$, and with $V_{Zout} \leq 0.1 V_L$ being the criterion of negligibility.

Solution

a) Applying (5.3.1), we compute

$$X_{Cmax} = \frac{1}{2\pi f_{max} C} = 2\,M\Omega$$

Since Z_{in} must be at least 10 times more than X_{Cmax}, we find

$$Z_{in} \geq 10 X_{Cmax} = 20\,M\Omega$$

Such a value can be achieved with an operational amplifier (op-amp) to be considered shortly.

b) The minimum impedance of the loading coil is equal to its ohmic resistance,

$$X_{Lmin} = R_L = 20\,\Omega$$

Thus, the output impedance of the separated component must be equal to or less than $2\,\Omega$; that is,

$$Z_{out} \leq 2\,\Omega$$

Since Z_{out} and X_L are connected in series, the voltage-divider rule produces

$$V_{Zout} = V_{out}\,[Z_{out}/(X_L + Z_{out})] = 2/22 V_{out} = 0.09 V_{out}$$

Hence, the criterion $V_{Zout} \leq 0.1 V_{out}$ has been met.

Discussion

All these calculations are straightforward and do not require any discussion.

5.3.1.2   Operational Amplifier

Fortunately, we have an electronic circuit with the required properties; it is an operational amplifier. The designations OP AMP, op-amp, and op amp are used interchangeably for this circuit;

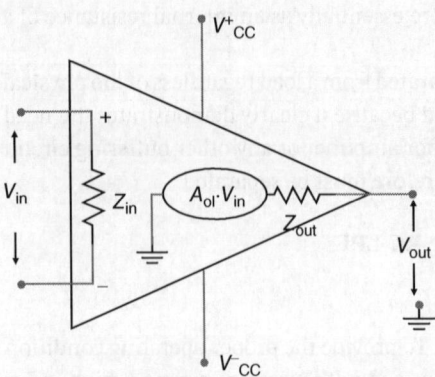

Figure 5.3.5 Equivalent circuit of an operational amplifier.

we will call it an *op-amp*. It is an integrated circuit, or chip; this is why we call it a device. The basics of an open-loop op-amp can be readily acquired through the analysis of its equivalent circuit shown in Figure 5.3.5: The input impedance, Z_{in} (Ω), is the resistance seen by the driving source (the input with respect to the op-amp circuit); the output impedance is the resistance seen by the load. The component shown as a voltage-controlled source, $A_{ol} \cdot V_{in}$, represents the essence of the op-amp operation: amplification of the input voltage, V_{in}, by open-loop gain, A_{ol}. (The term *open-loop op-amp* implies that an op-amp is used without any additional elements, whereas the term *closed-loop op-amp* means that the op-amp is wired with additional resistors providing negative feedback. We will examine a closed-loop op-amp shortly.)

An op-amp is an amplifier with three main distinguishing parameters: (i) very high input impedance, Z_{in} (Ω); (ii) very low output impedance, Z_{out} (Ω); and (iii) very high open-loop voltage gain, A_{ol}. To understand what values of the main characteristics of an op-amp we need, imagine working with an <u>ideal op-amp</u> which should have infinite input impedance, Z_{in} (Ω) $\to \infty$; zero output impedance, Z_{out} (Ω) $\to 0$; infinite open-loop gain, $A_{ol} \to \infty$; and infinite bandwidth, BW (Hz) $\to \infty$. These parameters of an ideal op-amp will help us to make an approximate analysis of the circuits based on this device. In reality, input impedance is between 10^5 and 10^8 Ω, output impedance is between 10 and 100 Ω, open-loop gain is between 10^5 and 10^8, and the bandwidth is quite limited.

This device has to be powered, which means it is an <u>active circuit</u>. We use the designations V_{CC}^+ and V_{CC}^- respectively, for the positive and negative terminals of a power supply. The supply voltage varies between 5 and 24 V for each V_{CC}, with ± 12 and ± 15 V being most typical. Power supplies are often ignored in schematics, yet we will follow this tradition.

The input with a positive sign is called the *noninverting input* and the input with a negative sign is called the *inverting input*. When a signal is applied to the noninverting input, it appears at the output (being amplified, of course) with the same polarity; clearly, the polarity of a signal is inverted when it is applied to the inverting input.

From your introductory electronics course, you no doubt recall how an op-amp works. If not, here are the basics: Consider a noninverting op-amp whose equivalent circuit is shown in Figure 5.3.5: An input voltage, V_{in}, is the difference between the noninverting (+) and the inverting (−) inputs; this is the differential input of an op-amp. This voltage sees a huge input impedance. As a result, almost no current flows between both inputs of an op-amp and between the input terminals and the ground. In other words, i^+ (A) $\to 0$ and i^- (A) $\to 0$. The input voltage, $V_{sig} = V_{in}$ (V), coming from an RC LPF, will be amplified and the output voltage, V_{out} (V), is given by

$$V_{out} (V) = A_{ol} \cdot V_{in} (V) \tag{5.3.1}$$

where A_{ol} is the gain of an open-loop op-amp. This V_{out} (V) is distributed between Z_{out} (Ω) and Z_L (Ω), where Z_L is an impedance of the loading coil and Z_{out} is a small output impedance of the op-amp (see Figure 5.3.4b). This explanation should clarify the op-amp basics illustrated in Figure 5.3.5.

Now we can appreciate all the advantages that an op-amp gives us when we need to work with a load like a voice coil. The op-amp essentially separates the signal source (in our case, an RC filter) from a load (in this consideration, a voice coil). Figure 5.3.6 depicts this situation.

An op-amp circuitry shown in Figure 5.3.6 is a closed *open-loop*. In practice, however, op-amps usually operate not in an open-loop configuration, but with *negative feedback*, as shown in Figure 5.3.7 for a noninverting amplifier. This circuitry is called a *closed-loop op-amp*. The reasons for using closed-loop op-amps are as follows:

The huge ($\geq 100\,000$) gain of open-loop op-amps has two major drawbacks: First, any small input will drive an op-amp to saturation. Here is why: Suppose you present 1 mV input to an op-amp whose $A_{ol} = 100\,000$; obviously, the output would be 100 V. An op-amp would not produce such a voltage because its *output can't be more than the power-supply voltage*, which typically is ± 12 V. This is why open-loop op-amps are used only in special applications, such as comparators. A closed-loop op-amp allows a circuit designer to avoid this problem by introducing controlled gain.

Second, suppose no input is presented to an op-amp; however, in practice, a small, unwanted voltage almost inevitably appears at the op-amp's input. This unwanted signal will be amplified by an open-loop op-amp and will appear at the amplifier's output.

In a closed-loop op-amp, the output voltage will be fed back to the input with a negative sign, thereby reducing the unwanted input to zero and thus eliminating the undesired effect.

In addition to these major advantages, a closed-loop op-amp allows for controlling the values of Z_{in} (Ω) and Z_{out} (Ω), stabilizing the gain and improving the frequency response (i.e. increasing the bandwidth).

To analyze the closed-loop op-amp shown in Figure 5.3.7, we need to note that the output voltage is distributed between resistors R_f (Ω) and R_i (Ω) according to the voltage divider rule. The voltage

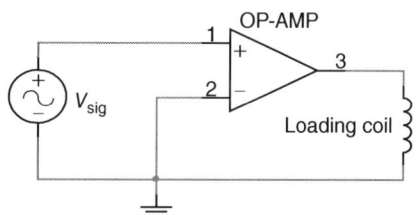

Figure 5.3.6 Operation of a noninverting op-amp with a signal source and a loading coil.

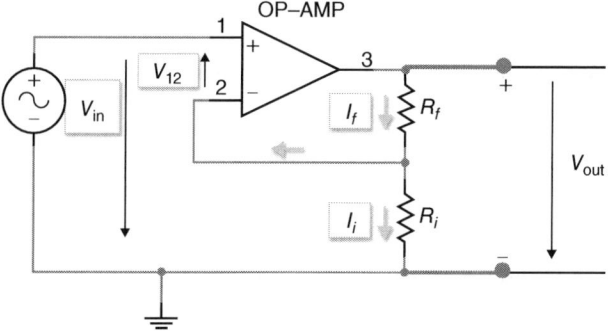

Figure 5.3.7 Noninverting op-amp with negative feedback.

drop across R_f, which is a part of the output voltage, is applied to the inverting input of the op-amp. Thus, this part of the output voltage provides *feedback*. This feedback is *negative* because the feedback voltage is subtracted from the signal voltage, as follows from the principle of an op-amp's operation.

We derive the formula of a closed-loop noninverting op-amp by using the basic mesh analysis.[2] First, the loop made of V_{out}, $I_f R_f$, V_{12}, and V_{in} gives us the equation,

$$V_{out} - I_f R_f - V_{12} - V_{in} = 0 \tag{5.3.2}$$

from which we can derive

$$I_f \text{ (A)} = \frac{V_{out} - V_{12} - V_{in}}{R_f} \tag{5.3.3}$$

The second loop – made of V_{in}, V_{12}, and $I_i R_i$ – gives us another equation,

$$V_{in} + V_{12} - I_i R_i = 0 \tag{5.3.4}$$

from which we obtain

$$I_i \text{ (A)} = \frac{V_{in} + V_{12}}{R_i} \tag{5.3.5}$$

Knowing that almost no current flows from input terminals of the op-amp because the input impedance is almost infinite, we can set $I_f \text{ (A)} = I_i$. (Follow the currents flow in Figure 5.3.7 to see why.) Also, there is no voltage drop across the input terminals of the op-amp. (Now you know why.) Therefore, $V_{12} = 0$. Hence, equating (5.3.3) and (5.3.5) enables us to find

$$\frac{V_{out} - V_{in}}{R_f} = \frac{V_{in}}{R_i} \tag{5.3.6}$$

Finally, simple algebraic manipulations with (5.3.6) allows us to obtain the formula for the ratio of the output and input voltages, which is the gain of a closed-loop op-amp, A_{cl},

$$A_{cl} = \frac{V_{out}}{V_{in}} = 1 + \frac{R_f}{R_i} \tag{5.3.7}$$

Equation (5.3.7) states that op-amp closed-loop gain, A_{cl}, is determined by the ratio of two external resistors, $\frac{R_f}{R_i}$.

To appreciate the typical values of closed-loop gain, consider the following example: $R_f = 47 \text{ k}\Omega$ and $R_i = 4.7 \text{ k}\Omega$. Then, $A_{cl} = 1 + \frac{R_f}{R_i} = 11$.

Using a closed-loop noninverting amplifier, we can stabilize the gain, since gain depends only on the values of $R_f (\Omega)$ and $R_i (\Omega)$. Also, by reducing the gain, we can significantly increase the op-amp bandwidth because, for a real op-amp, the gain-bandwidth product is a constant number.

We can now derive the formulas for the impedances of a negative-feedback op-amp by relying on the definition $Z = V/I$ and our analysis of a closed-loop op-amp; these impedances are given by

$$Z_{incl} (\Omega) = (1 + A_{ol} \cdot A_{cl}) Z_{inol} \tag{5.3.8}$$

and

$$Z_{outcl} (\Omega) = \frac{Z_{outol}}{1 + A_{ol} \cdot A_{cl}} \tag{5.3.9}$$

where and A_{ol} and A_{cl} are open-loop and closed-loop gains of an op-amp. Since $A_{ol} \cdot A_{cl} \gg 1$, the closed-loop input impedance, $Z_{incl} (\Omega)$, is much greater than the open-loop impedance,

---

2 See, for example, www.masteringelectronicsdesign.com.

Z_{inol} (Ω), whereas the closed-loop output impedance, Z_{outcl} (Ω), is much smaller than the open-loop impedance, Z_{outol} (Ω). In other words, *negative feedback significantly improves the values of impedances of a closed-loop op-amp compared with the values of an open-loop op-amp.*

Let us consider input-output and gain-frequency graphs, which are main op-amp's characteristics. They are shown in Figure 5.3.8.

When the input voltage varies between V_{CC}^+ and V_{CC}^-, the op-amp responds linearly so that the output voltage is proportional to the input, $V_{out} = A \cdot V_{in}$; however, when V_{in} is beyond these limits, the output voltage cannot be smaller than V_{CC}^- or greater than V_{CC}^+. This is where an op-amp comes to the saturation regions shown in Figure 5.3.8a.

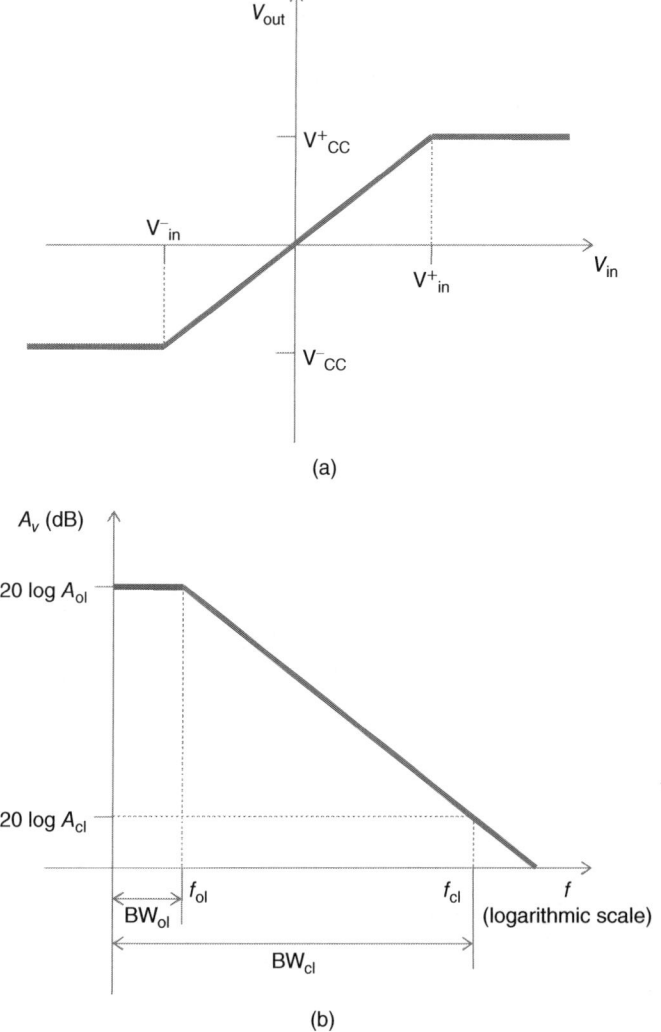

Figure 5.3.8 Operation of a real op-amp: (a) input–output characteristic of an op-amp and (b) gain-bandwidth product for open-loop and closed-loop amps.

The saturation in the input–output characteristic of an op-amp reflects a fundamental principle of operation of any amplifier: *An amplified (output) signal can draw power only from a power supply, whereas the input signal only controls the amount of power to be drawn.* (You may recall the operation of a common-emitter amplifier based on a bipolar transistor, where this concept can be clearly seen.) This phenomenon signifies a great advantage of a closed-loop op-amp because this op-amp can amplify meaningful values of an input signal. For example, if $A_{cl} = 20$ and $V_{in} = 1$ mV, the closed-loop output will be $V_{out(cl)} = 20$ mV. (What would be hypothetical $V_{out(ol)}$ in this example?) Putting it another way, the maximum input signal that an op-amp can amplify before reaching saturation is given by the input signal to make it equal to $V_{in(max)} = \frac{V_{CC}}{A_{cl}}$. In our example, $V_{in(max)} = 0.75$ V for either a positive or negative signal.

What is more, we can easily control this gain by changing the values of the feedback resistors. Practically speaking, the closed-loop gain is much more stable than the open-loop gain because A_{cl} depends only on the constant resistors' values, whereas A_{ol} depends on the properties of a semiconductor chip, which are greatly affected by the ambient temperature.

Figure 5.3.8b is, of course, built in logarithmic scale. It presents graphically the fact that *gain times bandwidth is constant for any given op-amp.* This figure illustrates that an open-loop op-amp has a huge gain but a small bandwidth, whereas the same op-amp in closed-loop configuration has a small gain but a big bandwidth. Thus, the area covered by $A \times BW$ is constant for any configuration. This important relationship helps us understand the existence of a trade-off between the gain and the bandwidth of an op-amp: *Controlling the gain of an op-amp, we control its bandwidth, too.*

Operational amplifiers are an important branch of electronics; there is a wealth of academic and industrial literature devoted to this subject. We, however, have to restrict ourselves by this brief review, which is necessary to support our further considerations.

5.3.1.3 Active Filters: Concept and Circuits

Assume that we want to build circuitry that will allow a passive RC filter to perform its task even when it is loaded with a simple coil. We know that an operational amplifier is a device that will allow us to achieve this goal. Thus, we can build a good LPF by combining a passive RC frequency selective circuit and an op-amp with negative feedback. Such a circuit, with a noninverting op-amp, is shown in Figure 5.3.9. Quite naturally, it is called an *active filter*.

The filter's cutoff frequency is still defined as f_C (Hz) $= \frac{1}{2\pi RC}$ because the main frequency's selective component here is an RC filter. We therefore assume that the op-amp's bandwidth covers the range of frequencies to be filtered; however, the inclusion of an op-amp changes other features of this filter drastically. In addition to separating an RC circuit from the load, another great advantage

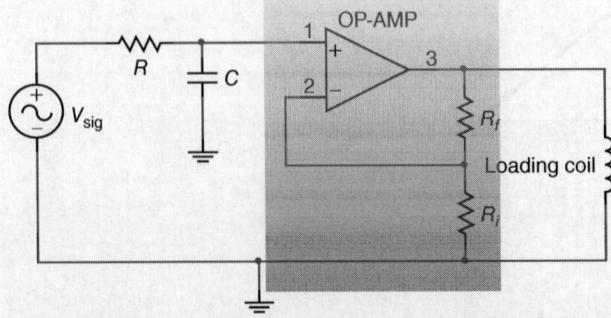

Figure 5.3.9 Schematic of an active low-pass filter with a noninverting op-amp.

of an active filter is that it enables us to easily manipulate its parameters to build the filter with the desired characteristics. In essence, active filters allow us to eliminate all the previously mentioned drawbacks of passive filters. To be sure, you can compare passive and active filters by simply referring to the passive filter's drawbacks listed above in Section 5.3.1.1. This comparison will be a good exercise for you.

Obviously, the need for an external power source – the only drawback of active filters – is the price we pay for having such good devices.

To analyze the performance of active filters, let us introduce their transfer functions.

5.3.1.4 Transfer Functions of an Active Filter: General View

Let us consider an active filter based on the inverting op-amp shown in Figure 5.3.10. Comparing this figure with Figure 5.3.6, we can readily see that here R and C components and feedback resistors R_i (Ω) and R_f (Ω) have been replaced by the complex impedances $Z_i(f)$ and $Z_f(f)$. These impedances provide frequency-selective properties to our circuit.

Let us denote the output and input voltages in frequency domain as $V_{out}(f)$ and $V_{in}(f)$. Then, by definition, a transfer function of this circuit is given by

$$H(f) = \frac{V_{out}(f)}{V_{in}(f)}$$

On the other hand, applying the approach we used in deriving (5.3.7), we can write for inverting op-amp the ratio output to input signals by replacing R_i (Ω) and R_f (Ω) with Z_i (Ω) and Z_f (Ω) as

$$H(f) = \frac{V_{out}(f)}{V_{in}(f)} = -\frac{Z_f}{Z_i} \tag{5.3.10}$$

Clearly, (5.3.10) gives us the formula for the transfer function of an active filter based on an inverting op-amp. Following the same pattern, we can write the formula for the transfer function of an active filter based on a noninverting amplifier (see Eq. (5.3.7)) as

$$H(f) = \frac{V_{out}(f)}{V_{in}(f)} = 1 + \frac{Z_f}{Z_i} \tag{5.3.11}$$

The transfer functions for specific types of filters are discussed below.

Analyzing (5.3.10) and (5.3.11), we can conclude that we are able to control the frequency response of an active filter by selecting the proper ratio of complex impedances.

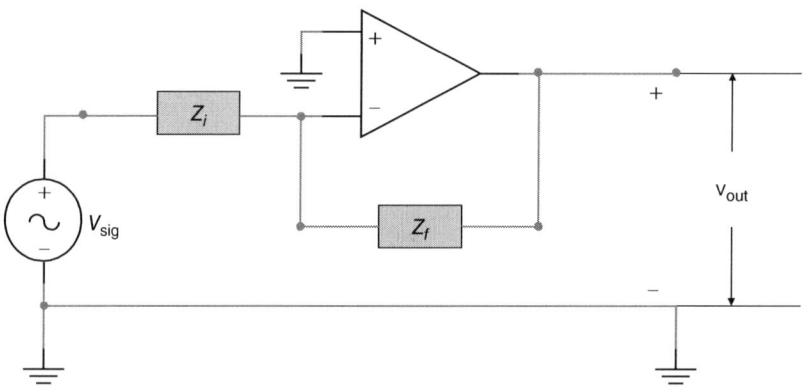

Figure 5.3.10 An active filter based on an inverting op-amp: general view.

5.3.1.5 Specific Types of Active Filters

Active LPF To build an *active LPF*, we need to use an RC circuit and an op-amp. Such a filter based on an inverting op-amp is shown in Figure 5.3.11.

The transfer function of this active LPF is given in (5.3.10). (Do you know why?) Figure 5.3.11 shows that we need to substitute R_i for Z_i and the expression $R_f \| (-jX_C)$ for Z_f; that is,

$$Z_f(f) = R_f \| (-jX_C) = \frac{R_f(-jX_C)}{R_f - jX_C} = \frac{R_f}{1 - \frac{R_f}{jX_C}} = \frac{R_f}{1 + j2\pi f C_f R_f} \quad (5.3.12)$$

(recall that $X_C\,(\Omega) = \frac{1}{2\pi f C}$). Therefore,

$$H_{LPF}(f) = \frac{V_{out}(f)}{V_{in}(f)} = -\frac{Z_f}{Z_i} = \frac{R_f}{R_i(1 + j2\pi f C_f R_f)} \quad (5.3.13)$$

This equation can be factored into two components: The first is the $\frac{R_f}{R_i}$ ratio, which is simply the gain of an RC LPF with inverting amplifier. It is clear that this gain does not depend on frequency; this is why it is sometimes called a *dc gain*:

$$A_{dc} = A_{cl0} = -\frac{R_f}{R_i} \quad (5.3.14)$$

We have to bear in mind that the real gain of an active filter changes with frequency because the op-amp's gain changes slightly, even within its bandwidth, and quickly drops to zero outside its low- and upper-frequency limits.

The second component is $\frac{1}{1+j2\pi f C_f R_f}$, which is the transfer function of a low-pass RC filter, as defined in (5.1.6) and (5.1.7):

$$H_{RCLPF} = \frac{1}{1 + j2\pi f C_f R_f} = \frac{-jX_C}{R_f - jX_C} \quad (5.3.15)$$

Therefore, the transfer function of an active LPF with an inverting op-amp is given by

$$H_{LPF}(f) = \frac{V_{out}(f)}{V_{in}(f)} = -\frac{Z_f}{Z_i} = -\frac{R_f}{R_i} \cdot \frac{1}{(1 + j2\pi f C_f R_f)} = A_{cl0} \cdot H_{RCLPF} \quad (5.3.16)$$

Equation (5.3.16) shows that this circuit performs low-pass filtering while simultaneously amplifying the input signal. This means that *the amplitude and phase of the frequency response of this active filter are simply an amplified version of the response of a passive low-pass RC filter*. Thus, if we take the amplitude part of the frequency response of an RC LPF shown in Figure 5.1.9, change

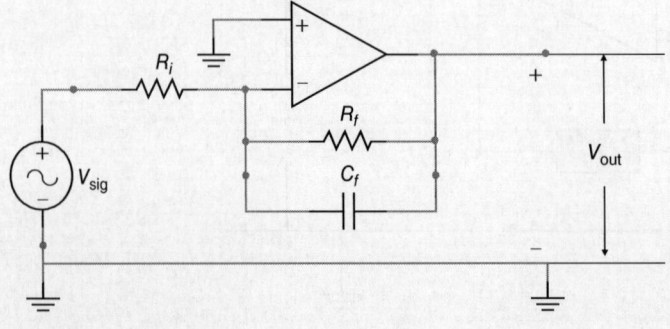

Figure 5.3.11 An RC active LPF based on an inverting op-amp.

5.3 Active and Switched-Capacitor Filters

"attenuation" to "amplitude response," and introduce gain value A_{cl0} instead of 1, we will obtain the graphs A_v vs. frequency and Θ vs. frequency of an active RC LPF.

It follows logically from this consideration that the cutoff frequency of this active filter is determined by the same equation as that applying to a passive filter (see Eq. (5.1.17)); that is,

$$f_C \text{ (Hz)} = \frac{1}{2\pi f C_f R_f} \tag{5.3.17}$$

To help you better grasp the idea of active filters, let us consider the following example.

Example 5.3.2 Designing an Active LPF Based on an Inverting op-amp

Problem

Using an inverting op-amp, design a low-pass active filter whose gain equals 10 and whose cutoff frequency is $f_C = 1000 \text{ Hz}$.

Solution

Consider the given quantities: $f_C \text{ (Hz)} = \frac{1}{2\pi C_f R_f} = 1000 \text{ Hz}$ and $A_{cl0} = -\frac{R_f}{R_i} = 10$. We have two equations and three unknown variables – R_f, R_i, and C_f; a typical situation in a design problem. Thus, we need to choose one quantity and compute the other two. Let us pick $C_f = 0.1 \text{ μF}$. Then,

$$R_f = \frac{1}{2\pi f_C C_f} = 1.59 \text{ k}\Omega$$

Taking the standard value of 1.6 kΩ for R_f, we can compute R_i as

$$R_i = -\frac{R_f}{A_{cl0}} = 0.16 \text{ k}\Omega, \text{ which is a standard value.}$$

Therefore, the active low-pass RC filter is designed. Figure 5.3.11 shows the schematic of such a filter.

Discussion

The design of this filter is a rather straightforward operation; we simply need to follow the definitions and choose a practical value of a capacitor. Also, we must not forget to pick the standard (not calculated) values of all the components.

The waveforms of the input and output signals obtained by simulation of this filter with given parameters and the amplitude response are shown in Figure 5.3.12a. The experimental setup of this simulation is the same as the filter's schematic shown in Figure 5.3.11; obviously, an oscilloscope has to be added.

Note that at low frequency, $f = 0.1 f_C = 100 \text{ Hz}$, the output amplitude is greater than the input amplitude, as the principle of an active LPF operation suggests. At the cutoff frequency, $f = f_C = 1000 \text{ Hz}$, the output amplitude is still greater than the input amplitude. We can estimate the output amplitude as

$$V_{out}|_{f=f_C} = 0.707 \times V_{in} \times A_{cl0} = 0.707 \times V_{in} \times \left(\frac{R_f}{R_i}\right) = 7.07 V_{in}$$

At high frequency, $f = 10 f_C = 10\,000 \text{ Hz}$, the output amplitude is smaller than the input amplitude, as expected, according to the LPF principle of operation. For your convenience, Figure 5.3.12b shows the simulated points at the amplitude part of the filter's frequency response. This figure visualizes the connection between the filter's operation in both the time and frequency domains.

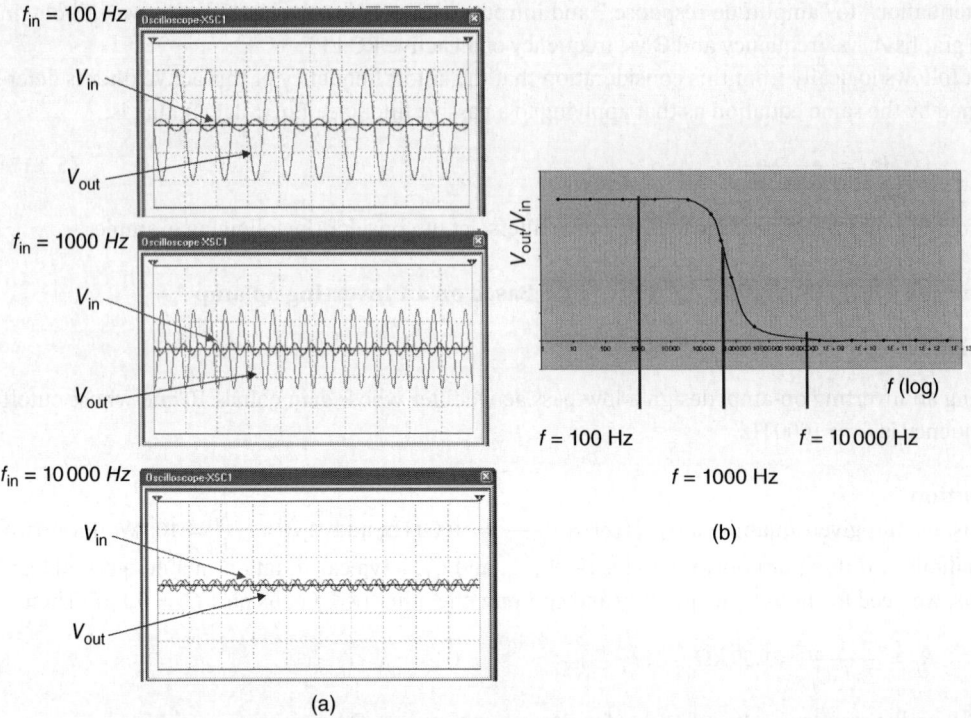

Figure 5.3.12 Operation of the active RC LPF given in Example 5.3.2: (a) waveforms of the input and output signals and (b) amplitude vs. frequency.

Active HPF We can use the approach considered in the preceding subsection to build and analyze *high-pass, band-pass, and band-stop active* filters. Consider Figure 5.3.13, where the schematic of an *RC active HPF* based on an inverting op-amp is shown. An analysis of this circuit can be easily done by referring to the HPF discussed in Section 5.2 (see Figures 5.2.2–5.2.6).

Note that in Figure 5.3.13 the input impedance, $Z_i(f)$, is given by components R_i (Ω), and C_i (F) connected in series:

$$Z_i(f) = R_i - jX_C = R_i + \frac{1}{j2\pi f C_i} \tag{5.3.18}$$

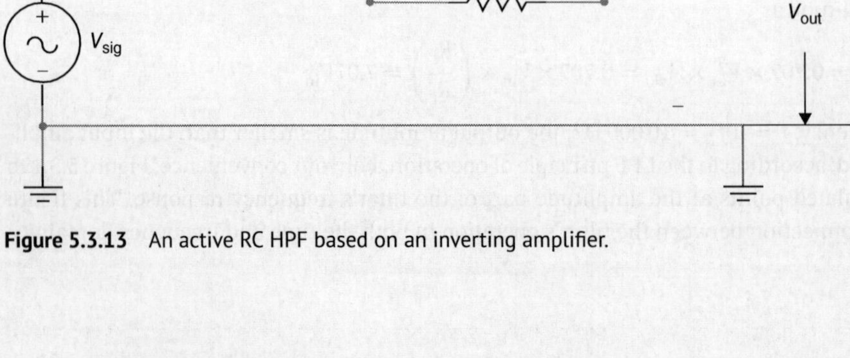

Figure 5.3.13 An active RC HPF based on an inverting amplifier.

The output impedance, Z_f, is simply equal to the feedback resistance, R_f:

$$Z_f(f) = R_f \tag{5.3.19}$$

Therefore, the transfer function of an active HPF is given by

$$H_{HPF}(f) = -\frac{Z_f}{Z_i} = -\frac{R_f}{R_i} = -\frac{R_f}{\left(R_i + \frac{1}{j2\pi C_i}\right)} = -\frac{j2\pi C_i R_f}{(1 + j2\pi C_i R_i)} \tag{5.3.20}$$

You are encouraged to recall the formula of the transfer function of a passive HPF and compare that formula with (5.3.20). The critical frequency of this filter is given by

$$f_{CHPF}\ (Hz) = \frac{1}{2\pi C_i R_i} \tag{5.3.21}$$

Consider the example of an active HPF with the following parameters: $R_i = 1.6\,k\Omega$, $R_f = 8\,k\Omega$, and $C_i = 0.1\,\mu F$. Its cutoff frequency is equal to $f_{CHPF} = 1000\,Hz$. Taking $V_{in} = 1\,V$, we can simulate the work of this filter using Multisim. The resulting waveforms and amplitude vs. frequency graph are shown in Figure 5.3.14.

Observe that at low frequency, $f = 0.1f_C = 100\,Hz$, the output amplitude is smaller than the input amplitude, as the principle of HPF operation requires. At cutoff frequency, $f = f_C = 1000\,Hz$, the output amplitude is about 3.5 times greater than the input amplitude. This is because the op-amp gain is equal to $A_{cl0} = \frac{R_f}{R_i} = 5$ and

$$V_{out}|_{f=f_C} = 0.707 \times \frac{R_f}{R_i} \times V_{in} = 3.5 V_{in}\ (V)$$

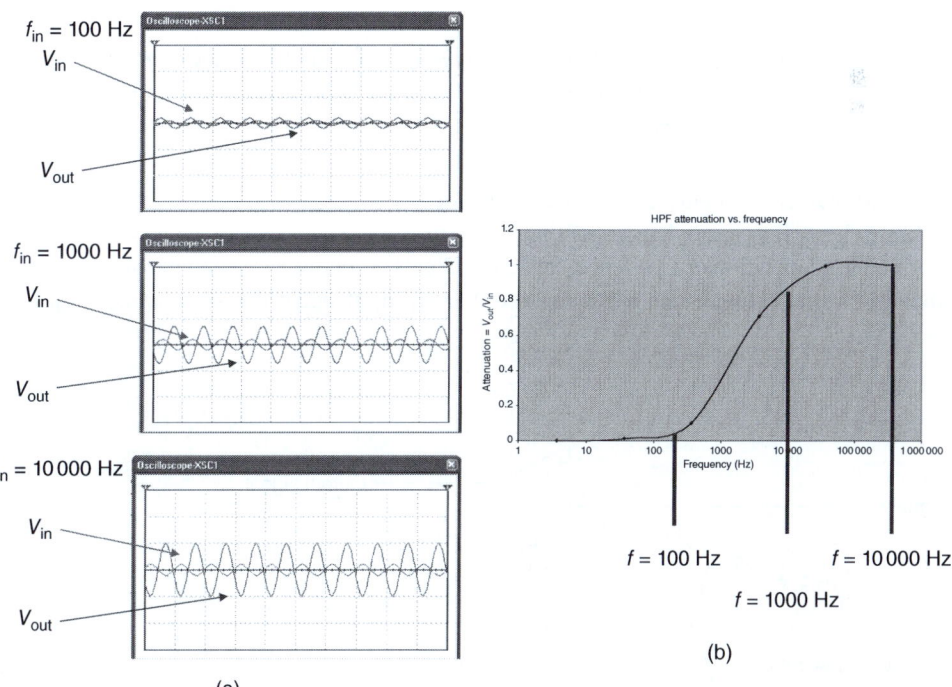

Figure 5.3.14 Active HPF: (a) The input and output waveforms at various frequencies and (b) amplitude vs. frequency.

At high frequency, $f = 10f_C = 10\,000$ Hz, the output amplitude is greater than the input amplitude, according to the principle of HPF operation. This concept is depicted in Figure 5.3.14b, where these three frequencies are shown in amplitude vs. frequency graph. Again, comparing Figures 5.3.14a and 5.3.14b allows us to better comprehend the filter's description in time domain and frequency domain.

Since BPF and BSF filters are the combinations of LPF and HPF filters, you will be able to devise and analyze active BPF and BSF filters, too. This will be a very good exercise. Appendix 5.3.A can help to perform this exercise.

5.3.1.6 Concluding Remarks on Active Filters
Our discussion of active filters most certainly helps us to learn the following main points:

- Active filters eliminate the problem associated with the effect of a load on the filter's operation.
- Active filters allow for amplification of the input signal, which is a big advantage of this circuitry. Indeed, the waveforms obtained by simulations show clearly that the output amplitudes can be greater than the input amplitudes, which is impossible for passive filters.
- Active filters are more practical and provide more flexibility in their design than passive filters. This is because active filters are integrated circuits and they easily allow for the construction of filters having several stages. However, they are just as restricted as passive filters in terms of their ability to change the shapes of their responses.

Most practical filters today are active because of advantages associated with this type of filter.

5.3.2 Switched-Capacitor Filters

Switched-capacitor filters are among the most important and ubiquitous types of active filters employed by today's electronic and communications industries because this is the only technology allowing for complete IC filter implementation.

5.3.2.1 Switched-Capacitor Filters: Concept and Circuits
Figure 5.3.15 shows the circuit diagram of this filter.

The principle of operation of a switched-capacitor filter is as follows: A clock-voltage source controls the switch that connects the capacitor to either an input-signal line or to an op-amp line.

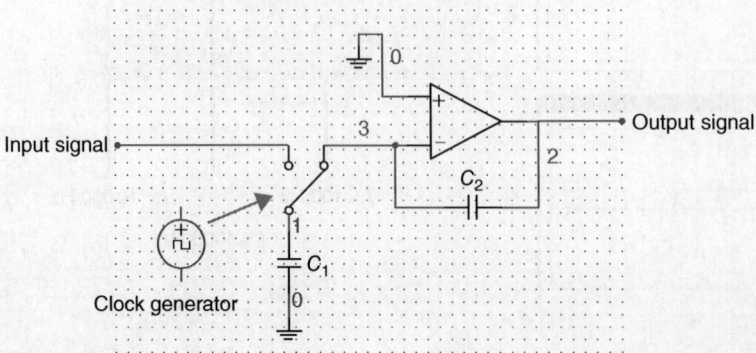

Figure 5.3.15 Circuit diagram of a switched-capacitor filter.

When the switch connects the capacitor to the input-signal line (voltage source), the capacitor is charging; when the switch connects the capacitor to the op-amp, the capacitor is discharging. Thus, the capacitor essentially presents its voltage to the op-amp. If the capacitor is connected to the op-amp for a long time, its voltage drops significantly below the input voltage; if, however, the capacitor is switched at a high frequency, it delivers voltage that is practically equal to an input signal. Thus, by controlling the clock frequency, f_{CLK}, we control the value of the voltage presented to the op-amp. (Recall the charging and discharging properties of a capacitor: It can be charged up to the applied voltage and can discharge to zero volts. Both processes take time whose duration depends on the capacitance, C (F), of the capacitor.)

We can describe the operation of a switched-capacitor filter, shown in Figure 5.3.15, by the following simple mathematical formulas (5.3.22).[3]

The charge, Q (C), of capacitor C_1 during one switching cycle is given by

$$Q(C) = C_1(F) \cdot V_{in}(V) \tag{5.3.22}$$

and its average current becomes

$$i_{C_1}(A) = \frac{Q(C)}{T(s)} = C_1(F) \, V_{in}(V) \times f_{CLK}(Hz) \tag{5.3.23}$$

Here, T (s) is the duration of a switching cycle, which is inversely proportional to the clock frequency, f_{CLK} (Hz). Observe that the current is proportional to the input voltage,

$$i_{C_1}(A) = (C_1(F) \cdot f_{CLK}(Hz)) \cdot V_{in}(V) = \frac{V_{in}(V)}{"R_{C_1}"} \tag{5.3.24}$$

in this regard, the switched capacitor, C_1, works as a resistor whose "resistance" is equal to

$$"R_{C_1}" = \frac{1}{C_1 \cdot f_{CLK}} \tag{5.3.25}$$

(We put R_{C_1} in quotation marks to stress that this is not a real resistance but the imitation of a resistance by a switched capacitor. In other words, "R_{C_1}" is the equivalent of a resistor that could be substituted for a switched capacitor.) Hence, the cutoff frequency of this active filter is equal to

$$f_C (Hz) = \frac{1}{2\pi"R_{C_1}"C_2} = \frac{C_1}{2\pi C_2} \cdot f_{CLK} \tag{5.3.26}$$

Equation (5.3.26) delivers the main point:

> The cutoff (critical) frequency of a switched-capacitor filter is proportional to the clock frequency, f_{CLK} (Hz), and depends on capacitances C_1 and C_2.

This means that by changing f_{CLK}, we can control (program) the cutoff frequency of a filter.

Example 5.3.3   Design of a Switched-Capacitor Filter (After Neamen 2010, pp. 1072–1073)

Problem

Consider an active LPF, such as the one reproduced below from Figure 5.3.11R. Design this filter with switched capacitors instead of resistors to meet the following specifications: dc gain is 5 and cutoff frequency, f_C, is 10 kHz.

---

[3] Maxim Integrated Products, Inc., *A Filter Primer*, Application Note 733, Sunnyvale, CA, 2016.

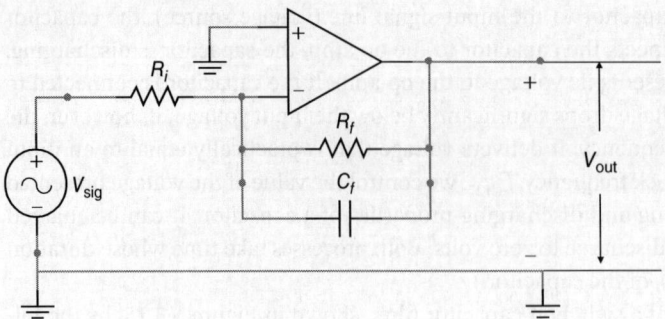

Figure 5.3.11R An RC active LPF based on an inverting op-amp.

Solution
The transfer function of this filter is given by

$$H_{\text{LPF}}(f) = \frac{V_{\text{out}}(f)}{V_{\text{in}}(f)} = -\frac{R_f}{R_i} \cdot \frac{1}{(1 + j2\pi f C_f R_f)} \tag{5.3.16R}$$

where the negative sign comes with inverting the op-amp and the cutoff frequency is given by

$$f_C \text{ (Hz)} = \frac{1}{2\pi f C_f R_f} \tag{5.3.17R}$$

Let us choose $C_f = 100\,\text{pF}$; then, the closest standard value, R_f must be

$$R_f = \frac{1}{2\pi f C_f f_C} = 160\,k\Omega$$

The switched-capacitor filter circuit is given in Figure 5.3.16; obviously, resistors R_i and R_f are replaced by switched capacitors C_1 and C_2, respectively. The transfer function of this filter remains

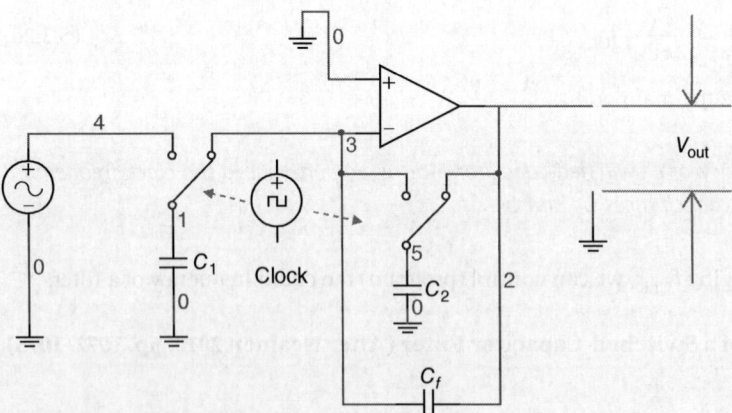

Figure 5.3.16 Switched-capacitor low-pass filter for Example 5.3.3.

the same as given in (5.3.16), but resistors R_i and R_f are replaced by their equivalents, R_{ieq} and R_{feq}, respectively. These equivalent resistors are can be found according to (5.3.25) as

$$R_{ieq} = \frac{1}{C_1 \cdot f_{CLK}}$$

and

$$R_{feq} = \frac{1}{C_2 \cdot f_{CLK}} \quad (5.3.27)$$

Hence, the transfer function of the switched-capacitor filter in question is given by

$$H_{LPF}(f) = -\frac{R_f}{R_i} \cdot \frac{1}{(1+j2\pi f C_f R_f)} = -\frac{R_{feq}}{R_{ieq}} \cdot \frac{1}{(1+j2\pi f C_f R_f eq)} = -\frac{C_1}{C_2} \cdot \frac{1}{1+j\frac{f}{f_c}} \quad (5.3.28)$$

Compare (5.3.28) with (5.3.16) and conclude that dc gain of this switched-capacitor filter is determined by the ratio of two capacitances, C_1 and C_2 because these capacitances determine the equivalent resistances by the principle of operation of the filter. To meet the required gain, we need to set

$$-\frac{C_1}{C_2} = -5$$

To find C_2, we refer to our predetermined value of $R_f = 160 \, k\Omega$ and the required value of $f_C = 10 \, kHz$. Choosing $f_{CLK} = 1 \, MHz$, which is 100 times greater than f_C, we can find C_2 as

$$C_2 = \frac{1}{R_{feq} \cdot f_{CLK}} = 6.25 \, pF$$

Taking the closest standard value, 6.2 pF, we can now find C_1 as

$$C_1 = 5C_2 = 31.25 \, pF$$

Again, replacing the actual value by the standard one, we find $C_2 = 32 \, pF$.

To compute the equivalent resistance, R_{ieq}, for this filter, we plug in the proper numbers into (5.3.27) and find

$$R_{ieq} = \frac{1}{C_1 \cdot f_{CLK}} = 31.3 \, k\Omega$$

Thus, the switched-capacitor filter is designed. You are encouraged to show the value of all parameters in Figure 5.3.16.

Discussion

- Focus especially on the derivation of the formulas and the computation of the equivalent resistances and the clock frequency for this type of filter. All other steps are mainly the same as for the passive and active filters discussed above in this section.
- Note that both C_1 and C_2 are much smaller than C_f; this relationship – when a switched capacitance is much smaller than a filtering capacitance – is typical for a switched-capacitor filter.

Though the schematic of a switched-capacitor filter looks complex, in reality it does not need any discrete resistors, capacitors or inductors. All its components are integrated in one chip and many ICs contain clocks as well.

A switched-capacitor filter can be put into operation without an op-amp. A switching circuit and a capacitor are the only components needed to make it work; however, in practice, this filter always contains an op-amp and therefore it belongs in the category of active filters. We, however, distinguish between active filters built with an op-amp, R and C components, and switched-capacitor filters because there is a fundamental difference between them:

> The active (and passive) filters considered so far process an analog input signal continuously. This is why they are called <u>continuous-time filters</u>. The switched-capacitor filters sample analog signals; consequently, they operate on a <u>discrete-time</u> basis.

A switched-capacitor filter is, in essence, a sampling device, which means that it converts a time-continuous signal into a time-discrete signal. Obviously, we must observe the requirement of the sampling (Nyquist) theorem, which necessitates keeping the sampling frequency, f_S, at least twice as high as the highest signal frequency, f_{max}. In a switched-capacitor filter, the sampling frequency is the clock frequency; i.e. $f_S = f_{CLK}$, and the filter's cutoff frequency, f_C, is considered the highest signal frequency, f_{max}. Typically, a switched-capacitor filter operates at

$$f_{CLK} \text{ (Hz)} > 100 f_C \text{ (Hz)}$$

which greatly facilitates aliasing problems.

All the above considerations prepare us to thoughtfully discuss the applications of the switched-capacitor filters.

5.3.2.2 Applications of Switched-Capacitor Filters

In addition to its great <u>advantage</u> – *easy programmable control of the cutoff frequency* – a switched-capacitor filter has another big advantage over other filter technologies: *It is a small, inexpensive silicon chip*. In fact, *this is the only technology that allows us to acquire an analog filter in completely integrated form*. The *advantages of switched-capacitor filters also include a highly accurate setting of the cutoff frequency and low sensitivity to temperature change*.

All these advantages determine the widespread usage of switched-capacitor filters in modern communication technology. However, the switched-capacitor filters have their *drawbacks* that affect their applications. These are the main ones:

a) *dc offset*: Since a switched-capacitor filter is an integrator, the filter exhibits a significant dc offset. The value of the filter's dc offset varies with the filter model and the configuration of the external resistors.
b) *Amplitude of input signal*: The amplitude of an input signal should be large enough to provide the best SNR but not too large to induce significant total harmonic distortion (to be considered in Chapter 6).
c) *Clock synchronization*: The clock of a switched-capacitor filter must be synchronized with the clocks of other circuits, such as DAC or ADC, working with this filter. (Refer to Chapter 4 for DAC and ADC consideration.)
d) *Aliasing*: As with any sampling device, aliasing – the topic discussed in Section 4.1.3 – is an inherent problem of a switched-capacitor filter.
e) *Clock feed-through*: The clock frequency and its harmonics can appear at the output pins even without an input signal; this phenomenon is called *clock feed-through*.

f) *Noise*: Filter's internal noise is generated by all its components, with the op-amps being the main culprits. Most of this noise lies within the filter's passband. This *noise is the primary weakness of switched-capacitor filters*.

Fortunately, *the shortcomings of the switched-capacitor filters are really minor concerns in their applications*. We listed them to make you aware of the small problems you might encounter when employing these filters because, again, the *switched-capacitor filters, due to their advantages, is today the main filter type used by the industry in the vast majority of applications*.

In addition to all their aforementioned advantages, the switched-capacitor filters possess the unique property called "universality."

Universality of Switched-Capacitor Filters By "universality," we mean the ability of the same chip to operate as an LPF, an HPF, a BPF, or a BSF. The switch to a specific standard function can be done by using several external resistors. What is more, many chips allow for the simultaneous availability of these functions. This universality is achieved by building every chip as a combination of two identical second-order building blocks; the implementation of every specific function can be done, then, by reconnecting outputs. But before we turn to the discussion of a universal filter, we need to pause to remind you of some basic information about an integrator:

An integrator is a circuit performing integration, as its name implies. It is usually built with R and C components and an op-amp. Thus, a switched-capacitor filter can work as an integrator. Integrators based on switched-capacitor filters are the building blocks of IC filters. This is why functional block diagrams of IC filters show integrators instead of detailed schematics of switched-capacitor filters.

Figure 5.3.17a gives an example of the block diagram of an industrial switched-capacitor filter. It shows the functional block diagram of a universal switched-capacitor filter consisting of two identical building blocks, A and B. Each block includes an op-amp, summation point, and two integrators. Here "X" stands for "A" or "B," *notch* means a BSF, and the numbers in parentheses indicate the pin numbers. This figure presents all the pins, whose descriptions are as follows: SHDN – shutdown input, V_{DD} – positive voltage supply, GND – ground, COM – common pin, CLK – clock input, EXTCLK – external/internal clock select input, INVX – inverting input of the filter summing X op-amp, NX/HPX – second-order notch/HPF output, SX – summing input, BPX – second-order BPF output, LPX – second-order LPF output.

For a detailed description of this diagram, refer to its source. You are encouraged to carefully review this block diagram to understand its functionality. The topology shown in Figure 5.3.17 allows a designer to build a filter that will perform LPF, BPF, HPF, and BSF functions simultaneously. It also allows for employing various filter prototypes, the topic to be discussed shortly.

An example of this capability with one block is shown in Figure 5.3.17b whose analysis reveals that this block provides, simultaneously, three out of four filter functions. Specifically, in one filter configuration, we can always get low-pass and band-pass outputs or, alternatively, either high-pass or band-stop (notch) outputs. The main parameters of these blocks are determined by the values of the clock frequency and the resistors.

The samples of the appropriate equations describing this block are as follows[4]:

---

4 Maxim Integrated Products, Specifications of *MAX7490/MAX7491, Dual Universal Switched-Capacitor Filters*, Sunnyvale, CA, 2009.

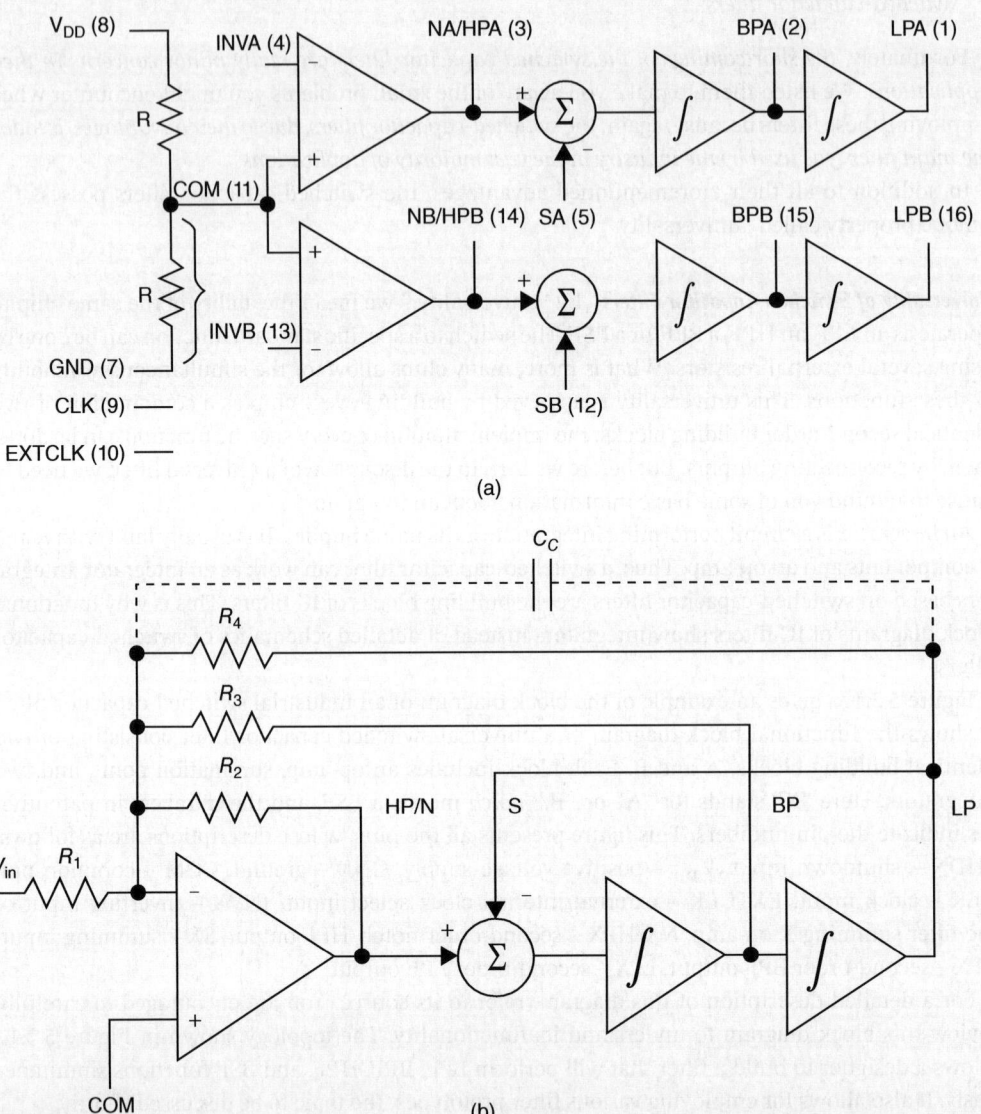

Figure 5.3.17 An example of the block diagram of an industrial universal switched-capacitor filter: (a) functional block diagram and (b) block diagram of a chip providing low-pass, high-pass, band-pass, and notch (band-stop) outputs. Source: Copyright Maxim Integrated Products (www.maximintegrated.com). Used by permission.

The center frequency, f_0, of the band-pass output measured at the peak frequency response, is given by

$$f_0 \text{ (Hz)} = \frac{f_{\text{CLK}}}{100} \cdot \sqrt{\left(1 + \frac{R_2}{R_4}\right)} \qquad (5.3.29)$$

which shows us that by changing the values R_2 and/or R_4, we can readily change the value of a center frequency.

The filter's quality factor, Q, is defined as the ratio of f_0 to the 3-dB bandwidth of a second-order BPF; it is given by

$$Q = \frac{R_3}{R_2} \cdot \sqrt{\left(1 + \frac{R_2}{R_4}\right)} \qquad (5.3.30)$$

The gains of the BPF (H_{OBP}), the LPF (H_{OLP}), the notch (band-stop) filter (H_{ON1} and H_{ON2}) are given by

$$H_{\text{OBP}} = -\frac{R_3}{R_1}$$

$$H_{\text{OLP}} = -\frac{R_2}{R_1} \cdot \frac{R_4}{R_2 + R_4}$$

$$H_{\text{ON1}} = -\frac{R_2}{R_1} \cdot \frac{R_4}{R_2 + R_4} \quad \text{at } f \to 0 \text{ Hz}$$

$$H_{\text{ON2}} = -\frac{R_2}{R_1} \quad \text{at } f \to (f_{\text{CLK}})/2 \qquad (5.3.31)$$

You have to bear in mind that these relationships, known as *design equations*, are different for different filter configurations even of the same filter type. The differences in the configurations are made by varying the connections of the external resistors and introducing new connections.

In general, there are more sophisticated modules containing customizable switched-capacitor filter array. Such *modules can be configured to perform virtually any filtering function without any external components.*

It is our hope that this brief introduction gives you an idea as to what switched-capacitor filters are, how they operate, and how they can be designed. For a more detailed discussion of this topic, consult the technical documentation and other sources listed in the book bibliography.

Questions and Problems for Section 5.3

- Questions marked with an asterisk require a systematic approach to finding the solution.
- Many questions and problems, including those marked with an asterisk, imply that you, in addition to reading the textbook, will do your research to find the answers. Consider such questions as mini-projects.

Active Filters – General View

1. Why are RC and RL filters called *passive*?

2. List the drawbacks of passive filters. Comment on each of them, explaining how serious each drawback is and what can be done to overcome each one.

3 *Consider the RC LPF shown in Figure 5.3.1:
   a) What is its output signal?
   b) Derive the formula for the output signal if an input signal is given by $v_{in}(t) = V_{in} \cos(\omega_{in} t + \theta_{in})$. (*Hint*: The output signal is the voltage drop across the loading coil. We can employ the regular circuit-analysis technique given that the capacitor and the inductor are connected in parallel.)

4 Explain why an RC LPF does not function properly when it is connected to a loading coil, as shown in Figure 5.3.1.

5 What can be done to ensure that an RC LPF operates properly even if connected to a loading coil?

6 Consider an RC LPF built with $R = 4.7\,k\Omega$ and $C = 40\,pF$. To provide the proper operating condition for this filter, it is necessary to separate the filter from a load with circuitry having high input impedance, Z_{in} (Ω), and low output impedance, Z_{out} (Ω). Calculate these impedances provided that they meet the following conditions:
   a) The input impedance must be Z_{in} (Ω) $> 10 X_{C max}$ (Ω) at $f_{max} = 4\,kHz$.
   b) The output impedance must be Z_{out} (Ω) $< X_{L min}$ (Ω), given that the ohmic resistance of the coil is $R_L = 20\,\Omega$.
   c) Prove that under the given output condition, most of the circuit's output voltage, V_{out} (V), will drop across the loading coil. (*Hint*: Refer to Figure 5.3.4 and Example 5.3.1.)

7 Using the Internet, the library, or any other research source, find out why an op-amp is called an *operational amplifier*, not simply an *amplifier*.

8 Sketch the equivalent circuit of an operational amplifier and comment on it, emphasizing the three main parameters of the op-amp.

9 What would be the input impedance, the output impedance, and the gain of an ideal op-amp?

10 Is an op-amp a passive or an active device? Explain.

11 There are noninverting op-amps and inverting op-amps. The inversion of what parameter of these types of op-amp are we referring to?

12 The gain of an open-loop op-amp is 300 000, the input voltage is 3 mV, and the power supply is ±5 V. Calculate the output voltage and explain how you arrived at your answer.

13 The same op-amp can operate in two modes: open loop and closed loop. Explain the difference and sketch the circuit diagrams of both op-amp versions.

14 Consider the closed-loop op-amp shown in Figure 5.3.7:
   a) Explain how it works.
   b) Explain why the feedback provided by resistors R_f and R_i is called negative?
   c) Derive the gain formula of a closed-loop noninverting op-amp.

15 *Based on the gain formula for a closed-loop noninverting op-amp, list three main advantages that a closed-loop op-amp has over an open-loop op-amp and emphasize what role these advantages play in op-amp applications. (*Hint*: You need to do some research into op-amp applications.)

16 *Consider the closed-loop inverting op-amp shown in Figure 5.3.P16.
   a) Explain how it works.
   b) Derive the gain formula of this op-amp.

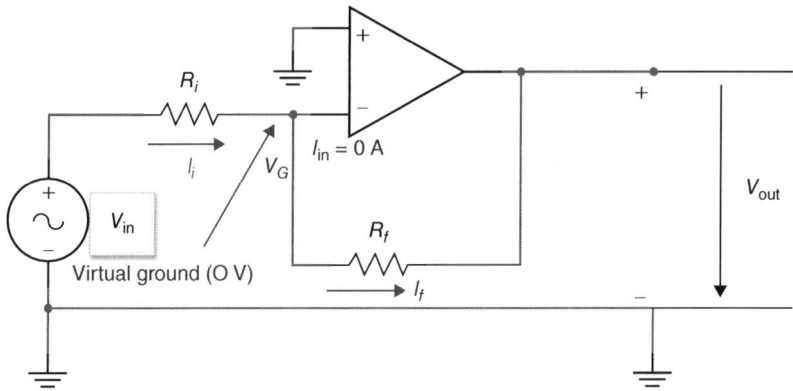

Figure 5.3.P16 Circuit diagram of a closed-loop inverting amplifier.

17 Compare closed-loop noninverting and inverting op-amps: Which amplifier would you prefer to use in your circuits? Why?

18 *Closed-loop noninverting and inverting op-amps are shown in Figures 5.3.7 and 5.3.P16. They include $R_f = 100\,k\Omega$, $R_i = 4.7\,k\Omega$, and $V_{CC} = \pm15$ V. Compare these op-amps by calculating the following parameters:
   a) Closed-loop gains.
   b) Output voltages if $V_{in} = 3$ mV and an open-loop gain is 200 000.
   c) Input and output impedances if $Z_{inol} = 8.2\,M\Omega$ and $Z_{outol} = 14.7\,\Omega$.
   d) Bandwidths if the 3-dB bandwidth of an open-loop op-amp is 3 Hz.
   e) Based on your calculations, discuss the advantages that each type of op-amp has.

19 *The gain of an open-loop op-amp is on the order of hundreds of thousand, whereas the gain of the same op-amp in a closed-loop configuration is on the order of tens. It seems that the greater the gain, the better the amplifier, but the industry works almost exclusively with closed-loop op-amps. Why?

20 The gains of an open-loop and a closed-loop op-amp are 100 000 and 24, respectively: What are these gains in dB?

21 Describe the advantage of using a logarithmic scale in modern communications.

22 Input voltage presented to a closed-loop op-amp is equal to 10 mV. The output voltage is equal to 250 mV. What is the gain in absolute number and in dB?

23 Consider an op-amp whose $A_{ol} = 10^5$, $A_{cl} = 20$, $BW_{ol} = 2\,Hz$, and $V_{CC} = \pm 5\,V$.
   a) What will be $V_{out(ol)}$ and $V_{out(cl)}$ if $V_{in} = 2\,mV$?
   b) Sketch the op-amp's V_{out}–V_{in} graphs similar to that shown in Figure 5.3.8a for both – open-loop and closed-loop – configurations.
   c) What will be closed-loop bandwidth, BW_{cl}, of this op-amp?
   d) Sketch the graph A_v-vs.-f in logarithmic scale. Refer to Figure 5.3.8b.

24 *The gain-bandwidth product of an op-amp is constant, implying that an open-loop op-amp has a very small bandwidth. How can we change the op-amp configuration to significantly increase its bandwidth? How can we control the op-amp's bandwidth?

25 What op-amp bandwidth – big or small – do we need? Why?

26 Sketch the circuits of a passive and an active filter, including the loading coil. List and comment on all the advantages and drawbacks of using an active filter rather than a passive filter.

27 The circuit of an active filter is shown in Figure 5.3.9:
   a) Explain the function of each component.
   b) What is its cutoff frequency if $R = 2.2\,k\Omega$ and $C = 0.002\,\mu F$, $R_f = 100\,k\Omega$, and $R_i = 4.7\,k\Omega$?

28 Consider the active filter shown in Figure 5.3.9:
   a) Explain why it is important to include a loading coil in this circuit.
   b) Explain why it is important to use a closed–loop op-amp in this application.
   c) What is the filter's gain if $R_f = 100\,k\Omega$ and $R_i = 4.7\,k\Omega$?

29 *Sketch the circuit diagram of an active LPF based on a closed-loop inverting op-amp with $R = 5\,k\Omega$, $C = 0.2\,nF$, $R_f = 100\,k\Omega$, and $R_i = 4.7\,k\Omega$:
   a) Compute its cutoff frequency.
   b) What will be the amplitude of an output signal, V_{out}, of this filter if its input is $v_{in}(t)$ (V) $= 6\cos(2\pi \times 0.1 \times 10^6 \times t)$?

30 Derive the formulas for transfer functions of active filters based on noninverting and inverting op-amps.

31 *Derive the transfer function of an active noninverting LPF like that shown in Figure 5.3.9.

32 *Consider an active noninverting LPF like that shown in Figure 5.3.9:
   a) Design a filter whose dc gain is 11 and whose cutoff frequency is 4 kHz.
   b) Compute its output voltages at $f = 0.4\,kHz$, $f = 4\,kHz$, and $f = 40\,kHz$.
   c) Compute its output phase shifts at $f = 0.4\,kHz$, $f = 4\,kHz$, and $f = 40\,kHz$.
   d) Sketch the graphs of the amplitude and phase frequency responses and show the computed points.

33 Derive the transfer function of an active inverting HPF.

34 Consider an active inverting HPF:
  a) Draw the circuit schematic of such a filter.
  b) Design an active inverting HPF with a dc gain of 11 and a cutoff frequency of 4 kHz.
  c) Compute its output voltages and phase shifts at $f = 0.4$ kHz, $f = 4$ kHz, and $f = 40$ kHz.
  d) Sketch the graphs of the amplitude and phase frequency responses and show the computed points.

35 Summarize all the advantages of active filters and explain how these advantages can be used in the filters' applications.

Switched-Capacitor Filters

36 Consider Figure 5.3.15:
  a) Explain the function of each component.
  b) Explain the principle of a switched-capacitor filter's operation.

37 Compare a switched-capacitor filter, shown in Figure 5.3.15, with an active first-order RC filter:
  a) We observe that the switched-capacitor filter includes two new components (a clock and a capacitor) instead of one passive resistor. Given that the clock is an active component, the switched-capacitor filter appears to be more sophisticated and more of a power-consuming unit than a simple RC filter. Why, then, does the text say that *"Switched-capacitor filters are among the most important and ubiquitous types of filters employed in today's electronics and communications industries?"*
  b) The switched-capacitor filter shown in Figure 5.3.15 includes two capacitors. What is the order of this filter?

38 The schematic of an active RC LPF with a dc gain of 10 and $f_C = 20$ kHz is shown in Figure 5.3.11. Design a switched-capacitor filter that meets the same specifications. Draw its schematic, showing the values of all components. (*Hint*: Follow Example 5.3.3.)

39 What is the order of the filter shown in Figure 5.3.16?

40 *Explain the main advantages of switched-capacitor filters over other analog filter technologies.

41 *Explain the main drawbacks of switched-capacitor filters.

42 A schematic of switched-capacitor filters typically contains op-amps. Are op-amps necessary components of these filters? Explain.

43 Consider a clock in a switched-capacitor filter:
  a) What is the role of the clock?
  b) How can we choose the clock's frequency?
  c) Is there any limit on the clock's frequency?

44 A cutoff frequency of a switched-capacitor filter is 20 kHz. What clock frequency would you recommend?

45 There are two categories of analog filters: continuous-time and discreet-time. To which category do switched-capacitor filters belong and why? Explain.

46 Electronic circuits that perform integration are called integrators. Explain the relationship between integrators and switched-capacitor filters. Prove your answer.

47 An example of the block diagram of an industrial universal switched-capacitor filter is shown in Figure 5.3.17:
  a) Explain how this circuit will operate in LPF mode.
  b) Will it be possible to use this circuit in HPF mode? Explain.
  c) If you use this circuit in BPF mode, how can you change f_0 from 10 to 100 kHz?
  d) If you use this circuit in BPF mode, how can you change its A_{cl} from 20 to 36?

5.3.A Active BPF and BSF

5.3.A.1 Active BPF

Another popular filter is the <u>band-pass active filter</u>. The schematic of such a filter with an inverting op-amp is shown in Figure 5.3.A.1. The analysis of this filter is also based on the analysis of a BSF, such as that given in Section 5.2. Just recall that a BPF is a combination of a LPF and a HPF (see Figures 5.2.1–5.2.6). The transfer function of this filter can be readily derived if we follow the patterns we have just developed for LPF and HPF active filters:

$$H_{\text{BPF}}(f) = -\frac{Z_f}{Z_i} \qquad (5.3.\text{A}.1)$$

where Z_f for LPF and Z_i for **HPF** can be obtained from (5.2.1) and (5.2.3) as

$$Z_f(f) = \frac{R_f}{1 + j2\pi f C_f} \text{ and } Z_i(f) = \frac{R_i}{R_i + \frac{1}{j2\pi f C_i}} \qquad (5.3.\text{A}.2)$$

(Remember that subscript in R_f and C_f refers to feedback components, not to frequency.)

We leave it for you to derive from (5.3.A.2) equations for the transfer function of a band-pass active filter with inverting op-amp and for the filter's amplitude and phase responses.

We should point out that this equation can be transformed to a more convenient form if we understand that an LPF sets a high cutoff frequency, f_{C2}, and an HPF sets a low cutoff frequency, f_{C1}. Hence, the values of these frequencies are given by

$$f_{C2} \text{ (Hz)} = \frac{1}{2\pi C_f R_f} \text{ and } f_{C1} \text{ (Hz)} = \frac{1}{2\pi C_i Ri} \qquad (5.3.\text{A}.3)$$

5.3.A Active BPF and BSF

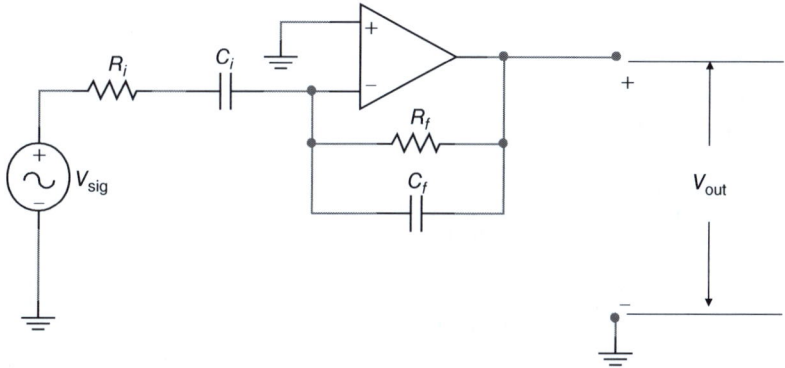

Figure 5.3.A.1 A band-pass active filter based on an inverting op-amp.

Referring to (5.1.20) and following that pattern, we can write

$$\frac{f}{f_{C1}} = \frac{R_i}{X_{Ci}}$$

and

$$\frac{f}{f_{C2}} = \frac{R_f}{X_{Cf}} \tag{5.3.A.4}$$

Now here is another exercise for you: Derive the formula for the HPF transfer function in terms of the f/f_{C1} and f/f_{C2} ratios.

The central frequency, f_0, is obviously given by

$$f_0 = \sqrt{(f_{C1} \times f_{C2})} \tag{5.3.A.5}$$

(See (5.2.11)).

The above discussion is sufficient to acquaint you with a band-pass active filter. We know that both passive and active versions of this filter are built with an LPF and an HPF connected in series. It is worth noting that the sequence of this connection – the LPF first and the HPF second or vice versa – is not important. This active BPF can be built in either sequence as long as the main requirement for building a BPF – $f_{C2} > f_{C1}$ – is met.

Figure 5.3.A.2a shows input and output waveforms at various frequencies; these were obtained by simulation of the circuit given in Figure 5.3.A.1 with $V_{in} = 1\,\text{V}$ and the following parameters: $R_i = 1.1\,\text{k}\Omega$, $C_i = 1.0\,\mu\text{F}$, $R_f = 10\,\text{k}\Omega$, and $C_f = 2.0\,\text{nF}$. Therefore, the critical frequencies are set as $f_{C1} = 144\,\text{Hz}$ and $f_{C2} = 1.06\,\text{kHz}$. The central frequency is given by $f_0 = 380\,\text{Hz}$. Figure 5.3.A.2b presents the amplitude vs. frequency graph of this filter.

Note that the input amplitude is equal to 1 V; this gives you a good reference point from which to evaluate the output amplitudes. When the frequency, $f = 50\,\text{Hz}$, is below the low cutoff frequency, f_{C1}, the output amplitude is small due to the work of the HPF section of our BPF. At $f_{C1} = 144\,\text{Hz}$, the output amplitude can be estimated as $0.707V_{in}$ times the BPF gain. When the frequency is equal to the central frequency, $f_0 = 380\,\text{Hz}$, the output amplitude reaches its maximum value. At the next frequency, $f = 760\,\text{Hz}$, the output amplitude is still high because this point is within the passband of our BPF. When the frequency is equal to the high cutoff frequency, $f_{C2} = 1.06\,\text{kHz}$, the output amplitude has the same value as it has at the low critical-frequency point; that is $V_{out} = 0.707V_{in} \times \text{gain}$.

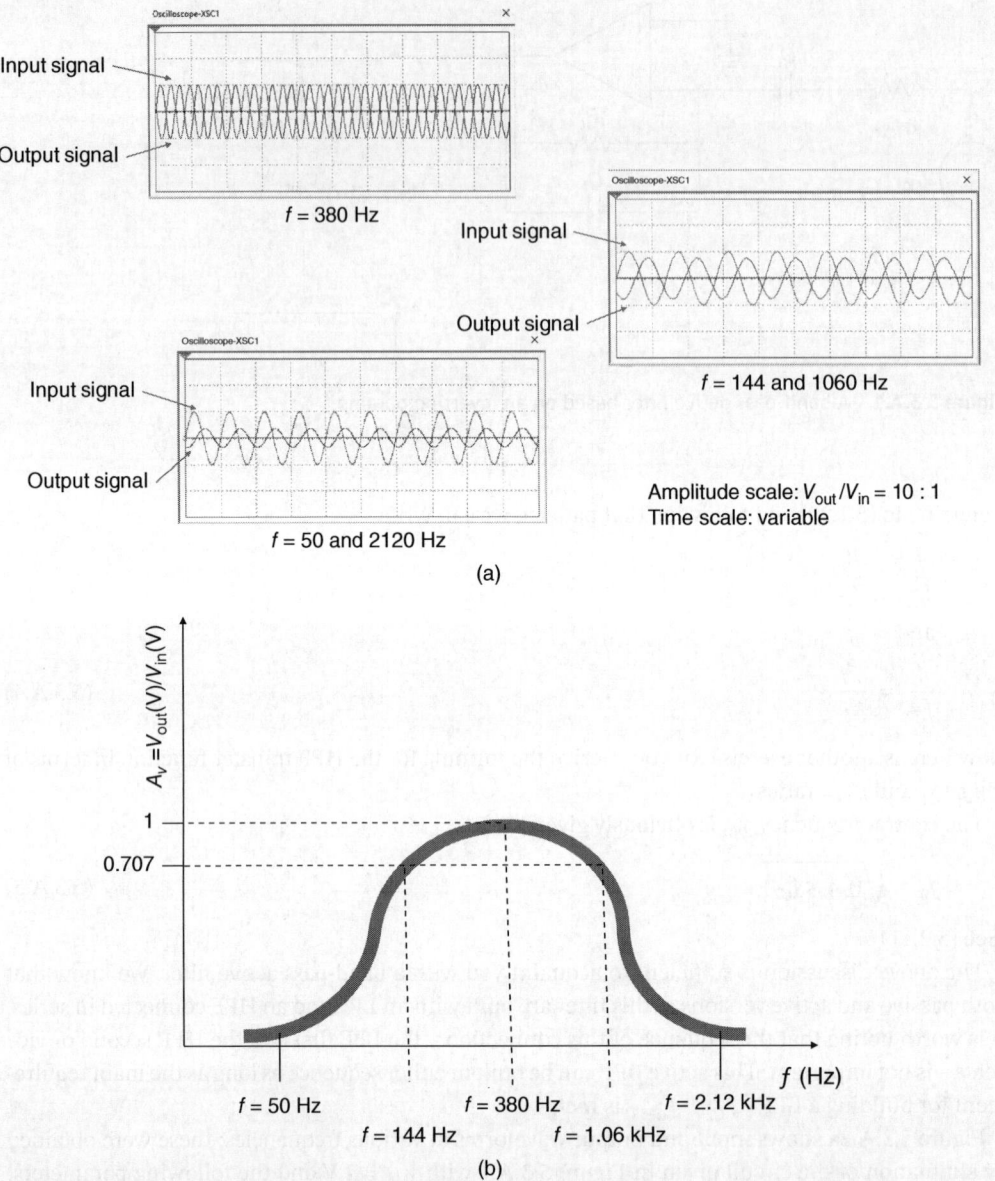

Figure 5.3.A.2 Active BPF: (a) Input and output waveforms at various frequencies and (b) amplitude response.

As the input frequency increases from $f = 2.12$ to 5.3 kHz, the output amplitude becomes increasingly smaller, as it must be according to the BPF principle of operation. Figure 5.3.A.2b illustrates these explanations in frequency domain. You can readily identify $\frac{V_{out}}{V_{in}}$ at every given frequency in this figure and in the corresponding waveform in Figure 5.3.A.2a.

Another version of an active BPF is shown in Figure 5.3.A.3. This filter includes three specific stages: An active LPF, which sets the upper critical frequency, $f_{C2} = 1/(2\pi C_f R_f)$; an active HPF,

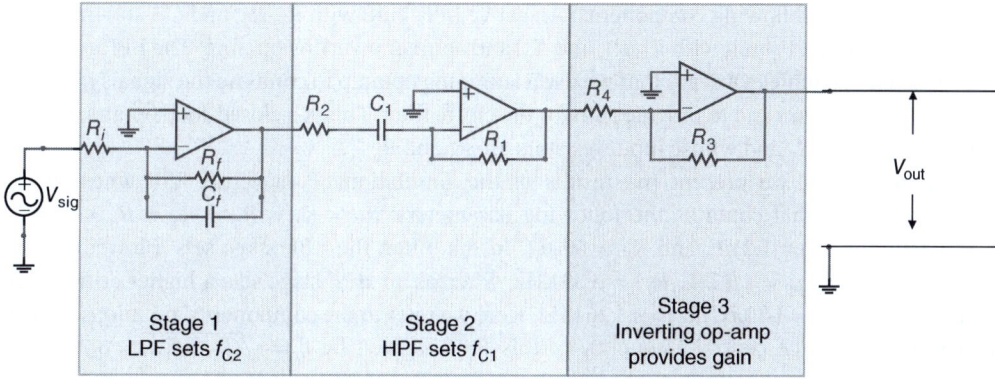

Figure 5.3.A.3 Schematic of an active three-stage BPF.[5]

which sets the lower critical frequency, $f_{C1} = 1/(2\pi C_1 R_2)$; and a closed-loop op-amp, which provides dc gain, $A_{dc} = -R_3/R_4$. This filter allows us to control the gain of the circuit, which is certainly an advantage in practical applications. You are encouraged to simulate its work by using Multisim or some other computerized tool to analyze the waveforms and the amplitude vs. frequency graph, as we have done with Figure 5.3.A.2.

5.3.A.2 Active BSF

The last type of filter we will discuss here is the *band-stop active filter*. Referring again to a passive BSF (see Section 5.2), we can conclude that building this type of filter requires that an LPF and an HPF be connected in parallel. The schematic of an active BSF is shown in Figure 5.3.A.4. The

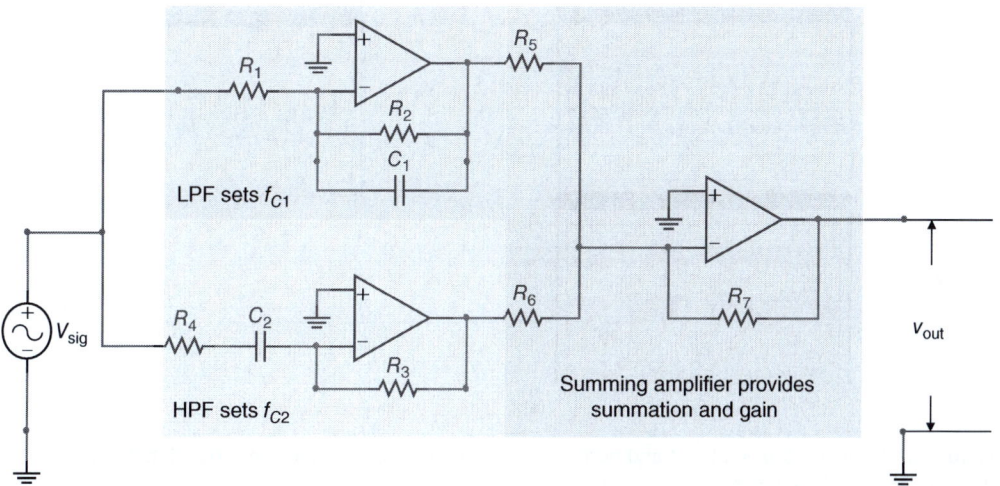

Figure 5.3.A.4 Schematic diagram of a band-stop active filter.[6]

---

5 After Alexander and Sadiku (2009, p. 644).
6 After Alexander and Sadiku (2009, p. 646).

filter consists of the following components: An active LPF, built with R_1, R_2, and C_1, and inverting op-amp; an active HPF, built with R_3, R_4, and C_2, and another inverting op-amp. The LPF and HPF are connected in parallel. Observe that we use a summing op-amp to combine the signals from the LPF and HPF branches and to provide gain for the entire filter. This is a closed-loop op-amp whose feedback resistor is R_7 and whose input resistors are R_5 and R_6.

In Figure 5.3.A.5, we present the results of the simulation of an active BSF whose $v_{in}(t)$ (V) = $\sin(\omega t)$ and that contains the following parameters: $R_1 = R_2 = R_4 = R_5 = R_6 = 10\,k\Omega$, $C_1 = 4.7\,nF$ and $C_2 = 1.2\,nF$, and $R_7 = 50\,k\Omega$. In this filter, the LPF stage sets a lower critical frequency, $f_{C1} = f_{CLPF} = 1/(2\pi C_1 R_2) = 3.39\,kHz$, whereas an HPF stage sets a higher critical frequency, $f_{C2} = f_{CHPF} = 1/(2\pi C_2 R_4) = 13.26\,kHz$. Here we refer to the components' notations shown in Figure 5.3.A.4. The central frequency of this BSF is given by $f_0 = \sqrt{(f_{C1} \times f_{C2})} = 6.70\,kHz$. The input and output waveforms at various input frequencies and amplitude responses at these frequencies are shown in Figure 5.3.A.5.

When the frequency of the input signal is lower than f_{C1} or greater than f_{C2}, the amplitude of the output signal is greater than the input amplitude; at the critical frequencies, the output amplitudes are about $0.707 \times A_{dc} \times V_{in}$; at $f = f_0$, the output amplitude reaches its smallest value. These observations can be easily traced at both waveforms (time domain) and on the amplitude vs. frequency graph (frequency domain) in Figure 5.3.A.5a,b.

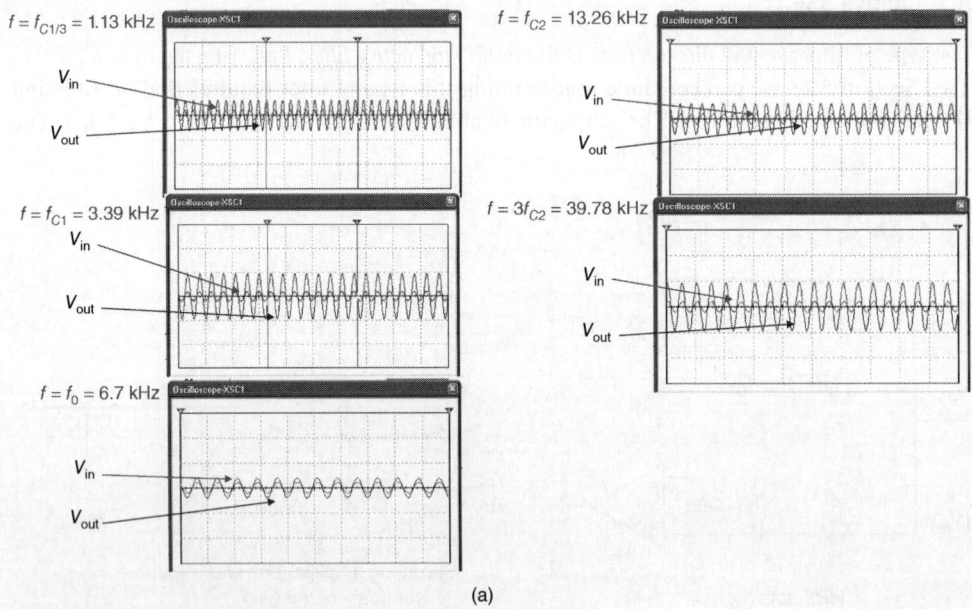

(a)

Figure 5.3.A.5 Active BSF: (a) Input and output waveforms at various frequencies and (b) amplitude responses at these frequencies.

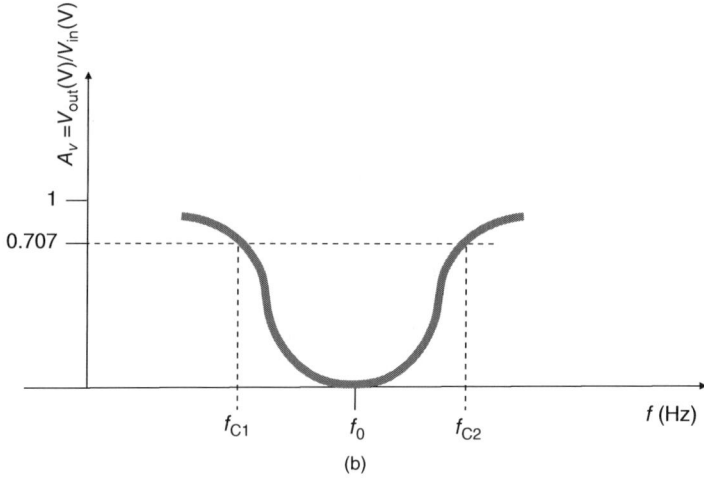

(b)

Figure 5.3.A.23 *(Continued)*

5.4 Filter Prototypes and Filter Design

Objectives and Outcomes of Section 5.4

Objectives

- To understand that filter design is a sophisticated operation and that the industry simplifies the design by developing typical filter-design procedures based on a set of established *filter prototypes,* each related to a specific application area.
- To learn that a filter prototype is determined by its transfer function, whose structure and coefficients are well known and tabulated.
- To study in depth the main filter prototypes (the Butterworth, Chebyshev 1, Chebyshev 2, elliptic, Bessel, and all-pass) and the areas of their application.
- To understand two main steps in filter design.
- To become familiar with MATLAB- and Multisim-based automated design options and to learn the traditional method of filter design based on developing a filter's transfer function.
- To study the concept of the design process exemplified by design of a second-order Butterworth filter.
- To understand the variety of existing filter technologies and to learn how to choose the proper technology for a specific task.

Outcomes

- Understand that filter design starts with a prototype, the model for the filter to be built. Recollect that an ideal filter – the first model that comes to a designer's mind – cannot be realized and

therefore cannot serve as a prototype in designing a filter. Recall that the solution to this problem lies in the design of a specific type of filter, the type that best approximates one characteristic of an ideal filter. Conclude that every one of this type constitutes a filter prototype.
- Realize that, specifically, we can build a filter with the best flatness in the passband or with the steepest rolloff in the transition band or with the best linearity of phase response; however, we cannot build a filter that would simultaneously have all these characteristics infinitely close to those of the ideal.
- Call to mind that the filter's characteristics are described by its transfer function; therefore, the design process boils down to constructing the proper transfer function and finding its coefficients. Realize that such an effort turns the design of every individual filter into a laborious, time-consuming project totally unacceptable in the modern industrial environment.
- Acquire knowledge that the filter prototypes eliminate the need to derive a transfer function of a new filter from scratch because each prototype has its "standard" transfer function (typically named after its developer) with numerical coefficients, the changing of which allows for designing a filter with the required characteristics.
- Bear in mind that all filter prototypes are standardized to a LPF and that all necessary information for designing HPF, BPF, and BSF can be obtained from the LPF prototype.
- Start learning the design process by considering the Butterworth filter prototype: Become familiar with its transfer function, $H(s) = b_0/B_n(s)$, where $s = \sigma + j\omega$ is a complex frequency, $b_0 = 1$ for a normalized ($\omega_C = 1$ rad/s) filter, and the nth order Butterworth polynomial has the form $B_n(s) = a_n s^n + a_{n-1} s^{n-1} + a_{n-2} s^{n-2} + \cdots + a_1 s + a_0$. Know that polynomial's coefficients are given in (5.4.5) and that they are tabulated in various forms.
- Learn that the amplitude response of an nth order Butterworth filter can be derived from its transfer function as $A_v = |H_n(\omega)| = 1/\sqrt{(1 + \omega^{2n})}$. Realize that here $\omega = \omega_x/\omega_C$ is the ratio of the current frequency, ω_x, to the cutoff frequency, ω_C.
- Understand that the formula of the Butterworth amplitude response shows that the higher the filter's order, the closer its amplitude response is to the ideal amplitude characteristic, as demonstrated in Figure 5.4.3. Comprehend, however, that the ideal graph could be attained only hypothetically at $n \to \infty$, which means that *building an ideal filer would require an infinite number of reactive components* because each n needs one C or L component.
- Realize that improvement in the Butterworth filter's amplitude response can be attained in practice only by increasing the number of RC cascaded filters, which is a price for this improvement.
- Be aware that the Butterworth filter's amplitude response can be obtained in logarithmic scale as A_v (dB) $\equiv |H(\omega)|_{dB}| = 10 \log(1 + \omega^{2n})$.
- Know that in a Butterworth filter, a 3-dB cutoff frequency – defined as the frequency at which the amplitude response of the Butterworth filter becomes equal to 0.707 – is the passband frequency, ω_P.
- Learn that the additional flexibility in designing a Butterworth filter is attained by introducing the design constant, ε, as $|H(\omega)| = 1/\sqrt{(1 + \varepsilon^2 (\omega_x/\omega_C)^{2n})}$ and its logarithmic counterpart, $R_{dB} = 10 \log(1 + \varepsilon^2)$. Understand that the design constant delimits the passband region.
- Remember that the main advantage of a Butterworth filter prototype is the flatness of the amplitude response in its passband and stopband, which is mainly required in audio transmission for preservation of a signal's amplitude. But know that the phase response of a Butterworth filter is far from linear, signifying another demonstration of the main principle of filter prototype usage: Only one characteristic of a given filter prototype can most closely approach that of an ideal filter.
- Learn that reversing the ratio of ω_x to ω_C allows for use of all the amplitude characteristics of a low-pass Butterworth prototype to find the amplitude characteristics of a high-pass Butterworth filter. Changing BW_x/BW_{3dB} to BW_{3dB}/BW_x enables us to determine the amplitude characteristics of band-pass and band-stop Butterworth prototypes.

5.4 Filter Prototypes and Filter Design

- Understand that the formula for a phase response of a Butterworth filter can be derived from its transfer function, $H(\omega)$, as Θ (rad) $= \arg(H(\omega)) = \tan^{-1}[\text{Im}\,(H(\omega))/\text{Re}(H(\omega))]$. Become familiar with the phase response of a second-order Butterworth filter, Θ_{But2} (rad) $= -\tan^{-1}[\omega\sqrt{2}/(1-\omega^2)]$, which is obtained by using the abovementioned derivation.
- Know that the time group delay of a Butterworth filter can be found as $T_D(s) = -d\Theta/d\omega$, which, for a second-order Butterworth filter, takes the form $T_{D\text{But2}}(s) = -d\Theta/d\omega = \sqrt{2}/[(1+\omega^2)/(1+\omega^4)]$.
- Learn that the higher order of a Butterworth filter – and therefore the closer a Butterworth filter's amplitude response is to that of the ideal graph – the larger the deviation of the phase shift and group delay from their ideal characteristics.
- Realize that the transfer function of a Butterworth filter can be presented in a factored form as $H(s) = A/(s-p_1)(s-p_2)\cdots(s-p_n) = A/\prod_{k=1}^{n}(s-p_k)$, where p_k are the poles and A is the filter's dc gain. Understand that knowing the poles allows for the complete recover of the transfer function.
- Become familiar with the technique of finding the pole positions on the complex plane, which provides an easy and convenient procedure for finding the filter's transfer function.
- Understand that the pole positions in the complex plane determine the general behavior of a Butterworth filter and that the values of a circuit's components are equal to twice the absolute values of the poles' real parts.
- Remember that all the frequency-response characteristics of the filters are important because they describe how well a filter passes the waveform (time-domain characteristic, that is) of an input signal.
- Understand that there are two main steps in filter design: Developing the filter's transfer function and implement this function in a filter circuit whose performance meets the required specifications.
- Become familiar with tools enabling us to automate filter design and comprehend that the general software tools, such as MATLAB and Multisim, facilitate the filter-design process.
- Learn that MATLAB can be used to design digital filters, whereas Multisim is capable of designing the circuits of analog filters; both tools can simulate filter operations and present the transfer function and responses of the designed filter.
- Be aware of the traditional approach to filter design through detailed design of such filter prototype as a second-order Butterworth filter.
- Understand that the filter-design handbooks usually contain tables of poles, but not the tables of polynomials. Understand that an experienced designer can reach a myriad of significant conclusions about the filter's properties by simply analyzing the locations of the poles on the real-imaginary plane.
- Develop a deep understanding of the design process by questioning some steps of the considered design process and finding the answers to these questions.
- Learn how to calculate the minimal order of a filter capable of meeting the required specifications.
- Realize that filter design is an art more than a science; however, it is an art based on a strong scientific foundation. Know that the design art aspect arises from the fact that in this process there is always one equation for two unknown variables and that this problem is resolved by assuming the value of one variable to find the second.
- Understand the problem in choosing the proper filter technology for a designed filter and get to know the filter's industrial classifications, enabling us to select the appropriate technology for a task.
- Review the available technology, such as passive filters, active filters, switched-capacitor filters, and digital filters.

5.4.1 Filter Prototypes

5.4.1.1 The Problem in the Filter Design – The Need for the Filter Prototypes

We start this discussion with two short notes: First, *filters* are, in fact, the *signal-processing* devices because they *modify the spectrum of an input signal*, enhancing or suppressing certain frequency-related features of this signal. Second, we have considered so far only the *fixed filters* with constant circuits and, consequently, constant characteristics. However, there are *adaptive filters* whose properties and characteristics change in response to incoming tasks; what is more, there are *reconfigurable filters*, whose schematic changes depending on the requests.

The approach to *designing a filter* seems to be quite obvious: Just design a filter with ideal characteristics. However, an *ideal filter* cannot serve as a design prototype for a real filter but, nonetheless, can still serve as an ultimate model in filter design. Why?

Let us recall the transfer function of a filter is introduced in Section 5.2.2 as

$$H(f) = \frac{V_{\text{out}}(f)}{V_{\text{in}}(f)} = |H(f)|e^{j\Theta} \tag{5.2.20R}$$

Considering a LPF as an example, we recall that the amplitude and the phase shift are given by

$$|H(f)| = A_v = \frac{1}{\sqrt{\left(1 + \left(\frac{f}{f_c}\right)^2\right)}} \tag{5.1.21R}$$

$$\Theta = -\tan^{-1}\left(\frac{f}{f_c}\right) \tag{5.1.22R}$$

An *ideal filter* is determined by the following conditions: (i) its amplitude, $|H(f)|_{\text{ideal}}$, is equal to 1 within the passband and is 0 outside the passband and (ii) its output phase shift, θ_{ideal} is proportional to the signal's frequency within the passband and is zero outside the passband. That is,

$$|\mathbf{H}(\mathbf{f})_{\text{ideal}}| = 1 \text{ for } f \in \text{passband}$$

and

$$|\mathbf{H}(\mathbf{f})_{\text{ideal}}| = 0 \text{ for } f \notin \text{passband} \tag{5.4.1a}$$

also,

$$\Theta_{\text{ideal}} = -t_d f \text{ for } f \in \text{passband}$$

and

$$\Theta_{\text{ideal}} = 0 \text{ for } f \notin \text{passband} \tag{5.4.1b}$$

These conditions are visualized in Figure 5.4.1.

It seems, again, that a designer just needs to develop the real circuitry that would have the characteristics of an ideal filter. There is, however, a fundamental obstacle to this straightforward approach, the obstacle formulated by the *Paley–Wiener criterion* for the realizability of a filter's amplitude response, $|H(\omega)|$. This criterion states that *the filter with the given amplitude response, $|H(\omega)|$, is physically realizable if and only if the following integral is finite*:

$$\int_{-\infty}^{+\infty} \left\{ \frac{|\ln|H(f)||}{1+f^2} \right\} df < \infty \tag{5.4.2}$$

An important clarification of this criterion is that

Figure 5.4.1 Amplitude and phase responses of an ideal filter.

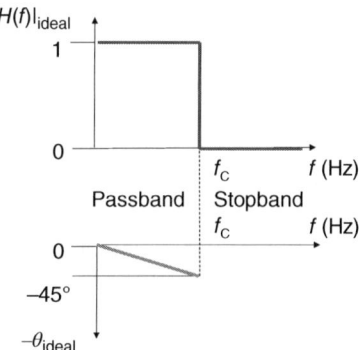

if $|H(f)|$ is zero over any finite band of frequency, then the member $|\ln|H(f)||$ goes to infinity and Criterion 5.4.2 is not satisfied. This means that the filter with the ideal amplitude response, like the one shown in Figure 5.4.1, is not physically realizable because its attenuation is zero in the stopband.

The Paley–Wiener criterion presents a necessary and sufficient condition for the physical realization of a filter with a given amplitude response, $|H(f)|$. This criterion is quite general and is applied to any type of filter – low-pass, high-pass, band-pass, and band-stop (notch).

Now, when we comprehend that ideal filters with zero-part amplitude responses cannot be built, we understand that the amplitude response of a physically attainable filter must *have a finite value over the entire range of frequencies*. This means, in particular, that $|H(f)|$ must be gradually changing and tending to – but never reaching – zero. Mathematically speaking, this means that $|H(f)|$ must be a continuous function over its range. Since the graph in Figure 5.4.1 has the point of discontinuity at f_C when it jumps from 1 to 0, a filter with such an amplitude response cannot be physically built.

In addition, as we will see shortly in our discussion of the Butterworth filter,

> an ideal filter cannot be built with a finite number of components; this is another way of saying that an ideal filter is a physical impossibility.

Thus, the problem of the filter design boils down to two contradictory requirements, which can be the best exemplified by considering a filter's amplitude response:

On the one hand, we can build filters only with gradually changing amplitude response; on the other hand, we need to maximally fit this response to ideal, step-like amplitude characteristic.

The industry, in its attempt to resolve this dilemma, has developed several standard filter prototypes *that can be used in the filter-design process. Why several?* Well, since we cannot build a filter whose frequency response matches the ideal characteristics completely, we can build one filter with the flattest amplitude response in passband, another filter with the steepest rectangular amplitude response in the transition band, and so on. In other words, each of these prototypes matches the ideal filter best in one or two parameters, but never completely. We will discuss these standard prototypes shortly.

5.4.1.2 Another Problem for Filter's Designer: Relationship Between Amplitude and Phase Responses

A thorough study of filters shows that in reality the *amplitude and phase parts of a filter's frequency response are not independent* and there is a trade-off between these two characteristics; that is, when building a filter with almost ideal amplitude response, we often compromise on the quality of its phase shift and vice versa. Fortunately, practical applications very often do not require that both of these characteristics be close to ideal. For example, the transmission of *audio signals* is very much insensitive to phase distortion, whereas such transmission requires a high-quality amplitude response. This is so because the human ear is sensitive to amplitude distortion and relatively insensitive to phase distortion. The requirements for *video transmission*, however, are just the opposite: Since the human eye is sensitive to phase distortion and can tolerate amplitude distortion, video needs almost ideal phase transmission and can afford significant amplitude distortion. Consider, for example, high-definition television (HDTV) transmission, which is the digital transmission of video signals: Distortion of a pulse waveform will result in a change in the brightness of object edges with respect to its center. Thus, an object would look bright at one side and dark at the other, not the picture you want to see on your TV screen. *Digital transmission* also requires minimum phase distortion to prevent pulse spreading (widening), which, in turn, leads to *intersignal interference, ISI*. Also, digital transmission needs the preservation of the signal waveform, which also requires minimum phase distortion.

Modern communications set much more demanding specifications for filters than that for just audio and video transmissions. Consider, for example, the filter requirements for the fifth generation of mobile communications known as 5G[7] (regarding 5G, refer to Section 1.2.3.3). This communications operates in frequency bands >6 and 20 GHz, which requires different filter technology than is used in 4G mobile devices. Also, the new filters must be reconfigurable and multiband devices to reduce the number of components in every unit. On the top of that, the size and cost of the filters must be significantly reduced. Complexity of new technology demands not only to reduce the complexity of the components but also to simplify the design process itself.

This discussion exemplifies the need for development of the filter prototypes.

5.4.1.3 Main Filter Prototypes – What and Why

The Graphs of Amplitude and Phase Responses of Standard Filter Prototypes So what are those filter prototypes? As introduced above, they are filters that have distinguished amplitude and phase responses. Thus, depending on the filter application, we can choose a specific standard prototype rather than inventing a filter from scratch. For example, if we are to design an audio system, we need a filter with the best possible amplitude characteristic in its passband. In this case, we' would pick the prototype called Butterworth filter because it meets the above requirement better than others. Figure 5.4.2 shows the amplitude and phase responses of the major filter prototypes; each prototype is named after its developer or a characteristic feature. Bear in mind that almost 90% of filters in use today are covered by these prototypes.

(In this subsection, we will interchangeably use radian frequency, ω (rad/s), and frequency, f (Hz), to shorten the writing in such expressions as a complex frequency, $s = \sigma + j\omega$ and transfer function, $H(\omega)$.) You can see that the amplitude response of the Butterworth filter has a smooth graph in all three regions – passband, transition band, and stopband. The Chebyshev filters have ripples either in the passband or the stopband, but their graphs are steeper than those of the Butterworth filter. The elliptic filter has the steepest amplitude response among all the prototypes, but its amplitude ripples at both the passband and the stopband. The Bessel filter exhibits an almost linear phase response.

Why Do We Need the Filter Prototypes? The *need for filter prototypes* can be explained as follows: Imagine we are filter designers and our task is to build a filter for a communication system. We

---

7 http://www.mwrf.com/systems/new-system-design-tools-must-5g-rf-front-ends?code=UM.

would like, of course, to build an ideal filter, but we know, from our discussion in Section 5.4.1.1, that an ideal filter cannot be physically realized. We also know from that discussion that *each physically realizable filter can best approximate the frequency response of an ideal filter in one parameter only*; that is, we can build a filter with the best flatness in the passband or with the steepest rolloff in the transition band or with the best linearity of phase response; what we cannot do is to build a filter that would have all these characteristics infinitely close to the ideal simultaneously. Since the filter's characteristics are described by its transfer function, we therefore choose the proper amplitude response, for example, by constructing the proper transfer function and finding its coefficients. But this is a formidable task because we do not know what this "proper" transfer function looks like, and so the only way we can do this work is by trial and error, that is, by deriving the transfer function, obtaining the frequency and time responses, and checking whether these responses satisfy our requirements. Obviously, with this approach, the design of every individual filter would become a laborious, time-consuming project that would be totally unacceptable in the modern industrial environment. This is where filter prototypes come into play.

To further explain the need for filter prototypes, we need to consider the general form of a transfer function:

$$H(s) = \frac{b_m s^m + b_{m-1} s^{m-1} + \ldots + b_1 + b_0}{a_n s^n + a_{n-1} s^{n-1} + \ldots + a_1 + a_0} \tag{5.4.3}$$

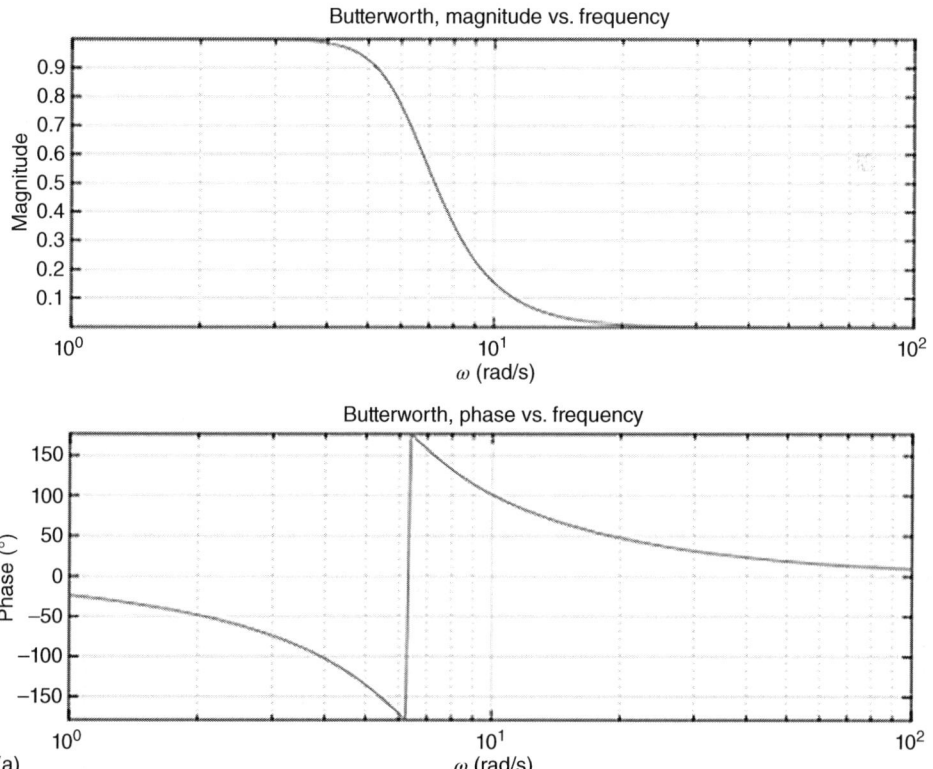

Figure 5.4.2 Amplitude and phase responses of the major filter prototypes: (a) Butterworth filter; (b) Chebyshev 1 filter; (c) Chebyshev 2 filter; (d) Cauer (elliptic) filter; and (e) Bessel filter.

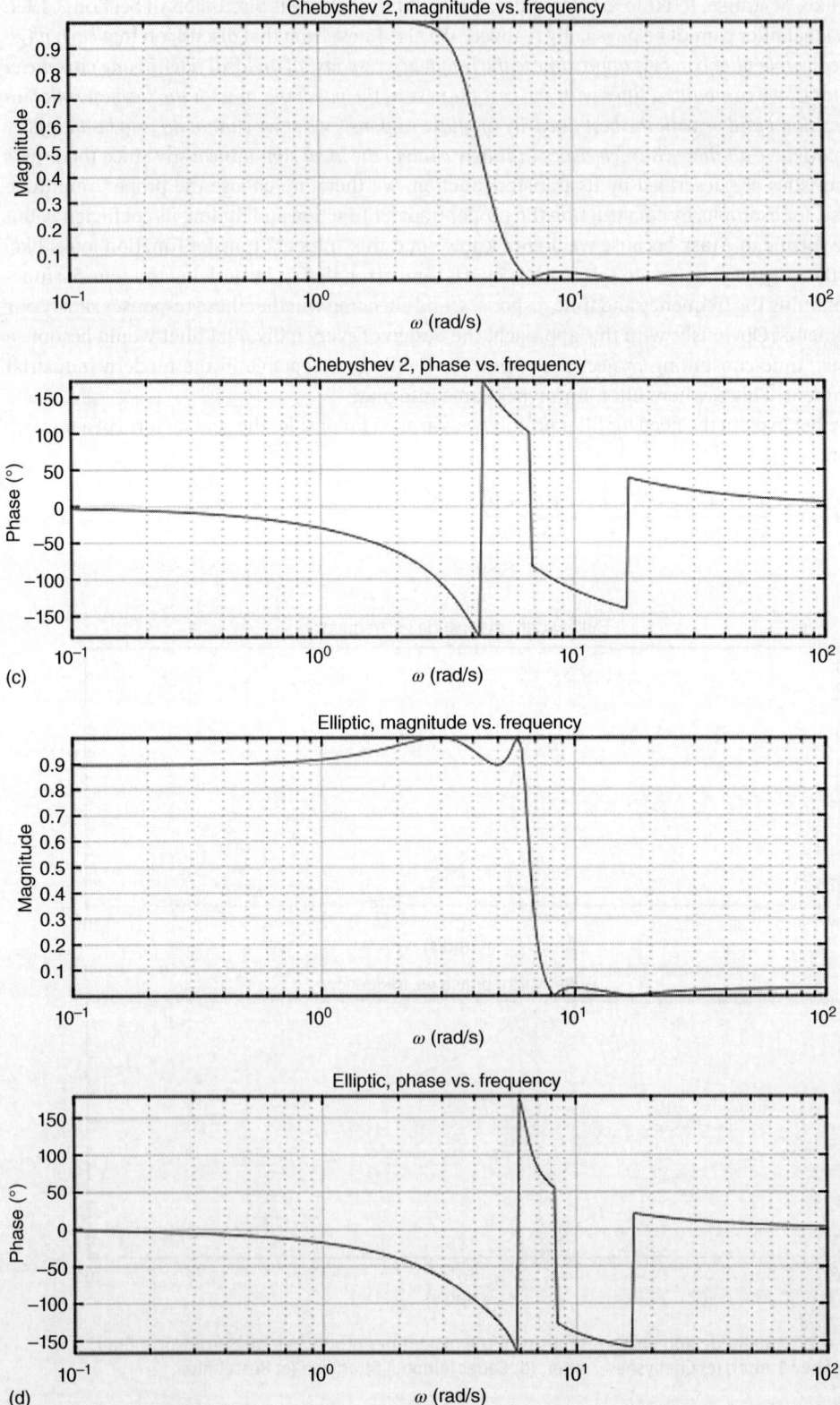

Figure 5.4.2 (Continued)

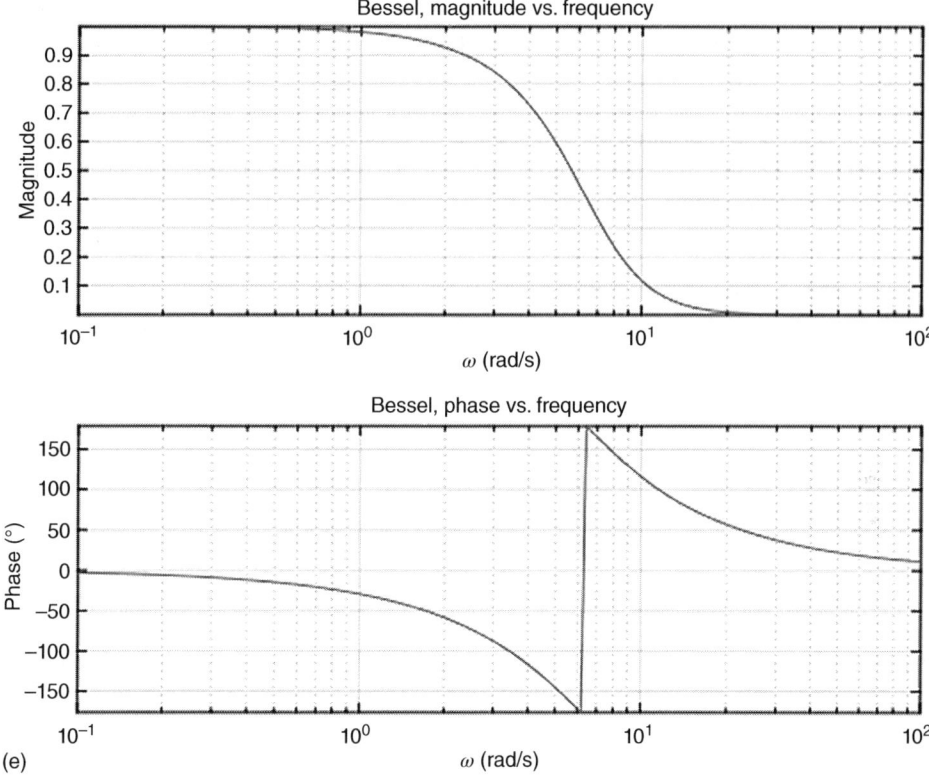

Figure 5.4.2 *(Continued)*

This transfer function is the ratio of two polynomials; this is why it is called a *rational function*. The transfer function has the following main features: First, the members of a transfer function depend on a complex frequency, $s = \sigma + j\omega$; therefore, we say that the transfer function is given in *s-domain*. We will, of course, discuss this point in the chapters that follow. Second, if we equate the *denominator* to zero, $(a_n s^n + a_{n-1} s^{n-1} + \cdots + a_1 s + a_0) = 0$, and solve this *characteristic equation*, we find the roots of this equation. These roots (solutions, that is) are called *poles*. The roots of a *numerator's characteristic equation* are called *zeros*. Based on this short introduction, we can consider the specific filter prototypes more rigorously than we did previously.

Each filter prototype has a unique transfer function with specific coefficients. We have already introduced the main filter prototypes in Figure 5.4.2 and emphasized that these prototypes differ in their amplitude, phase responses, and transient responses. For example, the Butterworth filter has a flat amplitude response in the passband, whereas the Bessel filter exhibits an almost linear phase response. No wonder the transfer function of the Butterworth filter differs from that of the Bessel filter.

Filter prototypes eliminate the need to derive a transfer function from scratch every time when we need to design a new filter. This is because most of this work has been done already; as a result, we have a set of "standard" transfer functions (most named after their developers), each of which allows us to optimize required filter characteristics.

It is important to know that all filter prototypes are standardized to a LPF. We can derive all necessary information for any filter type – HPF, BPF, and BSF – by applying the main results presented in this section.

5.4.1.4 Transfer Function of the Butterworth[8] Filter

The *Butterworth filter* is determined by a transfer function in *s*-domain, which should contain only poles and no zeros, and therefore its transfer function in general form can be deduced from (5.4.3) as

$$H_{\text{But}}(s) = \frac{b_0}{a_n s^n + a_{n-1} s^{n-1} + \ldots + a_1 + a_0} \quad (5.4.4)$$

Specifically, for the *normalized* Butterworth filter whose $\omega_C = 1$ rad/s and $b_0 = 1$, the coefficients of the denominator are

$$a_0 = 1$$
$$a_k = [\cos((k-1)\pi/2n)/\sin(k\pi/2n)] \times a_{k-1} \quad (5.4.5)$$

where $k = 1, 2, 3, \ldots, n$. For example, for $n = 4$ we find

$a_0 = 1,$
$a_1 = [\cos(0)/\sin(\pi/8)] \times 1 = 2.6131,$
$a_2 = [\cos(\pi/8)/\sin(\pi/4)] \times 2.613\,13 = 3.4142,$
$a_3 = [\cos(\pi/4)/\sin(3\pi/8)] \times 3.4142 = 2.6131,$
$a_4 = [\cos(3\pi/8)/\sin(\pi/2)] \times 2.6131 = 1.$

Now we can see *how the idea of a filter prototype is implemented through its transfer function*. Observe that the Butterworth's coefficients are symmetrical; that is, $a_0 = a_n$, $a_1 = a_{n-1}$, $a_2 = a_{n-2}$, and so on. (Can you prove this statement in general based on (5.4.5) and the fact that $a_0 = a_n$?) With these coefficients, the transfer function of a fourth order Butterworth filter can be written as

$$H_{\text{But}}^4(s) = \frac{1}{s^4 + 2.6131s^3 + 3.4142s^2 + 2.6131s + 1}$$

The denominator of the Butterworth transfer function is called a *Butterworth polynomial*. Clearly, if you know the Butterworth polynomial, you know the Butterworth transfer function. You do not need to compute all these coefficients by yourself; this work has been done many times by others and every book on filters contains a table listing the Butterworth polynomials. Surely, you can find them online. Remember that we need to know only half the coefficients thanks to their symmetrical properties. Also, remember $a_0 = 1$. This means that for $n = 4$, for example, the relationship among the coefficients is as follows: $a_0 = 1, a_1 = a_3, a_4 = a_0$; a_2 has no symmetrical member. Appendix 5.4.A contains the tables of the Butterworth polynomials in two formats.

Now we can write the transfer function of the normalized Butterworth filter of *n*th order as

$$H_{\text{But}n}(s) = \frac{1}{B_n(s)} \quad (5.4.6)$$

where $B_n(s)$ is a Butterworth polynomial whose coefficients are given by (5.4.5).

Another form of a Butterworth polynomial is given as a product of the first- and second-order polynomials. This form is popular because a filter can be easily realized as cascaded first-order and

---

8 Stephen Butterworth (1885–1958) was a British physicist and mathematician. His article, "On the Theory of Filter Amplifiers," *Wireless Engineer*, vol. 7, 1930, introduced his unique filter design, whose key feature is the flatness of its amplitude response, making it best suited for radio transmission, although the lack of amplitude sharpness could pose a problem in the quality of the sound.

second-order circuits. Table 5.4.A.2 presents the Butterworth polynomials in this format up to the eighth order.

5.4.1.5 Amplitude Response of the Butterworth Filter

Let us consider the relationship between Butterworth polynomials and the filter's amplitude response. As (5.4.5) shows, for $n = 1$, the Butterworth filter's transfer function is

$$H_{\text{But1}}(s) = \frac{1}{s+1} \tag{5.4.7}$$

In Section 5.2.2, we introduced a transfer function in frequency domain as $H(\omega)$. To relate our new s-domain presentation of a transfer function to its ω-domain form, we substitute $s = j\omega$ and find

$$H_{\text{But1}}(\omega) = \frac{1}{j\omega + 1} \tag{5.4.8}$$

Separating the real and imaginary parts, we get

$$H_{\text{But1}}(\omega) = \frac{1}{1+\omega^2} - j\frac{\omega}{1+\omega^2} \tag{5.4.9}$$

from which we determine

$$|H_{\text{But1}}(\omega)| = \frac{1}{\sqrt{(1+\omega^2)}} \tag{5.4.10}$$

If we assume that ω in (5.4.10) is the ratio of the current frequency, ω_x, to the cutoff frequency, ω_C, as

$$\omega \, (\text{rad/s}) = \frac{\omega_x}{\omega_C} \tag{5.4.11}$$

then we arrive at the form of the amplitude response of a first-order LPF presented in the Subsections 5.4.1.5 and 5.4.1.6:

$$|H_{\text{But1}}(\omega)| = \frac{1}{\sqrt{1 + \left(\frac{\omega_x}{\omega_C}\right)^2}} = \frac{1}{\sqrt{1 + \left(\frac{f_x}{f_C}\right)^2}} \tag{5.4.12}$$

Also, we need to remember that the cutoff frequency of a normalized filter's $\omega_C = 1$ rad/s, which allows us to omit ω_C in the formula for an amplitude response. Example 5.4.1 explains how to find the amplitude response of a second-order Butterworth filter.

Example 5.4.1 Amplitude Response of a Second-Order Butterworth Filter

Problem

Find the amplitude response of a normalized Butterworth filter of the second order.

Solution

By definition, the amplitude response is $|H(\omega)|$ and therefore we need to find the complex transfer function and determine its absolute value. Since the Butterworth filter's transfer function is given by (5.4.6), we find for $n = 2$

$$H_{\text{But2}}(s) = \frac{1}{B_2(s)} = \frac{1}{s^2 + 1.4142s + 1} = \frac{1}{s^2 + \sqrt{2}s + 1} \tag{5.4.13}$$

where the coefficients of a second-order Butterworth polynomial, $B_2(s)$, are given in Table 5.4.A.1 (see also Table 5.4.A.2). To present this transfer function in frequency domain, we substitute $s = j\omega$, and we get

$$H_{\text{But2}}(\omega) = \frac{1}{-\omega^2 + j\sqrt{2}\omega + 1} = \frac{1}{(1-\omega^2) + j\sqrt{2}\omega} \tag{5.4.14}$$

Separating the real and imaginary parts, we find

$$H_{\text{But2}}(\omega) = \frac{(1-\omega^2)}{(1-\omega^2)^2 + (\sqrt{2}\omega)^2} - j\frac{\sqrt{2}\omega}{(1-\omega^2)^2 + (\sqrt{2}\omega)^2} \tag{5.4.15}$$

from which we determine

$$|H_{\text{But2}}(\omega)| = \frac{1}{\sqrt{(1+\omega^4)}} \tag{5.4.16}$$

The problem is solved.

Discussion

- You are encouraged to explicitly perform all these manipulations.
- One point to stress is that reading any book on filters, we will encounter a table of the Butterworth polynomials such as Table 5.4.A.1 or 5.4.A.2. Unhappily, all too often such a table is presented without any explanation, so students remain confused about the role and the use of the table. We trust that our explanation and this example will provide you with a clear understanding of the relationship among the Butterworth filter's transfer function, its polynomials, and the amplitude response. Also, now we will be able to find the phase response from the Butterworth filter's transfer function given through its polynomials.

Generalizing the pattern developed in Example 5.4.1, we can find the amplitude response of an nth-order Butterworth filter as

$$|H_{\text{But}n}(\omega)| = \frac{1}{\sqrt{\left(1 + \left(\frac{\omega_x}{\omega_c}\right)^{2n}\right)}} = \frac{1}{\sqrt{\left(1 + \left(\frac{f_x}{f_c}\right)^{2n}\right)}} \tag{5.4.17}$$

where ω_x (f_x) is the current signal frequency, ω_C (f_C) is the cutoff frequency of a filter, and n is the filter's order ($n = 1, 2, 3, ..., N$).

When $n = 1$, we get (5.4.10), an amplitude response with which we are already familiar from our discussion throughout Sections 5.1 and 5.2. It is, of course, the amplitude response of an RC LPF, which is very far from an ideal LPF. If, however, we increase the filter's order, n, we will, as we proceed, continue to approach the ideal characteristic more closely. Figure 5.4.3 confirms this statement by showing an ideal magnitude response (in dashed shaded line) and the magnitude responses of the Butterworth filters of first, fourth, and eighth orders. Obviously, the ideal graph could be attained hypothetically at $n \to \infty$, which means that *we need an infinite number of reactive components to build an ideal filter*.

The Butterworth filter, as can be seen in Figure 5.4.3, provides a *flat amplitude response in passband and stopband*; in terms of this feature, this filter provides the best fit to the ideal characteristic. Therefore, the best *Butterworth filter's application* is in *audio* transmission, where this feature is necessary for the preservation of the amplitude of a signal. However, the lack of sharpness of the amplitude response can compromise the quality of audio transmission with the Butterworth filter.

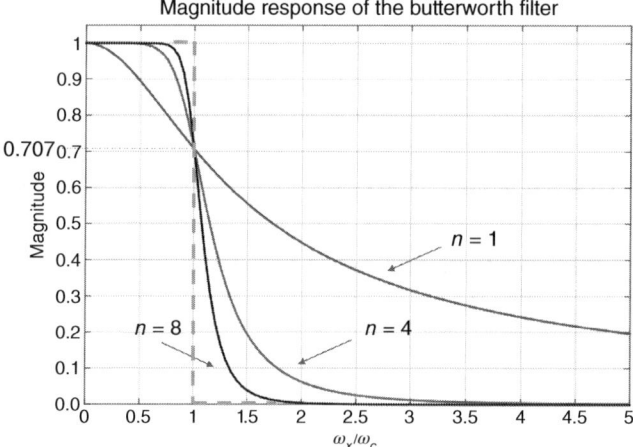

Figure 5.4.3 Amplitude response of a Butterworth filter of various orders.

We need to recall that the order of a filter is determined by the number of reactive components (C or L) in its circuitry. Theoretically, the high-order filters can be built either by including many reactive components in one circuit or by cascading several filters – or by combining both methods. From a practical standpoint, filter designers want to cascade first- and second-order filters rather than building a circuit with many reactive components. If we consider cascading the simple RC LPFs, we have to realize that building a Butterworth filter of the eighth order requires cascading eight RC LPFs; a 16th-order filter needs to cascade 16 RC LPFs, and so on. Of course, if we use the second-order filter, the number of units will be half that. Still, this number of circuits is the price paid for obtaining better amplitude response.

When $(\omega_x/\omega_C) = 1$ in Eq. (5.4.17), $|H(\omega)| = 0.707$, which means that for a Butterworth filter, a 3-dB critical frequency is the natural passband edge frequency. Indeed, since this filter does not show any ripples, we do not have to introduce here either ω_P or ω_S. Thus, *the cutoff frequency, ω_C, of the Butterworth filter defined as $|H(\omega)| = 0.707$ is still the main frequency-selectivity specification for this filter.*

5.4.1.6 Amplitude Response of the Butterworth Filter in Logarithmic Scale

When dealing with normalized filters, it is customary to present their amplitude response in logarithmic scale. Clearly, attenuation in dB is given by

$$A_v \text{ (dB)} = 20 \log \left(\frac{V_{\text{out}}}{V_{\text{in}}} \right) = 20 \log \frac{1}{\sqrt{\left(1 + \left(\frac{\omega_x}{\omega_C}\right)^{2n}\right)}}. \tag{5.4.18}$$

(See (5.1.S.1)). Remember that we refer to attenuation because we are talking about a passive filter; bear in mind, too, that our current explanation holds true even for active filters. Using (5.4.17), we can write

$$A_v \text{ (dB)} \equiv |H_{\text{But}n}(\omega)| = 20 \log \frac{1}{\sqrt{\left(1 + \left(\frac{\omega_x}{\omega_C}\right)^{2n}\right)}} = -10 \log \left(1 + \left(\frac{\omega_x}{\omega_C}\right)^{2n}\right) \tag{5.4.19a}$$

because $\log 1/\sqrt{x} = -1/2 \log(x)$. The negative sign in (5.4.19a) reflects the fact that for passive filters, the attenuation is always smaller than 1 and the base-10 logarithm of such a number is negative. The communication industry, however, wants to use attenuation as a positive number. This is why *the negative sign in the definition of attenuation is artificially introduced, so that*

$$A_v \text{ (dB)} \equiv |H_{\text{Butn}}(\omega)| = 10 \log \left(1 + \left(\frac{\omega_x}{\omega_C}\right)^{2n}\right) \tag{5.4.19b}$$

General Form of an Amplitude Response – The Design Constant, ε, and the Design Parameter, R_{dB} Let us introduce a new quantity, ε, called the *design constant*, so that (5.4.17) becomes

$$A_v \text{ (dB)} \equiv |H_{\text{Butn}}(\omega)| = \frac{1}{\sqrt{\left(1 + \varepsilon^2 \left(\frac{\omega_x}{\omega_C}\right)^{2n}\right)}} \tag{5.4.20}$$

The meaning of ε is quite clear:

In the Butterworth filter, the design constant, ε, delimits the passband region.

In fact, when $\frac{\omega_x}{\omega_C} = 1$, the amplitude becomes

$$|H_{\text{Butn}}(\omega)| = \frac{1}{\sqrt{(1+\varepsilon^2)}} \tag{5.4.21}$$

which determines the value of the amplitude response at the border of a passband. If $\varepsilon = 1$, then

$$|H_{\text{Butn}}(\omega)| = \frac{1}{\sqrt{(1+\varepsilon^2)}} = \frac{1}{\sqrt{2}} = 0.707$$

as we have seen many times before. If, however, we want to decrease the passband when a cutoff frequency, ω_C, is given, then we can simply assign ε equal to a desired value. For example, if $\varepsilon = 0.5$, then the new passband is limited by

$$|H_{\text{Butn}}(\omega)| = \frac{1}{\sqrt{(1+\varepsilon^2)}} = \frac{1}{\sqrt{1.25}} = 0.89$$

The structure of (5.4.20) shows that the minimum passband can be achieved at $\varepsilon = 0$.

The general amplitude response in logarithmic scale, according to (5.4.18) and (5.4.19b) is

$$A_v \text{ (dB)} \equiv |H_{\text{Butn}}(\omega)| = 10 \log \left(1 + \varepsilon^2 \left(\frac{\omega_x}{\omega_C}\right)^{2n}\right) \tag{5.4.22}$$

At $\omega_x/\omega_C = 1$, it gives

$$A_v(\text{dB})|_{\omega_x=\omega_C} \equiv R_{\text{dB}} = 10 \log(1+\varepsilon^2) \tag{5.4.23}$$

where the design parameter, R_{dB}, is the logarithmic equivalent of ε. For example, if $\varepsilon = 0.509$, then $R_{\text{dB}} = 1 \text{ dB}$. Some technical documents prefer using R_{dB} rather than ε; we have to be aware about this industry practice.

Figure 5.4.4 shows the Butterworth filter's amplitude response at various values of ε. It is clear that increasing the design constant results in increasing the steepness of the attenuation graph, but at the expense of decreasing the passband. The role of a design constant, ε, changes with the change of the filter prototype, as we will see in the next section devoted to the Chebyshev 1 filter.

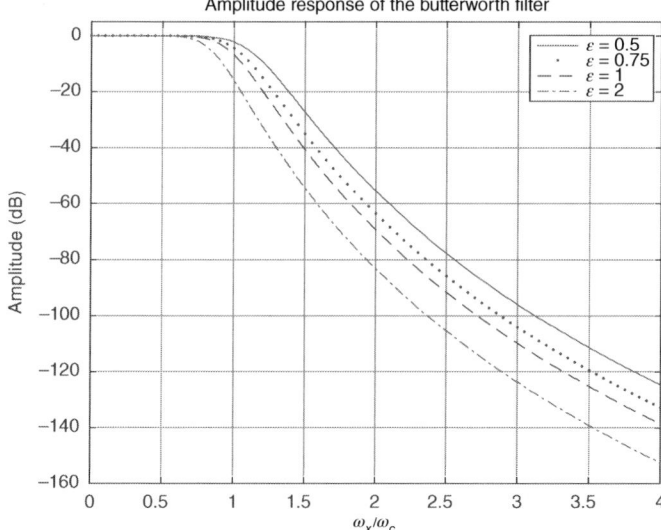

Figure 5.4.4 The Butterworth filter's amplitude response at various values of the design constant ε.

Table 5.4.1 How to find the amplitude response of any type of Butterworth filter.

Filter type	The ratio of frequencies or bandwidths
Low-pass	ω_x/ω_C
High-pass	ω_C/ω_x
Band-pass	$BW_x/BW_{3\,dB}$
Band-stop	$BW_{3\,dB}/BW_x$

Amplitude Response of Any Type of Butterworth Filter In the preceding subsections, we considered the amplitude response of a low-pass Butterworth filter. How can we apply this knowledge to describe the amplitude responses of a high-pass, band-pass, or band-stop Butterworth filters? To get the answer, let us introduce the following notations:

- $BW_{3\,dB}$ is 3-dB bandwidth.
- BW_x is the bandwidth of interest.
- ω_x is the frequency of interest (current frequency).
- ω_C is the cutoff frequency, as usual.

Table 5.4.1 shows how to apply (5.4.17), $|H_{\text{But}n}(\omega)| = \dfrac{1}{\sqrt{1+\left(\dfrac{\omega_x}{\omega_C}\right)^{2n}}}$, or (5.4.19b), $|H_{\text{But}n}(\omega)| = 10\log\left(1+\left(\dfrac{\omega_x}{\omega_C}\right)^{2n}\right)$, to any filter type – low-pass, high-pass, band-pass, and band-stop (Williams and Taylor 2006, p. 43.)

This approach simplifies the analysis and design of all types of Butterworth filters.

A final thought on this subject: Strictly speaking, we should say $\omega_C = 1$ rad/s, since we are working with a normalized filter; however, introducing ratio $\omega = \omega_x/\omega_C$ generalizes our discussion, making it more applicable to innumerable situations we may encounter.

5.4.1.7 Phase Response (Shift) and Time Group Delay of the Butterworth Filter

By definition, *the phase response* (shift) is given by

$$\Theta \text{ (rad)} = \arg(H(\omega)) = \tan^{-1} \frac{Im(H(\omega))}{Re(H(\omega))} \tag{5.4.24}$$

Considering, for example, the second-order Butterworth filter, we can find from (5.4.15) its phase shift as

$$\Theta_{But2} (°) = -\tan^{-1}\left(\frac{\sqrt{2}\omega}{1-\omega^2}\right) \tag{5.4.25}$$

For another phase shift examples, refer to Figure 5.4.5b. You are encouraged to derive the phase-shift formula for the Butterworth filter of any other order.

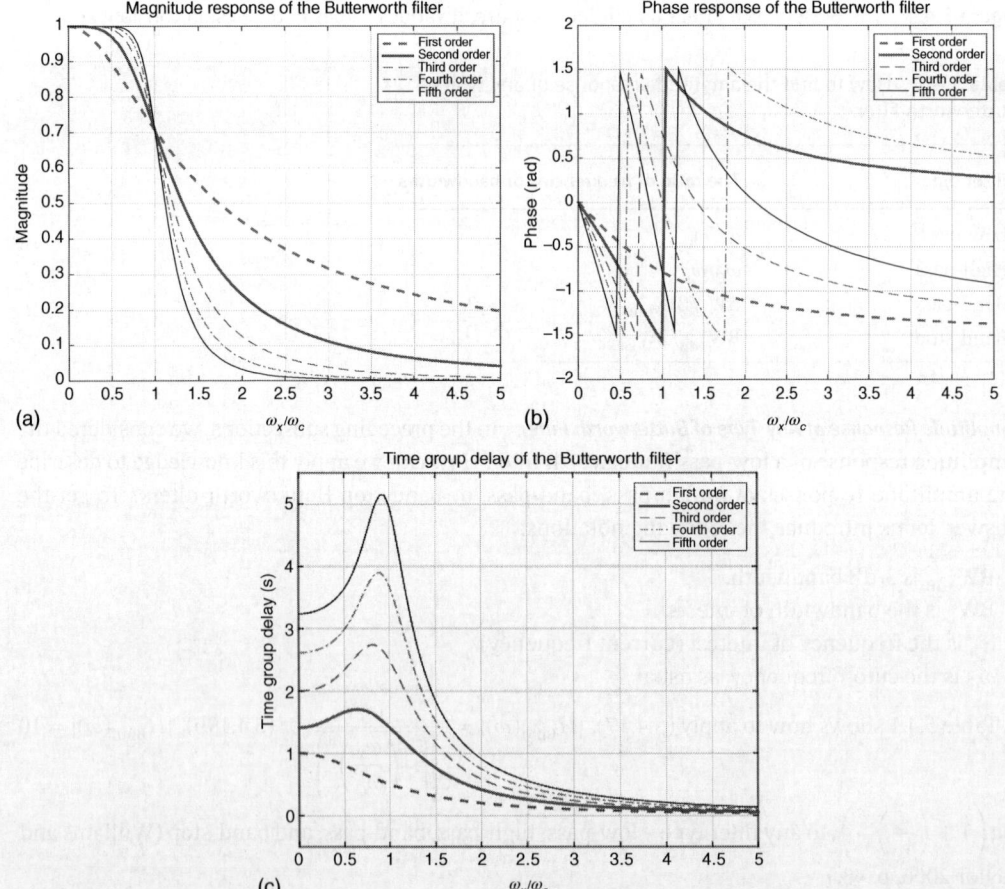

Figure 5.4.5 Responses of a Butterworth filter of various orders: (a) amplitude response; (b) phase response; and (c) time group delay.

The *time group delay* of a filter is defined in (5.1.27R) as $T_D(s) = -\frac{d\theta}{d\omega}$, which for the second-order Butterworth filter results in

$$T_{DBut2}(\omega) = -\frac{d\theta}{d\omega} = \sqrt{2} \cdot \frac{1+\omega^2}{1+\omega^4} \quad (5.4.26)$$

Figure 5.4.5 shows the responses of a Butterworth filter with orders from 1 to 5. We learned from Figure 5.4.3 that the amplitude response can be improved by increasing the filter's order. Figure 5.4.5a demonstrates this phenomenon again in a greater detail. Figure 5.4.5b shows the phase response of the Butterworth filter. Note that the third-, fourth-, and fifth-order filters exhibit two peaks within the chosen frequency range. Closely review these phase responses within the passband. Figure 5.4.5c visualizes the time group delay of this filter. We can see that this delay significantly increases around the cutoff frequency with an increase in the filter order. (This figure shows normalized group delay, the characteristic discussed in Section 5.1, Eq. (5.1.28)). In essence, whole Figure 5.4.5 clearly demonstrates how an increase in a Butterworth filter order improves the amplitude response and, simultaneously, causes the degradation of the phase response and time group delay. In other words,

the closer a Butterworth filter's amplitude response is to that of the ideal graph, the larger the deviation of the phase shift and group delay from their ideal characteristics.

5.4.1.8 Poles of the Butterworth Filter's Transfer Function

Consider the Butterworth filter's transfer function in the general form given in (5.4.6). Denoting the *poles* of this function as p_k, we can obtain the transfer function in the *factored form* as

$$H_{Butn}(s) = \frac{1}{B_n(s)} = \frac{1}{(s-p_1)(s-p_2)\cdots(s-p_k)} = \frac{1}{\prod_{k=1}^{n}(1-p_k)} \quad (5.4.27)$$

To find these poles, we need to set the denominator (the Butterworth polynomial, that is) equal to zero and solve this *characteristic equation*. However, there is a simpler way to find these poles. Consider the following squared amplitude response of the normalized ($\omega_C = 1$ rad/s) Butterworth filter transformed from (5.4.27):

$$(H_{Butn}(\omega))^2 = \frac{1}{1+\omega^{2n}} \quad (5.4.28)$$

Since we want to find the poles of $H(s)$, we substitute into (5.4.28) $\omega = \frac{s}{j}$ and obtain

$$(H_{Butn}(s))^2 = \frac{1}{1+\left(\frac{s}{j}\right)^{2n}} \quad (5.4.29)$$

The characteristic equation of the denominator, $1 + \left(\frac{s}{j}\right)^{2n} = 0$, yields $\left(\frac{s}{j}\right)^{2n} = -1$ or

$$(s^2)^n = -(j)^{2n} = e^{j\pi(2k-1)}(e^{j\pi/2})^{2n} = e^{j\pi(2k-1+n)} \quad (5.4.30)$$

because $e^{j\pi(2k-1)} = -1$ and $e^{j\pi/2} = j$. Here $k = 1, 2, 3, \ldots, n$. Therefore, an individual pole, p_k, is given by

$$p_k = e^{(j\pi(2k-1+n))/2n} \quad (5.4.31)$$

Incidentally, (5.4.31) for $k = 1$ results in $p_1 = e^{(j\pi(2-1+1)/2)} = e^{j\pi} = -1$ and therefore (5.4.27) produces $H_{But1}(s) = H_{Butn}(s) = \frac{1}{(s-p_1)(s-p_2)\cdots(s-p_k)} = \frac{1}{s+1}$.

It follows from (5.4.31) that the pole positions of a normalized Butterworth filter in the complex plane can be computed as (Williams and Taylor 2006, p. 43.)

$$\text{Pole position} \Rightarrow -\sin(\pi(2k-1)/2n) + j \cos(\pi(2k-1)/2n) \tag{5.4.32}$$

where, again, $k = 1, 2, 3, \ldots, n$. Indeed, $p_k = e^{j\pi\left(\frac{2k-1+n}{2n}\right)} = e^{j\pi\left(\frac{2k-1}{2n}\right)} e^{j\frac{\pi}{2}}$. Applying the Euler identity, we find $p_k = e^{(j\pi(2k-1))} e^{(j\pi/2)} = (\cos(\pi(2k-1)/2n) + j \sin(\pi(2k-1)/2n))(\cos \pi/2 + j \sin \pi/2)$, which yields (5.4.32) because $\cos \pi/2 = 0$, $j \sin \pi/2 = j$, and $j^2 = -1$.

Equation (5.4.32) shows that all poles have negative real parts and are located at the left half of the complex plane. Consider, for example, $n = 1$; we can easily perform all computations as follows:

For $k = 1$, $-\sin(\pi(2-1)/2) = -1$ and $j \cos(\pi(2-1)/2) = 0$

Thus, for a first-order filter, we have only one real pole, and it is located at point $(-1, 0)$ on the complex plane. (Depict it.) For $n = 2$, we find

For $k = 1$, $-\sin(\pi(2-1)/4) = -0.707$ and $j \cos(\pi(2-1)/4) = j0.707$
For $k = 2$, $-\sin(\pi(4-1)/4) = -0.707$ and $j \cos(\pi(4-1)/4) = -j0.707$

Therefore, for a second-order filter, we have two complex conjugate poles with negative real parts,

$$p_1 = -0.707 + j0.707 \text{ and } p_2 = -0.707 - j0.707$$

located at the circle with a radius of 1 rad/s. Now refer to Example 5.4.2 for a detailed explanation of this subject.

Example 5.4.2 Poles of a Fifth-order Butterworth Filter

Problem

Find the pole locations of a fifth-order Butterworth filter.

Solution

We need to compute and depict the poles. The results of computations based on (5.4.32) are presented in a tabular form:

K	$-\sin(\pi(2k-1)/2n)$	$j \cos(\pi(2k-1)/2n)$
	$n = 5$	
1	-0.309	$j0.951$
2	-0.809	$j0.588$
3	-1	0
4	-0.809	$-j0.588$
5	-0.309	$-j0.951$

The pole locations are shown in Figure 5.4.6. The diagram shows the locations of the real and imaginary parts of every pole. Note that the poles are located symmetrically with respect to a real axis because they are complex conjugates and have *negative real parts, which means that Butterworth filters of any order are stable*. The angle between two adjacent poles is equal to $\frac{180°}{n}$.

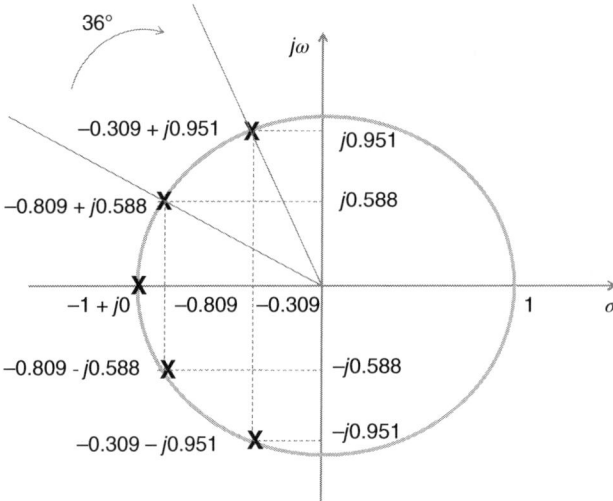

Figure 5.4.6 Pole locations of a normalized Butterworth filter for $n = 5$ in Example 5.4.2.

Discussion

The importance of knowing pole locations is threefold: First, if we know the pole locations, we can easily derive the filter's transfer function by inserting these poles in (5.4.27). Second, it helps us in designing a Butterworth filter because the values of a circuit's components are equal to twice the absolute values of the poles' real parts (Williams and Taylor 2006, pp. 43–44). Third – and what could be the most important point, to be further discussed shortly – we want to know the pole locations because these locations determine the general behavior of a filter.

We have presented a detailed account of the characteristics and functions of the Butterworth filter, one of the most popular prototypes. The other standard filter prototypes – Chebyshev 1 and 2, Cauer (elliptic), Bessel, and all-pass – can be analyzed in a similar manner; we leave this analysis to specialized literature.

5.4.2 Introduction to Filter Design

5.4.2.1 Two Main Steps in Filter Design

The discussion provided in this chapter has prepared you for a more detailed (and professional) look at a filter's operation. In general, we need to be able to *analyze* a given filter and *understand the process of designing* a filter with the required characteristics. Our ability to analyze and design filters gives us enough knowledge to read and understand *the technical documentation* developed by the industry. Though this textbook is not intended to be a guideline for a professional designer of filters, we nevertheless want to introduce the basic approach to this task and the basic concepts and terminology used in the design process.

> *Considering filter design, we can divide this process into two main parts:*
> 1. *Finding the filter's transfer function.*
> 2. *Implementing the transfer function of the filter being designed.*

The need to find the proper transfer function is explicitly discussed in Section 5.4.1: *The desired specifications for the filter are determined by its transfer function.* (Recall, for example, that a filter's amplitude response is simply the real part of its transfer function and the filter's phase response is an imaginary part of that transfer function.) The approach to finding the proper transfer function is also discussed in Section 5.4.1: We need to choose from the standard filter prototypes a transfer function that gives us a filter with the desired characteristics (i.e. flat amplitude response, steepest rolloff, and constant group delay). The implementation of this stage of filter design is further discussed in this section.

The second part of filter design – implementing the found transfer function – involves the following steps:

a) Choosing the appropriate filter technology – passive, active, switched-capacitor, and digital.
b) Designing a circuit.
c) Finding the values of the circuit's components.

These steps are greatly affected by the filter's application; we discuss them in this section.

5.4.2.2 Automated Design Options

Why study all these steps and processes – you may correctly ask – when automated tools should be available to facilitate filter design? Yes, there are several software packages, of course, that help us automate the design process and these will be considered briefly here. Apart from the specialized software, the general software tools, such as MATLAB and Multisim, facilitate the filter-design process and allow us to obtain the desired products.

With *MATLAB*,[9] for example, we can write our own MATLAB code to build and investigate a specific *analog filter*.

For designing *digital filters* (to be discussed in Section 5.5), MATLAB offers the Filter Design Toolbox, which requires entering the desired filter's specifications (that is, "create specification object," in MATLAB parlance), choosing the filter prototypes ("design methods," says MATLAB), and obtaining the result. To verify the design, we can plot the amplitude and phase responses of the designed filter. Of course, MATLAB offers you a variety of options, such as a default design method. The bottom line is that MATLAB calculates all the parameters of the designed *digital filter* and obtains the filter's specific responses. For an advanced design, MATLAB offers an additional package called a Filter Design and Analysis Tool (FDATool) and a Filter Visualization Tool (FVT). At the command "≫fdatool," the window shown in Figure 5.4.7a appears. Navigating this window is a straightforward task: After choosing the response type (e.g. Butterworth), the design method (e.g. infinite impulse response, IIR), the filter's order (e.g. the minimum order), and the frequency specifications (e.g. Units – Hz, $F_{pass} = 1000$ and $F_{stop} = 1800$), we need to press the "Design Filter" button. At the prompt "Designing Filter … Done" at the bottom left of the screen, we can observe all the responses and characteristics of the filter by clicking the "Analysis" button and choosing the parameter you want to see (e.g. amplitude response, phase response, time group delay, step response, pole/zero plot, and many others). Figure 5.4.7b shows some results of the filter design (**the amplitude (magnitude) and phase specifications**). Of course, MATLAB calculates the filter's transfer function; you can obtain the transfer-function coefficients from the "Analysis" drop-down menu.

Multisim,[10] in contrast to MATLAB, provides the tool for designing analog filters at the circuit level. From the main menu, first we need to choose "Tools," then "Circuits Wizard," and then

---
9 MATLAB®.
10 Multisim™ © 2019 National Instruments Corporation. All Rights Reserved.

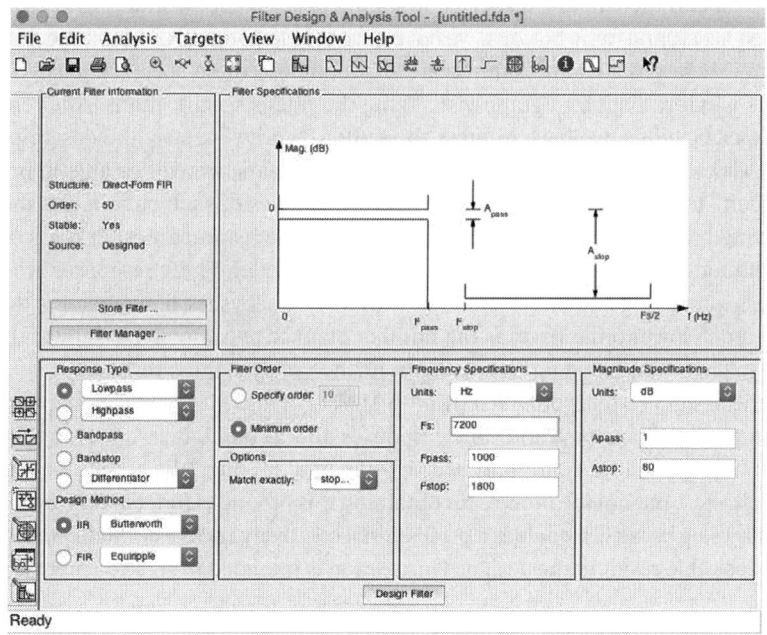

(a)

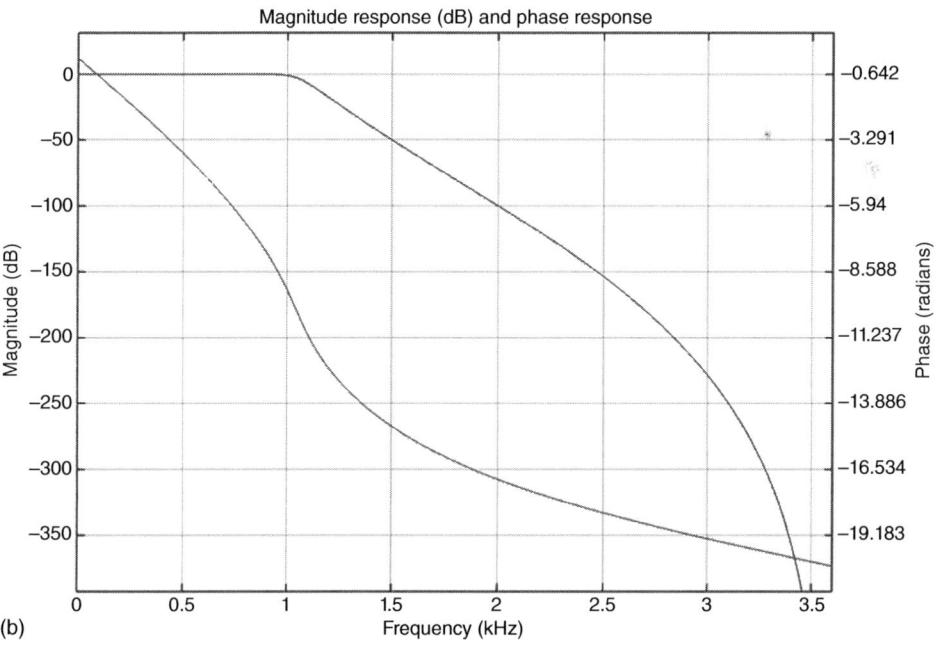

(b)

Figure 5.4.7 MATLAB filter-designing tool: (a) the window of the filter design and analysis tool in the initial state and (b) the amplitude and phase responses of the designed filter. Source: Reprinted with permission MATLAB®.

"Filter Wizard." We obtain the window displayed in Figure 5.4.8a. Here, working with the proper drop-down lists, text boxes, and radio buttons, we have to choose the desired type and specifications of our filter. Figure 5.4.8a gives us an example of how to choose these specifications. We need to observe that this window displays a graph visualizing the filter's specifications. After choosing all the necessary specifications, we have to press the button "Verify" to start the design process. After the prompt "Calculation was successfully completed" appears, we will be able to press the button "Build Circuit" to obtain the circuit's schematics; examples of such circuits are shown in Figure 5.4.8b. Having built each circuit, we can easily investigate its properties; in particular, we can build the amplitude and phase responses by using a Bode plotter. These responses are shown in Figure 5.4.8c. It is interesting and important to note that the Chebyshev filter achieves the same specifications at a much lower order (five, as the number of reactive components shows) than the Butterworth filter (eight). We leave it for the reader to further analyze these responses.

As we can see, the designer's tools provided by both MATLAB and Multisim can help us to find a filter's transfer function, frequency, and time responses and to build a circuit. However, most operations remain hidden, and we commonly see only the final product (the amplitude response, the suggested circuit, etc.), but not the process for obtaining this product. Understanding the design process is a necessary step because it enables a professional creatively correct or modify this process to achieve the best possible result in the design. This section is intended to give you an insight into the design process.

5.4.2.3 Design of a Second-order Butterworth Filter

Through several illustrations, we consider here one of the practical approaches to filter design developed by the industry.[11]

Suppose we need to design a LPF of a high order. We know that every high-order filter can be built by cascading the first-order and second-order filters: If we need a fourth-order filter to meet the specifications, we will cascade two second-order filters; if we need a fifth-order filter, we will cascade two second-order filters and one first-order filter, and so on. Since a first-order LPF is simple to design (we actually discuss this design in Sections 5.1 and 5.2), we start by designing a second-order filter such as that shown in Figure 5.4.9. The circuit of the active LPF (the shaded area) is wired to a source, to the oscilloscope, and to a Bode plotter. This is a well-know *Sallen–Key* circuit whose description can be found online or in any textbook on electronics, communications, or on filters (see, for example, Schaumann et al. 2010, pp. 175–176). Its transfer function is given by

$$H(s) = \frac{1}{(C_1 C_2 R_1 R_2)s^2 + C_2(R_1 + R_2)s + 1} \tag{5.4.33}$$

If we use conductances $G_1 = 1/R_1$ and $G_2 = 1/R_2$ and plug them into (5.4.33), we find another form of this transfer function:

$$H(s) = \frac{G_1 G_2}{(C_1 C_2)s^2 + C_2(G_1 + G_2)s + G_1 G_2} \tag{5.4.34}$$

The goal of this design is to find such values of circuit's resistors and capacitors that make this filter meet the given specifications.

Assuming, of course, that we know the desired filter specifications, we start designing a second-order filter by checking its performance against the requirements.

---

11 *Analog Electronic Filter Design Guide*, Frequency Devices, Inc., Ottawa, IL, 2009.

5.4 Filter Prototypes and Filter Design

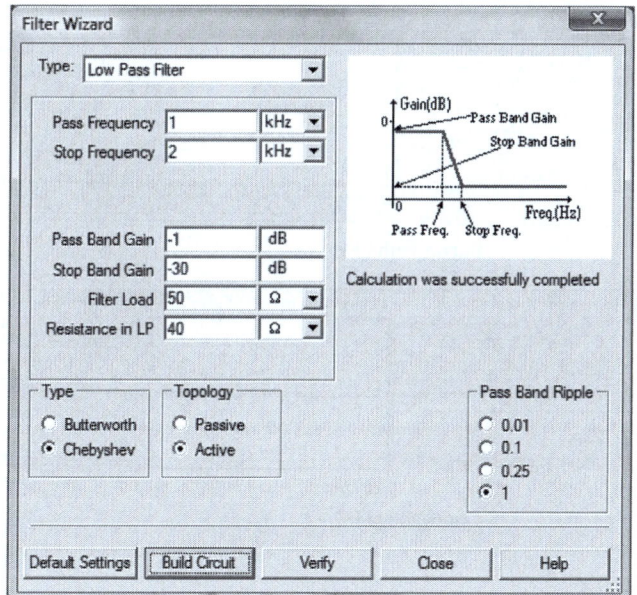

(a)

(b)

Butterworth LPF: V₂ 3, 750Ω, 1 Vrms source, 2 kHz, 0°; L_1 14.52 mH (14), L_2 12.16 mH (15), L_3 7.457 mH (16), L_4 1.552 mH (17); C_1 4.968 μF, C_2 5.503 μF, C_3 4.007 μF, C_4 1.838 μF; 50 Ω rload; XBP1.

Chebyshev LPF: V₁ 19, 750Ω source, 1 Vpk, 1 kHz, 0°; L_1 12.62 mH (20), L_2 14.63 mH (21), L_3 6.127 mH (22); C_1 5.684 μF, C_2 4.729 μF; 50 Ω rload (23); XBP2.

Figure 5.4.8 Designing an LPF with Multisim: (a) filter Wizard from Multisim; (b) the designed circuits of a low-pass filter for Butterworth and Chebyshev prototypes; and (c) the Bode plots of the amplitude and phase responses of the designed filters.

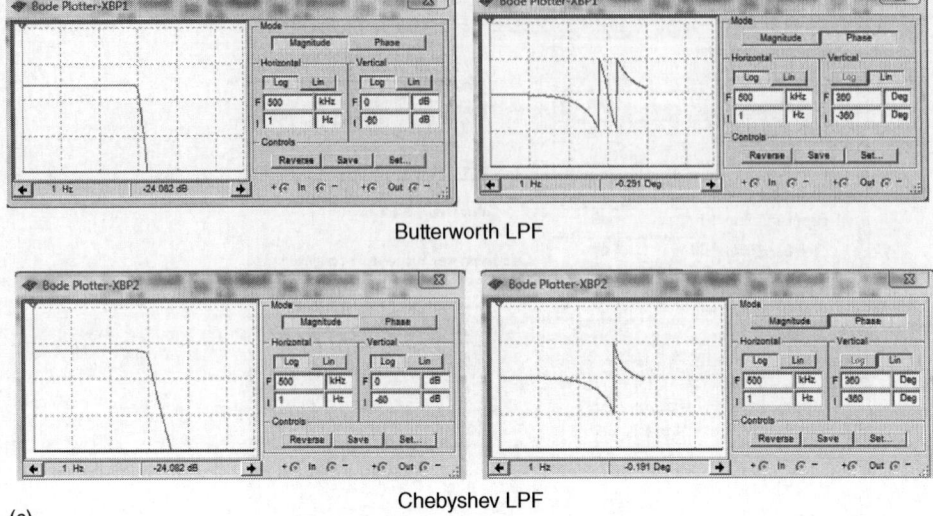

Butterworth LPF

Chebyshev LPF

(c)

Figure 5.4.8 (*Continued*)

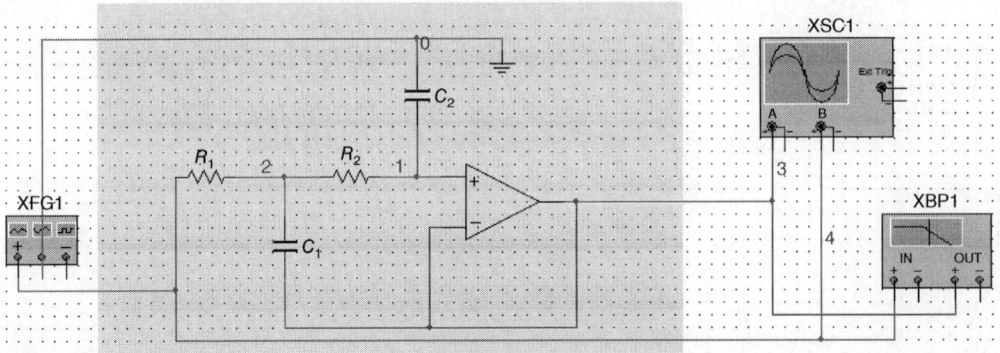

Figure 5.4.9 Circuit of a second-order active LPF with source and measuring instruments.

(Note that if we need to build a higher-order filter, we cascade new sections as necessary. This is one of the standard industry approaches, and it implies that *the filter's total order is determined in the design process.*)

We have to start the design process by choosing the filter's circuit (passive or active). Our next step is *choosing the filter prototype, which determines the filter's transfer function. Then we need to find the specific values of the transfer function's coefficients. These numbers allow us to compute the values of the circuit's components, the step that completes the design process.* To determine the coefficients, we have to use the tables of the transfer function's polynomials, which can be found in Appendixes 5.4.A.1 and 5.4.A.2 or in any filter-design handbook, such as Williams and Taylor (2006), or online.

A filter prototype, you recall, is defined by the filter's applications; since the applications of the filter being designed are not specified here, we can choose the Butterworth prototype as the simplest prototype to design. The Butterworth filter's transfer function has been obtained as

$$H_{\text{But}}(s) = \frac{b_0}{a_n s^n + a_{n-1} s^{n-1} + \ldots + a_1 + a_0} \tag{5.4.4R}$$

Comparing this equation with (5.4.33) and (5.4.34), we can see that

$$b_0 = G_1 G_2 = \frac{1}{R_1}\frac{1}{R_2} = 1$$
$$a_2 = C_1 C_2 = C_1 C_2 R_1 R_2$$
$$a_1 = C_2(G_1 + G_2) = C_2(R_1 + R_2)$$
$$a_0 = G_1 G_2 = 1. \tag{5.4.35}$$

Equation (5.4.35) reveals two important points here pertinent to the design process:

> One is that (5.4.35) shows that the values of the circuit's components can be found by comparing the transfer function of the specific filter circuit with the transfer function of a chosen prototype.

The most important point is that (5.4.35) clearly displays the main problem in filter design – the need for **normalization**:

> To present the filter's transfer function in explicit form, we must know all the polynomials' coefficient, but to know all the coefficients, we must have the values of all the components. On the other hand, we can't find the values of these coefficients until we complete the design process. The industry resolves this problem by assuming the values of one or several parameters to be equal to unity.

(The need for normalization stems from the situation discussed several times before: In design, we have more unknown variables than equations. Consider (5.4.35): From Table 5.4.A.1 presented in Appendix 5.4.A, we know that $a_0 = a_2 = 1$ and $a_1 = 1.4142$. But (5.4.35) contains four unknown variables, $C_1, C_2, R_1,$ and R_2. Therefore, an assumption for one value is needed and by normalization we close the gap.)

For example, we can assume that the cutoff frequency, ω_C, of our LPF is equal to 1 rad/s. Then, from the well-known formula, $\omega_C = \frac{1}{RC}$, we can find C (F) by taking any reasonable value – for example, 1 kΩ – of R.

> This process is called normalization.

This is how it works for our example: One of the popular industrial methods of normalization – which we will use now – entails setting the values of all resistors to 1 Ω; that is,

$$R_1 = R_2 = 1\,\Omega$$

Then, (5.4.33) becomes

$$H_{But2}(s) = \frac{1}{B_{But2}} = \frac{1}{(C_1 C_2)s^2 + 2C_2 s + 1} \tag{5.4.36}$$

From Table 5.4.A.1 we have

$$B_{But2}(s) = s^2 + 1.4142s + 1 \tag{5.4.37}$$

Comparing (5.4.36) and (5.4.37), we find $C_1 C_2 = 1$ and $2C_2 = 1.4142$ and therefore

$$C_2 = 0.7071\,\text{(F)}$$
$$C_1 = 1.4142\,\text{(F)} \tag{5.4.38}$$

> Thus, normalization equates the number of variables and equations and therefore makes the problem solvable.

Still, assuming all the resistors are equal to 1 Ω, does not look like a realistic approach. The help comes from careful analysis of (5.4.33) which reveals that

keeping the ratio of resistances to capacitances the same, we don't change the transfer function, that is, the filter's response.

Truly, let us multiply all capacitor values by factor M and divide all resistor values by the same factor. Plugging new values into (5.4.33) produces

$$H(s) = \frac{1}{\left(MC_1MC_2\left(\frac{R_1}{M}\frac{R_2}{M}\right)\right)s^2 + MC_2\left(\frac{R_1}{M} + \frac{R_2}{M}\right)s + 1} = \frac{1}{(C_1C_2R_1R_2)s^2 + C_2(R_1 + R_2)s + 1}$$

(5.4.39)

This is an important rule that applies to any filter's transfer function.

Therefore, instead of artificial values like $R_1 = R_2 = 1\,\Omega$, we can choose the realistic values of our resistors. If we take, for example, 1 kΩ, we need to divide the capacitor values by 1000 and obtain

$C_1 = 1.4142$ mF

$C_2 = 0.7071$ mF

It looks as though we have solved the problem but, in fact, *the designed second-order filter has a very special cutoff frequency value,*

$\omega_{C2} = 1$ rad/s

To understand how cutoff frequency is defined for a second-order filter, we simply extend the formula $\omega_{C1} = 1/RC$ to

$$\omega_{C2} = 1 \text{ rad/s} = \frac{1}{C_1C_2R_1R_2} \quad (5.4.40)$$

If we want to change this value while preserving the design, we can change, say, capacitor values but we must keep the resistor-to-reactance ratio the same. For example, if we change the cutoff frequency, f_{C2}, to 1 kHz rather than $\omega_{C2} = 1$ rad/s, we must divide the capacitor values by

$2\pi \times R = 2\pi \times 1000$

because f_C (Hz) $= \frac{1}{2\pi RC}$. Thus, we obtain by using (5.4.40)

$C_1 = 0.225$ μF and $C_2 = 0.1125$ μF

Choosing the closest standard values of resistors and capacitors,

$R_1 = R_2 = 1\,\text{k}\Omega, C_1 = 220$ nF, and $C_2 = 110$ nF

we can turn a general circuit like that in Figure 5.4.9 into the specific filter shown in Figure 5.4.10a.

Figure 5.4.10b presents the input and output waveforms at the critical frequency, whereas Figure 5.4.10c shows the amplitude and phase responses of this filter. Measurements of the values of the amplitude response at the Bode plot show that

$A_{pBut2} = -3.359$ dB at $f_p = 1.063$ kHz

$A_{sBut2} = -79.611$ dB at $f_s = 100$ kHz

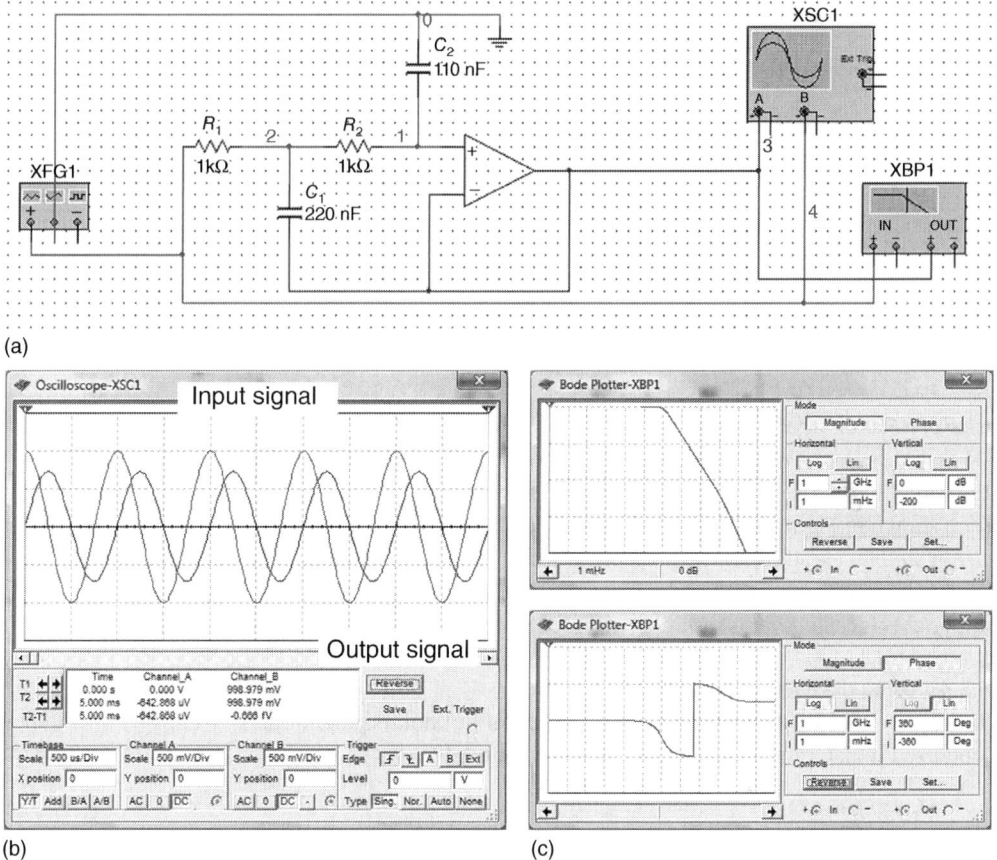

Figure 5.4.10 A designed Butterworth second-order LPF: (a) schematic of the filter; (b) waveforms of the input and output signals at $f = f_c = 1$ kHz; and (c) Bode plots of magnitude and phase responses.

Thus, applying (5.1.23), we can find the width of the transition band of this second-order Butterworth filter as

$$f_{TB-But2} = f_s - f_p \approx 99 \text{ kHz}$$

Bear in mind that Multisim's Bode plotter measures all values with some discreteness; hence, we cannot measure the amplitude at exactly the -3 and -80 dB attenuation levels.

(Note that we intentionally switch from radian frequency, ω (rad/s), to linear frequency, f (Hz), to make you feel comfortable with both measures.)

Thus, we complete the design of a second-order filter.

Again, pay close attention to the fact that we eventually divided the values of the capacitors by a scaling factor, $2\pi fR$.

By assigning realistic values of 1 kΩ to the resistors and 1 kHz (or to 1 krad/s) to the cutoff frequency and properly scaling the values of capacitors, we actually perform <u>denormalization</u>.

Thus, two main steps in filter design – normalization and denormalization – are clearly explained and demonstrated.

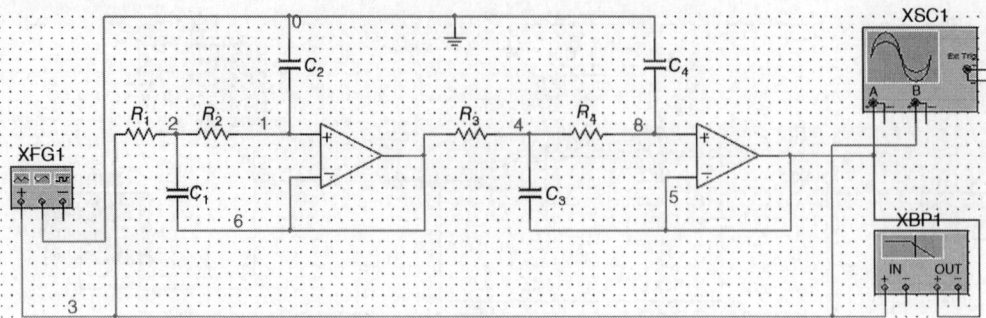

Figure 5.4.11 Fourth-order Butterworth LPF.

Design of a Fourth-order Butterworth LPF Suppose we are not happy with the steepness of an amplitude response in the transition area. We know, of course, that we can increase this steepness by increasing the filter's order (see Figures 5.4.3 and 5.4.5a). Thus, in general, we want to build the circuit shown in Figure 5.4.11. We have to understand, however, that we *cannot increase the filter's order by simply* cascading *two sections presented in Figure 5.4.10*; we *need to repeat this design procedure, but this time for a fourth-order filter*. We leave the problem of finding the values of all circuit's components (design, that is) as an exercise for you.

5.4.2.4 Using the Poles of a Transfer Function

What if we know the *poles* of the Butterworth filter but not its polynomials? Well, we know that a normalized transfer function can be given in a factored form:

$$H_{\text{Butn}}(s) = \frac{1}{B_n(s)} = \frac{1}{(s-p_1)(s-p_2)\cdots(s-p_k)} = \frac{1}{\prod_{k=1}^{n}(1-p_k)} \quad (5.4.27\text{R})$$

Hence, to obtain a transfer function as the product of linear and quadratic equations, we need to plug the poles into (5.4.27R). For instance, the poles of a fourth-order Butterworth LPF can be found as

$$p_{1,2} = -0.9239 \pm j\,0.3827$$
$$p_{3,4} = -0.3827 \pm j\,0.9239$$

Plugging these poles into the factored form of this filter's transfer function, we find

$$H(s) = \frac{1}{(s^2 + 0.7654s + 1)(s^2 + 1.8478s + 1)}$$

as in Table 5.4.A.2 given in Appendix 5.4.A.

The importance of using poles stems from the simple fact that filter-design handbooks usually contain tables of poles, but not the tables of polynomials. Thus, if we cannot find a table with the required polynomials, using the poles gives us an alternative approach to the same design procedure: We just compute the poles and then continue the design, as discussed here. (For computing the poles, refer to (5.4.31) and (5.4.32) and Example 5.4.2.)

An experienced designer can reach a myriad of significant conclusions about the filter's properties by simply analyzing the locations of the poles on the real-imaginary plane. This is why we made it a point to find and discuss the pole locations of the various filter prototypes.

5.4.3 The Design Process: Key Questions, Answers, and Salient Points

Our above discussion demonstrates the general approach taken by the industrial world to deal with such complex problems as filter design. In this subsection, we will examine this approach and draw some conclusions. This analysis will help you to understand the design process from a broader point of view and prepare you to work in any design-approach environment you might encounter.

5.4.3.1 Questions and Answers

Working at any project, it is very important to ask such questions that start with *why, what for, what if*, and so on. Specifically, analyzing the design process just demonstrated, you may want to ask the following questions:

Question: What filter prototype should we choose to start a design process? In general, are there any criteria or guidelines for choosing the right prototype for a specific design?

Answer: *Here, we chose Butterworth prototype just for demonstration purposes. In general, we choose the filter prototype based on the filter's application, as discussed in Sections 5.1, 5.2, and 5.3. Also, we need to bear in mind that a Butterworth filter is the easiest – and an elliptic the most difficult – prototype to design.*

Question: At what stage in the design process do we introduce the filter prototype?

Answer: *A specific prototype is realized over the same circuitry by simply finding the proper component values, which can be computed after we choose the proper polynomial (Butterworth, Chebyshev, Cauer, Bessel, etc.).*

Question: Where can we find the polynomials of the filters to be designed?

Answer: *We have four methods: Locate them in a filter-design handbook, compute them by using (5.4.31), find them from the filter transfer function in explicit form, and – of course – online.*

Question: Is there any method for finding the right filter order at the start of the design process but not the trial-and-error method during the process?

Answer: *Yes, see Example 5.4.3.*

These questions and their answers should clarify some points that remain obscure in the implementation of the design process.

Example 5.4.3 Calculating the Filter's Order

Problem

a) Consider the Butterworth filter with $f_C = 1.0$ kHz and $A_s = -80$ dB at $f_s = 100$ kHz. Find the required order of this filter.
b) Consider another Butterworth filter with $f_C = 1.0$ kHz and $A_s = -80$ dB at $f_s = 9.6$ kHz. Find the required order of this filter.

Solution

For a Butterworth prototype, the relationship between the filter's order, n, and the value of the filter's amplitude at any frequency, f_x (Hz), is mathematically given as

$$A_x = \frac{1}{\sqrt{\left(1 + \left(\frac{f_x}{f_c}\right)^{2n}\right)}} \quad (5.4.17\text{R})$$

Solving (5.4.17R) for n, we find, by referring to the stopband's amplitude, A_s, and frequency, f_s,

$$n = \frac{\log\left(\frac{1}{A_s^2} - 1\right)}{2\log\left(\frac{f_s}{f_c}\right)} \qquad (5.4.41)$$

Thus, we can find the solutions to problems (a) and (b) as follows:

a) For the Butterworth filter with $A_{sdB} = -80\,\text{dB}$ ($A_s = 0.0001$) at $f_s = 100\,\text{kHz}$ and $f_C = 1.0\,\text{kHz}$, we find, after plugging these numbers into (5.4.41), $n = 2$.
Thus, for these specifications, we need a Butterworth filter of the second order.
b) For another Butterworth filter, we find, after inserting the numbers $A_{sdB} = -80\,\text{dB}$ at $f_s = 9.6\,\text{kHz}$ and $f_C = 1.0\,\text{kHz}$, $n = 4.072 \Rightarrow 4$.

Therefore, here we need a Butterworth filter of the fourth order, which is the closest integer to the number actually obtained.

The problem is solved.

Discussion

First, compare Example 5.4.1 and Example 5.4.3, where we use the same prototypes with the same specifications: The filters' orders calculated in this example are exactly the same as those in the design example 5.4.1. This coincidence is the justification for the computational approach explored in this example to find a given filter's order.

Second, make a mental note that using this approach enables us to derive the formula for calculating the order of any filter prototype. Using Formula 5.4.41 and computing the filters' orders for specific amplitude specifications, we can find, for example, that to meet the same specifications, we need a seventh-order Butterworth filter, a fifth-order Chebyshev filter, or a third-order elliptic filter. These numbers further illustrate the main features of the filter prototypes.

Derivation of (5.4.41) enables us to find the filter's order in case of generalized Butterworth filter's amplitude response given in (5.4.20) as

$$A_x = \frac{1}{\sqrt{\left(1 + \varepsilon^2\left(\frac{f_x}{f_c}\right)^{2n}\right)}}$$

from which we can derive

$$n = \frac{\log\left(\frac{1-A_s^2}{\varepsilon^2 A_s^2}\right)}{2\log\left(\frac{f_s}{f_c}\right)} \qquad (5.4.42)$$

If we plug in the specifications given in point b of this example for $\varepsilon = 1$, we find $n = 4$.

5.4.3.2 Salient Points

As you can see from our examples and our discussion of filter design, there are no formulas or unique algorithm that we have to follow. *Filter design is an art more than a science; however, it is an art based on a strong scientific foundation* and there are many approaches to filter design dependent on this foundation.

Based on our extended discussion, we can now summarize the practical steps in filter design.

- To employ a simple, practical approach in finding the values of circuit components, we choose *free components* whose values we want to keep equal to one another in the entire circuit. We set their values to unity; this step is called *normalization*. Then we find the normalized transfer function of the filter being designed (for example, see Eqs. (5.4.33) and (5.4.36)).
- To compute the values of the other components, we compare the coefficients of the normalized transfer function with the polynomials of the filter prototype This comparison enables us to compute the required values (for example, see Eqs. (5.4.36) and (5.4.37)).
- To use normalization, we can employ two approaches: Either normalize the cutoff frequency or normalize the values of the free components.
- To make the values of the circuit components practical after finding the normalized values, we employ the *denormalization* process. This entails assigning realistic values to the free components and to the cutoff frequency and calculating the values of other components, keeping the ratio of component values the same.
- To implement the filter prototype (remember, the circuit is the same for any filter prototype), we assign to the circuit components the values recalculated after the denormalization.
- To choose the filter's technology (passive, active, switched-capacitor, etc.), we have to refer to the filter's application and to our experience and knowledge because there is no formal rule that could help us make this decision.
- To choose the main specifications for the filter being designed, we need to specify the cutoff (or passband) and stopband frequencies and passband and stopband amplitudes. For the generalized specifications, we need to add a design constant. These specifications allow us to compute the order of the required filter.

Though this process is specific for filters, the concept of design is very general and can be applied in many other areas of technology.

5.4.3.3 Choosing Filter Technology

Filter design results in the construction of a filter's circuit with the predetermined values of all the components. We have illustrated the design of passive and active filters, but we have not said a word up to now about when and why to choose either of these technologies. It is time to take up this subject.

How to Choose the Appropriate Filter First, we need to consider the main requirements in choosing filters for different applications. Though we have already covered this material, it is worth pausing here to summarize the main points we made. To do so, we – follow the hints of industrial world – have divided all the filter types into the following three categories[12]:

- *Filters providing high selectivity in frequency domain*: Filters in this category have to have a narrow transition band (or sharp rolloff, in industry parlance) of the amplitude response. This major requirement can be satisfied at the expense of other specifications, such as phase linearity. The main application areas for these filters include voice transmission, spectrum analysis, and modal analysis, which is the operation that determines the resonant frequencies of structures and objects.
- *Filters providing waveform preservation in time domain*: To preserve a signal's waveform, a filter must have the best possible linearity of its phase response and minimum harmonic distortion. It is a Bessel filter that can achieve this goal; this is why this prototype is mostly employed in video and data transmission.

---

12 *Analog Electronic Filter Design Guide*, Frequency Devices, Inc., Ottawa, IL, 2009.

- *Compromised filters*: Many applications require a narrow transition band along with high linearity of the phase response. One example of such an application is the task of determining the direction of a signal source by analyzing the waveforms of several receivers. Naturally, such filters are called compromised since, in their design, a proper balance between two interrelated specifications must be found.

In choosing the right filter technology for a specific task, we need to realize that today we have two main technologies: analog and digital. For analog filters, we can choose passive, active, or switched-capacitor filters. Digital filters offer design options that will be discussed later in the section that follows. For now, let us discuss the technologies available for *analog filters*.

Passive Analog Filters   These filters are based on discrete components – resistors, capacitors, and inductors – and *provide polynomial approximation of an ideal filter*. Their main advantages include a wide range of cutoff frequencies – from hundreds of hertz to tens of megahertz – and no need for external power. They do not generate any dc offset, and their design is a well-established procedure. However, passive filters suffer from many drawbacks discussed in Section 5.3.

Active Filters   Linear active filters have many features that eliminate or reduce the disadvantages of passive filters; however, active filters suffer from their own inherent drawbacks. Revisit Section 5.3 for detailed discussion of these issues.

All in all, this section gives you a glimpse into the complex but fascinating world of the design of analog filters. Using this knowledge, you will be able to learn and apply all the specifics of the design process in practice.

Questions and Problems for Section 5.4

- Questions marked with an asterisk require a systematic approach to finding the solution.
- Many questions and problems, including those marked with an asterisk, imply that you, in addition to reading the textbook, will do your research to find the answers. Consider such questions as mini-projects.

Filter Prototypes: What and Why

1   What are the filter prototypes?

2   Explain the need for developing filter prototypes.

3   List and explain the conditions that define an ideal filter.

4   Why do we need to develop various filter prototypes if – it seems – we can take an ideal filter as a prototype?

5   *We know that a filter is described by its transfer function. It seems, then, that we could construct any desired filter by developing the proper transfer function, but from a practical standpoint this approach does not work. Why not?

6   Why do we need to have several filter prototypes?

7 *What is the difference between a filter's transfer function in general and the transfer function of a filter prototype? Give an example.

8 *How do filter prototypes help in designing filters with the required characteristics?

9 For a normalized Butterworth filter, write its transfer function in general form.

10 The text says, "if you know the Butterworth polynomial, you know the Butterworth transfer function." Explain why this is so.

11 Considering the Butterworth filter as an example, explain how the concept of filter prototype is implemented through its transfer function.

12 Equation (5.4.5) gives a general formula for the coefficients of a Butterworth polynomial. This formula implies that these coefficients are the function of a variable. What variable?

13 Using (5.4.5), calculate the coefficients of the Butterworth polynomial of fifth order. Compare your results with that in Tables 5.4.A.1 and 5.4.A.2. Can you automate this work using MATLAB or MS Excel or any other computer tool?

14 *Referring to (5.4.5) and/or Tables 5.4.A.1 and 5.4.A.2, write the transfer functions of third-order and sixth-order Butterworth filters.

15 The text says that the Butterworth coefficients are symmetrical:
a) Prove this statement.
b) Using the symmetrical properties of the Butterworth coefficients, write the relationship between the first and the last coefficients for $n = 3$, $n = 4$, and $n = 5$.

16 *Consider the amplitude response of the third-order Butterworth filter:
a) Write the formula for this response.
b) Build graphs of this response in natural and logarithmic scales.

17 For the amplitude response of a Butterworth filter,
a) Write the formulas for $n = 2$, $n = 10$, and $n = 15$.
b) Build graphs of these responses in natural scale.
c) Prove that an ideal Butterworth filter cannot be built.

18 *What are the best applications for Butterworth filter? Why does a Butterworth filter fit these applications?

19 To improve the amplitude response of a Butterworth filter, we need to cascade basic RC (or RL) filters.
a) How many RC circuits do we need to build a fourth-order Butterworth filter? Explain why.
b) Practically speaking, the designer always wants to cascade first- and second-order filters to build the Butterworth filter of a required order. Sketch a block diagram of cascaded first- and second-order RC circuits to build a fourth-order Butterworth filter.

20 In filter specifications introduced in Section 5.1.4, there are three edge frequencies: The passband, ω_P; stopband, ω_S; and cutoff, ω_C. Which of them is the main frequency-selective specification for a Butterworth filter and why?

21 What is the passband of a Butterworth filter?

22 Compute the value of the amplitude response of a Butterworth filter in natural scale and logarithmic scale when its current frequency is equal to the cutoff frequency.

23 *Equation (5.4.20) generalizes the formula of the Butterworth filter's amplitude response by introducing the design constant, ε, as $|H(\omega)| = 1/\sqrt{(1 + \varepsilon^2(\omega_x/\omega_C)^{2n})}$. What is the meaning of the constant ε? Prove your answer.

24 The cutoff frequency of a Butterworth filter is 1 rad/s:
   a) What will be the filter's passband if $\varepsilon = 0.3$? $\varepsilon = 0.6$? $\varepsilon = 0.9$?
   b) How the changing in ε affects the steepness of the attenuation graph?
   c) How does the ε variation affects the passband length?

25 *Write the equations describing the amplitude response of a Butterworth filter of the following type: (a) low-pass, (b) high-pass, (c) band-pass, and (d) band-stop. (*Hint*: Refer to Table 5.4.1.)

26 Derive the phase-response formula for the Butterworth filter of *n*th order. (*Hint*: See (5.4.24) and (5.4.25).)

27 *Calculate the phase response (shift) and time group delay of a third-order Butterworth filter at $\omega = \omega_x/\omega_C = 1$. (*Hint*: You need to start with the transfer function of this filter, then find the phase response by using (5.4.24), and, after that, obtain the phase response and time group delay by applying (5.4.25) and (5.4.26). Verify your answers by examining Figure 5.4.5.)

28 Present the following transfer function in factored form: $H(s) = A/(s^2 + \sqrt{2}s + 1)$.

29 Why do we need to know the locations of a transfer function poles? List and explain three reasons for that. Use Figure 5.4.6 to exemplify your answer.

30 *Find the poles of the third-order Butterworth filter and show their locations on the complex plane. (*Hint*: Apply (5.4.32) and review Example 5.4.2.)

31 Consider Figure 5.4.2, where the amplitude and phase responses of all filter prototypes are shown:
   a) Which filter is the best for an audio transmission? Why?
   b) Which filter is the best for a video transmission? Why?

32 Based on the material presented in this section, summarize why we need to use filter prototypes in the design of required filters.

Introduction to Filter Design

33 What are the two main steps in filter design?

34 *Explain the role a filter's transfer function plays in filter design.

35 *Developing a filter's proper transfer function is a theoretical undertaking. List the steps that enable us to implement this theory into a tangible circuit.

36 Designing a filter's circuit based on a chosen prototype is a routine operation. Today, routine operations are mainly automated. Explain how filter design is automated.

37 Explain the difference in designing analog and digital filters with MATLAB.

38 Using MATLAB for the design of analog filters, build the amplitude and phase responses of a low-pass Butterworth filter with $n = 5$ and $\omega_C = 1$ rad/s.

39 If you need to design the circuit of an analog filter, which tool – MATLAB or Multisim – would you choose? Explain.

40 *Using Multisim, design the circuits and build the amplitude and phase responses of the Butterworth filter with $n = 5$ and $\omega_C = 1$ rad/s. Then build these graphs by using MATLAB and compare the graphs obtained by MATLAB and Multisim.

41 *Derive the transfer function of the filter shown in Figure 5.4.9. Compare your result with (5.4.33) and comment on any discrepancies.

42 *The text says that the standard industry approach is to determine the filter's order in the design process. Why cannot the order be determined before the design starts?

43 *Consider the design of a filter's circuit:
a) List all the steps needed to complete this design.
b) How are these steps related to the steps of implementing the filter's transfer function presented in the preceding subsection as (i) choosing the appropriate filter technology – passive, active, switched-capacitor, and digital, (ii) designing a circuit, and (iii) finding the values of the circuit's components?

44 Explain what the normalization process is in the filter design and why we need it.

45 *Consider designing a second-order Butterworth LPF:
a) What are the realistic values of the resistors, R_1 and R_2?
b) Using its standard transfer function, find the realistic values of its capacitors, C_1 and C_2.

46 *It follows from (5.4.33), $H(s) = \frac{1}{(C_1 C_2 R_1 R_2)s^2 + C_2(R_1 + R_2)s + 1}$, that keeping the ratio of resistances to capacitances constant, we do not change the filter's transfer function and therefore its frequency response. Prove this statement.

47 *Designing a normalized filter, we assume that $\omega_C = 1$ rad/s:
a) How can we change this assumed value of a cutoff frequency to a realistic desired value?
b) What are the initial parameters of the designed circuit and how they will change after changing ω_C?

48 Consider a second-oder Butterworth filter with $\omega_C = 1$ rad/s. Change its cutoff frequency to $\omega_C = 2$ rad/s and find the new values of its resistors and capacitors.

49 *Design a second-oder Butterworth filter with $\omega_C = 2$ rad/s:
  a) Find the values of its resistors and capacitors.
  b) Using Multisim, draw the circuit diagram of this filter.
  c) Using Multisim, find the waveforms of the input and output signals.
  d) Using Multisim, build the Bode plots of the amplitude and phase responses of this filter.
  e) Comment on the time-domain and frequency-domain graphs.

50 We use the normalization process to facilitate filter design. Why, then, do we need denormalization?

51 *Design a third-oder Butterworth filter with $\omega_C = 2$ rad/s:
  a) Find the values of its resistors and capacitors.
  b) Using Multisim, draw a circuit diagram of this filter. Identify the first-order filter and second-order filter.
  c) Using Multisim, find the waveforms of the input and output signals.
  d) Using Multisim, build Bode plots of the amplitude and phase responses of this filter.
  (*Hint*: Since the value of ε is not given, it implies that $\varepsilon = 1$.)

52 How many first-order and second-order filters do you need to cascade to build a 11th-order filter? Sketch a block diagram showing the resulting filter.

53 *Consider the poles role in the transfer function of a third-order Butterworth filter:
  a) Compute the poles of this filter.
  b) Write down the transfer function of this filter in factored form.
  c) Compare your results with the appropriate line in Table 5.4.A.2 and comment on discrepancies, if any.
  d) Show the locations of these poles in a complex plane.
  e) Based on the pole locations, draw a conclusion on the stability of this filter.

54 (Mini-project) Based on the analysis of the design process described in this section, develop the algorithm of a design process of analog filters. Using this algorithm,
  a) Draw a flowchart to illustrate this algorithm.
  b) Suggest modifications of the algorithm to streamline the design process.

55 (Mini-project) Design a second-order HPF from a given second-order active RC Butterworth LPF with $f_C = 10$ kHz:
  a) Build the circuit of the HPF using the same R and C components. What will be the cutoff frequency of this HPF?
  b) Derive the transfer function of the HPF from that of the LPF.
  c) Normalize the HPF transfer function based on a resistors' normalization.
  d) Find the polynomial of the filter in the design.
  e) Compute the values of the circuit components.
  f) Perform denormalization, obtain the realistic values of the components, and – using Multisim – draw the real circuit schematics.

g) Obtain the output waveforms and – using the Body plotter – the responses of the designed filter. Analyze your results.
(*Hint*: Refer to Table 5.4.A.1.)

The Design Process: Key Questions, Answers, and Salient Points

56 *Consider key questions and answers presented in Section 5.4.3:
   a) In designing a filter for video transmission, what prototype would you choose and why?
   b) How does the choice of a filter prototype affect the design process? Should we start the design by choosing a filter prototype?
   c) The first task in filter design is developing of the filter transfer function by using its polynomials. Where can we find these polynomials?
   d) What method other than trial and error can be applied in finding the filter order?

57 *Consider a Butterworth filter with $f_C = 10$ kHz and $A_s = -80$ dB at $f_s = 18$ kHz:
   a) Find the required order of this filter if $\varepsilon = 1$.
   b) What will be the order of this filter if $\varepsilon = 0.6$?
   c) Why the filter's order is important?
(*Hint*: Use Eq. (5.4.42).)

58 *Consider the statement that filter design is more an art than a science,
   a) Explain why.
   b) What artistic and scientific elements can you identify in filter design?

59 The text says that in filter design, we choose free components:
   a) What does this mean? Give an example.
   b) Are all the components free? Explain.

60 *Consider normalization in filter design:
   a) What two approaches can we use in the normalization process?
   b) Which approach do you prefer and why?

61 In filter design, how can we find the realistic values of circuit components if the normalized values are known?

62 The text says repeatedly that the circuit is the same for any filter prototype. How, then, can we design the filter of a specific prototype?

63 How can we choose the appropriate filter technology when designing a filter?

64 *In filter design, to meet the required specifications, we have to choose the filter prototype and calculate the values of its components. Suppose we did that for a second-order filter. Then, it appears that we need to design a fourth-order filter. The text says that we cannot cascade two designed second-order filers but start the design from scratch. Why?

65 The text says that from an application standpoint, we can sort all the filters into three categories: Filters providing high selectivity in frequency domain, filters providing waveform preservation in time domain, and compromised filters. But the text also says that we choose a filter prototype based on a given filter's application. Relate all the filter prototypes to one or more categories.

66 Consider passive filters:
   a) List their advantages and drawbacks.
   b) If you were a filter designer, would a passive filter always be your first choice of technology? Explain.

67 Consider active filters:
   a) List their advantages and drawbacks.
   b) If you were a filter designer, would an active filter always be your first choice of technology? Explain.

5.4.A Tables of the Butterworth Polynomials

Below are two tables of the coefficients of the Butterworth polynomials.

Another form of a Butterworth polynomial is given as a product of the first- and second-order polynomials. This form is popular because, in practice, a filter can be easily realized as cascaded

Table 5.4.A.1 Coefficients of the Butterworth polynomials.

Filter order, n	a_1	a_2	a_3	a_4	a_5	First and last coefficients
2	1.4142					$a_0 = 1$
3	$2.0000 = a_2$					$a_0 = a_3 = 1$
4	$2.6131 = a_3$	3.4142				$a_0 = a_4 = 1$
5	$3.2361 = a_4$	$5.2361 = a_3$				$a_0 = a_5 = 1$
6	$3.8637 = a_5$	$7.4641 = a_4$	9.1416			$a_0 = a_6 = 1$
7	$4.4939 = a_6$	$10.0978 = a_5$	$14.5918 = a_4$			$a_0 = a_7 = 1$
8	$5.1258 = a_7$	$13.1371 = a_6$	$21.8461 = a_5$	25.6883		$a_0 = a_8 = 1$
9	$5.7588 = a_8$	$16.5817 = a_7$	$31.1634 = a_6$	$41.9864 = a_5$		$a_0 = a_9 = 1$
10	$6.3924 = a_9$	$20.4317 = a_8$	$42.8020 = a_7$	$64.8824 = a_6$	74.2334	$a_0 = a_{10} = 1$

Table 5.4.A.2 Butterworth polynomials as products of the first- and second-order polynomials.

n	Butterworth polynomials $B_n(s)$
1	$(s+1)$
2	$s^2 + 1.4142s + 1$
3	$(s+1)(s^2 + s + 1)$
4	$(s^2 + 0.7654s + 1)(s^2 + 1.8478s + 1)$
5	$(s+1)(s^2 + 0.6180s + 1)(s^2 + 1.6180s + 1)$
6	$(s^2 + 0.5176s + 1)(s^2 + 1.4142s + 1)(s^2 + 1.9319s + 1)$
7	$(s+1)(s^2 + 0.4450s + 1)(s^2 + 1.2470s + 1)(s^2 + 1.8019s + 1)$
8	$(s^2 + 0.3902s + 1)(s^2 + 1.1111s + 1)(s^2 + 1.6629s + 1)(s^2 + 1.9616s + 1)$

first-order and second-order circuits. Table 5.4.A.2 presents the Butterworth polynomials in this format up to the eighth order. You are encouraged to verify these coefficients by multiplying the coefficients given in Table 5.4.A.1.

5.5 Digital Filters

Objectives and Outcomes of Section 5.5

Objectives

- To learn about digital filters, their principle of operation, and their different types.
- To compare analog and digital filters from the standpoint of their applications.

Outcomes

- Learn about digital filters, their principle of operation and their hardware.
- Revisit analog-to-digital conversion (ADC) and digital-to-analog conversion (DAC), which are the important parts of digital filters operations.
- Remember that digital filters are described by difference equations in contrast to analog filters, which are described by differential equations. Become familiar with discrete function, $v[n]$, which is a digital analog of a continuous function, $v(t)$. Learn from (5.5.5) how to determine the order of a difference equation and from (5.5.10) how to find its coefficients.
- Learn that there are two types of digital filters: Nonrecursive filters, which compute their output based on both the current and the previous inputs, and recursive filters, which compute their output using an input and the previous output.
- Become familiar with the impulse response of digital filters, which divides the filters into two classes: infinite impulse response (IIR) filters and finite impulse response (FIR) filters. Learn that the IIR filters are the recursive filters, and FIR filters are the nonrecursive filters.
- Master the fact that the transfer function of a digital filter – which can be derived from its difference equation – describes the digital filter in its entirety. Understand, too, that the order of a digital filter is the order of its transfer function, which is the order of either the numerator or the denominator, whichever is greater.
- Learn about adaptive filters, which are the new class of digital filters. Know that adaptive filters change their parameters (coefficients of their transfer function that is) over time to adapt their operation to the fluctuating characteristics of an input signal.
- Refer to Table 5.5.2 to comprehensively compare analog and digital filters and see which filter type works better in a specific application. Understand that, in short, analog filters are fast and have a large dynamic range in both amplitude and frequency scales; on the other hand, digital filters demonstrate superior performance characteristics, are readily programmable, can be made adaptive, are much easier to design, and are more readily automated in the design process than their analog counterparts. Realize that digital filters are the preferred tools in signal processing.
- Summarize the applications of various filter technology in practice.

5.5.1 What are Digital Filters?

5.5.1.1 Digital Filters – Principle of Operation

Our discussion of analog filters brings us to the point where we need to look at the other class of filters: digital. A digital filter comprises three main parts: An *analog-to-digital converter (ADC)*,

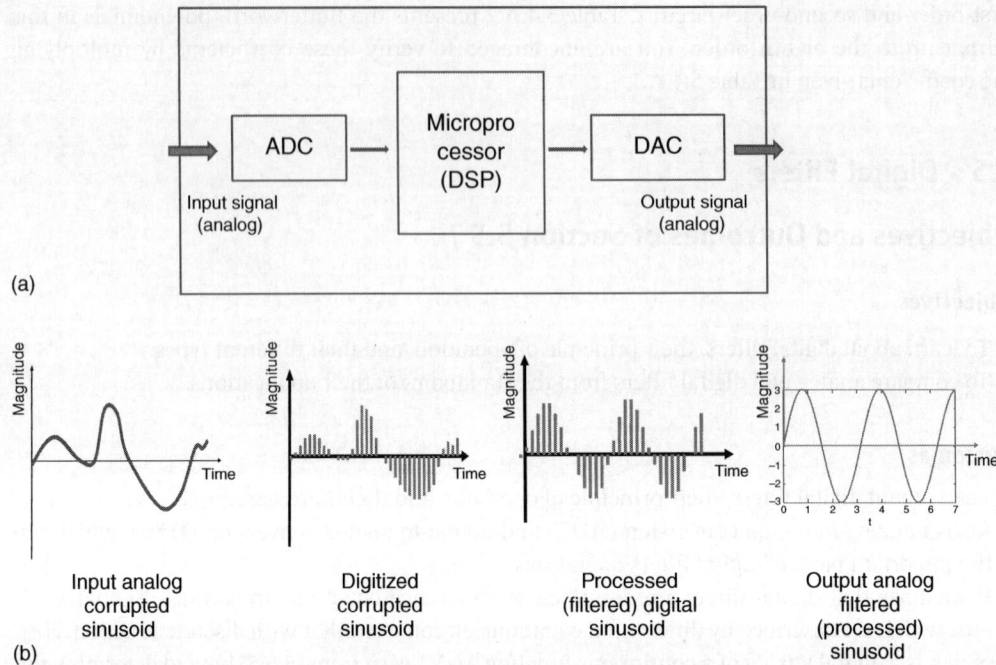

Figure 5.5.1 Digital filter: (a) a block diagram and (b) example of a digital filter operation.

a *digital-signal-processing (DSP) unit*, and a *digital-to-analog converter (DAC)*, as shown in Figure 5.5.1a.

An example of the operation of a digital filter is illustrated in Figure 5.5.1b: The input signal is an analog corrupted sinusoid. It is converted by an ADC into a set of numbers; these numbers are shown as a set of vertical lines of various amplitudes. This numerical presentation of a corrupted analog sinusoid is processed by a DSP unit and changed (filtered) as required. The output of the DSP unit is another set of numbers, shown as a set of vertical lines of various amplitudes. We can see that this output is a much improved (almost ideal, in fact) digital sinusoidal signal. This digital sinusoid is converted by a DAC into an output analog sinusoid from which most of the distortion has been filtered. *This is how a digital filter operates.*

It follows from the above explanation that the heart of a digital filter is the software executed by a DSP unit. This software processes an input signal converted into digital format; this is why a digital filter is considered part of a broader field called *digital signal processing (DSP). Since the main operation – the filtering of an input signal, that is – is performed by software, you can create a digital filter on any computing machine with sufficient computational ability.* An example is given in Section 5.4.2.2: There, we discussed the design of a digital filter performed by MATLAB, which was installed on a desktop computer. In reality, a DSP unit can be implemented as a general microprocessor, a specialized DSP processor, a field-programmable gate array (FPGA), or an application-specific integrated circuit (ASIC). Again, any processor constitutes the core of a digital filter. This description apprises you of the main *difference between digital and analog filters:*

Analog filters are replete with electronics (hardware, that is) and process an analog input signal directly by changing the parameters of the signal's waveform; digital filters, on the other hand, decompose an analog input signal into numbers (into digital format, of course), perform mathematical operations at these numbers, and then reconstruct the result of these operations into an analog output signal.

An important note: Digital-filter hardware includes two analog filters that are placed before ADC and after DAC units. The first one is called an *anti-alias* filter and the second one is called a *reconstruction* filter. The anti-alias filter removes from the input analog signal all frequency components greater than half of the sampling rate, whereas the reconstruction filter removes from the converted signal all frequency components above the sampling rate.

5.5.1.2 ADC and DAC Operations Revisited

These operations are fully discussed in Section 4.1. It should be, therefore, helpful (practically, necessary) to review this section to refresh your memory on all the details of ADC and DAC work, including the terminology used in this technology. We repeat here some information regarding ADC and DAC operations, focusing on the points needed to understand ADC and DAC application in digital-filter technology. (For example, we present a simplified explanation of a sampling operation though, in fact, it is a sample-and-hold operation, as discussed in Section 4.1.3.)

Digitizing an Analog Signal – Sampling Now we need to recall how to convert an analog signal into a digital signal and back. These operations are performed by the ADC and DAC units of a digital filter. An understanding of these operations is critical to the comprehension of the operation of a digital filter as a whole.

Consider the arbitrary analog signal shown in Figure 5.5.2a. To convert it to a digital format, we need first to *sample* it. We do this by taking measurements of the signal's magnitude (samples, a_0, a_1, a_2, a_3, ..., a_N) at sampling times (t_0, t_1, t_2, t_3, ..., t_N) separated by equal time intervals. This operation is called *sampling* and is shown in Figure 5.5.2b. Note that $t_0 = 0$, $t_1 = t_0 + \Delta t = \Delta t$, $t_2 = t_1 + \Delta t = 2\Delta t$, $t_3 = t_2 + \Delta t = t_1 + 2\Delta t = 3\Delta t$, etc. The final sampling time is given by $t_N = N\Delta t$. The period of this sampling operation is $T_s = \Delta t$. Thus, we can introduce the *sampling rate*, f_s, as

$$f_s \text{ (Hz)} = \frac{1}{T_s(s)} \tag{5.5.1}$$

After taking the samples, we can forget about the original analog signal and consider its representation by these samples, as shown in Figure 5.5.2c. Here we have a set of amplitudes (a_0, a_1, a_2, a_3, ..., a_N) corresponding to a set of sampling times (t_0, t_1, t_2, t_3, ..., t_N), which is the main goal of sampling:

We replace the real analog signal with a set of numbers (amplitudes a_0, a_1, a_2, a_3, ..., a_N); we know the location of each individual amplitude, a_n, because we know its sampling time, $t_n = n\Delta t$. Thus, we can simply replace the real analog signal by a table showing these amplitudes vs. sampling times.

Analyze Figure 5.5.2, which shows a sampling operation in sequence.

We can consider a sampling operation as the conversion of a continuous function, $v(t)$, into a discrete function, $v[n]$, where the discrete function is given as a set of numbers. Indeed, the analog

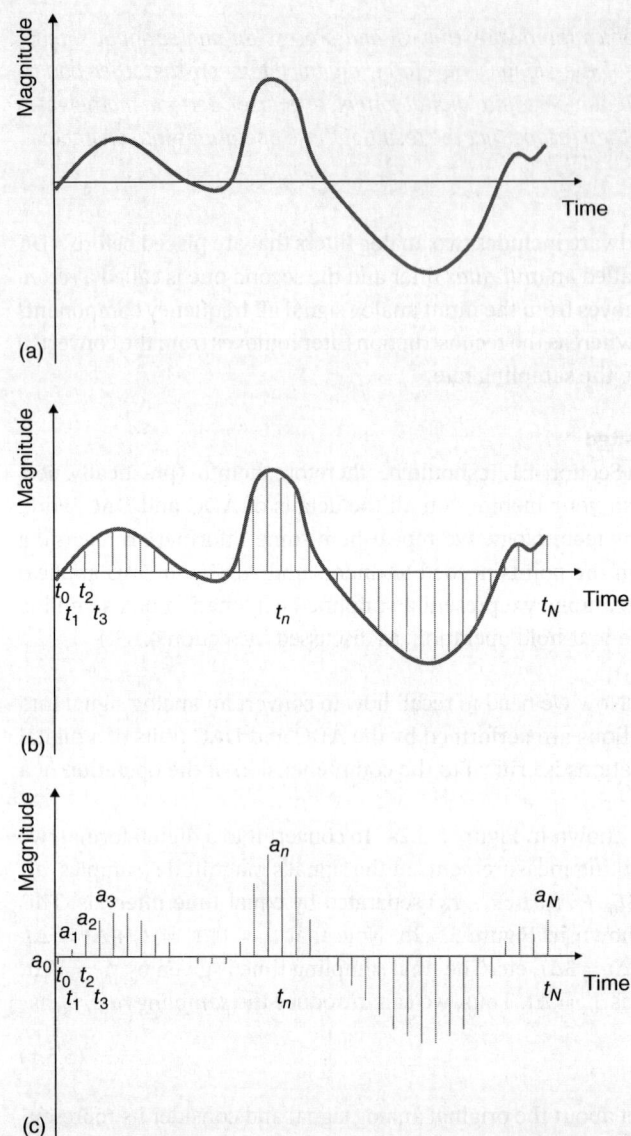

Figure 5.5.2 Sampling operation: (a) original analog signal; (b) sampling of the analog signal; and (c) samples representing the original analog signal.

signal, $v(t)$, shown in Figure 5.5.2a, is represented (after sampling) by a set of amplitudes, a_n. This set is a discrete function, $v[n]$, where $n = 0, 1, 2, 3, ..., N$. This function is presented in the form of Table 5.5.1, where the sampling amplitudes, a_n, are shown vs. sampling time, t_n (that is, vs. sampling number, n).

(Bear in mind that the amplitude values, a_n, shown in the table are approximate, not measured, values of the graph amplitudes.) Here, a_n is a set of numbers whose values depend on t_n, whereas t_ns are determined by the numbers n. All this means is that the set $a_n = f[n]$ is a *discrete function*,

Table 5.5.1 Discrete function, v[n], representing a sampled analog signal.

n (sampling number)	t_n (s) (sampling time)	a_n (V) (sampling amplitude)
0	0	−0.2
1	0.1	0.3
2	0.2	0.6
3	0.3	1.1
⋮	⋮	⋮
N = 156	15.6	0.4

v[n]. In our example, a_n is a set of decimal numbers, but, obviously, they can be converted into binary or hexadecimal numbers. In any event, we need to realize that

> *sampling enables us to take a major step toward implementing a digital filter, that is, moving from analog to digital format.*

Digitizing an Analog Signal – Nyquist (Sampling) Theorem (We recommend to revisit Section 4.1.3, where Nyquist (sampling) theorem is discussed from a different point of view.) The real question in sampling an analog signal is how many samples we want to take. Obviously, we need to consider not the total number of samples but the number of samples per second, that is, the sampling rate, f_S. If the sampling rate is too low, we can miss the important features of the signal being sampled. In this case, it will be difficult, if not impossible, to reconstruct the original signal. (Examine again Figure 4.1.4 to see the effect of a sampling rate on the result of sampling process.) The right choice of sampling frequency is determined by the fundamental rule called the *Nyquist*, or *sampling, theorem*, which states that

A sampling rate, f_S, must be at least twice the highest frequency, f_{MAX}, of the signal being sampled; that is,

$$f_S \text{ (Hz)} \geq 2f_{max} \text{ (Hz)} \tag{5.5.2}$$

see Eq. (4.1.1). When the sampling rate is equal to twice the maximum frequency, we call this rate the *Nyquist (sampling) frequency*:

$$f_{SA} = 2f_{max} \tag{5.5.3}$$

The Nyquist theorem is a fundamental premise that states that

> *we can completely restore an original signal from its samples provided that the sampling rate will be at least twice the highest signal frequency.*

Equation (5.5.2) shows that the Nyquist theorem determines the *minimum value* of a sampling frequency. But what if we take a sampling rate three times greater than f_{max}? If the sampling rate is too high, we will have more samples and this will allow us to better restore the original analog signal. The drawback of this approach, however, is that we will need more resources to collect additional information. With all the advances in modern digital technology, this requirement is not a problem anymore and so the trend is to sample analog signals at the highest possible rate. (Again, turn back to Chapter 4, where oversampling and undersampling, including the aliasing effect, are discussed.)

Since this theorem governs the conversion of signals from analog to digital format and back, it is not a surprise that it is called the *fundamental theorem of DSP*. We have discussed and will continue to discuss this theorem and its applications many times in this book.

This consideration briefly explains the operations performed by the ADC and DAC units shown in Figure 5.5.1. To complete our introduction to digital filters, we need to consider more closely the processing operations performed by a DSP unit.

5.5.1.3 Digital Filters – Difference Equation, Order, and Coefficients

A digital filter, as well as an analog filter, can be described by a mathematical equation in time domain. This approach brings us to the other major dissimilarity between digital and analog filters:

The mathematical tool used to describe an analog filter in time domain is <u>differential equations</u>, whereas the mathematical tool to describe a digital filter in time domain is, <u>difference equations</u>.

An analog filter can be described by the following general *differential equation*:

$$a_n \frac{d^n y(t)}{dt^n} + a_{n-1} \frac{d^{n-1} y(t)}{dt^{n-1}} + \cdots + a_1 \frac{dy(t)}{dt} + a_0 y(t) = b_m \frac{d^m x(t)}{dt^m} + b_{m-1} \frac{d^{m-1} x(t)}{dt^{m-1}}$$
$$+ \cdots + b_1 \frac{dx(t)}{dt} + b_0 x(t) \qquad (5.5.4)$$

Moving from differential to finite difference, we can rewrite this *differential* equation in the form of a *difference equation* as

$$a_0 y[n] + a_1 y[n-1] + \cdots + a_N y[n-N] = b_0 x[n] + b_1 x[n-1] + \cdots + b_M x[n-M] \qquad (5.5.5)$$

or

$$a_0 y[n] = b_0 x[n] + b_1 x[n-1] + \cdots + b_M x[n-M] - a_1 y[n-1] - \cdots - a_N y[n-N]$$
$$= \sum_{i=0}^{M} b_i x[n-i] - \sum_{j=1}^{N} a_j y[n-j] \qquad (5.5.6)$$

Here y and x are the *n-th* output and input signals, respectively, and b_i ($i = 0, 1, 2, 3, \ldots, M$) and a_j ($j = 1, 2, 3, \ldots, N$) are the coefficients. Difference equation (5.5.5) relates the output and input signals, whereas (5.5.6) shows how to compute the *n*-th output, $y[n]$, based on the previous outputs and all inputs. It is important to understand that, similar to an analog filter,

a difference equation completely describes the operation of a digital filter, and coefficients a_j and b_i completely specify the digital filter.

The main distinction between the differential (5.5.4) and difference (5.5.5) equations is that a differential equation is a function of a continuous variable (time), whereas a difference equation is a function of a discrete number (n).

For example, consider a digital filter that computes the average of the current and two previous input terms, that is, a *three-term average filter* (Lutovac et al. 2001, pp. 171–172): Its difference equation is given by

$$3y_n = x_n + x_{n-1} + x_{n-2} \qquad (5.5.7)$$

If you want to learn how to do these computations, write down (5.5.7) for each consecutive number:

$$3y_0 = x_0 + x_{-1} + x_{-2}$$
$$3y_1 = x_1 + x_0 + x_{-1}$$
$$3y_2 = x_2 + x_1 + x_0$$
$$3y_3 = x_3 + x_2 + x_1$$
$$\vdots$$
$$3y_N = x_N + x_{N-1} + x_{N-2}$$

where N is the number of the final output member. Here x_{-1} and x_{-2} are considered to be zero.

We need to understand that a difference equation implies that all input members needed for computations are stored in the processor's memory by the time the processor starts the calculations. This is why we set $x_{-1} = 0$ and $x_{-2} = 0$.

The *order* of a digital filter, in its simplest case, *is the number of the previous inputs used for calculating the current output*. Consider a three-term average filter: The member x_n is the current input, whereas the members x_{n-1} and x_{n-2} were the previous inputs; therefore, this is a second-order filter.

Consider another example: A *two-term difference filter* whose output is the difference between the current and the previous inputs; that is,

$$y_n = x_n - x_{n-1} \tag{5.5.8}$$

Since only one previous input is required for the computations, x_{n-1}, this is a first-order filter. However, if we consider the *central difference filter* given by

$$2y_n = x_n - x_{n-2} \tag{5.5.9}$$

we need to understand that, formally speaking, its difference equation should be written as

$$2y_n = x_n - 0 \cdot x_{n-1} - x_{n-2}$$

and, therefore, this is a second-order filter.

What if we need to use the previous outputs along with inputs to compute the current output? Consider, for example, a filter described by a difference equation:

$$3y_n = 2x_n - 5y_{n-1} - 4y_{n-2}$$

In this case, we have one input and two outputs, and the *filter's order is determined by either the input or the output number, whichever is the highest.*

In general, the order of a digital filter *described by the difference equation (5.5.5) is either N or M, whichever is greater.*

Coefficients, b_n, of digital filters can be easily found if we rewrite the previous difference equations (5.5.7), (5.5.8), and (5.5.9) in the following general form:

$$y_n = b_0 x_n + b_1 x_{n-1} + b_2 x_{n-2} + \cdots + b_N x_{n-N} \tag{5.5.10}$$

In (5.5.7), the coefficients are

$$b_0 = 1/3, b_1 = 1/3, b_2 = 1/3$$

the coefficients in (5.5.8) are

$$b_0 = 1, b_1 = -1$$

and the coefficients in (5.5.9) are

$$b_0 = 1/2, b_1 = 0, b_2 = -1/2$$

As was the case with analog filters, these coefficients specify the digital filters in terms of defining the output with given inputs.

5.5.1.4 Recursive (IIR) and Nonrecursive (FIR) Digital Filters and Their Difference Equations

Digital filters are divided into two classes: recursive and nonrecursive filters.

Nonrecursive filters compute their output based on both the current and the previous inputs. Recursive filters compute their output using an input and the previous output.

"Recursive" means "running back." The term refers to the fact that, with these filters, the previously calculated outputs are again used for calculating the current output. Nonrecursive filters do not need such feedback. The difference equations (5.5.7)–(5.5.9) are examples of nonrecursive filters because they compute their outputs based only on the current and the previous inputs. An example of the difference equation of a recursive filter is

$$y_n = x_n + y_{n-1} \tag{5.5.11}$$

Here the current output, y_n, is equal to the current input, x_n, and the previous output, y_{n-1}. The advantage of a recursive filter can be seen if we consider the computations described by (5.5.11) explicitly. Indeed, remembering that $y_{0-1} = 0$, we find

$$y_0 = x_0 + y_{n-1} = x_0$$
$$y_1 = x_1 + y_0$$
$$y_2 = x_2 + y_1$$
$$y_3 = x_3 + y_2$$
$$\vdots$$
$$y_{100} = x_{100} + y_{99}$$

Thus, a recursive filter can find any current output by using only two members: the current input and the previous output.

As for a nonrecursive filter, it needs to compute the sum of all previous inputs, so that, for example,

$$y_3 = x_3 + x_2 + x_1 + x_0$$

You can easily write the expression for y_{100}, the output for a nonrecursive filter, in order to understand the advantage of recursive filters. On the other hand, a recursive filter, in fact, sums up ("integrates") all previous inputs, too, but does so implicitly. Indeed, consider the third term of the above example on recursive calculations:

$$y_3 = x_3 + y_2$$

Plugging in the explicit expression for y_2, we find

$$y_3 = x_3 + y_2 = x_3 + x_2 + x_1 + x_0$$

which proves the statement.

From this discussion, we can deduce the general form of the *difference equations of a recursive filter* as follows:

$$a_0 y[n] + a_1 y[n-1] + \cdots + a_N y[n-N] = b_0 x[n] + b_1 x[n-1] + \cdots + b_M x[n-M] \quad (5.5.12a)$$

or

$$a_0 y[n] = b_0 x[n] + b_1 x[n-1] + \cdots + b_M x[n-M] - a_1 y[n-1] - \cdots - a_N y[n-N]$$

$$= \sum_{i=0}^{M} b_i x[n-i] - \sum_{j=1}^{N} a_j y[n-j] \quad (5.5.12b)$$

which are Eqs. (5.5.5) and (5.5.6) derived above.

Since the calculations of the output of a nonrecursive filter don't include the previous outputs, the *difference equation of the nonrecursive filter* takes the following general form

$$a_0 y[n] = b_0 x[n] + b_1 x[n-1] + \cdots + b_M x[n-M] = \sum_{i=0}^{M} b_i x[n-i] \quad (5.5.13)$$

5.5.1.5 Impulse Response of Digital Filters

The response of any system to an input signal can be found either in time domain through an impulse response or in frequency domain through a transfer function. The use of transfer function has been discussed in the preceding chapters; the impulse response will be considered in Chapters 6 and 7. Here we will introduce the basics of an impulse–response approach as applied to digital filters.

To refresh your memory, an impulse is a generalized function that is equal to zero everywhere except at the point where it exists, for example, at $t = 0$ second. The area under the impulse graph is equal to 1, which means that the impulse's amplitude tends to infinity. The impulse function, known as the Dirac delta function, is denoted as $\delta(t)$. It is a mathematical abstraction and cannot be realized practically. However, it is a very useful mathematical tool in everyday engineering work because some signals (very short pulses, for example) can be closely approximated by the delta function.

The importance of the impulse consideration stems from the fact that a filter's response to an individual impulse involves all the filter's dynamics and this response displays all the filter's properties. Though an impulse input is applied instantly and lasts theoretically for a zero second, the response of a filter to such an ultra-short input could last for some time; in fact, the duration of the impulse response demonstrates the important properties of the filter.

There are many important details of an impulse response in general. These brief remarks, nevertheless, are enough for understanding this simple fact: Impulse response enables us to comprehensively reconstruct the filter's total response in time domain.

Now, we can consider the basics of the impulse response of digital filters. Since recursive filters use the previous outputs for computing the current output, this feedback physically continues to supply energy into the filter's input, thus sustaining the impulse response of the filter for a long (theoretically, infinite) time. Thus, the *recursive filters are called IIR filters*. Certainly, the absence of such feedback is responsible for shortening the impulse response of *nonrecursive filters*, and these filters *are* called *FIR filters*. In the world of digital filters, these definitions are used interchangeably; that is, IIR means recursive filters and FIR means nonrecursive filters.

We should expect that, according to the terminology, IIR filters exhibit an impulse response that theoretically lasts forever, whereas FIR filters have an impulse response of finite duration. Figures 5.5.3 and 5.5.4 demonstrate the impulse response of IIR and FIR filters. We can readily see that for the same time interval shown in Figures 5.5.3 and 5.5.4, the response of an IIR filter still holds up significantly, whereas the response of a FIR filter is almost completely nullified.

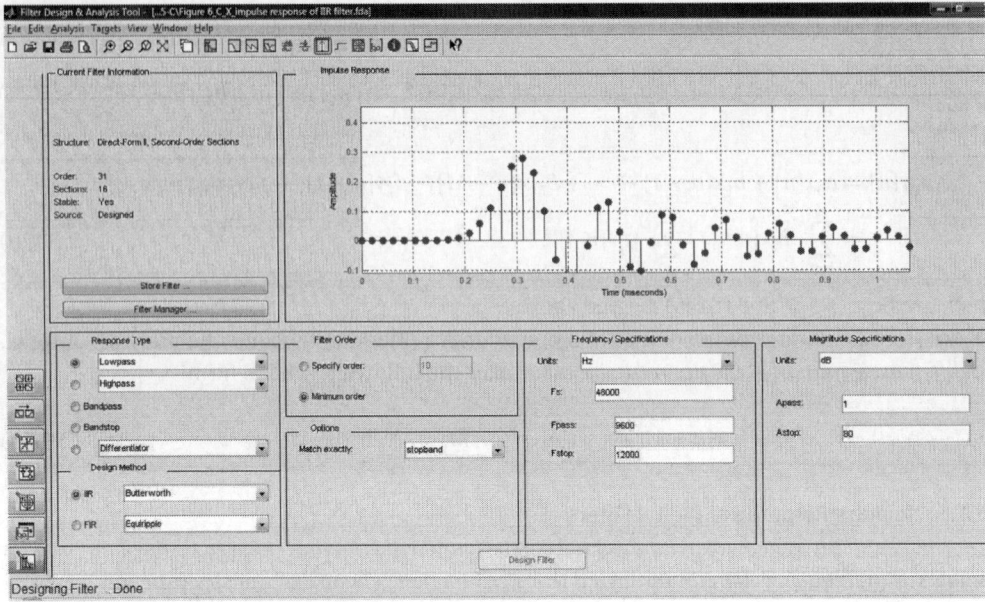

Figure 5.5.3 Impulse response of an IIR filter.

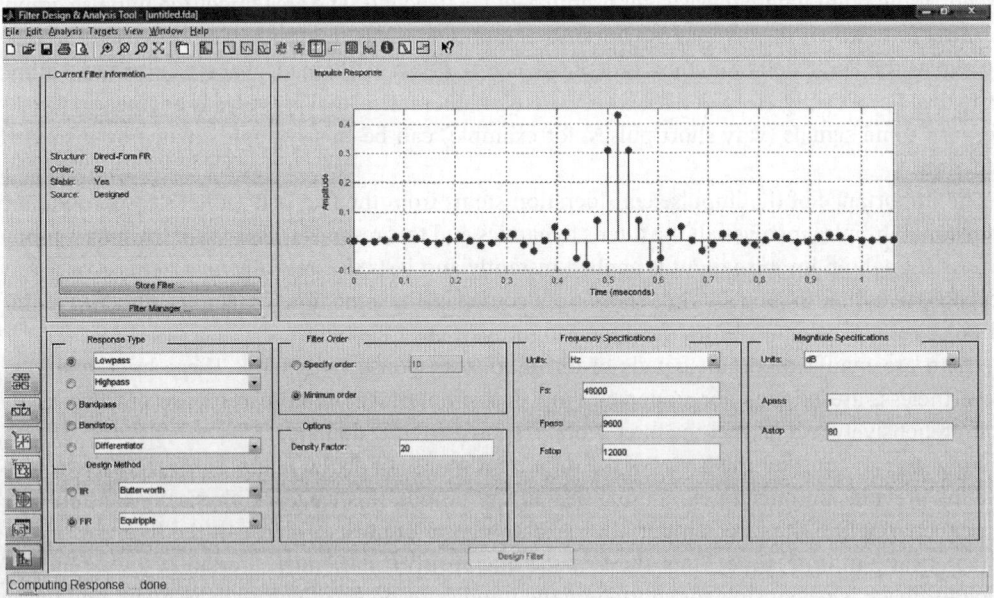

Figure 5.5.4 Impulse response of a FIR filter.

5.5.1.6 Transfer Function of a Digital Filter

In Section 5.4, the transfer function of an analog filter in general form has been introduced as

$$H(s) = \frac{b_m s^m + b_{m-1} s^{m-1} + \ldots + b_1 + b_0}{a_n s^n + a_{n-1} s^{n-1} + \ldots + a_1 + a_0} \tag{5.4.3R}$$

This transfer function in *s*-domain was obtained by applying the Laplace transform to the filter's differential equation. Here the complex frequency s is given by $s = \sigma + j\omega$.

Similarly, transforming a filter's difference equation into *z*-domain by using the *z*-transform, we find the transfer function of a digital filter to be

$$H(z) = \frac{b_0 z + b_1 z^{-1} + b_2 z^{-2} + \ldots + b_M z^{-M}}{a_0 z + a_1 z^{-1} + a_2 z^{-2} + \ldots + a_N z^{-N}} \quad (5.5.14)$$

We need to know that *z*-transform is a mathematical operation that transforms a discrete function given in time domain into a discrete function in *z*-domain. This transform is true for a linear, time-invariant digital filter. It also implies that the filter is causal. The former condition means that the principle of superposition is applied to the operation of a digital filter. The latter condition requires that the filter's response can appear only *after* the input is applied.

To better understand the origin of (5.5.14), let us take a simple approach (Lutovac et al. 2001, p. 176): Consider z^{-1} as a delay operator that introduces a delay (shift) by one sampling interval backward when applied to a sequence of samples (digital values). Applying operator z^{-1} to the value, x_n, of an input signal gives us the previous value of this signal, x_{n-1}; that is,

$$z^{-1} x_n = x_{n-1} \quad (5.5.15a)$$

For example, if we have an input signal as follows:

$$x_0 = 2$$
$$x_1 = 6$$
$$x_2 = -4$$
$$x_3 = 9$$

then

$$z^{-1} x_0 = x_{-1} = 0$$
$$z^{-1} x_1 = x_0 = 2$$
$$z^{-1} x_2 = x_1 = 6$$
$$z^{-1} x_3 = x_2 = -4 \quad (5.5.15b)$$

Also,

$$z^{-1}(z^{-1} x_3) = z^{-2} x_3 = x_1 = 6$$

and, in general,

$$z^{-m} x_n = x_{n-m} \quad (5.5.15c)$$

Obviously, this operation is true for an output signal $y(n)$; that is,

$$z^{-m} y_n = y_{n-m}$$

Now, consider the difference equation of a recursive filter (5.5.5a):

$$a_0 y[n] + a_1 y[n-1] + \cdots + a_N y[n-N] = b_0 x[n] + b_1 x[n-1] + \cdots + b_M x[n-M]$$

We understand that

$$y[n-k] = y_{n-k}$$

and

$$x[n-k] = x_{n-k}$$

Applying the delay operator z^{-m} to the input signal, we find

$$z^{-1}x_{n-1} = x_n$$
$$z^{-2}x_{n-2} = x_n$$
$$z^{-3}x_{n-3} = x_n$$
$$\vdots$$
$$z^{-M}x_{n-M} = x_n$$

Obviously, this procedure holds true for the output signal. Now the difference equation (5.5.12a) of a recursive filter,

$$a_0 y_n + a_1 y_{n-1} + \cdots + a_N y_{n-N} = b_0 x_n + b_1 x_{n-1} + \cdots + b_M x_{n-M}$$

takes the form

$$a_0 y_n + a_1 z^{-1} y_n + a_2 z^{-2} y_n + \cdots + a_N z^{-N} y_n = b_0 x_n + b_1 z^{-1} x_{n-1} + b_2 z^{-2} x_{n-2} + \cdots + b_M z^{-M} x_n \tag{5.5.16}$$

It is easy to rearrange this equation as follows:

$$(a_0 + a_1 z^{-1} + a_2 z^{-2} + \cdots + a_N z^{-N}) y_n = (b_0 + b_1 z^{-1} + b_2 z^{-2} + \cdots + b_M z^{-M}) x_n \tag{5.5.17}$$

which gives us a direct relationship between the input and output signals; that is, the *transfer function of a recursive (IIR) digital filter* can be expressed as

$$H(z) = \frac{y_n}{x_n} = \frac{b_0 + b_1 z^{-1} + b_2 z^{-2} + \cdots + b_M z^{-M}}{a_0 + a_1 z^{-1} + a_2 z^{-2} + \cdots + a_N z^{-N}} \tag{5.5.18}$$

Similarly, applying the z-transform to (5.5.12a), we can obtain the *transfer function of a nonrecursive (FIR) digital filter* as

$$H(z) = \frac{y_n}{x_n} = \frac{b_0 + b_1 z^{-1} + b_2 z^{-2} + \cdots + b_M z^{-M}}{a_0} \tag{5.5.19}$$

where usually $a_0 = 1$.

For example, the transfer function of a recursive filter described by the difference equation

$$a_0 y_n = b_0 x_n + a_1 y_{n-1}$$

is given by

$$H(z) = \frac{b_0}{a_0 - a_1 z^{-1}}$$

Using the concept of transfer function, we can redefine the definition of the order of a digital filter as

> *The order of a digital filter is the order of its transfer function, which is the order of either the numerator or the denominator, whichever is greater.*

This definition holds true for both recursive and nonrecursive filters. (Recall that the order [or degree] of a polynomial – both the numerator and the denominator of a transfer function are polynomials – is equal to the highest power of the polynomial variable.)

The transfer function of a second-order recursive filter in general is given by

$$H(z) = (b_0 + b_1 z^{-1} + b_2 z^{-2})/(a_0 + a_1 z^{-1} + a_2 z^{-2})$$

whereas the transfer function of a second-order nonrecursive filter in general is given by

$$H(z) = (b_0 + b_1 z^{-1} + b_2 z^{-2})/a_0$$

Consider, for example, the three-term average filter discussed previously in this section and write down its transfer function to fully understand this general expression.

Formally speaking, to obtain the transfer function of a digital filter, we need, of course, to perform a z transform of both sides of the digital filter's difference equation; this is similar to application of the Laplace transform to an analog filter's differential equation. The result is written as

$$H(z) = \frac{Y(z)}{X(z)} = \frac{b_0 + b_1 z^{-1} + b_2 z^{-2} + \cdots + b_M z^{-M}}{a_0 + a_1 z^{-1} + a_2 z^{-2} + \cdots + a_N z^{-N}} \tag{5.5.20}$$

which is, in essence, Eq. (5.5.18). Considering z as a frequency and following the procedure we use for analog fillers, we can find from (5.5.20) the frequency (amplitude and phase) response of a digital filter.

5.5.2 Conclusive Remarks on Digital and Analog Filters

5.5.2.1 Some Final Comments on Digital Filters

- We should always bear in mind that, in general, the *operation and description* of digital filters are analogous to that of analog filters.
- *The specifications of digital filters* are the same as those of analog filters, as discussed in Section 5.1.3.
- *The design of digital filters* is a special field requiring extensive research and laden with countless books. (Fortunately, there is no need for an end-user to learn design techniques. We can simply use these techniques in much the same manner as we use any software applications.)

5.5.2.2 Adaptive Filters

Thanks to advances in digital electronics and DSP, digital filters gave birth to a new class of filters – adaptive. These filters change their parameters (coefficients of its transfer function that is) over time to adapt their operation to the fluctuating characteristics of an input signal. Thus, to identify an unknown input signal or cancel its noise, an adaptive filter changes its coefficients to best meet the given filtering specifications. This is done by using the feedback mechanism, as shown in Figure 5.5.5. An adaptive filter first processes the input signal by its predetermined algorithm. The resulting output signal is compared with the desired signal at the point of summation and the difference (the error signal) modifies the algorithm by which the filter processes the input signal. In effect, the adaptive filter adjusts itself to the point where the output signal is equal to the desired signal regardless of the changes in the input signal.

Analyze Figure 5.5.5 closely. Describe the adaptive filter operation when, for example, additional noise is added to an input signal. Devise examples of the filter operation under other changes of the operational conditions.

(**Question**: *How is the feedback concept implemented in the adaptive filter shown in Figure 5.5.5?*)

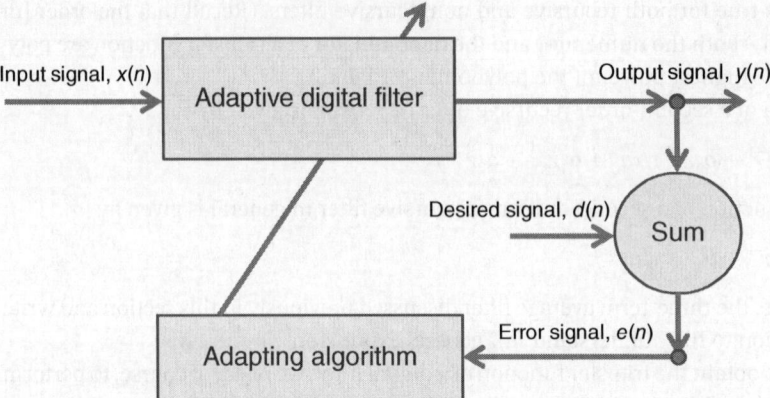

Figure 5.5.5 Block diagram of an adaptive filter.[11]

5.5.2.3 Comparison of Analog and Digital Filters

Finally, let us *compare* digital and analog filters. Table 5.5.2 briefly summarizes the main points of this comparison.

Analyze Table 5.5.2 closely: Recall every term and condition mentioned in this table. Such a review will be a good summary of all the discussion on filters given in this chapter.

In short, analog filters are inexpensive and fast and have a large dynamic range in both amplitude and frequency scales, whereas digital filters demonstrate superior performance characteristics, such as great steepness of the rolloff, small ripples, and high accuracy and stability of their parameters. In addition, they are readily programmable and can be made adaptive. Also, designing digital filters is much easier to do and is more automated than designing their analog counterparts, thereby reducing the total cost of filter applications. In general, digital filters are the preferred tools in signal processing unless you need simple and inexpensive filters for specific applications where analog filters are still in use. Finally, bear in mind that digital-filter hardware still cannot be built without anti-aliasing and reconstructive analog filters.

5.5.2.4 Summary of Applications of Various Filter Technologies

Now let us briefly summarize the key areas of applications for each technology discussed in this section:

Passive filters are typically used when

- External power is not available.
- Circuitry power lines need filtering.
- A high-frequency (multi-megahertz) range is to be covered.
- Active filters are typically used when
- Precise wide-range filters (from sub hertz to megahertz) are needed.
- A load problem has to be resolved.
- Gain of an input signal has to be provided.
- Switched-capacitor filters are typically used when
- Cost and space are paramount concerns.
- High accuracy with analog filters is required.
- A broad bandwidth (>10 kHz) is needed.

---

13 Smith (2007).

Table 5.5.2 Comparison of analog and digital filters.

Point of comparison	Analog filters	Digital filters	Winner
Speed[a]	Fast: The speed of operation is basically determined by the time needed for an electrical signal to propagate through a circuit	Slow: Speed of operation is determined by the time needed to process the tremendous number of samples; also, there are many delay elements (z^{-1}) in a filter	**Analog** (hardware still operates faster than software even though modern electronics effectively narrows this gap.)
Accuracy and stability	Low: The values of electronic components drift over time and with change in temperature. The errors caused by these drifts (changing the tolerances) are magnified with cascading filter stages, that is, with an increase of the filter's order	High: Digital filters depend much less (almost not at all) on electronic components. The values of their coefficients do not change because they are stored in the processor's memory	**Digital** (Have accuracy and stability that cannot be achieved by analog filters)
Dynamic range (amplitude and frequency)	A simple op-amp can operate with signals from µV to V (amplitude range of 1 million) and from 10^{-2} to 10^5 Hz (the frequency range is 10 million)	A 12-bit ADC has a saturation level of 4095 and a noise of 0.29 (the amplitude range is 14 000); sampling 10^5 Hz requires 10×10^6 samples to capture one cycle at 0.01 Hz, which means that it is almost impossible to achieve reasonable results. Also, this limitation results in cutting the passband of a high-pass digital filter	**Analog** (Much larger dynamic range in both amplitude and frequency)
Ripples	Even for a Butterworth filter, the ripples cannot be realistically made to less than 1%	Can be made as small as necessary (not more than 0.02%) for any filter prototype	**Digital**
Rolloff (Steepness)	Low: To achieve a shape factor of 1.1, an analog filter must be of the 30th order, which is impractical	High: Routinely achieve shape factor (ratio of passband to stopband frequencies) of 1.1	**Digital**
Stopband attenuation	Again, to achieve −100 dB stopband attenuation, we would need a thirtieth-order analog filter	Routinely achieve the level of −100 dB	**Digital** (Provides roll-off (steepness) hundreds and thousands of times better than analog)

(Continued)

Table 5.5.2 (Continued)

Point of comparison	Analog filters	Digital filters	Winner
Waveform preservation	Exhibit overshoot in step response	Also exhibit overshoots in step response	**None**
Noise[a]	Electronic components introduce thermal and other types of noise; this occurrence depends on time and temperature. In addition, an increase of the filter's order will result in an increase in the noise level	Noise is introduced by analog filters and ADC and DAC components; what is more, there is quantization noise. However, in total, digital filters introduce minimal noise and this does not change with time and temperature and does not depend on the filter's order	**Digital**

a) Note regarding the speed of operation: In Table 5.5.2, we refer to the speed of signal filtering. In communications, DSP is not just a filtering but a signal processing aiming to extract correct information from a received signal, which is typically deeply distorted by noise and other harmful effects during its transmission. Modern communication industry, by applying a masterful distribution of an entire task between hardware and software and using all the marvels of modern electronics, can build application specific integrated circuits, ASICs, capable to perform more than 10 trillion (!) operations per second.

○ Digital filters are typically used when
○ A high level of performance is needed and cannot be achieved with analog filters.
○ Programmable and adaptive filters have to be used.
○ A DSP unit already resides in the system.

Bear these points in mind when you need to choose the proper filter for your task.

Questions and Problems for Section 5.5

- Questions marked with an asterisk require a systematic approach to finding the solution.
- Many questions and problems, including those marked with an asterisk, imply that you, in addition to reading the textbook, will do your research to find the answers. Consider such questions as miniprojects.

What are Digital Filters?

1. Explain the difference between analog and digital filters.

2. Consider the hardware of a digital signal:
   a) Sketch the block diagram of a digital filter.
   b) What three main parts of its hardware do you need to show?
   c) Does this hardware include analog filters? Explain.

3 Explain the principle of operation of a digital filter. Give an example.

4 *The text says that a filter is a signal-processing device. Therefore, a digital filter is a digital signal processing, DSP, device. Figure 5.5.1, however, shows a separate DSP unit within the digital filter. Why?

5 Compare the hardware of an analog filter and of a digital filter: What is the heart of the analog-filter hardware and what is the heart of the digital-filter hardware. Explain.

6 Why is digital filtering considered as a part of digital signal processing, DSP?

7 Do we need to manufacture a special IC chip to make a digital filter? If your answer is *No*, how then can we produce a digital filter?

8 Can you use your laptop as a digital filter? If your answer is *Yes*, what conditions must be met?

9 When converting an analog signal into a digital one, the first operation is sampling. Explain this operation.

10 The text says that sampling is the conversion of a continuous function, $v(t)$, into a discrete function, $v[n]$:
   a) What is $v[n]$? Give an example.
   b) The continuous function is $v(t) = 3\cos(4t)$. What is its digital counterpart, $v[n]$?

11 The highest frequency that the human ear can hear is 20 kHz. What is the required sampling frequency? What is the required time interval, Δt?

12 Explain the similarity and dissimilarity between a differential equation and a difference equation. Give examples.

13 *Consider the difference equations for a four-term average filter, $4y_n = x_n + x_{n-1} + x_{n-2} + x_{n-3}$:
   a) Write down the equations for all consecutive numbers.
   b) What is the order of this filter?
   (*Hint*: Refer to (5.5.7) and its discussion.)

14 Consider the following example of a filter's difference equation: $3y[n] + 6y[n-1] + 12y[n-3] = 8x[n] + 4x[n-1] + 9x[n-2] + 2x[n-4]$. What is the order of this filter? Explain.

15 Consider two difference equations, (a) $y_n = x_n + y_{n-1}$ and (b) $y_n = x_n + x_{n-1} + x_{n-2} + \cdots + x_0$. Which one does describe a recursive filter? A nonrecursive? Explain.

16 Consider classifications of digital filters:
   a) What are the meanings of the terms "recursive" and "nonrecursive?"
   b) What filters are called recursive and nonrecursive?
   c) What filters are called infinite impulse response, IIR, and finite impulse response, FIR?
   d) What are the relationships between all four categories?

17 *Consider an impulse response of digital filters:
  a) What is an impulse (delta or Dirac) function?
  b) Explain the importance of studying the impulse response of digital filters.
  c) What is the difference between infinite impulse response (IIR) filters and finite impulse response (FIR) filters?
  d) What alternative terminology is used for IIR and FIR filters?
  e) An input digital signal is given by $\{x_0 = 3, x_1 = 6, x_2 = 2, x_3 = -7\}$. Apply operator z^{-1} and find the previous value of this signal, x_{n-1}.

18 The difference equation of a filter is given by $3y_n = 5x_n + 7y_{n-1}$:
  a) Which filter – recursive or nonrecursive – does this equation describe? Explain.
  b) Derive the transfer function of this filter.
  c) What is the order of this filter?

19 *Consider the transfer function of a nonrecursive (FIR) filter given in (5.5.19) as $H(z) = \frac{y_n}{x_n} = \frac{b_0 + b_1 z^{-1} + b_2 z^{-2} + \cdots + b_M z^{-M}}{a_0}$:
  a) Write the transfer function of a second-order filter.
  b) What is the output signal, y_n, of this filter if its input signal is $\{x_0 = 3, x_1 = 6, x_2 = 2, x_3 = -7\}$?

20 Describe the difference, if any, between the specifications of analog and digital filters.

21 *Design a digital filter low-pass IIR Butterworth filter of a minimum order with $F_{pass} = 10\,\text{kHz}$ and $F_{stop} = 12\,\text{kHz}$. Build the graphs of an amplitude response, a phase response, a time group delay, and a step response.
  (*Hint*: Use MATLAB or see Section 5.4.2.3.)

22 *Consider adaptive filters:
  a) What is an adaptive filter?
  b) Sketch an adaptive filter's block diagram and explain the principle of its operation.
  c) How can an adaptive filter be built?
  d) Can we build an analog adaptive filter? Explain.

23 Why do we need adaptive filters?

24 Consider Table 5.5.2, where analog and digital filters are compared from the standpoint of their applications:
  a) Which type of filter – analog or digital – would you prefer in general and why?
  b) What advantages do analog filters have over digital filters?
  c) What advantages do digital filters have over analog filters?
  d) Which type of filters – analog or digital – is the preferred one in DSP and why?

25 List main filter technologies.

26 We have a variety of filter technologies, namely, passive filters, active filters, switched-capacitor filters, and digital filters. Why do we need this variety instead of developing just one filter technology?

6

Spectral Analysis 1 – The Fourier Series in Modern Communications

"The study and measurement of the frequency content of a signal is called *spectral analysis*.[1]" The word *spectral* is an adjective signifying "related to" or "produced by" a spectrum. *Spectrum*, in turn, is the distribution of electromagnetic energy arranged in order of frequencies. In this chapter, we will study how to analyze the frequency content of a signal, that is, how to find the frequency components that make up this content. We will also study how to reconstruct the signal if we know its frequency content, an operation called *spectral synthesis*. In this chapter, we will consider periodic signals, which have a discrete spectrum; in Chapter 7, we will focus on nonperiodic signals, which have a continuous spectrum.

Objectives and Outcomes of Chapter 6

Objectives

- To learn the concept of spectral analysis and become comfortable in correlating the descriptions of all the processes given in time domain and in frequency domain.
- To gain a deeper understanding of spectral analysis and become familiar with the synthesis of periodic signals.
- To expand the student's knowledge of the spectral analysis of periodic signals.

Outcomes

- Distinguish between time domain and frequency domain.
- Understand how the Fourier series enables us to determine the spectrum of a periodic signal.
- Learn how to perform the spectral synthesis of a periodic signal.
- Become more familiar with, and feel more comfortable working with, the Fourier series by having learned the techniques for finding them for various types of periodic signals.
- Understand the effect of filtering on signals from the standpoint of spectral analysis.
- Gain insight into the concept of harmonic distortion.
- Learn the mathematical foundation of the Fourier series.
- Understand the power spectrum of periodic signals.

---

1 Taylor (1994).

6.1 Basics of Spectral Analysis

Objective and Outcomes for Section 6.1

Objective

To introduce the basics of spectral analysis and spectral synthesis.

Outcomes

After studying this section, you will be able to:

- Distinguish between time domain and frequency domain.
  - Differ between periodic and nonperiodic signals.
  - Explain the relationship between time domain and frequency domain.
  - Present cosine and sine signals in time domain and frequency domain.
  - Explain that each sinusoidal (cosine or sine) signal represents a single frequency; that is, the presentation of a signal through the cosine and sine functions shows the frequency content of the signal.
  - Distinguish between the waveform (the plot of a signal in time domain) and the spectrum (the graphical presentation of the signal's frequency content).
  - Experimentally verify the presentation of the same signal in time domain and in frequency domain.
- Apply the Fourier series to finding the spectrum of a periodic signal.
  - Demonstrate that the Fourier-series approach is based on the point that each sinusoidal signal in time domain represents a single frequency in frequency domain.
  - Explain that every periodic signal can be represented as a sum of cosines and sines and that this sum is the Fourier series.
  - Show that each cosine and sine of a Fourier series has a unique coefficient (amplitude) and a unique frequency.
  - Demonstrate calculations of the coefficients of the Fourier series.
  - Demonstrate developing signal waveform's formula into the Fourier series as the key operation in the spectral analysis.
- Understand how to perform the spectral synthesis of a periodic signal.
  - Show that the Fourier series allows for performing spectral synthesis, operation of the reverse of spectral analysis, that is, reconstruction of the signal's waveform from the given spectral components (harmonics) of the signal.
  - Ascertain that the waveform of a signal can be reconstructed from its spectrum by building the waveform of every spectral line (here the waveform is a cosine or sine) and then adding them up, as the Fourier series instructs.

6.1.1 Time Domain and Frequency Domain

6.1.1.1 Periodic and Nonperiodic Signals

All the electrical and optical signals we deal with in our professional life are divided into two categories: periodic and nonperiodic.

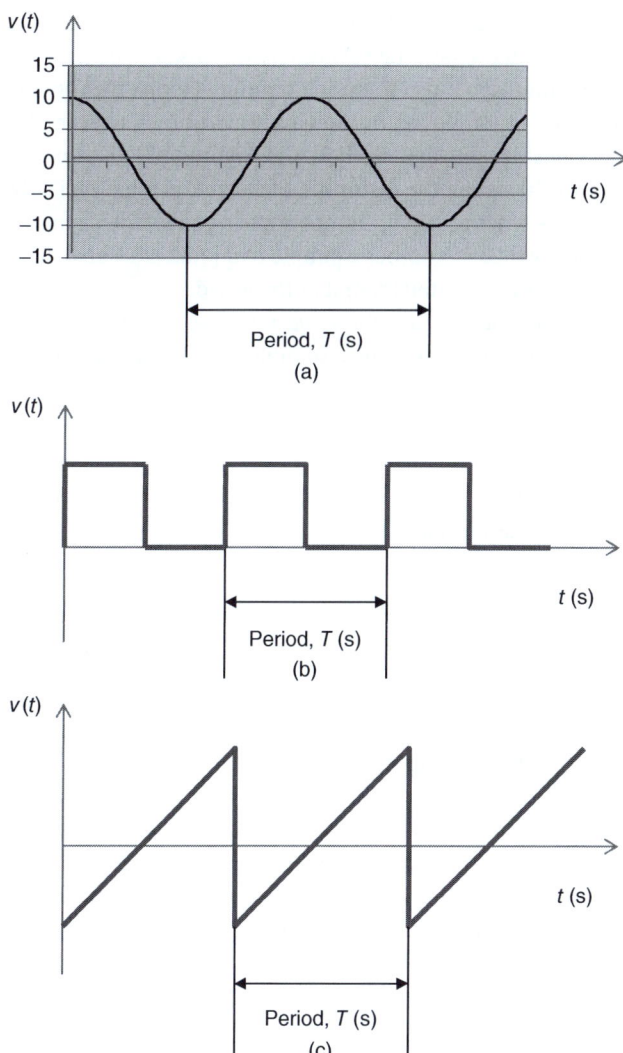

Figure 6.1.1 Periodic signals: (a) Sinusoidal signal; (b) square-wave signal; and (c) sawtooth signal.

A signal is called periodic if it repeats (copies) itself after a given time interval, which is called a period, T. All other signals are called nonperiodic (aperiodic).

Examples of periodic signals include sinusoidal signals (sine and cosine) and square-wave signals, which are shown in Figure 6.1.1a,b. (Study the signal in Figure 6.1.1c. Is it periodic? Can you identify the period of this signal?) It follows from the definition that periodic signals have constant amplitudes, periods, and phase shifts; indeed, how else could they reproduce themselves after each period?

Mathematically, periodic signals are described by the following formula:

$$v(t) = v(t + nT) \tag{6.1.1}$$

where $v(t)$ is the formula describing the signal's waveform, n is the integer ($n = 1,2,3, \ldots$), and T (s) is the period of the signal. For example, if $v(t) = A \cos(\omega t)$, then $A \cos(\omega t) = A \cos(\omega t + nT)$.

Pure periodic signals are special and ubiquitous; we use them in communications theory and in transmission, test, and measuring equipment employed in industry. In reality, however, the vast majority of signals are nonperiodic. Consider the signals shown in Figure 6.1.2. The signal in Figure 6.1.2a looks like a distorted sinusoidal signal; the distortion of a signal is what typically happens in electronic communications and is what results in nonperiodic signals. The signal in Figure 6.1.2b is the record of a human voice, which is another example of a practical signal. The signal in Figure 6.1.2c is noise, which is the most ubiquitous signal in the world.

From these examples, we can conclude that not only are the *majority of deterministic signals* nonperiodic but also that *all random signals* (and noise is a perfect example of a random signal) *are nonperiodic*.

6.1.1.2 Time Domain and Frequency Domain Revisited

We introduced the idea of time domain and frequency domain in Section 2.1 and now we need to delve into this topic in greater detail. Let us first agree that

> the time-domain description of a cosine signal, $v(t) = A\cos(2\pi ft) = A\cos(\omega t)$, is equivalent to a frequency-domain presentation of this signal by the line with an amplitude, A, and a frequency, f (or ω).

This concept is shown in Figure 2.1.9 and again in Figure 6.1.3 for three cosine signals with various amplitudes and frequencies.

You will recall that *time domain is a v–t space*, where the vertical axis, $v(t)$ (V), designates a signal's magnitude changing with time and measured in volts, whereas the horizontal axis, t (s), denotes time in seconds. *In time domain, we can show the signal's waveform, its amplitude, A (V), and its period, T (s)*. For example, all cosine signals shown at the left in Figure 6.1.3 (time domain) are described in general as

$$v(t) = A \cos(2\pi f_0 t) = A \cos(\omega_0 t) \quad (6.1.2)$$

where cos stands for cosine, A (V) is the amplitude, f_0 (Hz) is the frequency, and ω_0 (rad/s) $= 2\pi f_0$ is the radian (angular) frequency. Obviously, each cosine signal has its own amplitude and frequency, as exemplified by Figure 6.1.3. It is important to understand that *it is impossible to show the frequency value in a time-domain figure; thus, we can show only its period, T (s)*.

You are reminded that the relationship between the period, T (s), of a sinusoidal signal and its frequency, f_0 (Hz), is given by

$$T\,(\text{s}) = \frac{1}{f_0}\,(\text{Hz}) \quad (6.1.3)$$

> Equation (6.1.3) is a fundamental formula that governs the relationship between time and frequency domains.

Frequency domain *is an A–f space*, where the vertical axis, A (V), designates the amplitude of a signal (measured in volts in our example) and the horizontal axis, f_0 (Hz), shows the frequency in

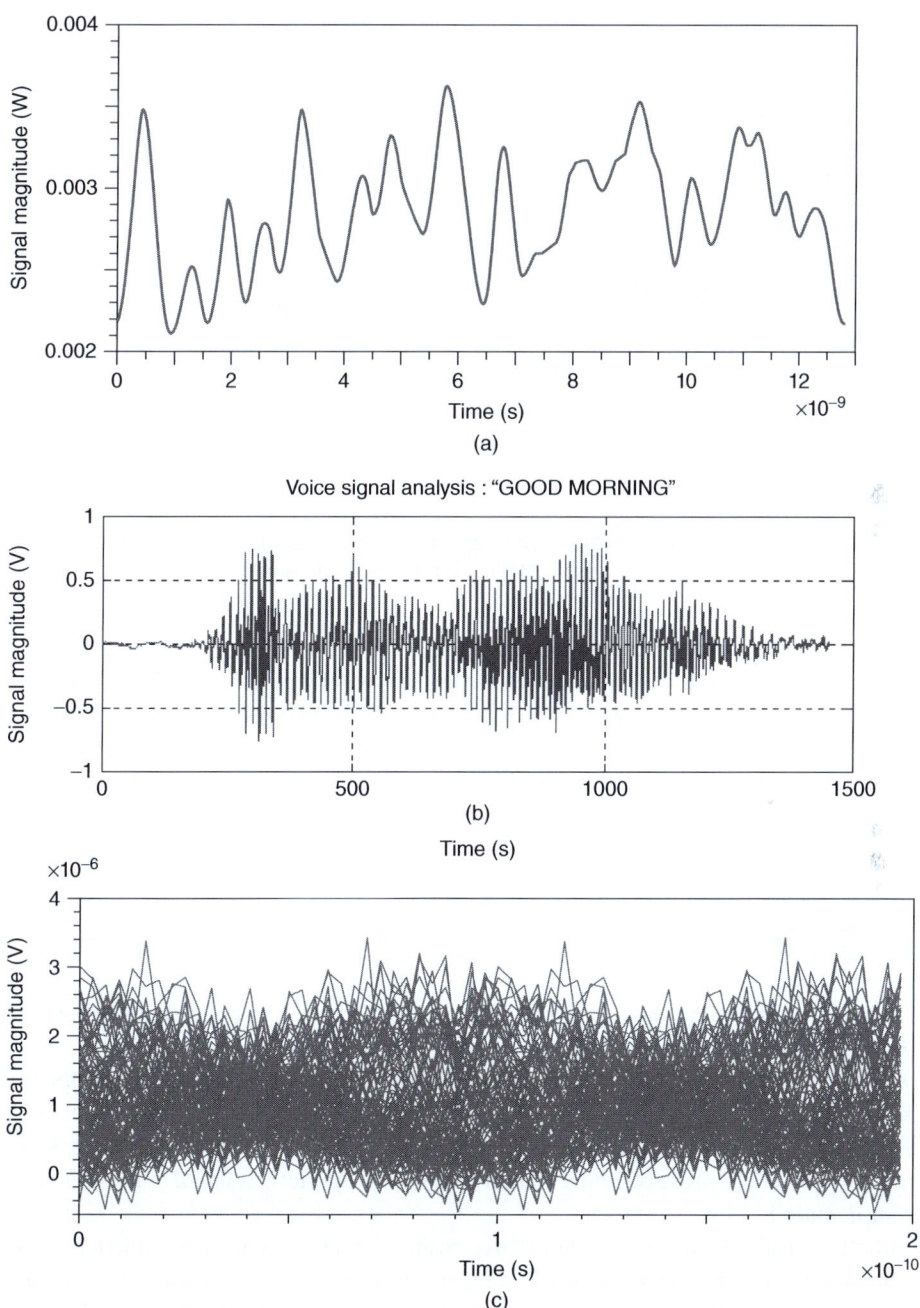

Figure 6.1.2 Nonperiodic signals: (a) Distorted sinusoidal signal; (b) record of human voice in time domain; and (c) noise. Source: (b) Professor Mohammed Kouar, New City College of Technology of the City University of New York. Reprinted with permission.

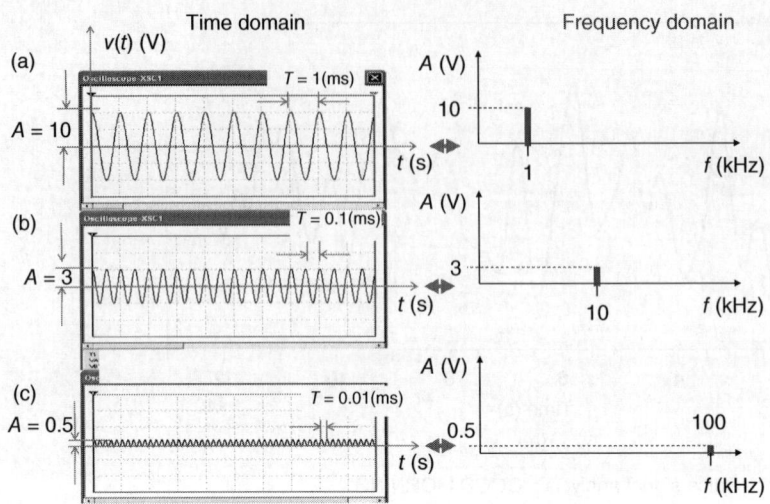

Figure 6.1.3 Cosine signals of (a) 1 kHz, (b) 10 kHz, and (c) 100 kHz with amplitudes of 10, 3, and 0.5 V shown in time domain (left) and frequency domain (right). (Graphs are not to scale.)

hertz. *In frequency domain, we can show the amplitude of a sinusoidal signal, A (V), and its frequency, f_0 (Hz).* The same signals shown in time domain (the left section of Figure 6.1.3) are presented in frequency domain (the right section of Figure 6.1.3); in time domain, we see the signals' waveforms, whereas in frequency domain, we see the lines whose heights represent the signals' amplitudes and whose locations along the f axis are specified by the values of the signals' frequencies. The point is:

> *A time-domain cosine signal, v(t) (V) = A cos(2πf₀t) = A cos(ω₀t), with its amplitude, A (V), and its period, T (s), is presented in frequency domain by the line with the amplitude, A (V), and the frequency, f_0 (Hz) = $\frac{1}{T(s)}$.*

Thus, the time-domain parameters of a cosine signal, A and T, are presented in frequency domain by A and f_0, where $f_0 = 1/T$. The only entity that we do not carry over from time domain to frequency domain is the signal's waveform. But we agree to represent a cosine signal by one line in frequency domain; thus, the signal's waveform is given by default. (You will recall that formula $v(t)$ (V) = $A \cos(\omega_0 t) = A \cos(2\pi f_0 t)$ shows that the waveform, $v(t)$, changes only because time, t (s), is varying since A (V) and f_0 (Hz) are constant; thus, a *waveform is the plot of a signal in the v–t space*, that is, in *time domain*.)

As an example, consider Figure 6.1.3, where three cosine signals are depicted on the left-hand side. To present our three signals in frequency domain, we show three lines with proper amplitudes and frequencies on the right-hand side. The corresponding formulas are shown in Table 6.1.1.

Note well: *When we refer to a waveform, we refer to the signal's representation in time domain.*

But what if a cosine signal has an *initial phase shift*, Θ? Well, in this case, the time-domain presentation of this signal has to be written as

$$v(t) = A \cos(2\pi f_0 t + \Theta) = A \cos(\omega_0 t + \Theta) \qquad (6.1.4)$$

and we obviously need the other coordinate plane in frequency domain to show the phase. Hence, this signal will be shown in frequency domain in two coordinate planes: One is A–f, as explained

Table 6.1.1 Formulas of cosine signals shown in time domain and in frequency domain in Figure 6.1.3.

Time domain	Frequency domain
$v_1(t)$ (V) $= 10 \cos(2\pi\ (1 \times 10^3)t)$	$A_1 = 10$ V at $f_1 = 1$ kHz
$v_2(t)$ (V) $= 3 \cos(2\pi\ (10 \times 10^3)t)$	$A_2 = 3$ V at $f_2 = 10$ kHz
$v_3(t)$ (V) $= 0.5 \cos(2\pi\ (100 \times 10^3)t)$	$A_3 = 0.5$ V at $f_3 = 100$ kHz

previously, and the other is Θ–f. To illustrate, we show the signal

$$v_1(t) = 10\ \cos(2\pi(1 \times 10^3)t + 30°)$$

in time and frequency domains in Figure 6.1.4a. We discussed this point in Section 2.2 and showed phase-shifted sinusoidal signals in both domains in Figure 2.2.3.

If we need to show a sine signal in frequency domain, we recall from Section 2.2 that

$$\sin \alpha = \cos(\alpha - 90°)$$

and therefore we have to show a phase shift of −90°. The sine signal, $v(t) = A \sin(2\pi f_0 t)$, in time and frequency domains is shown in Figure 6.1.4b. To compare time-domain and frequency-domain presentations of cosine and sine signals, see Figure 6.1.4c. This figure emphasizes that a *cosine wave* is a reference signal and, *by convention, its phase shift is zero,* provided that there is no initial phase shift. Therefore, it is depicted in frequency domain by only amplitude vs. frequency (A–f) line. On the other hand, a s*ine wave*, by definition, *has a −90° phase shift with respect to a cosine wave*. Therefore, it is depicted in frequency domain by both A–f and Θ–f lines.

Now let us summarize the relationship between time domain and frequency domain:

- In time domain, the v–t axes allow us to show the signals' waveforms, which contain all their parameters, namely, amplitudes, A, periods, T, and phase shifts, Θ.
- In frequency domain, the A–f axes allow us to depict amplitudes and phases of sinusoidal (cosine and sine) signals, A and θ, against their frequencies, $f_0 = 1/T$, as lines.
- A presentation with A–f axes (that is, without the phase shift) is sufficient only for a cosine signal, $v(t)$ (V) $= A \cos(\omega_0 t) = A \cos(2\pi f_0 t)$.
- For cosine signals with initial phase shifts and for sine signals, it is necessary to use the A–f and Θ–f axes to show both amplitudes and phases vs. frequency. See Figures 6.1.3 and 6.1.4.

It is worth noting that we have already described two different views – in time domain and in frequency domain – of the same process in our analysis of the operation of *filters*. We refer you to Section 5.1 and, particularly to our analysis of Figure 5.1.9a; we reproduce this figure for your convenience in Figure 6.1.5. Here, we present three waveforms (in time domain, obviously) measured at the filter's output at 0.1 kHz ($T = 10$ ms), 1 kHz ($T = 1$ ms), and 10 kHz ($T = 0.1$ ms) with Multisim simulation. In frequency domain, we show the V_{out} (V)-vs.-f (Hz) graph (filter's attenuation, in essence); we can measure the output amplitudes at the same frequencies. You can readily observe that the output amplitude, V_{out} (V) measured in both time domain and frequency domain, are the same.

The statement about the relationship between time domain and frequency domain looks like a hypothetical rule. How real is it? Well, we need to perform an experiment to confirm its validity. This experiment is shown in Figure 6.1.6. The same cosine signal is simultaneously presented to

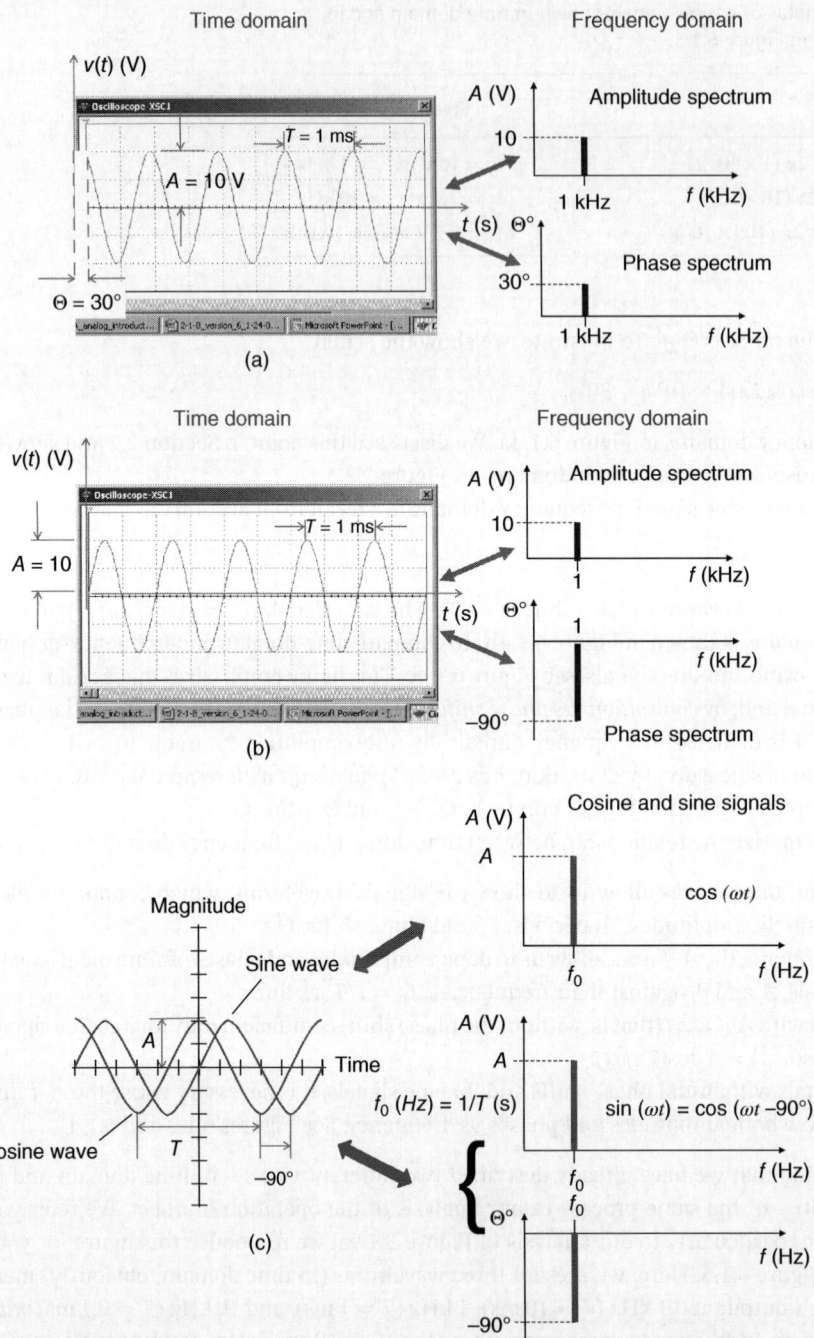

Figure 6.1.4 Sinusoidal signals with phase shifts in time and frequency domains: (a) Cosine signal with 30° phase shift; (b) sine signal; (c) comparison of cosine and sine signals in time domain and in frequency domain.

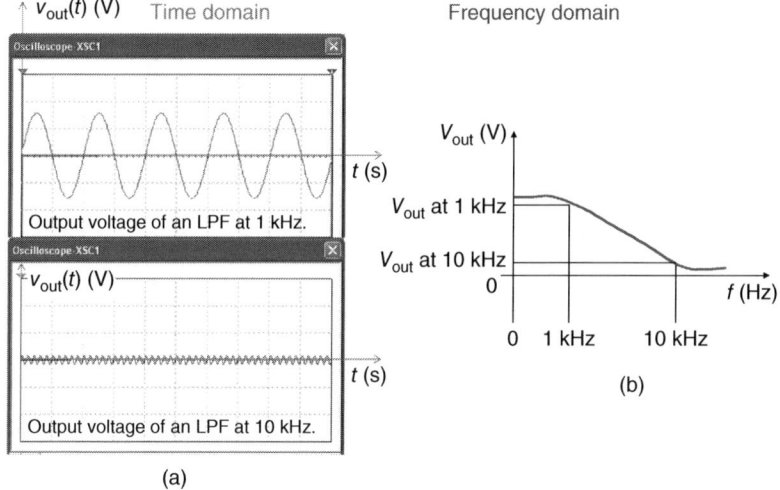

Figure 6.1.5 Output voltage of a low-pass filter presented in time (a) and frequency (b) domains.

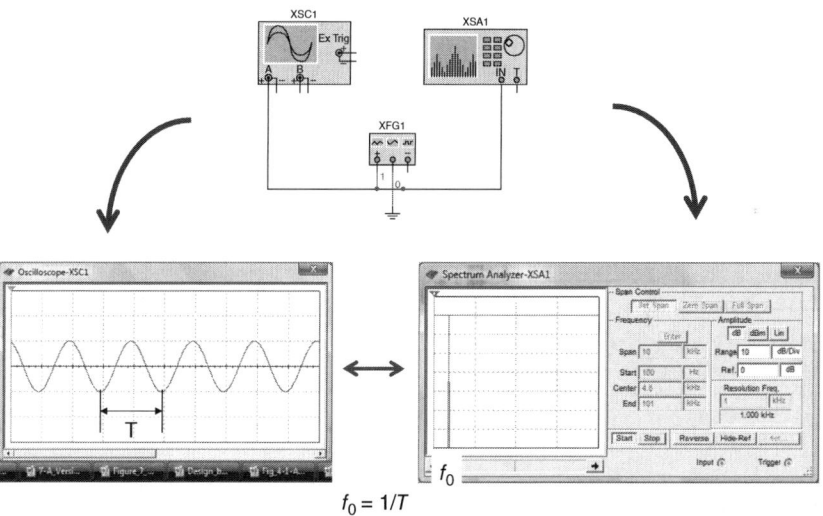

Figure 6.1.6 Experiment setup, the recorded waveform, and the spectrum of a cosine signal.

an oscilloscope and a spectrum analyzer. The signal's waveform and its spectrum are shown in Figure 6.1.6. You can clearly see that the cosine (a *waveform* in time domain) corresponds to a single line (a *spectrum* in frequency domain).

Next, let us do an experiment in time domain as shown in Figure 6.1.7: Consider that a cosine signal, $v(t)$ (V) $= A\cos(2\pi f_0 t)$, is presented to a band-pass filter whose central frequency is $f_{0BPF} = 1$ kHz and whose passband is 0.2 kHz. We expect that this signal will pass through the filter entirely provided that f_0 will be within a filter's passband. If, however, the signal's frequency will be out of the passband, then the filter's output should decrease. (The band-pass filter is designed with Multisim.[2])

---

[2] Multisim™ © 2019 National Instruments Corporation. All Rights Reserved.

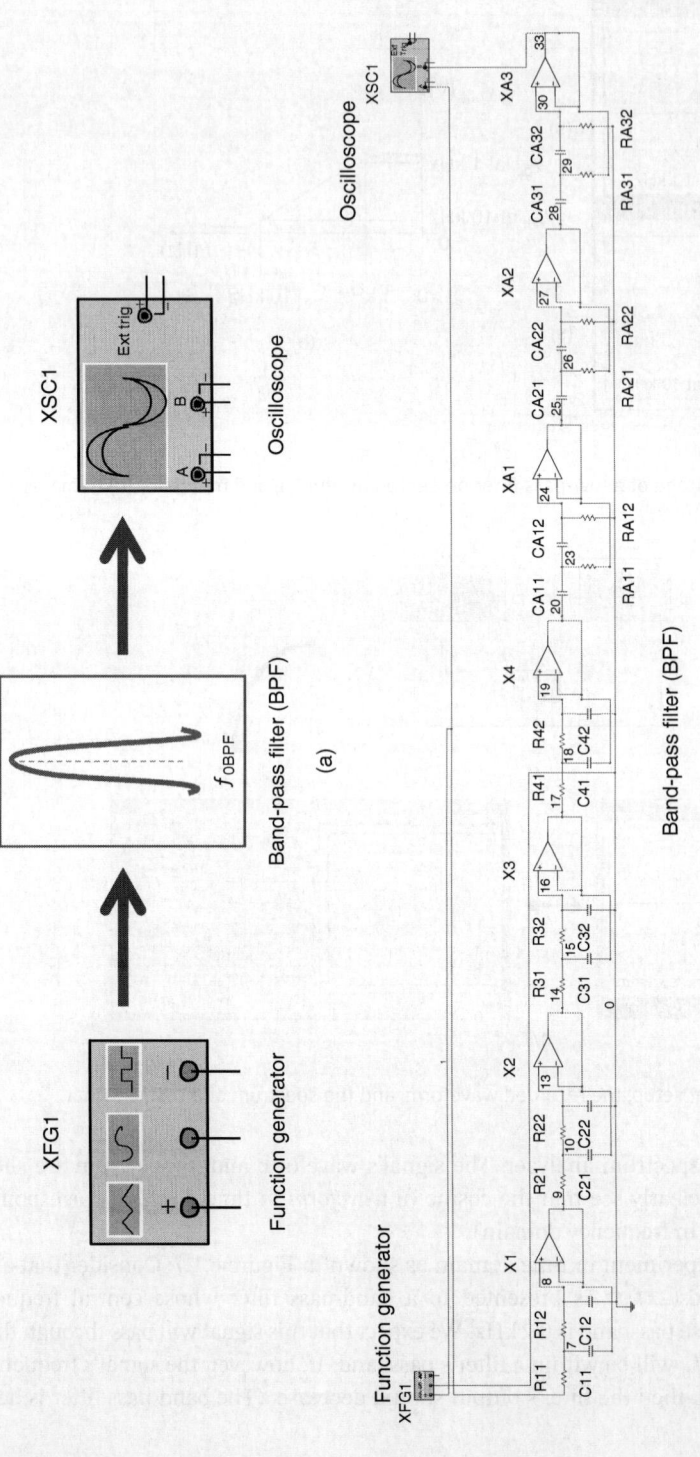

Figure 6.1.7 Experimental verification of the concept of time and frequency domains: (a) Concept of the experiment; (b) the experiment setup (consult Multisim for the standard form of a BPF circuitry); (c) the filter design and the filter's specifications; (d) measured waveforms; and (e) amplitudes of the input and output (filtered) signals.

Specifications of the BPF's components in Figure 6.1.7b

Resistances (Ω)		Capacitors (μF)		Op-amps	
R11	40	C11	11.83	X1	OPAMP_3T_VIRTUAL
R12	40	C12	8.403	X2	
R21	40	C21	13.95	X3	
R22	40	C22	2.492	X4	
R31	40	C31	20.88	XA1	
R32	40	C32	867.4 nF	XA2	
R41	40	C41	59.47	XA3	
R42	40	C42	227.6 nF		
RA11	69.27 mΩ	CA11	1 mF		
RA12	99.57 mΩ	CA12	1 mF		
RA21	50.71 mΩ	CA21	1 mF		
RA22	359.6 mΩ	CA22	1 mF		
RA31	18.55 mΩ	CA31	1 mF		

(b)

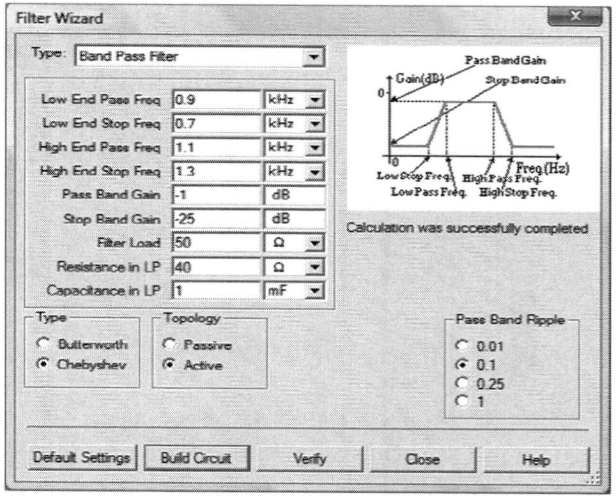

(c)

Figure 6.1.7 (Continued)

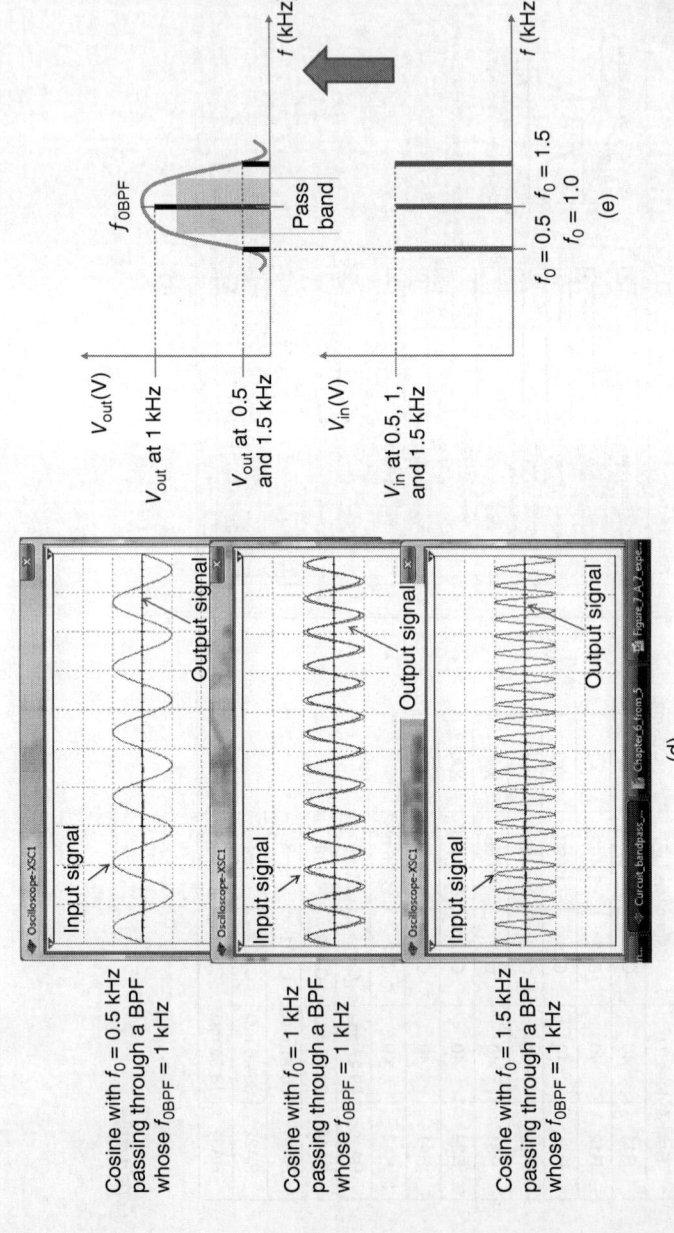

Figure 6.1.7 (*Continued*)

Figure 6.1.7a illustrates the concept of the experiment; Figure 6.1.7b,c shows accordingly the experiment setup, the filter design, and the filter's specifications; the waveforms of the input and output signals are shown in Figure 6.1.7d; the amplitudes of the input and output (filtered) signals are depicted in Figure 6.1.7e. We consider three signals whose frequencies are 0.5, 1, and 1.5 kHz. The output amplitudes of two cosine signals whose $f_0 = 0.5$ kHz and $f_0 = 1.5$ kHz are very small because their frequencies are outside the filter's passband. When, however, the cosine's frequency coincides with the BPF's central frequency, that is $f_0 = f_{0BPF}$, the filter's output is almost equal to the input because the signal's frequency is within the passband. Therefore, we can correctly interpret the results of this time-domain experiment only by resorting to the frequency-domain analysis, as Figure 6.1.7e shows.

6.1.1.3 Signal Spectrum

We need to realize that all periodic signals can be divided into two main classes: sinusoidal (harmonic) signals and nonsinusoidal signals. Sinusoidal signals, as we mentioned in Section 2.1, are either cosine or sine signals. All other periodic signals are nonsinusoidal. Square-wave and sawtooth signals, shown in Figure 6.1.1b,c, are examples of periodic but nonsinusoidal signals.

The concept of *spectrum* was introduced in Section 2.1, where we say that

> the spectrum of a signal shows what frequencies the given signal contains.

Thus, the spectrum of a single cosine is a single line in frequency domain. Figures 6.1.3, 6.1.4, and 6.1.6 show various cosine and sine signals (waveforms in time domain) and their spectra (lines in frequency domain). The important conclusion from this consideration is:

The spectrum of a sinusoidal signal (cosine or sine), $v(t) = A\cos(2\pi f_0 t + \Theta)$ or $v(t) = A\sin(2\pi f_0 t + \Theta)$, can be built as follows: First, we depict a line whose amplitude equals the signal's amplitude, A, and whose frequency equals the signal's frequency, f_0. This presentation is called amplitude spectrum. To complete the spectral description of a sinusoidal signal, a phase shift at the given frequency has to be added. This presentation is called a phase spectrum. The whole spectrum, which includes both amplitude and phase spectra, is called a frequency spectrum. Review Figures 6.1.3, 6.1.4, and 6.1.6 again. Remember that the term spectrum *refers to the signal presentation in frequency domain.*

We are quite clear now about the spectrum of a sinusoidal signal. *But what is the spectrum of a periodic nonsinusoidal signal?* In other words, what lines in frequency domain do we need to draw for a signal like a square wave? We present this question in visual form in Figure 6.1.8.

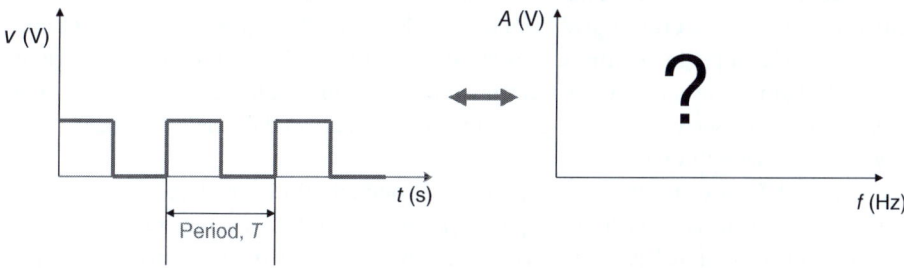

Figure 6.1.8 What is the spectrum of a nonsinusoidal periodic signal?

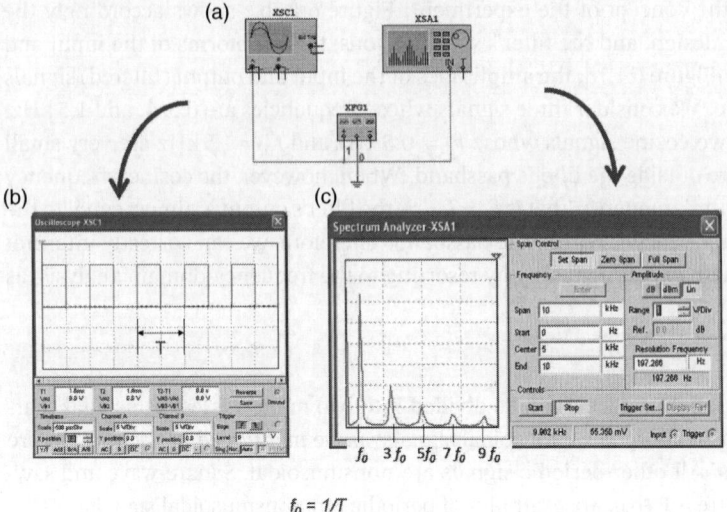

Figure 6.1.9 Experiment setup (a), the waveform (b), and amplitude spectrum (c) of a square wave.

We will answer this question with an experiment similar to that shown in aforementioned Figure 6.1.6. This time we present to both an oscilloscope and a spectrum analyzer a square-wave. The setup and the result are shown in Figure 6.1.9.

The spectrum analyzer shows that a square wave contains many frequencies; in other words, its spectrum is more sophisticated than the simple spectrum of a sinusoidal signal. Measurements show that this square wave contains only odd frequencies – f_0, $3f_0$, $5f_0$, $7f_0$, etc. Remember that f_0 (Hz) is the fundamental frequency equals f_0 (Hz) = $\frac{1}{T(s)}$, where T (s) is the signal period.

Do all nonsinusoidal signals have this spectrum? No! But

the spectrum of every nonsinusoidal signal contains many frequencies in contrast to the spectrum of a sinusoidal signal, which contains just one frequency.

We can confirm this conclusion by an experiment similar to that shown in Figure 6.1.9. Let us take a triangle signal and obtain its waveform and spectrum. The result of this experiment (a simulation with Multisim, actually) is shown in Figure 6.1.10. The experiment setup is shown in the center of Figure 6.1.10; the signal's waveform, copied from the oscilloscope screen, is shown at the left; the spectrum of the triangle, copied from the screen of the spectrum analyzer, is shown at the right, where the top right figure shows the entire spectrum and the bottom right figure shows the same spectrum in larger scale. Observe that a triangle – a periodic nonsinusoidal – signal contains many frequencies with various amplitudes.

From an analysis of Figures 6.1.9 and 6.1.10, we can conclude that *a periodic nonsinusoidal signal contains many frequencies, each with different amplitude*. What specific frequencies will any periodic nonsinusoidal signal contain? The answer is given by a theory called the *Fourier theorem*.

(**Question:** *Figures 6.1.9 and 6.1.10 show the amplitude spectra of periodic nonsinusoidal signals. The whole signal spectrum, however, must include a phase spectrum too. What do you think about the phase spectra of these signals?*)

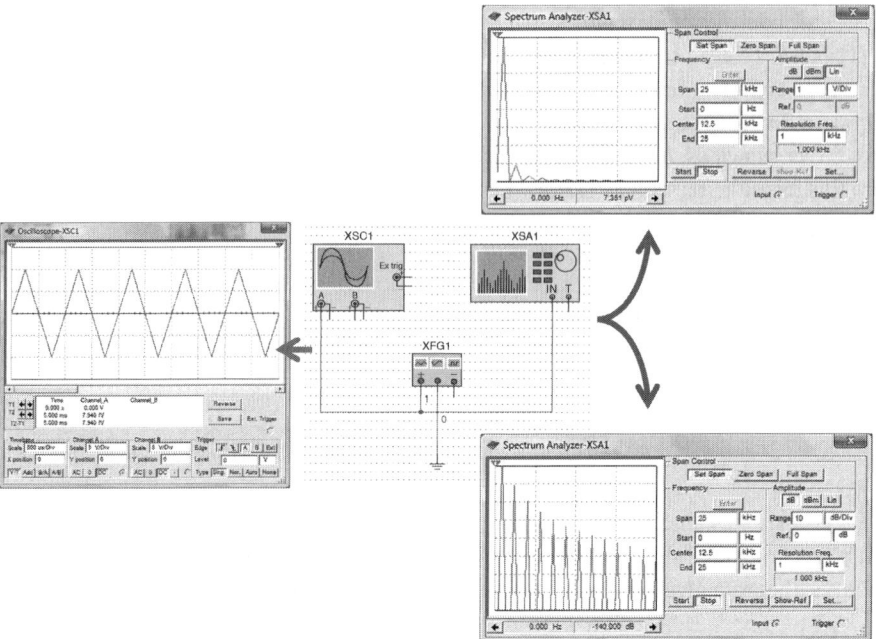

Figure 6.1.10 The experiment setup (center), the waveform of a triangle signal (left), and its spectrum (right). The entire spectrum (top right) and part of the spectrum in larger scale (bottom right).

6.1.2 The Fourier[3] Series

6.1.2.1 The Fourier Theorem

The problem we are trying to solve is this: What is the spectrum of a periodic nonsinusoidal signal? From our previous discussion, we can develop the following approach to solving this problem:

Since for each cosine and sine, we have a unique line in frequency domain, we need to represent a periodic nonsinusoidal signal as a combination of cosines and sines. Then, the set of lines representing individual sinusoidal signals determines the spectrum of the periodic nonsinusoidal signal.

But can we represent any periodic nonsinusoidal signal as a combination of cosines and sines? And if we can, how can we do this? The answer is given by the *Fourier theorem*:

> *Every periodic signal, v(t), can be represented as a sum of cosines and sines.*

How? The *Fourier formula* tells us:

$$v(t) \, (V) = A_0 + a_1 \cos 2\pi f_0 t + a_2 \cos 2\pi \, 2f_0 t + a_3 \cos 2\pi 3 f_0 t$$
$$+ \cdots + b_1 \sin 2\pi f_0 t + b_2 \sin 2\pi \, 2f_0 t + b_3 \sin 2\pi \, 3f_0 t + \cdots \tag{6.1.5a}$$

---

3 Jean Baptiste Joseph Fourier (1768–1830) was born in France into a family of tailors and orphaned at the age of 10. Despite this early hardship, he became a famous scientist and politician. Napoleon himself granted Fourier the title of baron. He studied mathematics in Paris under such famous mathematicians as Lagrange, Laplace, and Monge. In 1797, he became a chaired professor at the Ecole Polytechnique, the best technical university in France. He became involved in politics and served as a prefect of Grenoble; however, his interest in mathematics never waned and he discovered – while studying heat flow – the equation now bearing his name. To solve this equation, he developed an approach based on the expansion of function into the series named after him. This achievement came after many corrections he made to meet the criticisms from Lagrange and others; eventually he published the memoir entitled "The Analytical Theory of Hear" regarded as one of the key contributions to mathematics and engineering.

Here, $v(t)$ is a periodic nonsinusoidal signal with a period, T (s), and f_0 (Hz) = $\frac{1}{T(s)}$ is its fundamental frequency; A_0 (V) is a dc (constant) member, and the coefficients a_n (V) and b_n (V) are amplitudes of cosines and sines, respectively. Formula (6.1.5a) is a series consisting of members A_0, $a_n \cos 2\pi n f_0 t$, and $b_n \sin 2\pi n f_0 t$, where $n = 1,2,3, \ldots$.

Equation (6.1.5a) is known as the Fourier series of a signal, $v(t)$.

We can write this formula in a concise form:

$$v(t) = A_0 + \sum_{n=1}^{\infty} (a_n \cos 2\pi n f_0 t + b_n \sin 2\pi n f_0 t) \tag{6.1.5b}$$

where the summation goes from $n = 1$ to infinity.

Remember, for every cosine and sine signal, we have a unique line in frequency domain; therefore,

> as soon as we expand a periodic nonsinusoidal signal as the Fourier series, we obtain the frequency spectrum of the signal.

To represent a nonsinusoidal signal as a Fourier series, we need to know the coefficients A_0, a_n, and b_n. The Fourier theorem gives us the following *formulas for calculating these coefficients*:

$$A_0 = \frac{1}{T} \int_0^T v(t)\, dt \tag{6.1.6}$$

$$a_n = \frac{2}{T} \int_0^T v(t) \cos(2\pi n f_0 t)\, dt \tag{6.1.7}$$

$$b_n = \frac{2}{T} \int_0^T v(t) \sin(2\pi n f_0 t)\, dt \tag{6.1.8}$$

Since we know the formula, $v(t)$, of the signal and its period, T (and therefore its fundamental frequency, f_0), we can calculate A_0, a_n, and b_n by using (6.1.6), (6.1.7), and (6.1.8). As soon as we calculate these coefficients, we obtain the entire Fourier series of a given signal and therefore its spectrum.

(**Exercise**: Verify that the units on the right-hand side and the left-hand side of *(6.1.6), (6.1.7), and (6.1.8)* coincide.)

Example 6.1.1 illustrates the application of the Fourier theorem.

Example 6.1.1 The Spectrum of a Square-Wave Signal

Problem

Determine the spectrum of a square-wave signal shown in Figure 6.1.11a, whose $A = 10$ V and $T = 1$ ms.

Solution

To determine the signal's spectrum, we need to expand $v(t)$ into the Fourier series. To do so, we have to take the following steps:

Step 1: Define the signal $v(t)$.
Step 2: Calculate A_0.
Step 3: Calculate a_n and b_n.
Step 4: Obtain the Fourier series.
Step 5: Graphically build the signal's spectrum.

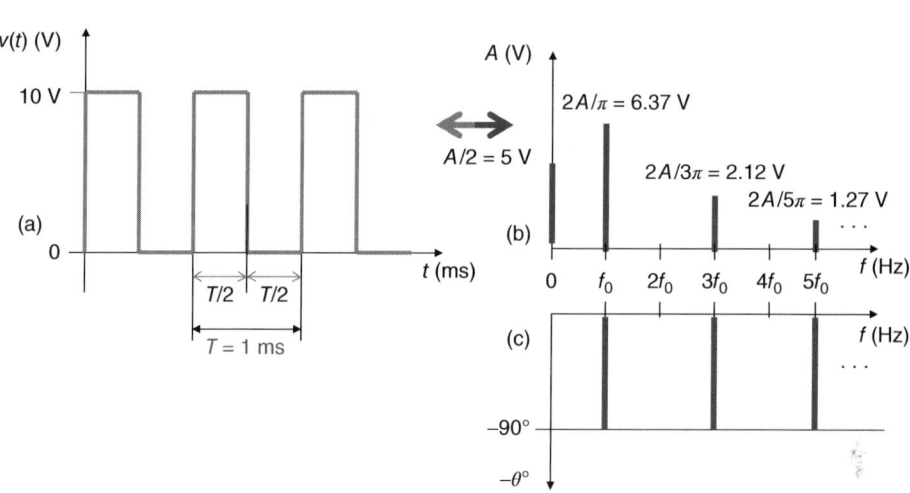

Figure 6.1.11 The square wave and its spectrum: (a) The waveform; (b) the amplitude spectrum; and (c) the phase spectrum.

Let us put this plan into effect:

Step 1: Examining Figure 6.1.11a, it is easy to see that within the time interval $0 \le t \le T/2$, $v(t) = A = 10\,V$; on the other hand, within interval $T/2 < t < T$, $v(t) = 0$. Therefore, the signal's waveform formula is defined as

$$v(t) = \begin{cases} A = 10\,V & \text{for } 0 \le t \le T/2 \\ A = 0\,V & \text{for } T/2 < t < T \end{cases} \tag{6.1.9}$$

Step 2 (see Sidebar 6.1.S.1 for more explanations on this step): Using (6.1.6), we find the dc member:

$$A_0 = \frac{1}{T}\int_0^T v(t)dt = \frac{1}{T}\left(\int_0^{T/2} A\,dt + \int_0^{T/2} 0\,dt\right)$$
$$= \left(\frac{1}{T}\right) A\big|_0^{T/2} = \left(\frac{A}{T}\right)(T/2 - 0) = \frac{A}{2} \tag{6.1.10}$$

Plugging in the given value of A, we obtain

$$A_0 = A/2 = 5\,V$$

Step 3: We use (6.1.7) for calculating a_n:

$$a_n = \frac{2}{T}\int_0^T v(T)v(t)\cos(2\pi nf_0 t)dt$$
$$= \frac{2}{T}\left(\int_0^{T/2} A\cos(2\pi nf_0 t)dt + \int_0^{T/2} 0\cos(2\pi nf_0 t)dt\right)$$
$$= \frac{2A}{T}\left(\frac{1}{2\pi nf_0}\right)\sin(2\pi nf_0 t)\big|_0^{T/2} = \frac{A}{T\pi nf_0}\sin(2\pi nf_0 T/2)$$

Recalling that $T = 1/f_0$ and $n = 1, 2, 3, \ldots$ we find

$$a_n = \frac{A}{\pi n}\sin(\pi n) = 0 \tag{6.1.11}$$

because $sin(\pi n)$ is equal to zero for every n equals 1, 2, 3, etc.

To calculate b_n, we apply (6.1.8):

$$\begin{aligned}
b_n &= \frac{2}{T} \int_0^T v(t) \sin(2\pi n f_0 t)\, dt \\
&= \frac{2}{T} \left(\int_0^{T/2} A \sin(2\pi n f_0 t) dt + \int_{T/2}^T 0 \sin(2\pi n f_0 t) dt \right) \\
&= -\frac{2A}{T} \left(\frac{1}{2\pi n f_0} \right) \cos(2\pi n f_0 t)\big|_0^{T/2} \\
&= -\frac{A}{\pi n}(\cos(2\pi n f_0 T/2) - \cos(0)) \\
&= -\frac{A}{\pi n}(\cos(\pi n) - 1)
\end{aligned}$$

If $n = 2, 4, 6, \ldots$, then $\cos(\pi n) = 1$ and $b_n = 0$; if $n = 1, 3, 5, \ldots$, then $\cos(\pi n) = -1$ and

$$b_n = -\frac{A}{\pi n}(-2) = \frac{2A}{\pi n} \tag{6.1.12}$$

Step 4: After performing these steps, we obtain the Fourier series for a given square wave by plugging the calculated coefficients a_n and b_n into (6.1.5a):

$$v(t) = \frac{A}{2} + \frac{2A}{\pi} \sin 2\pi f_0 t + \frac{2A}{3\pi} \sin 2\pi 3 f_0 t + \frac{2A}{5\pi} \sin 2\pi 5 f_0 t + \cdots \tag{6.1.13}$$

Or, in concise form:

$$v(t) = \frac{A}{2} + \sum_{n,\text{odd only}}^{\infty} \frac{2A}{\pi n} \sin(2\pi n f_0 t) \tag{6.1.14}$$

It is instructive to compare this result with the general formula for the Fourier series. To do so, we write the general equation, (6.1.5a), on top and the specific Fourier series for a square-wave signal (6.1.13), beneath it:

$$v(t) = A_0 + a_1 \cos 2\pi f_0 t + a_2 \cos 2\pi 2 f_0 t + \cdots + b_1 \sin 2\pi f_0 t + b_2 \sin 2\pi 2 f_0 t + b_3 \sin 2\pi 3 f_0 t + \cdots \tag{6.1.5R}$$

$\updownarrow \quad \updownarrow \quad \updownarrow \quad \updownarrow \quad \updownarrow \quad \updownarrow \quad \updownarrow$

$$v(t) = \frac{A}{2} \quad + 0 \quad + 0 \quad + \ldots + \frac{2A}{\pi} \sin 2\pi f_0 t \quad + 0 \quad + \frac{2A}{3\pi} \sin 2\pi 3 f_0 t + \tag{6.1.13R}$$

Thus, the spectrum of this type of square wave includes a constant member and odd sine harmonics only. The amplitudes of these harmonics are equal to $\frac{2A}{\pi n}$.

Step 5: Remembering that every sine and cosine has a unique line in frequency domain, we can now depict the *amplitude spectrum* of the given square-wave signal. This is shown in Figure 6.1.11b. Figure 6.1.11b gives us the answer to the question posed in Figure 6.1.8: Now we know what the amplitude spectrum of this square wave encompasses. Turning to Figure 6.1.4b, we recall that a sine signal has a $-90°$ phase shift, and therefore, the *phase spectrum* of the square wave must include this phase angle. This phase spectrum is shown in Figure 6.1.11c. *Thus, Figure 6.1.11b,c show the spectrum of the square wave whose waveform is presented in Figure 6.1.11a.*

Discussion

- First, we can conceptually compare the calculated and measured square-wave signal's spectra shown in Figures 6.1.9 and 6.1.11b. Now, we can readily understand why the measured spectrum includes many frequency lines.

- Second, we have to realize that the *dc member of the Fourier series, A_0, presents the average value of the signal over a period*. In our example, we clearly observe that $A_0 = \frac{A}{2}$, so we can come to this conclusion even without calculations.
- Third, we certainly understand that the Fourier series allows us to operate with periodic signals only. But what if we need to deal with a nonperiodic signal? The answer is that in such a case we would need to turn to the *Fourier transform*, not the *Fourier series*. A detailed discussion of this concept is given in Chapter 7.
- Interestingly, the example of a square wave has been used for many years to introduce the concept of a Fourier series; this example can be found in any textbook on communications, electronics, or signals and systems. We probably owe the popularity of this example to the simplicity in calculating the Fourier coefficients. In any event, we follow this academic tradition.

Sidebar 6.1.S.1 Calculating the Coefficients of a Fourier Series

For those of you who haven't taken the proper course in calculus and are not familiar with integration, here are some explanations on calculating the coefficients of a Fourier series:

Integration is a limit of summation when the independent variable goes to zero; that is,

$$\int_0^T v(t)dt = \lim_{t \to 0} \sum_{k=0}^{k=T} v_k(t)\Delta t \tag{6.1.S.1}$$

This definition has a clear interpretation:

A definite integral calculates the area under the curve described by the function v(t).

Consider Figure 6.1.S.1.1. The left-hand side shows that the area under the curve $v(t)$ can be *approximately* computed as a sum of the discrete rectangles with height, $v_k(t)$, and width, Δt. Now, imagine that this width, Δt, becomes smaller and smaller; as a result, these discrete rectangles will fill the area under the curve more and more accurately, and in the limit – when Δt goes to zero – we will measure the area *precisely*. This situation is shown on the right-hand side of Figure 6.1.S.1.1.

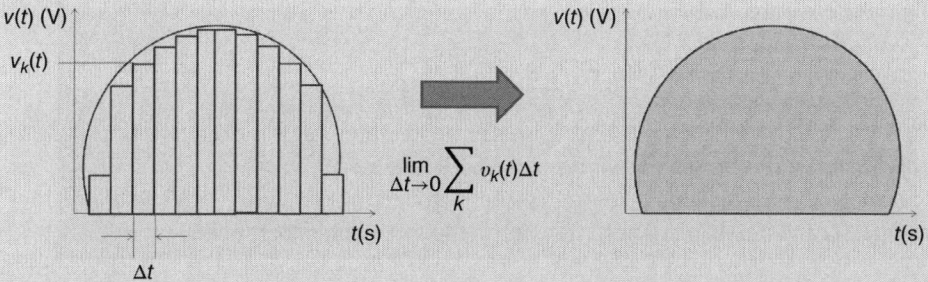

Figure 6.1.S.1.1 The concept of transition from summation to integration.

Now we can apply this knowledge to integration of a constant member. We refer specifically to 6.1.1. In this example, $v(t) = A$ for t between 0 and $T/2$ and $v(t) = 0$ for the second half of the signal's period. Therefore, the integral $\int_0^{T/2} A\,dt$ gives the area under the pulse, as shown in Figure 6.1.S.1.2. Performing the formal operations, we find

$$\int_0^{T/2} A\,dt = A \int_0^{T/2} dt = At\big|_0^{T/2} = \frac{AT}{2} - 0 = \frac{AT}{2} \tag{6.1.S.2}$$

(Continued)

Sidebar 6.1.S.1 (Continued)

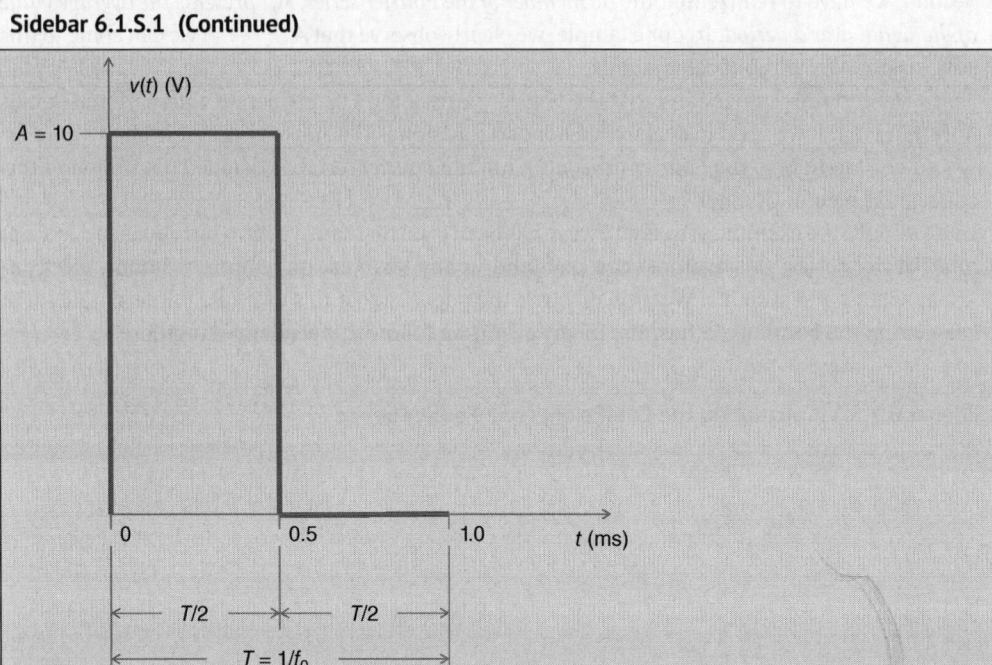

Figure 6.1.S.1.2 Area under the pulse of a square-wave signal.

On the other hand, the simple geometrical consideration (the area, S, of a rectangle is equal to the product of the width, W, and the height, H; that is, $S = W \times H$) gives $S_{rectangle} = \frac{T}{2} \times A = \frac{AT}{2}$, which confirms our calculations.

Let us now turn to the calculations of the Fourier series coefficients. Starting with a constant member, we need to recall that the Fourier formula determines A_0 as $1/T \int_0^{T/2} A\, dt$. Therefore, we find

$$A_0 = 1/T \int_0^{T/2} A\, dt = \frac{1}{T}\left(A \times \frac{T}{2}\right) = \frac{A}{2}$$

as *in* (6.1.10). Referring to Figure 6.1.S.1.2, we can clearly interpret this result: *A_0 is the average value of the square-wave amplitude, A, over the whole period, T.*

To calculate the coefficients a_n of the cosine members of the square-wave Fourier series (refer to (6.1.7) and (6.1.11)), we need to perform the following integration:

$$a_n = 2A/T \int_0^{T/2} \cos(2\pi n f_0 t)\, dt$$

For the first three coefficients, we have

$$\left.\begin{aligned} a_1 &= 2A/T \int_0^{T/2} \cos(2\pi f_0 t)\, dt \\ a_2 &= 2A/T \int_0^{T/2} \cos(2\pi\, 2f_0 t)\, dt \\ a_3 &= 2A/T \int_0^{T/2} \cos(2\pi\, 3f_0 t)\, dt \end{aligned}\right\} \quad (6.1.S.3)$$

The first three *harmonics* – $\cos(2\pi f_0 t)$, $\cos(2\pi 2 f_0 t)$, and $\cos(2\pi 3 f_0 t)$ – are shown in Figure 6.1.S.1.3. The pulse of the square wave is shown to remind that the integration is performed over the interval $T/2 \leq t \leq 0$. The integrals – $\int_0^{T/2} \cos(2\pi f_0 t) dt$, $\int_0^{T/2} \cos(2\pi 2 f_0 t) dt$, and $\int_0^{T/2} \cos(2\pi 3 f_0 t) dt$ – compute the areas under the proper curves. We can see in Figure 6.1.S.1.3 that the sum of the positive and negative areas in all three cases is equal to zero. This is why all a_n coefficients are equal to zero, as obtained in (6.1.11).

For calculating the sine coefficients, we need to perform the integration described in (6.1.8) and (6.1.12):

$$b_n = 2/T \int_0^{T/2} v(t) \sin(2\pi n f_0 t) dt$$

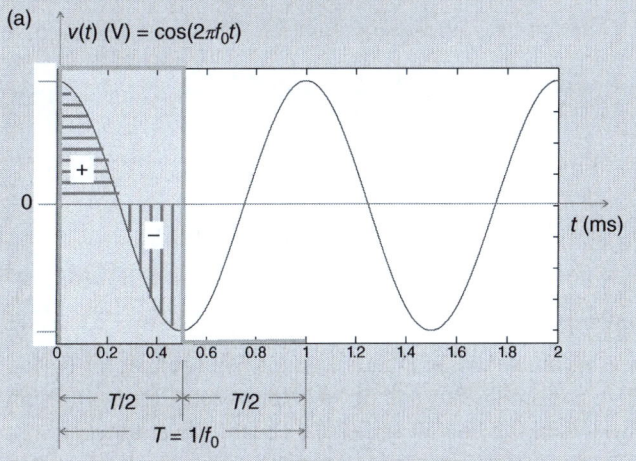

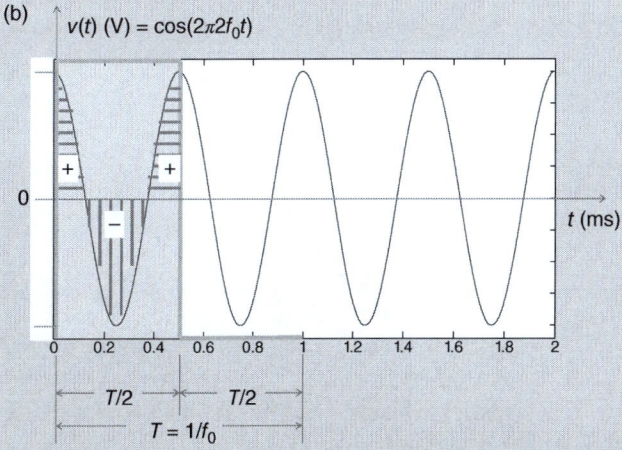

Figure 6.1.S.1.3 Calculating the cosine coefficients of the Fourier series of a square-wave signal: (a) First harmonic, b) second harmonic, and c) third harmonic. In all three harmonics, the sum of the positive and negative areas is equal to zero.

(Continued)

Sidebar 6.1.S.1 (Continued)

(c)

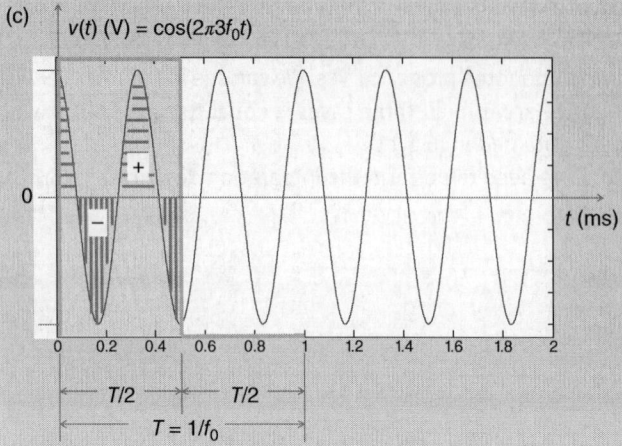

Figure 6.1.S.1.3 (Continued)

For the first three coefficients, we need to do the following integrations:

$$\left.\begin{array}{l} b_1 = 2/T \int_0^{T/2} v(t) \sin(2\pi f_0 t) dt \\ b_2 = 2/T \int_0^{T/2} v(t) \sin(2\pi\, 2f_0 t) dt \\ b_3 = 2/T \int_0^{T/2} v(t) \sin(2\pi\, 3f_0 t) dt \end{array}\right\} \quad (6.1.S.4)$$

The areas calculated by these integrals are shown in Figure 6.1.S.1.4. We see that for b_1 and b_3, the sum of positive and negative areas does not add up to zero, whereas for b_2 the sum does. This is why the odd b_n coefficients of the Fourier series of a square-wave signal are not equal to zero, but the even coefficients are. This consideration explains the result presented in (6.1.12).

(a)

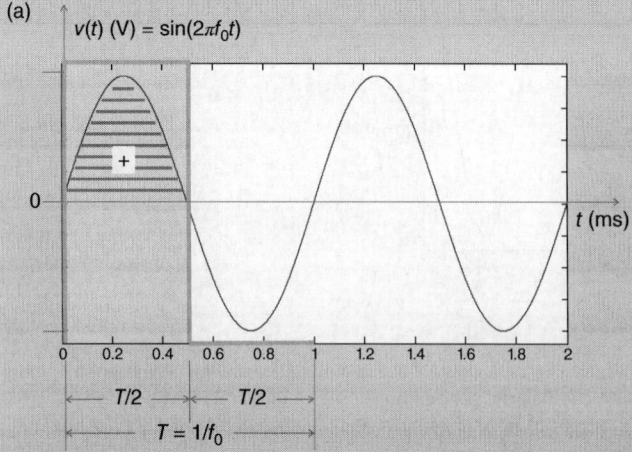

Figure 6.1.S.1.4 Calculating the sine coefficients of the Fourier series of a square-wave signal: (a) First harmonic, (b) second harmonic, and (c) third harmonic. In the first and third harmonics, the sum of the positive and negative areas is not equal to zero; in the second harmonic, this sum is equal to zero.

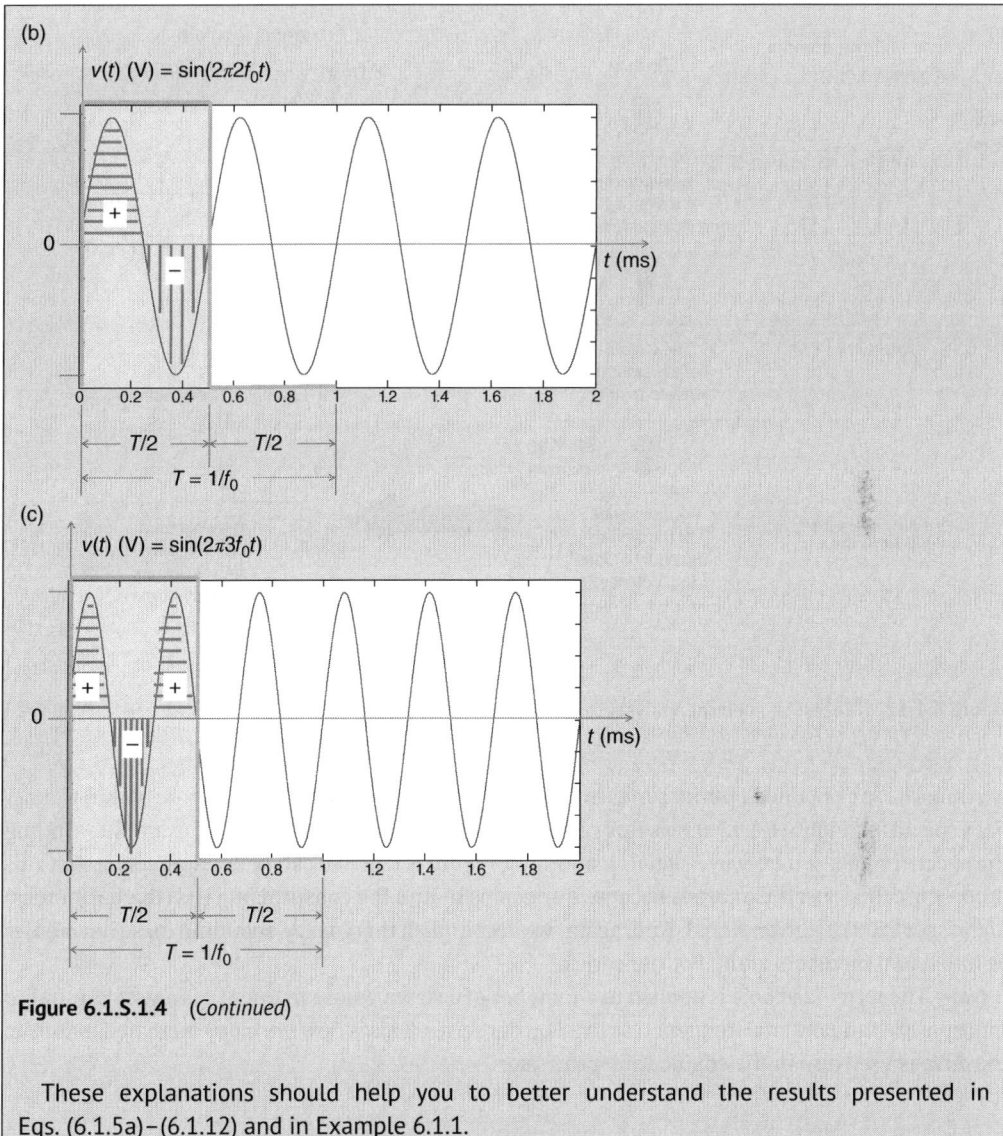

Figure 6.1.S.1.4 (Continued)

These explanations should help you to better understand the results presented in Eqs. (6.1.5a)–(6.1.12) and in Example 6.1.1.

6.1.2.2 Spectral Analysis – From the Whole to the Parts

The Fourier theorem and the Fourier series allow us to find out of what spectral components (harmonics, that is) a given signal consists. In other words, the Fourier series enables the spectral analysis of a given signal. We call this investigation an *analysis* because we are determining the components of a whole entity – the signal; we call it *spectral analysis* because we are finding the spectral components. Figure 6.1.12 presents an overview of the process of spectral analysis. In time domain, we start with a signal whose waveform, amplitude, and period are given. By analyzing the given signal, we develop the explicit formula of the signal's waveform, $v(t)$. Based on this expression, we calculate the coefficients of the specific Fourier series, which enable us to present $v(t)$ as a sum of specific cosines and sines. Since each cosine and sine has a unique line in frequency domain,

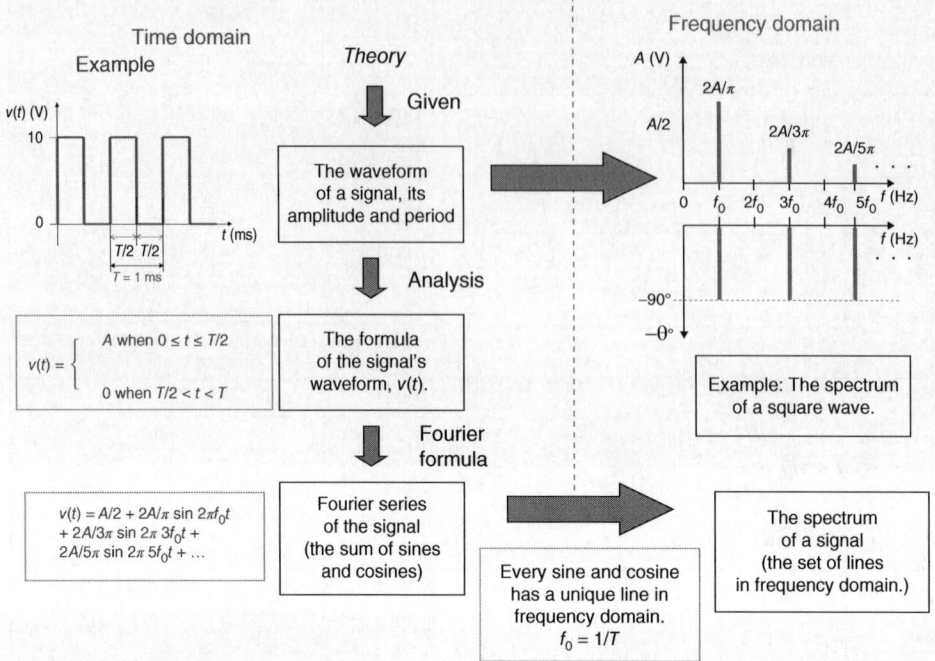

Figure 6.1.12 Process of spectral analysis.

we translate the obtained Fourier series into the lines in the A–f space and therefore obtain the signal's spectrum. Figure 6.1.12 shows this general process in conjunction with our example – finding the spectrum of a square-wave signal. Examining Figure 6.1.12, we can better appreciate that this process is called spectral analysis because our goal is to find the constituent parts (the harmonics) of the spectrum of a given signal. And, again, we accomplish this goal by obtaining these harmonics as individual members of the Fourier series.

Note: The term *harmonic* is applied to a member of a series whose frequency is a whole-number multiple of a fundamental frequency, as in a Fourier series. This is how *harmony* is created in music; the term is used also in the engineering profession.

6.1.3 Spectral Synthesis

6.1.3.1 Spectral Synthesis – From Parts to the Whole

Using spectral analysis, we solved the following problem: Given a signal's waveform, determine the signal's spectrum; that is, find the frequency components of the signal. But then again, we might encounter the opposite problem: Given the spectrum (frequency components) of a signal, build the signal's waveform. The process of finding the solution to this problem is called *spectral synthesis*. We visualize the spectral-synthesis problem in Figure 6.1.13.

To solve the synthesis problem, we need to remember the key point of spectral analysis: *Every line in frequency domain represents a harmonic signal.* For example, suppose we are given a spectral line with amplitude of 3 V and frequency of 60 Hz. It means that we are given a cosine signal, $3\cos(2\pi\, 60t)$. Figure 6.1.14, which is actually a partial reproduction of Figure 6.1.3, illustrates this explanation with specific values for the amplitude and the frequency.

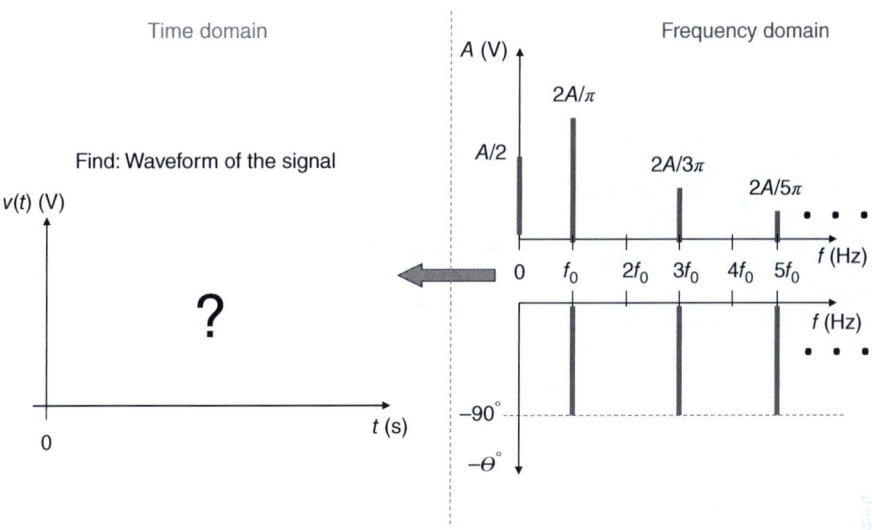

Figure 6.1.13 The waveform of a signal whose spectrum is given has to be found.

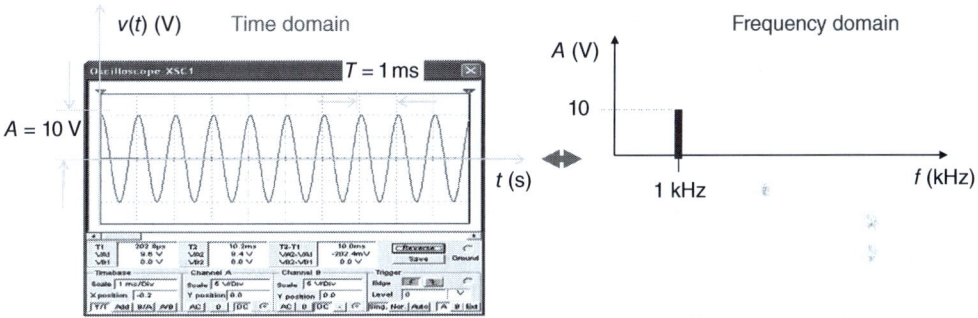

Figure 6.1.14 A single line in frequency domain having an amplitude, A, and a frequency, f_0, that correspond to a cosine signal in time domain with the same amplitude and the appropriate period, $T = 1/f_0$.

To complete the synthesis, we need to turn to the Fourier series,

$$v(t) = A_0 + a_1 \cos 2\pi f_0 t + a_2 \cos 2\pi 2f_0 t + a_3 \cos 2\pi 3f_0 t + \cdots + b_1 \sin 2\pi f_0 t$$
$$+ b_2 \sin 2\pi 2f_0 t + b_3 \sin 2\pi 3f_0 t + \cdots \quad (6.1.5aR)$$

This formula simply states that in time domain, the waveform, $v(t)$, is a sum of the cosine and the sine. Since each cosine and sine is a line in frequency domain, we arrive at the following conclusion:

> To build the waveform of a signal from its spectrum, we need to build the waveform of every spectral component (harmonic) and then add them up, as the Fourier series instructs.

An *example* of restoring the waveform of a square wave from its spectrum is shown in Figure 6.1.15. In Figure 6.1.15a, we take a dc member, A_0, in frequency domain and build the average value of the signal in time domain. Thus, from the entire Fourier series of a square-wave signal,

$$v(t) = \frac{A}{2} + \frac{2A}{\pi} \sin 2\pi f_0 t + \frac{2A}{3\pi} \sin 2\pi 3 f_0 t + \frac{2A}{5\pi} \sin 2\pi 5 f_0 t + \cdots \quad (6.1.13R)$$

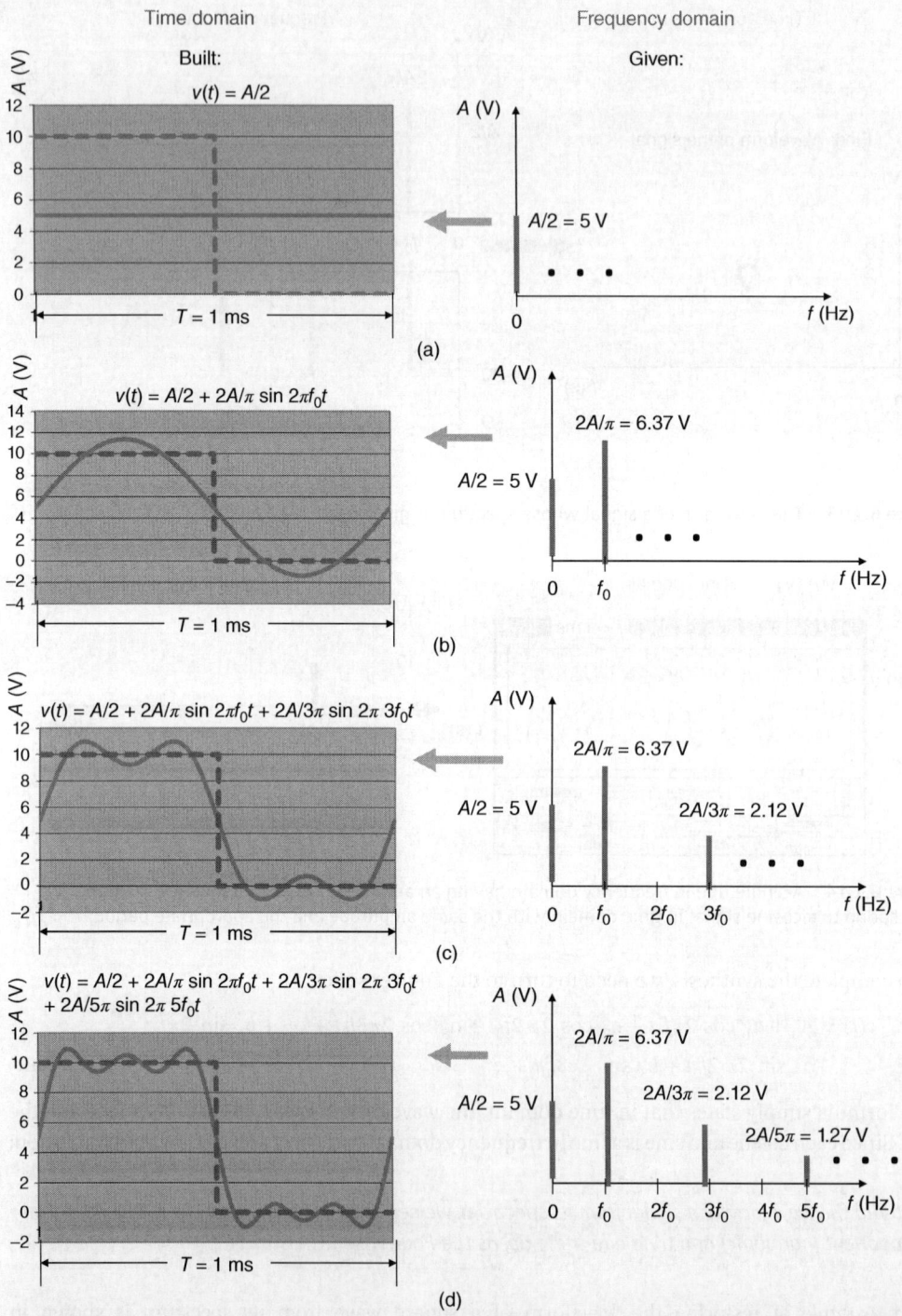

Figure 6.1.15 Synthesis of a square wave from its spectral components: (a) dc member; (b) sum of the dc member and the first harmonic; (c) sum of the first three members – dc member and first two harmonics; (d) sum of the first four members; and (e) sum of the first five members – dc member and the first, third, fifth, and seventh spectral components. The phase shift equals −90° for every spectral component is not shown.

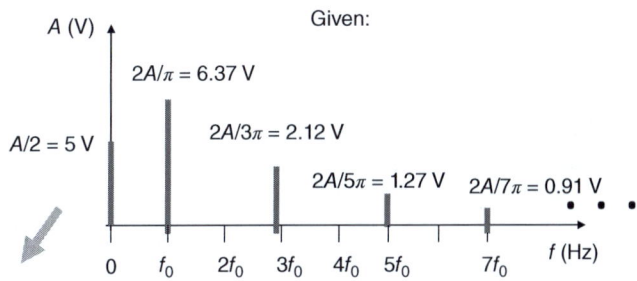

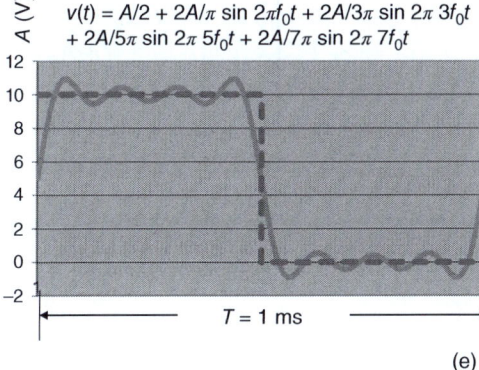

Figure 6.1.15 *(Continued)*

we take the first member in frequency domain and build it in time domain:

$$v(t) = \frac{A}{2} + \frac{2A}{\pi} \sin 2\pi f_0 t + \frac{2A}{3\pi} \sin 2\pi 3 f_0 t + \frac{2A}{5\pi} \sin 2\pi 5 f_0 t + \cdots$$

Note that the shape of this waveform has nothing in common with the square-wave waveform. In Figure 6.1.15b, we take the first two members of this Fourier series,

$$v(t) = \frac{A}{2} + \frac{2A}{\pi} \sin 2\pi f_0 t + \frac{2A}{3\pi} \sin 2\pi 3 f_0 t + \frac{2A}{5\pi} \sin 2\pi \, 5 f_0 t + \cdots$$

and add them up. This waveform does not resemble the square wave either, but at least it demonstrates the periodicity of the signal. Then, we add the third member of this Fourier series,

$$v(t) = \frac{A}{2} + \frac{2A}{\pi} \sin 2\pi f_0 t + \frac{2A}{3\pi} \sin 2\pi \, 3 f_0 t + \frac{2A}{5\pi} \sin 2\pi \, 5 f_0 t + \cdots$$

and, looking at the result shown in Figure 6.1.15c, we can readily visualize the shape of the waveform we are trying to reconstruct. We can also determine the period and amplitude of the signal in question. Figures 6.1.15d and 6.1.16e show the result of the summation of the first five and seven harmonics. Examining Figure 6.1.15, we can readily conclude that

the greater the number of harmonics (members of the Fourier series) we are summing up, the closer the reconstructed waveform to the original waveform.

At this point, we ask ourselves how many harmonics will we need to add to obtain the ideal (original) waveform. Analyzing Figure 6.1.15, we can promptly answer that this number goes to infinity. In fact, (6.1.5b) shows that summation in a Fourier series goes from $n = 1$ to infinity.

We can summarize the process of spectral synthesis in graphical format, as shown in Figure 6.1.16. The key point here is, again, that every line in frequency domain represents a cosine or sine with the proper amplitude and period in time domain. We then need to add all

6 Spectral Analysis 1 – The Fourier Series in Modern Communications

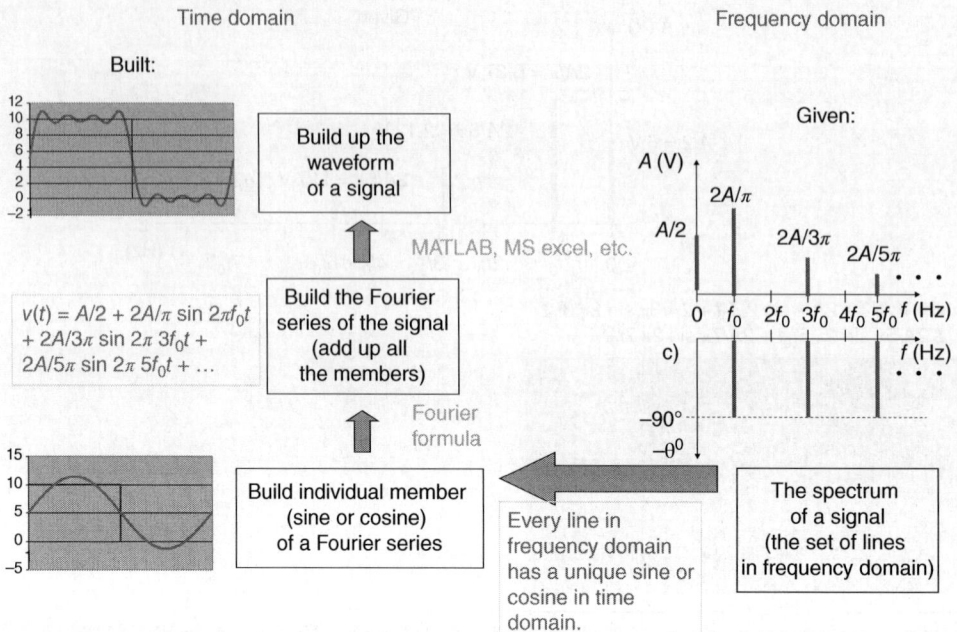

Figure 6.1.16 Process of spectral synthesis.

these individual waveforms to obtain the waveform of the signal in question. And, of course, we must not forget about the phase: In our example with a square wave, we do not show the phase explicitly. We acknowledge, however, existence of $\theta = -90^0$ because we know that every line in frequency domain corresponds to a sine signal in time domain for a square-wave signal.

Using the example with a square wave discussed previously, we can discern the relationship between spectral analysis and synthesis shown in Figure 6.1.17. We can see that

> *it is the Fourier series that bridges the time and frequency domains.*

Finally, Figure 6.1.18 presents the processes of spectral analysis and synthesis and their relationship in concise graphical form by using the example of a square wave. Here, we see unequivocally

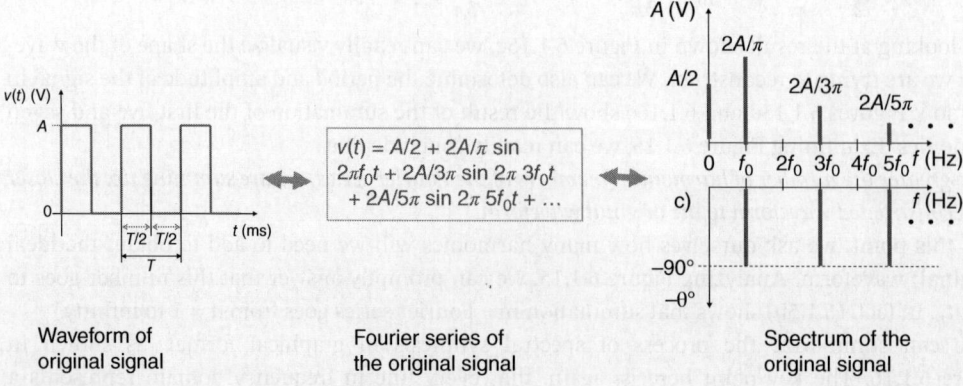

Figure 6.1.17 Spectral analysis and synthesis.

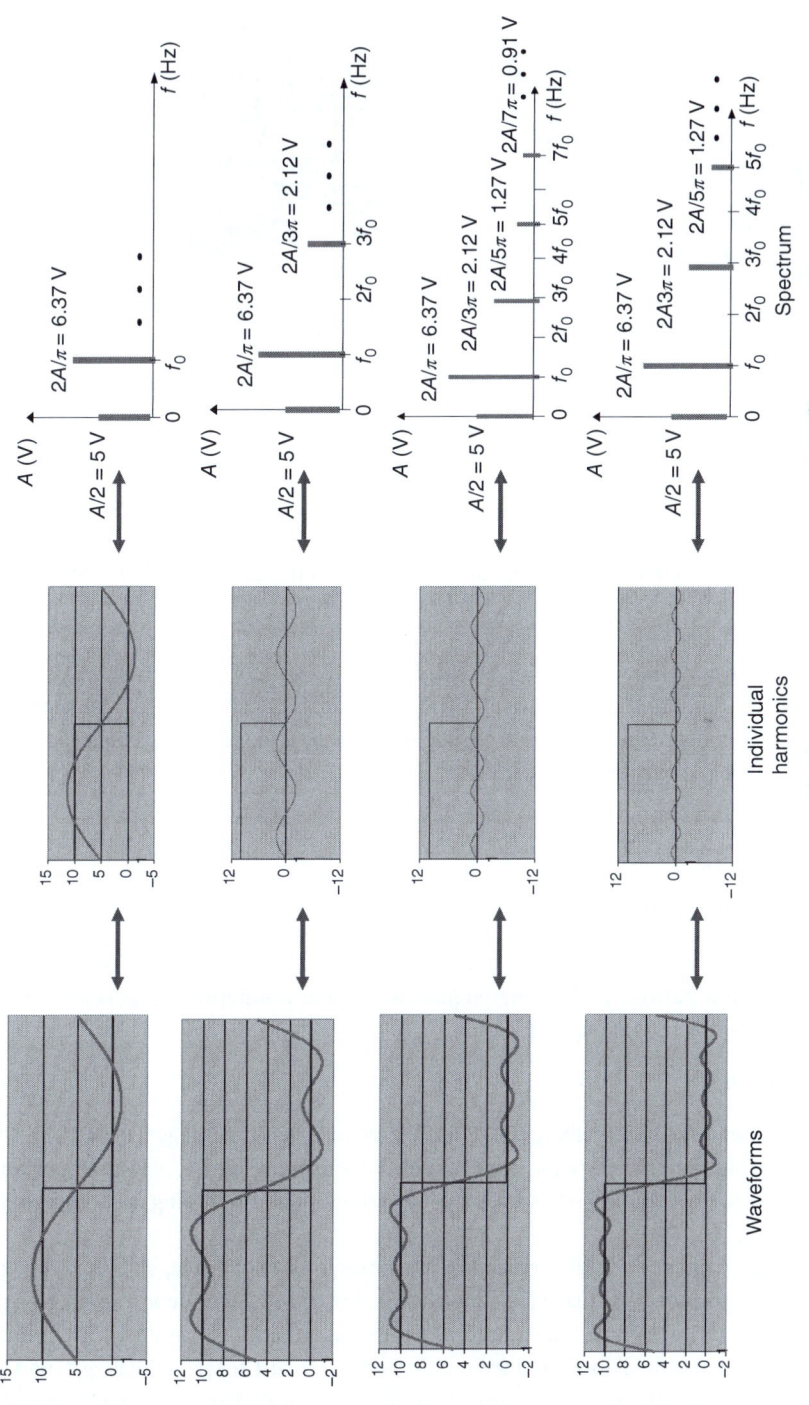

Figure 6.1.18 The role of individual harmonics and their summation in the processes of spectral analysis and synthesis. The phase shift, which equals −90° for every spectral component, is not shown.

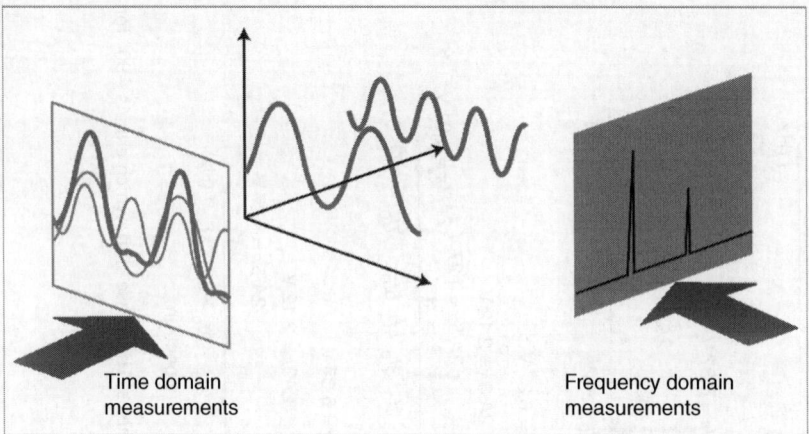

Figure 6.1.19 Presentations of two harmonics in time domain and frequency domain. Source: Reproduced with Permission of © Keysight Technologies, Inc., Courtesy of Keysight Technologies.

the roles played by the individual harmonics and their summation in our quest for a signal's waveform.

Let us consider the two harmonics of a signal and observe, simultaneously, their waveforms and their spectral lines. Such an approach, depicted in Figure 6.1.19, enables us to better understand the concept of spectral analysis and synthesis.

Figure 6.1.19 shows that the waveform in time domain is the sum of two constituent waveforms and that in frequency domain there are two spectral lines, one for each of the two harmonics.

Understanding the concept and technique of spectral analysis and synthesis is an important part of the education of every electrical engineer.

Summary of Section 6.1

Time Domain and Frequency Domain

- Periodic and nonperiodic signals: A signal is called periodic if it repeats (copies) itself after a given time interval called a period, T. All other signals are called nonperiodic (aperiodic). Periodic signals are described by the following formula:

$$v(t) = v(t + nT) \qquad (6.1.1R)$$

where $v(t)$ (V) is the formula describing the signal's waveform, n is the integer ($n = 1, 2, 3, \ldots$), and T (s) is the period of the signal. Periodic signals, such as sinusoidal signals, play an important role in communications. Nevertheless, the vast majority of real signals, including random signals, are nonperiodic.
- *Time domain and frequency domain revisited*: Time domain is a v–t space, where $v(t)$ (V) designates a signal's magnitude, whereas t (s) denotes time in seconds. Frequency domain is an A–f space, where the A (V) designates the amplitude of the signal and f (Hz) shows the frequency. The same cosine signal is described in time domain as $v(t) = A\cos(2\pi f_1 t) = A\cos(\omega_1 t)$ and is presented in frequency domain by the line with an amplitude A located at frequency f_1 (or ω_1). This statement is true only for a cosine signal. For a sine signal, we have to include the A–f and Θ–f axes to show the phase shift of $\theta = -90°$ (refer to Figures 6.1.3 and 6.1.4). This correspondence

between time domain and frequency domain holds true only for cosine and sine signals; its reality is confirmed experimentally.
- *Signal spectrum*: Presentation of a signal in frequency domain is called a signal's spectrum because it shows what frequencies the given signal contains. The spectrum of a sinusoidal signal (cosine or sine),

$$v(t) = A \cos(2\pi f_0 t + \Theta) \quad \text{or} \quad v(t) = A \sin(2\pi f_0 t + \Theta)$$

consists of a single frequency at f_0 (Hz), which is determined by the signal's period as $T(s) = 1/f_0$ (Hz). We depict this spectrum as a line with amplitude A in A–f space. This presentation is called *amplitude spectrum*. A signal's phase can be shown in Θ–f space as a line with the proper length, which is called a *phase spectrum*. Review Figures 6.1.3, 6.1.4, and 6.1.7.

The main question is, what is the spectrum of a periodic nonsinusoidal signal? The experiments show that a periodic nonsinusoidal signal contains many frequencies, each with different amplitude. To understand what specific frequencies any periodic nonsinusoidal signal will contain, we need to turn to the *Fourier series*.

Fourier Series
- *The Fourier theorem*: To find the spectrum of a periodic nonsinusoidal signal, we take the following approach: Since for each cosine and sine we can find a single line in frequency domain, we need to represent a periodic nonsinusoidal signal as a combination of cosines and sines. Then, the set of lines representing individual sinusoidal signals constitutes the spectrum of the signal. To implement this plan, we need to use the Fourier theorem, which states: Every periodic signal, $v(t)$, can be represented as a sum of cosines and sines as follows

$$v(t) \text{ (V)} = A_0 + a_1 \cos 2\pi f_0 t + a_2 \cos 2\pi 2 f_0 t + a_3 \cos 2\pi 3 f_0 t + \cdots + b_1 \sin 2\pi f_0 t$$

$$+ b_2 \sin 2\pi 2 f_0 t + b_3 \sin 2\pi 3 f_0 t + \cdots = A_0 + \sum_{n=1}^{\infty} (a_n \cos 2\pi n f_0 t + b_n \sin 2\pi n f_0 t)$$

(6.1.5R)

Here, $v(t)$ (V) is a periodic signal with a period, T (s), and f_0 (Hz) $= 1/T$ (s) is its fundamental frequency.

To repeat, as soon as we expand a periodic nonsinusoidal signal as the Fourier series, we obtain the spectrum of the signal. The coefficients A_0, a_n, and b_n of the Fourier series can be calculated by using (6.1.6)–(6.1.8).
- *Spectral analysis – from the whole to the parts*: The Fourier series enables the spectral analysis of a periodic nonsinusoidal signal, that is, finding all spectral components of the signal (refer to Figure 6.1.12). In spectral analysis, we find the constituent parts (the harmonics) of the spectrum of a given signal and we accomplish this by obtaining these harmonics, that is, individual members of the Fourier series.

Spectral Synthesis
- *Spectral synthesis – from parts to the whole*: In spectral synthesis, we need to solve the problem opposite to that in spectral analysis, namely: Given the spectrum (all frequency components) of a signal, build the signal's waveform. See Figure 6.1.13. To build the waveform of a whole signal from its spectrum, we need to build the waveform of every spectral component and then add them up, as the Fourier series instructs (refer to Figures 6.1.14–6.1.19).

Questions and Problems for Section 6.1

- Questions marked with an asterisk require a systematic approach to finding the solution.
- Many questions and problems, including those marked with an asterisk, imply that you, in addition to reading the textbook, will do your research to find the answers. Consider such questions as mini-projects.

Time Domain and Frequency Domain

1. Give definitions and examples of periodic and non-periodic signals.

2. Consider the two signals given in Figure 6.1.P2. Show which signal is periodic and which is nonperiodic and explain your reasoning.

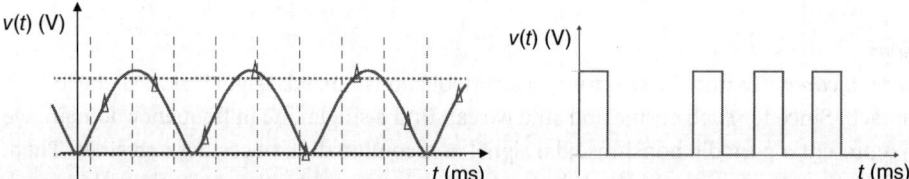

Figure 6.1.P2 Periodicity of the signals.

3. A majority of what type of signal – periodic or nonperiodic – exists in practice? Explain.

4. Define time domain and frequency domain.

5. If the waveform of a signal is given, to what domain – time domain or frequency domain – does this graph belong? Explain.

6. Consider the following signals: $v_1(t) = 12\cos(2\pi 60t)$, $v_2(t) = 6\cos(2\pi 120t)$, and $v_3(t) = 3\cos(2\pi 180t)$. Show these signals in time domain and frequency domain.

7. Since the same sinusoidal signal can be presented in time domain and frequency domain, there is a fundamental formula that relates one domain to the other. What is this formula? Give the equation and explain its meaning.

8. *Consider sinusoidal signal $v(t) = A\cos(2\pi f_0 t)$ with given amplitude A (V) and frequency f_0 (Hz). Can you show both of these parameters on a signal waveform? Explain.

9. *What parameter of a sinusoidal signal is shown in both time domain and frequency domain without alteration? Explain by using the figures in both domains.

10. *Depict the following signals in frequency domain:
    a) $v(t) = 8\cos(2\pi 60t)$.
    b) $v(t) = 8\cos(378t)$.

c) $v(t) = 8\cos(378t + 60°)$.
d) $v(t) = 8\sin(378t)$.

11 *Does the presentation of the same cosine signal in time domain and frequency domain represent a real phenomenon or is it an imaginary mental process developed for the convenience of discussion? Prove your answer.

12 *Can we measure signal parameters in time domain and frequency domain? If the answer is yes, what measuring instruments do we need to use in each domain? Explain.

13 *The presentations of the same signals and processes in time domain and frequency domain were introduced in the discussion of filter operations. Give examples of these presentations.

14 Define the spectrum of a signal.

15 Figure 6.1.P15 shows three periodic signals. Can you sketch the spectrum of each of these signals? Comment on your answers.

Figure 6.1.P15 Three periodic signals whose spectra are to be found.

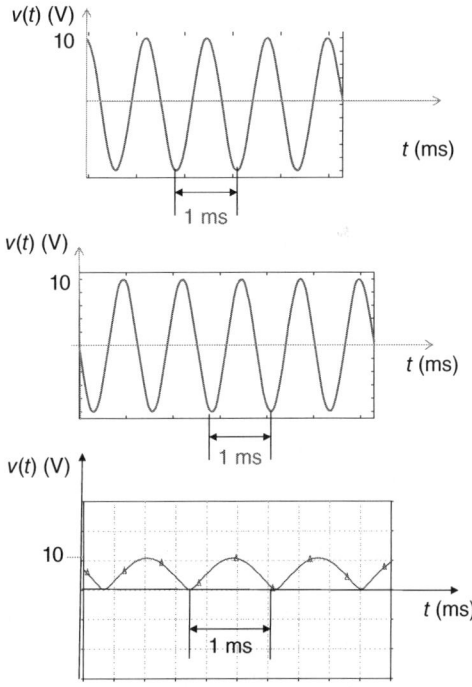

16 We distinguish between the amplitude spectrum and the phase spectrum of the same signal. Why? Explain and give an example.

17 The spectrum of a periodic nonsinusoidal signal consists of many frequencies, but only one of them is called *fundamental*. What is this frequency?

18 Refer to the signals shown in Figure 6.1.P15c. What are their fundamental frequencies?

The Fourier Series

19 The Fourier theorem states that *"every periodic signal, v(t), can be represented as a sum of cosines and sines."* Why do we need to represent a signal as a sum of cosines and sines? What can we achieve with this representation?

20 The Fourier series presents a given nonsinusoidal periodic signal as the following sum of cosines and sines: $v(t) = A_0 + a_1 \cos 2\pi f_0 t + a_2 \cos 2\pi 2 f_0 t + a_3 \cos 2\pi 3 f_0 t + \cdots + b_1 \sin 2\pi f_0 t + b_2 \sin 2\pi 2 f_0 t + b_3 \sin 2\pi 3 f_0 t + \cdots = A_0 + \sum_{n=1}^{\infty}(a_n \cos 2\pi n f_0 t + b_n \sin 2\pi n f_0 t)$.
  a) What do we know about the given periodic signal?
  b) What is f_0 and how can we compute it?
  c) What is the meaning of the number "n" and how does this number change in the Fourier series?
  d) What do we know about A_0, a_n, and b_n?

21 Consider the square-wave signal shown in Figure 6.1.P21. Calculate A_0, the three first cosine and three first sine members of its Fourier series.

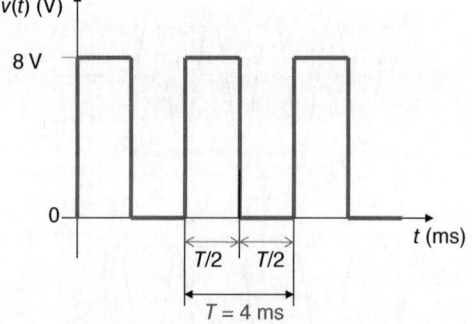

Figure 6.1.P21 A square-wave signal.

22 The waveform and the Fourier series of the bipolar shifted square wave signal are given in Figure 6.1.P22:
  a) In frequency domain, sketch the first three members and show all their parameters if $A = 2$ V and $T = 10$ ms.
  b) Sketch the waveform of the first harmonic of the signal and show its amplitude and period.

23 The waveform and the Fourier series of the sawtooth signal are given in Figure 6.1.P23:
  a) In frequency domain, sketch the first three members and show all their parameters if $A = 2$ V and $T = 10$ ms.
  b) Sketch the waveform of the first harmonic of the signal and show its amplitude and period.

24 The waveform and the Fourier series of the half-wave rectified signal are given in Figure 6.1.P24:
  a) In frequency domain, sketch the first three members and show all their parameters if $A = 2$ V and $T = 10$ ms.
  b) Sketch the waveform of the second harmonic of the signal and show its amplitude and period.

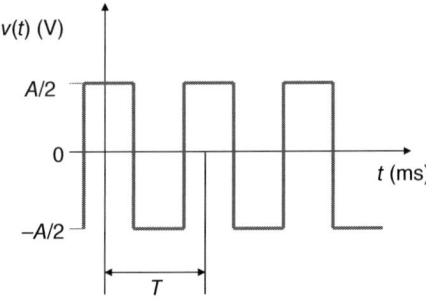

Fourier series

$v(t) = (2A/\pi)\cos 2\pi f_0 t - (2A/3\pi)\cos 2\pi (3f_0)t + (2A/5\pi)\cos 2\pi (5f_0)t + \ldots$

$= \sum_{n=1}^{\infty} (A\sin n\pi/2)/(n\pi/2)\cos 2\pi (nf_0)t$

Figure 6.1.P22 A bipolar shifted square-wave signal and its Fourier series.

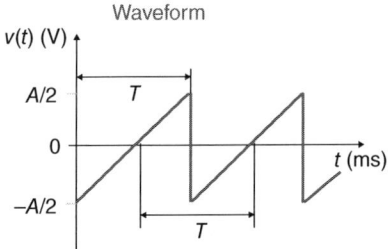

$v(t) = (A/\pi)\sin 2\pi f_0 t + (A/2\pi)\sin 2\pi (2f_0)t + (A/3\pi)\sin 2\pi (3f_0)t + (A/4\pi)\sin 2\pi (4f_0)t\ldots$

Figure 6.1.P23 A sawtooth signal and its Fourier series.

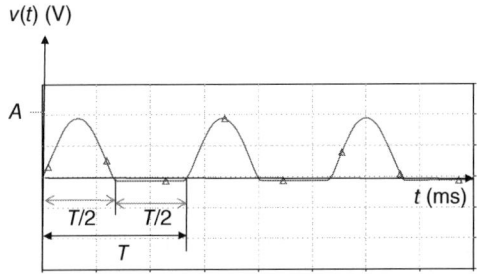

Fourier series

$v(t) = A/\pi + (A/2)\sin 2\pi f_0 t - (2A/3\pi)\cos 2\pi (2f_0)t - (2A/15\pi)\cos 2\pi (4f_0)t$
$\qquad - (2A/35\pi)\cos 2\pi (4f_0)t - \ldots$
$= A/\pi + (A/2)\sin 2\pi f_0 t - (2A/\pi)\sum_{n=2,4,6,}^{\infty}(\cos 2\pi (nf_0)t/(n^2-1))$

Figure 6.1.P24 A half-wave rectified signal and its Fourier series.

25 The waveform and the Fourier series of the digital signal are given in Figure 6.1.P25:
  a) In frequency domain, sketch the first three members and show all their parameters if $A = 2\,\text{V}$ and $T = 10\,\text{ms}$.
  b) Sketch the waveform of the third harmonic of this signal and show its amplitude and period.

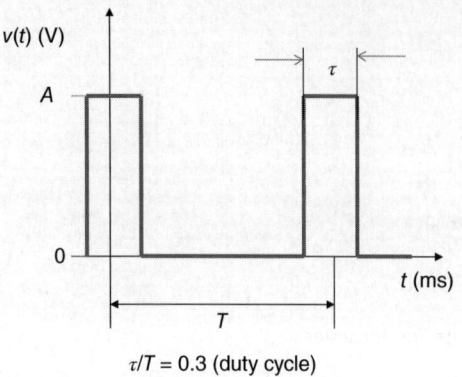

Figure 6.1.P25 A digital signal and its Fourier series.

$\tau/T = 0.3$ (duty cycle)

Fourier series

$v(t) = A\,\tau/T + \sum_{n=1}^{\infty} (2A/n\pi\,(\sin n\pi\,\tau/T))\cos 2\pi\,(nf_0)t$

26 *Find the spectrum of the bipolar square-wave signal shown in Figure 6.1.P26. Show all the calculations. Sketch the first four members of the spectrum.

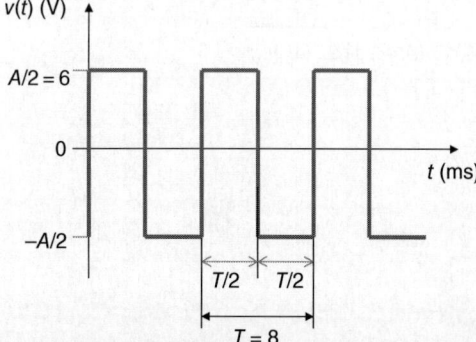

Figure 6.1.P26 A bipolar square-wave signal.

27 *Find the spectrum of the triangular signal shown in Figure 6.1.P27. Show all the calculations. Sketch the first four members of the spectrum.

28 Find the dc member, A_0, of the Fourier series for the signals shown in Figures 6.1.P21, 6.1.P26, and 6.1.P27.

29 It is said that the Fourier series enables the *spectral analysis* of a given signal:
  a) Explain the meaning of the word "analysis" in this context.
  b) What do we mean by "spectral analysis"?

30 *Describe the process of spectral analysis in conjunction with the meaning of the Fourier series (refer to Figure 6.1.12).

Figure 6.1.P27 The waveform of a triangular signal.

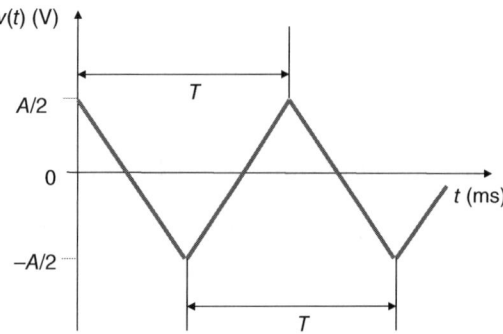

Spectral Synthesis

31 *Explain the difference between spectral analysis and spectral synthesis.

32 *The Fourier series of a signal is given by $v(t) = (A/\pi) \sin 2\pi f_0 t + (A/2\pi) \sin 2\pi(2f_0)t + (A/3\pi) \sin 2\pi(3f_0)t + (A/4\pi) \sin 2\pi(4f_0)t \cdots = \sum_{n=1}^{\infty}(A/n\pi) \sin 2\pi(nf_0)t$:
   a) Sketch the first four spectral components of this Fourier series if $A = 2$ V and $T = 1$ ms.
   b) Applying the spectral-synthesis technique, build the waveform of this signal based on the first four harmonics. (Use any computer tool you are comfortable with such as MS Excel or MATLAB.)

33 *The spectrum of a signal is shown in Figure 6.1.P33:
   a) Write down the expressions for the first three members of the Fourier series of this signal if $A = 2$ V and $f_0 = 1$ kHz.
   b) Applying the spectral-synthesis technique, build the waveform of this signal based on the first three harmonics. (Use any computer tool you are comfortable with such as MS Excel or MATLAB.)

Figure 6.1.P33 The spectrum of a signal.

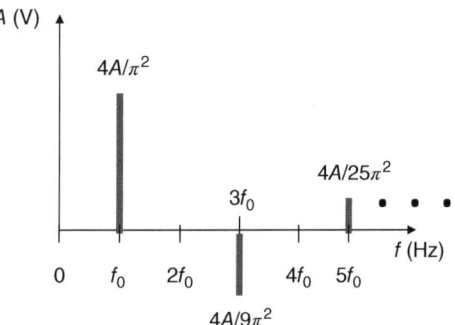

34 *Describe the process of spectral synthesis with reference to Figure 6.1.17.

35 *Describe the role of individual harmonics and their summation in the processes of spectral analysis and synthesis for a sawtooth signal (refer to Problem 23 in this section and Figure 6.1.19). Restrict the processes by three spectral components.

36 *Consider Figure 6.1.19, showing the presentation of two harmonics in time domain and frequency domain. Sketch a similar figure for two harmonics of the sawtooth signal discussed in Problem 35 in this section.

6.2 Introduction to Spectral Analysis

Objective and Outcomes of Section 6.2

Objective

- To gain a deeper understanding of the spectral analysis of a periodic signal.

Outcomes

After studying this section, you will be able to:

- Find the Fourier series for various types of periodic signals. Specifically, you will be able to
  o Derive the formulas for calculating the coefficients of the Fourier series.
  o Obtain the amplitude and the phase spectra of a periodic signal by calculating the coefficients of its Fourier series and, therefore, finding the Fourier series of the signal in an explicit form.
  o Simplify the aforementioned calculations by using the symmetry of periodic signals.
  o Use MATLAB to find the Fourier series of various signals.
- Demonstrate the effect of filtering on signals from the standpoint of spectral analysis. Specifically, you will be able to
  o Demonstrate that the effect of filtering on signals can be explained only from the spectral standpoint.
  o Develop the mathematical description of filtering of a single harmonic and present the results in time domain and frequency domain.
  o Develop the mathematical and graphical tools for describing the effect of filtering of a periodic signal in both time and frequency domains.
  o Explain the entire picture of signal filtering by relating the changes in the amplitudes and phases of a signal's spectral components to changes in the signal's waveform.
- Explain the concept of harmonic distortion. Specifically,
  o Show that the harmonic distortion is the phenomenon where an output signal contains more harmonics than an input signal.
  o Assess the level of harmonic distortion by calculating the ratio of the amplitudes of current harmonics to the fundamental harmonic amplitude and using the formula for the total harmonic distortion (THD).
  o Discern the relationship between changes in a signal's waveform and the signal's harmonic and phase distortions.

6.2.1 More About the Fourier Series

6.2.1.1 Coefficients of the Fourier Series

The Fourier theorem, presented in Section 6.1, states that any periodic signal can be represented as a trigonometric series, that is, as a sum of cosines and sines, as described in Eq. (6.1.5a),

$$v(t) = A_0 + a_1 \cos \omega_0 t + a_2 \cos 2\omega_0 t + a_3 \cos 3\omega_0 t + \cdots + b_1 \sin \omega_0 t$$
$$+ b_2 \sin 2\omega_0 t + b_3 \sin 3\omega_0 t + \cdots$$

6.2 Introduction to Spectral Analysis

Here, we use the notations

$$\omega_0 \text{ (rad/s)} = 2\pi f_0$$

to shorten the explanations. Equations (6.1.6)–(6.1.8) show how we can compute these coefficients. Sidebar 6.1.S.1 demonstrates the meaning of mathematical manipulations that enable the calculations of the coefficients. Now, it is time to understand where the formulas for the coefficients come from.

To derive these formulas, we first take an integral over one period from the left-hand and right-hand sides of the Fourier series, (6.1.5a):

$$\int_0^T v(t)dt = \int_0^T A_0 \, dt + \int_0^T a_1 \cos \omega_0 t \, dt + \int_0^T a_2 \cos 2\omega_0 t \, dt + \int_0^T a_3 \cos 3\omega_0 t \, dt + \cdots + \int_0^T b_1 \sin \omega_0 t \, dt + \int_0^T b_2 \sin 2\omega_0 t \, dt + \int_0^T b_3 \sin 3\omega_0 t \, dt + \cdots \quad (6.2.1)$$

We need to recall that

$$\int_0^T a_n \cos \omega_0 t \, dt = 0$$

and

$$\int_0^T a_n \sin \omega_0 t \, dt = 0 \quad (6.2.2)$$

because an average value – and the integral over a period does compute this average value – of a cosine and sine over one period is zero. Thus, from (6.2.1) we find

$$\int_0^T v(t) \, dt = \int_0^T A_0 \, dt = A_0 \int_0^T dt = A_0|_0^T = A_0 T$$

and therefore

$$A_0 = \int_0^T v(t) \rightarrow A_0 = 1/T \int_0^T v(t)$$

as in (6.1.6).

To derive the coefficients a_n, we multiply both sides of the Fourier series by $\cos \omega_0 t$ first, then by $\cos 2\omega_0 t$, then by $\cos 3\omega_0 t$, and so on, each time taking the integral over the period. For the fundamental harmonic, ω_0, we obtain

$$\int_0^T v(t) \cos \omega_0 t \, dt = \int_0^T A_0 \cos \omega_0 t \, dt + \int_0^T a_1 \cos \omega_0 t \cos \omega_0 t \, dt + \int_0^T a_2 \cos 2\omega_0 t \cos \omega_0 t \, dt + \int_0^T a_3 \cos 3\omega_0 t \cos \omega_0 t \, dt + \cdots + \int_0^T b_1 \sin \omega_0 t \cos \omega_0 t \, dt + \int_0^T b_2 \sin 2\omega_0 t \cos \omega_0 t \, dt + \int_0^T b_3 \sin 3\omega_0 t \cos \omega_0 t \, dt + \cdots \quad (6.2.3)$$

The dc term is obviously equal to zero:

$$A_0 \int_0^T \cos m\omega_0 t \, dt = 0$$

Generalizing the result of other products, we understand that we will have terms that contain either the same or various frequencies, that is,

$$\int_0^T a_m \cos m\omega_0 t \cos m\omega_0 t \, dt$$

or

$$\int_0^T a_n \cos n\omega_0 t \cos m\omega_0 t \, dt \, (n \neq m)$$

Integration of the terms containing the product of cosines of the same frequencies produces

$$\int_0^T \cos m\omega_0 t \cos m\omega_0 t \, dt = \int_0^T \cos^2 m\omega_0 t \, dt = \frac{T}{2} \qquad (6.2.4)$$

Integration of terms containing cosines of different frequencies produces

$$\int_0^T \cos n\omega_0 t \cos m\omega_0 t \, dt = 0 \quad (n \neq m) \qquad (6.2.5)$$

which you can easily verify by applying the trigonometric identity

$$\cos \alpha \cos \beta = \tfrac{1}{2}(\cos(\alpha+\beta) + \cos(\alpha-\beta))$$

Also, all members containing the product of the cosines and the sines will turn to zero,

$$\int_0^T \sin n\omega_0 t \cos m\omega_0 t \beta \, dt = 0 \qquad (6.2.6)$$

regardless of whether or not n equals m because of the orthogonality relationship of the sine and cosine. (The orthogonality relationship of the cosine and sine is another means to express this well-known fact: Their relative phase shift is equal to 90°.) With all these results in mind, we obtain from (6.2.3)

$$\int_0^T v(t) \cos \omega_0 t \, dt = a_n \int_0^T \cos m\omega_0 t \cos m\omega_0 t \, dt = a_n \int_0^T \cos^2 m\omega_0 t \, dt = a_n \frac{T}{2} \qquad (6.2.7)$$

Therefore,

$$a_n = \frac{2}{T} \int_0^T v(t) \cos n\omega_0 t \, dt \qquad (6.2.8)$$

as in (6.2.7).

The properties similar to those shown in (6.2.4)–(6.2.6) can be applied to a sine function. That is, from

$$\int_0^T A_0 \sin m\omega_0 t \, dt = 0$$

$$\int_0^T \sin m\omega_0 t \sin m\omega_0 t \, dt = \int_0^T \sin^2 m\omega_0 t \, dt = \frac{T}{2}$$

$$\int_0^T \sin n\omega_0 t \sin m\omega_0 t \, dt = 0, \quad (n \neq m) \text{ and}$$

$$\int_0^T \sin n\omega_0 t \cos m\omega_0 t \, dt = 0$$

we can derive the formula for b_n as

$$b_n = \frac{2}{T} \int_0^T v(t) \sin n\omega_0 t \, dt \qquad (6.2.9)$$

As we see, the real foundation of the spectral analysis of periodic signals is the Fourier series, given in (6.1.5a); the coefficients of this series can be obtained by plugging the mathematical description of a periodic signal into the Fourier formulas (6.1.6)–(6.1.8).

6.2.1.2 Amplitude and Phase Spectra

We have mentioned many times that the cosine signal is, by default, a reference signal with zero phase and therefore can be represented in the frequency domain by its amplitude only. The sine signal must include a $-90°$ phase angle in a frequency-domain presentation. If a specific cosine or sine has an initial phase shift, Θ, we need to show this phase in frequency domain, too, by introducing the Θ–f axes in addition to the A–f axes. This is why we have stated in Section 6.1 that, in general, we actually have the *amplitude spectrum* and *phase spectrum* of a signal; the former is presented on the A–f plane and the latter is presented on the Θ–f plane.

Now, we need to generalize our discussion and justify introducing the concept of amplitude and phase spectra, an introduction that leads to the amplitude-phase formula of the Fourier series. This formula will include both the amplitude and phase spectra of a signal.

Let us consider the general form of the Fourier series, (6.1.5b):

$$v(t) = A_0 + \sum_{n=1}^{\infty}(a_n \cos n\omega_0 t + b_n \sin n\omega_0 t) \qquad (6.2.10a)$$

The sum,

$$a_n \cos n\omega_0 t + b_n \sin n\omega_0 t$$

can be presented as

$$a_n \cos n\omega_0 t + b_n \sin n\omega_0 t = A_n \cos(n\omega_0 t + \Theta_n) \qquad (6.2.10b)$$

where

$$A_n = \sqrt{a_n^2 + b_n^2} \qquad (6.2.11a)$$

and

$$\Theta_n = -\tan^{-1}\left(\frac{b_n}{a_n}\right) \qquad (6.2.11b)$$

Formulas (6.2.11a) and (6.2.11b) can be obtained by applying the following trigonometric identity:

$$\cos(\alpha + \beta) = \cos \alpha \cos \beta - \sin \alpha \sin \beta \qquad (6.2.12)$$

Thus,

$$A_n \cos(n\omega_0 t + \Theta_n) = A_n \cos n\omega_0 t \cos \Theta_n - A_n \sin n\omega_0 t \sin \Theta_n$$

Plugging this expression into (6.2.10b),

$$a_n \cos n\omega_0 t + b_n \sin n\omega_0 t = A_n \cos(n\omega_0 t + \Theta_n) = A_n \cos \Theta_n \cos n\omega_0 t - A_n \sin \Theta_n \sin n\omega_0 t$$

we find

$$a_n = A_n \cos \Theta_n$$

and

$$b_n = -A_n \sin \Theta_n \qquad (6.2.13)$$

from which (6.2.11a) and (6.2.11b) immediately follow.

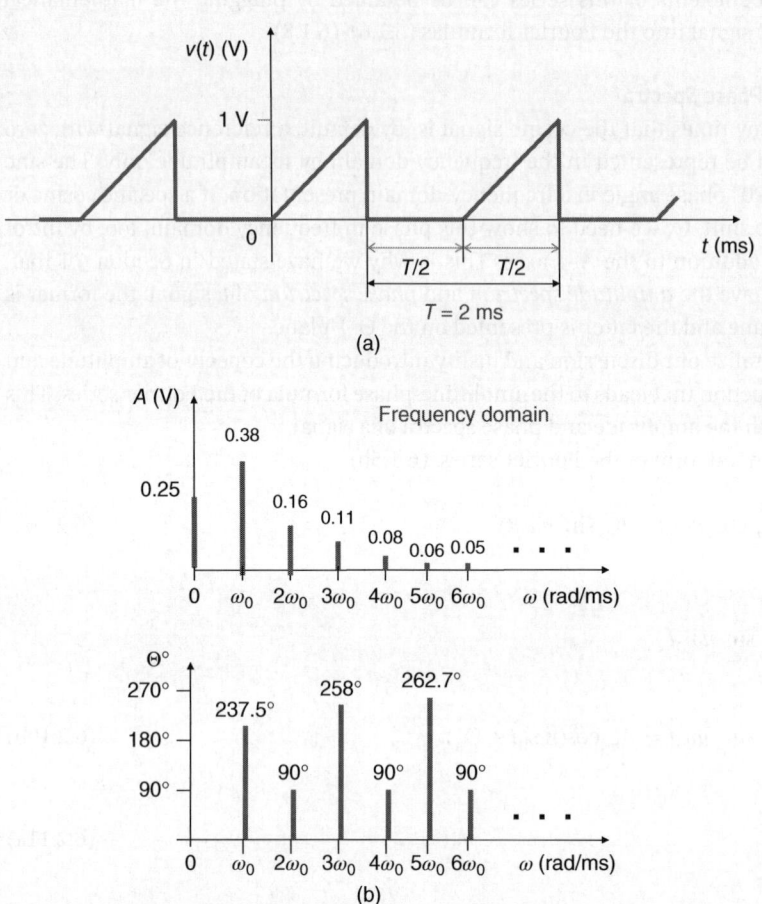

Figure 6.2.1 Delayed sawtooth signal and its spectrum for Example 6.2.1: (a) The signal's waveform; (b) the signal's amplitude and phase spectrum.

Thus, the Fourier series, given by (6.2.10a), can be presented in the following format:

$$v(t) = A_0 + \sum_{n=1}^{\infty}(a_n \cos n\omega_0 t + b_n \sin n\omega_0 t) = A_0 + \sum_{n=1}^{\infty} A_n \sin(n\omega_0 t + \Theta_n) \quad (6.2.14)$$

where A_n and Θ_n are given by (6.2.11a) and (6.2.11b).

This form clearly shows that we need to include both the amplitudes and phases in the spectrum of a signal. To ponder this consideration from a practical standpoint, let us study Example 6.2.1.

Example 6.2.1   The Spectrum of a Delayed Sawtooth Signal[4].

Problem

Find the spectrum of the periodic signal shown in Figure 6.2.1a if $A = 1$ V and $T = 2$ ms.

---

4 After Alexander and Sadiku (2009, pp. 763–764).

Solution

We need to follow the steps described in Example 6.1.1, that is, find the formula for this signal's waveform and expand it into the Fourier series.

Step 1: Define the signal $v(t)$: The signal's waveform formula, $v(t)$, can be defined from an observation of Figure 6.2.1a as

$$v(t) = \begin{cases} kt & \text{for } 0 \leq t \leq \frac{T}{2} \\ 0 & \text{for } \frac{T}{2} < t < T \end{cases} \quad (6.2.15)$$

The slope of this signal is given by

$$k = (A-0)\,(V)/\left(\frac{T}{2}-0\right)(s) = \frac{2A}{T}\,(V/s)$$

Given that $A = 1$ V and $\frac{T}{2} = 1$ ms, we compute $k = 1$ V/ms.

Step 2: Calculate A_0: The dc member of this signal can be calculated by using (6.1.6):

$$A_0 = \frac{1}{T}\int_0^T v(t)dt = \frac{1}{T}\left(\int_0^{T/2} kt\,dt + \int_{T/2}^T (0)dt\right)$$

$$= \frac{k}{T}\frac{t^2}{2}\bigg|_0^{T/2} = \frac{2A}{T^2}\left(\frac{\left(\frac{T}{2}\right)^2}{2} - 0\right) = \frac{A}{4} = \frac{1}{4}\,V$$

Steps 3: Calculate a_n and b_n: Applying (6.2.8), we compute a_n as

$$a_n = \frac{2}{T}\int_0^{T/2} v(t)\cos(n\omega_0 t)dt = \frac{2}{T}\left(\int_0^{T/2} kt\cos(n\omega_0 t)dt + 0\right)$$

$$= \frac{2k}{T}\int_0^{T/2} t\cos(n\omega_0 t)dt = \frac{4A}{T^2}t\cos(n\omega_0 t)dt$$

We need to recall that the integration of function $(t\cos n\omega_0 t)$ goes as follows:

$$\int_0^{T/2}(t\cdot\cos(n\omega_0 t))dt = \left(\frac{\cos(n\omega_0 t)}{(n\omega_0)^2} + \frac{t\sin(n\omega_0 t)}{n\omega_0}\right)\bigg|_0^{T/2}$$

$$= \frac{\cos\left(n\omega_0\frac{T}{2}\right) - 1}{(n\omega_0)^2} + \frac{\frac{T}{2}\sin\left(n\omega_0\frac{T}{2}\right)}{n\omega_0}$$

Since $\omega_0 = 2\pi/T$, we find

$$\frac{\cos\left(n\omega_0\frac{T}{2}\right) - 1}{(n\omega_0)^2} + \frac{\frac{T}{2}\sin\left(n\omega_0\frac{T}{2}\right)}{n\omega_0} = \frac{T^2}{4}\frac{(\cos(\pi n) - 1)}{(\pi n)^2} + \frac{T^2}{4}\frac{\sin(\pi n)}{\pi n}$$

If n is even ($n = 2, 4, 6, \ldots$), then $\cos(\pi n) = 1$; if n is odd ($n = 1, 3, 5, \ldots$), then $\cos(\pi n) = -1$; this result can be presented as

$$\cos(\pi n) = (-1)^n$$

Also, $\sin(\pi n) = 0$, regardless of whether n is even or odd. Thus, we obtain coefficients a_n in the following form:

$$a_n = \frac{T^2}{4}\frac{((-1)^n - 1)}{(\pi n)^2} \quad (6.2.16)$$

which means that

$$a_n = \begin{cases} 0 & \text{for } n = 2, 4, 6, \ldots \\ \dfrac{T^2}{4} \dfrac{-2}{(\pi n)^2} & \text{for } n = 1, 3, 5, \ldots \end{cases} \quad (6.2.17a)$$

Plugging our numbers into (6.2.16), we find

$$a_n \, (\text{V}) = \frac{(-1)^n - 1}{(\pi n)^2} \quad (6.2.17b)$$

Similarly, we can calculate coefficients b_n by using (6.2.9):

$$b_n = \frac{2}{T} \int_0^T v(t) \sin n\omega_0 t \, dt = \frac{2k}{T} \int_0^T t \sin n\omega_0 t \, dt$$

All the manipulations in finding b_n are exactly the same as they were for a_n except for the integration of a function ($t \sin n\omega_0 t$). In this case,

$$\int_0^{T/2} t \sin(n\omega_0 t) dt = \left(\frac{\sin(n\omega_0 t)}{(n\omega_0)^2} - \frac{t \cos(n\omega_0 t)}{n\omega_0} \Bigg|_0^{T/2} \right)$$

If we do all the required manipulations (and we leave that as an exercise for you), we should arrive at the following formula for b_n:

$$b_n = \frac{-\left(\frac{T}{2}\right)^2 \cos(\pi n)}{\pi n} = \frac{T^2}{4} \frac{(-1)^{n+1}}{\pi n} \quad (6.2.18a)$$

Thus,

$$b_n = \begin{cases} \dfrac{T^2}{4} \dfrac{(-1)}{\pi n} & \text{for } n = 2, 4, 6, \ldots \\ \dfrac{T^2}{4} \dfrac{1}{\pi n} & \text{for } n = 1, 3, 5, \ldots \end{cases} \quad (6.2.18b)$$

With $T = 2$ ms given in this example, we find

$$b_n = \frac{(-1)^{n+1}}{\pi n} \quad (6.2.18c)$$

Step 4: Obtain the Fourier series: Plugging into the Fourier series (6.2.10a) the obtained coefficients and given $A = 1$ V and $T = 2$ ms, we find

$$v(t) = A_0 + \sum_{n=1}^{\infty} (a_n \cos(n\omega_0 t) + b_n \sin(n\omega_0 t))$$

$$= \frac{1}{4} + \sum_{n=1}^{\infty} \left[\frac{(-1)^n - 1}{(\pi n)^2} \cos(\pi n t) + \frac{(-1)^{n+1}}{\pi n} \sin(\pi n t) \right] \quad (6.2.19)$$

It helps to present this result separately for odd and even orders, n:

$$v(t) = \frac{1}{4} + \sum_{\substack{n=1 \\ n \text{ even}}}^{\infty} \left[\frac{(-1)}{\pi n} \sin(\pi n t) \right] \quad \text{for } n = 2, 4, 6, \ldots \quad (6.2.20a)$$

and

$$v(t) = \frac{1}{4} + \sum_{\substack{n=1 \\ n \text{ odd}}}^{\infty} \left[\frac{(-2)}{(\pi n)^2} \cos(\pi n t) + \frac{1}{\pi n} \sin(\pi n t) \right] \quad \text{for } n = 1, 3, 5, \ldots \quad (6.2.20b)$$

Step 5: Graphically build the signal's spectrum: To build the spectrum of this signal, we need to present its Fourier series in amplitude-phase form, as in (6.2.14),

$$v(t) = A_0 + \sum_{n=1}^{\infty}(a_n \cos n\omega_0 t + b_n \sin n\omega_0 t) = A_0 + \sum_{n=1}^{\infty} A_n \cos(n\omega_0 t + \Theta_n)$$

Applying (6.2.11a),

$$A_n = \sqrt{(a_n^2 + b_n^2)}$$

and using coefficients a_n and b_n from (6.2.16) and (6.2.18a), we find

$$A_n = \sqrt{a_n^2 + b_n^2} = \frac{T^2}{4\pi n}\sqrt{\left[\left(\frac{((-1)^n - 1)}{\pi n}\right)^2 + ((-1)^{n+1})^2\right]} \quad (6.2.21a)$$

which results in

$$A_n \begin{cases} |b_n| = \dfrac{T^2}{4\pi n}\sqrt{((-1)^{n+1})^2} = \dfrac{T^2}{4\pi n} & \text{for } n = 2, 4, 6, \ldots \\[1em] \dfrac{T^2}{4}\sqrt{\left(\left(\dfrac{-2}{(\pi n)^2}\right)^2 + \left(\dfrac{1}{\pi n}\right)^2\right)} = \dfrac{T^2}{4(\pi n)^2}\sqrt{(4 + \pi n^2)} & \text{for } n = 1, 3, 5, \ldots \end{cases} \quad (6.2.21b)$$

Plugging $T = 2$(ms) into (6.2.21a) yields

$$A_n = \begin{cases} \dfrac{1}{\pi n} & \text{for } n = 2, 4, 6, \ldots \\[1em] \dfrac{1}{(\pi n)^2}\sqrt{(4 + \pi n^2)} & \text{for } n = 1, 3, 5, \ldots \end{cases} \quad (6.2.21c)$$

Using (6.2.11b), we find the phase spectrum of the given signal as

$$\Theta_n(°) = -\tan^{-1}\frac{b_n}{a_n} = -\tan^{-1}\left(\frac{(-1)^{n+1}}{(-1)^n - 1}\pi n\right) \quad (6.2.22a)$$

To simplify this expression, we consider the phase for odd and even orders and find

$$\Theta_n(°) = \begin{cases} -\tan^{-1}\left(\dfrac{-1}{1-1}\pi n\right) = 90° & \text{for } n = 2, 4, 6, \ldots \\[1em] -\tan^{-1}\left(-\dfrac{\pi n}{2}\right) = 180° + \tan^{-1}\left(\dfrac{\pi n}{2}\right) & \text{for } n = 1, 3, 5, \ldots \end{cases} \quad (6.2.22b)$$

Now, we can plot both the amplitude and phase spectra of the given signal. These plots are shown in Figure 6.2.1b.

Discussion

- First, you will recall that the dc member represents the average value of a signal: Observe the signal's waveform and notice that for the first half of the period, the average value is $A/2$, whereas for the second half it is zero. Thus, the average value of the signal for the entire period is equal to the computed value, $A_0 = A/4$.
- Secondly, note that, according to (6.2.22b), for n even, the phase shift is simply $90°$; for n odd, however, the phase spectrum is more sophisticated and the values of Θ must be computed. For example, for $\Theta_1 = 180° + \tan^{-1}\left(\frac{\pi}{2}\right) = 180° + 57.5° = 237.5°$ for $n = 1$. Also, understand that in this example the fundamental frequency, ω_0, is equal to $\omega_0 = 2\pi/T = \pi$ (rad/ms) because $T = 2$ ms. This is why the frequency axis can be labeled in multiples of π or simply in numbers as $\omega_0 = 3.14$ rad/ms, $2\omega_0 = 6.28$ rad/ms, $3\omega_0 = 9.42$ rad/ms, and so on.

Sidebar 6.2.S.1 Using the Signal's Symmetry for Finding the Fourier Series Coefficients

Some signals possess a symmetry that allows us to simplify calculating the coefficients of a Fourier series. The two main types of *symmetry* we may encounter in reality are *odd* and *even* ones:

A function is called *odd* if

$$v(-t) = -v(t) \tag{6.2.S.1.1}$$

The most familiar example of an *odd* function is a *sine*; other examples include t, t^3, and other odd powers of t. Some other examples of odd functions are shown in Figure 6.2.S.1.1. You can see that *these functions are symmetrical with regard to both the y and t axes*. The main property of the odd function is that

$$v(t)_{odd} = \int_{-T/2}^{T/2} v(t)dt = 0 \tag{6.2.S.1.2}$$

which stems from the fact that one half of an odd function is negative at one half of its period and the symmetrical half is positive at the other half of its period.

An *even* function is defined as

$$v(-t) = v(t) \tag{6.2.S.1.3}$$

The most familiar example of an even function is a *cosine*; other examples include t^2, t^4, and other even powers of t. Figure 6.2.S.1.1 shows other examples of even functions.

Study Figure 6.2.S.1.1a,b first: They are instructive examples of odd and even functions. The odd function shown in Figure 6.2.S.1.1a has 10 V at 1 s and −10 V at −1 s, whereas the even function in Figure 6.2.S.1.1b has 10 V at both +1 s and −1 s. To identify the *odd symmetry* by looking at the waveform, we need to flip the negative half of the waveform over the y axis and then flip this half over the t axis. The resulting graph must coincide with the positive half of the waveform. (Do this exercise with all waveforms in Figure 6.2.S.1.1.)

The *even symmetry* can be identified by simply rotating the negative half of the waveform around the y axis; the resulting waveform must match the positive half. (Do this for the all waveforms shown in Figure 6.2.S.1.1.)

If $v(t)$ of a given signal is an *odd* function, then using its property, we find that the Fourier coefficients are

$$A_0 = 0$$
$$a_n = 0$$

and

$$b_n = 4/T \int_0^{T/2} v(t)\sin(2\pi n f_0 t)dt \tag{6.2.S.1.4}$$

Equation (6.2.S.1.4) can be verified by plugging the given odd $v(t)$ into (6.2.6)–(6.2.8) as follows:

$$A_0 = 1/T \int_0^T v(t)dt = 1/T \int_{-T/2}^{T/2} v(t)dt = 0 \tag{6.2.S.1.5a}$$

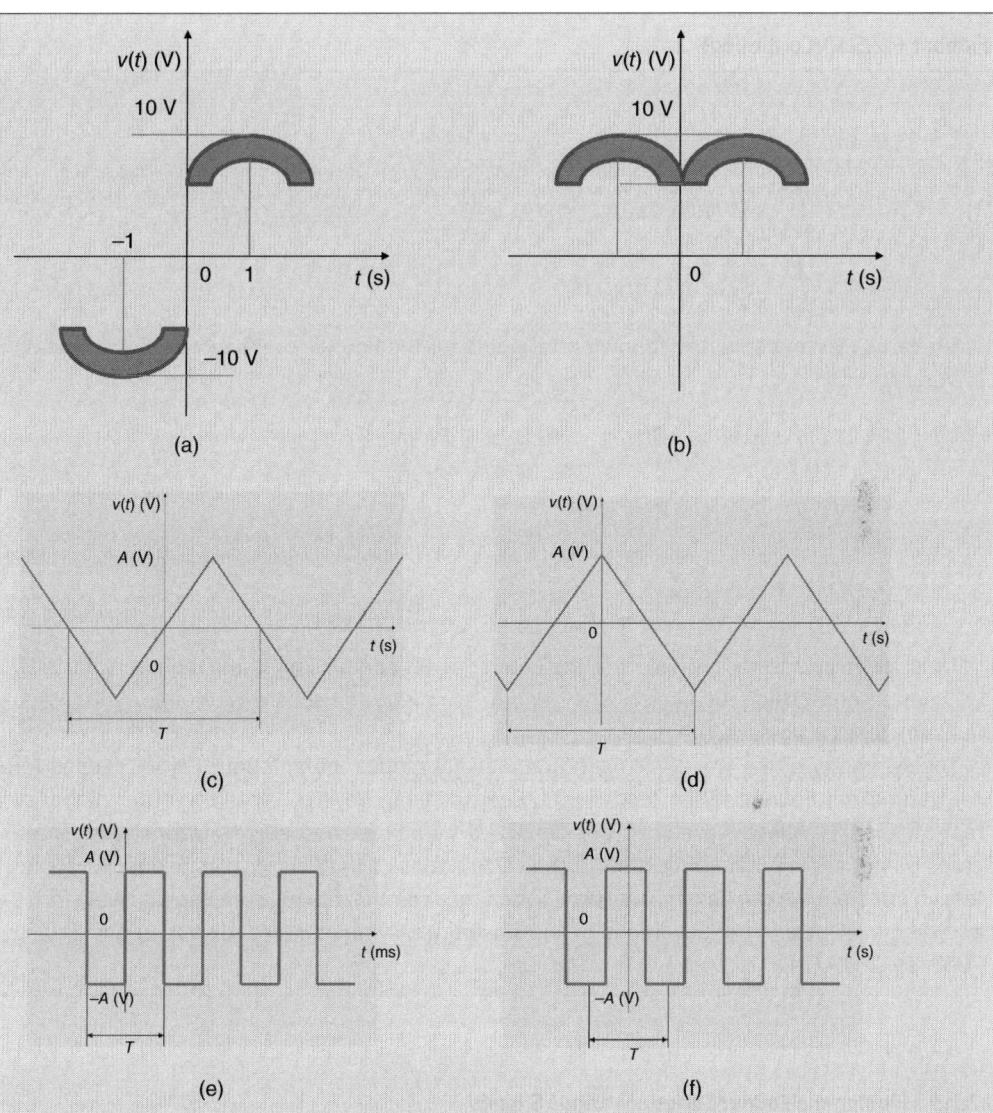

Figure 6.2.S.1.1 Examples of waveforms of odd and even functions: The waveforms of the odd functions are shown in (a), (c), and (e), whereas the even functions are shown in (b), (d), and (f).

$$a_n = 2/T \int_0^T v(t)\cos(2\pi n f_0 t)dt$$

$$= 2/T \int_{-T/2}^0 v(-t)\cos(2\pi n f_0 t)dt + 2/T \int_0^{T/2} v(t)\cos(2\pi n f_0 t)dt$$

$$= -2/T \int_{-T/2}^0 v(-t)\cos(2\pi n f_0 (-t))dt + 2/T \int_0^{T/2} v(t)\cos(2\pi n f_0 t)dt = 0 \quad (6.2.S.1.5b)$$

(Continued)

Sidebar 6.2.S.1 (Continued)

$$b_n = 2/T \int_0^T v(t)\sin(2\pi nf_0 t)dt$$

$$= 2/T \int_{-T/2}^0 v(-t)\sin(2\pi nf_0(-t))dt + 2/T \int_0^{T/2} v(t)\sin(2\pi nf_0 t)dt$$

$$= 4/T \int_0^{T/2} v(t)\sin(2\pi nf_0 t)dt \qquad (6.2.S.1.5c)$$

Similarly, we can derive the following formulas for the Fourier coefficients when $v(t)$ is an *even* function:

$$A_0 = 1/T \int_0^T v(t)dt = 2/T \int_0^{T/2} v(t)dt$$

$$a_n = 4/T \int_0^{T/2} v(t)\cos(2\pi nf_0 t)dt$$

and

$$b_n = 0 \qquad (6.2.S.1.6)$$

Thus, for an *odd* signal's waveforms, the Fourier series contains only sine members; for *even* waveforms, only cosine members. (What kind of symmetry the square-wave signal discussed in Example 6.1.1 possess?)

The symmetry of the waveforms can dramatically reduce the amount of work needed for calculating the Fourier series coefficients. Bear in mind, however, that there are plenty of real-life signals whose waveforms are described by neither odd nor even functions. These cases do not allow for simplification in the calculations of the Fourier coefficient. Nevertheless, in principle, we can develop any waveform into a combination of even and odd waveforms and then make use of the advantage of the waveform's symmetry (see Alexander and Sadiku 2009, pp. 764–774.).

6.2.1.3 Finding the Fourier Series of Various Signals

Typically, textbooks present tables with Fourier series of the most common waveforms to help students find the spectra of various periodic signals without actually performing calculations of their Fourier series. We, however, offer a different approach to this task: We present the MATLAB program, which enables us to *obtain* the Fourier series of various signals. Thus, you can obtain the Fourier series (and therefore its spectrum) of practically any periodic signal by simply following the instructions presented below. This approach is based on the following reasoning: To build a Fourier series of any waveform, we need to find the coefficients of this series. Thus, a program calculating such coefficients resolves the problem. To locate these coefficients, we have several options, which are discussed in this section. Our program can find coefficients A_0, a_n, and b_n of the trigonometric (cosine–sine) form of the Fourier series given by

$$v(t) = A_0 + \sum_{n=1}^{\infty}(a_n \cos 2\pi nf_0 t + b_n \sin 2\pi nf_0 t) \qquad (6.1.5bR)$$

6.2 Introduction to Spectral Analysis

This program also calculates the amplitude and phase of the Fourier series given in the amplitude-phase form, as

$$v(t) = A_0 + \sum_{n=1}^{\infty}(a_n \cos n\omega_0 t + b_n \sin n\omega_0 t) = A_0 + \sum_{n=1}^{\infty} A_n \sin(n\omega_0 t + \Theta_n) \qquad (6.2.14R)$$

The relationship between the coefficients of these two forms of the Fourier series – to remind you what we learned from (6.2.11a) and (6.2.11b) – are given by

$$A_n = \sqrt{a_n^2 + b_n^2} \quad \text{and} \quad \Theta_n = \tan^{-1} a_n/b_n$$

The program is presented in the MATLAB box 6.2.1.

MATLAB Box 6.2.1 *MATLAB script for finding the* **Fourier series of various periodic signals**[5]

```
function [] = fouriers()
clearall;        %clear all stored variables
symst;           %variable t
disp ('Fourier series'); disp('t-is the variable'); %display messages
T=input ('\n Enter Period \n');  %input period
w=2*pi/T;        %convert period to omega
d=input('Into how many parts do you want to divide your period? \n');  %input
%number for period division
T(d+1)=(2*pi)/w; % period final point
T(1)=0;          % period starting point
for s=2:1:(d+1)
    T(s)=(s-1)*(2*pi)/(w*d);    % period in-between points
end

for s=1:1:d
disp(['Enter the function from ', num2str(T(s)), ' to ',num2str(T(s+1)) ]);
%display messages
f(s)=input(")*(t/t);  %input function between indicated points of period
end
nm=input('Enter highest order \n'); %input top order
for n=1:1:nm
    s=0;     %reset variable "s"
for s=1:1:d
        a0(s)=(int(f(s),T(s),T(s+1)))/(T(d+1)-T(1));       %calculate DC
            %component for each function
        an(s)=2*(int(f(s)*cos(n*w*t),t,T(s),T(s+1)))/(T(d+1)-T(1));
%calculate %an for each function
        bn(s)=2*(int(f(s)*sin(n*w*t),t,T(s),T(s+1)))/(T(d+1)-T(1));
%calculate %bn for each function
end
    A0=sum(a0); %sum all DC component
a(n)=sum(an);    %sum all an components
b(n)=sum(bn);    %sum all bn components
end
for n=1:1:nm
f(n)=a(n)*cos(n*w*t)+b(n)*sin(n*w*t);  %construct function from %coefficients
```

(Continued)

---

[5] This MATLAB program was developed by Vitaly Sukharenko.

MATLAB Box 6.2.1 (Continued)

```
C(n)=sqrt(a(n)^2+b(n)^2);              %compute amplitude
theta(n)=atan(a(n)/b(n))/pi*180;       %compute phase in degrees
end

F=sum(f)+A0;     %sum function and DC component
disp('----------------------------------------');
disp('Answer: ');
disp(' ');
disp('DC component '); disp(A0);       %display DC component
disp(['Amplitude ', (C(1:1:nm))]);     %display Amplitude
disp(['theta ', (theta(1:1:nm))]);     %display phase
disp(['order ',num2str(1:1:nm)]);      %display order
disp(' ');
disp(['an ', (a(1:1:nm))]);            %display an's
disp(['bn ',(b(1:1:nm))]);             %display bn's

subplot(2,1,1);
ezplot(t,F); grid on;      %plot function
xlabel('Time');
ylabel('v(t)[V]');
title('Signal in time domain');

subplot(2,2,3);
stem(0,A0); grid on; hold on;
stem(1:1:nm,C(1:1:nm));    %plot Amplitude
xlabel('Normalized frequency (order) n = w/w0');
ylabel('A (V)');
title('Amplitude vs normalized frequency');

subplot(2,2,4);
stem(0,0);grid on; hold on;
stem(1:1:nm,theta(1:1:nm)); %plot theta
xlabel('Normalized frequency (order) n = w/w0');
ylabel('Phase in deg.');
title('Phase vs normalized frequency');

return;
```

We need to use the MATLAB version that includes the Symbolic Math Toolbox. Just run the program presented here and obtain the coefficients of a Fourier series for trigonometric and amplitude-phase forms. Based on this MATLAB program, we built Table 6.2.1, where all the necessary information for finding the Fourier series of the most common waveforms is given.

6.2.2 Effect of Filtering on Signals

6.2.2.1 Statement of the Problem

Suppose we present a periodic signal to a low-pass filter (LPF). What kind of output should we expect to see? The process leading to this question (with a square-wave signal shown as an example) is depicted in Figure 6.2.2.

Of course, we can perform an experiment. The setup of the experiment is shown in Figure 6.2.3a, where the output signal is still unknown. Figure 6.2.3b shows the output signal obtained for the specific input frequency.

Table 6.2.1 Fourier series of the most common signals.

Waveforms	Coefficients of Fourier series		
	A_0	a_n	b_n
Signal in time domain (square wave, ±1 V)	$A_0 = 0$	$a_n = 0$	$\dfrac{2A_{pp}}{\pi n}$ Only for odd functions
Signal in time domain (sawtooth, 0 to 1 V)	$\dfrac{A_{pp}}{2}$	$a_n = 0$	$-\dfrac{A_{pp}}{\pi n}$
Signal in time domain (triangular, 0 to 1 V)	$\dfrac{A_{pp}}{2}$	$a_n = 0$	$\dfrac{A_{pp}}{\pi n}$

(Continued)

Table 6.2.1 (Continued)

Waveforms	Coefficients of Fourier series		
	A_0	a_n	b_n
Signal in time domain (triangular wave)	$\dfrac{A_{pp}}{2}$	$-\dfrac{4A_{pp}}{(n\pi)^2}$ Only for odd functions	$b_n = 0$
Signal in time domain (full-wave rectified sine)	$\dfrac{2A_{pp}}{\pi}$	$\dfrac{-4A_{pp}}{\pi(4n^2-1)}$	$b_n = 0$
Signal in time domain (half-wave rectified sine)	$\dfrac{A_{pp}}{\pi}$	$\dfrac{-2A_{pp}}{\pi(n^2-1)}$ Only for even functions	$b_1 = \dfrac{A_{pp}}{2}$ All other $b_n = 0$

Signal in time domain	a_0	a_n	b_n
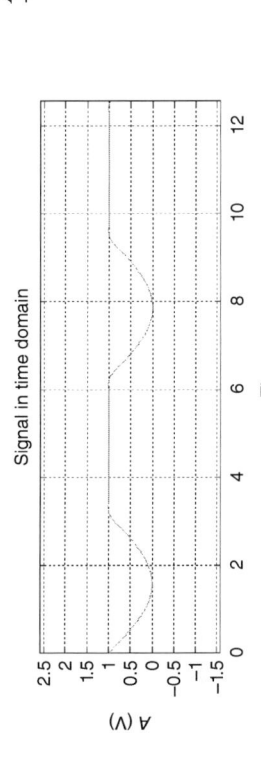	$\dfrac{A_{pp}(2\pi - 2)}{2\pi}$	$\dfrac{2A_{pp}}{\pi(n^2 - 1)}$ Only for even functions	$b_1 = -\dfrac{A_{pp}}{2}$ All other $b_n = 0$
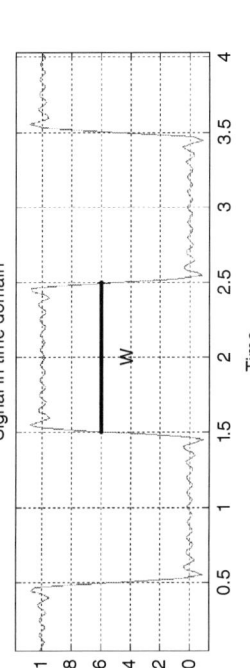	$\dfrac{wA_{pp}}{T}$	$\left((-1)^{\frac{n-1}{2}}\right)\dfrac{2A_{pp}}{n\pi}$ Only for odd functions	$b_n = 0$
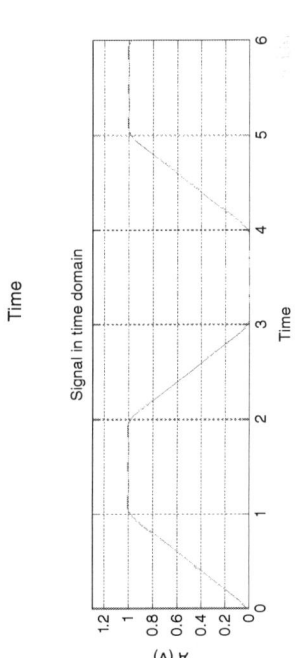	$\dfrac{A_{pp}}{2}$	$\dfrac{-4A_{pp}}{(n\pi)^2}$ Only for odd functions	$\left((-1)^{\frac{n-1}{2}}\right)\dfrac{4A_{pp}}{(n\pi)^2}$ Only for odd functions

(Continued)

Table 6.2.1 (Continued)

Waveforms	Coefficients of Fourier series		
	A_0	a_n	b_n
Signal in time domain (rotated plot, A (V) vs Time)	$\dfrac{A_{pp}(\pi + 2)}{2\pi}$	$\dfrac{-2A_{pp}}{(n^2 - 1)\pi}$ Only for even functions	$b_1 = A_{pp}\left(\dfrac{1}{2} - \dfrac{2}{\pi}\right)$ $b_n = \dfrac{-2A_{pp}}{n\pi}$ Only for odd functions
Signal in time domain (rotated plot, A (V) vs Time)	$A_0 = 0$	$a_n = 0$	$\dfrac{4nA_{pp}}{4n^2 - 1}$

Signal in time domain	A_0	a_n	b_n
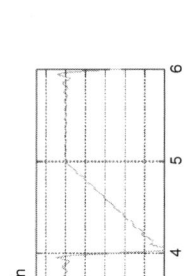	$A_0 = \dfrac{3}{4}A_{pp}$	$-\dfrac{2A_{pp}}{(n\pi)^2}$ Only for odd functions	$-\dfrac{A_{pp}}{n\pi}$
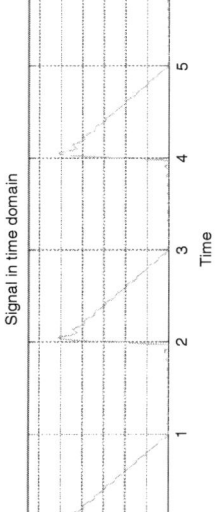	$A_0 = \dfrac{1}{4}A_{pp}$	$\dfrac{2A_{pp}}{(n\pi)^2}$ Only for odd functions	$\dfrac{A_{pp}}{n\pi}$

This table was prepared by Vitaly Sukharenko.

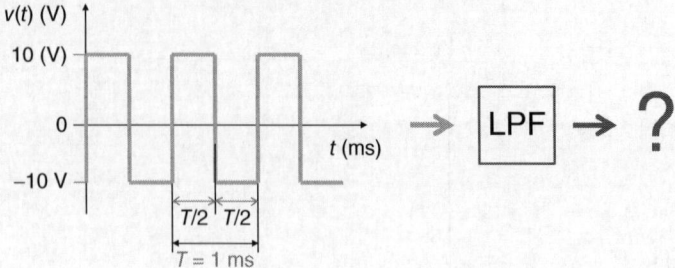

Figure 6.2.2 The output signal you would expect to see when a square-wave signal is presented to an LPF is an open question.

> Our previous discussions should convince us that the output signal depends mainly on the relationship between the signal's frequency, f_0, and the filter's cutoff frequency, f_C. This signal is also determined by the two other key filter specifications: the steepness of the roll-off (the width of the transition band) and attenuation of the stopband.

In the experiment shown in Figure 6.2.3, the input square wave has $A = 10\,\text{V}$ and $T = 1\,\text{ms}$, and therefore $f_0 = 1\,\text{kHz}$, whereas the filter's cutoff frequency is $f_C = 3.18\,\text{kHz}$.

6.2.2.2 Filtering a Single Harmonic

To understand why we obtain this output waveform but not any other, we need to revisit both the filtering process studied in Chapter 5 and the spectral analysis discussed in Section 6.1. Consider, for example, a single harmonic, $v_n^{in}(t) = A_n^{in} \sin(\omega_n t + \Theta_n^{in})$, presented to a low-pass RC filter, as shown in Figure 6.2.4a. The output harmonic, $v_n^{out}(t) = A_n^{out} \sin(\omega_n t + \Theta_n^{out})$, will have, of course, the same waveform (sine) and the same frequency, $\omega_n = 2\pi f_n$, but a different amplitude, A_n^{out}, and a phase shift, Θ_n^{out}. This is how we see the filtering process in time domain. In frequency domain, the same process is shown in Figure 6.2.4b. In both domains, we see the change in the amplitude and the phase shift of the output harmonic.

The aforementioned qualitative analysis and graphical presentation can be described mathematically in the following form:

Let $v_n^{in}(t) = A_n^{in} \sin(\omega_n t + \Theta_n^{in})$ be an nth input harmonic (with the amplitude, A_n^{in}, and the phase shift, Θ_n^{in}) presented to a filter. Let $v_n^{out}(t) = A_n^{out} \sin(\omega_n t + \Theta_n^{out})$ be the same nth harmonic that passed through the filter and now has amplitude A_n^{out} and output phase shift Θ_n^{out}; we call this the output harmonic. Then the output amplitude of the nth harmonic, A_n^{out}, is equal to the input amplitude of this harmonic, A_n^{in}, times the filter's attenuation, A_{vn}, computed at frequency ω_n:

$$A_n^{out} = A_n^{in} \times A_{vn} \tag{6.2.23a}$$

where

$$A_{vn} = \frac{1}{\sqrt{1 + \left(\frac{\omega_n}{\omega_c}\right)^2}} = \frac{1}{\sqrt{1 + \left(\frac{f_n}{f_c}\right)^2}} \tag{6.2.23b}$$

The output phase shift of the nth harmonic, Θ_n^{out}, is equal to

$$\Theta_n^{out} = \Theta_n^{in} + \Delta\Theta_n \tag{6.2.24a}$$

where the additional phase shift, $\Delta\Theta_n$, which the nth harmonic obtained after filtering, is

$$\Delta\Theta_n = -\tan^{-1}\left(\frac{\omega_n}{\omega_c}\right) = -\tan^{-1}\left(\frac{f_n}{f_c}\right) \tag{6.2.24b}$$

6.2 Introduction to Spectral Analysis | 553

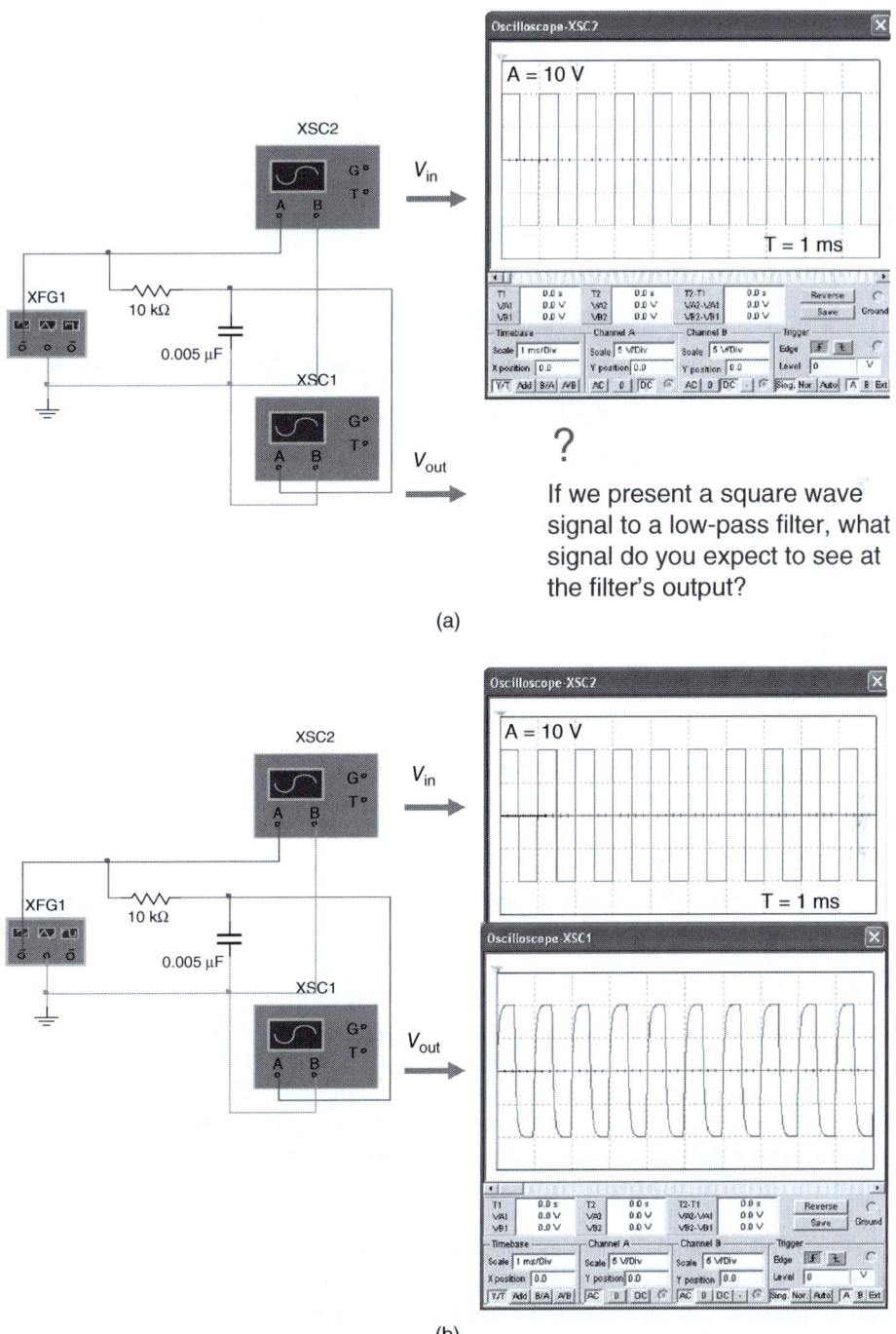

Figure 6.2.3 Time-domain presentation of filtering a square wave: (a) Experiment setup with an unknown output signal; (b) experiment setup with the sample of an output signal obtained.

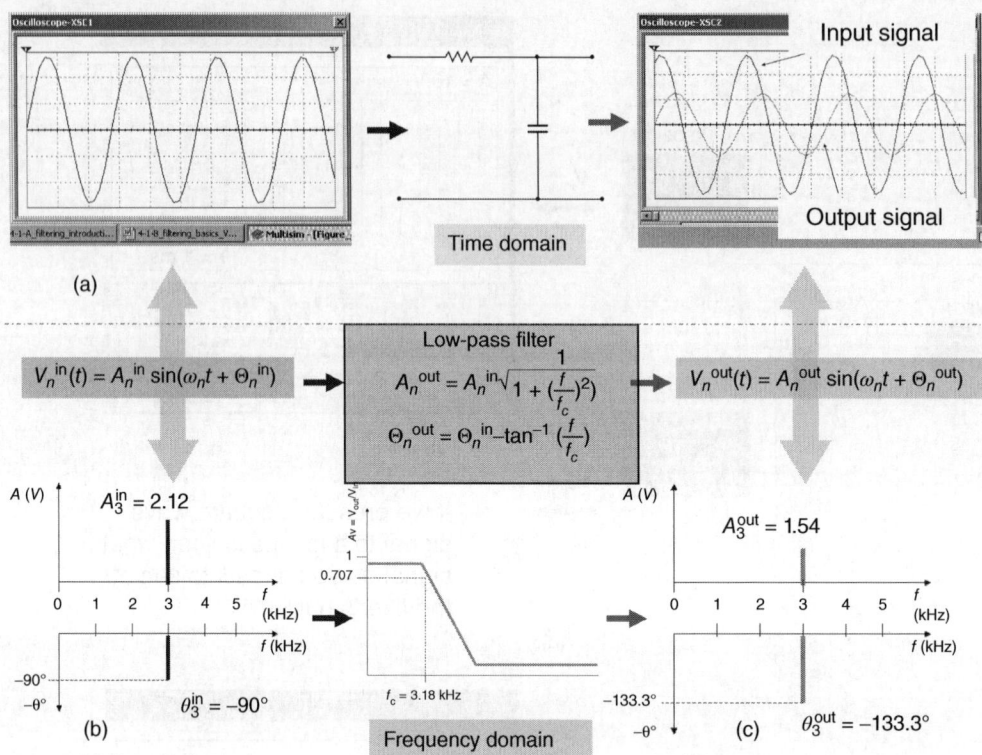

Figure 6.2.4 The (a) time-domain and (b) frequency-domain presentations of the filtering of a single harmonic: The output harmonic decreases in amplitude and undergoes an additional shift in phase.

Figure 6.2.4 shows – in both time and frequency domains – that the output harmonic experiences a decrease in amplitude and undergoes an additional phase shift with respect to the input harmonic. This point is discussed in Chapter 5 with regard to a single sinusoidal signal. Review and analyze Figure 6.2.4 closely: It gives you the key to understanding the effect of filtering on signals.

To exemplify the usage of (6.2.23a), (6.2.23b), (6.2.24a), and (6.2.24b), we have done the calculations for a square wave with $A = 10$ V and $T = 1$ ms that passes through the filter with $f_C = 3.18$ kHz. The results – the input amplitudes, the filter's attenuations, the output amplitudes, and the additional phase shifts of the first seven harmonics – are presented in Table 6.2.2:

Carefully analyze Table 6.2.2 to appreciate the effect of LPF on the signal. As a useful exercise, build the graph of every input and output harmonic on the same set of axes to visualize your explanations.

6.2.2.3 Filtering a Periodic Signal – Time and Frequency Domains

The next step in understanding the effect of filtering on signals is moving from the consideration of an individual harmonic to consideration of the entire periodic signal presented as a sum of many harmonics. Indeed, in the case of a square-wave signal (see Example 6.1.1), we know that this signal is presented as the following Fourier series (see Eq. (6.1.13)):

$$v(t) = A/2 + 2A/\pi \ \sin \ 2\pi f_0 t + 2A/3\pi \ \sin \ 2\pi \ 3f_0 t + 2A/5\pi \ \sin \ 2\pi 5f_0 t + \cdots$$

Therefore, *filtering of a square-wave signal is equivalent to filtering the sum of the harmonics shown in this Fourier series.*

Table 6.2.2 Example of the effect of filtering on signals: input amplitudes, filter attenuations, output amplitudes, and additional phase shifts of the first seven harmonics of a square-wave signal with $A = 10\,\text{V}$ and $T = 1\,\text{ms}$.

n	f_n (kHz)	A_n^{in} (V)	$A_{vn} = 1/\sqrt{(1+(f_n/f_C)^2)}$	A_n^{out} (V) $= A_n^{\text{in}} \times A_{vn}$	$\Delta\Theta_n = -\tan^{-1}(f_n/f_C)$
0 (dc)	$f_0 = 0$	$A_0^{\text{in}} = 5$	$A_{v0} = 1/\sqrt{(1+(f_1/f_C)^2)}$ $= 1/\sqrt{(1+(0/3.18)^2)}$ $= 1$	$A_0^{\text{out}} = A_0^{\text{in}} \times A_{v0}$ $= 5 \times 1 = 5$	$\Delta\Theta_0 = -\tan^{-1}(f_0/f_C)$ $= -\tan^{-1}(0/3.18)$ $= 0°$
1	$f_1 = 1$	$A_1^{\text{in}} = 6.37$	$A_{v1} = 1/\sqrt{(1+(f_2/f_C)^2)}$ $= 1/\sqrt{(1+(1/3.18)^2)}$ $= 0.95$	$A_1^{\text{out}} = A_1^{\text{in}} \times A_{v1}$ $= 6.37 \times 0.95$ $= 6.08$	$\Delta\Theta_1 = -\tan^{-1}(f_1/f_C)$ $= -\tan^{-1}(1/3.18)$ $= -17.5°$
3	$f_3 = 3$	$A_3^{\text{in}} = 2.12$	$A_{v3} = 1/\sqrt{(1+(f_3/f_C)^2)}$ $= 1/\sqrt{(1+(3/3.18)^2)}$ $= 0.73$	$A_3^{\text{out}} = A_3^{\text{in}} \times A_{v3}$ $= 2.12 \times 0.73$ $= 1.54$	$\Delta\Theta_3 = -\tan^{-1}(f_3/f_C)$ $= -\tan^{-1}(3/3.18)$ $= -43.3°$
5	$f_5 = 5$	$A_5^{\text{in}} = 1.27$	$A_{v5} = 1/\sqrt{(1+(f_5/f_C)^2)}$ $= 1/\sqrt{(1+(5/3.18)^2)}$ $= 0.54$	$A_5^{\text{out}} = A_5^{\text{in}} \times A_{v5}$ $= 1.27 \times 0.54$ $= 0.68$	$\Delta\Theta_5 = -\tan^{-1}(f_5/f_C)$ $= -\tan^{-1}(5/3.18)$ $= -57.5°$
7	$f_7 = 7$	$A_7^{\text{in}} = 0.91$	$A_{v7} = 1/\sqrt{(1+(f_7/f_C)^2)}$ $= 1/\sqrt{(1+(7/3.18)^2)}$ $= 0.41$	$A_7^{\text{out}} = A_7^{\text{in}} \times A_{v7}$ $= 0.91 \times 0.41$ $= 0.38$	$\Delta\Theta_7 = -\tan^{-1}(f_7/f_C)$ $= -\tan^{-1}(7/3.18)$ $= -65.6°$

This signal passed through a low-pass filter with $f_C = 3.18\,\text{kHz}$.

To obtain the entire output signal, we need to recall that our LPF is a *linear system* to which the *principle of superposition* can be applied. Application of the principle of superposition in our case means the following:

> *When filtering a periodic signal, we actually present to an LPF the sum of the harmonics; to find the filtered signal, we need to find the output of this sum. But following the principle of superposition, we can first find the result of the filtering of an individual harmonic and then obtain the entire output signal by summarizing the outputs of individual harmonics. Since the filtering of an individual harmonic is clearly explained (see Figure 6.2.4), the only step left to perform to obtain the entire filtered signal is summarizing the individual outputs.*

This process shown in Figure 6.2.5 requires your careful review and analysis.

Now, let us consider the filtering operation in frequency domain more closely. Look at Figure 6.2.6: It shows that if we present a square-wave signal to an *ideal* LPF, then the spectrum of an output signal will contain only undisturbed low-frequency harmonics, whose frequencies are less than the cutoff frequency of the filter. All harmonics with frequencies higher than the cutoff will be filtered out. Figure 6.2.6 shows an example of filtering the same-square signal ($A = 10\,\text{V}$ and $T = 1\,\text{ms}$) by the same LPF ($f_C = 3.18\,\text{kHz}$) that is presented in Figures 6.2.3–6.2.5.

Figure 6.2.6 introduces the idea of how a filtering works in frequency domain. In fact, studying a frequency-domain presentation is the best way to fully understand the effect of filtering on signals. (This is why the most direct relationship between input and output signals is given through

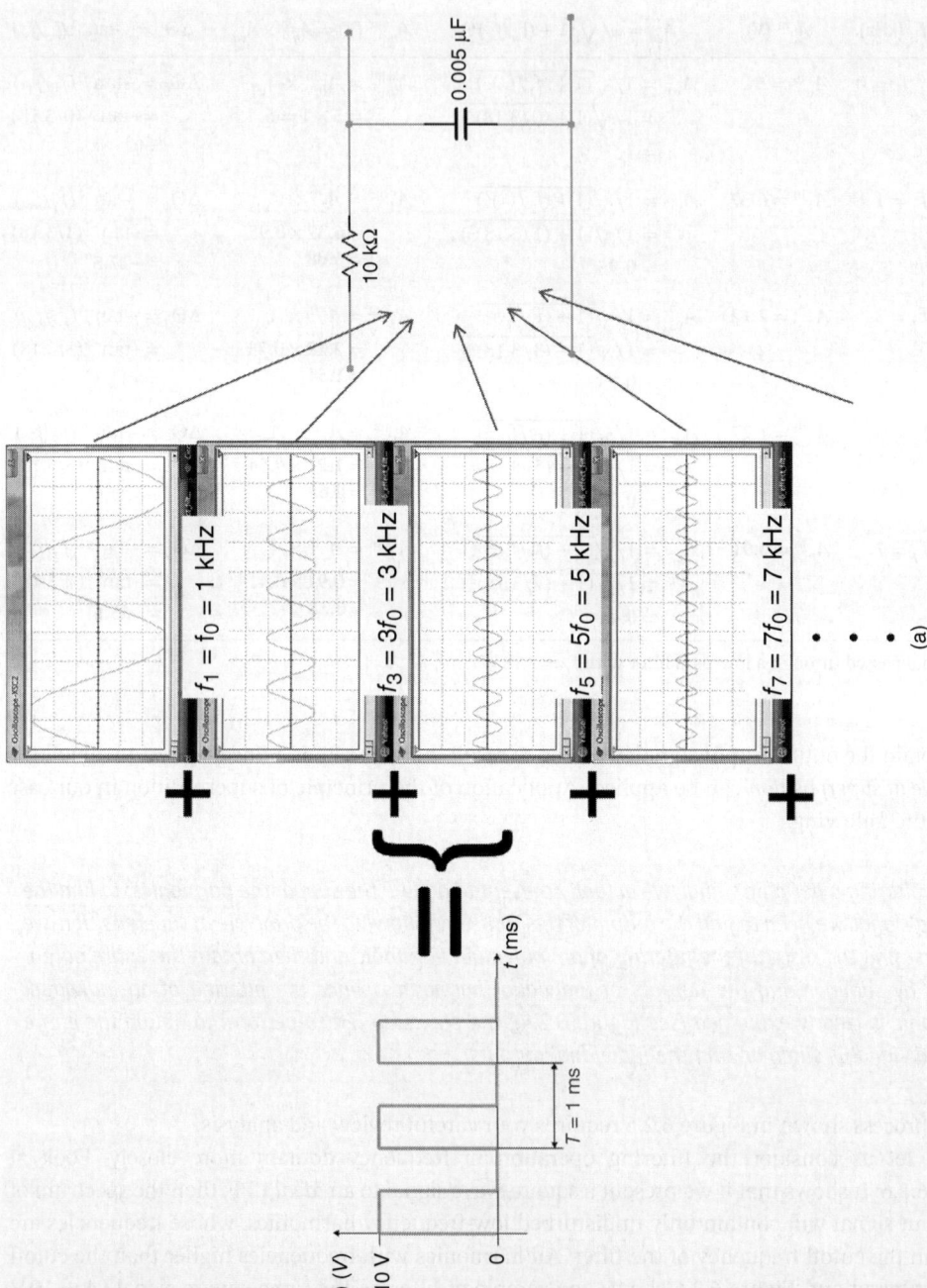

Figure 6.2.5 Process of filtering a square wave: (a) The square wave is expanded in the Fourier series (only four harmonics are shown) and each harmonic is presented to an LPF; (b) each harmonic is filtered by the LPF and the output of every harmonic is obtained; (c) all output harmonics are summarized, creating the entire output – the square-wave signal. The dc component is not shown.

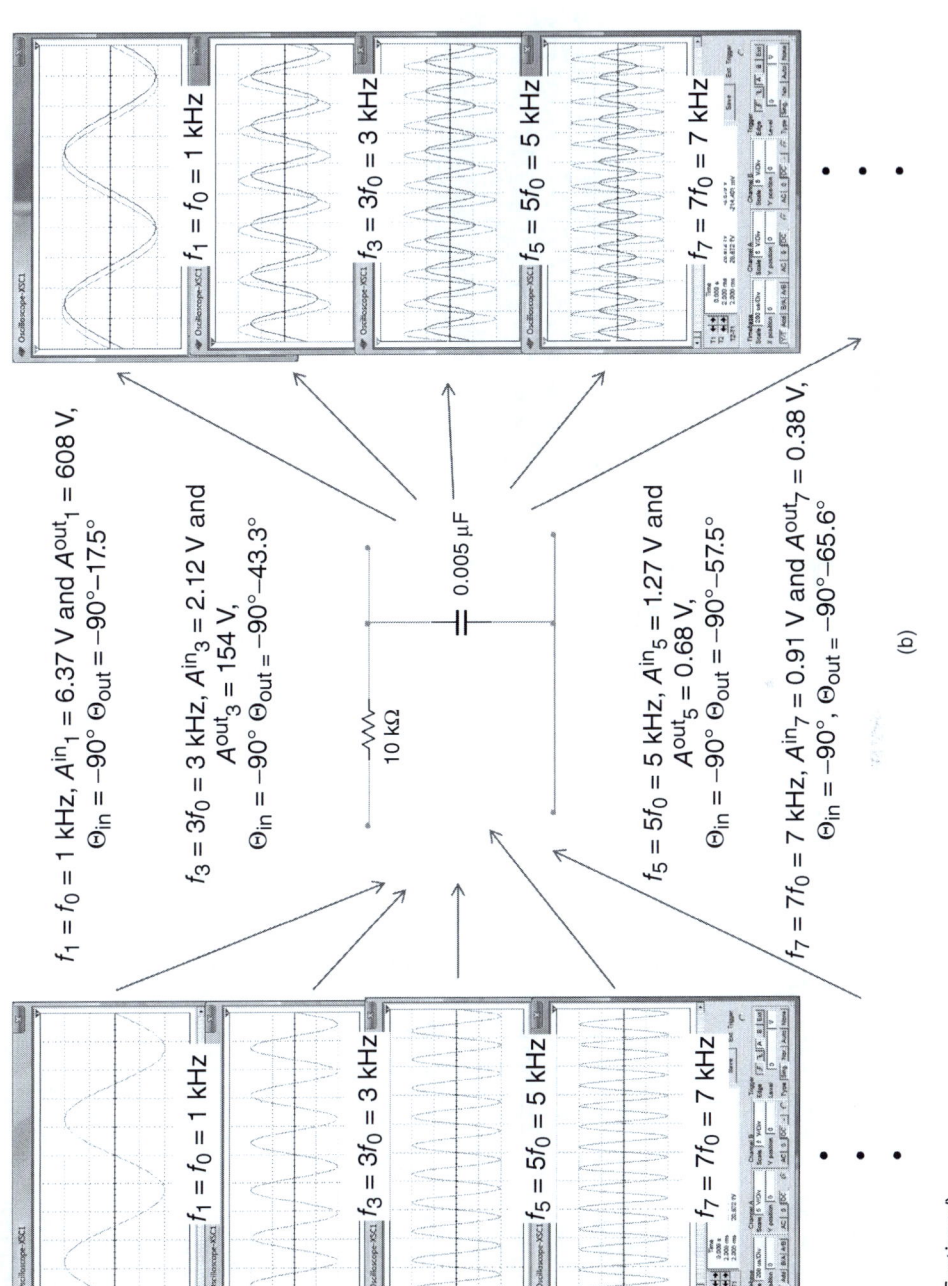

Figure 6.2.5 *(Continued)*

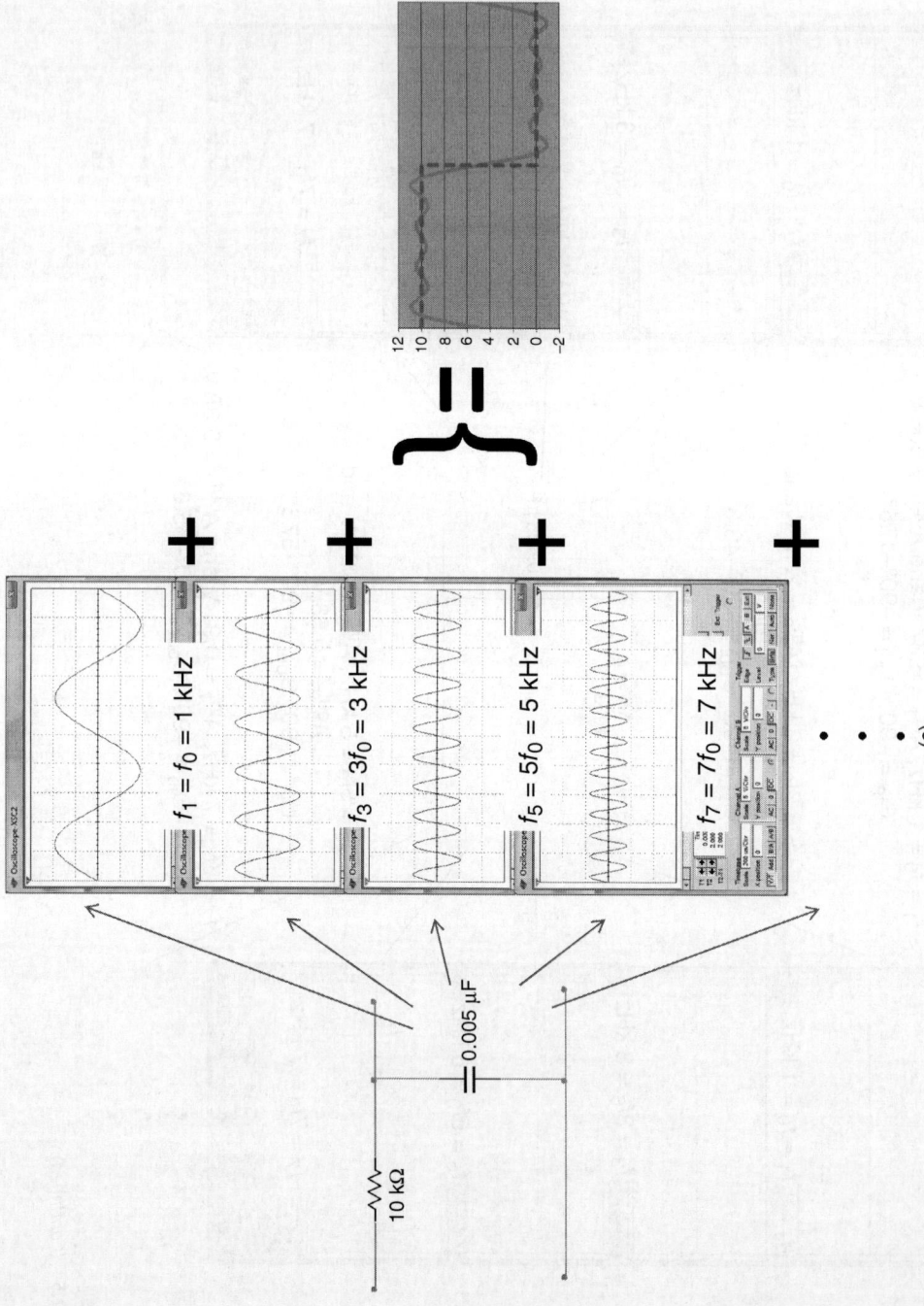

Figure 6.2.5 *(Continued)*

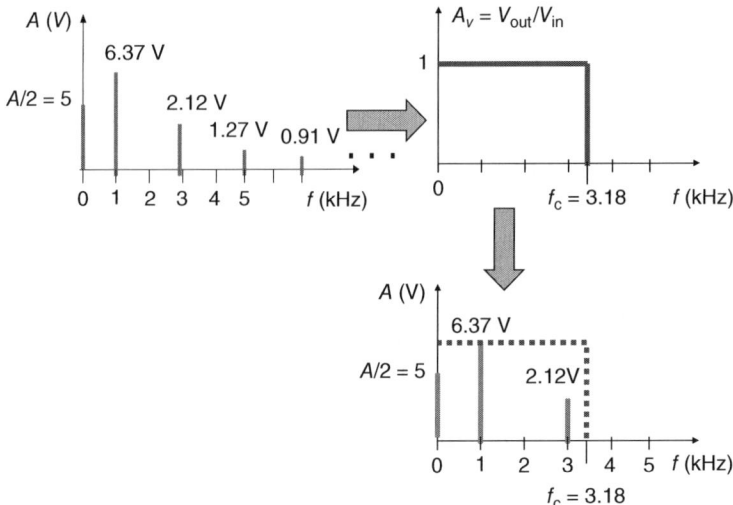

Figure 6.2.6 The square-wave signal passes through an *ideal* LPF: a frequency-domain presentation. (There are no additional phase shifts; the initial phases are not shown.)

the transfer function [frequency domain], not through differential equations [time domain]. We refer you to Chapter 5.) We know that the ideal filter cannot be realized. Applying, then, the signal-filtering concept to a real situation, as depicted in Figure 6.2.7, we can expect that presenting a square-wave signal to a real LPF will produce the following results:

- The low-frequency harmonics, whose frequencies are smaller than the filter's cutoff frequency, f_C, will pass with a small decrease in amplitudes and small values of the additional phase shift.
- The amplitudes of the intermediate-frequency harmonics, whose frequency is near f_C, will be significantly decreased (near 0.7 of their input values) by the filter and the additional phase shifts will be well pronounced.
- The high-frequency harmonics, whose frequency is greater than f_C, will be almost completely filtered out and the additional phase shifts will be close to the limit.

All these cases are shown in Figure 6.2.7.

Figure 6.2.7 shows an example of filtering, with a real LPF, the square-wave signal shown in previous figures. The top left of Figure 6.2.7a shows the input signal, the top right of the figure schematically shows the filter's attenuation, and the bottom of the figure shows both the input and the output signals superimposed on the filter's attenuation graph. The numbers show how the amplitudes of each harmonic get reduced after filtering; for example, the amplitude of the fundamental harmonic is reduced from 6.37 V at the input to 6.08 V at the output. In this example, $f_C = 3.18$ kHz; hence, the first harmonic ($f_0 = 1$ kHz) is still within the passband. The third harmonic ($3f_0 = 3$ kHz) is almost at the cutoff frequency, and its amplitude is reduced by approximately 0.707. The fifth harmonic ($5f_0 = 5$ kHz) is well beyond f_C; no wonder that its amplitude is reduced significantly (in fact, almost by half). See Table 6.2.2 for detailed calculations of these values. This example shows the concept of signal filtering as seen in frequency domain. Bear in mind that some of our figures do not show the phase shift induced by the filter; Figure 6.2.5b and Table 6.2.2, however, make this point crystal clear.

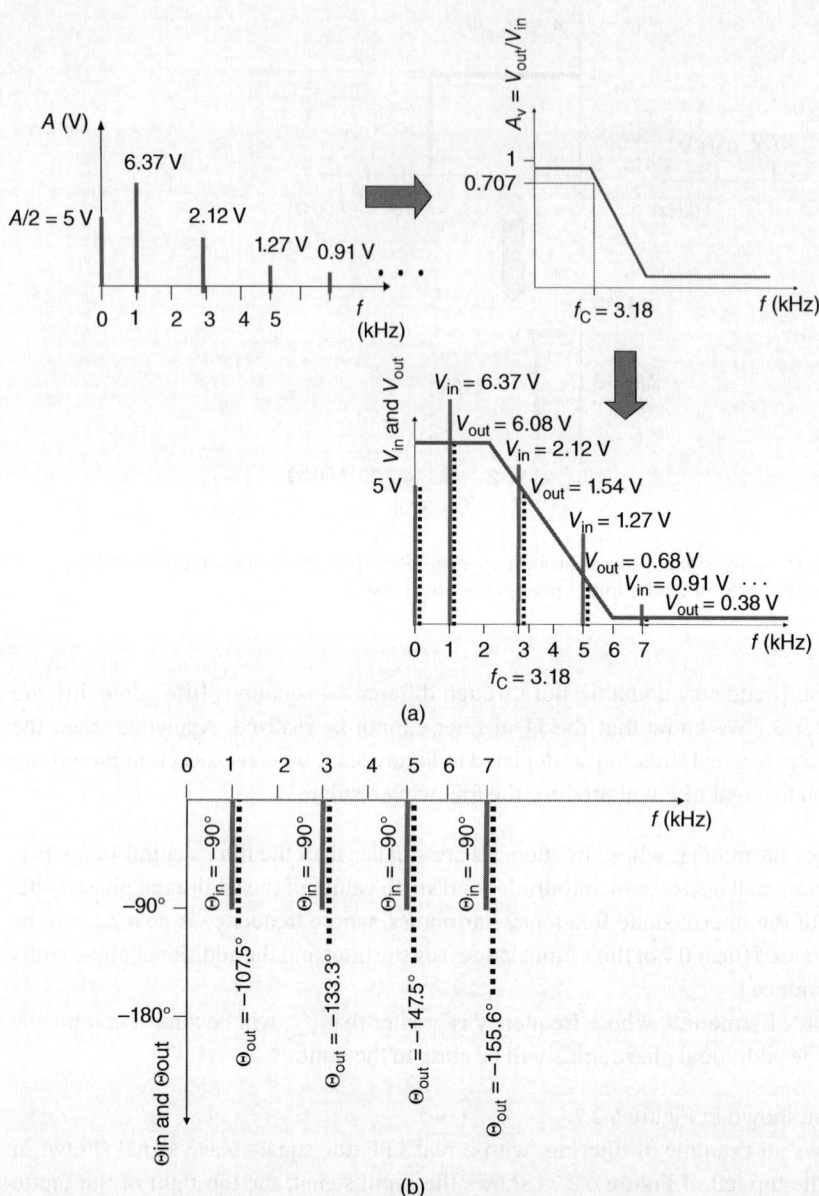

Figure 6.2.7 The frequency-domain presentation of filtering a square-wave signal by a real LPF: (a) Amplitude spectrum; (b) phase spectrum. (Not to scale.)

6.2.2.4 Filtering a Signal – The Entire Picture

After considering in detail the effect of filtering on signals, it is time to present this operation in its entirety. Figure 6.2.8a shows the experiment setup considered in Figure 6.2.3; this time we lowered the filter's cutoff frequency from 3.18 to 0.318 kHz while keeping the input signal the same. What waveform do you expect to see at the filter's output? Of course, the answer can be readily obtained by performing our experiment, the result of which is shown in Figure 6.2.8b.

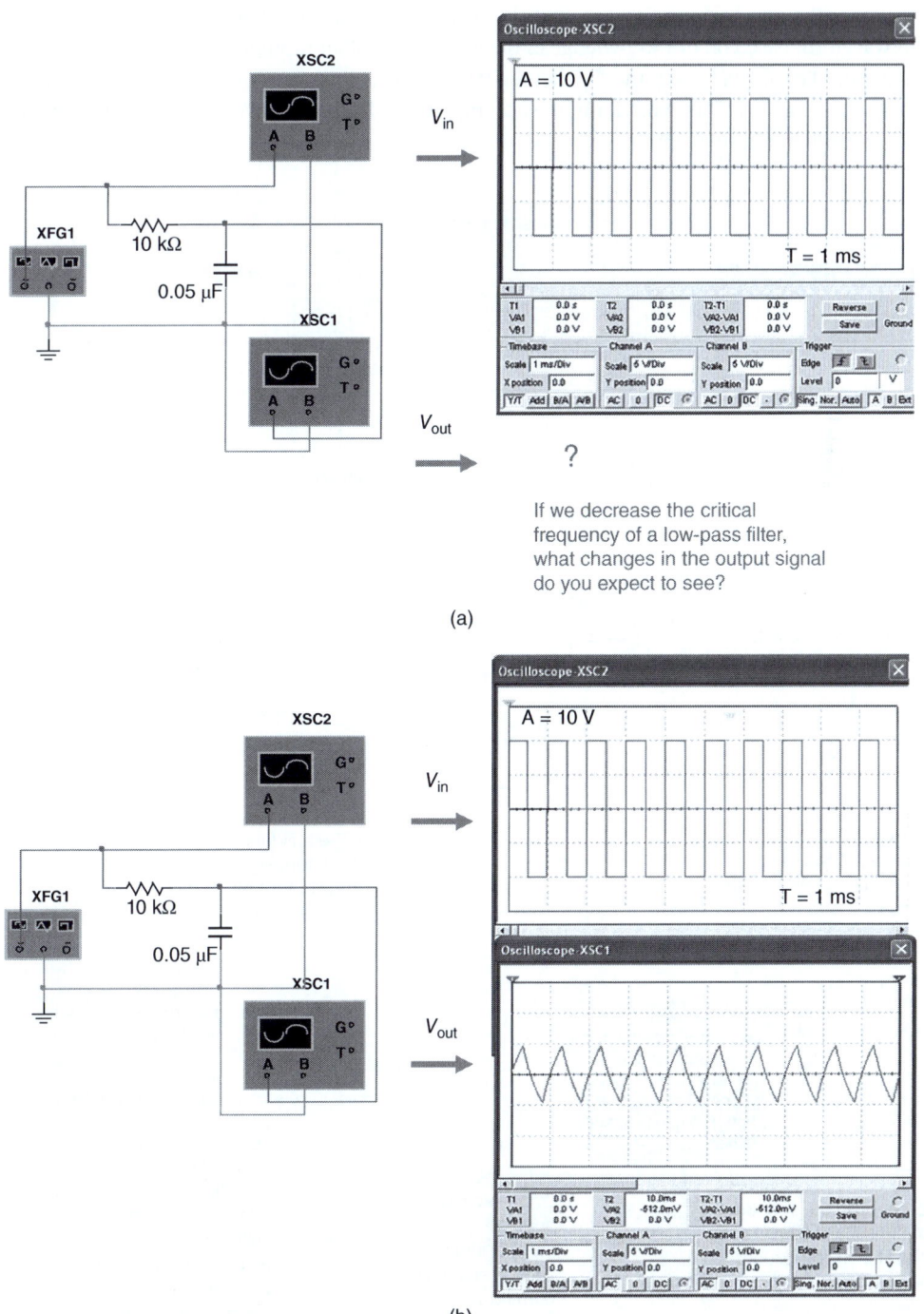

Figure 6.2.8 The square wave is presented to an LPF having a low (0.318 kHz) cutoff frequency: (a) What waveform should we expect to see at the output of the filter? (b) the waveform obtained by the experiment.

6 Spectral Analysis 1 – The Fourier Series in Modern Communications

Analyzing the output waveform in Figure 6.2.8b, we can say that the output signal closely resembles the sine (the one harmonic) wave. But the real question is, can we theoretically predict the output waveform based on our knowledge of the parameters of an input signal and the filter's cut-off frequency? The answer is yes and the method for arriving at this accurate prediction is presented in Figures 6.2.4–6.2.7, in Table 6.2.2 and the accompanying discussion. We visualize the summary of this approach in Figure 6.2.9.

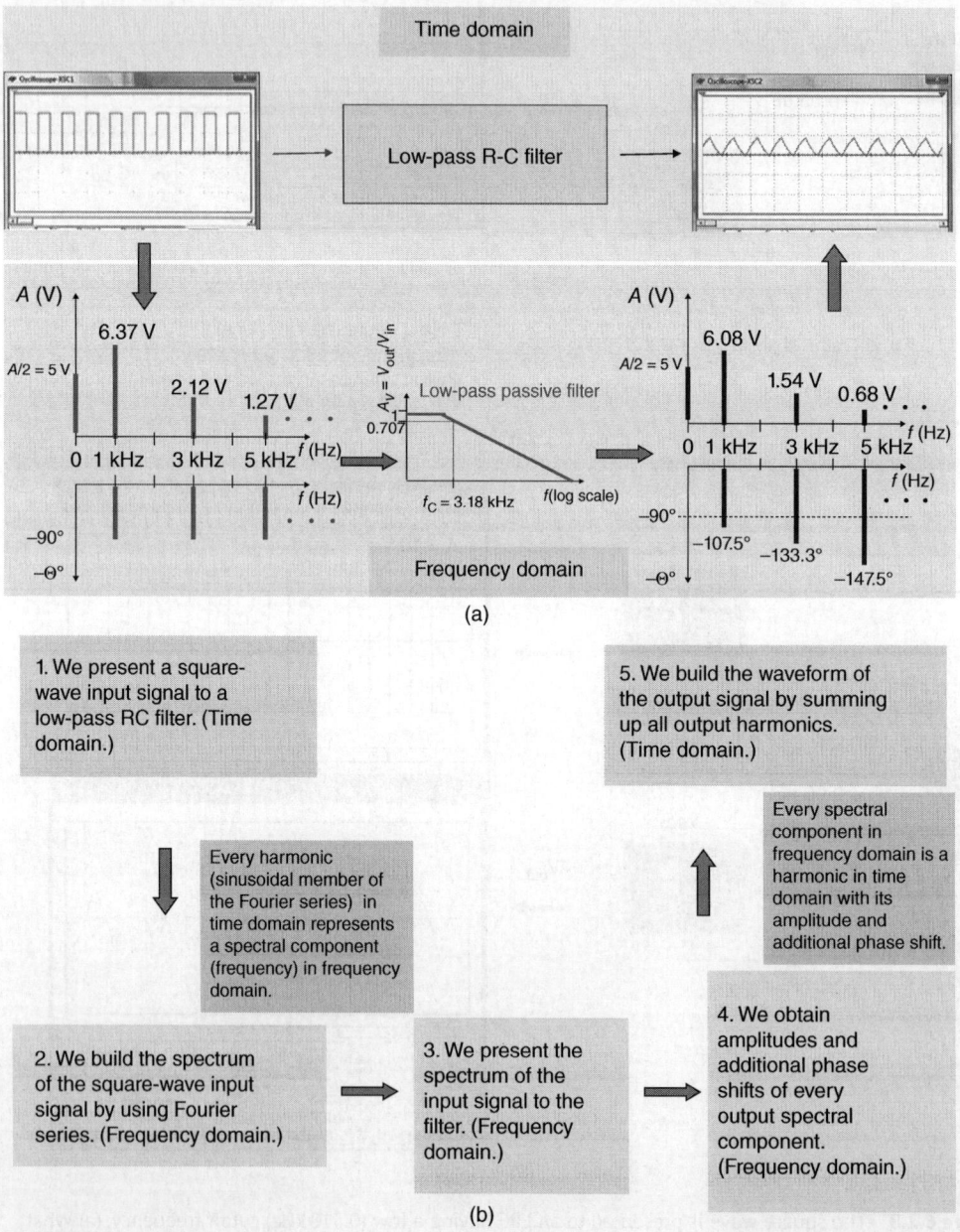

Figure 6.2.9 The entire picture of signal filtering: (a) Time- and frequency-domain presentations of filtering; (b) description of the sequence of steps in signal filtering.

Figure 6.2.9 shows that in time domain, we present a signal to an LPF and observe the output waveform; however, to predict this output, we need to find the spectrum of the input signal (see the "Spectral analysis" subsection in Section 6.1.), compute the amplitudes and phase shifts of each harmonic filtered by the given filter (see Table 6.2.2), and sum up all the output harmonics (see the "Spectral synthesis" subsection in Section 6.1) to obtain the waveform of the output.

Example 6.2.2 demonstrates the implementation of this approach.

Example 6.2.2 Effect of Filtering on Signals

Problem

A square-wave signal with $A = 10$ V and $T = 1$ ms is presented to an LPF with $f_C = 0.318$ kHz. Build the waveform of the output based on the seven ($7f_0$) output harmonics.

Solution

We need to follow the steps shown in Figure 6.2.9b:

First, the spectrum of this square wave has been found in Example 6.2.1 and is shown again in Figures 6.2.5–6.2.8. Secondly, performing calculations similar to those shown in Table 6.2.2, we find the output amplitudes and the additional phase shift for the first seven harmonics, as shown in Table 6.2.3:

Finally, we sum up the output harmonics and obtain the waveform of the output. Our summing up is done with MATLAB (see MATLAB box 6.2.2), but you can use MS Excel or any other software tool. The computed output waveform is presented in Figure 6.2.10.

Table 6.2.3 Input amplitudes, filter attenuations, output amplitudes, and additional phase shifts of the first seven harmonics for Example 6.2.2: $A_0^{in} = 10$ V, $T = 1$ ms, and $f_C = 0.318$ kHz.

n	f_n (kHz)	A_n^{in} (V)	$A_{vn} = 1/\sqrt{(1+(f_n/f_C)^2)}$	A_n^{out} (V) $= A_n^{in} \times A_{vn}$	$\Delta\Theta_n = -\tan^{-1}(f_n/f_C)$
0 dc	$f_0 = 0$	$A_0^{in} = 5$	$A_{v0} = 1/\sqrt{(1+(f_0/f_C)^2)}$ $= \sqrt{(1+(0/0.318)^2)}$ $= 1$	$A_0^{out} = A_0^{in} \times A_{v0}$ $= 5 \times 1 = 5$	$\Delta\Theta_0 = -\tan^{-1}(f_0/f_C)$ $= -\tan^{-1}(0/0.318)$ $= 0°$
1	$f_1 = 1$	$A_1^{in} = 6.37$	$A_{v1} = 1/\sqrt{(1+(f_1/f_C)^2)}$ $= 1/\sqrt{(1+(1/0.318))^2}$ $= 0.303$	$A_1^{out} = A_1^{in} \times A_{v1}$ $= 6.37 \times 0.303$ $= 1.93$	$\Delta\Theta_1 = -\tan^{-1}(f_1/f_C)$ $= -\tan^{-1}(1/0.318)$ $= -72.4° = -1.263$ rad
3	$f_3 = 3$	$A_3^{in} = 2.12$	$A_{v3} = 1/\sqrt{(1+(f_3/f_C)^2)}$ $= 1/\sqrt{(1+(3/0.318))^2}$ $= 0.105$	$A_3^{out} = A_3^{in} \times A_{v3}$ $= 2.12 \times 0.105$ $= 0.223$	$\Delta\Theta_3 = -\tan^{-1}(f_3/f_C)$ $= -\tan^{-1}(3/0.318)$ $= -83.9° = -1.465$ rad
5	$f_5 = 5$	$A_5^{in} = 1.27$	$A_{v5} = 1/\sqrt{(1+(f_5/f_C)^2)}$ $= 1/\sqrt{(1+(5/0.318))^2}$ $= 0.063$	$A_5^{out} = A_5^{in} \times A_{v5}$ $= 1.27 \times 0.063$ $= 0.081$	$\Delta\Theta_5 = -\tan^{-1}(f_5/f_C)$ $= -\tan^{-1}(5/0.318)$ $= -86.4° = -1.507$ rad
7	$f_7 = 7$	$A_7^{in} = 0.91$	$A_{v7} = 1/\sqrt{(1+(f_7/f_C)^2)}$ $= 1/\sqrt{((1+7/0.318)^2)}$ $= 0.045$	$A_7^{out} = A_7^{in} \times A_{v7}$ $= 0.91 \times 0.045$ $= 0.041$	$\Delta\Theta_7 = -\tan^{-1}(f_7/f_C)$ $= -\tan^{-1}(7/0.318)$ $= -87.4° = -1.525$ rad

6 Spectral Analysis 1 – The Fourier Series in Modern Communications

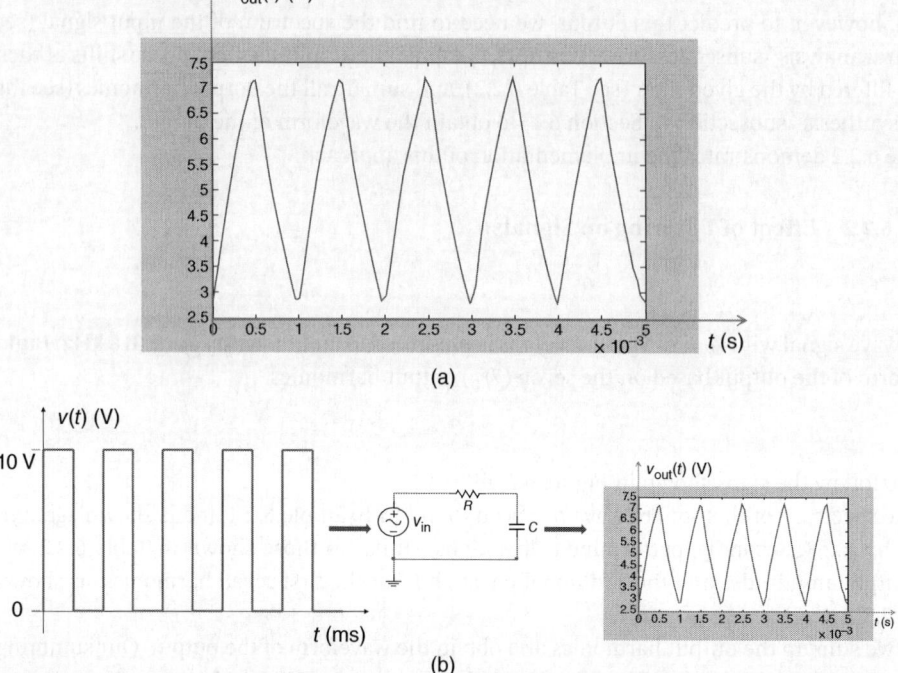

Figure 6.2.10 (a) The output waveform in Example 6.2.2; (b) the input and output waveforms in Example 6.2.2. (Not to scale.)

MATLAB Box 6.2.2 Synthesizing the Signal's Waveform in Example 6.2.2

Summing the dc member and the output harmonics of the filtered signal to obtain the output waveform.

```
»w=2*pi*1000;          %This is the value of fundamental frequency in rad/s

» A=5;                 %This is the value of dc member in volts

» F1=1.93*sin(w*t-1.263);     %This is the first output harmonic

» F3=0.223*sin(3* w*t-1.465);      % This is the third output harmonic

» F5=0.081*sin(5* w*t-1.507) % This is the fifth output harmonic

» F7=0.041*sin(7* w*t-1.525) %This is the seventh output harmonic

» S=A+F1+F3+F5+F7;     %This is the sum of the dc member and seven out-
put harmonics

»plot(t*1000,S)        %This is the command to MATLAB to plot the graph of this sum
                       %and choose the sampling rate based on the signal's period
```

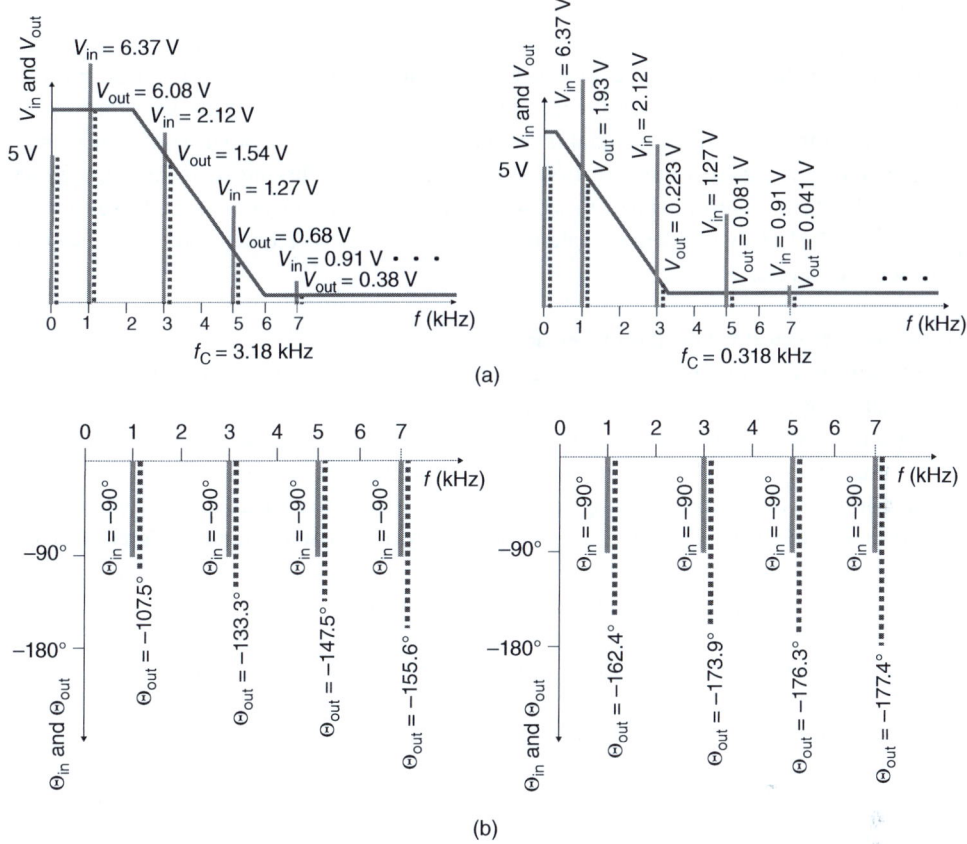

Figure 6.2.11 Spectra of input and output signals presented to an LPF with different cutoff frequencies: (a) Amplitude spectra (dc and the first four spectral components) of the square wave with $A = 10$ V and $T = 1$ ms presented to the LPFs with (left-hand side) $f_C = 3.18$ kHz and with (right-hand side) $f_C = 0.318$ kHz (right); (b) phase spectra of these signals. (Not to scale.)

Discussion

- First of all, this example demonstrates the techniques for making theoretical calculations of a filtered signal's output waveform. Compare the computed output waveform in Figure 6.2.10 with that in Figure 6.2.8b, which was obtained by simulation. You can see that these waveforms are very similar, though they differ in small details, which is explained by the small number of harmonics taken for the synthesis of the computed waveform. One practical piece of advice: *Don't forget about additional phase shifts when computing the output harmonics: You will obtain the wrong result otherwise.*
- Secondly, we need to compare the output waveforms shown in Figures 6.2.3 and 6.2.8: In these figures, we present the same square-wave signal with $A = 10$ V and $T = 1$ ms to the LPFs with different cutoff frequencies and obtain quite different output waveforms. To understand this result, we need to look at this operation from the frequency-domain perspective, as shown in Figure 6.2.11.

The values of the amplitudes shown in Figure 6.2.11 are taken from Tables 6.2.2 and 6.2.3. Figure 6.2.11a for LPF with $f_C = 3.18$ kHz shows that the amplitudes of the output harmonics still preserve significant values; this is why the output waveform shown in Figure 6.2.3 resembles an input signal – a square wave. In contrast, Figure 6.2.11b for LPF with $f_C = 0.318$ kHz shows that the amplitude of the first filtered harmonic is reduced more than three times, whereas the amplitudes of the following harmonics decrease 10 times and more. No wonder that the waveform of an output signal resembles a sinusoidal signal more than a square wave. In other words, decreasing the cutoff frequency10 times, from 3.18 to 0.318 kHz, results in filtering more high-frequency harmonics so that the output signal becomes closer to a single harmonic, that is, a sinusoid.

Figures 6.2.3–6.2.11 explicitly demonstrate how the filter's cutoff frequency affects the output spectrum and the waveform.

6.2.2.5 A Final Note on Effect of Filtering on Signals

We cannot overestimate the importance of signal filtering in modern communications. Refer to the recent history of communications: A legacy copper telephone line, which was approximately modeled by a low-pass RC filter, had the cutoff frequency of 4 kHz because the line was designed to deliver a human voice. When the Internet arrived, the telephone lines still remained the main transmission links available for providing access to the Internet from office or home computers. It became immediately clear that the direct transmission of a digital computer signal over the copper telephone wire is impossible. (Compare the output waveforms shown in Figures 6.2.3 and 6.2.10 to see how much a digital signal could get distorted. Imagine – and, better, simulate – the case when a digital signal of 4-MHz passes through an RC LPF with a 4-kHz cutoff frequency.) To solve the problem, engineers resorted to a modulation technique and a dial-up modem (*MOdulator-DEModulator*) appeared. Later, more sophisticated modems were developed to implement a digital subscriber line (DSL) that supported transmission at 1.5 Mb/s and higher over the same copper wire. But the demand for the higher transmission rates grew quickly, and today access to the Internet is mainly provided by an optical fiber capable to transmit hundreds of terabits per second.

Consider another example: modern wireless transmission technology called Wi-Fi (see Section 1.2). The best version of Wi-Fi transmits at 5-GHz carrier frequency and delivers data at the transmission rate of 54 Mb/s. These numbers tell us about the limitations of this technology caused, in particular, by the restriction of the channel's bandwidth.

Thus, regardless of the specific technology, all practical channels are bandlimited, as mentioned in Chapter 3. Whether we use Wi-Fi or fiber-optic technologies for transmission, the channels work as band-pass filters. This implies that the channels inevitably filter the signals being transmitted.

The analysis presented in this subsection equips us with the techniques for quantitative estimation of the distortion imposed by a signal filtering, which eventually enables us to control the quality of transmission. To learn more on this interesting and important topic, refer to specialized literature, for example, Oppenheim and Willsky (2015, pp. 437–440).

6.2.3 Harmonic Distortion

If we present a sinusoid (a single harmonic) to an amplifier and keep the amplifier in linear regime, then the output will be an amplified sinusoid, which is still a single harmonic. Remember that by *linear regime,* we mean that the output of the amplifier is proportional to the input. This situation is depicted in Figure 6.2.12, where both waveforms and the spectra of the input and output signals are shown.

If, however, the gain of an amplifier is too big, then the amplifier will be forced into nonlinear (saturation) regime. Here, the amplifier starts to distort the output signal, making it *nonsinusoidal*.

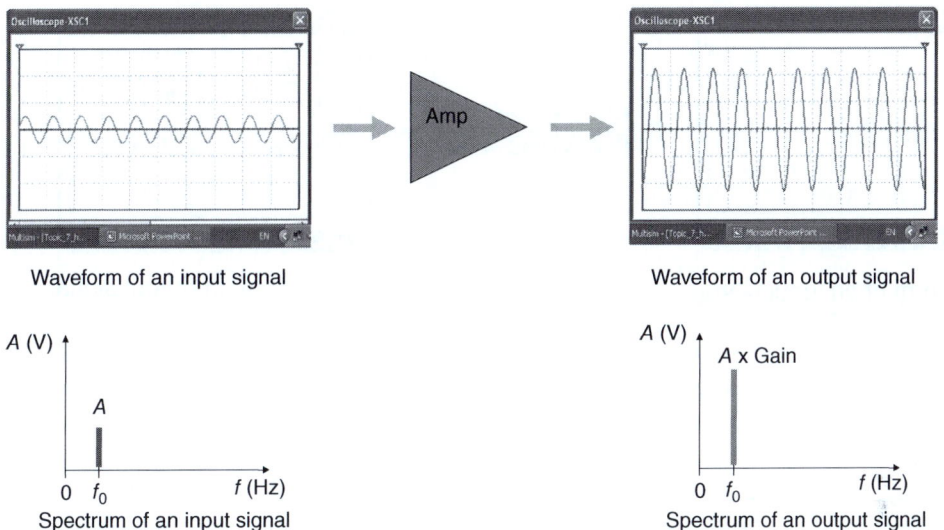

Figure 6.2.12 A sinusoidal signal presented to a linear amplifier: waveforms and spectra of the input and output signals.

For example, a cosine signal presented to an open-loop OP-AMP will become a square-wave signal due to clipping. We know that a sinusoidal signal has one harmonic, whereas the spectrum of a square wave contains many harmonics – in fact, an infinite number. This situation is shown in Figure 6.2.13.

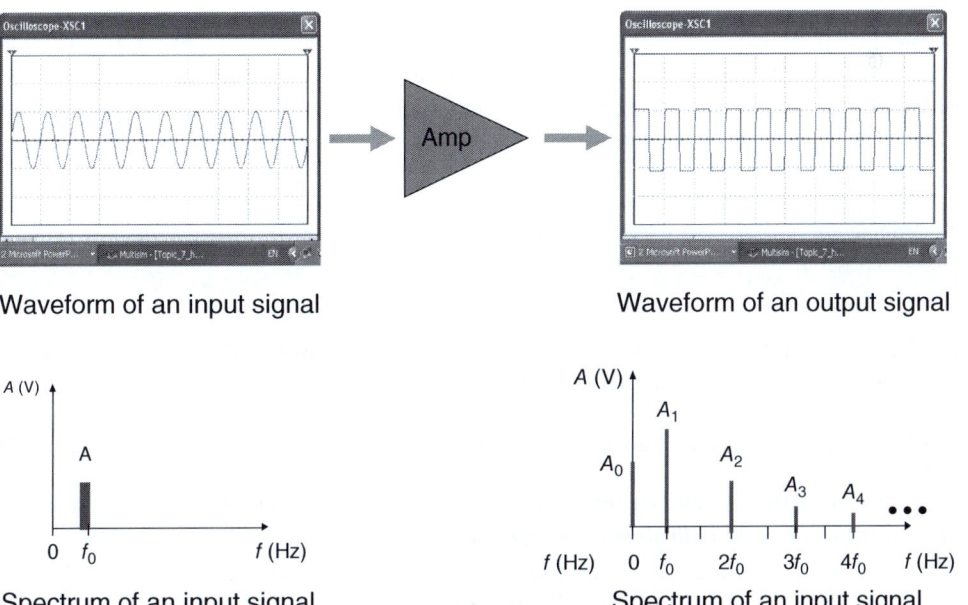

Figure 6.2.13 Example of harmonic distortion: The input sinusoidal signal becomes a square wave at the output of an amplifier operating in the deep nonlinear (saturation) regime; because of the change in its waveform, the output signal contains many harmonics in contrast to the input sinusoidal signal, which contains only one harmonic.

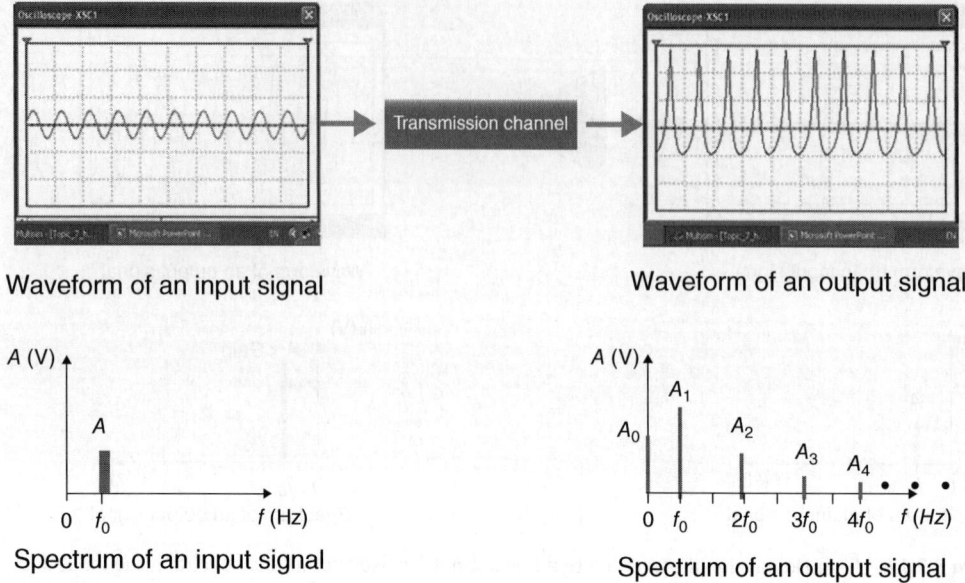

Figure 6.2.14 Harmonic distortion: The input sinusoidal signal after transmission becomes a periodic nonsinusoidal signal containing many harmonics.

Figure 6.2.14 illustrates the other example of harmonic distortion, the one in which the input sinusoidal signal after transmission becomes a periodic nonsinusoidal signal containing many harmonics.

The examples presented in Figures 6.2.13 and 6.2.14 demonstrate that in telecommunications we often meet a situation where the input signal is a sinusoid, but the output signal has a sophisticated waveform. From a frequency-domain standpoint, this means that at the input, we have a single-harmonic signal whereas the output signal contains many harmonics.

The phenomenon whereby an output signal contains more harmonics than an input is called harmonic distortion.

The measure of the harmonic distortion is the ratio of the amplitude of an individual output harmonic, A_i, to the output amplitude of the fundamental harmonic, A_1:

The harmonic distortion of an individual ith harmonic $= A_i/A_1$ \hfill (6.2.25)

To assess the overall distortion, we introduce total harmonic distortion, *THD*.
THD is the rms sum of the individual harmonic distortions; that is,

$$\text{THD} = \sqrt{[(A_2/A_1)^2 + (A_3/A_1)^2 + (A_4/A_1)^2 + (A_5/A_1)^2 + \cdots + (A_n/A_1)^2]} \times 100\%$$

$$= \sqrt{[(A_2)^2 + (A_3)^2 + (A_4)^2 + (A_5)^2 + \cdots + (A_n)^2]}/A_1 \times 100\% \hfill (6.2.26)$$

Example 6.2.3 demonstrates how to compute harmonic distortion.

Example 6.2.3 Total Harmonic Distortion

Problem

The amplitudes of the first five harmonics of the signal collected at the output of an amplifier are given by $A_1 = 5$ V, $A_2 = 3$ V, $A_3 = 1.6$ V, $A_4 = 0.7$ V, and $A_5 = 0.12$ V. Calculate the individual harmonic distortions and the THD.

Solution

Applying (6.2.25), we compute the individual harmonic distortions as

$$A_2/A_1 = 0.6, A_3/A_1 = 0.32, A_4/A_1 = 0.14, \text{ and } A_5/A_1 = 0.024$$

The THD is computed by using (6.2.26):

$$\text{THD} = \sqrt{[(A_2/A_1)^2 + (A_3/A_1)^2 + (A_4/A_1)^2 + (A_5/A_1)^2 + \cdots + (A_n/A_1)^2]} \times 100\%$$
$$= \sqrt{[(0.6)^2 + (0.32)^2 + (0.14)^2 + (0.024)^2]} \times 100\% = 69\%$$

Discussion

- The first point to consider is this: The obtained value of the THD is very big. This number tells us that an amplifier operates in a deep nonlinear regime. If we want to preserve the spectrum of an input signal, this operation is, obviously, not acceptable. If, however, we want to generate a set of harmonics – that is, the sinusoidal signals with frequencies equal to multiples of the fundamental frequency – then we can use this operation.
- A second point worth noting is that harmonic distortion is introduced not only by amplifiers but also by any electronic circuit and, of course, by any transmission channel. Real electronic filters, for example, inevitably introduce harmonic distortion. The wireless and wired transmission links distort the waveforms of the input signals, which change the spectrum of the output signal as compared with the input. We refer you to the preceding chapters – and to a rereading of this chapter, as necessary – where this point is discussed in detail.

If you analyze our discussion of harmonic distortion carefully, you will immediately realize that this distortion has been caused by distortion of the *signal amplitude*. In other words,

it is distortion of the amplitude of an input signal that causes harmonic distortion.

Indeed, (6.2.25) and (6.2.26) show that we evaluate the effect of harmonic distortion by considering the ratio of the amplitudes of the signals. But as we know from our discussion of electronic filters in Chapter 5, we can also encounter *phase distortion* of the output signal with respect to the input. *This phase distortion gives rise to distortion of an output waveform.* An example of such phase distortion is presented in Figure 6.2.15.

When a square-wave signal passes through a communications channel, the waveform of the output signal may become distorted due to a phase shift caused by the presence of reactive components. The result is a tilted waveform. We closely considered the reconstruction of the output waveform from the output spectrum in the preceding sections, and our discussion here enables us to perform the quantitative analysis of this phenomenon.

Changing the phase of the output signal with respect to the input is called phase distortion.

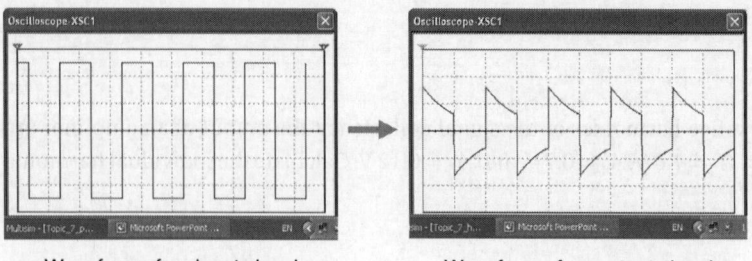

Waveform of an input signal　　　Waveform of an output signal

Figure 6.2.15 Change in the waveform of an output signal due to phase distortion.

The presence of phase distortion can be determined by observing the tilt (or sag) of the output waveform.

We cannot overestimate the importance of the signal's waveform for modern communications. Since time-domain and frequency-domain descriptions of the same signal are mutually dependent, changing a signal's waveform causes a change in the signal spectrum and vice versa, as harmonic distortion exemplifies. Optimization of a waveform for improving the communications quality is one of the trends in design of modern, specifically wireless (Bluetooth and Wi-Fi), communication systems.

Summary of Section 6.2

More About the Fourier Series

- *Coefficients of the Fourier series*: In Section 6.1, the formulas for calculating the coefficients of the Fourier series were introduced without explanations. In this section, these formulas have been derived, which gives us a deeper insight into the Fourier series application.
- *Amplitude and phase spectra*: The Fourier series can be presented in two trigonometric forms,

$$v(t) = A_0 + \sum_{n=1}^{\infty}(a_n \cos n\omega_0 t + b_n \sin n\omega_0 t)$$

and

$$v(t) = A_0 + \sum_{n=1}^{\infty} A_n \cos(n\omega_0 t + \Theta_n) \tag{6.2.14R}$$

where $A_n = \sqrt{a_n^2 + b_n^2}$ and $\Theta_n = -\tan^{-1} b_n/a_n$. The latter form clearly indicates the need for presenting two parts – amplitude spectrum and phase spectrum – of the whole signal's spectrum.
- *Using the signal's symmetry for finding the Fourier series coefficients*: Some signals possess a symmetry that allows us to simplify calculations of the Fourier coefficients. If, for example, the function describing the signal is odd, then the signal's Fourier series contains only sine members; if the function describing the signal is even, then the signal's Fourier series contains only cosine members.
- *Finding the Fourier series of various signals*: Understanding the nature of the coefficients of a Fourier series and learning various techniques for their calculation equip us with the tools for finding the Fourier series of various signals. Using MATLAB for automating the calculations makes this work easy and simple. An example of such a work is given in Table 6.2.1, where the Fourier series of the most common electrical signals calculated with MATLAB are presented.

Effect of Filtering on Signals

- *Statement of the problem*: After we have learned how to find the spectrum of a periodic signal by applying the Fourier series, we can tackle a more difficult problem: What happens with this signal when it passes through a filter? Experiments show that the signal preserves its period but experiences a change in the waveform. Hence, the problem is to develop a method for evaluating the changes and predicting the filtering output.
- *Filtering a single harmonic*: The key to assessing the result of filtering is to understand how a filter affects a single harmonic. We considered this problem very closely in Chapter 5 and know that the filtered harmonic preserves its waveform and frequency but experiences a decrease in amplitude and an addition in phase shift. We can calculate these changes by using the following equations:

The output amplitude of the *n*th harmonic, A_n^{out}, is given by

$$A_n^{out} = A_n^{in} \times A_{vn} \tag{6.2.23aR}$$

where

$$A_{vn} = \frac{1}{\sqrt{1 + \left(\frac{f_n}{f_c}\right)^2}} \tag{6.2.23bR}$$

The output phase shift of the nth harmonic, Θ_n^{out}, is equal to

$$\Theta_n^{out} = \Theta_n^{in} + \Delta\Theta_n \tag{6.2.24aR}$$

where

$$\Delta\Theta_n = -\tan^{-1}\left(\frac{f_n}{f_c}\right) \tag{6.2.24bR}$$

- *Filtering a periodic signal – time and frequency domains:* To understand the effect of filtering on an entire periodic signal, we need to recall that the signal is a sum of many harmonics, as the Fourier theorem states. Therefore, an output periodic signal obtained after the filtering is the sum of the individual harmonics constituting its Fourier series after this filtering. We can perform this summation thanks to the fact that a filter is a *linear system* to which the *principle of superposition* can be applied. Refer to Figures 6.2.5 and 6.2.6, where this process is presented.
- *Filtering a signal – the entire picture*: To comprehend the filtering operation in its entirety, we review the process as follows: We present a signal to an LPF and observe the output waveform in time domain; to predict this output, we consider the spectrum of the input signal. In frequency domain, we compute the amplitudes and phase shifts of each harmonic filtered by the given filter, which enables us to obtain the waveform of each harmonic in time domain. Finally, in time domain, we sum up all the output harmonics to obtain the waveform of the output signal. Figure 6.2.9 visualizes the filtering process. Analysis of this process shows that the output waveform depends on the ratio of the filter's cutoff frequency to the signal's fundamental frequency. It is this ratio that determines how much the amplitude and phase of an individual harmonic will change. The summation of individual harmonics in time domain requires the use of a computer.

Harmonic Distortion

- Harmonic distortion is the phenomenon in which the spectrum of an output signal contains more harmonics than the spectrum of an input signal. It is caused by the change of the waveform

of an output signal, for example, from a cosine to a square wave. The level of harmonic distortion is evaluated by calculating the ratio of the amplitudes of the individual harmonics to that of the fundamental harmonic.
- Phase distortion is the phenomenon in which only the phase spectrum of an output signal has changed. It results in slight changes in an output waveform, such as tilting or sagging.

Questions and Problems for Section 6.2

- Questions marked with an asterisk require a systematic approach to finding the solution.
- Many questions and problems, including those marked with an asterisk, imply that you, in addition to reading the textbook, will do your research to find the answers. Consider such questions as mini-projects.

More About the Fourier Series

1. Consider the sawtooth signal and its Fourier series shown in Figure 6.2.P1:
    a) Write down the formulas for the fifth, sixth, and seventh harmonics.
    b) Sketch the spectrum of this signal. Restrict the figure by the seventh spectral component.

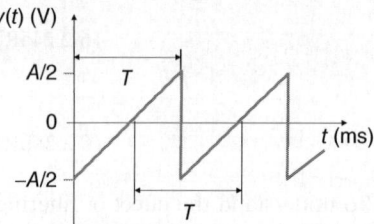

Fourier series:

$v(t) = (A/\pi) \sin 2\pi f_0 t + (A/2\pi) \sin 2\pi (2f_0)t + (A/3\pi) \sin 2\pi (3f_0)t + (A/4\pi) \sin 2\pi (2f_0)t \ldots$

$= \sum_{n=1}^{\infty} (A/n\pi) \sin 2\pi (nf_0)t$

Figure 6.2.P1 The waveform and the Fourier series of a sawtooth signal.

2. *Coefficients of the Fourier series were introduced in Section 6.1 as

$$A_0 = \frac{1}{T} \int_0^T v(t)\, dt \tag{6.1.6R}$$

$$a_n = \frac{2}{T} \int_0^T v(t) \cos(2\pi n f_0 t)\, dt \tag{6.1.7R}$$

and

$$b_n = \frac{2}{T} \int_0^T v(t) \sin(2\pi n f_0 t)\, dt \tag{6.1.8R}$$

Prove the correctness of these equations. (*Hint*: Refer to the Section 6.2.1.1 *Coefficients of the Fourier series*.)

3. A periodic signal, $v(t)$, can be presented in two different forms of the Fourier series:

$$v(t) = A_0 + \sum_{n=1}^{\infty} (a_n \cos n\omega_0 t + b_n \sin n\omega_0 t)$$

and

$$v(t) = A_0 + \sum_{n=1}^{\infty} A_n \cos(n\omega_0 t + \Theta_n)$$

where A_n and Θ_n are given by (6.2.11a) and (6.2.11b). Prove the identity of these forms. (*Hint*: Refer to (6.2.14).)

4 Why do we need to distinguish between the amplitude spectrum and the phase spectrum of the same signal?

5 Find the spectrum of the periodic signal shown in Figure 6.2.1a if $A = 3$ V and $T = 5$ ms. (*Hint*: Refer to Example 6.2.1.)

6 *Consider the shifted triangular signal shown in Figure 6.2.P6 with $A = 4$ V and $T = 0.2$ ms:
   a) Find the Fourier series of this signal.
   b) Calculate the coefficients of the series.
   c) Build the spectrum of this signal.

Figure 6.2.P6 A shifted triangular signal.

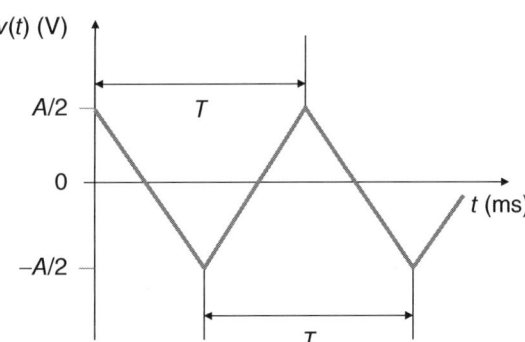

7 *Consider Table 6.2.1, where the waveforms of the most common signals are presented:
   a) Identify the odd and even signals in this table.
   b) Choose one odd and one even function from this table and show the calculations of the Fourier series coefficients for each signal by using the signal's symmetry.
   c) Compare your results with those given in this table.

8 Consider the waveform and the Fourier series of the sawtooth signal shown in Figure 6.2.P1. Why does this formula contain only sines? Explain.

9 Consider the waveform and the Fourier series of the full-wave rectified signal shown in Figure 6.2.P11. Why does this formula contain only cosines? Explain.

10 *Consider Table 6.2.1 and MATLAB box 6.2.1:
   a) Choose one odd and one even signal from Table 6.2.1, preferably the same one that you chose in Problem 6 (in this section).
   b) Find the Fourier series of these two signals by choosing one of the following three options:
      i) Use the MATLAB program given in MATLAB box 6.2.1.

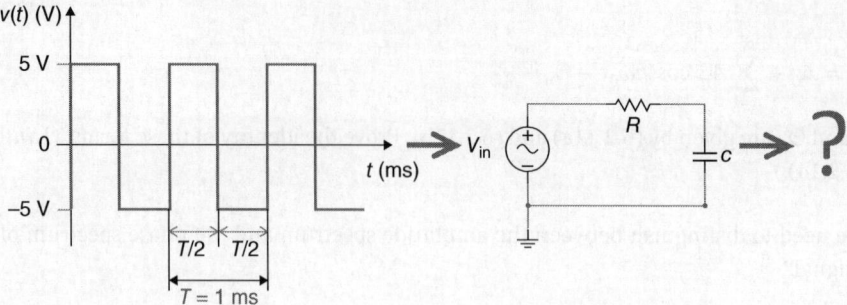

Figure 6.2.P11 Presenting a bipolar NRZ signal to a low-pass filter.

ii) Write a MATLAB script by yourself.
iii) Write your own computer program by using any programming language you are comfortable with.

Effect of Filtering on Signals

11 *Consider Figure 6.2.P11, where a bipolar NRZ signal is presented to a low-pass R–C filter:
  a) What can you tell about the waveform of a signal at the output of a low-pass filter without any additional investigation? Explain.
  b) Suggest a plan for finding the LPF's output signal. (*Hint*: Refer to Figures 6.2.4 and 6.2.5.)

12 *Consider a sawtooth signal and its Fourier series shown in Figure 6.2.P1:
  a) Show the second spectral component of this signal in frequency domain if $A = 2$ V and $T = 10$ ms.
  b) This signal is presented to a low-pass filter with a critical frequency of 0.2 kHz. Calculate the amplitude and phase of the second spectral component of the output signal and show it alongside the input second component. (*Hint*: Refer to Table 6.2.2.)
  c) Qualitatively sketch the waveform of the second output harmonic and show its parameters.

13 *The sawtooth signal shown in Figure 6.2.P1 is presented to another LPF with $f_C = 0.4$ kHz. Show the second harmonics of the input and output (filtered) signals in frequency domain and in time domain.

14 Assign $A = 3$ V and $T = 0.1$ ms to a full-wave rectified signal whose waveform and Fourier series are shown in Figure 6.2.P14 and perform the following tasks:
  a) Sketch the first four harmonics of the signal spectrum. Show all of the values.
  b) Sketch the waveform of the second harmonic of this signal and show its amplitude and period.
  c) This signal is presented to a low-pass filter with a critical frequency of 30 kHz. Calculate the amplitude and the phase shift of the output second harmonic and show both the input and output of the second harmonic on the same set of axes.
  d) On one set of axes, qualitatively sketch the waveforms of the second input (presented to the filter) and output (obtained after filtering) harmonics and show their parameters.

$v(t)$ (V)

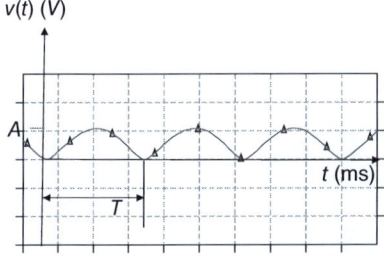

The Fourier series:

$v(t) = 2A/\pi + (4A/3\pi) \cos 2\pi (f_0)t - (4A/15\pi) \cos 2\pi (2f_0)t + (4A/35\pi) \cos 2\pi (3f_0)t - \ldots$

$= 2A/\pi - (4A/\pi) \sum_{n=1}^{\infty} (\cos 2\pi (nf_0)t/(4n^2 - 1))$

Figure 6.2.P14 Full-wave rectified signal presented to a low-pass filter.

15 *Refer to Questions 1, 13, and 14 and Figure 6.2.P1, which consider a sawtooth signal with its Fourier series and a low-pass filter with $f_C = 0.2$ kHz: Taking into account only the first three harmonics, sketch the spectra of the input and filtered signals. Show all the numbers.

16 Consider the sawtooth signal shown in Figure 6.2.P1 and assume $A = 6$ V and $T = 0.2$ ms. Present this signal to a real LPF with $f_C = 15$ kHz. Calculate the amplitudes of the first six harmonics and sketch the spectra of both the input and output signals. (*Hint*: See Figure 6.2.9.)

17 *How many harmonics do we need to sum up to obtain the waveform of a signal? Accompany your answer with detailed explanations.

18 Use any computerized tool (MS Excel, MATLAB, etc.) to do the following:
    a) Build the waveform of the filtered sawtooth signal discussed in Question 16 by using the first three harmonics.
    b) Repeat this process by using 6, 9, 12, and 15 harmonics.
    c) Build this waveform by using 33 and 66 harmonics.
    d) What conclusion regarding the quality of the waveform can you make based on your results obtained in all the operations of this problem?

19 *Return to Question 12: The answer is given in Figure 6.2.5, where the process of filtering a square wave is shown. Consider each step of this process and answer the questions in Table 6.2.P19.

20 *Compare Figures 6.2.3 and 6.2.8, which are reproduced in Figure 6.2.P20 for your convenience: Both figures show the same experiment, in which the same square-wave signal passes through low-pass RC filters. Why are the output signals so different? Explain with reference to Figure 6.2.5.

21 *If you aren't satisfied with the waveform of a signal that passed through a lowpass filter, what characteristic of the filter will you change first? In what way will you change this characteristic? Explain.

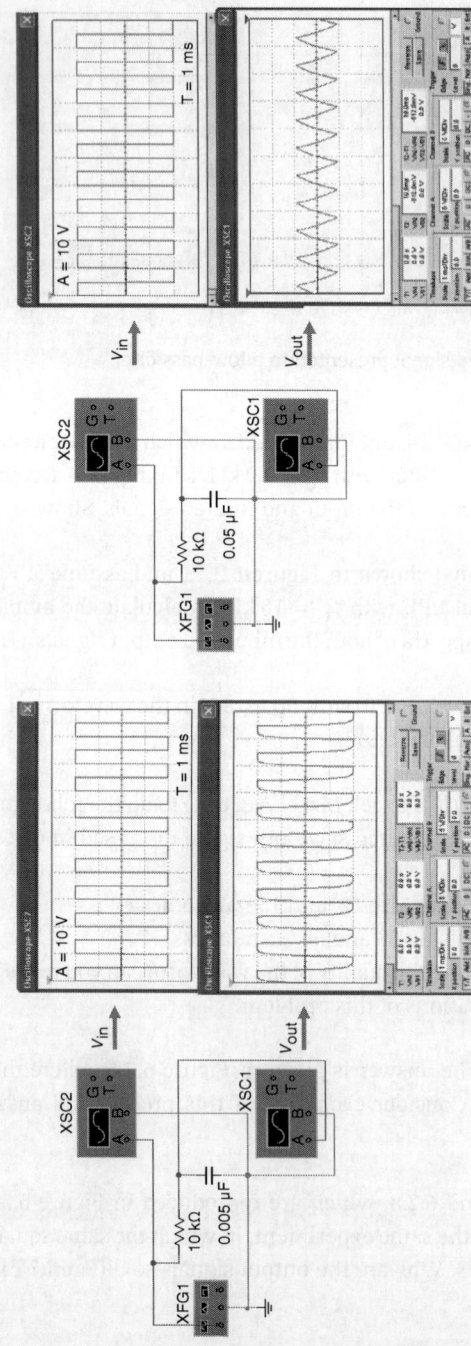

Figure 6.2.P20 Comparison of two experiments regarding the filtering of the same square wave signal.

Table 6.2.P19 Questions regarding the process of filtering the signals.

Designation in the caption of Figure 6.2.5	Step of filtering a signal shown in Figure 6.2.5	Question
6.2.5a	A square wave is expanded in the Fourier series (only four harmonics are shown) and each harmonic is presented to an LPF	6.2.5a. Q1: Is this expansion a real process or just a mathematical manipulation? 6.2.5a. Q2: If this is a real process, what device can perform this operation? Explain your answers
6.2.5b	Each harmonic is filtered by the LPF and the output harmonics are obtained	6.2.5b. Q1: Explain how we can find the output harmonics. (*Hint*: Refer to Table 6.2.2) 6.2.5b. Q2: How does an LPF "know" what harmonic to choose and how to filter it?
6.2.5c	All output harmonics are summarized, creating the entire output signal – the square wave	6.2.5c. Q1: Explain why we need to summarize the output harmonics 6.2.5c. Q2: Is this summation a real process or a mathematical manipulation? Explain

22 *We know that a waveform is a signal plot in time domain. Nevertheless, this subsection explains that we need to analyze frequency-domain processes to understand and control the filtered waveform. Explain why this is so. (*Hint*: Refer to Figure 6.2.9.)

23 *The text says that a cutoff frequency of a filter is the main parameter that determines the waveform of a filtered signal. However, a cutoff frequency is a frequency-domain parameter, whereas a waveform is a time-domain characteristic. How can a cutoff frequency determine the waveform?

Harmonic Distortion

24 *Consider the waveform and the spectrum of a cosine signal passing through an open-loop op-amp. Explain what harmonic distortion means by using this example. (*Hint*: Refer to Figure 6.2.13.)

25 To which phenomenon – amplitude-related or phase-related – does harmonic distortion refer? Explain.

26 *What devices or phenomena cause harmonic distortion? How do they do this?

27 A square wave signal is obtained by clipping a hugely amplified sinusoidal signal. How does this clipping affect the spectrum of the amplified signal? Explain by comparing the spectra of both sinusoidal and square wave signals.

28 The amplitudes of the first four harmonics of the Fourier series of an amplified signal are given by $A_1 = 6.37$ V, $A_2 = 2.12$ V, $A_3 = 1.27$ V, and $A_4 = 0.91$ V:
   a) Compute the individual harmonic distortions and the total harmonic distortion, THD.
   b) Does your result show a big or small THD?
   c) What would be an ideal value of a THD? Explain.

578 | 6 Spectral Analysis 1 – The Fourier Series in Modern Communications

29 What phenomenon is called phase distortion?

30 Compare two sets of the waveforms shown in Figure 6.2.P30:
   a) Determine in which case the waveform's distortion is caused by an amplitude-related phenomenon and in which case it is caused by a phase-related phenomenon.
   b) Give the reasoning for your answer.

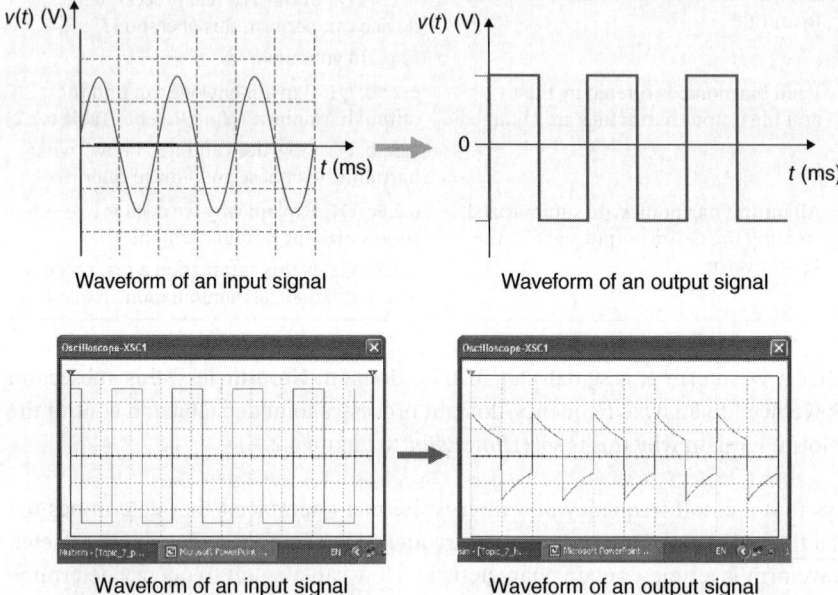

Figure 6.2.P30 Two sets of waveforms whose distortion was caused by two different phenomena.

6.3 Spectral Analysis of Periodic Signals: Advanced Study

Objectives and Outcomes of Section 6.3

Objectives

- To learn the mathematical foundations of the Fourier series.
- To get to know the power spectrum of periodic signals.

Outcomes

After studying this section, you will be able to:

- Explain the mathematical foundation of the Fourier series.
  o Present a sinusoidal signal in the polar and phasor forms.
  o Present a Fourier series in an exponential form and calculate its coefficients.
  o Show a Fourier series as a set of rotating phasors.
  o Explain the difference and relationship between two-sided and one-sided forms of a signal's spectrum.

o Demonstrate that the validity of the application of a Fourier series is limited by the convergence of the series to a given function and check the conditions of the convergence.
o Interpret the Gibbs phenomenon as an example of the limitation in the application of a Fourier series caused by its convergence.
- Explain the power spectrum of periodic signals.
o Distinguish between power and energy signals.
o Interpret the Parseval's theorem as a key formula for finding the distribution of a signal's power across its harmonics; that is, its power spectrum.
o Describe the relationship among the power, spectrum, and the waveform of a signal and transmission issues associated with this relationship.

6.3.1 Mathematical Foundation of the Fourier Series

6.3.1.1 The Fourier Series in Exponential and Phasor Forms

Polar and Phasor Forms of a Sinusoidal Signal Revisited From Section 2.3, we know that a sinusoidal signal can be presented in a polar form as

$$A_n \cos(n\omega_0 t + \Theta_n) \Rightarrow A_n \angle \Theta_n = A_n e^{j\Theta_n} \qquad (6.3.1a)$$

with $A_n = \sqrt{(a_n^2 + b_n^2)}$ and $\Theta_n = -\tan^{-1}(b_n/a_n)$.

You are reminded that a polar form, very much like a frequency-domain presentation, is a kind of agreement: In polar form, only the amplitude, A_n, and the phase, Θ_n, are shown, whereas the existence of a *sinusoidal* waveform and a frequency ($n\omega_0$) is just implied, but not presented explicitly. If we want to show the signal's frequency, we need to present the polar (which is a vector) as a rotating vector called a phasor. Thus, the *phasor form* of a sinusoidal signal is

$$A_n \cos(n\omega_0 t + \Theta_n) \Rightarrow A_n e^{j\Theta_n} e^{jn\omega_0 t} \qquad (6.3.1b)$$

You will recall that we obtained the *polar form* from a complex form of a signal presentation; that is, in our case,

$$A_n \angle \Theta_n = a_n + jb_n \qquad (6.3.2)$$

You will appreciate the importance of this form of the Fourier series shortly.

Fourier Series in an Exponential Form Refer again to Euler's formula,

$$e^{j\alpha} = \cos\alpha + j\sin\alpha$$
$$e^{-j\alpha} = \cos\alpha - j\sin\alpha \qquad (6.3.3a)$$

We can rearrange it as

$$\cos\alpha = \frac{(e^{j\alpha} + e^{-j\alpha})}{2}$$
$$\sin\alpha = \frac{(e^{j\alpha} - e^{-j\alpha})}{2j} \qquad (6.3.3b)$$

Formulas 6.3.3a and 6.3.3b give us the key to presenting the Fourier series in exponential form. Indeed, the harmonic members of a trigonometric form of the Fourier series given in (6.1.5a) and appearing again in (6.2.10a),

$$v(t) = A_0 + a_1 \cos \omega_0 t + a_2 \cos 2\omega_0 t + a_3 \cos 3\omega_0 t + \cdots$$
$$+ b_1 \sin \omega_0 t + b_2 \sin 2\omega_0 t + b_3 \sin 3\omega_0 t + \cdots$$
$$= A_0 + \sum_{n=1}^{\infty}(a_n \cos n\omega_0 t + b_n \sin n\omega_0 t)$$

can be presented as

$$\cos n\omega_0 t = \frac{(e^{jn\omega_0 t} + e^{-jn\omega_0 t})}{2}$$

$$\sin n\omega_0 t = \frac{(e^{jn\omega_0 t} - e^{-jn\omega_0 t})}{2j}$$

Hence, the Fourier series can be shown in an exponential form as

$$v(t) = A_0 + \sum_{n=1}^{\infty}(a_n \cos n\omega_0 t + b_n \sin n\omega_0 t)$$
$$= A_0 + \sum_{n=1}^{\infty}\left(a_n \frac{(e^{jn\omega_0 t} + e^{-jn\omega_0 t})}{2} + b_n \frac{(e^{jn\omega_0 t} + e^{-jn\omega_0 t})}{2j}\right)$$
$$= A_0 + \sum_{n=1}^{\infty}\left(\frac{a_n - jb_n}{2}e^{jn\omega_0 t} + \frac{a_n + jb_n}{2}e^{-jn\omega_0 t}\right) \quad (6.3.4)$$

If we now introduce new coefficients,

$$c_n = \frac{a_n - jb_n}{2}$$

$$c_{-n} = \frac{a_n + jb_n}{2}$$

$$c_0 = A_0 \quad (6.3.5)$$

and then take the sum from $-\infty$ to $+\infty$, we obtain the exponential form of the Fourier series as

$$v(t) = A_0 + \sum_{n=1}^{\infty}(a_n \cos n\omega_0 t + b_n \sin n\omega_0 t) = \sum_{n=-\infty}^{n=\infty}(c_n e^{jn\omega_0 t}) \quad (6.3.6)$$

In short,

$$v(t) = \sum_{n=-\infty}^{n=\infty}(c_n e^{jn\omega_0 t}) \quad (6.3.7)$$

Equation (6.3.7) is the Fourier series in exponential form.

Note that we do not need to show c_0 explicitly because the summation goes from minus to plus infinity and $n = 0$ is included.

Our next task is to define the coefficients, c_n. There are two ways to do this. First, from (6.3.5), we can see that c_n is a complex quantity,

$$c_n = \tfrac{1}{2}(a_n - jb_n)$$

and therefore can be presented as

$$c_n = |c_n| \angle \varphi \quad (6.3.8a)$$

where

$$|c_n| = \tfrac{1}{2}\sqrt{(a_n^2 + b_n^2)} \quad (6.3.8b)$$

and

$$\varphi_n = -\tan^{-1}(b_n/a_n) \text{ for positive } a_n \quad (6.3.8c)$$

Comparing (6.3.8c) with (6.3.2), (6.2.11a) and (6.2.11b),

$$A_n = \sqrt{(a_n^2 + b_n^2)} \quad \text{and} \quad \Theta_n = -\tan^{-1}(b_n/a_n)$$

we conclude that

$$A_n = 2|c_n| \quad \text{and} \quad \varphi_n = \Theta_n \quad (6.3.8d)$$

In the second approach, we can find the formula for the calculation of c_n without reference to a_n and b_n. Let us multiply both sides of (6.3.7),

$$v(t) = \sum_{n=-\infty}^{n=\infty} (c_n e^{jn\omega_0 t})$$

by $e^{-jm\omega_0 t}$ and take an integral over one period. Here, m changes from $-\infty$ to ∞ as n does, but they do not necessarily take the same value simultaneously. For example, n can be equal to 2 and m can be 5, but it is possible that $n = 3$ and $m = 3$. (This is the generalization of the approach we have explored in finding a_n and b_n in Section 6.2.) Thus, we find

$$\frac{1}{T}\int_0^T v(t)(e^{-jm\omega_0 t})dt = \sum_{n=-\infty}^{n=\infty} \left(c_n \left(\frac{1}{T}\int_0^T (e^{jn\omega_0 t} \cdot e^{-jm\omega_0 t})dt \right) \right) \quad (6.3.9)$$

All integrals where $n \neq m$ turns to zero (again, see (6.2.5) and (6.2.9)), which immediately produces

$$c_n = \frac{1}{T}\int_0^T v(t)(e^{-jn\omega_0 t})dt \quad (6.3.10)$$

Equation (6.3.10), along with formula $c_0 = A_0$, gives us the means for finding the coefficients of the Fourier series in exponential form.

Example 6.3.1 should clarify the treatment of the exponential form of the Fourier series.

Example 6.3.1   The Spectrum of a Pulse Train.

Problem

Consider a pulse train with the following parameters: Amplitude $V_{pt} = 10$ V, period $T = 8$ ms, and pulse duration $\tau = 1.6$ ms. The pulse train is shown in Figure 6.3.1a. Find the spectrum of this signal.

Solution

We still need to follow the steps presented in Example 6.1.1 and employed in Example 6.2.1.

Step 1. Define the signal $v(t)$: The signal's waveform formula, $v(t)$, can be defined from an observation of Figure 6.3.1 as

$$v(t) = \begin{cases} V_{pt} & \text{for } 0 \leq t \leq \tau \\ 0 & \text{for } \tau < t < T \end{cases} \quad (6.3.11)$$

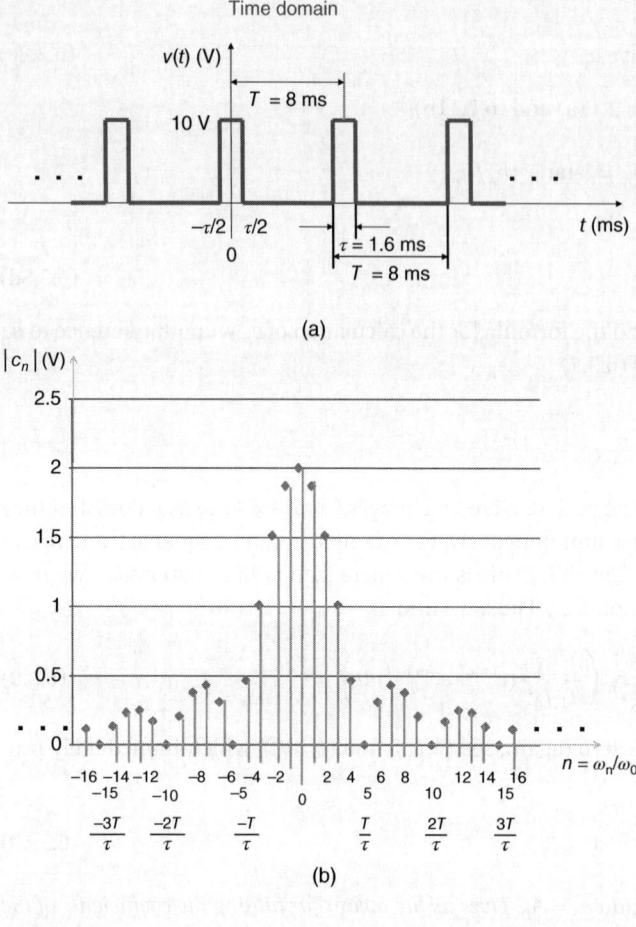

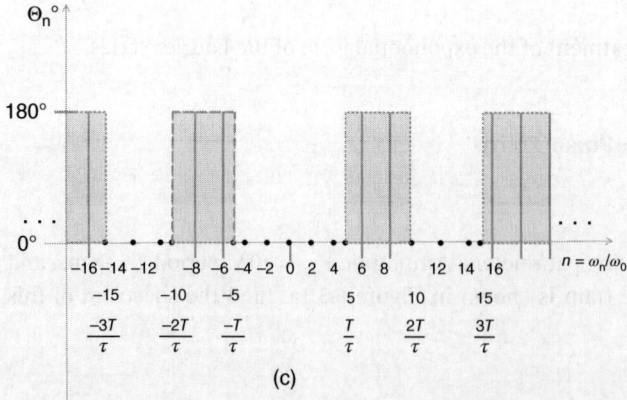

Figure 6.3.1 Spectral analysis of a pulse train: (a) Pulse train waveform; (b) amplitude spectrum; (c) phase spectrum; (d) amplitude spectrum with envelope; and (e) magnitude spectrum.

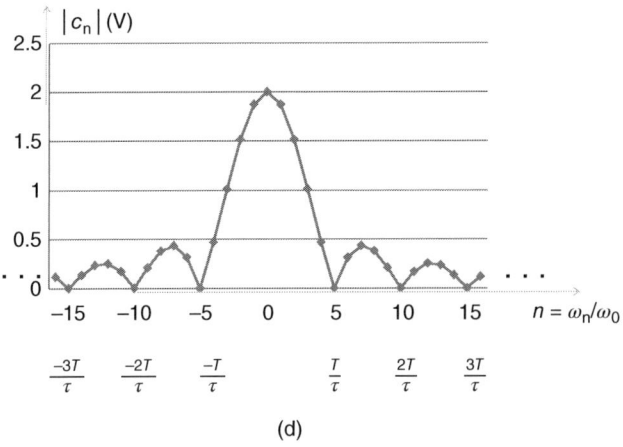

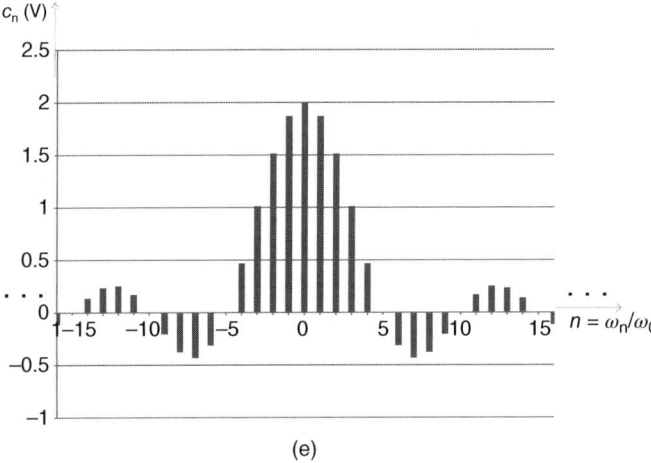

Figure 6.3.1 (Continued)

Step 2. Calculate $c_0 = V_0$: The dc member of this signal can be calculated by using (6.1.6) and (6.3.5),

$$c_0 = V_{pt0} = \left(\frac{1}{T}\right)\int_0^T v(t)\,dt = \left(\frac{1}{T}\right) V_{pt}t\big|_0^\tau = V_{pt}\,(\tau/T) \tag{6.3.12}$$

which results in

$$c_0 = V_{pt}(\tau/T) = 10\,\text{V} \times 0.2 = 2\,\text{V}$$

(Does this value make sense? Analyze the given pulse's parameters to find the answer.)

Step 3. Calculate c_n: We need to apply (6.3.10),

$$c_n = \frac{1}{T}\int_0^T v(t)\,(e^{-jn\omega_0 t})\,dt$$

Examining this integral, we understand that it will result in a difference between two exponential functions, $e^{-jn\omega_0 t_1} - e^{-jn\omega_0 t_2}$. Recalling that such a difference is the basis for (6.3.3b),

$$\sin\alpha = \tfrac{1}{2}j(e^{j\alpha} - e^{-j\alpha})$$

which we certainly want to explore, we need to change the integration limits from $\int_0^T$ to $\int_{-T/2}^{T/2}$. Figure 6.3.1a illustrates the equivalency of both sets of limits (from 0 to T and from $-T/2$ to $T/2$) for our integral. Then, we find that

$$c_n = \frac{1}{T}\int_0^T v(t)(e^{-jn\omega_0 t})dt = \frac{1}{T}\int_{-T/2}^{T/2} v(t)(e^{-jn\omega_0 t})dt$$

$$= \frac{V_{pt}}{T}\int_{-T/2}^{T/2}(e^{-jn\omega_0 t})dt = \frac{V_{pt}}{T}\cdot\frac{1}{-jn\omega_0}\cdot(e^{-jn\omega_0 t})\Big|_{-T/2}^{T/2}$$

$$= \frac{V_{pt}}{T}\cdot\frac{1}{-jn\omega_0}\cdot(e^{-jn\omega_0 t})\Big|_{-\tau/2}^{\tau/2} = \frac{V_{pt}}{T}\cdot\frac{e^{-jn\omega_0\frac{\tau}{2}} - e^{jn\omega_0\frac{\tau}{2}}}{-jn\omega_0}$$

because this integral is equal to zero everywhere except for the interval from $-\tau/2$ to $\tau/2$.

Applying (6.3.3b), we obtain

$$c_n = \frac{V_{pt}}{T}\cdot\frac{e^{-jn\omega_0\frac{\tau}{2}} - e^{jn\omega_0\frac{\tau}{2}}}{-jn\omega_0} = \frac{V_{pt}}{T}\cdot\frac{e^{jn\omega_0\frac{\tau}{2}} - e^{-jn\omega_0\frac{\tau}{2}}}{jn\omega_0}$$

$$= \frac{2V_{pt}}{Tn\omega_0}\cdot\frac{e^{jn\omega_0\frac{\tau}{2}} - e^{-jn\omega_0\frac{\tau}{2}}}{2j} = \frac{2V_{pt}}{Tn\omega_0}\sin\left(\frac{n\omega_0\tau}{2}\right) \qquad (6.3.13)$$

You will recall that $\omega_0 = 2\pi/T$ and therefore $n\omega_0 T = 2n\pi$. Using these formulas, we derive from (6.3.13)

$$c_n = \frac{2V_{pt}}{Tn\omega_0}\sin\left(\frac{n\omega_0\tau}{2}\right) = \frac{V_{pt}}{\pi n}\sin\left(\frac{\pi n\tau}{T}\right) \qquad (6.3.14)$$

Our sine depends on the ratio of the pulse duration, τ, to the signal period, T (s), called a *duty cycle*:

$$\text{Duty cycle} = \tau\,(s)/T\,(s) \qquad (6.3.15)$$

which was introduced in Section 3.2.2.

To make the coefficient c_n uniformly depend on a duty cycle, we rearrange (6.3.14) as follows:

$$c_n = \frac{V_{pt}}{\pi n}\sin\left(\frac{\pi n\tau}{T}\right) = \frac{\dfrac{V_{pt}\tau}{T}\sin\left(\dfrac{n\pi\tau}{T}\right)}{\dfrac{n\pi\tau}{T}} \qquad (6.3.16)$$

Equation (6.3.16) is a common form of the coefficients of an exponential Fourier series for a pulse train.

Plugging the given numerical values of the signal parameters into (6.3.16), we find

$$c_n = \frac{\dfrac{V_{pt}\tau}{T}\sin\left(\dfrac{n\pi\tau}{T}\right)}{\dfrac{n\pi\tau}{T}} = \frac{2\sin(0.2\pi n)}{0.2\pi n} \qquad (6.3.17a)$$

The function $\sin(x)/(x)$ is known as $\operatorname{sinc}(x)$; that is,

$$\sin(x)/(x) = \operatorname{sinc}(x)$$

Therefore, (6.3.17a) can be rewritten as

$$c_n = \frac{\dfrac{V_{pt}\tau}{T}\sin\left(\dfrac{n\pi\tau}{T}\right)}{\dfrac{n\pi\tau}{T}} = \frac{V_{pt}\tau}{T}\operatorname{sinc}\left(\frac{\pi n\tau}{T}\right) = 2\operatorname{sinc}(0.2n\pi) \qquad (6.3.17b)$$

Step 4. Obtain the exponential Fourier series: We obtain the Fourier series of the given pulse train by applying (6.3.7) and (6.3.16):

$$v(t) = \sum_{n=-\infty}^{n=\infty} (c_n e^{jn\omega_0 t}) = \frac{V_{pt}\tau}{T} \sum_{n=-\infty}^{n=\infty} \left(\frac{\sin\left(\frac{n\pi\tau}{T}\right)}{\frac{n\pi\tau}{T}} e^{jn\omega_0 t} \right)$$

$$= \frac{V_{pt}\tau}{T} \sum_{n=-\infty}^{n=\infty} \left(\frac{\sin\left(\frac{n\pi\tau}{T}\right)}{\frac{n\pi\tau}{T}} e^{jn\frac{2\pi}{T}t} \right) \quad (6.3.18)$$

(Again, c_0 is not shown here because $n = 0$ is included in summation.)

Step 5. Graphically build the signal's spectrum: To build the spectrum of this signal, we need to obtain the amplitude and phase of every harmonic. In our case, the amplitude is given by the absolute value of the coefficient c_n,

$$|c_n| = \frac{V_{pt}\tau}{T} \left| \frac{\sin\left(\frac{n\pi\tau}{T}\right)}{\frac{n\pi\tau}{T}} \right| = 2 \left| \frac{2\sin(0.2\pi n)}{0.2\pi n} \right| \quad (6.3.19)$$

The first positive and negative sixteen spectral lines of the *amplitude spectrum* are shown in Figure 6.3.1b vs. the *normalized frequency*, $n = \omega_n/\omega_0$. The normalized frequency, n, is derived from the definition of the Fourier series, where a current frequency, ω_n, is determined as a multiple of a fundamental frequency, ω_0; that is,

$$\omega_n = n\omega_0$$

The phase spectrum of a pulse train can be found from examining the complex coefficient, $c_n = \frac{1}{2}(a_n - jb_n)$, as

$$\varphi_n = \theta_n = -\tan^{-1}(b_n/a_n) \quad \text{for positive } a_n \quad (6.3.20a)$$

Since c_n in our case does not contain an imaginary part, $b_n = 0$, we have

$$\theta_n = 0 \quad \text{for positive } a_n$$

that is,

$$\theta_n = 0 \quad \text{for } \sin(n\pi\tau/T) > 0$$

Obviously,

$$\theta_n = 180° \quad \text{for } \sin(n\pi\tau/T) < 0$$

The condition

$$\sin(n\pi\tau/T) > 0$$

holds true only if

$$(n\pi\tau/T) < \pi$$

for one cycle, which means

$$(n\tau/T) < 1$$

or

$$n < T/\tau$$

If n is negative, then

$$-n > -T/\tau$$

In our example,

$\theta_n = 0$ for $n < 5$ and for $-n > -5$

and (6.3.20b)

$\theta_n = 180°$ for $n > 5$ and for $-n < -5$

Then, the cycle is repeated. The phase spectrum of a pulse train is shown in Figure 6.3.1c.

Figure 6.3.1d shows the continuous-frequency presentation of the amplitude spectrum of a pulse train. The significance of this graphic will become clear when we discuss the Fourier transform in Chapter 7.

Figure 6.3.1e demonstrates another form of a pulse train's amplitude spectrum. Since all coefficients of the Fourier series, c_n, in this case are real numbers, we can plot them directly. If we agree to depict c_n as a positive number when its phase is zero, then the c_n with a phase of 180° must be a negative number. Such a plot, called a magnitude spectrum, gives us both amplitude and phase information in one figure, which is the goal of building Figure 6.3.1e.

Discussion

- The spectrum of a pulse train is a very common example found in many textbooks listed in our bibliography. It is informative and – what's very important – enables us to bridge the methods for finding the spectra of periodic and nonperiodic signals. In addition, this example introduces several useful techniques, such as various approaches to determine the amplitude and phase spectra.
- It is crucial to understand that coefficients of the Fourier series of this example's waveform include $\sin(n\pi\tau/T)$, which means that the sine here is not a function but a specific number with given n and τ/T. For example, for duty cycle $\tau/T = 0.1$ and $n = 3$ (the third harmonic), we find $\sin(n\pi\tau/T) = \sin(3\pi \times 0.1) = \sin(0.3\pi) = 0.81$. We can completely avoid this confusion by defining the Fourier coefficients with the sinc(x) function, $c_n = \frac{V_{pt}\tau}{T}\text{sinc}\left(\frac{\pi n \tau}{T}\right)$, as in (6.3.17b), and performing calculations with a scientific calculator.
- A word of caution relating to *computing* c_0: We mentioned previously that this coefficient is automatically included in an exponential Fourier series, since the summation goes from $-\infty$ to ∞, as in (6.3.18); however, computation of this coefficient must be done separately from the calculation of other coefficients. Indeed, if we try to use the general formula for c_n,

$$|c_n| = \frac{V_{pt}\tau}{T}\left|\frac{\sin\left(\frac{n\pi\tau}{T}\right)}{\frac{n\pi\tau}{T}}\right|$$

we immediately encounter division by zero for $n = 0$! This is why we need to use a special formula (Eq. (6.3.12) in this example) to compute c_0.
- Consider the phase spectrum: We found that when

$$T/\tau = n = \pm 5$$

the phase changes from 0° to ±180° for the first cycle. For the second cycle, $2n$, the phase changes from ±180° to 0°; here,

$$2T/\tau = 2n = \pm 10$$

If we want to express this relationship with respect to $1/\tau$, as many authors do, we divide both sides of the given formula by T and find

$$1/\tau = n/T = \pm 5/T, 2/\tau = 2n/T = \pm 10/T, \text{etc.}$$

These points are shown in Figure 6.3.1b,c.

- We should remember that the spectrum of a pulse train includes an infinite number of positive and negative spectral lines, as (6.3.7) shows. Therefore, the number n goes from $-\infty$ to ∞, whereas any visual presentation of this spectrum includes only a finite number of harmonics. This is why all figures in Figure 6.3.1 contain ellipses.

Sidebar 6.3.S.1 The Other Forms of an Exponential Fourier Series

Let us summarize our findings in the preceding subsection:

- If $v(t)$ is a periodic signal, then it can be expanded into an exponential Fourier series, as defined in (6.3.7),

$$v(t) = \sum_{n=-\infty}^{n=\infty} (c_n e^{jn\omega_0 t}) \qquad (6.3.S.1.1)$$

- Coefficients c_n are defined by (6.3.10):

$$c_n = \frac{1}{T} \int_0^T v(t)(e^{-jn\omega_0 t}) dt$$

- As shown in (6.3.5), coefficients $c_n = a_n - jb_n$, where n changes from $-\infty$ to $+\infty$, are a complex quantity; their real and imaginary parts are given by

$$a_n = Re(c_n) \quad \text{and} \quad b_n = Im(c_n) \qquad (6.3.S.1.2)$$

Their arguments are defined as

$$\arg(c_n) = \tan^{-1}[Im(c_n)/Re(c_n)] = -\tan^{-1}(b_n/a_n) \qquad (6.3.S.1.3)$$

- Coefficient c_n can be presented in a polar form, as shown in (6.3.8a):

$$c_n = |c_n| \angle \varphi_n = |c_n| e^{j\varphi_n} \qquad (6.3.S.1.4)$$

where $|c_n| = \sqrt{(a_n^2 + b_n^2)}$ and $\arg(c_n) \equiv \varphi_n = -\tan^{-1}(b_n/a_n)$

The Fourier series as a series of phasors:

If $v(t)$ is a *real (noncomplex) function of time* – and in engineering that is always the case – then

$$(c_{-n}) = (c *_n) \qquad (6.3.S.1.5a)$$

where $(c *_n)$ is a complex conjugate to (c_n); i.e.

$$(c_{-n}) = (c *_n) = |c_n| e^{-j\varphi_n} \qquad (6.3.S.1.5b)$$

(Continued)

Sidebar 6.3.S.1 (Continued)

Equations (6.3.S.1.4) and (6.3.S.1.5b) show that the Fourier theorem can now be stated in the following form:

A periodic signal can be expanded as an infinite sum of phasors:

$$c_n e^{jn\omega_0 t} = |c_n| e^{j\varphi_n} e^{jn\omega_0 t} \quad \text{and} \quad c_{-n} e^{-jn\omega_0 t} = |c_n| e^{-j\varphi_n} e^{-jn\omega_0 t} \quad (6.3.S.1.6)$$

Since each pair of complex-conjugate phasors constitutes a real value, these phasors have to rotate in the opposite directions, as shown in Figure 6.3.S.1.1. (Don't confuse rotating phasors, $c_n e^{jn\omega_0 t}$, with their complex coefficients, c_n and $c_{-n} \equiv c^*_n$, presented in polar form.)

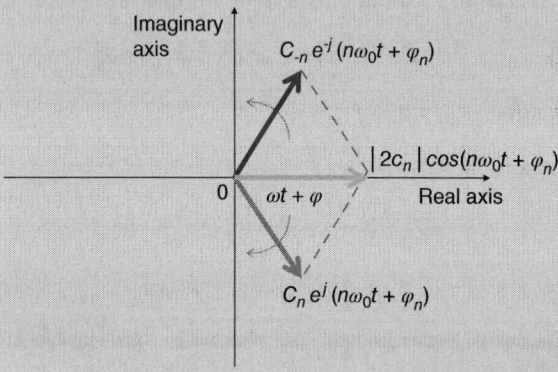

Figure 6.3.S.1.1 Rotating phasors of a Fourier series.

For a real signal, the Fourier series (6.3.S.1.6) must be composed as a sum of complex-conjugate pairs with its dc member, c_0, standing alone. Indeed,

$$v(t) = \sum_{n=-\infty}^{n=\infty} (c_n e^{jn\omega_0 t}) = c_0 + \sum_{n=-\infty}^{n=\infty} (c_n e^{jn\omega_0 t} + c_{-n} e^{-jn\omega_0 t})$$

$$= c_0 + \sum_{n=-\infty}^{n=\infty} ((|c_n| e^{j\varphi_n} e^{jn\omega_0 t}) + (|c_n| e^{-j\varphi_n} e^{-jn\omega_0 t}))$$

$$= c_0 + \sum_{n=1}^{n=\infty} 2|c_n| \cos(n\omega_0 t + \varphi_n) \quad (6.3.S.1.7)$$

where we use Euler's formula, given in (6.3.3b). The trigonometric (cosine–sine) form of the Fourier series follows immediately from (6.3.S.1.7); we leave this transformation as an exercise for you. (*Hint*: Review the manipulations shown in (6.3.4), (6.3.5), and (6.3.6).)

6.3.1.2 Two-Sided and One-Sided Spectra and Three Equivalent Forms of the Fourier Series

Presentation of the Fourier series in the exponential form (and especially as a sum of rotating phasors) leads us to an interesting and important conclusion: This presentation invokes both positive and negative frequencies, $n\omega_0$ and $-n\omega_0$. Therefore, the amplitude and phase spectra of a signal should be presented in both positive and negative half planes of the frequency domain. Indeed, for every positive phasor, $|c_n| e^{j\varphi_n} e^{jn\omega_0 t}$, we have a negative phasor, $|c_n| e^{-j\varphi_n} e^{-jn\omega_0 t}$. This spectrum is called *two-sided* for obvious reasons. It is shown in Figure 6.3.2a. If, however, we refer to the trigonometric form of the Fourier series, then we need to consider traditional amplitude and phase spectra, the spectra we have discussed in Sections 6.1 and 6.2. Such spectra are called *one-sided*

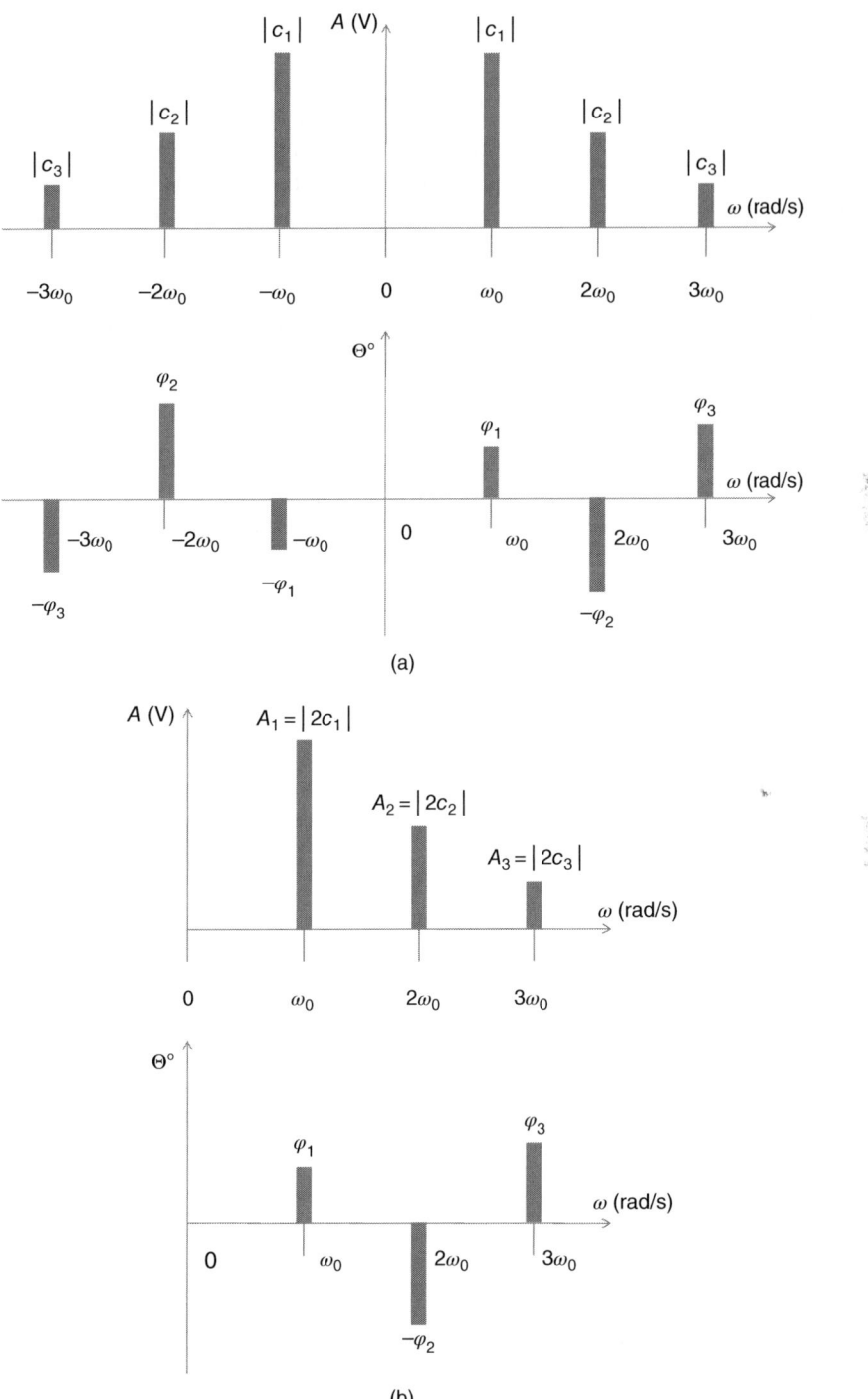

Figure 6.3.2 Examples of (a) two-sided and (b) one-sided amplitude and phase spectra of a signal.

because they refer to positive frequency only. We show an example of one-sided spectra again in Figure 6.3.2b. We referred to two-sided and one-sided spectra earlier in this book; this discussion should add to your knowledge of the topic.

It seems strange to consider negative frequencies because we know that in the real world there is no such thing as a negative frequency. This reference, however, is no more preposterous than the negative members of the Fourier series in a complex exponential form. Moreover, it is no more preposterous than the use of a complex number because we know that all numbers in the real world are real. We hope you get the point:

> *The complex exponential form of the Fourier series is a powerful and convenient tool that we use for sophisticated mathematical manipulations. The results of these manipulations must always be real numbers that we can measure and convert into tangible entities.*

Example 6.3.1, where we discuss the pulse train, brings us to its two-sided amplitude and phase spectra, shown in Figure 6.3.2. This is a typical situation when we use a complex exponential form of the Fourier series for the spectral analysis of a signal. When, however, we need to find the distributions of the voltage and power of this signal among its harmonics, we need to turn to one-sided spectrum because voltage and power are real entities. We will consider this point later in this section when we take up Parseval's theorem.

Compare (6.2.14) and (6.3.S.1.7) and find (6.3.21a):

$$v(t) = A_0 + \sum_{n=1}^{\infty} A_n \cos(n\omega_0 t + \Theta_n) = c_0 + \sum_{n=1}^{n=\infty} |2c_n| \cos(n\omega_0 t + \varphi_n) \tag{6.3.21a}$$

Therefore, the coefficients of two-sided and one-sided amplitude and phase spectra of the same signal relate to each other as follows:

$$A_0 = c_0$$

and

$$A_n = 2|c_n| \tag{6.3.21b}$$

Observe how this relationship is reflected in Figure 6.3.2.

To sum up, we present *the Fourier series in three equivalent forms*:

- Traditional trigonometric form, as in (6.2.1),

$$v(t) = A_0 + \sum_{n=1}^{\infty}(a_n \cos n\omega_0 t + b_n \sin n\omega_0 t)$$

- One-sided trigonometric form, as in (6.2.14),

$$v(t) = A_0 + \sum_{n=1}^{\infty} A_n \cos(n\omega_0 t + \Theta_n)$$

- Two-sided exponential form, as in (6.3.7),

$$v(t) = \sum_{n=-\infty}^{n=\infty} (c_n e^{jn\omega_0 t})$$

What form you choose depends on the task at hand, as our discussion and examples have shown and on which we elaborate shortly.

6.3.2 Conditions for Application of the Fourier Series

> **Sidebar 6.3.S.2 Convergence of the Fourier Series**
>
> The application of the Fourier series to spectral analysis is based on a key proposition, which we call the Fourier theorem: *Any periodic function can be expanded as a Fourier series.* This theorem states, in essence, that the Fourier series is an equivalent, valid presentation of the periodic function; in other words, *the Fourier series is simply another form of this function.* This is true, however, if – and only if – the Fourier series, as any series for that matter, converges to the given function.
>
> The mathematical study of the convergence of the Fourier series is a special subject and we, of course, will not delve into it here (see, for example, James 2015, pp. 285–292 for an interesting review of the mathematical development of the Fourier series and the Fourier transform.) Instead, we restrict ourselves to taking a practical approach to the problem by stating the *Dirichlet conditions of the convergence of a Fourier series*:
>
> Assume that $v(t)$ is a bounded periodic function that, in any period, has (i) a finite number of isolated maxima and minima and (ii) a finite number of points of finite discontinuity. If so, then its Fourier series expansion converges to $v(t)$ at all points where $v(t)$ is continuous and converges to the mean of discontinuity at points where $v(t)$ is discontinuous.
>
> These conditions require a point-by-point discussion:
>
> - *These are satisfactory conditions* – in fact, the necessary conditions for convergence of a Fourier series are not known. Fortunately, this fact does not cause a problem in practical applications. This is because almost all the signals we meet in engineering practice satisfy the Dirichlet conditions and therefore have a converged Fourier series.
> - The requirement of having signal $v(t)$ *bounded* means that $v(t)$ should be either absolutely integrable or square-integrable over its period; that is,
>
> $$\int_t^{t+T} |v(t)|\,dt \leq \infty \qquad (6.3.S.2.1a)$$
>
> or
>
> $$\int_t^{t+T} |v(t)|^2\,dt \leq \infty \qquad (6.3.S.2.1b)$$
>
> These requirements mean, practically speaking, that either $v(t)$ or $|v(t)|^2$ has a finite area over its period. (Some authors consider (6.3.S.2.1a) and (6.3.S.2.1b) the necessary conditions for existence of the Fourier series and call them the *weak Dirichlet conditions*.)
> - If $v(t)$ is absolutely integrable, then its Fourier series expansion exists and converges uniformly at any points where $v(t)$ is continuous. If $v(t)$ is square-integrable, then its Fourier series expansion converges in the mean; that is,
>
> $$\lim_{N \to \infty} \int_t^{t+T} |v(t) - v_N(t)|^2\,dt = 0 \qquad (6.3.S.2.2)$$
>
> where $v_N(t) = \sum_{n=-N}^{n=N}(c_n e^{jn\omega_0 t})$. Equation (6.3.S.2.2) says that the square mean difference between the signal, $v(t)$, and the partial sum of its Fourier series presentation, $v_N(t)$, goes to zero when the number of the sum members goes to infinity.

(Continued)

Sidebar 6.3.S.2 (Continued)

- The requirement for the signal, $v(t)$, to have a "*finite number of isolated maxima and minima*" is intuitively clear.
- The requirement for the signal, $v(t)$, to have "*finite points of finite discontinuity*" means that, in addition to isolated maxima and minima, the function can have interruptions in its continuity; however, the number of these interruptions must be finite and the discontinuities must be of a certain nature. The square wave pictured in Figure 6.3.S.2.1 is an example of a signal with a finite number of discontinuities (specifically, two). Indeed, we define this function as

$$v(t) = \begin{cases} A & \text{for } 0 \leq t \leq T/2 \\ 0 & \text{for } T/2 \leq t \leq T \end{cases}$$

This definition does not tell us how this function transitions from A to 0 and from 0 to A at $t = 0, t = T/2, t = T$, etc. Its graph implies that the signal discretely changes its value from A to 0 and back, which means that at the points $t = 0, t = T/2, t = T$, and $t = 3/2T$, the function $v(t)$ exhibits discontinuities. Such a presentation, however, is a mathematical model of a real signal. In reality, all signals we encounter in engineering applications exhibit a smooth transition from one point to another, which means they don't have such discontinuities. In other words, all practical signals meet the requirement to have the "*finite points of finite discontinuity*".

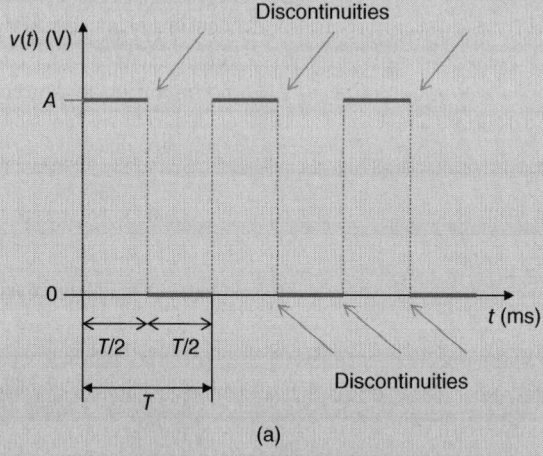

(a)

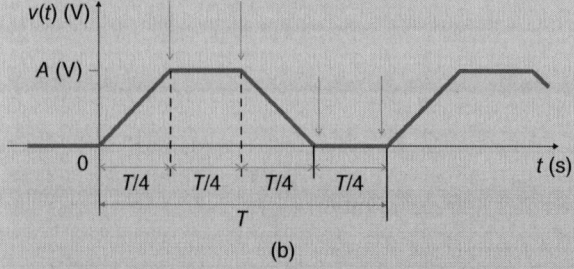

(b)

Figure 6.3.S.2.1 Signals with discontinuity: (a) A square wave with piecewise discontinuities; (b) a trapezoidal continuous signal with discontinuous first derivatives.

- The rate of convergence of the Fourier series *depends on the type of discontinuity of the given signal, v(t)*. In general, the smoother the signal, the higher the convergence rate of its Fourier series. *This rate is, obviously, determined by the pace with which the coefficients a_n and b_n decrease with an increase in n*. If a signal has a piecewise discontinuity, as does the square wave shown in Figure 6.3.S.2.1a, then the coefficients of its Fourier series expansion decrease as $1/n$. If a signal is continuous in itself but has discontinuous first derivatives (as does the trapezoidal signal shown in Figure 6.3.S.2.1b), then a_n and b_n decrease as $1/n^2$. If a continuous signal has first m continuous derivatives but has discontinuous $m + 1$ derivatives, then a_n and b_n decrease as $1/n^{m+1}$.

This discussion of the Dirichlet conditions provides us with a deeper understanding of the mathematical foundation (and limitations) of the Fourier series. Although the vast majority of practical signals allows for expansion into the Fourier series without any traps, knowing the limitations of this approach is certainly necessary to use this method professionally.

6.3.2.1 Gibbs[6] Phenomenon

If we review Figure 6.2.4 (the square wave) in Example 6.2.3, we certainly will note that the recovered waveform obtained by summing up the members of the Fourier series contains both undershoots and overshoots at the points of the signal's discontinuities. Undoubtedly, when the plot of a square wave jumps from 0 to A at $t = T^-$ *(left of $t = T$)*, we see an overshoot, and when this plot jumps from A to 0 at $t = T^+$ *(right of $t = T$)*, we see undershoot in its recovered waveform. Such spikes are characteristic of any waveform exhibiting piecewise discontinuity (see Table 6.2.1). The appearance of these spikes at the points of discontinuity is called the *Gibbs phenomenon*.

A close review of the Gibbs phenomenon reveals that the partial sum, $v_N(t) = \sum_{n=-N}^{n=N}(c_n e^{jn\omega_0 t})$, converges to the middle of this point (e.g. to $A/2$ in the example of a square wave with amplitude A) at the point of discontinuity. However, at the left and the right sides of this midpoint, the partial sum, $v_N(t)$, oscillates with the period $T/2N$ and an amplitude of about $0.09A$. These overshoot or undershoot amplitudes do not depend on N, and when $N \to \infty$, they collapse into spikes called "Gibbs' ears," which can be seen in Figure 6.2.4 and in Table 6.2.1.

The main point we draw from this discussion is this:

Reconstruction of the waveform of an ideal periodic digital signal with a piecewise discontinuity based on its expansion into the Fourier series never completely reproduces the waveform of this signal and will always exhibit overshoot and undershoot spikes at the points of discontinuity.

This phenomenon would seriously affect the operation of an ideal filter, if one existed, because it shows that such a filter (though never practically realizable) would always distort an input signal.

Fortunately, all practical signals, as mentioned previously, have a smooth transition from one point to another, which means that such signals are always continuous. Therefore, the *Gibbs*

---

6 **Josiah Willard Gibbs** (1839–1903) was an American physicist, chemist, and mathematician. He was the recipient of the first doctorate degree in engineering in the United States. He is recognized as a founder of chemical thermodynamics, physical chemistry, and vector analysis (independently of Oliver Heaviside). Though his work attracted some attention of the international scientific community in his lifetime, a full appreciation of the significance of his contributions only became evident during the first half of the twentieth century.

phenomenon is not a concern in engineering practice; we, however, must be familiar with it in order to understand why our theoretical considerations and practical calculations deliver different results.

To reflect the result of this discussion, we should use the sign of approximation instead of the equality sign when showing that a function, $v(t)$, with discontinuities is expanded into a Fourier series:

$$v(t) \approx \sum_{n=-\infty}^{n=\infty} (c_n e^{jn\omega_0 t}) \qquad (6.3.22)$$

This form must also be used when we take a partial sum of the Fourier series.

A final note on the mathematical foundation of the Fourier series: We always state that *only a periodic function can be expanded into a Fourier series*. This is true, however, if we need *to expand the signal, $v(t)$, over time interval that ranges from zero to infinity*. But we *can expand the nonperiodic signal into a Fourier series that converges over a finite time interval, $0 \leq t \leq \tau$*. This can be quite useful in applications where the signal is nonperiodic but is defined (or needs to be explored) over a finite time interval. More details on this technique can be found in (James 2015, pp. 293–298).

6.3.3 Power Spectrum of a Periodic Signal

6.3.3.1 Power and Energy Signals

We need to add another limitation to the use of the Fourier series:

The Fourier series can be applied only to a *power periodic signal*.

To understand the origin of this requirement, we need to recall that all signals are divided into two general categories: *energy and power*. The *total energy, E*, of a signal, $v(t)$, can be calculated as

$$E_T \text{ (J)} = \lim_{T \to \infty} \int_{-T}^{T} v^2(t) \, dt \qquad (6.3.23)$$

A signal, $v(t)$, is called an *energy signal* if the integral (6.3.23) exists and E is finite; i.e. $0 < E < \infty$.

The *total power, P*, of a signal, $v(t)$, can be calculated as

$$P_T \text{ (W)} = \lim_{T \to \infty} \left(\frac{1}{2T}\right) \int_{-T}^{T} v^2(t) \, dt \qquad (6.3.24)$$

A signal, $v(t)$, is called a *power signal* if the integral (6.3.24) exists and P is finite; i.e. $0 < P < \infty$.

If $v(t)$ is a periodic signal, then its average power per period is given by

$$P \text{ (W)} = \left(\frac{1}{T}\right) \int_{t}^{t+T} v^2(t) \, dt \qquad (6.3.25)$$

Equation (6.3.25) can be interpreted as average power dissipated in an electrical circuit over an 1-Ω resistor with $v(t)$ being either the voltage (v^2/R) or the current ($i^2 R$), as we will see shortly.

It should be noted that a signal can be either an energy signal or a power signal – or neither.

Let us compare the formula for a square-integrable signal over its period given in (6.3.S.2.1b),

$$\int_{t}^{t+T} |v(t)|^2 \, dt \leq \infty$$

with (6.3.25),

$$P \text{ (W)} = \left(\frac{1}{T}\right) \int_{t}^{t+T} v^2(t) \, dt$$

We can readily see that if the condition $\int_{t}^{t+T}|v(t)|^2\,dt \leq \infty$ is met, then the average power of a signal is finite; that is,

$$P(W) \leq \infty$$

This is why a Fourier series expansion of a periodic signal can be done only if the signal has finite average power. Fortunately, all practical periodic signals are power signals; thus, this requirement for the existence of a Fourier series is always met.

Keep always in mind that nonperiodic signals are energy signals; they carry finite total energy.

6.3.3.2 Parseval's Theorem

The Fourier series, as our discussion and the examples accompanying it demonstrate, shows the distribution of the signal's voltage across its harmonics. Indeed, the amplitudes of the harmonics are in volts and can be directly measured with a spectrum analyzer. If the signal's source produces current, then these amplitudes will be measured in amperes and show the distribution of the entire signal's current across the harmonics. Similarly, we can determine the distribution of the signal's power across the harmonics, thus obtaining the signal's power spectrum. We can confirm this finding by the following reasoning:

We know that the average power of a signal is given in (6.3.25) and that the periodic signal, $v(t)$, can be presented in the exponential form of the Fourier series, as in (6.3.7),

$$v(t) = \sum_{n=-\infty}^{n=\infty} (c_n e^{jn\omega_0 t})$$

Plugging $v(t)$ into (6.3.25), we find

$$P(W) = \left(\frac{1}{T}\right)\int_{t}^{t+T} v^2(t)\,dt = \left(\frac{1}{T}\right)\int_{t}^{t+T}\left[\sum_{n=-\infty}^{n=\infty}(c_n e^{jn\omega_0 t})\right]^2 dt$$

$$= \left(\frac{1}{T}\right)\int_{t}^{t+T}\left[\sum_{n=-\infty}^{n=\infty}(c_n e^{jn\omega_0 t})(c_n^* e^{-jn\omega_0 t})\right] dt$$

$$= \sum_{n=-\infty}^{n=\infty}\left[c_n c_n^* \left(\frac{1}{T}\right)\int_{t}^{t+T}(e^{jn\omega_0 t})(e^{-jn\omega_0 t})\,dt\right]$$

$$= \sum_{n=-\infty}^{n=\infty}[c_n c_n^*] = \sum_{n=-\infty}^{n=\infty}|c_n|^2$$

Thus, we have proved the following statement:

If $v(t)$ is a periodic signal, then its average power, $P(W)$, can be expressed as

$$P(W) = \left(\frac{1}{T}\right)\int_{t}^{t+T}[v(t)]^2 dt = \sum_{n=-\infty}^{n=\infty}(c_n c_n *) = \sum_{n=-\infty}^{n=\infty}|c_n|^2 \qquad (6.3.26)$$

where c_n is an nth coefficient of the signal's Fourier series and c_n^* is its complex conjugate.

Equation (6.3.26) is the mathematical form of *Parseval's theorem,* named after Marc-Antoine Parseval, an eighteenth century French mathematician.

Parseval's theorem, in essence, is a specific form of the principle of the conservation of energy (power); in fact, it says that *the power carried by a signal, v(t), is equal to the sum of the powers carried by all the signal's harmonics.* We can readily express Parseval's theorem through the real coefficients,

a_n and b_n, of a trigonometric Fourier series of the signal, $v(t)$, by using (6.3.7) and (6.3.8a):

$$P(\text{W}) = \left(\frac{1}{T}\right) \int_t^{t+T} [v(t)]^2 dt = \sum_{n=-\infty}^{n=\infty} |c_n|^2 = A_0^2 + \frac{1}{2}\sum_{n=1}^{\infty} A_n^2$$

$$= (A_0)^2 + \frac{1}{2}\sum_{n=1}^{n=\infty}(a_n^2 + b_n^2) \tag{6.3.27}$$

We will prove the validity of this statement shortly. In fact, in (6.3.27) we obtain the formula for the *root mean square (rms)* value, V_{rms}, of a periodic signal, $v(t)$, which is given by

$$V_{\text{rms}}^2 \,(\text{V}^2) = \left(\frac{1}{T}\right) \int_t^{t+T} v^2(t)\, dt \tag{6.3.28}$$

(Note that we do our integration from t to $t+T$ instead of limiting it from 0 to T because using the limits of integration, t and $t+T$, is the usual (and the most general) way to show integration over a period.)

Now, we can apply these general results to *the field of electrical circuits*. Let $v(t)$ (V) be the voltage across the resistor, $R\,(\Omega)$. The power then dissipated by this resistor is equal to

$$P_R\,(\text{W}) = V_{\text{rms}}^2 / R \tag{6.3.29}$$

where the *rms* value of $v(t)$ is given by (6.3.28). We need to use (6.2.14),

$$v(t)\,(\text{V}) = A_0 + \sum_{n=1}^{\infty} A_n \cos(n\omega_0 t + \Theta_n)$$

to obtain V_{rms}^2 through the harmonics of a $v(t)$ signal. Thus, we find

$$V_{\text{rms}}^2(\text{V}^2) = \left(\frac{1}{T}\right) \int_0^T v^2(t) dt$$

$$= \left(\frac{1}{T}\right) \int_0^T \left(A_0 + \sum_{n=1}^{\infty} A_n \cos(n\omega_0 t + \Theta_n) dt\right)^2 \tag{6.3.30}$$

Recalling that $(x+y)^2 = x^2 + 2xy + y^2$, we come to the following three terms of (6.3.30):

- $\left(\frac{1}{T}\right) \int_0^T A_0^2 dt = A_0^2$
- $\left(\frac{2}{T}\right) \int_0^T \left(A_0 \sum_{n=1}^{\infty} A_n \cos(n\omega_0 t + \Theta_n)\right) dt = 2A_0 \sum_{n=1}^{\infty} A_n \left(\frac{1}{T}\right) \int_0^T \cos(n\omega_0 t + \Theta_n) dt = 0.$
- $\left(\frac{1}{T}\right) \int_0^T \left(\sum_{n=1}^{\infty} A_n \cos(n\omega_0 t + \Theta_n)\right)^2 dt$

$$= \left(\frac{1}{T}\right) \int_0^T \left(\sum_{n=1}^{\infty} A_n \cos(n\omega_0 t + \Theta_n) \sum_{m=1}^{\infty} A_m \cos(m\omega_0 t + \Theta_m)\right) dt$$

$$= \left(\sum_{n=1}^{\infty} A_n \sum_{m=1}^{\infty} A_m \left(\frac{1}{T}\right) \int_0^T (\cos(n\omega_0 t + \Theta_n) \cdot \cos(m\omega_0 t + \Theta_m)) dt\right) = \frac{1}{2}\sum_{n=1}^{\infty} A_n^2$$

Since all members of this integral are equal to zero for $m \neq n$ and $\left(\frac{1}{T}\right)\int_0^T (\cos(n\omega_0 t + \Theta_n))^2 \, dt = \frac{1}{2}$, (6.3.30) reduces to the following equation:

$$V_{rms}^2 = \left(\frac{1}{T}\right)\int_0^T v^2(t)\,dt = \left(\frac{1}{T}\right)\int_0^T \left(A_0 + \sum_{n=1}^{\infty} A_n \cos(n\omega_0 t + \Theta_n)\right)^2 dt$$

$$= A_0^2 + \frac{1}{2}\sum_{n=1}^{\infty} A_n^2 = A_0^2 + \frac{1}{2}\sum_{n=1}^{n=\infty}(a_n^2 + b_n^2) \quad (6.3.31)$$

Equation (6.3.31) is the proof of (6.3.27). Therefore, power dissipated by the signal, $v(t)$ (V), over a resistor, R (Ω), is distributed across the signal's harmonics as

$$P_R \text{ (W)} = V_{rms}^2/R = 1/R\left(A_0^2 + \frac{1}{2}\sum_{n=1}^{\infty} A_n^2\right) \quad (6.3.32)$$

It is a common practice to consider power dissipated on the 1-Ω resistor; this power is called *normalized average power*, P_N (W). Thus,

$$P_N \text{ (W)} = V_{rms}^2/1\text{ (}\Omega\text{)} = A_0^2 + 1/2\sum_{n=1}^{\infty} A_n^2 = A_0^2 + \frac{1}{2}\sum_{n=1}^{\infty}(a_n^2 + b_n^2) \quad (6.3.33a)$$

Referring to (6.3.5) and (6.3.8b),

$$c_0 = A_0 \text{ and } |c_n| = \frac{1}{2}\sqrt{(a_n^2 + b_n^2)}$$

we find (6.3.33a) in the following form:

$$P_N = A_0^2 + 1/2\sum_{n=1}^{\infty}(a_n^2 + b_n^2) = c_0^2 + 2\sum_{n=1}^{n=\infty} c_n^2 \quad (6.3.33b)$$

Apparently, Eqs. (6.3.33a) and (6.3.33b) are the other forms of Parseval's theorem. Since c_0^2 is the dc power of a signal and $4c_n^2 = (a_n^2 + b_n^2)$ is the ac power carried by the nth harmonic, interpretation of Parseval's theorem for electric circuits is clear: *The normalized average power carried by a signal is the sum of its dc and ac power carried by all its harmonics.*

Equations (6.3.33a) and (6.3.33b) clearly show that Parseval's theorem in any form actually presents the *power spectrum* of a periodic signal:

> The power spectrum shows what portion of the signal's total power is carried by each harmonic.

We still need to remember that in all our considerations, we refer to normalized average power.

Example 6.3.2   The Power Spectrum of a Pulse Train

Problem

Consider the pulse train discussed in Example 6.3.1. Its waveform is shown in Figure 6.3.1, and its characteristics are as follows: amplitude, $V_{pt} = 10$ V; period, $T = 8$ ms; pulse duration, $\tau = 1.6$ ms; and duty cycle, $\tau/T = 0.2$. Find the power spectrum of this signal.

Solution
The solution to this problem looks rather straightforward: We need to use Parseval's theorem. However, before plugging in the coefficients of the Fourier series for the pulse train found in

Example 6.3.1, we need to decide which form of the Fourier series to use. Since power is a real entity, we need to use a one-sided form with real coefficients, given in (6.2.14); i.e.

$$v(t) = A_0 + \sum_{n=1}^{\infty} A_n \cos(n\omega_0 t + \Theta_n)$$

But we want to utilize the results of Example 6.3.1, where we built the two-sided spectrum of a pulse train. Therefore, we need to convert the two-sided spectrum of the periodic pulse train given in Example 6.3.1 into a one-sided spectrum. *The key to this conversion is Eq.* (6.3.21a),

$$A_0 = c_0 \quad \text{and} \quad A_n = 2|c_n|$$

Figure 6.3.3a compares a two-sided spectrum (reproduced from Figure 6.3.1) and a one-sided spectrum (computed here) of the given pulse train; the one-sided spectrum of a periodic pulse train is depicted in a larger scale in Figure 6.3.3b.

Now we can use (6.3.33a),

$$P_N \text{ (W)} = V_{\text{rms}}^2 = A_0^2 + 1/2 \sum_{n=1}^{\infty} A_n^2$$

to compute the power of individual harmonics and build the power spectrum of the pulse train. The power spectrum is shown in Figure 6.3.3c. The first 16 coefficients used for building the spectrum in Figure 6.3.3 are shown in Appendix 6.3.A.

The problem is solved.

Discussion

- Pay particular attention to the fact that $A_0 = c_0$, whereas $A_n = 2c_n$. Also, note that c_0 and c_n are computed by different formulas; for example, with a pulse train, we use (6.3.34) and (6.3.35).
- In this example, $|c_{-n}| = c_n = |c_n|$ because the coefficients are real.

6.3.3.3 A Signal's Bandwidth and Transmission Issues Associated with a Power Spectrum

Our discussion of the power spectrum of a periodic signal brings us to a very interesting question: What is the signal's bandwidth? You will recall the following definition:

> *A signal's bandwidth is the range of frequencies that the signal occupies; that is, the bandwidth of a given signal is the difference between its maximum and minimum frequencies.*

This definition implies, however, that the spectrum of a signal is finite, and the maximum and minimum frequencies of the signal are fixed, as shown in Figure 6.3.6. But in reality, we mostly deal with signals whose spectrum is infinite and whose maximum and/or minimum frequencies cannot be determined with certainty. The spectrum of the pulse train discussed in Examples 6.3.1 and 6.3.2 typifies such a situation (review Figure 6.3.1a,b). It is even more difficult to determine the bandwidth of a signal with continuous spectrum, such as that shown in Figure 6.3.1d. (We will study such signals shortly.)

The power spectrum is the means that helps to resolve this uncertainty. It delivers the meaningful definition of a signal's bandwidth with measurable characteristics. Indeed, when considering the transmission of a signal, one of the first things we want to determine is how much power the signal should carry. (The power/energy of a transmitted signal is an important metric in determining the quality and limitations of transmission.) If we want to transmit the total (maximum) power of a signal, we need to transmit all its harmonics, which means the signal occupies all frequencies from the maximum to the minimum.

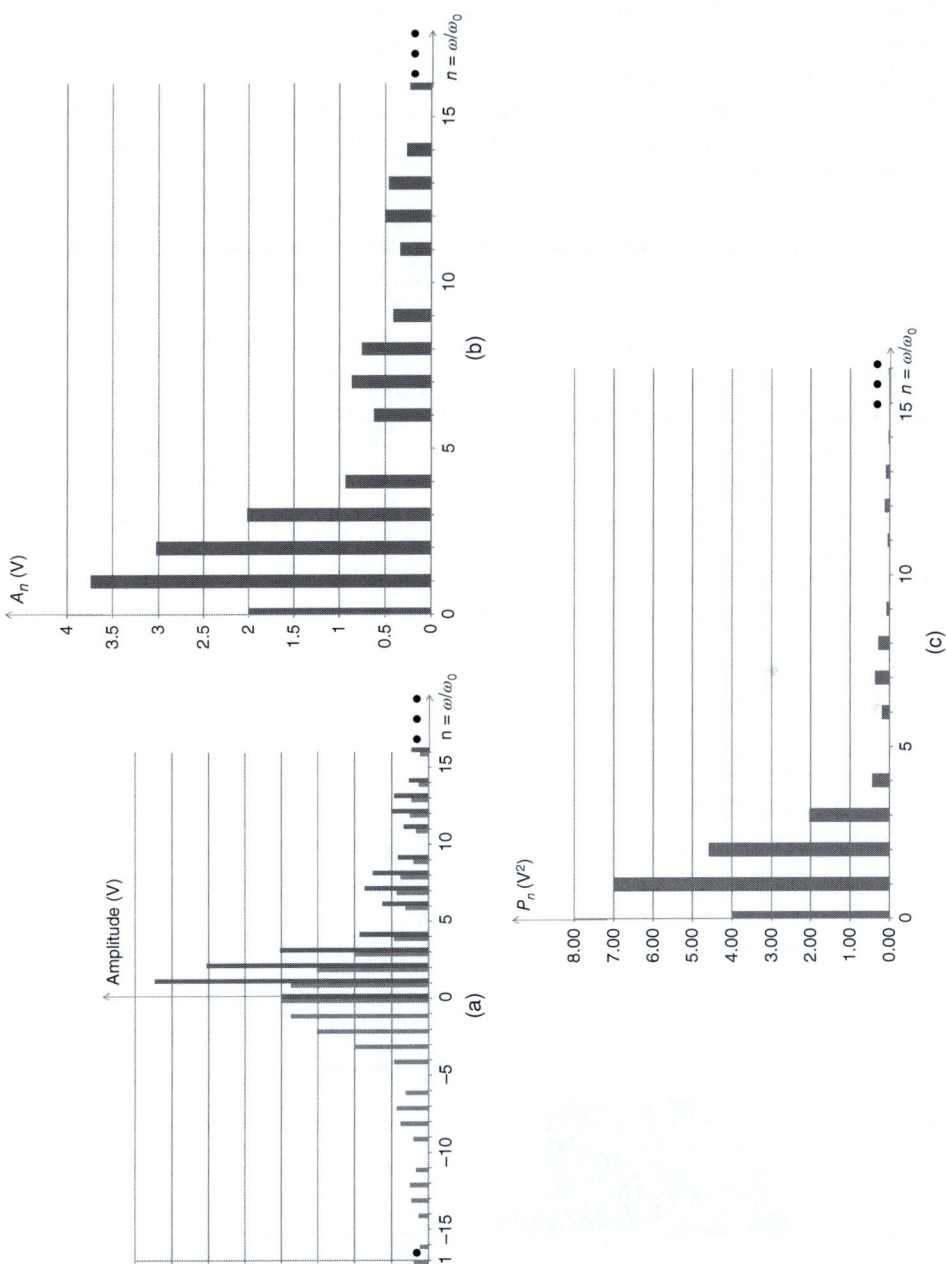

Figure 6.3.3 The spectra of the periodic pulse train for Example 6.3.2: (a) A two-sided and a one-sided spectrum; (b) a one-sided spectrum in a larger scale; and (c) a power spectrum.

> By determining the power of a signal being transmitted, we define the signal's bandwidth and vice versa.

In the case of a pulse train, for example, this means that for delivering the total power of the signal, we need to transmit an infinite number of spectral components; therefore, the signal's bandwidth is infinite. Such a result is, obviously, not practical and so we need to artificially restrict the signal's bandwidth. Such a restriction is logical and natural, but we have to introduce a quantitative measure; this means we need to develop and apply some criterion. And here is where Parseval's theorem comes into play:

> By deciding what percentage of the total power of the signal is sufficient for transmission, we determine the number of harmonics to transmit and, therefore, ascertain the signal's bandwidth.

Needless to say, the signal's bandwidth determines the required bandwidth of a transmission channel, which, as we stressed previously, is a commodity.

The relationship between the power and bandwidth of a signal is reciprocal: Knowing the signal's power, we define its bandwidth ; on the other hand, if we start by defining the bandwidth, we will determine its power (Figure 6.3.3b).

There is another important parameter to consider in regard to the role of the power spectrum in transmission: a signal's *waveform*. We refer to Figure 6.1.16, which shows that the more harmonics that are involved in the synthesis of a signal, the closer the signal's waveform is to its original (ideal) shape. Hence, we can say that the more harmonics we transmit, the better will be the waveform of a received signal. What is more, the better the waveform (shape) of a received signal, the higher the quality of transmission and the lower the probability of errors in deciphering the received information. But bear in mind that we want to restrict the number of harmonics to minimize the signal's bandwidth and its power; hence, these two requirements – receiving the best possible waveform and minimizing the signal's bandwidth and power – are, in fact, contradictory objectives, and so we need to find the optimal solution to this problem.

Regarding a signal's waveform, we need to remember that a transmission channel acts as a bandlimited filter, specifically, as a low-pass filter in a copper telephone line and as a bandpass filter in

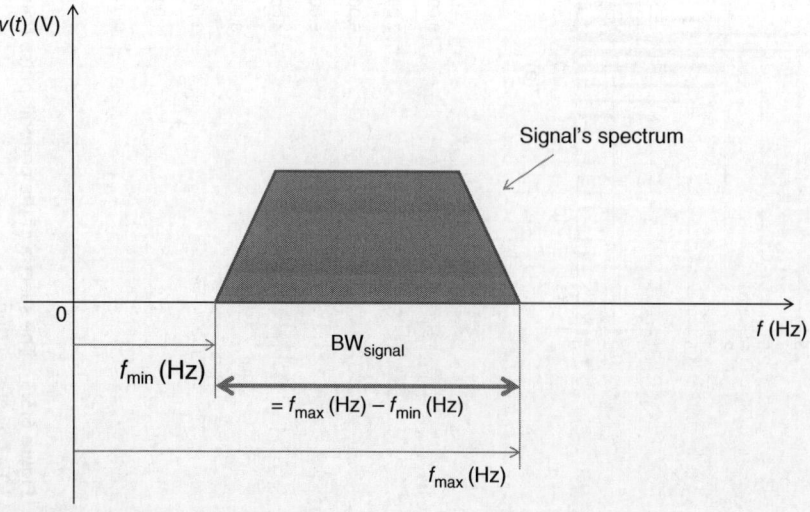

Figure 6.3.4 Definition of the bandwidth of a signal with specific maximum and minimum frequencies.

most transmission links. Such a filter introduces harmonic and phase distortions – the other cause of impairment of the signal's waveform and therefore the quality of transmission (see Sections 6.2.2 and 6.2.3).

We still need to consider another role that the power and bandwidth of a signal play in transmission: As we mentioned in Section 1.2 and will expand on shortly, the smaller the duration of a pulse, T_b, the wider its bandwidth (BW$_{signal}$); that is

$$\text{BW}_{signal}\,(\text{Hz}) \sim 1/T_b\,(\text{s}) \qquad (6.3.34)$$

But we need to remember, too, that the smaller the pulse duration, the greater the bit rate (BR):

$$\text{BR}\,(\text{b/s}) \sim 1/T_b\,(\text{s}) \qquad (6.3.35)$$

Thus, we have two contradictory requirements: to minimize the pulse duration in order to maximize the bit rate, but, at the same time, to maximize the bit duration in order to minimize the signal's bandwidth. This dilemma can be resolved by taking into account the third factor – the signal's power spectrum. We augment this discussion in Example 6.3.3.

Example 6.3.3 Power, Bandwidth, and Waveform of a Pulse Train

Problem

Consider the pulse train discussed in Example 6.3.1 and 6.3.2. What bandwidth does this signal occupy if we transmit 75% of its total power? 90%? 99%?

Solution

We transfer all the data from Example 6.3.2: $V_{pt} = 10\,\text{V}$, $T = 8\,\text{ms}$, $\tau = 1.6\,\text{ms}$, and $\tau/T = 0.2$. Also, we need to compute the fundamental frequency, $f_0 = 1/T = 125\,\text{Hz}$.

Step 1. Computing the total power: Since the power in question is given as a percentage of the total power, we need first to determine the total power. Conceptually, this can be done by applying Parseval's theorem; however, referring to (6.3.33a),

$$P_N\,(V^2) = A_0^2 + \tfrac{1}{2} \sum_{n=1}^{\infty} A_n^2$$

we immediately realize that we need to conduct a summation up to infinity. The explication comes from (6.3.27),

$$P\,(W) = \sum_{n=-\infty}^{n=\infty} |c_n|^2 = A_0^2 + \tfrac{1}{2} \sum_{n=1}^{\infty} A_n^2 = \left(\frac{1}{T}\right) \int_{t}^{t+T} [v(t)]^2\,dt$$

which clearly says how we can compute the total power of the signal in time domain. Refer to Figure 6.3.1a to follow our manipulations: Taking the initial time as $-\tau/2$ s, we find the upper limit of integration as $-\tau/2 + T$, which is presented in numbers as -0.0008 s and 0.0072 s. However, (6.3.33) defines our signal as

$$v(t) = V_{pt} = 10\,V\,(\text{for } 0 \leq t \leq \tau) \quad \text{and} \quad v(t) = 0\,(\text{for } \tau < t < T)$$

and therefore, this integral is not equal to zero only within the limits from $-\tau/2$ to $\tau/2$. Thus, we compute

$$P_{total} = \left(\frac{1}{T}\right) \int_{t}^{t+T} [v(t)]^2\,dt = \left(\frac{1}{T}\right) \int_{-\tau/2}^{\tau/2} [V_{pt}]^2\,dt = V_{pt}^2(\tau/T) = 20\,\text{W} \qquad (6.3.36)$$

We see again the importance of duty cycle, τ/T, in the description of a pulse train.

Table 6.3.1 Power and bandwidth of the pulse train in Example 6.3.3.

Normalized frequency, $n = \frac{\omega_n}{\omega_0}$	Normalized average P_0 (W) and the power of the individual harmonics, $P_n = \frac{1}{2}A_n^2$ (W)	Cumulative power by harmonics (W)	Percentage of total power by harmonics	Bandwidth of the pulse train by harmonics, $BW_{signal} = n \cdot f_0$ (Hz)
0	4.00	4.00	20	0
1	7.00	11.00	55	125
2	4.58	15.58	78	250
3	2.04	17.62	88	375
4	0.44	18.06	90	500
5	0.00	18.06	90	625
6	0.19	18.25	91	750
7	0.37	18.63	93	875
8	0.29	18.91	95	1000
9	0.09	19.00	95	1125
10	0.00	19.00	95	1250
11	0.06	19.06	95	1375
12	0.13	19.18	96	1500
13	0.11	19.29	96	1625
14	0.04	19.33	97	1750
15	0.00	19.33	97	1875
16	0.03	19.36	97	2000

Step 2. Computing power of harmonics and cumulative power: All these computations are straightforward; the results are shown in Table 6.3.1. We see that 75% of the total power is 15 W. From Table 6.3.1, we gather that a dc member and the first two harmonics give 15.58 W. Therefore, the bandwidth of a 75% signal power is equal to $2f_0 = 250$ Hz. Following this pattern, we calculate all the necessary numbers and collect them in Table 6.3.1. This table shows the normalized average power of each harmonic, the cumulative power by harmonics, the power in percentage by harmonics, and the bandwidth of the pulse train by harmonics. Thus, this table provides the solution to our problem.

Figure 6.3.5 depicts the information presented in Table 6.3.1:

- The power-spectrum column shows the power of the individual harmonics, P_n (W) vs. n.
- The cumulative-power column, P_{cum} (W) vs. n, demonstrates how the signal's power accrues with an increase in the normalized frequency.
- The percentage-power column, $\frac{P_{cum}}{P_{total}}$ (%) vs. n, illustrates what percentage of the total power corresponds to a specific value of the signal's frequency.
- Finally, the bandwidth-by-harmonics column, $BW_{signal} = nf_0$ (Hz), of a pulse train demonstrates the relationship between the signal's power and its bandwidth.

Let us consider an example using this table: for $n = 7$, $P_7 = 0.37$ W, $P_{cum} = 18.63$ W, $\frac{P_{cum}}{P_{total}} = 93\%$, and $BW_7 = 7 \cdot 125$ Hz $= 875$ Hz. You will recall that normalized frequency, $n = \frac{\omega_n}{\omega_0}$, allows us to compute the frequency of a given harmonic by using the formula $\omega_n = n \cdot 2\pi \cdot f_0$. For $n = 7$, for instance, $\omega_7 = 7\omega_0 = 7 \cdot 2\pi \cdot 125 = 5497.8$ rad/s.

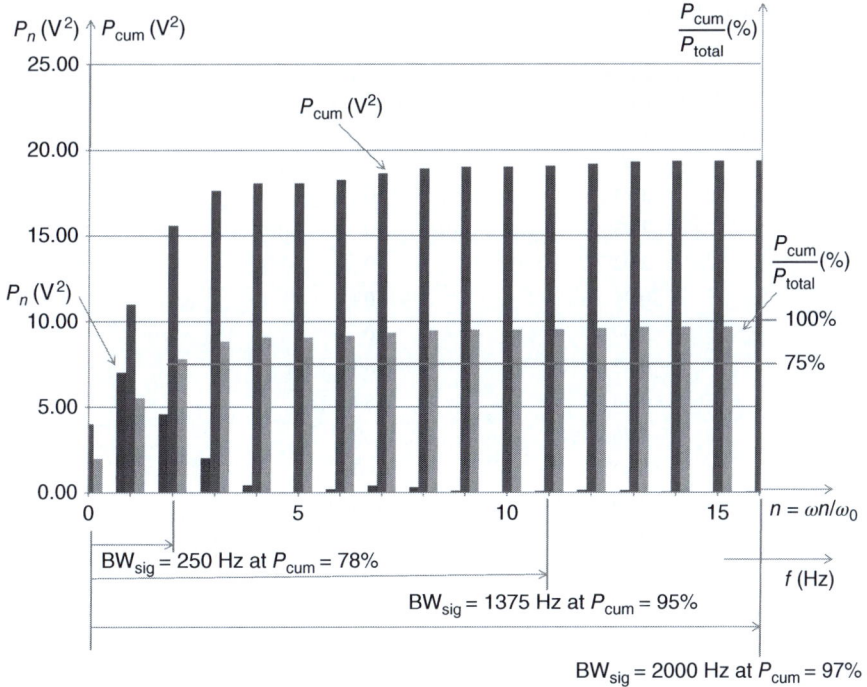

Figure 6.3.5 Pulse train for Example 6.3.2: power spectrum (P_n (W) vs. n), cumulative power (P_{cum} (W) vs. n), power in percentage ($\frac{P_{cum}}{P_{total}}$ (%) vs. n), and the signal's bandwidth (BW_{sig} (Hz) vs. P_{cum} (W)).

To illustrate the relationship between Table 6.3.1 and Figure 6.3.5, consider that the bandwidth of the pulse train equals $2f_0 = 250$ Hz: This bandwidth requires 78% of the signal's total power, which, in turn, necessitates inclusion of the dc component and the first two harmonics. If we need to transmit 95% of the total power, then we have to include 11 harmonics, and the signal's bandwidth then becomes $11f_0 = 1375$ Hz. Thus, Figure 6.3.5 and Table 6.3.1 present the solution to this problem by describing the relationship between the signal's power and its bandwidth.

Discussion

- As mentioned previously, the relationship between the signal's bandwidth and power is reciprocal; that is, we can specify the required power and receive the necessary bandwidth or we can specify the given bandwidth and receive the required power. Refer to Table 6.3.1 and Figure 6.3.6: If our transmission channel enables transmitting only 1500 Hz, then we have to set the bandwidth of our signal equal to this number. This setting means that the power delivered by this pulse train will be equal to 19.8 W, or 98% of the total power. Of course, we have to remember that *real bandwidth is the range of only positive frequencies*.
- In our discussion preceding this example, we recalled that reducing the number of a signal's harmonics changed the signal's waveform. We can illustrate this point using the results of our pulse-train example. Refer to (6.3.14),

$$v(t) = A_0 + \sum_{n=1}^{\infty} A_n \cos(n\omega_0 t + \Theta_n)$$

Table 6.3.2 Fourier coefficients for the pulse train in Example 6.3.2.

Normalized frequency, $n = \frac{\omega_n}{\omega_0}$	Amplitudes of harmonics, A_n (V)	Phases of harmonics, $\Theta_n{}^0$
0	2.00	0
1	3.74	0
2	3.03	0
3	2.02	0
4	0.94	0
5	0.00	0
6	0.62	180
7	0.86	180
8	0.76	180
9	0.42	180
10	0.00	180
11	0.34	0
12	0.50	0
13	0.47	0
14	0.27	0
15	0.00	0
16	0.23	180

to build the waveform of a pulse train by plugging in amplitudes A_n and phase shifts Θ_n of each harmonic.

○ To build the waveform with MS Excel, we collect A_n from Example 6.2.1, Θ_n from Example 6.1.1, and $f_0 = 125$ Hz from the problem's conditions and insert them in Table 6.3.2. Using this table, we can plot the waveforms of a pulse train. Such waveforms are shown in Figure 6.3.6. Comparing all the waveforms, we clearly see that the more powerful the signal, the better its waveform, that is, the closer it is to the ideal pulse train. This is because to boost a signal's power, we need to include more harmonics. Refer again to Section 6.3.1, where we discussed the synthesis of a square wave. The importance of a signal's waveform can be easily understood if we realize that to increase the bit rate – and we always want to increase the bit rate – we need to place the pulses as close to one another as possible. But the spacing between two adjacent pulses is determined by their shapes. Ask yourself, which pulses in Figure 6.3.6 – a or c – can be placed closer? Obviously, the pulses in Figure 6.3.6c allow for less spacing among them because their forms are closer to the rectangular than those in Figure 6.3.6a. Mentally, try to move the pulses in Figure 6.3.6a,c closer and closer to one another: In which figure do the pulses start to overlap sooner? Obviously, in Figure 6.3.6a.

> *You will recall that overlapping of pulses is called intersymbol interference (ISI), and it is one of the major obstacles to achieving high-quality digital communication.*

This discussion should explain the importance of a signal's power and bandwidth in electronic communication.

○ If we want to use MATLAB for this example, then we need to compute the Fourier coefficients, A_n. From (6.3.19) and (6.3.21a), we can derive the following formula for these computations for positive frequencies:

$$v_0 = c_0 = V * \tau/T$$

$$A_n = 2c_n = 4\sin(0.2n\pi)/(0.2n\pi)$$

Now we can write MATLAB code and plot the waveforms shown in Figure 6.3.6a–c. The MATLAB code for the 11 harmonics is presented in the following:

MATLAB Box 6.3.3 Building the Waveform of a Pulse Train from the Various Number of Harmonics in Example 6.3.3

```
V=10; %Amplitude [V]
T=0.008; %Period [sec]
tau=0.0016; %Duration [sec]
n=[-11:11]; %The number of harmonics
t=[0:T/512:0.018]; %Adjusting time
An=(V*tau/T).*sinc(n*tau/T);
w=n*(2*pi/T); %Define n frequencies for w [rad/sec]
w_t=w'*t; %Createbig matrix to do matrix multiplicationfor sum
vo=V*tau/T; %DC component [V]
v=vo+An*exp(-j*w_t); %DC and first 11 harmonics
plot(t,real(v)), grid, xlabel('Time [seconds]');
ylabel('Amplitude [V]');
title('DC and first 11 harmonics of the Fourier series')
```

- Can we use a two-sided spectrum to build a power spectrum? Yes. We leave the proof of this to you. It will be a good exercise toward a better understanding of the relationship between a two-sided and a one-sided spectrum.
- There is one more important point to mention: power budget. Consider a transmission system consisting of a transmitter, a transmission link with many components, and a receiver. Now we must determine how much power the transmitter, P_{TX} (W), has to launch into the transmission link in order to deliver enough power to the receiver. If P_{rec} (W) is the minimum power by which a receiver can recover the delivered information and P_{loss} (W) is the total losses a signal experiences on the end-to-end transmission, then the required launch power can be computed from this obvious formula:

$$P_{rec} \geq P_{TX} - P_{loss} \tag{6.3.37}$$

Simple rearrangement of (6.3.37) yields the requisite formula:

$$P_{TX} \geq P_{rec} + P_{loss} \tag{6.3.38}$$

For example, if $P_{rec} = 1$ mW and $P_{loss} = 99$ mW, then we must launch at least $P_{TX} = 100$ mW. Such a consideration is usually referred to as power-budget analysis. It imposes another restriction on the power in a transmission link and therefore on the bandwidth of the signal to be delivered.

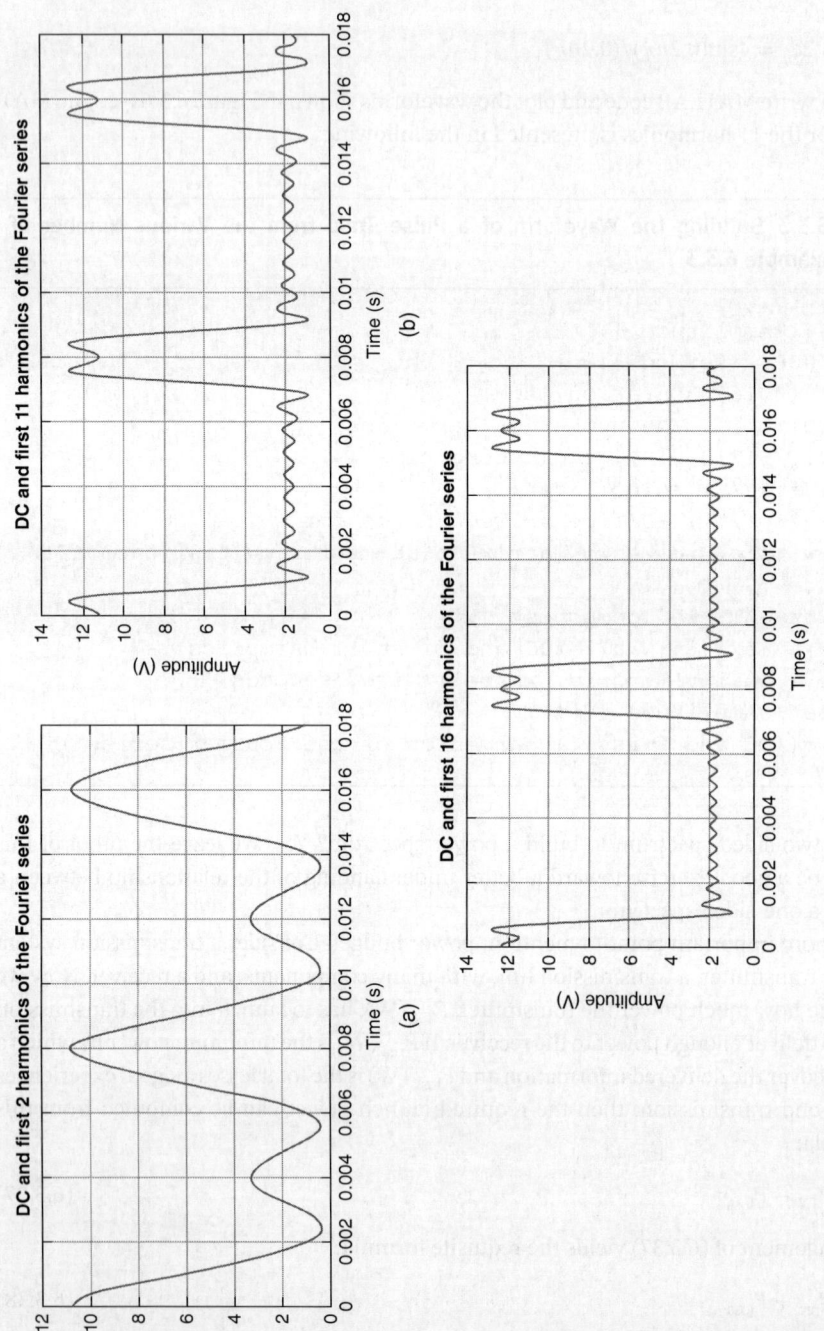

Figure 6.3.6 Waveforms of the pulse train for Example 6.3.2: (a) The pulse train at 75% of the total power with 250-Hz bandwidth (dc plus the first two harmonics); (b) the pulse train at 95% of the total power with 1375-Hz bandwidth (dc plus the first 11 harmonics); (c) the pulse train at 97% of the total power with 2000-Hz bandwidth (dc plus the first 16 harmonics).

- Finally, there is one other significant point to discuss: When reading books and papers on communications, we often encounter a reference to *power-bandwidth trade-off*. This *power-bandwidth trade-off is, in essence, the trade-off between the energy efficiency of the signal being transmitted and the bandwidth efficiency of the transmission channel*. This trade-off is a consequence of the fundamental Shannon theorem, which we discussed in Sections 1.2, 3.2, and 6.2. We will return to this subject in Chapter 8. For now, it is important to clearly understand that this phenomenon has nothing to do with *our discussion of the relationship between the POWER and BANDWIDTH of the same signal*.

At this juncture, we will pause *to summarize our discussion of the relationship between the transmitting power and the spectrum of a signal*:

- To determine a *signal's required power*, we calculate the number of harmonics to be transmitted to deliver this power and, in this way, define the signal's bandwidth. Conversely, in defining the signal's bandwidth, we determine the number of required harmonics and thereby ascertain the signal's power.
- In transmission, a signal's power has to meet two contradictory requirements: (i) The received power must be strong enough to be equal to the receiver's sensitivity and (ii) the transmitted power must be at minimum to reduce the transmission cost. The task is to find the optimal solution to this problem.
- The quality of the waveform of a received signal and therefore the quality of transmission increase with the number of harmonics being transmitted, which relates the quality of transmission to the signal's bandwidth and its power.
- A real transmission channel is always bandlimited and therefore acts as a filter, which causes harmonic and phase distortion – the additional source of degradation of transmission quality.
- In theory, the signal's bandwidth is the difference between the maximum and the minimum frequency of a signal; in practice, this bandwidth is the sum of harmonics delivering the required power.

Summary of Section 6.3

Mathematical Foundation of the Fourier Series

- *Polar and phasor forms of a sinusoidal signal revisited*: A sinusoidal signal can be presented in a polar form, which can be further developed into the phasor form (refer to (6.3.1a)). These forms enable us to present the Fourier series in an exponential form as

$$v(t) = \sum_{n=-\infty}^{n=\infty} (c_n e^{jn\omega_0 t}) \qquad (6.3.7R)$$

where complex coefficients, c_n, are defined by (6.3.8a). We can calculate the coefficients c_n by applying the following equation:

$$c_n = \left(\frac{1}{T}\right) \int_0^T v(t) (e^{-jn\omega_0 t}) dt \qquad (6.3.10R)$$

- *The other forms of an exponential Fourier series*: In general, the Fourier series can be presented in three equivalent forms:

Traditional trigonometric form, as in (6.2.1),

$$v(t) = A_0 + \sum_{n=1}^{\infty} (a_n \cos n\omega_0 t + b_n \sin n\omega_0 t)$$

One-sided trigonometric form, as in (6.2.14),

$$v(t) = A_0 + \sum_{n=1}^{\infty} A_n \cos(n\omega_0 t + \Theta_n)$$

Two-sided exponential form, as in (6.3.7),

$$v(t) = \sum_{n=-\infty}^{n=\infty} (c_n e^{jn\omega_0 t})$$

The last form allows for such additional variations as an infinite sum of phasors:

$$c_n e^{jn\omega_0 t} = |c_n| e^{j\varphi_n} e^{jn\omega_0 t} \quad \text{and} \quad c_{-n} e^{-jn\omega_0 t} = |c_n| e^{-j\varphi_n} e^{j-n\omega_0 t}) \qquad (6.3.S.1.6R)$$

and a sum of complex-conjugate pairs with its dc member, c_0, standing alone,

$$v(t) = c_0 + \sum_{n=1}^{n=\infty} 2|c_n| \cos(n\omega_0 t + \varphi_n) \qquad (6.3.S.1.7R)$$

What form you choose depends on the task at hand.

- *Two-sided and one-sided spectra*: Presentation of the Fourier series as a sum of rotating phasors invokes both positive and negative frequencies, $n\omega_0$ and $-n\omega_0$, which leads us to a *two-sided spectrum* of a signal. The trigonometric form of the Fourier series, however, allows for a *one-sided* spectrum because it refers to positive frequency only (refer to Figure 6.3.1). Despite introducing negative frequencies – the entities that do not exist in the real world – the complex exponential form of the Fourier series is a powerful and convenient tool that we use for sophisticated mathematical manipulations; results of these manipulations must always be real numbers that we can measure and convert into tangible entities.
- *Convergence of the Fourier series* The application of the Fourier series to the spectral analysis of a signal is based on a key proposition: The Fourier series is another form of this signal. This is true if – and only if – the Fourier series converges to the given function, for which the series must satisfy the *Dirichlet conditions*. *The rate of* convergence of the Fourier series depends on the type of discontinuity of the given signal. The smoother the signal, the higher the convergence rate of its Fourier series. This rate is determined by the pace with which the coefficients a_n and b_n decrease with an increase in n. Although the vast majority of practical signals allows for expansion into the Fourier series without any traps, knowing the limitations of this approach is certainly necessary to use this method professionally.
- *Gibbs phenomenon*: It is well known that the recovered waveform obtained by summing up the members of the Fourier series contains both undershoots and overshoots at the points of the signal's discontinuities, which is called the Gibbs phenomenon. This phenomenon shows that the Fourier series expansion of an ideal signal with a piecewise discontinuity never completely reproduces the waveform of this signal and will always exhibit overshoot and undershoot spikes at the points of discontinuity. Fortunately, all practical signals are always continuous, and the Gibbs phenomenon is not a concern in engineering practice.

Power Spectrum of a Periodic Signal

- *Power and energy signals*: You will recall that a signal can be either an energy signal or a power signal – or neither. The Fourier series can be applied only to a *power periodic signal*. Fortunately, all practical periodic signals are power signals; thus, this requirement for the existence of a Fourier series is always met.

- *Parseval's theorem*: The Fourier series shows the distribution of the signal's voltage across its harmonics. Similarly, we can determine the distribution of the signal's power across the harmonics, thus obtaining the signal's power spectrum. The average power P (W) of a periodic signal $v(t)$ can be expressed as

$$P\text{ (W)} = \left(\frac{1}{T}\right) \int_{t}^{t+T} [v(t)]^2 \, dt = \sum_{n=-\infty}^{n=\infty} (c_n c_n *) = \sum_{n=-\infty}^{n=\infty} |c_n|^2 \qquad (6.3.25\text{R})$$

where c_n is an nth coefficient of the signal's Fourier series in complex form and c_n* is its complex conjugate. This statement is called Parseval's theorem. It can be expressed through the real coefficients, a_n and b_n, of a trigonometric Fourier series of the signal, $v(t)$, as

$$P_N \text{ (W)} = (A_0)^2 + \tfrac{1}{2} \sum_{n=1}^{n=\infty} (a_n^2 + b_n^2) \qquad (6.3.33\text{aR})$$

where P_N (W) stands for normalized power. Equation (6.3.33a) shows the normalized power distribution across the signal's harmonics, which is another form of Parseval's theorem. We can find this theorem in still another form as

$$P_N = A_0^2 + 1/2 \sum_{n=1}^{\infty} (a_n^2 + b_n^2) = c_0^2 + 2 \sum_{n=1}^{n=\infty} c_n^2 \qquad (6.3.33\text{bR})$$

Here c_0^2 is the dc power of a signal and $4c_n^2 = (a_n^2 + b_n^2)$ is the ac power carried by the nth harmonic. Therefore, the normalized average power released by a signal over a 1-Ω resistor is the sum of its dc and ac power carried by all its harmonics. These equations demonstrate that Parseval's theorem in any form actually presents the *power spectrum* of a periodic signal, which shows what portion of the signal's total power is carried by each harmonic.

- *A signal's bandwidth and transmission issues associated with a power spectrum:* First, most real signals have infinite spectra, making it impossible to determine the bandwidth because its definition relies on the values of the signal's maximum and minimum frequencies. The power spectrum can resolve this uncertainty because we can determine what percentage of the total power of the signal is sufficient for transmission and – by applying Parseval's theorem – find the number of harmonics to transmit and therefore ascertain the signal's bandwidth. Thus, knowing the signal's power, we define its bandwidth and – reciprocally – by defining the bandwidth, we will determine its power. Second, we know that the more harmonics involved in the synthesis of a signal, the closer the signal's waveform is to its original (ideal) shape, the higher the quality of transmission, and the lower the probability of errors in deciphering the received information. Here, two requirements – receiving the best possible waveform and minimizing the signal's bandwidth and power – are contradictory objectives, and so we need to find the optimal solution to this problem. Third, the smaller the duration of a pulse, T_b, the greater the possible bit rate but, at the same time, the wider the signal's bandwidth. This consideration sets another restriction on the signal's power to be delivered at the receiver end (refer to Example 6.3.3 for a detailed discussion of these issues). In essence, this subsection introduces the concept of power-bandwidth trade-off in modern communications.

Questions and Problems for Section 6.3

- Questions marked with an asterisk require a systematic approach to finding the solution.
- Many questions and problems, including those marked with an asterisk, imply that you, in addition to reading the textbook, will do your research to find the answers. Consider such questions as mini-projects.

Mathematical Foundation of the Fourier Series

1. Why do we need to present a sinusoidal function in polar and phasor forms?

2. What is the difference between a polar form and a phasor form of a sinusoidal signal?

3. Present the signal $v(t) = 6\cos(180t + 1.23)$ in polar and phasor forms.

4. What is the key relationship that enables us to convert the Fourier series given in a trigonometric form into an exponential form? Explain.

5. *Consider the Fourier series of the square wave found in Example 6.1.1 as $v(t) = A/2 + 2A/\pi \sin 2\pi f_0 t + 2A/3\pi \sin 2\pi 3f_0 t + 2A/5\pi \sin 2\pi 5f_0 t + \cdots$ (6.1.13R). Present this Fourier series in an exponential form. Show all your mathematical manipulations. (*Hint*: Refer to equations from (6.3.1a) to (6.3.10).)

6. *Consider a pulse train with the following parameters: amplitude $V_{pt} = 2.8$ V, period $T = 4$ ms, and pulse duration $\tau = 0.8$ ms. Find the spectrum of this signal. (*Hint*: Refer to Example 6.3.1.)

7. *Consider the general form of the Fourier series of the pulse train given in (6.3.18), $v(t) = \frac{V_{pt}\tau}{T} \sum_{n=-\infty}^{n=\infty} \left(\frac{\sin\left(\frac{n\pi\tau}{T}\right)}{\frac{n\pi\tau}{T}} e^{jn\frac{2\pi}{T}t} \right)$:
   a) How will the spectrum of this pulse train change if the duty cycle, $\frac{\tau}{T}$, increases twice? Show all the mathematical manipulations along with your explanations.
   b) Is the expression $\sin(n\pi\tau/T)$ a function of time? Prove your answer.
   c) If a dc coefficient, c_0, is automatically included in an exponential Fourier series (6.3.18), why do we need to calculate it separately? How can we compute c_0? Give a mathematically based explanation.
   d) Phases of individual spectral members of this pulse train jump from 0° to 180°. Why don't they assume any other values? (*Hint*: Analyze Example 6.3.1.)

8. *Consider the Fourier series in the exponential form given in (6.3.20b) as $v(t) = \sum_{n=-\infty}^{n=\infty} (c_n e^{jn\omega_0 t})$:
   a) Present this series in a polar form and show how to find the amplitudes and phases of its members by using coefficients a_n and b_n of a trigonometric Fourier series.
   b) Equation (6.3.25) is accompanied by the statement, "A periodic signal can be expanded as an infinite sum of phasors: $c_n e^{jn\omega_0 t} = |c_n|e^{j\varphi_n}e^{jn\omega_0 t}$ and $c_{-n}e^{-jn\omega_0 t} = |c_n|e^{-j\varphi_n}e^{-jn\omega_0 t}$." Prove this statement.
   c) Regarding (6.3.25), the textbook says, "Since each pair of complex-conjugate phasors constitutes a real value, these phasors have to rotate in opposite directions, as shown in Figure 6.3.S.1.1." Why do these phasors have to rotate? Why in opposite directions?

9. *Presentation of the Fourier series in the exponential form requires the use of both positive and negative frequencies, $n\omega_0$ and $-n\omega_0$. Do we have negative frequencies in reality? How would you interpret the presence of negative frequencies in (6.3.20b), $v(t) = \sum_{n=-\infty}^{n=\infty} (c_n e^{jn\omega_0 t})$?

10 *The Fourier series of a real signal enables us to find the signal's spectrum. However, we can present this spectrum as a two-sided one or a one-sided spectrum:
  a) Are these spectra equivalent? Prove your answer.
  b) If these spectra are equivalent, why do we need two, not just one?

11 The Fourier series can be presented in three forms: traditional trigonometric form, (6.3.1a), one-sided trigonometric form, (6.2.14), and two-sided exponential form, (6.3.7). Which form would you choose for your practical application?

12 The Fourier theorem states that any periodic function can be expanded into a Fourier series. What is the necessary condition for the validity of this expansion?

13 The textbook explains that there are Dirichlet conditions of the convergence of a Fourier series. Briefly explain the meaning of these conditions.

14 *Figure 6.3.P14 shows an ideal pulse (a), a piecewise pulse (b), and a real pulse (c):
  a) Which version satisfies Dirichlet conditions? Explain.
  b) If the Fourier series of these pulses converge, which one has the highest and which one has the slowest convergence rate? Why?

15 The discussion of the Gibbs phenomenon in the textbook results in the following conclusion: "The Fourier series expansion of an ideal periodic digital signal with a piecewise discontinuity never completely reproduces the waveform of this signal and will always exhibit overshoot and undershoot spikes at the points of discontinuity." But the Fourier theorem states that any periodic signal can be developed into the Fourier series, which implies that the reconstruction of the signal (spectral synthesis) is possible. What is your opinion on this contradiction? Explain whether this contradiction restricts the practical use of the Fourier series for spectral analysis and synthesis.

Power Spectrum of a Periodic Signal

16 Explain the difference between power and energy signals. Give examples of both types.

17 The text says that "… all practical periodic signals are power signals; thus, this requirement for the existence of a Fourier series is always met." Explain what requirement this statement refers to. Also, explain how this statement relates to convergence of the Fourier series.

18 Consider a periodic electrical signal, $v(t)$ (V):
  a) What is its normalized average power?
  b) How does this power relate to the signal's rms value, V_{rms}?

19 *Consider a pulse train with the following parameters: amplitude $V_{pt} = 2.8\,\text{V}$, period $T = 4\,\text{ms}$, and pulse duration $\tau = 0.8\,\text{ms}$. Find the power spectrum of this signal. (*Hint*: Refer to Example 6.3.2; also, see Problem 6 and Example 6.3.1.)

20 How does the power of a signal affect its bandwidth? Explain.

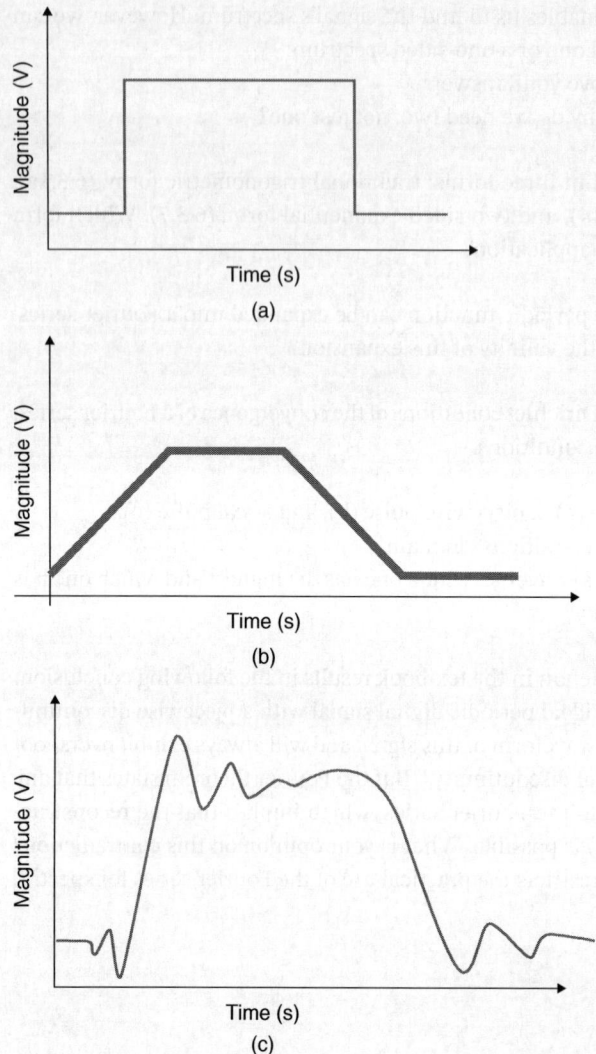

Figure 6.3.P14 Three waveforms of a pulse: ideal (a), piecewise (b), and real (c).

21 *It appears that a signal's power affects its waveform. Explain how it occurs. Give an example. (Refer to Sections 6.3.3.3 and 6.1.3.)

22 *Consider a pulse train with the following parameters: amplitude $V_{pt} = 2.8$ V, period $T = 4$ ms, and pulse duration $\tau = 0.8$ ms. What bandwidth does this signal occupy if we transmit 75% of its total power? 90%? 99%? (*Hint*: Refer to Examples 6.3.1–6.3.3; also, see Problems 6 and 19.)

23 Consider Table 6.3.1:
   a) How many harmonics can you transmit to deliver 97% of the signal's total power?
   b) What transmission bandwidth is required to transmit 97% of the signal's total power?
   c) Reconstruct the waveform of the signal carrying 97% of the signal's total power. (*Hint*: See Example 6.3.3.)

24 Explain the phenomenon called "intersymbol interference, ISI." How does this phenomenon depend on the number of harmonics being transmitted? Does ISI depend on the power of the signal being transmitted? Explain.

25 Consider the pulse train discussed in Example 6.3.2 and build the power spectrum of this signal from its two-sided spectrum.

26 Consider a simple point-to-point optical communications link shown in Figure 1.2.8. If a receiver's minimum power is $P_{rec} = 1\ \mu W$ and total power loss is $P_{loss} = 980\ \mu W$, what power must the transmitter generate?

27 Explain what calculations the industry calls the power budget.

28 What is the power-bandwidth trade-off? Explain.

6.3.A Fourier Coefficients of a Two-sided and a One-sided Spectrum of the Periodic Pulse Train for Example 6.3.2.

Normalized frequency, $n = \frac{\omega_n}{\omega_0}$	Fourier coefficients of a two-sided spectrum, c_n (V)	Fourier coefficients of a one-sided spectrum, $A_n = 2|c_n|$ (V)	Normalized average P_N (W) and the power of the individual harmonics, P_n (W) = ½ A_n^2 (V²)
−16	−0.12	0.00	0.00
−15	0.00	0.00	0.00
−14	0.13	0.00	0.00
−13	0.23	0.00	0.00
−12	0.25	0.00	0.00
−11	0.17	0.00	0.00
−10	0.00	0.00	0.00
−9	−0.21	0.00	0.00
−8	−0.38	0.00	0.00
−7	−0.43	0.00	0.00
−6	−0.31	0.00	0.00
−5	0.00	0.00	0.00
−4	0.47	0.00	0.00
−3	1.01	0.00	0.00
−2	1.51	0.00	0.00
−1	1.87	0.00	0.00
0	2.00	2.00	4.00
1	1.87	3.74	7.00
2	1.51	3.03	4.58
3	1.01	2.02	2.04

Normalized frequency, $n = \frac{\omega_n}{\omega_0}$	Fourier coefficients of a two-sided spectrum, c_n (V)	Fourier coefficients of a one-sided spectrum, $A_n = 2|c_n|$ (V)	Normalized average P_N (W) and the power of the individual harmonics, P_n (W) $= \frac{1}{2} A_n^2$ (V^2)
4	0.47	0.94	0.44
5	0.00	0.00	0.00
6	−0.31	0.62	0.19
7	−0.43	0.86	0.37
8	−0.38	0.76	0.29
9	−0.21	0.42	0.09
10	0.00	0.00	0.00
11	0.17	0.34	0.06
12	0.25	0.50	0.13
13	0.23	0.47	0.11
14	0.13	0.27	0.04
15	0.00	0.00	0.00
16	−0.12	0.23	0.03

7

Spectral Analysis 2 – The Fourier Transform in Modern Communications★

Objectives and Outcomes of Chapter 7

Objectives

- To become familiar with the Fourier transform and its application to the spectral analysis of various types of signals.
- To study the application of the Fourier transform in determining the response of circuits and systems.
- To learn about the application of the Fourier transform to digital signal processing in communications.

Outcomes

- Understand that a continuous nonperiodic signal can be obtained from a periodic signal by approaching its period to infinity, which results in reforming the discrete spectrum of the periodic signal to a continuous spectrum of the nonperiodic signal. Consequently, develop the Fourier transform from the Fourier series by tending the signal's period to infinity, thus developing a mathematical tool for finding the spectrum of a nonperiodic signal.
- Comprehend that the Fourier transform is a mathematical tool for the spectral analysis of a nonperiodic signal.
- Get to know that mathematically the Fourier transform is the integral operation that transforms a time-domain function into the frequency-domain function. Realize that these integrations of the main time-domain functions are collected in the table of the Fourier transform pairs. Understand that the Fourier transform of each time-domain function presents the spectrum of this function.
- Learn how to use the Fourier transform to spectral analysis of specific communication signals by applying its properties.
- Recall that nonperiodic signals are the energy signals and know that the Rayleigh energy theorem enables us to find the energy spectrum of a nonperiodic signal.
- Know that the Fourier transform of a system's impulse response, $h(t)$, produces the system's transfer function, $H(\omega)$. Realize that due to the convolution property of the Fourier transform, we can obtain the system's output in frequency domain, $V_{out}(\omega)$, simply by multiplying the system's input, $V_{in}(\omega)$, and the transfer function, $H(\omega)$; that is, $V_{out}(\omega) = V_{in}(\omega) \cdot H(\omega)$. Recall that the system's output in time domain is determined by applying the inverse Fourier transform, $v_{out}(t) = \mathbf{F}^{-1}\{V_{out}(\omega)\}$.

★In this chapter, we assume that the pulse duration, T_b (s), and pulse width, T_W (s), coincide; we denote this parameter τ.

Essentials of Modern Communications, First Edition. Djafar K. Mynbaev and Lowell L. Scheiner.
© 2020 John Wiley & Sons, Inc. Published 2020 by John Wiley & Sons, Inc.

- Know that in regard to the time-axis and magnitude-axis properties, signals can be classified as continuous periodic, continuous nonperiodic, discrete periodic, and discrete nonperiodic.
- Learn that there are four types of the Fourier transformations, one for each type of signal, enabling us to determine the spectrum of each type of signal. These Fourier transforms are: continuous-time Fourier series (FS), continuous-time Fourier transform (FT), discrete Fourier transform (DFT), and discrete-time Fourier transform (DTFT).
- Comprehend that in modern computerized technology, we have to use discrete signals and, therefore, employ either the discrete Fourier transform (DFT) or the discrete-time Fourier transform (DTFT).
- Grasp that one of the most ubiquitous operations in modern electronic technology is digital signal processing (DSP); it enables us to achieve high-quality transmission in electronic communication by changing the spectrum of the received signal in a desired way. Understand that obtaining the spectrum of a signal is a necessary condition for performing DSP.
- Get to know that DSP faces two major problems: supporting the required high processing speed and dealing with modern discrete and digital signals. Be aware that that only DFT meets these challenges.
- Learn that DFT is a mathematical operation that transforms one set of numbers, representing a discrete time-domain signal, into another set of numbers, representing this signal in frequency domain. Know that DFT is a discrete analogy of the continuous Fourier series, which would imply that DFT works only with periodic signals.
- Study that DFT can be used for finding the spectra of nonperiodic signals by limiting (truncating or windowing) these signals and thus making them finite.
- Become familiar with the problem that arises by the signal truncating: the more detailed the waveform (the plot of a signal in time domain), the less detailed the plot of its spectrum and vice versa. Know that this problem is often referred to as *the principle of uncertainty of Fourier analysis*.
- Be aware that we always have to truncate any signal presented to DFT and, therefore, we will always have an approximate, not exact, spectrum of a signal.
- Get to know the relationship among the Fourier transforms, namely: The Fourier transform can be derived from the Fourier series; the discrete-time Fourier transform (DTFT) can be derived from the Fourier transform; the discrete Fourier transform (DFT) can be derived from the discrete-time Fourier transform.
- Comprehend that DFT enables to perform its task only after the algorithm for fast execution of the spectral analysis – the fast Fourier transform (FFT)—had been developed. Know that there are many versions of FFT and the search for even faster algorithms to perform DFT continues.

7.1 Basics of the Fourier Transform

Objectives and Outcomes of Section 7.1

Objectives
- To understand the difference between periodic and nonperiodic signals.
- To become familiar with the Fourier transform.
- To start learning how to apply the Fourier transform to the spectral analysis of nonperiodic signals.

Outcomes

After studying this section, you will be able to

- Understand that a nonperiodic signal can be obtained from a periodic signal by tending its period to infinity, which results in reforming the discrete spectrum of the periodic signal to a continuous spectrum of the nonperiodic signal.
- Develop the Fourier transform by considering the Fourier series of a periodic signal in the limit when the period is being tended to infinity, thus developing a mathematical tool for finding the spectrum of a nonperiodic signal.
- Understand the concept of the Fourier transform and learn how to apply its mathematics to a simple example of a decaying exponential function. Relate the mathematical operation to the table of the Fourier transform pairs.
- Apply the Fourier transform to finding the spectra of a rectangular pulse and an exponential function.
- Understand the Rayleigh energy theorem as a mathematical tool for finding the energy spectrum of a nonperiodic signal.

7.1.1 The Fourier Transform in Spectral Analysis

7.1.1.1 From a Periodic to a Nonperiodic Signal

Let us consider a pulse train, the signal discussed in Example 6.1.1. The main characteristic of this signal is its duty cycle, defined in (6.3.15) as

$$\text{Duty cycle} = \frac{\tau}{T}$$

where τ (s) is the pulse duration and T (s) is the signal's period.

We can change the duty cycle by changing either τ or T. In both cases, the value of the duty cycle changes, but the effect of these changes on the signal spectrum will be quite different.

Changing the Pulse Duration, τ Let us start by *changing τ while keeping T constant*. The amplitude and phase spectra of a pulse train with various τ are shown in Figure 7.1.1.

MATLAB Box 7.1.1 The Waveforms and Spectra of a Pulse Train for $\tau = T/8$ in Figure 7.1.1[1]

```
tau=T/8:
V=input('Enter the amplitude [V]');
T=input('Enter the pulse period [sec]');
N=12; tau=T/8; %Pulse period [sec]
fs=1./T; %Sampling frequency [Hz]
D=tau./T; %Duty cycle
N=N/2; t=-0.02:T/512:0.02; %Adjusting time
pulseperiods=[-N:N]*T; %Pulse period
%The function of the pulse train:
```

[1] Developed by Ina Tsikhanava.

(Continued)

> **MATLAB Box 7.1.1 (Continued)**
>
> ```
> v=V.*pulstran(t,pulseperiods,@rectpuls,tau);
> subplot(3,1,1), plot(t,v), grid
> axis([-0.02 0.02 -1 V+1]);
> title('The waveform of a pulse train with tau=T/8');
> xlabel('Time [sec]');
> ylabel('Amplitude [V]');
> %Amplitude spectrum of the pulse train:
> a0=V*D; %The DC component a0
> n=-12:12; %Adjusting the number of harmonics
> an=V*sin(pi*n*D)./(n*pi); %The an coefficients
> Phase=atan2(0,an); %Phase in radians
> An=sqrt(an.*an); %The bn coefficients are all zero
> fn=n./T; %Adjusting frequency
> subplot(3,1,2), stem([0 fn],[a0 An]), grid
> xlabel('Frequency [Hz]'), ylabel('Amplitude [V]')
> title('Amplitude spectrum of the pulse train with tau=T/8')
> subplot(3,1,3), stem([0 fn],[0 Phase]); grid
> title('Phase spectrum of the pulse train with tau=T/8');
> xlabel('Frequency [Hz]'), ylabel('Phase [rads]')
> ```

When decreasing the duration of the pulse while keeping the period constant, we are trying to create the *ideal pulse train*, in which the duration of the pulse would be negligible with respect to the period.

The following discussion needs a brief reminder: The first zeros of the amplitude spectrum of a pulse train in Figure 7.1.1 are at the frequencies $\pm 1/\tau$ (s). Thus, *the frequency range, from $-1/\tau$ to $1/\tau$, can be considered as a measure of the bandwidth of the pulse train because most of the signal's power is concentrated here*. For a one-sided spectrum, a signal's bandwidth can be estimated as

$$\text{BW (Hz)} = 1/\tau \text{ (s) or BW (rad/s)} = 2\pi/\tau \tag{7.1.1}$$

See Section 6.3.3.2, particularly Figure 6.3.3.

An important observation to draw from Figure 7.1.1 is that the narrower the pulse (that is, the smaller the pulse duration, τ), the broader the signal's amplitude bandwidth (BW(Hz) = $1/\tau$ (1/s)) and vice versa. This well-known phenomenon is called *reciprocal spreading*; it holds true for the signals of any waveforms, not only for pulses. The physics behind this phenomenon is that *the faster the time variations of a signal, the greater the number of frequencies its spectrum contains and vice versa*. The pulse perfectly exemplifies this point: The pulse's width, τ (s), shows us how fast the pulse is changing (the smaller the τ, the faster the change); correspondingly, a narrower pulse has a greater number of frequencies in its spectrum than a wider one. Exemplifying this statement, Figure 7.1.1 shows that for $\tau = T/4$, $1/\tau \approx \text{BW} = 500\,\text{Hz}$ and for $\tau = T/32$, $1/\tau \approx \text{BW} = 4000\,\text{Hz}$. Therefore, *$1/\tau$ (Hz) is a measure of a pulse's bandwidth*.

It follows from the above observation that an ideal pulse would have a uniformly distributed amplitude spectrum with equal amplitudes for every harmonic. Surely, as $\tau \to 0$, the points of the first intersection of the amplitude spectrum with frequency axis tend to infinity, $1/\tau \to \infty$. (Compare these points for $\tau = T/2$ and $\tau = T/64$ in Figure 7.1.1 and imagine what happens with a further decrease of τ.) We remember, of course, that the summation of harmonics in a Fourier series extends to infinity. But according to Parseval's theorem, such a signal would carry infinite power, which, obviously, is not possible. Thus, *the train of the ideal pulses (with $\tau \to 0$) can't be built*.

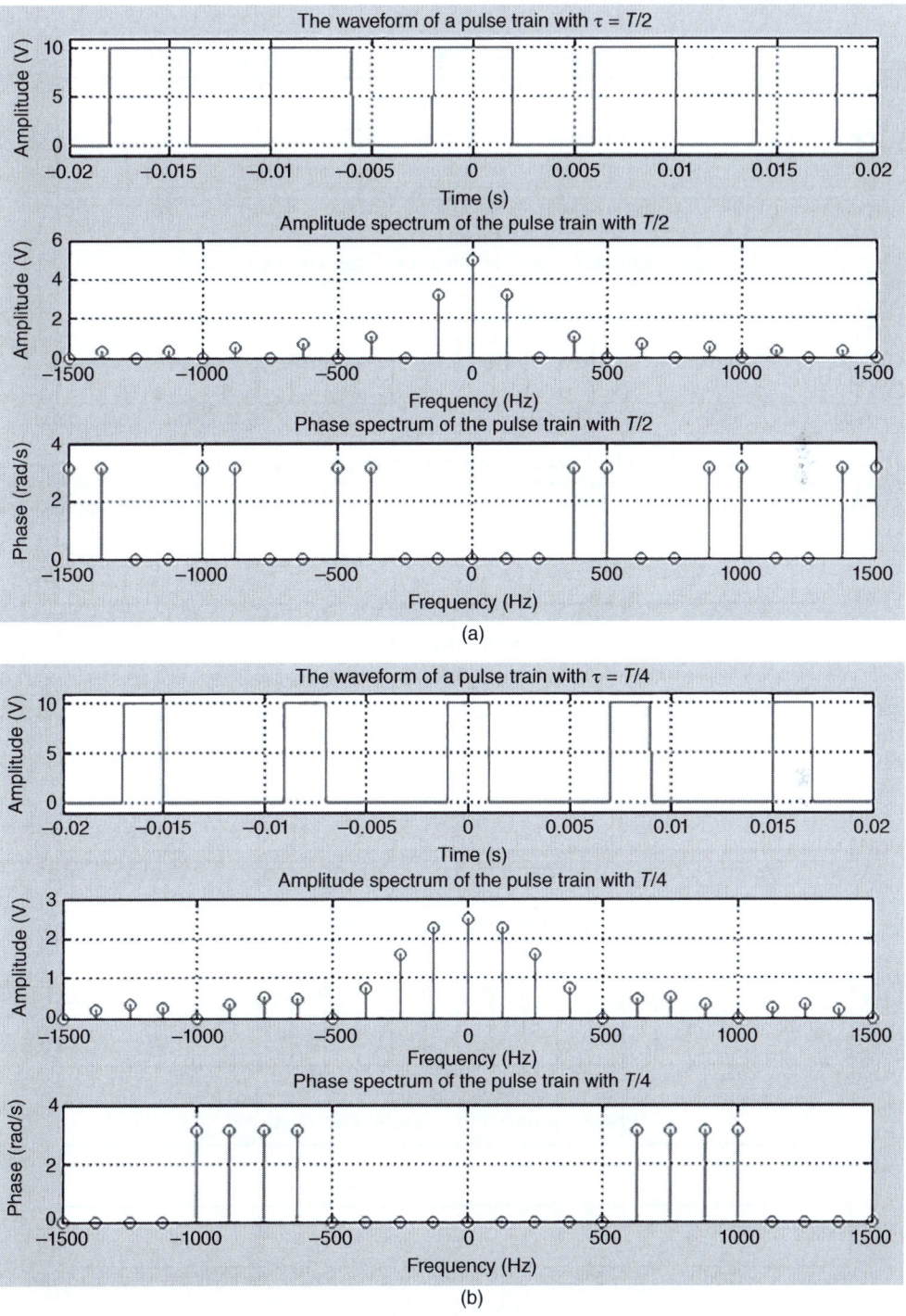

Figure 7.1.1 The waveforms and spectra of a pulse train with various τ and constant T. The MATLAB code for building one of the graphs is presented in MATLAB box 7.1.1. (a) Pulse train with $\tau = T/2$. (b) Pulse train with $\tau = T/4$. (c) Pulse train with $\tau = T/8$. (d) Pulse train with $\tau = T/16$. (e) Pulse train with $\tau = T/32$; time and frequency scales are the same as in the previous figures. (f) Pulse train with $\tau = T/32$; time and frequency scales are changed to demonstrate the whole picture of the waveform and the spectra. (g) Pulse train with $\tau = T/64$; time and frequency scales are the same as in the figures with a range from $\tau = T/2$ to $\tau = T/16$.

620 | *7 Spectral Analysis 2 – The Fourier Transform in Modern Communications*

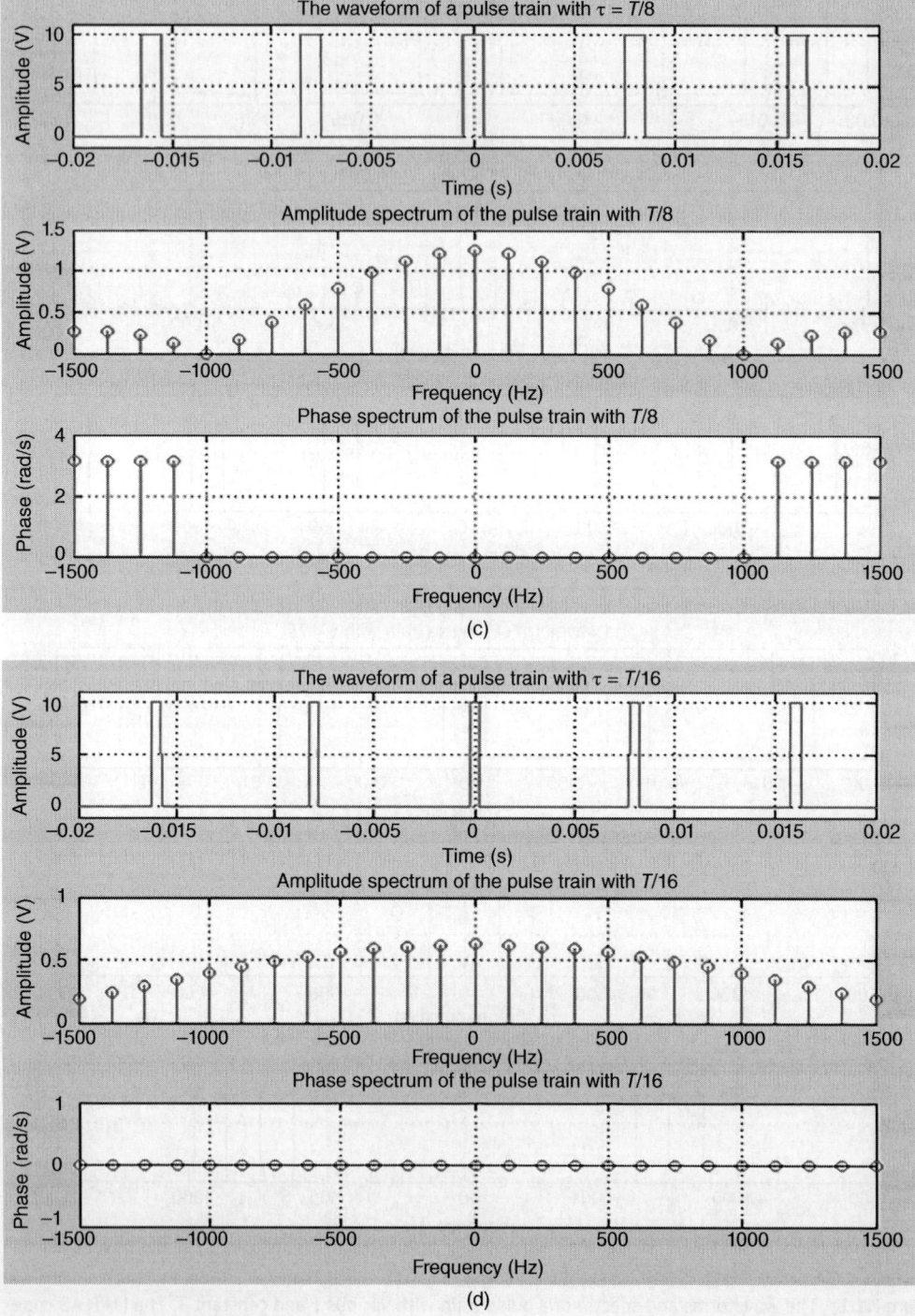

Figure 7.1.1 (*Continued*)

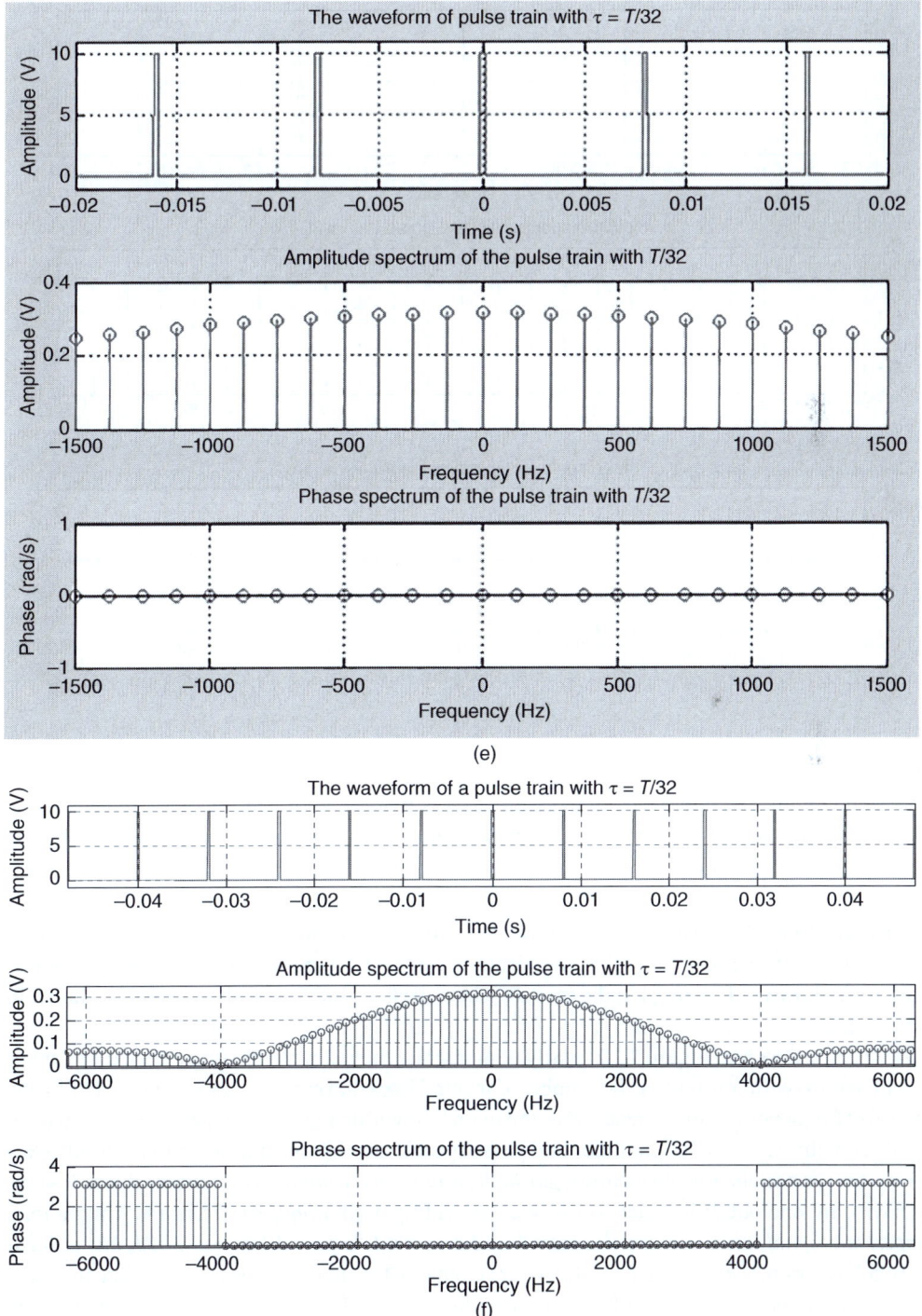

Figure 7.1.1 (*Continued*)

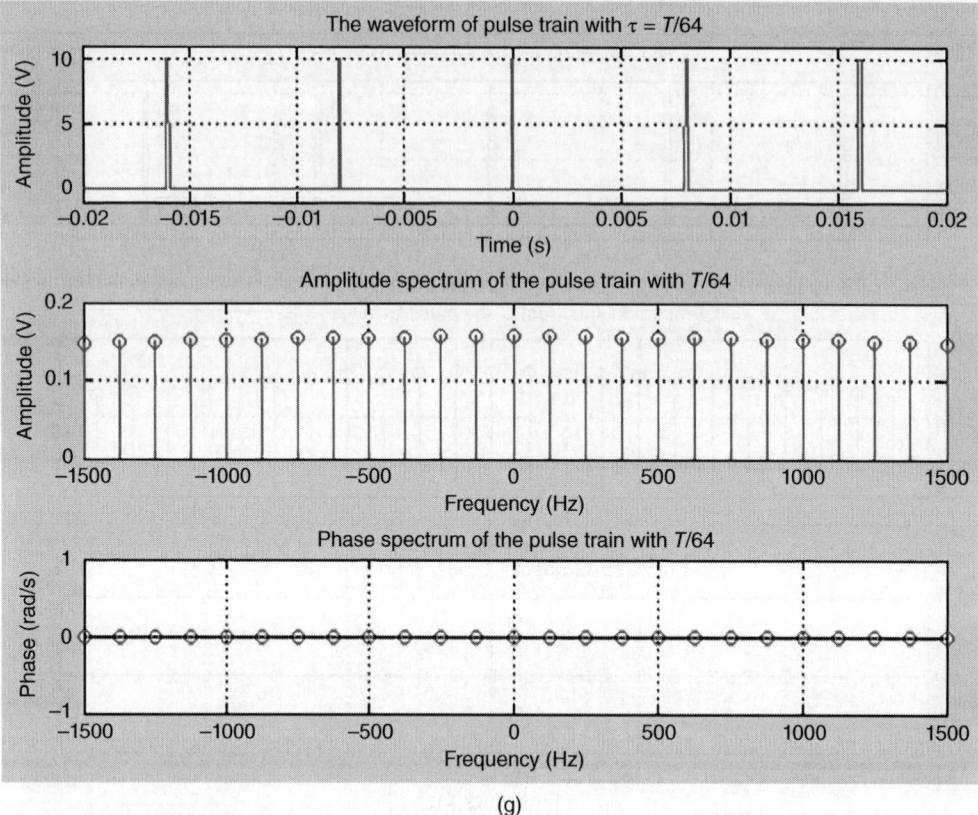

(g)

Figure 7.1.1 (*Continued*)

A second point to be noted regarding Figure 7.1.1 is that the narrower the pulse, the closer the adjacent spectral lines. Therefore, the ideal pulse train would have a continuous spectrum, but, of course, it will never occur.

Changing the Period, T (s), of a Pulse Train Now let us vary the duty cycle by *changing T and keeping τ constant*. We have two versions of this investigation: The first, where we keep the number of pulses constant, is shown in Figure 7.1.2; the second, where we keep the time scale constant, is shown in Figure 7.1.3.

The spectra shown in Figure 7.1.2 clearly demonstrate the main trend: As the period of a pulse train increases (with constant τ), the number of spectral lines that can be accommodated within one interval of frequency becomes greater. (See the main lobe within the $\pm 1/\tau$ interval.) This means that the adjacent lines in both the amplitude and phase spectra move closer and closer. In the limit, *when $T \to \infty$, the spectral lines become indistinguishable and form a continuous spectrum*. But at $T \to \infty$, we have only one pulse; in other words, we are dealing with a *nonperiodic signal*. Therefore, *a nonperiodic signal has a continuous spectrum, in contrast to a periodic signal, which has a linear (discrete) spectrum*. (Since in Figure 7.1.2 we are changing the time scale to accommodate the same number of pulses, the duration of pulses seems to decrease, but this is only the visual effect; in reality, τ remains constant.)

Figure 7.1.3 demonstrates our main point in an even more convincing way: The greater the period of a pulse train, the closer the spectrum comes to being continuous. (Figure 7.1.3f,g shows the same

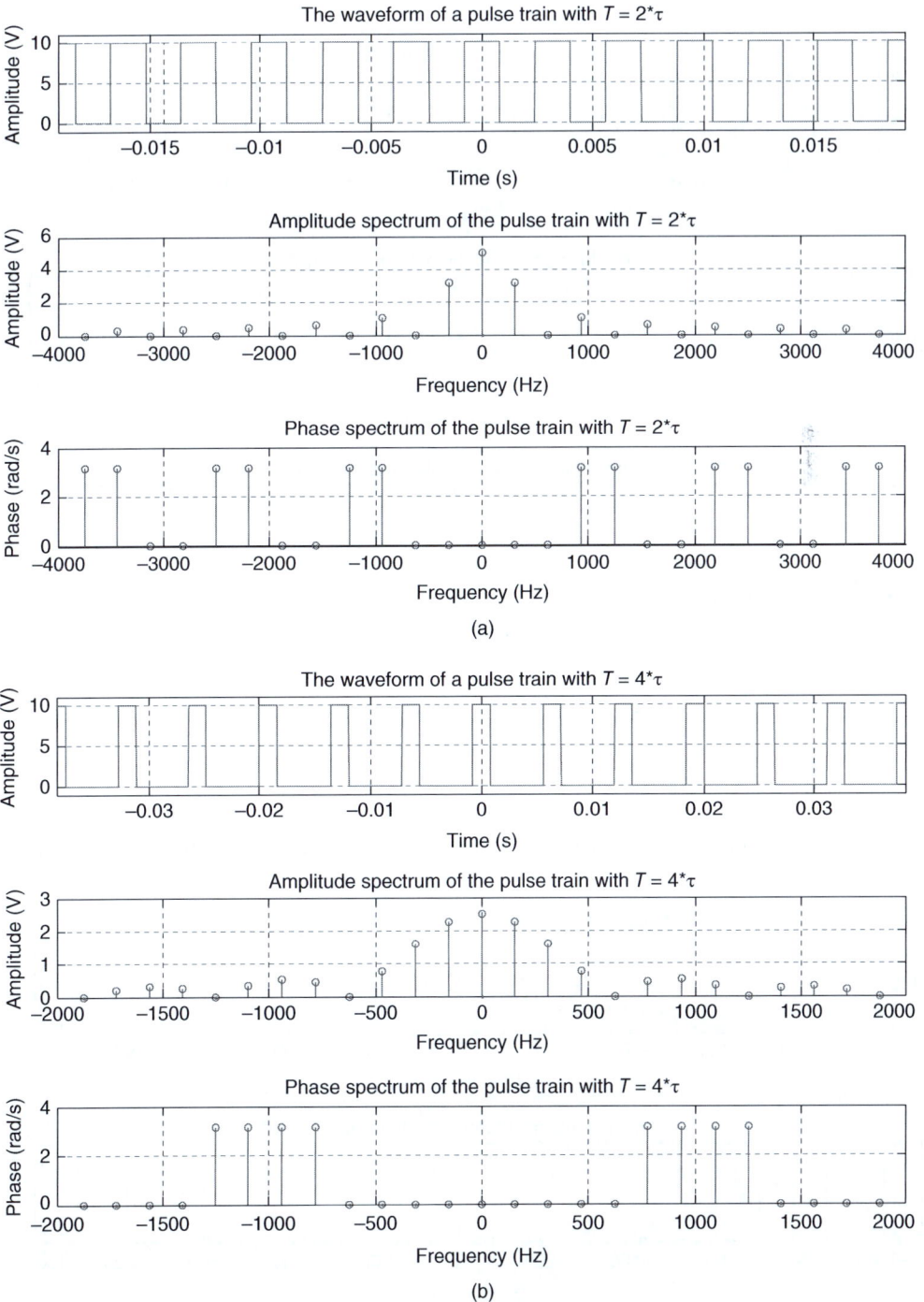

Figure 7.1.2 The waveforms and spectra of a pulse train with constant τ and variable T. Here, the number of pulses is constant, but the time scale is changing. ((a) Pulse train with $T = 2\tau$. (b) Pulse train with $T = 4\tau$. (c) Pulse train with $T = 8\tau$. (d) Pulse train with $T = 16\tau$. (e) Pulse train with $T = 32\tau$. (f) Pulse train with $T = 64\tau$.)

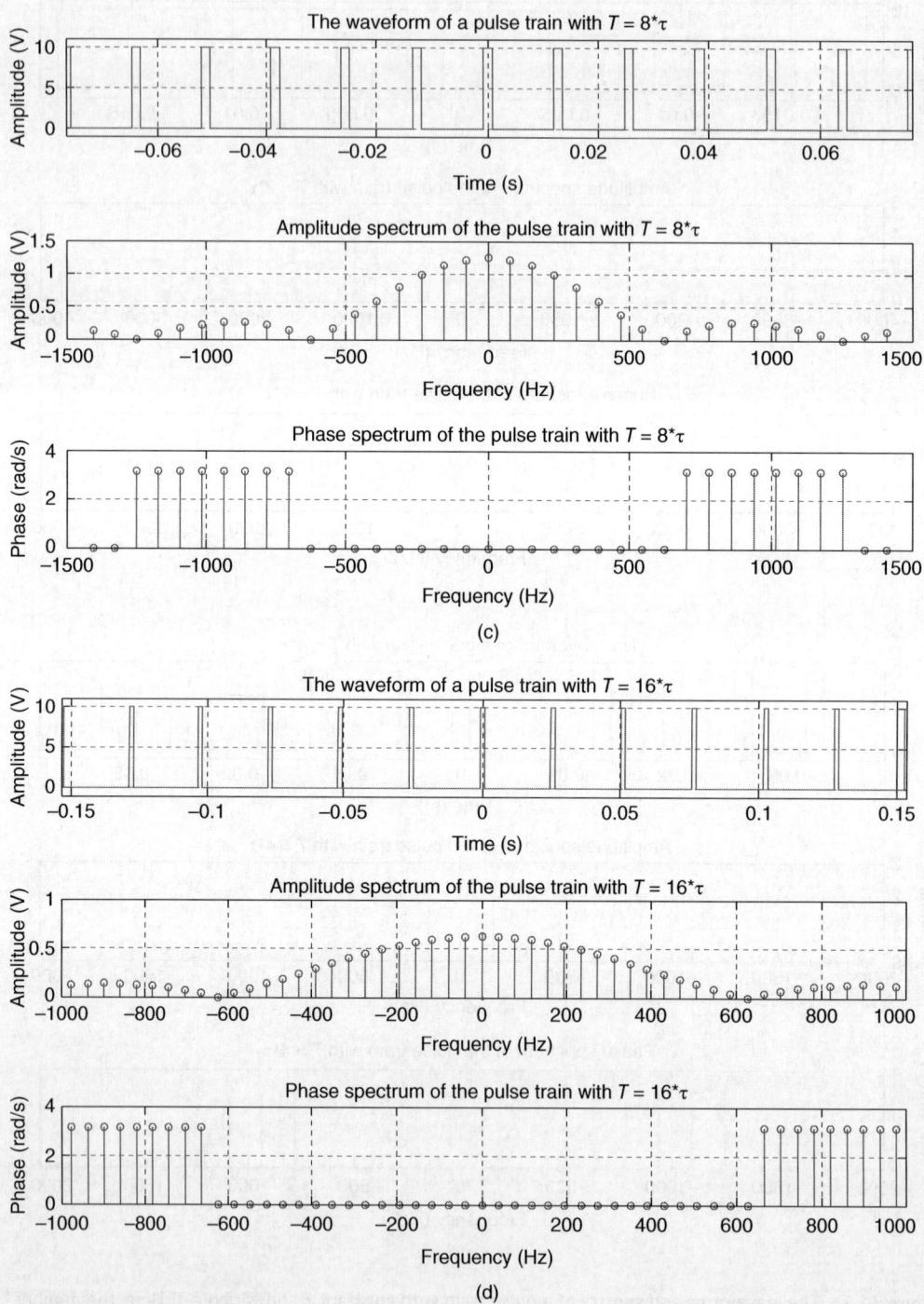

Figure 7.1.2 (Continued)

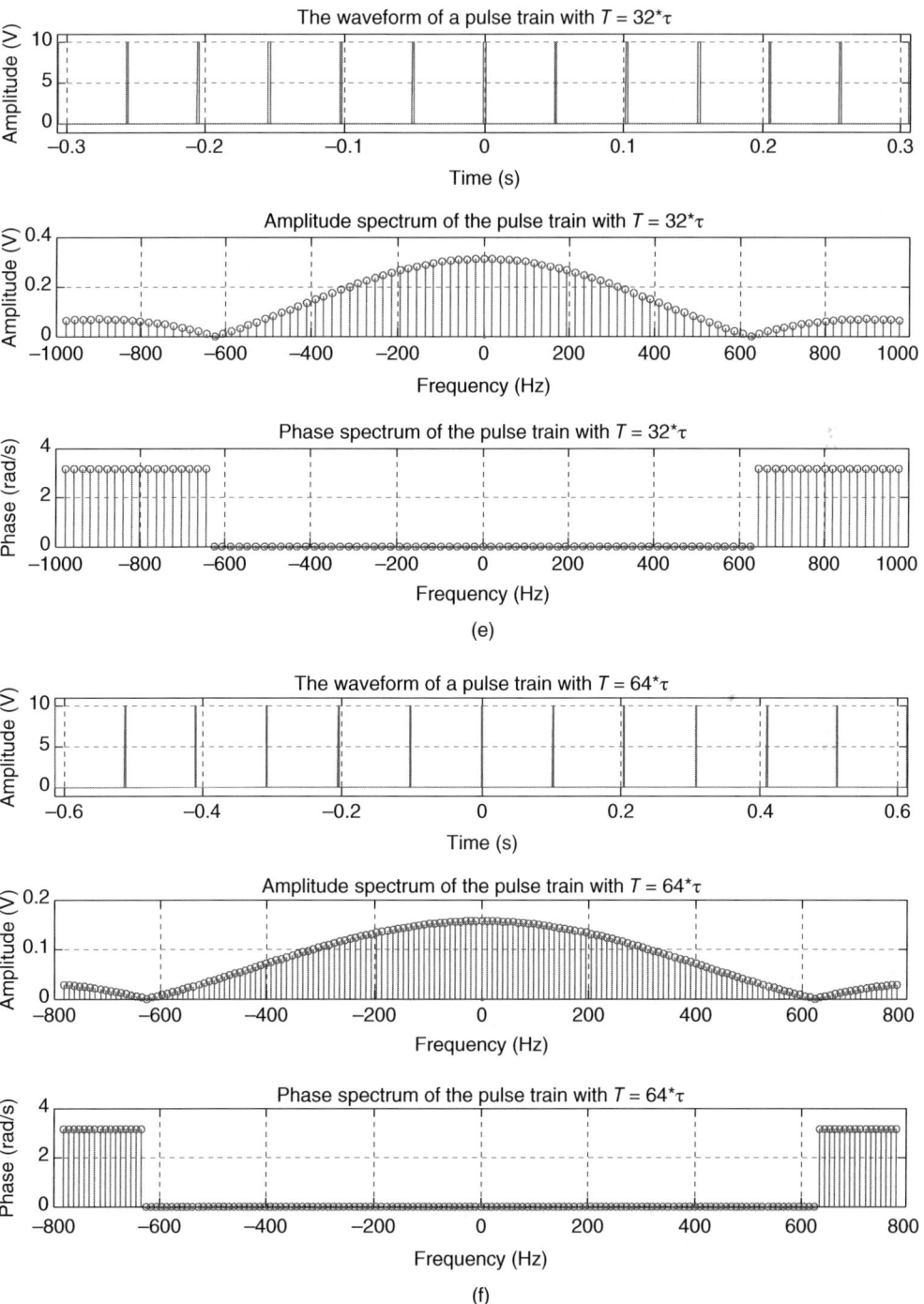

Figure 7.1.2 (Continued)

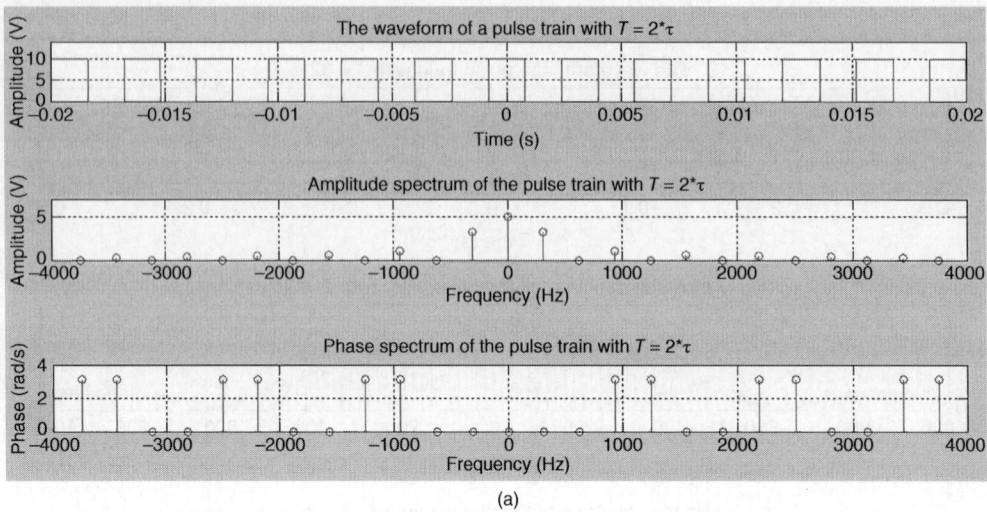

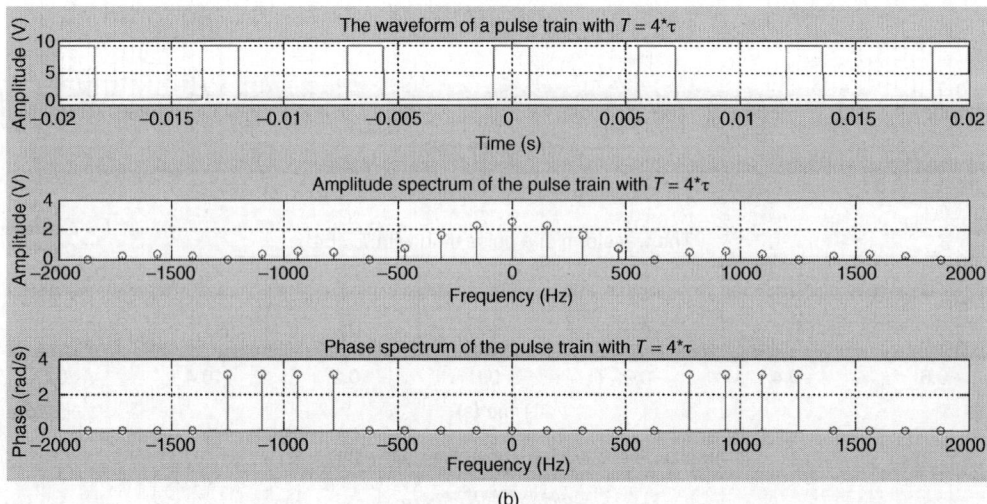

Figure 7.1.3 The waveforms and spectra of a pulse train with constant τ and variable T. Here, the time scale is constant, and therefore the number of pulses is changing. ((a) $T = 2\tau$. (b) $T = 4\tau$. (c) $T = 8\tau$. (d) $T = 16\tau$. (e) $T = 32\tau$. (f) $T = 64\tau$; the time and frequency scales are the same as in the previous figures. (g) $T = 64\tau$; the time and frequency scales changed to demonstrate the whole picture of the waveform and the spectra of the pulse train.)

case, $T = 64\tau$, but at different frequency scales to visually emphasize that the spectrum approaches the continuous state.) Also, we see that the greater the T, the fewer the number of pulses that can be accommodated within the same time interval; as T goes to infinity, the number of pulses within any given time interval tends to one.

Pay particular attention to this important fact: *The shape of the amplitude and the width of the main lobe (the bandwidth, that is) do not change when we vary T(s).* This is because both characteristics are determined by τ, not by T, and here τ is constant.

Compare Figure 7.1.1 with Figures 7.1.2 and 7.1.3 and find the similarities and differences. Doing this exercise will enable you to fully understand the material that follows.

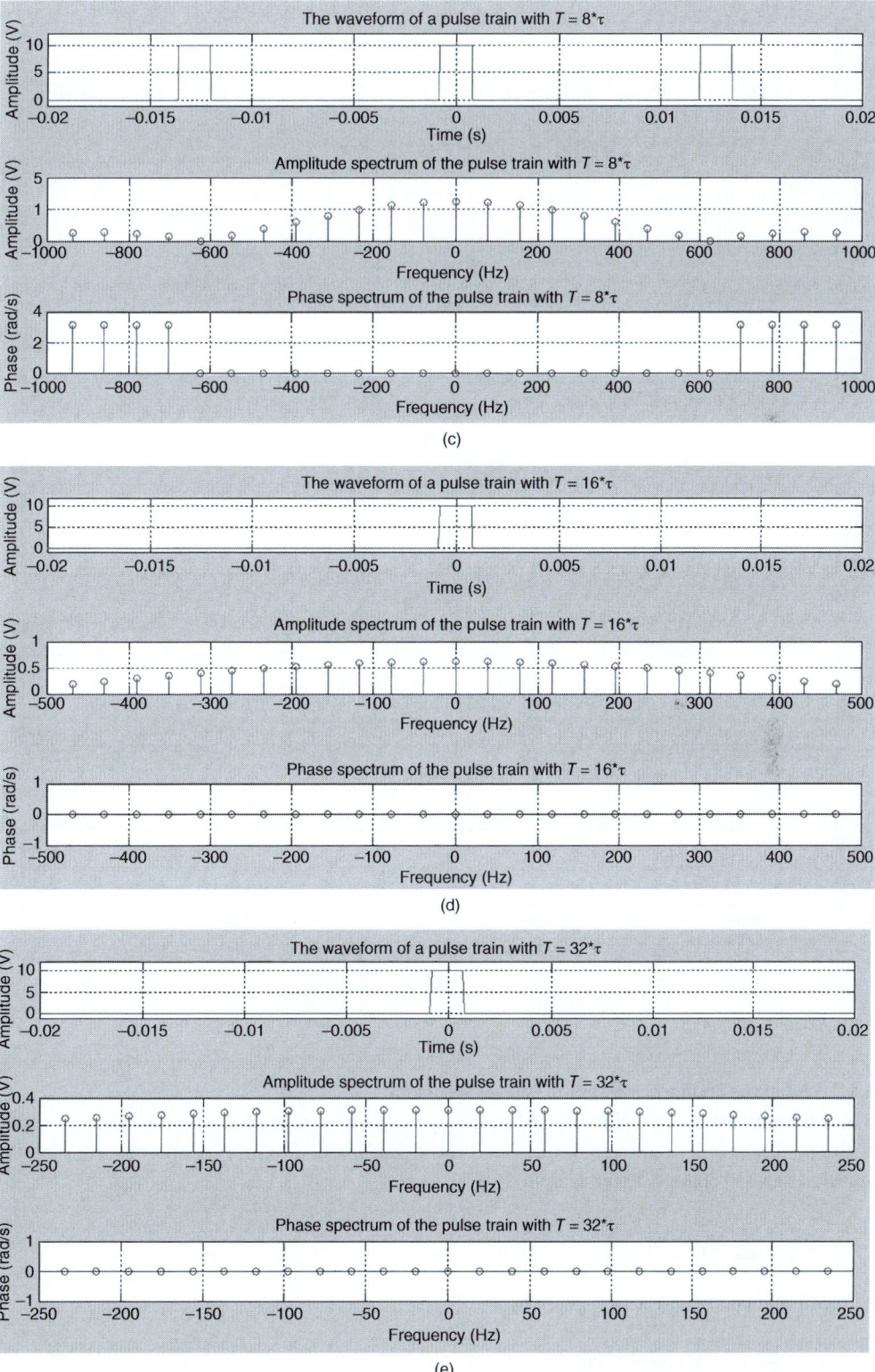

Figure 7.1.3 (*Continued*)

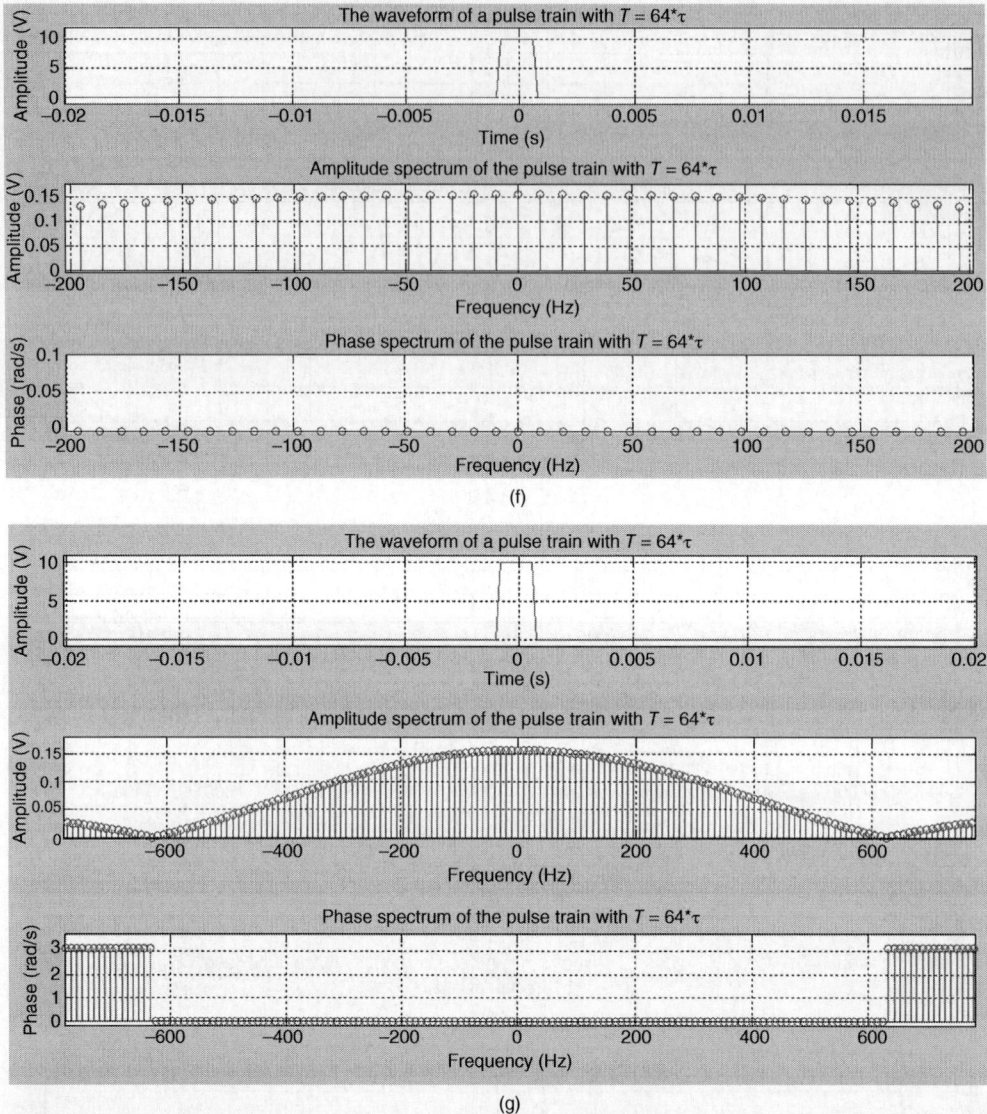

Figure 7.1.3 (Continued)

7.1.1.2 From the Fourier Series to the Fourier Transform

To find the *mathematical tool for describing the spectrum of a nonperiodic signal*, we start with a Fourier series for a periodic signal, then tend the signal's period to infinity and observe how the Fourier series will change. Refer again to (6.3.7),

$$v(t) = \sum_{n=-\infty}^{n=\infty} (c_n e^{jn\omega_0 t})$$

and recall that coefficients c_n are given by

$$c_n = \frac{1}{T} \int_{t}^{t+T} v(t)(e^{-jn\omega_0 t}) dt$$

Plugging the coefficients into the Fourier series, we find

$$v(t) = \sum_{n=-\infty}^{n=\infty} \left[\left(\frac{1}{T} \int_t^{t+T} v(t)(e^{-jn\omega_0 t}) dt \right] \cdot e^{jn\omega_0 t} \right) \quad (7.1.2)$$

The interval between two adjacent spectral lines can be found as

$$\Delta\omega = (n+1)\omega_0 - (n)\omega_0 = \omega_0 = 2\pi/T \quad (7.1.3)$$

Now we can reformat (7.1.2) with explicit reference to the spacing between the adjacent spectral lines, $\Delta\omega$, as

$$v(t) = \sum_{n=-\infty}^{n=\infty} \left[\left(\frac{1}{2\pi} \int_t^{t+T} v(t)(e^{-jn\omega_0 t}) dt \right] \cdot e^{jn\omega_0 t} \right) \Delta\omega \quad (7.1.4)$$

As the period increases, the spectral lines draw closer and closer, which means that *in the limit the incremental spacing, $\Delta\omega$, turns to differential, $d\omega$*. In other words,

when $T \to \infty$, $2\pi/T \to 0$ and $\Delta\omega \to d\omega$.

This condition also entails changing from discrete harmonics, $n\omega_0$, to continuous frequency, ω, and replacing the summation by integration. Therefore, at $T \to \infty$, (7.1.4) becomes

$$v(t) = \frac{1}{2\pi} \int_{-\infty}^{\infty} \left[\int_{-\infty}^{\infty} v(t)(e^{-j\omega t}) dt \right] e^{j\omega t} d\omega \quad (7.1.5)$$

Let us denote the core of this integral as

$$F(\omega) = \int_{-\infty}^{\infty} v(t)(e^{-j\omega t}) dt \quad (7.1.6a)$$

and call it the *Fourier transform*. That is, the Fourier transform, $F(\omega)$, of the time-domain function $v(t)$ is given by

$$\boldsymbol{F}\{v(t)\} = F(\omega) = \int_{-\infty}^{\infty} v(t)(e^{-j\omega t}) dt \quad (7.1.6b)$$

Here $\boldsymbol{F}$ is the Fourier-transform operator and $F(\omega)$ is the result (the Fourier transform.

Plugging this Fourier transform into (7.1.5), we determine the *inverse Fourier transform* of $F(\omega)$:

$$\boldsymbol{F}^{-1}\{F(\omega)\} = \frac{1}{2\pi} \int_{-\infty}^{\infty} F(\omega) e^{j\omega t} d\omega = v(t) \quad (7.1.7)$$

Obviously, an inverse Fourier transform produces $v(t)$.

7.1.1.3 The Fourier Transform Briefly Explained

To understand what the Fourier transform means and does, let us apply the definition of the Fourier transform given in (7.1.6a) to specific time-domain functions.

Example 7.1.1 The Fourier Transform of a Decaying Exponential Time-Domain Function

Problem

Find the Fourier transform of the decaying exponential function $v(t) = e^{-\alpha t} u(t)$.

Solution

First, we need to recall that the multiplication of any function by a unit-step (Heaviside) function $u(t)$ simply means that the given function exists only on the positive half of axis t. Second, we must refer to the definition of the Fourier transform given in (7.1.6a):

$$F(\omega) = \int_{-\infty}^{\infty} v(t)(e^{-j\omega t}) dt$$

Now, plugging the given time-domain function into (7.1.6a), we find

$$F(e^{-\alpha t} u(t)) = \int_{-\infty}^{\infty} v(t)(e^{-j\omega t}) dt = \int_{-\infty}^{\infty} e^{-\alpha t} u(t)(e^{-j\omega t}) dt$$

$$= \int_{0}^{\infty} e^{-\alpha t}(e^{-j\omega t}) dt = \int_{0}^{\infty} e^{-(\alpha+j\omega)t} dt = \left[-\frac{e^{-(\alpha+j\omega)t}}{\alpha + j\omega}\right]_{0}^{\infty} = -\frac{e^{-\infty} - e^{0}}{\alpha + j\omega} = \frac{1}{\alpha + j\omega}$$

Thus,

$$F(e^{-\alpha t} u(t)) = \frac{1}{\alpha + j\omega} \tag{7.1.8}$$

The problem is solved.

Discussion

Most important, these manipulations show that the Fourier transform of time-domain functions results in obtaining the frequency-domain functions, as demonstrated in (7.1.8).

(a) We can perform these mathematical manipulations by using the Symbolic Math Toolbox. For the time-domain functions considered in this example, we use the following MATLAB code to determine the required Fourier transform:

The Fourier transform of $v(t) = e^{-\alpha t} u(t)$ with $\alpha = 3$:

```
%Fourier transform
syms w t a
a=3;
H=sym('heaviside(t)');
F=fourier(H*exp(-3*t));
pretty(F)
```

Answer: $F = \frac{1}{wi+3}$

(Note that a unit-step (Heaviside) function $u(t)$ is denoted by "H" in this MATLAB script.)

(b) As an illustration of how to obtain the inverse Fourier transform, $F^{-1}\{F(\omega)\}$, consider the following MATLAB script for $F(\omega) = 1/(j\omega + 3)$ from this example:

```
%Inverse FT
syms w t a
a=3; %Defining a
v=ifourier(1/(w*i+a),w,t);
v=subs(v,'heaviside(t)','H') %replacing heaviside(t) by H for function v
```

Answer: $v(t) = H*\exp(-3*t)$, as expected.

Thus, MATLAB makes finding the direct and inverse Fourier transforms of given functions an easy operation.

(Exercise: Build the graph of $v(t) = H*\exp(-3*t)$. (*Hint*: Use the MATLAB plotting ability.))

Example 7.1.1 demonstrates that

the Fourier transform is a mathematical operation that transforms a time-domain function into a frequency-domain function.

Indeed, the function $v(t) = e^{-\alpha t}$, which depends only on t, is transformed by the Fourier integral into the function $F(\omega) = 1/(\alpha + j\omega)$, which depends only on ω. Conceptually, this operation is depicted in Figure 7.1.4a. The idea of the inverse Fourier transform is shown in Figure 7.1.4b.

If we continue applying the Fourier integral 7.1.6a to various time-domain functions, we will obtain their Fourier transforms. The results of such manipulations are typically tabulated; Table 7.1.1 presents several samples of the Fourier transform pairs.

The first glance at Table 7.1.1 reveals that the left column presents the given time-domain functions and the second right columns exhibits the corresponding frequency-domain functions obtained by applying the Fourier transforms to the given functions. These transformations from time domain to frequency domain are reciprocal; that is, a time-domain function can be obtained by applying the inverse Fourier transform to the corresponding frequency-domain function. For

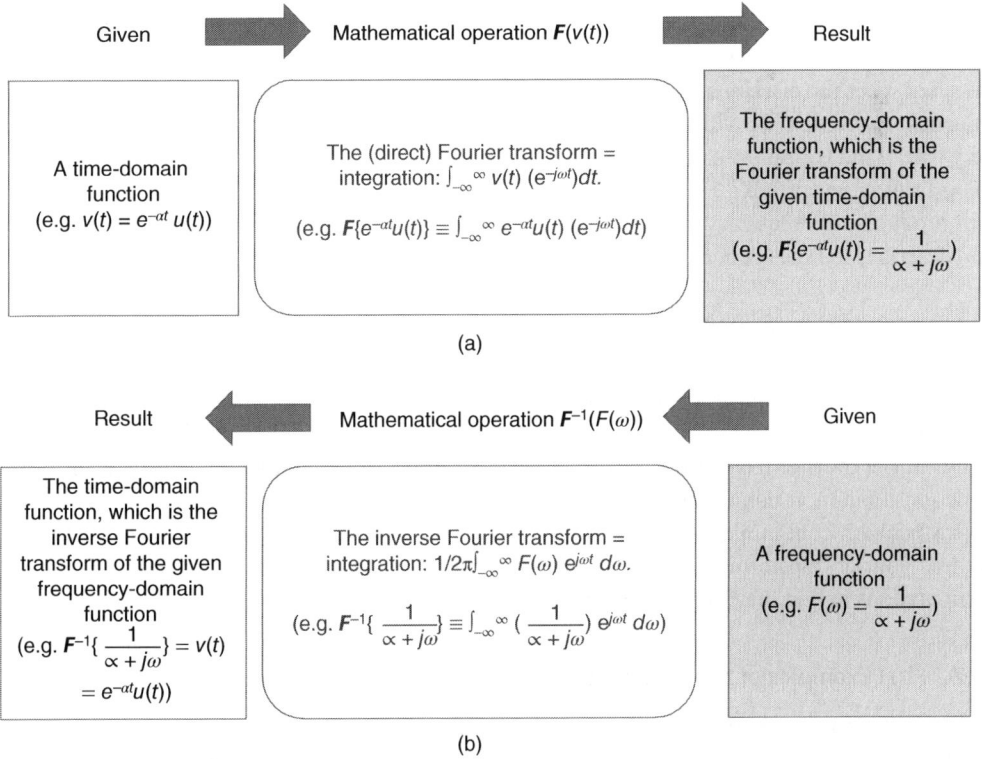

Figure 7.1.4 Conceptual visualization of (a) the (direct) Fourier transform according to (7.1.6b) and (b) the inverse Fourier transform according to (7.1.7).

Table 7.1.1 The Fourier transforms.

Time-domain function, $v(t) = F^{-1}(F(\omega))$	Fourier transform ($\rightarrow$), $F^{-}(v(t))$ Inverse Fourier transform ($\leftarrow$), $F^{-1}(F(\omega))$	Frequency domain function, $F(\omega) = F^{-}(v(t))$	Conditions and comments
$p(t) = \begin{cases} A \text{ for } \|t\| < \tau \\ 0 \text{ for } \|t\| > \tau \end{cases}$	$\leftarrow \rightarrow$	$A\tau \cdot \text{sinc}\left(\dfrac{\omega\tau}{2}\right)$	τ-duration of rectangular pulse, $p(t)^a$
$e^{-\alpha t}u(t)$	$\leftarrow \rightarrow$	$\dfrac{1}{\alpha + j\omega}$	$\alpha > 0$, constant
$t^n \cdot e^{-\alpha t}u(t)$	$\leftarrow \rightarrow$	$\dfrac{n!}{(\alpha + j\omega)^{n+1}}$	$\alpha > 0$, constant
1	$\leftarrow \rightarrow$	$2\pi\delta(\omega)$	$\delta(\omega)$ – delta function
$u(t)$	$\leftarrow \rightarrow$	$\pi\delta(\omega) + \dfrac{1}{j\omega}$	$u(t)$ – step function
$e^{j\omega_0 t}$	$\leftarrow \rightarrow$	$2\pi\delta(\omega - \omega_0)$	
$\cos(\omega_0 t)$	$\leftarrow \rightarrow$	$\pi[\delta(\omega - \omega_0) + \delta(\omega + \omega_0)]$	
$\sin(\omega_0 t)$	$\leftarrow \rightarrow$	$\pi[\delta(\omega + \omega_0) - \delta(\omega - \omega_0)]$	

a) See Section 7.1.2.1.

instance, applying the Fourier transform to $v(t) = e^{-\alpha t}u(t)$, results in $F(\omega) = F(e^{-\alpha t}u(t)) = \dfrac{1}{\alpha + j\omega}$. Inversely, $F^{-1}\left(\dfrac{1}{\alpha + j\omega}\right) = e^{-\alpha t}u(t)$. This result is demonstrated in Example 7.1.1.

The delta function in frequency domain, $\delta(\omega_0)$, is depicted as a vertical line located at ω_0. Thus, the Fourier transform of a sinusoidal signal results in two vertical lines located at $\pm\omega_0$, which is a two-sided spectrum of a sinusoidal signal. The Fourier transform of a rectangular pulse is discussed in the next subsection; we will elaborate on this topic in Section 7.2. (You are encouraged to analyze every pair of the Fourier transforms shown in Table 7.1.1.)

This brief review of Table 7.1.1 brings us to the fundamental conclusion: Since the Fourier transform produces the frequency-domain function, $F(\omega)$, of a given time-domain function, $v(t)$, and $F(\omega)$ describes the spectrum of $v(t)$, we understand that

> the Fourier transform enables us to find the spectrum of a nonperiodic time-domain function.

In other words, the Fourier transform is a main tool in the spectral analysis of nonperiodic (that is, vast majority of) signals.

The Fourier transform is thoroughly discussed in Section 7.2. We can find the Fourier transforms of the most common nonperiodic signals by using the tables, similar to Table 7.1.1, available in any textbook on communications and online. Of course, MATLAB also enables us to obtain the required Fourier transforms. Thus, it is extremely seldom that we need to resort to mathematical manipulations to find the Fourier transform of a specific time-domain function.

7.1.2 First Examples of the Fourier Transform Applications

7.1.2.1 A Rectangular Pulse

We need to discuss specifically a rectangular pulse and its Fourier transform because this pulse plays a significant role in consideration of nonperiodic signals and digital transmission. Such a pulse is designated as $p(t)$ or rect(t).

There are variations in the presentation of a rectangular pulse. Let us start by considering a pulse with amplitude A (V) and width (duration) τ (s) described as

$$p(t) \equiv \text{rect}\left(\frac{t}{\tau}\right) = A\left[u\left(t+\frac{\tau}{2}\right) - u\left(t-\frac{\tau}{2}\right)\right] \tag{7.1.9}$$

in which $u(t)$ is a unit-step function. This definition means that a rectangular pulse is constructed from two unit-step functions, as shown in Figure 7.1.4a, where the pulse exists only from the $-\tau/2$ to the $+\tau/2$ intervals. *This definition is used in MATLAB to calculate a rectangular pulse.*

To find the Fourier transform of $p(t)$, we need to apply (7.1.6b):

$$\mathbf{F}\{p(t)\} = P(\omega) = \int_{-\infty}^{\infty} p(t)(e^{-j\omega t})dt = A\int_{-\tau/2}^{\tau/2}(e^{-j\omega t})dt \tag{7.1.10}$$

The limits of integration have been changed because the pulse exists only from $-\tau/2$ to $\tau/2$; also, we should remember that the numerical value of a unit-step function is equal to one. By continuing the integration, we find

$$\mathbf{F}\{p(t)\} = \int_{-\tau/2}^{\tau/2} A(e^{-j\omega t})dt = -\left.\frac{Ae^{-j\omega t}}{j\omega}\right|_{-\tau/2}^{\tau/2} = -\frac{A\left(e^{-\frac{j\omega\tau}{2}} - e^{\frac{j\omega\tau}{2}}\right)}{j\omega}$$

Now we need to resort to the Euler formula, which we have already used many times:

$$e^{jx} = \cos x + j\sin x \text{ and } e^{-jx} = \cos x - j\sin x$$

and its corollary

$$e^{-jx} - e^{jx} = -2j\sin x$$

With the latter formula, the Fourier transform of a rectangular pulse takes the following form:

$$\mathbf{F}\{p(t)\} = -\frac{A\left(e^{-\frac{j\omega\tau}{2}} - e^{\frac{j\omega\tau}{2}}\right)}{j\omega} = \frac{2A\sin\left(\frac{\omega\tau}{2}\right)}{\omega} = \frac{A\tau\sin\left(\frac{\omega\tau}{2}\right)}{\frac{\omega\tau}{2}} = A\tau\,\text{sinc}\left(\frac{\omega\tau}{2}\right)$$

where $\text{sinc}\left(\frac{\omega\tau}{2}\right) = \dfrac{\sin\left(\frac{\omega\tau}{2}\right)}{\left(\frac{\omega\tau}{2}\right)}$, as we have discussed in this section.

Therefore,

$$\mathbf{F}\{p(t)\} = \frac{A\tau\sin\left(\frac{\omega\tau}{2}\right)}{\frac{\omega\tau}{2}} = A\tau\,\text{sinc}\left(\frac{\omega\tau}{2}\right) \tag{7.1.11}$$

as shown in Table 7.1.1

The other form of a rectangular pulse also used in this book is given by

$$\text{rect}(t) \equiv p(t) = \begin{cases} A & \text{for } -\tau/2 \leq t \leq \tau/2 \quad \text{(that is, for } (|t| \leq \tau)) \\ 0 & \text{otherwise} \quad\quad\quad\quad\quad\quad \text{(that is, for } (|t| > \tau)) \end{cases} \tag{7.1.12}$$

From Figure 7.1.5a,b, we can see that this rectangular pulse is identical to that given in (7.1.9). Therefore, the Fourier transform of this pulse is described by (7.1.11),

$$\text{rect}(\omega) = \mathbf{F}\{\text{rect}(t)\} = P(\omega) = \frac{A\tau\sin(\omega\tau/2)}{\omega\tau/2} = A\tau\,\text{sinc}(\omega\tau/2) \tag{7.1.13}$$

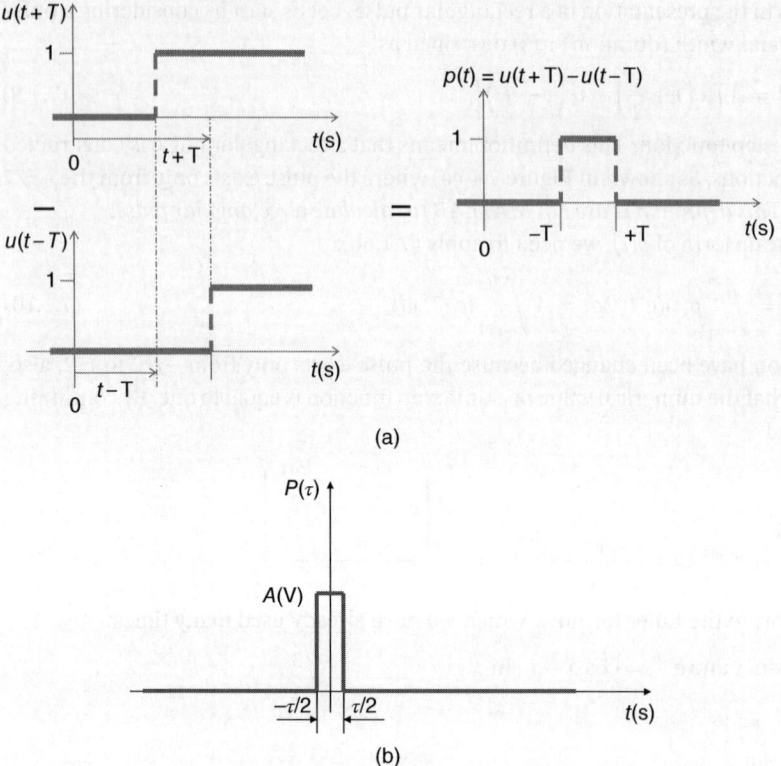

Figure 7.1.5 (a) Rectangular pulse constructed from two unit-step functions; (b) ideal rectangular pulse with amplitude A and duration τ.

Both pulses are shown in Figure 7.1.5a,b.

Finding the Fourier transform of a rectangular pulse at $\omega = 0$ seems impossible because it requires dividing by zero. We can, however, apply L'Hopital's rule in the limit and obtain

$$\frac{A\tau \sin(\omega\tau/2)}{\omega\tau/2} = A\tau \left|\frac{A\tau \sin(\omega\tau/2)}{\omega\tau/2}\right| = A\tau \lim_{\omega \to 0} \left|\frac{A\tau \sin(\omega\tau/2)}{\omega\tau/2}\right|$$

$$= A\tau \lim_{\omega \to 0} \left|\frac{\frac{d}{d\omega}\sin(\omega\tau/2)}{\frac{d}{d\omega}\omega\tau/2}\right| = A\tau \lim_{\omega \to 0} |\cos(\omega\tau/2)| = A\tau(\text{V s}) \qquad (7.1.14)$$

where A (V) is the amplitude and τ (s) is the width (duration) of the rectangular pulse.

Calculating the Fourier transforms of a decaying exponential function and a rectangular pulse as well as an analysis of Table 7.1.1 yields the following *conclusions*:

- Finding the Fourier transform of a time-domain function by integration is not always easy, particularly if the time-domain function is complicated. Fortunately, we need not perform this integration because most practical Fourier transforms – Fourier pairs, that is – are tabulated.
- If the required Fourier transform cannot be found in a table, we can obtain it by using MATLAB. Finding the Fourier transform of an exponential decaying function and a rectangular pulse in the examples given previously demonstrated the advantage of this approach.

7.1.2.2 Basics of the Spectral Analysis of a Nonperiodic Signal

It is instructive to compare the Fourier series given in (6.3.7) with the inverse Fourier transform given in (7.1.7):

$$\left. \begin{array}{l} v(t) = \sum_{n=-\infty}^{n=\infty}(c_n e^{jn\omega_0 t}) \\ v(t) = 1/2\pi \int_{-\infty}^{\infty} F(\omega) e^{j\omega t} d\omega \end{array} \right\} \quad (7.1.15)$$

Both equations allow us to equate a given signal to its presentation in frequency domain through either the discrete coefficients, c_n, or a continuous Fourier transform, $F(\omega)$. Therefore, the rule is quite clear:

To find the spectrum of a signal, we need to use a Fourier series for a periodic signal and a Fourier transform for a nonperiodic signal.

But, from a practical standpoint, how can we find the spectrum of a nonperiodic signal by using its Fourier transform? The answer can be found in an analogy between the Fourier series and the Fourier transform presented in (7.1.15). In the case of a periodic signal, its amplitude spectrum is given by the absolute values of the coefficients of its Fourier series, $|c_n|$, and its phase spectrum is given by the arguments, $\arg(c_n)$, of these coefficients (see Eq. (6.3.8a)). Extending this logic, we can say that the amplitude spectrum of a nonperiodic signal is given by the absolute value of its Fourier transform, $|F(\omega)|$; the phase spectrum is provided by the argument, $\arg[F(\omega)]$, of this transform. Thus,

If $F(\omega)$ is the Fourier transform of a nonperiodic signal, $v(t)$, then the amplitude spectrum of $v(t)$ is given by $|F(\omega)|$ and the phase spectrum of $v(t)$ is provided by $\arg[F(\omega)]$.

Obviously, to use this rule, we need to find the Fourier transform of a specific signal.

Armed with this fount information, we can now turn to an example of finding the spectrum of a nonperiodic signal.

Example 7.1.2 The Spectra of Nonperiodic Signals

Problem

Find the amplitude and phase spectra of

(a) The causal exponential decaying signal with amplitude $A = 1$ V and decay constant $\alpha = 3$ (1/s),

$$v(t) = Ae^{-\alpha t} u(t) \quad \text{for} \quad \alpha > 0$$

(b) A rectangular pulse with amplitude $A = 1$ V and duration $\tau = 2$ s,

$$v(t) = p(T)$$

Solution

(a) As a reminder, the waveform of a causal decaying exponential function is shown in Figure 7.1.6a. The Fourier transform of this signal is presented in (7.1.6a) and can also be found in Table 7.1.1 as

$$\mathbf{F}\{e^{-\alpha t} u(t)\} = \frac{1}{\alpha + j\omega}$$

This is a complex function of frequency; its absolute value (modulus) gives us the amplitude spectrum, whereas its argument gives us the phase spectrum. Hence, we need to present $F(e^{-\alpha t} u(t))$ as

$$\frac{A}{\alpha + j\omega} = A \left(\frac{\alpha}{(\alpha^2 + \omega^2)} - \frac{j\omega}{(\alpha^2 + \omega^2)} \right) \quad (7.1.16)$$

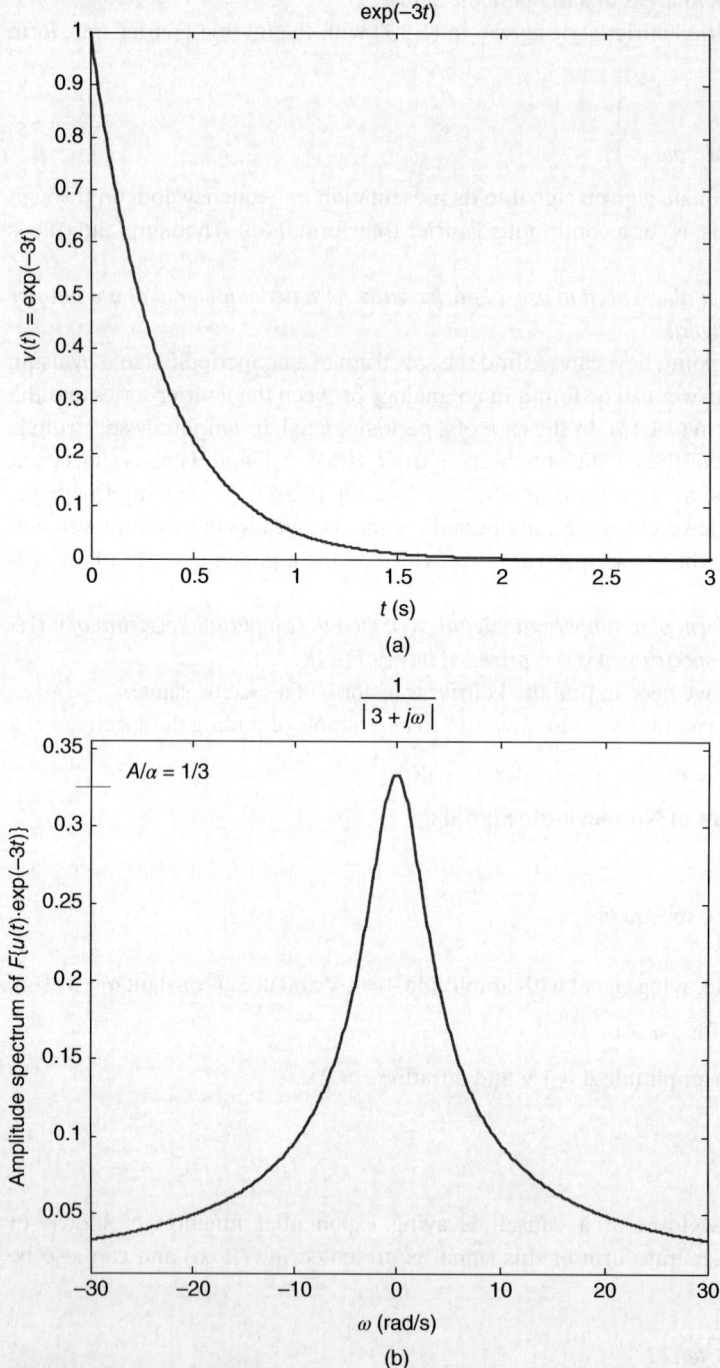

Figure 7.1.6 Causal decaying exponential signal $Ae^{-\alpha t} \cdot u(t)$: (a) Waveform; (b) amplitude spectrum; (c) phase spectrum.

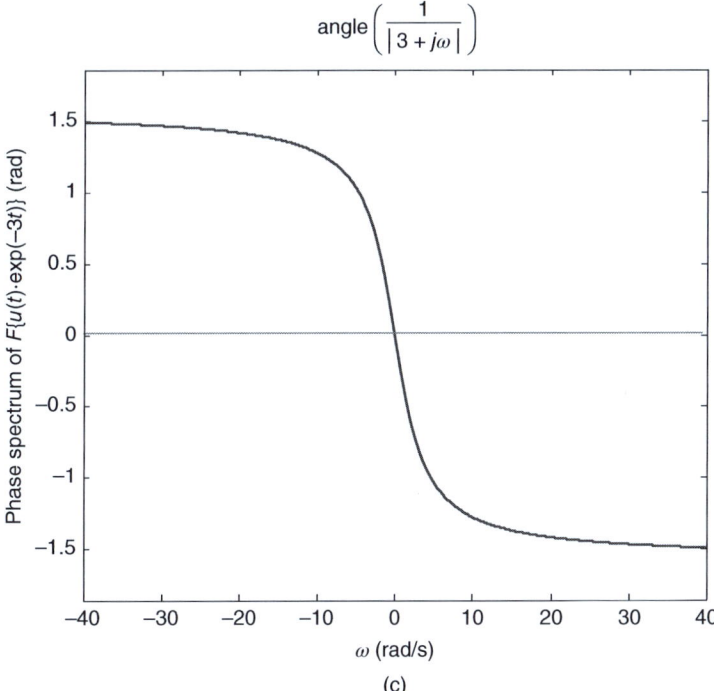

Figure 7.1.6 (Continued)

The modulus of this complex number is equal to

$$| F(e^{-\alpha t}u(t)) | = A\left(\sqrt{\left[\left(\frac{\alpha}{(\alpha^2 + \omega^2)}\right)^2 + \left(\frac{(-j\omega)}{(\alpha^2 + \omega^2)}\right)^2\right]}\right) = \frac{A}{\sqrt{(\alpha^2 + \omega^2)}} \quad (7.1.17)$$

and its argument is given by

$$\Theta = \arg F(e^{-\alpha t}u(t)) = -\tan^{-1}\left(\frac{\omega}{\alpha}\right) \quad (7.1.18)$$

Let us analyze the amplitude spectrum first. When $\omega = 0$, the amplitude equals

$$| F(e^{-\alpha t}u(t)) | = \frac{A}{\alpha} \quad \text{for} \quad \omega = 0$$

When $\omega \to \pm\infty$, the amplitude goes to zero. Thus, we should expect the amplitude spectrum, $|F(e^{-\alpha t}u(t))|$, to have a maximum value equal to A/α at $\omega = 0$ and to approach zero as ω goes to plus-minus infinity. Interestingly, the amplitude of this graph, as follows from (7.1.17), equals 0.707 at $\omega = \alpha$. The graph in Figure 7.1.6b shows the amplitude spectrum of signal $e^{-\alpha t}u(t)$ with given $A = 1$ V and $\alpha = 3$ (1/s).

The course of the phase spectrum is clear from (7.1.18): When ω goes from zero to $-\infty$, the phase angle increases from 0° to 90° and stays positive; when ω goes from zero to $+\infty$, the phase angle decreases from zero to $-90°$ and stays negative. At $\omega = \alpha$, the phase shift equals 45°. Figure 7.1.6c visualizes our analysis; the angles are shown in radians.

Figure 7.1.6 is built with the following MATLAB script:

```
%Waveform of the causal decaying exponential function, v(t) = e^-3t·u(t):
» syms w t
» v = exp(-3*t);
» ezplot(v,[0,3])

(1) %Amplitude spectrum of the Fourier transform of the causal decaying
exponential function, |F{v(t) = e^-3t·u(t)}|
    » syms w t
    » H=sym('heaviside(t)');
    » F=fourier(H*exp(-3*t));
    » AF=abs(F);
    » ezplot(AF,[-30,30])

(2) %Phase spectrum of the Fourier transform of the causal decaying exponential
function, arg[F{v(t) = e^-3t·u(t)}]
    » syms w t
    » H=sym('heaviside(t)');
    » F=fourier(H*exp(-3*t));
    » ANF=angle(F);
    » ezplot(ANF,[-40,40])
```

The problem in part (a) of this example is solved.

(b) The waveform of a rectangular pulse, $p(t)$, can be found in Figure 7.1.7a. The Fourier transform of this signal is given in (7.1.13) and in Table 7.1.1 as

$$\mathbf{F}\{p(t)\} = P(\omega) = \frac{A\tau \sin\left(\frac{\omega\tau}{2}\right)}{\frac{\omega\tau}{2}} = A\tau \operatorname{sinc}\left(\frac{\omega\tau}{2}\right)$$

Focus especially on the amplitude of the Fourier transform, $A\tau$, whose units are V s = V/Hz; this is because this amplitude represents a *spectral density* of the signal.

Therefore, the amplitude spectrum of a single pulse is given by

$$|P(\omega)| = \left| \frac{A\tau \sin\left(\frac{\omega\tau}{2}\right)}{\frac{\omega\tau}{2}} \right| = \left| A\tau \operatorname{sinc}\left(\frac{\omega\tau}{2}\right) \right| \qquad (7.1.19a)$$

and its phase spectrum is determined as

$$\arg(P(\omega)) = \arg\left(\frac{A\tau \sin\left(\frac{\omega\tau}{2}\right)}{\frac{\omega\tau}{2}} \right) = \arg\left(A\tau \operatorname{sinc}\left(\frac{\omega\tau}{2}\right) \right) \qquad (7.1.19b)$$

Let us set $A = 1$ V and $\tau = 2$ s and analyze the *amplitude spectrum*, $\left| \frac{A\tau \sin\left(\frac{\omega\tau}{2}\right)}{\frac{\omega\tau}{2}} \right|$. For this example, the amplitude is equal to $A\tau = 2$ V/Hz. Obviously, the plot of an *amplitude spectrum*, $\left| \frac{A\tau \sin\left(\frac{\omega\tau}{2}\right)}{\frac{\omega\tau}{2}} \right|$, reaches zero every time the argument of sine equals $\pi, 2\pi, 3\pi$, etc. Thus, the first zero will be reached at $\omega_1(\tau/2) = \pi$; that is, $\omega_1 = \frac{\pi}{\tau/2}$. For this example, $\tau = 2$ s and we find that $\omega_1 = \pi$. Hence, $\omega_2 = 2\pi$, $\omega_3 = 3\pi$, etc. If we want to see these points on a frequency scale, we need to divide ω by 2π. This is what the graph shows in Figure 7.1.7b.

The *phase spectrum* of this signal is given as 0° at intervals
- from $\omega = 0$ to $\omega_1 = \frac{\pi}{\tau/2}$
- from $\omega_2 = \frac{2\pi}{\tau/2}$ to $\omega_3 = \frac{3\pi}{\tau/2}$, etc.

and as 180° at intervals
- from $\omega_1 = \frac{\pi}{\tau/2}$ to $\omega_2 = \frac{2\pi}{\tau/2}$
- from $\omega_3 = \frac{3\pi}{\tau/2}$ to $\omega_4 = \frac{4\pi}{\tau/2}$, etc.

A similar distribution of the phase angles holds true for negative ω. This spectrum is shown in Figure 7.1.7c.

The MATLAB script used for building Figure 7.1.7 is as follows:

(a)
```
%The waveform of a rectangular pulse:
» syms w t
» H=sym('heaviside(t+1)-heaviside(t-1)');
» ezplot(H, [-10,10])
```

(b)
```
%Amplitude spectrum of the rectangular pulse:
» syms w t
»H=sym('heaviside(t+1)-heaviside(t-1)');
» F=fourier(H);
» AF=abs(F);
» ezplot(AF, [-10,10])
```

(c)
```
%Phase spectrum of the rectangular pulse:
» syms w t
»H=sym('heaviside(t+1)-heaviside(t-1)');
» F=fourier(H);
» ANF=angle(F);
» ezplot(ANF, [-30,30])
```

It is very important to compare these spectra with those shown in Figure 6.3.1: We can readily see how discrete spectra turn to continuous ones when the signal transforms from a periodic pulse train to a single pulse. We can also plot the magnitude spectrum of a rectangular pulse similar to that shown in Figure 6.3.1c.

7.1.2.3 Rayleigh Energy Theorem

Another fact to bear in mind is that a *nonperiodic signal is an energy signal*. Of course, the integral

$$P = \lim_{T \to \infty} \left(\frac{1}{T} \int_{-\tau/2}^{\tau/2} v^2(t) dt \right) \tag{6.3.25R}$$

cannot be determined for a nonperiodic signal because its limits are the half-periods. Thus, we need to turn to (6.3.23),

$$E = \int_{-\infty}^{\infty} [v(t)]^2 dt$$

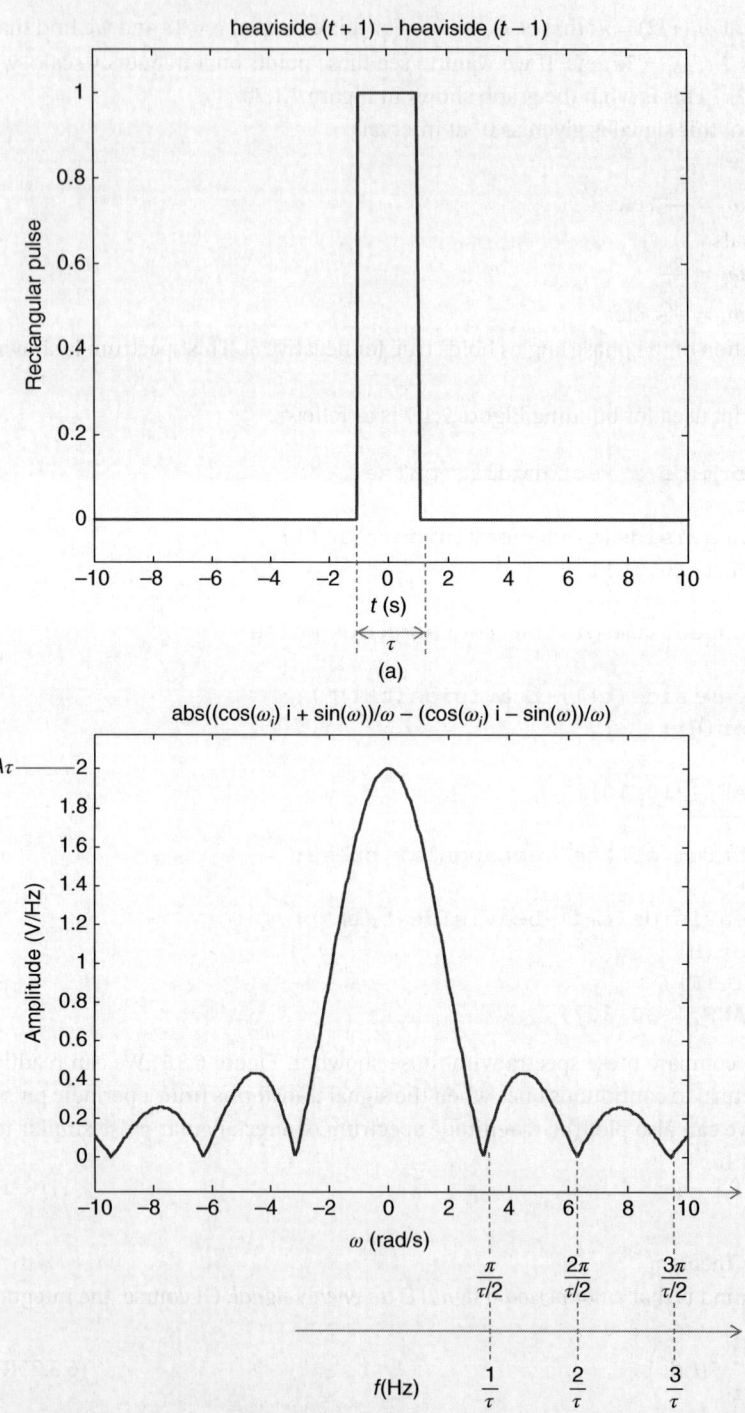

Figure 7.1.7 Rectangular pulse $p(t)$: (a) Waveform; (b) amplitude spectrum; (c) phase spectrum.

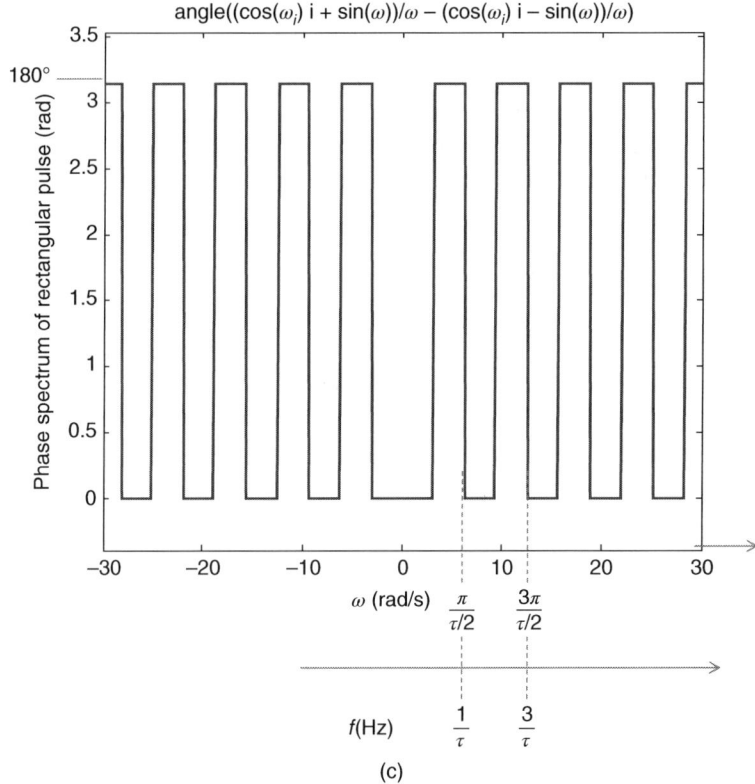

Figure 7.1.7 *(Continued)*

which defines the energy of a nonperiodic signal. Since this signal is related to its Fourier transform, as

$$v(t) = \frac{1}{2\pi} \int_{-\infty}^{\infty} F(\omega)e^{j\omega t}\, d\omega \tag{7.1.7R}$$

we can express the signal's energy through its Fourier transform. So plugging (7.1.7) into (7.3.33), we find, after some manipulations (James and Dyke 2018, p. 370),

$$E = \frac{1}{2\pi} \int_{-\infty}^{\infty} |F(\omega)|^2\, d\omega \tag{7.1.20}$$

Examining (7.1.20), we immediately recognize that it is analogous to Parseval's theorem, presented in (6.3.35). In fact, some textbooks refer to (7.1.20) as Parseval's theorem for nonperiodic signals; on the other hand, most books call it the Rayleigh theorem.

> *The energy distribution of a nonperiodic signal along its spectrum is described by the function $|F(\omega)|^2$, called the <u>energy spectral density</u>; the portion of this energy concentrated in a differential frequency band, $d\omega$, is given by*
>
> $$dE = |F(\omega)|^2\, d\omega \tag{7.1.21}$$

If we need to compute the total energy of a nonperiodic signal, we now have two options: Eq. (6.3.23) or Eq. (7.1.20). Practically speaking, integrating in time domain by using (6.3.23) is often less difficult than manipulating with (7.1.20) in frequency domain, but only the Rayleigh theorem provides us with an option to find the frequency distribution of the signal's energy.

As a brief example, consider the amplitude spectrum of the rectangular pulse shown in Figure 7.1.7b. The energy concentrated in the main lob of this signal is (Carlson and Crilly 2010, pp. 50–52.)

$$E_{\text{main}} = 1/2\pi \int_{-2/\tau}^{2/\tau} |F(\omega)|^2 \, d\omega = 1/2\pi \int_{-2/\tau}^{2/\tau} (A\tau)^2 \, \text{sinc}^2(\omega\tau) \, d\omega = 0.92 A^2 \tau \qquad (7.1.22)$$

whereas the total energy of the pulse is $A^2\tau$. Thus, more than 90% of the pulse's energy is concentrated in the main lobe of its spectrum, that is in the frequency band between $-1/\tau$ and $1/\tau$.

As is true for Parseval's theorem, the Rayleigh energy theorem delivers the _energy spectrum_ of a nonperiodic signal.

Summary of Section 7.1

- The Fourier transform in spectral analysis:
- The vast majority of real signals are nonperiodic and their spectra can be found only with the Fourier transform. This subsection presents basics of finding the spectrum of a nonperiodic signal. We discuss how to transit from a periodic to a nonperiodic signal, from a Fourier series to a Fourier transform, and how to use the Fourier transform and its inverse version. The important highlights of these considerations are as follows:
- A nonperiodic signal has a continuous spectrum, in contrast to a periodic signal, which has a linear (discrete) spectrum.
- To find the spectrum of a signal, we need to use a Fourier series for a periodic signal and a Fourier transform for a nonperiodic signal.
- The Fourier transform is a mathematical operation that transforms a time-domain function into a frequency-domain function. Refer to Figure 7.1.4.
- If $F(\omega)$ is the Fourier transform of a *nonperiodic signal*, $v(t)$, then the amplitude spectrum of $v(t)$ is given by $|F(\omega)|$ and the phase spectrum of $v(t)$ is provided by $\arg(F(\omega))$.
- Finding the Fourier transform of a time-domain function by integration is not an easy task. Fortunately, most practical Fourier transforms – Fourier pairs, that is – are tabulated. If the required Fourier transform cannot be found in a table, we can obtain it by using MATLAB.
- *Rayleigh energy theorem*: A nonperiodic signal is an energy signal. To find the frequency distribution of the signal's energy, we need to apply the Rayleigh theorem, given as

$$E = \frac{1}{2\pi} \int_{-\infty}^{\infty} |F(\omega)|^2 d\omega \qquad (7.1.20\text{R})$$

The energy distribution of a nonperiodic signal along its spectrum is described by the function $|F(\omega)|^2$, called _energy spectral density_. The portion of this energy that is concentrated in a differential frequency band, $d\omega$, is given by

$$dE = |F(\omega)|^2 d\omega \qquad (7.1.21\text{R})$$

As is true for Parseval's theorem, the Rayleigh energy theorem delivers the _energy spectrum_ of a nonperiodic signal.

Questions and Problems for Section 7.1

- Questions marked with an asterisk require a systematic approach to finding the solution.
- Many questions and problems, including those marked with an asterisk, imply that you, in addition to reading the textbook, will do your research to find the answers. Consider such questions as mini-projects.

The Fourier Transform in Spectral Analysis

1. Explain the difference between a periodic and a nonperiodic signal. Give examples.

2. What is a duty cycle of a nonperiodic signal? Explain.

3. We can change the duty cycle of a pulse train by changing either the pulse duration, τ (s), or the signal's period, T (s).
   a) How does the change of τ affect the spectrum of this pulse train?
   b) How does the change of T affect the pulse train's spectrum?

4. Consider a pulse of $\tau = 3$ ms width (duration):
   a) What is its bandwidth?
   b) What will be the pulse bandwidth if its width reduces to 0.3 ms?
   c) Why does the pulse's bandwidth change with the change of its width?

5. Consider a pulse as a carrier of information (a bit). Its power is proportional to the area under the pulse. In transmission, we want to minimize power-per-bit parameter; thus, it would seem that an ideal pulse should be of zero width. Can an ideal pulse be built in practice? Explain

6. Consider a periodic pulse train:
   a) What parameters of the periodic signal should we change to turn this periodic signal into a nonperiodic one?
   b) What kind of nonperiodic signal can we obtain by changing the parameters of the periodic pulse train?

7. Compare Figures 7.1.1–7.1.3 and describe their similarities and differences.

8. What mathematical tools do we use to find the spectra of periodic and nonperiodic signals?

9. Define the Fourier transform, $F(\omega)$. Explain the meaning of each member of $F(\omega)$.

10. What is the difference between the Fourier series and the Fourier transform? (*Hint*: Start with the definitions of each of these mathematical tools.)

11. Examine Figure 7.1.4:
    a) Why do we need the inverse Fourier transform?
    b) Applying the Fourier transform to a time-domain function, $v(t)$, we obtain $F(\omega)$. If we then apply the inverse Fourier transform to frequency-domain function, $F(\omega)$, we receive the original $v(t)$. Why do we need to do these seems to be mutually exclusive operations?

12. Using the definition of the Fourier transform, find the Fourier transform of $v(t) = 3e^{-5t}$. Show all your manipulations.

13 Using MATLAB, find the Fourier transform of $v(t) = 3e^{-5t}$. To verify your answer, find the inverse Fourier transform of $F^{-1}(\omega) = F(e^{-5t})$. Show your code and the MATLAB answers.

14 Using the table, find the Fourier transform of
   a) $v(t) = t^3 \cdot e^{-2t} u(t)$.
   b) $p(t) = 6[u(t+5/2) - u(t-5/2)]$

First Examples of the Fourier Transform Applications

15 Equation (7.1.9) describes a rectangular pulse as $p(t) \equiv \text{rect}(t) = A[u(t+\tau/2) - u(t-\tau/2)]$, where τ is a pulse duration (width) and $u(t)$ is a unit-step function. What is the meaning of this description? Sketch a figure to support your explanations.

16 *Consider a rectangular pulse $p(t) \equiv \text{rect}(t) = 5[u(t+3) - u(t-3)]$:
   a) Sketch this pulse.
   b) Find its Fourier transform.
   c) Show all your manipulations.

17 *Find the amplitude and phase spectra of $v(t) = 3e^{-5t}$.

18 *Find the amplitude and phase spectra of $p(t) = 15\,\text{sinc}(1.5\omega)$.

19 Can we apply Parseval's theorem to finding the power spectrum of a nonperiodic signal? Explain your reasoning.

20 *We know that a periodic signal is a power signal and a nonperiodic signal is an energy signal. We can apply Parseval's theorem to find the power spectrum of a periodic signal by calculating the power carried by each individual harmonic. A nonperiodic signal, however, has a continuous spectrum, where no individual harmonic exists. How, then, can we find the energy spectrum of a non-periodic signal? Explain.

21 *Consider the rectangular pulse given in Problem 16 as $p(t) = 5[u(t+3) - u(t-3)]$:
   a) Find the energy of this pulse concentrate in the main lob, that is, in the frequency band between $-1/\tau$ and $1/\tau$.
   b) What percentage of the pulse's total energy does this main-lob energy constitute? Show your calculations.
   c) What is the bandwidth of this pulse? Justify your choice.

7.2 Continuous-Time Fourier Transform: A Deeper Look

Objectives and Outcomes of Section 7.2

Objectives

- To become familiar with the concept and existence of the Fourier transform.
- To learn the main Fourier transform pairs and the main properties of the Fourier transform.
- To study examples of applications of the Fourier transform to the spectral analysis of various nonperiodic signals and systems.

Outcomes

- Get to know that mathematically the Fourier transform is the integral operation that transforms a time-domain function into the frequency-domain function. Learn that the Fourier transform exists only if the Fourier integral, $\int_{-\infty}^{\infty} v(t)\, e^{-j\omega t}\, dt$, converges, which, in turn, puts a set of Dirichlet's conditions on the $v(t)$ function. Know that the signals we encounter in our practice always satisfy Dirichlet's conditions and the Fourier transforms of the realizable signals always exist.
- Recall that a function is a rule that assigns to each element of an input set of numbers a unique element in the output set of numbers. Remember that if y is the function of x, $y = f(x)$, x *may or may not* be an inverse function of y denoted as $x = f^{-1}(y)$. Learn that a *transform* is also a mathematical operation, but this operation produces the *function* in response to an entry *function*. Know that a transform is a reversible operation; that is, the original function can be obtained from the transformed-by function by the inverse transform.
- Understand that the Fourier transform can be presented symbolically as $\boldsymbol{F}\{v(t)\}$ and its results is denoted as $F(\omega)$; that is $\boldsymbol{F}\{v(t)\} = F(\omega)$. It can also be shown in the form of a table. Realize that the inverse Fourier transform is designated as $\boldsymbol{F}^{-1}\{F(\omega)\} = v(t)$.
- Become familiar with Dirac delta function, also called *impulse function* and denoted as δ-function. Get to know that the *impulse function* is the mathematical depiction of an impulse (an instant impact) often used to describe a real excitation of an engineering unit. Learn that the Fourier transform of a delta function shows its spectrum: A single impulse contains all frequencies from $-\infty$ to ∞ with a constant amplitude.
- Study the Fourier transform pairs shown in Table 7.2.1. Realize that this table is a collection of integrations of the time-domain functions performed with the Fourier integral. Understand that the Fourier transform of each time-domain function presents the spectrum of this function. For example, $\boldsymbol{F}\{\cos(\omega_0 t)\} = \pi[\delta(\omega - \omega_0) + \delta(\omega + \omega_0)]$, which means that two-sided spectrum of a cosine function consists of two lines located at ω_0 and $-\omega_0$, as we know from the preceding sections.
- Learn the main properties of the Fourier transform collected in Table 7.2.2. Follow the discussion of this properties accompanying the table. Study, for instance, how the modulation property helps to obtain the spectrum of an amplitude-modulated signal. Know that all these properties are proved by mathematical manipulations based on the Fourier transform definition.
- Know that the Fourier transform of a system's impulse response, $h(t)$, produces the system's transfer function, $H(\omega)$. Learn how to find a system's impulse response by using the system's differential equations. Become familiar with the examples of such an application of the Fourier transform to finding $H(\omega)$ of an RC LPF, which enables us to obtain the amplitude and phase responses of this filter. Realize that due to the convolution property of the Fourier transform, we can obtain the system's output in frequency domain, $V_{out}(\omega)$, simply by multiplying the system's input, $V_{in}(\omega)$, and the transfer function, $H(\omega)$; that is, $V_{out}(\omega) = V_{in}(\omega) \cdot H(\omega)$. Recall that $v_{out}(t) = \boldsymbol{F}^{-1}\{V_{out}(\omega)\}$.

7.2.1 Definition and Existence of the Fourier Transform

The formal definition of the Fourier transform is as follows:
If $v(t)$ is a time-domain function, then its Fourier transform is given by

$$\boldsymbol{F}\{v(t)\} = \int_{-\infty}^{\infty} e^{-j\omega t} v(t) dt = F(\omega) \qquad (7.2.1)$$

The inverse Fourier transform of $F(\omega)$ obviously produces $v(t)$ and is given by

$$\mathbf{F}^{-1}\{F(\omega)\} = \frac{1}{2\pi}\int_{-\infty}^{\infty} e^{j\omega t} F(\omega)d\omega = v(t) \qquad (7.2.2)$$

(These two equations are the replicas of (7.1.6b) and (7.1.7).) You are also reminded that $v(t)$ is a given function in time domain, $\mathbf{F}$ is an operator depicting the Fourier transform, and $F(\omega)$ is the result of this operation; that is, $\mathbf{F}\{v(t)\} = F(\omega)$.

Now we can formally define the requirement for the existence of the Fourier transform:

The Fourier transform exists if and only if the Fourier integral $\int_{-\infty}^{\infty} v(t)e^{-j\omega t} dt$ converges,

which, in turn, puts a set of conditions on the $v(t)$ function known as *Dirichlet's conditions*:

1. The time-domain function, $v(t)$, must be absolutely integrable,

$$\int_{-\infty}^{\infty} |v(t)|dt < \infty \qquad (7.2.3)$$

   which means that this integral must be finite.
2. The function $v(t)$ must have a finite number of maxima and minima and a finite number of discontinuities in any finite interval.

The second of Dirichlet's conditions comprises an environment sufficient for the existence of the Fourier integral (7.2.1). It is interesting to note that (7.2.3) implies that $F(\omega)$ exists; indeed, (Langton 2008, pp. 12–13.)

$$|F(\omega)| \le \int_{-\infty}^{\infty} f(t)e^{-j\omega t}dt = \int_{-\infty}^{\infty} |f(t)| |e^{-j\omega t}| dt$$

and

$$|F(\omega)| \le \int_{-\infty}^{\infty} |f(t)| dt$$

since $|e^{-j\omega t}| = 1$.

Fortunately, the signals we encounter in our practice always satisfy Dirichlet's conditions and the Fourier transforms of the realizable signals always exist. (If you sense that you have already encountered these conditions in this text, you are right: We introduced them in Section 7.1 with respect to the Fourier series. Here we apply them to the Fourier transform and provide a detailed explanation.)

Examples of finding the Fourier transforms of an exponential decaying function and a rectangular pulse are given in Section 7.1; reviewing them will help you to better comprehend the current material.

7.2.2 The Concept of Function and the Transform

In order to understand the term "transform," we need to remind ourselves the meaning of the term "function" discussed in Section 2.2:

A *function is a rule* that assigns to each element of an input set of numbers a *unique element* in the output set of numbers. The input set is called the *domain* and the output set is called the *range of the function*.

This definition is supported by Figure 2.2.S.2.2 reproduced here.

If x is an arbitrary element of the input set X and y is its unique assigned element in the output set Y, then we say that *y is the function of x* and denote it as (James 2015, p. 3.)

$$y = f(x)$$

We must recall that if y is the function of x, $y = f(x)$, x *may or may not* be an inverse function of y denoted as $x = f^{-1}(y)$. Again, refer to Sidebar 2.2.S.2 *Function and Signal* to refresh your memory on this topic.

Figure 7.2.1a visualizes the concept of function from an application standpoint: Function is a mathematical operation that produces the set of the output numbers in response to a set of the input numbers. Consider, for example, function, $y = x^2$, shown in Figure 7.2.1a: It produces output y equals square of x for every x presented to the function.

(**Question**: *Does $y = x^2$ has an inverse function? [Hint: Apply the definition of a function to find the correct answer.]*)

A *transform* is also a mathematical operation, but this operation produces the *function* in response to an entry *function*, as Figure 7.2.1b demonstrates. For instance, the Fourier transform produces frequency-domain function $F(\omega) = 1/(\alpha + j\omega)$ from the time-domain function $v(t) = e^{-\alpha t}$. This example can be shown symbolically as $\boldsymbol{F}\{e^{-\alpha t}\} = 1/(\alpha + j\omega)$, where $\boldsymbol{F}\{\}$ stands for the Fourier operator. In general, when we present to a transform a set of functions defined in one domain, the transform produces the corresponding set of functions in the other domain. Example of such general transform is shown in the bottom portion of Figure 7.2.1b where a set of *time-domain* functions is transformed into the set of functions in *frequency domain*.

In contrast to function, a transform is a reversible operation; that is, the original function can be obtained from the transformed-by function by the inverse transform. For example, the inverse Fourier transform of frequency-domain function, $F(\omega) = 1/(\alpha + j\omega)$ produces $\boldsymbol{F}^{-1}(1/(\alpha + j\omega)) = e^{-\alpha t}$ in time domain. This is why Figure 7.2.1b shows double-arrow lines connecting the functions being transformed. For further explanations, see, for example, (Roden 2003, pp. 474–475).

A transform can be presented symbolically as $\boldsymbol{F}\{v(t)\}$ in the example of the Fourier transform; it can also be shown in form of table, as Table 7.1.1 demonstrates. (Recall that a function can also be presented as a table, as demonstrated in Figure 2.2.S.2.1b.)

In engineering, the most used transforms are the Laplace transform, the Fourier transform, and the *z*-transform. The Laplace transform converts continuous time-domain signals into

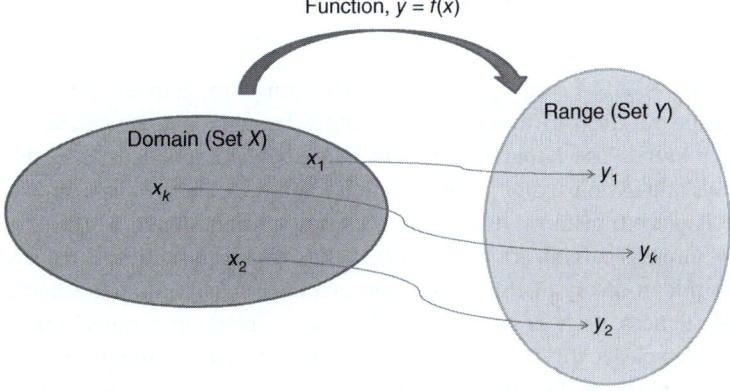

Figure 2.2.S.2.2R Function is a rule assigning to each element from a domain set a unique element from a range set.

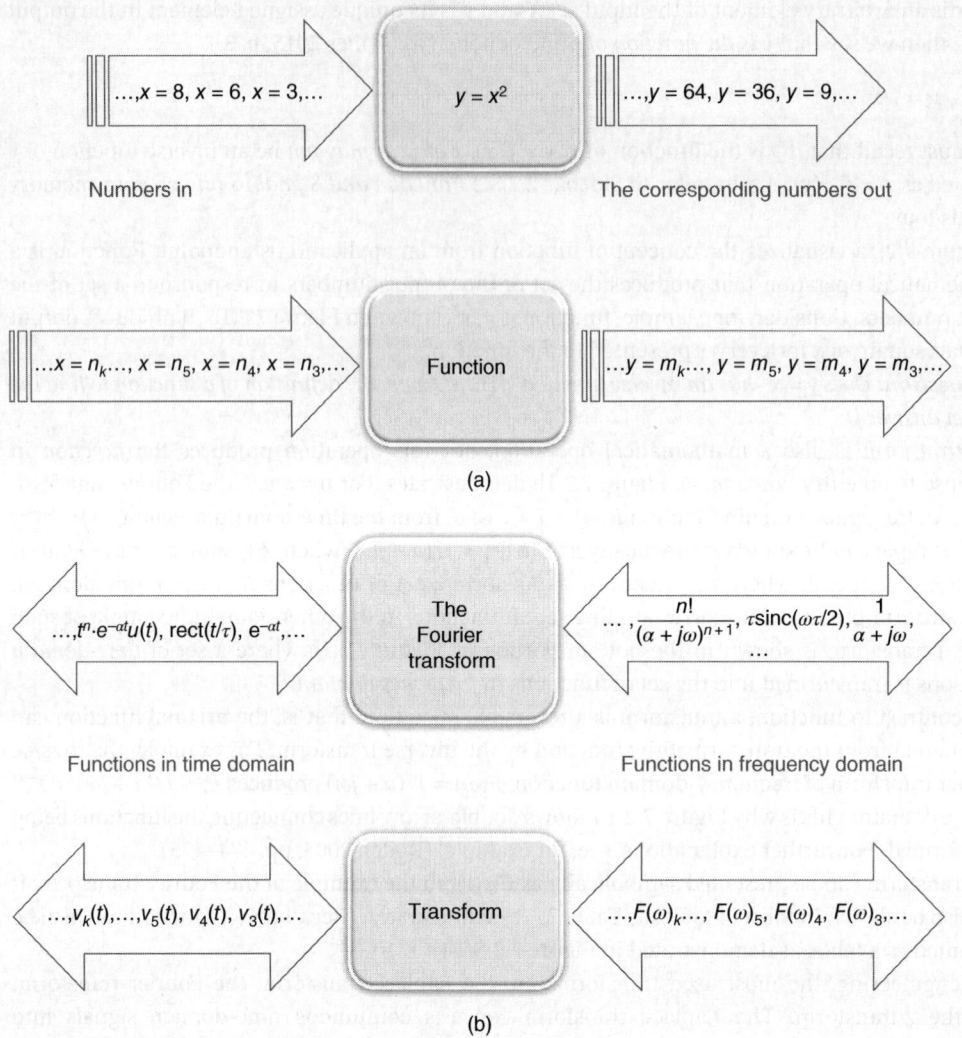

Figure 7.2.1 (a) The concept of function and (b) the transform.

s-domain functions, where $s = \sigma + j\omega$ is a complex frequency. This transform converts differential equations – that are, obviously, given in time domain – into s-domain algebraic equations whose solutions can be easily found. The Laplace transform finds its main applications in the areas of control theory, signals and systems, and advanced circuit analysis. The z-transform is a discrete-time analogy to the Laplace transform; its application, clearly, lies in digital world. The Fourier transform, as we have already learned, is a main mathematical tool in the spectral analysis and, therefore, plays a major role in theory of communications. (A deeper look into this subject reveals, of course, myriad of nuances, such as an active application of the Fourier transform in control theory. Unhappily, we must leave this interesting subject to the specialized literature.)

Let us turn back to the Fourier transform. Before presenting additional examples of how to obtain the Fourier transforms of time-domain functions, we need to introduce a *delta*, or *Dirac function*, also known as an *impulse function* $\delta(t)$. This function is equal to zero everywhere except at one point, $t = a$, where a is a specific value of t. Graphically, it looks like a spike with zero width and infinite amplitude. See Sidebar 7.2.S.1 "Dirac Delta Function" for a detailed description of this function and its properties.

Sidebar 7.2.S.1 Dirac[2] Delta Function

Formally speaking, the *Dirac delta function*, $\delta(t)$, is a generalized, not a common, mathematical function with quite unusual properties. However, the rigorous definition of this function exists, and this function finds many useful applications in engineering analysis. For one, the delta function defines the Fourier transforms of such functions as 1, $u(t)$, $e^{j\omega t}$, and $\cos(\omega t)$ and $\sin(\omega t)$, as we can see in Table 7.2.1. This function can be thought of as a very sharp, narrow pulse that is presented graphically as a straight line – refer to Figure 7.2.S.1.1. Note that the Dirac delta function is also called the *impulse function* and is often simply denoted as δ-*function*.

The Dirac function can be defined in time, space, or in any other variable. To define the delta function, $\delta(x)$, consider the rectangular pulse centered on a and having the width, W, and the amplitude, A/W, as shown in Figure 7.2.S.1.1a:

$$P(x) = \begin{cases} A/W & \text{for } a - W/2 < x < a + W/2 \\ 0 & \text{otherwise} \end{cases} \qquad (7.2.S.1.1)$$

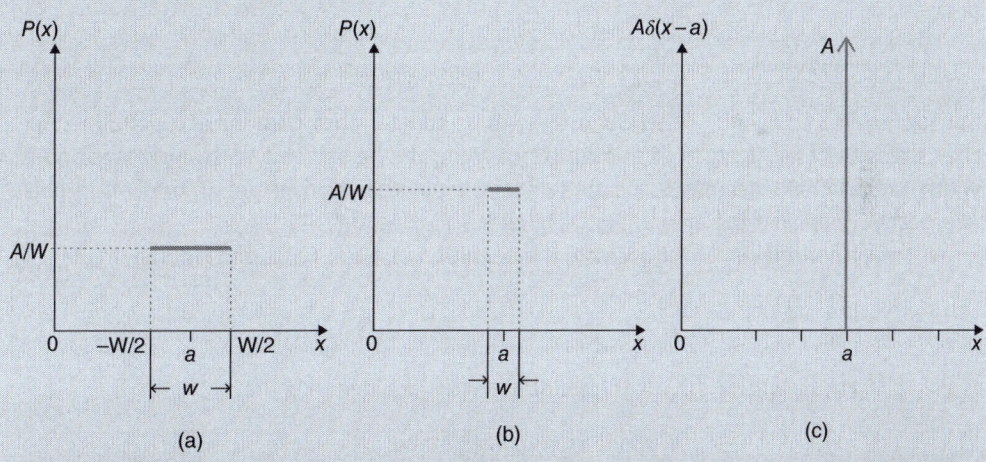

Figure 7.2.S.1.1 (a) A rectangular pulse with width, W, amplitude, A/W, and area, A, (b) the same pulse with decreased width but the same area, and (c) a symbolic presentation of the Dirac delta function, $A\delta(x - a)$, as the limit of the process $W \to a$.

(Continued)

---

2 Paul Dirac (1902–1984) was a British physicist who made fundamental contributions to the development of quantum mechanics in its early years. From 1932 to 1969, he was Lucasian Professor of Mathematics at Cambridge University – the position once occupied by Isaac Newton. He is most famous for the introduction of the wave equation, named after him, reconciling quantum mechanics and the theory of relativity and the prediction of anti-matter, specifically, the proton – an elementary particle identical to the electron but carrying a positive charge. This prediction was experimentally confirmed and he shared the Nobel Prize for Physics in 1933. He spent his later years as a professor of physics in the United States. In 1995, a memorial to him was erected in the Westminster Abbey in London.

Sidebar 7.2.S.1 (Continued)

The area under this pulse is A; indeed,

$$\int_{-\infty}^{\infty} P(x)dx = \int_{-W/2}^{W/2} \left(\frac{A}{W}\right) dx = A \tag{7.2.S.1.2}$$

Let us now make this pulse increasingly narrower while keeping its area constant. To meet this condition, the height of this pulse must increase proportionally, as shown in Figure 7.2.S.1.1b.
In the limit, when W tends to a single point, a, we find

$$\lim_{w \to a} \int_{-\infty}^{\infty} P(x)dx = \lim_{w \to a} \int_{-W/2}^{W/2} \left(\frac{A}{W}\right) dx = A \int_{-\infty}^{\infty} \delta(x-a)dx = A \tag{7.2.S.1.3}$$

because $\int_{-\infty}^{\infty} \delta(x-a)dx = 1$ by definition.

If we consider a unit pulse, then the multiplicative constant $A = 1$ and (7.2.S.1.3) become

$$\int_{-\infty}^{\infty} \delta(x-a)dx = 1 \tag{7.2.S.1.4}$$

If a is located at the origin of the coordinate plane, that is, $a = 0$, then

$$\int_{-\infty}^{\infty} \delta(x)dx = 1 \tag{7.2.S.1.5a}$$

or

$$\int_{-\infty}^{\infty} \delta(x)dx = A \tag{7.2.S.1.5b}$$

Equations (7.2.S.1.5a) and (7.2.S.1.5b) are the most popular presentations of a delta function; they stress the main feature of this function: While the height of a pulse increases and its width proportionally decreases, the area under the pulse keeps constant. In the limit, a pulse approaches a line, but the area under this line is still constant.

Two important properties follow from the definition of a $\delta(x)$-function centered on point a:

$$\delta(x-a) = 0 \quad \text{for } x \neq a \tag{7.2.S.1.6}$$

$$\int f(x)\delta(x-a)dx = f(a) \text{ if the point } x = a \text{ is included into the range of integration and}$$

$$\int f(x)\delta(x-a)dx = 0 \text{ if not} \tag{7.2.S.1.7}$$

Here is an example: The integrand of the Fourier integral (7.2.1),

$$\mathbf{F}\{v(t)\} = F(\omega) = \int_{-\infty}^{\infty} e^{-j\omega t} v(t) dt$$

includes the complex exponential function $e^{-j\omega t}$. But what is the Fourier transform of this function itself? As we will soon learn from Table 7.2.1, it is given by

$$\mathbf{F}\{e^{j\omega_0 t}\} = \int_{-\infty}^{\infty} e^{j\omega_0 t} dt = 2\pi\delta(\omega - \omega_0) \tag{7.2.S.1.8}$$

If we plug $2\pi\delta(\omega-\omega_0)$ into the inverse Fourier transform, we should obtain $e^{j\omega_0 t}$. (Mesiya 2013, p. 57.) Indeed,

$$\int_{-\infty}^{\infty} e^{j\omega_0 t} 2\pi\delta(\omega-\omega_0)\, d\omega = e^{j\omega_0 t} \tag{7.2.S.1.9}$$

according to (7.2.S.1.7). Thus, (7.2.S.1.9) is the proof of (7.2.S.1.8). Since the Fourier transform gives the spectrum of a nonperiodic signal, as was discussed in Section 7.1, we see that the spectrum of a complex exponential function consists of one line centered at ω_0, similar to that shown in Figure 7.2.S.1.1c. If this line is centered at $\omega_0 = 0$, we get

$$\mathbf{F}\{e^{j0t}\} = \mathbf{F}\{1\} = 2\pi\delta(\omega) \tag{7.2.S.1.10}$$

This is an important result: The Fourier transform of 1 is $2\pi\delta(\omega)$. Clearly, the Fourier transform of a constant A is $2\pi A\delta(\omega)$. Obviously, the inverse Fourier transform of $2\pi\delta(\omega)$ is 1, but what is your guess as to the Fourier transform of $\delta(t)$? Try to determine it.

Equations (7.2.S.1.8 – 7.2.S.1.10) are, in fact, examples of how to find the Fourier transforms. It is interesting to note that the delta function is the derivative of the Heaviside or unit-step function, $u(x)$, and therefore can be defined as

$$u'(x) = \delta(x) \tag{7.2.S.1.11}$$

The term *impulse function* emphasizes the engineering application of this function: It is the mathematical depiction of an impulse (an instant impact) often used to describe a real excitation of an engineering unit.

Now we are ready for examples.

Example 7.2.1 Finding the Fourier Transform of a Single Impulse[3]

Problem

Find the Fourier transform of a single impulse.

Solution

As the definition of the Fourier transform states, we must have an explicit formula of a time-domain function, $v(t)$, to be transformed into the frequency domain. Indeed, refer to (7.2.1),

$$\mathbf{F}\{v(t)\} = \int_{-\infty}^{\infty} v(t) e^{-j\omega t}\, dt = F(\omega)$$

A single impulse is described by a delta function, as Figure 7.2.S.1.1c in Sidebar 7.2.S.1 shows. Thus, a single pulse located at $t = 0$ is described as $A\delta(t)$, where the multiplicative constant A is included to make this problem a general one. Next, we can perform integration, as (7.2.1) requires:

$$\mathbf{F}\{v(t)\} = \int_{-\infty}^{\infty} v(t) e^{-j\omega t}\, dt = \int_{-\infty}^{\infty} A\delta(t) e^{-j\omega t}\, dt \tag{7.2.4}$$

Using the property of the delta function given in 7.2.S.1.7,

$$\int f(x)\delta(x-a)\, dx = f(a)$$

---

[3] After Mesiya (2013, pp. 56–57).

we find the Fourier transform of the delta function as

$$F\{A\delta(t)\} = A\int_{-\infty}^{\infty} \delta(t)e^{-j\omega t}dt = A \qquad (7.2.5)$$

This is because our delta function is defined at the origin of the t-axis; that is, $\delta(t) = \delta(t-0)$. Thus, the Fourier transform of a delta function, $A\delta(t)$, is given by

$$F\{A\delta(t)\} = A$$

The problem is solved.

Discussion

- Equation (7.2.5) states that a single impulse contains all frequencies from $-\infty$ to ∞ with a constant amplitude, A. If we consider a single *unit* pulse, then, obviously, $A = 1$. The graph of the Fourier transform of a delta function, $A\delta(t)$, is shown in Figure 7.2.2. This figure graphically demonstrates the same statement: When $\omega = 0$ rad/s, then the spectral density, SD, is equal to A; when $\omega = 10$ rad/s, SD $= A$; when $\omega = -30$ rad/s, SD $= A$, and so on. Regardless of the value of the frequency, the spectral density of the Fourier transform of $A\delta(t)$ is always the same. This is why

> *when a single impulse is presented to a system as an input signal, it is presenting to the system a signal containing all frequencies!*

This single-pulse feature is widely used in testing and in describing a system's behavior.
- The MATLAB code for this example is as follows:

```
Fs=1000; %Sampling frequency [Hz]
A=1; %The amplitude of the delta function [V]
x=-100:100; %Specify index x
delta=(x==0); %Define the delta sequence
d=delta*A; %Delta function
subplot(2,1,1), stem(x,d) %Plot delta function in time domain
xlabel('Time [sec]'), ylabel('Amplitude [V]');
title('Delta function in time domain with A=1V');
axis([-100 100 -0.25 A+0.25]), grid;
D=fft(d,512); %Transform the original signal into frequency domain
f=(0:255)/256*(Fs/2); %Scale the frequency vector
subplot(2,1,2), plot(f,abs(D(1:256))); %Delta function in freq domain
xlabel('Frequency [Hz]'), ylabel('Amplitude [V]'), grid;
title('Spectrum of a delta function with A=1V');
```

The graphs of a delta function and its spectrum by MATLAB are shown in Figure 7.2.2b.

Let us consider another example.

Example 7.2.2   Finding the Fourier Transform of a Cosine Function
Problem

Find the Fourier transform of the cosine function, $\cos(\omega_0 t)$, of a specific frequency, ω_0.

Solution

The solution is rather straightforward: Apply the Fourier integral (7.2.1) and find

$$F\{v(t)\} = \int_{-\infty}^{\infty} v(t)e^{-j\omega t}dt = \int_{-\infty}^{\infty} \cos(\omega_0 t)e^{-j\omega t}dt \qquad (7.2.6)$$

7.2 Continuous-Time Fourier Transform: A Deeper Look | 653

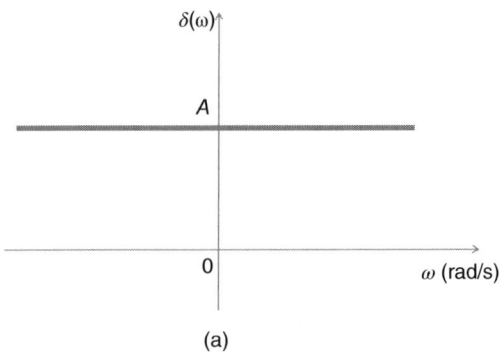

(a)

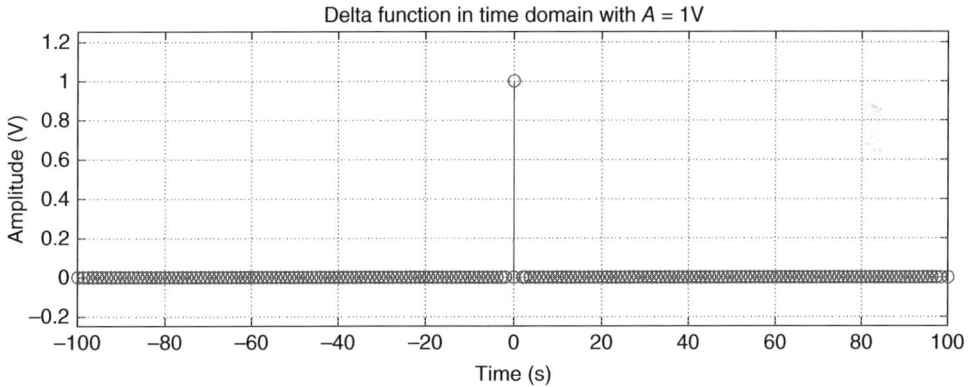

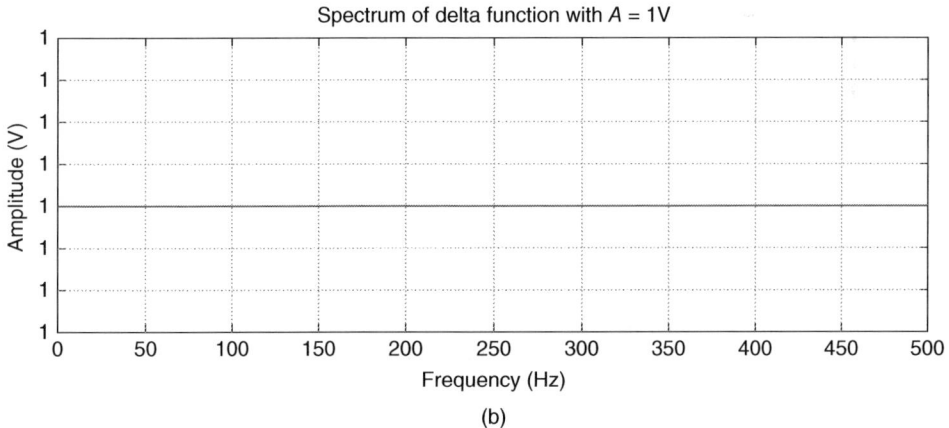

(b)

Figure 7.2.2 Graphs for Example 7.2.1: (a) The spectrum of the delta function built manually; (b) the waveform and the spectrum of the delta function built with MATLAB.

Recalling Euler's formula,

$$e^{jx} = \cos x + j \sin x \quad \text{and} \quad e^{-jx} = \cos x - j \sin x$$

we present the cosine function as a combination of exponential functions:

$$\cos(\omega_0 t) = \tfrac{1}{2}(e^{j\omega_0 t} + e^{-j\omega_0 t})$$

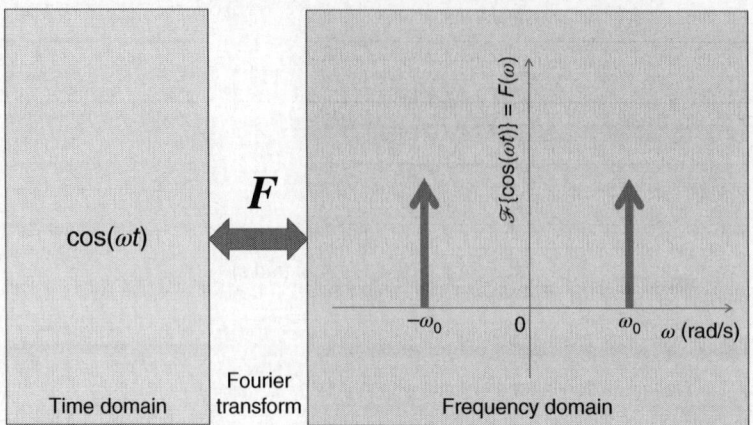

Figure 7.2.3 Time-domain function cos(ωt) and its Fourier transform in Example 7.2.2.

Plugging this expression into (7.2.6) yields

$$F\{\cos(\omega_0 t)\} = \int_{-\infty}^{\infty} \cos(\omega_0 t) e^{-j\omega t} dt = 1/2 \int_{-\infty}^{\infty} (e^{j\omega_0 t} + e^{-j\omega_0 t}) e^{-j\omega t} dt$$

$$= \frac{1}{2} \int_{-\infty}^{\infty} (e^{-j(\omega+\omega_0)t} + e^{-j(\omega-\omega_0)t}) dt$$

Now we can use the Fourier transform obtained in (7.2.S.1.8), which results in

$$F\{\cos(\omega_0 t)\} = \frac{1}{2} \int_{-\infty}^{\infty} (e^{-j(\omega+\omega_0)t} + e^{-j(\omega-\omega_0)t}) dt = \pi\delta(\omega + \omega_0) + \pi\delta(\omega - \omega_0) \qquad (7.2.7)$$

Equation (7.2.7) is the Fourier transform of the cosine function; that is, it is the solution to this problem.

Discussion

Equation (7.2.7) shows that the spectrum of a cosine function consists of two lines located at ω_0 and $-\omega_0$. This is because $\delta(\omega + \omega_0)$ and $\delta(\omega - \omega_0)$ are equal to zero everywhere except at points ω_0 and $-\omega_0$, respectively, according to the property of the delta function, (7.2.S.1.6). Obviously, this transform gives us a two-sided spectrum of a cosine function. For a one-sided spectrum, we need to show one line at ω_0 only. Refer to Section 6.1, where we depicted the spectrum of a cosine as a single line. A graphical presentation of (7.2.7) is shown in Figure 7.2.3.

7.2.3 Table of the Fourier Transform

Examples 7.2.1 and 7.2.2 bring us to consideration a table of the Fourier-transform pairs. The abbreviated version of this table is given in Section 7.1. That Table 7.1.1 is accompanied by some explanations on its usage. Here, the expanded version of such a table is presented.

There are several pertinent comments we should make with regard to Table 7.2.1:

- Study closely the pairs of Fourier transforms starting from $1 \leftrightarrow \delta(\omega)$. They all involve the delta (Dirac) function, $\delta(x)$. In fact, this function can be introduced naturally using the Fourier integral as follows:

$$\delta(t) = \frac{1}{2\pi} \int_{-\infty}^{\infty} e^{-j\omega t} dt \qquad (7.2.8)$$

Table 7.2.1 Fourier transform pairs.

Time-domain function, $v(t) = F^{-1}(F(\omega))$	Fourier transform ($\rightarrow$), $F(v(t))$ Inverse Fourier transform ($\leftarrow$), $F^{-1}(F(\omega))$	Frequency domain function, $F(\omega) = F(v(t))$	Conditions and comments
1	$\leftarrow \rightarrow$	$2\pi\delta(\omega)$	$\delta(\omega)$ – delta function of ω
$\delta(t)$	$\leftarrow \rightarrow$	1	$\delta(t)$ – delta function of t
$\delta(t - t_0)$	$\leftarrow \rightarrow$	$e^{-j\omega t_0}$	$\delta(t)$ – delta function of t; t_0 – constant
$u(t)$	$\leftarrow \rightarrow$	$\pi\delta(\omega) + \dfrac{1}{j\omega}$	$u(t)$ – unit step function
$|t|$	$\leftarrow \rightarrow$	$\dfrac{-2}{\omega^2}$	
$\mathrm{sgn}(t)$	$\leftarrow \rightarrow$	$\dfrac{2}{j\omega}$	sign(t) – sign (signum) function of t
$e^{j\omega_0 t}$	$\leftarrow \rightarrow$	$2\pi\delta(\omega - \omega_0)$	ω_0 – real constant
$p(t) = \begin{cases} 1 \text{ for } |t| < \tau \\ 0 \text{ for } |t| > \tau \end{cases}$ or $\mathrm{rect}(\dfrac{t}{\tau})$	$\leftarrow \rightarrow$	$\tau \cdot \mathrm{sinc}\left(\dfrac{\omega\tau}{2}\right)$	τ – duration of rectangular pulse, $p(t)^a$
$e^{-\alpha t} u(t)$	$\leftarrow \rightarrow$	$\dfrac{1}{\alpha + j\omega}$	$\mathrm{Re}(\alpha) > 0$, constant
$t^n \cdot e^{-\alpha t} u(t)$	$\leftarrow \rightarrow$	$\dfrac{n!}{(\alpha + j\omega)^{n+1}}$	$\mathrm{Re}(\alpha) > 0$, constant
$e^{-\alpha|t|}$	$\leftarrow \rightarrow$	$\dfrac{2\alpha}{\alpha^2 + \omega^2}$	$\mathrm{Re}(\alpha) > 0$, constant
$|t| \cdot e^{\alpha|t|}$	$\leftarrow \rightarrow$	$\dfrac{2(\alpha^2 - \omega^2)}{\alpha^2 + \omega^2}$	$\mathrm{Re}(\alpha) > 0$, constant
$\cos(\omega_0 t)$	$\leftarrow \rightarrow$	$\pi[\delta(\omega - \omega_0) + \delta(\omega + \omega_0)]$	ω_0 – real constant
$\sin(\omega_0 t)$	$\leftarrow \rightarrow$	$j\pi[\delta(\omega + \omega_0) - \delta(\omega - \omega_0)]$	ω_0 – real constant
$\cos(\omega_0 t) \cdot u(t)$	$\leftarrow \rightarrow$	$\dfrac{\pi}{2}[\delta(\omega - \omega_0) + \delta(\omega + \omega_0)]$	ω_0 – real constant
$\sin(\omega_0 t) \cdot u(t)$	$\leftarrow \rightarrow$	$j\dfrac{\pi}{2}[\delta(\omega + \omega_0) - \delta(\omega - \omega_0)]$	ω_0 – real constant
$e^{-\alpha t}\cos(\omega_0 t) \cdot u(t)$	$\leftarrow \rightarrow$	$\dfrac{\alpha + j\omega}{(\alpha + j\omega)^2 + \omega_0^2}$	ω_0 – real constant
$e^{-\alpha t}\sin(\omega_0 t) \cdot u(t)$	$\leftarrow \rightarrow$	$\dfrac{\omega_0}{(\alpha + j\omega)^2 + \omega_0^2}$	ω_0 – real constant

a) See Section 7.1.2.1.

- Observe, too, the pairs $1 \leftrightarrow \delta(\omega)$ and $\delta(t) \leftrightarrow 1$. They demonstrate the Fourier-transform property called *duality*; we will refer to this property many times throughout this book.
- If you think that the Fourier transforms of such simple functions as sine and cosine look very complex, well, you are right. They *are*. This is the very nature of the Fourier transform. However, the meaning of these transforms is clear: See (7.2.7), Example 7.2.2, and Figure 7.2.3.
- As soon as the exponential function, $e^{-\alpha t}$, appears, the real part of α must be positive. We need the exponent to be raised to a negative real power, that is, to $e^{-|\alpha|t}$. This is simply because the Fourier transform of the function $e^{+\alpha t}$, where $\mathrm{Re}(\alpha) > 0$, does not exist.
- You can prove these Fourier transforms by applying the definitions given in (7.2.1) and (7.2.2). Bear in mind, however, that many of these transforms cannot be obtained by direct integration and so you will need to use the limiting forms of the proper functions.

7.2.4 Properties of the Fourier Transform

Let us now turn to a discussion of the main properties of the Fourier transform, listed in Table 7.2.2.

In Table 7.2.2, all time-domain functions are assumed causal and proper Fourier transforms are one-sided.

The properties of the Fourier require some discussion that follows.

7.2.4.1 Units

Time is measured in seconds and frequency is measured in the number of cycles per second. The units of any Fourier transform, as follows from (7.2.1), are determined by the units of its integrand. We reproduce here the Fourier transform integral (7.2.1) with units shown for every member of the integrand:

$$F(\omega) = F\{v(t)\} = \int_{-\infty}^{\infty} e^{-j\omega t} v(t) dt$$

$$= \int_{-\infty}^{\infty} [\text{dimensionless}] \cdot [\text{units of time-domain function}] \cdot [\text{seconds}] \tag{7.2.9}$$

Table 7.2.2 Main properties of the Fourier transform.

Property	Time domain $v(t)$	Frequency domain $F(\omega) = F\{v(t)\}$
Signal representation	$v(t)$	$F(\omega)$
Units	Voltage, $v(t)$ [V]	$F\{v(t)\} = V(\omega)$ [V s]
	Current, $i(t)$ [A]	$F\{i(t)\} = I(\omega)$ [A s]
Linearity	$a_1 v_1(t) + a_2 v_2(t)$, a_1 and a_2 – constants	$F\{a_1 v_1(t) + a_2 v_2(t)\} = a_1 F_1(\omega) + a_2 F_2(\omega)$
Duality	$v(t)$	$F\{v(t)\} = F(\omega)$
	$F(t)$	$F\{F(t)\} = 2\pi v(-\omega)$
Time scaling	$v(kt)$, k – constant	$F\{v(kt)\} = \frac{1}{|k|} F\left(\frac{j\omega}{k}\right)$
Frequency scaling	$v(t) = F^{-1}\{F(k\omega)\} = \frac{1}{|k|} v\left(\frac{t}{k}\right)$	$F(k\omega)$
Time shift	$v(t - \tau)$	$F\{v(t - \tau)\} = F(\omega) \cdot e^{-j\omega\tau}$
Frequency shift	$F^{-1}\{F(\omega - \omega_0)\} = v(t) e^{j\omega_0 t}$	$F(\omega - \omega_0)$
Modulation (frequency translation)	$v(t) \cos(\omega_0 t)$	$F\{v(t) \cos(\omega_0 t)\} = \frac{1}{2}[F(\omega + \omega_0) + F(\omega - \omega_0)]$
Time differentiation	$v'(t)$ $\vdots$ $v^n(t)$	$F\{v'(t)\} = j\omega F(\omega) \; \vdots \; F\{v^n(t)\} = (j\omega)^n F(\omega)$
Frequency differentiation	$F^{-1}\left\{(j)^n \frac{d^n F(\omega)}{d\omega^n}\right\} = t^n v(t)$	$(j)^n \frac{d^n F(\omega)}{d\omega^n}$
Integration	$\int_{-\infty}^{t} v(t) dt$	$F\left\{\int_{-\infty}^{t} v(t) dt\right\} = \frac{F(\omega)}{j\omega} + \pi F(0) \delta(\omega)$
Convolution in time domain	$v_1(t) \ast v_2(t)$	$F_1(\omega) \cdot F_2(\omega)$
Convolution in frequency domain	$v_1(t) \cdot v_2(t)$	$F_1(\omega) \ast F_2(\omega)$

Therefore, if $v(t)$ (V) is the voltage signal, its Fourier transform results in $F(\omega)$ (V s), as shown in Table 7.2.2. You are encouraged to analyze the units of all other signals and parameters used in communications. Bear in mind that units of the Fourier transform are intermediate results; eventually for practical applications, we need to invert all Fourier transforms to time-domain functions with their units.

7.2.4.2 Linearity

Table 7.2.2 shows that the Fourier transform exhibits both *homogeneity* and *superposition (additivity)* properties, which constitute its linearity. Know that here homogeneity means that an increase in the $v(t)$ amplitude will result in a proportional increase in the $F(\omega)$ amplitude, whereas superposition means that the sum of two time-domain signals will result in the sum of two corresponding frequency-domain signals. Review Example 7.2.2 again and find how the linearity property of the Fourier transform is applied in finding the spectrum of a cosine function. This property enables us to break down a sophisticated problem into small pieces, find the Fourier transforms of these small problems, and obtain the total solution by combining the individual results.

7.2.4.3 Duality

To visualize this property, refer to the example in Figure 7.2.1 which shows that the waveform (time-domain function) of a delta function is a vertical line and its spectrum (frequency-domain function) is a horizontal line. Now, if the waveform of a function is a horizontal line, then the function's spectrum is a vertical line by the duality property. (Build these two graphs.) Table 7.2.1 shows that if $v(t) = 1$, then its Fourier transform is $\boldsymbol{F}\{1\} = 2\pi\delta(\omega)$; reversely, if $v(t) = \delta(t)$, then $\boldsymbol{F}\{\delta(t)\} = 1$. Table 7.2.2 states that if $\boldsymbol{F}\{v(t)\} = F(\omega)$, then replacing ω by t in the latter function, we find its Fourier transform as $\boldsymbol{F}\{F(t)\} = 2\pi v(-\omega)$. For example, from Table 7.2.1 we know that if $v(t) = \text{rect}\left(\frac{t}{\tau}\right)$, then $\boldsymbol{F}\left\{\text{rect}\left(\frac{t}{\tau}\right)\right\} = \tau\cdot\text{sinc}\left(\frac{\omega\tau}{2}\right)$. Using the duality property, we find that for $v(t) = \tau\cdot\text{sinc}\left(\frac{t\tau}{2}\right)$ the Fourier transform is

$$\boldsymbol{F}\left\{\tau\cdot\text{sinc}\left(\frac{t\tau}{2}\right)\right\} = 2\pi v(-\omega) = 2\pi\,\text{rect}\left(-\frac{\omega}{\tau}\right)$$

(**Exercise**: Using Table 7.2.1, we find that $\boldsymbol{F}\{e^{-3|t|}\} = \frac{6}{3^2+\omega^2}$. Applying the duality property, find the Fourier transform of $F(t) = \frac{6}{3^2+t^2}$.)

7.2.4.4 Modulation

This property is also called *frequency translation* or *frequency shifting*. In communication theory, it is one of the most important properties of the Fourier transform. It enables us to find the spectra of amplitude-modulated signals. Table 7.2.2 shows this property as $\boldsymbol{F}\{v_M(t)\cos(\omega_0 t)\} = \frac{1}{2}[F(\omega + \omega_0) + F(\omega - \omega_0)]$. Here, $v_M(t)$ is a message (information) signal and $\cos(\omega_0 t)$ is a carrier signal. In words, this formula states that the Fourier transform of a sinusoidal signal with variable (modulated) amplitude, $\boldsymbol{F}\{v(t)\cos(\omega_0 t)\}$, results in two bands of frequencies centered at $\pm\omega_0$. Figure 7.2.4 presents this property graphically. It shows that the spectrum of a message signal centered at ω (rad/s) = 0 and restricted by its bandwidth, BW. The spectrum of an amplitude-modulated signal, however, contains two bands of frequencies centered at $\pm\omega_0$. Each of these bands is restricted by the bandwidth of a message signal, $\pm$BW. Triangle is a typical figure to show a frequency band of an unknow shape. (Why do the magnitude axis in Figure 7.2.4 is labeled as an absolute value of a function being analyzed?)

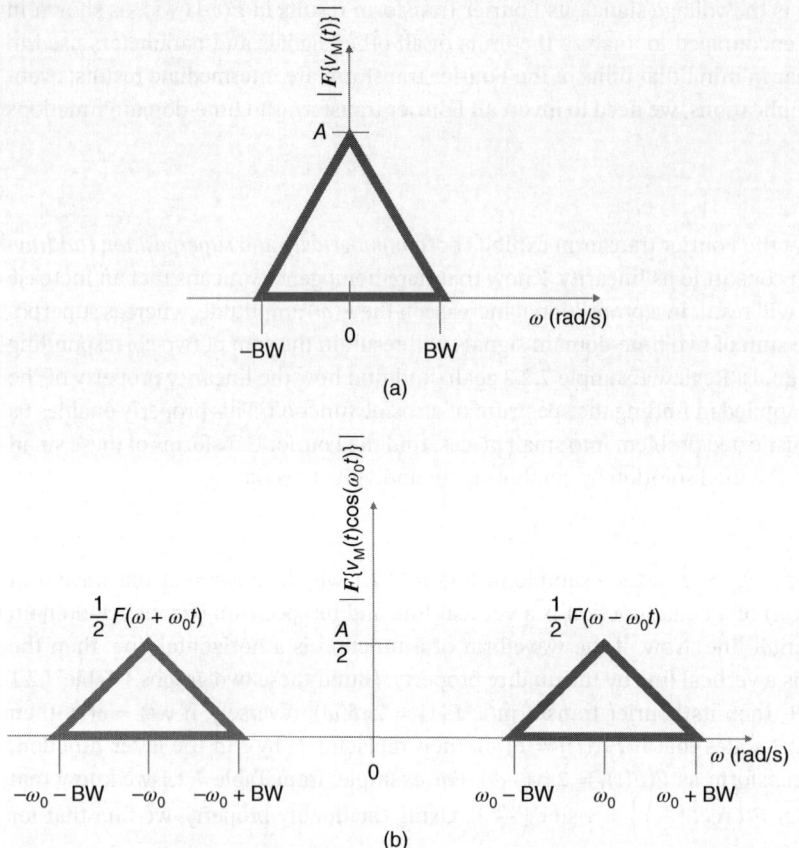

Figure 7.2.4 Application of the Fourier transform to finding the spectrum of an amplitude-modulated signal: (a) The spectrum of a message signal, $v_M(t)$; (b) the spectrum of an AM signal, $v_M(t)\cos(\omega_0 t)$.[4]

Since $\cos(\omega_0 t) = \frac{1}{2}(e^{j\omega_0 t} + e^{-j\omega_0 t})$, multiplication of a time-domain signal $v(t)$ by an exponential, $e^{j\omega_0 t}$, results in the shifting of the signal's frequency, ω, by ω_0 in frequency domain. This is why multiplication of the function $\cos(\omega_0 t)$ by a signal, $f(t)$, results in the translation of the signal's frequency, ω, by $\pm\omega_0$. This result is also known as <u>*modulation theorem*</u>, which we will use in the chapters that follow. This standpoint gives you a hint for the proof of the modulation property.

(**Question**: *Consider a single tone AM signal, $v_{AM}(t) = (A_C + A_M \cos\omega_M t)\cos\omega_C t$. Here, A_C and ω_C are the amplitude and frequency of a carrier and A_M and ω_M are the amplitude and frequency of a message (information) signal. What is its one-side spectrum? Draw the figure. Hint: Apply the Fourier transform to $v_{AM}(t)$.)*

We will thoroughly discuss amplitude modulation in Chapter 8.

7.2.4.5 Convolution in Time and in Frequency and a Transfer Function

The importance of these properties becomes clear if we recall that we are looking for the system's output when the system and its input (excitation) are given. In time domain, the system's output

---

4 After Alexander and Sadiku 2009, p. 819.

can be found by using the *convolution integral* defined as

$$v_{\text{out}}(t) = \int_{-\infty}^{\infty} v_{in}(\tau)h(t-\tau)d\tau = v_{\text{in}}(t) * h(t) \tag{7.2.10}$$

Here $h(t)$ is the system's response to an impulse modeled by $\delta(t)$) and ✡ is the symbol of convolution operation. In practice, application of the convolution integral for finding the system's output is a difficult mathematical operation (though recently it becomes practically achievable thanks to advances in mathematics and computer science). But in frequency domain, the convolution reduces to simple multiplication, as Table 7.2.2 shows. Hence, applying the Fourier transform to $v_{\text{in}}(t)$ ✡ $h(t)$, yields

$$V_{\text{in}}(\omega)\cdot H(\omega)$$

where $V_{\text{in}}(\omega) = \boldsymbol{F}\{v_{\text{in}}(t)\}$ and

$$H(\omega) = \boldsymbol{F}\{h(t)\} \tag{7.2.11}$$

Thus, the Fourier transform of the whole convolution integral (7.2.10) results in

$$V_{\text{out}}(\omega) = V_{\text{in}}(\omega)\cdot H(\omega) \tag{7.2.12}$$

You will recall that H(ω) is the transfer function of a system, which means that the Fourier transform of a system's impulse response is the system's transfer function in the frequency domain, as (7.2.11) shows. Since the spectrum of delta function includes all possible frequencies, as Figure 7.2.1 demonstrates, the impulse response and, consequently, H(ω) are comprehensively describe the system's response to any input frequency.

Therefore, the convolution property enables us to obtain the system's transfer function and, consequently, find the system's output in frequency domain. The output in time domain, $v_{\text{out}}(t)$, can be determined by applying the inverse Fourier transform as

$$\boldsymbol{F}^{-1}\{V_{\text{out}}(\omega)\} = v_{\text{out}}(t) \tag{7.2.13}$$

This consideration emphasizes the importance of the convolution property of the Fourier transform.

7.2.4.6 Time Differentiation

Understand that differentiation of a function in time domain is equivalent to multiplying the Fourier transform of this function by $j\omega$. That is, $\boldsymbol{F}\{v'(t)\} = j\omega F(\omega)$. This property helps in transmuting differential equations of communications systems into frequency domain.

7.2.4.7 Other Properties of the Fourier Transform

At this stage of reading this book, you are well equipped to consider all other properties of the Fourier transform shown in Table 7.2.1 by following the patterns presented above. Bear in mind that all these properties can be proved by relatively simple manipulations based on the Fourier transform definition. A detailed consideration of these properties along with their proofs can be found, for example, in Alexander and Sadiku (2009, pp. 816–825.)

7.2.5 Example of Using the Fourier Transform

We are now ready to apply the Fourier transform to the analysis of communication circuits.

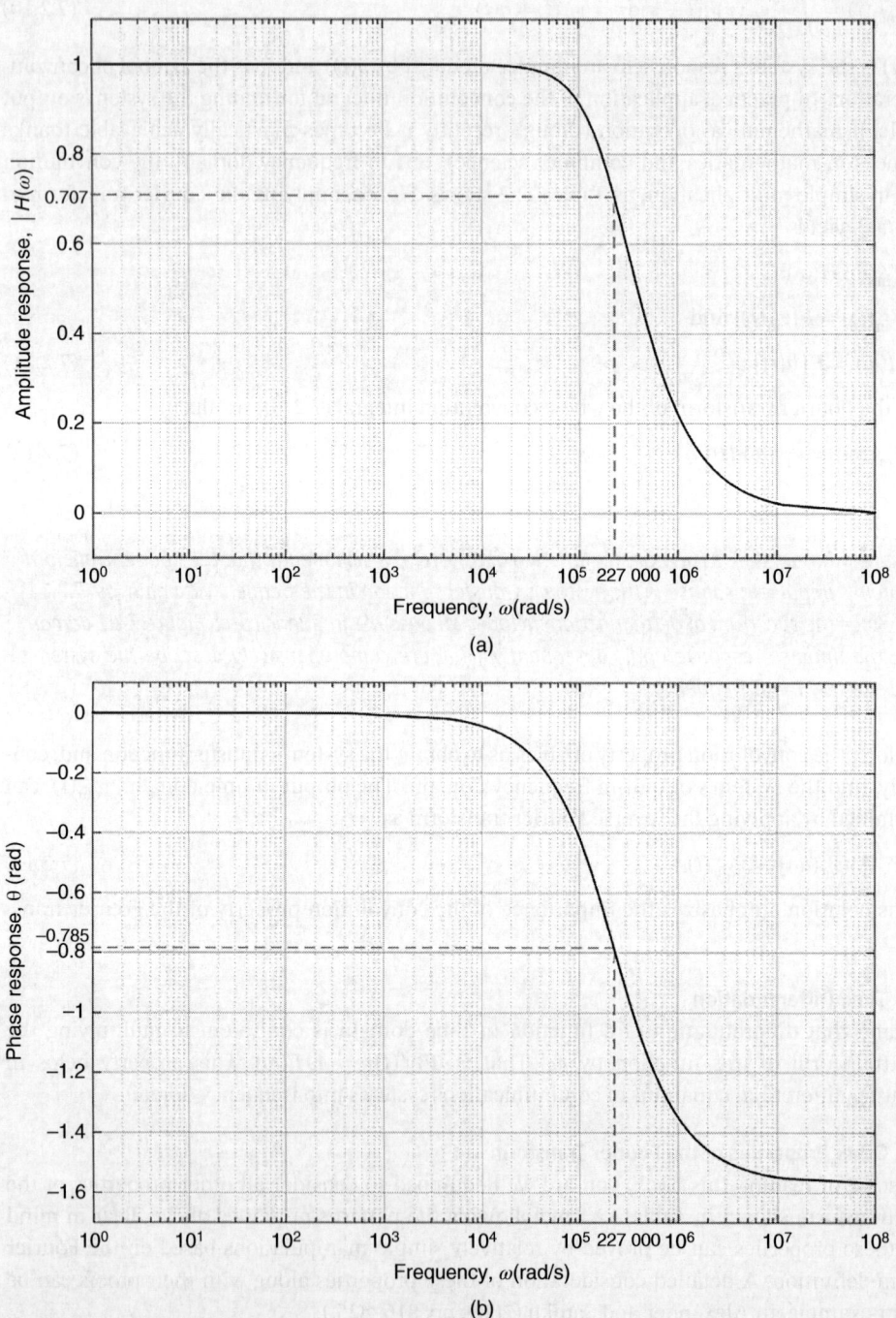

Figure 7.2.5 Amplitude and phase responses of the RC LPF in Example 7.2.3.

Example 7.2.3 Amplitude and Phase Responses of an RC Low-Pass Filter (LPF)

Problem

Using the Fourier transform, find the amplitude and phase frequency responses of the series RC LPF with $R = 2.2\,\text{k}\Omega$ and $C = 0.002\,\mu\text{F}$. (This filter is discussed in Example 5.1.5. Its decaying constant and cutoff frequency are $\alpha = \omega_C\,(\text{rad/s}) = \frac{1}{RC} \approx 227.27\,\text{krad/s}$.)

Solution

We need to find the filter's transfer function, $H(\omega)$, from which the amplitude response will be determined as its absolute value, $|H(\omega)|$, and the phase response will be obtained as an argument of this function, $\arg H(\omega) = \Theta(\omega)$.

A system's transfer function is given as the Fourier transform of the impulse response, $h(t)$,

$$H(\omega) = \mathbf{F}\{h(t)\} \qquad (7.2.11\text{R})$$

The RC LPF's impulse response is given by

$$h(t) = \alpha e^{-\alpha t} u(t) \qquad (7.2.14)$$

(See Sidebar 7.2.S.2 "The impulse response of an RC LPF" for derivation of (7.2.14).
Therefore,

$$H(\omega) = \mathbf{F}\{h(t)\} = \mathbf{F}\{\alpha e^{-\alpha t} u(t)\} = \frac{\alpha}{\alpha + j\omega} = \frac{\alpha^2}{\alpha^2 + \omega^2} - \frac{j\alpha\omega}{\alpha^2 + \omega^2} \qquad (7.2.15)$$

is the transfer function of an RC LPF.

The amplitude response is found as

$$|H(\omega)| = |\mathbf{F}\{\alpha e^{-\alpha t} u(t)\}| = \left|\frac{\alpha^2}{\alpha^2 + \omega^2} - \frac{j\alpha\omega}{\alpha^2 + \omega^2}\right| = \frac{\alpha}{\sqrt{(\alpha^2 + \omega^2)}} \qquad (7.2.16\text{a})$$

and the phase response is obtained as

$$\arg H(\omega) = \arg\left(\frac{\alpha^2}{\alpha^2 + \omega^2} - \frac{j\alpha\omega}{\alpha^2 + \omega^2}\right) = \Theta(\omega) = -\tan^{-1}\left(\frac{\omega}{\alpha}\right) \qquad (7.2.16\text{b})$$

Thus, by using the Fourier transform, we found the magnitude and phase characteristics of an RC lowpass filter, a result we already knew and discussed many times in Sections 5.1 and 5.4.

The graphs of both responses are shown in Figure 7.2.5.

Discussion

- Sidebar 7.2.S.2 "Finding the impulse response of an RC LPF" shows the various approaches to derivation of (7.2.14) and discusses implications of its use.
- It is necessary to realize that a system's impulse response enables us to find the response to any input. As an example, let us consider the response of an RC LPF to an input rectangular pulse based on the filter's impulse response. The rectangular pulse is given as,

$$v_{\text{in}}(t) = \text{rect}(t) = \begin{cases} A(\text{V}) & \text{for } 0 \leq t \leq T \\ 0(\text{V}) & \text{otherwise} \end{cases}$$

(Notice that a rectangular pulse given in Table 7.2.1 is centered at $t = 0$ and restricted $\pm \tau/2$; here the rectangular pulse is centered at $t = T/2$ and bounded by 0 and T.)

In solving this problem, we also outline the Fourier-transform technique in the following three steps:

1) Find the Fourier transform of the input signal by referring to Table 7.2.1:
$$V_{in}(\omega) = F\{v_{in}(t)\} = F\{\mathbf{rect}(t)\} = A\,\sin(\omega T)/(\omega) = AT\,\text{sinc}(\omega T) \qquad (7.2.17)$$

2) Find the transfer function of the circuit as the Fourier transform of an impulse response:
$$H(\omega) = F\{h(t)\} = F\{\alpha e^{-\alpha t} u(t)\} = \frac{\alpha}{\alpha + j\omega} = \frac{\alpha^2}{\alpha^2 + \omega^2} - \frac{j\alpha\omega}{\alpha^2 + \omega^2} \qquad (7.2.15R)$$

3) Using (7.2.17) and (7.2.15), find the response of the RC LPF in frequency domain as
$$V_{out}(\omega) = V_{in}(\omega) \times H(\omega) = [2AT\,\text{sinc}(\omega t)] \times \frac{\alpha}{\alpha + j\omega} \qquad (7.2.18)$$

The problem is solved. We can, additionally, determine the circuit's output in time domain as
$$v_{out}(t) = F^{-1}\{V_{out}(\omega)\} = F^{-1}\left\{[2AT\,\text{sinc}(\omega t)] \times \frac{\alpha}{\alpha + j\omega}\right\} \qquad (7.2.19)$$

To find $v_{out}(t)$ in explicit form, we need to perform this inverse Fourier transform.

- We can now better interpret the meaning of the circuit's parameters. We know that the frequency at which the amplitude value is equal to $1/\sqrt{2} = 0.707$ is called the *cutoff frequency, ω_c*. Looking at (7.2.16a), we find that

$$|H(\omega_c)| = \frac{1}{\sqrt{\left(1 + \left(\frac{\omega}{\alpha}\right)^2\right)}} = 1/\sqrt{2}$$

when $\alpha = \omega$; that is, the cutoff value of the angular frequency, ω_c (rad/s), is determined as

$$\omega_c\,(\text{rad/s}) = \alpha = 1/RC \text{ or } f_C\,(\text{Hz}) = 1/2\pi RC$$

as we well know.

We also know from our previous discussion that the phase shift is equal to -45^0 at $\omega = \omega_c$. All these values are shown in Figure 7.2.5.

From the definition, discussion and examples of the Fourier transform, we can again highlight the important **conclusion**:

> The Fourier transform, $F\{v(t)\} = \int_{-\infty}^{\infty} f(t) e^{-j\omega t}\,dt = F(\omega)$, of a time-domain function, v(t), gives us the spectrum of this function. This spectrum can be presented in a polar form,
>
> $F(\omega) = |F(\omega)| \angle \theta(\omega)$
>
> where $|F(\omega)|$ is the magnitude (amplitude) spectrum and $\theta(\omega)$ is the phase spectrum.

If the time-domain function, $v(t)$, is the *response of a given device to an impulse input*, then its Fourier transform produces the device's amplitude and phase responses, as shown above.

> **Sidebar 7.2.S.2 The Impulse Response of an RC LPF**
>
> *7.2.S.2.1 The Impulse Response of a First-Order System*
> What exactly is an impulse response, h(t)? Well, formally speaking, *an impulse response, h(t), is the solution to the system's differential equation, whose right-hand side is the delta function.*

To put these words into a formula, we refer to the general system's differential equation,

$$a_n y^n(t) + a_{n-1} y^{n-1}(t) + \cdots + a_1 y'(t) + a_0 y(t) = b_m x^m(t) + b_{m-1} x^{m-1}(t)$$
$$+ \cdots + b_1 x'(t) + b_0 x(t)$$

If we replace the input signal, $x(t)$, by a delta function, $\delta(t)$, then we must replace the output, $y(t)$, by an impulse response, $h(t)$; that is,

$$a_n h^n(t) + a_{n-1} h^{n-1}(t) + \cdots + a_1 h'(t) + a_0 h(t) = b_0 \delta(t) \tag{7.2.S.2.1}$$

Solving this equation, we find the impulse response, $h(t)$.

Consider, for example, the response of a first-order system to an impulse, $\delta(t)$. From (7.2.S.2.1), the required differential equation can be found as

$$a_1 h'(t) + a_0 h(t) = b_0 \delta(t) \tag{7.2.S.2.2}$$

This equation can be written in a classical form as

$$h'(t) + \alpha h(t) = \alpha \delta(t) \tag{7.2.S.2.3}$$

where the damping constant, α, is given by $\alpha = a_0/a_1 = b_0/a_1$ under the assumption $a_0 = b_0$.

To find the impulse response of a first-order system, we have to solve this differential equation. To do so, we need to (1) assume the form of a trial solution, (2) determine the constant, and (3) verify the solution. Let us perform these steps:

1. The form of a *trial solution* can be found based on the following reasoning: As any real system, our system is <u>causal</u>, which means that $h(t)$ *must be equal to zero for* $t < 0$. On the other hand, for the duration $t > 0$, no excitation is applied to the circuit because the impulse is applied only at $t = 0$ (see Figure 7.2.S.1.1a and Eq. (7.2.S.1.6)). Therefore, for $t < 0$ and $t > 0$, our system is undriven, and *the solution to* (7.2.S.2.3) *should have the form of an RC LPF's free response, which is* $v^*_{out}(t) = Ae^{-\alpha t}$, as discussed in Section 5.1. Since our system is *causal*, $h(t)$ must be multiplied by the unit-step function. Thus, the trial solution to (7.2.S.2.3) is

$$h(t) = Ae^{-\alpha t} u(t) \tag{7.2.S.2.4}$$

2. *Constant A* can be found by determining response $h(t)$ at $t = 0$, as we did in Section 5.1. Since (7.2.S.2.3) must be satisfied at all times, and the excitation is applied only at $t = 0$, we can obtain the solution at $t = 0$ by integrating this equation from $t = 0^-$ into $t = 0^+$, where 0^- and 0^+ are infinitesimal times just before and after the zero instant, $t = 0$; that is,

$$\int_{0-}^{0+} h'(t) dt + \alpha \int_{0-}^{0+} h(t) dt = \alpha \int_{0-}^{0+} \delta(t) dt \tag{7.2.S.2.5}$$

You will recall that $h'(t) = dh/dt$ and $\int_{0-}^{0+} h'(t) dt = \int_{0-}^{0+} (dh(t)/dt) dt = \int_{0-}^{0+} dh(t) = (h(0^+) - h(0^-))$. Performing the integration and remembering that $\int_{0-}^{0+} \delta(t) dt = 1$ according to (7.2.S.1.5a), we obtain

$$(h(0^+) - h(0^-)) + \alpha \int_{0-}^{0+} h(t) dt = \alpha \tag{7.2.S.2.6}$$

Here, $h(0^-) = 0$ because this system is causal. The integral $\int_{0-}^{0+} h(t) dt$ must be equal to zero because it shows the area covered by the finite-value function, $h(t) = Ae^{-\alpha t} u(t)$, over an infinitesimal increment, $(0^+ - 0^-)$. Hence, (7.2.S.2.6) reduces to

$$h(0^+) = \alpha \tag{7.2.S.2.7}$$

(Continued)

Sidebar 7.2.S.2 (Continued)

Given that $h(t) = Ae^{-\alpha t}u(t)$, $h(0^+)$ becomes

$$h(0^+) \equiv (Ae^{-\alpha 0^+})u(0^+) = \alpha \qquad (7.2.S.2.8)$$

Since $e^0 = 1$ and $u(0^+) = 1$ by definition, the constant A is determined as

$$A = \alpha \qquad (7.2.S.2.9)$$

3. Finally, the *trial solution* to (7.2.S.2.3) takes the form

$$h(t) = \alpha e^{-\alpha t}u(t) \qquad (7.2.S.2.10)$$

To verify *our solution*, we substitute (7.2.S.2.10) into the original equation, (7.2.S.2.3). The *left-hand side* of (7.2.S.2.3) becomes

$$h'(t) + \alpha h(t) = [\alpha e^{-\alpha t}u(t)]' + \alpha[\alpha e^{-\alpha t}u(t)]$$
$$= \alpha[-\alpha e^{-\alpha t}u(t) + e^{-\alpha t}u'(t) + \alpha e^{-\alpha t}u(t)] = \alpha e^{-\alpha t}(u(t))' \qquad (7.2.S.2.11)$$

Thus, (7.2.S.2.3) turns into

$$\alpha e^{-\alpha t}u'(t) = \alpha\delta(t) \qquad (7.2.S.2.12)$$

This holds true because $u'(t) = \delta(t)$, and the derivative $u'(t)$ is zero for $t < 0$ and $t > 0$; therefore, the product $e^{-\alpha t}u'(t)$ reduces to $e^0\delta(t) = \delta(t)$. This means that the assumed solution, $h(t) = \alpha e^{-\alpha t}u(t)$, is the true solution to (7.2.S.2.3) and *is the impulse response of a first-order system*.

Example 7.2.S.2.1 Finding the Impulse Response of an RC Low-Pass Filter

Problem

Find the impulse response of the RC circuit shown in Figure 7.2.S.2.1 (middle part) with $R = 2\,k\Omega$ and $C = 2\,\mu F$. (Figure 7.2.S.2.1 shows the input impulse and expected impulse response based on (7.1.4) and (7.2.4).)

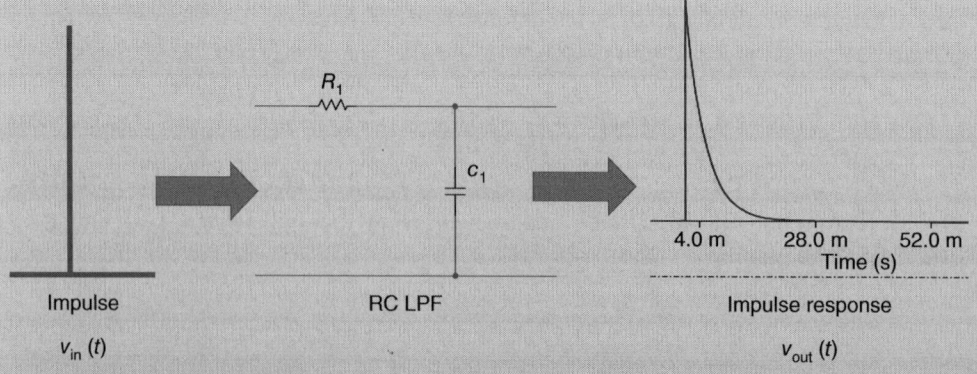

Figure 7.2.S.2.1 Symbolic picture of finding the impulse response of the RC LPF in Example 7.2.S.2.1.

Solution

To start, we need to derive the differential equation of an RC LPF. Applying Kirchhoff's voltage law, we find:

$$v_{in}(t) = v_R(t) + v_C(t) \tag{7.2.S.2.13}$$

Expressing $v_R(t)$ through $v_C(t) \equiv v_{out}(t)$ as

$$v_R(t) = i_R(t)R \quad \text{and} \quad i_C(t) = RCv'_C(t)$$

results in

$$v_R(t) = i_R(t)R = RCv'_C(t) = RCv'_{out}(t) \tag{7.2.S.2.14}$$

Combining (7.2.S.2.13) and (7.2.S.2.14), we obtain

$$RCv'_{out}(t) + v_{out} = v_{in}(t) \tag{7.2.S.2.15a}$$

With $\alpha = 1/RC$, the first-order differential equation describing this RC LPF becomes

$$v'_{out}(t) + \alpha v_{out} = \alpha v_{in}(t) \tag{7.2.S.2.15b}$$

Since in this example $v_{out} = h(t)$ and $v_{in} = \delta(t)$, (7.2.S.2.15b) becomes (7.2.S.2.3)

$$h'(t) + \alpha h(t) = \alpha \delta(t)$$

The solution to this equation is

$$h(t) = \alpha e^{-\alpha t} u(t)$$

as in (7.2.S.2.10).

Referring to the *decay constant*, $\alpha = 1/RC$ (s^{-1}), or the *time constant*, $\tau = RC$ (s), we present (7.2.S.2.10) as

$$h(t) = \alpha e^{-\alpha t} u(t) = \left(\frac{1}{\tau}\right) e^{-t/\tau} u(t) = \left(\frac{1}{RC}\right) e^{-(1/RC)t} u(t) \tag{7.2.S.2.16}$$

Plugging $R = 2\,k\Omega$ and $C = 2\,\mu F$ into (7.2.S.2.16), we compute

$$h(t) = (250)\, e^{-250t} u(t)$$

The graph of this response is shown in Figure 7.2.S.2.2. This impulse response, obviously, is a one-sided decayed exponential function whose initial value is $h(0)u(t) = 250\,s^{-1}$ and whose decay constant is $\alpha = 250\,s^{-1}$.

Discussion

There are several points to consider here

- If we analyze our manipulations from the start (the circuit's voltage law given in (7.2.S.2.13)) to the result (the circuit's impulse response, (7.2.S.2.16)), we understand that this example demonstrates the concept of finding an *impulse response* : Solve the circuit's differential equation *for h(t) when the input is* $\delta(t)$. Figure 7.2.S.2.3 illustrates this definition.
- Does this *form* of the impulse response of an RC LPF – the decaying exponential function – look reasonable to you? We think it makes sense because (i) the input is applied only at $t = 0$, which means that this input creates the initial condition and (ii) after the initial impulse is applied, there is no input signal here, which means that the RC circuit is undriven. This is why we should expect to see the impulse response of this first-order

(Continued)

Sidebar 7.2.S.2 (Continued)

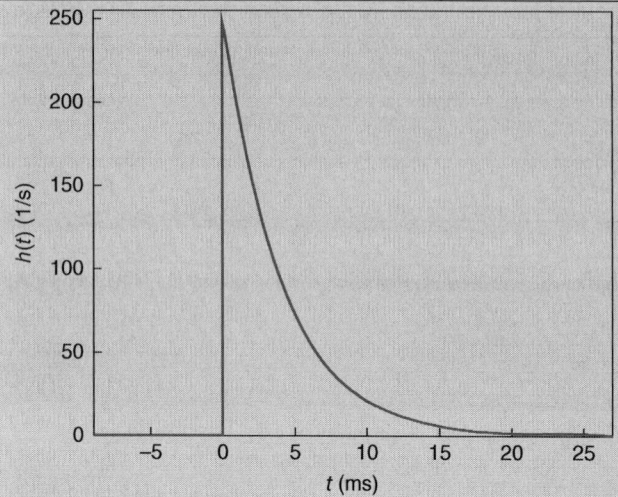

Figure 7.2.S.2.2 Impulse response of an RC LPF.

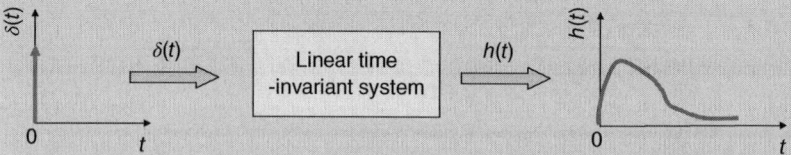

Figure 7.2.S.2.3 Conceptual view of the impulse response: The hypothetical system's response to an individual impulse.

system in the *form of a system's free response*, which is a decaying exponential function, as we learned in Section 5.1. Nevertheless, the impulse response is not a real output signal, as we point out below.

- To discuss the meaning of the impulse response, let us examine its dimension. From (7.2.S.2.16), $h(t) = (1/RC)e^{-(1/RC)t}u(t)$, we find

$$h(t) = (1/RC)\,[1/\text{s}]e^{-(1/RC)t}[1]\,u(t)[1] \to h(t)\,[1/\text{s}] \tag{7.2.S.2.17}$$

where units are given in brackets.

Why could the response of an *RC* circuit, which must be the voltage across the capacitor, carry dimension 1/s? The answer is that $h(t)$ is not a real physical output signal, which can be measured by an instrument like an oscilloscope. It is just a *mathematical model of the real process* where the real input is replaced by an impulse and the real output is modeled by the impulse response. The real output signal, $v_{\text{out}}(t)$, is given by

$$v_{\text{out}}(t)\,(\text{V}) \approx \sum_{m=0}^{\infty}(V_{\text{in}}(m\Delta T))\,h(t - m\Delta T))\Delta T \to \int_{0}^{\infty} v_{\text{in}}(\tau)\,h(t-\tau)d\tau \tag{7.2.10R}$$

which shows that $v_{\text{out}}(t)$ is the result of the integration of the product of the input amplitude, V_{in} (V), the impulse response, $h(t)$ (s^{-1}), and dt (s). Thus, the *dimension of* $v_{\text{out}}(t)$ is given by

$$[\text{V}] \bullet [1/\text{s}] \bullet [\text{s}] = [\text{V}]$$

(***Exercise:*** Analyze the dimension of the entire differential equation, (7.2.S.2.3).)

This concludes our discussion of Example 7.2.S.2.1.

(**Question**: Based on this example, can you find the impulse response of an RLC circuit by using (7.2.S.2.1)?)

Sidebar 7.2.S.3 Alternative Methods of Finding a Transfer Function

We have found the circuit's transfer function by applying the Fourier transform to the impulse response. We can achieve this goal, however, by using the circuit's differential equations or the Fourier integral. These methods will be exemplified by considering the transfer function of a series RC LPF.

1. *Finding the RC LPF's transfer function by using the circuit's differential equation*
   This equation was obtained in (7.2.S.2.15b) as

   $$v'_{out}(t) + \alpha v_{out}(t) = \alpha v_{in}(t)$$

   where $\alpha = 1/RC$. Using the Fourier transform, we transmute this equation into the frequency domain

   $$j\omega V_{out}(\omega) + \alpha V_{out}(\omega) = \alpha V_{in}(\omega)$$

   or

   $$(j\omega + \alpha)V_{out}(\omega) = \alpha V_{in}(\omega) \quad (7.2.S.3.1)$$

   Hence, the RC LPF transfer function in frequency domain is given by

   $$H(\omega) = \frac{V_{out}(\omega)}{V_{in}(\omega)} = \frac{\alpha}{\alpha + j\omega} = \frac{\alpha^2}{\alpha^2 + \omega^2} - \frac{j\alpha\omega}{\alpha^2 + \omega^2}$$

   This is the result we already obtained in (7.2.15). Observe that in frequency domain the transfer function is obtain by simple mathematical (algebraic, in fact) manipulations.

2. *Finding the RC LPF's transfer function by applying the Fourier transform integral*
   The RC LPF's transfer function is defined in (7.2.15) as

   $$H(\omega) = \mathbf{F}\{h(t)\} = \mathbf{F}\{\alpha e^{-\alpha t} u(t)\}$$

   If we apply the definition of the Fourier transform given in (7.2.1)–(7.2.15), we find for an RC LPF

   $$H(\omega) = \int_{-\infty}^{\infty} \alpha e^{-\alpha t} u(t) e^{-j\omega t}\, dt \quad (7.2.S.3.2)$$

   Performing the integration, we obtain

   $$H(\omega) = \mathbf{F}\{\alpha e^{-\alpha t} u(t)\} = \frac{\alpha}{\alpha + j\omega} = \frac{\alpha^2}{\alpha^2 + \omega^2} - \frac{j\alpha\omega}{\alpha^2 + \omega^2} \quad \text{for} \quad \alpha > 0$$

   We would like, however, to present here a slightly different way to calculate the Fourier integral (Smith 2003, pp. 253–255.) Referring to Euler's formula,

   $$e^{-j\omega t} = \cos \omega t - j \sin \omega t$$

   we can separate the real and imaginary parts of the Fourier integral,

   $$\int_{-\infty}^{\infty} \alpha e^{-\alpha t} u(t) e^{-j\omega t}\, dt = \int_0^{\infty} \alpha e^{-\alpha t} \cos \omega t\, dt - j \int_0^{\infty} \alpha e^{-\alpha t} \sin \omega t\, dt \quad (7.2.S.3.3)$$

(Continued)

> **Sidebar 7.2.S.3 (Continued)**
>
> Note that the lower integration limit has changed from $-\infty$ to 0 due to the step (Heaviside) function, $u(t)$. Now, let us perform the integration:
>
> $$\text{Re}[H(\omega)] = \text{Re}[\boldsymbol{F}\{\alpha e^{-\alpha t}u(t)\}] = \alpha \int_0^\infty e^{-\alpha t} \cos \omega t \, dt = \frac{\alpha e^{-\alpha t}}{\alpha^2 + \omega^2}$$
>
> $$(-\alpha \cos \omega t + \omega \sin \omega t)\big|_{t=0}^\infty = \frac{\alpha^2}{\alpha^2 + \omega^2} \quad \text{for} \quad \alpha > 0 \qquad (7.2.\text{S}.3.4)$$
>
> Similarly,
>
> $$\text{Im}[H(\omega)] = \text{Im}[\boldsymbol{F}\{\alpha e^{-\alpha t} u(t)\}] = -j\alpha \int_0^\infty e^{-\alpha t} \sin \omega t \, dt = -\frac{\alpha e^{-\alpha t}}{\alpha^2 + \omega^2}$$
>
> $$(-\alpha \sin \omega t - \omega \cos \omega t)\big|_{t=0}^\infty = \frac{-\omega \alpha}{\alpha^2 + \omega^2} \quad \text{for} \quad \alpha > 0 \qquad (7.2.\text{S}.3.5)$$
>
> Here, we use the facts that at $\omega \to \infty$, the exponential function $e^{-\alpha t}$ goes to zero and at $t = 0$, the $\sin(\omega t)$ equals zero. Thus, the Fourier transform of an impulse response of an RC LPF is given by
>
> $$H(\omega) = \boldsymbol{F}\{\alpha e^{-\alpha t}u(t)\} = \text{Re}[F(\omega)] + j \, \text{Im}[F(\omega)] = \frac{\alpha^2}{\alpha^2 + \omega^2} - j\frac{\omega \alpha}{\alpha^2 + \omega^2} \quad \text{for} \quad \alpha > 0$$
>
> as in (7.2.15).

Questions and Problems for Section 7.2

- Questions marked with an asterisk require a systematic approach to finding the solution.
- Many questions and problems, including those marked with an asterisk, imply that you, in addition to reading the textbook, will do your research to find the answers. Consider such questions as miniprojects.

1. Equation (7.2.1) reads $\boldsymbol{F}\{v(t)\} = \int_{-\infty}^{\infty} e^{-j\omega t}v(t)dt = F(\omega)$. Explain the meaning of each member.

2. The text says that the Fourier transform exists if and only if the Fourier integral $\int_{-\infty}^{\infty} v(t)e^{-j\omega t} \, dt$ converges. Why is convergence of the Fourier integral necessary? What happens if the Fourier integral does not converge?

3. It follows from the text statements that the Fourier integral converges if the function $v(t)$ meets the Dirichlet's conditions:
   a) Does a $\sin(\omega t)$ meet the Dirichlet's conditions? Prove your answer.
   b) What about a rectangular pulse?

4. *The text says that the Fourier transforms of the realizable signals always exist. Why? Prove this statement. (*Hint*: See definition of a realizable signal.)

5. Compare a function and a transform:
   a) What is the difference between a function and a transform?
   b) What do these two mathematical entities have in common?

c) Define two mathematical operations symbolically denoted by arrows: $\{4 \to 64\}$ and $\{3e^{-2t} \to 3/(2+j\omega)\}$. Explain your reasoning.

d) What will be the results of the reverse operations of $\{4 \to 64\}^{-1}$ and $\{3e^{-2t} \to 3/(2+j\omega)\}^{-1}$?

6 A delta function is defined as a vertical line whose area is finite. It would seem that a line, by definition, does not have width. Then, what area can exist under a line?

7 It seems that a delta function is a pure mathematical abstraction. How can we use this abstraction in engineering practice?

8 Table 7.2.1 shows that $F\{\cos(\omega_0 t)\} = \pi[\delta(\omega - \omega_0) + \delta(\omega + \omega_0)]$. What is the role of delta functions in this formula? How can you present this formula graphically?

9 *Table 7.2.1 shows the Fourier transform pairs for the most popular functions used in engineering. Each entry of Table 7.2.1 has been obtained by applying the Fourier integral to a given time-domain function. To confirm this statement, prove the following result:

a) $F\{\delta(t)\} = 1$.

b) $F\{e^{j\omega_0 t}\} = 2\pi\delta(\omega - \omega_0)$.

c) $F\{\cos(\omega_0 t) \cdot u(t)\} = \frac{\pi}{2}[\delta(\omega - \omega_0) + \delta(\omega + \omega_0)]$.

10 *Table 7.2.1 shows that $F\{\delta(t - t_0)\} = e^{-j\omega t_0}$.

a) What is the meaning of this result?

b) Depict the amplitude spectrum of $\delta(t - t_0)$.

11 Find the Fourier transform of

a) $v(t) = 3\cos(12t)$;

b) $v(t) = \delta(t - 4)$;

c) $v(t) = \sin(\omega_0 t)$.

12 Table 7.2.1 shows that $F(\cos(\omega_0 t))$ and $F(\cos(\omega_0 t)u(t))$ are different? Why is this difference?

13 What and why is the difference between $e^{-\alpha t}u(t)$ and $e^{-\alpha|t|}$?

14 The Fourier transforms containing $e^{-\alpha t}$, $\cos(\omega_0 t)$ or $\sin(\omega_0 t)$ include conditions "$\text{Re}(\alpha) > 0$ constant" or "ω_0 – real constant." Why do these conditions have to be applied?

15 *Consider the properties of the Fourier transform:

a) If $v(t)$ is an energy signal, what will be the unit of its Fourier transform?

b) What is the Fourier transform of $v(t) = 5\cos(3t) + 2\sin(7t)$?

c) What is the Fourier transform of $v(t) = e^{j6t}$?

d) What is the Fourier transform of $v(t) = (5 + 4\cos(3t)) \cdot \cos(33t)$?

e) What is the Fourier transform of $v(t) = \frac{d(2e^{6t})}{dt}$?

16 Determine the impulse response of a series RC LPF whose $R = 3.0\,\text{k}\Omega$ and $C = 0.006\,\mu\text{F}$.

17 Consider an impulse response of a linear time-invariant communication system:

a) How can we find it? What is the difference between impulse response and free response of an RC LPF? What do these responses have in common?
b) Is an impulse response a real signal or a mathematical model? Explain.

18 Consider a series RC LPF whose $R = 3.0\,\text{k}\Omega$ and $C = 0.006\,\mu\text{F}$. Using the Fourier transform, find the amplitude and phase frequency responses of this filter.

19 *Find the amplitude and phase responses of an RC LPF whose $R = 2.2\,\text{k}\Omega$ and $C = 0.002\,\mu\text{F}$ to a step signal $v(t) = 3u(t)$.

20 What methods of finding the transfer function of a communication system do you know? Give an example.

7.3 The Fourier Transforms and Digital Signal Processing

Objectives and Outcomes of Section 7.3

Objectives

- To learn that there are four types of signals – continuous periodic, continuous nonperiodic, discrete periodic, and discrete nonperiodic – and, correspondingly, four types of the Fourier tools: Continuous-time Fourier series (FS), continuous-time Fourier transform (FT), discrete Fourier transform (DFT), and discrete-time Fourier transform (DTFT).
- To become familiar with digital signal processing (DSP) and the two major problems in application of the Fourier transforms to this technology: processing speed and digital nature of communications signals.
- To understand why only DFT can be employed for DSP and how DFT can be applied for the spectral analysis of both discrete periodic and discrete nonperiodic signals.

Outcomes of Section 7.3

- Realize that the main application of the Fourier transform is finding the spectrum of a signal and the frequency response of a communication system (see Sections 7.2 and 7.3).
- Know that in regard to the time-axis and magnitude-axis properties, signals can be classified in four types or categories: continuous periodic, continuous nonperiodic, discrete periodic, and discrete nonperiodic (see Figure 7.3.1).
- Learn that there are four types of the Fourier transformations, one for each type of signal, enabling us to determine the spectrum of each type of signal. These Fourier transforms are continuous-time Fourier series (FS), continuous-time Fourier transform (FT), discrete Fourier transform (DFT), and discrete-time Fourier transform (DTFT) (see Figure 7.3.2).
- Comprehend that in modern computerized technology, we have to use discrete signals and, therefore, employ either the discrete Fourier transform (DFT) or the discrete-time Fourier transform (DTFT).
- Grasp the point that one of the most ubiquitous operations in modern electronic technology is digital signal processing (DSP); it enables us to achieve high-quality transmission in electronic communication. DSP is a fully computerized operation that changes the spectrum of the received signal in a desired way. Understand that obtaining the spectrum of a signal is a necessary condition for performing DSP.

- Get to know that DSP faces two major problems: supporting the required high-processing speed and dealing with modern discrete and digital signals. Realize that the analysis of the DSP requirements concludes that only DFT can be used for this work.
- Learn that DFT is a mathematical operation that transforms one set of numbers, representing a discrete time-domain signal, into another set of numbers, representing this signal in frequency domain. Be aware that, in essence, DFT is a discrete analogy of the continuous Fourier series, which implies that DFT can work only with periodic signals.
- Observe the problem in employing DFT for finding the spectrum of a nonperiodic signal: On the one hand, DFT can be applied only to a finite signal; on the other hand, a nonperiodic signal is not finite. Be aware that the solution lies in limiting (truncating or windowing) a nonperiodic signal and making it finite.
- Become familiar with problem that arises by the signal truncating: the more detailed the waveform (the plot of a signal in time domain), the less detailed the plot of its spectrum and vice versa. Know that this problem is often referred to as *the principle of uncertainty of Fourier analysis*.
- Be aware that since we need a finite signal for the DFT operation, we always have to truncate any signal presented to DFT and, therefore, we will always have an approximate, not exact, spectrum of a signal.
- Get to know the relationship among the Fourier transforms, namely: The continuous Fourier transform can be derived from the Fourier series; the discrete-time Fourier transform (DTFT) can be derived from the continuous Fourier transform; the discrete Fourier transform (DFT) can be derived from the discrete-time Fourier transform (see Figure 7.3.8).
- Comprehend that DFT enables to perform its task only after the algorithm for fast execution of the spectral analysis – the fast Fourier transform (FFT) – had been developed. Know that there are many versions of FFT and the search for even faster algorithms to perform DFT continues.

In this section, we will turn to an important application of the Fourier transform – *signal processing*. Refer to the general block diagram of a communications system (Figure 1.1.2 reproduced here) and concentrate on the receiver end. A receiver, of course, accepts a transmitted signal and produces an output signal, which delivers the intended information. But what is critical in this function is for the receiver to produce an output signal of the highest possible quality – think about a picture on your screen or a voice from your speaker – from the received signal, distorted by transmission flaws. The receiver achieves this goal through a set of operations called *signal processing*. Since most of signals today are digital, this operation is called *digital signal processing, DSP*. The Fourier transforms play the major role in this important operation. (For brief introduction to DSP see the Sidebar 7.3.S.1 "A word about DSP" in this section.)

7.3.1 Signals and the Fourier Transformations

Comparing the Fourier series and the Fourier transform, we referred to two types of signals – *periodic and nonperiodic*. We certainly noticed that regardless of the periodicity of the signals, we distinguished between *continuous and discrete signals*. Thus, we need to understand that, in general,

> we can meet four types of signals derived from the following combinations: periodic and nonperiodic, continuous and discrete.

Figure 7.3.1 pictures these signals. We recall that <u>continuous</u> and <u>discrete</u> are the characteristics of signals describing their continuity along the <u>time axis</u>. The terms <u>analog</u> and <u>digital</u> refer to the

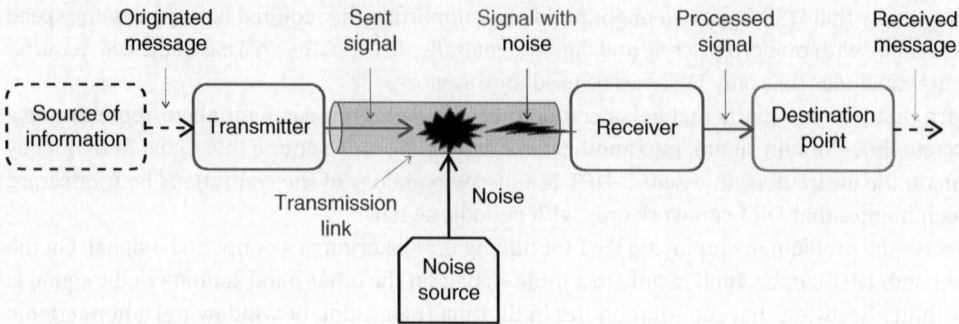

Figure 1.1.2R General block diagram of a communication system.

continuity of signals along the magnitude axis. See Chapter 2 for an introduction to classification of signals, where we stressed that a signal is called *continuous if we can draw its waveform without lifting a pen off the paper*. We can observe the four types of signals – continuous periodic, continuous nonperiodic, discrete periodic, and discrete nonperiodic – in Figure 7.3.1a–d, respectively.

You will recall that we can obtain a *discrete signal* from a continuous one by *sampling* the continuous signal. Sampling, however, gives us the discrete amplitudes with infinite precision. (Figure 7.3.1e visualizes these explanations: It shows several sampling points and amplitudes.) Clearly, such amplitudes cannot be processed by computers. Thus, these amplitudes must be approximated to finite values to obtain *digital signals*. Figure 7.3.1f shows how the sampled amplitudes have truncated and presented with finite (digital) values. Revisit Chapter 3 for explanation of analog-to-digital conversion, ADC. Refer to Smith (2003, pp. 1–67) and Manolakis and Ingle (2011, pp. 2–9) for introduction to the signal classification.

We learned in Chapter 6 and Section 7.1 that the Fourier series enables us to find the spectra of continuous periodic signals and the Fourier transform allows for obtaining the spectra of continuous nonperiodic signals.

But how can we find the spectra of other types of signals – discrete nonperiodic and periodic? The answer: There are two other types of Fourier transforms that perform these operations: the discrete-time Fourier transform (DTFT) and the discrete Fourier transform (DFT).

While the names of these two transforms are almost identical, their operating methods are, in fact, quite different. We will discuss these transforms in detail shortly; for now, we need to understand their place in the overall category of *Fourier transformations*. The relationship among the type of signals and type of Fourier transformations is symbolically pictured in Figure 7.3.2, which is self-explanatory. Examine this figure very closely and pay attention to the characteristics of signals, correspondence of the signals to the spectra, and the type of Fourier transform used in each line. For example, the Fourier transform applied to a time-continuous nonperiodic signal delivers its frequency-continuous spectrum, whereas discrete Fourier transform (DFT) applied to a periodic time-discrete signal allows for finding the discrete-frequency spectrum of this signal. (In developing these classifications, we follow Smith (2003, pp. 141–146) and McClellan et al. (2003, p. 392).)

The only question that might arise is why the sampled sinusoidal signal in Figure 7.3.2d is called nonperiodic. The answer is because this signal is time-windowed (time-restricted). The underlining point of this explanation is as follows:

A sinusoidal signal is periodic (and its spectrum contains only one line) if and only if it runs from negative infinity to positive infinity. As soon as the sinusoidal signal is time-restricted (windowed), as in Figure 7.3.2d, it is not periodic anymore and its spectrum contains many frequencies.

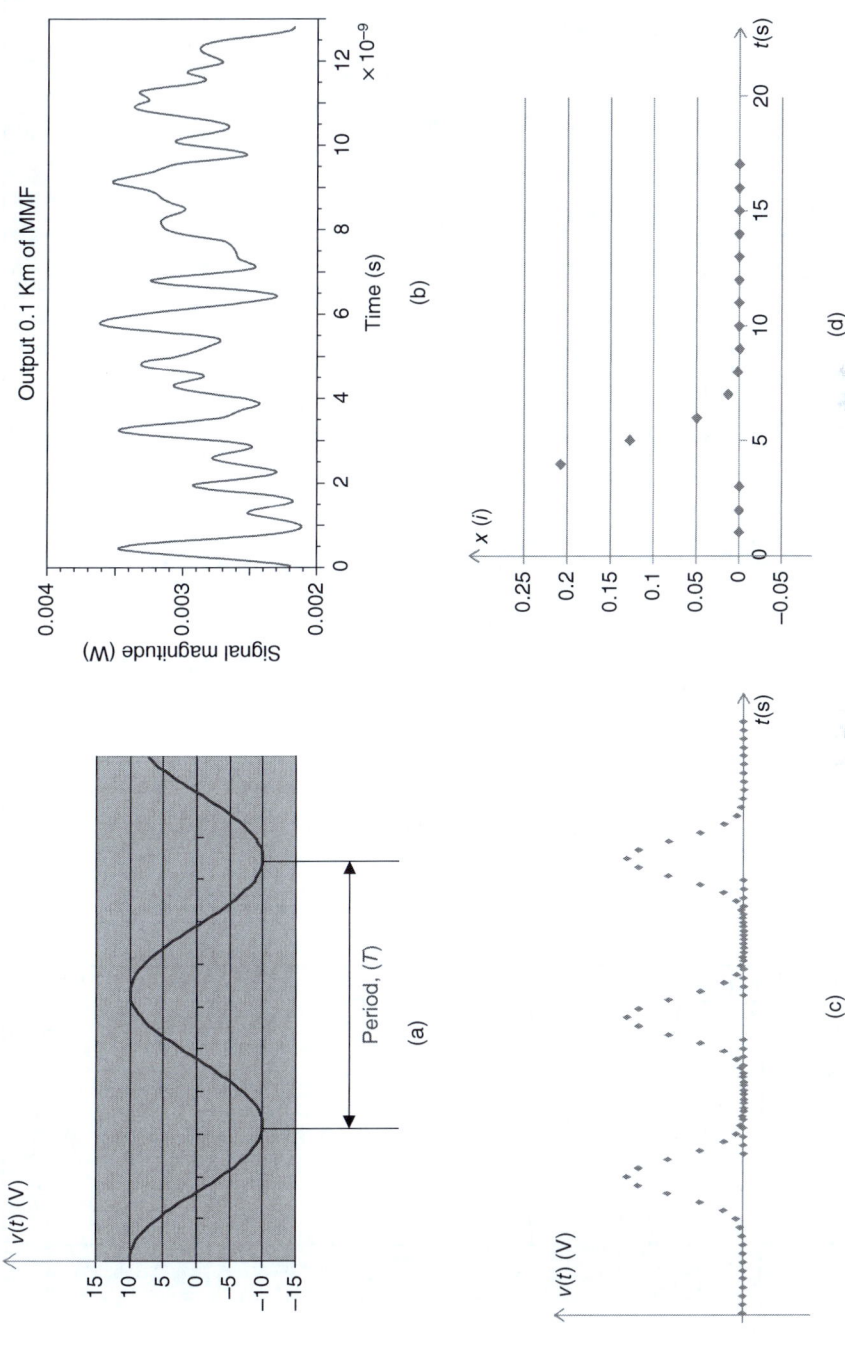

Figure 7.3.1 Signals: (a) Continuous periodic; (b) continuous nonperiodic; (c) discrete periodic; (d) discrete nonperiodic; (e) sampling of a continuous signal; (f) digitizing a sampled signal.

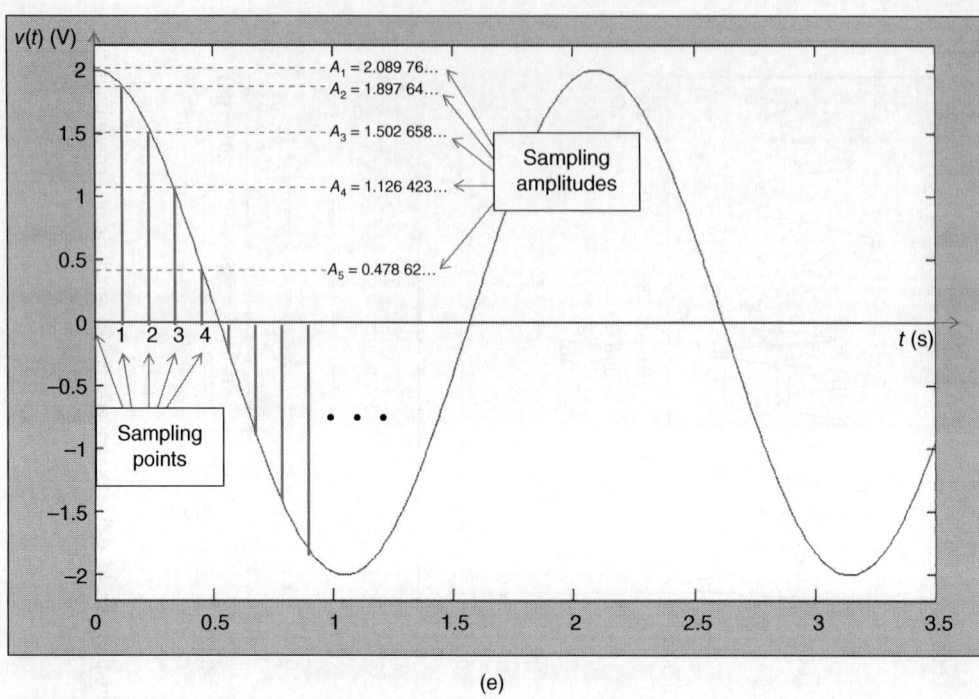

(e)

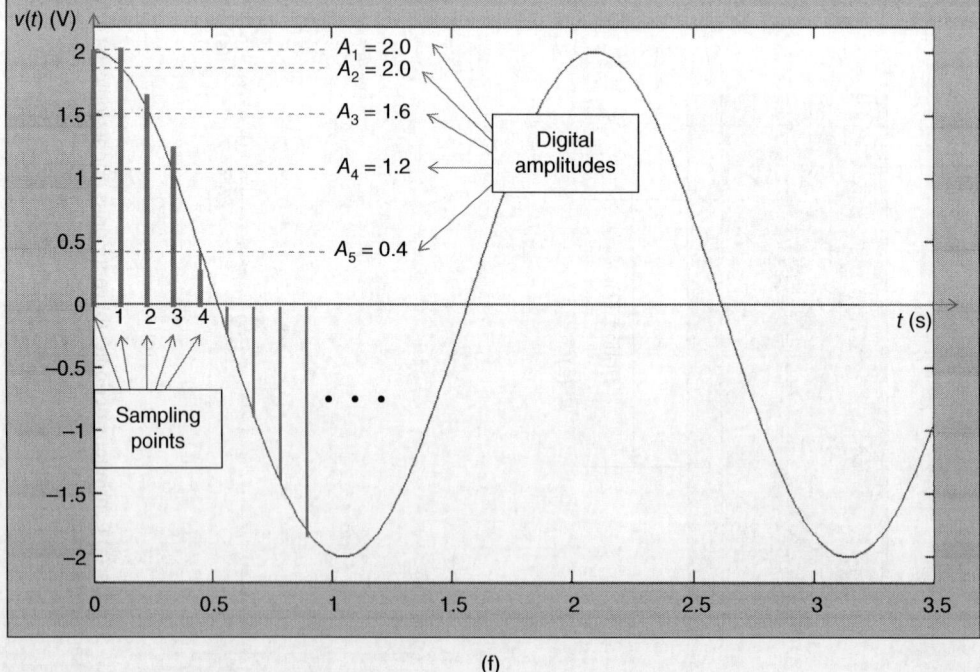

(f)

Figure 7.3.1 (*Continued*)

7.3 The Fourier Transforms and Digital Signal Processing | 675

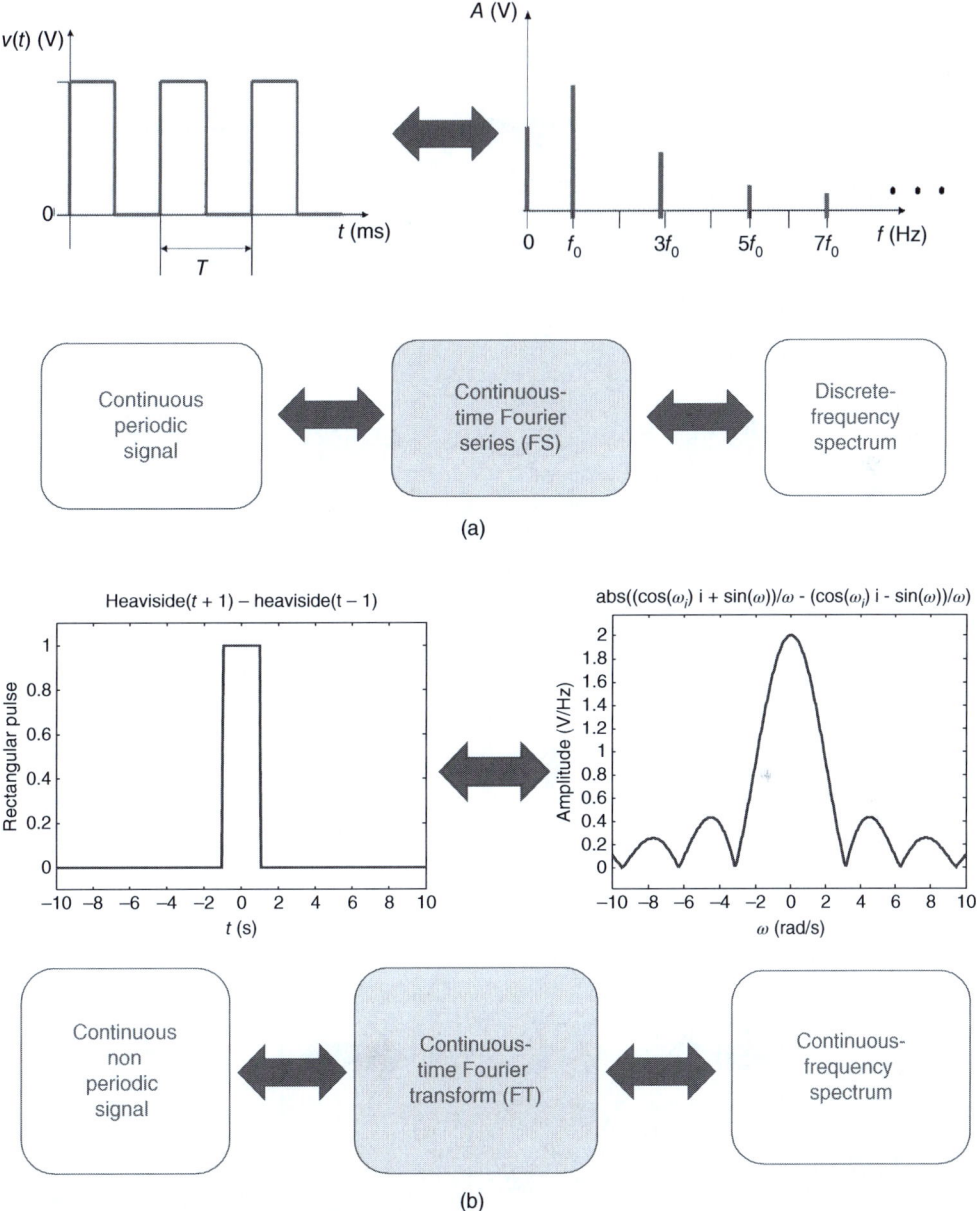

Figure 7.3.2 Signals, spectra, and the Fourier transformations: (a) The spectrum of a continuous periodic signal found with a continuous-time Fourier series (FS) is the discrete-frequency spectrum; (b) The spectrum of a continuous nonperiodic signal found with a continuous-time Fourier transform (FT) is the continuous-frequency type; (c) The spectrum of a discrete periodic signal found with a discrete Fourier transform (DFT) is the discrete-frequency kind; (d) the spectrum of a discrete nonperiodic signal found with the discrete-time Fourier transform (DTFT) is the continuous-frequency spectrum.

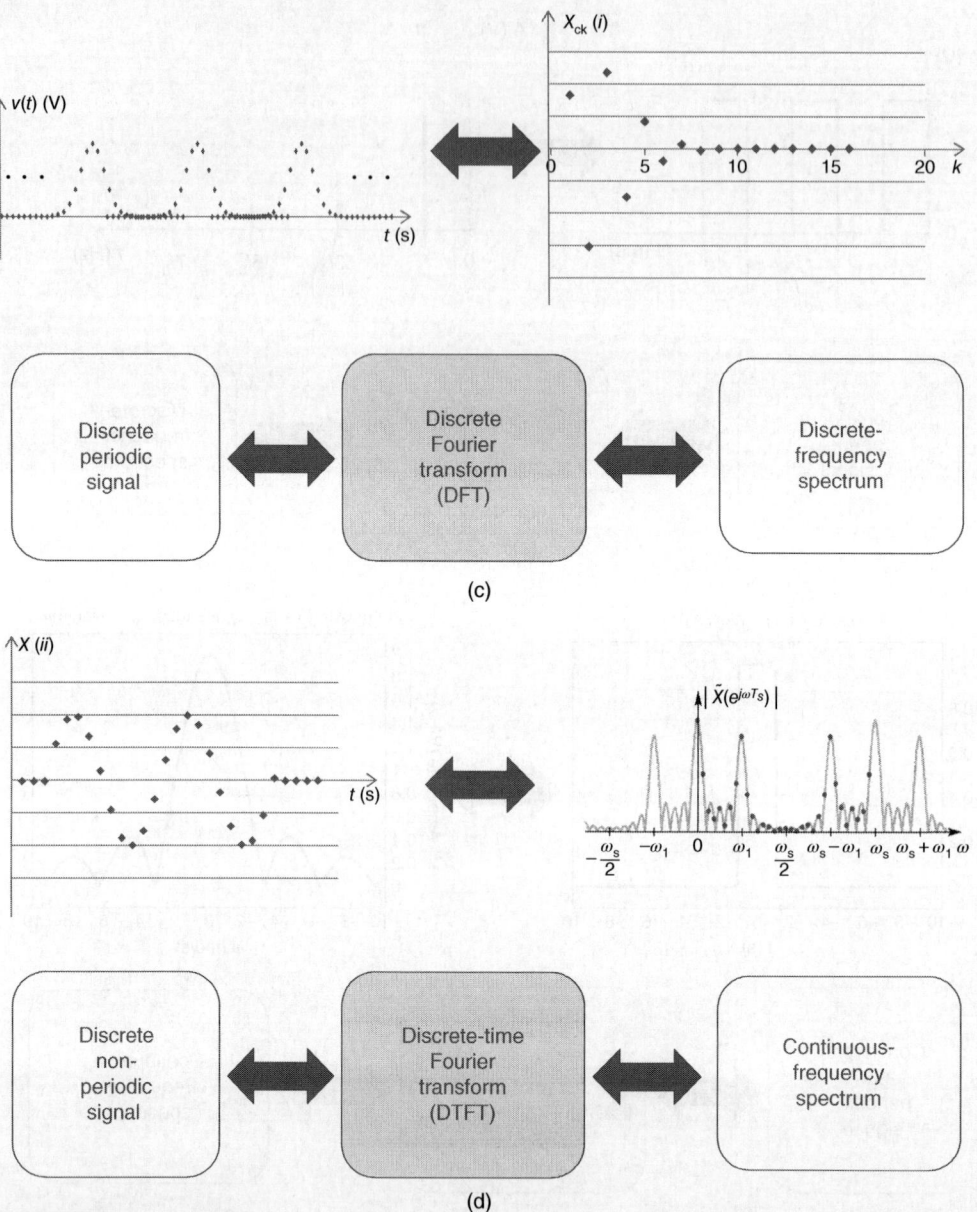

Figure 7.3.2 (Continued)

This explanation should clarify why the spectrum in Figure 7.3.2d is so rich.

What is important to bear in mind is that DFT plays the major role in DSP, as we will soon see.

As Figure 7.3.2 shows, the Fourier series (FS) and Fourier transform (FT) enable us to find the spectra of continuous signals, whereas the DFT and DTFT allow us to obtain the frequency content of discrete signals.

Do not be confused by the fact that a square wave looks like a discrete signal; it is, in fact, a time-continuous signal, but with two discontinuities per period. It is not differentiable at the corners, but this is another matter. (See our discussions of Dirichlet's condition in Sections 6.3 and 7.2.) Due to this feature, FS provides only an *approximate* decomposition of a square wave.

In this regard, it is interesting to point out that in 1807 Joseph Fourier submitted for publication his manuscript about heat propagation in solid bodies. It proposed the decomposition of a periodic signal into a sum of cosine and sine functions (the Fourier series). His paper, however, met an objection from the outstanding French mathematician Joseph-Louis Lagrange, who claimed that such decomposition was impossible for signals with corners, such as a square wave. The Fourier paper was finally published in 1822. It gave birth to a new direction in mathematics and engineering – spectral analysis – and was widely recognized as being accurate. It took another hundred years before Josiah Gibbs proved that Lagrange had been right: The Fourier series cannot *exactly* reconstruct a square wave. See our discussion of the Dirichlet conditions; also, see the subsection "The Gibbs phenomenon" in Section 6.3.2.1.

From our current discussion, we might conclude that FS and FT along with DFT and DTFT provide the comprehensive tools needed for finding the spectra of any type of signal we know today. Figure 7.3.2 specifies the use of the Fourier transforms depending on the continuity or discreteness of the signals.

It seems that the problem of finding the spectra of all existing signals can be readily solved. But we need to realize that our goal is not just to find these spectra but to use them to change the signals in a desired way. In other words, finding the spectra is only an intermediate step in achieving the ultimate goal – building the desired signals. Refer to Sidebar 7.3.S.1.

Sidebar 7.3.S.1 A Word About DSP

Consider any electronic communication device: Each and every one of them needs to process signals to provide high-quality communication. Today, we enjoy excellent quality of voice, video and data transmission regardless of distance; this quality is achieved thanks to the deep, fast processing of the signals that deliver the information. Since the processing is done digitally, it is called, appropriately, digital signal processing (DSP).

Removing or reducing the noise from the received signal by filtering is a typical example of how signal processing works. When we use digital filters for processing digital signals (and this is exactly what we do in modern electronic communication because all signals today are digitized for transmission and processing), we are engaging in *digital signal processing, DSP*. Refer to Section 5.5, to gain a broader perspective on this topic.

In general, *digital signal processing can be done in time domain, spatial domain, wavelet domain, and frequency domain*. Processing in time domain means working with one-dimensional signals. When referring to spatial domain, we imply performing the operations with multidimensional signals. Wavelets are truncated sinusoidal-like signals. Processing signals in wavelet domain means decomposing a signal on wavelets. This is similar to decomposing a periodic signal on sinusoidal signals in the Fourier series. Since wavelets carry information about both time and frequency, this decomposition allows for both time and frequency analysis. We will concentrate, however, on the frequency-domain analysis (spectral analysis, that is) because this approach is mainly used in DSP for communications.

Examples of digital signal processing are given in Figure 7.3.S.1.1. Demonstration of the result of a DSP operation is presented in Figure 7.3.S.1.1a. Two images of the same picture show the quality of a picture without DSP (left) and after DSP (right). The difference is seen clearly.

(Continued)

Sidebar 7.3.S.1 (Continued)

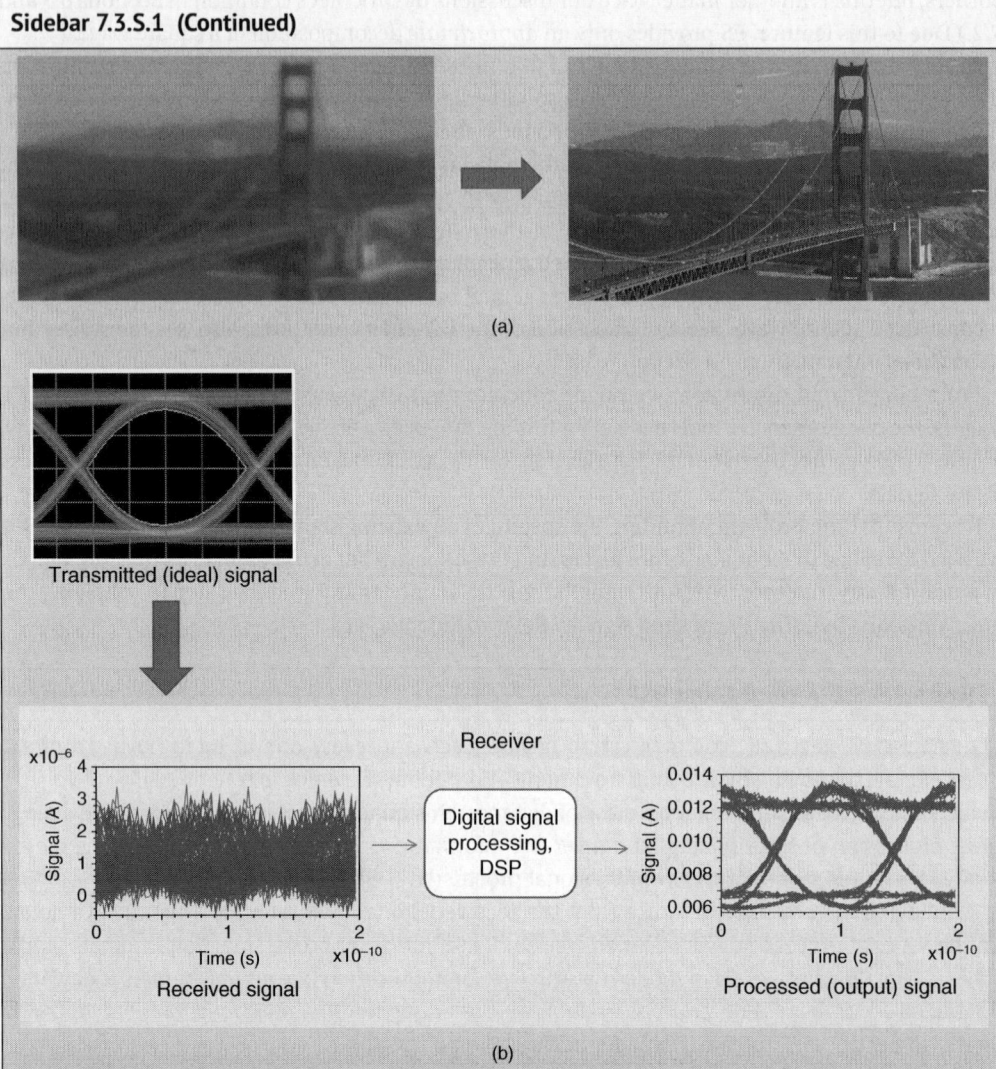

Figure 7.3.S.1.1 Examples of digital signal processing: (a) The example of a DSP operation; (b) sent, received, and processed signals in on-off keying (OOK) modulation format; (c) sent, received, and processed signals in quadrature phase-shift keying (QPSK) modulation format. Source: Figure 7.3.S.1.1a is courtesy of Mr. Wolf Perlov. Reprinted with permission.

Considering DSP at the signal level, we need to know that when a communications system uses a simple on–off keying (OOK) modulation scheme, transmission quality is evaluated by an eye diagram, which is a combination of many 1s and 0s. *The wider the "eye," the better the received signal and vice versa.* Figure 7.3.S.1.1b shows the DSP operation for such a signal (counterclockwise from the upper left corner):

- An ideal signal sent from the transmitter;
- The signal, buried in noise due to transmission flaws, arrived at the receiver;

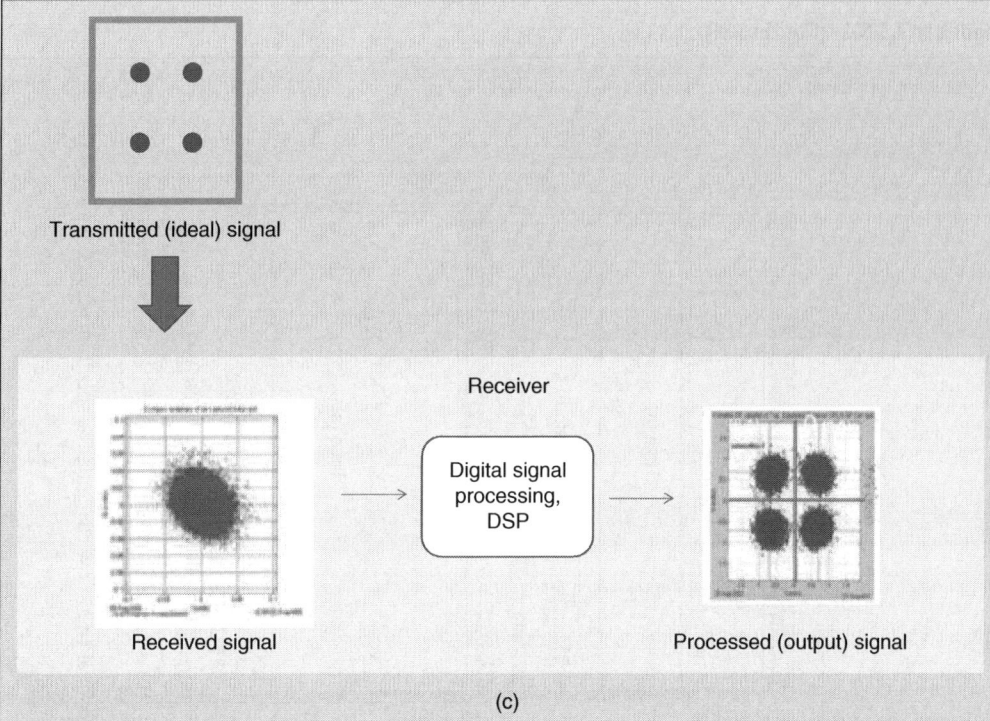

(c)

Figure 7.3.S.1.1 (Continued)

- The signal, buried in noise, is presented to a DSP system;
- The processed (DSP output) signal, close to the ideal signal, enables the reliable recovery of the information.

Figure 7.3.S.1.1c shows the signal with a more sophisticated modulation format, quadrature phase-shift keying (QPSK). In this case, the quality of transmission is evaluated by the quality of a constellation diagram, where each dot represents a symbol carrying two bits. The sequence of steps is the same as in the previous figure (counterclockwise from the upper left corner):

- An ideal signal sent from the transmitter;
- At the receiver input, this signal is represented by a large spot, which is – in fact – four merged dots due to deterioration of the signal during transmission;
- The signal, buried in noise, is presented to a DSP system;
- After being processed by the DSP system, the output signal, now close to the ideal signal (it shows four clearly distinguishable dots), enables reliable restoration of the information.

Modulation formats and associated topics are discussed in Chapters 8–10.

But what is the role of the spectra of these signals in this operation? Eventually, *we need to obtain the desired waveform of an output signal by properly changing the spectrum of the received signal*. (Refer to Chapter 6, where we consider the relationship between the waveform and the spectrum of a signal; also, see Section 6.2.2, which is particularly pertinent to this discussion.) Thus, removing, reducing, or increasing certain spectral components will modify the output signal in a desired way. This is exactly what DSP is doing. Therefore,

(Continued)

Sidebar 7.3.S.1 (Continued)

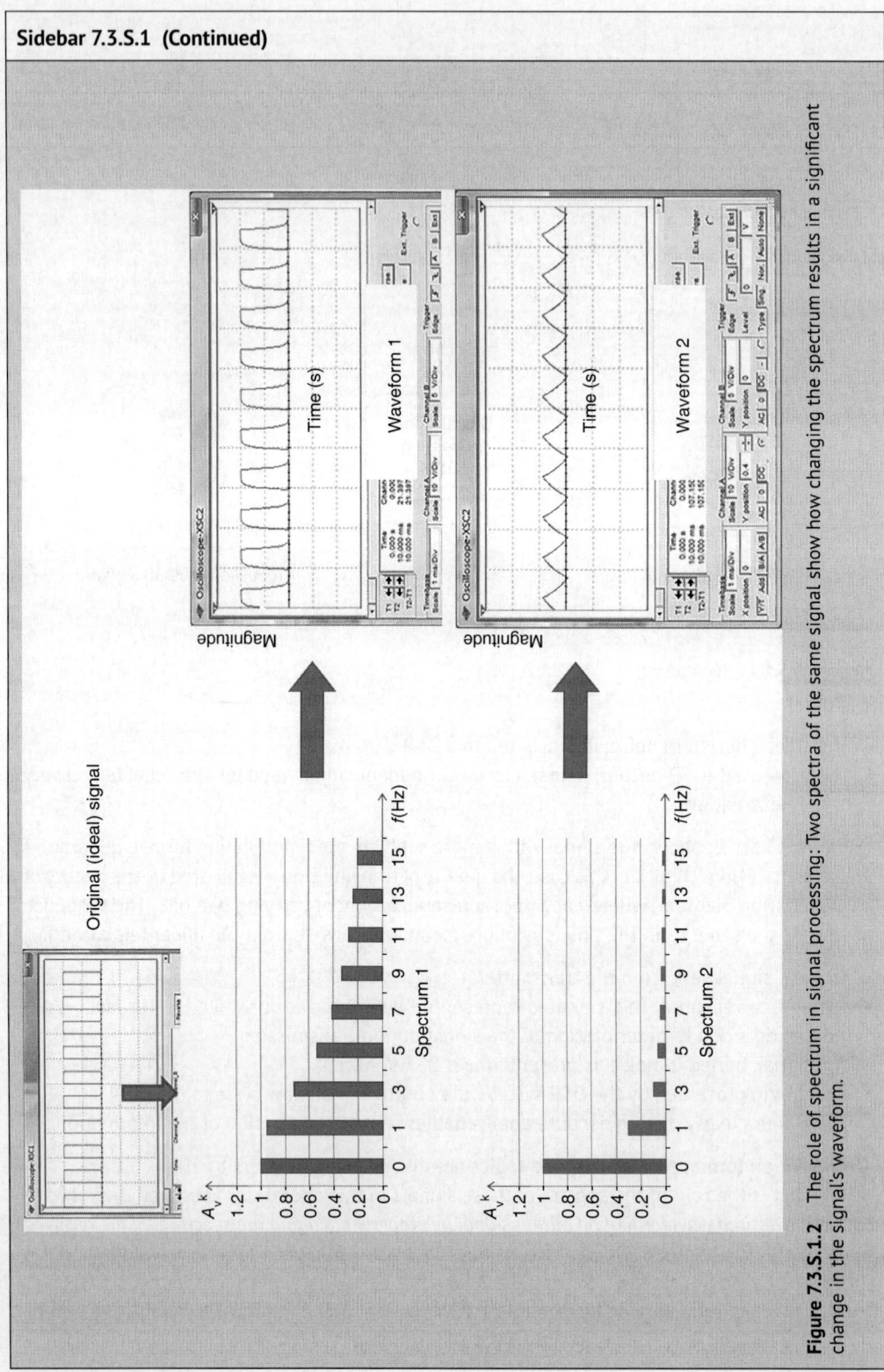

Figure 7.3.S.1.2 The role of spectrum in signal processing: Two spectra of the same signal show how changing the spectrum results in a significant change in the signal's waveform.

> finding the spectra of signals is a necessary condition for digital signal processing.
> Figure 7.3.S.1.2 illustrates this concept. It pictures the waveform of an original (ideal) square wave and two versions of its processed spectrum. The difference is clear: In spectrum 2, the amplitudes of the harmonics (spectral components) are significantly smaller than the proper amplitudes in spectrum 1. The result of such a change of the spectrum is seen in the waveforms. This simple example shows how we can control a signal's waveform (and the quality of transmission in general) by changing its spectrum.
> This brief introduction to DSP is intended to highlight the role of spectral analysis in this application.

7.3.2 Determining the Fourier Transform Required for DSP

Let us reiterate:

Digital signal processing involves *two significant steps*:

- *Obtaining the spectrum of a signal*
- *Changing this spectrum in a desired way*.

We know that to find the spectrum of a signal, we can use either the Fourier series or the Fourier transform; unfortunately, however, *there is a major problem in the practical application of these methods in DSP: processing speed*.

Modern communications delivers a huge volume of information and, therefore, it must operate at high transmission speed (transmitting capacity). Mobile communications at 1 Gb/s = 10^9 b/s and wired optical communications at 1 Tb/s = 10^{12} b/s and even at 1 Pb/s = 10^{15} b/s have become the norm, and *the signal processing of this information must be done in real time*. Indeed, a surgeon performing an emergency operation cannot wait a couple of hours to consult with a specialist who needs time to review the patient's MRI image. Even in a social situation, you do not want to wait a couple of hours to see, for example, the picture your friend has just placed on Instagram. Bottom line: *Signal processing speed must be in the same order of magnitude as the transmission speed*. To understand what kind of time scale we are talking about, we need to remind ourselves that light – the signal carrier in optical communication, which is the linchpin of modern communications – has a 200-THz frequency; that is, its period equals 5×10^{-15} s, or 5 fs! We cannot process such as signal in real time by performing the direct calculation of the Fourier series or integration of the Fourier integral, as (6.1.5a) and (7.2.1) require.

A second key problem is the *nature of modern signals*: The Fourier series and the Fourier transform were developed for working with *continuous and analog signals*, whereas today almost all signals used in electronic communication are *discrete and digital*. Therefore, we need the new versions of the Fourier operation to find the spectrum of these signals. It seems, as Figure 7.3.2c,d show, that for discrete signals we have the tools needed in the forms of the discrete-time Fourier transform (DTFT) and the discrete Fourier transform (DFT). But a closer look reveals that, in reality, only DFT can do the job. Why? Let us consider this closely.

7.3.3 Digital Signal Processing (DSP) and Discrete Fourier Transform (DFT)

7.3.3.1 The Problem: Choosing the Best Type of FT for DSP

Let us look again at the sequence of problems needed to be addressed:

- To provide high-quality electronic communication.
  ○ Obviously, this entails processing the received signals.

- To process the signals.
  - This relies on modifying the signals' spectra in a desired way, which calls for obtaining the spectra.
- To obtain the required spectra to process the signals.
  - This operation must be done at modern transmission speeds, that is, at 10^{12} b/s or even at 10^{15} b/s.
- To ensure that the signals are discrete and digital.
  - This is absolutely necessary because signal processing at high speed can be done only by using modern computerized equipment.
- To obtain the spectra of discrete signals.
  - This requires using only the DTFT and DFT types of the Fourier transforms.

Thus, we arrive at a simple *problem*: determine which type of the Fourier transforms – DTFT or DFT – works better for DSP. The solution to this problem is the subject of this subsection.

Figure 7.3.2 shows that the discrete-time Fourier transform (DTFT) works with nonperiodic time-discrete signals. Formally, DTFT is given by (McClellan et al. 2003, p. 392.)

$$X(e^{j\omega}) = \sum_{i=-\infty}^{i=\infty} x[i]e^{-j\omega i} \qquad (7.3.1)$$

where $x[i]$ is a discrete-time signal and $X(e^{j\omega})$ is a complex amplitude. It is important to stress that $X(e^{j\omega})$ is a function of *continuous* frequency ω; it is always periodic and its period always equals 2π.

It follows from this definition that an infinite number of sinusoids is required for an exact decomposition of the signals. (This is not a surprise: Just recall that the summation in a Fourier series and the limits in the Fourier-transform integral also range from negative infinity to positive infinity.) The point is that *no computer can implement an algorithm that includes an infinite number of members*. Hence, DTFT cannot be used in practical DSP systems.

Therefore, we come to the conclusion that only DFT can be employed for DSP. But, you may wonder, where can we use the FS, FT, and DTFT transforms? The answer: in our theoretical works, mathematical research, and academic exercises for a deeper understanding of the nature of the Fourier transforms and for the discovery of their unique properties.

We have chosen DFT simply by eliminating the other three types of Fourier transforms, but we need to prove the correctness of our selection. Let us briefly consider the DFT operation. Our intention is to introduce the concept of DFT, leaving the detailed description of this Fourier transform to the many specialized texts. Among them, for a clear and practical explanation, we recommend (Smith 2003, pp. 141–185.) Also, references (Miao 2007, pp. 243–256), (McClellan et al. 2003, pp. 399–402), (Chaparro 2015, pp. 437–458), (Stern and Mahmoud 2004, pp. 52–60), and (Manolakis and Ingle 2011, pp 392–398) showed be of great help in studying this topic.

7.3.3.2 How Discrete Fourier Transform (DFT) Works

Reminder About the Fourier Series Before discussing DFT, it will be helpful to briefly revisit the *Fourier series* considered in Section 6.1. There, the Fourier series given in (6.1.5b),

$$v(t) = A_0 + \sum_{n=1}^{\infty}(a_n \cos 2\pi n f_0 t + b_n \sin 2\pi n f_0 t)$$

stated that any periodic signal can be decomposed into cosines and sines with specific amplitudes and frequencies. Since each cosine and sine function represents a single frequency, *the Fourier series shows the content of frequencies of which the signal, v(t), is composed* – that is, the *spectrum* of the signal. This is what all the Fourier transforms do.

In the example with a square wave, we found that the spectrum of this time-domain signal consists of a set of sinusoids (graphically shown as lines in frequency domain) with specific amplitudes and located at the multiples of the fundamental frequency, ω_0 (rad/s) = $2\pi/T$, where T (s) is the signal's period. The amplitudes – the coefficients of the members of the Fourier series – are determined by the rules given in Eqs. (6.1.6)–(6.1.8):

$$A_0(V) = 1/T \int_0^T v(t)dt$$

$$a_n = 2/T \int_0^T v(t)\cos(2\pi n f_0 t)dt$$

$$b_n = 2/T \int_0^T v(t)\sin(2\pi n f_0 t)dt$$

In Figure 6.1.16, we reconstructed the one period of a square wave in time domain by plugging the spectral components – cosines and sines with their amplitudes – into the Fourier series (6.1.5a). In Section 6.3, we presented the Fourier series in an exponential form,

$$v(t) = \sum_{n=-\infty}^{n=\infty} (c_n e^{jn\omega_0 t}) \tag{6.3.7R}$$

The complex amplitudes, c_n, relate to the amplitudes a_n and b_n as

$$c_n = \tfrac{1}{2}(a_n - jb_n)$$

thanks to Euler's formula,

$$e^{-jx} = \cos(x) - j\sin(x)$$

In general, c_n is determined by

$$c_n = 1/T \int_0^T v(t)(e^{-jn\omega_0 t})dt \tag{6.3.10R}$$

The purpose of this reminder is to facilitate your understanding of the DFT operation, as will become apparent shortly.

DFT Explained The operation of the discrete Fourier transform (DFT) is described by (Smith (2003, pp. 152–153) and McClellan et al. (2003, pp. 392–394)):

$$X_k = \sum_{i=0}^{N-1}\left(x[i]\cdot e^{-j2\pi\frac{k}{N}i}\right) \tag{7.3.2}$$

The inverse discrete Fourier transform (IDFT) is presented by:

$$x[i] = \frac{1}{N}\sum_{k=0}^{N-1}\left(X_k e^{j2\pi\frac{k}{N}i}\right) \tag{7.3.3}$$

The inverse DFT can also be shown as a sum of real, X_{ck}, and imaginary, X_{sk}, parts:

$$x[i] = \sum_{k=0}^{N/2}\left(X_{ck}\cos\left(2\pi\frac{k}{N}i\right)\right) + \sum_{k=0}^{N/2}\left(X_{sk}\sin\left(2\pi\frac{k}{N}i\right)\right) \tag{7.3.4}$$

Here $x[i]$ is a discrete time-domain signal represented by a set of numbers, $x_0, x_1, x_2, \ldots, x_{N-1}$, which means that index i runs from 0 to $N-1$ with the increment equal to 1. These numbers are the sample values obtained by sampling the continuous signal, $x(t)$, at the equal intervals $\Delta t = 1/fs$, with fs being the sampling rate and N the total number of samples. (Refer to Section 3.2, where digitizing a continuous signal is discussed; also, see Figure 7.3.2c, where the discrete signal and the discrete spectrum with DFT are shown.) The complex amplitude, X_k, has index k running from 0 to $N-1$ and with the step equal to 1. It is broken down into the real, X_{ck}, and imaginary, X_{sk}, discrete amplitudes of the cosine and sine members, with index k running from 0 to $N/2$ and with the step equal to 1.

Equations (7.3.2)–(7.3.4) are the discrete analogies of the continuous equations of the Fourier series (6.1.5b) and (6.3.7), reproduced above. In fact, DFT is so analogous to FS that some authors call DFT a *discrete Fourier series*. We will turn to this analogy shortly.

The discrete cosine and sine here are called basis functions:

$$c_k[i] = \cos\left(2\pi \frac{k}{N} i\right) \tag{7.3.5a}$$

and

$$s_k[i] = \sin\left(2\pi \frac{k}{N} i\right) \tag{7.3.5b}$$

They are not traditional continuous functions but, rather, a set of numbers because the indexes k, i, and N are the actual numbers and, therefore, the arguments of cosines and sines are fractions of 2π. For example, for $i = 2$, $k = 5$, and $N = 32$, we have $\cos\left(2\pi \frac{5}{32} 2\right) = -0.38$ and $\sin\left(2\pi \frac{5}{32} 2\right) = 0.92$.

Hence, the right-hand side of (7.3.2) is a set of numbers, too. Thus, (7.3.2) states that

DFT is a mathematical operation that transforms one set of numbers, representing a discrete time-domain signal, into another set of numbers, representing this signal in frequency domain.

An important point should be kept in mind after reviewing (7.3.2):

> As is the case with the Fourier series, DFT decomposes the discrete time-domain signal into a set of cosine and sine members. Since each cosine and sine represents a single frequency, such decomposition means that, applying DFT, we find the frequency content of the given signal; that is, we find its spectrum.

The description of the DFT operation must be completed by presenting the amplitudes X_{ck} and X_{sk}. They are given by (Smith 2003, pp. 153–160)

$$X_{ck} = \frac{1}{N/2} \sum_{i=0}^{N-1} \left(x[i] \cos\left(2\pi \frac{k}{N} i\right)\right) \tag{7.3.6a}$$

and

$$X_{sk} = -\frac{1}{N/2} \sum_{i=0}^{N-1} \left(x[i] \sin\left(2\pi \frac{k}{N} i\right)\right) \tag{7.3.6b}$$

Here, index i runs from 0 to $N-1$ and index k runs from 0 to $N/2$, as in (7.3.4). Amplitude X_{ck} is calculated differently at two extreme points, $k = 0$ and $k = N/2$, namely

$$X_{c0} = \frac{1}{N} \sum_{i=0}^{N-1} \left(x[i] \cos\left(2\pi \frac{0i}{N}\right)\right) = \frac{1}{N} \sum_{i=0}^{N-1} (x[i]) \tag{7.3.7a}$$

$$X_{cN/2} = \frac{1}{N} \sum_{i=0}^{N-1} \left(x[i] \cos\left(2\pi \frac{Ni}{2N}\right)\right) = \frac{1}{N} \sum_{i=0}^{N-1} (x[i] \cos(\pi i)) \tag{7.3.7b}$$

Table 7.3.1 Comparison of the Fourier series and the discrete Fourier transform.

The Fourier series (FS)		The discrete Fourier transform (DFT)	
Time domain	Frequency domain	Time domain	Frequency domain
Independent variable – time, t (s)		Independent index – sampling number, i	
	Independent variable – frequency, f (Hz) or ω (rad/s)		Independent index – frequency index, k
Signal – continuous-time periodic, $v(t)$ [V]		Signal – discrete-time periodic, $x[i]$	
	Transform – sum, including $f = 0$, of cosines and sines with specific amplitudes and frequencies		Transform – sum, including $k = 0$, of discrete cosines and sines with specific amplitudes and indexes i and k
	Frequencies – multiples (nf_0) of the fundamental frequency, $f_0 = 1/T$, where T is the signal's period		Frequencies – discrete frequencies, $f_k = \frac{k}{N} f_s$, where f_s is the sampling frequency

Tip: Analyze this table, reread the discussions of Eqs. (7.3.1)–(7.3.3), and try to work out by yourself answers to questions that may arise in studying this topic.

Amplitude X_{sk} turns to zero at these two points.

It is instructive to compare the Fourier series (7.1.5) with the discrete Fourier transform (7.3.2). The similarity is that in both equations, the left-hand side is the time-domain signal being processed and the right-hand side is the set of cosines and sines onto which the given time-domain signal is decomposed. Calculations of the coefficients – a_n, b_n and X_{ck}, X_{sk} – are conceptually similar, too. The main difference is that (7.1.5) describes the continuous signal, whereas (7.3.2) deals with the discrete signal. Also, there is a difference in a constant (dc) member and in the range of summation. Nevertheless, it is useful to refer to the analogy with the Fourier series when trying to understand the DFT concept. Table 7.3.1 compares these two Fourier transformations.

(**Question:** *We have shown that only DFT can work for DSP. However, DFT works only with periodic signals, whereas the majority of real-life signals are non-periodic. How is this contradiction resolved?*)

Let us consider an example of finding the DFT of a specific time-domain signal.

Example 7.3.1 Finding the DFT of a Train of Gaussian Pulses

Problem

Find the DFT of the train of Gaussian pulses given as

$$v(t) = e^{-2.77\left(\frac{t-b}{c}\right)^2} \tag{7.3.8}$$

where $b = T/2$ and $c = 0.5$. This function is defined on one period, T, and repeats itself thereafter; it is shown in Figure 7.3.3a.

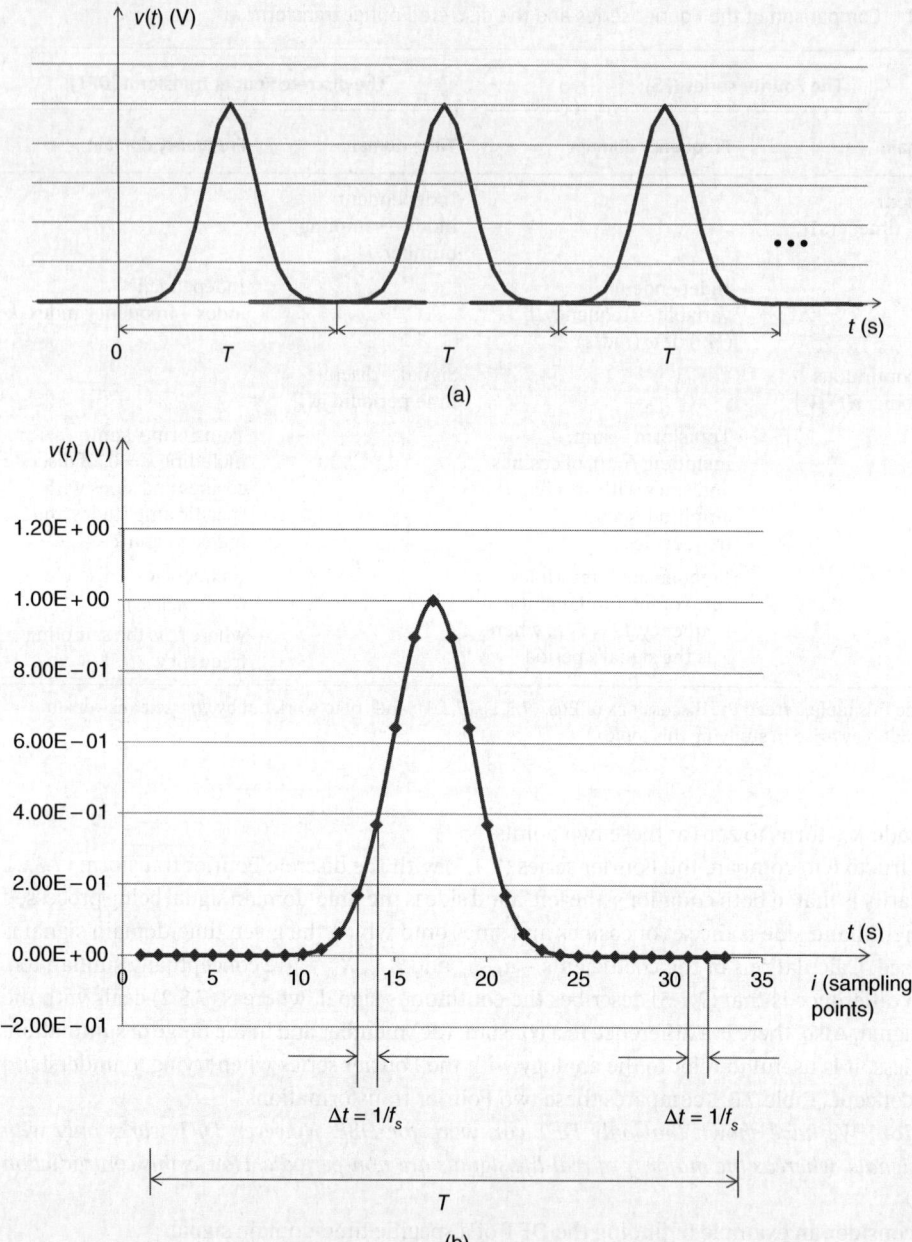

Figure 7.3.3 (a) Gaussian pulse train and (b) sampling of a Gaussian pulse.

Solution

Follow these steps:

1. Sample this continuous signal over one period: Set the sample numbers (sample points) at equal intervals, $\Delta t = 1/f_s$, and measure the amplitudes at these points.

- This sampling is done with MS Excel. The sampled signal with $N = 32$ sample points is displayed in Figure 7.3.3b; hence, *index i (the sample numbers) runs here from 0 to 31*.

- Collect the sample values (amplitudes at the sampling points) obtained when sampling.
- The set of these numbers, along with their indexes i's (sample numbers), is shown in the time-domain column of Table 7.3A.1. We must realize that instead of a continuous time-domain signal, shown in Figure 7.3.3a, we now have the set of numbers – data, that is – shown in Table 7.3A.1.
- Apply the DFT given in (7.3.2) to the discrete time-domain signal (data) given in Table 7.3A.1. To do so,
  (a) Calculate amplitudes X_{ck} and X_{sk} for each k, according to (7.3.6a) and (7.3.6b), where *index k runs from 0 to 16 and index N from 0 to 32*.
  (b) Calculate basis functions $c_k[i] = \cos\left(2\pi \frac{k}{N} i\right)$ and $s_k[i] = \sin\left(2\pi \frac{k}{N} i\right)$ for each index, i.
  (c) Multiply each basis function by its corresponding amplitude, such as $A_1 \cdot c_1[i]$, $A_2 \cdot c_2[i]$, ..., $B_1 \cdot s_1[i]$, $B_2 \cdot s_2[i]$, ...

This is how we execute the above calculations:

Calculations for Point 3a: According to (7.3.6a), we need to find the set of amplitudes X_{ck}^i for one k corresponding to every i from 0 to $N-1$, which is equal to 31 in this example. We then sum them up to obtain X_{ck}. For example, for $k = 0$, we find $X_{c0}^0, X_{c0}^1, X_{c0}^2, \ldots, X_{c0}^{31}$, sum them up, and obtain X_0. We repeat the procedure for all other ks and find $X_{c1}, X_{c2}, X_{c3}, \ldots, X_{c15}$. The following are examples of calculation of X_{c1}^1 and X_{c1}^2:

$$X_{c1}^1 = x[1] * \cos\left(2\pi \frac{1}{32} 1\right) = (1.49 \times 10^{-11} \times 0.98) = 1.46 \times 10^{-11} \quad \text{for } k = 1 \text{ and } i = 1$$

$$X_{c1}^2 = x[2] * \cos\left(2\pi \frac{1}{32} 2\right) = (3.70 \times 10^{-01} \times 0.92) = 3.42 \times 10^{-10} \quad \text{for } k = 1 \text{ and } i = 2$$

Similarly, we can find amplitudes $X_{s0}^0, X_{s0}^1, X_{s0}^2, \ldots, X_{s0}^{15}$ and $X_{s1}, X_{s2}, X_{s3}, \ldots, X_{s15}$.

Table 7.3A.2 shows the values of amplitudes X_{c0}^i, X_{c1}^i, and X_{s1}^i, X_{s2}^i, for all *is* along with the values of X_{c0}, X_{c1} and X_{s1}, X_{s2}. Figure 7.3.4 visualizes the distribution of all X_{ck} and X_{sk} amplitudes vs. frequency index k.

Remember: Amplitudes X_{c0} and $X_{cN/2}$ must be calculated differently, according to (7.3.7a) and (7.3.7b), and amplitudes X_{s0} and $X_{sN/2}$ are equal to zero. Samples of these calculations are presented in Table 7.3A.2.

Calculations for Point 3b: Examples of three calculated basis functions vs. sampling number i for $k = 1$, $k = 2$, and $k = 8$ are shown in Figure 7.3.5.

Calculations for Point 3c: Examples of these multiplications for $k = 1$, $k = 2$, and $k = 8$ are shown in Figure 7.3.5.

4. These steps completed, the problem is solved.

Discussion

1. *Calculations:* A word to the wise: To understand how these calculations work, perform some of the ones described in this example. They are easy to do it with MS Excel, but, of course, you can use any computer tool. Such an exercise would also help you to better comprehend the meaning of all the DFT operations.
2. *Amplitudes X_{ck} and X_{sk}*: Review Figure 7.3.4, where amplitudes X_{ck} are the points (numbers) at the specific location of the frequency axis represented here by index k. Compare this figure with the spectrum of a square wave shown in Figure 6.1.9 and observe their similarity. Note, too, that amplitudes X_{sk} in this example are equal to zero. We met similar situations, you will recall, in discussing the Fourier series of some periodic functions in Section 7.1.

688 | *7 Spectral Analysis 2 – The Fourier Transform in Modern Communications*

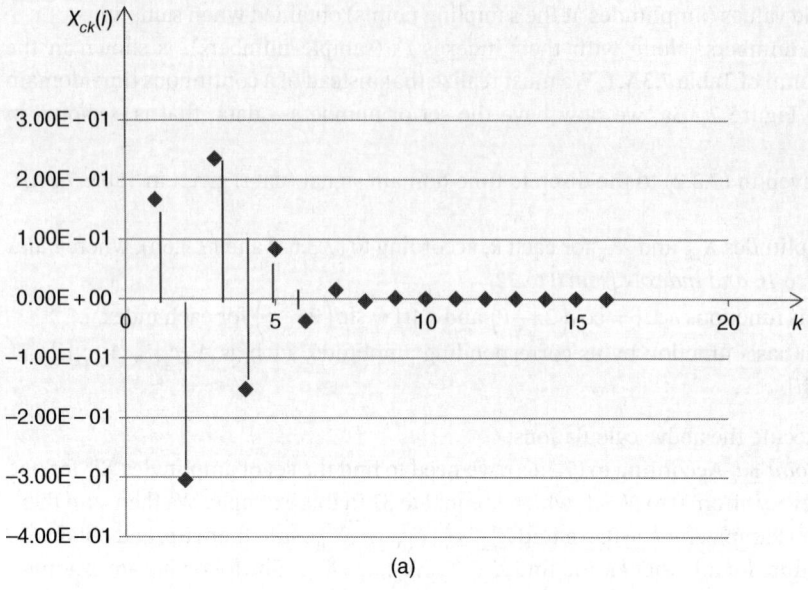

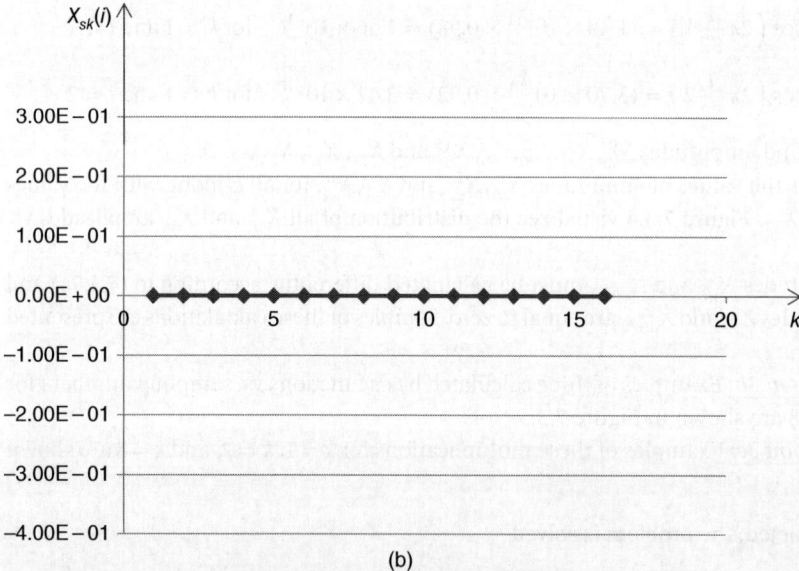

Figure 7.3.4 (a) Amplitudes X_{ck} and (b) X_{sk} vs. index k.

3. *Basis functions and index k:* Pay particular attention to Figure 7.3.5 because it shows the role of the *frequency index, k*: This index determines the number of cycles of a cosine or sine. Thus, by presenting a signal as a combination of these basis functions, we show the frequencies of this signal; that is, we perform a Fourier (spectral) analysis of the signal. To discern why index k represents the frequency domain, we need to be able to explain why the parameter $k/N = f/f_s$ (with N being the total number of sampling points and f_s being the sampling frequency) is sometimes called *frequency*, though it is dimensionless.

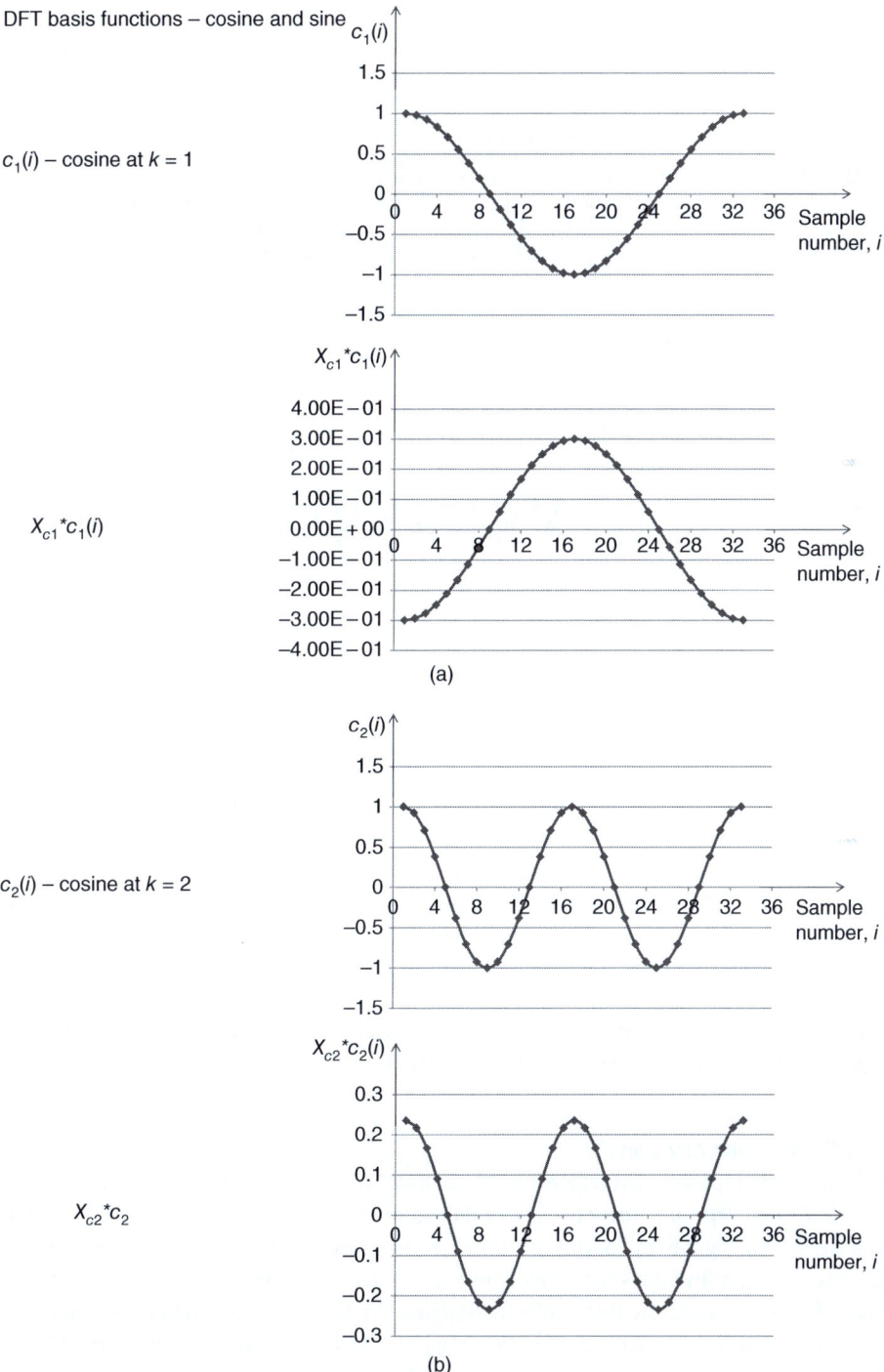

Figure 7.3.5 Some basis cosine functions, $c_k[i]$, and cosine members, $A_k \cdot c_k[i]$, of the discrete Fourier transform of the Gaussian pulse in Example 7.3.1: (a) $k = 1$; (b) $k = 2$; (c) $k = 8$.

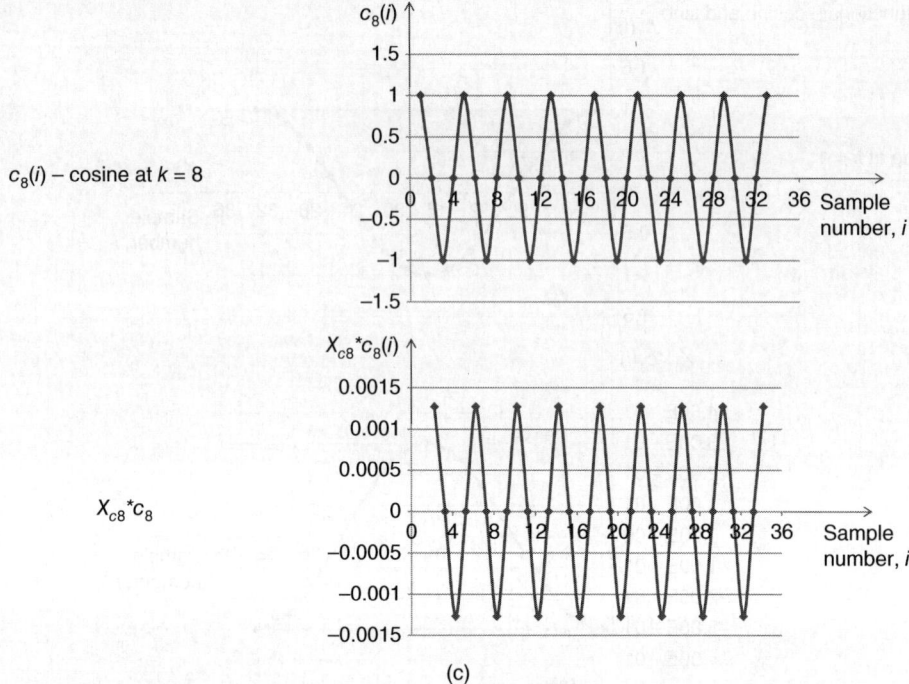

Figure 7.3.5 (Continued)

4. *Take special note*: We must understand that we work in discrete space; therefore, strictly speaking, all the graphs in Figure 7.3.5 must be shown as a set of dots (numbers, that is). The *continuous lines connecting these dots are shown only to illustrate* that these numbers represent cosine and sine functions. Figure 7.3.6 demonstrates this point, showing both versions – a set of dots and the same set connected by continuous lines – for comparison.
5. *Synthesis of a time-domain signal*: If we want to reconstruct (synthesize) the time-domain signal from its spectrum (from its DFT, that is), we need to plug into the right-hand side of (7.3.8) all the found members and perform summation. This would be a good exercise to undertake in order to gain a fuller understanding of the nature of DFT.

7.3.3.3 Can DFT Work with Any Signal?

It seems we have a serious problem here: *We have stated that DFT can work only with periodic signals and that DFT is the only type of Fourier transform that can be used in DSP*. But we know, too, that the vast majority of real signals in communications are nonperiodic. Can DFT work with them? The answer: Yes, DFT can find the spectrum of any type of signal and here is how:

First, we need to understand the difference in applying the Fourier transforms to periodic and nonperiodic signals. Let us start with a periodic signal to which the Fourier series and DFT are applied. We recall that *a periodic signal repeats itself after each period and ranges from negative infinity to positive infinity*. Consider the Fourier series that operates over a periodic signal: The summation of its harmonics (spectral components) is performed from $-\infty$ to ∞, whereas *calculations of the amplitudes of these harmonics are limited by the period of the signal*. (See Eqs. (7.1.5)–(7.1.8), reproduced at the beginning of this section.) This is where the periodicity of a signal comes into

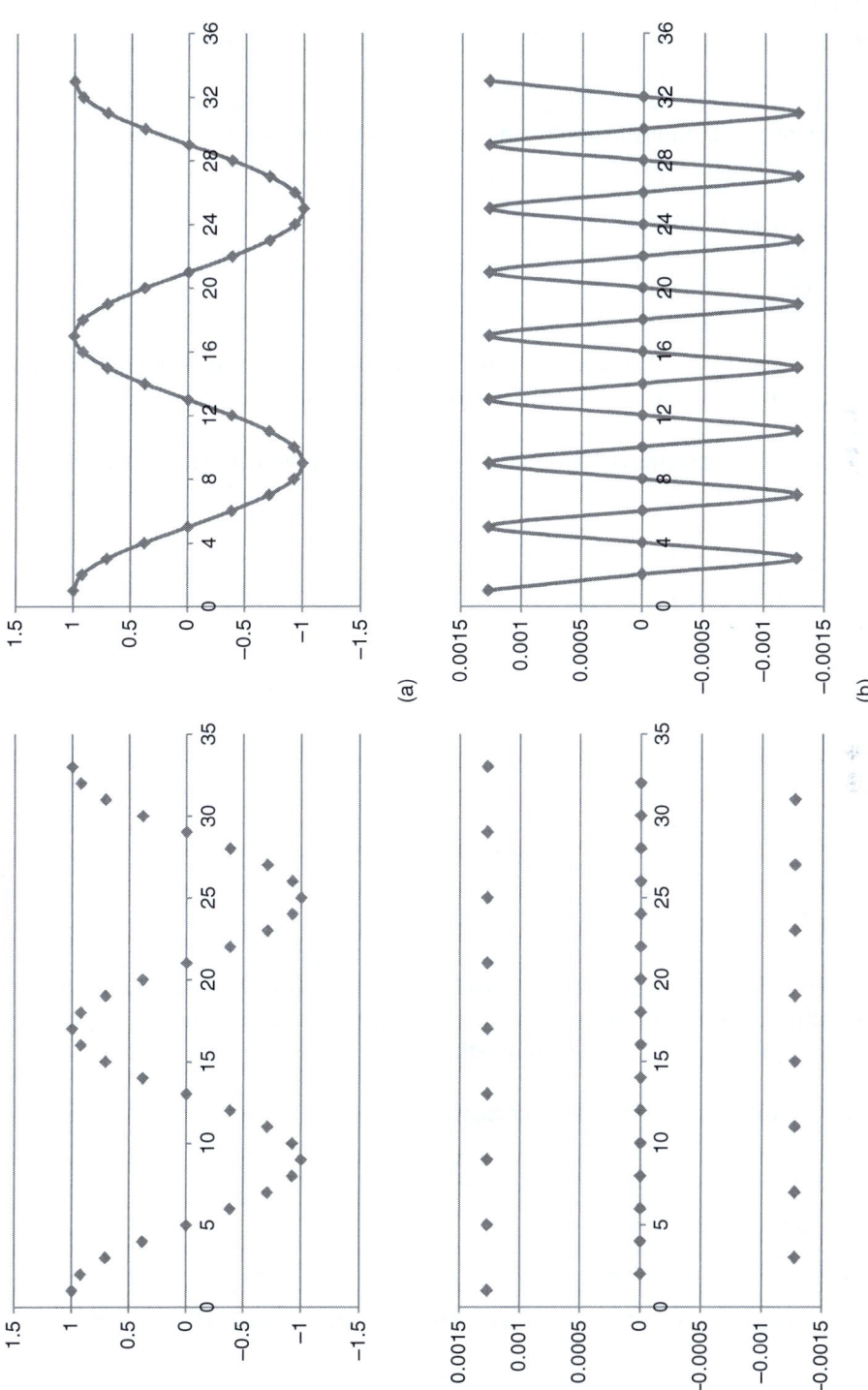

Figure 7.3.6 Basis cosine functions, $c_k[i]$, presented as sets of dots (left-side charts) and sets of dots connected by solid lines (right-side charts) for (a) $k = 2$ and (b) $k = 8$.

a play! In DFT, as (7.3.6a) and (7.3.6b) show, the summation in calculations of amplitudes also is performed over a finite range of samples. This is analogous to a finite time interval. Thus, *the periodicity of a signal results in calculations of spectral components over a finite interval.*

Finding the spectrum of a nonperiodic continuous signal calls for calculations over infinite time intervals, as (7.2.1) shows for the Fourier transform. Similarly, for a discrete signal, the sampling is performed from negative infinity to positive infinity, as (7.3.1) shows for DTFT. These equations are duplicated here for your convenience:

$$F(\omega) = \int_{-\infty}^{\infty} v(t)(e^{-j\omega t})dt \tag{7.2.1R}$$

$$X(e^{j\omega}) = \sum_{i=-\infty}^{i=\infty} x[i]e^{-j\omega i} \tag{7.3.1R}$$

This infinity appears because *a nonperiodic signal does not have any natural limitations*, such as a period for a periodic signal. This is why DFT cannot be directly used for finding the spectrum of a nonperiodic signal.

In short, the problem of applying DFT to finding the spectrum of a nonperiodic signal boils down to the following contradiction:

> On the one hand, DFT can be applied only to a finite signal; on the other hand, a nonperiodic signal is not finite. So here is the solution: Since we cannot change DFT, let us limit a nonperiodic signal and make it finite.

How can we impose limitations on a nonperiodic signal? In some cases, it is almost natural. Consider Example 7.3.1 and imagine that the given Gaussian pulse is a single one. Then, the interval T shown in Figure 7.3.1 seems to be a natural limitation of this signal, though – mathematically speaking – such a single pulse goes from $-\infty$ to ∞. Thus, we truncate a signal and consider the parts left over negligible. The same is true for a single rectangular pulse, such as that considered in Example 7.1.2. In other cases, the situation is more complicated. For instance, for the signal shown in Figure 7.3.1b, we cannot find any natural limits. In this case, we simply need to decide how to truncate a signal to preserve most of its content. Remember, we cannot work with infinite signals under any circumstances; hence, we must truncate them to process them. Do we change the spectrum of a signal by truncating the signal? Yes, of course, but we can save the main features of the signal's spectrum. How much? This depends on how much of the signal's content we preserve. The following example provides some insight into this issue.

Example 7.3.2 Truncating (Windowing) a Signal[5]

Problem

1. Given a sinusoidal signal,

    $$v(t) = Vo + V_1 \cos(\omega_1 t + \theta)$$

    where $Vo = 0.4$ V, $V_1 = 1$ V and $\omega_1 t = 4\pi \approx 12.6$ rad/s, find the best way to truncate it.

---

[5] After McClellan et al. (2003, pp. 393–396).

Solution

Since we were not given any rules or guidelines to determine which truncation method is best, we need to try various versions of truncation and then decide which one to choose.

To truncate a sinusoidal signal means to limit it from the left and the right, which is equivalent to multiplying the sinusoidal signal and a rectangular window. The rectangular window and its amplitude spectrum – calculated with MATLAB – along with the sinusoidal signal and its amplitude spectrum are shown in Figure 7.3.7a,b, respectively. To solve the problem, we will window (truncate) the *given* sinusoidal signal by multiplying the signal by windows of various widths and we will find its amplitude spectrum by applying the Fourier transform. These operations are executed with MATLAB. An example of the MATLAB code is given in the discussion section.

Analyzing Figure 7.3.7c–f, we can now conclude that there is no such thing as the best truncation method; in fact, we can choose any windowed signal, depending on the requirement of the signal's application. Problem solved!

Discussion

Let us pause here to highlight the following important points:

1. By truncating a signal – and we always operate with truncated signals in computerized world – we inevitably change the form of the signal's spectrum. Instead of lines (δ-functions, that is) of the spectrum of an ideal signal, as in Figure 7.3.7b, we have the bell-curve peaks of various widths, as shown in Figure 7.3.7c–f for the windowed signals.
2. The implication of Point 1 is that by windowing a signal, we introduce some uncertainty in determining its spectrum. Indeed, the spectrum of an ideal (infinite) sinusoidal signal can be determined precisely by three numbers: -4π, 0, and 4π. The spectra of all truncated signals can be determined with various levels of accuracy but only approximately.
3. Figure 7.3.7 gives us a hint as to what truncating window width we should choose. If we truncate a long piece of sinusoid, as shown in Figure 7.3.7c with $T_W = 16$ s, then its spectrum will resemble the ideal spectrum of Figure 7.3.7b with its well-pronounced peaks; the drawback of this spectrum, however, is that it does not show the small spectral components that can be seen in the other versions of this spectrum shown in Figure 7.3.7d–f. If we window a short piece of the sinusoid, as shown in Figure 7.3.7c with $T_W = 1$ s, then its spectrum does not resemble the ideal spectrum at all, but we can see all the details of the spectrum. We can conclude, therefore, that *the spread of a function in time domain and the spread of its spectrum (Fourier transform) are inversely proportional*
and
the more detailed the waveform (the plot of a signal in time domain), the less detailed the plot of its spectrum and vice versa.
Our ability to determine the details of a plot is often called *resolution*. Hence, point 3b states that the higher the time-domain resolution of a signal, the lower its frequency-domain resolution, and vice versa. We mentioned already that this point is often referred to as the *principle of uncertainty of Fourier analysis*.
4. Since we need a finite signal for the DFT operation, we always have to truncate any signal presented to DFT. Therefore, *when using a computer, we will always have an approximate, not exact, spectrum of a signal*. In Figure 7.3.7a, the spectrum of an ideal sinusoidal signal is drawn manually – and it is precise; the spectra of all windowed signals, shown in Figure 7.3.7c–f, are calculated by computer, and they are all approximate.

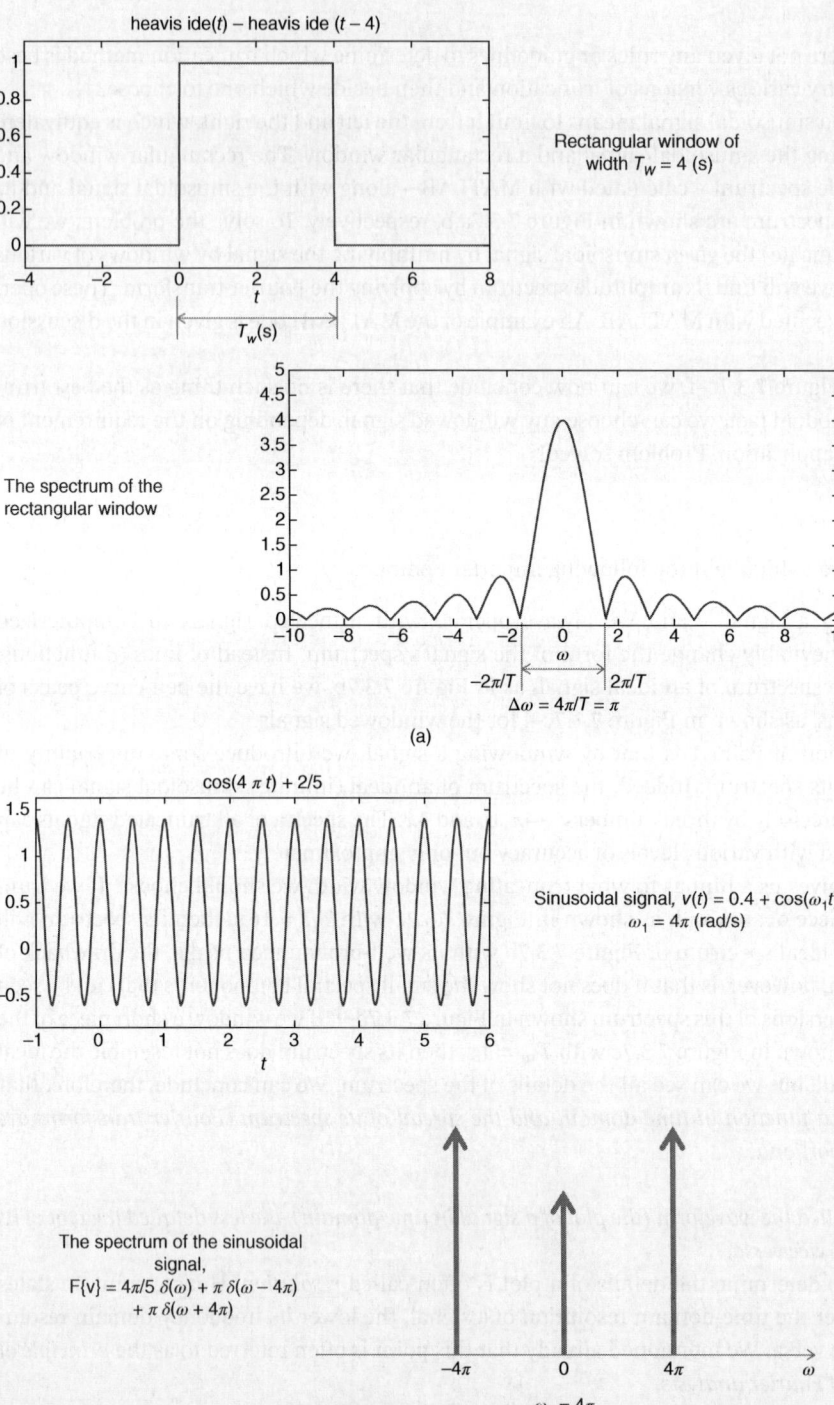

Figure 7.3.7 The waveforms and the spectra of (a) a rectangular window; (b) a given sinusoidal signal; (c) a windowed sinusoidal signal with window width $T_W = 16$ s; (d) a windowed sinusoidal signal with $T_W = 4$ s; (e) a windowed sinusoidal signal with $T_W = 1$ s; (f) a windowed sinusoidal signal with $T_W = 0.2$ s.

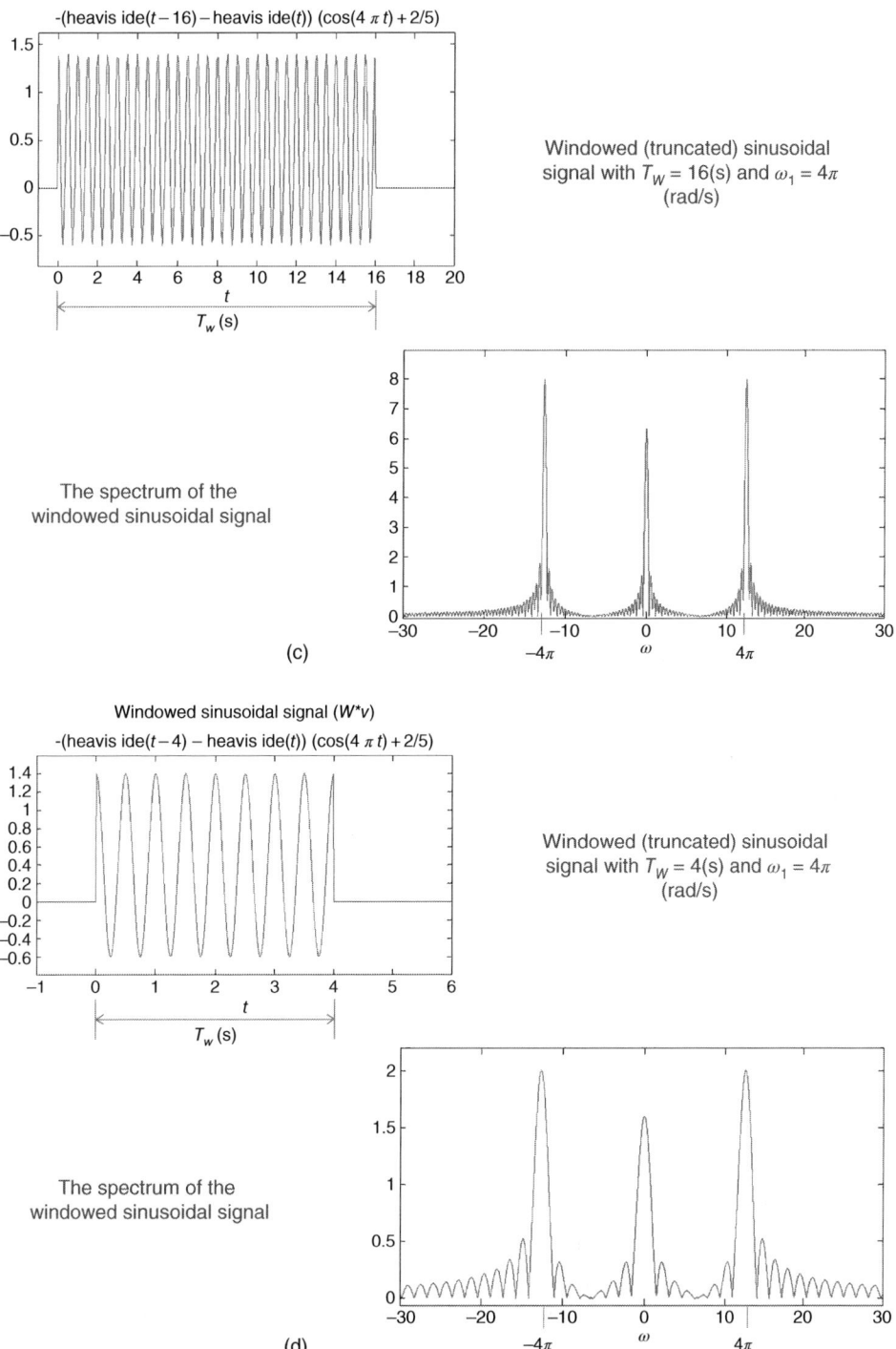

Figure 7.3.7 (Continued)

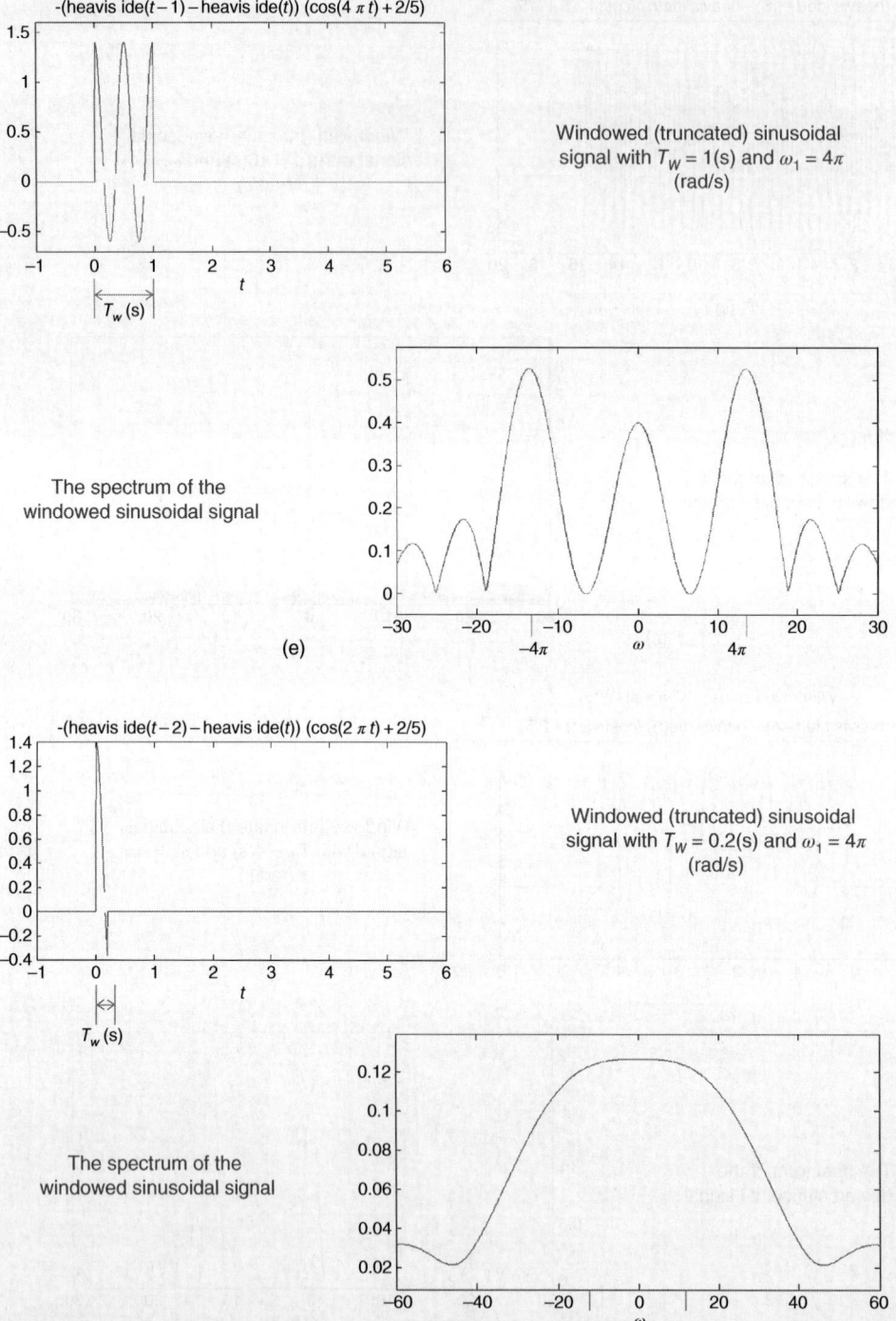

(e) Windowed (truncated) sinusoidal signal with $T_W = 1$(s) and $\omega_1 = 4\pi$ (rad/s)

The spectrum of the windowed sinusoidal signal

(f) Windowed (truncated) sinusoidal signal with $T_W = 0.2$(s) and $\omega_1 = 4\pi$ (rad/s)

The spectrum of the windowed sinusoidal signal

Figure 7.3.7 (*Continued*)

5. An example of MATLAB code is presented beneath. You can change this code to calculate the various versions of waveforms and their spectra. You can also select the parts of the code to generate separately the waveforms and the signals.

```
≫syms w t                              %start symbolic toolbox
≫ v= 0.4 + cos(4*pi*t) ;               %build a sinusoidal signal
≫W=sym('heaviside(t)-heaviside(t-4)'); %build a rectangular window of width 4
≫WV=W*v;                               %window truncate) the sinusoidal signal
≫F=fourier(WV);                        %find the Fourier transform of the truncated signal
≫AF=abs(F);                            %find the amplitude spectrum of F
≫ ezplot(AF,[−1,6])                    %plot the obtained spectrum
```

7.3.4 Relationship Among All Fourier Transforms

Review Figure 7.3.2 again, which shows that for every type of signal there is a proper Fourier transform. But all the signals, though looking different, are in fact deeply connected. Indeed, we show in Section 7.1 that a nonperiodic single rectangular pulse can be obtained from a periodic pulse train by tending the period to infinity. Also, it is not difficult to understand that a discrete signal becomes a continuous one when the time interval between the adjacent points approaches zero. If the signals have underlying connections, do their Fourier transforms relate to one another? Yes, they do and here is how:

Let us recall that in Section 7.1 we showed that the Fourier transform can be derived from the Fourier series by performing the following manipulations:

- Start with the Fourier series given in (6.3.7) as $v(t) = \sum_{n=-\infty}^{n=\infty} (c_n e^{jn\omega_0 t})$.
- Plug the coefficients, $c_n = \frac{1}{T} \int_t^{t+T} v(t)(e^{-jn\omega_0 t})dt$, into the Fourier series, (6.3.7) and find

$$v(t) = \sum_{n=-\infty}^{n=\infty} \left(\frac{1}{T} \int_t^{t+T} v(t)(e^{-jn\omega_0 t})dt \cdot e^{jn\omega_0 t} \right) \qquad (7.1.2R)$$

- Rewrite (7.1.2) with reference to the interval between two adjacent spectral lines,

$$\Delta\omega = (n+1)\omega_0 - (n)\omega_0 = \omega_0 = 2\pi/T \qquad (7.1.3R)$$

to find

$$v(t) = \sum_{n=-\infty}^{n=\infty} \left(\frac{1}{2\pi} \int_t^{t+T} v(t)(e^{-jn\omega_0 t})dt \cdot e^{jn\omega_0 t} \right) \Delta\omega \qquad (7.1.4R)$$

- Keep increasing period T so that the incremental spacing, $\Delta\omega$, turns to differential, $d\omega$; that is, when $T \to \infty$, then $\Delta\omega \to d\omega$. In the limit, *discrete harmonics, $n\omega_0$, change to continuous frequency, ω, and the summation is replaced by integration.*
- Turn (7.1.4) and (7.1.5) by tending $T \to \infty$,

$$v(t) = \frac{1}{2\pi} \int_{-\infty}^{\infty} \left[\int_{-\infty}^{\infty} v(t)(e^{-j\omega t})dt \right] e^{j\omega t} d\omega \qquad (7.1.5R)$$

whose core is called the *Fourier transform*

$$F\{v(t)\} = F(\omega) = \int_{-\infty}^{\infty} v(t)(e^{-j\omega t})dt \qquad (7.1.6aR)$$

698 | *7 Spectral Analysis 2 – The Fourier Transform in Modern Communications*

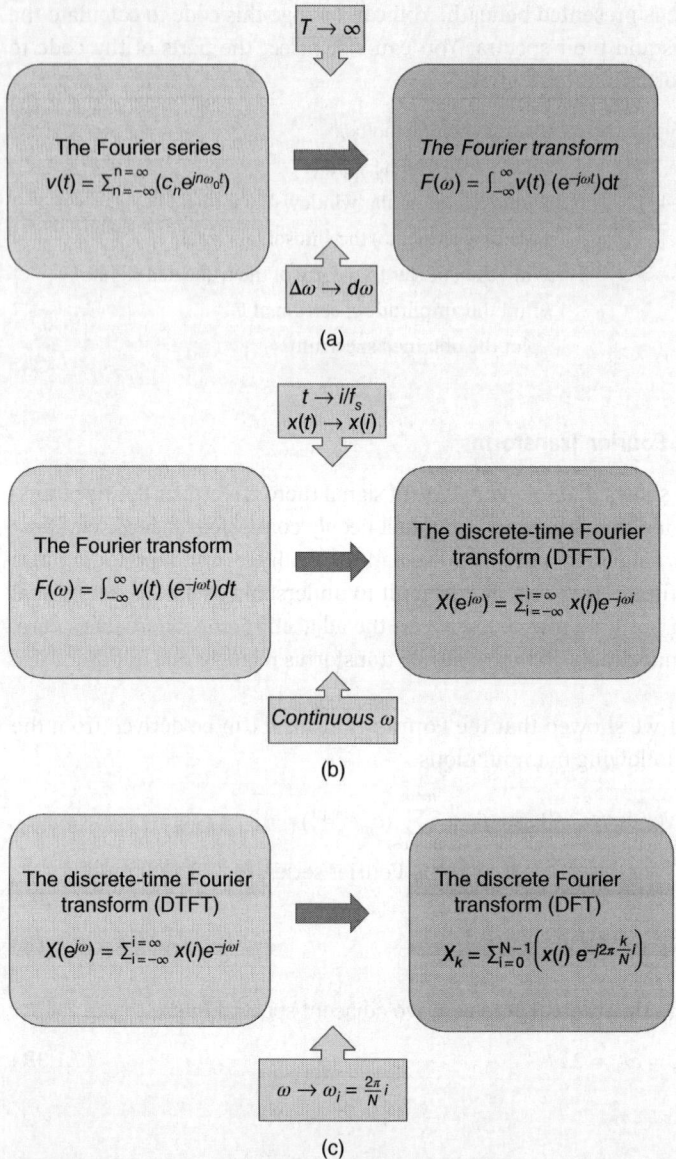

Figure 7.3.8 Relationship among the Fourier transforms (a) Derivation of the Fourier transform from the Fourier series; (b) derivation of the discrete-time Fourier transform (DTFT) from the Fourier transform; (c) derivation of the discrete Fourier transform (DFT) from the discrete-time Fourier transform.

This derivation of the Fourier transform from the Fourier series is schematically shown in Figure 7.3.8a, where the main transformations in time domain, $T \to \infty$, and frequency domain, $\Delta\omega \to d\omega$, are highlighted.

Figure 7.3.8 shows schematically all the other connections among the Fourier transforms. The discrete-time Fourier transform (DTFT) can be derived from the Fourier transform by sampling a continuous signal, $x(t)$, at discrete points i over the equal intervals $\Delta t = 1/f_s$, where f_s is the sampling frequency. As a result of this operation, the discrete intervals, i/f_s, replace the continuous time, t, and the time-discrete signal, $x[i]$, takes the place of the time-continuous signal, $x(t)$. Notice that in DTFT the frequency remains continuous. In the same venue, the discrete Fourier transform (DFT) can be obtained from DTFT by changing the continuous frequency, ω, to discrete, equally spaced frequencies, $\Delta\omega_i = \frac{2\pi}{N} i$.

We leave the formal derivation of these transformations to you as a challenging exercise; you can, of course, find these derivations in specialized textbooks, such as McClellan et al. (2003, pp. 392–399) and Manolakis and Ingle (2011, pp. 236–241).

7.3.5 Fast Fourier Transform (FFT)

Now that we become familiar how to work with the discrete Fourier transform, it seems the problem of computing the spectra of discrete signals for DSP is solved. Well, not quite. There is still a "small" issue left: writing the proper computer program and executing it. But we know, of course, that before writing a computer program, it is necessary to develop the algorithm (the sequence of logical steps) that this program will execute. And here is where a new problem arises. If we follow the definition of DFT given in (7.3.2),

$$X_k = \sum_{i=0}^{N-1} \left(x[i] \cdot \exp\left[-j2\pi \frac{k}{N} i\right] \right)$$

we need to perform N multiplications, $x[i] * e^{-j2\pi \frac{k}{N} i}$, for each i, and N additions of members, $x[i] * e^{-j2\pi \frac{k}{N} i}$, for every i running from 0 to $N-1$. (Refer to Example 7.3.1 to see the specific manipulation.) All these manipulations are totaled in N^2 operations. For $N = 32$, as in Example 7.3.1, we will need to perform 1024 operations; for a computer that performs each operation in one microsecond, this will take $2048 \times 10^{-6} = 0.002$ s. It looks like a short execution time, but we need to realize that in practice N is extremely large. For $N = 10^3$, for example, the time needed to execute such a program is equal to $t = 2N^2 \times 10^{-6}$ s $= 2$ s; however, for $N = 10^6$, the execution time is approximately 23 days. Today, to process an optical signal delivering information at $1T$ (b/s) $= 10^{12}$ b/s, we are dealing with N being well above 10^{12}. You can easily imagine, therefore, how long it would take to calculate only one sample by this conventional method. Thus, *DFT would remain just another academic curiosity had it not been for a revolutionary change in the computational approach.*

This change came in 1965, when Cooley and Tukey (1965)[6] introduced the algorithm called a *fast Fourier transform* (FFT). This algorithm enables reducing the number of operations for an N-point DFT from N^2 to $N\log_2 N$. To demonstrate how significant this advantage is, let us take the numbers from the above example and compute: For $N = 10^3$, the execution time with the FFT technique becomes equal to $10^3 \log_2 10^3 \times 10^{-6}$ s ≈ 0.01 s, or 200 times faster than with the DFT approach. But with $N = 10^6$, FFT executes the required operations in $N^2/(N \cdot \log_2 N) = 20\,000\,000$ times faster than does conventional DFT. If we plug any large number N into the formula $N^2/(N \cdot \log_2 N)$, its quotient allows us to really appreciate the advantage of FFT. In reality, the time saved executing the DFT operations with FFT is even greater than what our simple examples show. *Thus, the greater the number N, the more pronounced the advantage of using FFT instead of conventional DFT.* What is more, FFT calculations are more precise than DFT's because execution of the FFT algorithm requires fewer steps and, therefore, introduces fewer round-off errors.

---

6 Many experts believe that the modern information, DSP-based age began with this work.

From this consideration, we can deduce that

(1) *the* fast Fourier transform (FFT) is the computational algorithm *for executing the discrete Fourier-transform (DFT) concept,*
(2) FFT reduces the number of operations for an N-point DFT at least by the ratio N^2 to $N \log_2 N$,
(3) DFT becomes a practical tool for modern DSP but only thanks to the speed and precision of the FFT algorithm.

We do not intend to discuss FFT algorithms in detail, let alone to consider FFT programs, because this book addresses the user of this method, not the developer of these algorithms or programs. If you need more information on this topic, turn to specialized literature. We recommend (Smith 2003, pp. 225–242) and (McClellan et al. 2003, pp. 278–286) to start. What follows is a brief introduction to the basic ideas of FFT.

There are three main steps in the FFT algorithm:

(1) Decomposition of a single N-point time-domain discrete signal into N signals, each of which is a single-point individual signal.
(2) Calculation of the frequency spectrum for each individual single-point time-domain signal.
(3) Synthesis of the individual frequency spectra into one spectrum for the entire signal.

Example 7.3.1 should help in understanding of some elements of these calculations. For instance, in the current example, we consider a 32-point signal. FFT decomposes this signal into 32 individual signals and performs operations similar to those we did in Example 7.3.1.

There is an underlying assumption in the original Cooley–Tukey FFT algorithm: N has to be the power of two in order for the decomposition to occur recursively. (It is called *radix-2* algorithm.) The subsequent developments, however, brought about the new variants of the FFT algorithm that enable us to operate with N being prime numbers.

There are many variations of FFT today; recently, it was announced that scientists of MIT and Harvard developed a new version of FFT that works even significantly faster than traditional FFT.

In MATLAB, the command for executing FFT is *fft*. We do not really need to know how it is executed; as users, we need only to know how to apply it. Bear in mind that all computerized calculations associated with spectral analysis are done with FFT. For instance, in Example 7.3.2 we found the spectrum of a windowed sinusoid by writing the MATLAB command *fourier(WV)*, and this command was actually executed with FFT. As a result, we obtain the spectrum of the signal almost instantly.

This short introduction should help you appreciate the role and place of FFT in spectral analysis in particular and in DSP in general.

It is worth noting that, meeting the demand of the fast-advancing new field of quantum communications, a *quantum Fourier transform (QFT)* has been developed.

Questions and Problems for Section 7.3

- Questions marked with an asterisk require a systematic approach to finding the solution.
- Many questions and problems, including those marked with an asterisk, imply that you, in addition to reading the textbook, will do your research to find the answers. Consider such questions as miniprojects.

1 What is digital signal processing? At what part of a communication system this operation is performed?

2 List four types of signal employed in modern communication, qualitatively sketch their waveforms, and explain their peculiarities.

3 Consider classification of the signals employed in modern communications:
   a) Are analog and continuous signals the same? Explain.
   b) Is there any difference between discrete and digital signals?
   c) What is antonym to *continuous* in the signal classification?
   d) What is antonym to *digital* in the signal classification?

4 How can we obtain a digital signal from a continuous signal?

5 Consider Table 7.3.P5: Put the corresponding Fourier operation against each type of communication signal. Explain your reasoning.

Table 7.3.P5 Communication signals and the Fourier transformations.

Communication signal	Correct placement of the Fourier transformation	The Fourier transformation
Continuous periodic		Continuous-time Fourier transform (FT)
Continuous nonperiodic		Discrete-time Fourier transform (DTFT)
Discrete periodic		Continuous-time Fourier series (FS)
Discrete non-periodic		Discrete Fourier transform (DFT)

6 Is a square wave discrete or continuous signal? Explain.

7 Recall the Gibbs phenomenon: What mathematical problem underlies this phenomenon?

8 Consider digital signal processing, DSP:
   a) Briefly explain this operation.
   b) Is there any difference between DSP and filtering?
   c) Give examples of DSP you are familiar with.
   d) Whether your smartphone performs DSP on any input signal? Explain.

9 In determining what Fourier transform will work for DSP we refer to processing speed as a major problem to be resolved.
   a) Why is this so?
   b) What Fourier transform meets this requirement better than all others? How?

10 Since almost all signals used in electronic communication are *discrete and digital*, it would seem that both DTFT and DFT can be employed for DSP. In reality, only DFT can do the job. Why?

11 The text says that "DFT is a mathematical operation that transforms one set of numbers, representing a discrete time-domain signal, into another set of numbers, representing this signal in frequency domain." With reference to (7.3.2), interpret the meaning of this statement.

12 Compare the Fourier series and the discrete Fourier transform:
a) What do they have in common?
b) What is the difference between them?

13 *Find the DFT of the train of Gaussian pulses given as $v(t) = e^{-2.77\left(\frac{t-b}{c}\right)^2}$, where $b = T/4$ and $c = 0.25$.

14 Figure 7.3.P14 shows the same signal in different versions:
a) What is difference between these versions?
b) Which version is the correct presentation of the intended operation? Why?

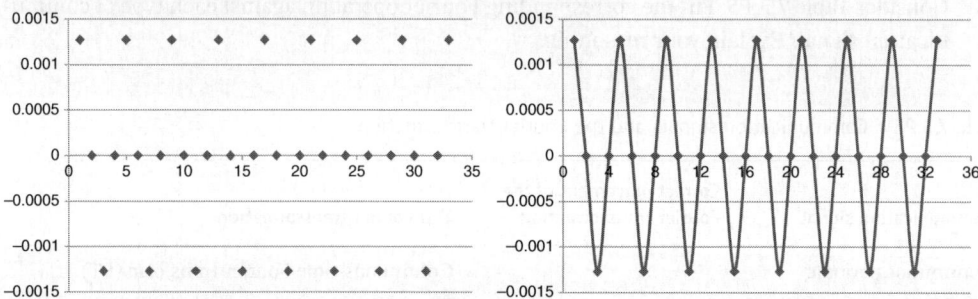

Figure 7.3.P14 The same signal presented in two versions.

15 *In analog world, the Fourier series decomposes an original time-domain signal into a set of cosines and sines; since each sinusoidal signal represents a single frequency, this decomposition delivers the signal's spectrum. How can we determine the spectrum of a digital signal?

16 Consider a DFT operation: What are the basis functions? Why do we need them? Give an example.

17 DFT works only with discrete periodic signals, the text says. We know, however, that the most practical signals are nonperiodic. How, then, can DFT be used for spectral analysis of these signals?

18 The text says that applying DFT to finding the spectrum of a nonperiodic signal boils down to the following contradiction: On the one hand, DFT can be applied only to a finite signal; on the other hand, a nonperiodic signal is not finite. How can this contradiction be resolved? Explain.

19 Consider two sinusoidal signals: $v_1(t) = 0.5 + 2\cos(25.2t)$ and $v_2(t) = v_1(t) \cdot rect(t/6)$, where $rect(t/6)$ is a rectangular pulse defined as $rect(t/6) = u(t+3) - u(t-3)$:
a) Qualitatively sketch the waveforms of these signals.
b) Qualitatively sketch the spectrum of each of these signals and explain your reasoning.
c) Using MATLAB (or any other computerized tool), determined the spectra of both signals and compare your results with those obtained in Problem 19a.
d) How will the spectrum of signal $v_3(t) = v_1(t) \cdot rect(t/2)$ qualitatively change compare to signal $v_1(t)$ and $v_2(t)$?

20 The text says that by truncating a signal we inevitably change the form of the signal's spectrum. Why?

21 There is the principle of uncertainty of Fourier analysis. Explain what it is. Give an example.

22 We can determine the signal's spectrum either analytically (calculating manually by formula) or by using a computerized tool. Which method produces more accurate result? Explain.

23 *Figure 7.3.8 shows the relationship among the Fourier transforms:
   a) Figure 7.3.8a shows that the Fourier transform (FT) can be derived from a Fourier series (FS). Provide detailed explanation of this derivation.
   b) Repeat the same task for Figure 7.3.8b,c.

24 Consider the fast Fourier transform (FFT):
   a) Why is it called "fast?"
   b) If N^2 is the total number of operations needed for completing a DFT task and each operation takes 0.1 μs to perform by a computer, how long will it take to perform 10^8 operations with traditional DFT and with FFT?
   c) What is the solution to the problem to inherently low processing speed of DFT?

25 How long will it take to perform on task whose $N = 10^{12}$ by using (a) traditional DFT algorithm and (b) FFT algorithm? What if $N = 10^{15}$?

26 The text says that FFT is not only much faster than a traditional DFT algorithm but it is also more accurate. How is it possible?

27 The original version of the FFT algorithm required that N has to be the power of two in order for the decomposition to occur recursively. Is it an advantage or a drawback of the FFT algorithm? Explain.

7.3.A Data and Samples of Calculations for Example 7.3.1

Table 7.3A.1 Data for the DFT in Example 7.3.1.

Time domain		DFT	Frequency domain		
Sample number, i	Sample value, $x[i]$		Frequency index, k	Amplitude A_k	Amplitude B_k
0	4.80E−13	Forward DFT	0	1.66E−01	0.00
1	1.49E−11		1	−3.05E−01	2.48E−17
2	3.70E−10		2	2.35E−01	−2.58E−17
3	7.37E−09	Inverse DFT	3	−1.52E−01	2.86E−17
4	1.18E−07		4	8.27E−02	−4.19E−17
5	1.50E−06		5	−3.78E−02	−2.18E−17

(Continued)

Table 7.3A.1 (Continued)

Time domain		DFT	Frequency domain		
Sample number, i	Sample value, $x[i]$		Frequency index, k	Amplitude A_k	Amplitude B_k
6	1.54E−05		6	1.45E−02	−3.99E−18
7	1.27E−04		7	−4.69E−03	−1.17E−17
8	8.32E−04		8	1.27E−03	6.05E−18
9	4.39E−03		9	−2.90E−04	1.01E−17
10	1.85E−02		10	5.55E−05	−9.04E−17
11	6.27E−02		11	−8.93E−06	−2.84E−16
12	1.70E−01		12	1.21E−06	−4.86E−17
13	3.69E−01		13	−1.37E−07	2.61E−16
14	6.42E−01		14	1.31E−08	4.52E−19
15	8.95E−01		15	−1.06E−09	−2.28E−16
16	1.00E+00		16	1.66E−01	7.27E−18
17	8.95E−01				
18	6.42E−01				
19	3.69E−01				
20	1.70E−01				
21	6.27E−02				
22	1.85E−02				
23	4.39E−03				
24	8.32E−04				
25	1.27E−04				
26	1.54E−05				
27	1.50E−06				
28	1.18E−07				
29	7.37E−09				
30	3.70E−10				
31	1.49E−11				

Table 7.3A.2 Samples of calculations of A_k and B_k for Example 7.3.1.

		Cosine amplitude $A_0 = \frac{1}{32} * \sum_{i=0}^{31}(A_0^{\ i})$ $= 1.66E-01$	Cosine amplitude $A_1 = \frac{1}{16} * \sum_{i=0}^{31}(A_1^{\ i})$ $= -3.05E-01$	Sine amplitude $B_1 = -\frac{1}{16} * \sum_{i=0}^{31}(B_1^{\ i})$ $= 2.48E-17$	Sine amplitude $B_2 = -\frac{1}{16} * \sum_{i=0}^{31}(B_2^{\ i})$ $= -2.58E-17$
Index, i	Members, $x[i]$	Amplitudes, $A_0^{\ i} = x[i]$	Amplitudes, $A_1^{\ i} = x[i]$ $* \cos\left(2\pi \frac{1}{N} i\right)$	Amplitudes, $B_1^{\ i} = x[i]$ $* \sin\left(2\pi \frac{1}{N} i\right)$	Amplitudes, $B_2^{\ i} = x[i]$ $* \sin\left(2\pi \frac{2}{N} i\right)$
0	4.80E−13	4.80E−13	4.80E−13	0.00E+00	0.00E+00
1	1.49E−11	1.49E−11	1.46E−11	2.91E−12	5.70E−12
2	3.70E−10	3.70E−10	3.42E−10	1.42E−10	2.62E−10
3	7.37E−09	7.37E−09	6.13E−09	4.10E−09	6.81E−09

Table 7.3A.2 (Continued)

		Cosine amplitude $A_0 = \frac{1}{32} * \sum_{i=0}^{31}(A_0^i)$ $= 1.66E - 01$	Cosine amplitude $A_1 = \frac{1}{16} * \sum_{i=0}^{31}(A_1^i)$ $= -3.05E - 01$	Sine amplitude $B_1 = -\frac{1}{16} * \sum_{i=0}^{31}(B_1^i)$ $= 2.48E - 17$	Sine amplitude $B_2 = -\frac{1}{16} * \sum_{i=0}^{31}(B_2^i)$ $= -2.58E - 17$
Index, i	Members, $x[i]$	Amplitudes, $A_0^i = x[i]$	Amplitudes, $A_1^i = x[i]$ $* \cos\left(2\pi\frac{1}{N}i\right)$	Amplitudes, $B_1^i = x[i]$ $* \sin\left(2\pi\frac{1}{N}i\right)$	Amplitudes, $B_2^i = x[i]$ $* \sin\left(2\pi\frac{2}{N}i\right)$
4	1.18E−07	1.18E−07	8.32E−08	8.32E−08	1.18E−07
5	1.50E−06	1.50E−06	8.36E−07	1.25E−06	1.39E−06
6	1.54E−05	1.54E−05	5.90E−06	1.42E−05	1.09E−05
7	1.27E−04	1.27E−04	2.47E−05	1.24E−04	4.84E−05
8	8.32E−04	8.32E−04	5.10E−20	8.32E−04	1.02E−19
9	4.39E−03	4.39E−03	−8.56E−04	4.30E−03	−1.68E−03
10	1.85E−02	1.85E−02	−7.09E−03	1.71E−02	−1.31E−02
11	6.27E−02	6.27E−02	−3.48E−02	5.21E−02	−5.79E−02
12	1.70E−01	1.70E−01	−1.20E−01	1.20E−01	−1.70E−01
13	3.69E−01	3.69E−01	−3.07E−01	2.05E−01	−3.41E−01
14	6.42E−01	6.42E−01	−5.93E−01	2.46E−01	−4.54E−01
15	8.95E−01	8.95E−01	−8.78E−01	1.75E−01	−3.43E−01
16	1.00E+00	1.00E+00	−1.00E+00	1.23E−16	−2.45E−16
17	8.95E−01	8.95E−01	−8.78E−01	−1.75E−01	3.43E−01
18	6.42E−01	6.42E−01	−5.93E−01	−2.46E−01	4.54E−01
19	3.69E−01	3.69E−01	−3.07E−01	−2.05E−01	3.41E−01
20	1.70E−01	1.70E−01	−1.20E−01	−1.20E−01	1.70E−01
21	6.27E−02	6.27E−02	−3.48E−02	−5.21E−02	5.79E−02
22	1.85E−02	1.85E−02	−7.09E−03	−1.71E−02	1.31E−02
23	4.39E−03	4.39E−03	−8.56E−04	−4.30E−03	1.68E−03
24	8.32E−04	8.32E−04	−1.53E−19	−8.32E−04	3.06E−19
25	1.27E−04	1.27E−04	2.47E−05	−1.24E−04	−4.84E−05
26	1.54E−05	1.54E−05	5.90E−06	−1.42E−05	−1.09E−05
27	1.50E−06	1.50E−06	8.36E−07	−1.25E−06	−1.39E−06
28	1.18E−07	1.18E−07	8.32E−08	−8.32E−08	−1.18E−07
29	7.37E−09	7.37E−09	6.13E−09	−4.10E−09	−6.81E−09
30	3.70E−10	3.70E−10	3.42E−10	−1.42E−10	−2.62E−10
31	1.49E−11	1.49E−11	1.46E−11	−2.91E−12	−5.70E−12

8

Analog Transmission with Analog Modulation

Objectives and Outcomes of Chapter 8

Objectives

- To understand the need for modulation by studying the difference between baseband and broadband transmissions.
- To become familiar with analog modulation, in which a carrier is always a sinusoidal signal and a message is always an analog signal.
- To realize that modulation is imposing the changes in a message signal onto one of the parameters (amplitude, frequency, or phase) of the carrier wave.
- To learn the mathematical analyses and become familiar with the practical aspects of amplitude, frequency, and phase modulations.
- Master that analog transmission is based on analog modulation; therefore, quality of analog transmission can be improved by using various techniques of analog modulation and demodulation.

Outcomes

- Learn that in baseband transmission the same signal contains information and transmits this information, which implies that baseband transmission occurs in low-frequency range of the EM spectrum because information signals typically occupy this range.
- Understand that baseband transmission, though being simple and inexpensive, has significant drawbacks and that the solution to the problems of baseband transmission is broadband transmission, in which an information signal is superimposed onto a sinusoidal carrier signal. Deduce that broadband transmission occurs at the band of frequencies broader and higher than the frequencies of an information signal. Realize that broadband transmission greatly exceeds its baseband counterpart in transmission rate and distance.
- Grasp the fact that superimposing an information signal upon a carrier is called modulation, which means changing one of the sinusoidal carrier's parameters (amplitude, frequency, or phase) according to variations in the information signal.
- Master the fact that in analog modulation, we can achieve amplitude modulation (AM), frequency modulation (FM), or phase modulation (PM).
- Gain knowledge of the mathematical analysis of a tone amplitude and frequency modulations, in which information is presented by a sinusoidal signal. Realize that in reality, a sinusoidal signal does not bear any information; nevertheless, this analysis enables us to find all the necessary characteristics and parameters of AM and FM signals in time domain and frequency domain.

- Study how to extend the results obtained in the analysis of sinusoidal AM and FM to a general case, such as when information is presented (borne) by an arbitrary signal.
- Become familiar with practical aspects of the operations of AM and FM transmission systems, such as increasing their power and bandwidth efficiencies and improving their signal-to-noise ratios (SNRs) and noise immunity.
- Realize that applying various modulation and demodulation techniques, we can improve quality of analog transmission. Learn that an FM transmission system with preemphasis and de-emphasis circuits exemplifies such techniques.

8.1 Basics of Analog Modulation

Objectives and Outcomes of Section 8.1

Objectives

- To understand the concept of analog modulation.
- To learn the basics of amplitude modulation.
- To acquire knowledge of the basics of frequency modulation.
- To become familiar with the concept of phase modulation.

Outcomes

- Recall the operation of a communication system by revisiting its block diagram.
- Learn that a transmission is called baseband if the same signal contains information and transmits information. Comprehend that information signals typically occupy the low-frequency range of the EM spectrum, which means that the baseband transmission occurs in this range.
- Understand that baseband transmission, though being a simple and inexpensive, has significant drawbacks, which preclude this transmission mode from becoming the mainstream in modern communication.
- Realize that the solution to the problems of baseband transmission is broadband transmission, in which an information signal is superimposed onto a carrier sinusoidal signal. Deduce that broadband transmission occurs at the band of frequencies broader and higher than the frequencies of an information signal.
- Grasp the fact that superimposing an information signal upon a carrier is called modulation, which is the method for creating broadband transmission. Learn that modulation means changing one of the sinusoidal carrier's parameters (amplitude, frequency, or phase) according to variations in the information signal.
- Use Table 8.1.1 to compare the properties of baseband and broadband transmissions and to learn the specific advantages of broadband transmission.
- Master the fact that in analog modulation a carrier is a sinusoidal signal and that by modulating one of its parameters we can achieve AM, FM, or PM.
- Gain knowledge of the basics of amplitude modulation by considering a tone AM, where a message signal is a sinusoidal one. Grasp the concept of a tone AM and its modulation index and learn how to compute the modulation index based on the received amplitude-modulated signal. Get to know that the frequency of a carrier must be at least 10 times greater than that of a message signal.

- Derive the formula for a tone AM by inserting the equation for a message signal into the amplitude of a carrier. Use this formula for calculating an instantaneous value of a tone AM signal and for deriving its formulas for envelopes.
- Learn that the spectrum of a tone AM signal can be determined by presenting the AM signal's formula as a sum of three sinusoidal members; find out that this spectrum consists of a carrier frequency, an upper sideband (USB), and a lower sideband (LSB).
- Come to know that the total power borne by an AM signal is distributed among its carrier, and the USB and LSB components and that the carrier typically consumes most of the transmitting power without delivering any information. Learn how to compute the power distribution and understand that low power efficiency is one of the major drawbacks of AM transmission.
- Learn that the main reason to switch from amplitude modulation to frequency modulation is to avoid using an amplitude to carry information because a signal's amplitude is easily distorted by noise, which results in receiving erroneous information.
- Understand the concept of FM, which states that the frequency of an FM signal changes as the message (modulating, information, or intelligence) signal requires.
- Realize that the amplitude of an FM signal is constant, which implies that the average power of an FM signal does not depend on the message being transmitted and therefore remains constant.
- Deduce that the FM frequency, $f_{FM}(t)$ (Hz), is given by $f_{FM}(t) = f_C$ (Hz) $+ k_{FM} \cdot v_M(t)$, where f_C (Hz) is the constant carrier's frequency, $v_M(t)$ is the message signal, and k_{FM} (Hz/V) is a modulation (deviation) constant also called the frequency sensitivity of a demodulator. Derive that for a single-tone message signal, $v_M(t) = A_M \cos(2\pi f_M t)$, the FM frequency becomes $f_{FM}(t) = f_C$ (Hz) $+ k_{FM} A_M \cos(2\pi f_M t)$.
- Understand that in a single-tone FM, its frequency, f_{FM} (Hz), continually varies between its maximum, $f_C + k_{FM} A_M$, and minimum, $f_C - k_{FM} A_M$, values and therefore must be referred to as *instantaneous* FM frequency, meaning that there is no period of an FM signal. Realize that time intervals between zero crossings of an FM waveform contain information about a message signal.
- Be aware that k_{FM} (Hz/V) is the input–output characteristic of a voltage-controlled oscillator (VCO), the device performing frequency modulation in practice.
- Derive the equation of a single-tone FM signal as $v_{FM}(t)$ (V) $= A_C \cos(2\pi f_C t + \beta \cos(2\pi f_M t))$, where $\beta = k_{FM} A_M$ (Hz)$/f_M$ (Hz) is the modulation index. Understand that this equation enables us to compute the instantaneous value of an FM signal.
- Learn that β is proportional to the amplitude of a message signal and remember that, in contrast to AM, the FM modulation index is typically greater than 1.
- Gather that the greater the FM modulation index, β, the greater the difference between the maximum and minimum FM frequency values, the farther the waveform of an FM signal from a sinusoidal signal, and the greater the number of frequencies in the FM signal. Understand that this means that the greater the β, the greater the number of harmonics (side frequencies) in the spectrum of an FM signal.
- Comprehend that the greater the β, the greater the bandwidth of an FM signal defined in (8.1.25) as BW_{FM} (Hz) $= f_n^+$ (max) $- f_n^-$ (min). Know that theoretically the FM bandwidth tends to infinity with any β but in practice it is estimated by (8.1.26) as $BW_{FM} \approx 2nf_M$ or approximated by Carson's rule as $BW_{FM} \approx 2(\beta + 1)f_M$, where n is the number of harmonic and f_M (Hz) is the message signal's frequency.
- Gain the knowledge that in FM transmission the amplitude and frequency of the carrier must be greater (and the frequency must be much greater) than the amplitude and frequency of the message signal.

- Learn that phase modulation is similar to frequency modulation and these two are often referred to as angle modulation.
- Realize that in phase modulation, the message signal changes (modulates) the phase of the carrier and this procedure is described in (8.1.30) as $v_{PM}(t) = A_C \cos(\omega_C t + k_\theta A_M \cos(\omega_M t))$, where k_θ (rad/V) is the phase modulation (deviation) constant.
- Comprehend that (8.1.30) enables us to plot the waveform of a PM signal and compute its instantaneous value.
- Understand that the analysis of phase modulation is similar to that of frequency modulation. Be aware that in PM, a *maximum phase deviation*, $\Delta_\theta = k_\theta A_M$, cannot exceed 2π to avoid ambiguity; that is, $-\pi < \Delta_\theta \leq +\pi$, as shown in (8.1.31).

8.1.1 Why We Need Modulation: Baseband and Broadband Transmission

8.1.1.1 Baseband Transmission and Its Major Problems

Let us recall the basics of communications discussed in Chapter 1: Suppose we are assigned to develop a communication system that would transmit information in any required format. How can we do it? What devices do we need? The solution to this problem, based on our knowledge and common sense, can be found in this sequence of steps illustrated in Figure 8.1.1:

- Convert original information into electrical format by using a *transducer*.
- Prepare a signal for transmission and send it by employing a *transmitter (Tx)*.
- Deliver the signal over the distance desired through a *transmission link* (which can be implemented with various transmission media, such as copper wire, air, or optical fiber).
- Recover the information from a received electrical signal with a *receiver (Rx)*.
- Convert the recovered electrical information signal into its original form (e.g. voice, video, or data) with another *transducer*.

Note especially that a signal can be transmitted over a link in various forms, such as electromagnetic waves via air or light through optical fiber; in any case, it must be converted into electrical form first.

Let us consider, as an example, the operation of your smart phone. When a microphone, which is a transducer here, converts a voice to an electrical signal, it produces an electrical signal whose magnitude changes according to variations in the sound waves. Figure 8.1.2 shows such a record. Figure 8.1.2a shows the plot of the electrical signal representing a human voice in time domain, that is, as a *waveform*. Figure 8.1.2b shows what frequencies this signal contains, that is, the *spectrum* of this signal.) This electrical signal is delivered via a transmission link to a receiver. (What is the transmission link for your cell phone? What is a receiver?) There, the electrical signal is converted back into sound; as you have no doubt surmised, the transducer at the receiver end is a speaker.

This is how – conceptually, of course – Alexander Bell transmitted the first ever voice message in 1876. The main feature of Bell's system – from the standpoint of this discussion – is that the

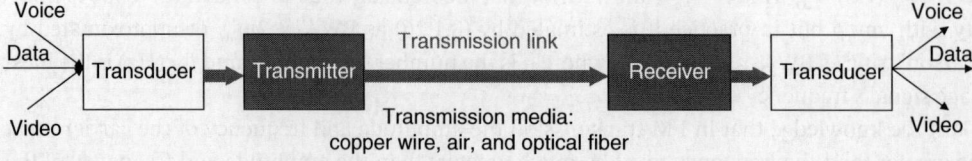

Figure 8.1.1 Detailed block diagram of a communication system.

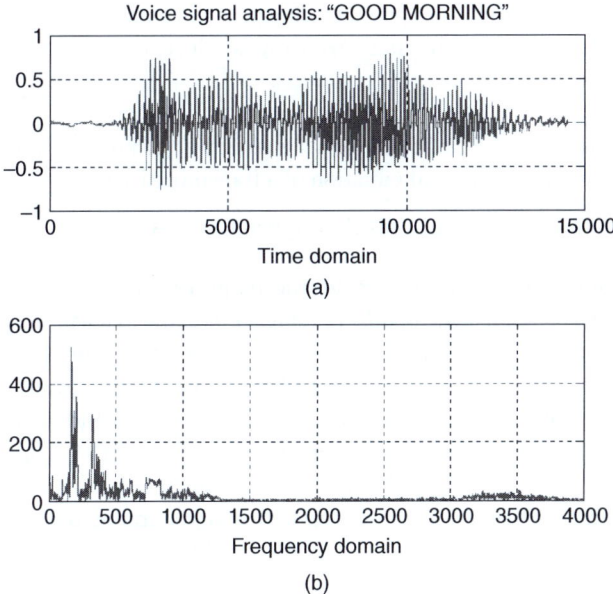

Figure 8.1.2 The record of a human voice in time (a) and frequency (b) domains. Source: Reproduced with permission from Prof. Mohammed Kouar, New York City College of Technology of the City University of New York.

signal transmitted to a receiver was the voice message directly converted into an electrical wave. This first voice transmission is an example of so-called *baseband transmission* because this type of transmission delivers information *at the base band of frequencies*, the band (range) at which the information is generated. Usually, the base band (range) of frequencies of the original information signal is concentrated in a low-frequency range. Figure 8.1.2 exemplifies this by showing that the voice signal is concentrated in the band of frequencies from 300 to 3300 Hz. Another example is music as we can hear it: Its frequency range is from 300 to 20 kHz.

If we ponder this description, we realize that there is another important feature of this transmission system: Here, *the same signal that contains the information to be delivered also carries this information.*

> *A transmission is called baseband if the same signal contains and transmits information.*

The consequence of this definition is that we need a low-pass channel to deliver a baseband signal; in other words, both the signal and the link are in the same low-frequency band. Digital transmission also uses this approach; examples include digital subscriber lines, digital magnetic recording, and digital optical recording.

The *main advantage of analog baseband transmission* is its simplicity and, therefore, low cost. It seems that baseband transmission is a natural way to deliver information by means of an analog signal, but, in reality, it is not the main transmission method today. Why? Because baseband transmission has the following drawbacks:

- Baseband transmission occurs at inherently low transmission speed (rate).
- Baseband transmission allows for transmission of only one signal at a time.

- *Analog* baseband transmission, where information is imposed by changing the *amplitude* of the signal, delivers a distorted signal, and it is difficult, sometimes even impossible, to extract correct information from the received signal.
- Baseband transmission can deliver a signal over a very limited transmission distance.
- Baseband transmission cannot control either the frequency of a signal or a transmission link, and therefore, this transmission cannot minimize the attenuation of a transmitted signal.

The origins and mechanisms of these drawbacks are discussed in Appendix 8.1.A.

8.1.1.2 Solution to the Problems of Baseband Transmission – Broadband Transmission

Broadband transmission occurs at the band of frequencies broader and higher than the frequencies of an information signal. How can we transmit an information signal at frequencies other than those within the signal's own band? We need to use a *carrier* signal, one whose band of frequencies is higher and broader than those of the base band. Radio transmission and all other broadband transmission technologies use this principle: They take an information signal – voice, for example – and superimpose it upon a carrier signal. Since a carrier signal is not restricted by any band of frequencies or magnitudes, as is an information signal, we can generate this signal in any band of frequencies we desire and at any required power range.

> *The general scheme of broadband transmission is as follows: At the transmitter end, an information signal is superimposed upon a carrier signal by a modulator; the modulated signal is then sent over a transmission link and, at the receiver end, the information signal is extracted from the received modulated signal by a demodulator.*

This concept is pictured in Figure 8.1.3. Pay special attention to the modulated signal. This is a carrier sinusoidal signal whose amplitude changed to deliver the message (in this example, 1s and 0s) presented by the information signal. This is how the information signal is superimposed upon the carrier signal in this example.

Superimposing an information signal upon a carrier signal is called modulation.

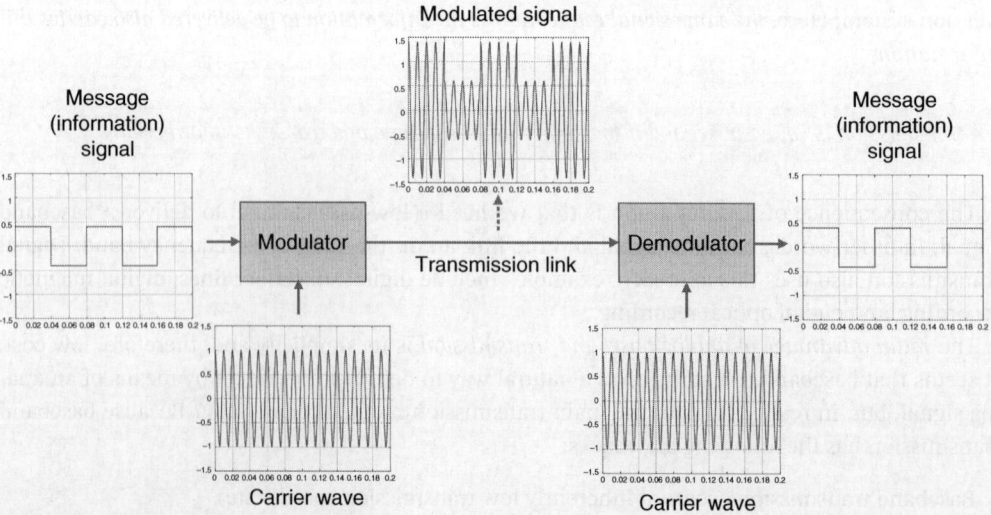

Figure 8.1.3 Concept of broadband transmission (modulation).

Since we want to make the transmission that should occur at a frequency higher than that of the baseband, we need to shift the frequency range of a transmitting signal at the higher end. This approach to modulation is called modulation by *frequency translation*.

What would be the best candidate for a *carrier signal*? A sinusoidal signal. The main advantage of this signal is that it occupies minimal (theoretically – zero) bandwidth. (Refer to Chapters 2 and 7, where this point is discussed.) This advantage is critical, as we will see in the following discussions. In addition, a sinusoidal signal is easy to manipulate (amplify and change).

> *The main advantage of broadband transmission is that it enables us to eliminate or mitigate the drawbacks of baseband transmission.*

Let us contrast baseband transmission's drawbacks listed previously and in Appendix 8.1.A to the advantages of broadband transmission:

- Low transmission rate of baseband transmission is replaced by the high transmission rate of broadband transmission simply because the former transmits information by a high-frequency carrier and BR (b/s) $\sim f_{CARRIER}$ (Hz). Refer to (1.3.17). Consequently, broadband transmission systems can transmit a high volume of information per second because H (bit) = BR (b/s) $\times T_{tr}$ (s), as (1.3.1) states.
- Baseband's restriction in transmission to only one signal at a time is not a problem for broadband transmission. Figure 8.1.4 shows how broadband communication allows for the transmission of several signals simultaneously: It is done by superimposing different information signals, even if they are in one frequency range, onto various carriers of different frequencies. For example, several radio stations can broadcast their talk shows (voice signals) over the same geographical area without interference by using different frequencies for their carrier signals. And, of course, we have telephone, cable TV, and Internet connections simultaneously over a single optical fiber in and from our homes because all these information signals use carriers of different frequencies. The technique of sending many signals over the same transmission link simultaneously is called *multiplexing*; Figure 8.1.4 shows an example of frequency-division multiplexing, FDM. (see Appendix 10.A for discussion of multiplexing.)
- In *analog* baseband transmission, it is difficult, sometimes even impossible, to extract correct information from the received signal. Not so in broadband transmission. Even in the analog form of transmission, broadband communications has the flexibility to deliver information not only

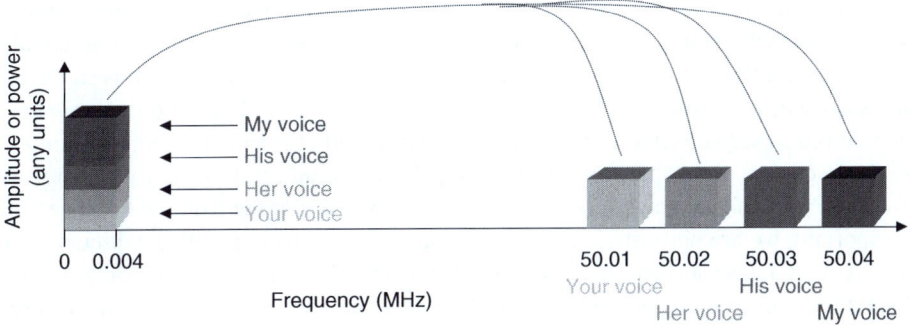

Figure 8.1.4 Broadband transmission allows for delivering many signals simultaneously. (The frequency values are for illustrative purpose only.)

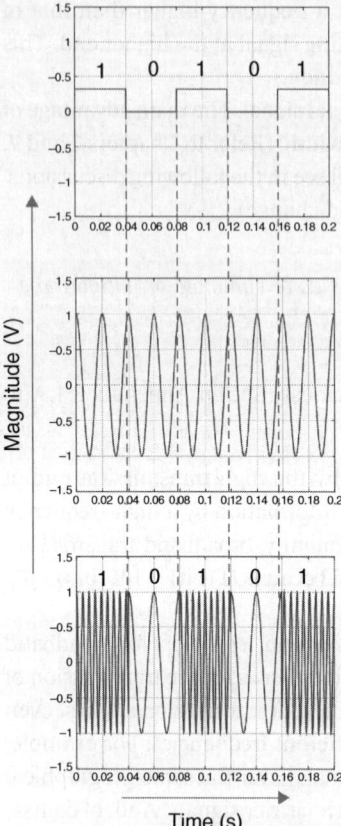

Figure 8.1.5 Broadband transmission with frequency modulation does not depend on the amplitude of a received signal.

by virtue of the signal's amplitude – which makes it difficult to recover information from the distorted signal, as shown in Figure 8.1.A.2 – but also by changing the frequency of the carrier. This is not possible in baseband transmission because the frequency of a signal being transmitted (information signal, that is) is fixed by its very nature. Figure 8.1.5[1] demonstrates the concept of FM in which changes in the signal's amplitude, in contrast to baseband transmission, do not affect the information delivered by this signal because information is delivered by frequency variations. Thus, broadband systems can deliver information much more reliably than baseband can.

- Recall that baseband transmission can deliver a signal over a very limited transmission distance. In contrast, a broadband system can transmit information over long distances. Since in broadband we are transmitting, in essence, a modulated carrier signal, we can make this signal as powerful as needed, thus avoiding the distance limitation inherent in baseband transmission. Recalling that the power delivered by the wave is proportional to the square of both the amplitude and the frequency, we realize that by increasing the amplitude and frequency of a carrier signal, we can create a signal powerful enough to be transmitted over the required distance.

Equally important, we are not restricted in the size of antennas any more. As we discussed in Appendix 8.1.A, to deliver a voice signal in the range below 4 kHz, we need a 75-km antenna. For 300-MHz FM transmission, however, we would need an antenna of only 1 m. What is more, we saw that in baseband transmission, to radiate signals for voice (4 kHz), music (20 kHz), and

---

1 MATLAB code for this and all other figures and examples in this section are prepared by Ms. Ina Tsikhanava.

analog TV (6.5 MHz), we need antennas of different sizes (7.5 km, 0.7 km, and 4.6 m, respectively) that would vary more than a thousand times. Thus, the ratio of the biggest to the smallest sizes in baseband transmission could be about 6.500 kHz/4.0 kHz = 1625. In contrast, in broadband transmission, with a 200-MHz carrier, the ratio of the biggest to the smallest antenna sizes for analog TV and voice would be around 206.5 Hz/200.004 Hz ≈ 1.03. In other words, an antenna of the same size would serve for both voice and video transmissions.

- Baseband transmission cannot minimize the attenuation of a transmitted signal, but broadband can thank to its ability to choose the correct frequency range to achieve the best transmission conditions in general and the least attenuation in particular. The most striking example is in the comparison of radio, satellite, and optical transmissions: For radio transmission, we use the megahertz range, for satellite the gigahertz range, and for optical the terahertz range. If we were restricted to only one low frequency, as we were in the baseband mode, then all these transmissions would not be possible because signals in radio frequencies and microwave ranges cannot propagate through an optical fiber.

Table 8.1.1 summarizes this discussion, concisely comparing baseband and broadband transmissions.

As Table 8.1.1 shows, broadband transmission outperforms baseband transmission by all measures. The price we pay for these advantages is the spectrum needed for broadband transmission. Since spectrum is a commodity in wireless communications, the communications industry sometimes resorts to narrowband transmission in specific applications.

As we remember from Chapters 1 to 3, there are analog and digital signals in electronic communications. Consequently, we need to use modulation for both of them. We call the *modulation an analog if both an information signal and a carrier have analog waveforms*. In this chapter, we consider analog modulation and we turn to digital modulation in Chapters 9 and 10.

This brief introduction to broadband transmission should convince us of the need to modulate a carrier. Now we can consider the modulation techniques more closely.

8.1.2 Basics of Amplitude Modulation

8.1.2.1 What Type of Analog Modulation Can We Have?

To reiterate, modulation means superimposing an information signal onto a *carrier*. How can we do this? As we learned in the previous subsection, a carrier signal must be a simple sinusoid, which is governed by the following formula:

$$v(t) = A\,\cos(\omega t + \theta) = A\,\cos(2\pi f t + \theta) \tag{8.1.1}$$

Superimposing an information signal onto a carrier simply means changing some parameter of the carrier to enable it to carry the information. Equation (8.1.1) clearly shows that we can change either the amplitude, A, or the frequency, f, or the phase shift, θ. Consequently, we can arrange either *AM* or *FM* or *PM*. Figure 8.1.6 illustrates this point. Since frequency is nothing more than changing phase shift with respect to time, the two latter forms of modulation are often combined in one type, which is called *angle (or angular) modulation*. The term *angle* or *angular*, refers to the entire argument $(2\pi f t + \theta)$ of a sinusoidal function, which is an angle.

8.1.2.2 What Is Amplitude Modulation (AM)

As the term suggests, AM, is changing the amplitude of a carrier signal to deliver information. Consider Figure 8.1.7, where an example of AM is demonstrated: The digital signal (a) carries information in the form of 1s and 0s; the carrier signal (b) is a cosine; the modulated signal (c) is a carrier

Table 8.1.1 Comparison of baseband and broadband transmissions.

Parameter	Baseband	Broadband
Bit rate (transmission rate) is proportional the carrier frequency, $BR\ (b/s) \sim f_{CARRIER}\ (Hz)$. (Bit is a measure of volume of information; thus, bit-per-second (b/s) is transmission rate.)	Very low because a carrier is the information signal and it is in the low-frequency range	Very high because we can use a carrier signal at any frequency range, including optical
Number of signals transmitted simultaneously	One signal at a time because all information signals are in one frequency range and they interfere (see Figure 8.1.1)	Many information signals simultaneously because each of them is superimposed onto its own carrier with a different frequency (see Figure 8.1.4)
Reliability	Low because the information is delivered by an original signal, which can be easily distorted by noise during the transmission	Higher than in baseband transmission because information is delivered by a modulated signal, which provides more means to deliver correct information despite distortion of the signal being transmitted
Transmission distance	Very limited because the power of the transmitted signal cannot be made as big as needed. In addition, antennas of different sizes are required for each type of transmission (voice, music, and TV)	Less limited because the carrier signal can be made as powerful as needed. In addition, the antenna sizes are almost the same for the different types of transmission
Choosing the right frequency for specific transmission conditions	The frequency of a transmitted signal is predetermined by the nature of the information to be delivered and therefore cannot be adjusted for specific transmission links (copper wire, air, or optical fiber)	The frequency of a transmitted signal is the frequency of the carrier and can be adjusted for specific transmission links

signal whose amplitude changed in such a way that this signal carries information (those 1s and 0s). At the destination point, the receiver can decipher the information (the sequence of 1s and 0s) that had been sent with the received modulated signal.

This specific type of amplitude modulation is called *amplitude-shift keying, ASK*, which was introduced many years ago. In fact, the operation of the "ancient" dial-up modem was based on this principle because ASK enables the delivery of digital (computer-originated) information by a sinusoidal carrier over a copper-wire telephone line designed for the transmission of analog signals.

A word about terminology: In amplitude modulation we have two input signals – an information signal and a carrier. An information-bearing signal is also called a message, modulating, baseband, or intelligent signal. The result of modulation (combining these two signals) is an amplitude-modulated signal.

8.1 Basics of Analog Modulation | 717

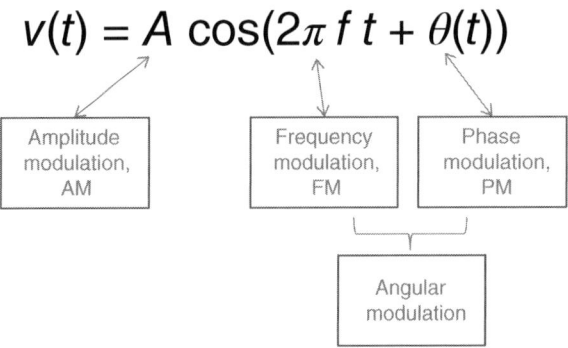

Figure 8.1.6 By changing a parameter of a sinusoidal carrier signal, we can obtain amplitude, frequency, or phase modulation.

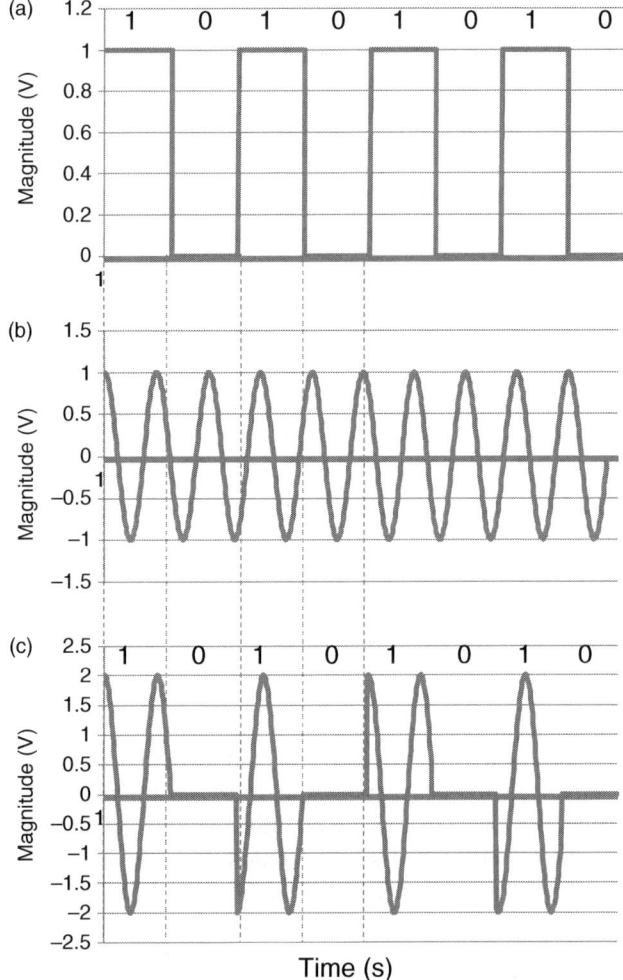

Figure 8.1.7 Example of amplitude modulation: amplitude-shift keying (ASK).

8 Analog Transmission with Analog Modulation

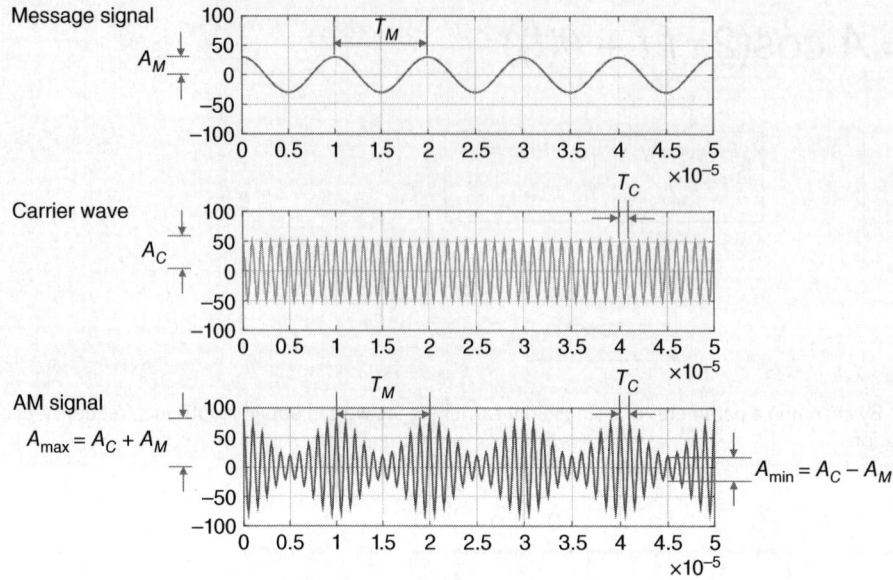

Figure 8.1.8 Tone (sinusoidal) amplitude modulation. Message signal (top), Carrier wave (middle), and AM signal (bottom).

Amplitude modulation[2] was the first type of modulation used in practice thanks to its simplicity; its inherent advantages keep this type of modulation current and, until recently, it was used in TV transmission, although today the area of its direct application reduces to just radio transmission. Nevertheless, amplitude modulation is a part of the most modern modulation formats used in digital transmission.

When considering the basics of AM, all textbooks traditionally start the discussion with a situation where the message (information or modulating) signal is a sinusoidal. This type of AM is called **tone modulation** because a pure sinusoid represents a single tone. Following this tradition, we present this AM in Figure 8.1.8.

In Figure 8.1.8, the *message signal* (a) that modulates the carrier has an amplitude, A_M (V), and a period, T_M (s), so its frequency is given by

$$f_M \text{ (Hz)} = \frac{1}{T_M(s)} \tag{8.1.2a}$$

Since $\omega_M = 2\pi f_M$, the angular frequency of a modulating (information) signal is defined as

$$\omega_M \text{ (rad/s)} = \frac{2\pi}{T_M(s)} \tag{8.1.2b}$$

The *carrier* in Figure 8.1.8b is characterized by its amplitude, A_C (V), and its period, T_C (s), so that its frequency is f_C (Hz) = $1/T_C$ (s) and its radian frequency is ω_C (rad/s) = $2\pi/T_C$ (s).

Since both constituents of the AM signal are sinusoidal waves described by their amplitudes and periods, it seems that the basic parameters of the AM signal should also be its amplitude and frequency (its period). Still, by definition, the amplitude is the maximum value of a sinusoid's

---

2 Invention of amplitude modulation is attributed to Reginald Fessenden. His first AM transmission system was demonstrated in December 1906, when this system broadcast voice and music from the first station in Brant Rock, Massachusetts. The signal was received several hundred miles away by the ships equipped with radio.

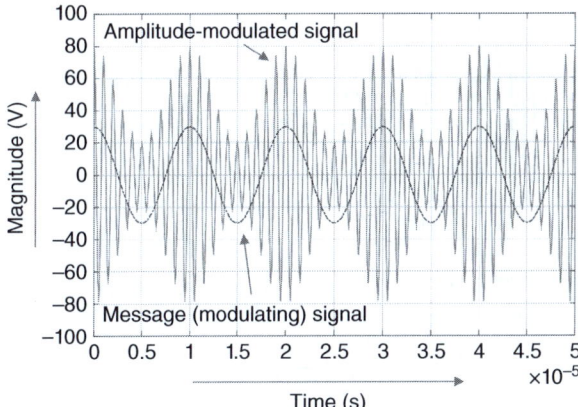

Figure 8.1.9 Concept of amplitude modulation.

magnitude, and the period is a specific time interval. In other words, both amplitude and frequency are specific numbers for any given sinusoidal signal; that is, they are *constants*. However, the AM signal shown in Figure 8.1.8c has a variable "amplitude" and two periods, which means that neither the "amplitude" nor the "period" of this AM signal can be termed this way because they do not satisfy the definitions of these parameters.

To describe an AM signal with measurable parameters, we introduce the following four quantities: A_{max} (V) and A_{min} (V), T_C (s), or f_C (Hz), and T_M (s), or f_M (Hz). Here, A_{max} (V) is the maximum amplitude and A_{min} (V) is the minimum amplitude of a modulated signal; they are equal to the sum and the difference of the amplitudes of the modulating, A_M (V), and carrier, A_C (V), signals:

$$A_{max}(V) = A_C(V) + A_M(V) \tag{8.1.3a}$$

and

$$A_{min}(V) = A_C(V) - A_M(V) \tag{8.1.3b}$$

These formulas clearly emphasize the fact that a modulated signal is a combination of both the message and the carrier signals. Obviously, when $A_{min} = A_{max}$, there is no amplitude modulation, since there are no variations in the carrier amplitude.

The waveform of a modulated signal features two periods: the small period of a carrier signal, T_C (s), and the big period of a modulating signal, T_M (s). Thus, we can distinguish between the high frequency of a carrier, f_C (Hz) = $1/T_C$ (s) and the low frequency of a modulating signal, f_M (Hz) = $1/T_M$ (s).

Remember, it is only the AM signal (Figure 8.1.8c) that we can see on the screen of an oscilloscope if we look into the transmission link of an AM transmission system; this is the signal the receiver obtains. Figure 8.1.9 should help us comprehend the concept of amplitude modulation: It shows the information (modulating) signal and the carrier signal, whose amplitude changed accordingly to impress the information on it.

8.1.2.3 Modulation Index
Review Figure 8.1.10: What are the differences among the three AM signals shown there? The top AM signal has $A_{min} = 0$ V and $A_{max} = 100$ V, the middle one has $A_{min} \approx 20$ V and $A_{max} \approx 80$ V, and

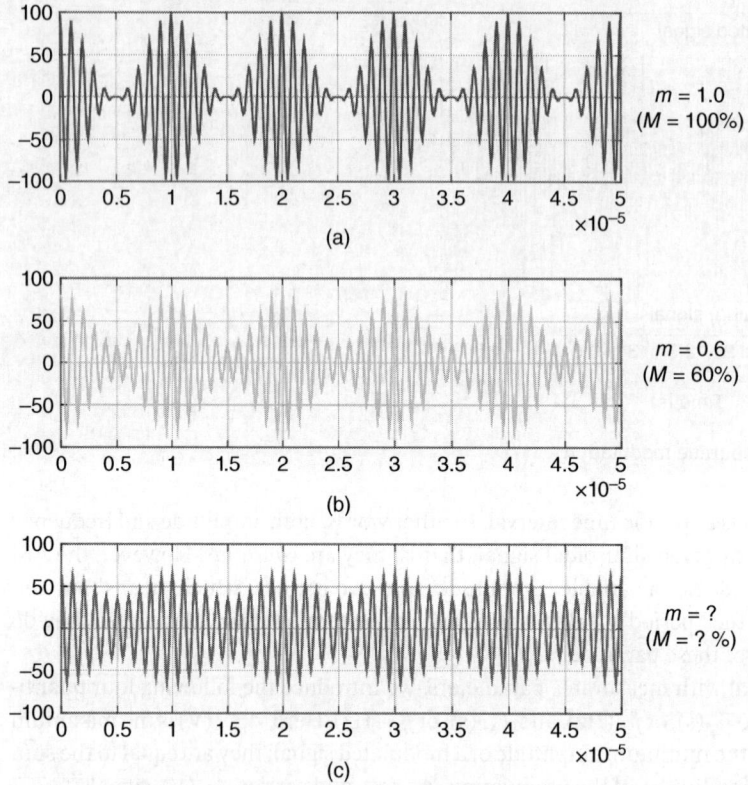

Figure 8.1.10 Modulation indexes of AM signals.

the bottom one has $A_{\min} \approx 35$ V and $A_{\max} \approx 65$ V. Thus, all three signals have a different ratio of $A_{\min}$ and $A_{\max}$. From a visual perspective, the top signal is modulated more deeply than the middle one and even more deeply than the bottom one. The parameter that describes the *depth of modulation* is called a **modulation index, m**.

Formally, the modulation index is defined as the ratio of the amplitudes of the message and carrier signals,

$$m = \frac{A_M(\text{V})}{A_C(\text{V})} \tag{8.1.4}$$

We can define a modulating index in percentages, as

$$M(\%) = m \times 100 = \frac{A_M(\text{V})}{A_C(\text{V})} \times 100$$

However, at the receiver end, only the modulated signals, such as those shown in Figure 8.1.10 can be obtained. In this case, (8.1.4) does not help directly because an AM signal does not exhibit A_M and A_C. Nevertheless, using (8.1.3a) and (8.1.3b), we can relate A_M and A_C to $A_{\max}$ and $A_{\min}$ as follows: Adding up $A_{\max}(\text{V}) = A_C(\text{V}) + A_M(\text{V})$ and $A_{\min}(\text{V}) = A_C(\text{V}) - A_M(\text{V})$ results in

$$A_C(\text{V}) = \frac{A_{\max} + A_{\min}}{2} \text{ (V)} \tag{8.1.5a}$$

8.1 Basics of Analog Modulation

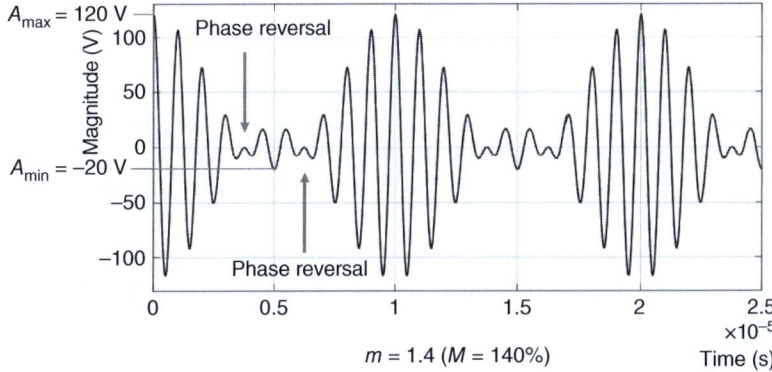

Figure 8.1.11 Overmodulated AM signal.

and subtracting A_{min} (V) from A_{max} (V) gives

$$A_M \text{ (V)} = \frac{A_{max} - A_{min}}{2} \text{ (V)} \quad (8.1.5b)$$

Now we can use (8.1.4) and define the modulation index as

$$m = \frac{A_{max} - A_{min} \text{ (V)}}{A_{max} + A_{min} \text{(V)}} \quad (8.1.6)$$

It is (8.1.6) that we can use to calculate a modulation index in practice.

We can thus comprehend the meaning of the numbers shown at the right side of each graph in Figure 8.1.10: They are the modulation indexes of these signals. The top AM signal is modulated at the deepest level and, therefore, its modulation index is the greatest of the three. The middle AM signal is modulated deeper than the bottom one but less so than the top; it is quite reasonable, therefore, that its modulation index should be in between those two indexes. Obviously, the modulation index of the bottom signal should be the smallest. Computing these modulation indexes, we find that the top signal has $A_{min} = 0$ V and $A_{max} = 100$ V, which gives $m = \frac{A_{max} - A_{min} \text{ (V)}}{A_{max} + A_{min} \text{(V)}} = 1$; the middle signal, with $A_{min} \approx 20$ V and $A_{max} \approx 80$ V, has $m = \frac{60 \text{ V}}{100 \text{ V}} = 0.6$; the bottom one (c), with $A_{min} \approx 35$ V and $A_{max} \approx 65$ V, features $m = \frac{30 \text{ V}}{100 \text{ V}} = 0.3$. Hence, our conjecture about the values of these modulating indexes was correct.

It follows from (8.1.6) that the minimum value of m is zero. True, when $A_{max} = A_{min}$, there is no modulation and, naturally, $m = 0$. What is the maximum value of m? It seems that it should be at $A_{min} = 0$, which gives $m = 1$. In fact, many textbooks consider $m = 1$ as the maximum value. Can the modulation index be greater than 1? Yes, as Figure 8.1.11 shows. This type of signal is called *overmodulated*. To compute its modulation index, we need to understand that there are phase reversals at the minimum amplitude, which means that A_{min} changes its sign. In the example shown in Figure 8.1.11, the minimum amplitude is given as $A_{min} = -20$ V, whereas $A_{max} = 120$ V. Applying (8.1.6), we compute

$$m = \frac{A_{max} - A_{min} \text{ (V)}}{A_{max} + A_{min} \text{(V)}} = \frac{120 - (-20) \text{ (V)}}{120 + (-20)(V)} = 1.4$$

Bear in mind, however, that overmodulation is not the norm and, typically, the value of an AM modulation index is between 0 and 1.

8 Analog Transmission with Analog Modulation

Example 8.1.1 Computing the AM Modulation Index

Problem

The AM signal shown in Figure 8.1.9 is received at the receiver end of an AM transmission line: (a) compute its modulation index in fraction and percent; (b) find the amplitudes of the modulating (information), A_M, and the carrier, A_C, signals; (c) estimate the frequencies of both signals.

Solution

a) Observation of Figure 8.1.9 yields $A_{max} = 80$ V and $A_{min} = 20$ V. We apply (8.1.6) and find

$$m = \frac{A_{max} - A_{min} (V)}{A_{max} + A_{min}(V)} = \frac{60 \text{ V}}{100 \text{ V}} = 0.6$$

from which, $M = 60\%$.

b) To find A_M and A_C, we use (8.1.5a) and (8.1.5b):

$$A_C (V) = \frac{A_{max} + A_{min}}{2} (V) = 50 \text{ V}$$

and

$$A_M (V) = \frac{A_{max} - A_{min}}{2} (V) = 30 \text{ V}$$

c) From Figure 8.1.9, we estimate the period of a modulating signal as $T_M \approx 10$ μs and its frequency as f_M (Hz) $= 1/T_M$ (s) $= 0.1$ MHz. Also, we count 10 cycles of a carrier signal within 10 μs, which gives $T_C = 1$ μs or $f_C = 1$ MHz.

Thus, all the problems in this example are solved.

Discussion

The ability to analyze experimental data in general and the graphs in particular is an important skill of an engineer. You should develop this ability over the course of your studies.

8.1.2.4 Relationship Between Frequencies of Information and Carrier Signals

According to (8.1.4), $m = \frac{A_M(V)}{A_C(V)}$. Therefore, we know that when $A_M = A_C$, $m = 1$, and when $A_M = 0$, $m = 0$. All this means that we know how the *ratio of the amplitudes*, $\frac{A_M(V)}{A_C(V)}$, affects an AM signal: When this ratio is near zero, the AM signal is slightly modulated, and when this ratio approaches 1, the AM signal is modulated deeper. But what about frequencies? How does the relationship between the frequencies of an information signal and a carrier signal affect an AM signal?

Figure 8.1.12 demonstrates three AM signals generated by the same circuit with various values of f_M and f_C. When f_C is 20 or 10 times greater than f_M, the AM signal is of good quality with well-pronounced features of amplitude modulation; however, when the values of frequencies f_M and f_C become closer or equal, the AM signal loses its characteristic features and we cannot even call it amplitude-modulated. Based on this observation, we come to the following conclusion:

The carrier frequency, f_C (Hz) must be at least 10 times greater than the frequency of the message signal, f_M (Hz); that is, $\frac{f_C(\text{Hz})}{f_M(\text{Hz})} \geq 10$.

(**Question**: What do you think about the ratio $\frac{A_M(V)}{A_C(V)}$? Do any restrictions or recommendations apply to this ratio?)

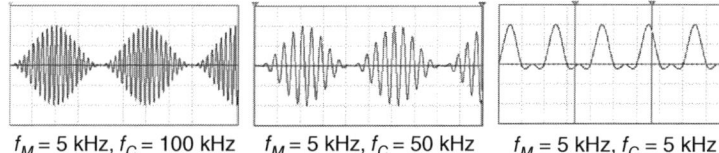

$f_M = 5$ kHz, $f_C = 100$ kHz $f_M = 5$ kHz, $f_C = 50$ kHz $f_M = 5$ kHz, $f_C = 5$ kHz

Figure 8.1.12 AM signals with various values of frequencies of message, f_M, and carrier, f_C, signals: $f_C \gg f_M$ (left), $f_C > f_M$ (middle), and $f_C = f_M$ (right).

8.1.2.5 The Formula for an AM Signal and It Instantaneous Value

We must be able to compute the value of an AM signal at any given instant, an *instantaneous value*. The path to do this may be found in (8.1.1), which enables us to find the instantaneous value of any cosine signal by plugging in the values of the amplitude, A, and the frequency, f. Thus, we first need to derive the formula describing an AM signal in its entirety.

The Formula of an Amplitude-Modulated Signal Suppose we need to deliver a message signal, $v_M(t) = A_M \cos(\omega_M t)$, using amplitude modulation. Obviously, we need to employ a carrier signal, $v_C(t) = A_C \cos(\omega_C t)$. Amplitude modulation means changing the amplitude of a carrier signal to superimpose the message on it. Hence, we need to *insert the message signal into the carrier amplitude* as

$$v_{AM}(t) = (A_C + v_M(t)) \cos(\omega_C t) \tag{8.1.7}$$

Though $v_{AM}(t)$ in (8.1.7) looks like a regular sinusoidal signal with radian frequency ω_C and amplitude $(A_C + v_M(t))$, but it is not. *What distinguishes this signal from a regular cosine is that its amplitude varies as the formula for $v_M(t)$ requires.* Based on (8.1.7), we can derive the formula for a *tone* amplitude-modulated signal as follows

$$v_{AM}(t) = (A_C + v_M(t)) \cos(\omega_C t) = (A_C + A_M \cos \omega_M t) \cos(\omega_C t) \tag{8.1.8a}$$

Using a modulation index, $m = \frac{A_M(V)}{A_C(V)}$ yields

$$v_{AM}(t) = A_C \left(1 + \frac{A_M(V)}{A_C(V)} \cos \omega_M t \right) \cos \omega_C t = A_C(1 + m \cos \omega_C t) \cos \omega_C t \tag{8.1.8b}$$

To further explore the formula for an AM signal (8.1.8b), we need to use the well-known trigonometric identity: $\cos A \times \cos B = \frac{1}{2}(\cos(A-B) + \cos(A+B))$ and find

$$v_{AM}(t) = (A_C + A_M \cos \omega_M t) \cos \omega_C t = A_C \cos \omega_C t + \frac{1}{2}(A_M \cos \omega_M t \times \cos \omega_C t)$$
$$= A_C \cos \omega_C t + \frac{A_M}{2} \cos(\omega_C - \omega_M)t + \frac{A_M}{2} \cos(\omega_C + \omega_M)t \tag{8.1.9}$$

You will recall that $A_M = mA_C$. Substituting for A_M in the aforementioned expression, we obtain the explicit formula for a sinusoidal amplitude-modulated signal:

$$v_{AM}(t) = A_C \cos \omega_C t + m \frac{A_C}{2} \cos(\omega_C - \omega_M)t + m \frac{A_C}{2} \cos(\omega_C + \omega_M)t \tag{8.1.10}$$

This formula allows us to calculate an amplitude-modulated signal in its entirety and compute its instantaneous value.

Instantaneous Value of an AM Signal Applying (8.1.9) or (8.1.10), we can compute the instantaneous value of an AM signal. For example, for the AM signal shown in Figure 8.1.9 with $A_M = 30$ V, $A_C = 50$ V, $f_M = 0.1$ MHz, and $f_C = 1$ MHz at $t = 3$ μs we compute

$$v_{AM}(t)|_{t=3\,\mu s} = A_C \cos \omega_C t + \frac{A_M}{2} \cos(\omega_C - \omega_M)t + \frac{A_M}{2} \cos(\omega_C + \omega_M)t = -77.9 \text{ V}$$

Return to Figure 8.1.9, find the point $t = 3$ μs, and identify the magnitude value of the signal at this instant.

Envelope of an AM Signal An envelope is an imaginary line within which the signal is contained. Review Figure 8.1.9 and ask yourself what line or lines delimit an AM signal. The answer is given in Figure 8.1.13, where these two lines (that is envelopes) are shown. The top line is a *positive envelope* and the bottom line is a *negative envelope*. To derive the formulas for these envelopes, we recall that (8.1.8a) gives the general formula for an AM signal, $v_{AM}(t) = (A_C + A_M \cos \omega_M t) \cos(\omega_C t)$. The maximum values of $v_{AM}(t)$ are reached when $\cos(\omega_C t) = 1$, and the minimum values are at $\cos(\omega_C t) = -1$; therefore, $v_{AM}(t)$ is delimited between these extremes. Consequently, the positive envelope is given as

$$v_{ENV}^+(t) = (A_C + A_M \cos \omega_M t) \quad (8.1.11a)$$

and the negative envelope is

$$v_{ENV}^-(t) = -(A_C + A_M \cos \omega_M t) \quad (8.1.11b)$$

Figure 8.1.13 illustrates the concept of an envelope with reference to the AM signal shown in Figure 8.1.9.

(**Exercise**: Build two envelopes of an AM signal by taking the numbers form Example 8.1.1.)

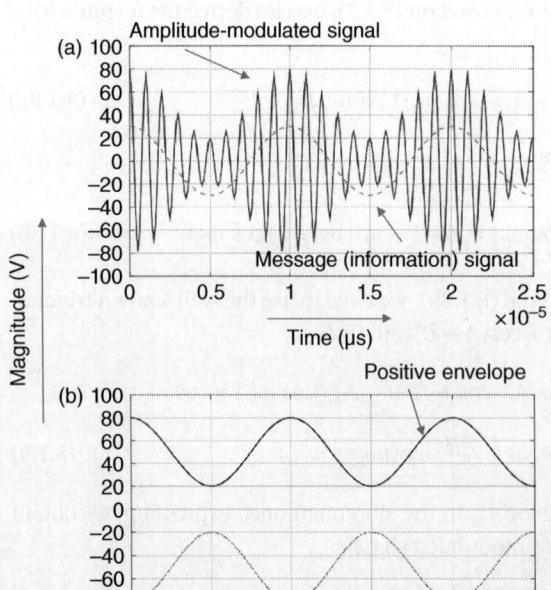

Figure 8.1.13 Envelopes of an AM signal: (a) An AM signal (reproduction of Figure 8.1.9) and (b) positive and negative envelopes of the AM signal.

8.1.2.6 The Spectrum of an AM Signal

What frequencies does an AM signal consist of? We know that there are two input signals – information and carrier – that form the resulting AM signal; however, these signals are not simply added but combined in a more sophisticated manner, as (8.1.9) shows. Referring to this equation, we can see that an AM signal consists of three sinusoidal signals:

$$A_C \cos \omega_C t \equiv A_C \cos 2\pi f_C t$$
$$\frac{A_M}{2} \cos(\omega_C - \omega_M)t \equiv \frac{A_M}{2} \cos 2\pi(f_C - f_M)t$$
$$\frac{A_M}{2} \cos(\omega_C + \omega_M)t \equiv \frac{A_M}{2} \cos 2\pi(f_C + f_M)t$$

You will recall from our discussions in Section 2.1 and Chapter 7 that every cosine signal can be represented by a unique line in frequency domain. Following this rule, we show these three cosines in frequency domain in Figure 8.1.14. This figure, in fact, presents *the spectrum* of an AM signal, which has three spectral components:

- *carrier signal* with frequency f_C and amplitude A_C,
- *lower sideband (LSB)* with frequency $f_C - f_M$ and amplitude $A_M/2$,
- *upper sideband (USB)* with frequency $f_C + f_M$ and amplitude $A_M/2$.

Figure 8.1.14 also shows the information signal with frequency f_M and amplitude A_M to give the scale of the figure (remember, $\frac{f_C(\text{Hz})}{f_M(\text{Hz})} \geq 10$). From examining Figure 8.1.14, we come to the following conclusion:

> *In an AM signal, the information is delivered by the sideband signals, whereas a carrier spectral component contains no information.*

(Can you justify this conclusion? *Hint*: Observe the amplitudes and frequencies of the spectral components.)

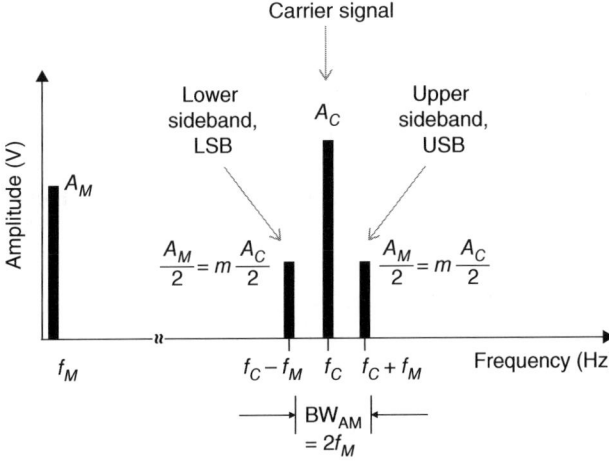

Figure 8.1.14 Spectrum of an AM signal.

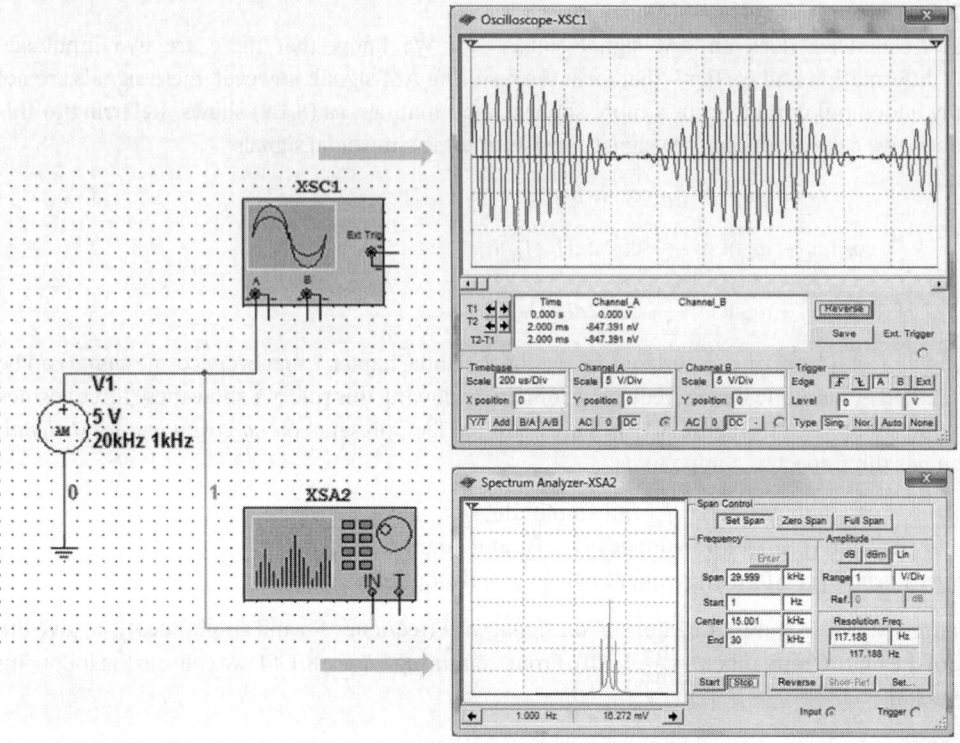

Figure 8.1.15 The setup of the experiment (left) and the waveform (top right) and the spectrum (bottom right) of the same AM signal.

Having the spectrum, we can determine the <u>bandwidth</u> *of an AM signal*. The bandwidth of a signal, by definition, is the difference between the signal's maximum and minimum frequencies. For an AM signal, the maximum frequency is $f_C + f_M$ and the minimum is $f_C - f_M$. Therefore,

$$\text{BW}_{AM} \text{ (Hz)} = (f_C + f_M) - (f_C - f_M) = 2f_M \tag{8.1.12}$$

How realistic are our theoretical considerations of the concept and spectrum of a tone AM signal? Figure 8.1.15 demonstrates the Multisim simulation, where the same AM signal is simultaneously presented to the oscilloscope, which shows the signal's waveform in time domain, and the spectrum analyzer, which shows the signal's spectrum in frequency domain. These figures confirm our theory of an AM signal. It is important to master the time- and frequency-domain presentations; engineers rely on this dual view of a signal in their practical work.

Example 8.1.2   Finding the Spectrum of an AM Signal

Problem

Compute and depict the frequency spectrum of a sinusoidal AM signal with the following parameters: $A_C = 50$ V, $A_M = 30$ V, $\omega_C = 6.28 \times 10^6$ rad/s, and $\omega_M = 6.28 \times 10^3$ rad/s.

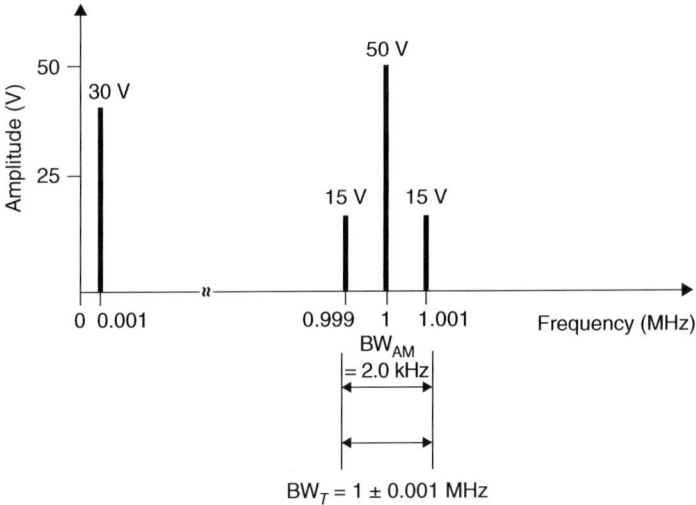

Figure 8.1.16 Spectrum of the AM signal in Example 8.1.2.

Solution

We need to calculate all the parameters shown in Figure 8.1.14. They are

Given: f_C (Hz) $= \omega_C/2\pi = 1 \times 10^6$ Hz, f_M (Hz) $= \omega_M/2\pi = 1 \times 10^3$ Hz, $A_C = 50$ V, and $A_M = 30$ V
Computed: $f_C - f_M$ (Hz) $= (\omega_C - \omega_M)/2\pi = 0.999 \times 10^6$ Hz and $f_C + f_M$ (Hz) $= 1.001 \times 10^6$ Hz

$$m = 0.6 \Rightarrow m\frac{A_C}{2} = \frac{A_M}{2} = 15 \text{ V}$$

The frequency spectrum of this signal is shown in Figure 8.1.16. Study this figure carefully.

Discussion

- We compute and depict the bandwidth of the AM signal, BW_{AM} (Hz), which is the *band of frequencies occupied by the AM signal*; it is equal, in this example, to $2f_M = 2$ kHz. For transmission of this signal, however, we need the *bandwidth of the transmission link*, BW_T, which is the same band of frequencies but centered on f_C; that is, $BW_T = 1 \pm 0.001$ MHz in this example. This is to say that not all 2 kHz values are created equal from the transmission standpoint. For instance, 2 kHz from 0 to 2 kHz is easy and inexpensive to transmit, but 2 kHz from 999 to 1001 kHz is more difficult and thus more expensive to transmit.
- In Figures 8.1.14 and 8.1.16, we depict the spectrum as a set of lines, but the screen of the spectrum analyzer in Figure 8.1.16 shows not lines but pulse-like shapes. Why is this? The fact is that only an ideal *infinite* sinusoidal signal has a single frequency and is represented in frequency domain by a single line. In reality, any sinusoidal signal is *finite* and therefore contains from a few to many frequencies (refer to Section 7.3). All these frequencies are concentrated around a central frequency to form a continuous spectrum. Such a spectrum takes the shape of a peak because the central frequency has the biggest amplitude, the next two symmetrical peripheral frequencies have smaller amplitudes, the third peripheral frequency pair has even smaller amplitudes,

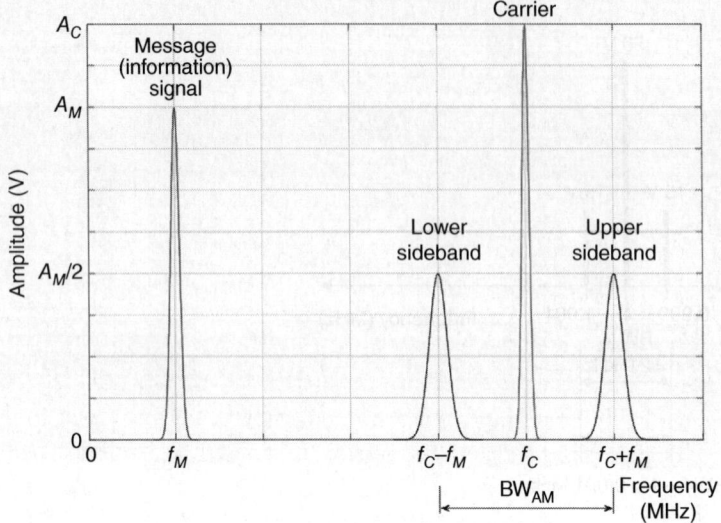

Figure 8.1.17 The real spectrum of an AM signal: Each spectral component includes many frequencies that form a continuous local spectrum around its central frequency; here, the central frequencies in hertz are $f_M, f_C - f_M, f_C,$ and $f_C + f_M$.

and so on. Examine Figure 8.1.17 and observe how the above explanation applies to the spectrum of a modulating signal with a central frequency, f_M, and to the LSB ($f_C - f_M$), carrier (f_C), and to USB ($f_C + f_M$) spectral components of an AM signal.

A close look at Figures 8.1.14–8.1.17 shows that amplitude modulation results in shifting the low frequency of an information signal to the vicinity of the high frequency of a carrier. This is why this type of modulation is often called modulation by *frequency translation*, or *frequency conversion*. It is mathematically described by the multiplication of a baseband signal by a sinusoidal carrier, as demonstrated in Figure 8.1.10. Bear this term in mind when studying any other types of analog modulation (frequency and phase modulations), which we will be doing shortly.

8.1.2.7 Power Distribution in an AM Signal

To analyze the power distribution in an AM signal, refer to (8.1.10), $v_{AM}(t) = A_C \cos \omega_C t + m\frac{A_C}{2} \cos(\omega_M - \omega_C)t + m\frac{A_C}{2} \cos(\omega_M + \omega_C)t$. As you will recall from your circuit-analysis course, the average power of a sinusoid delivered to a load with impedance R is equal to

$$P\,(\text{W}) = \frac{(V_{\text{rms}})^2}{R} = \frac{\left(\frac{V_{\text{pk}}}{\sqrt{2}}\right)^2}{R} = \frac{(V_{\text{pk}})^2}{2R} \tag{8.1.13}$$

where V_{rms} is the rms voltage and V_{pk} is the peak voltage. For the spectral components of the AM signal shown in Figure 8.1.16, V_{pk} is equal to A_C and $A_M/2$, respectively. Thus,

$$P_C\,(\text{W}) = \frac{(A_C)^2}{2R} \tag{8.1.14a}$$

and

$$P_{USB}(W) = P_{LSB} = P_{SB} = \frac{\left(\frac{A_M}{2}\right)^2}{2R} = \frac{m^2}{4}\frac{(A_C)^2}{2R} = \frac{m^2}{4}P_C \qquad (8.1.14b)$$

The total power, P_T, of an AM signal is equal to

$$P_T(W) = P_C + P_{USB} + P_{LSB} = P_C + 2P_{SB} = P_C + 2\frac{m^2}{4}P_C = \left(1 + \frac{m^2}{2}\right)P_C \qquad (8.1.14c)$$

To clearly interpret this formula, let us consider a case where $m = 1$. Then the power of each sideband is equal to $P_{SB} = 0.25P_C$ and the total power is given by $P_T(W) = P_C + 2P_{SB} = 1.5P_C$. Thus, $P_C(W) \approx 0.67P_T$ and $2P_{SB}(W) = 0.5P_C \approx 0.33P_T$. In other words, *almost 67% of the total power is consumed by the carrier signal, which delivers no information, whereas only 33% of the total power is concentrated in the information signals.* This is the price we pay for the advantages of AM transmission.

To look at this phenomenon from a different angle, we introduce the *power efficiency*, η, of AM transmission, defined as

$$\eta = \frac{P_M(W)}{P_T(W)} \qquad (8.1.15a)$$

where $P_M = 2P_{SB}$ is the sum of the power of two message signals and P_T is the total power. Using (8.1.14a)–(8.1.14c), we obtain

$$\eta = \frac{P_M(W)}{P_T(W)} = A_M^2/(A_C^2 + 2A_M^2) \qquad (8.1.15b)$$

Considering, as in the previous example, $m = 1$, which means that $A_M = A_C$, we find

$$\eta = P_M/P_T = \frac{A_M^2}{A_C^2 + 2A_M^2} = 0.33$$

This result shows again the power inefficiency of amplitude modulation.

Example 8.1.3 Power Distribution in an AM Signal

Problem

Calculate (1) the total power and (2) the power ratio of the sideband signals and the carrier signal to the total power of the amplitude-modulated signal if $A_C = 50$ V, $A_M = 30$ V, and $R = 50\,\Omega$.

Solution
(1) The solution is straightforward: Plug in the values of the amplitudes of the carrier and sideband signals in (8.1.14a) and (8.1.14b) and compute:

$$P_C = \frac{(A_C)^2}{2R} = 25\text{ W}$$

$$P_{USB} = P_{LSB} = \frac{\left(\frac{A_M}{2}\right)^2}{2R} = 2.25\text{ W}$$

To find the required total power by (8.1.15b), we compute the modulation index as $m = \frac{A_M}{A_C} = 0.6$ and calculate

$$P_T = \left(1 + \frac{m^2}{2}P_C\right) = 1.18 P_C = 29.5 \text{ W}$$

We can check our solution by summing up all the constituent powers as

$$P_T = P_C + P_{USB} + P_{LSB} = 25 + 2.25 + 2.25 = 29.5 \text{ W}$$

(2) The power ratio of the carrier and sideband signals to the total power is given as

$$\frac{P_C}{P_T} = \frac{25 \text{ (W)}}{29.5 \text{ (W)}} = 0.85$$

and

$$\frac{P_{SB}}{P_T} = \frac{2.25 \text{ (W)}}{29.5 \text{ (W)}} = 0.08$$

The problem is solved.

Discussion

- Thus, the sideband containing the information consumes only 8% of the total power of the transmitted AM signal. Again, this is a major drawback to the use of a carrier signal. However, the picture is not totally bleak. We need to maintain a proper perspective on this matter; that is, we need to recall, too, all the advantages of broadband transmission. In addition, there are transmission techniques that help to cope with this problem. These will be discussed in the section that follows.
- Note that for calculating the *ratio* of power of all constituent signals, we do not need to know the value of a load resistor.

8.1.2.8 AM Modulation and Demodulation

Now, that we have become familiar with the basics of amplitude modulation and its signal, we need to consider modulation and demodulation operations in AM transmission. Refer to Figure 8.1.1 to refresh your memory of the block diagram of an AM transmission system.

At the sending end, an *AM transmitter* performs *modulation*, which is a straightforward operation: A transducer converts an information signal (e.g. voice) into an electrical signal and presents it to a modulator. An oscillator generates a carrier sinusoid and presents its signal to the modulator too. This modulator performs the operations described by the formula for an AM signal, $v_{AM}(t) = A_C \cos\omega_C t + A_M \cos\omega_M t \cos\omega_C t$, shown in (8.1.8a). After additional amplification, this AM signal is sent out and broadcast by an antenna.

(**Exercise**: Based on the given description, sketch the block diagram of an AM transmitter and the circuit of an AM modulator.)

At the destination point, the receiver must extract the message (information) signal from the received AM signal; that is, the receiver must perform *demodulation*. Figure 8.1.18 shows the principle of operation of an AM demodulator, called an *envelope detector*. The structure of this detector can be determined by knowing that an *information signal is the positive envelope of an AM signal*.

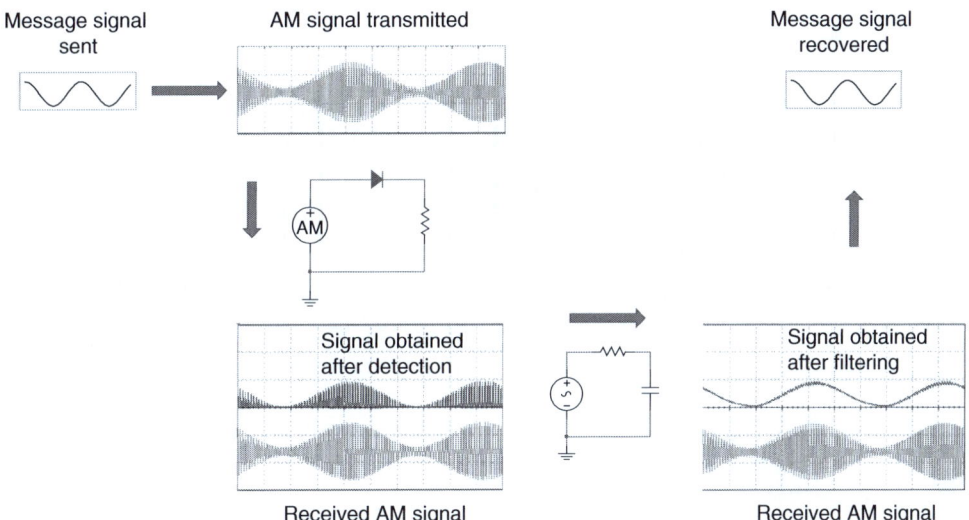

Figure 8.1.18 Principle of operation of an AM envelope detector (demodulator).

Therefore, the first step in demodulation should be removing the negative half of an AM wave; this is accomplished by a diode. Nonetheless, the positive half of the AM wave still contains both the high-frequency oscillations and the low-frequency information signal. Therefore, we need to employ a low-pass filter (LPF) whose output is the peak variations (envelope) of the received AM signal. This is how the message signal can be recovered.

(**Exercise**: The value of the cutoff frequency, f_C, of a LPF is critical for quality recovery of a message signal. If f_C is too small, the high-frequency components will appear at the output – observe in Figure 8.1.18 the ripples at the signal obtained after filtering. What if f_C is too big? Build the circuit of an AM envelope detector and simulate its operation with various values of f_C.)

Figure 8.1.18 illustrates the demodulation of a single AM signal. Our radio set, however, receives many AM signals from different stations and we want to choose one of them. We do this by *tuning*, that is, making our receiver choose the carrier frequency of a desired station, as shown in Figure 8.1.19a. This operation is done by a superheterodyne AM detector, whose block diagram is shown in Figure 8.1.19b. The detector contains a local oscillator whose frequency, f_{LO}, can be tuned as needed. The main point is that changing the RF tuning for a desired station (that is, f_C), we simultaneously vary the frequency f_{LO}, so that the difference, called intermediate frequency, $f_{IF} = f_C - f_{LO}$, remains constant. This simultaneous tuning is called *ganged tuning*. Thanks to the ganged tuning, we obtain an AM signal with a constant carrier frequency f_{IF} from which a demodulator can recover a message signal as discussed previously in conjunction with Figure 8.1.18. Automatic gain control (AGC) provides the constant gain for incoming signals with various power.

(*Question: In general, our radio set receives simultaneously many AM signals of various carrier frequencies. We tune a receiver in search of the carrier frequency of the desired radio station, which means we keep changing the receiver's frequency. The text says, however, that the superheterodyne detector processes the AM signal with a constant frequency. How could that be?*)

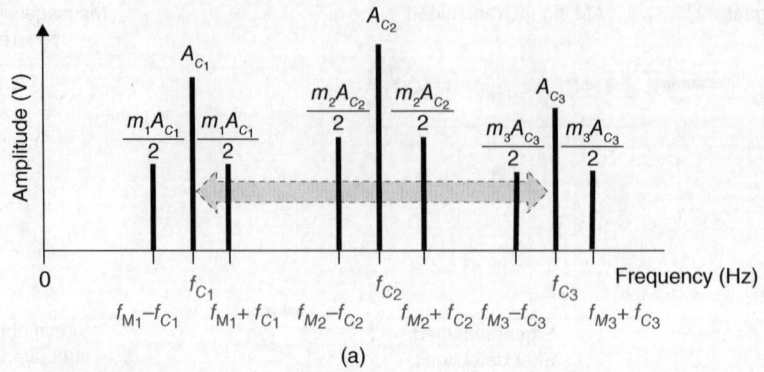

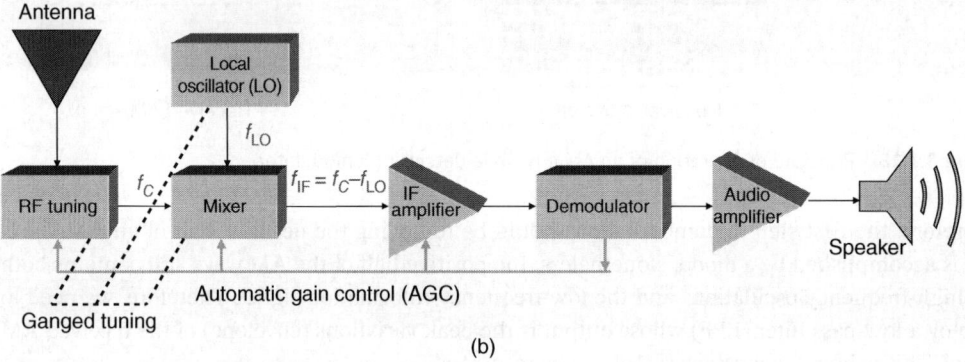

Figure 8.1.19 Principle of operation of a superheterodyne AM detector: (a) tuning in frequency domain and (b) block diagram of an AM super heterodyne detector.

8.1.2.9 The Main Drawback of Amplitude Modulation

You will recall that an AM signal carries information by its amplitude. Noise distorts the amplitude of any signal. Thus, the amplitude of a received AM signal will be distorted and the received information might be wrong. This drawback is caused by the very nature of the AM signal, and it cannot be eliminated by any technological advances. Figure 8.1.20 shows the AM signal sent by a transmitter and the received AM signal distorted by noise.

The ultimate solution to this problem is to avoid using an amplitude for carrying information. Figure 8.1.6 shows that the other carrier parameter that can be used for modulation is the carrier signal's frequency. This is how we arrived at the need for the FM which is considered in the next subsection.

This concludes the basics of amplitude modulation.

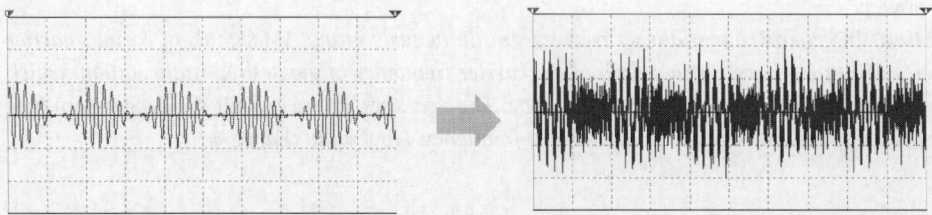

Figure 8.1.20 A clear AM signal sent (left) and a noise-distorted AM signal received (right).

8.1.3 Basics of Frequency Modulation (FM)

8.1.3.1 Frequency Modulation: Why and What

Frequency modulation, FM, was invented to overcome the main problem of amplitude modulation namely, to make a modulated signal less susceptible to noise. To understand how, Figure 8.1.5, where the concept of FM was introduced, is reproduced here for your convenience.

As Figure 8.1.5R shows, the main feature of a frequency-modulated signal is this:

> *The frequency of an FM signal changes as the message (modulating, information, or intelligence) signal requires.*

Figure 8.1.5R The concept of frequency modulation.

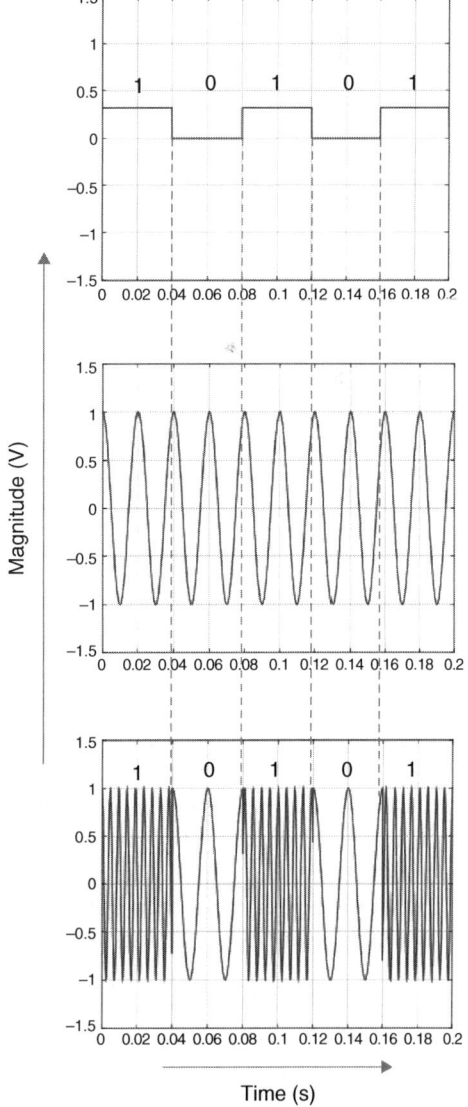

Note also the following distinguishing features of an FM signal:
- The amplitude of an FM signal is constant; this is in contrast to that of an AM signal, whose amplitude varies according to changes in the message signal.
- The waveform of an FM signal does not resemble an information signal in any way; again, this is in contrast to an AM signal, whose envelope repeats the changes in the information signal.

The main advantage of an FM signal, obviously, is that the distortion of the signal's amplitude does not directly affects the transmission quality. Your own experience with AM and FM radio, we are sure, confirms that FM provides a higher fidelity transmission.

The constancy of an FM signal's amplitude has a simple consequence: The average power of an FM signal does not depend on the message being transmitted; it therefore remains constant. To prove this statement, refer to (8.1.14a):

$$P\,(\text{W}) = \frac{(A)^2}{2R}$$

where P (W) is the average signal power, A (V) is the signal amplitude, and R (Ω) is the resistance over which this power is developed. Since an FM signal has constant amplitude, A_C, its average power stays constant.

8.1.3.2 The Frequency of an FM Signal

To make the discussion of the frequency of an FM signal more specific, we turn to the simplest case of an FM transmission in which the message (modulating, or information, or intelligence) signal is sinusoidal. (Remember, in analog modulation, a carrier is always a sinusoidal signal.) Since a sinusoidal signal represents a single acoustical tone, this type of frequency modulation is called a *single-tone*. The basic parameters of a single-tone FM signal are similar to those of an AM tone signal; that is, they are amplitudes and frequencies of the message and carrier signals and their combinations. Omitting phases, as we did in our discussion of amplitude modulation, we can write the following expressions for the message (modulating) signal,

$$v_M(t) = A_M \cos(\omega_M t) = A_M \cos(2\pi f_M t)$$

and for the carrier signal,

$$v_C(t) = A_C \cos(\omega_C t) = A_C \cos(2\pi f_C t)$$

(**Question**: *Before proceeding with this discussion, ask yourself which signal – carrier or modulating – you want to change and which parameter of the signal you want to change.*)

We need to realize that frequency modulation is the process of changing the carrier frequency in accord with changes in the message signal. Specifically, for a sinusoidal FM, we want the following:

When the magnitude of a message signal increases to the maximum value, A_M, the frequency of the FM signal also increases to the maximum, $(f_{FM})_{max}$; when the magnitude of a modulating (information) signal decreases to the minimum value, $-A_M$, the frequency of the FM signal decreases to its minimum value, $(f_{FM})_{min}$, too. Figure 8.1.21 visualizes this explanation

Therefore, the frequency of an FM signal must be the sum of the carrier frequency, f_C, and the information signal, $v_M(t) = A_M \cos(2\pi f_M t)$, with the appropriate coefficient, as

$$f_{FM}(t)\,(\text{Hz}) = f_C + k_{FM} A_M \cos(2\pi f_M t) \tag{8.1.16}$$

Here k_{FM} (Hz/V) is a *modulation (deviation) constant,* also called the *frequency–sensitivity factor* of a demodulator. We elaborate on the meaning of this coefficient shortly.

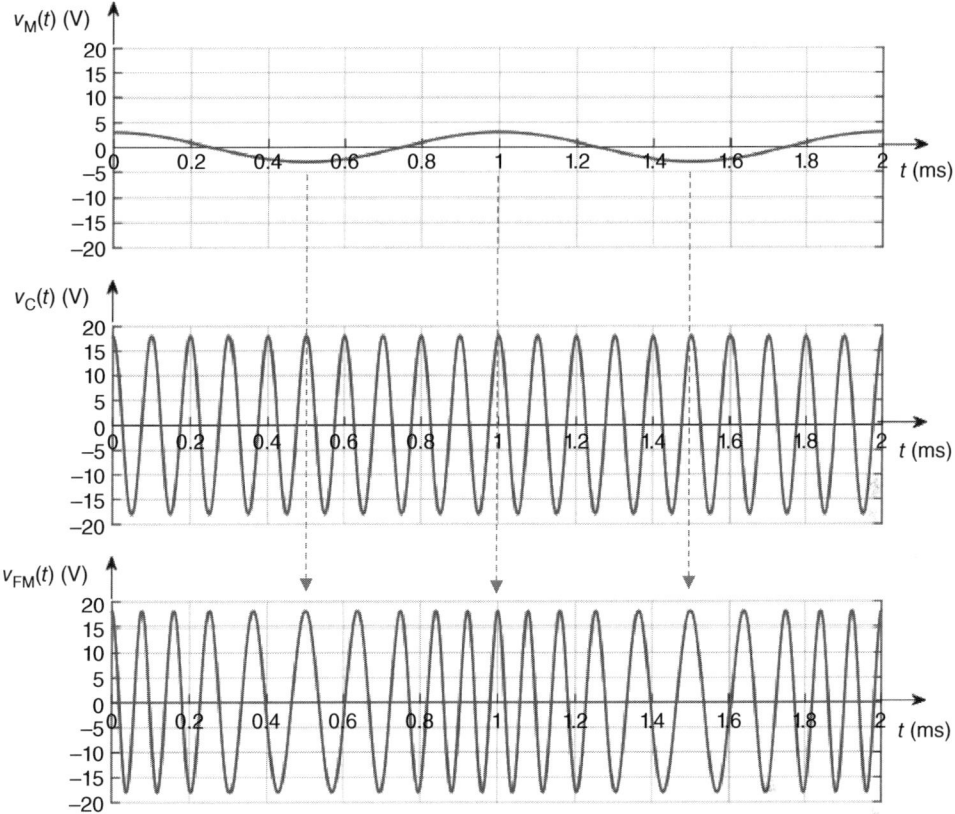

Figure 8.1.21 Single-tone FM signal: Message signal (top), carrier wave (middle), and FM signal (bottom).

As we have already learned, the frequency of a regular sinusoidal signal is a constant (number). In FM, however, $f_{FM}(t)$ is always changing, and therefore, (8.1.16) gives us an *instantaneous value* of an FM frequency.

Equation (8.1.16) explicitly presents the idea of frequency modulation:

> *The frequency of an FM signal follows the variations of a message signal that carries information.*

Figure 8.1.22 illustrates this concept.

The frequency of an FM signal, $f_{FM}(t) = f_C + k_{FM} A_M \cos(2\pi f_M t)$, deviates from the frequency of a carrier, f_C, by $k_{FM} A_M \cos(2\pi f_M t)$. This deviation is proportional to the amplitude of the message signal, A_M (V), and depends on its frequency, f_M (Hz) and on the modulation (deviation) constant, k_{FM} (Hz/V). Since the value of a cosine is restricted as $-1 \leq \cos(2\pi f_M t) \leq 1$, the *peak (maximum) deviation of an FM frequency from a carrier frequency* is given by

$$\Delta f \text{ (Hz)} = k_{FM} A_M \tag{8.1.17}$$

so that f_{FM} (Hz) varies between $f_C + k_{FM} A_M$ and $f_C - k_{FM} A_M$; that is,

$$f_C - k_{FM} A_M \leq f_{FM} \leq f_C + k_{FM} A_M \tag{8.1.18}$$

8 Analog Transmission with Analog Modulation

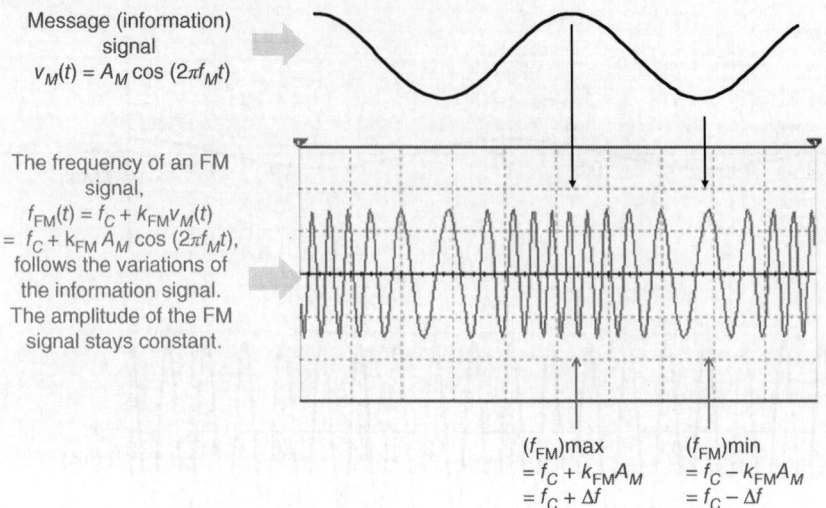

Figure 8.1.22 The frequency of an FM signal.

In other words,

$$(f_{\text{FM}})_{\max} \text{ (Hz)} = f_C + k_{\text{FM}} A_M = f_C + \Delta f \quad (8.1.19a)$$

and

$$(f_{\text{FM}})_{\min} \text{ (Hz)} = f_C - k_{\text{FM}} A_M = f_C - \Delta f \quad (8.1.19b)$$

Equations (8.1.19a) and (8.1.19b) introduce two new parameters of an FM signal: the maximum and minimum values of the frequency of a modulated signal.

At this point, you might rightly ask what is the mystical $k_{\text{FM}}(t)$ (Hz/V)? The answer should be sought in an analysis of the units of (8.1.17) and (8.1.18). All members of these equations are frequencies measured in hertz. Since the product, $k_{vM}(t) A_M$, must be in hertz, too, whereas A_M is in volts, the dimension of $k_{vM}(t)$ must be (Hz/V); that is,

$$\Delta f \text{ (Hz)} = k_{\text{FM}}(t) \text{ (Hz/V)} A_M(\text{V}) \quad (8.1.20)$$

We will learn shortly that this coefficient, $k_{vM}(t)$ (Hz/V), carries very real function: It is the input/output characteristic of a VCO, the device performing frequency modulation in practice. It also characterizes the sensitivity of an FM demodulator to a change in FM frequency.

Example 8.1.4 The Peak Frequency Deviation of an FM Signal

Problem

Given: Message (information) signal, $v_M(t)$ (V) = $3\cos(6.28 \times 10^3 t)$, carrier signal, $v_C(t)$ (V) = $18\cos(6.28 \times 10^4 t)$, and the modulation constant, k_{FM} (Hz/V) = 1000.

Calculate: Peak (maximum) frequency deviation and maximum and minimum values of the FM frequency.

Plot: An FM signal showing all the computed values on the figure.

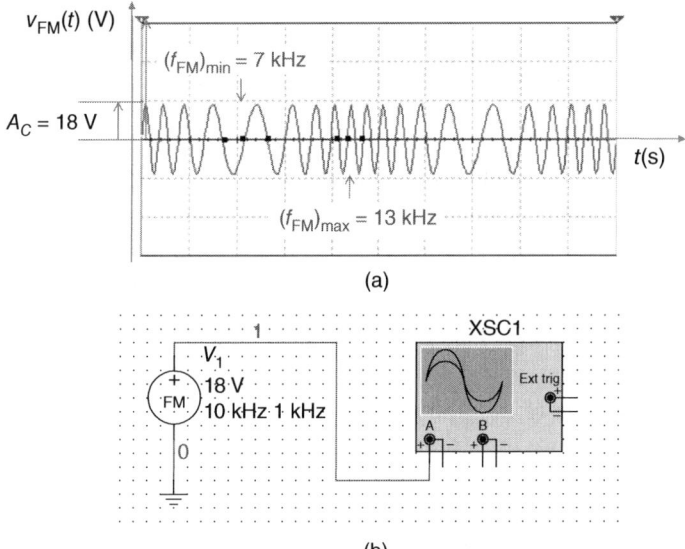

Figure 8.1.23 The plot of the FM signal (a) and the simulation circuit (b) for Example 8.1.4.

Solution

Since (8.1.16)–(8.1.22) refer mostly to the spectral (cyclic) frequencies, we will convert the given radian frequencies into frequencies, as follows:

$$\omega_M = 6.28 \times 10^3 \text{ rad/s} \Rightarrow f_M = \omega_M/2\pi \approx 1 \text{ kHz}$$

and

$$\omega_C = 6.28 \times 10^4 \text{ rad/s} \Rightarrow f_C = \frac{\omega_C}{2\pi} \approx 10 \text{ kHz}$$

Applying (8.1.18) for peak frequency deviation, we compute

$$\Delta f = k_{FM} A_M = 3 \text{ kHz}$$

The use of (8.1.19a), (8.1.19b), and (8.1.20) yields

$$(f_{FM})_{max} = f_C + \Delta f = 13 \text{ kHz}$$

and

$$(f_{FM})_{min} = f_C - \Delta f = 7 \text{ kHz}$$

Figure 8.1.23 shows the plot of the FM signal (a) and the simulation circuit (b) for this example. The figure displays the regions of the signal's waveform with maximum and minimum frequencies, which correspond to the maximum and minimum values of the information signal's amplitude.

Discussion

- We usually want $f_C \gg f_M$ and $A_C > A_M$, and in this example we follow these recommendations. Such a relationship is quite similar to those required in amplitude modulation.

- Again, pay particular attention to the fact that the dimension of k_{FM} is Hz/V, which provides the dimension of Δf in Hz. As we have mentioned many times, checking the dimensions of formula members is the first step in verifying the accuracy of the mathematical operations.
- *Timely tip*: In considering a signal's frequency, we obviously should consider its period as well; however, the frequency of an FM signal varies because it is a function of time. Therefore, the general concept of a period cannot be applied to an FM signal and cannot be shown on its waveform. This explains why in FM we always refer to *instantaneous frequency*, which is a derivative of a phase. (Recall the definition of a derivative.) There is, however, no such quantity as an "instantaneous period." This is why we do not refer to a "period" in discussing an FM signal's waveform.
- To perceive the effect of an information signal on an FM waveform, refer to the points where this waveform crosses the time axis. These points are called, obviously, *zero crossings*, several of which are shown as black dots in Figure 8.1.23. There we can clearly see how the time intervals between the zero crossings in the area of $(f_{FM})_{max}$ differ from the time intervals between the zero crossings in the area of $(f_{FM})_{min}$. Since these differences are caused by a variation in the information signal, we can say that *the zero crossings of an FM signal contain information about a message signal*.

8.1.3.3 Modulation Index of an FM Signal

Since in AM operation we vary the magnitudes, we need to know the swing in amplitudes; that is, $A_{max} = A_C + A_M$ and $A_{min} = A_C - A_M$. And since in FM operation we vary the frequencies, we need to know how frequencies fluctuate. Recall that Δf (Hz) is the peak deviation of the frequency of an FM signal; hence, the greater or smaller the Δf, the greater or smaller the fluctuations of f_{FM}. This means that Δf (Hz) *indicates the depth of the frequency modulation*. In our discussion of amplitude modulation, we introduced a *modulation index* as a measure of the depth of modulation. Following the same logic, we can introduce the modulation index for frequency modulation; this index must be proportional to Δf (Hz). Typically, the term *modulation index* suggests that this should be a dimensionless number and to meet this requirement, the industry chooses the ratio of Δf (Hz) to f_M (Hz). The Greek letter β (beta) or another notation, m_f, is used to designate the *modulation index of an FM signal*:

$$\beta = m_f = \frac{\Delta f (\text{Hz})}{f_M (\text{Hz})} = \frac{(k_{FM} A_M)(\text{Hz})}{f_M (\text{Hz})} \qquad (8.1.21)$$

Note well:

> *In FM, the modulation index is proportional to the amplitude of an information signal.*

The value of the modulation index of an FM system varies from small (<0.25) to big (>0.25). If $\beta < 0.25$, FM is called *narrowband*; if $\beta > 0.25$, FM is called *wideband*. Typically, the FM modulation index, β (m_f), is greater than 1. This is in contrast to amplitude modulation, where m is usually no greater than 1. Examine Figure 8.1.24 to visualize the idea of FM modulation index.

To build the graphs in Figure 8.1.24, we used the following formula for a single-tone FM signal

$$v_{FM}(t) = A_C \cos(2\pi f_C t + \beta \sin(2\pi f_M t)) \qquad (8.1.22)$$

The derivation of this formula is presented in Section 8.2.

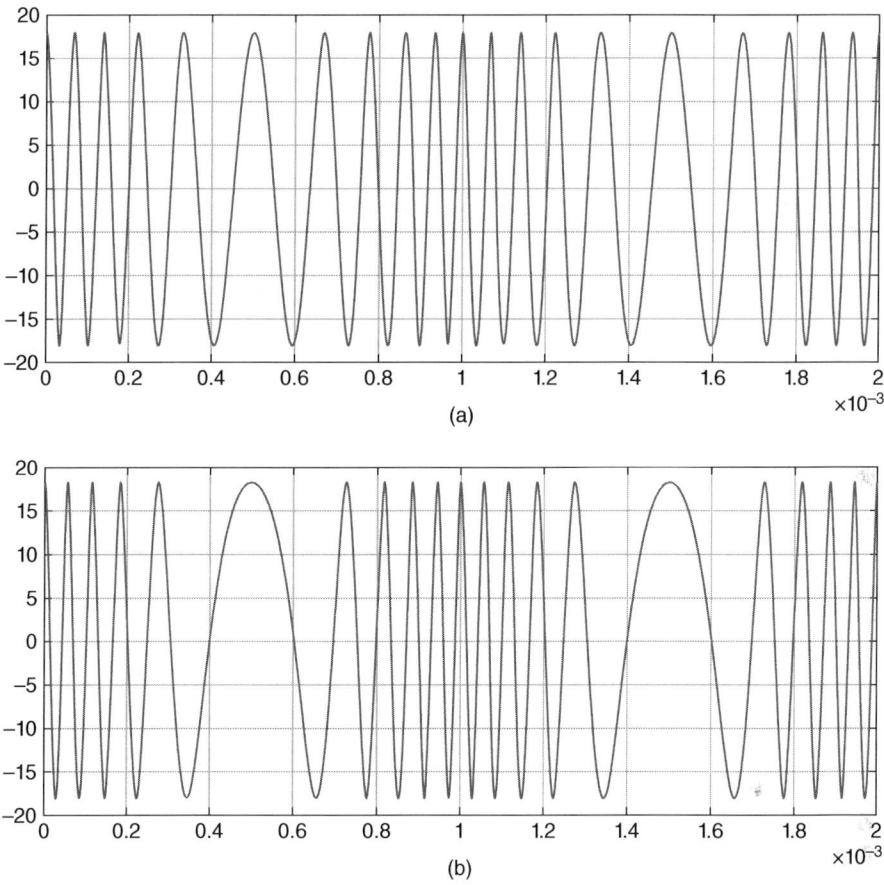

Figure 8.1.24 FM modulation index: This is a measure of the depth of modulation and it is proportional to the amplitude of a modulating (information) signal, A_M. All parameters of these two signals are the same except for A_M, which changes from $A_M = 5$ V (a) to $A_M = 8$ V (b). The modulation index of the top signal is $\beta = 5$. Can you surmise the modulation index of the bottom signal?

Example 8.1.5 Calculating the FM Modulation Index

Problem

Calculate the modulation index for the message (information) signal, $v_{M1}(t)$ (V) $= 5\cos(6.28 \times 10^3 t)$, carrier signal, $v_C(t)$ (V) $= 18\cos(6.28 \times 10^4 t)$, and the modulation constant, k_{FM} (Hz/V) $= 1000$. What will be the modulation index for a new message signal $v_{M2}(t)$ (V) $= 8\cos(6.28 \times 10^3 t)$?

Solution

The modulation index is given by (8.1.21) as

$$\beta = \frac{\Delta f(\text{Hz})}{f_M(\text{Hz})} = \frac{(k_{FM} A_M)(\text{Hz})}{f_M(\text{Hz})}$$

Plugging in the numbers, we compute $\beta_1 = 5$ for $v_{M1}(t)$ and $\beta_2 = 8$ for $v_{M2}(t)$. The problem is solved.

Discussion

- To emphasize the concept of modulation index, we plot in Figure 8.1.24 two FM signals discussed in this example. Can you determine which signal is $v_{M1}(t)$ and which is $v_{M2}(t)$?
- Analyzing Figure 8.1.24, we can easily see the difference: The greater the depth of modulation, the greater the deviation of the FM frequency, $\Delta f = k_{FM} A_M$, from a nonmodulated carrier frequency, f_C, and the greater the modulation index, $\beta = \frac{\Delta f (\text{Hz})}{f_M (\text{Hz})} = \frac{(k_{FM} A_M)(\text{Hz})}{f_M (\text{Hz})}$. Since the frequency of a message signal, f_M, is constant, changing the modulation index, β, means changing the peak frequency deviation, Δf. This is why *some books refer to Δf (Hz) as a modulation index* per se. Be aware of these two definitions of the *modulation index* when working out the problems.
- Observe also that the amplitudes of both FM signals are constant and equal to A_C; it is the change in A_M that affects the frequency deviation, Δf, and consequently the modulation index, β.

We need to remember that all the aforementioned considerations are made for a single-tone FM signal in which a message signal is a single sinusoid.

8.1.3.4 The Spectrum and Bandwidth of an FM Signal

Performing the spectral analysis of a frequency-modulated signal in a formal, mathematical way is a rather sophisticated undertaking. We pursue this topic in Section 8.2. At this stage of our study, we cannot obtain a simple formula that would describe the frequency spectrum of an FM signal. Therefore, we will take a qualitative, intuitive approach to the subject.

The Spectrum of an FM Signal Frequency spectrum, by definition, is the set of frequencies contained in a given signal. From our study of spectral analysis, we know that a pure sinusoidal signal contains one frequency, but the signals of other waveforms contain many. This study should convince us that *the further a signal's waveform deviates from the pure sinusoidal form, the more sophisticated its spectrum*.

Figure 8.1.25 illustrates this point:

- The pure cosine contains one frequency (Figure 8.1.25a).
- A combination of two sinusoidal signals – an AM signal – contains three frequencies (Figure 8.1.25b).
- A nonsinusoidal signal – a square wave – contains an infinite number of frequencies (Figure 8.1.25c).

Based on this reasoning, let us pose a question: How many and what kinds of frequencies does an FM signal contain? The answer can be found by analyzing previous figures: *The closer an FM signal is to a pure cosine, the fewer the number of the frequencies it contains and vice versa*. Also, *the closer an FM signal is to a pure cosine, the smaller its modulation index*. Indeed, Figure 8.1.26 shows that the top FM signal is a pure cosine, which means that there is no modulation and consequently $\beta = 0$. The second signal from the top shows slight variations in its frequency. Its modulation index is $\beta = 1$; this signal contains nine harmonics. The third signal from the top in Figure 8.1.26 reveals substantial frequency deviations; no wonder its modulation index is $\beta = 3$, and the spectrum of this signal contains 15 harmonics. The bottom FM signal deviates significantly from a pure cosine.

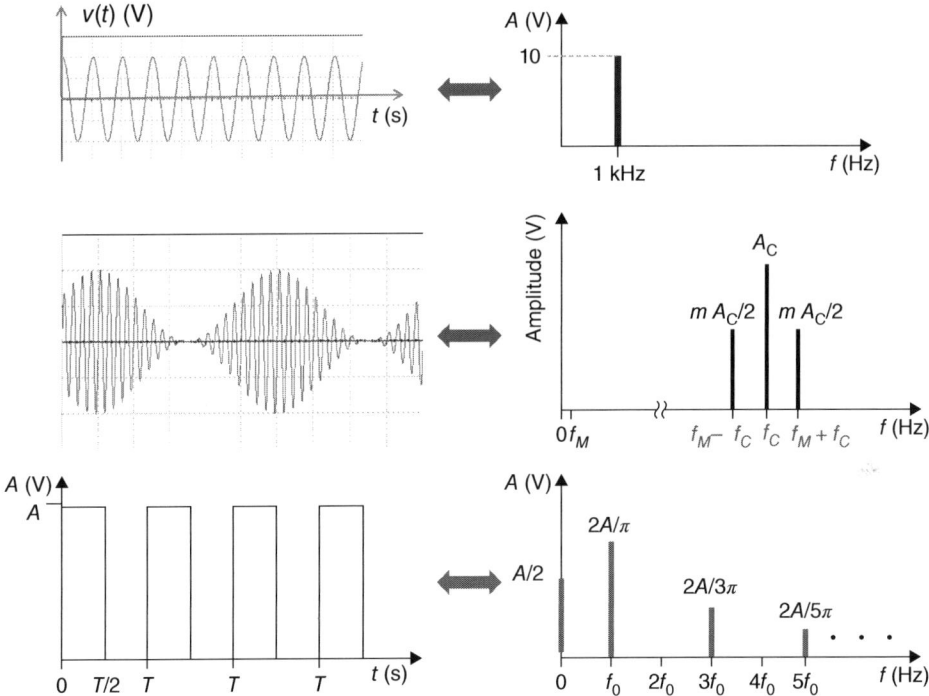

Figure 8.1.25 Waveforms of signals and their spectra: The farther a signal's waveform deviates from the sinusoidal form, the more sophisticated its spectrum. (Top – a sinusoidal signal, middle – an AM signal, and bottom – a square wave signal).

This is why its modulation index is the greatest of the all shown ($\beta = 6$), and the spectrum of this signal should contain 21 frequencies. Thus, we arrive at this conclusion:

The farther an FM signal is from a pure cosine, the greater the number of the frequencies it contains, and the higher its modulation index is.

Figure 8.1.26 shows that all the lines of the side spectral components are evenly spaced and separated by the same frequency, f_M. Thus, the ascending frequency of the first order, $n = 1$ (which is adjacent to the central frequency, f_C) is equal to f_1^+ (Hz) $= f_C + f_M$ and the descending frequency of the first order is equal to f_1^- (Hz) $= f_C - f_M$. The second order frequencies are given by $f_2^\pm$ (Hz) $= f_C \pm 2f_M$. Thus, the highest order side frequencies are equal to

$$f_n^\pm \text{ (Hz)} = f_C \pm nf_M \tag{8.1.23}$$

where n is the number of visible (significant) side frequencies.

The normalized amplitude of an unmodulated carrier signal, $A_C = 1$, is shown in the top graph of Figure 8.1.26 at $\beta = 0$. To estimate the relative value of the amplitudes of all spectral components (harmonics), the vertical axis shows the value 1 in other plots in this figure. Note that, in contrast to an AM spectrum, the amplitudes of the FM harmonics do not descend in a regular order from the highest at f_C to the lowest at f_n. Thus, the amplitudes of $f_C \pm 2f_M$ harmonics could be greater than those of $f_C \pm f_M$, as the spectrum of the FM signal with $\beta = 3$ demonstrates in Figure 8.1.26.

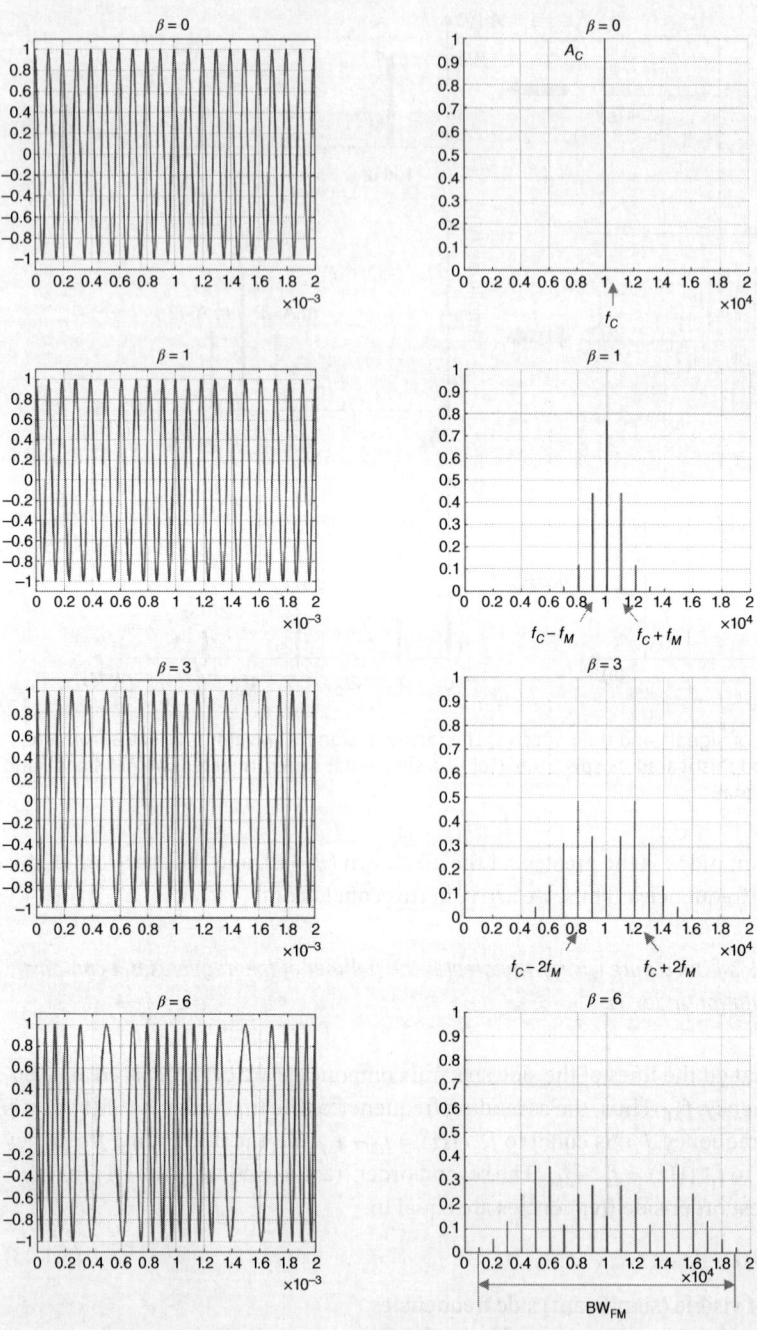

Figure 8.1.26 FM signals with various spectra and their modulation indexes.

Bandwidth of an FM Signal Bandwidth, by definition, is the range of frequencies a given signal occupies. That is,

$$\text{BW}_{\text{FM}} \text{ (Hz)} = f_n^+ \text{ (max)} - f_n^- \text{ (min)} \tag{8.1.24}$$

This approach does not work here because, theoretically, any FM signal contains an infinite number of harmonics; in other words, number n runs from $-\infty$ to ∞. In practice, engineers always use the bandwidth that contains only a finite number of harmonics; this approach is realized by considering only the sidebands with significant amplitudes (SAs). There is no standard or theoretical definition of "significant amplitude"; hence, it is up to an engineer to determine at what fraction of unity the relative amplitude should be considered *significant*. For example, we can state that the amplitude is significant if its harmonic has a relative amplitude equal to 1% of one, that is, 0.01. In this way, we determine the *number of one-side harmonics, n, of a given FM signal*. As soon as the number n is established, calculating the FM bandwidth becomes a minor task. Indeed, since the highest order frequency, according to (8.1.23), is equal to f_n^+ (Hz) $= f_C + nf_M$ and the lowest one is given as f_n^- (Hz) $= f_C - nf_M$, *the estimated bandwidth of a sinusoidal FM signal* can be calculated as

$$\text{BW}_{\text{FM}} \approx f_C + nf_M - (f_C - nf_M) = 2nf_M \tag{8.1.25}$$

Equation (8.1.25) uses the approximation sign instead of the equality sign to stress that it calculates only the part of the infinite FM bandwidth delimited by the *significant harmonics*.

From Figure 8.1.26, you can see that for $\beta = 1$, we have four side frequencies on each side; that is, $n = 4$. Hence, in this case, $\text{BW}_{\text{FM}} = 2 \times 4f_M = 8f_M$. Clearly, *the deeper the modulation, the more sophisticated the waveform of an FM signal, the greater the number of side frequencies, and the wider the bandwidth.*

Note that in contrast to the bandwidth of an AM signal, the bandwidth of an FM signal depends indirectly on the modulation index, $\beta = \Delta f/f_M = k_{\text{FM}}A_M/f_M$, and therefore on a message signal. However, *it is the peak frequency deviation, Δf (Hz) $= k_{\text{FM}}A_M$, that determines the FM bandwidth.*

(**Question**: How do you interpret this statement?)

Two other timely tips regarding the *FM bandwidth*:

1. Higher order side frequencies usually have very small amplitudes. Thus, in reality, we overestimate the bandwidth needed for practical use. If we neglect those sidebands with small amplitudes and still want to transmit not less than 98% of the total signal power, we can use the following approximate formula:

$$\text{BW}_{\text{FM}}^{\text{CBR}} \approx 2(\beta + 1)f_M = 2(\Delta f + f_M) = 2(k_{\text{FM}}A_M + f_M) \tag{8.1.26}$$

This formula is called *Carson's bandwidth rule, CBR*.[3] Carson's formula neglects the harmonics whose amplitudes are smaller than 5% of A_C.

2. Formula $\text{BW}_{\text{FM}} \approx 2(\Delta f + f_M)$ emphasizes the important advantage of an FM transmission: With variations of f_M, the FM bandwidth does not vary as much as an AM bandwidth. (You will recall that $\text{BW}_{\text{AM}} = 2f_M$.) Consider the following example: If $\Delta f = 100$ kHz and we will use

---

[3] John Renshaw Carson(1886–1940) was an American mathematician and electrical engineer. He graduated from Princeton University and worked most of his professional life at AT&T's Bell Telephone Laboratories. He invented single-sideband amplitude modulation, which we will study in Section 8.2 and made important contributions to the theory of frequency modulation. He pointed out the disadvantages of narrowband ($\beta < 0.25$) FM and introduced the bandwidth rule now bearing his name.

three different message signals – $f_{M1} = 0.2$ kHz, $f_{M2} = 2$ kHz and $f_{M3} = 20$ kHz – the bandwidth will change as follows:

$$BW_{FM1} = 2(\Delta f + f_M) = 2(100 + 0.2) \text{ kHz} = 200.4 \text{ kHz},$$

$$BW_{FM2} = 2(100 + 2) \text{ kHz} = 204 \text{ kHz},$$

$$BW_{FM3} = 2(100 + 20) \text{ kHz} = 240 \text{ kHz}$$

Thus, when the modulating frequency, f_M, changes a 100-fold, the FM bandwidth changes less than 20% (from 200.4 to 240 kHz in our example). This is the other advantage of FM over AM. Indeed, consider the AM signal: If we change f_M a 100-fold (e.g. from 0.2 to 20 kHz), the AM bandwidth would change by 90% (from 0.4 to 40 kHz in this example). Observe, however, that the bandwidth of an FM signal is much greater (>200 kHz) than that of an AM signal (≤40 kHz) for the same message signal.

The following example should clarify all the points of our discussion of the FM spectrum and bandwidth.

Example 8.1.6   Spectrum and Bandwidth of an FM Signal

Problem

Compute all the side frequencies and the bandwidth of an FM signal if its carrier and message signals and the modulation constant are given as $v_M(t)$ (V) $= 3\cos(6.28 \times 10^3 t)$, $v_C(t)$ (V) $= 15\cos(6.28 \times 10^4 t)$, and $k_{FM} = 1000$ Hz/V, respectively.

Solution

Making use of the calculations performed in Examples 8.1.4 and 8.1.5, we can easily obtain $f_M = 1$ kHz, $f_C = 10$ kHz, and $\beta = 3$. To compute the side frequencies, we need to apply (8.1.24), which yields the results shown in Table 8.1.2. This table presents the values of all the side frequencies.

The bandwidth can be computed by using its definition, (8.1.24), as

$$BW_{FM} = 2nf_M = 12 \text{ kHz}$$

If we use the approximate CBR formula, we will obtain

$$BW_{FM}^{CBR} \approx 2(\beta + 1)f_M = 2(3 + 1) \times 1 \text{ kHz} = 8 \text{ kHz}$$

The problem is solved.

Table 8.1.2   Side frequencies of the FM signal in Example 8.1.6.

Order, n	Ascending order	Descending order
1	$f_1^+ = f_C + f_M = 10$ kHz $+ 1$ kHz $= 11$ kHz	$f_1^- = f_C - f_M = 10$ kHz-1 kHz $= 9$ kHz
2	$f_2^+ = f_C + 2f_M = 10$ kHz $+ 2$ kHz $= 12$ kHz	$f_2^- = f_C - 2f_M = 10$ kHz-2 kHz $= 8$ kHz
3	$f_3^+ = f_C + 3f_M = 10$ kHz $+ 3$ kHz $= 13$ kHz	$f_3^- = f_C - 3f_M = 10$ kHz-3 kHz $= 7$ kHz
4	$f_4^+ = f_C + 4f_M = 10$ kHz $+ 4$ kHz $= 14$ kHz	$f_4^- = f_C - 4f_M = 10$ kHz-4 kHz $= 6$ kHz
5	$f_5^+ = f_C + 5f_M = 10$ kHz $+ 5$ kHz $= 15$ kHz	$f_5^- = f_C - 5f_M = 10$ kHz-5 kHz $= 5$ kHz
6	$f_6^+ = f_C + 6f_M = 10$ kHz $+ 6$ kHz $= 16$ kHz	$f_6^- = f_C - 6f_M = 10$ kHz-6 kHz $= 4$ kHz

Discussion

- Looking at Figure 8.1.26 and Table 8.1.2, we can compute the bandwidth immediately by subtracting the lowest frequency from the highest. The highest frequency is 16 kHz and the lowest is 4 kHz; therefore, the range of frequencies occupied by this signal (bandwidth, that is), 16 kHz − 4 kHz = 12 kHz, is the same number obtained by formal computations. Applying the approximate formula results in a different bandwidth because this formula neglects lower-order frequencies, whose amplitude is less than 5% of A_C. Hence, this result shows that, for practical use, we can restrict ourselves to a narrower bandwidth (8 kHz). How important is this result? Well, remember that the calculated bandwidth, BW_{FM}, determines the required transmission bandwidth of a link that will transport this FM signal, and of course, the higher the needed transmission bandwidth, the more expensive will be this link. Therefore, we do not want to use a 12-kHz link if an 8-kHz link will do. On the other hand, (8.1.26) can give us a reasonable means of determining the required transmission bandwidth; then it is up to us, the experts in the field, to make the appropriate decision.
- To conclude our discussion of the waveform, spectrum, and bandwidth of an FM signal, consider the circuit and the result of simulation shown in Figure 8.1.27. In this circuit, the same FM signal is presented to an oscilloscope to obtain the waveform of the signal and to a spectrum analyzer to obtain the spectrum of the signal. The parameters of the signal are as follows: $A_C = 5\,V, f_C = 10\,kHz, f_M = 1\,kHz, \beta = 6$. Since $\beta = \frac{\Delta f (Hz)}{f_M (Hz)}$, then $\Delta f = \beta f_M = 6\,kHz$. Figure 8.1.27 shows that the spectrum of this FM signal contains nine one-side frequencies; thus, $n = 9$ and $BW_{FM} = 2nf_M = 18\,kHz$. The use of Carson's rule, CBR, yields $BW_{FM}^{CBR} \approx 2(\beta + 1)f_M = 14\,kHz$.

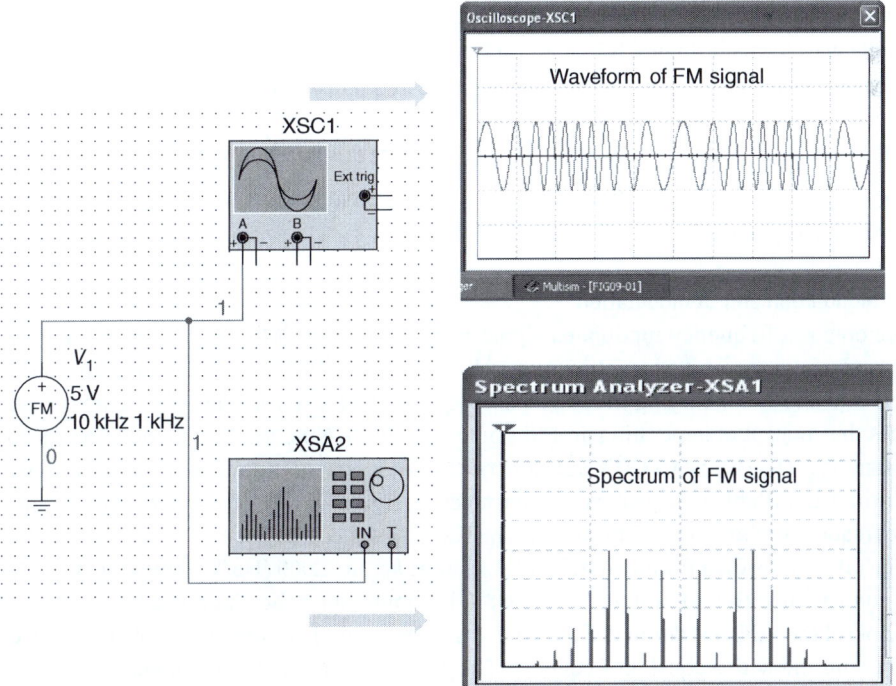

Figure 8.1.27 Waveform and spectrum of an FM signal.

8.1.3.5 Relationship Between Parameters of Message and Carrier Signals in FM Transmission

In considering AM modulation, we studied the relationship between the parameters of message and carrier signals. For amplitudes, we found that the ratio $\frac{A_M(V)}{A_C(V)}$ determines the modulation index, $m = \frac{A_M(V)}{A_C(V)}$; for frequencies, we found that f_C must be at least 10 times greater than f_M to obtain an AM signal of decent quality.

In FM modulation, there is no direct relationship between A_M and A_C; in general, however, we want to create a powerful carrier signal, which means that its amplitude should be big enough to make long-distance transmission possible. Thus, typically, we want $A_C > A_M$.

As for the relationship between f_C and f_M, we refer to Figure 8.1.28, which clearly shows the role of $\frac{f_C(Hz)}{f_M(Hz)}$ in shaping the FM waveform. These waveforms are built with $\frac{f_C(Hz)}{f_M(Hz)} = 16.6$ (a), $\frac{f_C(Hz)}{f_M(Hz)} = 8.3$ (b), and $\frac{f_C(Hz)}{f_M(Hz)} = 1$ (c). Modulation index in all three cases remains the same, $\beta = 6$. Of course, when $\frac{f_C(Hz)}{f_M(Hz)} = 1$, the term "FM" cannot be applied to this signal. Thus, we conclude that in FM, f_C must be greater than f_M.

An analysis of Figure 8.1.28 raises the obvious question of how much greater f_C must be than f_M and why do these waveforms change the way they do there. The answer lies in the mathematical description of their frequencies:

$$f_{FM}(t)\,(Hz) = f_C + k_{FM}A_M\,\cos(2\pi f_M t) \qquad (8.1.16R)$$

This equation shows that the amplitude of a message signal, A_M, determines the difference between the maximum and minimum values of f_{FM} (that is, the depth of modulation), whereas the message signal's frequency, f_M, determines the *periodicity* of these maxima and minima. The smaller the $\frac{f_C(Hz)}{f_M(Hz)}$ ratio, the fewer the number of cycles which the carrier signal, $v_C(t)$, makes within one period of the message signal, $v_M(t)$. (Count the numbers of cycles between two adjacent minima in the top and middle waveforms in Figure 8.1.28 and relate these numbers to the $\frac{f_C(Hz)}{f_M(Hz)}$ ratios given for these waveforms.) When $f_C = f_M$, we want to change the frequency within its own period, which is impossible.

(**Question**: Do you want $\frac{f_C(Hz)}{f_M(Hz)} = 16.6$ or $\frac{f_C(Hz)}{f_M(Hz)} = 8.3$? Putting this question succinctly, do you want $\frac{f_C(Hz)}{f_M(Hz)} \gg 1$ or just $\frac{f_C(Hz)}{f_M(Hz)} > 1$? To find the answer, you need to develop a criterion for determining the desired ratio between f_C and f_M.)

8.1.3.6 FM Modulation and Demodulation

How can we obtain a frequency modulated signal in practice? For *FM modulation*, fortunately, we have a wonderful device called a voltage-controlled oscillator (VCO). It generates a sinusoidal carrier signal and allows for varying the carrier's frequency by an external voltage. Figure 8.1.29a demonstrates the concept of generating an FM signal with VCO. Note that the modulation (deviation) constant, k_{FM} (Hz/V), is a VCO's transfer coefficient; it is a quantitative measure of how a VCO transforms the variable voltage of an input (message) signal, $v_M(t)\,(V) = A_M\cos(2\pi f_M t)$, into the variable frequency of an output FM signal, $f_{FM}\,(Hz) = f_C + \beta\cos(2\pi f_M t)$.

(**Exercise**: Take all the data from Example 8.1.4, change k_{FM} to 2.000 (Hz/V), build the new FM signal, and compare it with that built in Example 8.1.4. Comment on the difference.)

The principle of the VCO operation is clarified in Figure 8.1.29b. If dc voltage is applied, the VCO generates the sinusoid with a constant frequency. The greater the dc voltage, the higher the output frequency – compare the two top figures. When the input voltage changes, then that the frequency of a generated signal changes accordingly – see the bottom figure. FM modulation with a VCO is a straightforward operation, and the VCO is an example of a *direct FM modulator*.

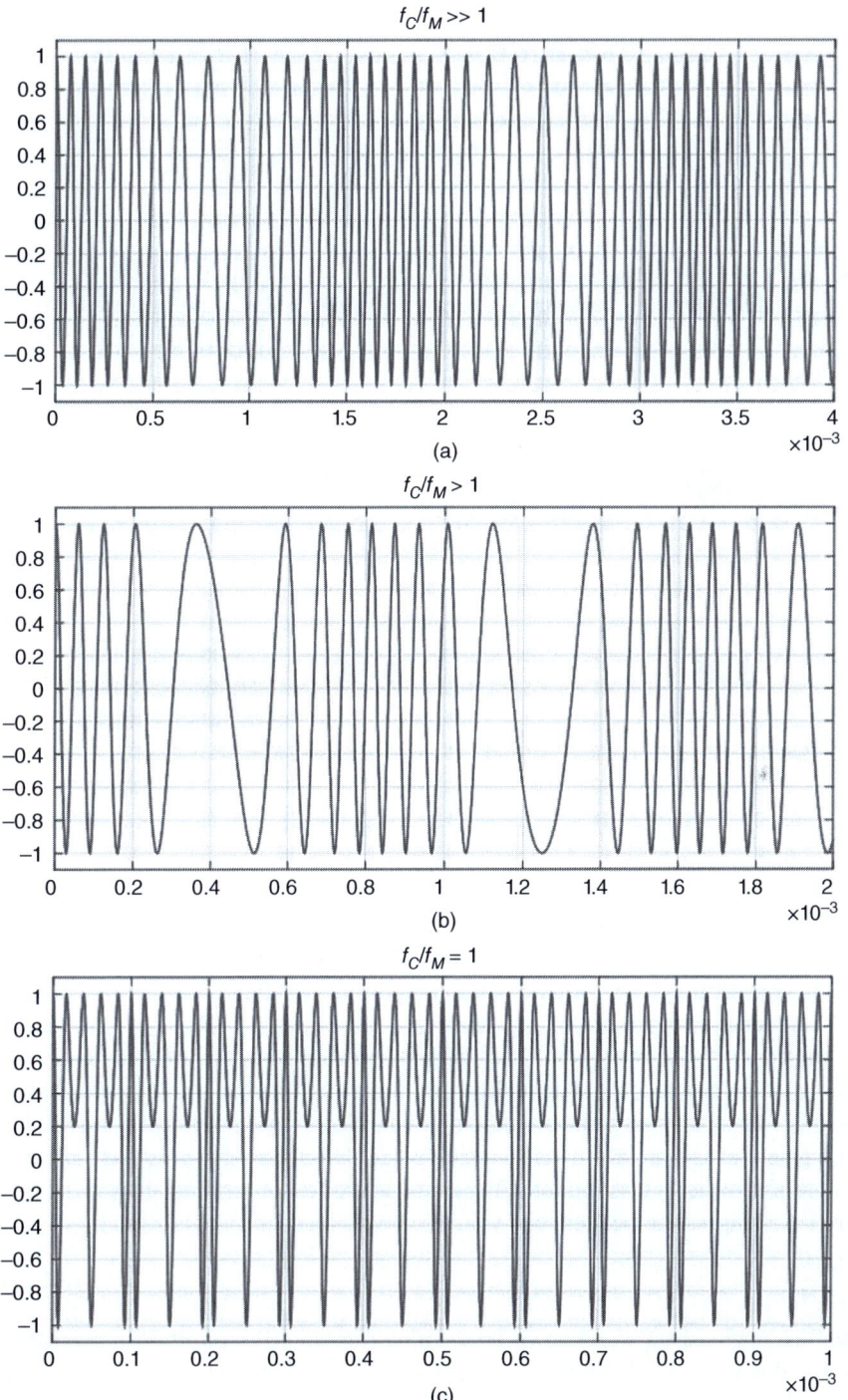

Figure 8.1.28 FM signals with different f_C and f_M ratios: $\frac{f_C(\text{Hz})}{f_M(\text{Hz})} \gg 1$ (a), $\frac{f_C(\text{Hz})}{f_M(\text{Hz})} > 1$ (b), and $\frac{f_C(\text{Hz})}{f_M(\text{Hz})} = 1$ (c). The modulation index is constant, $\beta = \frac{k_{FM}A_M}{f_M} = 6$.

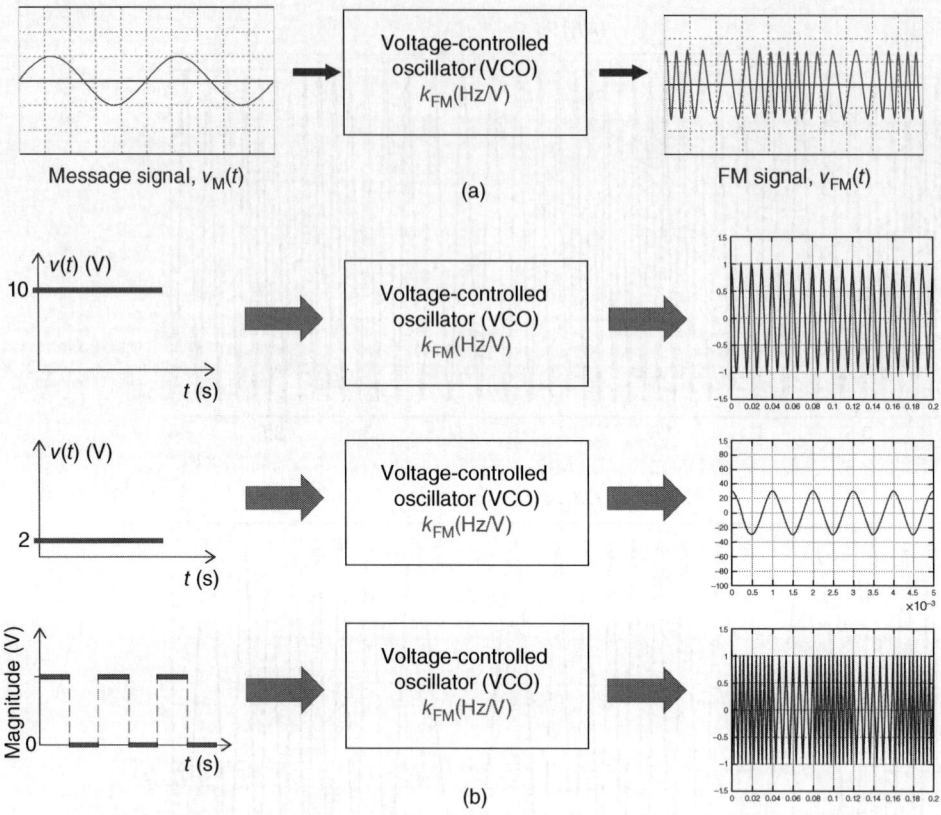

Figure 8.1.29 Operation of a voltage-controlled oscillator: (a) concept of FM modulation with VCO and (b) principle of VCO operation.

The task of *FM demodulation* is to recover the original (information) signal from the incoming FM signal, as shown in Figure 8.1.30a. There are two approaches to this task. First, we can use a *resonant circuit* to convert frequency variations into amplitude variations (see Figure 8.1.30b). Here, the circuit's frequency swing from $(f_{FM})_{max}$ (Hz) $= f_C + \Delta f$ to $(f_{FM})_{min}$ (Hz) $= f_C - \Delta f$ results in current variations from I_{max} (A) to I_{min} (A). Thus, the problem boils down to the demodulation of amplitude variations, a task covered in AM demodulation. Since this principle of operation is based on using the slope of a resonant curve, the proper circuit is called a *slope detector*. Figure 8.1.30c shows the linear segment of a resonant curve in a greater scale. This figure shows that a resonance circuit is an example of a *frequency discriminator*, the circuit is used to convert frequency variations into voltage (current) changes. The frequency-to-voltage conversion coefficient is given by

$$\frac{1}{k_{FM}} = \frac{\Delta v(V)}{\Delta f(Hz)} \tag{8.1.27}$$

where k_{FM} (Hz/V) is, obviously, a well-known *modulation (deviation) constant*. Because $1/k_{FM}$ (V/Hz) determines the slope of the linear segment of a slope detector, k_{FM} is called the *frequency-sensitivity factor* of this type of FM demodulator.

The second approach, which is the main type of FM demodulation today, uses a *phase-locked loop* (PLL) device. A PLL is one of the most widely used electronic components, of course, in an

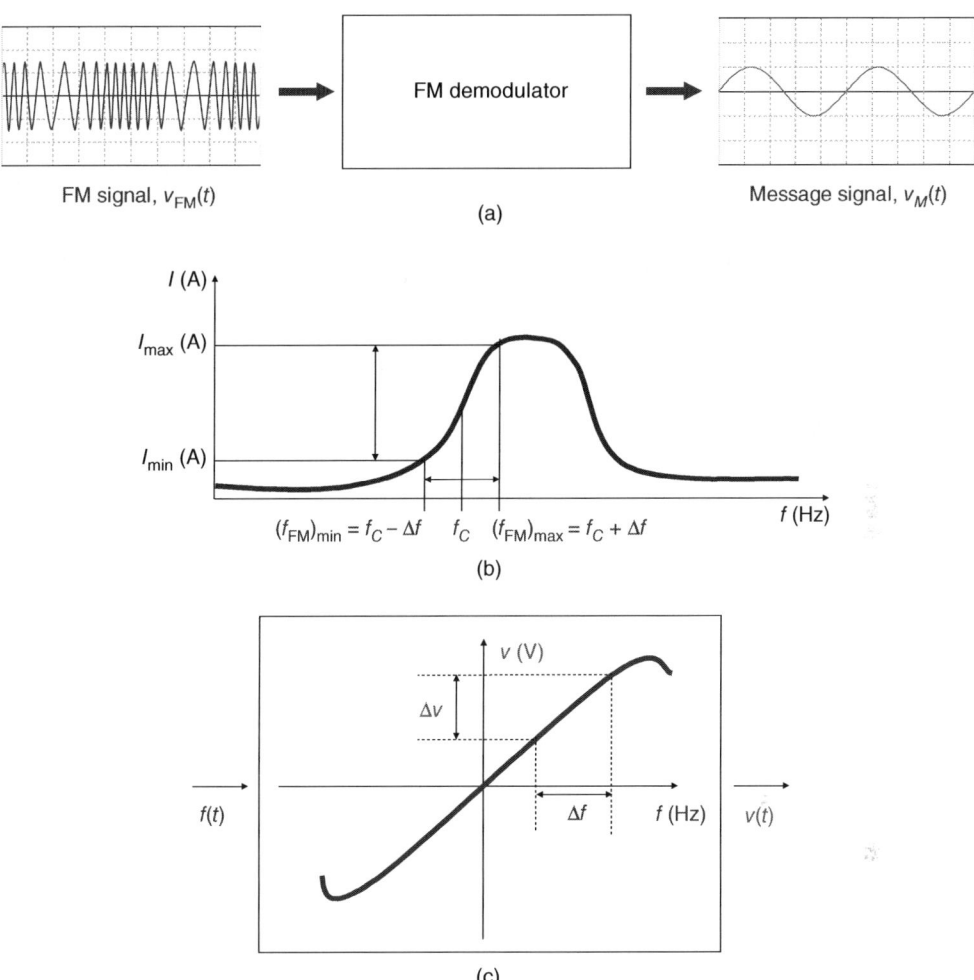

Figure 8.1.30 FM demodulation: (a) the concept; (b) the principle of FM demodulation with a slope detector; and (c) the slope detector is a frequency discriminator.

integrated-circuit (IC) form. PLL is a classic example of a closed-loop (negative-feedback) system. The block diagram of a PLL is shown in Figure 8.1.31a.

A PLL operates as follows: A phase detector constantly compares the $v_{FM}(t)$ and $v_{VCO}(t)$ signals. If a difference exists between them, the phase detector generates a voltage V_{PD} proportional to $v_{FM}(t) - v_{VCO}(t)$. After filtering, this voltage becomes *the detected message (information) signal*, that is, the demodulator output. This voltage also forces the VCO to generate a signal that will reduce the difference between $v_{in}(t)$ and $v_{VCO}(t)$ to zero, that is, track an input FM frequency. Since the input FM frequency keeps changing, the PLL dynamically follows these changes. Therefore, to keep the difference between input FM and VCO generated signals equal to zero, the VCO must generate an exact copy of the input FM signal. But to make VCO generate this exact copy, its driving voltage, $v_{out}(t)$, must be the exact copy of the message signal. And this voltage, $v_{out}(t)$, is the output of the PLL FM demodulator.

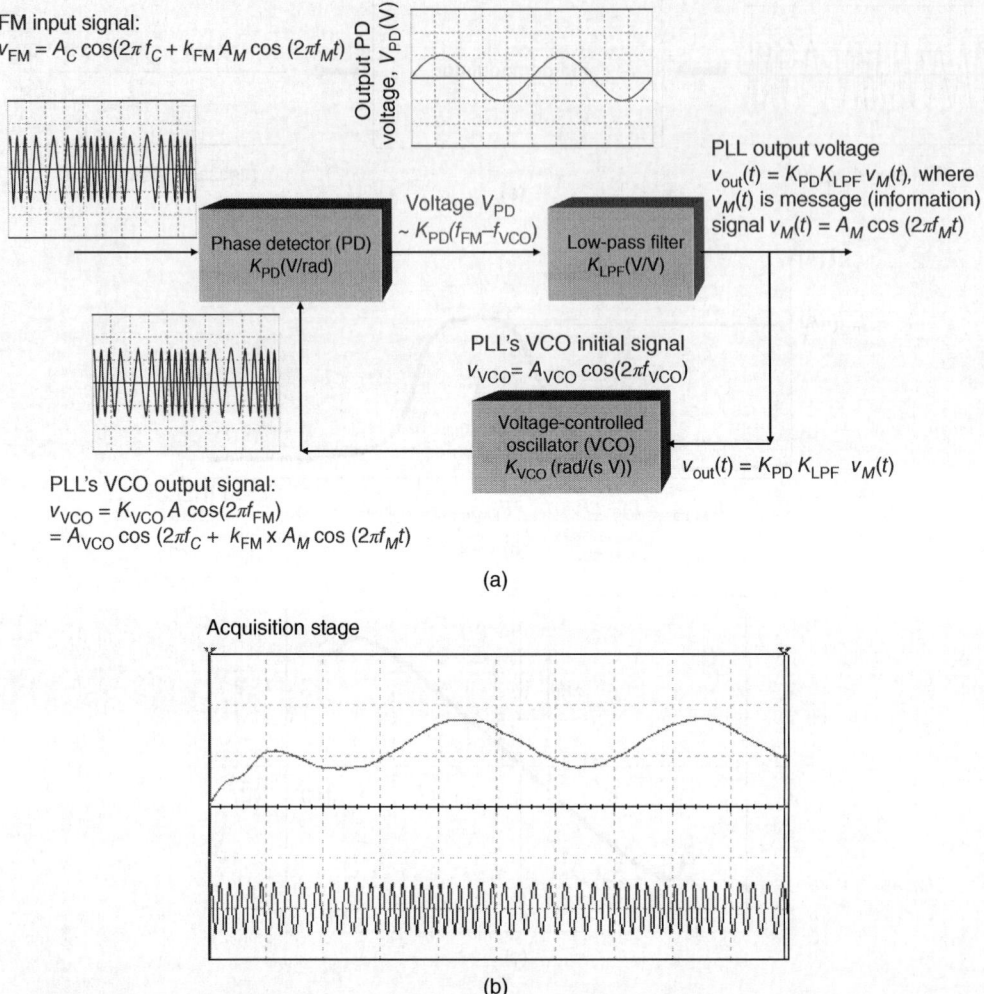

Figure 8.1.31 FM demodulation with PLL: (a) the PLL principle of operation and (b) the acquisition stage and the recovered message signal with PLL.

Figure 8.1.31b shows the result of a Multisim simulation of a PLL operation. Observe that PLL needs a short acquisition interval to adjust VCO's signal to an input FM signal. After this stage, the PLL produces a replica of the sent information signal.

Interestingly, a PLL principle is used to build an *FM modulator* in systems where high precision of the total operation is required.

We proceed with a more meticulous discussion of FM in Section 8.2.

8.1.4 Basics of Phase Modulation (PM)

8.1.4.1 How to Generate a Phase-Modulated Signal

Phase and *frequency* are two closely related parameters of a sinusoidal signal, as we learned in Chapter 2. Therefore, phase modulation is simply a version of frequency modulation, and they are usually combined under one term: *angle (angular) modulation*. The word *angle* refers to an argument $(2\pi ft + \theta)$ (rad) of a cosine function, $A\cos(2\pi ft + \theta)$. (As we will learn in Section 8.2, this angle is also called *instantaneous phase*, $\psi(t)$ (rad) $= 2\pi ft + \theta$, whereas θ is called *initial phase* or *phase shift*. Do not confuse these two meanings of phase.)

As the term *phase modulation* suggests, we need to vary the phase, θ, of a cosine signal to deliver information. This operation can be described by the following formula

$$\theta_M(t)\,(\text{rad}) = k_\theta v_M(t) \tag{8.1.28}$$

where $v_M(t)$ (V) is the message signal and k_θ (rad/V) is the *phase modulation (deviation) constant*.

Figure 8.1.32 illustrates phase modulation of a binary message signal: When the amplitude of an information signal is HIGH, the phase shift of the modulated signal maintains one value, specifically $\theta_M(t) = 0$ rad; when the amplitude is LOW, the phase of the modulated signal changes to another value, such as $\theta_M(t) = \pi$ in our example. Therefore, when a modulating signal switches between two states, the modulated signal changes its phase between two values, too. Thus, the main feature of a phase-modulated signal is this:

> The changes in the phase shift of a PM signal follow the variations in the message signal.

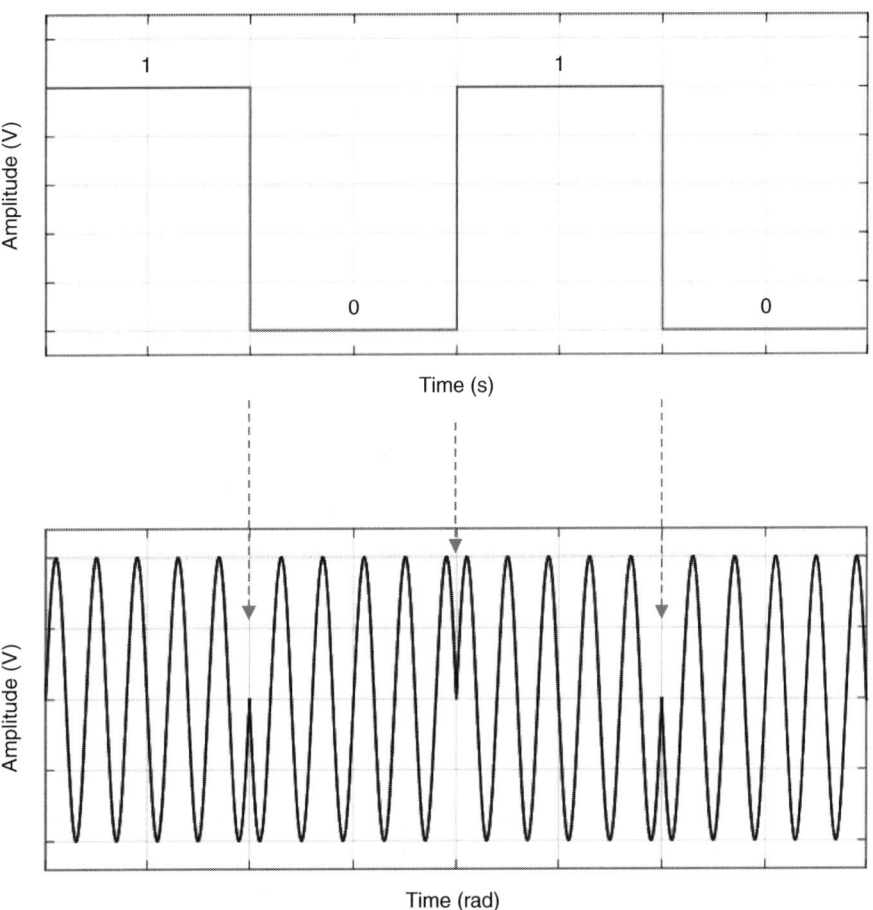

Figure 8.1.32 Example of phase modulation: phase-shift keying (PSK).

The signal shown in Figure 8.1.32 is called a *phase-shift keying (PSK)* signal. Compare the ASK, FSK, and PSK signals shown in Figures 8.1.5, 8.1.7, and 8.1.27 to see how the same concept – modulation – can be implemented in different ways.

Let us consider phase modulation in a case where both the modulating and carrier signals are sinusoidal. Since $v_M(t) = A_M \cos(2\pi f_M t)$, (8.1.28) yields

$$\theta_M(t) \text{ (rad)} = k_\theta \text{ (rad/V)} \, v_M(t) \text{ (V)} = k_\theta A_M \cos(2\pi f_M t) = k_\theta A_M \cos(\omega_M t) \quad (8.1.29)$$

If we plug $\theta_M(t)$ into the formula for a carrier signal, $v_C(t)$ (V) $= A_C \cos(2\pi f_C t + \theta(t))$, we obtain

$$v_{PM}(t) \text{ (V)} = A_C \cos(2\pi f_C t + \theta_M(t)) = A_C \cos(\omega_C t + k_\theta A_M \cos(\omega_M t)) \quad (8.1.30)$$

From (8.1.30) we can readily derive that the frequency of a PM signal is given by

$$\omega_{PM}(t) \text{ (rad/s)} = \omega_C + k_\theta A_M \cos(\omega_M t) \quad (8.1.31)$$

Compare (8.1.31) with (8.1.16): You can see the similarity of PM and FM signals. The formal difference between them is a coefficient for the member "$\cos(\omega_M t)$." For an FM signal, this coefficient is a *maximum frequency deviation*, Δ_f (Hz) $= k_{FM} A_M$, and for a PM signal, this coefficient is a *maximum phase deviation*, Δ_θ (rad) $= k_\theta A_M$. The real difference between these deviations is that Δ_f has no formal limitations, whereas Δ_θ cannot exceed 2π to avoid ambiguity; more accurately,

$$-\pi < \Delta_\theta \leq +\pi \quad (8.1.32)$$

Pay attention to the dimension of k_θ, which is rad/V, so the units of $k_\theta A_M$ are radian too.

The *maximum phase deviation*, Δ_θ (rad) $= k_\theta A_M$, is also referred to as the *phase modulation* index, β_{PM} (rad). In contrast to AM and FM, where the modulation indexes are dimensionless, this modulation index carries a dimension, rad. Also, a phase modulation index does not depend on the frequency of a modulating signal, f_M.

(**Question**: Can we introduce a dimensionless phase modulation index? If not, why not? Hint: Consider how the FM modulation index was introduced.)

The following example elaborates on this brief introduction to phase modulation.

Example 8.1.7 Plotting a Sinusoidal PM Signal

Problem

Plot a PM signal if the carrier, message signal, and the modulation constant are given as $v_M(t)$ (V) $= 3\cos(6.28 \times 10^5 t)$, $v_C(t)$ (V) $= 15\cos(62.8 \times 10^6 t)$, and $k_\theta = 3$ rad/V, respectively.

Solution

The solution is simple: Apply (8.1.30), $v_{PM}(t) = A_C \cos(2\pi f_C t + k_\theta A_M \sin 2\pi f_M t)$, and calculate the signal. The result is presented in Figure 8.1.33.

Discussion

Analyze Figure 8.1.33: Observe how the phase changes with a change in the amplitude of a message signal. The arrows indicate particular points where the message signal crosses the *time* axis as it descends. Obviously, these particular points are no more than examples; you can come up with many others on your own. However, these examples clearly illustrate how phase modulation occurs.

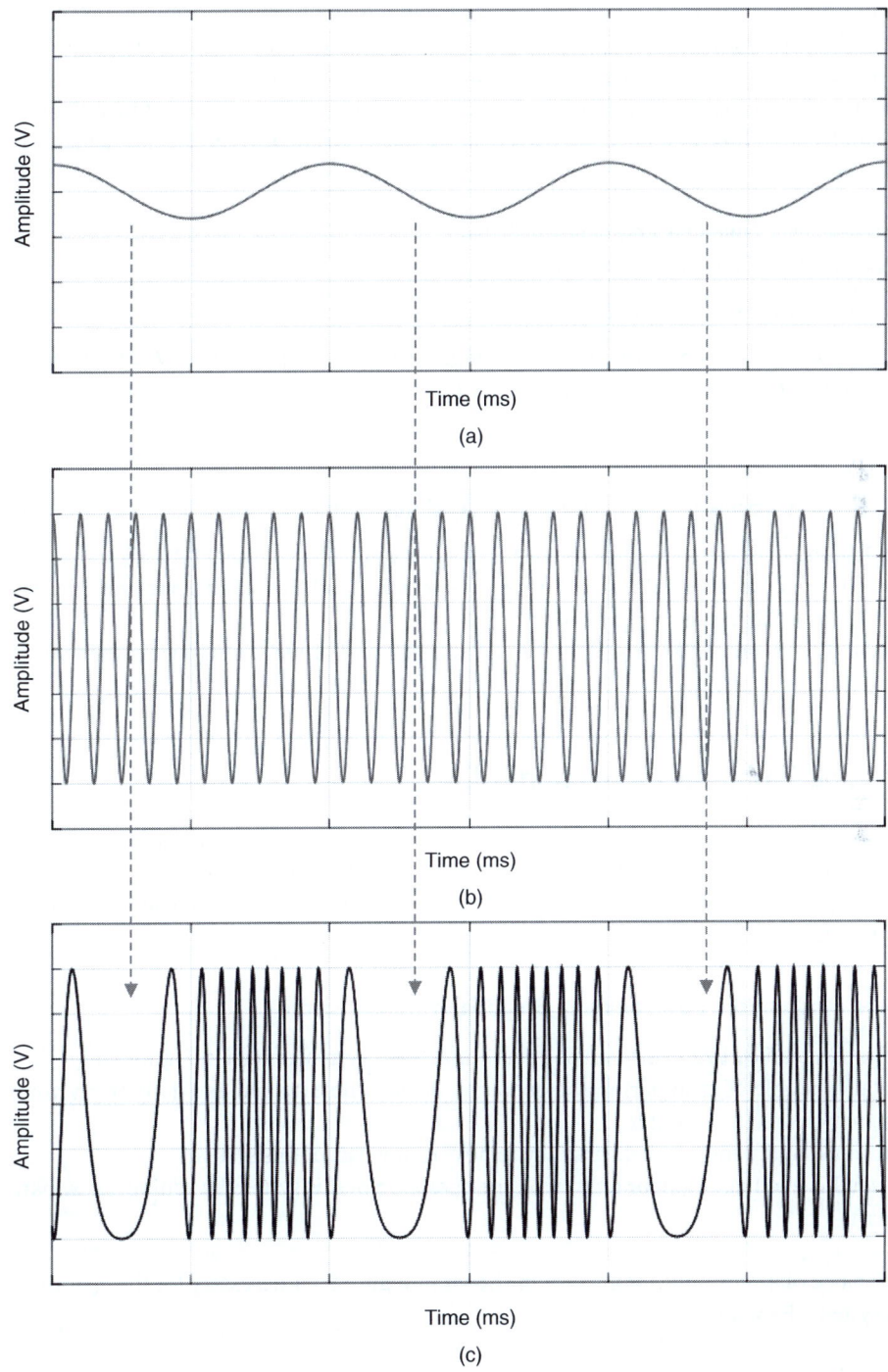

Figure 8.1.33 Sinusoidal phase modulation for Example 8.1.7: (a) message signal; (b) carrier signal; and (c) phase-modulated signal.

Compare this figure with Figure 8.1.22, where a sinusoidal frequency modulation is shown. As you can see, it is almost impossible to distinguish one from the other. In other words, these two figures visually confirm the similarity between FM and PM. But do not be mistaken: Even though PM and FM appear to be very similar in many ways, they are quite different types of modulation. However, to understand this difference, we need to delve into their theory more deeply, which we will do in Section 8.2.

8.1.4.2 Instantaneous Value of a Sinusoidal PM Signal

Let us rewrite (8.1.30) in a more concise form:

$$v_{PM}(t)\,(V) = A_C\,\cos(\omega_C t + \Delta_\theta\,\cos(\omega_M t)) \tag{8.1.33}$$

Here, the argument of the cosine is the sum of two angles, $(2\pi f_C t)$ (rad) and $(\Delta_\theta \cos(2\pi f_M t))$ (rad). Applying the simple trigonometric formula, $\cos(A + B) = \cos A \cos B - \sin A \sin B$, to (8.1.33), we obtain

$$\begin{aligned}v_{PM}(t)\,(V) &= A_C\,\cos(\omega_C t + \Delta_\theta\,\cos(\omega_M t)) \\ &= A_C\,\cos(\omega_C t)\,\cos(\Delta_\theta\,\cos(\omega_M t)) - A_C\,\sin(\omega_C t)\,\sin(\Delta_\theta\,\cos(\omega_M t))\end{aligned} \tag{8.1.34}$$

Equation (8.1.34) enables us to calculate the instantaneous value of a sinusoidal PM signal.

(**Exercise**: Find the value of a PM signal at $t = 5\,\mu s$ by taking all the signal's parameters from Example 8.1.7. Show this value on Figure 8.1.33.)

We continue discussion of PM modulation in Section 8.2 with a more rigorous mathematical approach.

Questions and Problems for Section 8.1

- Questions marked with an asterisk require a systematic approach to finding the solution.
- Many questions and problems, including those marked with an asterisk, imply that you, in addition to reading the textbook, will do your research to find the answers. Consider such questions as mini-projects.

Why We Need Modulation: Baseband and Broadband Transmission

1. Consider the detailed block diagram of a communication system shown in Figure 8.1.1:
   a) Explain the role of each part in the operation of the whole system and accompany your explanation with an example.
   b) Can you simplify this block diagram by removing any part of this system?
   c) Do you have a suggestion for improving the operation of this system by adding a new part or parts?

2. Is every block of a communications system shown in Figure 8.1.1 necessary for the operation of this system? Explain.

3. Consider the operation of your cell phone: Name all the parts of this communication system (refer to Figure 8.1.1).

4. Consider the spectrum of the human voice shown in Figure 8.1.2: What feature of this spectrum is a general one for any human voice and what feature is specific to this record only?

5   Why is the direct transmission of a human voice over a communication system called baseband transmission?

6   Give as many examples of baseband transmission as you can.

7   When you turn on your radio, you can hear a human voice. Is this baseband transmission? Explain.

8   Baseband transmission suffers from a number of drawbacks, but it is still in use. Why?

9   Explain how broadband transmission works using broadcast radio as an example.

10 Why is modulation a main feature of broadband transmission?

11 In analog broadband transmission, a carrier is always a sinusoidal signal. Why? Can we use any other analog signal as a carrier? Explain.

12 Broadband transmission provides a higher transmission rate than baseband. Why? Explain. Give an example.

13 Why does broadband transmission enable us to deliver several messages simultaneously whereas baseband does not?

14 Why does broadband transmission enable us to use frequency modulation but baseband does not? Why is this important?

15 *Compare the transmission distances and the attenuation of the signals in baseband and broadband transmissions. (*Tip*: Review Section 1.3.2 and particularly Figure 1.3.3.)

16 Contrast baseband and broadband transmissions by comparing all the important characteristics of each type.

17 Broadband transmission is superior to baseband by all measures. But we do incur a cost for all these benefits. What is this cost?

Basics of Amplitude Modulation

18 *Describe all the types of analog modulation we can create by using a sinusoidal carrier.

19 Qualitatively explain how we can create an amplitude modulation. Give an example.

20 How does AM radio work? Explain by referring to Figures 8.1.2, 8.1.3, and 8.1.7.

21 *Name all three constituents of an AM signal and show them in Figure 8.1.7.

22 Compute the modulation index of the AM signal shown in Figure 8.1.8

23 Can we modulate the carrier $v_C(t) = 30\cos(500t)$ by a message signal $v_M(t) = 3\cos(600t)$? Explain.

24 *A full tone AM signal is composed from the carrier $v_C(t) = 12\cos(6280t)$ and the message signal $v_C(t) = 3\cos(314t)$.
   a) Derive the formula for this AM signal.
   b) Build its waveform and show T_C (s) and T_M (s) on this graph.
   c) Compute its modulation index.
   d) Compute the instantaneous value of this AM signal at $t = 3$ ms.

25 *Consider an AM signal whose carrier is $v_C(t) = 12\cos(6283t)$ and whose message is $v_M(t) = 3\cos(314t)$:
   a) Derive the formulas for the envelopes of this AM signal.
   b) Build the graphs of these envelopes.
   c) Explain why an AM signal has two envelopes.

26 *Consider the AM signal whose carrier is $v_C(t) = 12\cos(6283t)$ and the message is $v_M(t) = 3\cos(314t)$:
   a) Derive the formula for determining the spectrum of this signal.
   b) Find the spectrum of this signal.
   c) Build a figure showing this spectrum. Indicate all given and computed quantities.

27 *Consider Figure 8.1.14, where the spectrum of an AM signal is shown:
   a) Why are side-frequency components called "upper" and "lower"?
   b) Figure 8.1.14 shows **lines** for USB and LSB. Why, then, are these components called side**bands**?

28 *Figure 8.1.17 shows that the spectrum of each spectral components of an AM signal occupies not a single line (frequency) but a band of frequencies:
   a) Why?
   b) Do you want this band be small or large?
   c) What is an ideal band for each of the three spectral components?

29 *Consider an AM signal whose carrier is $v_C(t) = 12\cos(6283t)$ and whose message is $v_M(t) = 3\cos(314t)$:
   a) Calculate the power distribution in this signal.
   b) Compute its power efficiency.
   c) How can we improve the power efficiency of this signal?

Basics of Frequency Modulation

30 AM seems to be the most natural and simple type of modulation. Why do we need to develop FM?

31 Describe the concept of frequency modulation. Give an example.

32 *What is the main advantage of FM over AM?

33 *Calculate the power distribution in an FM signal whose carrier is $v_C(t) = 12\cos(6283t)$ and whose message is $v_M(t) = 3\cos(314t)$.

34 *Consider the FM signal whose carrier is $v_C(t) = 12\cos(6283t)$, whose message is $v_M(t) = 3\cos(314t)$, and whose k_{FM} (Hz/V) $= 800$:
a) Derive the formula for this FM signal.
b) Build its waveform and show how variations in $v_M(t)$ are reflected in changes in $f_{FM}(t)$.
c) Calculate the peak frequency deviation.
d) Compute the modulation index of this FM signal.
e) Compute the instantaneous value of this FM signal at $t = 3$ ms and show the computed signal's magnitude in the figure built for this problem.

35 *Equation (8.1.16), describing the frequency of an FM signal, contains a modulation (deviation) constant k_{FM} (Hz/V), carrying unusual units. Does this constant have a practical meaning or is it just mathematical sleight of hand?

36 *Consider the waveform of the FM signal shown in Figure 8.1.21c. Can you show frequencies and periods of a carrier and of a message signal on this graph? Explain.

37 *The text discusses only the case of frequency modulation whose message signal is a sinusoid. Can we produce an FM signal when the message is an arbitrary signal? Consider as an example the waveform of a human voice, given in Figure 8.1.2a, to prove your answer.

38 *The text says that "the zero crossings of an FM signal contain information about a message signal." What information? How can you read this information?

39 Calculate the modulation index for an FM signal whose message (information) signal is $v_{M1}(t)$ (V) $= 8\cos(6.28 \times 10^3 t)$, the carrier signal is $v_C(t)$ (V) $= 18\cos(6.28 \times 10^4 t)$, and the modulation constant is k_{FM} (Hz/V) $= 1000$:
a) What will be the modulation index for a new message signal, $v_{M2}(t)$ (V) $= 16\cos(6.28 \times 10^3 t)$?
b) Build the graph of the new FM signal and compare it with the waveforms given in Figure 8.1.24.

40 *Consider the spectra of FM signals shown in Figure 8.1.26:
a) How is a signal's waveform related to the number of a signal's side frequencies?
b) How is the number of a signal's side frequencies related to the value of a signal's modulation index?
c) How is a signal's waveform is related to a signal's bandwidth?

41 *How does the estimated bandwidth of a sinusoidal FM signal depend on the signal's modulation index?

42 *There are three values of an FM signal's bandwidth: the exact, the estimated, and the approximated (Carson's rule). How do they differ?

43 Consider the bandwidths of three FM signals whose $\Delta f = 200$ kHz and message signals are $f_{M1} = 3$ kHz, $f_{M2} = 30$ kHz, and $f_{M3} = 300$ kHz:

a) Compute the FM bandwidth for each of the message signals.
b) Compare the changes in three FM bandwidths in absolute numbers and in percent.
c) Compare the changes in FM bandwidths with those in AM bandwidths modulated by the same message signals. Based on this comparison, which modulation type – FM or AM – do you prefer and why?

44 The text says that the change f_M 100-fold causes a change in the AM bandwidth by 90% but in the FM bandwidth by only 20%. Why is this? (*Hint*: Consider the formulas for AM and FM bandwidths.)

45 What parameters of an FM signal do you need to know to compute its bandwidth?

46 *Consider the following message signal $v_M(t)$ (V) = $4\cos(6.28 \times 10^3 t)$? Give the order of magnitude of the amplitude and the frequency of a carrier appropriate to this message signal.

47 One of the most popular FM modulators is a voltage-controlled oscillator, VCO:
a) Explain its operating principle.
b) What is the role of a modulation (deviation) constant, k_{FM} (Hz/V), in a VCO operation?
c) Consider Figure 8.1.29a, showing a VCO's input and output signals. Qualitatively sketch the new output signal that would be produced if k_{FM} (Hz/V) of this VCO would be doubled.

48 We can use a resonant circuit as an FM demodulator:
a) Explain the operating principle of this demodulator.
b) What is the role of a modulation (deviation) constant, k_{FM} (Hz/V), in its operation?
c) Consider Figure 8.1.30a: Qualitatively sketch the new demodulated message signal if the FM demodulator is a slope detector and its k_{FM} (Hz/V) would be doubled.

49 Describe the principle of operation of a phase-locked loop, PLL, demodulator.

50 The text says that a PLL output signal serves simultaneously as a demodulated message signal and as an error-correcting feedback signal. It sounds if the detected message contains an error, which obviously should not be. Explain this seeming contradiction.

51 A PLL FM demodulator requires an acquisition time before it starts to demodulate a message signal correctly:
a) Is this an advantage or a drawback of this type of FM demodulator? Explain.
b) Go online and find out what is a typical acquisition interval for modern PLL FM demodulators.

Basics of Phase Modulation

52 We know the reasons for the development of AM and FM types of analog modulation. What do we want to achieve with PM?

53 Write down two formulas: One for the frequency of an FM signal, (8.1.16), and another for the frequency of a PM signal, (8.1.31). List similarities and differences between them.

54 *Describe mathematically a PSK signal shown in Figure 8.1.32.

55 *Consider the PM signal whose carrier is $v_C(t) = 12\cos(6283t)$, whose message is $v_M(t) = 3\cos(314t)$, and whose $k_\theta = 5$ rad/V:
   a) Derive the formula for this PM signal.
   b) Build its waveform and show how variations in $v_M(t)$ are reflected in changes in $f_{FM}(t)$.
   c) Calculate the peak phase deviation, Δ_θ. What is the range of change for this deviation?
   d) Compute the modulation index of this PM signal.
   e) Compute the instantaneous value of this PM signal at $t = 3$ ms and show the computed signal's magnitude in the figure built for this problem.

56 *The *maximum phase deviation*, Δ_θ (rad/V) $= k_\theta A_M$, is also referred to as the *phase modulation index*, β_{PM}. Why does this modulation index carry a dimension, whereas modulation indexes of AM and FM are dimensionless?

57 *FM and PM types of modulation are very similar. Are they basically one and the same or still different types? Why don't we combine them into one type?

8.1.A Drawbacks of Baseband Transmission

In this appendix, the drawbacks of baseband transmission are considered in detail.

The major problem of baseband transmission is an inherently low transmission speed (rate).

This drawback stems from the very nature of baseband transmission: The carrier of information in this transmission is the original signal (the source of the information), which is always in a low-frequency band.

You will recall from Sections 1.2 and 1.3 that transmission rate, or bit rate, BR (b/s), is proportional to bandwidth, BW (Hz), which, in turn, is proportional to the frequency of a carrier, $f_{CARRIER}$ (Hz); that is,

$$\text{BR (b/s)} \sim \text{BW (Hz)} \sim f_{CARRIER} \text{ (Hz)} \tag{8.1.A.1}$$

In other words, the higher the frequency of a carrier signal, the greater the transmission rate. It is vitally important to have BR (b/s) as high as possible because, as Hartley's law (introduced in Section 1.2) states,

$$H \text{ (bit)} = \text{BR (b/s)} \times T_{tr} \text{ (s)} \tag{8.1.A.2}$$

where H (bit) is the volume of information and T_{tr} (s) is the transmission time. A simple example: Suppose we need to download a program of 336 Mbit in volume. If we would use an archaic dial-up modem with a maximum transmitting capacity of $C = 56$ kb/s, we would need T_{tr} (s) $= H$ (bit)/BR (b/s) $= 6000$ seconds $= 100$ minutes $= 1.67$ hour, but with a modern downstream transmission rate of BR $= 600$ Mb/s, it will take only T_{tr} (s) $= 0.56$ seconds. Refer to Sections 1.2 and 1.3, where this principle is discussed in detail.

Again, in baseband transmission, the carrier frequency is low by the very nature of this transmission; therefore, the bit rate is very limited.

In addition to the major drawback of baseband transmission, there are other problems associated with this mode:

- *Problem*: We can *transmit only one signal at a time*. Indeed, when you talk over the landline phone with your friend, no one else in your home can use this telephone line. This is because the

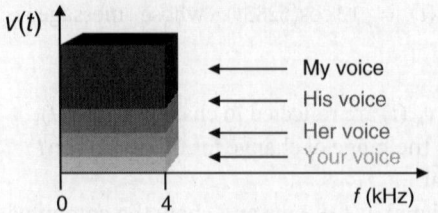

Figure 8.1.A.1 Baseband transmission can deliver only one signal at a time.

frequencies of all voices are concentrated in the same band and therefore cannot be distinguished when transmitted simultaneously. Figure 8.1.A.1 visualizes this explanation.

Another example is radio transmission: Imagine several radio stations directly transmitting the voices of different people over the same area by using only a microphone-amplifier-speaker system. Will you be able to distinguish among them? Of course, not. This is an example of the *interference* of various signals. We are quite familiar with the phenomenon. Suffice to refer to a crowded room where everybody speaks loudly; quite obviously we cannot distinguish our conversation from another for all the voices are within the same frequency range and interfere with one another.

- *Problem*: In an *analog* baseband transmission, information is affected by changing the *amplitude* of the signal. (Remember, the frequency of an information signal is determined by the signal's nature and therefore cannot be changed.) As we discussed in Chapter 1, any signal in this world suffers from distortion during transmission and an analog signal makes it difficult to recover the information from a distorted received signal. Consider the example shown in Figure 8.1.A.2, where the changes in amplitude (better to say *magnitude*, as discussed in Chapter 2) deliver the information.

Now imagine you phone your friend and say, "Good morning." Let us assume that a magnitude of 5 V represents the sound "G" and a magnitude of 2 V represents the sound "O"; thus, the magnitudes of the signal in the left graph of Figure 8.1.A.2 at instants 1 and 2 represent the sounds "G" and "O" of your greeting. Now look at the magnitudes of the received signal shown in the right graph of Figure 8.1.A.2: At the same instants, these magnitudes are quite different and represent different letters (sounds). What sounds would your friend hear? "B" and "U" or "J" and "Y"? Extending the two examples to the point of absurdity, with such a transmission, your friend may hear "Bad evening" instead of "Good morning." Remember, your recipient does not know what message you sent. He or she has only the received signal and cannot understand by that alone what was sent. As Figure 8.1.A.2 shows, the received amplitudes at instants 1 and

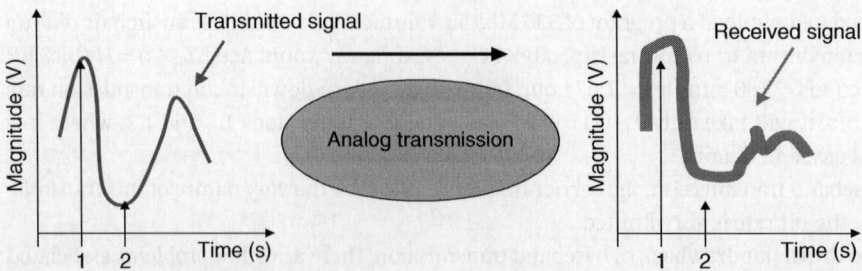

Figure 8.1.A.2 Signal distortion during analog transmission.

2 are quite different from the sent amplitudes. Therefore, your recipient will obtain the wrong information, not what was actually sent.

Therefore, *analog baseband transmission delivers a distorted signal to the destination point, and it is difficult, sometimes even impossible, to extract correct information from the received signal.*

Ask your grandparents to compare the quality of the old-fashioned (analog) telephone transmission with today's digital voice delivery.

- *Problem*: Baseband transmission has *very limited transmission distance*.

Indeed, it is intuitively clear that the longer the transmission distance, the more powerful must be the transmitted signal. Since in baseband transmission one signal performs both operations (carrying information and transmitting it), we have no choice except to increase the power of this signal to transmit it over a long distance. (Remember that the power of an electrical signal is proportional to the square of its amplitude and its frequency.) To understand the problem better, let us return to the example of telephone transmission: If we were to use baseband transmission throughout a long-distance call, we would need a microphone that generates a huge amount of power. The size of such a microphone would be much more than, for example, a telephone system in the first days of telephony. In fact, we would not even call such a hypothetical device a *micro*phone; it would be called a "jumbo phone." Imagine the myriad inconveniences associated with the use of such a giant apparatus, not to mention the tremendous amount of power this device would consume. And the use of such "technology" could never be justified from an economic standpoint. You certainly do not want to pay thousands of dollars to have your voice delivered to another city by this method.

Besides the economic issue, there is a strong technological limitation to the power of an analog baseband signal: The power of a signal is proportional, in part, to the square of its frequency. Since the frequency of an original signal is low, its power is low, too. To provide sufficient power for a baseband signal, a designer would have to overcome unresolvable problems. For example, in radio or any wireless transmission, the size of the antenna, the final segment of a transmitter, must be about $\lambda/10$, where λ (m) is the wavelength of a radiating signal. The frequency-wavelength relationship is given by the well-known formula $\lambda f = c$, where $c = 3 \times 10^8$ m/s is the speed of light in vacuum. Therefore, voice transmission ($f \leq 4\,\text{kHz} \Rightarrow \lambda \geq 0.75 \times 10^5$ m) would require an antenna of 7.5 km. For transmitting music ($f \leq 20\,\text{kHz} \Rightarrow \lambda \geq 0.7 \times 10^4$ m), an antenna of 0.7 km would be required. These numbers not only show unacceptable antenna sizes but they also mean that a music antenna would not work for voice transmission. (It is worth noting that there are a variety of antenna designs and, correspondingly, a variety of formulas describing these antennas and their sizes. All these formulas, however, rely on a general rule: Antenna size is inversely proportional to its frequency or directly proportional to its wavelength. Formula $\lambda/10$, for estimating the order of an antenna size, is accepted by the industry.)

- Last, but not least, there are still other problem: Baseband transmission is done at the frequency determined by an information signal, whereas the transmission media (copper wire, optical fiber, air, etc.) has minimum attenuation at the frequencies determined by their properties. (Review Section 1.3.2 and pay special attention to Figure 1.3.3.)

Hence, we *don't have any means to minimize signal attenuation*, which is another reason for shortening a transmission distance.

This discussion provides the detailed explanations to the list of drawbacks of the baseband transmission given in this section.

8.2 Analog Modulation for Analog Transmission – An Advanced Study

Objectives and Outcomes of Section 8.2

Objectives

- To develop a generalized view on analog modulation for analog transmission.
- To learn in depth the types and techniques of amplitude modulation.
- To acquire knowledge of rigorous mathematical analysis of time-domain and frequency-domain presentations of angular (frequency and phase) modulation.
- To master the main aspects of analog transmission.

Outcomes

- Learn that there are two basic types of modulation – analog and digital. Realize that in analog modulation, both message and carrier signals are analog, whereas in digital modulation, the picture is mixed.
- Recall that analog modulation is implemented by changing the amplitude, frequency, or phase of a sinusoidal carrier, which gives rise to AM, FM, and PM, respectively.
- Realize that the type of amplitude modulation discussed in Section 8.1 is, in fact, called full, or double-sideband transmitted carrier (DSB-TC) AM and that its title reflects the main features of this type of AM.
- Learn how to develop the mathematical descriptions and graphical presentations of the waveforms and spectra of full (DSB-TC) AM signals modulated by sinusoidal and arbitrary message signals.
- Realize that the full AM transmission suffers from low efficiencies in the power and bandwidth utilization.
- Understand that power efficiency is drastically improved by using a double-sideband suppressed carrier (DSB-SC) AM type and both power and bandwidth efficiency are maximized by employing a single-sideband suppressed carrier (SSB-SC) AM transmission. Learn how to determine the waveforms and spectra of these types of AM with various message signals.
- Grasp the concept of filtering and phasing methods for generating an SSB-SC AM signal and become familiar with the advantages and drawbacks of these methods.
- Compare DSB-TC, DSB-SC, and SSB-SC types of AM transmissions from the standpoint of their applications and learn that a system-view approach requires taking into consideration not only transmission issues but also modulation and demodulation processes. Get to know that the demodulation quality improves with utilizing a vestigial sideband (VSB) AM type.
- Realize that application of classic analog AM transmission today is shrinking dramatically, yet the concept of amplitude modulation is actively being used in the most modern digital modulation methods.
- Review the concept of angular modulation that comprises both frequency and phase modulation and learn that terms *instantaneous angle* and *instantaneous phase* of a sinusoidal function are synonyms.
- Recall that PM is the changes in an instantaneous phase that incorporate the message-signal variations. Study the mathematical description of a PM signal, its SNR, and the practical limitations of analog phase modulation. Understand that, nevertheless, the PM concept is widely used in modern digital modulation schemes.

- Acquire knowledge of how to derive the most general formula for FM presented in (8.2.25).
- Conclude that it is a *modulated parameter* of a sinusoidal carrier – the amplitude in AM, the frequency in FM, and the phase in PM – that depends linearly on a message signal and that this property is the important feature of analog modulation. Recall also that an *AM signal* is a linear function and that *FM and PM signals* are nonlinear functions of a message signal.
- Learn that a single-tone frequency modulated signal, $v_{FM}(t)$, presented in (8.2.36) as $v_{FM}(t) = A_C \cdot \sum_{n=-\infty}^{\infty} J_n(\beta) \cdot \cos(\omega_C + n\omega_M)t$, is a series of harmonics $\cos(\omega_C \pm n\omega_M)t$ with amplitudes $A_C \cdot J_n(\beta)$. In this formula, A_C and ω_C are the parameters of a carrier, ω_M is a message frequency, and $J_n(\beta)$ are the Bessel functions of the nth order with modulation index β. Realize that (8.2.36) gives the spectrum of a single-tone (sinusoidally modulated) FM signal.
- Acquire the knowledge of how to find and sketch the spectrum of a single-tone FM signal by utilizing MATLAB-generated graphs and the Table of the Bessel function.
- Master the know-how of determining the bandwidth of a single-tone FM signal by meeting one of two criteria, namely either (i) by the values of significant amplitudes or (ii) by the required power of an FM signal.
- Realize that Carson's rule, $BW_{FM}^{CBR} = 2(\beta + 1)f_M$, estimates the FM bandwidth within which not less than 98% of the total signal power is transmitted, whereas the general formula, BW_{FM} (Hz) $= 2nf_M$, estimates the FM bandwidth determined by the number, n, of transmitted harmonics.
- Generalize the approach to analysis of a single-tone FM signal to an arbitrary FM and learn how to perform practical calculations for dual-tone FM.
- Study how noise affects an FM transmission and learn how to estimate the FM SNR in (8.2.71) and the FM figure of merit in (8.2.68). Realize that FM systems exhibit a unique feature: the improvement of SNR after the demodulation of an FM signal. Learn that this feature, however, takes into effect only after the SNR_{in} threshold has been overcome.
- Understand that the higher frequency components of a demodulated FM signal will suffer from noise more than the lower-frequency components, resulting in the distortion of the signal presented to the user. Learn that this problem is solved by pre- and de-emphasis technique. Study in (8.2.74) how to calculate the factor of improvement provided by this method and in Figure 8.2.25 the block diagram of an FM system implementing this technique.

8.2.1 Classification of Modulation Revisited

Basically, we have two modulation types: analog and digital. In analog modulation, both the message (information) and carrier are the analog signals. Digital modulation, however, presents a mixed picture. There are types of digital modulation that transmit the analog information signal with a digital carrier and – what might be even more surprising – types of modulation that impose a digital information signal upon an analog carrier. *This section considers only analog modulation*; digital modulation is the subject of Chapters 9 and 10.

You will recall that in analog modulation, we use a sinusoidal signal as a carrier, as it occupies the smallest (theoretically – zero) bandwidth. This is a big advantage because, remember, the less spectrum (band of frequencies) we use, the smaller our expenses.

Analog modulation is implemented by changing the amplitude, frequency, or phase of a sinusoidal carrier signal, as was introduced in Section 8.1 and shown in reprised Figure 8.1.6R.

The examples of analog AM, FM, and PM are shown in Figures 8.2.1–8.2.3, respectively. In these examples, the carrier is a sinusoidal signal, and the message (information, modulating, or intelligent) signals are various analog signals.

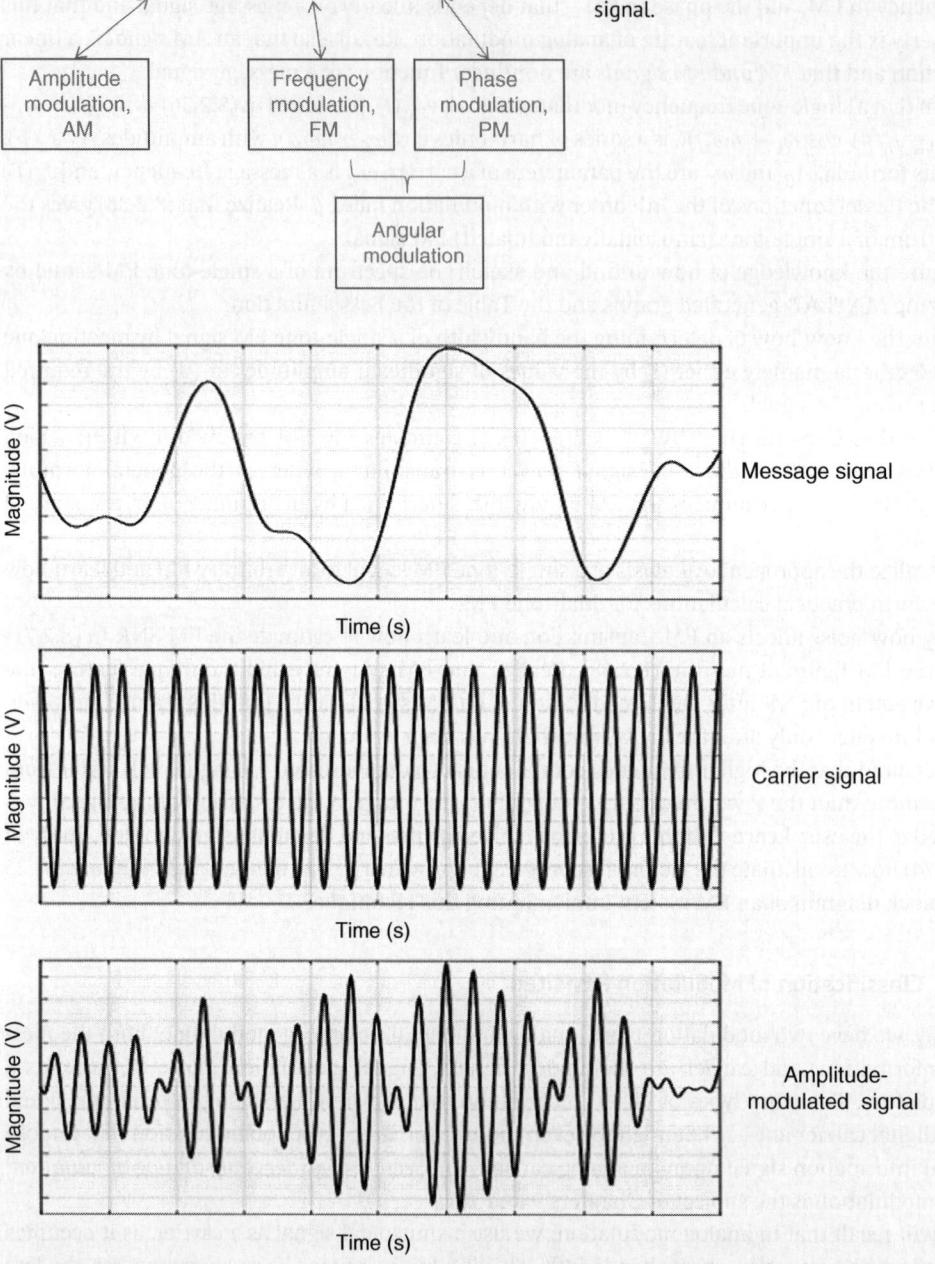

Figure 8.1.6R Amplitude, frequency, and phase modulations of a sinusoidal carrier signal.

Figure 8.2.1 Example of analog amplitude modulation. Source: Reproduction of Figure 2.3.2 (see Figures 8.1.7–8.1.9, where the concept of AM is introduced).

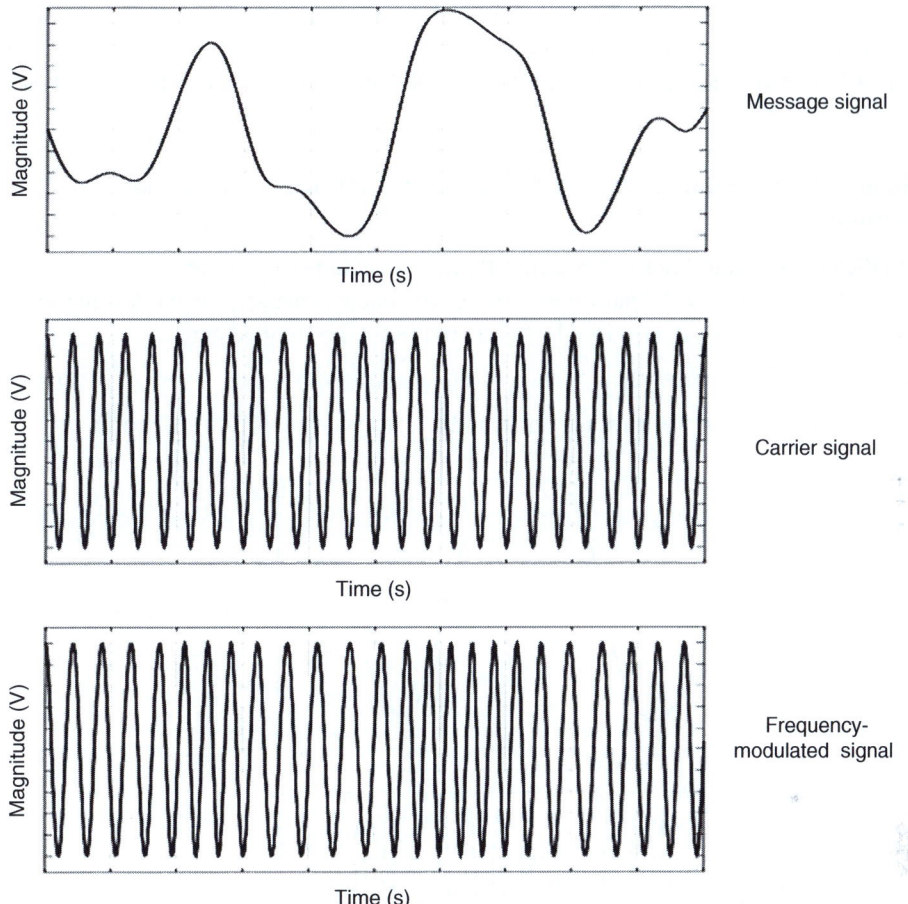

Figure 8.2.2 Example of analog frequency modulation of non-periodic information signals (see Figure 8.1.7).

In Figure 8.2.1, we can clearly see that the amplitude-modulated (AM) signal is the carrier signal, whose amplitude changes according to changes in the information (message) signal. To visualize this process, we can virtually superimpose the message signal onto the carrier signal and observe the resulting AM signal.

Figure 8.2.2 shows how the frequency of a sinusoidal carrier varies in response to changes in the magnitude of an information signal. When the magnitude of an information signal reaches its maximum, the frequency of a frequency modulation (FM) signal is at maximum, too; on the other hand, when the amplitude of an information signal reduces to its minimum, the frequency of an FM signal is at its minimum as well.

Figure 8.2.3a explicitly demonstrates how the information signal changes the phase of a carrier to obtain the PM signal. Compare FM and PM signals in Figure 8.2.3b and observe the difference between them. These pictures are worth a thousand words of explanation.

8 Analog Transmission with Analog Modulation

This review should help refresh your memory on the basic types of analog modulation. We now begin advanced study of amplitude modulation.

(**Question**: *Which modulation method of analog modulation – AM, FM, and PM – is the best? Explain your reasoning.*)

8.2.2 Advanced Consideration of Amplitude Modulation, AM, and Its Application in Analog Transmission

8.2.2.1 Full (Double-Sideband Transmitted Carrier, DSB-TC) Amplitude Modulation

Referring to Section 8.1, we recall that a *tone AM signal*, where a message signal is sinusoidal (demonstrated in Figure 8.2.4), is described by (8.1.7) and (8.1.8a) combined here as

$$v_{AM}(t) = (A_C + v_M(t))\cos(\omega_C t) = (A_C + A_M \cos(\omega_M t))\cos(\omega_C t) \tag{8.2.1}$$

where $v_M(t)$ (V) $= A_M \cos(\omega_M t)$ is a message signal and $v_C(t)$ (V) $= A_C \cos(\omega_C t)$ is a carrier.

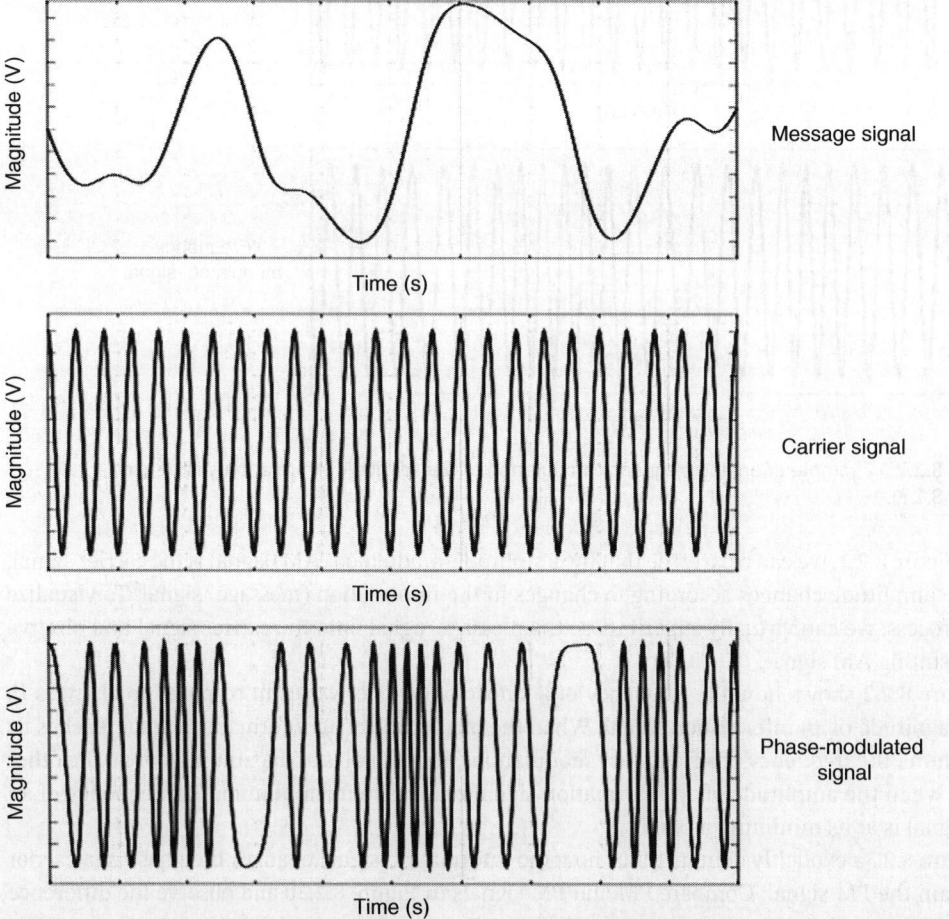

Figure 8.2.3 Example of analog phase modulation (a) with nonperiodic information signal and (b) comparison of frequency modulation (left) and phase modulation (right) of the same nonperiodic information signal (refer to Figure 8.1.28).

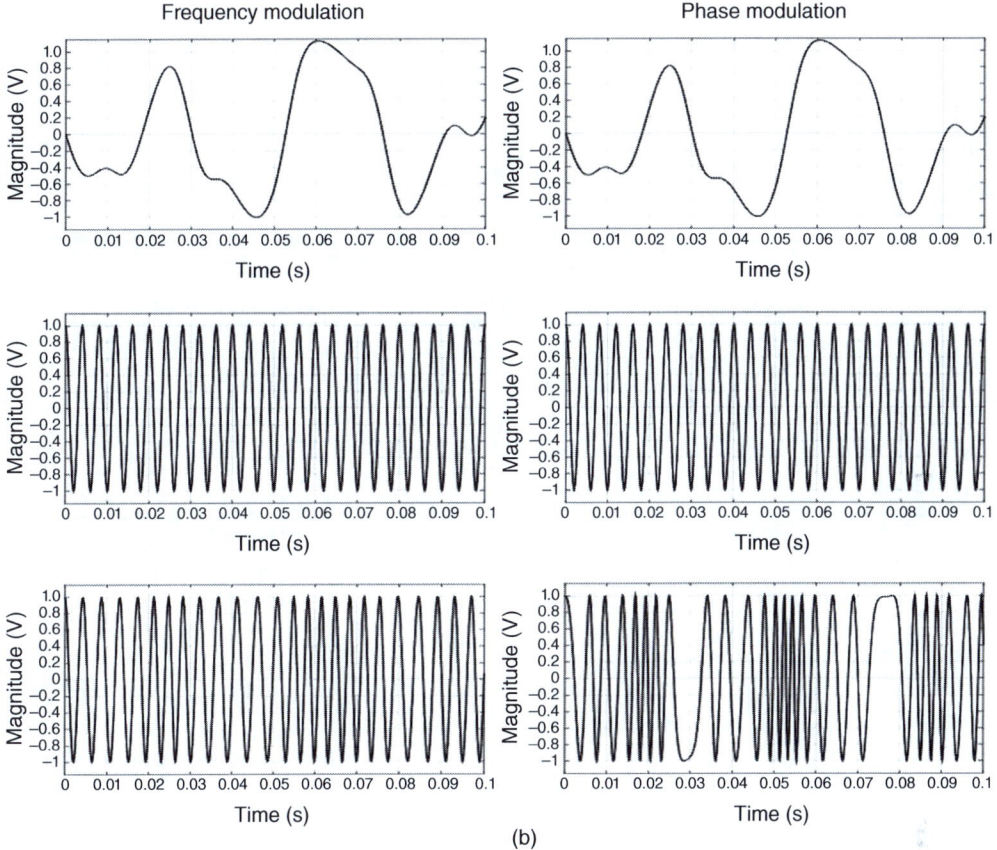

Figure 8.2.3 *(Continued)*

By simple manipulations, we obtained (8.1.9), reproduced below as

$$v_{AM}(t) = A_C \cos(\omega_C t) + \frac{A_M}{2}\cos(\omega_C + \omega_M)t + \frac{A_M}{2}\cos(\omega_C - \omega_M)t \qquad (8.2.2)$$

Equations (8.2.1) and (8.2.2) describe the *waveform* and the *spectrum* of a tone AM, discussed in Section 8.1. A summary of that discussion is presented in Figure 8.2.4. Examining that figure, we notice that modulating a high-frequency sinusoidal carrier by a low-frequency sinusoidal information signal results in an AM signal with the spectrum concentrated around the carrier frequency, $\omega_C(f_C)$, and consisted of three components: $\omega_C(f_C)$, $\omega_C + \omega_M(f_C + f_M)$, and $\omega_C - \omega_M(f_C - f_M)$. (You will recall that we can use either a two-sided or a one-sided spectrum of a signal, as discussed in Section 6.3. Here, in Figure 8.2.4 one-sided spectrum is used.)

Now, we want to consider a more general situation, one in which a message signal, $v_M(t)$, is an analog but not a sinusoidal. In this case, the spectrum of $v_M(t)$ is not a single line but a continuum of frequencies from 0 to f_M, as Figure 8.2.5a shows; this is why such a nonsinusoidal analog signal is called a *band-limited signal*. (Some authors call it *bandpass* signal.) The form of the spectrum shown in Figure 8.2.5a is purely illustrative; it does not represent any real signal. To see an example of the spectrum of a real signal, refer to Figure 8.1.2, where the spectrum of a human voice is demonstrated.

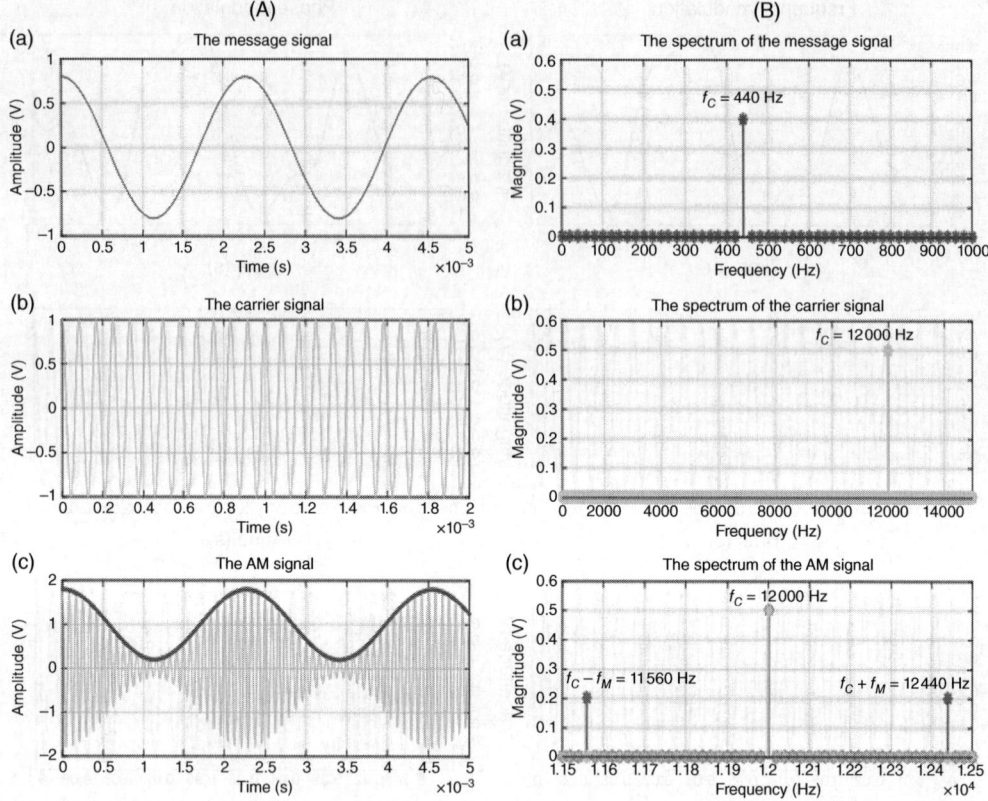

Figure 8.2.4 The waveforms (A) and one-sided spectra (B) of a tone AM signal and its components: (A) The waveforms of the message signal (a), the carrier wave (b), and the AM signal (c); (B) the spectra of the message signal (a), the carrier wave (b), and the AM signal (c).

The spectrum of an arbitrary AM signal carrying an analog but nonsinusoidal message signal can be found by simply extending the logic applied to the tone AM: The spectrum will be concentrated around the carrier frequency, $\omega_C(f_C)$, and will have two sidebands – an USB and a LSB, as qualitatively shown in Figure 8.2.5a.

(**Reminder**: The spectrum of a periodic signal was obtained previously by using the Fourier series (refer to Section 6.1). To find the spectrum of a nonperiodic signal, we need to apply the Fourier transform (introduced in Section 7.1); this is how we can find the continuous spectrum of an arbitrary AM signal. Remember, too, that the spectrum strength of a nonperiodic signal is represented by a *spectral density*, $S(f)$ (V/Hz). (As we already learned, the spectral density of a signal, $S(f)$ (V/Hz), is the average voltage of a signal in 1 Hz bandwidth around frequency f (Hz). Thus, $S(f) = E/\text{BW}$ (V/Hz), where E (V) is the signal's rms magnitude in volts). Figure 8.2.5a illustrates this key point.)

(**Exercise**: Using the online sources, find the order of magnitude of E (V) and BW (Hz) of your cell phone and compute its $S(f)$ (V/Hz).)

To make this analysis more specific, we will also consider a *dual-tone AM*, in which the message signal is a sum of two sinusoidal signals, i.e.

$$v_M(t) = v_{M1} + v_{M2} = A_{M1} \cos(\omega_{M1}t) + A_{M2} \cos(\omega_{M2}t) = A_{M1} \cos(2\pi f_{M1}t) + A_{M2} \cos(2\pi f_{M2}t)$$

The waveforms and spectra of this signal and its components are shown in Figure 8.2.5b. Generalizing (8.2.1) and (8.2.2), we can describe amplitude modulation mathematically as

$$v_{AM}(t) = A_{AM}(t) \cos(\omega_C t) = (A_C + v_M(t)) \cos(\omega_C t) = (A_C \cos(\omega_C t) + v_M(t) \cos(\omega_C t) \quad (8.2.3)$$

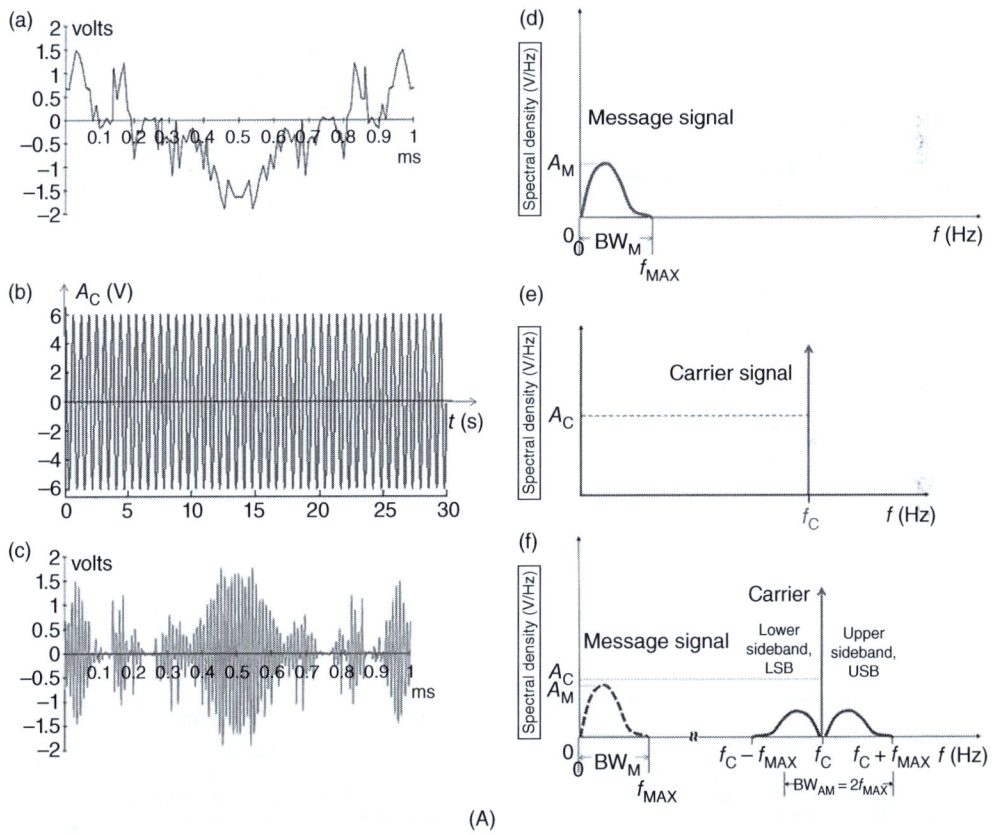

(A)

Figure 8.2.5 Examples of AM modulation: (A) the waveforms of a message signal (a), the carrier wave (b), and the AM signal (c), and the one-sided spectra of a message signal (d), the carrier wave (e), and an arbitrary full AM signal (f). (B) The waveforms and one-sided spectra of a dual-tone full AM signal and its components, where the left side shows the waveforms of the message signal (a), the carrier wave (b), and the AM signal (c) and the right side shows the spectra of the message signal (d), the carrier wave (e), and the AM signal (f). (C) Sinusoidal and dual-tone message signals and corresponding AM signals in greater scale. (Graphs in Figure 8.2.5a are not to scale and the spectra in this part of the figure are shown qualitatively.)

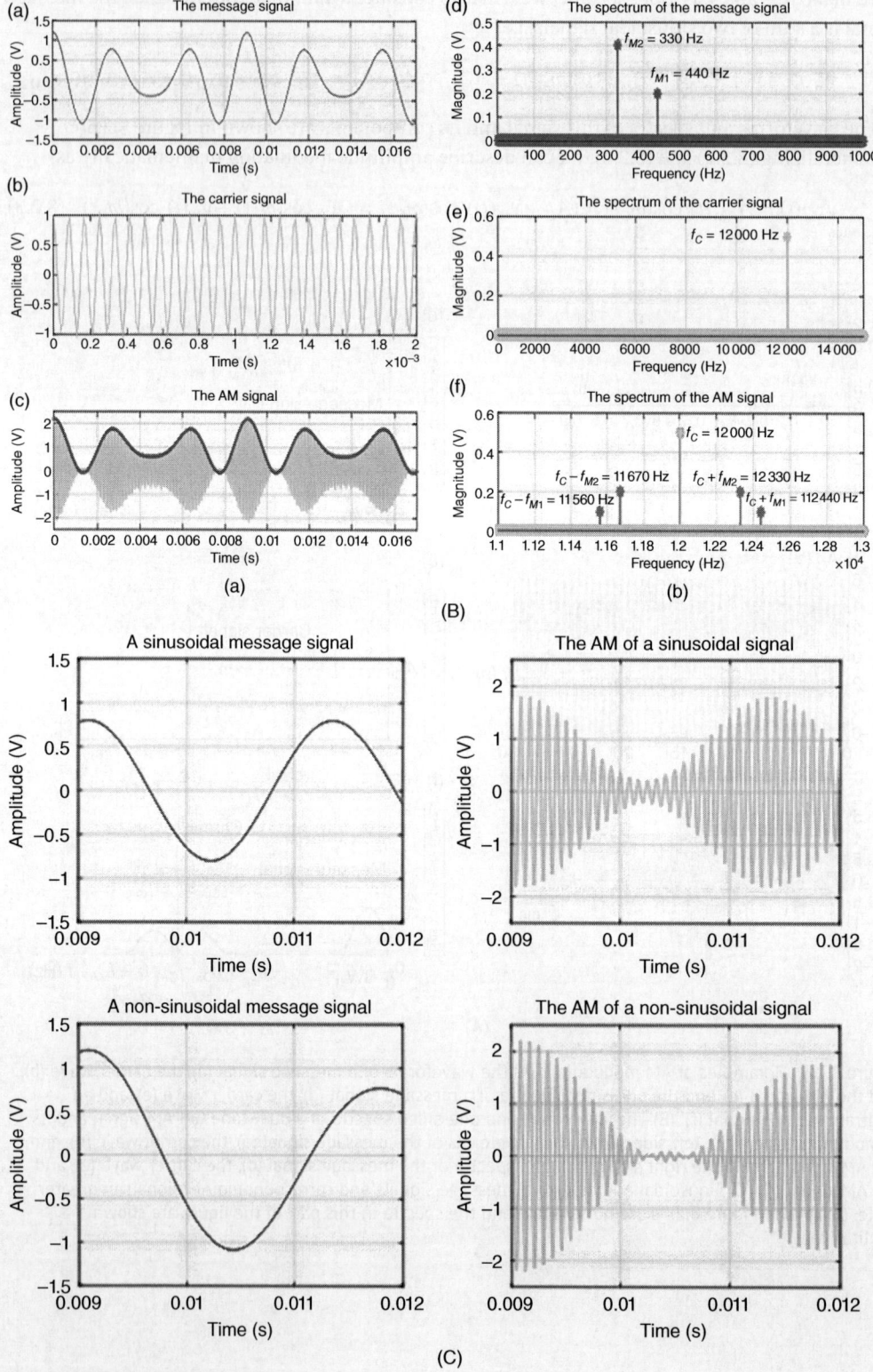

Figure 8.2.5 (*Continued*)

where $v_M(t)$ is a message (information) signal and $A_C \cos(\omega_C t)$ is a carrier. Note that (8.2.3) shows that the amplitude of an AM signal, $A_{AM}(t) = (A_C + v_M(t))$, is a linear function of a message signal, $v_M(t)$).

An amplitude modulation with a sinusoidal carrier and an arbitrary message signal is called a *full*, or *double sideband, DSB-TC;* (8.2.3) is the general formula for this signal.

Now we can describe the spectrum shown in Figure 8.2.5 mathematically by recalling that the Fourier transform of $v_{AM}(t)$ produces the spectrum of this signal (refer to Section 7.1). The Fourier transform of (8.2.3) takes the form

$$F\{v_{AM}(t)\} = F\{(A_C \cos(\omega_C t) + v_M(t) \cos(\omega_C t)\} = F\{(A_C \cos(\omega_C t)\} + F\{v_M(t) \cos(\omega_C t)\} \tag{8.2.4}$$

According to Table 7.1.1, Fourier Transforms, the Fourier transform of the first member of (8.2.4) is

$$F\{(A_C \cos(\omega_C t)\} = A_C \pi \delta(\omega - \omega_C) + A_C \pi \delta(\omega + \omega_C) \tag{8.2.5}$$

Here, $\delta(\omega)$ is the *delta (Dirac or impulse) function*. The delta function is equal to zero everywhere except at one point, which is ω_C in (8.2.5). This function is depicted as a spike surmounted by an arrow, as Figure 8.2.5a (the middle image) demonstrates. It is customary to show the multiplicative constant as its amplitude. Thus, (8.2.5) shows that the two-sided spectrum of an $A_C \cos(\omega_C t)$ consists of two lines, one is at ω_C and the other is at $-\omega_C$. The Fourier transform of the second member of (8.2.5) can be found by applying *modulation property* $F\{f(t) \cos(\omega_0 t)\} = \frac{1}{2}(F(\omega - \omega_0) + F(\omega + \omega_0))$, given in Table 7.2.2, Main Properties of the Fourier Transform, as

$$F\{v_M(t) \cos(\omega_C t)\} = \frac{1}{2}F(\omega - \omega_C) + \frac{1}{2}F(\omega + \omega_C) \tag{8.2.6}$$

Therefore, the Fourier transform of the AM signal is given by

$$F\{v_{AM}(t)\} = A_C \pi \delta(\omega - \omega_C) + A_C \pi \delta(\omega + \omega_C) + \frac{1}{2}F(\omega - \omega_C) + \frac{1}{2}F(\omega + \omega_C) \tag{8.2.7}$$

Equation (8.2.7) describes the two-sided spectrum of the AM signal whose waveform is given by (8.2.3). Each side of this two-sided spectrum includes a line stemming from the sinusoidal carrier, $v_C(t) = A_C \cos(\omega_C t)$, and two bands of frequency – USB and LSB – caused by the message signal, $v_M(t)$. Leaving only a one-sided positive spectrum, we obtain

$$F\{v_{AM}(t)\} = A_C \pi \delta(\omega - \omega_C) + \frac{1}{2}F(\omega - \omega_C) + \frac{1}{2}F(\omega + \omega_C) \tag{8.2.8}$$

as shown in Figure 8.2.5a. Since the spectrum of this AM signal contains a carrier frequency and two bands of side frequencies, this type of amplitude modulation is called *full AM* or *DSB-TC AM*.

If $v_M(t)$ is a sinusoidal signal, then (8.2.1) becomes $v_{AM}(t) = (A_C + A_M \cos(\omega_M t)) \cos(\omega_C t)$; its Fourier transform takes the form

$$F\{v_{AM}(t)\} = A_C \pi \delta(\omega - \omega_C) + A_C \pi \delta(\omega + \omega_C) + \frac{1}{2}A_M \pi \delta(\omega - \omega_C) + \frac{1}{2}A_M \pi \delta(\omega + \omega_C) \tag{8.2.9}$$

and the frequency sidebands collapse into lines, as in Figure 8.2.4 for a one-sided spectrum.

Example 8.2.1 The Waveform and Spectrum of a Full (DSB-TC AM) Signal

Problem

Consider an amplitude-modulated signal with $v_M(t) = \text{sinc}(3t)$ as the message and $v_C(t) = \cos(30t)$ as the carrier. Find the waveforms and the spectra of $v_M(t)$, $v_C(t)$ and $v_{\text{DSB-TC}}(t) \equiv v_{AM}(t)$ signals.

Solution

Using (8.2.3), we find

$$v_{AM}(t) = (A_C + v_M(t))\cos(\omega_C t) = (1 + sinc(3t))\cos(30t)$$

The one-sided waveforms and the spectra of all three signals are presented in Figure 8.2.6. They are built with MATLAB, and the code is given in the following Discussion section. The spectra are found, of course, by applying the Fourier transform discussed here and in Section 7.1.

Discussion

- The MATLAB code for building Figure 8.2.6, with comments pertaining to this example, is presented below:

The MATLAB code for 8.2.1:

```
syms t w
vm=sin(3*t)./(3*t); % The message signal is sinc(3t)
figure(1);ezplot(vm,[0,7,-0.5,1.2]) % The waveform of the
message signal is plotted %in Figure 9.B-6a
xlabel('Time [seconds]');ylabel('vm(t)');grid;
Fm = fourier(vm); % The Fourier transform of the message signal
figure(2);ezplot(Fm,[0,50,-0.2,1.2]) % The plot of the one-sided
spectrum of the %message signal shown in Figure 8.2.6b
xlabel('\omega [rad/s]');ylabel('Fm(\omega)');grid;
vc=cos(30*t); % The carrier signal is cos(30t)
figure(3);ezplot(vc,[0,5]) % The waveform of the carrier is plotted in %Figure 8.2.6a
xlabel('Time [seconds]');ylabel('vc(t)');grid;
FC = fourier(vc); % The Fourier transform of the carrier
figure(4);ezplot(FC+30,w,[0,50,0,10]); % The plot of the one-sided
spectrum of the %carrier shown in Figure 8.2.6b
xlabel('\omega [rad/s]');ylabel('Fc(\omega)');grid;
vam=vc+vm*vc % The waveform of the AM signal
figure(5);ezplot(vam,[0,7,-2.5,2.5]) % The waveform of the AM signal is plotted in %Figure 8.2.6a
xlabel('Time [seconds]');ylabel('vam(t)');grid;
F = fourier(vam) % The Fourier transform of the AM signal
figure(6);ezplot(F,[0,50]) % The one-sided spectrum of the AM signal is %plotted in Figure 8.2.6b
xlabel('\omega [rad/s]');ylabel('F(\omega)');grid;
```

- Comparison of Figure 8.2.6 with Figure 8.2.5: We can clearly see how the general concept of amplitude modulation with a band-limited analog message signal is applied in a particular case. We urge you to compare every single plot and see, for instance, that the spectrum of the message signal, sinc(3t), in this example is centered on $\omega_M = 0$ rad/s, whereas the spectrum of the AM signal – though preserving its shape – is centered on $\omega_C = 30$ rad/s.

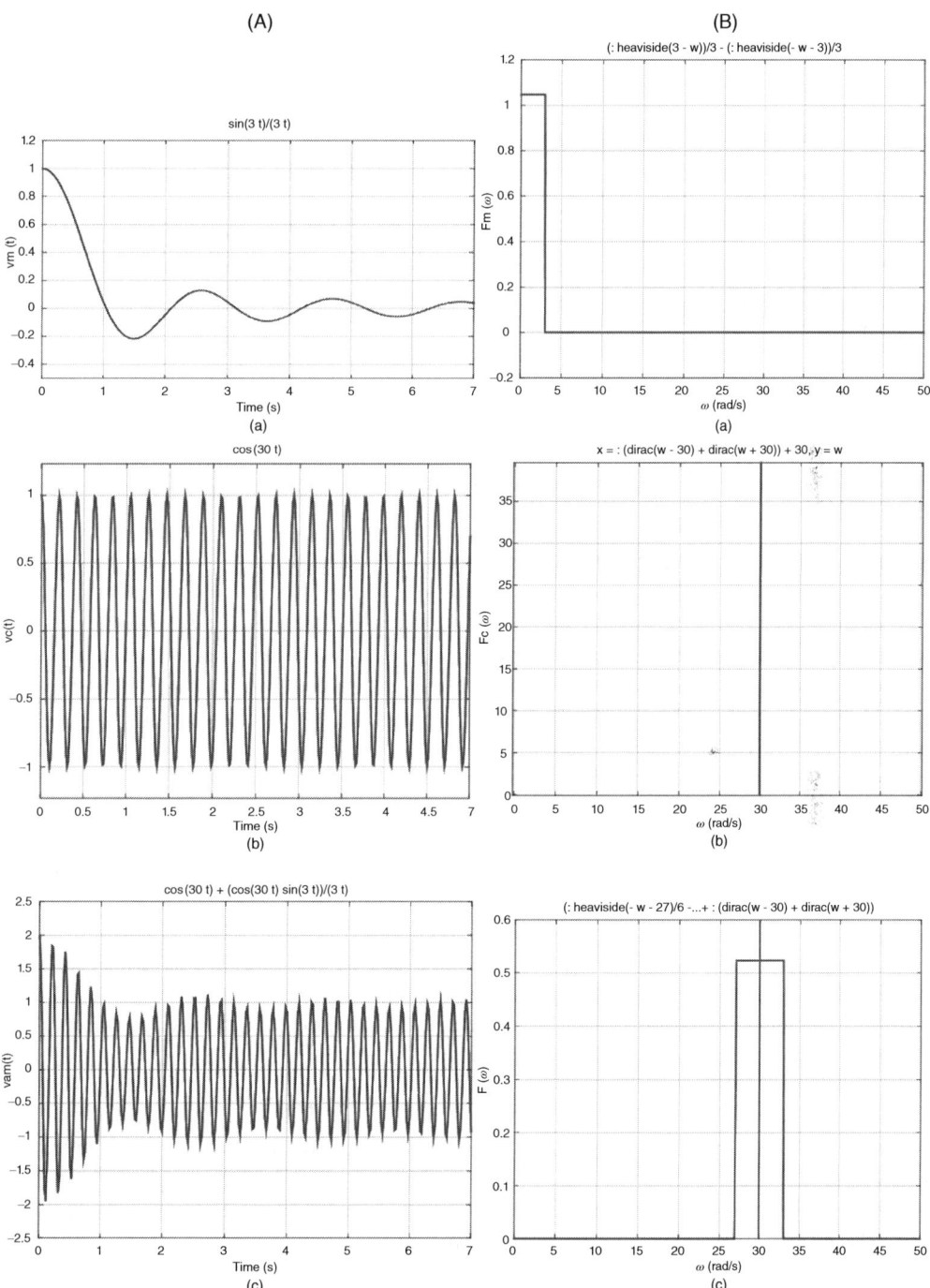

Figure 8.2.6 The one-sided waveforms and spectra of the full AM signal in Example 8.2.1: (A) the waveforms of the message signal (a), the carrier wave (b), and the AM signal (c). (B) The spectra of the message signal (a), the carrier wave (b), and the AM signal (c). (Figures are not to scale.)

8.2.2.2 Problems of Full AM Transmission

There are two main problems with the transmission of a full AM signal: *waste of both power and bandwidth*.

Power Efficiency Analysis of the *power* of a tone AM signal given in Section 8.1 – see (8.1.14b) and (8.1.15b) – shows that a carrier, which contains most of the total AM power, delivers no information. We can extend this analysis to a full AM signal by considering the spectral components – the carrier and two sidebands – of the DSB-TC AM signal shown in Figure 8.2.6. Referring to the power of the carrier, P_C (W), to the power of the message signal containing the power of the two sidebands, P_M (W) = $2P_{SB}$, and to the total as P_{AM} (W) $\equiv P_T = P_C + P_M$, we introduce in (8.1.15a), the *power efficiency*, as

$$\eta = \frac{P_M(W)}{P_T(W)} \tag{8.2.10}$$

Referring to the tone AM discussed in Section 8.1, we know that for $m = 1$, $\eta = 0.33$. The conclusion is that the full AM system wastes power by transmitting the carrier signal, which delivers no information.

Bandwidth Efficiency The *bandwidth* analysis in Section 8.1 – see (8.1.12) – shows that a *tone* full AM signal delivers two frequency sidebands containing the same information, which amounts to unnecessary duplication. The same is true for an *arbitrary* full AM, as Figure 8.2.6 shows. Indeed, the bandwidth of a full AM signal is given by BW_{AM} (Hz) = $2f_M$, which means that the full AM signal occupies twice the amount of bandwidth necessary to deliver information carried by the $v_M(t)$ signal. As we have indicated many times, bandwidth is a commodity and saving it is an important task of the designer of a communication system. Therefore, *from both the power and bandwidth standpoints, DSB-TC AM is a very inefficient means of delivering information*.

8.2.2.3 Double-Sideband Suppressed Carrier (DSB-SC) AM

To overcome the power inefficiency of a full AM transmission, we need simply **not** transmit a carrier, which can be done by removing this signal from a full AM.

(*Question*: Modulation is changing the parameters of a carrier signal. How, then, can we make a modulated signal if we remove a carrier? Doing so would seem to contradict the whole idea of a broadband (modulated) transmission, but DSB-SC modulation exists. Explain.)

An AM transmission without a carrier is called a DSB-SC *AM or DSSC* or simply *DSB AM*. The formula for a DSB signal can be derived as follows:

$$v_{AM-DSB}(t) = v_{AM}(t) - v_C(t) = (v_C(t) + v_M(t) \cdot v_C(t)) - v_C(t) = v_M(t) \cdot v_C(t) \tag{8.2.11}$$

(**Exercise**: Sketch the block diagram of a DSB-SC modulator that performs the operation described by Eq. (8.2.11).)

To find the spectrum of this type of AM, we consider a *tone AM* because of its simplicity. Thus, for $v_M(t) = A_M \cos(\omega_M t)$ and $v_C(t) = A_C \cos(\omega_C t)$, we find

$$v_{AM-DSB}(t) = v_M(t) \cdot v_C(t) = A_M \cos(\omega_M t) \cdot A_C \cos(\omega_C t)$$

$$= \tfrac{1}{2} A_M \cdot A_C \cos(\omega_C + \omega_M)t + \tfrac{1}{2} A_M \cdot A_C \cos(\omega_C - \omega_M)t \tag{8.2.12}$$

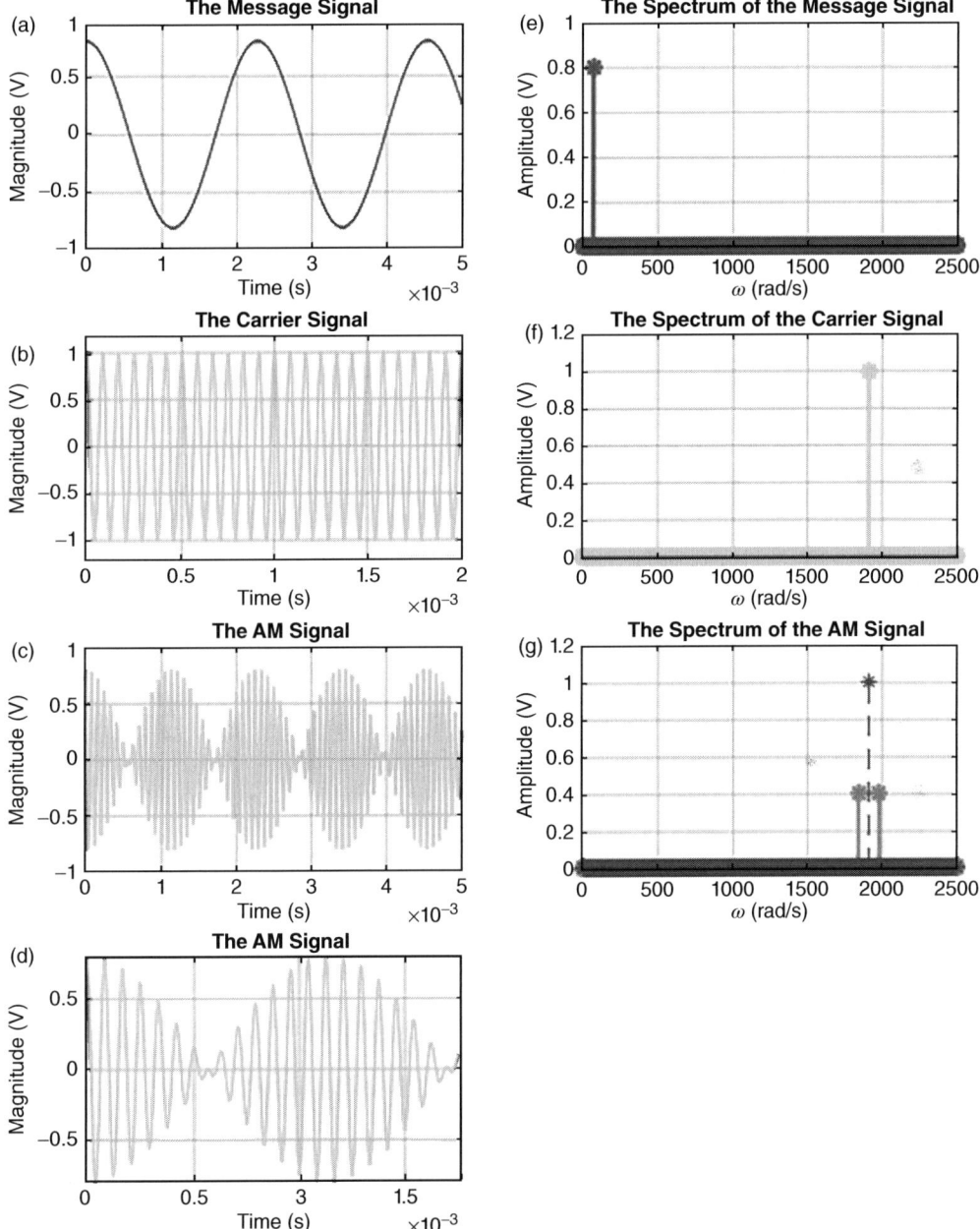

Figure 8.2.7 The waveforms and spectra of a double-sideband (DSB) tone AM signal and its components: The left column: the waveforms of the message signal (a), the carrier signal (b), the DSB AM signal (c), and the DSB AM signal in greater scale (d); the right column: the spectra of the message signal (e), the carrier signal (f), and the DSB AM signal (g). (Figures are not to scale.)

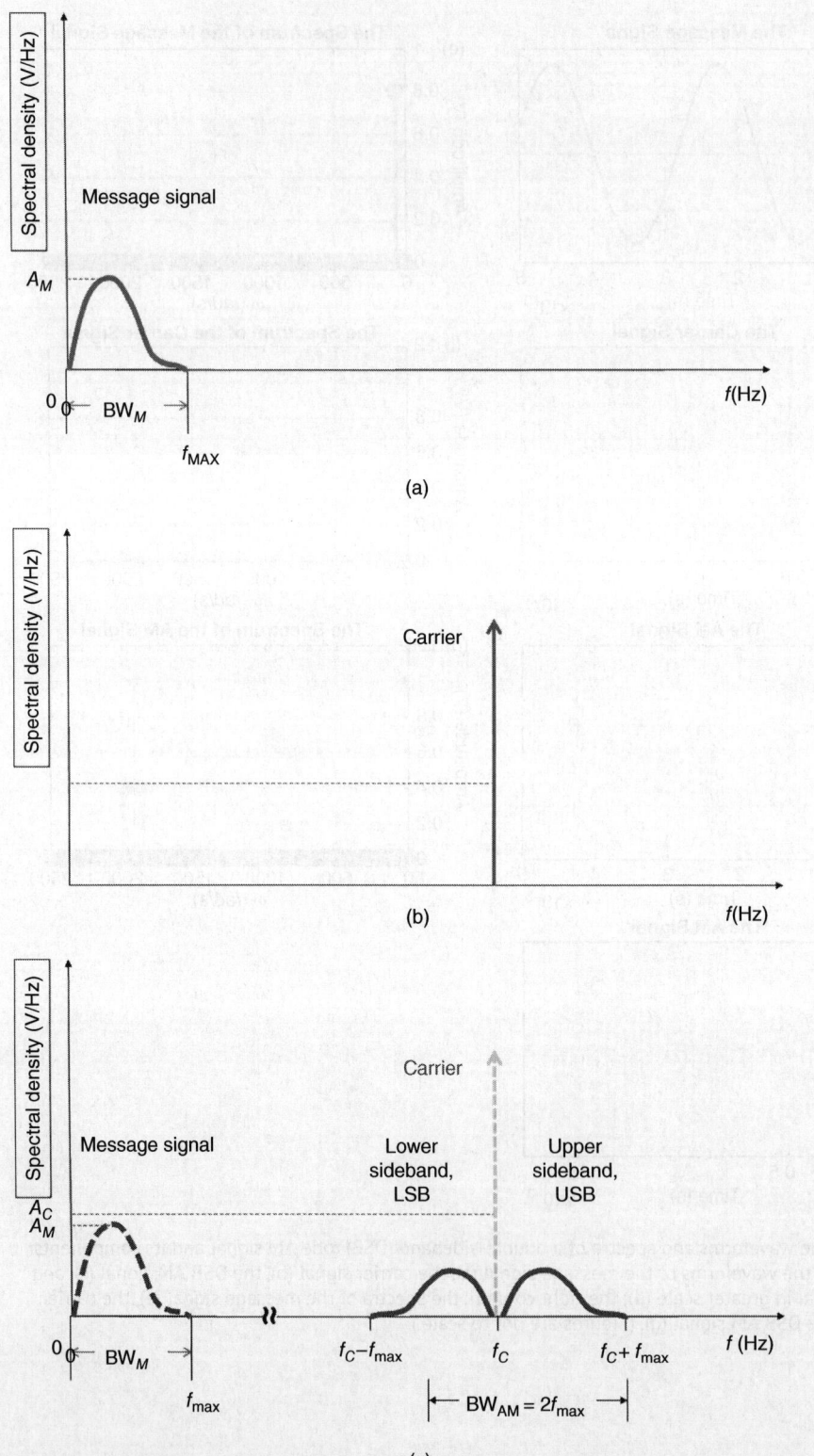

Figure 8.2.8 The spectra of an arbitrary DSB AM signal and its components: The spectrum of a message signal (a), the spectrum of a carrier signal (b), and the spectrum of a DSB AM signal (c). (Figures are not to scale, and the spectra in this figure are shown qualitatively.)

Compare this formula with (8.1.11a) and observe the similarity and difference between the DSB AM and full AM signals. Obviously, the spectrum of a DSB AM should not contain a carrier line but must contain, as (8.2.12) shows, USB and LSB. Refer to Figure 8.2.7, where the waveforms and the spectra of a DSB AM signal and its components are shown. Compare this figure with Figure 8.2.4, consider the main features of a DSB AM and full AM signals, and answer the following questions by referring to (8.2.11) and (8.2.12):

- Amplitudes of DSB and full AM signals are different. By how much? Why?
- The envelope period of a full AM signal is equal to the period of a message signal T_M (s), whereas the envelope period of a DSB AM signal is equal to $T_M/2$. Why?
- The waveform of a DSB AM signal shown in a greater scale in Figure 8.2.7 demonstrates the 180° phase change at a node but the waveform of a full AM signal does not exhibit any phase change. Why?

If the message signal is an analog but not a sinusoidal, DSB AM still can be described by using (8.2.11) as

$$v_{\text{AM-DSB}}(t) = v_M(t) \cdot v_C(t) = v_M(t) \cdot A_C \cos(\omega_C t). \tag{8.2.13}$$

We can readily observe the similarity of this equation to (8.2.3). Following the analysis presented in (8.2.3)–(8.2.9), we can determine the spectrum of a DSB AM signal by using the modulation property of the Fourier transform as

$$F\{v_{\text{AM-DSB}}(t)\} = F\{v_M(t) \cdot A_C \cos(\omega_C t)\} = \frac{A_C}{2} F(\omega - \omega_C) + \frac{A_C}{2} F(\omega + \omega_C) \tag{8.2.14}$$

Therefore, the spectrum of a DSB AM signal contains two bands of frequency due to presence of the message signal, $v_M(t)$, and it does not include a carrier signal. The one-sided spectrum of an arbitrary DSB AM signal is shown in Figure 8.2.8.

The total power of DSB-TC given by $P_{\text{AM}} \equiv P_T = P_C + P_M$. Since $P_C = 0$ in a DSB AM signal, $P_T = P_M$. Thus, *a DSB AM signal uses 100% of its transmitting power to deliver the information signal.*

Example 8.2.2 The Waveform and Spectrum of a DSB AM Signal

Problem

Consider an amplitude-modulated signal with $v_M(t) = \text{sinc}(3t)$ as the message and $v_C(t) = \cos(30t)$ as the carrier. Find the waveforms and the spectra of the $v_M(t)$, $v_C(t)$, and $v_{\text{AM-DSB}}(t)$ signals.

Solution

We follow the pattern of Example 8.2.1. Using (8.2.12), we find $v_{\text{AM-DSB}}(t) = v_M(t) v_C(t) = \text{sinc}(3t) \cdot \cos(30t)$. The waveform and the spectrum of the obtained DSB AM signal are shown in Figure 8.2.9a.

Compare the waveform and spectrum of $v_{\text{AM-DSB}}(t)$ shown in Figure 8.2.9a and that of full AM, $v_{\text{AM}}(t)$, shown in Figure 8.2.9b and list all the differences you observe. Highlight the advantage that DSB AM signal exhibits over a full AM.

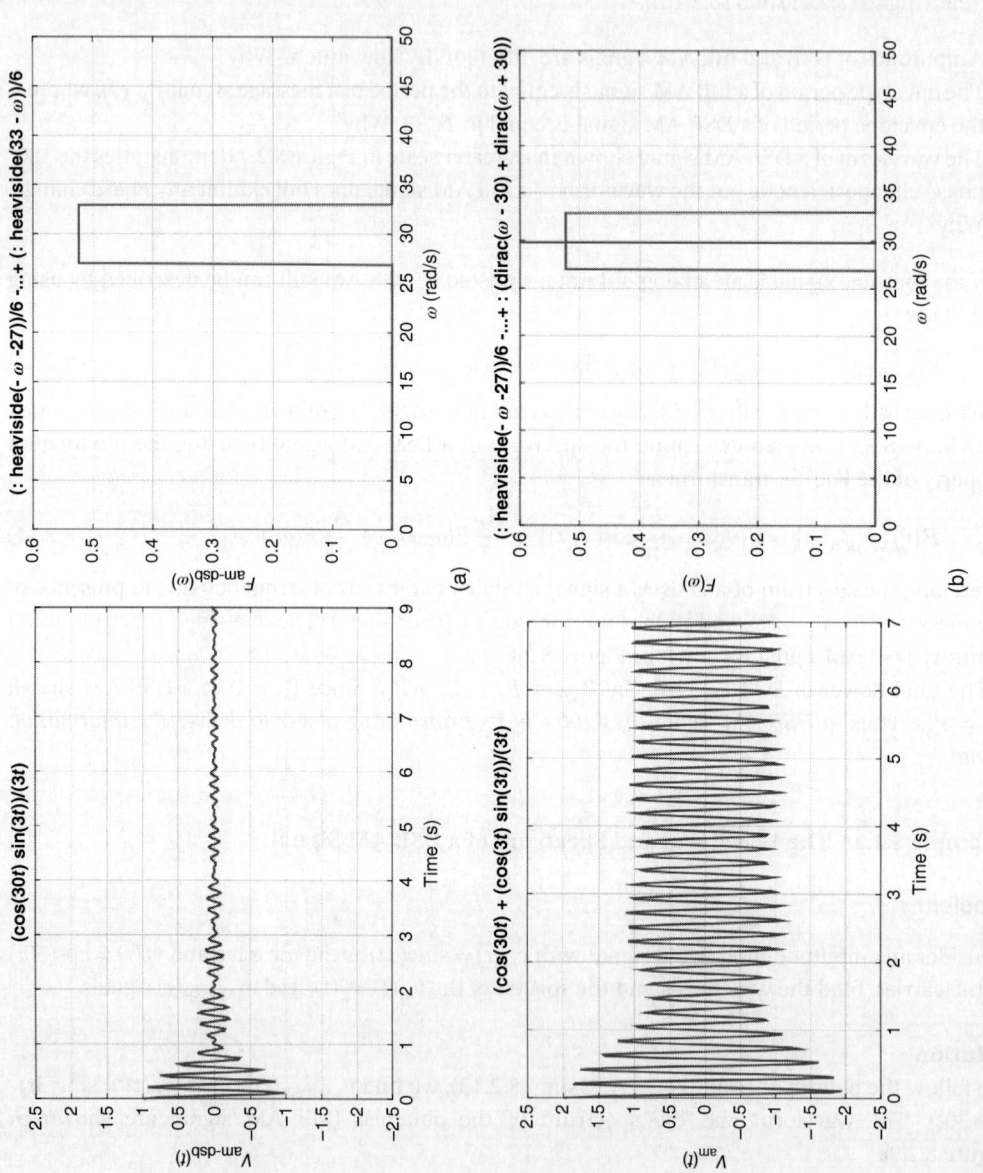

Figure 8.2.9 Waveform and spectrum of (a) DSB AM signal and (b) that of full AM signal. (The latter figures are reproduced from Figure 8.2.6.)

Discussion

- Realize that in a DSB AM system, all the transmitted power is distributed between the sidebands carrying the information; this is why the DSB AM system is 100% power efficient.
- Though a DSB signal is certainly more power-efficient than a full AM signal, its transmission still suffers from a lack of bandwidth efficiency because both sidebands carry the same information.
- Since both full AM and DSB AM signals are quite similar, the MATLAB codes in Example 8.2.1 and this example are also similar. We leave the development of the former code as an exercise for you to work out.

8.2.2.4 Single-Sideband Suppressed Carrier (SSB-SC) AM

To overcome the bandwidth inefficiency of a full AM and a DSB AM, we need to simply block the transmission of the carrier and one of the sidebands, which in principle can be done by suppressing (filtering) them in transmission. The new signal is called a *single-sideband suppressed carrier (SSB-SC or SSB) AM*. We will use the simpler notation, *SSB AM*, which includes the other popular designation of this type of AM – *suppressed-sideband AM*.

To obtain SSB AM, we can subtract one of the sidebands from a DSB signal as

$$v_{AM-SSB}(t) = v_M(t) \cdot v_C(t) - v_{USB}(t) \tag{8.2.15a}$$

For example, for a tone AM, we find

$$
\begin{aligned}
v_{AM\text{-}SSB}(t) &= v_M(t) \cdot v_C(t) - v_{USB}(t) = A_M \cos(\omega_M t) \cdot A_C \cos(\omega_C t) - v_{USB}(t) \\
&= \frac{1}{2} A_M \cdot A_C \cos(\omega_C + \omega_M)t + \frac{1}{2} A_M \cdot A_C \cos(\omega_C - \omega_M)t - \frac{1}{2} A_M \cdot A_C \cos(\omega_C + \omega_M)t \\
&= \frac{1}{2} A_M \cdot A_C \cos(\omega_C - \omega_M)t = v_{LSB}(t)
\end{aligned}
\tag{8.2.15b}
$$

This approach to building an SSB AM transmitter – called the *filtering method* – includes a bandpass (sideband) filter; this filter must eliminate a carrier and one sideband. Equation (8.2.15b) exemplifies removing a USB but, in principle, either one can be removed. Figure 8.2.10a,b shows the block diagrams of USB and LSB transmitters and the corresponding spectra. These figures demonstrate that to achieve the ideal results, we must use not just a bandpass filter (BPF) but an *ideal bandpass filter*. Such a filter must have *step-like magnitude* characteristics and *cutoff frequencies* $\omega_{ch} = \omega_C$ and $\omega_{cl} = \omega_C - \omega_M$ if we want to transmit a USB and cutoff frequencies $\omega_{ch} = \omega_C + \omega_M$ and $\omega_{cl} = \omega_C$ if we want to transmit LSB, as shown in Figure 8.2.10a,b.

The main drawback of this method is obvious: We do not have an ideal BPF; thus, we cannot completely eliminate one sideband and the carrier. This is why the approach called the *phasing method* has been developed for generating an SSB AM signal. The idea is to subtract one sideband from a DSB signal at the circuit level. This concept can be mathematically demonstrated with a tone AM as follows: Let us take the DSB AM signal obtained in (8.2.12):

$$
\begin{aligned}
v_{AM\text{-}DSB}(t) &= v_M(t) \cdot v_C(t) = A_M \cos(\omega_M t) \cdot A_C \cos(\omega_C t) \\
&= \frac{1}{2} A_M \cdot A_C \cos(\omega_C + \omega_M)t + \frac{1}{2} A_M \cdot A_C \cos(\omega_C - \omega_M)t
\end{aligned}
$$

Recalling that

$$\sin \alpha \cdot \sin \beta = \frac{1}{2} \cos(\alpha - \beta) - \frac{1}{2} \cos(\alpha + \beta)$$

8 Analog Transmission with Analog Modulation

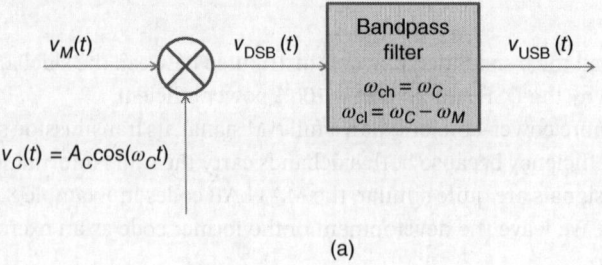

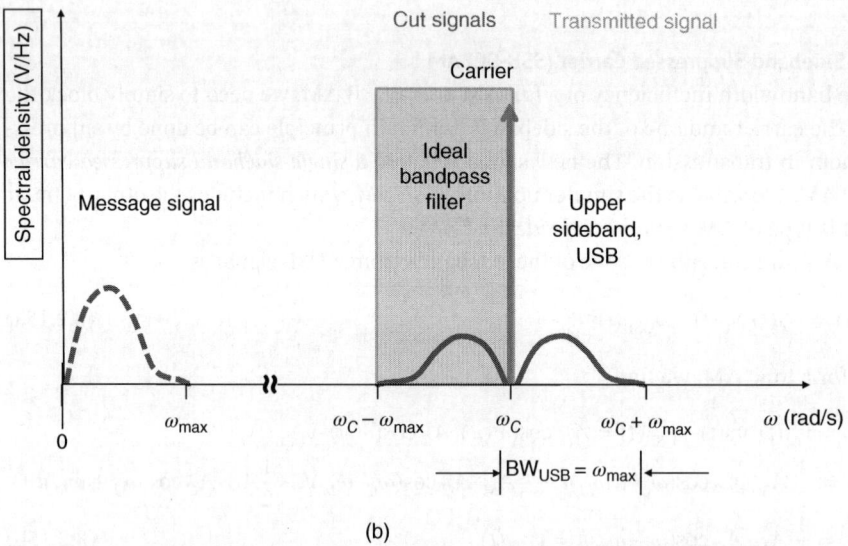

Figure 8.2.10 Filtering method for generating an SSB AM signal: (a) The block diagram of a USB transmitter with an ideal bandpass filter and the spectrum of a cut LSB band and a transmitted USB band and (b) similar figures for an LSB AM transmission. (Figures are not to scale.)

we realize that by adding or subtracting the $A_M \sin(\omega_M t) \cdot A_C \sin(\omega_C t)$ signal to or from $v_{\text{AM-DSB}}(t)$, we can obtain one of the sidebands. Indeed,

$$v_{\text{USB}}(t) = A_M \cos(\omega_M t) \cdot A_C \cos(\omega_C t) + A_M \sin(\omega_M t) \cdot A_C \sin(\omega_C t)$$
$$= \frac{1}{2} A_M \cdot A_C \cos(\omega_C + \omega_M)t + \frac{1}{2} A_M \cdot A_C \cos(\omega\, \omega_M)t + \frac{1}{2} A_M \cdot A_C \cos(\omega_C + \omega_M)t$$
$$- \frac{1}{2} A_M \cdot A_C \cos(\omega_C - \omega_M)t$$
$$= A_M \cdot A_C \cos(\omega_C + \omega_M)t \quad (8.2.16a)$$

Similarly,

$$V_{\text{LSB}}(t) = A_M \cos(\omega_M t) \cdot A_C \cos(\omega_C t) - A_M \sin(\omega_M t) \cdot A_C \sin(\omega_C t)$$
$$= A_M \cdot A_C \cos(\omega_C - \omega_M)t \quad (8.2.16b)$$

These operations can be accomplished with a circuit whose block diagram is shown in Figure 8.2.11. A message signal, $v_M(t)$, and a carrier, $v_C(t)$, go directly to the cosine DSB generator, which produces a $v_{\text{DSB}}^C(t)$ signal. The same signals, $v_M(t)$ and $v_C(t)$, passing through -90° phase-shift units, go to the

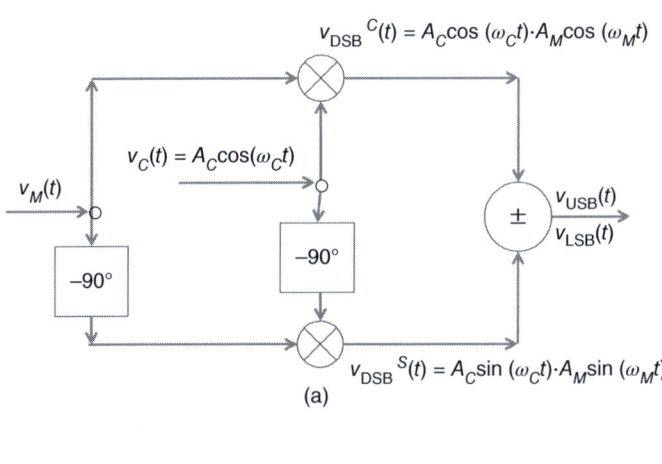

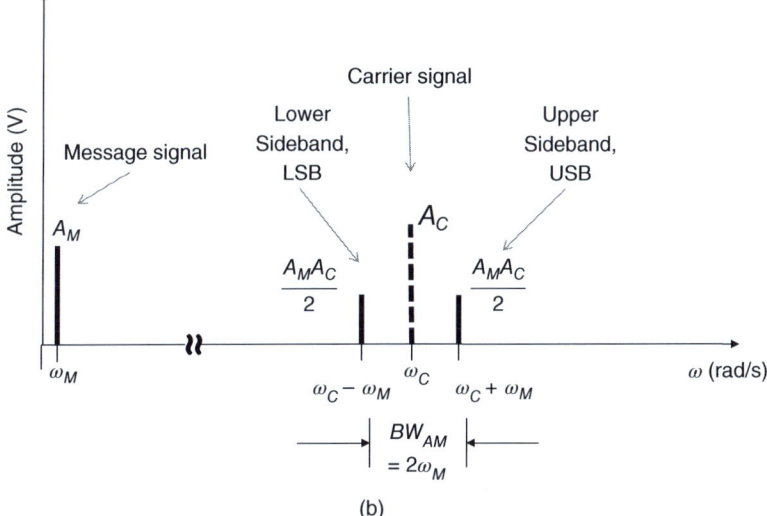

Figure 8.2.11 Phasing method of generating an SSB AM signal: (a) The block diagram of an SSB generator and (b) the spectra of DSB, USB and LSB tone AM signals. (Figures are not to scale.)

sine DSB generator, which produces a $v_{DSB}^S(t)$ signal. At the output, these cosine DSB and sine DSB signals are either added to give a $v_{USB}(t)$ signal or subtracted to yield a $v_{LSB}(t)$ signal.

*(**Question**: What are the waveforms of the USB and LSB signals shown in Figure 8.2.11? Sketch these waveforms and show their amplitudes and periods.) (Hint: Examine Figure 8.2.11 and remind yourself of the fundamental relationship between the time and frequency domains.)*

*(**Question**: Phasing method of generating SSB AM signal is the example of implementing the mathematical formula in a real circuit. Do you know another example of such approach?)*

If the message signal is nonsinusoidal, we need to turn to the *Hilbert transform* for a mathematical description of an SSB AM signal, but this transform is outside the scope of our book. We can logically extend the phasing method used previously to an arbitrary message signal and qualitatively find the spectrum of such an SSB AM; this spectrum is shown in Figure 8.2.12. The conclusion we draw from observing this figure is clear: *SSB transmission reduces the bandwidth of an AM signal to the minimum value, the bandwidth of a message signal*.

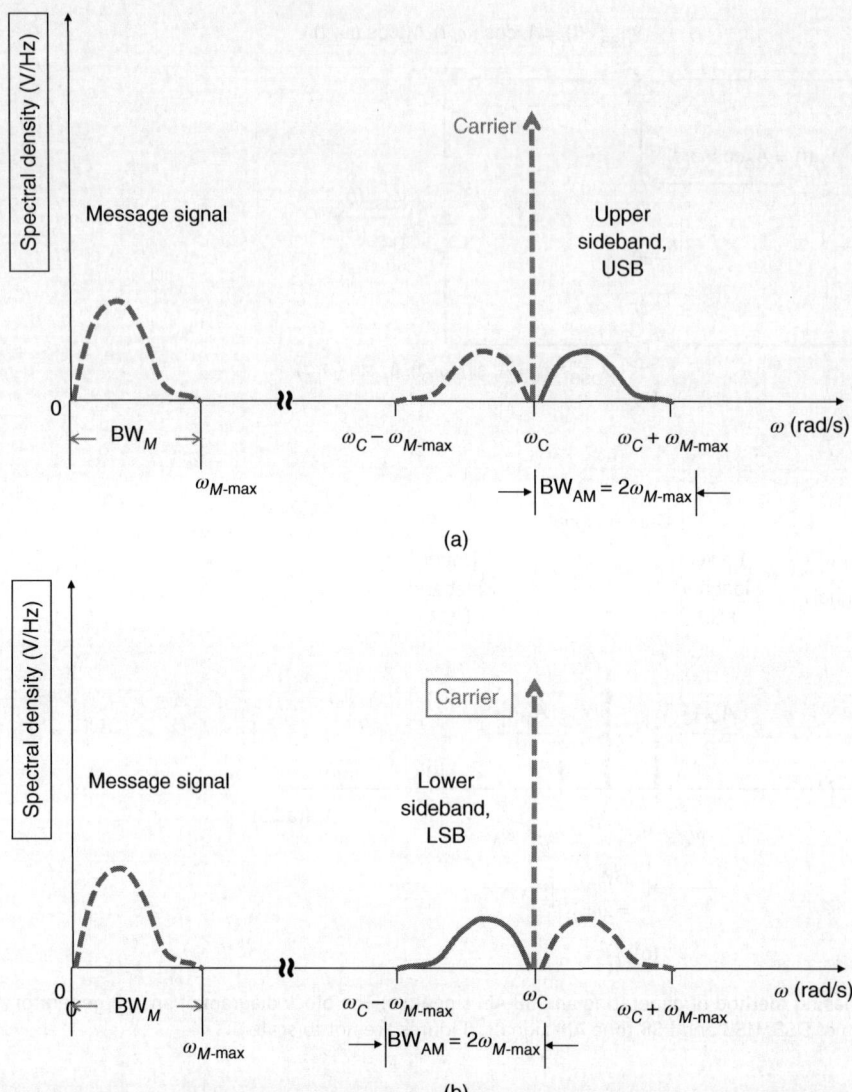

Figure 8.2.12 The spectrum of (a) a USB signal and (b) an LSB signal generated from a DSB AM signal by the phasing method. (Figures are not to scale.)

8.2.2.5 Full AM, DSB, or SSB – Which Type to Choose?

Our review of the three types of amplitude modulation shows that full AM uses both power and bandwidth inefficiently, DSB saves power but does not improve the bandwidth efficiency, and SSB efficiently uses transmitter's power and saves transmission bandwidth. We should elaborate on all the advantages of SSB transmission:

- The highest power efficiency in SSB transmission is achieved by channeling the total power that would normally be used for transmitting a full AM into the transmission of just one signal. Thus, this signal – SSB – can be made stronger and be able to travel farther.

- SSB efficiently uses power generated by a transmitter not just because it delivers only one sideband but also because SSB transmits only during the modulation process, that is, only when someone speaks into a microphone or when music is transmitted. Without the modulation, there is no transmission.
- Since SSB uses a smaller bandwidth, this AM type enables the transmission of more signals within a given range of spectrum, thus increasing the spectral efficiency of the transmission. (See Section 1.2 to brush up on spectral efficiency and the importance of this aspect of modern communications.)
- The smaller bandwidth of an SSB results, as we will learn shortly, in reducing noise impact on transmission.
- SSB transmission has no *selective fading*. As Figure 8.2.5 shows, the full AM signal includes many signals with various frequencies; in radio transmission, these signals are affected by the atmosphere quite differently and they arrive at the receiver with various phase shifts. These phase shifts can lead to the mutual cancellation of some signals, which is called selective fading.

It seems that the conclusion is obvious: We should always use only SSB. In practice, however, all three types are used, each finding its specific niche. Why? The answer lies in the fact that power and bandwidth efficiency are not the only criteria for choosing a type of AM transmission. The other equally important issues are the *modulation* and *demodulation* processes. We leave the consideration of modulation methods and techniques to specialized technological manuscripts because modulation does not impose any restrictions on the type of AM transmission. *Demodulation*, on the other hand, does, and here are the main points to bear in mind about it:

1. At the destination point of an AM transmission, a carrier wave serves as a reference signal because it is a pure sinusoidal signal. Thus, the carrier provides a receiver with sufficient information needed for demodulation – the first and most important of all being time. Cutting off a carrier in both DSB and SSB transmissions makes the complete recovery of information difficult. One solution is to *generate a similar signal at the receiver side and reinsert it into the received signal*, but this reinserted signal must have exactly the same frequency and phase as the real carrier. In practice, this requirement is very difficult to meet. Another solution is transmitting a small portion of a carrier to preserve the benefits of DSB or SSB transmissions, yet still having a carrier at the receiver side. The price paid: transmitting a small amount of extraneous power. This type of transmission is called the *pilot carrier*.
2. Some types of transmission meet restrictions in bandwidth. For example, the Federal Communications Commission (FCC) requires that an analog broadcast TV signal should not exceed 6 MHz. Video information occupies up to 4.2 MHz; to provide the best quality and reliability of transmission, it needs to use full AM. Thus, the full, DSB-TC AM signal occupies $2 \times 4.2\,\text{MHz} = 8.4\,\text{MHz}$. To cut the bandwidth of this signal and preserve the quality of transmission, a special type of AM called *vestigial-sideband (VSB)* has been developed. In VSB, a portion of an LSB, along with a carrier and the whole USB, is transmitted. In TV transmission, VSB provides the AM bandwidth that meets the 6-MHz restriction and reliably delivers all the signals needed for complete demodulation. The spectrum of a VSB signal is shown in Figure 8.2.13. Though the days of analog TV are history, the VSB technique is still in use in some modern digital TV systems.

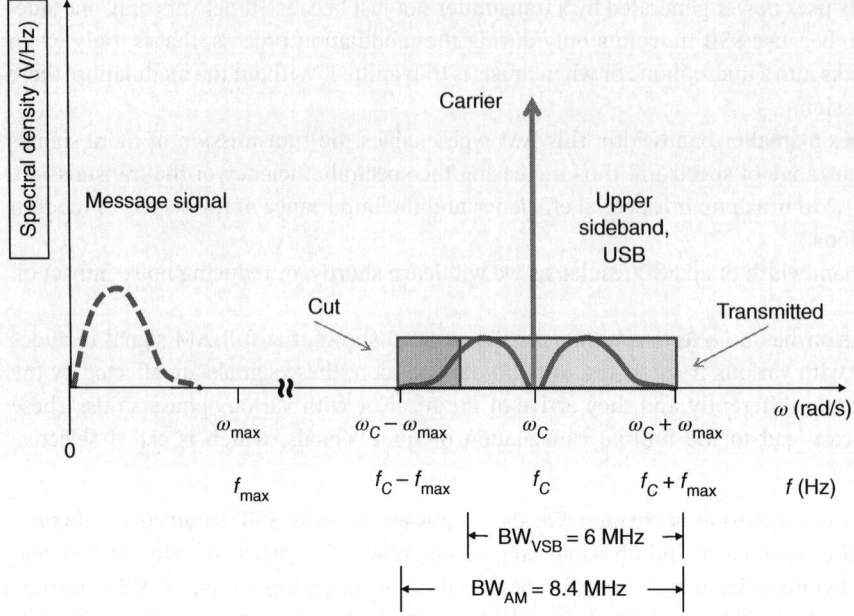

Figure 8.2.13 The spectrum of a vestigial sideband, VSB, signal. (Figures are not to scale.)

8.2.2.6 Applications of AM Transmission

As we are well aware, the heyday of AM transmission lies in the distant past; with the recent retirement of analog TV, the other big segment of AM application disappeared. Did this declare the end of AM transmission once and for all? Well, not quite. This type of transmission still finds its applications, the main and best known, of course, being AM radio and broadcast TV. They use DSB with a pilot carrier for the transmission of stereo signals and – for TV only, of course – the delivery of the color aspects of a picture. What's more, SSB transmission is used in two-way radios, which find application in military, marine, and ham (amateur radio) communication. Even in digital TV, AM is still used but in VSB format with modern multilevel coding schemes.

The point is that the concept of amplitude modulation is a must-know subject for any communication professional because it is widely used in modern coding formats applied in digital modulation, as we will learn in Chapters 9 and 10.

8.2.3 Advanced Consideration of Angular (Phase and Frequency) Modulation and Its Application in Analog Transmission

8.2.3.1 Angular Modulation

What follows is an in-depth discussion of angular modulation.

Angle (Phase) of a Sinusoidal Function You will recall that *angular modulation* is the common term covering both FM and PM (see the short reminder at the beginning of this section and refer to Section 8.1). To fully understand the use of the term *angular*, we need to enhance the basic definitions (refer to Chapter 2). A regular sinusoidal carrier,

$$v_C(t) = A_C \cos(\psi_C(t)) = A_C \cos(2\pi f_C t + \theta_C) = A_C \cos(\omega_C t + \theta_C) \qquad (8.2.17)$$

includes the following parameters:

- A_C (V) is the signal's *amplitude* (a constant).
- $\psi_C(t)$ (rad) = $\omega_C t + \theta_C = 2\pi f_C t + \theta_C$ is the *instantaneous phase*, often simply called the *phase*, or *angle* (a variable, a function of time).
- f_C (Hz) is the *linear (cyclic) frequency*, often simply called the *frequency* (a constant).
- ω_C (rad/s) = $2\pi f_C$ is the *radian (angular) frequency* (a constant).
- θ_C (rad) is the *initial phase* (phase shift or phase offset), often also called – confusingly – the phase. This parameter is obtained from $\psi_C(t)$ at $t = 0$ and it is a constant.

At any given instant, $\psi_C(t)$ takes a specific value, which therefore gives a specific value to $v_C(t)$ because, of course, the value of a cosine (or a sine) is determined by the value of its angle. For example, consider a carrier, $v_C(t) = A_C \cos(\omega_C t + \theta_C)$, with $A_C = 1$ V, $\theta_C = 0$, and $\omega_C = 3.14$ rad/s; then for $t = 1/6$ seconds, we find $\psi_C(t) = \pi/6$ rad = 30° and $v_C(t) = 0.866$ V. This example explains that modulation of either radian frequency, ω_C, or of the initial phase, θ_C, will result in changing the angle of a cosine (or a sine), justifying the term *angular modulation*. Pay close attention to the units: the units of $\psi_C(t)$ must be radians (or degrees).

Phase Modulation Here, we generalize our introduction to PM, given in Section 8.1.

Phase modulation, PM, is the operation that changes an <u>instantaneous</u> phase of the carrier sinusoidal signal, $\psi_C(t)$, according to variations of the message signal, $v_M(t)$.

This is done by adding a message signal to the carrier's phase as

$$\psi_{PM}(t) = \psi_C(t) + k_\theta v_M(t) = \omega_C t + \theta_C + k_\theta v_M(t) \qquad (8.2.18)$$

Here, k_θ (rad/V) is the *phase modulation (deviation) constant*. The initial phase, θ_C, which is a constant, plays no role in the dynamic process of changing $\psi_{PM}(t)$ and can therefore be omitted. Thus,

$$\psi_{PM}(t) = \psi_C(t) + k_\theta v_M(t) = \omega_C t + k_\theta v_M(t) \qquad (8.2.19)$$

A PM signal is described by

$$v_{PM}(t) = A_C \cos \psi_{PM}(t) = A_C \cos(\omega_C(t) + k_\theta v_M(t)) \qquad (8.2.20)$$

which is (8.1.29) in a slightly modified form. Thus, we arrive at the form of a PM signal, which has been used in Section 8.1 and shown in Figure 8.1.8. This is why from this point all formulas and considerations presented in Section 8.1 can be applied without alterations.

(**Question**: *Figure 8.1.8, reproduced at the beginning of this section, shows that phase modulation is achieved by varying phase Θ in $v(t) = A \cos(\omega t + \theta)$. In discussion of (8.2.18), $\psi_{PM}(t) = \omega_C t + \theta_C + k_\theta v_M(t)$, however, the text says that θ_C is a constant and can be neglected. Do you see contradiction between these two statements? Explain.*)

The given generalization is important because it shows that

in PM, the instantaneous phase is a linear function of the message signal, $v_M(t)$, as highlighted by (8.2.20).

We need to recall from (8.1.30) that the induced phase variations are restricted as

$$-\pi < k_\theta v_M(t) \leq +\pi \text{ or} |k_\theta v_M(t)| \leq \pi \tag{8.2.21}$$

The restrictions, delimited by (8.2.21), enable us to avoid ambiguity imposed by the cycling nature of an angle.

The importance of this generalization stems from the fact that the SNR of a demodulated PM is proportional to the square of the amplitude of the message signal, A_M, and to the square of k_θ. The proof of this statement and the derivation of the following formula are beyond the scope of this book, but the implications are of concern to us:

$$\text{SNR}_{\text{PM}} = \frac{(k_\theta \gamma A_M)^2 P_s}{2 N_0 f_{\max}} \tag{8.2.22}$$

Here,

- k_θ (rad/V) is the phase modulation (deviation) constant (as above).
- γ is the attenuation constant.
- A_M (V) is the amplitude of a sent message signal (and thus γA_M is the amplitude of a received message signal).
- P_s (W) is the average normalized power of a message signal.
- N_0 (V²/Hz) is the power spectral density of noise within the bandwidth of a BPF in a receiver.
- $f_{\max}$ (Hz) is the maximum frequency of the output LPF in the receiver.

Equation (8.2.22) shows several ways to increase the SNR_{PM}, but all of them come with a price. If we want to increase the A_M, we need to increase the transmitted power, which is practically always restricted; what's more, in wireless communications, this is simply impossible due to the limited battery power in mobile devices. Increasing the phase-modulation constant, k_θ, will result in an increase in the required transmission bandwidth caused by increasing the frequency deviation. Also, increasing k_θ is restricted, as (8.2.21) shows.

Due to these practical limitations, *an analog PM is not a popular form of analog modulation*. This is why from here on we will concentrate on frequency modulation. Nevertheless, *PM plays an important role in digital modulation*, as we will learn in Chapters 9 and 10. Since both types of modulation are very similar, studying FM will provide enough information for us to understand both phase and frequency modulation.

Frequency Modulation Based on the general definition given in (9.2.20), we can write the general formula for a frequency-modulated signal as

$$v_{\text{FM}}(t) = A_C \cos(\psi_{\text{FM}}(t)) \tag{8.2.23}$$

We want to derive the general formula for $\psi_{\text{FM}}(t)$. We know that an instantaneous FM frequency is a derivative of an instantaneous FM phase; that is,

$$\omega_{\text{FM}}(t) = \frac{d(\psi_{\text{FM}}(t))}{dt} \tag{8.2.24}$$

We also know from (8.1.16) that the FM frequency is the sum of a carrier frequency and a variable member imposed by a message signal,

$$\omega_{\text{FM}}(t) = \omega_C + k_{\text{FM}} v_M(t) \tag{8.2.25}$$

Now we perform the following steps:

- From (8.2.24), we find $d(\psi_{FM}(t)) = \omega_{FM}(t)dt$.
- Taking the integrals from both sides, $\int_0^t d[\Psi_M(t)] = \int_0^t \omega_{FM}(t)dt$, yields $\psi_{FM}(t) = \int_0^t [\omega_C + k_{FM} v_M(t)]dt$ or $\psi_{FM}(t) = \int_0^t \omega_C dt + \int_0^t k_{FM} v_M(t)dt$.
- Finally, we obtain

$$\psi_{FM}(t) = \omega_C t + k_{FM} \int_0^t v_M(t)dt \qquad (8.2.26)$$

Plugging $\psi_{FM}(t)$ into (8.2.26), we find

$$v_{FM}(t) = A_C \cos(\psi_{FM}(t)) = A_C \cos\left(\omega_C t + k_{FM} \int_0^t v_M(t)dt\right) \qquad (8.2.27)$$

Equation (8.2.27) is the most general formula for a frequency-modulated signal.

You can find this equation in almost any textbook on communications. The significance of this formula will become apparent shortly; at this point, we need to stress that the instantaneous FM frequency can be obtained from (8.2.27) as

$$\omega_{FM}(t) = \frac{d(\Psi_{FM}(t))}{dt} = \frac{d\left(\omega_C t + k_{FM} \int_0^t v_M(t)dt\right)}{dt} = \omega_C + k_{FM} \cdot v_M(t) \qquad (8.2.28)$$

Equation (8.2.28) shows that the instantaneous FM frequency is a linear function of the message signal.

Refer to (8.2.22), $\psi_{PM}(t)$ (rad) $= \omega_C t + k_\theta \cdot v_M(t)$, to recall that in PM, the *phase of a PM signal is also a linear function of the message signal*. It is worth recalling too that, according to (8.2.1), the *amplitude of an AM signal is a linear function of the message signal*. Therefore, it is a modulated parameter of a sinusoidal carrier – the amplitude in AM, the frequency in FM, and the phase in PM – that depends linearly on a message signal. This property is an important feature of analog modulation.

What is also important to stress is the different nature of AM, FM, and PM modulations. An AM signal depends *linearly* on the message signal, $v_M(t)$,

$$v_{AM}(t) = A_{AM}(t) \cos(\omega_C t) = (A_C + v_M(t)) \cos(\omega_C t) \qquad (8.2.29)$$

whereas the FM and PM signals depend *nonlinearly* on the message signal,

$$v_{FM}(t) = A_C \cos(\psi_{FM}(t)) = A_C \cos\left(\omega_C t + k_{FM} \int_0^t v_M(t)dt\right) \qquad (8.2.27R)$$

and

$$v_{PM}(t) = A_C \cos(\psi_{PM}(t)) = A_C \cos(\omega_C(t) + k_\theta v_M(t)), \qquad (8.2.20R)$$

This is why *AM is called linear modulation, and FM and PM are called nonlinear modulations*. (Are you confused with the linearity of PM and FM modulations? Reread the above explanations to understand that their modulated parameters – the phase and the instantaneous frequency – are the linear functions of the message signals, whereas the PM and FM modulated signals depend on the message signal nonlinearly.)

The consequence of this difference is that the analysis (specifically, the spectral analysis) of an AM signal is much simpler than that of FM and PM signals.

(**Question**: With the introduction of $\psi_{PM}(t)$ (rad), the term phase becomes ambiguous, particularly because we also refer to initial phase or phase shift. How would you suggest we designate these entities to avoid ambiguity?)

8.2.3.2 Sinusoidal (Single-Tone) Frequency Modulation (FM)

Modulation Index and Instantaneous Frequency of a Single-Tone FM Signal We will now discuss frequency modulation with a sinusoidal message signal, which is also called *single-tone FM* (see Section 8.1). Such a case is of general importance because it leads to obtaining significant general results; its practical importance lies in the audio application of FM. Thus, a message signal is given by

$$v_M(t) = A_M \cos(\omega_M t)$$

Plugging the formula for $v_M(t)$ into (8.2.27), we find

$$v_{FM}(t) = A_C \cos\left(\omega_C t + k_{FM} \int_0^t A_M \cos(\omega_M t) dt\right)$$

$$= A_C \cos\left(\omega_C t + \frac{k_{FM} A_M}{\omega_M} \sin(\omega_M t)\right) = A_C \cos(\omega_C t + \beta \sin(\omega_M t)) \quad (8.2.29)$$

Now we can see that the FM modulation index,

$$\beta = m_f = \frac{\Delta f(\text{Hz})}{f_M(\text{Hz})} = \frac{(k_{FM} A_M)(\text{Hz})}{f_M(\text{Hz})} \quad (8.1.21R)$$

introduced in Section 8.1 without explanation, *stems, in fact, from the general definition of an FM signal*. Also, we can fully comprehend the meaning of a modulation index: It normalizes the frequency deviation, Δf (Hz) = $k_{FM} A_M$, to the frequency of a message signal, f_M (Hz), the signal that causes this deviation. Equation (8.2.29) also shows the origin of (8.1.22),

$$v_{FM}(t) = A_C \cos(2\pi f_C t + \beta \sin(2\pi f_M t)) \quad (8.1.22R)$$

which too was introduced in Section 8.1 without derivation.

Referring to (8.2.23), we can rewrite the instantaneous phase of a sinusoidal FM signal as

$$\psi_{FM}(t) = \omega_C t + \frac{k_{FM} A_M}{\omega_M} \sin(\omega_M t) = \omega_C t + \beta \sin(\omega_M t) \quad (8.2.30)$$

Taking the derivative of $\psi_{FM}(t)$, we obtain *the instantaneous frequency of a single-tone FM signal* as

$$\omega_{FM}(t) = \frac{d(\psi_{FM}(t))}{dt} = \omega_C + k_{FM} A_M \cos(\omega_M t) \quad (8.2.31)$$

We introduced this instantaneous frequency in (8.1.17) without derivation. Now, we understand the origins of the modulation index and the formula for the instantaneous frequency of a single-tone FM signal. Just remember that coefficient k_{FM} carries the dimension rad/s/V and represents the transfer coefficient of a VCO, the device generating a frequency-modulated signal.

Finding the Spectrum of a Single-Tone FM Signal Equation (8.2.29) enables us to take a further step in analyzing a single-tone FM signal. Specifically, using $\cos(A + B) = \cos A \cdot \cos B - \sin A \cdot \sin B$, we can rewrite this equation in the following form:

$$v_{FM}(t) = A_C \cos\left(\omega_C t + \frac{k_{FM} A_M}{\omega_M} \sin(\omega_M t)\right) = A_C \cos(\omega_C t + \beta \sin(\omega_M t))$$

$$= A_C \cos(\omega_C t) \cos(\beta \sin(\omega_M t)) - A_C \sin(\omega_C t) \sin(\beta \sin(\omega_M t)) \quad (8.2.32)$$

In (8.2.32), we have strange members: $\cos(\beta \sin(\omega_M t))$ and $\sin(\beta \sin(\omega_M t))$. It is difficult to figure out what the cosine of a sine function can be. Fortunately, this relationship is well known in mathematics and it has been proved that each of these members can be presented in the form of a series, namely

$$\cos(\beta \sin(\omega_M t)) = J_0(\beta) + 2J_2(\beta) \cos(2\omega_M t) + 2J_4(\beta) \cos(4\omega_M t) + \cdots$$
$$= J_0(\beta) + J_{-2}(\beta) \cos(2\omega_M t) + J_2(\beta) \cos(2\omega_M t) + J_{-4}(\beta) \cos(4\omega_M t) + J_4(\beta) \cos(4\omega_M t) + \cdots$$
$$= \sum_{n=-\infty}^{\infty} J_n(\beta) \cos(n\omega_M t) \quad \text{for } n \text{ even} \tag{8.2.33a}$$

Similarly,

$$\sin(\beta \sin(\omega_M t)) = 2J_1(\beta) \sin(\omega_M t) + 2J_3(\beta) \sin(3\omega_M t) + \cdots$$
$$= \sum_{n=-\infty}^{\infty} J_n(\beta) \sin(n\omega_M t) \qquad \text{for } n \text{ odd} \tag{8.2.33b}$$

Here the coefficients $J_n(\beta)$ are called *Bessel*[4] *functions of the first kind and order n*. The derivations of these equations are well beyond our scope, but if you are interested in the details, consult specialized mathematics books, such as Riley et al. (1998), pp. 458–470 or Bowman (2012), pp. 147–159.

With Bessel functions, (8.2.32) takes the form

$$v_{FM}(t) = A_C \cos(\omega_C t + \beta \sin(\omega_M t)) = A_C \cos(\omega_C t) \cdot \cos(\beta \sin(\omega_M t))$$
$$- A_C \sin(\omega_C t) \cdot \sin(\beta \sin(\omega_M t))$$
$$= A_C \cos(\omega_C t) \sum_{n=-\infty}^{\infty} J_n(\beta) \cos(n\omega_M t) - A_C \sin(\omega_C t) \sum_{n=-\infty}^{\infty} J_n(\beta) \sin(n\omega_M t) \tag{8.2.34}$$

In this equation, we can factor out the common term A_C and set the summation, $\sum_{n=-\infty}^{\infty} J_n(\beta)$, over all the sinusoidal terms; then (8.2.34) can be rewritten as follows:

$$v_{FM}(t) = A_C \sum_{n=-\infty}^{\infty} J_n(\beta)[\cos(\omega_C t) \cdot \cos(n\omega_M t) - \sin(\omega_C t) \sin(n\omega_M t)] \tag{8.2.35}$$

Applying $\cos(A + B) = \cos A \cdot \cos B - \sin A \cdot \sin B$, we can collapse the members in brackets back into one cosine function, $(\cos(\omega_C t) \cdot \cos(n\omega_M t) - \sin(\omega_C t) \cdot \sin(n\omega_M t)) = \cos(\omega_C t + n\omega_M t)$, and obtain

$$v_{FM}(t) = A_C \sum_{n=-\infty}^{\infty} J_n(\beta) \cdot \cos(\omega_C + n\omega_M)t \tag{8.2.36}$$

Analysis of this equation reveals the main point of these manipulations:

Equation (8.2.36) presents a single-tone frequency-modulated signal, $v_{FM}(t)$, as a series of harmonics $\cos(\omega_C \pm n\omega_M)t$ with amplitudes $A_C J_n(\beta)$.

---

[4] **Friedrich Wilhelm Bessel** (1784–1846) was a German astronomer and mathematician. Despite the lack of a formal university education, he nevertheless made such a significant contribution to the field of astronomy that, in 1811, at the age of 27, he was appointed as director and professor at the Koenigsberg Observatory, Prussia. His observations and calculations at this institution made him one of the most prominent astronomers of his day. During his research in Koenigsberg, he developed the Bessel functions in the form we use them today. These functions have found a variety of applications in mathematics and engineering, and they have made Bessel's name familiar to every professional working in these fields.

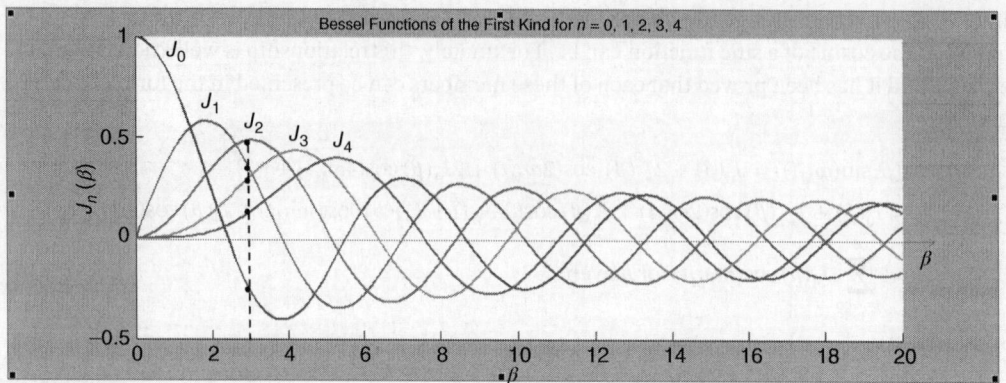

Figure 8.2.14 Bessel function of the first kind for various orders (*n*) and modulation indexes (β), built with MATLAB.

We remember, of course, that each cosine function represents a single frequency and is shown as a single line with its proper amplitude in frequency domain. Therefore,

Equation (8.2.36) gives the spectrum of a single-tone FM signal.

One question remains, however: What are those Bessel functions that makeup the coefficients $J_n(\beta)$ in (8.2.43)? The presentation of these functions in a closed form requires the introduction of other mathematical functions and operations that are outside the scope of this textbook. Fortunately, though, we do not need to be familiar with all the mathematical details because everything that we need to know is that the Bessel functions is presented in tabulated and graphical forms. Graphs of the Bessel functions of the first kind and order *n* are shown in Figure 8.2.14; these graphs are plotted with MATLAB. The numerical values of these functions are listed in Table 8.2.1, which is also built with MATLAB.

The MATLAB code for building Table 8.2.1 is given in the appendix to this section.

The graphs shown in Figure 8.2.14 and the values of the Bessel functions given in Table 8.2.1 enable us to analyze the spectrum of a single-tone FM signal. To do so, we need to explore the properties of the Bessel functions.

8.2.3.3 The Spectrum of a Single-Tone FM Signal, the Main Properties of the Bessel Functions, and Narrowband and Wideband FM

The Spectrum of a Single-Tone FM Signal Equation (8.2.36) shows that an FM signal can be presented as a series, that is, as a sum of the harmonics $\cos(\omega_C \pm n\omega_M)t$ with their amplitudes, $A_C \cdot J_n(\beta)$. For example, a series of $\beta = 3$ and $n = 4$ takes the form

$$\begin{aligned}
v_{FM}(t) = &\cdots A_C \cdot J_{-4}(3) \cdot \cos(\omega_C - 4\omega_M)t + A_C \cdot J_{-3}(3) \cdot \cos(\omega_C - 3\omega_M)t \\
&+ A_C \cdot J_{-2}(3) \cdot \cos(\omega_C - 2\omega_M)t + A_C \cdot J_{-1}(3) \cdot \cos(\omega_C - \omega_M)t + A_C \cdot J_0(3) \cdot \cos(\omega_C)t \\
&+ A_C \cdot J_1(3) \cdot \cos(\omega_C + \omega_M)t + A_C \cdot J_2(3) \cdot \cos(\omega_C + 2\omega_M)t + A_C \cdot J_3(3) \cdot \cos(\omega_C + 3\omega_M)t \\
&+ A_C \cdot J_4(3) \cdot \cos(\omega_C + 4\omega_M)t + \cdots
\end{aligned} \quad (8.2.37)$$

We can readily see the similarity between (8.2.37) and the Fourier series discussed in Chapter 6.

Table 8.2.1 Values of Bessel function of the first order for various orders and modulation indexes built with MATLAB.

	beta=1	beta=2	beta=3	beta=4	beta=5	beta=6	beta=7	beta=8	beta=9
n=0	0.7652	0.2239	-0.2601	-0.3971	-0.1776	0.1506	0.3001	0.1717	-0.0903
n=1	0.4401	0.5767	0.3391	-0.0660	-0.3276	-0.2767	-0.0047	0.2346	0.2453
n=2	0.1149	0.3528	0.4861	0.3641	0.0466	-0.2429	-0.3014	-0.1130	0.1448
n=3	0.0196	0.1289	0.3091	0.4302	0.3648	0.1148	-0.1676	-0.2911	-0.1809
n=4	0.0025	0.0340	0.1320	0.2811	0.3912	0.3576	0.1578	-0.1054	-0.2655
n=5	2.4976e-04	0.0070	0.0430	0.1321	0.2611	0.3621	0.3479	0.1858	-0.0550
n=6	2.0938e-05	0.0012	0.0114	0.0491	0.1310	0.2458	0.3392	0.3376	0.2043
n=7	1.5023e-06	1.7494e-04	0.0025	0.0152	0.0534	0.1296	0.2336	0.3206	0.3275
n=8	9.4223e-08	2.2180e-05	4.9344e-04	0.0040	0.0184	0.0565	0.1280	0.2235	0.3051
n=9	5.2493e-09	2.4923e-06	8.4395e-05	9.3860e-04	0.0055	0.0212	0.0589	0.1263	0.2149
n=10	2.6306e-10	2.5154e-07	1.2928e-05	1.9504e-04	0.0015	0.0070	0.0235	0.0608	0.1247
n=11	1.1980e-11	2.3043e-08	1.7940e-06	3.6601e-05	3.5093e-04	0.0020	0.0083	0.0256	0.0622
n=12	4.9997e-13	1.9327e-09	2.2757e-07	6.2645e-06	7.6278e-05	5.4515e-04	0.0027	0.0096	0.0274
n=13	1.9256e-14	1.4949e-10	2.6591e-08	9.8586e-07	1.5208e-05	1.3267e-04	7.7022e-04	0.0033	0.0108
n=14	6.8854e-16	1.0729e-11	2.8802e-09	1.4362e-07	2.8013e-06	2.9756e-05	2.0520e-04	0.0010	0.0039
n=15	2.2975e-17	7.1830e-13	2.9076e-10	1.9479e-08	4.7967e-07	6.1917e-06	5.0590e-05	2.9260e-04	0.0013
n=16	7.1864e-19	4.5060e-14	2.7488e-11	2.4717e-09	7.6750e-08	1.2019e-06	1.1612e-05	7.8006e-05	3.9330e-04
n=17	2.1154e-20	2.6593e-15	2.4435e-12	2.9469e-10	1.1527e-08	2.1872e-07	2.4945e-06	1.9422e-05	1.1202e-04

- For each order, n, we have two harmonics, $\cos(\omega_C \pm n\omega_M)t$, with two amplitudes, $A_C J_{\pm n}(\beta)$. For example, for $f_C = 100\,\text{kHz}$, $f_M = 15\,\text{kHz}$, $A_C = 10\,\text{V}$, $\beta = 6$, and $n = 3$, we have

$$A_C \cdot J_{-3}(6) \cdot \cos(\omega_C - 3\omega_M)t \text{ and } A_C \cdot J_3(6) \cdot \cos(\omega_C + 3\omega_M)t;\text{ that is,}$$

$$A_C \cdot J_{-3}(6) \cdot \cos(2\pi(f_C - 3f_M)t) \text{ and } A_C \cdot J_3(6) \cdot \cos(2\pi(f_C + 3f_M)t);\text{ that is,}$$

$$|-1.15| \cdot \cos(2\pi \times 55 \times 10^3)t) \text{ and } (1.15) \cdot \cos(2\pi(145 \times 10^3)t)$$

Since Table 8.2.1 does not show negative values of n, we need to find $J_{-n}(\beta)$ by the rule

$$J_{-n}(\beta) = (-1)^n J_n(\beta) \tag{8.2.38}$$

which stems from the properties of the Bessel functions. What's more, *though we find certain values of $J_{-n}(\beta)$ being negative, we always depict an FM spectrum with positive amplitudes.* This is because in engineering practice, when such measurements are taken with a spectrum analyzer, the amplitudes always appear positive. To comply with this convention when writing the equations, we show the absolute values of negative amplitudes, as written above.

- The frequencies of the harmonics $\omega_C \pm n\omega_M$ are located symmetrically around the carrier frequency, ω_C (or f_C); they are spaced ω_M (or f_M) apart. An example of the spectrum of an FM signal with the sinusoidal message for $\omega_C = 40\,\text{rad/s}$, $\omega_M = 4\,\text{rad/s}$, $A_C = 12\,\text{V}$, $\beta = 3$, and $n = 4$ is shown in Figure 8.2.15 (see also Figure 8.1.26).

Main Properties of the Bessel Functions The amplitudes, $A_C \cdot J_n(\beta)$, of spectral components (sidebands or side frequencies) are determined by the Bessel functions, $J_n(\beta)$, which are often called relative amplitudes. We have mentioned some properties of these functions previously (see, for instance,

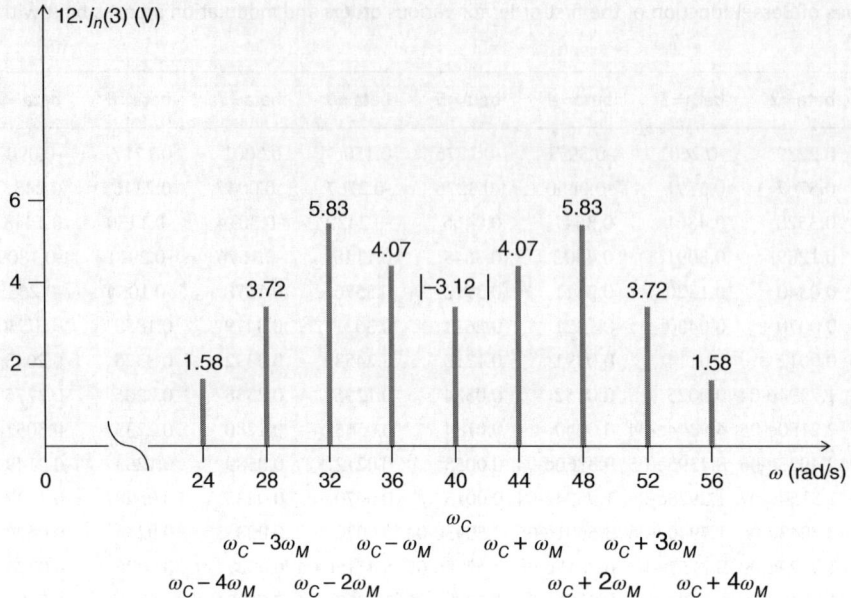

Figure 8.2.15 Example of the spectrum of an FM signal with a sinusoidal message for $\omega_C = 40$ rad/s, $\omega_M = 4$ rad/s, $A_C = 12$ V, $\beta = 3$, and $n = 4$.

Eq. (8.2.38)). That notwithstanding, the following main properties of the Bessel functions (and therefore the amplitudes of the spectral components of an FM spectrum) shown in Figure 8.2.14 and Table 8.2.1 must be stressed in particular:

- The Bessel functions oscillate around zero and diminish with an increase in β.
- They are oscillatory, but not periodic.
- All of them, except for $J_0(\beta)$, start at zero; $J_0(\beta)$ has an initial value of unity; therefore, the maximum amplitude of an FM spectrum is the amplitude of the unmodulated carrier, $A_C J_0(\beta)$ at $\beta = 0$, which simply equals A_C.

The properties of the Bessel functions discussed above can be readily seen in Figure 8.2.14. Indeed, the graphs show that these functions become smaller (i) as the modulation index, β, increases and (ii) as the order of the functions, n, increases. We can find the values of the Bessel functions of various orders by measuring these values at every value of β. For example, consider the line drawn at $\beta = 3$ in Figure 8.2.14. Observe how it intersects the graphs of the Bessel functions of various orders. We can estimate the approximate values of the Bessel functions as $J_0(3) \approx -0.26$, $J_1(3) \approx 0.34$, $J_2(3) \approx 0.49$, and $J_4(3) \approx 0.31$. But, of course, the values of the Bessel functions can be more easily and more precisely found in the table. Just compare our graphical estimates with the numbers given in Table 8.2.1.

- The amplitudes of a single-tone FM spectrum are symmetrical with respect to $\omega_C (n = 0)$. They do not decrease or increase monotonically with a change of n or β, as Table 8.2.1 and Figure 8.2.15 show.
- The following property, which is important to know for the discussion that follows shortly on FM power and bandwidth, concerns the square of $J_n(\beta)$:

$$\sum_{n=-\infty}^{\infty} J_n^2(\beta) = 1 \tag{8.2.39}$$

8.2 Analog Modulation for Analog Transmission – An Advanced Study

This property is intuitively plausible, but we leave its proof to specialized textbooks.

Narrowband and Wideband FM Table 8.2.1 shows that the greater the modulation index, β, the greater the number of Bessel functions with significant amplitudes and therefore the greater the number of visible spectral components. For example, when $\beta \ll 1$, a single-tone FM signal has only two sidebands, but when $\beta \gg 1$, there are many sidebands, as seen in Figures 8.1.26 and 8.2.18. When $\beta \ll 1$, FM is called narrowband; when $\beta \gg 1$, FM is called wideband. (Can you explain why?) John Carson (see Section 8.1) stressed the advantages of wideband FM over narrowband many years ago; today, FM transmission occurs only in the wideband mode.

8.2.3.4 The Bandwidth of a Single-Tone FM Signal

How to Determine the Bandwidth of a Single-Tone FM Signal Theoretically, the *bandwidth of an FM signal* lasts from $-\infty$ to ∞ because $J_n(\beta)$ is not equal to zero for any n and β. But in engineering practice, we must have a finite value for bandwidth. Fortunately, in a single-tone FM, the values of $J_n(\beta)$ become negligibly small with an increase in n, as Table 8.2.1 shows. Thus, we can choose some criterion by which to restrict the value of n and therefore determine a finite FM bandwidth. In choosing the value of the FM bandwidth, we have to remember that, on the one hand, we want a large bandwidth to provide high-quality transmission but, on the other hand, a large bandwidth is associated with high transmission cost. Bandwidth also could be restricted by regulations. Bearing this trade-off in mind, we can choose an n value by taking one of two possible approaches: (a) the value of the significant amplitudes or (b) the required power of the FM signal.

a) *Bandwidth by the value of significant amplitudes, SAs*: First, we need to define what amplitudes are considered "significant." There are no formal criteria for choosing a significant value. Thus, we need to refer to engineering practice, in which the values of significant amplitudes for

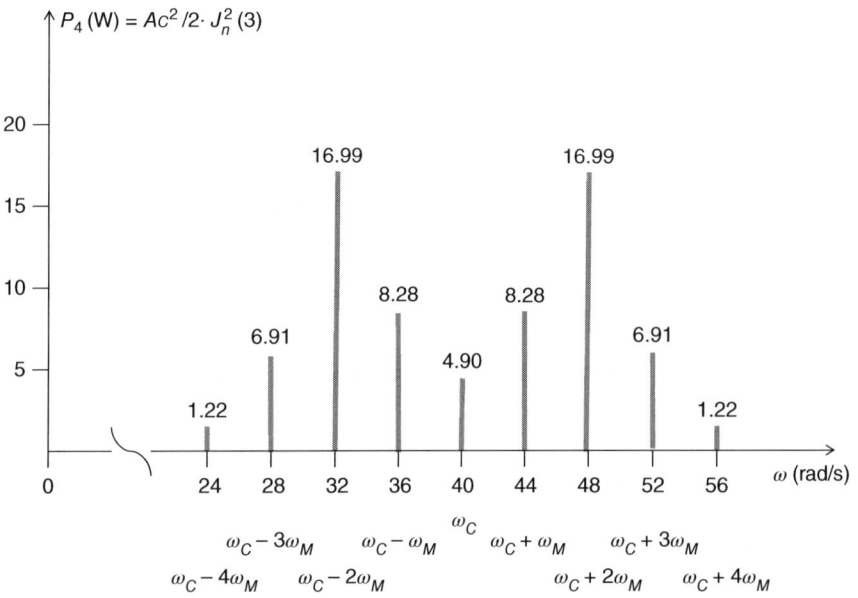

Figure 8.2.16 Example of the power spectrum for a single-tone FM signal with $\omega_C = 40$ rad/s, $\omega_M = 4$ rad/s, $A_C = 12$ V, $\beta = 3$, and $n = 4$.

high-fidelity transmission are set at 1% (−40 dB) of the amplitude of an unmodulated carrier, i.e. $J_0(0) \cdot A_C$. We usually consider a normalized spectrum, for which $A_C = 1$. This is justified by the fact that A_C simply multiplies all the Bessel functions evenly and therefore the only *relative values* of amplitudes, $J_n(\beta)$, are important. Since $J_0(0) = 1$, we should take unity as a reference. Thus, 1% gives 0.01. When referring to the amplitudes in volts, we obtain a significant amplitude of 1% in decibels, as SA = 20 log(1/100) = −40 dB. For voice transmission, acceptable quality can be achieved by setting the significant amplitudes at 10% = 0.1(−20 dB). For example, for $\beta = 6$, high-quality transmission requires that the significant relative amplitudes, $J_{\pm n}(6)$, must be no less than 0.01, which gives 11 sidebands (see Table 8.2.1). For voice transmission, the significant relative amplitudes must be no less than 0.1, which gives nine sidebands.

b) *Bandwidth by the required power of an FM signal*: We need to recall that both the power and bandwidth of a signal are determined by the number of its transmitted harmonics. Therefore, by choosing a signal's power, we determine its bandwidth and vice versa. The relationship between a signal's power and its bandwidth is discussed in Chapter 7 in reference to Parseval's theorem. You will recall that the power of a signal is determined by the well-known formula P (W) = $A^2/2R$, where A (V) is the signal's amplitude and R (Ω) is the load resistor over which the power is released. For an ac signal, this equation gives an average power and for $R = 1\,\Omega$, we have *average normalized power*, which we will consider from here on. Since the amplitude of an individual kth harmonic of a single-tone FM signal is given by $A_C \cdot J_k(\beta)$, the *maximum power* of this signal is shown by

$$P_{\max} = \frac{A_C^2}{2} \cdot \sum_{k=-\infty}^{\infty} J_k^2(\beta) \tag{8.2.40}$$

(Since $\sum_{k=-\infty}^{\infty} J_k^2(\beta) = 1$, the maximum average normalized power of an FM signal is a constant and is given by $P_{\max} = \frac{A_C^2}{2}$, as mentioned in Section 8.1.)

The power of a sinusoidal FM signal containing n harmonics is therefore presented as

$$P_n = \frac{A_C^2}{2} \sum_{k=-n}^{n} J_k^2(\beta) \tag{8.2.41}$$

What we want to know is what percentage (fraction) of maximum power we need to transmit to achieve the required transmission quality. This measure is obtained from (8.2.40) and (8.2.41) as

$$\frac{P_n(W)}{P_{\max}(W)} = \frac{\frac{A_C^2}{2} \sum_{k=-n}^{n} J_k^2(\beta)}{\frac{A_C^2}{2} \sum_{k=-\infty}^{\infty} J_k^2(\beta)} = \sum_{k=-n}^{n} J_k^2(\beta) \tag{8.2.42}$$

because $\sum_{k=-n}^{n} J_k^2(\beta) = 1$, according to (8.2.39). Calculations show that by choosing

$$n = \beta + 1 \tag{8.2.43}$$

we achieve

$$\frac{P_n(W)}{P_{\max}(W)} = \sum_{k=-\beta+1}^{\beta+1} J_k^2(\beta) \geq 0.98 \tag{8.2.44}$$

Thus, when we define FM bandwidth as

$$BW_{FM} = 2(\beta + 1)f_M \tag{8.2.45}$$

we transmit not less than 98% of the total signal power. We recognize, of course, (8.2.45) as *Carson's rule*, introduced in Section 8.1. This discussion, however, emphasizes the importance of this rule as a power-based criterion for estimating BW_{FM} (Hz).

The FM Bandwidth – A Discussion Consider the values of the relative amplitudes, $J_n(\beta)$, given in Table 8.2.1: When $\beta = 1$, the value of $J_n(1)$ equals 0.1149 at $n = 2$ and 0.0196 at $n = 3$; thus, it approximately equals 0.1 at $n > 2$. Similarly, for $\beta = 2$, $J_n(2) \approx 0.1$ at $n = 3$; for $\beta = 3$, $J_n(3) \approx 0.1$ at $n = 4$, etc. In other words, $J_n(\beta)$ becomes small (~ 0.1) when $n = \beta + 1$. If we substitute this value of n into (8.1.25), we obtain

$$BW_{FM} \text{ (Hz)} \approx 2(\beta + 1) \cdot f_M \text{ (Hz)}$$

This is another justification for Carson's rule. Now we understand that this rule relies on the property of the Bessel functions.

- We need to remember that theoretically the FM bandwidth ranges from $-\infty$ to ∞; therefore, *all the above formulas define the approximate value of the FM bandwidth*. In fact, in this section, we determine the criteria for the bandwidth approximation.
- Important note regarding power distribution in the spectrum of an FM signal: Consider, for example, the spectrum of the single-tone FM signal shown in Figure 8.2.15. Based on (8.2.41), we can write the power spectrum of this signal (with $\omega_C = 40$ rad/s, $\omega_M = 4$ rad/s, $A_C = 12$ V, $\beta = 3$, and $n = 4$) in an expanded form as

$$P_4(W) = A_C^2/2 \sum_{k=-4}^{4} J_k^2(3)$$
$$= A_C^2/2 \cdot J_{-4}^2(3) + A_C^2/2 \cdot J_{-3}^2(3) + A_C^2/2 \cdot J_{-2}^2(3) + A_C^2/2 \cdot J_{-1}^2(3) + A_C^2/2 \cdot J_0^2(3)$$
$$+ A_C^2/2 \cdot J_1^2(3) + A_C^2/2 \cdot J_2^2(3) + A_C^2/2 \cdot J_3^2(3) + A_C^2/2 \cdot J_4^2(3)$$
$$= 1.22 + 6.91 + 16.99 + 8.28 + 4.90 + 8.28 + 16.99 + 6.91 + 1.22 \qquad (8.2.46)$$

The sum of these power components is equal to

$$\frac{A_C^2}{2} \sum_{k=-4}^{4} J_3^2(3) = 71.7$$

which gives 99.6% of the total power $P_{total} = A_C^2/2 = 72$ W. This power spectrum is depicted in Figure 8.2.16. Understand that the total signal power, determined by the amplitude of the carrier, is distributed among the sidebands after the modulation, as (8.2.46) and Figure 8.2.16 exemplify. *This is why the amplitude of the ω_C component, which is the amplitude of a modulated carrier, is not – in contrast to the spectrum of an AM signal – the greatest amplitude of an FM spectrum.*

Spectrum and Bandwidth of a Single-Tone FM Signal To complete our discussion of the spectrum and bandwidth of an FM signal with a sinusoidal message, consider an individual harmonic of such a signal:

$$A_C J_n(\beta) \cos(\omega_C + n\omega_M)t = A_C J_n(\beta) \cos(n\omega_{FM}t) \qquad (8.2.47)$$

We know from Section 7.1 that we can find the spectrum of a time-domain function by taking its Fourier transform. Let us apply this rule to an individual harmonic (8.2.47): Consulting Table 7.1.1, we find

$$\mathcal{F}\{A_C J_n(\beta) \cos(n\omega_{FM}t)\} = A_C J_n(\beta)\pi(\delta(\omega + n\omega_{FM}) + \delta(\omega - n\omega_{FM})] \qquad (8.2.48)$$

To apply this manipulation to the whole single-tone FM signal, we first rewrite this signal as

$$v_{FM}(t) = A_C \sum_{n=-\infty}^{\infty} J_n(\beta) \cos(n\omega_{FM} t) = A_C \sum_{n=-\infty}^{\infty} J_n(\beta) \cos(\omega_C + n\omega_M t) \tag{8.2.49}$$

Now we can obtain the Fourier transform of the whole $v_{FM}(t)$ by performing the Fourier transform over each individual harmonic and summing up the results:

$$F\{v_{FM}(t)\} = F\left\{A_C \cdot \sum_{n=-\infty}^{\infty} J_n(\beta) \cdot \cos(n\omega_{FM} t)\right\}$$

$$\Rightarrow V_{FM}(f) = A_C \cdot \pi \sum_{n=-\infty}^{\infty} J_n(\beta)[\delta(\omega + n\omega_{FM}) + \delta(\omega - n\omega_{FM})]$$

$$= A_C \pi \sum_{n=-\infty}^{\infty} J_n(\beta)[\delta(\omega + (\omega_C + n\omega_M)) + \delta(\omega - (\omega_C + n\omega_M))] \tag{8.2.50}$$

Recalling that the delta function is depicted as a single line limited by its multiplicative constant, we now understand the meaning of a graphical presentation of a single-tone FM signal as being a set of separate lines with given amplitudes.

This analysis provides a guideline to finding and analyzing the spectrum of an FM signal. Let us consider an example of such a quest.

Example 8.2.3   Spectrum of an FM Signal

Problem

The FCC assigned specific frequencies for FM radio broadcasting. For example, Channel 201 (in the FCC's designation) has the central (carrier) frequency $f_C = 87.9$ MHz; the central frequency for Channel 202 is $f_C = 88.1$ MHz, etc. The FCC also set aside the 200-kHz band for an individual channel. Channel 201, for instance, occupies 87.9 ± 0.1 MHz; that is this channel is assigned the band from 87.8 to 88.0 MHz. (For additional details on the FCC regulations, visit its website.) Each FM channel is allowed a ± 75 kHz frequency deviation. Typically, the maximum frequency of a message signal is $f_M = 15$ kHz, which provides decent transmission quality. (You will recall that the human ear can distinguish sound frequency up to 20 kHz, but the most upper frequencies carry very low energy.) **Based on the information given, determine the spectrum and the bandwidth of an FM signal transmitted by Channel 201**. (Strictly speaking, Channels 201 to 301 are not assigned for commercial broadcasting, but we still use the parameters of Channel 201 as an example.)

Solution

Our discussion shows that we must determine at what order, n, we obtain the required finite value of a bandwidth. We know that there are two approaches to finding this number: (1) By the value of the significant amplitudes and (2) by the required power of an FM signal. Let us consider both.

(1) Finding the FM Spectrum by the Value of the Significant Amplitudes

Since the problem refers to $f_{max} = 15$ kHz of a message signal, this transmission must deliver more than a simple voice. (Recall that almost all the energy from a voice is concentrated between 300 and 3300 Hz.) Therefore, we have to refer to high-fidelity transmission and require that the SAs must satisfy the 1% criterion; that is, this transmission must have at least SA ≥ 0.01, as previously noted. Now we can find the required spectrum and bandwidth as follows:

Table 8.2.2 Frequencies, relative amplitudes, and relative power of the FM signal in Example 8.2.3.

N	$f_c \pm nf_M$ (MHz)	BW_{FM} (MHz) $= f_{max} - f_{min}$ $= 2nf_M$ (MHz)	$J_n(5)$ (V)	$J_n^2(5)$ (V²) $= (-1)^{2n} \cdot J_{-n}^2(5) + J_n^2(5)$ $= 2 \cdot J_n^2(5)$
0	87.9	0	−0.1776	$J_0^2(5) = 0.0315$
1	87.9 + 0.015 = 87.915 87.9 − 0.015 = 87.885	0.030	−0.3276	$2 \cdot J_1^2(5) = 0.2146$
2	87.9 + 0.030 = 87.930 87.9 − 0.030 = 87.870	0.060	0.0466	$2 \cdot J_2^2(5) = 0.0043$
3	87.9 + 0.045 = 87.945 87.9 − 0.045 = 87.855	0.090	0.3648	$2 \cdot J_3^2(5) = 0.2662$
4	87.9 + 0.060 = 87.960 87.9 − 0.060 = 87.840	0.120	0.3912	$2 \cdot J_4^2(5) = 0.3061$
5	87.9 + 0.075 = 87.975 87.9 − 0.075 = 87.825	0.150	0.2611	$2 \cdot J_5^2(5) = 0.1363$
6	87.9 + 0.090 = 87.990 87.9 − 0.030 = 87.810	0.180	0.1310	$2 \cdot J_6^2(5) = 0.0343$
7	87.9 + 0.105 = 88.005 87.9 − 0.105 = 87.795	0.210	0.0534	$2 \cdot J_7^2(5) = 0.0057$
8	87.9 + 0.120 = 88.020 87.9 − 0.120 = 87.780	0.240	0.0184	$2 \cdot J_8^2(5) = 0.0007$
9	87.9 + 0.135 = 88.035 87.9 − 0.135 = 87.765	0.270	0.0055	$2 \cdot J_9^2(5) = 0.0001$

- Determine the modulation index as $\beta = \frac{\Delta f(\text{Hz})}{f_M(\text{Hz})} = \frac{\Delta f(\text{Hz})}{f_{max}(\text{Hz})} = \frac{75 \text{ kHz}}{15 \text{ kHz}} = 5$.
- Consult Table 8.2.1 and find that $J_9(5) = SA_9 = 0.0055$.
- Obtain $n = 9$ and compute $BW_{FM} = 2 \times nf_M = 2 \times 9 \times 15 \text{ kHz} = 270 \text{ kHz}$.

The spectrum of FM Channel 201, computed by the significant-amplitudes criterion, is given in Table 8.2.2 and shown in Figure 8.2.17. Clearly, for $n = 9$, the spectrum does not fit into the allotted band and must be adjusted. Based on the equation $BW_{FM} = 2nf_M$, we have to choose $n = 6$ to get

$$BW_{FM} = 2nf_M = 2 \times 6 \times 15 \text{ kHz} = 180 \text{ kHz}$$

For $n = 6$, we find $J_6(5) = SA_6 = 0.1310$, which is much greater than the desired 0.01 value. What's more, it even exceeds the 10% criterion for voice transmission. How much the transmission quality will suffer as a result of limiting the number of sidebands is yet to be determined, but we must remember that there is a trade-off between transmission quality and bandwidth restrictions. In this case, we have no choice but to adhere to the bandwidth restrictions.

(2) Finding the FM Spectrum by the Required Power of an FM Signal

If we want to transmit no less than 98% of a signal's total power, the solution is straightforward: Use (8.2.42),

$$n = \beta + 1$$

and build the spectrum. Since $\beta = 5$ in this example, we compute $n = 6$. The application of Carson's rule yields, of course, the same bandwidth that we computed in Point (1) above in this example:

$$BW_{FM} = 2(\beta + 1)f_{max} = 2 \times 6 \times 15 \text{ kHz} = 180 \text{ kHz}$$

The problem is solved.

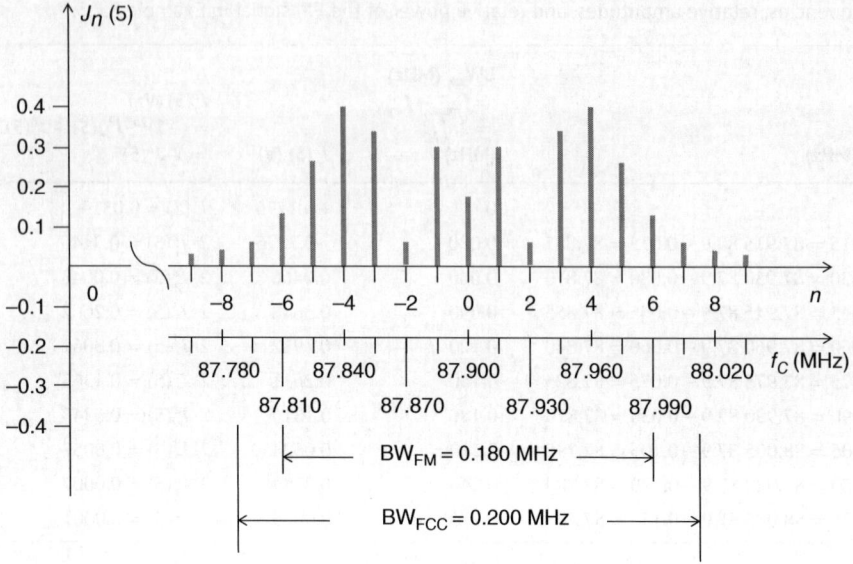

Figure 8.2.17 The amplitude spectrum and bandwidth of a single-tone FM signal for Example 8.2.3.

Discussion

- Examine Figure 8.2.17: First, observe that all the spectral lines are shown as positive quantities, whereas Table 8.2.2 shows that many of them are negative. This is because, as we already mentioned, in engineering practice, these lines are always considered positive. To emphasize this point, we label the vertical axis as an absolute value of a relative amplitude, $|J_n(\beta)|$. Secondly, study closely the horizontal axis: We label it with the order of a Bessel function, n, and the frequency, f, of each spectral component (sideband). In this way, we can clearly show the bandwidths – one allotted by the FCC and the other obtained by calculations – in two measures.
- To avoid confusion, we need to clarify the following: The order of a Bessel function, n, is equal to the number of sidebands (side frequencies), but the total number of harmonics includes the fundamental harmonic, $J_0(\beta)$, and therefore is greater than n by 1. Thus, to calculate the spectrum of a single-tone FM signal and its bandwidth, n is the correct number to use. Refer, for example, to (8.2.43), $BW_{FM} = 2nf_M$, where n is the order of a Bessel function and the number of sidebands. However, to calculate the power of any FM signal by adding up the power of all harmonics, we need to include the fundamental harmonic, $J_0(\beta)$. For example, for $n = 6$, we transmit seven harmonics and for $n = 3$, we transmit four harmonics.
- We stressed previously that using Carson's rule, we obtain the approximate bandwidth within which the FM signal carries *not less than* 98% of its total power. How good is this approximation? The exact number can be obtained by computing the power carried by all harmonics within the assigned bandwidth, as (8.2.41) describes. In this example, calculated values of $J_0^2(5) + 2J_n^2(5)$ for n running from 1 to 9 are given in Table 8.2.2. For $n = 9$, we compute that 99.9% of the total power is transmitted; for $n = 6$, the transmitted percentage of the total power is 99.3%; for $n = 5$, we compute 95.9%. Thus, for $n = 6$, our transmission delivers 99.3% of total signal power, which is greater than the 98% projected by Carson's rule.
- *To sum up*: The problem in this example is solved by setting $n = 6$, as a result of which our transmission fits in the allotted bandwidth and delivers 99.3% of the total signal power. The only remaining issue is the quality of the signal, but that issue will be discussed in Chapter 10.

- *Important note*: In this example, we computed the FM bandwidth, $\text{BW}_{\text{FM}} = 2(\beta + 1)f_{\max}$, and the modulation index, $\beta = \Delta f/f_M = \Delta f/f_{\max}$, based on the maximum frequency of a message signal, $f_{\max}$. This approach contains a contradiction: On the one hand, formulas of both BW_{FM} and β were derived for a single-tone modulation; on the other hand, the characteristic $f_{\max}$ implies existence of more than one sinusoidal signal. This contradiction is discussed later in this section.

Calculations performed in Example 8.2.3 can be automated, and it can be done with MATLAB. What's more, such automation enables us to solve many various problems related to finding the spectrum of an FM signal, problems we often meet in practice (see Appendix 8.2.A.).

8.2.3.5 General Case of an FM Signal (An Arbitrary Message Signal)

Estimating the Transmission Bandwidth of an FM Signal Modulated by an Arbitrary Message When a message signal is arbitrary (not a single-tone sinusoidal), the complete analysis of an FM signal becomes a formidable task. In such a case, as always in engineering, we should break down the whole task into small pieces. To start, we ask ourselves why we need to analyze the FM signal. As our discussion in this section shows, we need this analysis in order to (i) find the spectrum of the FM signal and (ii) determine the bandwidth required for transmission of this signal. For practical purposes, estimation of the FM bandwidth is the primary task and, fortunately, we can estimate a transmission bandwidth without finding the spectrum of the FM signal. It can be done by using Carson's rule, where, instead of the frequency of a sinusoidal message, f_M, we substitute the maximum frequency of the message signal, $f_{\max}$. We did this in Example 8.2.3 as

$$\text{BW}_{\text{FM}} = 2(\beta + 1)f_{\max} \tag{8.2.51}$$

Some textbooks refer to this formula as *generalized Carson's rule*. By pursuing this approach, we can generalize the evaluation of an FM transmission bandwidth. Two steps are involved:

Step 1: We recall the following definitions and manipulations used in an FM single-tone modulation:
- An FM bandwidth is defined in (8.1.27) as $\text{BW}_{\text{FM}} \approx 2nf_M$, with n being the number of sidebands and f_M the frequency of a sinusoidal message signal.
- We determine in (8.2.42) that $n = \beta + 1$. Substituting it into (8.1.27), we obtain the Carson's rule as

$$\text{BW}_{\text{FM}} \approx 2(\beta + 1)f_M \tag{8.2.44R}$$

- The modulation index is given in (8.1.20) by $\beta = \Delta f/f_M$, where $\Delta f = k_{\text{FM}} \cdot A_M$ is the frequency deviation with A_M being the amplitude of the message signal and k_{FM} being a modulation constant.
  Now, Carson's rule for a tone modulation can be presented in the form

$$\text{BW}_{\text{FM}} \approx 2(\beta + 1)f_M = 2(\Delta f + f_M) \tag{8.2.52}$$

Step 2: We can apply Carson's rule in a general way as follows:
- Introduce the maximum frequency deviation, $\Delta f_{\max}$, as

$$\Delta f_{\max} = k_{\text{FM}} \cdot A_{\max} \tag{8.2.53}$$

where A_{max} is the maximum amplitude of an information signal. This definition relies on the fact that a modulation constant, k_{FM}, does not depend on the nature of an information signal but remains the same for both sinusoidal and arbitrary signals.
- Define modulation index, β, as

$$\beta = \frac{\Delta f_{max}(Hz)}{f_{max}(Hz)} = \frac{k_{FM} A_{max}(Hz)}{f_{max}(Hz)} \quad (8.2.54)$$

- Generalize (8.2.51) as

$$BW_{FM}(max) = 2(\beta + 1)f_{max} = 2(\Delta f_{max} + f_{max}) \quad (8.2.55)$$

Equation (8.2.55) enables us to estimate the transmission bandwidth of an arbitrary-modulated FM signal. The accuracy of this estimation is as good as the accuracy of Carson's rule, which, in general, underestimates the FM bandwidth.

Multitone Modulation – An Analytical Approach To approach the finding of the spectrum of an arbitrary FM signal, we consider as an example an FM signal modulated by the sum of two sinusoidal signals. Thus, we have

$$v_M(t) = A_{M1} \cos(\omega_{M1} t) + A_{M2} \cos(\omega_{M2} t) \quad (8.2.56)$$

From (8.2.27), (8.2.29), and (8.2.30), we find

$$v_{FM}(t) = A_C \cos\left(\omega_C t + k_{FM} \int_0^t v_M(t) dt\right) = A_C \cos(\omega_C t + \beta_1 \sin(\omega_{M1} t) + \beta_2 \sin(\omega_{M2} t)) \quad (8.2.57)$$

Here $\beta_i = \frac{\Delta f_i(Hz)}{f_{Mi}(Hz)} = \frac{k_{FMi} A_{Mi}(Hz)}{f_{Mi}(Hz)}$ and $i = 1, 2$. Introducing notations

$B_1 = \beta_1 \sin(\omega_{M1} t)$ and $B_2 = \beta_2 \sin(\omega_{M2} t)$

we rewrite (8.2.57) as

$$v_{FM}(t) = A_C \cos(\omega_C t + B) \quad (8.2.58)$$

where $B = B_1 + B_2$. Now we need to apply the rules $\cos(A + B) = \cos A \cdot \cos B - \sin A \cdot \sin B$ and $\sin(A + B) = \sin A \cdot \cos B + \cos A \cdot \sin B$ to (8.2.57). This yields

$$v_{FM}(t) = A_C \cos(\omega_C t + B) = A_C \cos(\omega_C t) \cos(B_1 + B_2) - A_C \sin(\omega_C t) \sin(B_1 + B_2)$$
$$= A_C [\cos(B_1) \cos(B_2) - \sin(B_1) \sin(B_2)] \cos(\omega_C t)$$
$$- A_C [\sin(B_1) \cos(B_2) - \cos(B_1) \sin(B_2)] \sin(\omega_C t) \quad (8.2.59)$$

(**Exercise**: Explicitly derive (8.2.59) from (8.2.58).)

Plugging back $B_1 = \beta_1 \sin(\omega_{M1} t)$ and $B_2 = \beta_2 \sin(\omega_{M2} t)$, we get

$$v_{FM}(t) = A_C [\cos(\beta_1 \sin(\omega_{M1} t)) \cos(\beta_2 \sin(\omega_{M2} t)) - \sin(\beta_1 \sin(\omega_{M1} t)) \sin(\beta_2 \sin(\omega_{M2} t))] \cos(\omega_C t)$$
$$- A_C [\sin(\beta_1 \sin(\omega_{M1} t)) \cos(\beta_2 \sin(\omega_{M2} t))$$
$$- \cos(\beta_1 \sin(\omega_{M1} t)) \sin(\beta_2 \sin(\omega_{M2} t))] \sin(\omega_C t) \quad (8.2.60)$$

Now we have the same cosine-of-sine and sine-of-cosine members that we had in (8.2.32). Then, following the procedure employed in (8.2.32)–(8.2.36), we can present $v_{FM}(t)$ in the form of a

Fourier series with the Bessel functions being its coefficients as follows:

$$v_{FM}(t) = A_C \sum_{n=-\infty}^{\infty} J_n(\beta_1) \sum_{m=-\infty}^{\infty} J_m(\beta_2) \cdot \cos(\omega_C + n\omega_{M1} + m\omega_{M2})t \qquad (8.2.61)$$

Compare this equation with (8.2.36) to see their similarity.

Equation (8.2.61) presents the spectrum of an FM signal modulated by two sinusoidal signals.

Note that we can obtain (8.2.61) more clearly if we use the exponential form of a Fourier series, as we did in Section 6.2. To proceed, we need to recall Euler's identity, $e^{j\theta} = \cos\theta + j\sin\theta$, from which we find $\cos\theta = Re(e^{j\theta})$. Then, we apply this formula to the FM signal to derive

$$v_{FM}(t) = A_C \cos(\omega_C t + \beta_1 \sin(\omega_{M1} t) + \beta_2 \sin(\omega_{M2} t))$$
$$= A_C Re[e^{j(\omega_C t + \beta_1 \sin(\omega_{M1} t) + \beta_2 \sin(\omega_{M2} t))}] = A_C Re[e^{j(\omega_C t)} e^{j(\beta_1 \sin(\omega_{M1} t))} e^{j(\beta_2 \sin(\omega_{M2} t))}] \qquad (8.2.62)$$

Here we need to introduce the following forms of presenting a sinusoidal signal through a series with the Bessel functions:

$$e^{j\beta_1 \sin(\omega_{M1} t)} = \sum_{n=-\infty}^{\infty} J_n(\beta_1) e^{jn\omega_{M1} t}$$

$$e^{j\beta_2 \sin(\omega_{M2} t)} = \sum_{m=-\infty}^{\infty} J_n(\beta_2) e^{jm\omega_{M2} t} \qquad (8.2.63)$$

To find the explicit formula for $v_{FM}(t)$, we plug (8.2.63) into (8.2.62):

$$v_{FM}(t) = A_C Re[e^{j(\omega_C t)} e^{j\beta_1 \sin(\omega_{M1} t)} e^{j\beta_2 \sin(\omega_{M2} t)}]$$
$$= A_C Re\left[e^{j(\omega_C t)} \sum_{n=-\infty}^{\infty} J_n(\beta_1) e^{jn\omega_{M1} t} \sum_{m=-\infty}^{\infty} J_n(\beta_2) e^{jm\omega_{M2} t}\right] \qquad (8.2.64)$$

A simple rearrangement produces

$$v_{FM}(t) = A_C Re\left[e^{j(\omega_C t)} \sum_{n=-\infty}^{\infty} J_n(\beta_1) e^{jn\omega_{M1} t} \sum_{m=-\infty}^{\infty} J_n(\beta_2) e^{jm\omega_{M2} t}\right]$$
$$= A_C \sum_{n=-\infty}^{\infty} J_n(\beta_1) \sum_{m=-\infty}^{\infty} J_n(\beta_2) e^{jm\omega_{M2} t} \cos(\omega_C + n\omega_{M1} + m\omega_{M2})t \qquad (8.2.65)$$

Equation (8.2.61) shows that the FM signal contains the following frequencies:

1. The carrier frequency at ω_C.
2. The components at $\omega_C \pm n\omega_{M1}$ due to the modulating signal ω_{M1}.
3. The components at $\omega_C \pm m\omega_{M2}$ due to the modulating signal ω_{M2}.
4. The components at $\omega_C \pm n\omega_{M1} \pm m\omega_{M2}$ for $n, m = 1, 2, 3, \ldots, \infty$ due to the beating among all the involved signals caused by the nonlinear nature of the frequency modulation.

This analysis enables us to understand and predict the spectrum of an FM signal modulated by the sum of two sinusoidal messages. To find out more specifics about this spectrum (that is, its amplitudes and corresponding frequencies), we need to perform the explicit calculations in (8.2.65) (see the Appendix 8.2.A).

Another approach to finding the spectrum of a multitoned-modulated FM signal is based on obtaining the Fourier transform of the given signal, $v_{FM}(t)$. Implementing this approach analytically would be a difficult undertaking but, with the use of MATLAB or any other computerized tool, this task becomes readily achievable.

Dual-Tone Frequency Modulation – Practical Calculations For practical calculations of a dual-tone FM spectrum, we take a straightforward approach: Find the Fourier transform of a given FM signal using MATLAB. Example 8.2.4 demonstrates this approach.

Example 8.2.4   Finding the Spectrum of an FM Signal Modulated by a Dual-Tone Message

Problem

Apply the Fourier transform and use MATLAB to find the spectrum of an FM signal modulated by a dual-tone message, $v_{FM}(t)\,(V) = A_C \cos(\omega_C t + \beta_1 \sin(\omega_{M1}t) + \beta_2 \sin(\omega_{M2}t))$, with the following parameters: $A_C = 1\,V$, $\omega_C = 2\pi 12000\,\text{rad/s}$, $\beta_1 = 5$ and $\beta_2 = 10$.

Consider two cases:

1. The modulating sinusoids are harmonically related and have the following frequencies: $\omega_{M1} = 2\pi 440\,\text{rad/s}$ and $\omega_{M2} = 2\pi 880\,\text{rad/s}$.
2. The modulating sinusoids are harmonically unrelated and have the following frequencies: $\omega_{M1} = 2\pi 440\,\text{rad/s}$ and $\omega_{M2} = 2\pi 330\,\text{rad/s}$.

Solution

Case 1: In this case, (8.2.57) takes the form

$$v_{FM}(t)\,(V) = \cos(2\pi 12000 t + 5 \sin(2\pi 440 t) + 10 \sin(2\pi 880 t))$$

The MATLAB code for solving this case is given below; the results are shown in Figures 8.2.18 and 8.2.19.

MATLAB code and spectra of the FM signal for **Example 8.2.4**:

```
t=0:0.0000001:0.009;
x = cos(2*pi*12000*t+5*sin(2*pi*440*t)+10*sin(2*pi*880*t));
subplot(2,1,1);
plot(t,x);
title('Frequency Modulated Dual Tone Signal in Time Domain'); grid;
xlabel('Time [seconds]'); ylabel('Amplitude [V]');
axis([0 0.005 -1 1]);
Fs = 80000;
t = 0:1/Fs:1-(1/Fs);
x = cos(2*pi*12000*t+5*sin(2*pi*440*t)+10*sin(2*pi*880*t));
xdft = (1/length(x))*fft(x);
f = -40000:(Fs/length(x)):40000-(Fs/length(x));
subplot(2,1,2);
plot(f,abs(fftshift(xdft)));
title('Spectrum of Frequency Modulated Dual Tone Signal'); grid
xlabel('Frequency [Hz]'); ylabel('Amplitude [V]');
```

To view the spectrum in detail, we zoom in on the area around the carrier frequency; the result is shown in Figure 8.2.19.

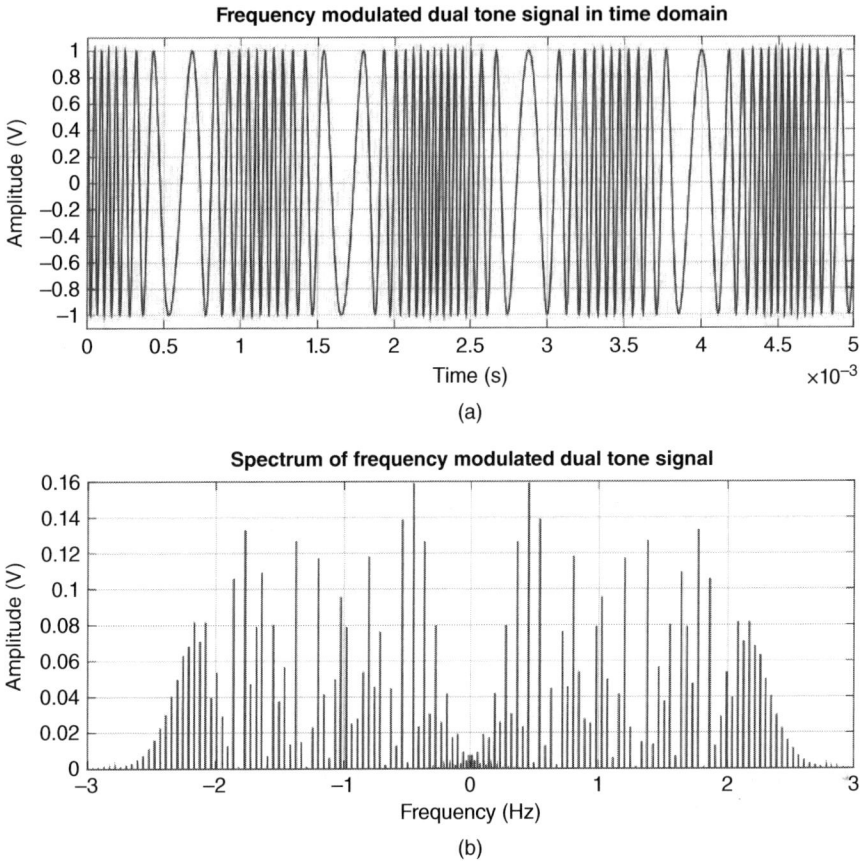

Figure 8.2.18 The waveform of the sinusoidal dual-tone FM signal (a) and its spectrum (b) with harmonically related modulating signals for Example 8.2.4.

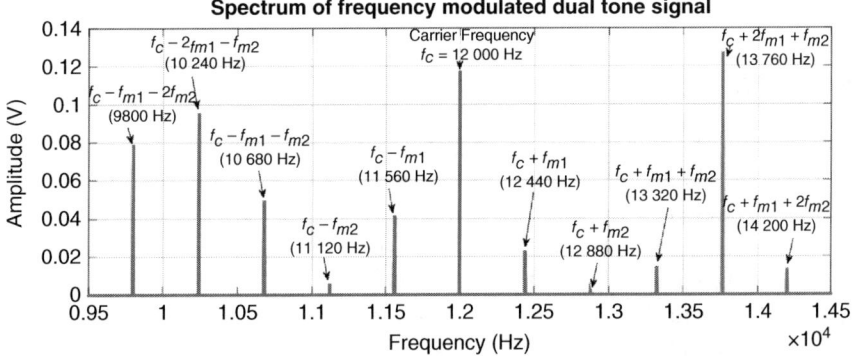

Figure 8.2.19 The detailed spectrum of the FM signal modulated by two harmonically related sinusoidal messages for Example 8.2.4.

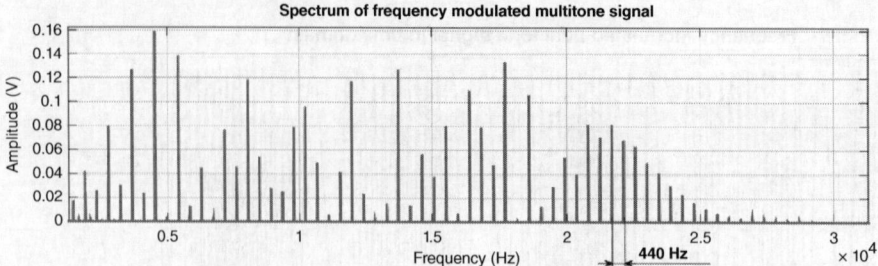

Figure 8.2.20 The sidebands' separation in the FM signal modulated by two harmonically related sinusoidal messages for Example 8.2.4.

Analyze Figure 8.2.19 carefully; specifically identify each sideband in terms of its relationship to the general rule: $f_C \pm n f_{M1} \pm m f_{M2}$. For example, the sideband at frequency 13320 Hz is the result of summing $f_C = 12\,000$ Hz and $f_{M1} = 440$ Hz and $f_{M2} = 880$ Hz. Compare this spectrum with that obtained by the single-tone modulation shown in Figures 8.2.A.2 and 8.2.A.3 in Appendix 8.2.A. Observe how adding the second modulating signal changes the FM spectrum compared with the single-tone spectrum. Note that the spectrum contains all sidebands according to the rule $f_C \pm n f_{M1} \pm m f_{M2}$ with $n, m = 1, 2, 3, \ldots, \infty$. Finally, we gather from Figure 8.2.20 that all sidebands are separated by $f_{M1} = 440$ Hz. This is a *feature of FM signals modulated by harmonically related sinusoidal messages*.

Case 2: In this case, we consider two harmonically unrelated modulating sinusoids with radian frequencies $\omega_{M1} = 2\pi 440$ rad/s and $\omega_{M2} = 2\pi 333$ rad/s. Now (8.2.57) takes the form $v_{FM}(t) = \cos(2\pi 12000 t + 5\sin(2\pi 440 t) + 10\sin(2\pi 333 t))$. You are encouraged to modify the MATLAB code given in this example for Case 1 to obtain the FM spectrum for this case.

Compare Figures 8.2.18 and 8.2.21. Observe the changes in the waveform and general structure of the spectrum of the two FM signals. These differences clearly demonstrate the variety of waveforms and spectra of FM signals modulated by even slightly different message signals. Thus, *it is impossible to perform the general spectral analysis of an FM signal modulated by an arbitrary message*.

Figure 8.2.22 demonstrates sidebands caused by various combinations of carrier, f_C, messages, f_{M1} and f_{M2}, and frequencies $f_C \pm f_{M1} \pm f_{M2}$. Identify, for example, the following spectral components:

$f_C = 12\,000$ Hz,
$f_C + f_{M1} = 12\,440$ Hz and $f_C - f_{M1} = 11\,560$ Hz,
$f_C + f_{M2} = 12\,333$ Hz and $f_C - f_{M2} = 11\,667$ Hz,
$f_C + f_{M1} + f_{M2} = 12\,773$ Hz and $f_C - f_{M1} - f_{M2} = 11\,227$ Hz,
$f_C + 2f_{M1} + f_{M2} = 13\,213$ Hz and $f_C - 2f_{M1} - f_{M2} = 10\,787$ Hz,
$f_C + f_{M1} + 2f_{M2} = 13\,106$ Hz and $f_C - f_{M1} - 2f_{M2} = 10\,894$ Hz,
$f_C + 2f_{M1} + 2f_{M2} = 13\,546$ Hz and $f_C - 2f_{M1} - 2f_{M2} = 10\,454$ Hz,
$f_C + 3f_{M1} + 2f_{M2} = 13\,986$ Hz and $f_C - 3f_{M1} - 2f_{M2} = 10\,014$ Hz,
$f_C + 2f_{M1} + 3f_{M2} = 13\,879$ Hz and $f_C - 2f_{M1} - 3f_{M2} = 10\,121$ Hz,
$f_C + 3f_{M1} + 3f_{M2} = 14\,319$ Hz and $f_C - 3f_{M1} - 3f_{M2} = 9681$ Hz.

You can continue this exercise for $f_C \pm n f_{M1} + m f_{M2}$ where $n, m = 3, 4, 5, \ldots$.

Figure 8.2.23 shows even more detailed spectrum of the FM signal modulated by two harmonically unrelated sinusoids. As we can see, the first pair of sidebands are

$$f_C - 3f_{M1} + 4f_{M2} = 12\,012 \text{ Hz and } f_C + 3f_{M1} - 4f_{M2} = 11\,988 \text{ Hz}$$

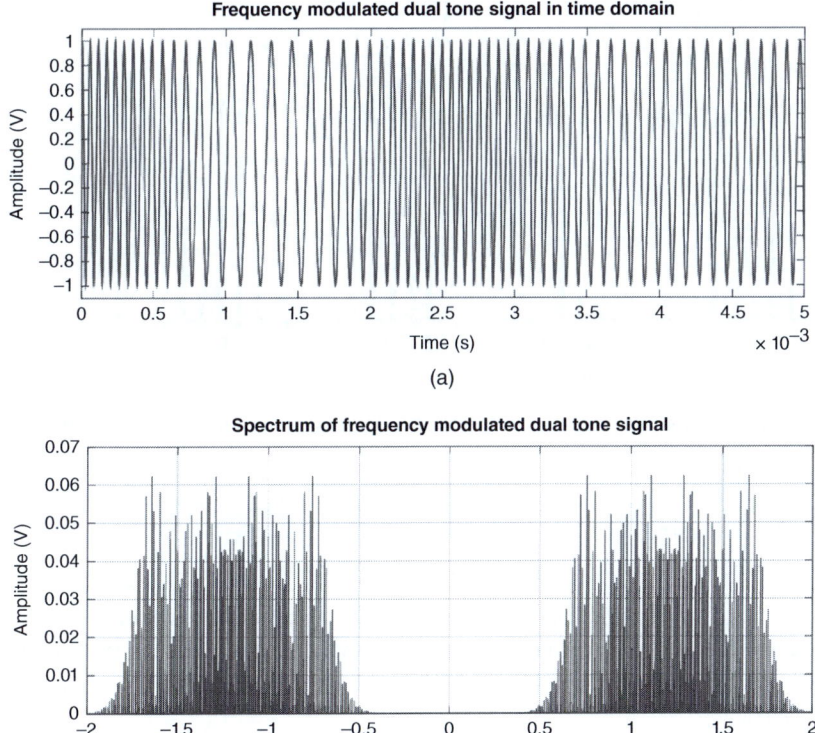

Figure 8.2.21 The waveform of the sinusoidal dual-tone FM signal (a) and its two-sided spectrum (b) with harmonically unrelated modulating signals for Example 8.2.4.

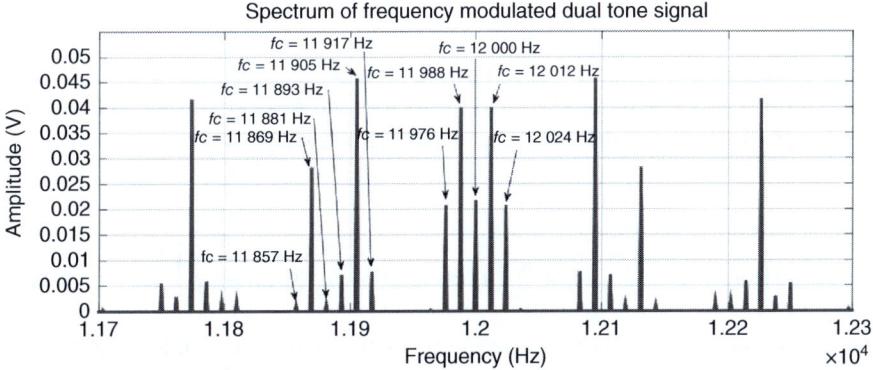

Figure 8.2.22 The detailed spectrum of a sinusoidal dual-tone FM signal with harmonically unrelated modulating signals for Example 8.2.4.

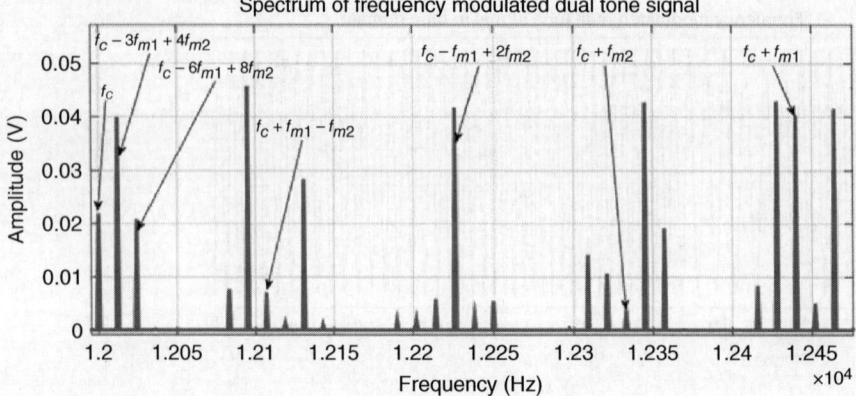

Figure 8.2.23 The sidebands around the carrier frequency of the sinusoidal dual-tone FM signal with harmonically unrelated modulating signals for Example 8.2.4.

and the second pair are

$$f_C - 6f_{M1} + 8f_{M2} = 12\,024 \text{ Hz and } f_C + 6f_{M1} - 8f_{M2} = 11\,976 \text{ Hz}$$

These examples demonstrate how sophisticated the combinations of frequencies could be even in the relatively simple case of frequency modulation. Again, this sophistication is due to the nonlinear nature of the FM.

The problem is solved.

Discussion

We can extend this approach to the FM multitone modulation produced by many sinusoidal message signals. What's more, we can perform the spectral analysis of an arbitrary modulated FM signal provided that we will be able to describe its waveform analytically. This is the key to enabling the application of the Fourier transform.

Finding the Spectrum of an FM Signal Modulated by an Arbitrary Message If a modulating (information, or message) signal is periodic, it can be developed in the Fourier series, as we learned in Chapter 6 (see Equation (6.1.5a)). Hence, the waveform of this signal can be written as

$$v_{\text{FM}}(t) = A_C \cos\left(\omega_C t + k_{\text{FM}} \int_0^t v_M(t)dt\right)$$

$$= A_C \cos\left(\omega_C t + k_{\text{FM}} \int_0^t \left[A_0 + \sum_{n=1}^\infty (a_n \cos n\omega_0 t) + \sum_{n=1}^\infty (b_n \sin n\omega_0 t)\right] dt\right) \quad (8.2.66)$$

From here we can proceed by extending the procedure discussed in the previous subsection on two-tone modulation. How effective the process and how productive the result will depend on the message signal's waveform and, therefore, on its Fourier series.

If a message signal is nonperiodic, we must use (8.2.27), $v_{\text{FM}}(t) = A_C \cos\left(\omega_C t + k_{\text{FM}} \int_0^t v_M(t)dt\right)$, which is the most general form of an FM signal. Whether we will be able to obtain the FM spectrum analytically depends on the waveform of the message signal. In any event, by performing the integration in (8.2.27), we will obtain the formula of the signal's waveform, from which we can find the Fourier transform of this signal by using MATLAB.

Figure 8.2.24 Diagram of an FM transmission and signal-to-noise ratio, SNR, of the received and demodulated FM signals.

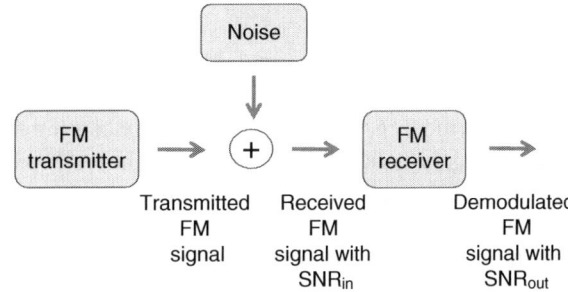

Finally, we need to understand that, in the most general case, a message signal is a random one, and its analysis requires the use of the theory of probabilities; however, we leave the discussion of this approach to specialized textbooks.

8.2.3.6 Effect of Noise on an FM Signal

A solid mathematical investigation of the effect of noise on FM transmission quality involves many branches of mathematics and engineering. Such an approach is based on various assumptions regarding the nature of an FM signal and various models of an FM transmission system, that is, its transmitters, receivers, and propagation links. The result of such investigations is various metrics for assessing this effect; however, such an approach is beyond the scope of our study, and so we provide here just a brief review of how noise affects an FM signal.

Signal-to-Noise ratio (SNR) and the Figure of Merit (FoM) in FM Consider Figure 8.2.24: An FM signal encounters noise in the course of its transmission. We refer to noise that adds its power to the FM signal; *such noise is called additive.* The FM noise-distorted signal is received at the input of a receiver; its signal-to-noise ratio is denoted as SNR_{in}. In recovering the information from the received signal, the receiver performs several stages of signal processing. The result is a demodulated FM signal with its signal-to-noise ratio termed as SNR_{out}. We are interested, of course, in comparing SNR_{in} with SNR_{out}. The ratio SNR_{out}/SNR_{in} shows how much the quality of the FM signal improves (if SNR_{out} becomes greater than SNR_{in}) or degrades (if SNR_{out} becomes smaller than SNR_{in}). This ratio is usually considered a figure of merit of an FM transmission, FoM_{FM}; that is,

$$FoM_{FM} = \frac{SNR_{out}}{SNR_{in}} \tag{8.2.67}$$

A strict mathematical derivation of the formula for FoM_{FM} entails a very advanced procedure and depends on the receiver's model, but the main point is that FoM_{FM} is proportional to β^2 for $\beta > 1$ as

$$FoM_{FM} = \left(\frac{SNR_{out}}{SNR_{in}}\right)_{FM} \approx \frac{3\beta^2}{2} \tag{8.2.68}$$

Thus, for commercial FM broadcasting with $\beta = 5$ discussed in Example 8.2.3, we find $FoM_{FM} = \frac{SNR_{out}}{SNR_{in}} \approx \frac{3\beta^2}{2} = 17.5$. We usually measure FoM_{FM} in decibels:

$$FoM_{FM} \text{ (dB)} = 10 \log \left(\frac{SNR_{out}}{SNR_{in}}\right)_{FM} = 12.4 \text{ dB} \tag{8.2.69}$$

Incidentally, the figure of merit for AM transmission is given by (Leung 2009, p. 232)

$$FoM_{AM} = \left(\frac{SNR_{out}}{SNR_{in}}\right)_{AM} \approx \frac{1}{3} \tag{8.2.70}$$

Think about this: $\text{FoM}_{\text{FM}} = \frac{\text{SNR}_{\text{out}}}{\text{SNR}_{\text{in}}} \approx \frac{3\beta^2}{2}$. The signal-to-noise ratio of a demodulated FM signal is $1.5\beta^2$ times greater than the input, SNR_{in}, which means that demodulation dramatically improves the output, SNR_{out}. What's more, the FM's figure of merit rapidly increases with an increase of the modulation index, β, giving us a powerful tool for improving FM transmission quality.

Improving the SNR_{out} is very unusual in the world of electronics. For instance, compare (8.2.69) and (8.2.70): We see that FM demodulation unusually increases (that is improves) its SNR_{out}, whereas AM demodulation reduces its SNR_{out}, an expected and typical result. Also, you will recall that the noise figure, NF, discussed in Section 4.2 in regard to an operational amplifier, is the ratio of SNR_{out} to SNR_{in}, and SNR_{out} is always smaller (worse) than SNR_{in}. This is because an operational amplifier adds its own noise to the input noise. Thus, decreasing SNR_{out} is a typical event because any electronic communication module generates its own noise, which of course adds to the input noise. FM transmission is an exception because of the nonlinear nature of this modulation. It "hides" the information signal from the additive noise. We refer to (8.2.29), $v_{\text{FM}}(t)\,(V) = A_C \cos(\omega_C t + \frac{k_{\text{FM}} A_M}{\omega_M} \sin(\omega_M t))$, which shows that the message signal, $A_M \sin(\omega_M t)$, is placed "inside" the carrier wave, $A_C \cos(\omega_C t)$.

Though (8.2.68) is derived for a single-tone FM signal, this result holds true in the general case of frequency-modulation transmission. This is why FM transmission always provides better transmission quality than AM, as we know from our experience with radio broadcasting.

The improvement in the SNR_{out} comes with a price, however, as can be seen from (8.2.68): The increase in β – which significantly improves the FoM_{FM} – results in an increase in transmission bandwidth, as Carson's rule says:

$$\text{BW}_{\text{FM}}\,(\text{Hz}) = 2(\beta + 1)f_M \qquad (8.2.45\text{R})$$

Therefore, in FM, there is a trade-off between transmission quality, measured by FoM_{FM}, and FM bandwidth. We pointed out that a similar trade-off exists in phase modulation. FM, however, offers a significant advantage over PM, as the following consideration demonstrates: The SNR of a PM system is given in (8.2.25) as $\text{SNR}_{\text{PM}} = \frac{(k_\theta \gamma A_M)^2 P_s}{2 N_0 f_{\text{max}}}$. The SNR of an FM system, under similar assumptions, can be expressed as

$$\text{SNR}_{\text{FM}} = \frac{3}{2} \left(\frac{k_{\text{FM}} \gamma A_M}{2\pi} \right)^2 \left(\frac{P_s}{N_0 f_{\text{max}}^3} \right), \qquad (8.2.71)$$

where

- k_{FM} (rad/V) is the frequency modulation (deviation) constant.
- γ is the attenuation constant.
- A_M (V) is the amplitude of a sent message signal.
- γA_M is the amplitude of a received message signal.
- P_s is the average normalized power of a message signal.
- N_0 (V^2/Hz) is the power spectral density of noise within the bandwidth of a BPF in a receiver.
- f_{max} (Hz) is the maximum frequency of the output LPF in the receiver.

Discussing a PM system, we considered the methods of improving its SNR along with the price we pay for this improvement – see (8.2.25). As a reminder, we concluded that the increase in the modulation constant, k_θ, was the best way to improve the SNR, but this benefit caused an increase in the transmission bandwidth. All these considerations can also be applied to an FM system. However, an

FM system offers a major advantage over a PM system: Increasing the phase modulation constant, k_θ, in PM is restricted by

$$|k_\theta v_M(t) \, (\text{rad})| \leq \pi \tag{8.2.21R}$$

whereas increasing the frequency modulation constant, k_{FM}, is restricted by the following condition:

$$k_{FM} v_M(t) \leq f_C \, (\text{Hz}) \tag{8.2.72}$$

The meaning of this restriction becomes clear when we recall that the frequency of an FM signal is given in (8.2.25) as $f_{FM}(t)$ (Hz) $= f_C + k_{FM} v_M (t)$. Thus, (8.2.72) requires that a member, $k_{FM} v_M (t)$ (Hz), modulating an FM frequency, $f_{FM}(t)$, must always be smaller than or equal to a carrier frequency, f_C. Since f_C is always high, this restriction is not as severe as (8.2.21); therefore, FM allows much more flexibility in the trade-off between signal quality and transmission bandwidth. This is why analog radio broadcasting relies on FM, not on PM, transmission.

SNR_{in} Threshold in FM The improvement in the SNR of an FM signal after demodulation holds true only if the SNR_{in} is high; otherwise, the FM receiver simply does not work properly. The value of the SNR_{in} at and below which the FM receiver is unable to correctly extract information from a received signal is called the FM threshold, or the SNR threshold in FM. Typically, the threshold, (SNR_{in}) threshold in FM broadcasting is about 12 dB for mono and 17 dB for stereo transmission. These numbers give us an idea as to the level of restrictions imposed on a received signal in an FM system. There is, of course, a theory describing this phenomenon and enabling its evaluation, but we will leave these topics for specialized study.

Preemphasis and De-emphasis in FM Transmission Examine again Figure 8.2.24. Noise causes still another problem in FM transmission: The power of noise in a detected FM signal is proportional to the square of the signal's frequency,

$$P_{noise} \, (\text{W}) \sim f^2 \tag{8.2.73}$$

(Can you justify this statement? *Hint*: We consider additive noise.)

Thus, the higher frequency components of the demodulated FM signal will suffer from noise more than the lower frequency components, resulting in distortion of the signal presented to the user. To realize how harmful this phenomenon can be, consider commercial FM stereo broadcasting. The stereo effect is achieved by transmitting simultaneously two signals, called *left* and *right*. The right signal is imposed on a 38 to 53-kHz band of the message-signal (baseband) bandwidth and the left signal occupies the range from 38 kHz down to 23 kHz of this spectrum. Therefore, according to (8.2.73), noise will affect the right signal more than the left signal, and the listener will receive a distorted message.

It looks as though this problem could be easily resolved by suppressing the higher frequency components of the demodulated FM signal by placing an additional LPF at the output of the FM receiver. But this approach, in addition to decreasing the noise power, would distort the message signal by suppressing its higher frequency portion of the spectrum. To reduce the detrimental effect of this additional filtering, an FM signal is usually predistorted at the transmission end by inserting a circuit that artificially amplifies the higher frequency band of the message-signal spectrum. Such an operation is called *preemphasis* of an FM signal, whereas decreasing the amplitudes of the

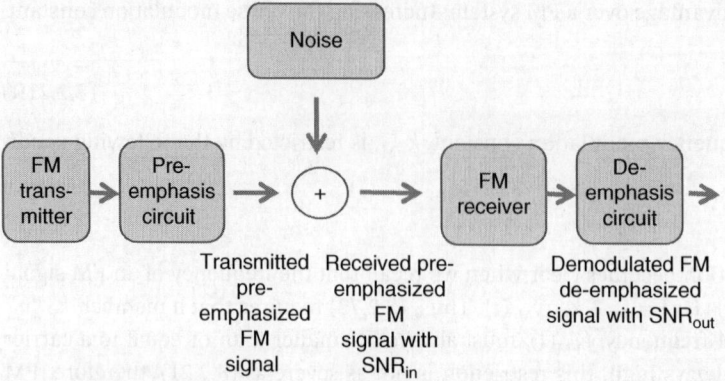

Figure 8.2.25 Block diagram of an FM transmission system with preemphasis and de-emphasis circuits.

higher-frequency portion of the received FM signal is called *de-emphasis*. A block diagram of an FM system with these additional circuits is shown in Figure 8.2.25.

Clearly, the preemphasis and de-emphasis circuits must match in regard to increasing and suppressing the higher frequency components of the FM signal. If this condition is met, then the operation of such an FM system is described as follows:

The preemphasis circuit artificially increases the amplitudes of the higher frequency components of a transmitted FM signal; the de-emphasis circuit suppresses the higher frequency components of a demodulated FM signal along with the noise embedded in this signal. As a result, the noise in the output FM signal is reduced, and the message signal returns to its original form.

As you might surmise, the preemphasis and de-emphasis circuits are high-pass and low-pass filters, respectively. The factor of improvement achieved with this operation can be presented as

$$I = \frac{\left(\frac{f_M}{f_C}\right)^3}{3\left[\left(\frac{f_M}{f_C}\right) - \tan^{-1}\left(\frac{f_M}{f_C}\right)\right]} \qquad (8.2.74)$$

Here, f_M is the frequency of a message signal and f_C is the 3-dB cutoff frequency of the preemphasis and de-emphasis filters. For example, in a commercial FM broadcast, $f_M = 15$ kHz and $f_C = 2.1$ kHz. Thus, $I = 22$ or 13 dB. This is a significant improvement in the quality of FM transmission.

Questions and Problems for Section 8.2

- Questions marked with an asterisk require a systematic approach to finding the solution.
- Many questions and problems, including those marked with an asterisk, imply that you, in addition to reading the textbook, will do your research to find the answers. Consider such questions as mini-projects.

Classification of Modulation Revisited

1. What is modulation? Why do we need modulation? What two main types of modulation do you know?

2. Explain the difference between analog and digital modulation.

3   The text says that analog modulation is implemented by changing the amplitude, frequency, or phase of a sinusoidal carrier:
    a) Why does it refer only to a sinusoidal carrier?
    b) Can another analog signal be a carrier in analog modulation? Explain.

4   With references to Figures 8.2.1, 8.2.2, and 8.2.3, explain how AM, FM, and PM can be obtained.

Advanced Consideration of Amplitude Modulation

5   Explain what is called a tone AM signal. Give an example of its waveform and spectrum.

6   A full (DSB-TC) AM signal is composed from the carrier, $v_C(t) = 10\cos(2\pi 3333t)$, and the message signal is $v_M(t) = 3\cos(2\pi 40t)$:
    a) Derive the formula for this AM signal.
    b) Sketch its waveform and one-sided spectrum. Show all the available numbers on both figures.
    c) Compute the modulation index of this signal.

7   Consider a dual-tone full AM signal with the following parameters: $v_C(t) = 10\cos(2\pi 12\,000t)$ and $v_M(t) = 0.4\cos(2\pi 440t) + 0.8\cos(2\pi 330t)$:
    a) Use any computerized tool to build the waveforms and spectra of this AM signal and its components.
    b) Compare your results with the tone AM signal discussed in Problem 6.

8   Why is AM called a linear modulation?

9   *Consider a full AM with a carrier, $v_C(t) = A_C \cos(\omega_C t)$, and arbitrary message signal, $v_M(t)$:
    a) Write the formula for the waveform of this signal.
    b) Write the formula for the spectrum of this signal.
    c) What kind of AM spectrum – discrete or continuous – do you expect to see? Explain.
    d) Qualitatively sketch the spectrum of this AM signal.

10 Name and explain two main problems of a full AM signal.

11 Compute the power efficiency of a dual-tone full AM signal with the following parameters: $v_C(t) = 10\cos(2\pi 12\,000t)$ and $v_M(t) = 0.4\cos(2\pi 440t) + 0.8\cos(2\pi 330t)$.

12 Compute the bandwidth of a dual-tone full AM signal with the following parameters: $v_C(t) = 10\cos(2\pi 12\,000t)$ and $v_M(t) = 0.4\cos(2\pi 440t) + 0.8\cos(2\pi 330t)$. How efficient is the use of this bandwidth in AM transmission?

13 *Consider a double-sideband suppressed carrier (DSB-SC or simply DSB) AM signal:
    a) Explain why it is called double-sideband? Suppressed carrier?
    b) Derive the formula for a DSB AM signal.
    c) Qualitatively sketch the waveform and the spectrum of a DSB AM signal.

14 A carrier signal, $v_C(t) = 10\cos(2\pi 3333t)$, and a message signal, $v_M(t) = 3\cos(2\pi 40t)$, are used to compose a DSB AM signal:
   a) Write the formula for this AM signal.
   b) Build the waveform of this signal.
   c) Sketch the spectrum of this signal.
   d) Don't forget to show all the numbers in the figures with the waveform and spectrum.

15 *Assume a message signal is a square wave with $A = 3\,\text{V}$ and $T = 25\,\text{ms}$ and its carrier is $v_C(t) = 10\cos(2\pi 3333t)$. They form a DSB AM signal.
   a) Write the formula for this AM signal.
   b) Build the waveform of this signal.
   c) Sketch the spectrum of this signal.
   d) Don't forget to show all the numbers in the figures with the waveform and spectrum.

16 What is the main drawback of a DSB AM signal? Explain.

17 Consider a single-sideband suppressed-carrier (SSB-SC or simply SSB) AM signal:
   a) Why do we need this signal?
   b) How can we create such a signal in principle?

18 *An SSB AM signal is composed from the carrier $v_C(t) = 10\cos(2\pi 3333t)$ and the message signal $v_M(t) = 3\cos(2\pi 40t)$.
   a) Write the formula for this AM signal.
   b) Build the waveform of this signal.
   c) Sketch the spectrum of this signal.
   d) Repeat this problem if the message signal is a square wave with $A = 3\,\text{V}$ and $T = 25\,\text{ms}$ and the carrier is the same as above.
   e) Don't forget to show all the numbers in the figures with the waveform and spectrum.

19 We can generate an SSB AM signal by filtering one of the sidebands using an ideal BPF, as Figure 8.2.10 shows. In reality, we must consider the application of a real filter: Qualitatively sketch a figure similar to Figure 8.2.10 if a real BPF with f_0, f_{C_1}, and f_{C_2} is used (see Section 5.2 for BPF).

20 *Figure 8.2.11 shows the use of the phasing method for generating an SSB AM signal with a sinusoidal message signal. To generalize this approach, consider an arbitrary message signal: In principle, suggest the circuit for such a case and qualitatively sketch the spectrum of the new SSB AM signal this circuit has generated.

21 *If you are a designer of a new AM communication system, which type of amplitude modulation – full, DSB, or SSB – would you choose? Explain your reasoning.

22 *Consider the demodulation of an AM signal:
   a) Refer to the waveform of the tone full AM signal in Figure 8.2.4a, explain its demodulation process, and show the resulting waveform.

b) Compare the waveforms of the tone full AM signal in Figure 8.2.4 and the tone DSB signal in Figure 8.2.7 and explain which signal can be easier and more accurately demodulated and why.

23 Explain why a special type of AM called *vestigial-sideband (VSB)* has been developed. Why is this AM used even in digital transmission?

24 *Using online and library sources, list all the applications of AM transmission in modern communication systems.

Advanced Consideration of Angular (Phase and Frequency) Modulation

25 Why are FM and PM modulations often combined under the term *angular modulation*?

26 *In describing a sinusoidal function, we use the term *phase* to denote two different parameters.
   a) What are those parameters?
   b) How can we distinguish between them?
   c) What is the relationship between these two parameters?

27 What role does an analog PM play in modern communications?

28 *Referring to (8.2.20) and (8.2.28), the text says that the instantaneous phase in PM and instantaneous frequency in FM are linear functions of a message signal. However, the text later states that both PM and FM are nonlinear modulations. How can you reconcile these apparently contradictory statements? Explain in detail.

29 Why is AM called linear modulation and why are FM and PM called nonlinear modulations? Explain how this difference affects the spectral analysis of these signals.

30 *A single-tone FM signal is given by $v_{FM}(t) = 10\cos(80t + 2\sin(4t))$.
   a) Write the formula presenting this signal as a sum of harmonics, that is, the formula for the signal's spectrum.
   b) Sketch the spectrum of this signal.

31 *Determine the spectrum of an FM signal with $v_{FM}(t) = 5\cos(628t + 3\sin(6.28t))$.

32 Considering the properties of the Bessel functions, the text says that "they are oscillatory, but not periodic." Explain the meaning of this statement with reference to Figure 8.2.14.

33 *The amplitudes of harmonics of a single-tone FM signal change differently with an increase in n. Can you mathematically describe these changes? Can you predict the value of an individual amplitude? Explain and give an example.

34 What are the bandwidths of two single-tone FM signals with $\beta_1 = 0.3$ and $\beta_2 = 3$? Which one is narrowband FM and which is wideband FM?

35 What is the bandwidth of a single-tone FM signal with $\beta = 5$ and $f_M = 12\,\text{kHz}$ if

a) SA = −20 dB?
b) the required transmitting power is no less than 98% of the total power?
c) Sketch the spectrum of this signal and show all the relevant numbers.

36 Estimate the transmission bandwidth of an arbitrary-modulated FM signal with $k_{FM} = 30$ kHz/V, $A_{max} = 2$ V, and $f_{max} = 10$ kHz. How accurate is this estimation?

37 *Find the spectrum of an FM signal modulated by a dual-tone message with the following parameters: $A_C = 2$ V, $\omega_C = 2\pi 8000$ rad/s, $\beta_1 = 3$, and $\beta_1 = 8$. Consider two cases: (a) $\omega_{M1} = 2\pi 240$ rad/s and $\omega_{M2} = 2\pi 480$ rad/s and (b) $\omega_{M1} = 2\pi 240$ rad/s and $\omega_{M2} = 2\pi 420$ rad/s.
a) Build the FM spectra of both cases.
b) Show the difference in these spectra.

38 Compute the figure of merit, FoM, for a single-tone FM with $\beta = 6$.
a) Is the computed FoM good or bad?
b) What would be an ideal FoM?

39 *Refer to (8.2.68): How can we improve the FoM of a single-tone FM? How practical would be our recommendation?

40 Compare the SNR_{PM} given in (8.2.25) with the SNR_{FM} given in (8.2.71) by assuming that all the common parameters are the same for both transmission types:
Which SNR is better?
By how much?

41 What is the SNR threshold in FM? Explain.

42 Consider the preemphasis and de-emphasis operations in FM transmission:
a) What do we want to achieve with these operations?
b) How can we quantify this achievement? Give an example by finding online the message frequency and 3-dB cutoff frequencies of the filters.

8.2.A Finding the Spectrum of an FM Signal with MATLAB

The FCC rules for FM broadcasting were explained in Example 8.2.3. We consider the specific FM signal with the following parameters: $f_c = 87.9$ MHz, $\Delta f = 75$ kHz, and $f_M = 15$ kHz. The spectrum of this signal was found in Example 8.2.3 by using MS Excel and manually plotting the proper figures. In this appendix, we want to show how this work can be automated with MATLAB and what advantages this tool provides.

The **task** is to determine the spectrum of the FM signal given in Example 8.2.3 by considering the following criteria:

1. The value of the significant amplitudes of the sidebands.
2. The value of the given bandwidth.
3. The values of the significant amplitudes AND the bandwidth simultaneously.

   Criterion 1: In this case, the selection criterion is SA ≥ 0.01. The sample of MATLAB code is given at the end of this appendix. Three samples of the FM spectrum built by MATLAB are shown in

8.2.A Finding the Spectrum of an FM Signal with MATLAB

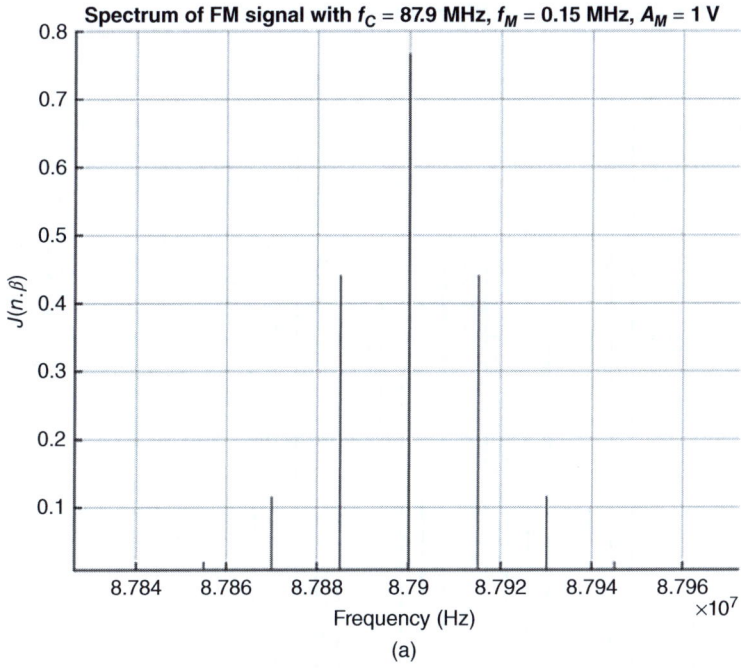

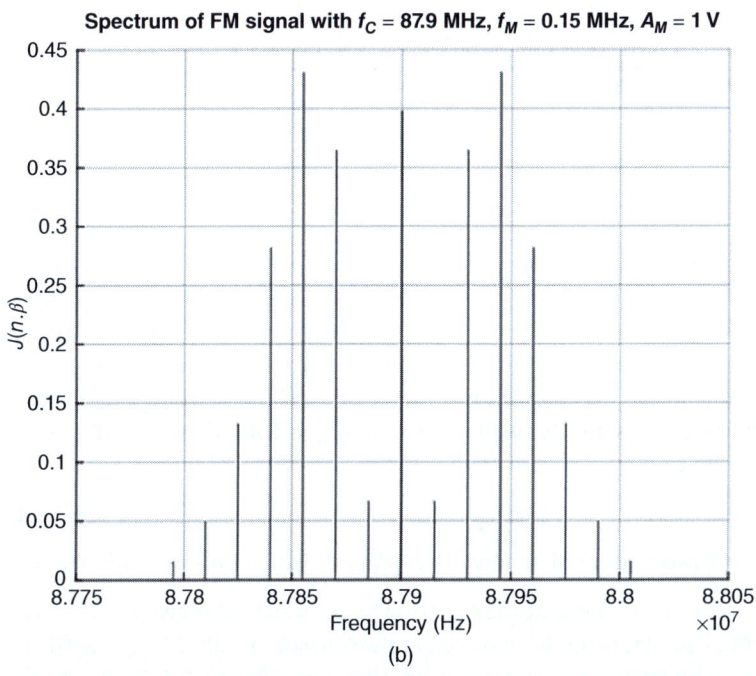

Figure 8.2.A.1 The spectra of an FM signal with $f_C = 87.9$ MHz, $f_M = 0.15$ MHz, $A_M = 1$ V, and various β selected by Criterion 1 – the value of the significant amplitudes, $SA \geq 0.01$; here $\beta = (k_{FM}A_M)/f_M$ is the modulation index, n is the number of sidebands, BW_{FM} (kHz) is the FM bandwidth. (a) $\beta = 1$, $n = 3$, and $BW_{FM} = 90$ kHz; (b) $\beta = 4$, $n = 7$, and $BW_{FM} = 210$ kHz; and (c) $\beta = 9$, $n = 13$, and $BW_{FM} = 390$ kHz.

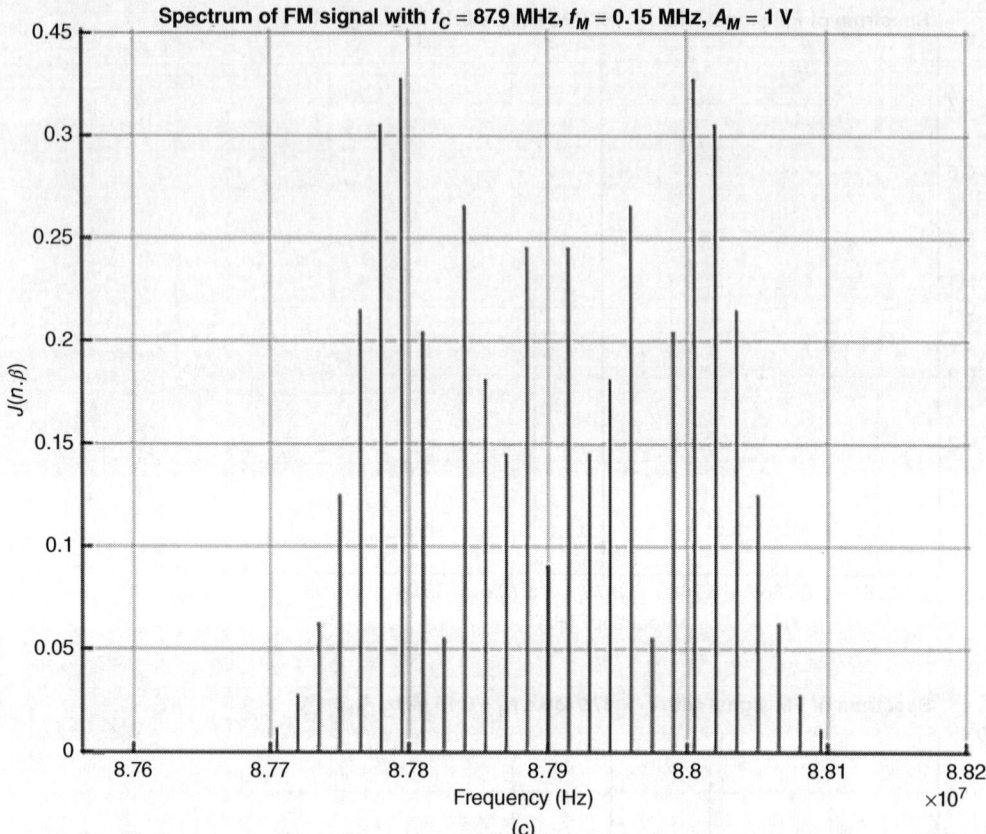

Figure 8.2.A.1 (Continued)

Figure 8.2.A.1. The code allows for the substitution of any value of β; you are encouraged to build all the FM spectra for β that vary from 1 to 9. Obviously, the MATLAB result calculated for $\beta = 5$ must coincide with that in Example 8.2.3; you are encouraged to verify this. Analyzing Figure 8.2.A.1a–c, we can see that the increase in β causes an increase in the number of sidebands, n, and, consequently, in the bandwidth of the signal, as (8.1.23), $BW_{FM} = 2nf_M$, predicts. Since Carson's rule estimates an FM bandwidth as $BW_{FM} = 2(\beta + 1)f_M$, we can find an approximate FM bandwidth as

$$BW_{FM} \approx 2\beta f_M \qquad (8.2.A.1)$$

for $\beta \gg 1$. This equation emphasizes that FM bandwidth is proportional to the modulation index.

1. *Criterion 2*: In this case, we need to determine the spectrum for the same FM transmission, but by using a different criterion – limitation of the allowed bandwidth; specifically, $BW_{FM} \leq 200$ kHz. The spectra of this FM signal for various values of β are shown in Figure 8.2.A.2. Clearly, the spectra for β from 1 to 5 are the same in both Figures 8.2.A.1 and 8.2.A.2. For $\beta > 5$, the FM

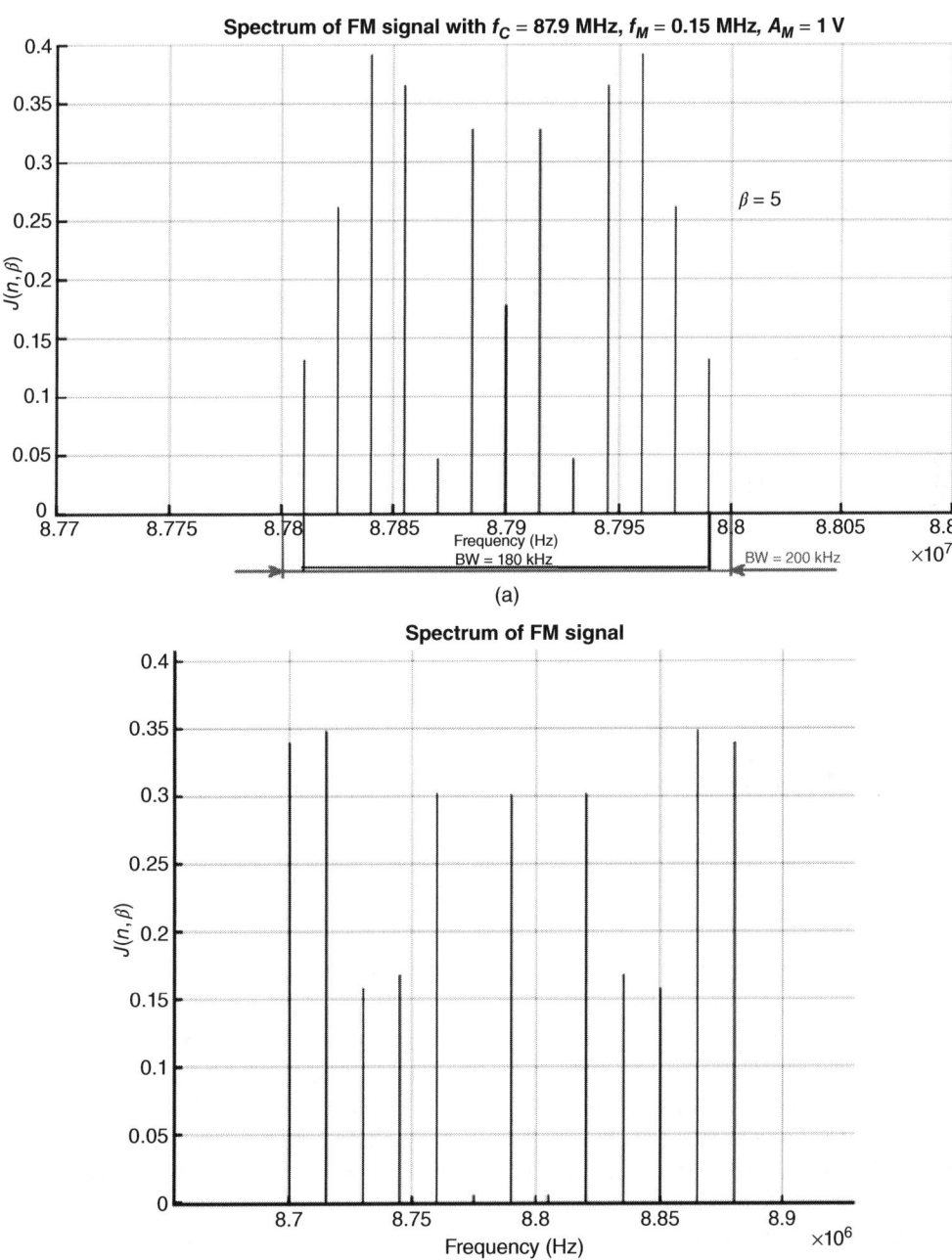

Figure 8.2.A.2 The spectra of an FM signal with $f_C = 87.9$ MHz, $f_M = 0.15$ MHz, $A_M = 1$ V, and various β selected by Criterion 2 – the value of the required bandwidth, $BW_{FM} \leq 200$ kHz. (a) $\beta = 5$, (b) $\beta = 7$, and (c) $\beta = 9$.

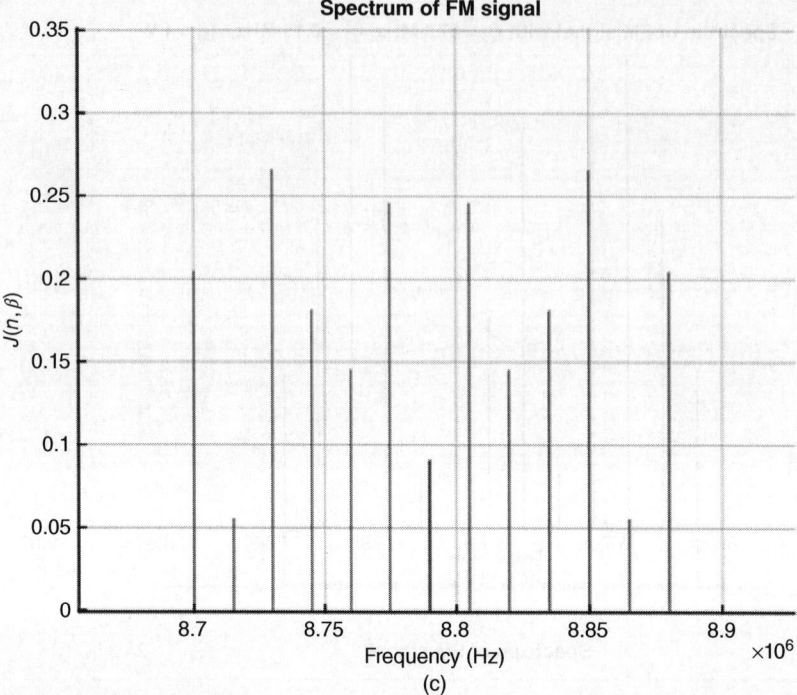

Figure 8.2.A.2 (*Continued*)

bandwidth exceeds the required 200 kHz and MATLAB automatically cutsoff all sidebands beyond this limit, presenting the FM spectrum within the required bandwidth. Interestingly, the number of sidebands for any β from 6 to 9 is the same because of the restriction in bandwidth, but the amplitudes of these sidebands change significantly from one value of β to another. We can obtain these spectra, of course, from Figure 8.2.A.1 by manually restricting the number of sidebands to $n = 13$, but the approach presented here *does this work automatically* and does not require any prior knowledge of the FM spectrum. It is instructive to compare this figure with Figures 8.2.17 and 8.2.A.1.

2. *Criterion 3*: Here, we use, simultaneously, two criteria for determining the FM spectra: the value of the significant amplitudes and the required bandwidth. The MATLAB code, which exemplifies our approach for SA ≥ 0.01 and BW$_{FM} \leq 200$ kHz, is given below. You are encouraged to modify this code to find the FM spectra meeting the two previous criteria. Figure 8.2.A.3 shows the spectra. We leave the detailed analysis of this figure to you; just notice that though the restricted bandwidth is the same for any β, the content of the spectrum varies from one value of a modulation index to another. This implies that the power delivered by these various signals also changes. In addition, study Figure 8.2.A.3c: It shows the actual and the required bandwidth, indicating which sidebands are cut due to the restriction in bandwidth. To clearly understand this, two-criteria example, compare this figure with Figures 8.2.A.1 and 8.2.A.2. This example demonstrates the power of the automated approach, which enables us to find an FM spectrum that satisfies not just one criterion but a combination of various criteria.

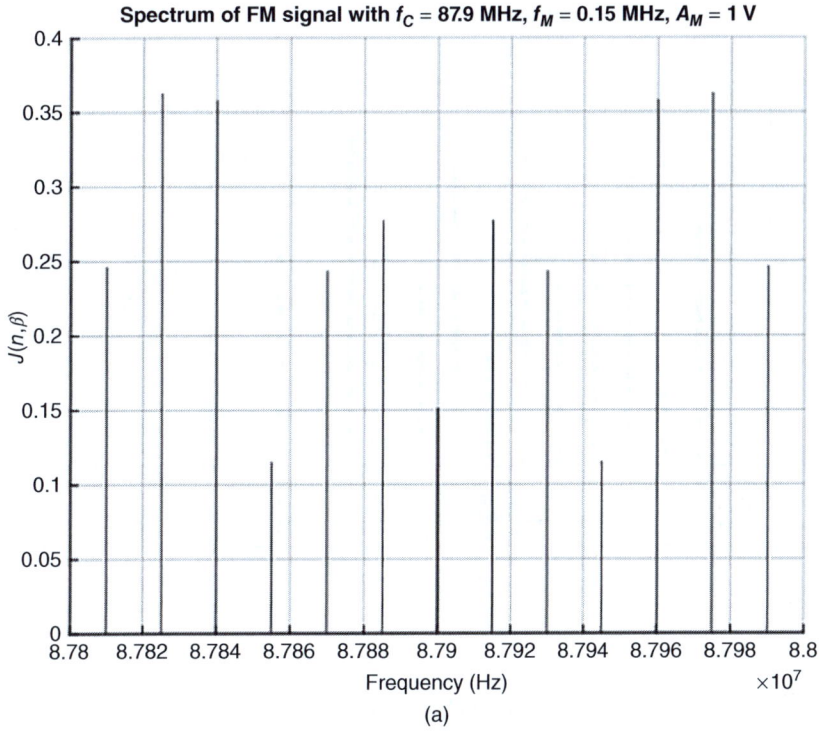

(a)

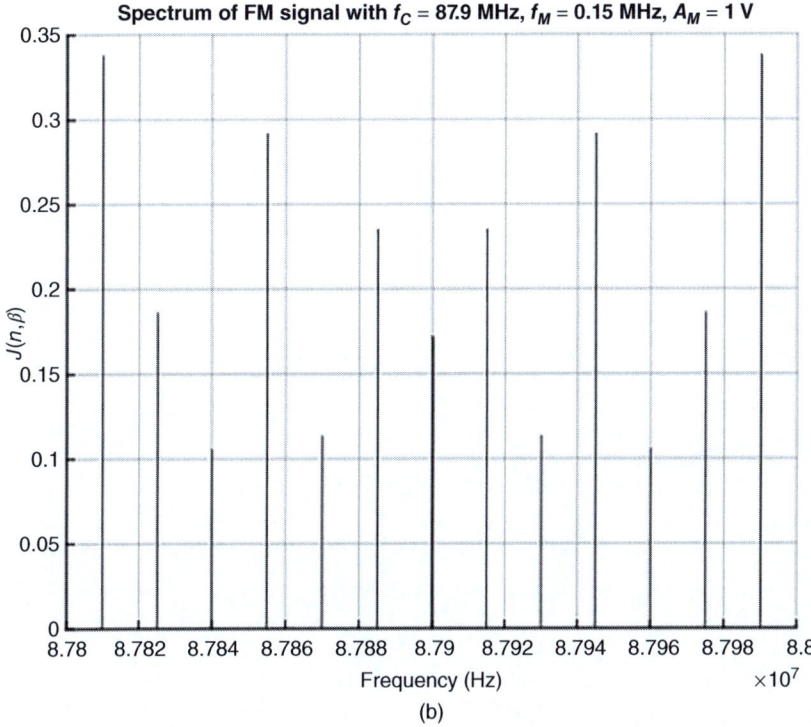

(b)

Figure 8.2.A.3 The spectra of an FM signal with $f_C = 87.9$ MHz, $f_M = 0.15$ MHz, $A_M = 1$ V, and various β selected by Criterion 3 – the value of the significant amplitudes AND the required bandwidth; that is, SA ≥ 0.01 and $BW_{FM} \leq 200$ kHz. (a) $\beta = 6$, actual $n = 9$ and BW = 270 kHz, restricted (shown) BW = 180 kHz, (b) $\beta = 8$, actual $n = 11$ and BW = 330 kHz, restricted (shown) BW = 180 kHz, and (c) $\beta = 9$, actual $n = 13$ and BW = 390 kHz; shown are the required BW = 180 kHz and actual BW = 390 kHz.

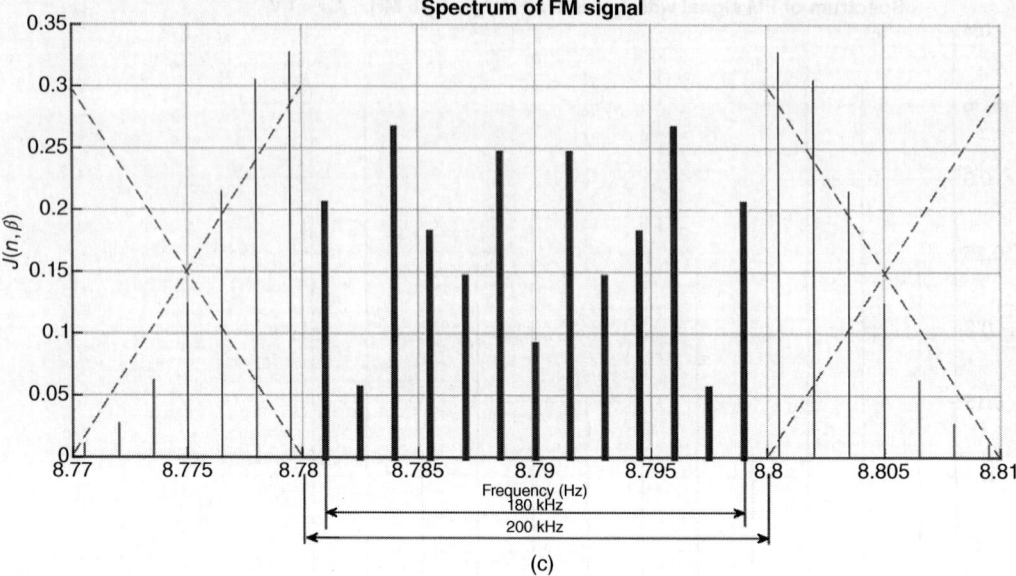

Figure 8.2.A.3 (*Continued*)

The MATLAB code for Criterion 3, SA ≥ 0.01 AND BW$_{FM}$ $\leq$ 200 (kHz).

```
beta=input('Enter beta: ')
Am=1;fm=15000;Ac=1;fc=87900000;SA=0.01;
BW_fcc=200000; %Maximum bandwidth must be less than 0.2MHz
m=0:1:17;f=m*fm;
J=zeros(length(m),1);
n=-1; %it is not 0 because we don't count fc itself
for (i=1:1:length(m))
    J(i)=Ac*besselj(m(i),beta); %J(i) starting at fm=0
if abs(J(i))>SA
n=n+1;
elseif i<(beta+1)
n=n+1;
end
end
n %Number of sidebands >0.01
BW_fm=2*fm*(beta+1)
if BW_fm>BW_fcc
n=fix(BW_fcc/(2*fm)) %Number of sidebands to satisfy BW<0.2MHz
BW_fm=2*n*fm
end
xlabel('Frequency [Hz]'); ylabel('J(n,beta)');
title('Spectrum of FM signal with fc=87.9MHz, fm=0.15MHz, Am=1V')
grid;
for (i=1:1:n+1)
```

```
    line([fc+f(i) fc+f(i)], [0 abs(J(i))]);
    line([fc-f(i) fc-f(i)], [0 abs(J(i))]);
end
```

The advantage of using MATLAB (or any other computerized tool, for that matter) is not limited by only the automation of calculations. Another advantage is that we do not need to refer to a table of the Bessel functions; even more important, we can solve sophisticated problems – as shown in this example – problems that could be difficult, if not impossible, to solve manually.

The analysis of this Appendix, along with Example 8.2.3, should give us deeper insight into the main features of the FM spectrum.

9

Digital Transmission with Binary Modulation

Objectives and Outcomes of Chapter 9

Objectives

- To understand that digital modulation is the technique in which the discrete changes in a message signal are imposed onto parameters (amplitude, frequency, or phase) of a sinusoidal carrier wave.
- To summarize the essentials of digital transmission and learn how to assess the quality of a digital transmission by revisiting the mathematical foundation for calculating the probability of error, employing the bit error rate (BER) metric, and acquiring knowledge of the eye diagram.
- To introduce binary amplitude shift-keying (ASK), frequency shift-keying (FSK), and phase shift-keying (PSK) modulations, to provide their mathematical description, and to describe modulation and demodulation principles of ASK, FSK, and PSK.
- To assess the transmission quality of ASK, FSK, and PSK.

Outcomes

- Revisit the basics of digital transmission, including the operation and characteristics of a communication system and its modules; learn classification and description of noise and the methods of reducing its harmful effect; grasp the idea of noise figure (NF) and its importance for the operation of a communication system.
- Review the Gaussian probability density function (bell curve) and the basics of calculating the probability of finding the random variable beyond a certain value (threshold), which lays the foundation for computing the error probability in digital transmission.
- Learn how to assess the quality of digital transmission by using BER and other quality metrics and realize that BER is the probability of error.
- Become familiar with eye diagram, the visual technique enabling us to qualitatively and quasi-quantitatively assess the transmission quality. Learn how to relate the observable parameters of an eye diagram to the parameters of the Gaussian curve and estimate the error probability based on this relationship.
- Get to know binary shift keying, the digital modulation technique widely employed in modern communications.
- Study binary ASK, FSK, and PSK modulations, their generation, detection, characteristics, advantages, and shortcomings and become familiar with the main areas of their applications.

Essentials of Modern Communications, First Edition. Djafar K. Mynbaev and Lowell L. Scheiner.
© 2020 John Wiley & Sons, Inc. Published 2020 by John Wiley & Sons, Inc.

- Compare quality of transmission in ASK, FSK, and PSK systems based on the required bandwidth and BER characteristics and learn that each type of shift-keying modulation has its own advantages and shortcomings.

9.1 Digital Transmission – Basics

Objectives and Outcomes of Section 9.1

Objectives

- To summarize the essentials of digital transmission.
- To revisit the mathematical foundation for calculating the probability of error by using the Gaussian (bell) curve.
- To learn how to assess the quality of a digital transmission by employing the BER metric.
- To acquire knowledge of the eye diagram and its application to estimating the transmission quality.

Outcomes

- Revisit the block diagram of a communication system to understand that a communication system must support a high bit rate, deliver information of high quality, be noise immune, be resilient to any internal and external disruptions, expend minimum energy per transmitted bit, incur minimal installation and operational costs, be flexible and reconfigurable in meeting ever-changing customer demands, and possess a plethora of other important properties.
- Concentrate on the transmission aspects of the operation of a communication system: the transmission (bit) rate and the quality of delivered information.
- Realize that to meet the quality demand, all components of a communication system – transmitter, link, and a receiver – have to operate in sync, for which their main characteristics must be verified.
- Get to know that a transmitter, Tx, has two main characteristics: slope efficiency, SE, and modulation bandwidth, BW_{mod}. Understand that SE is a Tx's input/output (I/O) characteristic and BW_{mod} is a measure of Tx's ability to transmit a high-speed digital signal. Be aware that BW_{mod} is determined by the rise time of a Tx. Realize that a Tx must also meet the requirements related to its spectral parameters.
- Learn that a receiver, Rx, is also described by its input/output and modulation bandwidth characteristics; however, these characteristics for various Rx types require specific forms. Get to know that an Rx needs an additional characteristic, which specifies the minimum values of the received signal. Be aware that in an optical communication system, the Rx's I/O is a photodiode's responsivity, and the minimum received power is called receiver sensitivity.
- Master the fact that a transmission link is described by its attenuation (the loss of signal strength per unit length), bandwidth (the measure of its capacity to transmit information flow), latency (the measure of signal delay), and signal-to-noise ratio (SNR).
- Understand that for controlling the SNR, we need to know the nature and characteristics of noise. Learn that noise is classified as external and internal and that there are three main types of internal noise – flicker, shot, and thermal – of which thermal noise is omnipresent. Get to know that the harmful effect of noise can be reduced by various measures, starting with its filtering. Come to know such an important characteristic as NF.

9.1 Digital Transmission – Basics

- Compare the characteristics of a Tx, a channel, and an Rx in order to verify whether these components would support the operation of a given communication system. For example, be aware that in an optical system knowing the signal power launched into an optical fiber by a Tx and the loss introduced by a channel (link) enables us to calculate the received power and measure it against the Rx's sensitivity. Perform similar verification for modulation, spectral and other characteristics of all components.
- Remind yourself that Shannon's law assumes the noise model called additive white Gaussian noise (AWGN).
- Revisit basics of probability theory, in particular the definition and meaning of Gaussian probability distribution (density) function. Be aware that the Gaussian probability distribution function (PDF) is widely used as a model of real stochastic processes, including noise in digital transmission. Remember that the Gaussian PDF is depicted as a bell curve and is characterized by its mean value μ and standard deviation σ.
- Recall that the probability of finding a random Gaussian variable Z within any given interval is equal to the area under the Gaussian (bell) curve for this interval. Calculate this probability by using a standard normal PDF form given either as Integral 9.1.20 or as Table 9.1.1.
- Learn that finding the probability of errors in digital transmission requires calculating the probability of finding Z in the Gaussian tail, i.e. $Z > a$. Understand that it can be done by deriving Integral 9.1.22, denoting it as Q-function, and tabulating the results of integration as exemplified by Table 9.1.2. Be aware that for calculating the error probability, the research and industrial communities also use the complementary error function, $\text{erfc}(x)$, directly related to Q-function. Remember that all these results are valid only for a standard normal Gaussian PDF.
- Understand that errors in a digital transmission are caused by noise. Get to know that the assumed mathematical model of noise is a Gaussian PDF with $\mu = 0$ and σ being a measure of noise intensity. Realize that a received signal is a sum of transmitted signal and noise; this sum is described by the Gaussian PDF with means V_1 and V_0 and standard deviations σ_1 and σ_0, where V_1 and V_0 are the mean values of bit 1 and bit 0 and σ_1 and σ_0 are the standard deviations of noise of bit 1 and bit 0.
- Learn that in a digital transmission, the receiver makes the decision as to which bit – 1 or 0 – it receives by comparing the mean values V_1 and V_0 of the received signal with the threshold value, V_{TH}, which is typically set as $V_{TH} = (V_1 - V_0)/2$. Be aware that if $V_1 > V_{TH}$, then the decision is that bit 1 is most likely; if $V_0 < V_{TH}$, then the decision is that bit 0 is most likely.
- Come to know that noise changes the actual values of received pulses so that the sampled (measured) value of a real bit 1 could be smaller than V_{TH} and the sampled value of a real bit 0 could be greater than V_{TH}, which significantly increases the probabilities of erroneously identifying bit 1 for bit 0 and vice versa (see Figure 9.1.16).
- Master the fact that the erroneous detection of a received noisy signal is mathematically described by the error probability, which is calculated as the probability of finding the sampled value of bit 1 smaller than V_{TH} and/or the value of bit 0 greater than V_{TH}. Recall that this problem is resolved by (9.1.22), and actual calculations are reduced to finding the value of Q-function or erfc function from a Table 9.1.2. Remember that this explanation is visualized in Figure 9.1.17, which shows that the probability that a receiver erroneously identifies bit 1 as bit 0 is equal to the area under the tail of curve PDF_1 in the interval from V_{TH} to infinity, and the probability that bit 0 will be identified as bit 1 is equal to the area under the tail of curve PDF_0 in the interval from minus infinity to the threshold.
- Learn that the average probability of erroneously identifying a received bit is, in fact, the bit error ratio (rate), BER, which is one of the main quality metrics in digital transmission. Be aware that

this fact is presented in (9.1.34b) as $\text{BER} = P_e = Q\left(\frac{V_1 - V_0}{\sigma_1 + \sigma_0}\right) = \frac{1}{2}\,\text{erfc}\left(\frac{V_1 - V_0}{\sqrt{2}(\sigma_1 + \sigma_0)}\right)$. Get to know that, on the other hand, BER is defined in (9.1.33) as $\text{BER} = \dfrac{\text{Number of received erroneous bits}}{\text{Total number of received bits}}$ and that both definitions are correct because of the definition of probability, as shown in (9.1.35) and (9.1.36).

- Realize that one acronym, BER, encompasses two definitions: (1) *bit error **rate***, which is the number of erroneous bits per unit of time, as the word "rate" implies and (2) *bit error **ratio***, which is the ratio of the number of erroneous bits to the total number of bits and know that both definitions mean the same thing.
- Grasp the point that the BER has its limitation because it measures exclusively the probability of mistakenly taking bit 1 for bit 0 and vice versa; that is, BER is a measure of *bit flips* only. Realize that bit flips are caused by noise; this is why the BER depends solely on SNR. Be aware that in reality the errors in digital transmission might occur due to various phenomena: for example, due to signal distortion caused by such impairments as dispersion and intersymbol interference and simply due to missing or inserting bits.
- Get to know that there are several metrics for the assessment of transmission quality. Cognize that the BER is considered the most conclusive quality measure; the others include SNR, eye diagram, and the quality factor (to be discussed shortly). Be aware that additionally, the *error vector magnitude (EVM)* is often used for assessing the quality of a digital transmission.
- Familiarize yourself with the eye diagram, which is a visual device that enables us to qualitatively and quasi-quantitatively assess the transmission quality. Understand that an *eye diagram* (or *eye pattern*) is formed by overlaying thousands and even millions of transmitted pulses. Take a glance at the complete eye diagram shown in Figures 9.1.20 and 9.1.21 and see how much this pattern resembles the human eye.
- Realize that since an eye diagram is formed from millions of pulses, it is a statistical device by its very nature. Consider an example shown in Figure 9.1.20B where the upper level of an eye diagram, called *eye level one*, is formed by summing up the upper levels of all pulses that contributed to this eye pattern. Understand that if upper levels of all contributing pulses were equal in values, then an ideal eye level one would be a single line. Examine that in reality the heights (amplitudes) of contributing pulses vary, and these variations (caused mainly by noise) make the eye level a wide strip. Understand that the noisier the signal, the wider the level. Comprehend that everything said about the upper level of an eye diagram is applied to its lower level called *eye level zero*.
- Understand that measuring the samples of a received signal allows us to build a histogram that could then be approximated by a continuous curve. Get to know that all experiments confirm that usually such curves are closely described by the Gaussian PDF.
- Examine Figure 9.1.21b exhibiting the snapshot of an actual eye diagram with fuzzy lines and a narrow opening due to noise. Understand that the wider the eye opening, the better the received signal and vice versa. Conclude that an eye diagram can be used to rate transmission quality.
- Learn that an eye pattern enables us to evaluate the transmission process not only qualitatively but also quantitatively even though the quantitative assessment of transmission quality acquired with an eye diagram is only a crude estimation of the actual results. Realize that using the eye diagram is still a convenient and time-saving procedure compared with obtaining the BER.
- Understand that the frequencies of measured signal values of levels one and zero in an eye diagram are described by two Gaussian PDFs when the number of samples goes to infinity. See Figure 9.1.21b to visualize this description. Remember that the means of the PDF_1 and PDF_0, V_1 and V_0, determine logic 1 and logic 0, respectively. Realize that calculating the width of levels one

and zero presents a problem caused by the random nature of an eye diagram. Comprehend that there are no numbers that delimit the widths of these levels, but there are only the probabilities that the sampled values of these levels will not exceed $\pm 3\sigma$ limit.
- Observe that an eye diagram's parameters related to signal strength are measured along eye's vertical axis, and the signal's time-related parameters are measured along the horizontal axis. Specifically, the strength parameters include a signal amplitude, an eye diagram's height, and threshold, and the timing parameters are a sampling point, jitter, and rise and fall times.
- Come to know that for evaluating transmission quality, we need to create and scrutinize an eye diagram instead of simply investigating a single pulse because a single pulse is a snapshot of a whole pulse train, but an eye diagram, formed by overlaying millions of pulses, contains the statistical characteristics of the entire transmission process.
- Learn that measuring eye diagram's parameters and processing the results yields in the calculation of quality factor, $\text{QF} = \frac{V_1 - V_0}{\sigma_1 + \sigma_0}$. Study in (9.1.40c) that QF is a measure of SNR. Get to know, through (9.1.41), that the QF enables us to calculate the BER as $\text{BER} = \frac{1}{2}\,\text{erfc}\left(\frac{\text{QF}}{\sqrt{2}}\right) = Q(\text{QF}) = Q\left(\frac{V_1 - V_0}{\sigma_1 + \sigma_0}\right)$, where we can use either Q-function or erfc function.
- Finally, examine Figure 9.1.22 and Example 9.1.2 to conclude how we can estimate the quality of digital transmission by using an eye diagram.

9.1.1 Essentials of Digital Transmission Revisited

We discussed the various aspects of digital transmission in Chapters 1 and 3, where the basic concepts and techniques of digital communications were covered. In this section, we review some aspects of that material and significantly expand our knowledge of digital transmission. This discussion will put in perspective the information presented in the abovementioned chapters and enhance our knowledge of digital transmission, which is needed for studying the more advanced material to be introduced in Section 9.2 and Chapter 10.

9.1.1.1 Block Diagram of a Communication System

Let us look again at the block diagram of a communication system originally presented and discussed in Sections 1.1 and 1.2. (Interestingly, such a block diagram was prepared by Claude Shannon for his seminal work "A Mathematical Theory of Communication.") Reviewing that sections will refresh your memory of this block diagram; here we want to take several additional steps toward understanding the operation of a communication system.

What do we want from a communication system? First and foremost, it must deliver a maximum volume of information over a minimum transmission time, a characteristic called the transmission (bit) rate. *But besides a high bit rate, a communication system must also deliver information of high quality, be noise immune, be resilient to any internal and external disruptions, expend minimum energy per transmitted bit, incur minimal installation and operational costs, be flexible and reconfigurable in meeting ever-changing customer demands, and possess a plethora of other important properties.* We concentrate in this section, however, only on the transmission aspects of the operation of a communication system: the transmission rate(and the quality of the delivered information.

To meet the quality demand, all components of a communication system have to operate in sync; to be certain of this, we need to verify the components' characteristics. Consider a simple example: If a transmitter has a modulation bit rate of 50 Gb/s but a receiver can demodulate signals at only 50 Mb/s, this system either underutilizes its potential by maintaining the Rx rate or does not deliver

9 Digital Transmission with Binary Modulation

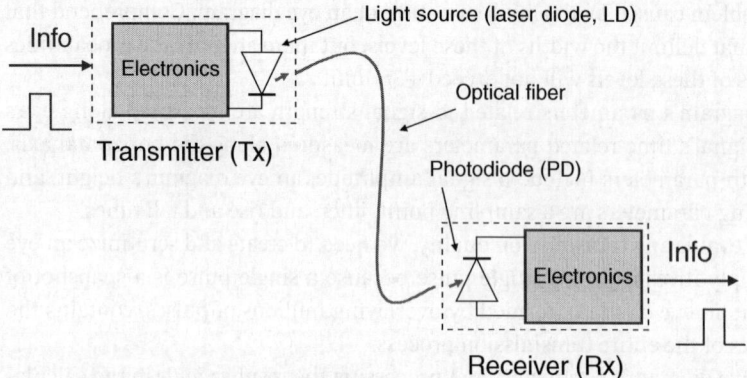

Figure 9.1.1 (Reproduction of Figure 1.1.8.) Block diagram of an optical (fiber-optic) communication system.

correct information by operating at the Tx rate. (Can you explain why the system, if operating at the Tx modulation rate, does not deliver correct information?)

This consideration brings us to the need to discuss briefly the main characteristics of a transmitter, a receiver, and a transmission link (channel). We exemplify this discussion by studying the optical communication system shown in Figure 9.1.1.

9.1.1.2 Characteristics of a Transmitter, Tx

Let us start with a transmitter: What characteristics are needed to describe it? The first one, naturally, is the transmitter's input/output (I/O) relationship. For a laser diode (LD), shown in Figure 9.1.1, the input is a forward driving current, I_F (mA), and the output is the power of light, P_{light} (mW). Their ratio, called the *slope efficiency* (SE), is the LD's I/O characteristic. SE is given by

$$\text{SE (mW/mA)} = \frac{\Delta P_{light}(\mu W)}{\Delta I_F (mA)} \tag{9.1.1}$$

where ΔI_F (mA) is the increment of the forward current and ΔP_{light} (mW) is the increment of light power caused by ΔI_F. Obviously, SE is calculated within the linear part of an $I_F - P_{light}$ graph so that P_{light} (mW) = SE (mW/mA) $\cdot I_F$ (mA). To put this characteristic in perspective, compare typical numbers of low-cost LEDs and LDs: An LED produces 300 µW at 20 mA, and an LD gives 600 µW at 7 mA. In other words, the LD produces twice the LED's power by consuming a nearly three times less current.

The second characteristic of a transmitter is its ability to transmit a high-speed digital signal. This ability is described by a modulation bandwidth, BW_{mod} (Hz). (Regarding the industrial usage of the term "bandwidth" for digital transmission, refer to Chapter 3.) Figure 9.1.2 demonstrates the problem described by the modulation bandwidth: Even if a driving current, I_F (mA), presents

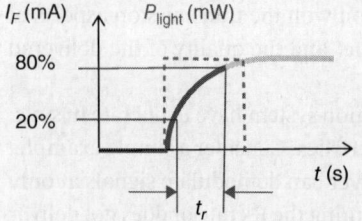

Figure 9.1.2 Rise time and modulation bandwidth.

an ideal, step-like electrical pulse to an LD, the LD needs some time – called *rise time*, t_r (s) – to produce the optical pulse. (Refer to Section 3.2 for a review of the subject of rise time.) Clearly, the greater the rise time, the fewer the pulses per second an LD can produce. This is why a modulation bandwidth is calculated as

$$\text{BW}_{\text{mod}} \text{ (b/s)} = \frac{k}{t_r \text{ (s)}} \tag{9.1.2}$$

Here, k is a constant whose value, along with the value of t_r, is usually specified by the manufacturer. Comparing again an LED and an LD, we find that an LD's t_r (s) is typically from 10 to 100 times smaller than an LED's. Today's LDs can be modulated at a rate exceeding 50 Gb/s, but this is not enough for a modern communication system operating at a bit rate exceeding tens of terabits per second. The attempts to directly modulate LDs at such bit rates face fundamental physical limitations stemming from the fact that a transmitter's LD must simultaneously perform two tasks: generate a light beam and modulate this beam. The solution to the problem is found in separating these tasks: In modern transmitters, an LD generates only the CW light, while an external modulator modulates this beam.

The two aforementioned transmitter characteristics – input/output and modulation bandwidth – are typical for a transmitter of any type of communication system. Think about our smart phones, for example: We want them to generate maximum output power by consuming the minimum power of a battery, and we want them to transmit the maximum bit rate by using the simplest circuitry. These characteristics are very much similar to those attributed to a Tx of an optical communication system, specifically to its LD.

The other transmitter characteristics for different types of communication systems can vary. For instance, we want an ideal LD to generate a single wavelength (frequency), but we want a cell phone's transmitter to be able to generate all the audible frequencies from 20 Hz to 20 kHz.

9.1.1.3 Characteristics of a Receiver, Rx

Regarding a receiver, we certainly want to describe it by the first two characteristics that we use for a transmitter: input/output and modulation bandwidth. In the case of an optical system, the photodiode, which is the front end of the receiver (see Figure 9.1.1), must produce maximum photocurrent, I_P (µA), in response to the power, P_{rec} (µW), of a received optical signal. Thus, a PD's input/output characteristics, called *responsivity*, R (µA/µW), is given by

$$R \text{ (µA/µW)} = \frac{\Delta I_P (\text{µA})}{\Delta P_{\text{rec}(\text{µW})}}. \tag{9.1.3}$$

We must remember that the received optical signal is typically weak, which results in a weak photocurrent; this is why the units of both I_P and P_{rec} are typically in the micro range.

As for a PD's modulation bandwidth, it is specified in the same way as for an LD, i.e. by (9.1.2). The requirement for a PD's spectral characteristic is just the opposite of that for an LD: An ideal LD should generate one wavelength but an ideal receiver must be able to operate with all the possible wavelengths that can be transmitted by its communications system. For example, an ideal PD must cover the entire range of today's transmission wavelengths, i.e. from 800 to 1600 nm. Fortunately, such PDs exist. The wideband spectral characteristic of a receiver is desirable in general because it makes possible the use of one type of Rx in a variety of communication systems. This property simplifies the supply and lowers the operational cost of a network.

In contrast to a Tx, an Rx must be additionally specified by the minimum strength values of the received signal. For example, the specification for a PD's minimum received power, called *receiver sensitivity,* is defined as follows:

Receiver sensitivity is the minimum received power required to sustain the given BER.

Typical values of PD sensitivity vary from -15 to -50 dBm for a BER from 10^{-9} to 10^{-12}. (We urge you to recalculate these power values in µW.) Concentrate on the meaning of "receiver sensitivity:" This characteristic requires not just minimum received power, which causes a PD's response with minimum photocurrent, but the value of the received power, which provides PD operation at the required quality level.

Receiver sensitivity is a typical Rx characteristic regardless of the physical nature of a communication system, but there are, of course, a number of receiver-specific characteristics used in specific communications systems.

Comparing the characteristics of a given Tx and Rx, we can check whether they would work for the same communication system. For example, knowing the maximum transmission power generated by a Tx and the loss introduced by a channel (link), we can calculate the received signal power and measure it against the Rx's sensitivity. Such calculations, called a *power budget* (or link budget), are performed by designers of optical communications systems. (See, for example, Mynbaev and Scheiner 2001, pp. 284–286 for a detailed discussion of the power-budget approach.)

9.1.1.4 Characteristics of a Transmission Channel (Link)

To complete the discussion of an entire communication system, we need to characterize a channel (transmission link) that connects a Tx to an Rx.

Channel Attenuation To do this, we first consider their input/output relationship. It might sound strange, but a *channel* also has its own *input/output relationship*. The input is the strength of the signal that has been sent for transmission, and the output is this strength at the channel end. For example, in an optical communication system (see Figure 9.1.1), the input is the light power launched into an optical fiber, P_{in} (mW), and the output is the power emerging from the optical fiber at the endface of a PD, P_{out} (mW). The I/O of a channel is characterized by an output-to-input ratio. Since a channel is a passive device, the output signal is always weaker than the input, and the channel is characterized by a *loss* measured in decibels (dB). The loss is given by

$$\text{Loss (dB)} = 10 \, \log\left(\frac{P_{out}(\text{mW})}{P_{in}(\text{mW})}\right) \tag{9.1.4}$$

For optical fiber, if $P_{in} = 1$ mW and $P_{out} = 1$ µW, then the loss $= -30$ dB. The loss in dB is always a negative number because $P_{out} < P_{in}$. (Prove the last statement by referring to the property of 10-base logarithm.) If you compare two optical fibers supplied by different manufacturers and find that one optical fiber's loss_1 is -30 dB and the other's loss_2 is -50 dB, can you tell which fiber is better? In reality, you cannot. It is intuitively clear that the longer the transmission link, the greater the loss it introduces. This is true for copper wire, coaxial cable, optical fiber, and air; that is, it is true regardless of the physical nature of the link. This means that the loss value depends not only on the quality of a link but also on its length. Therefore, the loss is not a comprehensive I/O characteristic of a transmission channel's quality. What characteristic would you suggest? In such a characteristic, we must include the loss but exclude the length. This can be done by measuring the loss per unit of length. Thus, we arrive at *attenuation*, A (dB/km), a channel's characteristic obtained by dividing the loss by the length; that is

$$A \text{ (dB/km)} = \frac{\text{Loss (dB)}}{\text{Length (km)}} \tag{9.1.5}$$

Bear in mind that the communication industry wants *attenuation to be a positive number*, which can be achieved either by introducing an artificial negative sign for the loss (dB) in (9.1.5) or by calculating the loss as $10\log\left(\frac{P_{in}(mW)}{P_{out}(mW)}\right)$ in (9.1.4). Typical attenuation of modern fiber-optic cable (which is optical fiber placed in a protective surrounding) is about 0.3 dB/km. This means that the loss introduced by the 100-km long optical fiber is -30 dB; that is, the signal becomes a thousand times weaker at the output. In contrast, the attenuation of copper wire is about 40 dB/km, which means that a signal's loss is -40 dB after 1 km of transmission. Compare these two links: A 1-mW input signal is still 1-µW strong after 100 km of transmission through optical fiber, but a 1-mV input signal becomes 0.1-µV weak after only 1 km of transmission via copper wire. No wonder that an optical fiber blankets the world to deliver global communication traffic and copper wire remains in use only for signal transmission over very short distances (meters and centimeters). In this comparison, we refer to transmission of an optical signal at 200 THz via optical fiber vs. transmission an electrical signal at 100 kHz via copper-wire cable.

(*Important*: Do all these calculations to ensure you understand every number and every statement in the above discussion.)

To sum up:

Attenuation, A (dB/km), describes a decrease in signal's strength per unit of length of a transmission link and this is one of the most important characteristics of the transmission link's quality and its input/output relationship.

It must be noted that in *wireless communications,* the assessment of a channel's loss is not straightforward. This is because in wireless transmission, the electromagnetic waves of various frequencies travel through air along various paths. Therefore, this communication technology is not restricted to one transmission line. As a result, the term "channel" in wireless communications becomes ambiguous since, in contrast to wireline transmission, one communication line can be established wirelessly through several pathways. For example, a *Wi-Fi* system can theoretically deliver a message through 13 channels (pathways) occupying the range from 2412 to 2484 MHz and separated by increments of 5 MHz. Loss in a wireless channel may vary from one pathway to another, which means that a channel's *attenuation,* defined in (10.1.5), can be directly applied to an individual pathway but not to the entire transmission line. Nevertheless, the total loss of a received signal can be evaluated although its value can change randomly due to the fluctuations in propagation conditions; this channel characteristic is called *fading.*

Channel Bandwidth The second characteristic of any unit of a communication system is its ability to transmit high-speed digital transmission. For Tx and Rx, we introduce a *modulation bandwidth* determined by the rise time given in (9.1.2). We can do similar calculations for optical fiber and other transmission media. Indeed, since a communication system must support the required *transmission rate* and since the system includes the components of different physical makeup (e.g. electronic, optical, acoustic, mechanical, etc.), the only common characteristic in computing the system's rate is the rise time. The total rise time of a system, t_{rs}, is given by

$$t_{rs}(s) = \sqrt{(t_{r1}^2 + t_{r2}^2 + t_{r3}^2 + \cdots + t_{rn}^2)} \tag{9.1.6}$$

where t_{rn} is the rise time of an *n*-component. Then the system's modulation bandwidth can be computed as in (9.1.2); that is, $BW_{mod-s} = k/t_{rs}$.

Calculating a system's modulation bandwidth enables us, however, to compute only the distribution of $t_{rs}(s)$ and find out which component lowers the system BW_{mod-s}. To determine the link's ability to transmit digital traffic at the required speed, we need to turn once more to Shannon's law,

$$C(\text{b/s}) = \text{BW (Hz)} \log_2 (1 + \text{SNR}) \tag{1.3.2R}$$

You will recall that (1.3.2) states that the *maximum transmission rate (speed), C (b/s),* that a communication channel can reliably provide is proportional to the channel's available *bandwidth*, BW (Hz), and depends on the *SNR*. What this means is that *C (b/s) plays the role of a modulation bandwidth for a communication channel.* Shannon's law explicitly shows that the key member in limiting the *C (b/s)* is the *channel's bandwidth*, BW (Hz), that can be effectively used for transmission. Therefore, Shannon's law, which was derived from information theory, not from technological aspects, can be applied only to a *bandlimited* channel (link). Therefore, we can say that the use of Shannon's law implies that we are considering a bandlimited channel. (What if BW (Hz) in (1.3.2) is zero? Or tends to infinity? Can these situations occur in practice? *Hint*: See, for example, Proakis and Salehi 2014, p. 306.)

All transmission channels (links), in reality, are composed of tangible media; therefore, they are bandlimited. Consider this well-known example: Copper wire can transmit ac signals but cannot transmit optical signals, although both are simply electromagnetic (EM) waves of different frequency ranges, with ac being in kHz and light in THz bands. On the other hand, optical fiber is a perfect insulator for ac and even for radio-frequency signals but can conduct light. How can we determine the bandwidth of copper wire and optical fiber? Up to what frequencies does copper wire conduct EM waves and beginning at what frequencies does optical fiber start conducting light?

To answer these questions, we need to find out by what criterion we determine whether a channel conducts or insulates. Obviously, a channel insulates a signal if the signal has been launched in the input but cannot be detected at the channel's output. But why does this signal disappear and to where? Our preceding discussion shows that any channel introduces loss; therefore, in an insulator, a signal diminishes after a very short distance because of the high level of loss. In contrast, a conductor exhibits low loss in its specific frequency range.

Since the loss property of a transmission medium is defined by its attenuation, we can determine whether a channel will conduct or insulate a signal by observing the attenuation vs. frequency graph of the channel. Such a graph is called *spectral attenuation*, and Figure 9.1.3 shows these graphs for copper-based cable and optical fiber.

Studying Figure 9.1.3 shows that attenuation of copper-based cable grows exponentially with an increase in frequency, whereas optical fiber has a minimum attenuation at a certain range of frequencies (wavelengths). Thus, the bandwidths of twisted pair and coaxial cable are restricted at the megahertz frequency range, where their attenuation is at an acceptable (but still very high) level. The bandwidth of optical fiber is also restricted to a "short" terahertz frequency range, where optical fiber's attenuation reaches its minimum. Bottom line: *The available bandwidth of a transmission channel is determined by its attenuation.*

To elaborate, consider the following example: In general, optical fiber can conduct EM waves over a wide range of terahertz frequencies; for instance, we are all familiar with the use of optical fiber for the transmission of visible light ranging from 400 to 700 nm. However, today's *transmission* optical fiber (which overlays the globe to deliver Internet traffic) has minimum attenuation in the range of $1562 - 1530 = 32$ nm, which corresponds approximately to $195.9 - 191.9 = 4$ THz. (We learned that since λ (m) $\cdot f$ (Hz) $= c = 3 \times 10^8$ m/s, $\lambda = 1550$ nm approximately corresponds to $f = 193.4$ THz.) It is optical fiber's high attenuation outside the range from 191.9 to 195.9 THz that limits optical fiber's bandwidth usable for communications. We need to plug this bandwidth value, 4 THz, into Shannon's law (1.3.2) to calculate the maximum transmission rate. (Be aware that today the optical

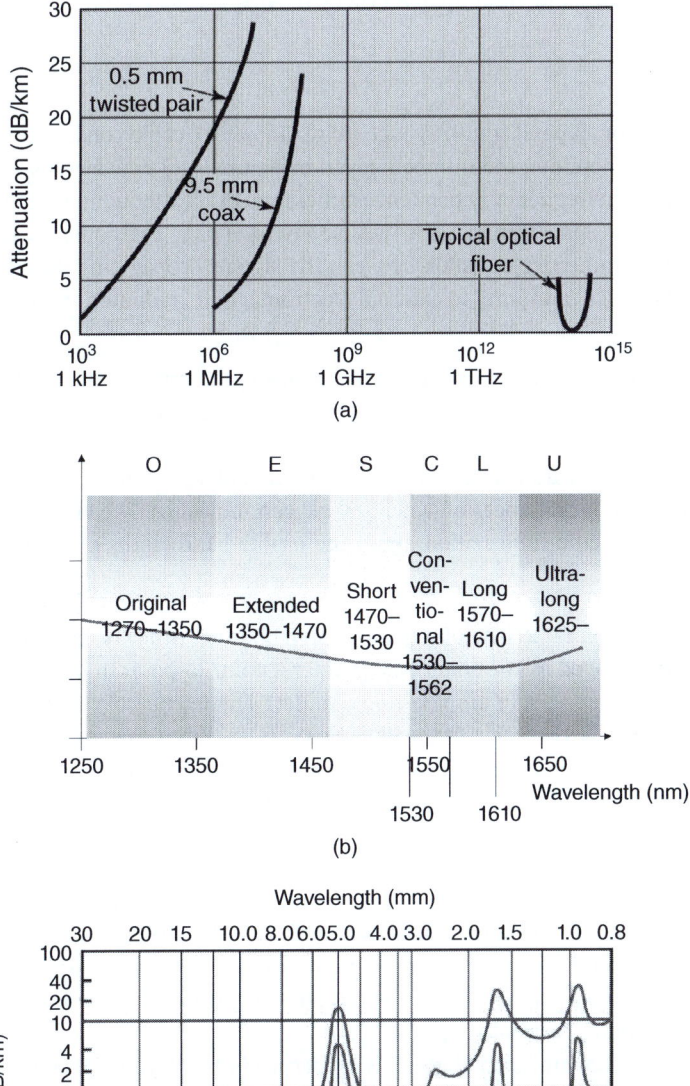

Figure 9.1.3 Spectral attenuation of various transmission media: (a) copper-based cable and optical fiber (general view); (b) optical fiber for long-distance transmission (detailed view); and (c) the Earth's atmosphere (general view: curve A – at sea level; curve B – at 4 km above the sea level). (Figure 9.1.3a,c is reprised Figure 1.3.3a,b. See permission information.)

communications industry expands this bandwidth up to 8 THz by moving the signal transmission into 1600-nm range.)

Briefly

- For copper-based cable, attenuation depends on the gauge (wire diameter), cable construction, test methods, and other factors. Therefore, the numbers given in Figure 9.1.3 convey only the order of magnitude and the trends in frequency dependence rather than exact attenuation values.
- In reference to copper wire used for signal transmission, we use a twisted pair: Two copper wires are twisted around each other to minimize crosstalk between the signals propagating through each wire and to maximize the pair's immunity to external electromagnetic radiation in differential signaling.

The above point can be applied to wireless transmission, too. However, the spectral attenuation of air exhibits much greater variation than that of wireline channels because each possible path experiences various power losses due to a variety of obstacles. As general guidelines, Figure 9.1.3c shows spectral attenuation of the Earth's atmosphere, which can be considered "free space." The causes of peaks in attenuations are the resonances of water and oxygen molecules. We encourage you to use online sources to find specific information about the usable bandwidths of wireless channels.

Channel Latency Another important characteristic of a transmission channel (link) is its *latency*, the measure of a signal's delay introduced by the channel. Consider a point-to-point link without any embedded networking modules such as an amplifier or a router: The latency of this link is determined by the time needed for a carrier wave to travel from Tx to Rx. From this standpoint, the minimum latency is introduced by *air*, where EM waves travel at the speed closest to that in vacuum, $c = 3 \times 10^8$ m/s. It would seem that the second fastest transmission channel would be *optical fiber* because photons travel at the speed of light. However, within optical fiber, photons (light) travel at the speed $v = c/n \approx 2 \times 10^8$ m/s, with n being optical fiber's refractive index, $n \approx 1.5$. *Coaxial cable*, delivering high-frequency EM waves, causes a delay and, from the propagation-speed standpoint, is similar to optical fiber. The EM wave travels within the cable at velocity $v = V_F \cdot c$, where V_F is a *velocity factor* whose value varies from approximately 0.6 to 0.99. V_F depends on the cable's properties. Formally, $V_F = 1/\sqrt{\mu_r \varepsilon_r}$, where ε_r is the relative permittivity (a dielectric constant) and μ_r is relative permeability. Normally, $\mu_r = 1$.

You might expect that the slowest transmission channel is *copper wire* (a twisted pair) because the electrons travel at a much slower velocity than photons, but the real picture is not that simple. Yes, electrons travel several meters per hour, but this so-called *drift speed* is not the signal velocity. A signal within a conductor is delivered by the fluctuations of the electromagnetic field, which couples to electrons and propagates down a copper wire. Thanks to this effect, the signal velocity within the wire is close to the speed of light in vacuum; in fact, the velocity factor of copper wire is about 0.95.

Therefore, our latency discussion brings a surprise result: *Air precipitates the least delay, whereas optical fiber and the copper-based channel introduce intermediate delays; coaxial cable may cause an even much longer delay than optical fiber.*

It should be noted that the time delay depends on the ambient conditions; for example, the temperature variations cause the fluctuations in signal delay in optical fiber.

(**Question**: *Optical fiber and coaxial cable are the transmission media that conduct EM waves. The propagation speed is determined by refractive coefficient, n, in optical fiber and by velocity factor, V_F, in cable. Should there be any relationship between n and V_F? If yes, what is it?*)

9.1 Digital Transmission – Basics

The delay caused by a transmission channel and discussed above is called a *propagation delay*. If we need to know the duration of time needed to relay a message from origin to destination through a whole network, we have to take into account other types of delay, such as *transmission delay*, *processing delay*, and *queuing delay*.

Channel Signal-to-Noise Ratio, SNR Shannon's law reveals that a channel's maximum transmission rate also depends on the *SNR*. The greater the SNR, the greater the C (b/s). Refer to Section 1.3 for a discussion of the SNR factor. Obviously, we cannot control noise power; thus, the only way to increase the SNR is to increase the signal power. But this straightforward approach has its limitation. In optical fiber, for example, we cannot increase the signal's power beyond its linear maximum because optical fiber becomes a nonlinear channel. Thus, the SNR and, therefore, the transmission rate cannot be increased indefinitely by increasing the signal power; this restriction in optical communications is called the *nonlinear Shannon limit*.

Mobile communications, too, cannot significantly increase the signal power due to limitations in the size and weight of mobile-device batteries. Therefore, the channel's maximum transmission rate, as determined by Shannon's law, is always limited.

Finally, we need to know that transmission-channel theory is based on the models of these channels. In a simple case, such as a traditional twisted-pair telephone line, a low-pass filter closely models the real communication channel. (Refer to Chapter 4, where filters are discussed.) In sophisticated cases, such as wireless transmission through a number of reflecting and deflecting points and randomly changing environmental conditions, building a model and providing a comprehensive mathematical description of a transmission channel can be a challenge.

To discuss a channel's SNR more thoroughly, we need to consider noise in detail.

9.1.1.5 The Model of Noise in Shannon's Law

In his law, Shannon includes noise, as (1.3.2) and the block diagram of a communication system in Figure 1.1.2R demonstrate. But what kind of noise did he consider? He relied on the noise model called *AWGN*. In this term, the word *additive* refers to the assumption that noise simply adds to a signal, as Figure 1.1.2R shows. Another assumption is found in the word *white*, which relates to white light. This light is modeled by a composition of all wavelengths (frequencies) uniformly spread from zero to infinity. All these wavelengths (frequencies) are of the same amplitudes (power spectral densities). The word Gaussian pertains to the PDF, called the *bell* or *Gaussian* function. (This topic is discussed in the next subsection.) Therefore, Shannon's law – one of the most fundamental equations governing communication theory – holds true only for a specific noise model. Does this fact undermine the generality of Shannon's law? Is the AWGN model universal? It has

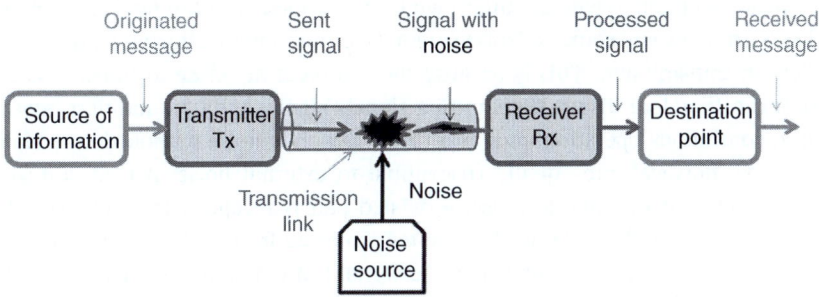

Figure 1.1.2R Block diagram of a communication system.

9 Digital Transmission with Binary Modulation

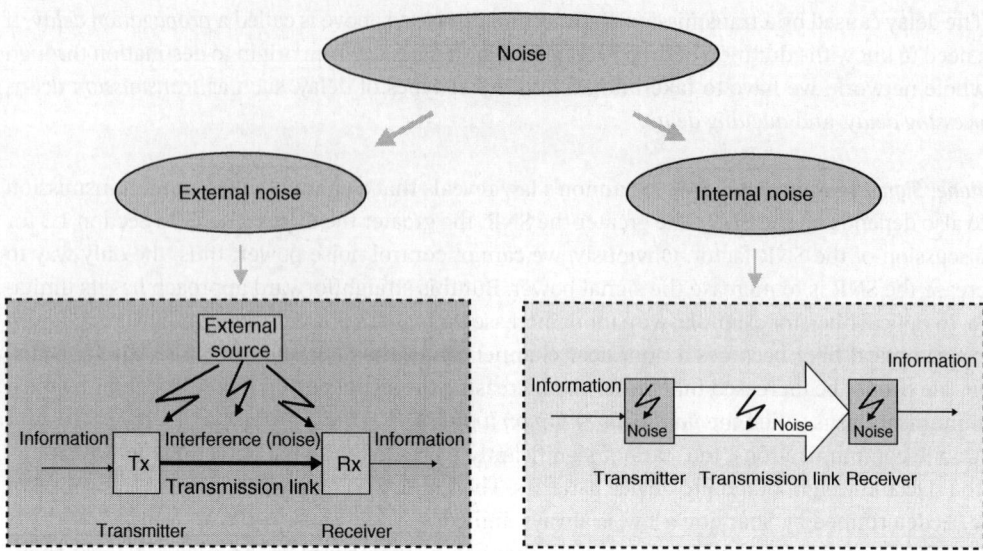

Figure 9.1.4 External and internal noise.

been, and still is, the most widely used noise model. It covers many practical cases, but it does not apply to all the situations we encounter in practice. To analyze this model closely, we need to present a short primer on noise.

What is *noise* in modern communication systems? It is a *harmful EM radiation that distorts the information signal*. Noise originates from two different causes: (i) external, such as natural or man-made and (ii) internal, such as electronic circuits and devices that make up the system. Thus, we distinguish between *external (extrinsic) noise* and *internal (intrinsic) noise*, as shown in Figure 9.1.4.

External Noise This is any radiation from outside sources that interferes with a transmitting signal and distorts it. Lightning is one such source of stray electrical signals. Other natural sources of such disturbances are the sun and other cosmic bodies that generate electrical and optical radiation, which may interfere with transmitted signals.

The main external sources of interfering signals, however, are not natural but man-made. For example, EM waves radiated by radio and TV broadcasting stations interfere with communication signals being transmitted wirelessly. Electrical motors in elevator systems and even the electronic equipment in your college's laboratory generate their own external noise. You can no doubt come up with your own list of troublesome sources. Noise originating from manufacturing equipment is especially a problem in transmission. This is because these sources are close to transmission lines and their signals can therefore distort transmitting signals significantly. In addition, most of these sources are in continuous operation and, as a result, generate noise without letup. Not all transmission channels, however, are equally susceptible to external noise. Air, of course, is the transmission channel least immune to noise, a twisted pair is receptive to EM waves of kilohertz and megahertz ranges, and coaxial cable is better protected from EM radiation than a twisted pair but is still vulnerable. The most immune to external EM radiation channel is optical fiber. A well-designed fiber-optic cable protects optical fiber from any EM radiation from zero to X-ray.

External noise can be significantly reduced or even virtually eliminated by careful design, such as the screening of a device or link. Examples of screening include coaxial cable and a metallic case for any measuring devices, such as oscilloscopes, function generators, etc.

Internal Noise This noise is generated by the components of electronic circuits. These include passive components (conductors, resistors, inductors, and capacitors) and active components (transistors and ICs). Ideal reactive components (L and C) do not generate noise. In contrast to external noise, internal noise cannot be eliminated by simple protective measures. However, careful circuit design helps to reduce internal noise substantially.

There are three main types of **internal noise**: *thermal, shot, and flicker*.

Thermal noise is the *deviation of an instantaneous number of electrons from their average value because of temperature change*. Thus, any conductor, including resistors and semiconductor devices, is the source of thermal noise, which is also referred to as *Johnson* or *white noise*. Its cause is as follows: Flowing electrons (which constitute current) interact with the lattice atoms of a conductor. This random interaction results in a variation of the instantaneous value of electrons traversing a given cross section of the conductor, that is, in a variation of the instantaneous value of the current (or voltage). The higher the temperature, the greater the motion of the lattice atoms, impeding the electron flow and increasing the variation of the instantaneous number of electrons. Therefore, thermal noise (i) is a random process by its very nature and (ii) increases with an increase in temperature.

The *average power of thermal noise*, P_{th}, is given by

$$P_{th} (W) = kTBW \qquad (9.1.7)$$

where $k = 1.38 \times 10^{-23}$ W/(K × Hz) or (J/K) is the Boltzmann[1] constant; T (K) is absolute temperature in kelvins, K [T (K) = 273 °C + X(°C)]; and BW (Hz) is the bandwidth in which the measurement is made. Equation (9.1.7) shows that this noise is called "thermal" because its average power is directly proportional to temperature. Note that thermal noise power is independent of circuitry resistance, but it is proportional to the bandwidth. For example, a resistor at room temperature (27 °C) whose measured BW = 1 MHz generates P_{th} = 0.0041 pW.

Figure 9.1.5 shows the results of an industrial measurement of thermal noise. Figure 9.1.5 demonstrates that thermal noise power is evenly distributed across the spectrum, which means that thermal noise can be modeled by white noise. It also shows that an increase in noise measured bandwidth will cause the increase in thermal noise power.

As noted previously, two other *types of internal noise are shot noise* and *flicker noise*:

Shot noise *is a deviation of the number of generated electrons from the average number of electrons*. It is generated by *active components*, such as transistors and ICs. The physical cause of shot noise is simply the finiteness of the charge quantum, which results in statistical fluctuation of the current. For example, a 1-A dc current actually has 57-nA rms fluctuations measured in 10-kHz bandwidth, which means this current fluctuates by approximately 0.000 006%. The shot-noise model assumes

---

[1] **Ludwig Eduard Boltzmann**, (1844–1906), an Austrian physicist who made fundamental contributions to modern science. He is acclaimed for relating thermodynamic and other properties of materials to their atomic structure, the concept known today as statistical mechanics. Using this approach, he developed our current interpretation of the second law of thermodynamics, proving that this law is a statistical statement about probabilities. Based on the statistical method, he devised his well-known formula, $S = k \ln W$, where S is entropy, k is the Boltzmann constant (through which most engineering students know his name), and W is the number of ways the atoms or molecules of a thermodynamic system can be arranged. This equation is engraved on his tombstone in Vienna, Austria.

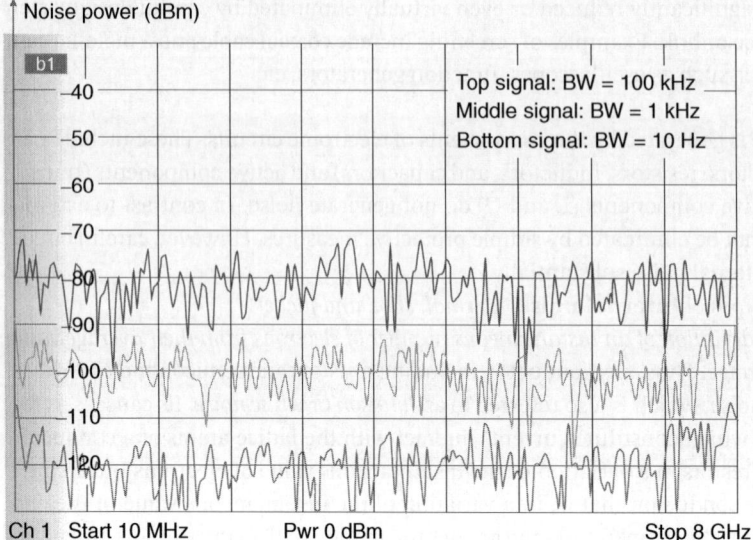

Figure 9.1.5 The measurements of thermal noise at various bandwidths. Source: Reprinted with permission of Rohde & Schwarz (Fundamentals of Vector Network Analysis (Primer), http://www.rohde-schwarz.com).

that charge carriers making up the current act independently as they cross the potential barrier (p–n junction). In this model, shot-noise power, represented by the square of the rms value of average noise current, $(I_{rms}^2)_{shot}$, is given by the Schottky formula

$$(I_{rms}^2)_{shot} = 2qI_{dc}\text{BW} \tag{9.1.8}$$

Here $q = 1.6 \times 10^{-19}$ coulombs is the charge of an electron, I_{dc} (A) is the bias current, and BW (Hz) is the bandwidth in which the measurement is made. Therefore, shot noise (i) is a random process by its very nature, (ii) increases with an increase in the bias current, and (iii) is flat over the frequency spectrum; that is, it is white like thermal noise.

Both thermal and shot noises are generated by electronic components according to their physical principles. We *cannot* reduce them by an improved design or through better manufacturing of the devices and the materials.

Electronic components and devices are additionally the sources of **flicker noise**, which is also referred to as *excess* or *pink* or *1/f noise*. This noise is caused by imperfections in electronic devices, such as fluctuations in the resistance of resistors and bias current in transistors. Despite the variety of sources, flicker noise's essential power is concentrated in the low-frequency range (<1 kHz typically) and is always inversely proportional to frequency. There is no precise formula for noise power since this noise is device-specific. However, if we represent noise power by the square of the average rms value of the current, $(I_{rms}^2)_{flicker}$, then the general formula is given by:

$$(I_{rms}^2)_{flicker} = (kI^m \text{BW})/f^n \tag{9.1.9}$$

Here I (A) is the current in the device, k is a constant (specific for a particular device), BW (Hz) is the bandwidth of a measurement, f (Hz) is an operating frequency, and m (between 0.5 and 2) and n (near 1) are the coefficients. So far, we have concentrated on finding the power of all types of internal noise and have mentioned their spectral distribution. To complete this explanation, we present in Figure 9.1.6 the whole picture of the spectrum of internal noise.

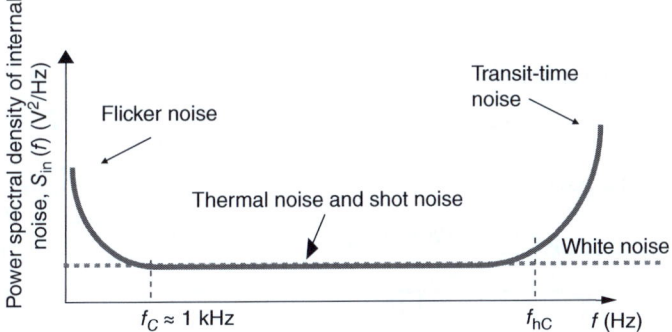

Figure 9.1.6 The spectrum of internal noise.

The power of flicker noise is significant only in the low-frequency range; it lessens with an increase in frequency and typically can be neglected beyond $f_C \approx 1$ kHz. The power of thermal noise and of shot noise is relatively small and spreads uniformly across the spectrum (theoretically, from 0 to infinity). At high frequencies, starting from the device-specific high-frequency cutoff, f_{hC} (Hz), the power of internal noise rises again because of the transit-time effect when the transit time of the charge carriers crossing the p–n junction becomes comparable to the signal's period.

Can we eliminate or at least decrease internal noise? It would seem that since this noise is generated by the same circuitry that is used for processing a signal, the answer is no. However, analysis of (9.1.7)–(9.1.9) show that although we cannot eliminate this noise power by making the right-hand sides of these equations equal to zero, we can control one parameter: the bandwidth, BW (Hz), involved in signal transmission. Minimizing BW significantly reduces the noise power being added to the signal power and therefore improves the SNR.

Examining the internal-noise spectrum in Figure 9.1.6, we understand, for example, that by using a high-pass filter, we can remove flicker noise from the signal being transmitted provided that the message's spectrum is above 1 kHz. (This is illustrated in Figure 9.1.7a.) The harmful effect of thermal and shot noise can be reduced by using a band-pass filter (BPF) as shown in Figure 9.1.7b. The BPF can save the whole transmitting signal and reject the noise components outside the filter's passband. (See Chapter 5 to refresh your memory on filters.) This operation can significantly improve the SNR for the signal being transmitted. For instance, Figure 9.1.7b shows the spectrum of a DSB AM signal (reproduced from Figure 8.2.9) and the BPF's amplitude response. It can be seen that the BPF enables the signal to pass with some portion of thermal and shot noise. However, most of the internal noise whose spectrum is outside the BPF's passband, $f_{C2} - f_{C1}$, is rejected; that is, the noise power contributing to this SNR becomes significantly less. This is how filtering can decrease the harmful effect of internal noise.

Figure 9.1.7b demonstrates an important point: It is not the power of signal or power of noise that plays a vital role in the operation of a communication system; it is the ratio of those powers known as SNR that determines transmission quality.

9.1.1.6 An Amplifier in a Transmission Channel: Internal Noise, SNR, and Noise Figure

Examine Figure 9.1.4 again: It shows that besides a transmitter and a receiver, internal noise is generated by a transmission link. A bare transmission link made of copper wire (a twisted pair), coaxial cable, optical fiber, or air is a passive device. What kind of internal noise can it generate? For copper-based links it is, obviously, thermal noise. For optical fiber, there are a number of internal phenomena negatively affecting signal transmission that can therefore be considered noise. Air

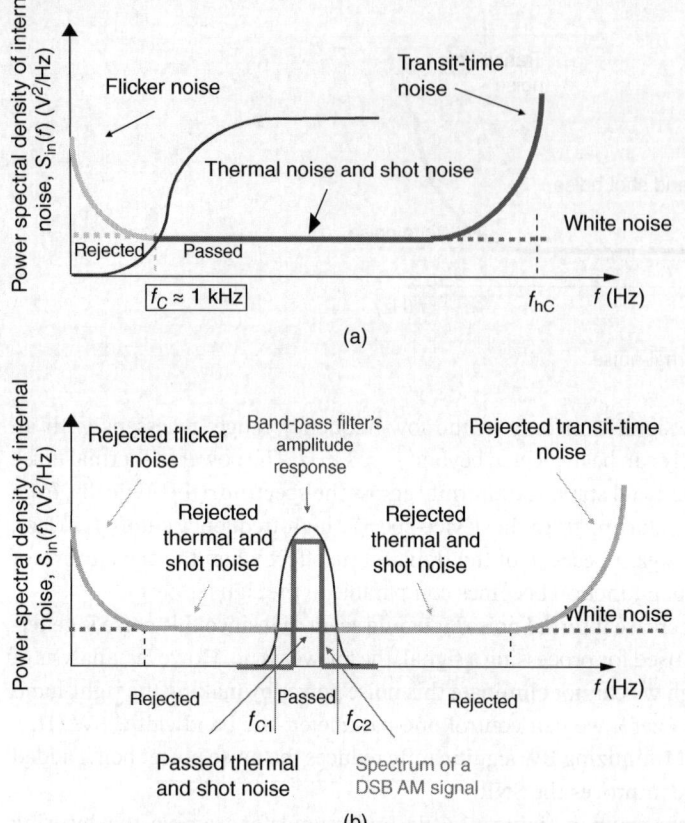

Figure 9.1.7 Reducing the harmful effect of noise by filtering: (a) Using a high-pass filter to eliminate flicker noise; (b) using a band-pass filter to increase the signal-to-noise ratio of a DSB AM signal.

does not generate internal noise per se with respect to EM transmission; all signal disturbances in wireless communications are caused by external sources.

In addition to a bare transmission link, a communication channel in general includes other modules supporting the operation of a communication network. The first such module is of course an *amplifier*, which is a part of almost all communication channels.

Any amplifier, regardless of its makeup, does one thing: *It transfers energy, or power, supplied by an external source into the signal being amplified.* Take, for example, the operation of a basic common-emitter amplifier by sketching its schematic: The base (input) variable signal simply changes the resistance of the p–n junction, thus enabling variable current to flow from V_{CC} to ground, resulting in greater variable output voltage. If the transistor gain, $A = 300$, and the input signal is $V_{in} = 0.1$ V, what will be the output voltage? Students typically compute $V_{out} = V_{in}A = 30$ V, which is wrong. The correct answer depends on the value of the supply voltage, V_{CC}. If $V_{CC} = 5$ V, then the correct answer is $V_{out} = 5$ V because an amplifier cannot produce more than the supplied voltage; in other words, $V_{out} \leq V_{CC}$. All this applies to an op-amp, whose maximum output cannot exceed its V_{CC}. (See Section 5.3 for our analysis of the op-amp.)

While transferring the voltage from V_{CC} to V_{in}, an electronic amplifier generates thermal, shot, and flicker internal noise as the preceding discussion explains. Generating internal noise is an inherent (and unfortunate) part of the operation of an electronic amplifier (and any types of amplifier, for that matter).

Another example is an optical amplifier (OA) which boosts an optical signal directly in optical domain. OA amplifies the input signal by stimulating emission. Here is how it works: An input photon stimulates radiation of another photon of the same wavelength (frequency). This duo travels in the same direction and the two oscillate in sync. They then stimulate radiation of another two photons; this action produces four photons, and so on. Remember that light is a stream of photons; therefore, this chain operation results in light amplification or light generation. Stimulated emission is central to a laser's operation. Because of stimulated emission, a laser beam is monochromatic (i.e. one bright saturated color), exceptionally well directed (focused and narrow), and coherent (all photons oscillate in phase). (The term laser is the acronym for Light Amplification by Stimulated Emission of Radiation.)

An OA amplifies an incoming transmitted optical signal by adding stimulated photons to the signal. Indeed, the greater the number of photons in a given stream, the more powerful this stream (the signal being transmitted) becomes. However, in addition to stimulated emission, the same OA produces stray photons, whose wavelengths are close to but different from stimulated ones and whose traveling directions and phases of oscillations are unpredictable. This phenomenon is called spontaneous emission. The part of spontaneous emission that contributes to signal distortion (called amplified spontaneous emission, ASE) is the internal noise generated by an OA. The worst part of this phenomenon is that the same mechanism that provides signal amplification also produces ASE, which means that the greater the OA gain, the higher the power of spontaneous emission.

How well does the internal noise of an op-amp and an OA fit the AWGN model incorporated into Shannon's law? Well, it is certainly additive because it adds to the transmitting signal, but its spectral characteristics are far from those of a pure white-noise model. For one thing, electronic and OAs have limited bandwidths. An op-amp's spectral-range limitation stems from the bandwidth-gain trade-off; in an OA, ASE bandwidth is limited due to the process of its generation. In general, the limitations in noise bandwidths do not change their white-noise nature, but we should be aware of these restrictions in the application of Shannon's law.

Now you may correctly ask, how does the internal noise of an amplifier affect the operation of a communication channel? Let us take an optical-communication channel to answer this question. Study Figure 9.1.8, where the NF of an OA is introduced. The SNRs at the input, SNR_{in}, and the output, SNR_{out}, of an OA are given as

$$SNR_{in} \text{ (dB)} = 10 \log \left(\frac{P_{signal}^{in} \text{ (mW)}}{P_{noise}^{in} \text{ (mW)}} \right)$$
$$SNR_{out} \text{ (dB)} = 10 \log \left(\frac{A \cdot P_{signal}^{in} \text{ (mW)}}{A \cdot P_{noise}^{in} \text{ (mW)} + P_{ASE} \text{ (mW)}} \right)$$

(9.1.10)

Here A is the OA's gain. If an OA were an ideal device, it would amplify the input signal and input noise evenly, and SNR_{out} (dB) would be equal to SNR_{in} (dB). In reality, however, the OA generates its internal noise, ASE, whose power is added to the power of the amplified noise. Now SNR_{out}

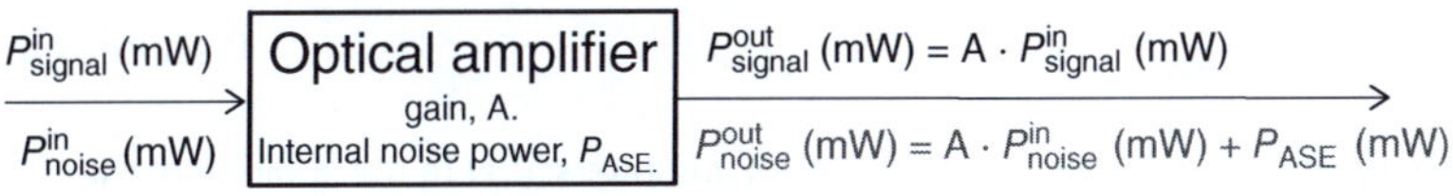

P_{signal}^{in} (mW) $\rightarrow$ P_{noise}^{in} (mW)	Optical amplifier gain, A. Internal noise power, P_{ASE}	P_{signal}^{out} (mW) = A · P_{signal}^{in} (mW) $\rightarrow$ P_{noise}^{out} (mW) = A · P_{noise}^{in} (mW) + P_{ASE} (mW)

Figure 9.1.8 Noise figure of an optical amplifier.

(dB) is not equal to SNR_{in} (dB), as (9.1.10) shows, and so the greater the P_{ASE} (mW), the greater the difference. Therefore, the difference between SNR_{out} (dB) and SNR_{in} (dB) is the measure of how much of the noise power is added by the amplifier itself; this measure is called the *NF* (dB) and it is given by

$$NF\,(dB) = 10\,\log\left(\frac{SNR_{in}}{SNR_{out}}\right) = SNR_{in}\,(dB) - SNR_{out}\,(dB) \tag{9.1.11}$$

Consider an example: If $SNR_{in} = 30$ dB and $SNR_{out} = 25$ dB in an OA, then the NF of this OA is NF = 5 dB, which is a typical value for today's OA. Bear in mind that (9.1.10) and (9.1.11) hold true for any type of amplifier regardless of its makeup; simply substitute $P_{internal\text{-}noise}$ for P_{ASE} in (9.1.10) or use voltage instead of power.

Note that the NF is measured by the *ratio of SNR_{in} to SNR_{out}* in absolute numbers, but NF in decibels is given by the *difference of SNR_{in} (dB) and SNR_{out} (dB)* as (9.1.11) shows.

The detrimental effect of an amplifier's internal noise in modern communications can be understood clearly by examining the fiber-optic transmission link shown in Figure 9.1.9. An initial $SNR_{initial}$ (dB) degrades due to an increase in the total power of the internal noise, to which every in-line OA contributes. NF is a measure of how the SNR deteriorates after each amplification. Assuming that NF is the same for every OA, we find that SNR_{final} (dB) = $SNR_{initial}$ (dB) $- n \cdot NF$ (dB), where n is the number of OAs. On the other hand, the communication industry determines the required value of the SNR_{final} based on the BER needed for a given transmission. Typically, for an optical link operating at BER = 10^{-12}, the minimum SNR_{final} must be 24 dB. Using all this data, we can assess the performance of a transmission channel. For example, If $SNR_{initial} = 38$ dB, NF = 6 dB, and $n = 3$, then the $SNR_{final} = 20$ dB, which means that this optical channel cannot support the industrial requirements for the BER because of the high NF value. In essence, we see that *internal noise generated by an amplifier restricts the channel's transmission distance*. The above points can be applied not only to an optical channel but to a transmission link of any type.

(Question: *In the above example, what NF value will be necessary to meet the 24-dB requirement for the SNR_{final}?)*

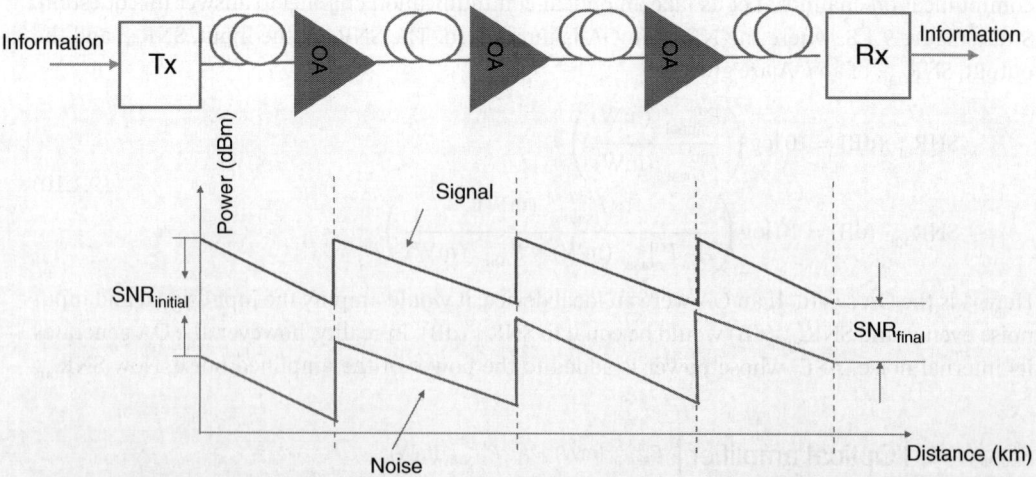

Figure 9.1.9 Fiber-optic communication link with several optical amplifiers: Due to the addition of the internal noise from each amplifier, the final signal-to-noise ratio becomes unacceptable after a specific transmission distance.

A summing up: The above deliberation gives us deeper insight into the operation of a communication system and prepares us to study the methods of assessing transmission quality.

9.1.2 Assessing the Quality of Digital Transmission: The Gaussian[2] (Bell) Curve and the Probability Value

The following subsections discuss how to evaluate the quality of digital transmission. To do so, we need to consider three points: the application of normal probability distribution presented by the Gaussian (bell) curve in finding certain probabilities, the use of an eye diagram in assessing transmission quality, and the probabilistic approach to BER calculations based on the two above mentioned points. We want to explore the formula for BER and explain how it relates to the practical measurements performed in assessing digital-transmission quality.

9.1.2.1 Gaussian (Bell) Normal Probability Distribution

Let us start with a review of the basic terms and definitions:

> A random variable, X, which can take an infinite number of values designated x that correspond to the sampling points on any line interval, is called a continuous random variable.

Incidentally, a line interval with its sampling points is called a *sample space*. This definition says that every time we measure X, it assumes one of the x values, say x_k, or x_m, or x_n. For example, if a sample space is the line interval from $-\infty$ to ∞, sampling X might result in $x_k = 0.47$ or $x_m = -12.86$ or $x_n = 34.92$. Since X is a random variable, we do not know the specific value that it might take; we can only find the *probability* that X will be within a specific interval, say from $x_m = -12.86$ to $x_n = 34.92$.

Now let us consider the Gaussian (bell-shaped or, simply, bell) curve shown in Figure 9.1.10. This curve is an example of the probability distribution (or density) function, PDF, denoted as $f(x)$, where x can run from $-\infty$ to ∞. The Gaussian PDF is often called *normal* distribution. This is a very important example because the Gaussian curve can be met in many engineering, financial, biological, military, and other applications. In fact, there is what's called the *central limit theorem*, which states – loosely put – that if a random continuous variable X represents a set of many independent events (which is often the case), its PDF is a Gaussian distribution.

A Gaussian PDF is described by the following formula:

$$f(x) = \frac{1}{\sigma\sqrt{2\pi}} e^{-\frac{(x-\mu)^2}{2\sigma^2}} \qquad (9.1.12)$$

where μ is the *mean* (expected value) and σ^2 is the *variance* so that σ is the *standard deviation*. These values are calculated by the following formulas

$$\mu = \frac{\sum X}{N} \qquad (9.1.13a)$$

---

[2] The curve was named after Karl Frederick Gauss (1777–1855), a world-renowned German mathematician often referred to as the Prince of Mathematics. He published his work on this curve in 1809. History, however, tells us that the curve was first mentioned in 1733 in the work of Abraham De Moivre (1667–1754), a French mathematician who lived most of his life in England. He was a friend of Sir Isaac Newton and a fellow of the Royal Society. De Moivre made a significant contribution to the early development of the theory of probability. In 1774, Pierre Simon Laplace (1749–1827), one of the greatest of French scientists, published a paper rigorously investigating this curve. Laplace's work in probability and statistics laid the foundation for the modern theory of probability.

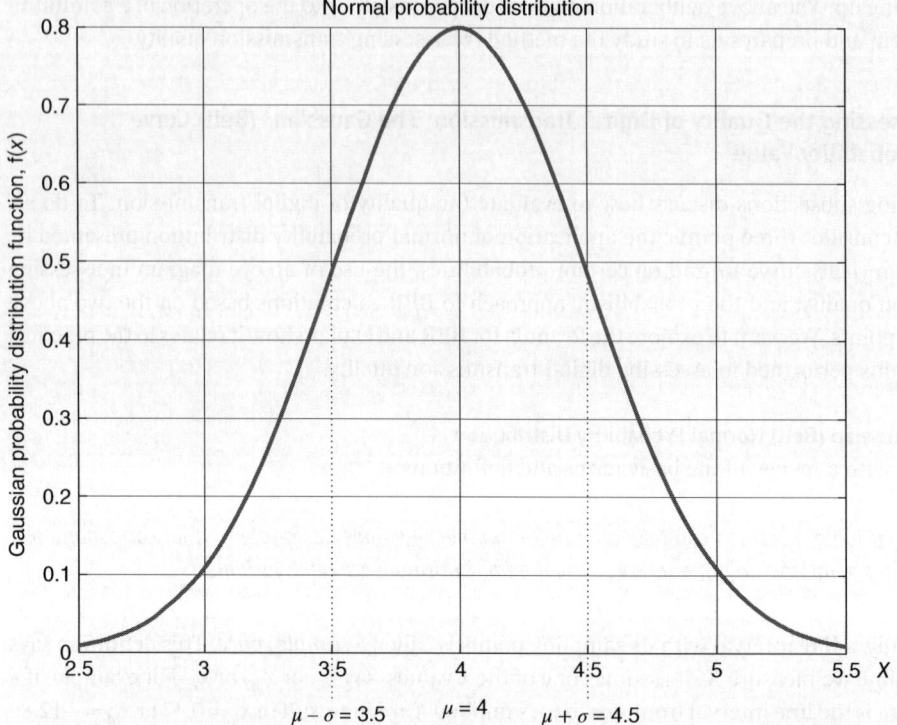

Figure 9.1.10 Gaussian (bell) curve, which is the graphical representation of the Gaussian probability distribution function with mean $\mu = 4$ and standard deviation $\sigma = 0.5$. The figure shows the central position, $\mu = 4$, and the inflection points, $\mu \pm \sigma = 4 \pm 0.5$.

and
$$\sigma = \sqrt{\frac{\sum (X - \mu)^2}{N}} \tag{9.1.13b}$$

Here N is the total number of samples (cases).

The parameters μ and σ are shown in Figure 9.1.10. (Note that at points $\mu + \sigma$ and $\mu - \sigma$, the curvature of the graph $f(x)$ changes from concave downward to concave upward or vice versa. In other words, when $x = \mu + \sigma$ and $x = \mu - \sigma$, the sign of the curvature [concavity] changes. These points, $x = \mu + \sigma$ and $x = \mu - \sigma$, are called *inflection* points.)

9.1.2.2 Finding the Probability Value with the Bell Curve

A PDF, such as the $f(x)$ curve shown in Figure 9.1.10 and given in (9.1.12), does not provide us with the probability of anything; it shows only the concentration of the possible measured values of X along the real axis. To compute the probability of finding X between points $x_a = a$ and $x_b = b$, we need to calculate the area under the curve $f(x)$ within this interval. In short, *the probability that X takes a value within the interval from a to b,*

$$P(b < X < a) \tag{9.1.14}$$

is the area under the Gaussian curve $f(x)$ from a to b.

Figure 9.1.11a shows such an area for the interval from 0 to −5; the area for the interval from 6 to ∞ is also demonstrated there. *The probability that variable x assumes any possible value in the*

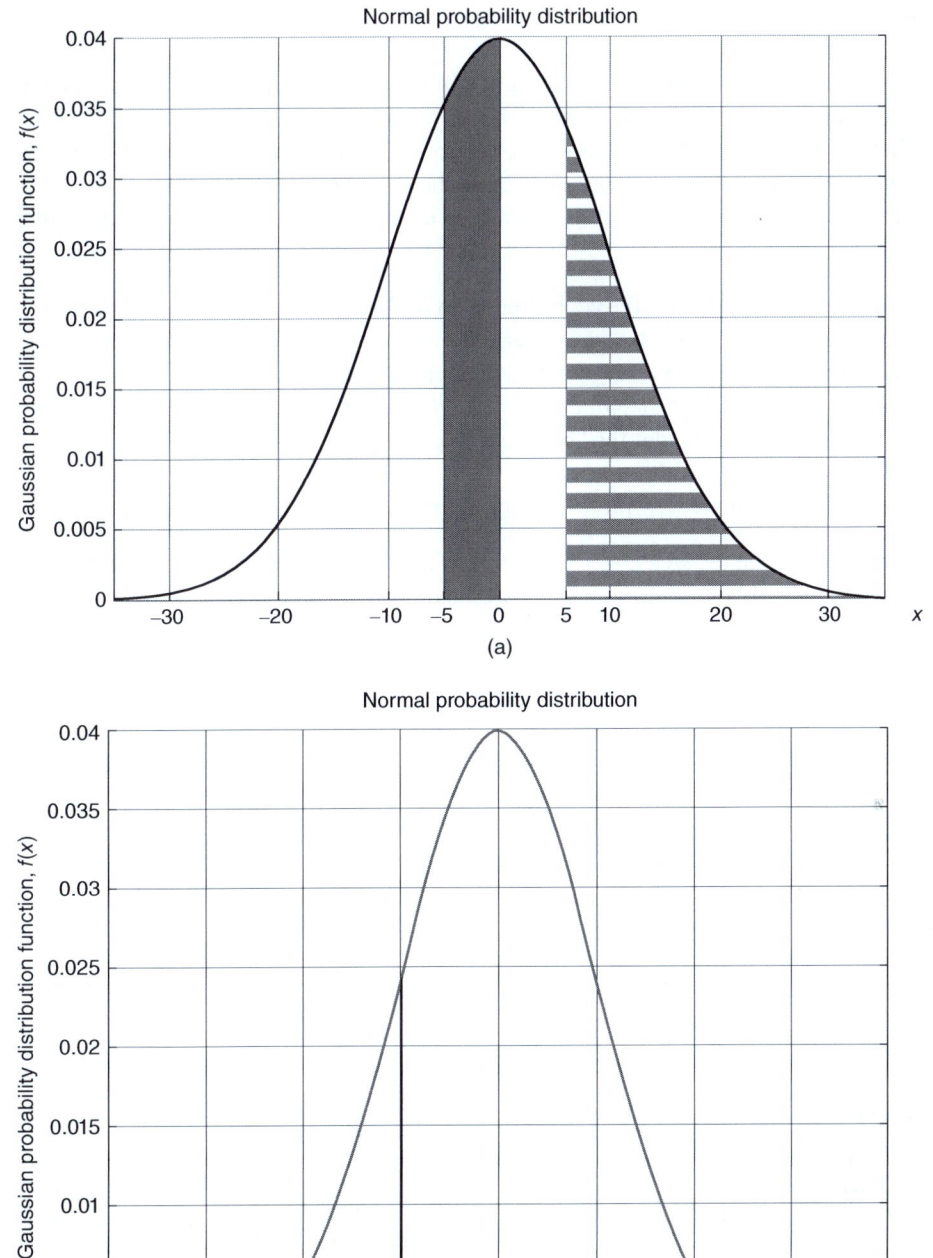

Figure 9.1.11 A Gaussian (bell-shaped) curve representing normal probability distribution: (a) The area under the curve, $f(x)$, for the interval from 0 to -5 is equal to the probability of finding the value of x within this interval. The same is true for the interval from 6 to ∞. (b) The probability of finding X around a specific value, $x = -10$, is equal to $f(x)\, dx$ at $x = -10$.

interval from $-\infty$ *to* ∞ *is equal to one* because this is a sure event, implying that the area under the whole curve is equal to 1.

Can we determine the probability of finding X not within an interval but, rather, being equal to a specific value, say, $x = -10$? No, because the probability is an area under the curve and the area under the line $x = -10$ is zero. We can, however, determine the probability of finding X *around* $x = -10$, that is, within an infinitesimal interval, dx, as shown in Figure 9.1.11b.

The Gaussian curve presented in (9.1.12) describes *normal probability distribution* and has the following key properties:

- The value of the *mean*, μ, defines the central position of the Gaussian curve. In Figure 9.1.10, $\mu = 4$, and in Figure 9.1.11, $\mu = 0$. If μ assumes a different value, the curve will move as a whole to a position with a different center. Figure 9.1.12a demonstrates this point.
- The value of the *variance*, σ^2, or the *standard deviation*, σ, determines the spread (width) of the curve. The greater the σ, the wider the "bell" and vice versa. Figure 9.1.12b reveals that σ is a measure of a Gaussian curve width.
- The Gaussian curve is symmetric with respect to its central position, i.e. to the mean, μ.

This property implies that the area under the curve for an arbitrary interval, $\mu + a$, is equal to the area for the interval $\mu - a$. This, in turn, means that the probability of finding X between μ and a is equal to the probability of finding X between μ and $-a$. In formula form, this becomes

$$P(\mu \leq X \leq \mu + a) = P(\mu - a \leq X \leq \mu) \tag{9.1.15}$$

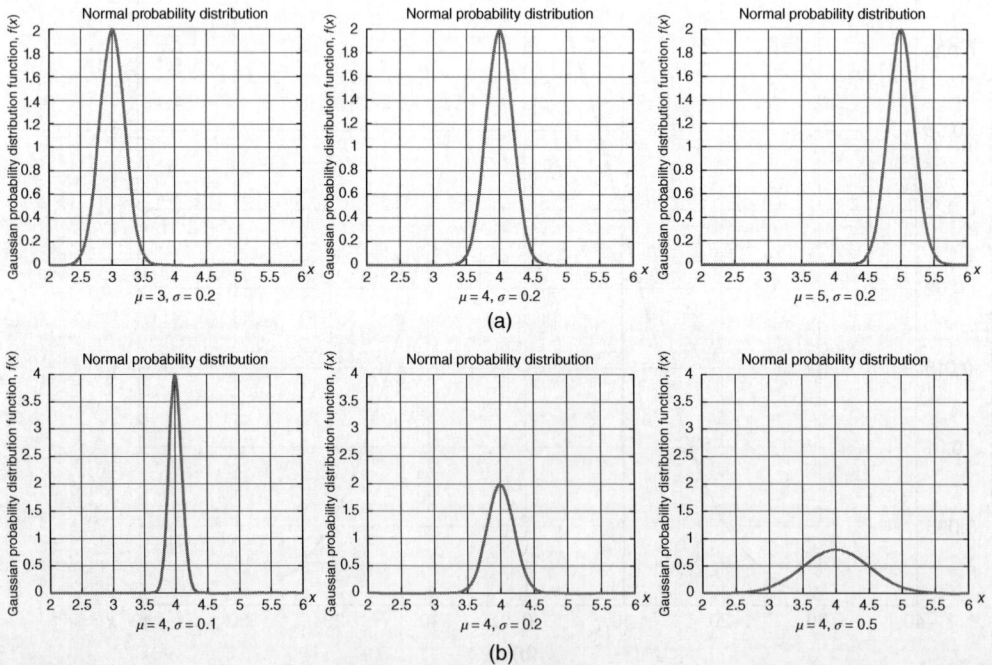

Figure 9.1.12 Gaussian curves with various mean and standard deviation values: (a) The mean value changes from $\mu = 3$ (left) to $\mu = 4$ (middle) to $\mu = 5$ (right), whereas $\sigma = 0.2$ stays constant. (b) The mean $\mu = 4$ stays constant but the standard deviation changes from $\sigma = 0.1$ (left) to $\sigma = 0.2$ (middle) to $\sigma = 0.5$ (right).

Equation (9.1.12) shows that $f(x)$ depends only on two parameters, μ and σ^2. This is why mathematicians use the notation $\boldsymbol{X \sim N(\mu, \sigma^2)}$ to indicate that *a continuous random variable X has a normal distribution with mean μ and variance σ^2*.

Note how the heights of the curves change in Figure 9.1.12b: the greater the σ, the lower the height. This is because the whole area under the curve must be equal to 1 regardless of the values of the curve's parameters. Hence, when the curve becomes wider (σ becomes greater), the height becomes lower to keep the whole area under the curve constant.

9.1.2.3 Standard Normal Probability Distribution

It is clear from the subsection 9.1.2.2 that to determine the probability of finding a continuous random variable between values a and b, we need to calculate the area under the Gaussian curve at this interval. In general, the area under a curve for the interval from a to b described by the function $f(x)$ can be computed by the integration of this function as

$$\text{Area under the curve } f(x) \text{ for } a < x < b = P(a < X < b) = \int_a^b f(x) dx \qquad (9.1.16)$$

In the case of normal probability distribution, we need to plug the $f(x)$ given by (9.1.12) into (9.1.16), which yields

$$P(a < X < b) = \frac{1}{\sigma\sqrt{2\pi}} \int_a^b e^{-\frac{(x-\mu)^2}{2\sigma^2}} dx \qquad (9.1.17)$$

Unfortunately, integral (9.1.17) cannot be evaluated analytically (in closed form). In addition, as follows from (9.1.17), we need to plug in specific values of a, b, μ, and σ to compute the probability for each specific data set. Thus, we must relegate this work to a computer, but it is still a tedious, time-consuming process. The solution could be a table of calculated values, but we would need so many tables of so many values of a, b, μ, and σ that such an undertaking would make this solution impractical. Thus, to solve this computational problem, *a standard normal probability distribution function (a standard normal PDF)* is introduced by bringing in a new variable, z, as follows:

$$z = \frac{x - \mu}{\sigma} \qquad (9.1.18)$$

This transform (9.1.12) into the form

$$f(z) = \frac{1}{\sqrt{2\pi}} e^{-\frac{z^2}{2}} \quad \text{for } -\infty < z < \infty \qquad (9.1.19)$$

Equation (9.1.19) shows that this PDF has the mean value zero and the standard deviation equals 1; i.e. $\mu_s = 0$ *and* $\sigma_s = 1$. This is why the mathematical notation for a standard normal PDF is $\boldsymbol{Z \sim N(0,1)}$.

Based on (9.1.17) and (9.1.19), we can say that calculating the probability of finding Z between sampling points a and b now comes down to evaluating the following integral:

$$P(a < Z < b) = \frac{1}{\sqrt{2\pi}} \int_a^b e^{-\frac{z^2}{2}} dz \qquad (9.1.20)$$

The results of these calculations are tabulated, and *we can apply an obtained table for any μ and σ* by using (9.1.20). In other words, *the real-world probability problems with the Gaussian curve must be converted into standard normal PDF form*. Doing so facilitates our calculations by using the widely available standard table and simple data transformation.

The Table 9.1.1 with the calculated probabilities of finding a random variable Z within any given interval is presented below. You can find such a table in any telecommunications textbook and online.

Using this Table 9.1.1 is quite simple: We set the interval from $\mu = 0$ to the required value of z and determine from the table the probability of finding z in this interval. For example, let us calculate the probability of finding Z within interval $0 < Z < 1.45$. We need to find $z = 1.45$ in the Table 9.1.1 by looking in the column farthest to the left for the first decimal, $z = 1.4$, and in the top row for the second decimal, $z = 0.05$. At the intersection of the proper column and row, we find the number 0.4265. This is the area under the Gaussian curve for the given interval. It means that the probability of finding Z between 0 and 1.45 is equal to 0.4265, or 42.65%. Figure 9.1.13 illustrates these calculations.

We know that the Gaussian curve is symmetrical; thus, we do not need a table for negative values of Z. Indeed, in our example, the probability of finding z between 0 and −1.45 is exactly the same as for the same positive interval ($0 < Z < 1.45$), that is, 0.4265.

For this section, *we need to calculate the probability of finding the variable Z at a tail of the Gaussian curve; that is, $Z > a$ or $Z < a$.* Consider Figure 9.1.14: As an example, we want to determine the probability of finding Z in the right tail of the bell curve for Z from 2.56 to infinity, i.e. for $Z \geq 2.56$. The problem is that the standard table gives us the probability for Z in the interval from a given value to the center. But since the probability of finding Z anywhere under the curve is 1 (a sure event) and the Gaussian curve is symmetrical, the probability of finding Z from 0 (the center position) to

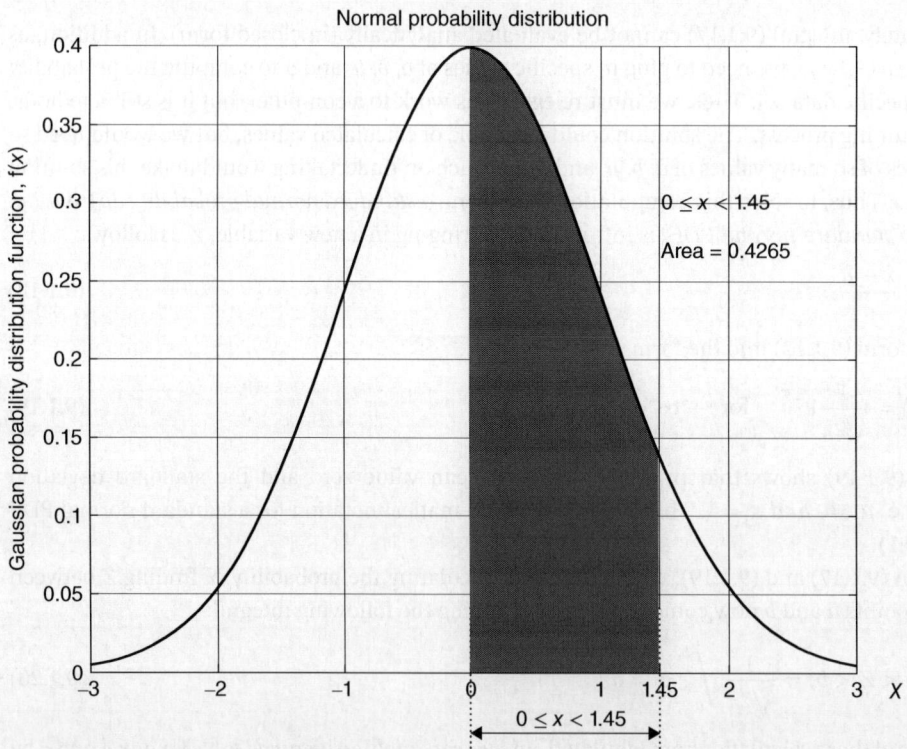

Figure 9.1.13 Calculating the probability of finding random variable z within the interval $0 < Z < 1.45$: This probability is equal to the area under the Gaussian PDF for the given interval.

Table 9.1.1 Standard normal probability distribution function.

	0.00	0.01	0.02	0.03	0.04	0.05	0.06	0.07	0.08	0.09
0.0	0	0.0040	0.0080	0.0120	0.0100	0.0199	0.0239	0.0279	0.0319	0.0359
0.1	0.0398	0.0438	0.0478	0.0517	0.0557	0.0596	0.0636	0.0675	0.0714	0.0753
0.2	0.0793	0.0832	0.0871	0.0910	0.0948	0.0987	0.1026	0.1064	0.1103	0.1141
0.3	0.1179	0.1217	0.1255	0.1293	0.1331	0.1368	0.1406	0.1443	0.1480	0.1517
0.4	0.1554	0.1591	0.1628	0.1664	0.1700	0.1736	0.1772	0.1808	0.1844	0.1879
0.5	0.1915	0.1950	0.1985	0.2019	0.2054	0.2068	0.2123	0.2157	0.2190	0.2224
0.6	0.2257	0.2291	0.2324	0.2357	0.2389	0.2422	0.2454	0.2486	0.2517	0.2549
0.7	0.2580	0.2611	0.2642	0.2673	0.2704	0.2734	0.2764	0.2794	0.2823	0.2852
0.8	0.2881	0.2910	0.2939	0.2967	0.2995	0.3023	0.3051	0.3078	0.3106	0.3133
0.9	0.3159	0.3186	0.3212	0.3238	0.3264	0.3289	0.3315	0.3340	0.3365	0.3389
1.0	0.3413	0.3438	0.3461	0.3485	0.3508	0.3531	0.3554	0.3577	0.3599	0.3621
1.1	0.3643	0.3665	0..3686	0.3708	0.3729	0.3749	0.3770	0.3790	0.3810	0.3830
1.2	0.3849	0.3869	0.3888	0.3907	0.3925	03944.	0.3962	0.3980	0.3997	0.4015
1.3	0.4032	0.4049	0.4066	0.4082	0.4099	0.4115	0.4131	0.4147	0.4162	0.4177
1.4	0.4192	0.4207	0.4222	0.4236	0.4251	0.4265	0.4279	0.4292	0.4306	0.4319
1.5	0.4332	0.4345	3.4357	0.4370	0.4382	0.4394	0.4406	0.4418	0.4429	0.4441
1.6	0.4452	0.4463	0.4474	0.4484	0.4495	0.4505	0.4515	0.4525	0.4535	0.4545
1.7	0.4554	0.4564	0.4573	0.4582	0.4591	0.4599	0.4608	0.4616	0.4625	0.4633
1.8	0.4641	0.4649	0.4656	0.4664	0.4671	0.4678	0.4686	0.4693	0.4699	0.4706
1.9	0.4713	0.4719	0.4726	0.4732	0.4738	0.4744	0.4750	0.4756	0.4761	0.4767
2.0	0.4772	0.4778	0.4783	0..4788	0.4793	0.4798	0.4803	0.4808	0.4812	0.4817
2.1	0.4821	0.4826	0.4830	0.4834	0.4838	0.4842	0.4846	0.4850	0.4854	0.4857
2.2	0.4861	0.4864	0.4868	0.4871	0.4875	0.4878	0.4881	0.4884	0..4887	0.4890
2.3	0.4893	0.4896	0.4898	0.4901	0.4904	0.4906	0.4909	0.4911	0.4913	0.4916
2.4	0.4918	0.4920	0.4922	0.4925	0.4927	0.4929	0.4931	0.4932	0.4934	0.4936
2.5	0.4938	0.4940	0.4941	0.4943	0.4945	0.4946	0.4948	0.4949	0.4951	0.4952
2.6	0.4953	0.4955	0.4956	0.4957	0.4959	0.4960	0.4961	0.4962	0.4963	0.4964
2.7	0.4965	0.4966	0.4967	0.4968	0.4969	0.4970	0.4971	0.4972	0.4973	0.4974
2.8	0.4974	0.4975	0.4976	0.4977	0.4977	0.4978	0.4979	0.4979	0.4980	0.4981
2.9	0.4981	0.4982	0.4982	0.4983	0.4984	0.4984	0.4985	0.4985	0.4986	0.4986
3.0	0.4987	0.4987	0.4987	0.4988	0.4988	0.4989	0.4989	0.4989	0.4990	0.4990

∞ is 0.5. Therefore, the probability of finding Z in the interval from 2.56 to ∞ is simply a difference between the probability of Z being in the right half of the curve, which is $P(0 < Z < \infty) = 0.5$, and the probability of Z being between the center and 2.56, which is $P(0 < Z < 2.56) = 0.4948$ (see Table 9.1.1). In formula,

$$P(Z \geq 2.56) = P(0 < Z < \infty) - P(0 < Z < 2.56) = 0.5 - 0.4948 = 0.0052 \qquad (9.1.21)$$

Thus, the chance of finding Z in the given tail of the curve is only 0.52%. These calculations are illustrated in Figure 9.1.14.

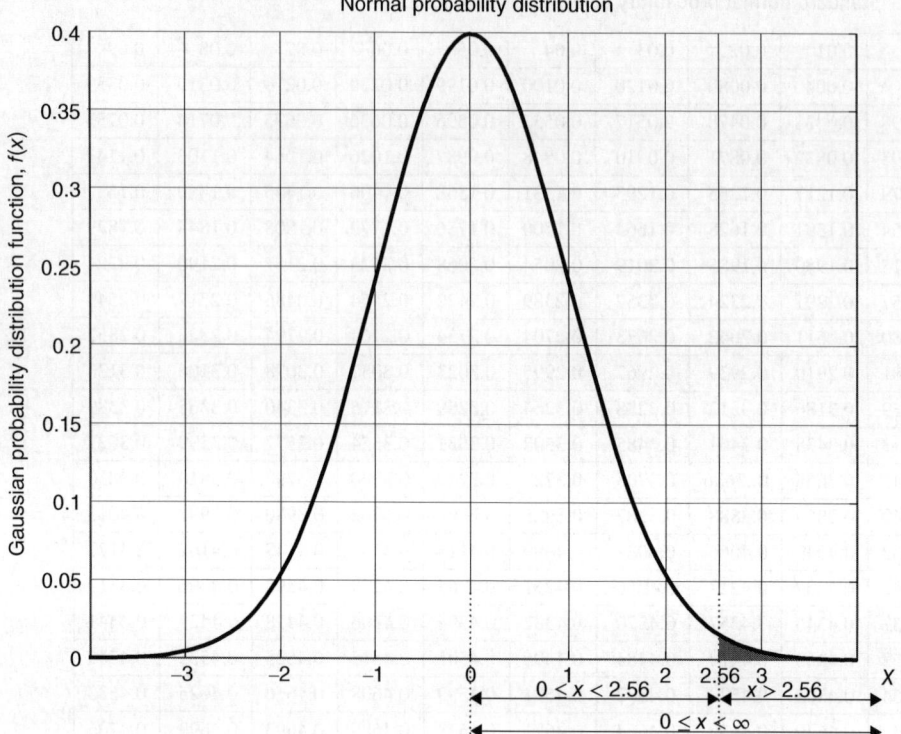

Figure 9.1.14 Calculating the probability of finding Z in the tail of the Gaussian curve.

9.1.2.4 The Gaussian Curve and Q-Function

We achieved our goal to determine the probability of finding Z in the Gaussian tail; i.e. $Z > a$. However, our approach involved two steps, whereas the definition of finding the required probability (9.1.19) suggests that there is a straightforward one-step method to calculate this probability. Indeed, if we properly change the limits of the integral (9.1.20), we immediately attain our goal. Thus, instead of integrating from a to b, as before,

$$P(a < Z < b) = \frac{1}{\sqrt{2\pi}} \int_a^b e^{-\frac{z^2}{2}} dz \qquad (9.1.20R)$$

we will now integrate from a to infinity, as shown in (9.1.22):

$$P(Z > a) = \frac{1}{\sqrt{2\pi}} \int_a^\infty e^{-\frac{z^2}{2}} dz \qquad (9.1.22)$$

Integral (9.1.22) still cannot be evaluated in a closed form, but the results of the numerical integration are tabulated, as shown below.

Let us denote (9.1.22) as $Q(a)$ and call it the *Q-function*; i.e.

$$P(Z > a) = Q(a) = \frac{1}{\sqrt{2\pi}} \int_a^\infty e^{-\frac{z^2}{2}} dz \qquad (9.1.23)$$

We see that the Q-function is simply a designation of Integral (9.1.23), which calculates the probability of finding variable Z greater than given constant a. Now we understand the meaning of this function (calculating the probability value) and the restriction of its application (to the Gaussian

Table 9.1.2 MATLAB code and the probabilities of finding $P(Z > a)$ for a standard normal gaussian PDF (Q-function).

```
a=0:10;
q=qfunc(a); %Numerical values of the standard. norm. distr.
Q=(exp((-a.^2)./2)./(a.*sqrt(2.*pi)));
P=[a;q;Q];
f = figure('Position',[550 800 300 250]);
rnames = {' ',' ',' ',' ',' ',' ',' ',' ',' ',' ',' '};
cnames = {'a','P(Z>a)=Q(a)','Q(a)'};
% Create the uitable
t = uitable(f,'Data',P','ColumnName',cnames,'RowName',rnames);
% Set width and height
t.Position(3) = t.Extent(3);
t.Position(4) = t.Extent(4);
```

	a	Accurate values by MATLAB $P(Z > a) = Q(a)$	Approximate values by (9.1.24) $Q(a)$
	0	0.5000	Inf
	1	0.1587	0.2420
	2	0.0228	0.0270
	3	0.0013	0.0015
	4	3.1671E−05	3.3458E−05
	5	2.8665E−07	2.9734E−07
	6	9.8659E−10	1.0126E−09
	7	1.2798E−12	1.3050E−12
	8	6.2210E−16	6.3153E−16
	9	1.1286E−19	1.1422E−19
	10	7.6199E−24	7.6946E−24

curve only). The MATLAB code and Q-function values calculated for the integer values of $a = 1, 2, 3, \ldots, 10$ are given in Table 9.1.2. For example, if $a = 3$, then $P(Z > a) = Q(a) = 0.0013$. Remember that you can execute these calculations directly with MATLAB, without resorting to Tables 9.1.1 or 9.1.2.

For $a > 3$, the Q-function can be computed approximately by using the following equation:

$$Q(a) \approx \frac{1}{a\sqrt{2\pi}} e^{-\frac{a^2}{2}} \quad \text{(for } a > 3\text{)} \tag{9.1.24}$$

Table 9.1.2 shows the results of precise calculations of the Q-function with MATLAB (left-hand column) and the approximate values of $Q(a)$ (right-hand column) computed by using (9.1.24). The graph shown in Figure 9.1.15 helps to find the Q-function when its argument, a, is given.

In calculating the probability of finding the variable Z greater than a given constant, a, many textbooks refer to the *error function*, erf(a), instead of a Q-function. The error function is defined as

$$\text{erf}(a) = \frac{2}{\sqrt{\pi}} \int_0^a e^{-\frac{z^2}{2}} dz \tag{9.1.25a}$$

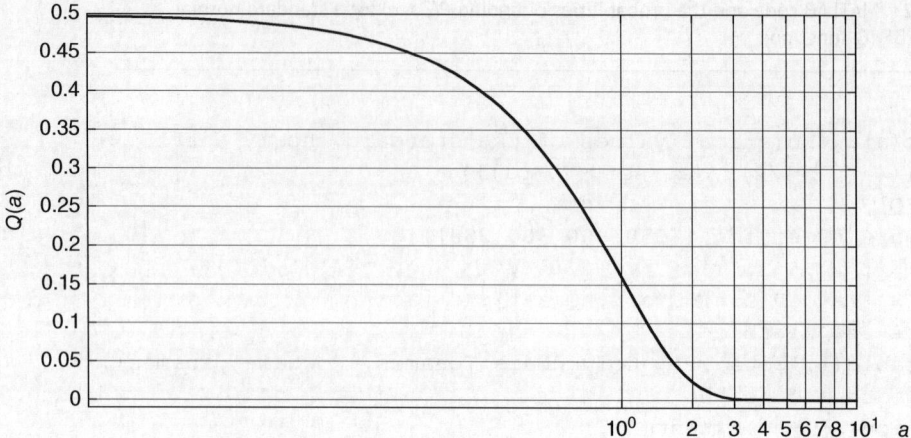

Figure 9.1.15 Graph showing $P(Z > a)$ vs. a, that is, $Q(a)$.

An error function relates to Gaussian PDF as

$$\int_{-x}^{x} G(x)dx = \text{erf}\left(\frac{x}{\sigma\sqrt{2}}\right) \tag{9.1.25b}$$

where we use x instead of a. To calculate quality metrics in communications, research and industrial communities use *complementary error function*, erfc(x), defined as

$$\text{erfc}(x) = 1 - \text{erf}(x) \tag{9.1.25c}$$

This function is directly related to Q-function as

$$\text{erfc}(x) = 2Q(x\sqrt{2}) \tag{9.1.25d}$$

or inversely

$$Q(x) = \tfrac{1}{2}\,\text{erfc}\left(\frac{x}{\sqrt{2}}\right) \tag{9.1.25e}$$

This short detour into the theory of probability is necessary for our understanding of how we can assess transmission quality.

9.1.3 Assessing the Quality of Digital Transmission: Bit Error Rate and More

Why do we need to calculate the probability of finding the variable Z greater than a given constant, a? For evaluation of transmission quality. How? This is the subject of the current subsection, which discusses various approaches to assessing transmission quality and the metrics used for this assessment.

9.1.3.1 Decision-Making Procedure in the Presence of Noise

First, we need to remind you that

> *in digital transmission, a receiver makes the decision as to which bit – 1 or 0 – it received by comparing the value of the received signal with the threshold value, V_{TH}.*

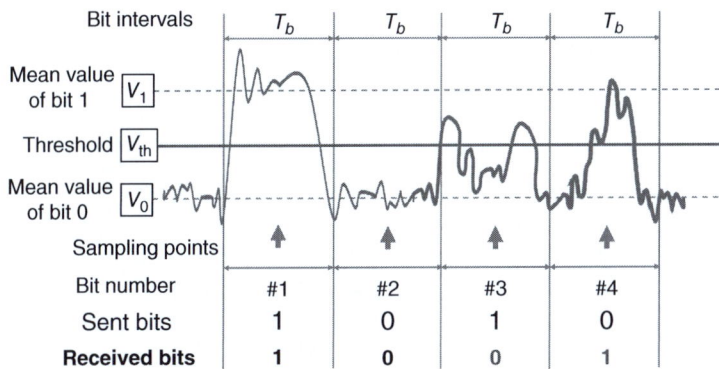

Figure 9.1.16 Decision procedure in digital transmission: Bits #1 and #2 have their signal levels, respectively, above and below V_{TH}, and so the receiver most likely identifies these signals as logic 1 and logic 0. Bits #3 and #4 erroneously have their signal levels below and above V_{TH}, respectively, due to noise; consequently, the receiver might incorrectly identify these signals as logic 0 and logic 1.

Figure 9.1.16 illustrates the decision procedure. Typically, V_{TH} is set in the middle between the mean values of bit 1, V_1, and bit 0, V_0.

Without noise, this approach works well. When noise distorts the received signal, however, a receiver can encounter either of the two situations shown in Figure 9.1.16. First, the received signal levels could be clearly above or below the threshold voltage, as seen in bits #1 and #2. In these cases, the probability that a receiver correctly deciphers bit #1 as logic 1 and bit #2 as logic 0 is very high. However, when noise distorts bit #3 carrying logic 1 so badly that its sampled (measured) level drops below the threshold, the receiver might incorrectly detect this bit as logic 0. When the sampled level of bit #4 is distorted above the threshold, this bit might be misinterpreted by the receiver as logic 1 instead of true logic 0; in other words, the probability of the correct decision on this bit is very low.

This example shows the origin of errors caused by the distortion of received signal by noise. Given that noise is a random process, it is apparent that the associated errors are also random events and they must be evaluated by using the theory of probability. (It is important to realize that Figure 9.1.16 is a *snapshot* of a digital transmission and that such sampling is performed on trillions of bits that are being transmitted every second over modern communication networks.)

To calculate the probability of erroneously detecting 1 instead of 0 and 0 instead of 1, we need to refer to Section 9.1.2, where all the necessary terms and calculations were introduced. To apply that theory, we assume that *noise is a random process with the Gaussian PDF described by* (9.1.12) *with mean $\mu = 0$ and standard deviation σ*. Since the received signal is a sum of transmitted signal and noise, this sum is described by the Gaussian PDF with means V_1 and V_0 and standard deviations σ_1 and σ_0, where V_1 and V_0 are the mean values of bit 1 and bit 0 and σ_1 and σ_0 are the standard deviations of noise at bit 1 and bit 0.

Therefore, *the sampled signal for bit 1 is a random continuous variable described by a Gaussian PDF with mean V_1 and standard deviation σ_1*. Similarly, bit 0 is described by V_0 and σ_0. Figure 9.1.17a shows how the Gaussian curves describe noise at levels 1 and 0. Thus, we, in fact, assume *that a transmitted digital signal is accompanied by AWGN* the model presumed by Shannon when he derived his law.

The mean value of PDF_1 obviously differs from the mean of PDF_0; for instance, the ideal V_1 would be equal to 5 V and the ideal V_0 would be equal to 0 V in non-return-to-zero (NRZ) encoding. The standard deviations σ_1 and σ_0, by their very nature, describe the signals' distribution around the

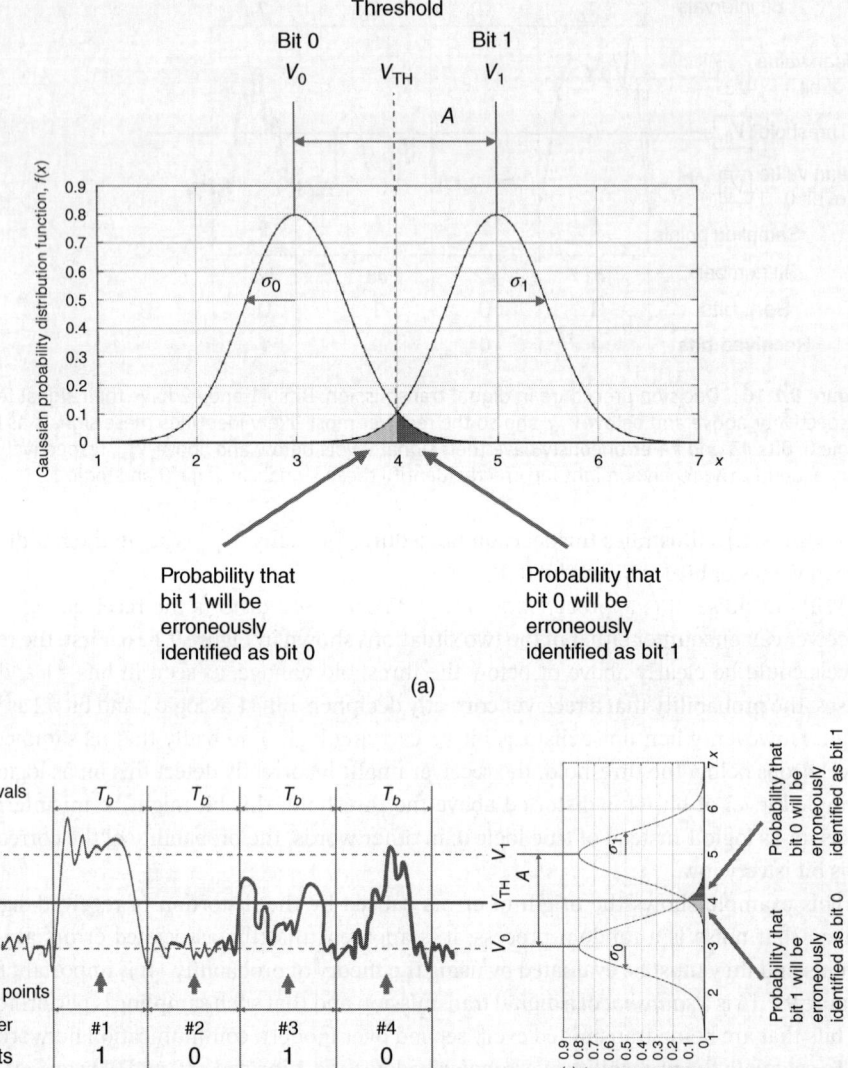

Figure 9.1.17 Concept of the error probability in digital transmission: (a) Gaussian curves as models for noisy bit 1 and bit 0; (b) digital transmission with noise and Gaussian curves.

means V_1 and V_0. Since we have assumed that all variations in signal levels are caused by noise, then, obviously, without noise, σ_1 and σ_0 would hypothetically be equal to zero.

The assumption that this probability distribution is described by the Gaussian curve is quite plausible because, as discussed in Section 9.1.2, the Gaussian curve is a good model for many real processes typified by *white* noise. It describes perfectly the most common type of noise – *thermal* (refer to Section 9.1.1).

Figure 9.1.17a shows *how to apply the Gaussian PDF to the calculation of* error probability in digital transmission. When noise is significant, σ_1 and σ_0 are big enough to make PDF_1 intersect PDF_0. Thus, the left tail of PDF_1 crosses the threshold line and enters the area that belongs to PDF_0.

Hence, the probability that a receiver erroneously identifies bit 1 as bit 0 becomes equal to the area under the tail of curve PDF_1 in the interval from minus infinity to V_{TH}. Likewise, the right tail of PDF_0 crosses the threshold and enters the area of PDF_1; then the probability that bit 0 will be identified as bit 1 is equal to the area under PDF_0 in the interval from the threshold to minus infinity.

It is intuitively clear that the bigger the noise, the wider the PDF (that is, the greater the σ) and the greater the probability to erroneously identify bits 1 and 0. This is because with the increase in the noise level, the larger part of PDF_1 will be in the area of PDF_0 and vice versa. What if the noise power becomes very small? Clearly, the PDF becomes narrow (that is, σ becomes small), so that the smaller part of PDF_1 will penetrate the area of PDF_0 and vice versa. In this case, the error probability tends to zero. We must realize, however, that the tails of a PDF go to infinity and therefore the probability of error will never be equal to zero in the presence of noise (that is, as long as σ does not become zero).

(**Exercise:** Sketch two figures similar to Figure 9.1.17a with (i) big noise power, that is, great σ and (ii) small noise power, that is, small σ. Compare all three figures to visualize the above explanations.)

Figure 9.1.17b combines the origin of errors in the digital transmission of noisy signal with calculations of error probabilities by using Gaussian PDF. Figure 9.1.17b shows how the physical process – the harmful effect of noise on a signal – is described mathematically. Your thoughtful analysis of Figure 9.1.17b is the key to a full understanding of the error calculations and assessing transmission quality based on these calculations.

9.1.3.2 The Probability of Error in Detecting the Received Signal: Bit Error Rate (Ratio)

Our goal is to derive the general formula for BER. But before starting the mathematical manipulations, we need to further clarify how detection operates in a digital transmission. Let us focus on the receiver part of a communications system shown in Figure 1.1.2R, which is reproduced at the beginning of this section. The block diagram of the detector of a digital signal is presented in Figure 9.1.18. Here, the sent signal affected by noise is received by a receiver frontend. The linear filter (integrator) smooths the pulse train over, thus producing a stream of numbers, that is a variable $z(t)$. The sampler (timer) samples this stream within the bit interval T_b and produces the set of samples $z(T_b)$. The decision circuit compares the value of each sample with the given threshold, V_{TH}. Based on this comparison, the circuit decides which bit – 1 or 0 – is received. (Refer to Figure 9.1.16 to see sampling and decision procedures. Bear in mind that the receivers for specific transmission systems will be discussed in the subsequent Section 9.2.)

With this operation in mind, we can turn to the mathematical description of the probability of error. (In the coming manipulations, we follow Hsu 2003, pp. 226–229.) First, let us understand that the variable $z(T_b)$ represents the value of the received signal, which is the sum of the sent signal,

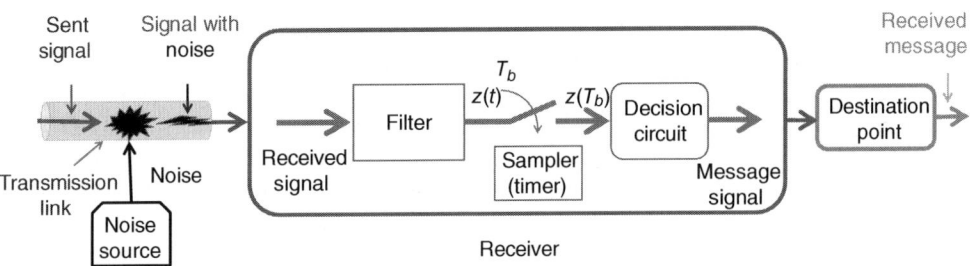

Figure 9.1.18 The receiver part of a communication system: detection of a digital signal.

s_i, and noise, n; that is,

$$z = s_i + n \qquad (9.1.26)$$

where $i = 0, 1$ so that s_0 stands for bit 0 and s_1 stands for bit 1. The structure of (9.1.26) reflects the additive property of noise.

Secondly, as stated in this subsection previously, we consider additive, white, and Gaussian noise (AWGN) with zero mean, $\mu = 0$, and standard deviation, σ. It is described by the Gaussian PDF given in (9.1.12), which in this case takes the form

$$f(x) = \frac{1}{\sigma\sqrt{2\pi}} e^{-\frac{(x)^2}{2\sigma^2}} \qquad (9.1.27)$$

Third, combining (9.1.26) and (9.1.27), we obtain the Gaussian PDF of a received signal with noise as

$$f(s_1) = \frac{1}{\sigma_1\sqrt{2\pi}} e^{-\frac{(z-V_1)^2}{2\sigma_1^2}} \qquad (9.1.28a)$$

and

$$f(s_0) = \frac{1}{\sigma_0\sqrt{2\pi}} e^{-\frac{(z-V_0)^2}{2\sigma_0^2}} \qquad (9.1.28b)$$

Reminder: Here V_1 and V_0 are the mean values and σ_1 and σ_0 are standard deviations of PDF_1 and PDF_0, respectively (see Figure 9.1.17).

The fourth and key point:

> The probability that bit 1 will be erroneously identified as bit 0 is equal to the area under the tail of PDF_1 which penetrates into the "territory" under PDF_0 and extends from $-\infty$ to the threshold, V_{TH}.

(Examine Figures 9.1.17a,b and reread the discussion of these figures.) This area and, consequently, the error probability for bit 1 is given by

$$P_{1e}(z < V_{TH}) = \int_{-\infty}^{V_{TH}} f(s_1) dz = \frac{1}{\sigma_1\sqrt{2\pi}} \int_{-\infty}^{V_{TH}} e^{-\frac{(z-V_1)^2}{2\sigma_1^2}} dz \qquad (9.1.29a)$$

See Eqs. (9.1.20)–(9.1.22) and Figure 9.1.14.
Likewise,

$$P_{0e}(z > V_{TH}) = \int_{V_{TH}}^{\infty} f(s_0) dz = \frac{1}{\sigma_0\sqrt{2\pi}} \int_{V_{TH}}^{\infty} e^{-\frac{(z-V_0)^2}{2\sigma_0^2}} dz \qquad (9.1.29b)$$

The threshold, V_{TH}, has been defined as $V_{TH} = \frac{V_1+V_0}{2}$. It is often called the *optimum threshold*.

The fifth and last step: Let us introduce a new variable, y, as

$$y = \frac{z - V_0}{\sigma_0} \qquad (9.1.30)$$

The new limit for this variable can be found as follows: When $z = V_{TH}$, then $y = \frac{V_{TH}-V_0}{\sigma_0} = \frac{V_1-V_0}{2\sigma_0}$. Also, $dz = \sigma_0 \, dy$. Making these substitutions into (9.1.29b), we obtain

$$P_{0e}\left(y > \frac{V_1 - V_0}{2\sigma_0}\right) = \frac{1}{\sqrt{2\pi}} \int_{\frac{V_1-V_0}{2\sigma_0}}^{\infty} e^{-\frac{y^2}{2}} dy \qquad (9.1.31)$$

Now, compare this equation with (9.1.23),

$$P(Z > a) = Q(a) = \frac{1}{\sqrt{2\pi}} \int_a^\infty e^{-\frac{z^2}{2}} dz \qquad (9.1.23R)$$

and, making the assumption that $\sigma_0 = \sigma_1 = \sigma$, find

$$P_e = \frac{1}{\sqrt{2\pi}} \int_{\frac{V_1 - V_0}{2\sigma}}^\infty e^{-\frac{y^2}{2}} dy = Q\left(\frac{V_1 - V_0}{2\sigma}\right) = \tfrac{1}{2}\operatorname{erfc}\left(\frac{V_1 - V_0}{2\sqrt{2}\sigma}\right). \qquad (9.1.32)$$

This equation enables us to calculate the probability of error based on the parameters of a received signal.

Now, it is time to turn to one of the main topics of this chapter – the metric of digital transmission quality called *bit error ratio (rate), BER*. It is defined as

$$\text{BER} = \frac{\text{Number of received erroneous bits}}{\text{Total number of received bits}} \qquad (9.1.33)$$

It will be shown shortly that BER *is* the probability of error and therefore the formula for calculating BER is given by

$$\text{BER} = P_e = Q\left(\frac{V_1 - V_0}{2\sigma}\right) = \tfrac{1}{2}\operatorname{erfc}\left(\frac{V_1 - V_0}{2\sqrt{2}\sigma}\right). \qquad (9.1.34a)$$

If the standard deviations of PDF_1 and PDF_0 are different, that is $\sigma_0 \neq \sigma_1$, then (9.1.34a) takes the form

$$\text{BER} = P_e = Q\left(\frac{V_1 - V_0}{\sigma_1 + \sigma_0}\right) = \tfrac{1}{2}\operatorname{erfc}\left(\frac{V_1 - V_0}{\sqrt{2(\sigma_1 + \sigma_0)}}\right) \qquad (9.1.34b)$$

Significance of Eqs. (9.1.34a) *and* (9.1.34b) *in digital communications cannot be overstated because it is the general formula for calculating one of the most important transmission quality metrics – the BER.*

Example 9.1.1 Calculating the Probability of Error (BER) in Digital Transmission

Problem

Given $V_1 = 3.1$ V, $V_0 = -2.9$ V, $\sigma_1 = 0.6$ V, and $\sigma_0 = 0.4$ V:

(a) Calculate the average probability of error (BER) in this received signal.
(b) What would this probability (BER) be if σ_0 became 0.6 V?

Solution

Calculations can be done by using (9.1.34a).

(a) The distance between level one and level zero is $V_1 - V_0 = 3.1 - (-2.9) = 6.0$ V. The sum of the standard deviations is $\sigma_1 + \sigma_0 = 1.0$ V. Thus, $\frac{V_1 - V_0}{\sigma_1 + \sigma_0} = 6$. Inspection of Table 9.1.2 yields $Q(6) = P_e = \text{BER} = 9.9 \times 10^{-10}$.
(b) If $\sigma_0 = 0.6$ V, then $\frac{V_1 - V_0}{\sigma_1 + \sigma_0}$ becomes equal to 5. Then $Q(5) = P_e = \text{BER} = 2.9 \times 10^{-7}$.

The problem is solved.

Discussion

- To fully understand this example, visualize every step of the performed calculations by reviewing Figure 9.1.17b. It helps to realize that the distance $V_1 - V_0$ between the means of level one and level zero is the *signal amplitude, A*. Figure 9.1.17b also visualizes the relationship between the signal amplitude and the values of standard deviations: When standard deviations are fixed, an increase in $V_1 - V_0$ results in decreasing the error probability (BER); that is, increasing signal amplitude improves transmission quality.
- An increase in σ reflects the increase in noise intensity, which results in increasing (worsening) the probability of error (BER). This means that the increase in a standard deviation shows the decrease in SNR provided that the signal amplitude does not change.
- Understanding that $A = V_1 - V_0$ describes the signal strength and that $\sigma_1 + \sigma_0$ describes the noise intensity leads to the conclusion that the ratio $(V_1 - V_0)/(\sigma_1 + \sigma_0)$ reflects the level of SNR. This means that the probability of error (BER) directly depends on SNR.

Pay attention to the definition of a signal amplitude, $V_1 - V_0$. When V_0 is a negative number, this difference turns to a sum of V_1 and V_0; that is, $V_1 - (-V_0) = V_1 + V_0$. When V_0 is a positive number, this difference remains the difference between V_1 and V_0. In any event, it shows the distance between the mean values of levels one and zero, that is, the signal amplitude.

9.1.3.3 BER: A Discussion

- What is the meaning of BER? From our discussion in this section, we can conclude that the *bit error ratio is nothing more than the average probability of mistaking received bit 1 for bit 0 and vice versa*. Why? The definition of BER, which is the ratio of the number of erroneous bits to the total number of bits, implies that we know how many erroneous bits we received, but *that is not true*. In fact, we can only know the total number of received bits, but we cannot know how many of them are erroneous due to the nature of the transmission process: A receiver does not know what message a transmitter sends and therefore cannot determine how many bits in error it has detected. It would seem, then, that we can precisely determine BER afterward by arranging a test in which we obtain the sent message and compare it with the received one. But this approach does not work either because transmission is a random process and one snapshot does not fully represent its statistical parameters. We would need to repeat this test thousands of times and to process the obtained data for calculating the main statistical parameters. This is what a bit-error-rate tester (BERT) does. As a result, we can determine only the *average* number of erroneous bits received. Thus, the definition of BER in fact means the following:

$$\text{BER} = \frac{\text{Average number of erroneous bits detected}}{\text{Total number of bits received}} \quad (9.1.35)$$

It is worth comparing this definition with the basic definition of probability, which says

$$\text{Probability} = \frac{\text{Number of actual occurrences of a specific event}}{\text{Total number of possible occurrences}} \quad (9.1.36)$$

Thus, BER is simply the *probability* of determining the actual occurrences of erroneous bits.

In a real transmission, however, a receiver cannot count the average number of erroneous bits because it cannot identify them. *The receiver can only measure the samples of the received signal and, based on these measurements, compute that signal's statistical parameters, such as mean values V_1 and V_0 and the noise levels σ_1 and σ_0, and, eventually, calculate the probability of receiving a bit in error (that is, the expected value of the bit error ratio).* All of this is done based on (9.1.34b), $\text{BER} = Q\left(\frac{V_1 - V_0}{\sigma_1 + \sigma_0}\right)$.

- To fully interpret (9.1.34b), we need to refer to Table 9.1.2 and Figure 9.1.15. They show that Q-function (BER, that is) decreases (improves, that is) with an increase in its argument, $\frac{V_1-V_0}{\sigma_1+\sigma_0} = \frac{V_1-V_0}{2\sigma}$. Therefore, BER improves with an increase in distance $A = V_1 - V_0$ between the mean values of bit 1 and bit 0 (that is, with an increase in signal strength or amplitude) and with a decrease in standard deviation σ (that is, with a decrease in the noise level).
- Strictly speaking, one acronym, BER, encompasses two definitions: (i) *bit error **rate***, which is the number of erroneous bits per unit of time, as the word "rate" implies, and (ii) *bit error **ratio***, which is the ratio of the number of erroneous bits to the total number of bits. However, the communication industry often ignores this nuance and uses the acronym interchangeably in both instances. This is because the two definitions, in essence, mean the same thing. How so? Well, if we know the number of erroneous bits during a time interval and also the system's bit rate (the number of bits during the same interval), then we immediately know the bit error ratio. Indeed, let N_e (errbit) be the *number of erroneous bits per time interval, T* (s); then the *rate* of erroneous bits is

$$\text{BER}_e \text{ (errbit/s)} = \frac{\text{Number of erroneous bits, } N_e \text{ (errbit)}}{\text{Time (s)}} \quad (9.1.37a)$$

Similarly, the *total bit rate*, BR_T (bit/s), *is the total number of bits,* N_T (totbit), *during the same time interval, T* (s),

$$\text{BR}_T \text{ (bit/s)} = \frac{\text{Total number of bits, } N_T \text{ (totbit)}}{\text{Time (s)}}$$

Then the *ratio BER* is

$$\text{BER} = \frac{\text{BER}_e \text{ (errbit/s)}}{\text{BR}_T \text{(bit/s)}} = \frac{N_e \text{ (errbit)}/T \text{ (s)}}{N_T \text{(totbit)}/T \text{ (s)}} = \frac{N_e \text{ (errbit)}}{N_T \text{ (totbit)}} \quad (9.1.37b)$$

For example, if we receive 3 bits in error for 100 seconds and the bit rate of our system is 5 Gb/s, then BER_e (errbit/s) $= \frac{3 \text{ errbits}}{100 \text{ s}}$ and BR_t (bit/s) $= \frac{500 \times 10^9 \text{ bit}}{100 \text{ s}}$, which results in BER $= \frac{\frac{3}{100} \text{ (errbit/s)}}{\frac{500 \times 10^9}{100} \text{ (bit/s)}} = \frac{3 \text{ errbit}}{500 \times 10^9 \text{ (totbit)}} = 6 \times 10^{-12}$. And remember, we are talking about received (detected) bits.
- The BER is rightly considered a most comprehensive quantitative measure of digital transmission quality. It has, however, a *limitation*: the BER measures exclusively the probability of mistakenly taking bit 1 for bit 0 and vice versa; that is, BER is a measure of *bit flips* only. The bit flips are caused by noise; this is why BER depends solely on SNR. In reality, errors in digital transmission might occur due to other phenomena; for example, due to signal distortion caused by such impairments as dispersion and intersymbol interference, or simply due to missing or inserting bits.
- The BER is just a measure. Even if we know its value, how can this knowledge help us to improve transmission quality? There is a practical benefit to knowing BER: we can develop forward error correction (FEC) methods to enhance transmission reliability and to control errors despite the presence of noise.
- It is worth mentioning that the BER for every modulation format is described by a specific formula, which will be demonstrated in Sections 9.2 and 10.1.
- There are several *metrics for the assessment of transmission quality* (Freude et al. 2012). The *BER* is considered the most conclusive quality measure; it is, however, sometimes difficult and time-consuming to obtain its precise value. The *SNR* is also a popular measure of transmission quality; it can be obtained by direct measurements. The SNR, nevertheless, is an intermediate

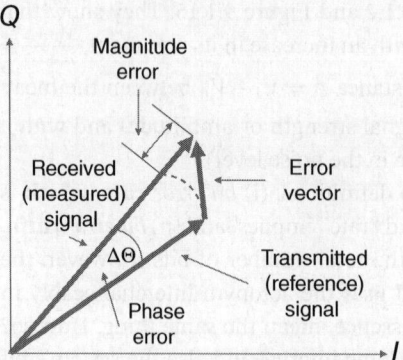

Figure 9.1.19 Error vector magnitude (EVM).

quality measure and needs additional processing for arriving at a final conclusion on a transmission quality. An eye diagram (to be discussed shortly) and the quality factor (QF) which is a measure of its shape, are also the measures of SNR and indicators of the general state of a digital transmission.

- Additionally, the quality metric called *EVM* is often used in modern communications. To introduce the EVM, we need to consider a *signal space* shown in Figure 9.1.19. This space is created by two axes, I (in phase) and Q (quadrature). Three signals – transmitted (reference), received (measured), and error – are presented by their vectors. The error vector reflects *all the impairments* that cause the difference between the referenced (ideal) signal and the measured (real or actual) signal. This difference is measured by the magnitude error and phase error. If amplitudes of all vectors are measured in rms values then the EVM is defined as

$$\text{EVM (dB)} = 10 \log \left(\frac{A_{\text{error}}}{A_{\text{reference}}} \right) \quad (9.1.38)$$

Of course, we refer to average values of all vectors because all of them are random by the nature of communication process. The EVM carries a wealth of information on signal impairments and, in this regard, is a more general quality metric than the BER. The EVM is actively used in wireless transmission and modern optical communications, the technologies that employ the sophisticated modulation formats. (These formats are discussed in Section 9.2 and Chapter 10.) The detailed description of EVM theory, measurements, and applications can be found in literature; some references, including MATLAB, are given in the bibliography section of this book. (See all the bibliography entries that start with "EVM.")
- In general, the quantitative assessment of systems performance, including operation of modern communication systems, becomes a separate branch of informatics and attracts considerable interest from a research community (Balsamo et al. 2018).

We will continue to explore the BER subject in the sections that follow.

9.1.4 Eye Diagram

9.1.4.1 Eye Diagram: The Concept

This discussion is restricted to the digital transmission delivering bits 1 and 0. Specifically, we concentrate on NRZ digital encoding, where bit 1 is carried by a pulse with amplitude A *(V or W)*, and bit 0 is represented as a zero-voltage (or zero-power) signal. These pulses are demonstrated in Figure 9.1.17b. (Refer to Figure 3.1.10 to brush up on transmission codes.)

An eye diagram (pattern) is a visual device that enables us to qualitatively and quasi-quantitatively assess transmission quality.

The concept of forming an eye diagram by overlaying bit 1 on bit 0 is shown on the left-hand side of Figure 9.1.20A. Figure 9.1.20Ab shows eight possible combinations of three pulses and all the possible transitions among those bits. In reality, an eye diagram is formed by overlaying thousands and even millions of transmitted pulses. For a test, the stream of pulses is generated by a pseudo-random binary sequence (PRBS) generator built into a bit error rate tester (BERT). A first glance at the complete eye diagram on Figure 9.1.20Ab reveals how much this pattern resembles the human eye.

Since an eye diagram is formed from millions of pulses, it is a statistical device by its very nature. Figure 9.1.20b shows how the upper and lower levels of an eye diagram, called *eye level one* and *eye level zero*, are formed by summing up the upper levels of all pulses contributing to this eye pattern. Ideally, if the values of all contributing pulses (their heights, that is) were equal, the eye levels would be tiny lines. In reality, the heights (amplitudes) of contributing pulses vary as shown on Figure 9.1.20Ba. These variations make the eye levels look not like single lines but like wide strips – see Figure 9.1.20B. They are caused mainly by noise; therefore, the noisier the signal,

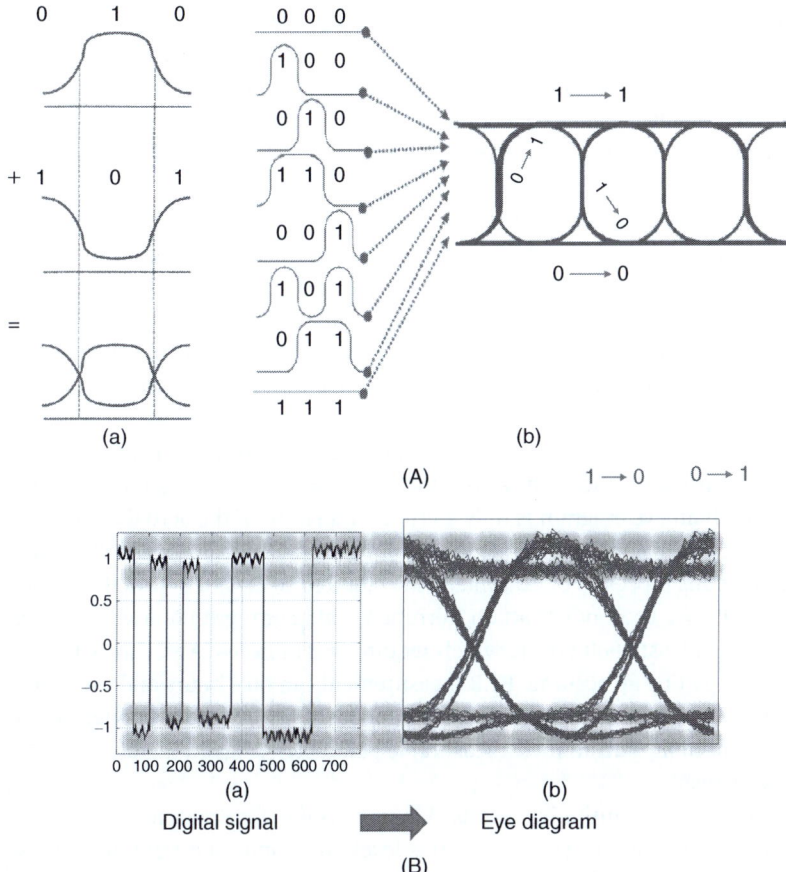

Figure 9.1.20 Eye diagram: (A) The concept of formation and (B) the composition of eye levels from individual pulses.

the wider the levels one and zero of the eye diagram. Bear in mind that the variations in pulse amplitudes are random events; therefore, there is no one-number correspondence between a pulse amplitude and the width of an eye diagram level. This is why the dashes that relate the pulse amplitudes and the eye diagram levels are given in Figure 9.1.20B as fuzzy lines. To understand how these considerations can be turned into measurable parameters, we need to consider histograms.

(Figure 9.1.20B focuses on the variations in levels [amplitudes] of the pulses and shows the ideal transitions [vertical lines] from one level to another. In reality, due to rise and fall times, these transitions are the ascending and descending curves that look like those shown in Figure 9.1.20A and in Figure 9.1.20Bb. To recall the rise and fall time phenomenon, see Figures 9.1.2 and 3.2.4.)

A histogram shows the frequencies of occurrence of various values of a continuous variable by using bars of different heights. Figure 9.1.21a demonstrates the concept of a histogram. In our application, a histogram shows how often upper levels of contributing pulses take on specific values. For example, Table 1 in Figure 9.1.21 shows that the values of an eye's level one (top of a pulse) between -2 and -1 appear three times, values between -1 and 0 appear four time, values between 0 and 1 appear five times, and values between 1 and 2 appear three times. Histogram 1 presents this information in a graphical format. In general, a histogram shows that some values of top pulse levels appear frequently whereas other values appear very seldom. The frequencies of value occurrences are represented by the heights of the corresponding bars. The set of histograms in Figure 9.1.21a shows that when the number of individual independent events (upper levels of streaming pulses in our case) increases, the histogram describing this case assumes a bell-like shape. Clearly, when the number of individual events goes to infinity, the discrete histogram tends to continuous probability distribution. Since all individual events (upper pulse levels in our example) are independent, their distribution will be the Gaussian PDF. The eye level zero is formed and described in similar manner.

Figure 9.1.21b exhibits a snapshot of the oscilloscope screen showing an actual eye diagram. It can be seen that a real eye diagram is formed by fuzzy lines and has a narrow opening due mainly to noise. The greater the noise level, the wider (thicker) these fuzzy lines and the narrower the eye diagram opening. An ideal eye diagram, in contrast, should look like the ones shown on Figure 9.1.20Ab: wide open and with clear delineating lines. In general, *the wider the eye opening, the better the received signal, and vice versa*. Thus, the eye diagram can be used to evaluate transmission quality.

The eye pattern enables us to evaluate the transmission process not only qualitatively but also quantitatively. It must be clearly understood, however, that the quantitative assessment of transmission quality acquired with an eye diagram is only a crude estimation of the actual results. So, why do we still need to use an eye diagram? Consider the example: Obtaining the precise BER value is a difficult and time-consuming process. For example, if the required BER is 10^{-12} and the bit rate is 100 Gb/s, we need $10^{12}/10^{11} = 1.1$ seconds to attain 1 erroneous bit. For representative statistical results, it is necessary to get at least 1000 errors, which require 18.3 minutes. And this is for one test only. By some estimates, getting an accurate BER measurement can take hours and even days. In contrast, an eye diagram can be obtained almost instantaneously. What's more, modern oscilloscopes display eye diagrams alongside all their measurable parameters. Again, the price for this convenience is approximate results.

How can an eye diagram deliver quantitative results? Let us consider, for example, the parameters of eye level one shown in Figure 9.1.21b. To form this level, thousands and millions of pulse measurements are used. The histogram built from these numbers is approximated by a Gaussian PDF, which is shown in Figure 9.1.21b on the right-hand side of the upper level. We know that a Gaussian PDF is described by its mean value and standard deviation. The *mean value of eye level*

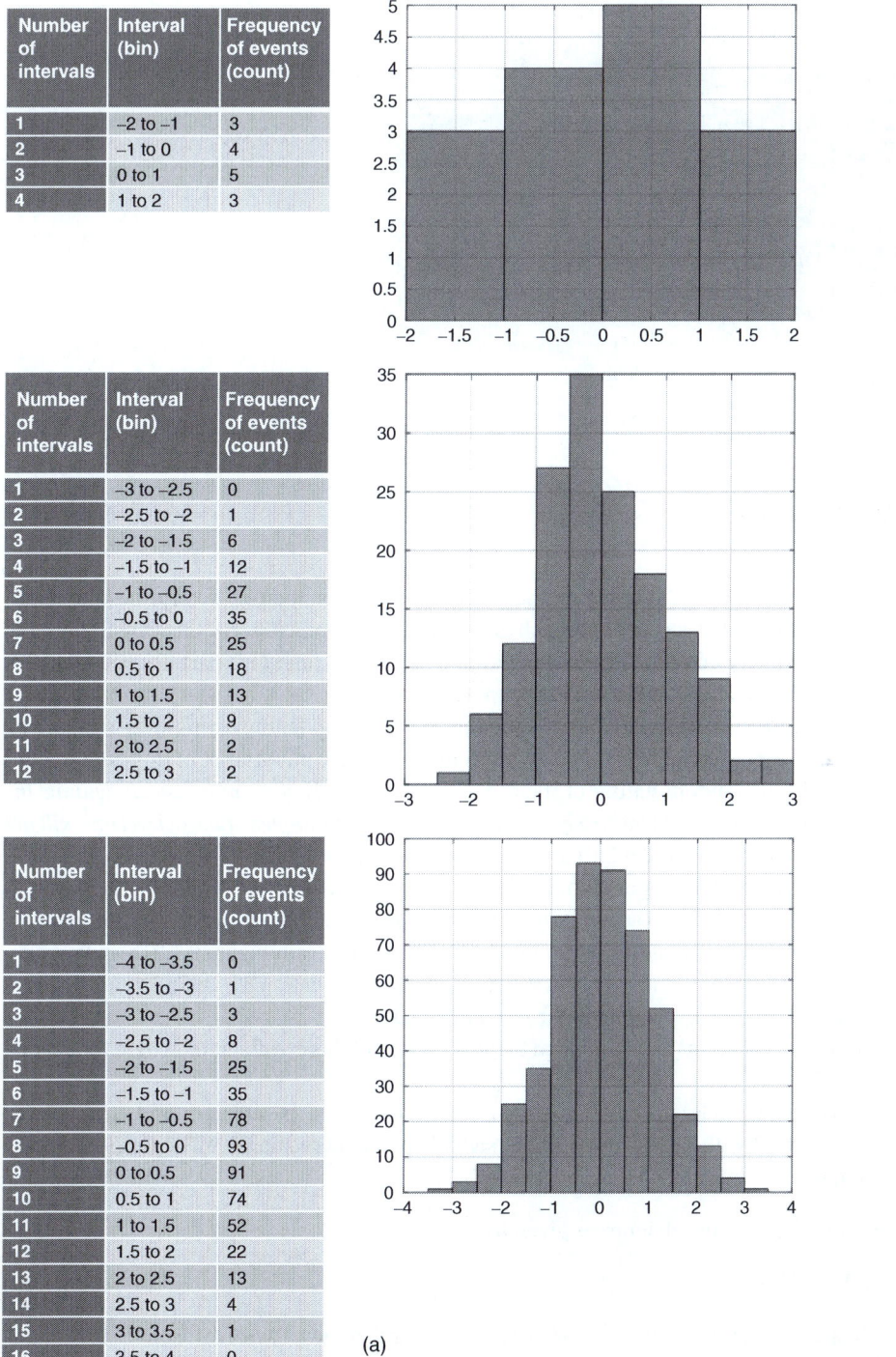

(a)

Figure 9.1.21 Eye diagram and its parameters: (a) The set of histograms showing how Gaussian PDF is formed and (b) the snapshot and all the parameters of an eye diagram. Source: (b) "Agilent Bit Error Ratio Testers" 5988-9514EN, 2003. © Keysight Technologies, Inc. Courtesy of Keysight Technologies.

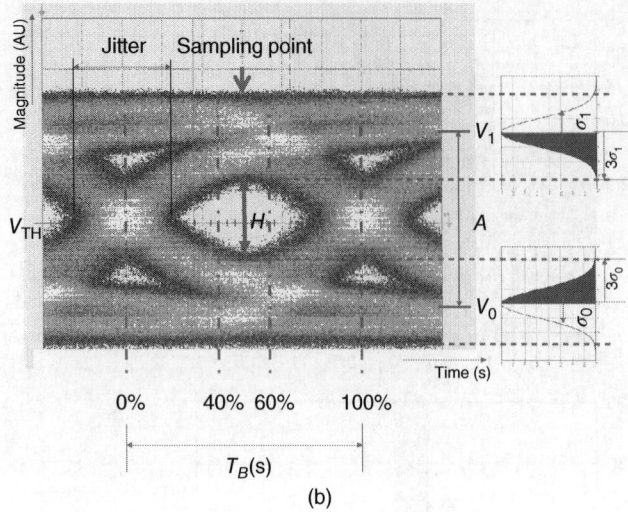

Figure 9.1.21 (Continued)

one, V_1, is calculated by (9.1.13a) based on the measured numbers; then V_1 determines the nominal value of logic 1 (bit 1). A similar approach is used for V_0. (In practice, for this calculation, the communication industry takes data from the middle part of level one. This part lies between 40% and 60% of the bit interval, T_b, as shown in Figure 9.1.21b.)

But how can we calculate the width of eye level one? (Since the widening of this level is caused by noise, this width is a measure of noise effect, which emphasizes the importance of this calculation.) The answer lies in the random nature of the eye diagram: *There is no number which delimits the level one from its top or bottom; there is only the **probability** that a sampled value of level one will not exceed a certain limit.* (See Figure 9.1.13 and its discussion.) If we set this limit as $\pm 3\sigma$ deviations from the mean value μ, then this probability can be calculated as $P(\mu - 3\sigma < Z < \mu + 3\sigma) \approx 0.997$. This implies that the range of $\pm 3\sigma$ accounts for 99.7% of all values of the Gaussian PDF. 3-σ criterion is used by many industries, including communications. Examine again Figure 9.1.21b and observe the Gaussian PDF of eye level one, its mean value V_1, standard deviation σ_1, and $\pm 3\sigma_1$ areas under this bell curve. For this level, the area under -3σ is shaded.

All the above considerations can be applied to the level zero of this eye diagram, where the shaded area is under $+3\sigma$.

Observe that an eye diagram exhibits signal's strength-related parameters and time-related parameters. Specifically, the main measurable signal's parameters shown in Figure 9.1.21b are defined as follows:

- *Amplitude, A (V)*, of an eye diagram is given by

$$A(V) = V_1 - V_0 \tag{9.1.39a}$$

where V_1 and V_0 are the mean values of eye level one and eye level zero, respectively.

- Height of an eye diagram, H (V), the measure of the eye pattern opening, is calculated as

$$H(V) = (V_1 - 3\sigma_1) - (V_0 + 3\sigma_0) \tag{9.1.39b}$$

where $\pm 3\sigma_1$ and $\pm 3\sigma_0$ are the estimates of cumulative noise in levels one and zero, respectively.

- The optimum threshold value, V_{TH}, is typically determined as

$$V_{TH} = \frac{V_1 + V_2}{2} \tag{9.1.39c}$$

- *Sampling point* is the instant at which the eye height is measured; it should be at the point where the eye is most open, which is typically at the center of the eye diagram.
- *Jitter* is the variation in bit interval value, T_b, measured at the threshold crossing points. It affects the position of the sampling point and therefore the accuracy of height measurements (see Appendix 9.2.A)
- *The ascending curve from bit 0 to bit 1 and the descending curve from bit 1 to bit 0* are the bases for measuring the rise and fall times, respectively.

(Know that modern oscilloscopes are capable of directly measuring and displaying most of these parameters.)

9.1.4.2 Estimating Transmission Quality with an Eye Diagram

Why do we need to create and scrutinize an eye diagram instead of simply investigating a single pulse to evaluate transmission quality? Because a single pulse is just a snapshot of a whole pulse train. Today's optical networks, operating at the bit rate of 1 Tb/s, transmit 10^{12} bits each second. Obviously, we need to assess this transmission statistically, and one pulse is not a representative measure of the whole process. Since an *eye diagram* is formed by overlaying millions of pulses, *it shows the statistical characteristics of the transmission process*, whereas a pulse represents only an individual event. And remember: All parameters of an eye pattern are determined after sampling millions of bits and processing the results. Our discussion of the formation and the parameters of an eye diagram provides a solid foundation for assessing the transmission quality.

Figure 9.1.21b shows that eye diagram's height, H (V), is much smaller than the eye amplitude, A (V), due to noise. Therefore, the eye diagram is an indicator of SNR of a pulse train being transmitted. It would seem that we can use the ratio H/A or another eye parameter as a measure of SNR but, in fact, the situation is not that simple. Remember, an eye diagram is nothing more than a visual representation of the result of processing a stochastic transmission process. In reality, collecting, processing, and obtaining parameters of this process, including building an eye diagram, is not a simple task; it requires a special equipment and hours of processing. Scientists and engineers work hard to develop faster, more timely ways of performing this procedure. (See, for example, Ippei Shake et al. 2004, pp. 1296–1302.) Typically, such a procedure results in measuring the *quality factor, QF*, defined as

$$QF = \frac{V_1 - V_0}{\sigma_1 + \sigma_0} \tag{9.1.40a}$$

Here, as before, V_1 and V_0 are the mean values of eye level one and eye level zero, respectively and σ_1 and σ_0 are the standard deviations of these levels. (Be aware that many textbooks, research papers, and technical documents denote quality factor as Q. We designate it QF to avoid confusion with the Q-function.) The QF definition (9.1.40a) is supported by Recommendation ITU-T G.976 of the Telecommunications Sector of the International Telecommunication Union. In decibels,

$$QF\,(dB) = 20\,\log(QF) \tag{9.1.40b}$$

In reference to eye diagram parameters, the QF can be generally seen as a direct measure of the SNR, namely

$$QF = SNR = \frac{V_1 - V_0}{\sigma_1 + \sigma_0} \tag{9.1.40c}$$

In optical communications, however, relationship between QF and OSNR (optical SNR) depends on the photodetector optical bandwidth, BW_{op}, and electrical bandwidth of receiver electronics, BW_{el}. In logarithmic scale, it is given by

$$QF\,(dB) = OSNR\,(dB) + (BW_{op}\,(dB) - BW_{el}\,(dB)) \qquad (9.1.40d)$$

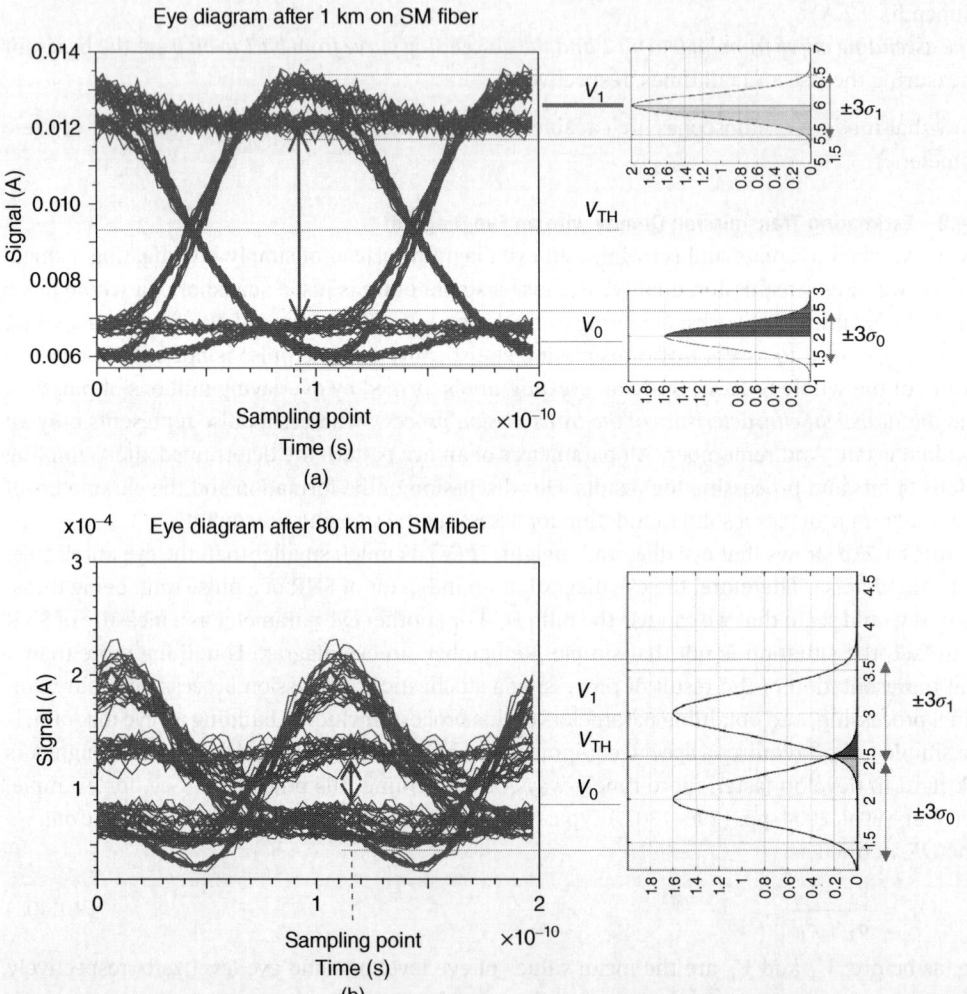

Figure 9.1.22 Eye diagram and the probability of errors: (a) the eye diagram is wide open, the two bell curves are far apart, and the tails barely intersect, meaning that the probability of deciphering bit 1 as bit 0 and vice versa is very low; (b) the eye diagram is open only slightly, the two bell curves are close to each other, and the tails intersect significantly; therefore, the probability of erroneously detecting bit 1 as bit 0 and vice versa is significant; and (c) the eye diagram is almost completely closed, the two bell curves are overlapping to a larger degree, and the tails intersect substantially, meaning that the probability of incorrectly identifying bit 1 as bit 0 and vice versa is high (not to scale). (Eye diagrams in Figure 9.1.22a–c is simulated with RSoft's LinkSim program.)

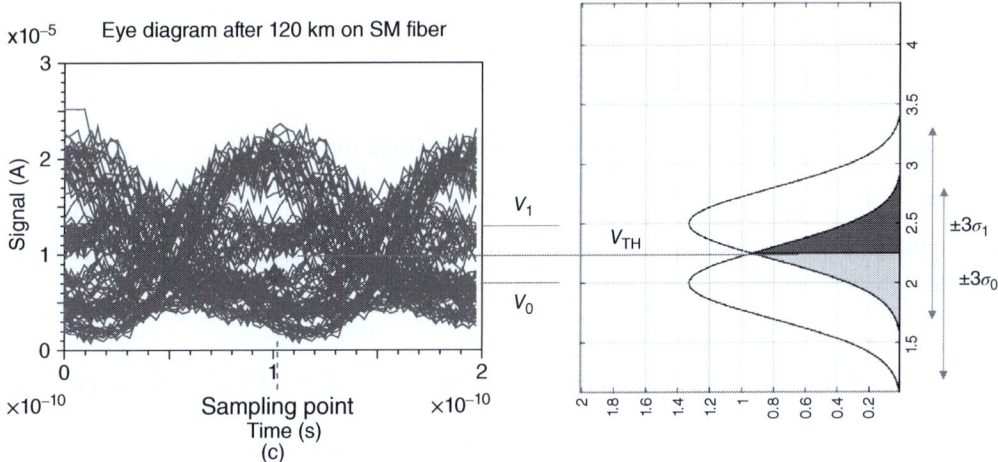

Figure 9.1.22 (Continued)

The quality factor is a measurable parameter which enables us to calculate the BER as

$$\text{BER} = \tfrac{1}{2}\,\text{erfc}\left(\frac{QF}{\sqrt{2}}\right) = Q(QF) = Q\left(\frac{V_1 - V_0}{\sigma_1 + \sigma_0}\right) \tag{9.1.41}$$

where Q-function and $erfc$ function are defined in (9.1.23) and (9.1.25a), respectively. Refer to (9.1.34a) where BER is introduced. Relate the BER derived from probability theory in (9.1.34a) and the BER determined by a measurable parameter – the quality factor – in (9.1.41).

The importance of Eq. (9.1.41) cannot be overestimated because it shows that (i) QF is the factor that determines BER; (ii) BER, in fact, depends exclusively on the SNR; (iii) BER can be calculated based on results of a measurement.

(**Exercise**: In Figure 9.1.21b, identify all the eye diagram parameters needed to calculate (9.1.41).)

To further explore how an eye diagram can be used for BER estimation, let us examine Figure 9.1.22. Figure 9.1.22 shows that when an eye diagram is wide open, the Gaussian PDF_1 and PDF_0 are set far apart, and the probability of erroneously identifying bit 1 as bit 0 and vice versa is low. But when the eye is closed, PDF_1 and PDF_0 are significantly overlapped, and the probability of error is very high. (Revisit Figures 9.1.20 and 9.1.21 to fully understand this explanation.) Such a qualitative assessment of the state of a digital transmission can be turned, as we have learned in this subsection, into an approximate quantitative analysis for computing the BER with (9.1.41).

Thus, Figure 9.1.22 as a whole demonstrates how a physical process – increasing the noise level – is reflected in an eye diagram and related to the Gaussian PDF. This demonstration provides the key to estimating the probability of error, that is, the BER.

(**Question:** *Can you estimate the BER for each of the three eye diagrams shown in Figure* 9.1.22?)

To put this consideration onto solid mathematical footing, refer to (9.1.29a) and (9.1.29b), which calculate the probability of identifying bit 1 as bit 0 when the tail of PDF_1 penetrates the "territory" of PDF_0, and vice versa.

We are concentrating only on those parameters of an eye diagram that are directly related to measuring and calculating BER, whereas the eye diagram, in fact, contains a wealth of other useful information. We encourage you to closely investigate the eye diagram information by reading specialized publications from companies producing measuring instruments.

To conclude this discussion, let us turn to an example:

Example 9.1.2 Estimation of BER by Using an Eye Diagram

Problem

The eye diagram of a received digital signal is shown in Figure 9.1.23. Estimate BER of this transmission.

Solution

We need to calculate the quality factor, QF, for which we have to find all of its parameters. Rough estimation of these parameters can be obtained from the given eye diagram. Let us place a sampling point at $t = 0$ ps, The mean value of eye level one is approximately equal to $V_1 \approx (0.5 + 0.2)/2 = 0.35$ V. Crude estimation of the mean value of eye level zero gives $V_0 \approx -0.35$ V. Thus, $V_1 - V_0 \approx 0.7$ V.

Full-width noise of eye level one is about $0.5 - 0.2 = 0.3$ V. The rule of thumb is that the standard variation of noise PDF is 1/6 of this width; that is, $6\sigma_1 \approx 0.3$ V. Assuming that standard variations are the same for both eye levels, we compute $\sigma_1 = \sigma_0 \approx 0.05$ V.

Applying (9.1.40a), the quality factor is computed as

$$\text{QF} = \frac{V_1 - V_0}{\sigma_1 + \sigma_0} = \frac{0.7 \text{ V}}{0.1 \text{ V}} = 7$$

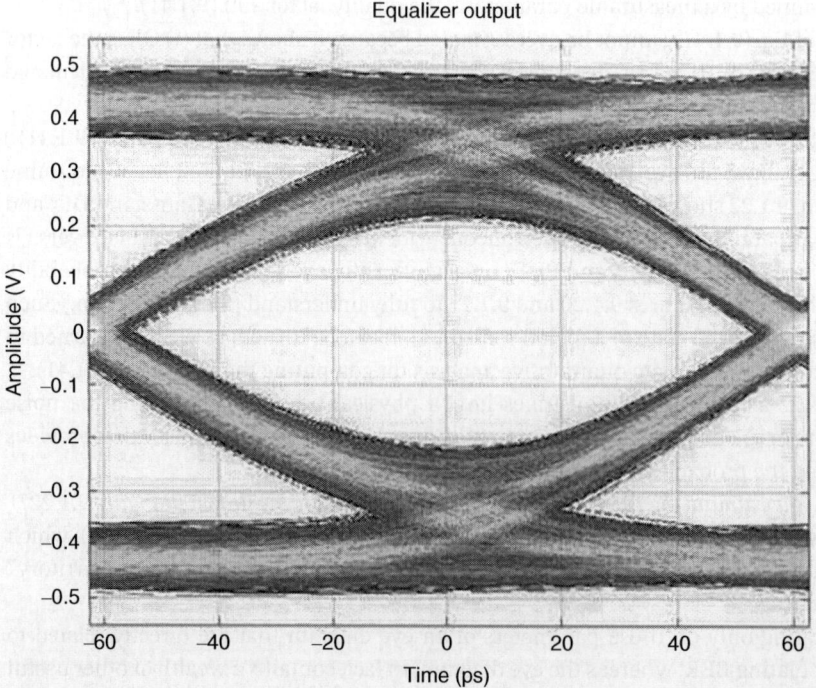

Figure 9.1.23 An eye diagram for Example 9.1.2. Source: Reprinted with permission Navraj Nandra https://blogs.synopsys.com/theeyeshaveit/2012/11/11/reducing-pci-express-3-0-fuzz-with-multi-tap-filters/.

Consulting Table 9.1.2, we find

$$\text{BER} = Q(\text{QF}) = Q(7) \approx 1.3 \times 10^{-12}$$

The problem is solved.

Discussion

- This example is significant, despite its simplicity, because it shows how to use eye diagram parameters to estimate transmission quality. But we still need to bear in mind that this is only a rough guess of the value of the BER.
- Though height of the eye diagram is 2.5 times smaller than the eye amplitude (approximately 0.4 V vs. 1.0 V), nevertheless, the eye is wide opened and information could be reliably deciphered from the transmitted message. This example demonstrates that the eye height, though a noteworthy specification, is not a decisive parameter in the assessment of transmission quality.
- The calculated result looks plausible because the eye is wide opened and its contour is clearly delimited; thus, $\text{BER} \approx 10^{-12}$, which is a standard for modern optical communications, looks reasonable.

Questions and Problems for Section 9.1

- Questions marked with an asterisk require a systematic approach to finding the solution.
- Many questions and problems, including those marked with an asterisk, imply that you, in addition to reading the textbook, will do your research to find the answers. Consider such questions as mini-projects.

Basics of Digital Transmission Revisited

1 This chapter is devoted to digital transmission. What other types of transmission do you know? Explain.

2 We discussed several versions of the block diagram of a communication system (see Figure 9.1.1). Why is the block diagram shown in Figure 1.1.2R referred to as Shannon's model?

3 *Figure 1.1.2R shows a block diagram of a communication system with noise:
   a) What kind of noise – external or internal – is this? Explain.
   b) What measures can be taken to reduce the harmful effect of this noise?
   c) Is this noise additive or multiplicative? Explain.
   d) Does noise affect the transmitted signal only in a communication channel?

4 *The communication system shown in Figure 1.1.2R is designed to deliver information from a source to a destination point. List the main characteristics this system must possess.

5 *Figure 1.1.2R shows a set of components connected by links. Why is this set called a *system*?

6 Compare Figures 1.1.2R and 9.1.1: Describe how each component shown in Figure 1.1.2R is presented in Figure 9.1.1.

7 *What characteristics of the Tx, the Rx, and the transmission link of a communication system must be in accord with one another?

8 List and explain the two main characteristics of a transmitter.

9 What is the input/output characteristic of a transmitter?

10 *The example of industrial specifications of a laser diode (LD) includes $P_{light} = 600\,\mu W$ at $I = 7\,mA$ and typical graph of normalized output power vs. forward current shown in Figure 9.1.P10:
   a) Calculate the slope efficiency of this LD.
   b) Explain why this graph does not start at the zero point. (*Hint*: Do your research.)

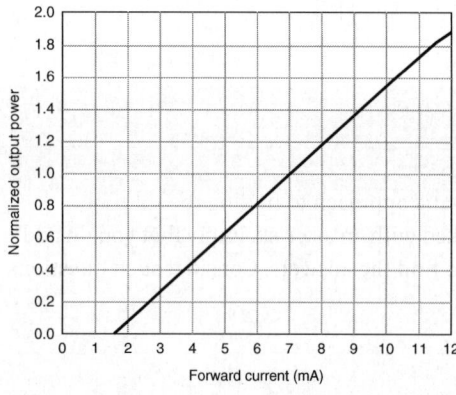

Figure 9.1.P10 Example of an industrial specification for the input–output characteristic of an LD.

11 Consider the modulation bandwidth, BW_{mod} (Hz), of a transmitter, Tx:
   a) If a Tx's rise time is $t_r = 1$ ps and $k = 0.48$, what is its BW_{mod} (Hz)?
   b) Compare two transmitters: $t_{r1} = 1$ ps and $t_{r2} = 0.01$ ps. Which do you prefer? Why?
   c) Explain why t_r determines the Tx modulation bandwidth.

12 *What requirement for Tx's spectral bandwidth do you know? What would be an ideal spectral bandwidth of a Tx?

13 What is the input/output characteristic of a receiver?

14 If an Rx's rise time is $t_r = 2$ ps and $k = 0.35$, what is its BW_{mod} (Hz)?

15 Compare the spectral bandwidths of an optical Tx and an optical Rx: What do they have in common and how do they differ?

16 We attribute two main characteristics to a Tx, but three main characteristics to an Rx. What is the Rx's third main characteristic?

17 The receiver sensitivity of a photodiode (PD) is -23 dBm for BER $= 10^{-12}$:
   a) Explain what is meant by *receiver sensitivity*. What is the value of the given receiver sensitivity in absolute numbers?
   b) Why is receiver sensitivity related to BER?
   c) If the required BER changes to 10^{-15}, in what way – an increase or a decrease – must receiver sensitivity be changed? Give your estimate.

18 *Consider the communication system shown in Figure 9.1.1:
   a) If the Tx signal power is 0 dBm and the link loss is -25 dB, what signal power will reach the Rx?
   b) If the Rx sensitivity is -23 dBm, will this system be operational? Explain.

19 Explain the difference between loss and attenuation of a communication link.

20 If the signal power launched into a communication link is $P_{in} = -5$ dBm and the output power is $P_{out} = -35$ dBm, what is the link loss in dB? In absolute numbers?

21 If loss is defined by Eq. (9.1.4), Loss (dB) $= 10 \log\left(\frac{P_{out}(mW)}{P_{in}(mW)}\right)$, then this number will always be negative. Why?

22 Assume that two manufacturers offer optical fibers. The first manufacturer claims that its fiber exhibits -34 dB losses and the second advertises -45 dB losses. Can you determine whose fiber is better and explain why?

23 Explain why attenuation, A (dB/km), is a comprehensive characteristic of link quality but loss (dB) is not.

24 Compute the loss and attenuation of an optical fiber if $P_{in} = 5$ mW, $P_{out} = 5\,\mu$W, and $L = 100$ km?

25 What will be the P_{out} (mW) for an optical fiber if $P_{in} = 0$ dBm, $L = 120$ km, and $A = 0.2$ dB/km?

26 Find the maximum transmission distance for an optical fiber whose attenuation is $A = 0.3$ dB/km if the power launched in is 4 mW and the receiver sensitivity is $4\,\mu$W.

27 Using attenuation, A (dB/km), we can compare achievable transmission distances of such different transmission media as air, copper wire, coaxial cable, and optical fiber. Why is this possible?

28 Attenuation of a transmission medium is the ratio of signal strength's loss to medium's length. Does it depend on any other parameter? If so, what is this parameter and how does attenuation depend on it?

29 Consider channel bandwidth, BW_{mod-s}:
   a) Modulation bandwidth of a Tx and an Rx determines transmission rate of these devices. It can be calculated as k/t_r, where t_r (s) is rise time. Can we determine channel transmission capacity by using a similar equation? Explain.

b) Equation (9.1.6) enables us to calculate channel rise time. Does this allow us to determine the channel transmission rate? Explain.

30 Review Shannon's law, C (b/s) = BW (Hz) $\log_2(1 + \text{SNR})$:
a) What is the role of channel bandwidth, BW (Hz), in determining channel maximum transmission rate, C (b/s)?
b) Find the C (b/s) of a channel if its BW (Hz) = 4 THz and its SNR = 10^3.
c) According to the text, Shannon's law implies that we are considering a bandlimited channel. How does the limitation in channel bandwidth follow from Shannon's law?

31 Explain what characteristic of a transmission channel determines its available bandwidth. Give examples.

32 Review Figure 9.1.3 that shows how attenuation of various transmission media depends on frequency. Based on Figure 9.1.3, determine approximately the bandwidth value for each transmission medium.

33 Signal traveling through a transmission channel experiences delay. What are the reasons for this delay?

34 Consider propagation delay of all four transmission media (air, twisted pair, coaxial cable, and optical fiber):
a) Why do these media introduce propagation delays?
b) Refractive indexes of air and optical fiber are given by $n_{\text{air}} = 1.001$; $n_{\text{of}} = 1.5$. Velocity factors of twisted pair and coaxial cable are equal to $V_{\text{Ftp}} = 0.96$ and $V_{\text{Fcoax}} = 0.94$. Calculate the propagation delays for a 1 km transmission through each transmission medium.
c) Compare all the calculated delays and interpret the result.

35 Shannon's law shows that channel maximum transmission rate, C (b/s), depends on the channel bandwidth, BW (Hz) and its signal-to-noise ratio (SNR). Explain which factor affects C (b/s) more significantly and why.

36 What should be an ideal value of SNR?

37 What is SNR = 36 dB in absolute number?

38 Noise determining SNR in Shannon's law is modeled as *additive white Gaussian noise (AWGN)*. Explain the meaning of each term in this model.

39 Examine the role of noise in transmission:
a) What is noise?
b) What kinds of noise do you know?
c) What are the sources of external noise and internal noise?
d) Which kind of noise – external or internal – can be easily reduced and how?

40 In a communication system, signal power is 30 mW and noise power is 0.3 µW. What will be transmission capacity of this system if BW = 200 MHz?

41 Consider internal noise in a communication system:
  a) What types of internal noise do you know?
  b) How do you distinguish each type of internal noise?
  c) Which type of internal noise is the most difficult to reduce and why?

42 Focus on thermal noise:
  a) Calculate the average power of thermal noise, P_{th}, generated by a resistor at room temperature and bandwidth 3 MHz.
  b) Figure 9.1.5 shows that if noise bandwidth increases, the power of thermal noise increases too. Prove this statement by examining Figure 9.1.5. Why is this so?
  c) List and explain all the measures that can be taken to reduce the average power of thermal noise.
  d) Can we eliminate thermal noise completely? Explain your answer.

43 Does a bare transmission link generate internal noise? Explain for each transmission medium.

44 If a transmission link includes an amplifier, will its internal noise change? If yes, in what way?

45 Explain the meaning of such amplifier characteristic as noise figure (NF). In what units is NF measured?

46 Why does SNR_{out} of an amplifier differ from SNR_{in}?

47 Consider an optical communication system: Compute optical signal-to-noise ratios $OSNR_{in}$ and $OSNR_{out}$ and the noise figure (NF) of an optical amplifier (OA) if $P_{in} = 0.3\,\mu W$, $P_{out} = 30\,\mu W$, $P_{noise}^{in} = 2\,nW$, and $P_{internal\text{-}noise} = 0.4\,\mu W$.

48 Signal-to-noise ratio of the input signal $SNR_{in} = 10 \times 10^3$. This signal is presented to an amplifier with noise figure NF = 5.2 dB.
  a) What will SNR_{out} be?
  b) What should be the ideal value of NF in dB and what would SNR_{out} be in this case? Explain.

49 We include an amplifier in a transmission link to compensate for the loss in a signal's strength and be able to transmit the signal over longer distance. It would seem that we can include as many amplifiers as we want to transmit the signal over any desired distance. Is this true? Prove your answer.

50 Consider a transmission link that includes five series amplifiers. If SNRs at the start and at the end of transmission link are given by $SNR_{in} = 60\,dB$ and $SNR_{out} = 36\,dB$, what is a noise figure of each amplifier provided that they all have the same NF? Show your calculations with comments.

Assessing the Quality of Digital Transmission: The Gaussian (Bell) Curve and the Probability Value

Problems

51 *This section is devoted to assessing the quality of digital transmission. This means that there are quantitative parameters that enable us to evaluate the transmission quality in numbers. What are those parameters?

52 A sampling space of random variable X is the line interval from -10 to 10. We measure X 200 times. Can we predict which value X will take on 199th measurement? On any specific measurement? Explain.

53 Figure 9.1.10 shows that when $x = 4.5$, $f(x) = 0.5$. Explain what it means. Why does $f(x)$ becomes equal to 0.5 too when $x = 3.5$?

54 The area under the whole bell-shaped curve in Figure 9.1.10 is equal to 1 when x runs from $-\infty$ to ∞. What is the probability of finding x anywhere within its sampling space? Explain.

55 The text says *the probability that X takes any value within the interval from a to b, is the area under the Gaussian curve f(x) from a to b*. How can probability be equal to area?

56 Consider Gaussian PDF of random variable X: What is the probability of finding $x = 3.2$? Explain.

57 Sketch three Gaussian PDFs with the same σ but whose $\mu_1 = -1$, $\mu_2 = 0$, and $\mu_3 = 3$ and comment on their differences.

58 Sketch three Gaussian PDFs with the same μ but whose $\sigma_1 = 0.01$, $\sigma_2 = 0.10$, and $\sigma_3 = 0.99$ and comment on their differences.

59 Take two Gaussian PDFs with $\mu_1 = \mu_2 = 0$ but whose $\sigma_1 = 0.1$, $\sigma_2 = 0.9$. If you were to calculate the probabilities P_1 and P_2 of finding x between 0 and 0.1, which probability would you expect to be greater? Explain.

60 We learned that to determine the probability of finding a continuous random variable between values a and b, we need to calculate the area under the Gaussian curve at this interval. For Gaussian PDF, this calculation boils down to evaluating (9.1.17),

$$P(a < X < b) = \frac{1}{\sigma\sqrt{2\pi}} \int_a^b e^{-\frac{(x-\mu)^2}{2\sigma^2}} dx \tag{9.1.17R}$$

Explain why we cannot use this integral for calculations in practice. What is a *standard normal probability distribution function (a standard normal PDF)* and why do we need it?

61 Both Integrals (9.1.17), $P(a < X < b) = \frac{1}{\sigma\sqrt{2\pi}} \int_a^b e^{-\frac{(x-\mu)^2}{2\sigma^2}} dx$, and (9.1.20), $P(a < Z < b) = \frac{1}{\sqrt{2\pi}} \int_a^b e^{-\frac{z^2}{2}} dz$, calculate the probability of finding X or Z between sampling points a and b. Why the former integral cannot be used in practice and the latter can?

Questions and Problems for Section 9.1

62. Using Table 9.1.1, determine the probability of finding Z within interval $-1.36 < Z < 0$. Sketch the Gaussian PDF curve and show the area corresponding to your calculations.

63. Calculating the probability of finding random variable Z in a tail of the Gaussian curve
    a) Calculate $P(Z \geq 1.84)$, that is, the probability of finding Z from 1.84 to infinity.
    b) Repeat the calculations for $Z \leq -1.84$.
    c) Sketch the bell curve and show the results of your calculations for both cases.

64. For $a = 6$, determine the probability of finding $Z > 6$ by using the $Q(a)$ function values given in Table 9.1.2 and computed by using (9.1.24). Compare the results and explain why they differ.

65. Find the relationship between Q-function, $Q(a)$, and complementary error function, erfc(a).

66. Suppose you build two graphs, Q-function vs. a and erfc function vs. a, where $a = 1,2,3, \ldots, 10$. Do you expect them to be identical or different? Build these graphs to prove your answer.

67. Explain how a receiver determines which logic – bit 1 or bit 0 – is obtained in a noisy digital signal. Sketch a figure supporting your explanation.

68. The nominal values for level one and level zero are 3.3 and 0.5 V, respectively. The receiver's threshold is 1.9 V. The measured values of received pulses at sampling points are 2.1 V and 1.8 V. What logic bits will the receiver decipher? Explain.

69. Figure 9.1.17a shows that the probability to erroneously identify bit 1 as bit 0 is the area under the tail of the PDF1 penetrating beyond the threshold. How would this probability change if σ_1 increases two times?

70. Refer to Figure 9.1.17 and Equation (9.1.34b): Given $V_1 = 2.1$ V, $V_0 = -1.8$ V, $\sigma_1 = 0.7$ V, and $\sigma_0 = 0.6$ V, calculate the average probability of error in this received signal. What would be this probability if σ_0 became 1.25 V?

71. The text makes an assumption that "… noise is a random process with the Gaussian probability distribution function (PDF) described by $f(x) = \frac{1}{\sigma\sqrt{2\pi}} e^{-\frac{(x-\mu)^2}{2\sigma^2}}$, where μ is the mean (expected value) and σ is the standard deviation." How important is this assumption? What if noise were described by any other probability distribution function?

72. Calculate the average probabilities, P_e, of error in digital transmissions whose $V_1 - V_0 = 2.5$ V and (1) $\sigma = 0.1$ V, (2) $\sigma = 0.4$ V, and (3) $\sigma = 0.7$ V. What trend in P_e do you see?

73. Examine Figure 9.1.18: It shows that a receiver presents to a decision circuit samples of an incoming signal taken at every bit interval, T_b:
    a) The incoming signal is a stream of pulses carrying bits 1 and 0. Why do we need to take samples of these pulses?
    b) How important is it to take samples every T_b? Can we take the samples at any other instants?

74 Every textbook and industrial document defines BER as the ratio of the number of erroneous bits to the total number of bits, but our text states that bit error ratio is nothing more than the average probability of taking received bit 1 for bit 0 and vice versa:
   a) How can the *ratio* be equal to the *probability*?
   b) How can these two standpoints at the same entity be compliant?
   c) Which of the above definitions can be used in practice? Why? How?

75 We know that BER is the ratio of the number of erroneous bits to the total number of bits. Suppose you received 10^{10} total number of bits for one second:
   a) Can you determine how many received bits are erroneous?
   b) Can you determine which of the received bits are erroneous?
   c) Explain.

76 The bit error rate, BER, is just a probability of error occurrence:
   a) How can such an entity be one of the most important metrics of transmission quality?
   b) BER depends on $\frac{V_1 - V_0}{\sigma_1 + \sigma_0}$. What is the physical meaning of the numerator? Denominator? The entire ratio?

77 Explain how the same acronym, BER, is defined as a bit error *ratio* or as a bit error *rate*.

78 *The text says that "the bit error rate is rightly considered the most comprehensive quantitative measure of digital transmission quality:"
   a) Does the BER account for all the possible errors in digital transmission? Explain.
   b) Does knowing the BER have any productive output?

79 *Consider an error vector magnitude, EVM:
   a) Explain how it can be used to assess the transmission quality.
   b) Compare the BER and the EVM: Which metric would you prefer to use in your practice and why?

80 Discuss the concept of eye diagrams:
   a) Explain what eye diagram is and how it is formed.
   b) How many pulses (bits) are used to form an eye pattern?
   c) Why do we need an eye diagram in addition (or instead of) a signal's waveform?
   d) Why is it called "an eye" diagram (pattern)?

81 Explain how an eye diagram can be used to qualitatively assess transmission quality.

82 *Consider the quantitative assessment of transmission quality with an eye diagram:
   a) List all the measurable parameters of an eye diagram that can be used to quantitatively evaluate the quality of transmission.
   b) Describe the use of these parameters.
   c) Do these parameters find the deterministic values of transmission quality or do they calculate the probabilities of finding those values within a range? Explain.

83 *Discuss the formation of an eye diagram based on Figure 9.1.20A,B:

a) Sketch an imaginable eye diagram that would be built from a thousand identical pulses. What will change if you take a million identical pulses?
b) Why do upper and lower levels of individual pulses (bits) vary in reality?
c) Show how eye level zero will be formed in reality.

84 Consider the pulse train containing a hundred pulses (bits). The amplitude nominal value of its zero level is 0 V. The measured amplitudes and the frequencies of their appearance are given in Table 9.1.P84:
a) Build the histogram of this measurement.
b) If you approximate this histogram by a Gaussian PDF, what will be its mean, μ, and standard deviation, σ?

Table 9.1.P84 Measured amplitudes of the lower level of a pulse train.

The measured values of zero level (V)	Frequency of appearance
−0.2	10
−0.1	15
0	46
0.1	17
0.2	12

85 Examine the eye diagram of a noisy signal shown in Figure 9.1.21b:
a) There are two eye measurements, A (V) and H (V). What is the difference between them? Can you quantitatively estimate this difference?
b) The eye level one (upper level) is fuzzy and wide (thick). Why so?
c) Are there strict boundaries of level one? Can you estimate the width of this level?
d) What are the similarity and the difference between eye level zero (lower level) and level one of this eye diagram?

86 *Examine eye diagrams in Figures 9.1.20B and 9.1.21b:
a) What are the sampling points over there? Why choosing the sampling point is important in the analysis of an eye diagram?
b) What phenomenon causes jitter? What would be the jitter value in an ideal eye diagram? Why?
c) How can jitter, which results from variations in time measurements, affect the eye height, which depends on variations in amplitude measurements?
d) Show how we can measure rise and fall times in the eye diagram presented in Figure 9.1.20A,B.

87 *The text says that an eye diagram is an indicator of the transmission SNR and gives (9.1.40c) to estimate the SNR:
a) Estimate SNR by using the eye pattern shown in Figure 9.1.21b.
b) The text says that eye diagram opening is the general indicator of the transmission's quality; which means that eye height, H (V), should be the measure of this quality. However, the formula for SNR, (9.1.40c), does not include H (V). Why?

88 Based on the measured samples and built histograms, the following parameters of eye level one and level zero, respectively, have been calculated: $V_1 = 3.1\,\text{V}$, $V_0 = 0.3\,\text{V}$, $\sigma_1 = 0.2\,\text{V}$, and $\sigma_0 = 0.2\,\text{V}$. What is the quality factor (QF) in absolute number and in decibels and what is the bit error rate (BER) of this transmission?

89 Why is Eq. (9.1.41) so important in assessing the quality of digital transmission?

90 Explain how the opening of an eye diagram can be related to the probability of erroneously identifying bit 1 for bit 0 and vice versa. Sketch the figures to show the low probability, the medium probability, and the high probability of error.

91 Figure 9.1.P91 shows an eye diagram with the following measured values: $V_1 = 79\,\text{mV}$, $V_0 = -79\,\text{mV}$, and $6\sigma_1 = 6\sigma_0 = 96\,\text{mV}$. Estimate the BER of this transmission. Calculate roughly the height of eye diagram, H (V).

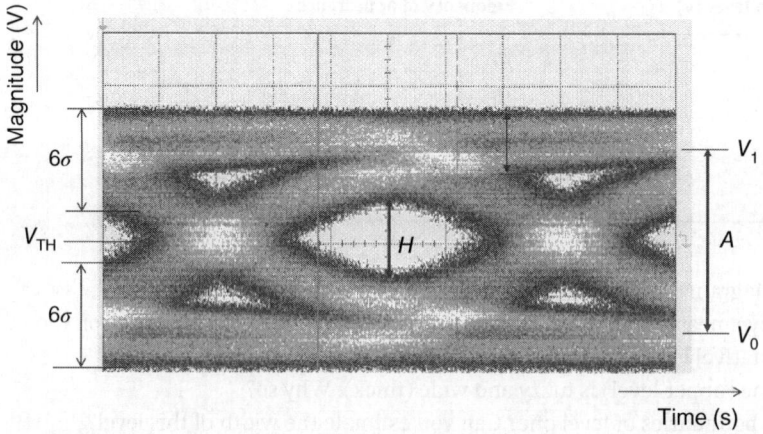

Figure 9.1.P91 Eye diagram for Problem 91 (Section 9.1).

9.2 Introduction to Digital Transmission – Binary Shift-Keying Modulation

Objectives and Outcomes of Section 9.2

Objectives

- To introduce binary shift-keying modulation in digital transmission.
- To learn the concept of ASK, FSK, and PSK modulations.
- To provide the mathematical description and introduce modulation and demodulation principles of ASK, FSK, and PSK.
- To assess the transmission quality of ASK, FSK, and PSK.
- To consider advantages, shortcomings, and applications of ASK, FSK, and PSK.

Outcomes

- Learn that this section discusses transmission of a digital signal by using a sinusoidal carrier with three types of shift keying modulations. Get to know that these types of modulation are ASK, FSK, and PSK. Understand that the binary-encoded message signal causes changes in a sinusoidal carrier amplitude in ASK, in a carrier frequency in FSK, and in a carrier phase in PSK modulation.
- Focus on the concept of producing a binary amplitude shift-keying (BASK) signal depicted in Figure 9.2.2a showing that a modulator allows the carrier to go through when the bit message is 1 and blocks partly or entirely the carrier from passing when the bit message is 0. Understand how an entire ASK transmission system operates by reviewing Figure 9.2.2b.
- Realize that mathematically an ASK signal is a set of two sinusoidal signals with the same frequencies and phases but different amplitudes as (9.2.2) shows. Understand that the consequence of this mathematical description of ASK is that most of the general properties of amplitude modulation discussed in Sections 8.1 and 8.2 are applicable to ASK modulation.
- Comprehend that the spectrum of an ASK signal depends on the message signal. Be aware that, for example, the spectrum of an ASK signal with a square wave message is merely the message spectrum shifted by f_C and symmetrically placed around the carrier frequency whose amplitudes are twice as small as the amplitudes of the message spectrum.
- Remember that the ASK signal's bandwidth is the sum of the frequencies of harmonics delivering the required power. See Example 9.2.2 to become familiar with the bandwidth of an ASK signal with a square wave message. Come to know that the bandwidth of an ASK with a pulse as a message signal is the range of frequencies between first nulls (zero crossings) of the graph *power spectral density vs. frequency*, according to common industry practice.
- Be aware that, in an ASK transmission, the probability of receiving an error signal with a given digital SNR, $\frac{E_b}{N_0}$, is $\text{BER}_{\text{ASK}} = Q\left\{\sqrt{\frac{E_b}{N_0}}\right\}$, where E_b (J) is the average bit energy and N_0 (W/Hz) is the one-sided noise power spectral density. Get to know that this formula enables us not only to assess the quality of the ASK transmission but also to compute the received power necessary to support the required BER, as Example 9.2.3 demonstrates.
- Learn that the primary advantage of ASK modulation is its simplicity, but this modulation has all the drawbacks inherent in amplitude modulation, the main one being noise susceptibility. Come to know that, however, ASK is one of the main modulation formats in optical communications, the linchpin of the modern communications. Study in Example 9.2.4 how the ASK modulation works in optical communications today.
- Realize that an ASK detector (demodulator) cleans a received signal from the sinusoidal carrier, samples it at every bit interval, T_b, and compares the sampled value with the threshold to determine which bit – 1 or 0 – is received. Be aware that there are two basic types of ASK demodulators: coherent and incoherent. Review Figure 9.2.9 to understand the principle of operation of a coherent ASK demodulator.
- Get to know the FSK modulation in which bit 1 of a message signal is represented by one frequency of a carrier wave and bit 0 by another frequency.
- Understand that a binary frequency shift-keying (BFSK) modulation employs two sinusoidal carrier waves with different frequencies and the same amplitudes and phases, and its signal can be mathematically described as the product of carrier waves and message signals. Grasp the fact

that the carrier frequency, ω_C, is fixed, and introducing the frequency deviation, $\Delta\omega$, we can present a BFSK signal as a combination of two carrier waves that differ only in frequency; that is, $\omega_{Ci} = \omega_C \pm \Delta\omega$, where $i = 1, 2$, as in (9.2.27). Realize that this conclusion means that a BFSK signal can be treated as a combination of two ASK signals.

- Comprehend that the spectrum and bandwidth of a BFSK signal depends, of course, on a message signal. Find that the spectrum of the BFSK signal with a square wave message shown in (9.2.30) is formed from the spectra of two ASK signals, one of which is carried by ω_{C1} and the other by ω_{C2}. Observe that each spectral component is shifted by a multiple of the fundamental (message) frequency, ω_M (Hz) $\equiv \omega_0$ (Hz), frequency.

- Grasp the fact that the power bandwidth of a BFSK signal, considered as a combination of two ASK signals, is merely the sum of two power spectra of the ASK signals separated by doubled frequency deviation $2\Delta\omega$. as shown in Figure 9.2.12. Learn that this statement is true regardless of whether a BFSK carries a periodic message, such as a square wave, or a nonperiodic message, such as a pulse train. Compare Figures 9.2.12 and 9.2.13 to visualize this fact and see that the BFSK bandwidth sets the limitation on the bit rate through (9.2.33), $\Delta\omega_{min}/2\pi$ (Hz) = BR (b/s).

- Learn that the BER of an FSK modulation is governed by the same formula as that of an ASK due to a simplified strategy of treating FSK as two ASK signals. Realize that in reality, however, FSK is a *nonlinear modulation*, which requires a different technique for more accurate analysis.

- Understand that there are two types of FSK modulation: a *discontinuous-phase* FSK (DPFSK) shown in Figure 9.2.10 and a continuous-phase FSK (CPFSK) shown in Figure 9.2.15. Be aware that in DPFSK, the phases of two different frequencies change abruptly at the transition from bit 1 to bit 0 and back, which causes an increase in the signal's bandwidth. Come to know that the DPFSK problem can be solved by employing a voltage-controlled oscillator (VCO) which generates an FSK signal with two different frequencies while keeping the phases of these signals continuous during the transitions. Learn that this system is called a CPFSK and that the CPFSK is the method by which an FSK signal is generated today because a CPFSK signal occupies less bandwidth than a DPFSK signal.

- Master the point that a carrier frequency, $\omega_C/2\pi$ (Hz), must be a multiple of the bit rate, BR (b/s), and the value of the frequency deviation, $\Delta\omega/2\pi$ (Hz), must be a multiple of BR(b/s)/2 in order to generate the CPFSK signal and facilitate detecting this signal. Examine Section 9.2.3.7 and Example 9.2.5 to fully comprehend this point.

- Study coherent and incoherent FSK detections and learn that a coherent detection is definitely more accurate than an incoherent one.

- Conclude that FSK has the same inherent advantage over ASK as FM does over AM: It is less susceptible to noise because changes in signal amplitude do not directly affect transmission quality. Comprehend that, however, the FSK needs a greater bandwidth to transmit its signal. Learn that FSK, thanks to its high noise immunity, finds a wide range of applications running from many mobile devices to the Global System for Mobile Communications to the Internet of Things (IoT).

- Learn that a PSK modulation delivers a digital message by changing the phase of a sinusoidal carrier. Understand that in binary PSK, BPSK, changes from bit 1 to bit 0 and back are presented by a phase change of 180° and that a BPSK modulation can be considered as two ASK modulations whose carrier waves have equal amplitudes and frequencies but a 180° difference in phases. Realize that this simplified approach enables us to comprehend BPSK main features and develop its mathematical description.

- Come to know that the spectral characteristics of a PSK signal are similar to those of an ASK signal, and therefore, the bandwidths of PSK signals with a square wave message and with a pulse train are the same as the ASK bandwidths.
- Find that coherent BPSK demodulation is similar to that of the ASK but requires that a local oscillator must know the phase of a carrier wave, which is impossible without obtaining additional information. Learn that differential phase shift-keying (DPSK) resolves this issue by keeping the carrier phase without change when bits 1 are transmitted and changing the carrier phase by 180° for every bit 0. Be aware that this DPSK advantage comes with a price: DPSK requires more bit energy than BPSK to sustain the same BER.
- Observe from (9.2.64a), $\text{BER}_{\text{BPSK}} = Q\left(\sqrt{\frac{2E_b}{N_0}}\right)$, that the BER_{PSK} is greater than that of ASK and FSK and therefore a PSK modulation is more advantageous than its ASK and FSK counterparts because, in addition to a better BER, its bandwidth is as small as that of ASK and smaller than that of FSK by $\Delta\omega/2\pi$ (Hz).
- Comprehend that the aforementioned advantage makes PSK one of the main modulation formats not only in high-speed wireless communications but also in another major field – optical communications, and therefore PSK is the basis for most modern modulation techniques.
- Finally, examine Table 9.2.4 to compare the pulse-train bandwidth and BER characteristics of ASK, FSK, and PSK modulations and see the advantages and shortcomings of each type of shift-keying modulation.

This section considers the types of modulation that enable digital transmission, specifically transmission of a digital signal by using an analog, sinusoidal carrier. To increase the efficiency of your study, we recommend a quick review of Chapter 3, where the basics of digital transmission are discussed, and Section 8.1, where the basics of analog modulation are considered. Digital modulation is a main branch of telecommunications with research, education, and the industry underpinning it (see the bibliography). Here, we will present an introduction to the specific topic, *shift-keying modulation*, the introduction that gives you the foundation and guidelines for further study.

9.2.1 Digital Signal over a Sinusoidal Carrier – Binary Shift-Keying Modulation

This subsection considers *binary* modulations enabling transmission of a digital signal. In considering three versions of this type of modulation – ASK, FSK, and PSK – we will follow a format that makes their comparison easy. A general view of all three types of modulations is given in Figure 9.2.1.

Observing Figure 9.2.1, we see that the message signal causes changes in a carrier amplitude in ASK, in a carrier frequency in FSK, and in a carrier phase in PSK modulation. The following subsections consider each of these modulations.

9.2.2 Binary Amplitude-Shift Keying (ASK)

9.2.2.1 ASK Concept and Waveform

ASK has been briefly mentioned in Section 8.1 as an example of amplitude modulation (see Figure 8.1.7). It is, indeed, the version of AM with a sinusoidal carrier wave and a digital message (information) signal. Figure 9.2.2a shows the concept of producing an ASK signal: The carrier wave is presented to a multiplier (a modulator, in fact) controlled by the message signal. The message here is a digital binary signal; this is why this type of ASK is called *binary ASK and often referred to as BASK*. In this section, we restrict our discussion to the binary type of ASK only.

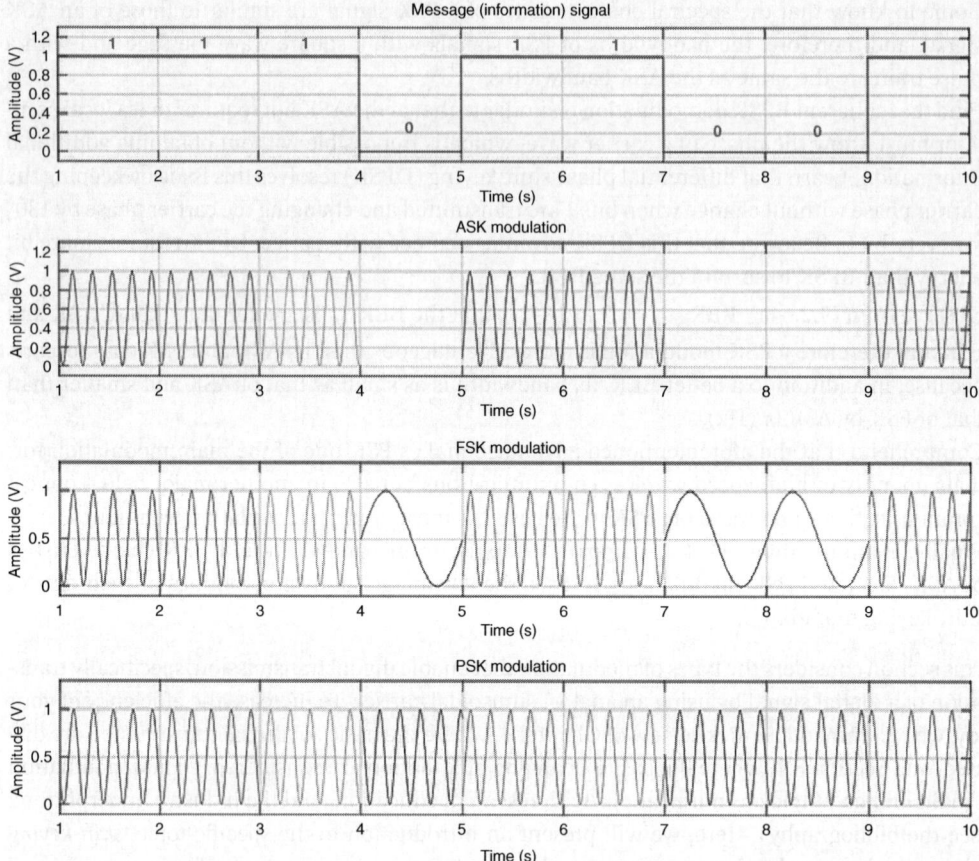

Figure 9.2.1 Three types of binary shift keying modulation: (a) The pulse train, which is a digital binary message signal; (b) amplitude-shift keying (ASK); (c) frequency-shift keying (FSK); and (d) phase-shift keying (PSK).

Figure 9.2.2a demonstrates that when the message is "1," the product of the message and carrier signals is one and the modulator allows the carrier to go through; when the message is "0," the product of two signals is zero, which blocks the carrier from passing. In reality, the modulator, of course, is a digital electronic device that does not simply pass or block the carrier but controls its amplitude. It assigns the maximum amplitude to the carrier pulse delivering bit 1 and any smaller value, including zero, to the carrier pulse delivering bit 0. Examples of waveforms of an ASK signal are shown in Figure 9.2.2a,b. The ASK signal implements the general concept of modulation discussed in Sections 8.1 and 8.2; that is, we change one of the parameters of a carrier wave (amplitude, frequency, or phase) to deliver the message (information) signal.

Figure 9.2.2b demonstrates an entire ASK transmission system, which consists of a transmitter, a transmission link, and a receiver. The transmitter generates the ASK-modulated signal, the transmission link carries the modulated signal, and the receiver recovers the information signal. In this example, the modulated signal has two levels of amplitude: the maximum value for bit 1 and a smaller, nonzero value for bit 0. The information signal shown in Figure 9.2.2b is a periodic unipolar square wave. Of course, no information signal can be a periodic one; Figure 9.2.2b is just an illustration of the ASK concept, a graphic that we will elaborate on shortly.

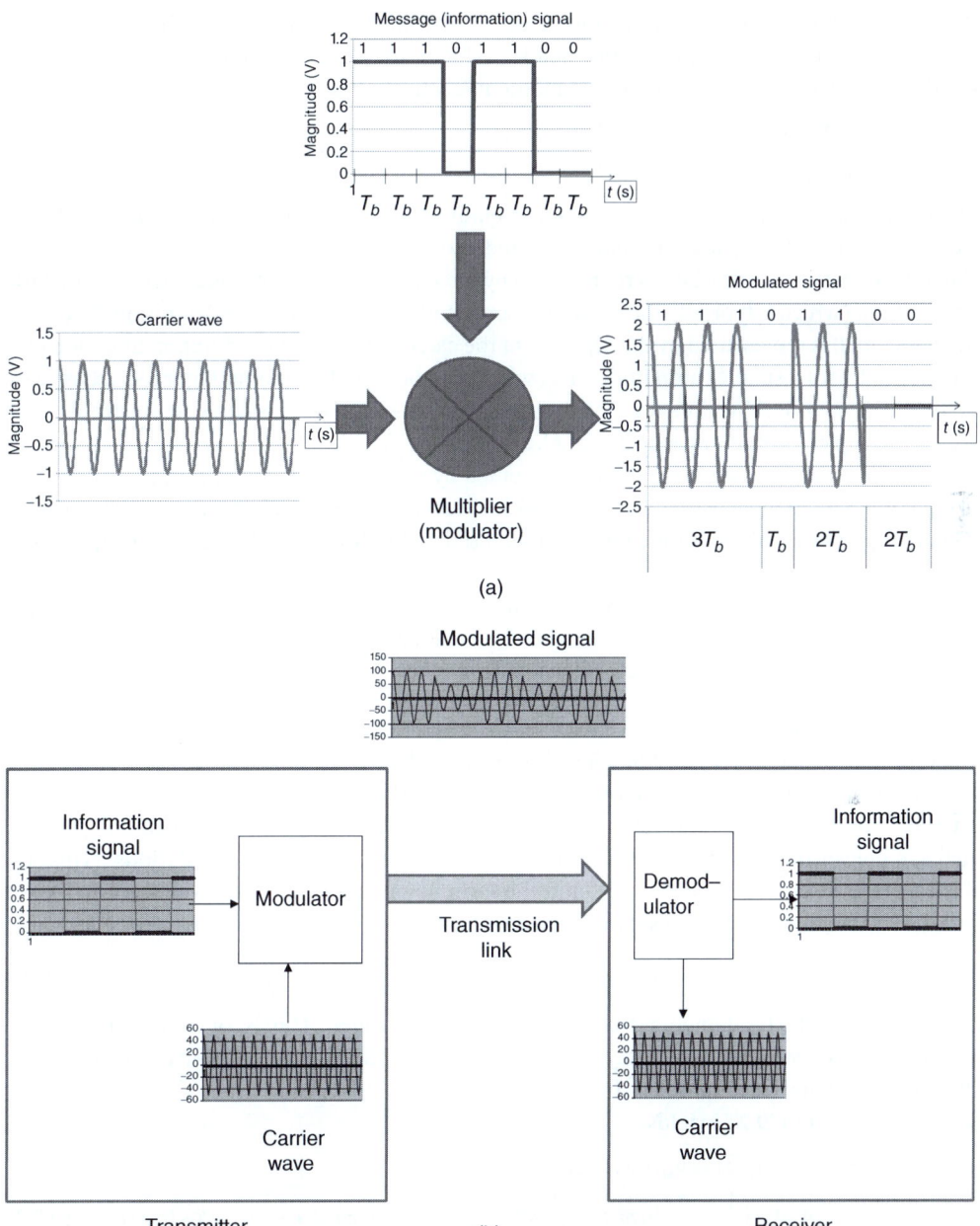

Figure 9.2.2 Amplitude-shift keying (ASK) modulation: (a) The modulation concept and (b) an ASK transmission system.

9.2.2.2 Mathematical Description of ASK

Assume that a carrier wave is given by $v_C(t) = A_C \cos(\omega_C t)$. Then the ASK modulation depicted in Figure 9.2.2a can be mathematically described as the product of the carrier and message signals; that is,

$$v_{ASK}(t) = v_M(t) \cdot v_C(t) = v_M(t) \cdot A_C \cos(\omega_C t) \tag{9.2.1}$$

When the message signal is bit 1, the modulator allows the maximum value of the carrier signal to go through; when bit 0 is applied, the smaller value of the amplitude of the carrier wave goes through. These explanations are presented mathematically in (9.2.2).

$$\text{Bit } 1 \to A_{C1} \cos(\omega_C t) \quad \text{for } t = T_b$$
$$\text{Bit } 0 \to A_{C0} \cos(\omega_C t) \quad \text{for } t = T_b \tag{9.2.2}$$

Here, T_b (s) is the duration of a bit, as shown in Figure 9.2.2a; A_{C1} and A_{C0} are the amplitudes of the carrier, with $A_{C1} > A_{C0}$; ω_C is the carrier frequency.

Equation (9.2.2) shows that bit 1 is represented by the segment of the sinusoidal carrier wave with amplitude A_{C1} (which is typically the maximum amplitude), frequency ω_C, and duration T_b; on the other hand, bit 0 is represented by the segment of the carrier with the same duration and the same frequency but with a smaller amplitude, A_{C0}, which can be zero. Examining Figure 9.2.2b reveals that for several identical bits in a row, the duration of a sinusoidal wave is equal to the proper sum of T_b; for example, for three 1s, the sinusoid lasts $3T_b$.

Since changes in a message signal are represented by various values of the carrier's amplitude, this type of modulation is justifiably called ASK. Also, we can reasonably expect that the most of the general properties of amplitude modulation, discussed in Sections 8.1 and 8.2, are applicable to ASK modulation.

It is interesting to note that the term *shift keying* refers to the Morse telegraph, where an electrical key is used to deliver DOT and DASH signals. The modulator shown in Figure 9.2.2a operates for binary ASK modulation, in fact, exactly like a telegraph key.

9.2.2.3 ASK Spectrum

To determine the spectrum of an ASK signal, we refer to the mathematical description of the ASK operation given in (9.2.1). The question is, how do we describe a message signal? Let us consider the periodic unipolar square wave shown in Figure 9.2.2b as a message signal. (Again, no periodic signal can deliver information, but this example allows us to simplify the calculations.) This signal is carefully discussed in Section 6.1, and the formula of its waveform is given in (6.1.13). We reproduce below this equation in slightly modified form

$$v_M(t) = A_M \left(\frac{1}{2} + \frac{2}{\pi} \sin \omega_0 t + \frac{2}{3\pi} \sin 3\omega_0 t + \frac{2}{5\pi} \sin 5\omega_0 t + \cdots \right) \tag{6.1.13R}$$

where $\omega_0 \equiv \omega_M$ is the fundamental frequency of a message signal. This is, of course, the Fourier series of our message signal, which says that summing up all these harmonics gives the waveform of the message signal (see Figure 6.1.16).

Inserting $v_M(t)$ into (9.2.1) yields

$$v_{ASK}(t) = v_M(t) \cdot v_C(t) = v_M(t) \cdot A_C \cos(\omega_C t)$$
$$= \left(A_C A_M \left[\frac{1}{2} + \frac{2}{\pi} \sin \omega_0 t + \frac{2}{3\pi} \sin 3\omega_0 t + \frac{2}{5\pi} \sin 5\omega_0 t + \cdots \right] \right) \cdot \cos(\omega_C t) \tag{9.2.3}$$

Referring to the well-known trigonometric identity $\sin A \cdot \cos B = (\sin(A+B) + \sin(A-B))/2$, we can rewrite (9.2.3) as follows:

$$v_{ASK}(t) = \frac{A_C A_M}{2} \cdot \cos(\omega_C t)$$
$$+ \frac{A_C A_M}{\pi} \cdot [(\sin(\omega_C + \omega_0)t + \sin(\omega_C - \omega_0)t)$$
$$+ \frac{1}{3}(\sin(\omega_C + 3\omega_0)t + \sin(\omega_C - 3\omega_0)t)$$
$$+ \frac{1}{5}(\sin(\omega_C + 5\omega_0)t + \sin(\omega_C - 5\omega_0)t) + \ldots] \tag{9.2.4}$$

Equation (9.2.4) says that the spectrum of the ASK signal consists of

- a carrier component with amplitude $\frac{1}{2}(A_C \cdot A_M)$ and frequency ω_C and
- harmonics with amplitudes $(A_C \cdot A_M)/n\pi$ and frequencies $(\omega_C + n\omega_0)$ and $(\omega_C - n\omega_0)$, where $n = 1, 3, 5, 7, 9, \ldots$

The spectrum of this signal is shown in Figure 6.1.16, where the phase shift of $-90°$ is omitted for simplicity.

Example 9.2.1 expands these explanations.

Example 9.2.1 The Spectrum of an ASK Signal with a Periodic Square Wave Message

Problem

Determine the spectrum of an ASK signal with a periodic square wave message signal if the parameters of the signals are given as $A_C = A_M = 1$ V, $\omega_C = 2\pi \times 2 \times 10^6$ rad/s, and $\omega_0 = 2\pi \times 1 \times 10^3$ rad/s. To simplify the calculations, we (i) assume that the amplitudes of message and carrier signals are equal to one and (ii) take the frequency of the message signal, ω_0, from Example 6.1.1.

Solution

We need to apply (9.2.4) and perform the calculations, which yield

$$\begin{aligned}
v_{ASK}(t)(V) =\ & \frac{A_C A_M}{2}\cos(\omega_C t) + \frac{A_C A_M}{\pi}[(\sin(\omega_C + \omega_0)t + \sin(\omega_C - \omega_0)t) \\
& + \frac{1}{3}(\sin(\omega_C + 3\omega_0)t + \sin(\omega_C - 3\omega_0)t) \\
& + \frac{1}{5}(\sin(\omega_C + 5\omega_0)t + \sin(\omega_C - 5\omega_0)t) + \cdots] \\
=\ & 0.5\cos(2\pi \cdot 2000 \cdot 10^3 t) + 0.32[\sin(2\pi \cdot 2001 \cdot 10^3 t) + \sin(2\pi \cdot 1999 \cdot 10^3 t) \\
& + \frac{1}{3}(\sin(2\pi \cdot 2003 \cdot 10^3 t) + \sin(2\pi \cdot 1997 \cdot 10^3 t)) + \frac{1}{5}(\sin(2\pi \cdot 2005 \cdot 10^3 t) \\
& + \sin(2\pi \cdot 1995 \cdot 10^3 t)) + \cdots]
\end{aligned} \qquad (9.2.5)$$

The required spectrum is shown in Figure 9.2.3.

Figure 9.2.3a reminds us of the waveform and the spectrum of a unipolar square wave, which was considered in Section 6.1. This square wave is the message signal in our example. Figure 9.2.3b demonstrates the spectrum of the ASK signal along with the spectrum of the message signal. We can readily see that the ASK spectrum is merely the message spectrum shifted by the value of f_C and symmetrically placed around the carrier frequency. We also observe that the amplitudes of the ASK spectrum are twice as small as the amplitudes of the message spectrum, as Eqs. (6.1.13R) and (9.2.3) require.

Discussion

Note that neither graphic in Figure 9.2.3 shows the phase shifts needed for sine components; this is why the figure is called *the **amplitude** spectrum*. We omitted the phases to concentrate on the main point of this discussion. Refer to Figure 6.1.11 to recall the total spectrum of a square wave.

What will be *the spectrum of a real, nonperiodic message signal*? We know from Section 7.1 that in this case, we need to use the Fourier transform and the result will be a continuous spectrum of the signal. Since ASK is intended to deliver a rectangular pulse train, let's consider this case:

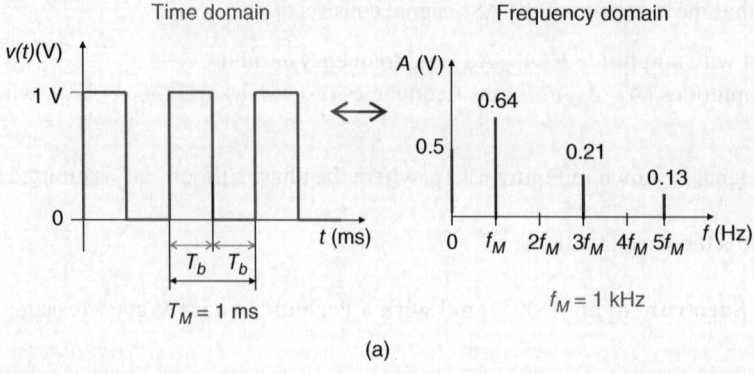

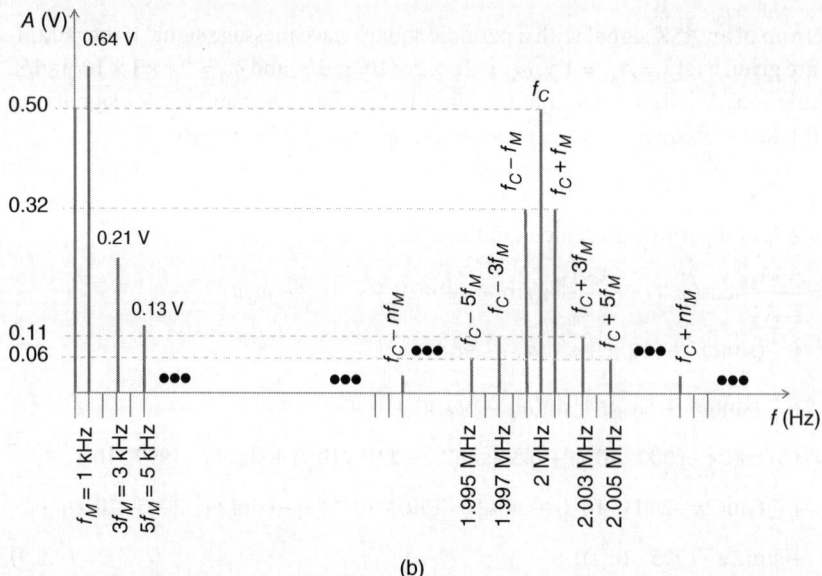

Figure 9.2.3 The amplitude spectrum of the ASK signal for Example 9.2.1: (a) The waveform (left) and the spectrum (right) of the message signal and (b) the spectrum of the ASK signal. Here f_M (Hz) $\equiv f_0$ (Hz) $= 1/T$ (s) is the fundamental frequency and T is the period of the message signal. Note that in Figure 9.2.3a the phase shift of $-90°$ is omitted.

First, we need to remember that a rectangular pulse has been discussed in Section 7.1. The formula of the Fourier transform of a rectangular function, $p(t)$, is given there as

$$F\{p(t)\} = AT_b \, \text{sinc}\left(\frac{\omega T_b}{2}\right) \tag{7.1.13R}$$

Its waveform and amplitude spectrum, shown in Figure 7.1.7R, are reproduced below. (Note that many textbooks use Greek letter τ to denote bit duration; that is, T_b (s) $= \tau$ (s).)

Now, we can better understand the formation of a pulse-train spectrum. Let us consider Figure 9.2.4, where the example of ASK spectrum is shown.

The left-hand side of Figure 9.2.4 shows the waveforms of a rectangular pulse train, $v_M(t)$ (V), a carrier wave, $v_C(t)$ (V) $= \cos(2\pi f_C t)$, and the transmitted ASK signal, $v_{\text{ASK}}(t)$ (V) $= v_M(t) \cdot \cos(2\pi f_C t)$; the right-hand side shows the three corresponding spectra, $F\{v_M(t)\}$ (V/Hz), $F\{\cos(2\pi f_C t)\}$ (V/Hz),

9.2 Introduction to Digital Transmission – Binary Shift-Keying Modulation

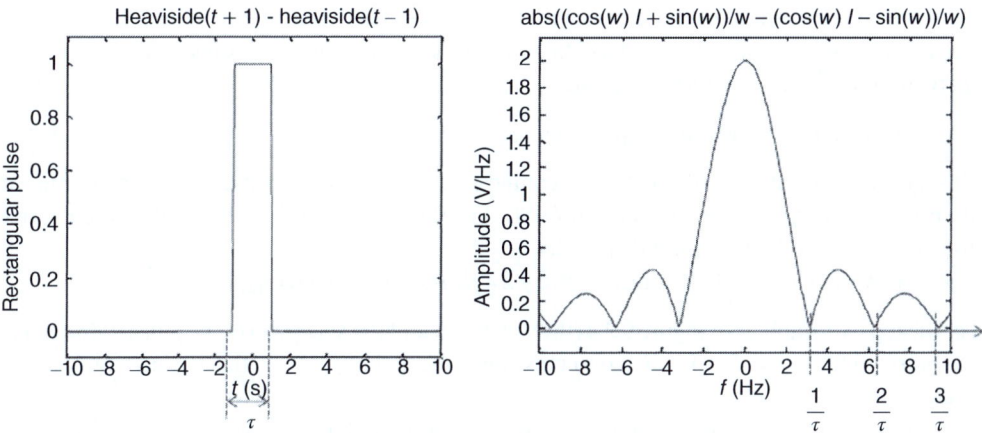

Figure 7.1.7R The waveform (a) and amplitude spectrum of a rectangular pulse (b) (remember, $\tau = T_b$).

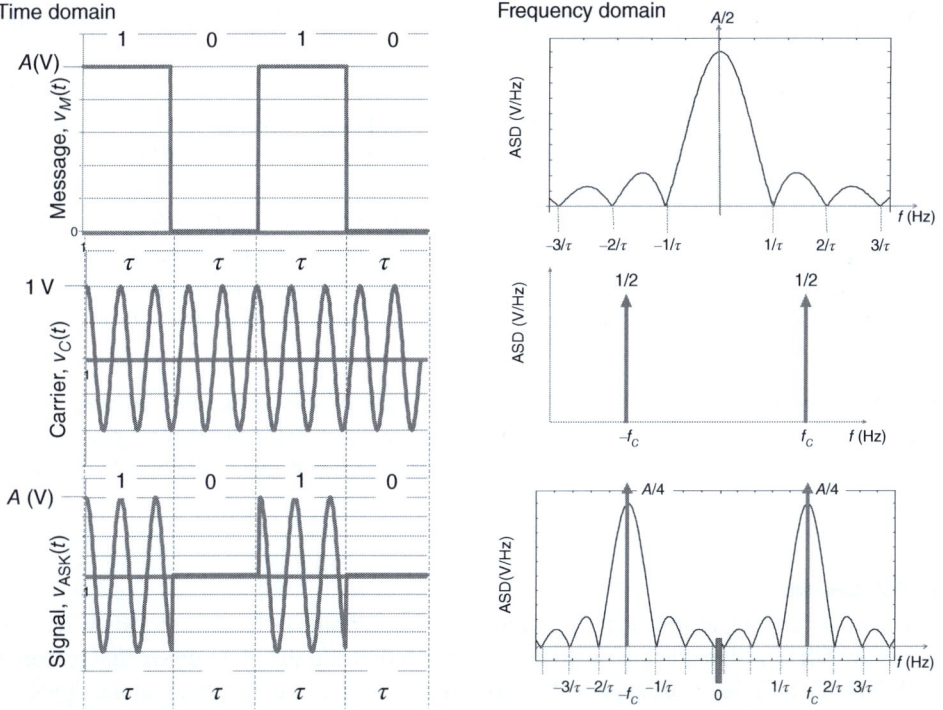

Figure 9.2.4 The waveforms and spectra of a rectangular pulse train, sinusoidal carrier, and an ASK signal (figures are not to scale).[3]

and $F\{v_{ASK}(t)\}$ (V/Hz). The time-domain presentation demonstrates that the ASK signal is obtained by multiplication of the message signal (here, the rectangular pulse train) and the carrier sinusoid. The frequency-domain presentation shows that (i) the constituents' spectra are the result of the Fourier transforms of the message and carrier time-domain functions and (ii) the ASK spectrum

---

3 After Stern and Mahmoud (2004, p. 220).

is the result of convolution of the message and carrier spectra. Referring to Table 7.2.2 we recall that in a case where the carrier is a sinusoidal function, such convolution results in the modulation property of the Fourier transform that states

$$F\{f(t)\cos(\omega_0 t)\} = \tfrac{1}{2}[F(\omega + \omega_0) + F(\omega - \omega_0)] \tag{9.2.6}$$

This is how the resulting ASK spectrum is composed. We also need to recall that the magnitude of each spectrum is *amplitude spectral density (ASD)* in V/Hz or V s, which is the result of the Fourier transform of time-domain magnitude in volts.

After reviewing Figure 7.1.7R, study Figure 9.2.4 to grasp the following points:

- For the top two graphics of Figure 9.2.4 (the waveform and the spectrum of the pulse train): Since the pulse width is $\tau = T_b$, the first zero crossing is located at $f_1 = 1/\tau$, the second – at $f_2 = 2/\tau$, and so on.
- For the middle two graphics of Figure 9.2.4 (the waveform and the spectrum of the carrier wave): Since the carrier is given by $\cos(2\pi f_C t)$, its Fourier transform results in two delta (Dirac) functions centered at $2\pi(f - f_C)$ and $2\pi(f + f_C)$ (see Table 7.1.1).
- For the bottom two graphics of Figure 9.2.4 (the waveform and the spectrum of the ASK signal): The waveform shows that each bit 1 contains $\tau/T_C = \tau \cdot f_C = 3$ cycles of the carrier wave.
- Analyzing the ASK signal's spectrum shown in Figure 9.2.4, we find that the graphs have only two side lobes toward each other before they start overlapping. Since the half of the main lobe also occupies $1/\tau$ (Hz), the total frequency range from 0 Hz to $\pm f_C$ (Hz) is equal to $3/\tau$ (Hz). In other words, the range from 0 (Hz) to $\pm f_C$ (Hz) can accommodate only three $1/\tau$ (Hz) bands. We will see the importance of this fact shortly when we analyze bandwidth and power spectrum of an ASK signal. For now, we can conclude that this ASK spectrum demonstrates one important fact regarding choosing the right value of a carrier frequency: *The greater the carrier frequency, the greater the number of spectral components that can be used for transmission; this means that the received signal will better reproduce the original message.*

{*Question*: Do you know why "the received signal will better reproduce the original message?"}

(*Exercise*: Let $\tau = 15 \times 10^{-15}$ s, $f_C = 200$ THz, and $A = 3$ V. Insert all the numbers in places of the letters shown in Figure 9.2.4. How many cycles of the carrier are within a single bit 1? Prove your answer.)

9.2.2.4 ASK Bandwidth

As soon as the spectrum of an ASK signal is found, we can determine its bandwidth. You will recall that the bandwidth of a signal is the range of frequencies that the signal occupies; that is, bandwidth, BW (Hz), is the difference between the maximum and minimum frequency of the signal, as discussed in Chapters 6 and 7. Thus,

$$\text{BW (Hz)} = f_{\max} - f_{\min} \tag{9.2.7}$$

Theoretically, however, the spectrum of both periodic and nonperiodic signals tends to infinity and bandwidth cannot be determined by using the above formula. As we learned in Section 6.2, the practical solution to this problem lies in determining what portion of the signal's total power we want to transmit. For a periodic signal, this criterion boils down to a simple fact: *The signal's bandwidth is the sum of the frequencies of harmonics delivering the required power.*

Example 9.2.2 clarifies this point.

Example 9.2.2 Determining the Bandwidth of an ASK Signal with a Square Wave Message

Problem

Determine the bandwidth of the ASK signal with a square wave message given in Example 9.2.1 ($A_C = A_M = 1$ V, $\omega_C = 2\pi \times 2 \times 10^6$ rad/s, and $\omega_M = 2\pi \times 1 \times 10^3$ rad/s) if the delivered power will be 75% of the total power. Solve this problem for delivering 95% and 98% of total power.

Solution

The general approach to solving such a problem entails three steps, as Example 6.3.3 demonstrates:

1. Find the signal's total power by using Eq. (6.3.36R),

$$P_T \text{ (W)} = \frac{1}{T} \int_t^{t+T} [v(t)]^2 dt \tag{6.3.36R}$$

2. Derive the formula describing the power distribution along the individual harmonics of the ASK signal.
3. Compute the power of the individual harmonics and the cumulative power and determine the required bandwidth by calculating the number of needed harmonics, as shown in Table 6.3.2.

We carry out this plan as follows:

1. To compute the total power, we need to define $v_{ASK}(t)$ for this example by using (9.2.1) as

$$v_{ASK}(t) = \begin{cases} A_M \cdot A_C \cos(\omega_C(t)) & \text{for } 0 < t \leq T/2 \\ 0 & \text{for } T/2 < t \leq T \end{cases} \tag{9.2.8}$$

This is because the square wave delivers power only during half of its period, $T/2 = T_b$. Thus, (6.3.36) takes the form

$$P_T(\text{W}) = \frac{1}{T} \int_t^{t+T} [v_{ASK}(t)]^2 dt = \frac{1}{T} \int_t^{t+T} [A_M \cdot A_C \cos(\omega_C t)]^2 dt$$

$$= \frac{(A_M \cdot A_C)}{T} \int_t^{t+T} \cos^2(\omega_C t) dt = \frac{(A_M \cdot A_C)}{T} \left[\frac{t}{2} + \frac{1}{4\omega_C} \sin(2\omega_C t) \right] \Big|_0^{T/2}$$

$$= \frac{(A_M \cdot A_C)}{T} \left[\frac{T}{4} + \frac{1}{4\omega_C} \sin\left(2\left(\frac{2\pi}{T}\right)\left(\frac{T}{2}\right)\right) \right] = \frac{(A_M \cdot A_C)^2}{4} \tag{9.2.9}$$

Here we apply the rule

$$\int_0^{T/2} \cos^2(\omega_C t) dt = \left[\frac{t}{2} + \frac{1}{4\omega_C} \sin(2\omega_C t) \right] \Big|_0^{T/2}$$

and the formula

$$\omega_C = 2\pi/T$$

2. We derive the equation to calculate power carried by individual harmonics by following the procedure used for derivation of (6.3.42a). Specifically, the voltage carried by an ASK signal

composed from n harmonics presented in (9.2.4) is given as

$$v_{ASK}(t) = \frac{1}{2}(A_C \cdot A_M)\cos(\omega_C t) + (A_C \cdot A_M)/\pi[(\sin(\omega_C + \omega_M)t + \sin(\omega_C - \omega_M)t)$$
$$+ \frac{1}{3}(\sin(\omega_C + 3\omega_m)t + \sin(\omega_C - 3\omega_M)t)$$
$$+ \frac{1}{5}(\sin(\omega_C + 5\omega_M)t + \sin(\omega_C - 5\omega_M)t) + \cdots]$$
$$= (A_C \cdot A_M/2)\cos(\omega_C t) + \sum_{n=-\infty}^{\infty}(A_C \cdot A_M)/\pi(\cos(\omega_C + n\omega_M)t), n \text{ odd only} \quad (9.2.10)$$

Since dissipated power is given by

$$P_R\,(W) = \frac{V_{rms}^2}{R} \quad (6.3.38R)$$

we need to calculate V_{rms}^2 as

$$V_{rms}^2 = \frac{1}{T}\int_0^T v^2(t)dt \quad (6.3.39R)$$

Plugging $v_{ASK}(t)$ from (9.2.10) into (6.3.39), we find

$$V_{rms}^2 = \frac{1}{T}\int_0^T v_{ASK}^2(t)dt$$
$$= \frac{1}{T}\int_0^T \left[\left(\frac{A_C \cdot A_M}{2}\right)\cos(\omega_C t) + \sum_{n=-\infty}^{\infty}\left(\frac{A_C \cdot A_M}{n\pi}\right)\cos(\omega_C + n\omega_M)t\right]^2 dt \quad (9.2.11)$$

Notice that integration is performed from zero to T, as in (6.3.40), and n is odd only. Since $(x+y)^2 = x^2 + 2xy + y^2$, we found the following three members of (9.2.11):

$$\frac{1}{T}\int_0^T \left(\left(\frac{A_C \cdot A_M}{2}\right)\cos(\omega_C t)\right)^2 dt = \frac{(A_C \cdot A_M)^2}{8}$$

$$\frac{1}{T}\left(\frac{A_C \cdot A_M}{2}\right)\cos(\omega_C t) \cdot \sum_{n=-\infty}^{\infty}\left(\frac{A_C \cdot A_M}{n\pi}\right)(\cos(\omega_C + n\omega_M)t) = 0$$

$$\frac{1}{T}\int_0^T \sum_{n=-\infty}^{\infty}\left(\frac{A_C \cdot A_M}{n\pi}\right)[\cos(\omega_C + n\omega_M)t)]^2 dt$$

$$= \frac{1}{2}\sum_{n=-\infty}^{\infty}\left(\frac{A_C \cdot A_M}{n\pi}\right)^2 = \sum_{n=1}^{\infty}\left(\frac{A_C \cdot A_M}{n\pi}\right)^2 \quad (9.2.12)$$

The manipulations leading to the results of this integration are similar to those we used in deriving (6.3.40). Do not forget, we assume $R = 1\,\Omega$ to obtain a normalized power.

3. Equation (9.2.12) enables us to compute the power of individual harmonics. Then we need to sum them up one by one, find the required power, and – consequently – the bandwidth. The results are presented in Table 9.2.1. (For details on this operation, refer to Section 6.3 and see Table 6.3.2.)

Figure 9.2.5a shows the power spectrum of the signal discussed here and Figure 9.2.5b demonstrates the cumulative power delivered by the signal's harmonics. Observe the correlation between Figure 9.2.5a,b: As the higher-order harmonics become smaller and smaller, they contribute less and less to the cumulative power, as the curve in Figure 9.2.5b shows.

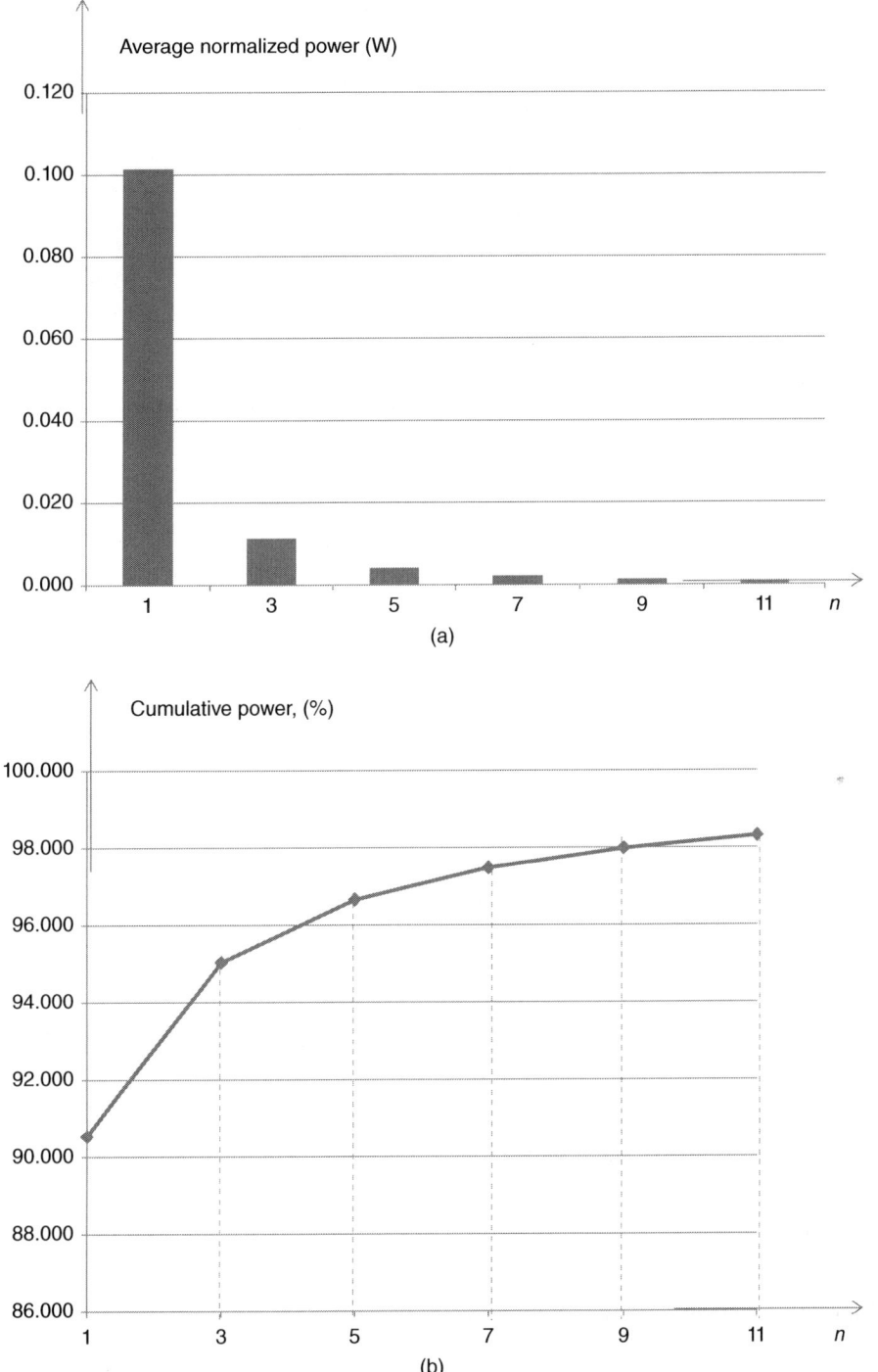

Figure 9.2.5 (a) Power spectrum and (b) cumulative power in percentage of the total power of the ASK signal in Example 9.2.2.

Table 9.2.1 Power and bandwidth of the ASK signal in Example 9.2.2.

Harmonic number, n	Normalized average power of the individual harmonics, $P_n(W) = \left(\frac{A_c \cdot A_M}{n\pi}\right)^2$	Cumulative power by harmonics, $P_c(W) = \frac{1}{8} + P_n$	Percentage of total power ($P_T = 0.25\,V^2$) by harmonics	Bandwidth of the ASK signal by harmonics, $BW_{signal} = 2n \cdot f_M$ (kHz)
1	0.101	0.226	90.528	2
3	0.011	0.238	95.032	6
5	0.004	0.242	96.653	10
7	0.002	0.244	97.480	14
9	0.001	0.245	97.980	18
11	0.001	0.246	98.315	22

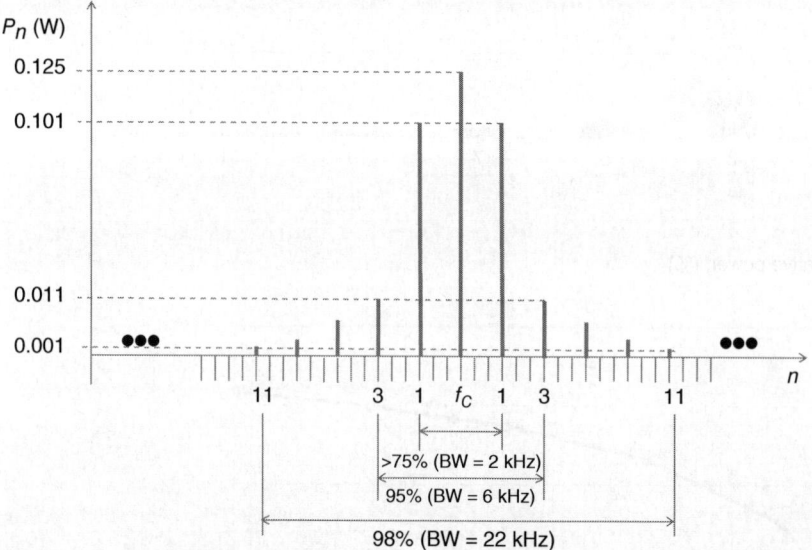

Figure 9.2.6 Power spectrum, percentage of the total power, and bandwidth of the binary ASK signal in Example 9.2.2.

Thus, the solution to this problem can be gathered from Table 9.2.1 as follows:
The bandwidth of the given ASK signal with a square wave message is equal to

2 kHz for 75% power,
6 kHz for 95% power,
22 kHz for 98% power.

Refer to Figure 9.2.6, where this solution is visualized.

Discussion

- Note that the relationship between percentage of delivered total power, the required number of harmonics, and the ASK bandwidth is shown in Figure 9.2.6. We see that to deliver 75% of the total power, we need only one harmonic. This implies that the receiver will "see" only a single

sinusoid at the frequency $f_C + f_M$, enabling it to recognize only 1s and 0s and the frequency of the message signal, f_M. For the 95% and 98% levels, we need to transmit 3 and 11 harmonics, respectively; in the latter case, the receiver will be able to reconstruct the pulse significantly closer to the original pulse's shape. We discussed this issue in Chapter 6; refer, for example, to Figure 6.3.6.
- If a message signal is nonperiodic, its spectrum, as we know, will be continuous, such as that shown in Figure 9.2.4. The bandwidth of an ASK signal with a nonperiodic message can be found in a way similar to that presented in this example, but we need to use the Fourier transform of the signal, as we did in Section 7.1. We will learn shortly how to find the bandwidth of an ASK signal with a nonperiodic message.
- This problem provides a good lesson in the application of spectral analysis to the characterization of various signals; that is, finding their spectrum, bandwidth, and other key features. We will use this example as a guide for characterizing other binary shift keying signals.

9.2.2.5 Bandwidth and Bit Rate of ASK

The relationship between the bandwidth and the bit rate of a signal is an important issue in the consideration of a transmission system.

> *When we transmit a stream of pulses, we describe its bit rate in bits per second, b/s; but to design a transmission channel correctly, we need to know what bandwidth in hertz this stream occupies.*

An ASK transmission is a good example of this problem: We have a digital message characterized by a bit rate, BR (b/s), whereas transmission occurs through a communications channel (link) with limited bandwidth, BW_{ch} (Hz). Hence, we must know the bandwidth of the transmitted ASK signal, BW_{ASK} (Hz), to properly design a transmission system; that is, to ensure that BW_{ch} accommodates BW_{ASK}.

To be sure, for an ASK signal, the relationship between BW and BR depends on the message signal. For the *square wave message*, as in Figure 9.2.2 and Example 9.2.1, this relationship is simple: The period of a square wave message, T_M, is twice the width of the pulse, T_b, i.e. $T_M = 2T_b$, which means

$$f_M \text{ (Hz)} = 1/T_M \quad \text{and} \quad BR \text{ (b/s)} = 1/T_b \Rightarrow f_M \text{ (Hz)} = \frac{BR}{2} \text{(b/s)}$$

Since the bandwidth of an ASK signal is given in multiples of f_M, we can write for the square wave message,

$$BW_{ASK} \text{ (Hz)} = \frac{BR_{ASK}}{2} \text{(b/s)} \qquad (9.2.13)$$

(Does this formula look familiar to you? It should be. See (1.3.2) and (1.3.3).)

In general, however, this relationship should be found through a more sophisticated method: We need to derive the formula of an ASK signal with an arbitrary message, $v_{ASK}(t)$, use the Fourier transform to find its spectrum and, with additional manipulations, determine the power spectrum. From this we can find the needed bandwidth based on the required received power. Such operations are similar to those we performed in Section 7.1 and also here in Example 9.2.2. Finding the bandwidth of an arbitrary ASK signal would be a good exercise for you. We leave this problem to more specialized textbooks (such as Stern and Mahmoud 2004, Page 218 and Carlson and Crilly 2010, Page 651) and simply present the main results in concise form.

The average normalized power spectral density, $S(f)$, of a pulse train, which is a one-sided spectrum of the train, has a $\text{sinc}^2(\pi ft)$ shape:

$$S(f) \sim \text{sinc}^2(\pi ft) \; (\text{V}^2/\text{Hz}) \tag{9.2.14}$$

Such a graph is shown in Figure 9.2.7. The first zeros occur at $f\,(\text{Hz}) = 1/\tau\,(\text{s})$, where $\tau\,(\text{s}) = T_b$ is the pulse width (duration). Since the bit rate is inversely proportional to the pulse width,

$$\text{BR (b/s)} = \frac{1}{\tau(s)} = \frac{1}{T_b(s)} \tag{9.2.15}$$

we can say that the first zeros occur at BR (b/s). The area under the main lobe provides 90% of the signal's total power; if we include both second lobes, we obtain 95%, and so on. Therefore, if we want the channel bandwidth to deliver 90% of the total power, then the required bandwidth is equal to double the bit rate; i.e.

$$\text{BW}_{\text{ASK}}\,(\text{Hz}) = (f_C + \text{BR}) - (f_C - \text{BR}) = 2\,\text{BR (b/s) at 90\% power} \tag{9.2.16a}$$

If we need to deliver 95% of the total power, then we find

$$\text{BW}_{\text{ASK}}\,(\text{Hz}) = 4\text{BR (b/s) at 95\% power} \tag{9.2.16b}$$

and so on.

It is common industry practice to consider pulse bandwidth as the range of frequencies between first nulls (zero crossings) of the graph *power spectral density vs. frequency*. In Figure 9.2.7, these nulls occur at $fc\,(\text{Hz}) = \text{BR}$, which provides 90% of the signal power. Be careful, however, not to infer that the first-null bandwidth is the rule; typically, bandwidth is accompanied by notations clarifying how it is defined in each specific case. Such an approach is similar to that used in defining rise/fall time, where we always have to refer to which rule is being applied: 90% to 10% or 80% to 20%.

We need to recall that, theoretically, the bandwidth of a train of ideal rectangular pulses tends to infinity. Transmitting such a hypothetical signal through a band-limited channel will result in cutting the signal's bandwidth and therefore distortion of the waveforms of the pulses. Figure 9.2.8 (top) shows that a receiver will obtain curve-shaped pulses instead of rectangular ones. If, however, we send curve-shaped pulses (ones without discontinuities at the corners), then such pulses will

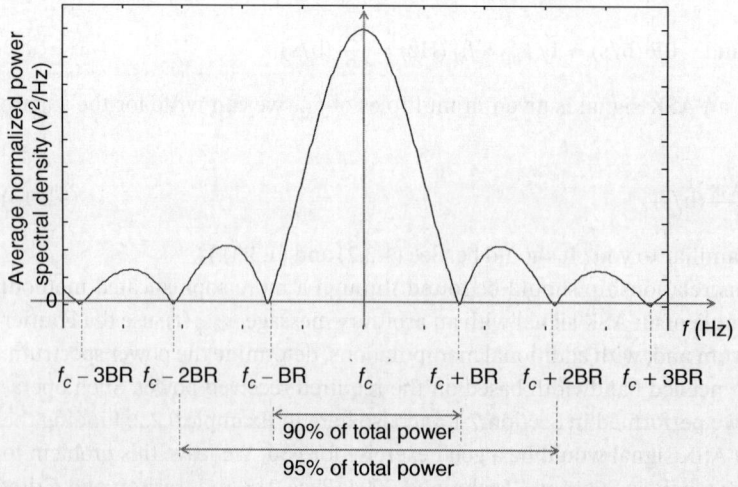

Figure 9.2.7 The power spectrum and bandwidth of a rectangular pulse train.

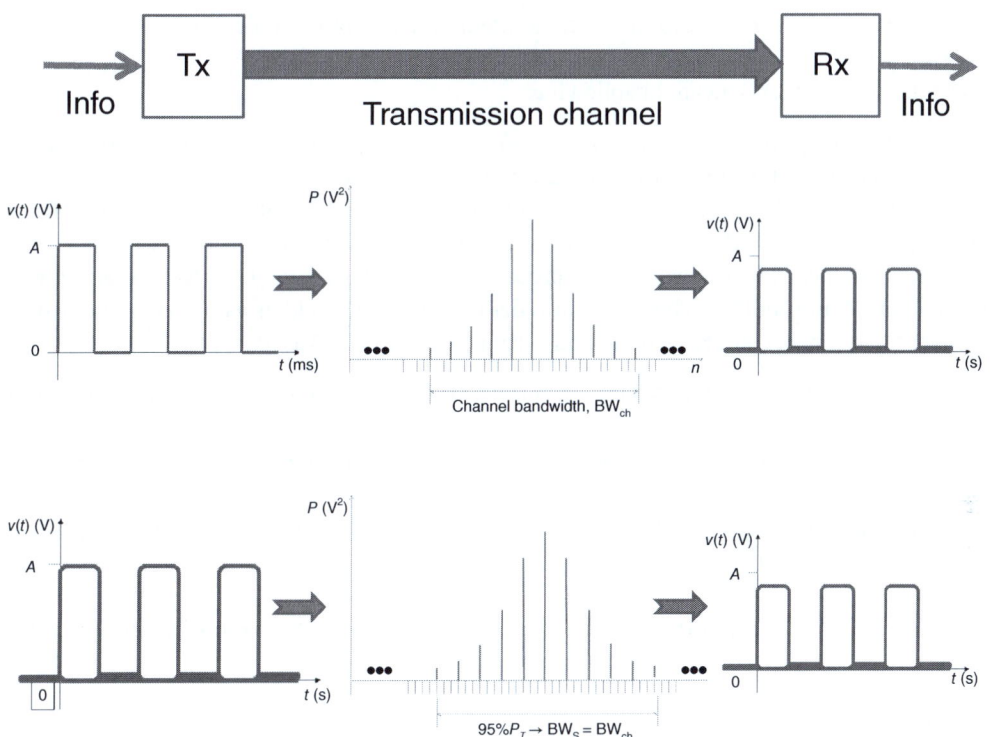

Figure 9.2.8 Pulse waveforms and their changes due to channel bandwidth: If rectangular pulses are sent, then their big (theoretically infinite) bandwidth will be cut by a bandlimited channel to a smaller one, which will result in a smother shape of the output pulses (top); if smooth, curved pulses are sent and their bandwidth fits the channel's bandwidth, then the shape of the pulses will be substantially preserved (bottom).

require a limited bandwidth. *By properly shaping the pulses before transmission, we can make the signal's bandwidth fit the channel bandwidth*, which results in minimal distortion of a transmitted ASK signal. This situation is shown in the bottom section of Figure 9.2.8. Sending curve-shaped pulses is a popular approach to reducing the required bandwidth for ASK transmission.

9.2.2.6 Bit Error Ratio, BER, of ASK System

In Section 9.1, we introduce and discuss BER. Here we want to elaborate on this important topic and apply the BER basic concept to specific types of shift-keying digital transmission.

A reminder: *The bit error ratio, BER, is the ratio of the number of erroneous bits to the number of total bits transmitted.* For instance, if a system transmitted 10^{12} bits and one bit was in error, then BER $= 1/10^{12} = 10^{-12}$. BER is one of the most important measures of transmission quality. Obviously, the ideal BER is equal to zero; today's optical transmission routinely requires BER $= 10^{-12}$ and this requirement is rising to 10^{-15}.

To discuss the BER of ASK, we need to obtain a formula for its calculation. Fortunately, this formula can be found in any textbook on communications; in our notations, it takes the form

$$\text{BER}_{\text{ASK}} = Q\left\{\sqrt{\frac{E_b}{N_0}}\right\} \tag{9.2.17}$$

(The derivation of (9.2.17) can also be found in literature; see, for example, Haykin and Mosher, 2007, Page 419.)

The notations in (9.2.17) mean the following:

- E_b (J) is the average bit energy given in (9.1.38) as E_b (J) = P_b (W) · T_b (s), where P_b (W) and T_b (s) are the power and the interval of an individual bit, respectively.
- N_0 (V²/Hz), or N_0 (W/Hz), is the one-sided noise power spectral density. It can be derived from (1.3.12) as N_0 (W/Hz) = P_N (W) · T_b (s), where P_N (W) the noise power.
- Q is the Q-*function*, which calculates the probability of finding a variable x greater than a given value a for a Gaussian PDF. This function is defined in (9.1.22), and its values are presented in Table 9.1.2. It can, of course, be found and calculated in MATLAB as *qfunc*.

In essence, the ratio E_b (J)/N_0 (W/Hz) = E_b (J)/N_0 (J) is the SNR of a received digital signal (see Eq. (9.1.40)). Therefore (9.2.17) tells us the probability of receiving an error signal with a given digital SNR. You are urged to review Sections 9.1.3 and 9.1.4 to refresh your memory on these topics.

These considerations are very important in solving the practical problems of modern transmission, as Example 9.2.3 can attest.

Example 9.2.3 Bit Error Ratio and the Required Received Power (after Mesiya 2013, p. 601)

Problem

A binary ASK signal is transmitted via a fiber-optic communication system. The required BER_{ASK} is 1.3×10^{-12}. The noise added by the link is characterized by its power spectral density, $N_0 = 1.6 \times 10^{-18}$ W/Hz. What signal power must be delivered to a receiver to sustain the given BER_{ASK} at BR = 100 Gb/s?

Solution

We have learned that the strength of an optical signal is characterized by power in watts. Also, we know that BR = 100 Gb/s is a regular transmission speed in modern optical communications and, in fact, even much higher speeds are commercially available.

How can we approach to solving this problem? We need to relate BER to signal power. This relationship can be found through (9.2.17) that refers BER to the average bit energy, E_b (J), by using which and giving timing we can obtain P_{rec} (W). Thus, we write

$$\text{BER}_{\text{ASK}} = 1.3 \times 10^{-12} = Q\left\{\sqrt{\frac{E_b}{N_0}}\right\}$$

Using Table 9.1.2 of Q-function values, we find the argument of Q-function as

$$Q\left\{\sqrt{\frac{E_b}{N_0}}\right\} = 1.3 \times 10^{-12} \Rightarrow \sqrt{\frac{E_b}{N_0}} = 7.0$$

(In the discussion section of this example, you can find the MATLAB code for finding the argument of Q-function.)

Therefore,

$$E_b = 49 N_0 = 49 \times 1.6 \times 10^{-18} \text{ J} = 78.4 \times 10^{-18} \text{ J}$$

From (9.1.38) we get

$$P_b (W) = P_{rec} (W) = \frac{E_b(J)}{T_b(s)} = E_b (J) \cdot BR (b/s) \tag{9.2.18}$$

That is,

$$P_{rec} (W) = 78.4 \times 10^{-8} \times 100 \times 10^9 = 7.8 \times 10^{-6} \, W = 7.8 \, \mu W$$

The communication industry typically uses logarithmic measures for power; hence, converting milliwatts into dBm as

$$P (dBm) = 10 \cdot \log(P (mW/1 \, mW)$$

yields

$$P_{rec} (dBm) = 10 \, \log (0.0078/1) = -21.1 \, dBm$$

For modern optical communications, this is the typical received power required for BER $= 10^{-12}$. The problem is solved.

Discussion

- Note that we can obtain the formula for direct calculation of $P_{rec} (W) = P_b (W)$ by dividing both the numerator and the denominator of (9.2.17) by T_b (s); this manipulation results in

$$BER = Q\left\{\sqrt{\frac{E_b/T_b}{N_0/T_b}}\right\} = Q\left\{\sqrt{\frac{P_b(W)}{P_N(W)}}\right\} \tag{9.2.19}$$

The noise spectral density, N_0 (W/Hz), is related to the noise power, P_N (W), as

$$P_N (W) = \frac{N_0(W/Hz)}{T_b(s)} = N_0 (W/Hz) \cdot BR (b/s) \tag{9.2.20}$$

(This result has been introduced in (9.1.39b).)
Now, remember that

$$Q\left\{\sqrt{\frac{P_b(W)}{P_N(W)}}\right\} = BER = 1.3 \times 10^{-12}$$

The argument of Q-function is found from Table 9.1.2 to be equal to 7, which means

$$\sqrt{\frac{P_b(W)}{N_0(W/Hz) \cdot BR(b/s)}} = 7.0$$

Direct computations yield

$$P_{rec} (W) = 49 N_0 \cdot BR (W) = E_b \, BR (W) = 7.8 \, \mu W$$

as above.
- We are aware that the minimum received power that enables a communications system to sustain the required BER is called *receiver sensitivity*, P_{rec} (W).

- Since the received power is proportional to the bit rate, the greater the bit rate, the greater the amount of power required at the destination point.
- This example shows that the smaller the noise power spectral density, the lower the amount of power required at the receiver.
- The importance of this example becomes immediately apparent as we turn to the discussion of the application of ASK modulation in optical communications.

MATLAB Code for Working with *Q*-function in Example 10.2.3

The following MATLAB code finds the argument of *Q*-function $Q(x) = 1.3*10^{(-12)}$:

```
for x = 1:1:10
Q = qfunc(x);
if Q<1.3e-12
if Q>1.2e-12
x
end
end
end
```

9.2.2.7 ASK Advantages, Drawbacks, and Applications

The main advantage of ASK modulation is its simplicity, which is why this modulation was used in the first application of digital binary signals: teletype service. However, this modulation has all the drawbacks inherent in amplitude modulation, the main one being noise susceptibility. Because of this shortcoming, ASK was not actively used for years in electronic communications; in fact, many textbooks covered ASK simply for the sake of completeness, explaining that it was out of favor. That was true until the advent of optical communications about 30 years ago. From those days and until recently, this industry has used binary ASK as a main modulation format. Today, optical communications use more sophisticated modulation formats (see Chapter 10); nevertheless, BASK is still the first among the modulation formats of optical – and therefore all modern – communications.

Optical communications – in a very basic form – delivers bit "1" when a light pulse passes through optical fiber and bit "0" when a receiver "sees" no light. The light pulses are formed from the segments of a continuous-wave light. On the other hand, light, as we all know, is a form of electromagnetic radiation, and continuous-wave light is of course an electromagnetic wave, too. Therefore, light pulses are the segments of a sinusoidal wave, as shown in Figure 9.2.2; that is, ON–OFF optical modulation is a binary ASK. What distinguishes optical ASK from conventional BASK are the characteristics of light and optical transmission, primarily light frequency and transmission speed. To clarify this point, let us turn to Example 9.2.4.

Example 9.2.4   ASK Parameters for Wireless and Optical Communications

Problem

How many cycles of a carrier wave does one pulse of binary ASK transmission contain if ASK is applied (1) for a Wi-Fi transmission supported by a 5-GHz wave and operating at 1 Gb/s transmission rate and (2) for an optical-communication link supported by a 200-THz wave and operating at 1 Tb/s?

Solution

To find the number of cycles within one pulse, we need to know the duration (width) of the pulse, T_b, and the interval (period), T, of one cycle. We know that the pulse's duration can be computed as

$$T_b(\text{s}) = \frac{1}{\text{BR(b/s)}}$$

The period of an EM wave can be found from the fundamental relationship among the wavelength, λ (m), the frequency, f (Hz), and the speed of light in vacuum, c (m/s):

$$\lambda(\text{m}) \cdot f(\text{Hz}) = c(\text{m/s}) \qquad (9.2.21)$$

as

$$T_{\text{EM}}(\text{s}) = \frac{\lambda(\text{m})}{c(\text{m/s})} \qquad (9.2.22)$$

Employing these formulas, we can perform the required calculations:

(1) For Wi-Fi transmission, we compute

$$T_{b1} = \frac{1}{\text{BR}_1(\text{b/s})} = 1 \times 10^{-9} = 1 \text{ ns}$$

and

$$T_{\text{EM1}} = \frac{1}{f_1(\text{Hz})} = 0.2 \times 10^{-9} = 0.2 \text{ ns}$$

The number of carrier-wave cycles within one pulse, n_1, is given by

$$n_1 = \frac{T_{b1}(\text{s})}{T_{\text{EM1}}(\text{s})} = 5$$

Thus, for Wi-Fi transmission, each BASK pulse contains 5 cycles of an EM wave.

(2) Similar calculations for optical communications result in the following numbers:

$$T_{b2} = 1 \times 10^{-12} = 1 \text{ ps}$$

and

$$T_{\text{EM2}} = 5 \times 10^{-15} = 5 \text{ fs}$$

The number of cycles is equal to

$$n_2 = \frac{T_{b2}(\text{s})}{T_{\text{EM2}}(\text{s})} = 200$$

Therefore, for optical communications, each BASK pulse carries 200 cycles of an optical wave.

Discussion

- The 5-GHz frequency of the Wi-Fi carrier is classified as the *super-high frequency (SHF)* range of the radio-frequency (RF) band. See Section 1.2.3.
- Though the bit rate of optical communications is 1000 times greater than that of Wi-Fi, the number of cycles accommodated by one optical pulse is only 40 times greater than that of Wi-Fi. Why is this?
- The greater the number of cycles per bit, the more reliable the transmission. This is because the corruption of several individual cycles will not corrupt the entire pulse.

- This example shows the importance of having the carrier frequency, f_C, high enough to perform high-quality modulation. Imagine that T_C (s) > T_b; that is, f_C (Hz) < BR (b/s); then ASK modulation would be simply impossible. Refer to conclusion of our discussion of Figure 9.2.4: the higher the carrier frequency, the greater the number of harmonics that can be accommodated within the range from 0 to f_C. Indeed, if f_C (Hz) = $1/T_b$, then, as Figure 9.2.4 shows, the two graphs of the ASK spectrum will overlap and we would not be able to decipher the message from the resulting signal.
- It is necessary to know that optical pulse experiences various distortions during transmission through optical fiber, including a decrease in its amplitude due to losses and an increase in its duration (width) caused by dispersion. For detailed discussion of this topic, refer to specialized texts, such as Mynbaev and Scheiner (2001, pp. 49 and 57.)

9.2.2.8 Detection (Demodulation) of an ASK Signal

We will consider, as an example, a coherent demodulation of an ASK signal. The term *coherence* refers to correlation between the phases of two waves. *Coherent demodulation* implies *beating an input signal with a locally generated harmonic, $2\cos(\omega_c t)$, of the same frequency and phase as a carrier wave*. The typical block diagram of a coherent ASK receiver is shown in Figure 9.2.9. A bandpass filter, BPF, enables the passing of an ASK signal and rejects out-of-band noise. After multiplication, the input pulse train, noise, and the product $\cos(\omega_c t) \times 2\cos(\omega_c t) = 2\cos^2(\omega_c t) = 1 + \cos(2\omega_c t)$, are presented to an integrator. Here, the integrator is a special type of a filter with an ability to maximize the SNR for an output signal. It can do this because the waveform of an input signal is known and the filter's impulse response is matched to this waveform. For this property, it is called a *matched filter*. It presents to the timer the maximum value of its output signal at the sampling instance, which improve the probability to correctly detect the received bits.

The integrator additionally removes the high-frequency component, $\cos(2\omega_c t)$, and sends the digital pulse train down the line. Hence, coherent demodulation improves the process of eliminating a carrier wave from a received signal and, in general, increases the accuracy of recovering information. (Removing a carrier wave is also achieved with incoherent demodulation – a demodulation type that is still in use today – by direct filtering a sinusoidal component from the modulated input signal.) A timer (called a *sampler*) controls the detection timing process by taking a sample every bit interval. A decision circuit is essentially a comparator that compares the sampled pulse value with a threshold and decides which bit – 1 or 0 – is obtained. (See Section 9.1.3 for full explanation of the decision process in digital transmission.)

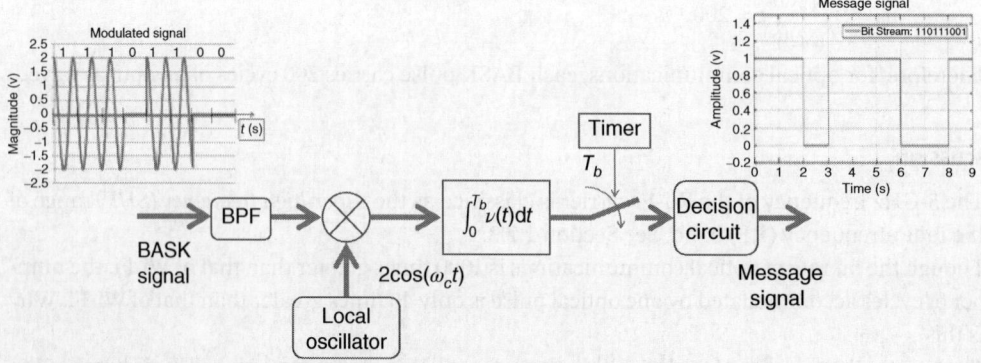

Figure 9.2.9 Typical configuration of a coherent BASK receiver.

9.2 Introduction to Digital Transmission – Binary Shift-Keying Modulation

To conclude our discussion of binary ASK modulation, it is necessary to mention that there are modulation formats that contain several bits in one symbol (pulse). These formats are called M-ary or multilevel; they are mentioned in Section 3.2 and discussed in Chapter 10. Therefore, *there are ASK modulations other than the binary type*. In fact, modern optical communications has moved from binary to such modulations as quadrature-amplitude modulation (QAM) and quadrature phase-shift keying (QPSK) concepts that we will take up in next chapter. This means that modern communications actively explores sophisticated ASK modulations. We would not consider these types of ASK here, but the basic knowledge you have acquired from this discussion provides you with a solid foundation for the study of all current and future modifications of ASK modulation.

9.2.3 Binary Frequency-Shift Keying (FSK)

9.2.3.1 FSK Concept and Waveform

As the name suggests, this is the shift keying modulation in which bit 1 of a message signal is represented by one frequency of a carrier wave and bit 0 by another frequency. The principle of operation of a *binary FSK* (often designated BFSK) transmission system is shown in Figure 9.2.10.

The principle of FSK modulation is similar to that of frequency modulation discussed in Section 9.1: The message signal changes the carrier frequency to enable it to carry information. Figure 9.2.10 shows that an FSK transmission system operates similarly to any broadband

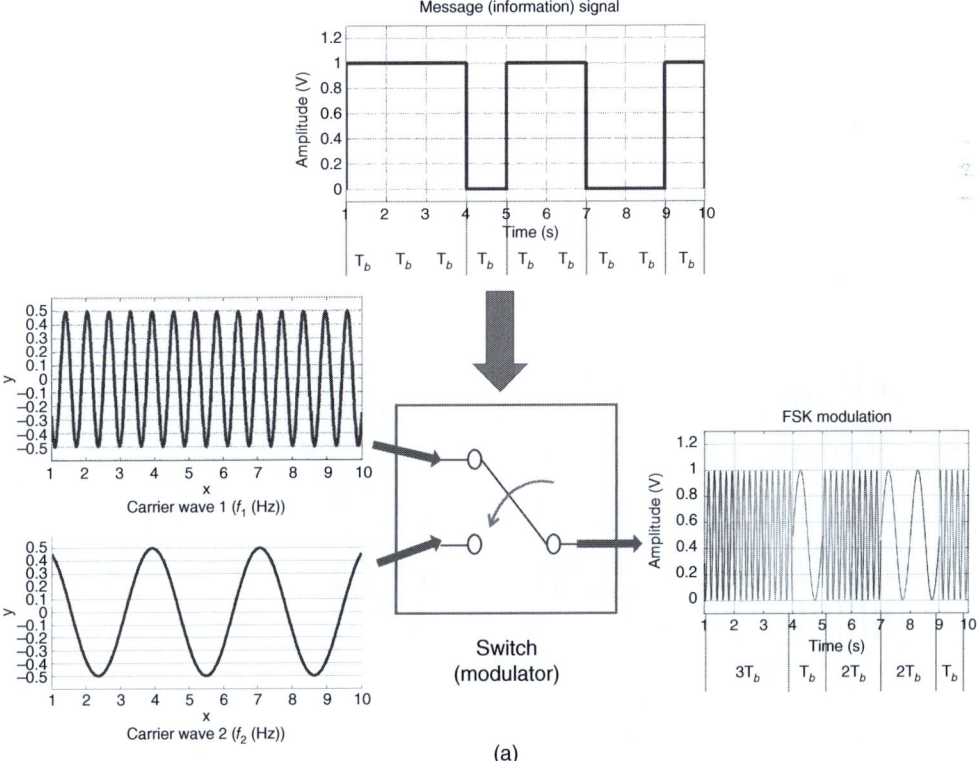

Figure 9.2.10 Binary frequency-shift keying (FSK) modulation: (a) The modulation concept; (b) a MATLAB/Simulink simulation of FSK modulation; (c) a binary FSK signal built with MATLAB; and (d) an FSK transmission system.

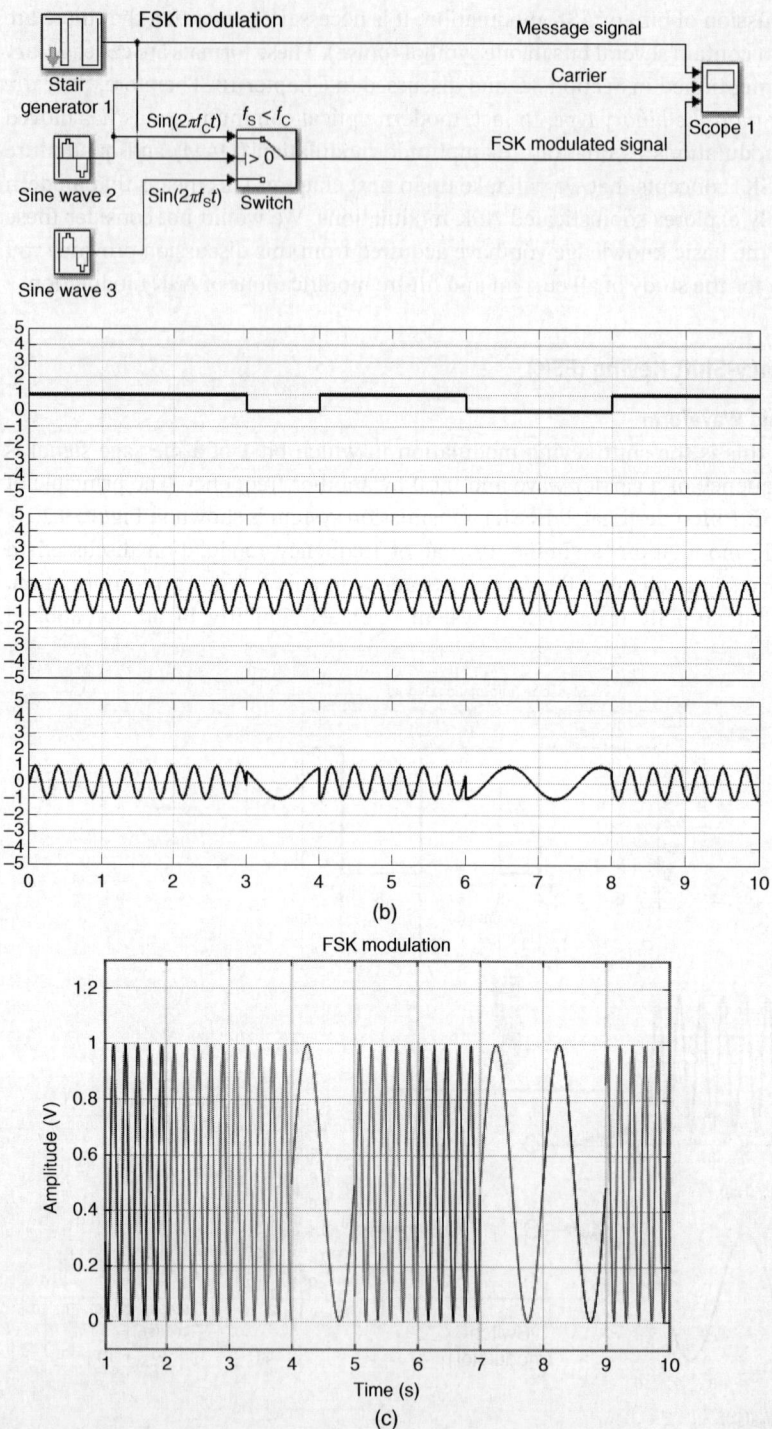

Figure 9.2.10 (Continued)

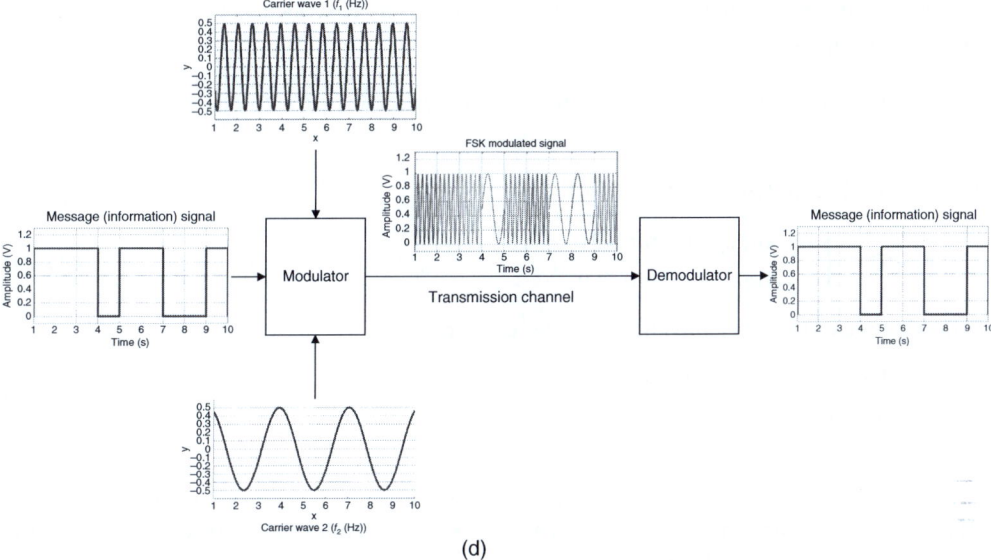

(d)

Figure 9.2.10 *(Continued)*

modulation system: A message signal (the binary signal in this case) changes a carrier wave (which uses two frequencies in a BFSK system) to deliver information. A modulator (a switch in Figure 9.2.10a) assigns f_{C1} to carry bit 1 and f_{C2} to carry bit 0. The FSK modulated signal travels through a communication channel (link). When the FSK signal arrives at its destination, the receiver demodulates the signal, which means it discards the carriers and extracts the message (the information). In essence, *a BFSK system functions as two ASK systems with different carrier frequencies but the same amplitudes.*

Figure 9.2.10b demonstrates the circuit for MATLAB/Simulink simulation of an FSK modulation and the waveforms of a message (top), carrier (middle), and the FSK (bottom) signals. Figure 9.2.10c shows the FSK signal built with MATLAB by applying (9.2.24). We can clearly see in Figure 9.2.10b,c that the change from f_{C1} to f_{C2} and back causes the signal phases change abruptly. This is because the switch shown in Figure 9.2.10a does not control the phases. This simple modulation method is called *DPFSK*. Figure 9.2.10d presents an entire FSK transmission system.

9.2.3.2 Mathematical Description of FSK

Two sinusoidal carrier waves with different frequencies and the same amplitudes used in BFSK are given by

$$v_{C1}(t) = A_C \cos(\omega_{C1} t)$$

and

$$v_{C2}(t) = A_C \cos(\omega_{C2} t)$$

Therefore, a BFSK-modulated signal can be mathematically described as the product of carrier waves and message signals; that is,

$$v_{\text{FSK}}(t) = v_{M1}(t) \cdot v_{C1}(t) + v_{M2}(t) \cdot v_{C2}(t) = v_{M1}(t) \cdot A_C \cos(\omega_{C1}t) + v_{M2}(t) \cdot A_C \cos(\omega_{C2}t) \tag{9.2.23}$$

As a result, BFSK is described by the formula

$$v_{\text{FSK}}(t) = \begin{cases} A_C \cos(\omega_{C1}t) & \text{for bit 1} \\ A_C \cos(\omega_{C2}t) & \text{for bit 0} \end{cases} \tag{9.2.24}$$

Following the format presented in (9.2.2), we can write

$$\begin{aligned} \text{Bit } 1 &\Rightarrow A_C \cos(\omega_{C1}t) \quad \text{for } t_1 = T_b \\ \text{Bit } 0 &\Rightarrow A_C \cos(\omega_{C2}t) \quad \text{for } t_0 = T_b \end{aligned} \tag{9.2.25}$$

where T_b (s) is the duration of the bit, ω_{C1} (rad/s) and ω_{C2} (rad/s) are the frequencies of the carrier with $\omega_{C1} > \omega_{C2}$, and A_C (V) is the carrier's amplitude. Observe again that the amplitude of the carrier waves is the same; we expect this because FSK delivers information by changing only the carrier frequency. Formula (9.2.24) shows that a BFSK signal is the combination of two carrier waves that differ only in frequency.

Assuming that the base carrier frequency, ω_C, is fixed, two carrier frequencies can be determined as

$$\omega_{C1} = \omega_C + \Delta\omega$$
and
$$\omega_{C2} = \omega_C - \Delta\omega \tag{9.2.26}$$

Here, $\Delta\omega$ is the *frequency deviation*, the term we used in introducing the concept of frequency modulation in Section 9.1. Now an FSK signal can be described as

$$v_{\text{FSK}}(t) = \begin{cases} A_C \cos(\omega_{C1}t) = A_C \cos(\omega_C + \Delta\omega)t & \text{for bit 1} \\ A_C \cos(\omega_{C2}t) = A_C \cos(\omega_C - \Delta\omega)t & \text{for bit 0} \end{cases} \tag{9.2.27}$$

9.2.3.3 FSK Spectrum and Bandwidth with Square Wave Message

First, let us consider an FSK signal with a periodic square wave message signal. This message signal, as we mentioned in our ASK discussion, can be developed into the following Fourier series:

$$v_{M1}(t) = A_{M1}\left(\frac{1}{2} + \frac{2}{\pi}\sin\omega_0 t + \frac{2}{3\pi}\sin 3\omega_0 t + \frac{2}{5\pi}\sin 5\omega_0 t + \cdots\right) \tag{6.1.13R}$$

But in BFSK we need to find the Fourier series for two message signals, $v_{M1}(t)$ and $v_{M2}(t)$. To do so, we note from Figure 9.2.10a that the second signal, $v_{M2}(t)$, is simply a complement of the first, $v_{M1}(t)$; that is, $v_{M2}(t) = 1 - v_{M1}(t)$. For example, if $v_{M1}(t)$ is bit 1, then $v_{M2}(t)$ is bit 0; likewise, if $v_{M1}(t)$ is bit 0, then $v_{M2}(t)$ is bit 1. Since the period, T (s), of a square wave is given, the fundamental (message) frequency, f_0 (Hz) $\equiv f_M$ (Hz) $= 1/T$ (s), for both message signals is the same. Thus,

$$v_{M2}(t) = 1 - v_{M1}(t) = \left(1 - \frac{A_{M1}}{2}\right) - A_{M1}\left(\frac{2}{\pi}\sin\omega_0 t + \frac{2}{3\pi}\sin 3\omega_0 t + \frac{2}{5\pi}\sin 5\omega_0 t + \cdots\right) \tag{9.2.28}$$

In finding the spectrum of a BFSK, we follow the approach used in deriving (9.2.4) for the ASK spectrum:

- Take the main formula of a BFSK signal, (9.2.23), in the following form

$$v_{BFSK}(t) = v_{M1}(t) \cdot v_{C1}(t) + v_{M2}(t) \cdot v_{C2}(t) = v_{M1}(t) \cdot A_C \cos(\omega_{C1} t) + v_{M2}(t) \cdot A_C \cos(\omega_{C2} t)$$
(9.2.29)

- Plug (6.1.13R) and (9.2.28) into (9.2.29) and use the trigonometric identity $\sin A \cdot \cos B = (\sin(A+B) + \sin(A-B))/2$ to obtain

$$v_{FSK}(t)(V) = \frac{A_C A_M}{2} \cos(\omega_{C1} t)$$

$$+ \frac{A_C A_M}{\pi}[(\sin(\omega_{C1} + \omega_0)t + \sin(\omega_{C1} - \omega_0)t)$$

$$+ \frac{1}{3}(\sin(\omega_{C1} + 3\omega_0)t + \sin(\omega_{C1} - 3\omega_0)t)$$

$$+ \frac{1}{5}(\sin(\omega_{C1} + 5\omega_0)t + \sin(\omega_{C1} - 5\omega_0)t) + \cdots]\}$$

Segment of FSK spectrum centered at ω_{C1}

$$+ A_C \left(1 - \frac{A_{M1}}{2}\right) \cos(\omega_{C2} t)$$

$$- \frac{A_C A_M}{\pi}[(\sin(\omega_{C2} + \omega_0)t + \sin(\omega_{C2} - \omega_0)t)$$

$$+ \frac{1}{3}(\sin(\omega_{C2} + 3\omega_0)t + \sin(\omega_{C2} - 3\omega_0)t)$$

$$+ \frac{1}{5}(\sin(\omega_{C2} + 5\omega_0)t + \sin(\omega_{C2} - 5\omega_0)t) + \cdots]\}$$

Segment of FSK spectrum centered at ω_{C2} (9.2.30)

Equation (9.2.30) describes the spectrum of a BFSK signal, and Figure 9.2.11 depicts this spectrum. We can see that the spectrum of a BFSK signal is formed from the spectra of two ASK signals, one of which is carried by ω_{C1} and the other by ω_{C2}. This result is expected and obvious because we treat the FSK signal as two ASK signals with different frequencies. Observe that in Figure 9.2.11

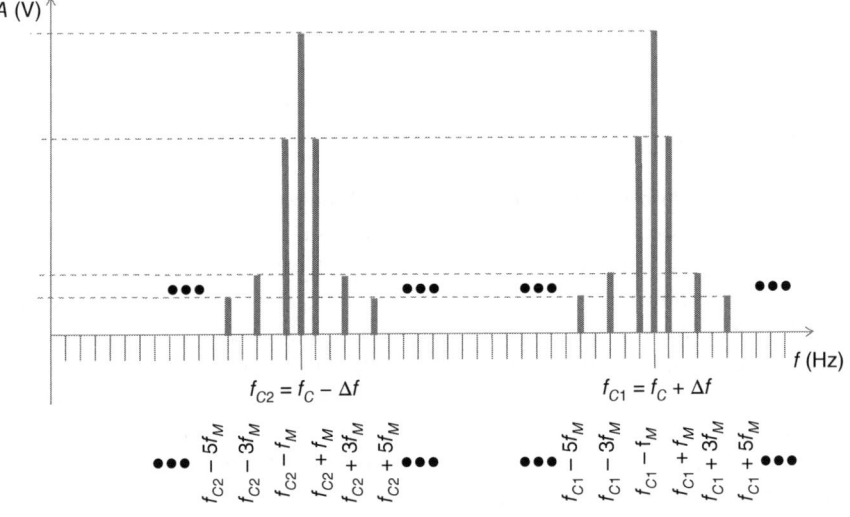

Figure 9.2.11 Spectrum of a BFSK signal with a periodic message.

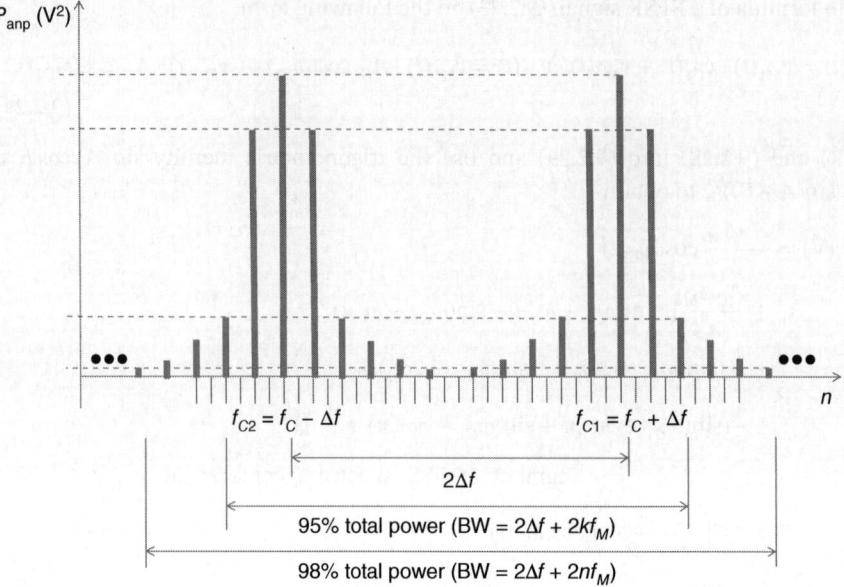

Figure 9.2.12 The power bandwidth of a binary FSK signal with a periodic message signal.

we (1) refer to f (Hz) rather than to ω (rad/s), as in (9.2.30), and (2) we use a different notation for the fundamental (message) frequency, f_M (Hz) $\equiv f_0$ (Hz), to stress that each spectral component is shifted by a multiple of this frequency.

Since a BFSK signal can be considered a combination of two ASK signals, its *bandwidth* can be found based on this presumption. For an FSK with a periodic message, the *power spectrum* is merely two power spectra of an ASK signal separated by doubled frequency deviation $2\Delta f$. To find the specific numbers, we need to perform calculations similar to those we did in Example 9.2.2. (We urge you to carry out these calculations.) We present a general view of such a spectrum and FSK bandwidth in Figure 9.2.12. Figure 9.2.12 shows that there are two identical spectra centered at $f_{C1} = f_C + \Delta f$ and at $f_{C2} = f_C - \Delta f$. The bandwidth is determined by the percentage of the total power, which in this case means by the number of harmonics. For example, Figure 9.2.12 shows that we need $n = 16$ harmonics to obtain 95% and $n = 22$ harmonics to obtain 98% of the total signal power. (See Example 9.2.2, where the bandwidth of an ASK signal is found and all manipulations and results are discussed in detail.)

9.2.3.4 FSK Spectrum and Bandwidth with a Rectangular Pulse-Train Message

FSK spectrum and bandwidth are a combination of the spectra and bandwidths of two ASK signals even when a message signal is nonperiodic. Refer to the spectrum and bandwidth of an ASK with the pulse-train message shown in Figure 9.2.7. Now, we can easily understand that the FSK spectrum is a combination of two ASK spectra centered at $f_{C1} = f_C + \Delta f$ and $f_{C2} = f_C - \Delta f$ and separated by the double-frequency deviation $f_{C1} - f_{C2} = 2\Delta f$. Thus, the FSK bandwidth is equal to the ASK bandwidth plus $2\Delta f$. Figure 9.2.13 depicts this. Figure 9.2.13 shows that the first null crossing occurs at $1/\tau$ (s) = BR (b/s), within which the FSK signal delivers 90% of the total power, and the second zero crossing occurs at $2/\tau$ (s) = 2 BR (b/s), within which the FSK signal delivers 95% of the total power. (Remember, τ (s) = T_b (s).) Thus, the FSK bandwidth is given by

$$BW_{FSK} \text{ (Hz)} = 2\Delta f + 2BR \text{ (b/s) at 90\% power} \qquad (9.2.31a)$$

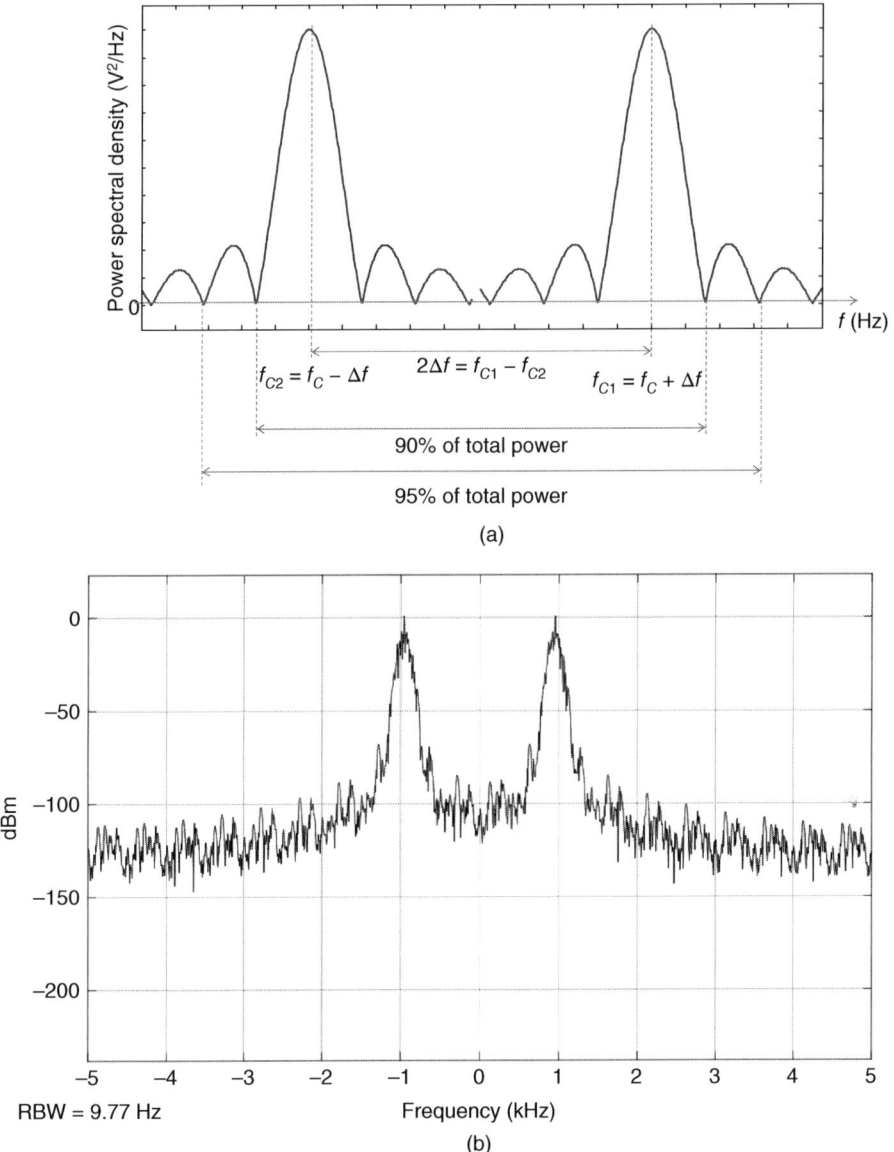

Figure 9.2.13 Power spectrum and bandwidth of an FSK signal with pulse train message: (a) Calculated power spectrum and percentage of total power and (b) example of a simulated spectrum of an FSK signal with overlapping main lobes.

and

$$BW_{FSK} \text{ (Hz)} = 2\Delta f + 4\,BR \text{ (b/s)} \text{ at } 95\% \text{ power} \tag{9.2.31b}$$

As we already learned, (9.2.11) states BW_{ASK} (Hz) = 2BR (b/s) at 90% power and BW_{ASK} (Hz) = 4 BR (b/s) at 95% power. Remember, too, that

$$BW_{FSK} \text{ (Hz)} = BW_{ASK} \text{ (Hz)} + 2\Delta f \tag{9.2.32}$$

regardless of the nature of the message signal and the percentage of the total power taken into account.

You should ask at this point, what must the value of $2\Delta f$ be? Examining Figure 9.2.13a, we observe that if $2\Delta f$ is less than $1/\tau$ (s) = BR (b/s), then the two graphs overlap before their first zero crossings; in other words, they cannot fully build even their main lobes. Thus, the requirement is

$$2\Delta f \geq \frac{2}{\tau(s)} = 2\text{BR (b/s)}$$

or

$$\Delta f_{\min}\,(\text{Hz}) = \frac{1}{\tau(s)} = \text{BR (b/s)}. \quad (9.2.33)$$

For example, if the bit rate is equal to 100 Mb/s, then the minimum frequency deviation must be 100 MHz. The FSK spectrum, obtained by Simulink and shown in Figure 9.2.13b, confirms this: We can see overlapping of two main lobes.

9.2.3.5 Bit Error Ratio, BER, and Remarks on our BFSK Discussion

The bit error ratio of a BFSK signal is evaluated using the same formula that is used for ASK; that is

$$\text{BER}_{\text{FSK}} = \text{BER}_{\text{ASK}} = Q\left(\sqrt{\frac{E_b}{N_0}}\right) \quad (9.2.17\text{R})$$

This equation and its derivation can be found in many textbooks on communications (see, for example, Sklar 2001, p. 213).

All discussions regarding BER, including the calculations given in Example 9.2.3, can be directly applied to an FSK signal.

The following example should help you to become more familiar with BER calculations.

Example 9.2.5 **BER and Receiver Sensitivity for BFSK transmission**

Problem

An industrial graph "BER vs. P_{rec}" is presented in Figure 9.2.14, where P_{rec} is the minimum power enabling the receiver to sustain the required BER, that is, the *receiver sensitivity* (see Example 9.2.3). For this transmission, find the allowed noise spectral density, N_0 (W/Hz), to provide BER = 10^{-12} and BR = 10 Gb/s.

Solution (See Example 9.2.3)

First, we need to gather all the given data: We know the BER and we can determine the P_{rec} from Figure 9.2.14. Secondly, we know that the allowed noise spectral density, N_0, is related to BER as

$$\text{BER}_{\text{FSK}} = Q\left(\sqrt{\frac{E_b}{N_0}}\right) \quad (9.2.17\text{R})$$

Therefore, the plan for the solution is simple: Express N_0 (J/Hz) through the given quantities in the sequence of manipulations.

We are given that

$$Q\left(\sqrt{\frac{E_b}{N_0}}\right) = 10^{-12}$$

Using the MATLAB code given in Example 9.2.3, we find $\sqrt{\frac{E_b}{N_0}} = 6.7$. Thus, $N_0 = \frac{E_b}{44.9}$.

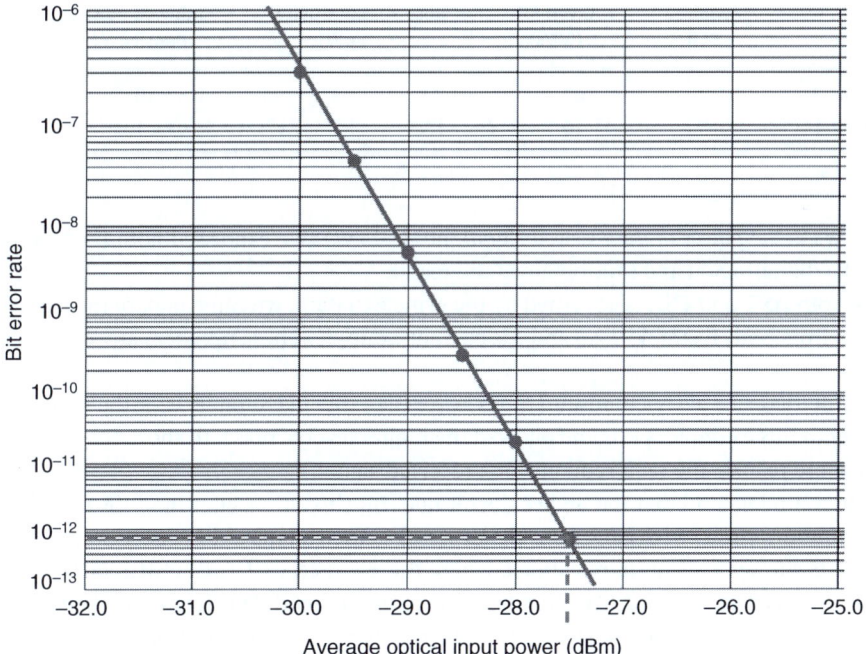

Figure 9.2.14 Industrial graph of BER vs. received power: Measured results for BR = 10 Gb/s. Source: Reprinted with permission of Discovery Semiconductors, Inc., Ewing, NJ, USA.

Now, the problem boils down to expressing E_b through P_{rec}. Recalling (9.2.19),

$$P_{rec} (W) = 2E_b \cdot BR \; (J \times b/s)$$

we find $E_b \; (J) = \frac{P_{rec}}{2BR}$ (W s).

Examining Figure 9.2.14, we gather that the needed power for BER = 10^{-12} is approximately -27.6 dBm, which corresponds to $1.73 \, \mu W$.

Therefore,

$$E_b = \frac{P_{rec}}{2BR} = \frac{1.73 \times 10^{-6} \; (W)}{2 \times 10 \times 10^9 \; \left(\frac{1}{s}\right)} = 0.087 \times 10^{-15} \; J$$

Finally,

$$N_0 = \frac{E_b}{44.9} = 1.94 \times 10^{-18} \; W/Hz$$

This is the answer to our problem.

Discussion

- We need to realize that all manipulations here and in Example 9.2.3 are based on a straightforward idea: BER is a function (in this case, a Q-function) of the digital SNR. That is:

$$BER = Q\sqrt{(P_{signal}(W)/P_{noise}(W))} \; \text{digital}.$$

Table 9.2.2 Comparison of BER data for ASK and FSK transmissions in Examples 9.2.3 and 9.2.4.

	Required BER	Given BR (Gb/s)	Receiver sensitivity, P_{rec} (μW)	Allowed noise spectral density, N_0 (W/Hz)
ASK – Example 9.2.3	10^{-12}	100	7.8	0.8×10^{-18}
FSK – Example 9.2.4	10^{-12}	10	1.73	1.94×10^{-18}

Here, P_{signal} (W) is *the power of the received signal*, i.e. P_{rec} (W). We expressed both powers in parameters of a digital signal and thus obtained the result.

- Since BER of both ASK and FSK is governed by the same equation, we can compare the significant numbers from both Example 9.2.3 and the current example. These numbers are shown in Table 9.2.2.
Since the BER in this example is a thousand times lower than that in Example 9.2.3, the required received power in this example is less and allowed the higher level of noise. In short, the smaller the BER, the more relaxed the requirements for a received signal.
- It is important to understand that the higher the bit rate, BR (b/s), the better the BER must be. In mobile communications, where BR is on the order of 1 Gb/s, a typical BER is between 10^{-5} and 10^{-7}, whereas in optical communications, where the BR exceeds 1 Tb/s, a BER of 10^{-12} is typical and 10^{-15} becomes required. This trend is easy to comprehend: The definition of bit error ratio – *the ratio of the number of erroneous bits to the number of total bits transmitted* – implies that two defining numbers are measured over the same time interval. Therefore, the greater the BR (b/s), the greater the number of total transmitted bits, and – if the number of erroneous bits is kept constant – the better BER is required. (If we refer to the definition of BER – the number of erroneous bits per unit time – then we arrive to the given conclusion in a more straightforward manner.)

We need to realize that a simplified strategy that considers FSK as two ASK signals implies that FSK is *linear modulation* because the ASK modulation is indeed a linear type. However, in reality, FSK is a type of frequency modulation and as such, it is a *nonlinear modulation*, which requires the different technique in its analysis. Such a rigorous approach, however, is used in more advanced courses. Fortunately, this simplified analysis provides plausible results with reasonable accuracy. A more sophisticated analysis is necessary when a high level of accuracy is required. We leave such an undertaking to advanced textbooks and continue to follow the simplified method.

9.2.3.6 Discontinuous-Phase FSK (DPFSK) and Continuous-Phase FSK (CPFSK)

Our preceding analysis treats an FSK signal as a direct combination of two ASK signals. What's more, we discussed only one type of FSK modulation, the *DPFSK* shown in Figure 9.2.10. Its main feature is that the phases of two different frequencies change abruptly at the transition from bit 1 to bit 0 and back. As we have learned in the preceding subsection, DPFSK is appealing thanks to its simplicity in both implementation and analysis, but it has a major drawback: The phase interruptions cause an increase in the signal's bandwidth. This is because DPFSK delivers individual segments of sinusoids, and such segments have extended bandwidths, as shown in Figure 7.3.7.

This DPFSK problem can be solved by replacing a simple switch with a VCO in an FSK modulator. In this case, the message signal causes the VCO to generate an FSK signal with two different frequencies; the VCO does this while keeping the phases of these signals continuous, without abrupt interruptions. Such a signal is called a CPFSK.

CPFSK is the method by which an FSK signal is generated today.

Since the phases of a modulated carrier change smoothly at the transitions from bit 1 to bit 0 and back, a CPFSK signal occupies less bandwidth than a DPFSK signal. This is equivalent to saying that the side lobes of a CPFSK spectrum contain less energy (they are smaller in amplitudes) than similar lobes of a DPFSK (see Figure 9.2.13). Thus, using CPFSK, we can save on the bandwidth of a channel through which this signal is transmitted.

The circuit for generating a CPFSK signal with a VCO is shown in Figure 9.2.15a. The waveforms of a message signal and corresponding CPFSK signal are demonstrated in Figure 9.2.15b,c, which show the transitions from bit 1 to bit 0 and back on a larger scale. The latter figure demonstrates that two different frequencies smoothly transit from one to the other without undergoing phase changes at the transition points. Figure 9.2.15d contrasts the waveform of DPFSK (top) and CPFSK (bottom) signals to demonstrate the difference in changing the phases of DPFSK (discontinuous-phase FSK, remember?) and CPFSK signals.

9.2.3.7 Mathematical Description of a CPFSK Signal

How can we mathematically describe a CPFSK signal? As we will learn shortly, to better detect a BFSK signal, we need to make signals v_{M1} and v_{M2} *orthogonal* in a mathematical sense. *Two signals are called orthogonal over the interval T_b if*

$$\int_0^{T_b} v_{M1}(t) \cdot v_{M2}(t) dt = 0 \quad (9.2.34)$$

Based on (9.2.27), we can present two FSK signals with different frequencies, $\omega_1 = \omega_C + \Delta\omega$ and $\omega_2 = \omega_C - \Delta\omega$, and different phases, θ_1 and θ_2, as

$$v_{M1}(t) = A_C \cos(\omega_C t + \Delta\omega t + \theta_1)$$

and

$$v_{M2}(t) = A_C \cos(\omega_C t - \Delta\omega t + \theta_2) \quad (9.2.35)$$

To achieve the <u>continuous-phase transition</u>, we need to make the initial phases of $v_{M1}(t)$ and $v_{M2}(t)$ equal; that is,

$$\theta_1 = \theta_2 = \theta \quad (9.2.36)$$

(Explain how this statement relates to (9.2.35).) Plugging v_{M1} and v_{M2} under Condition (9.2.36) into (9.2.34), we find

$$\int_0^{T_b} v_{M1}(t) \cdot v_{M2}(t) dt = \int_0^{T_b} A_C \cos(\omega_C t + \Delta\omega t + \theta) \cdot A_C \cos(\omega_C - \Delta\omega t + \theta) dt \quad (9.2.37)$$

Recalling the trigonometric formula $\cos A \cdot \cos B = \frac{1}{2} \cos(A+B) + \frac{1}{2} \cos(A-B)$, we can rewrite (9.2.37) as

$$\int_0^{T_b} A_C \cos(\omega_C t + \Delta\omega t + \theta) \cdot A_C \cos(\omega_C - \Delta\omega t + \theta) dt$$

$$= \frac{A_C^2}{2} \int_0^{T_b} \cos(2\omega_C t + 2\theta) dt + \frac{A_C^2}{2} \int_0^{T_b} \cos(2\Delta\omega t) dt \quad (9.2.38)$$

(In the succeeding manipulations, we follow Mesiya, 2013, pp. 612–614.) Let us evaluate the first integral:

$$\frac{A_C^2}{2}\int_0^{T_b}\cos(2\omega_C t+2\theta)dt = \frac{A_C^2}{4\omega_C}(\sin(2\omega_C t+2\theta))\big|_0^{T_b}$$

$$= \frac{A_C^2}{4\omega_C}[\sin(2\omega_C T_b+2\theta)-\sin(2\theta)] \qquad (9.2.39)$$

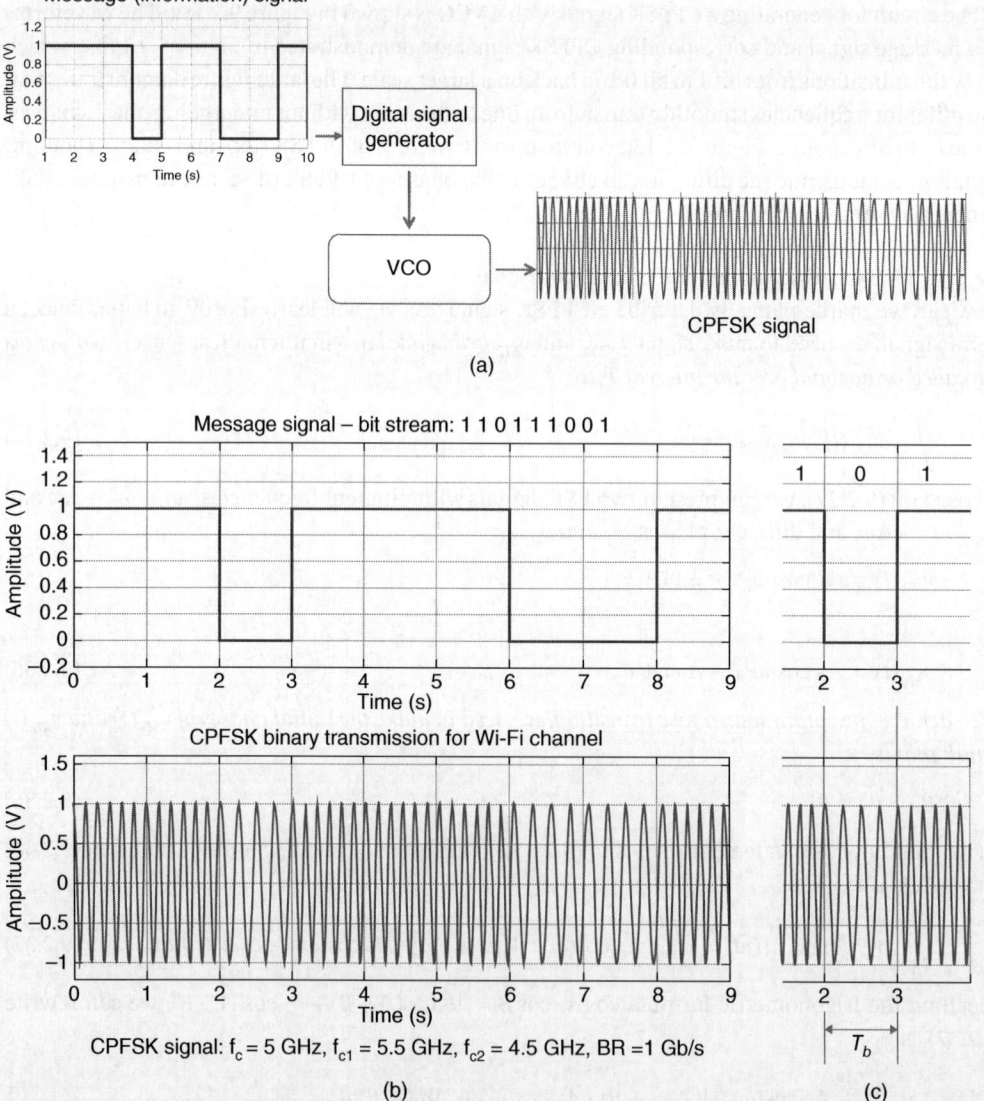

Figure 9.2.15 A continuous-phase FSK (CPFSK) signal: (a) Generating a CPFSK signal; (b) the waveform of a CPFSK signal; (c) the waveform of a CPFSK signal at the transition from 1 to 0 and from 0 to 1 on a larger scale; and (d) comparison of FSK and CPFSK waveforms by MATLAB/Simulink simulation.

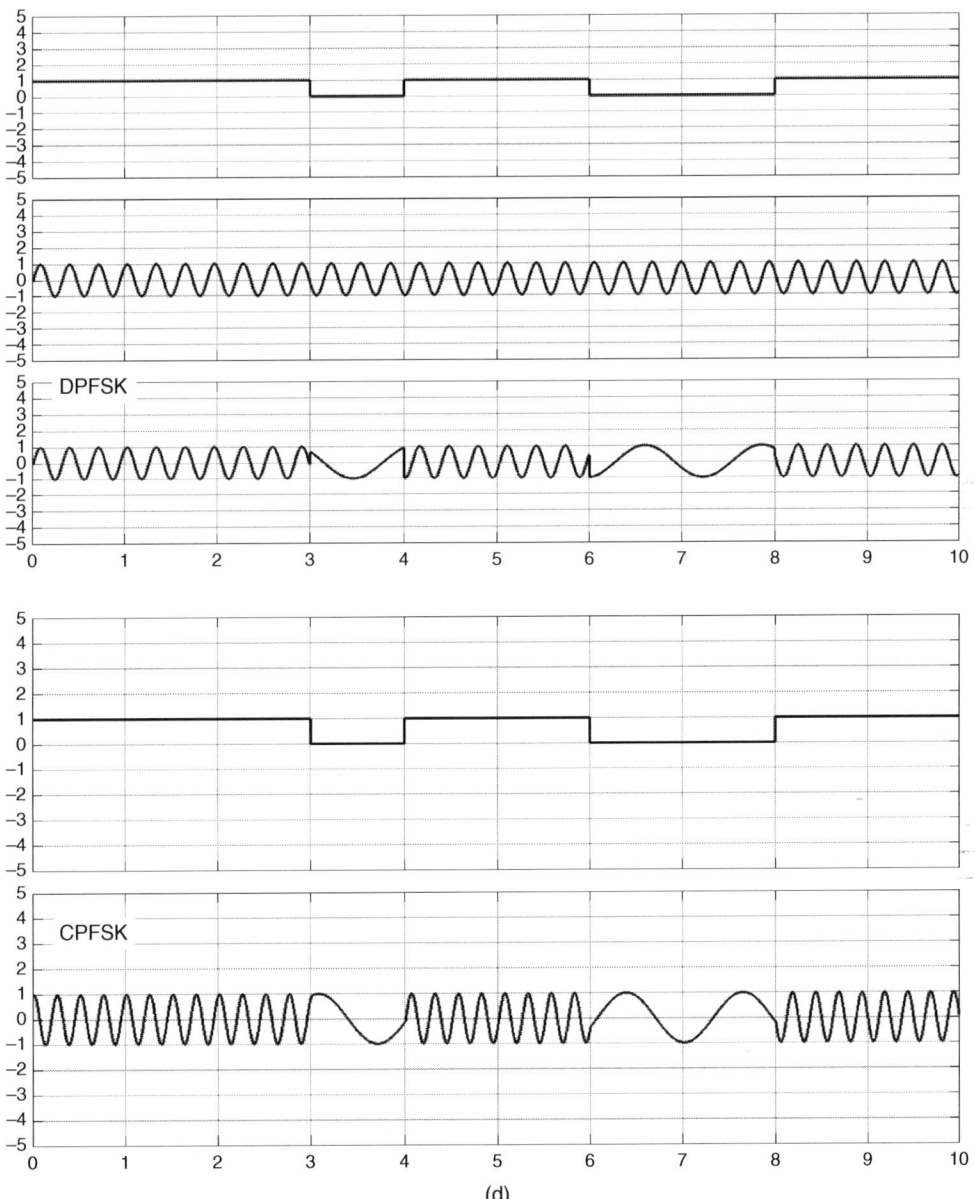

Figure 9.2.15 (Continued)

Now let us set the requirement:

$$f_C \cdot T_B = \frac{f_C}{\text{BR}} = \frac{T_b}{T_C} = n \qquad (9.2.40)$$

where $n = 1, 2, 3, \ldots$. (You should check the units in this formula to verify that n is a dimensionless constant.) Condition (9.2.40), as you can see, requires that a carrier frequency, f_C (Hz), must be a multiple of the bit rate, BR (b/s), or, equivalently, it must be a whole number of the cycles of the carrier wave within the bit duration, T_b (s).

(Refer to Example 9.2.3 to refresh your memory that the ratio $\frac{T_b}{T_C}$ determines the number of cycles of a carrier wave per one-bit duration, T_b.) The physics underlying (9.2.40) is obvious: *This condition requires that the phase of a carrier frequency must be a multiple of 2π within T_b, thereby enabling a VCO to know when a bit ends or starts.*

Under (9.2.40), we find

$$2\omega_C T_b = 4\pi f_C T_b = 4\pi n$$

and, therefore, (9.2.39) becomes

$$\frac{A_C^2}{4\omega_C}[\sin(2\omega_C Tb + 2\theta) - \sin(2\theta)] = \frac{A_C^2}{4\omega_C}[\sin(4\pi n + 2\theta) - \sin(2\theta)] \quad (9.2.41)$$

Now let us resort to familiar trigonometric formula, $\sin(A+B) = \sin A \cdot \cos B + \cos A \cdot \sin B$, through which we can obtain from (9.2.41) the following result:

$$\frac{A_C^2}{4\omega_C}[(\sin(4\pi n + 2\theta) - \sin(2\theta)]$$

$$= \frac{A_C^2}{4\omega_C}[(\sin(4\pi n) \cdot \cos(2\theta) + \cos(4\pi n) \cdot \sin(2\theta) - \sin(2\theta)] = 0 \quad (9.2.42)$$

This is because $\sin(4\pi n)=0$ and $\cos(4\pi n)=1$ for any n, which reduces (9.2.42) to a simple difference $(\sin(2\theta) - \sin(2\theta))$.

Thus, we have found that by meeting Condition (9.2.40), we turned integral $\frac{A_C^2}{2}\int_0^{T_b} \cos(2\omega_C t + 2\theta)dt$ to zero.

Now we will evaluate the second integral in (9.2.39):

$$\frac{A_C^2}{2}\int_0^{T_b} \cos(2\Delta\omega t)dt = \frac{A_C^2}{2}\left.\frac{\sin(2\Delta\omega t)}{2\Delta\omega}\right|_0^{T_b}$$

$$= T_b \frac{A_C^2}{2}\frac{\sin(2\Delta\omega Tb)}{2\Delta\omega Tb} = \frac{A_C^2}{2}T_b\frac{\sin(2\pi\Delta fTb)}{2\pi\Delta fTb}$$

$$= \frac{A_C^2}{2}T_b \operatorname{sinc}(2\Delta fT_b) \quad (9.2.43)$$

As we learned in Section 6.3, $\operatorname{sinc}(x)$ turns to zero when $x = 1, 2, 3, \ldots$. Hence, it must be

$$(2\Delta f T_b) = k \quad (9.2.44)$$

where $k = 1, 2, 3, \ldots$. Condition (9.2.44) sets the requirement on frequency deviation; namely

$$\Delta f \text{ (Hz)} = \frac{k}{2T_b(s)} = k\frac{BR(b/s)}{2} \quad (9.2.45)$$

These derivations can be summarized as follows:

- To facilitate the detection of an FSK signal, as we will learn shortly, its two constituent signals must be mathematically orthogonal as

$$\int_0^{T_b} v_{M1}(t) \cdot v_{M2}(t)dt = 0 \quad (9.2.34R)$$

- To minimize the bandwidth of an FSK signal, we need to generate a continuous-phase version of the FSK signal, CPFSK, which requires that the initial phases of two FSK frequencies must be equal at the transition points as

$$\theta_1 = \theta_1 = \theta \tag{9.2.36R}$$

- Orthogonality of the constituent FSK signals plus Condition (9.2.36) lead to two requirements for generating the CPFSK signal:
  - The relationship between the carrier frequency and the bit rate must meet the following conditions:

$$f_C \cdot T_b = \frac{f_C}{BR} = \frac{T_b}{T_C} = n \quad \text{for } n = 1, 2, 3, \ldots \tag{9.2.40R}$$

  - The value of the frequency deviation must be determined by the following formula:

$$\Delta f \text{ (Hz)} = \frac{k}{2T_b} = k\frac{BR(b/s)}{2} \tag{9.2.45R}$$

These considerations are discussed in Example 9.2.6.

Example 9.2.6   Generating a CPFSK Signal

Problem

A BFSK transmission for a Wi-Fi communication relies on carrier frequency $f_C = 5$ GHz and operates at $BR = 1$ Gb/s.

a. Find the conditions for generating the CPFSK signal for this transmission.
b. Comment on your choice of constants n and k.
c. Build the waveforms of this signal.

Solution
a) To generate the CPFSK signal, we need to satisfy two conditions:
  (1) $f_C \text{ (Hz)} = n \cdot BR \text{ (b/s)}$.
  (2) $\Delta f \text{ (Hz)} = k\frac{BR(b/s)}{2} = \frac{k}{2T_b(s)}$.
  The first condition is satisfied by the given values of $f_C = 5$ GHz and BR = 1 Gb/s, which give $n = 5$. To satisfy the second condition, we need to choose the constant k. Let us take the minimal value of k, $k = 1$, which results in the minimal value of $\Delta f_{min} = BR/2 = 0.5$ GHz.
b) In choosing constant n, we have restrictions on both sides, f_C and BR: The carrier frequency, f_C (Hz), is typically given either by spectral regulations, such as those imposed by the FCC, and/or by the properties of a transmission medium, such as air, optical fiber, or coaxial cable. On the other side, the value of bit rate, BR (b/s), is typically required, which eliminates any freedom in its choosing. Fortunately, there are no restrictions on the value of this constant, and so n can be very large. For instance, if we were to use CPFSK for optical transmission with typical parameters of $f_C = 200$ THz and $BR = 1$ Tb/s, we would find $n = 200$.
Constant k determines the value of the frequency deviation when the bit rate is given, as (9.2.45) shows. On the one hand, the greater the k, the greater the separation between the two frequencies, $f_{C1} = f_C + \Delta f$ and $f_{C2} = f_C - \Delta f$, and the easier and more accurate the detection of the BFSK signal. (Modern signal-processing technology, however, practically eliminates the problem of distinction between two frequencies.) On the other hand, the smaller the k, the less bandwidth

a BFSK signal occupies. The communication industry, of course, chooses the minimal bandwidth, which is minimal Δf, and therefore $k = 1$. This choice is even more justified by advanced analysis, which takes into account the effect of noise.

c) The required waveform is shown in Figure 9.2.16a; the corresponding MATLAB code is given below.

Discussion

- Though in general $\Delta f = k \cdot BR/2 = k/2T_b$ and integer k can hypothetically take any value from 1 to infinity, we should know that, typically, $k = 1$. Such a CPFSK is called *minimum shift keying, MSK*. It is also called *Sunde's FSK* after the engineer who introduced it. Figure 9.2.16b shows the FSK signal with the same parameters as in Figure 9.2.16a but with $k = 4$. And, of course, when k is not an integer, we do not have a CPFSK signal, as Figure 9.2.16c, where $k = 1.3$, attests.
- Bear in mind that the ratio of frequency deviation to bit rate is called in BFSK the modulation index, h_{BFSK}. Thus,

$$h_{\text{BFSK}} = \frac{\Delta f \text{ (Hz)}}{\text{BR (b/s)}} \qquad (9.2.46)$$

- Analyze Figure 9.2.16: Here we see that the number of cycles per T_b is indeed an integer and the phases change continuously at the transition from bit 1 to bit 0 and back. (See Figure 9.2.16c, where these transitions are shown in greater detail.)

This concludes Example 9.2.6.

9.2.3.8 Detection (Demodulation) of an FSK Signal

Basically, we have two methods to detect an FSK signal: *incoherent (noncoherent)* and *coherent*.

Incoherent BFSK Detection The *incoherent* demodulator treats the FSK signal as a combination of two ASK signals and detects only received frequencies to recover the bits. This demodulator works with a CPFSK signal; it does not detect phase changes. Let us consider the principle of operation of an incoherent BFSK detector in detail.

The block diagram of an *incoherent BFSK detector* is presented in Figure 9.2.17. The detector includes two bandpass filters, BPF1 and BPF2, two envelope detectors, a comparator, and a decision circuit. BPF1 allows for the passing of frequency ω_{C1} whereas BPF2 passes frequency ω_{C2}.

When a CPFSK signal arrives, it is directed into two branches. In Branch 1, the BPF1 allows the sinusoidal signal with frequency ω_{C1} delivering bit 1 to pass. *Envelope detector* 1 removes high-frequency oscillations and recovers the envelope of the signal corresponding to bit 1. The simplest envelope detector consists of a diode which is followed by a capacitor in parallel with a resistor. Its output, voltage V_+, enters the noninverting input of a comparator. At the same time, BPF2 does not allow the signal carrying bit 1 to pass because frequency ω_{C1} does not match its passband; therefore, envelope detector 2 senses no signal, and its output signal is equal to zero. Therefore, voltage V_-, presented to the inverted comparator input, is zero. The comparator compares two inputs and directs the result to a decision circuit. Since in this case $V_+ > V_-$, the circuit decides that bit 1 is delivered.

When bit 0 arrives, it is detected by BPF2 and envelope detector 2, respectively, while Branch 1 passes no signal. (See Chapter 5 to refresh your memory on filters.) Since this time $V_+ = 0$ and $V_- > 0$ and therefore $V_+ < V_-$, the decision circuit determines that bit 0 is delivered. As a result,

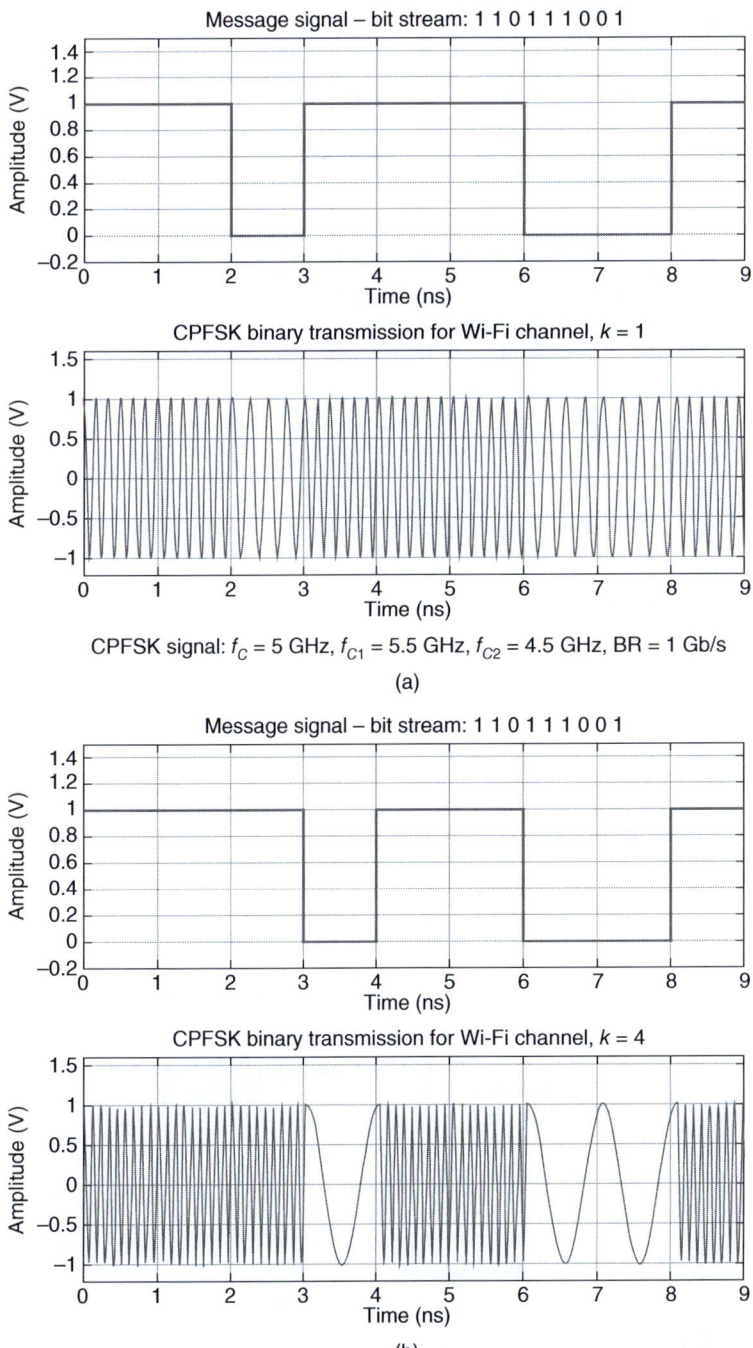

Figure 9.2.16 The waveforms of the CPFSK signal in Example 9.2.6: (a) The message signal and the waveform of the FSK signal for $k = 1$; (b) the same signals for $k = 4$; and (c) the same signals for $k = 1.3$.

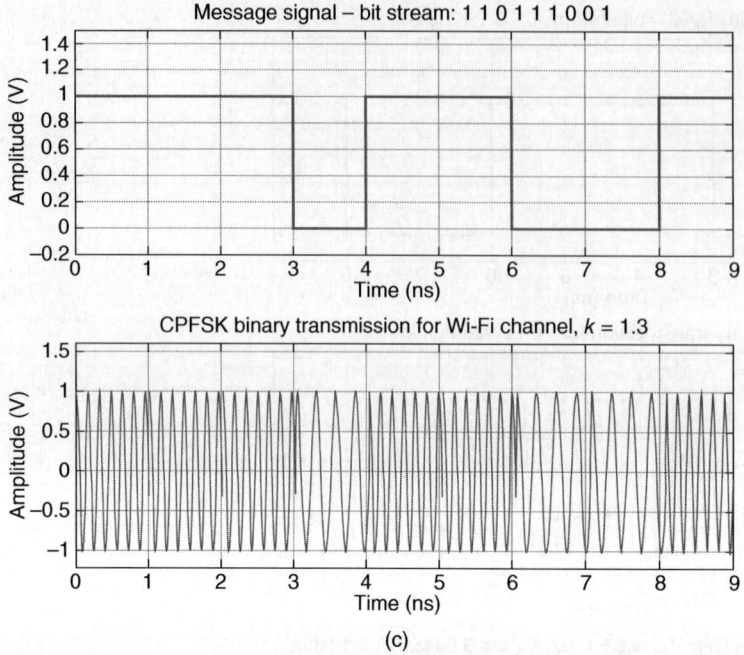

Figure 9.2.16 (Continued)

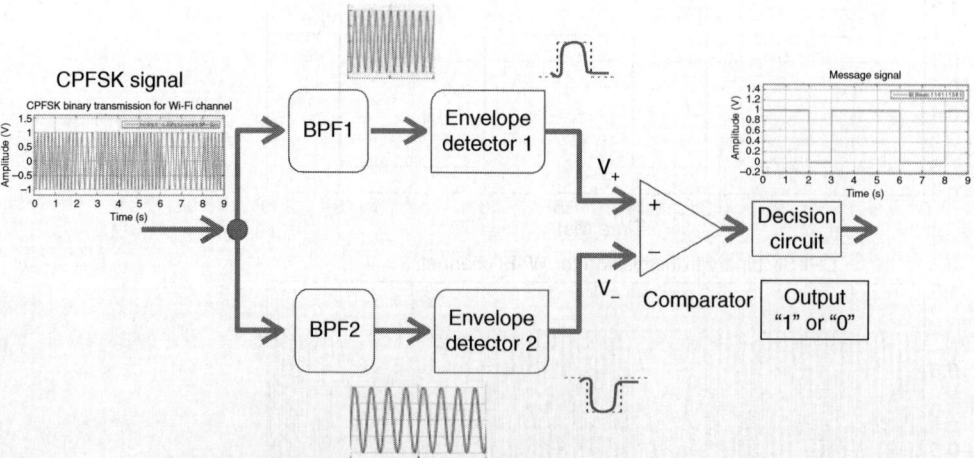

Figure 9.2.17 The block diagram of an incoherent BFSK detector.

the message signal is reconstructed. It is important to understand that the bits arrive in sequence; thus, at each single bit interval, either bit 1 or bit 0 is detected.

Understand that the envelope detector operates as a rectifier; therefore, the uncovered amplitude of an arrived signal is a smooth version of an original rectangular pulse, as Figure 9.2.17 shows. Naturally, this signal is called *bandlimited*, whereas the decision circuit produces rectangular pulses with theoretically unlimited bandwidth.

9.2 Introduction to Digital Transmission – Binary Shift-Keying Modulation

This simplified presentation does not take into account the presence of noise but merely explains the concept of incoherent BFSK detection.

Incoherent detection, with all its simplicity, fundamentally lacks accuracy for this reason: *To precisely detect the logic of a received bit, the detector must know the real duration of a bit, that is, the starting and ending points of each bit. But all FSK signals theoretically have equal amplitudes; hence, the only parameters by which the starting and ending points can be accurately detected are the signals' phases. Since the incoherent demodulator cannot detect these phases, it cannot accurately determine the real duration of a bit. The result: a high probability of error.*

See the Appendix 9.2.A *Jitter* to this section for more on the problem of incoherent detection.

Coherent BFSK Detection A *coherent detector*, as its name suggests, can accurately detect the phases of received signals and thus determine the start and the end of each bit interval. (See Section 9.2.2.6 for the meaning of the term *coherent detector*.) This fact justifies its preferred status as a more accurate FSK receiver. The block diagram of a coherent CPFSK detector is presented in Figure 9.2.18. Each branch of the detector contains a local oscillator, a multiplier, and an integrator. Common to both branches are a timer, a summation circuit, and a decision circuit. The incoming CPFSK signal is directed to two branches simultaneously. In Branch 1, the input signal is multiplied by $2\cos(\omega_1 t)$, which is generated by local oscillator 1. The product of this multiplication is integrated over a T_b interval. The resulting signal is algebraically summed over T_b with the signal from Branch 2 taken at the same time and their difference is presented to a decision circuit, which decides what bit has arrived. The timer supplies the bit duration to all the needed components to synchronize their operation.

The coherent detection is based on integration of the product of the locally oscillated signals, $2\cos(\omega_i t)$, and the incoming signals, $v_{Mi} = A_C \cos(\omega_{Ci} t)$, where $i = 1, 2$. As mentioned previously, $\omega_1 = \omega_{C1}$ and $\omega_2 = \omega_{C2}$ (see Eq. (9.2.35)). We neglected the initial phases in v_{M1} and v_{M2} since they are equal to each other and can be equal to zero.

This is how the operation of coherent detection can be described mathematically. (The following derivation is based on Haykin and Moher, 2007, pp. 414–416.) When bit 1 is transmitted, both

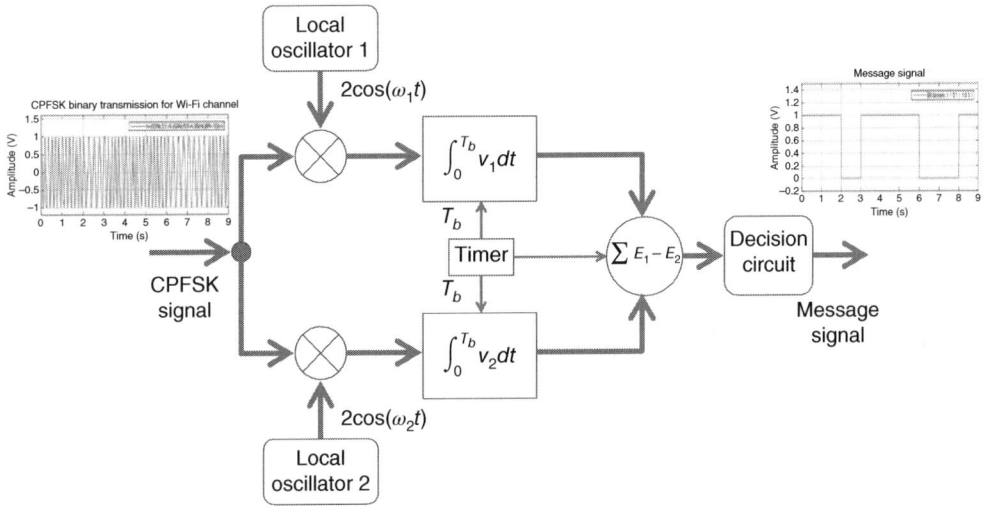

Figure 9.2.18 The block diagram of a coherent BFSK detector.

branches receive signal $A_C \cos(\omega_{C1} t)$. The signal in Branch 1, after multiplication by the local signal, becomes

$$v_1(t) = A_C \cos(\omega_{C1} t) \cdot 2\cos(\omega_{C1} t) \tag{9.2.47}$$

The integral of this product takes the form

$$E1 = \int_0^{T_b} A_C \cos(\omega_{C1} t) \cdot 2\cos(\omega_{C1}) dt \tag{9.2.48}$$

The use of the identity $2\cos^2 \alpha = 1 + \cos^2 \alpha$ enables us to evaluate (9.2.48) as

$$E1 = \int_0^{T_b} A_C \cos(\omega_{C1} t) \cdot 2\cos(\omega_{C1} t) dt = A_C \int_0^{T_b} (1 + \cos(2\omega_{C1} t)) dt$$

$$= A_C t \big|_0^{T_b} + 0 = A_C T_b \tag{9.2.49}$$

This is because the integral of $\cos(2\omega_{C1} t)$ is zero, as shown in (9.2.50):

$$A_C \int_0^{T_b} (\cos(2\omega_{C1} t)) dt = \frac{A_C}{2\omega_{C1}} \sin(\omega_{C1} T_b) = \frac{A_C}{2\omega_{C1}} \sin(2\pi f_{C1} T_b)$$

$$= \frac{A_C}{2\omega_{C1}} \sin(2\pi \cdot n) = 0 \quad (n = 1, 2, 3, \ldots) \tag{9.2.50}$$

Here, we refer to (9.2.40), where the condition for obtaining CPFSK signal $f_{C1} \cdot T_b = n$ was introduced.

Thus, for bit 1, the signal presented for comparison from Branch 1 is given as

$$E1 = A_C T_b \tag{9.2.51}$$

At the same time, the signal of bit 1 processed in Branch 2 is given by

$$E2 = \int_0^{T_b} A_C \cos(\omega_{C2} t) \cdot 2\cos(\omega_{C1}) dt = 0 \tag{9.2.52}$$

This integral yields zero because signals $\cos(\omega_{C1} t)$ and $\cos(\omega_{C2} t)$ are orthogonal (See Eq. (9.2.34), where we refer to the need to make the two constituent FSK signals orthogonal for better detection.) Therefore, for bit 1, Branch 2 produces signal

$$E2 = 0$$

The easiest way to compare these signals is to take their difference, $E1 - E2$, which for bit 1 results in

$$E1 - E2 = A_C T_b - 0 > 0 \tag{9.2.53}$$

This difference is obtained by a summation circuit and presented to the decision circuit, which chooses bit 1 because the difference, $E1 - E2$, is positive.

If bit 0 is transmitted, the operation performed by both branches will be symmetrical to the previous one, and the outputs of the branches will be

Branch 1 $\to E1 = 0$

Branch 2 $\to E2 = A_C T_b$ \qquad (9.2.54)

Consequently, the difference between the two output signals,

$$E1 - E2 = 0 - A_C T_B < 0 \qquad (9.2.55)$$

will be negative, and so the decision circuit chooses bit 0.

When analyzing this discussion, notice how a coherent detector precisely determines the start and end of each individual bit: As soon as bit 1 starts, the integral in Branch 1, $E1 = \int_0^{T_b} A_C \cos(\omega_{C1} t) \cdot 2\cos(\omega_{C1} t) dt$, becomes equal to $A_C T_b$, and the proper integral in Branch 2, $E2 = \int_0^{T_b} A_C \cos(\omega_{C1} t) \cdot 2\cos(\omega_{C2} t) dt$, becomes equal to zero. Therefore, the difference, $E1 - E2$, immediately becomes positive. Symmetrically, when bit 0 starts, the difference, $E1 - E2$, immediately becomes negative. Thus, the decision circuit can make a well-justified selection.

Is a coherent detection really more accurate than incoherent detection? The answer is a definite yes. More rigorous analysis, which we skip here, shows that under the same conditions, the probability of error of coherent detection is more than four times less (read, better) than that of the incoherent. (Roden, 2003, pp. 386–387.)

To complete our discussion of FSK detection, we need to note that there are a number of versions of incoherent and coherent detectors designed to perform specific tasks. Discussing even several of them could easily be the subject of a special textbook. We believe, however, that the knowledge of the principles of FSK demodulation that you acquire from this section will allow you to understand any version of an FSK demodulator that you might encounter in your professional career.

9.2.3.9 BFSK: Advantages, Drawbacks, and Applications

FSK has the same inherent advantage over ASK as FM does over AM: It is less susceptible to noise because changes in signal amplitude do not directly affect transmission quality. But there is a price for this advantage: the need for greater bandwidth to transmit an FSK signal.

The areas of FSK application are determined by its advantage. Since wireless (radio) transmission is much more affected by noise than wired transmission, FSK has found its main applications in this area. In the early days of telecommunications, FSK was widely used in such system as a radioteletype. Today, various versions of FSK modulation are used in many types of mobile devices. Examples of these applications include medical devices for patient monitoring, wireless peripherals (such as mice, keyboards, speakers, and headphones), smart home and smart building devices, remote meter reading, garage door-opening systems, tire-pressure monitoring systems, railway-temperature monitoring, crane control at construction sites, to mention a few.

FSK is used in the Global System for Mobile Communications (GSM), which has been the de-facto standard for mobile communications. As a result, FSK modulation is embedded into many devices and systems supporting the *IoT*. Figure 9.2.19 demonstrates some of the FSK applications employed in contemporary devices and systems; several measuring instruments are also shown. (Some of these modules, by the way, also use ASK modulation.) As we can see, FSK modulation technology is involved in an array of applications that range from small personal and household devices to huge industrial systems. Rapid expansion of wireless technology in creating a smart environment (think about the development of smart cities) makes FSK one of the most ubiquitous modulation techniques. (See Section 1.1.5 for a brief introduction to the IoT.)

Therefore, mastering the fundamentals of FSK technology given here, you will certainly acquire very useful knowledge and become well prepared for further study of all the material needed for success in your professional career.

This concludes the discussion of FSK modulation.

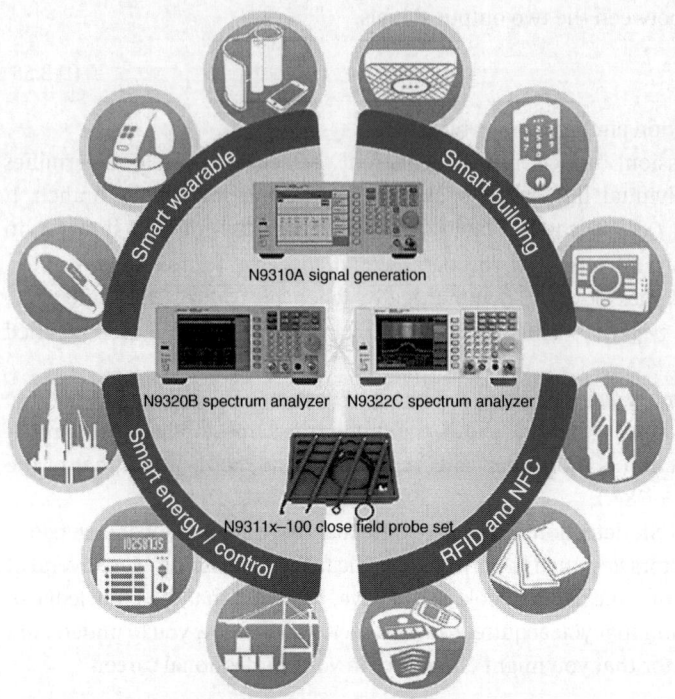

Figure 9.2.19 Examples of applications of FSK modulation technology in the Internet of Things. Source: © Keysight Technologies, Inc. Courtesy of Keysight Technologies.

9.2.4 Binary Phase-Shift Keying (PSK)

9.2.4.1 PSK Concept and Waveform

PSK delivers a digital message by changing the phase of a sinusoidal carrier. In binary PSK, sometimes called BPSK, changes from bit 1 to bit 0 and back are presented by a phase change of 180°, or π. Figure 9.2.20 shows the concept of PSK modulation.

A BPSK modulation can be considered as two ASK modulations; both modulations have carrier waves with equal amplitudes and frequencies but with the phase difference equal to 180°, or π. Such a simplified approach, used in discussing FSK modulation, will allow us to quickly understand the main features and the spectrum of PSK. This approach is shown in Figure 9.2.20a. A detailed view of the phase changes computed with MATLAB is seen in Figure 9.2.20b, where we can note that at every transition from 1 to 0 and back, the phase of a carrier wave changes by 180°. Simulations with MATLAB/Simulink, shown in Figure 9.2.20c, demonstrate the block diagram of a simulation setup and the waveforms of the message and PSK signals. Here, the phase changes at transitions from 1 to 0 and from 0 to 1 are clearly seen. The transition from bit 1 to bit 0 of a message signal and the corresponding phase change of a carrier wave on a larger scale, obtained with MATLAB/Simulink, are presented in Figure 9.2.20d. Finally, Figure 9.2.20e demonstrates how an entire PSK transmission system works: The modulator assigns carrier wave 1, with phase shift $\theta = 0$, to bit 1 of the message signal and carrier wave 2, with phase shift $\theta = \pi$, to bit 0. The PSK modulated signal is delivered to the destination point via a transmission channel; the demodulator recovers the message from the received PSK signal and discards the carrier.

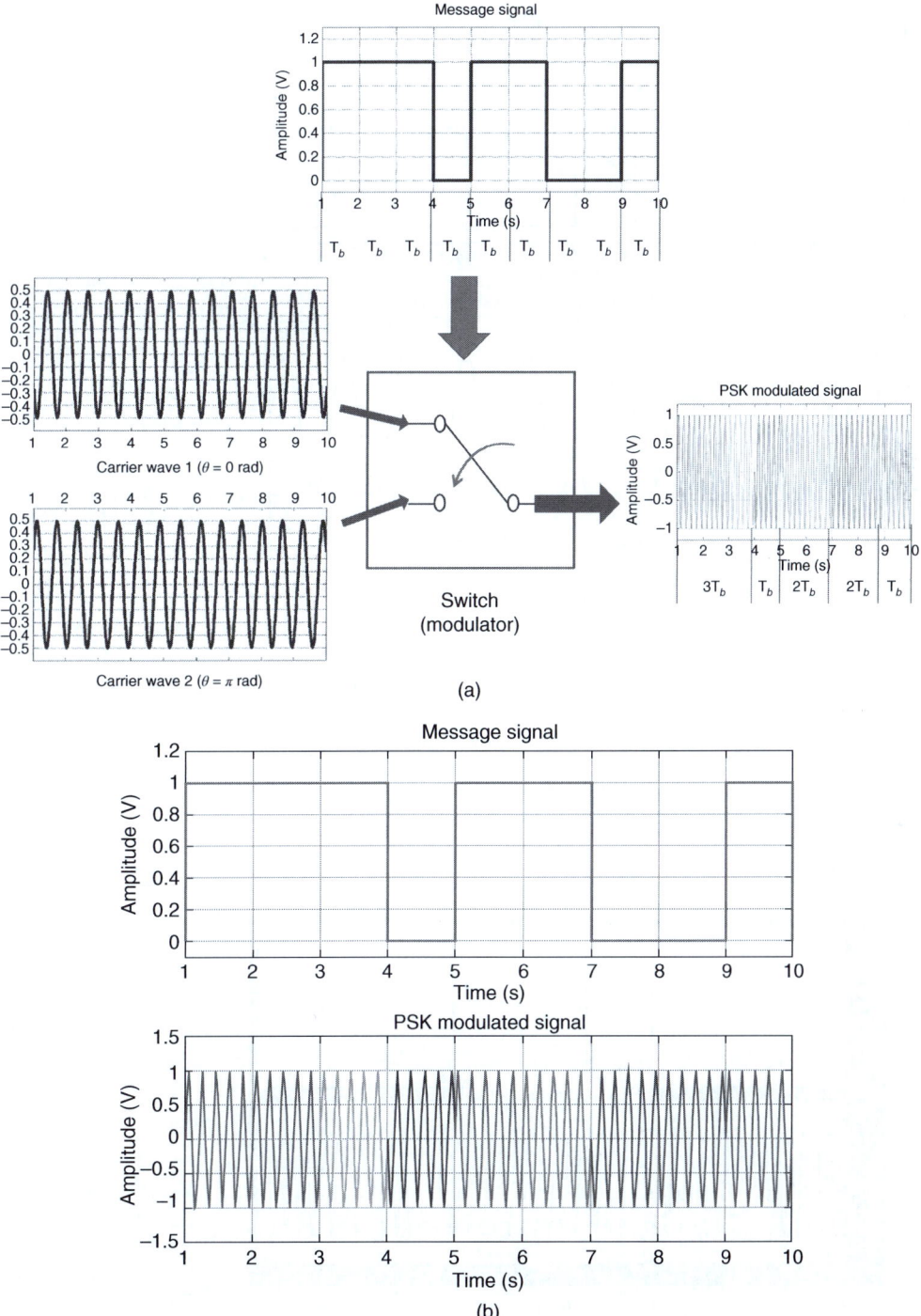

Figure 9.2.20 Concept of binary phase-shift keying (PSK) modulation: (a) The modulation setup; (b) a binary PSK signal built with MATLAB; (c) the binary PSK signal built with MATLAB on a larger scale; (d) MATLAB/Simulink simulation of PSK modulation; and (e) a PSK transmission system.

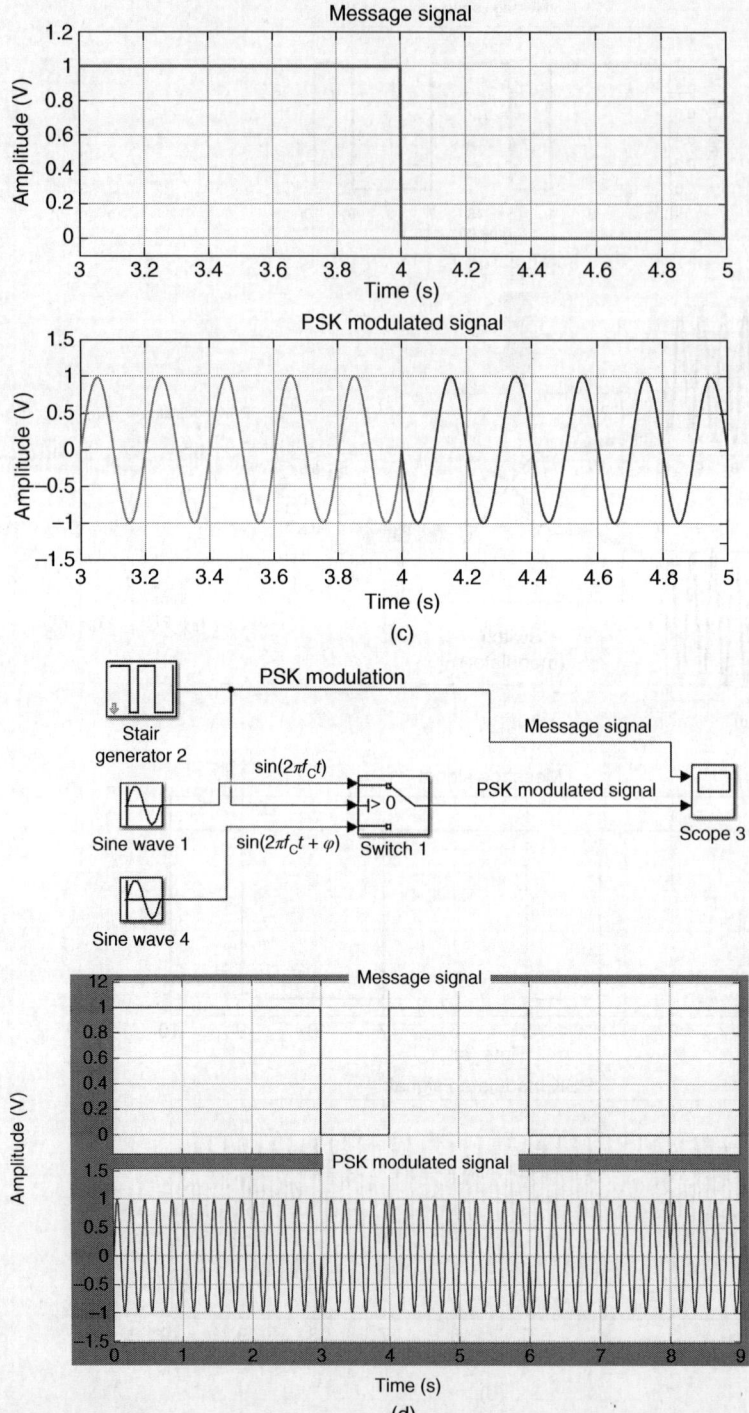

Figure 9.2.20 (Continued)

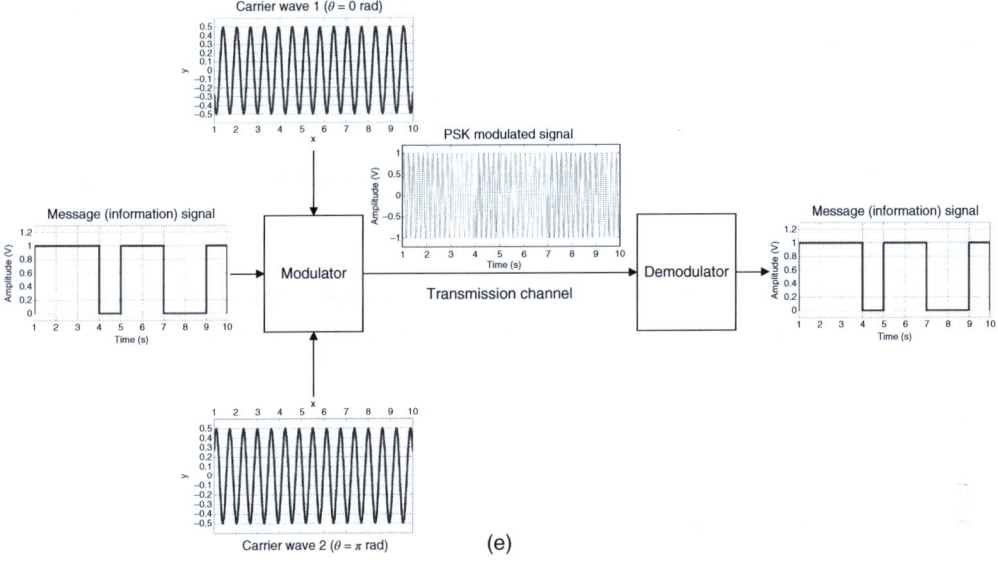

Figure 9.2.20 (Continued)

9.2.4.2 PSK Mathematical Description; PSK Spectrum and Bandwidth with a Square Wave Message

Two *PSK carrier waves* are given by

$$v_{C1}(t) = A_C \cos(\omega_C t + 0)$$

and

$$v_{C2}(t) = A_C \cos(\omega_C t + \pi) = -A_C \cos(\omega_C t) \tag{9.2.56}$$

Thus, a binary PSK signal is described by the formula

$$v_{PSK}(t) = \begin{cases} A_C \cos(\omega_C t) & \text{for bit 1} \\ -A_C \cos(\omega_C t) & \text{for bit 0} \end{cases} \tag{9.2.57a}$$

or, following the format used in this section and first presented in (9.2.2), we can write

$$\begin{aligned} \text{Bit } 1 &\to A_C \cos(\omega_C t) \quad \text{for } t_1 = T_b \\ \text{Bit } 0 &\to -A_C \cos(\omega_C t) \quad \text{for } t_0 = T_b \end{aligned} \tag{9.2.57b}$$

where T_b is the duration of a bit, ω_C is the carrier's frequency, and A_C is the carrier's amplitude. Again, (9.2.57a) and (9.2.57b) show that a binary PSK signal is a combination of two carrier waves whose amplitudes and frequencies are equal but whose phases differ by 180°.

To explore the procedure used for an FSK modulation, let us assume that a PSK signal includes two message signals, $v_{M1}(t)$ and $v_{M2}(t)$, where the second message signal, $v_{M2}(t)$, is a complement of the first, $v_{M1}(t)$; that is, $v_{M2}(t) = 1 - v_{M1}(t)$. Then a *BPSK-modulated signal* becomes the product of carrier waves and message signals as

$$v_{PSK}(t) = v_{M1}(t) \cdot v_{C1}(t) + v_{M2}(t) \cdot v_{C2}(t) = v_{M1}(t) \cdot A_C \cos(\omega_C t) - v_{M2}(t) \cdot A_C \cos(\omega_C t) \tag{9.2.58}$$

Let us start, as we do in this section, with considering a PSK signal with a *periodic square wave message signal*. This message signal, as we learned before, can be developed into the following Fourier series:

$$v_{M1}(t) = A_{M1} \left(\frac{1}{2} + \frac{2}{\pi} \sin \omega_0 t + \frac{2}{3\pi} \sin 3\omega_0 t + \frac{2}{5\pi} \sin 5\omega_0 t + \ldots \right) \quad (6.1.13R)$$

Now, we can find the PSK spectrum by implementing the method developed for an FSK signal in the previous subsection:

Use (9.2.58) and write

$$v_{PSK}(t) = v_{M1}(t) \cdot v_{C1}(t) + v_{M2}(t) \cdot v_{C2}(t) = v_{M1}(t) \cdot A_C \cos(\omega_C t) - v_{M2}(t) \cdot A_C \cos(\omega_C t)$$
$$= (v_{M1}(t) - v_{M2}(t)) \cdot A_C \cos(\omega_C t) = (v_{M1}(t) - 1 + v_{M1}(t)) \cdot A_C \cos(\omega_C t)$$
$$= -A_C \cos(\omega_C t) + 2v_{M1}(t) \cdot A_C \cos(\omega_C t) \quad (9.2.59)$$

Assume $A_{m1} = 1$, plug (6.1.13R) into (9.2.59), and obtain

$$v_{PSK}(t) = -A_C \cos(\omega_C t) + 2v_{M1}(t) \cdot A_C \cos(\omega_C t) = -A_C \cos(\omega_C t)$$
$$+ 2 \left(\frac{1}{2} + \frac{2}{\pi} \sin \omega_0 t + \frac{2}{3\pi} \sin 3\omega_0 t + \frac{2}{5\pi} \sin 5\omega_0 t + \cdots \right) \cdot A_C \cos(\omega_C t) \quad (9.2.60)$$

Apply trigonometric identity we are already familiar with,

$$\sin A \cdot \cos B = (\sin(A+B) + \sin(A-B))/2$$

and obtain, after simple manipulations,

$$v_{PSK}(t) = 2A_C[(\sin(\omega_C + \omega_0)t + \sin(\omega_C - \omega_0)t) + \frac{1}{3}(\sin(\omega_C + 3\omega_0)t + \sin(\omega_C - 3\omega_0)t)$$
$$+ \frac{1}{5}(\sin(\omega_C + 5\omega_0)t + \sin(\omega_C - 5\omega_0)t) + \ldots] \quad (9.2.61)$$

(We did similar manipulations in the subsection on ASK modulation, when we derived (9.2.4) from (9.2.3).)

Find that the spectrum of a PSK signal is the same as an ASK spectrum with a suppressed carrier and a double amplitude of each harmonic.

Refer to Examples 9.2.1 and 9.2.2 and Figures 9.2.3 and 9.2.6 to construct the amplitude spectrum, power spectrum, and bandwidth of a PSK signal from those characteristics of an ASK signal.

Bottom line:

The bandwidth of a PSK signal with a square wave message is the same as the bandwidth of an ASK signal.

Likewise, the bandwidth of a PSK signal for a pulse train is the same as the bandwidth of the ASK signal given in Figure 9.2.7.

Therefore, the spectral characteristics of a PSK signal are similar to those of an ASK signal.

9.2.4.3 Demodulation of a Binary PSK Signal

Coherent demodulation of a BPSK signal is similar to that of the BASK. Compare the block diagram of the coherent BPSK receiver shown in Figure 9.2.21 to that of BASK presented in Figure 9.2.9 and observe that they are almost identical. (Remind yourself what other similarities between BPSK and BASK are discussed in this subsection.) This is why the process of BPSK demodulation is similar to that of BASK. This process includes the following consecutive steps:

- A bandpass filter removes out-of-band noise.
- A multiplier combines the input signal and a locally generated sinusoid to produce a new signal, whose frequency contains a double carrier-frequency component, $2\omega_C$.

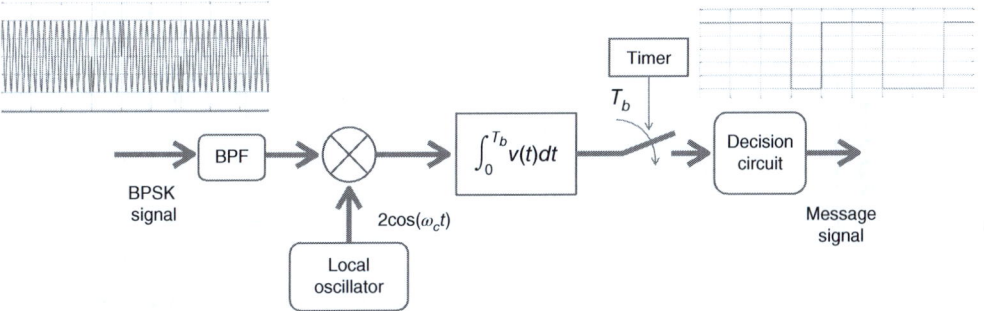

Figure 9.2.21 Block diagram of a coherent BPSK receiver.

- The combined signal containing a message signal and the double-frequency harmonic is presented to an integrator (a lowpass matched filter) that removes the double-frequency harmonic with high accuracy.
- The message signal cleared from the sinusoidal component reaches a decision circuit that recovers the pulses under timing control by comparing the value of an obtained pulse with a threshold.

See Figure 9.2.9 and accompanying discussion of the BASK demodulation for comparison with the above BPSK demodulation. Note that noise, though bandlimited by the BPF and the integrator, passes through the system regardless of whether bit 1 or bit 0 is detected. Fortunately, however, this demodulation process does not increase the level of the passing noise as an FSK coherent detector does (see below).

Differential BPSK Coherent BPSK demodulation requires, in fact, that a local oscillator generate the sinusoid of the same frequency and *phase* as a carrier wave, as was stressed in the discussion of an ASK coherent demodulator. This is necessary to obtain a double-frequency signal after multiplying an input BPSK signal by a locally generated sinusoid, which has been shown mathematically as $\cos(\omega_c t) \times 2\cos(\omega_c t) = 2\cos^2(\omega_c t) = 1 + \cos(2\omega_c t)$. Again, this equation does not work if the two sinusoids have different phases. But how, you may ask, can a receiver know the phase of a carrier being sent by a transmitter? If the frequency $\omega_c t$ is set in advance for both a transmitter and a receiver, the phase of the carrier changes during the transmission and is unknown to the receiver. Thus, coherent demodulation needs additional information to perform properly. Note that the ASK and the FSK signals deliver information about a carrier phase by virtue of their envelopes: They have different envelopes for bit 1 and bit 0. In contrast, the PSK envelopes are the same for both bits.

To resolve this issue, DPSK, was invented. This is how a transmitter modulates a DPSK signal in encoding signified NRZ-S:

- When bits 1 are transmitted, then the carrier sinusoid keeps its phase without change.
- When bit 0 is transmitted, then the carrier changes its phase by 180° for every bit.

An NRZ-S modulation compared with a regular BPSK modulation is shown in Figure 9.2.22. A pulse train delivering a message signal is shown in Figure 9.2.22a. BPSK and DPSK signals are shown in Figure 9.2.22b,c, respectively. In a BPSK signal, the carrier phases change at every transition from bit 1 to bit 0 and vice versa, as the arrows indicate. In a DPSK signal, the carrier phases change only at every bit 0 (see arrows in Figure 9.2.22c). Note that DPSK transmission requires an initial segment of a carrier wave equal to the bit duration to set the reference phase. Clearly, the opposite encoding, through which the carrier phase changes only when bit 1 is transmitted, exists and is denoted as NRZ-M.

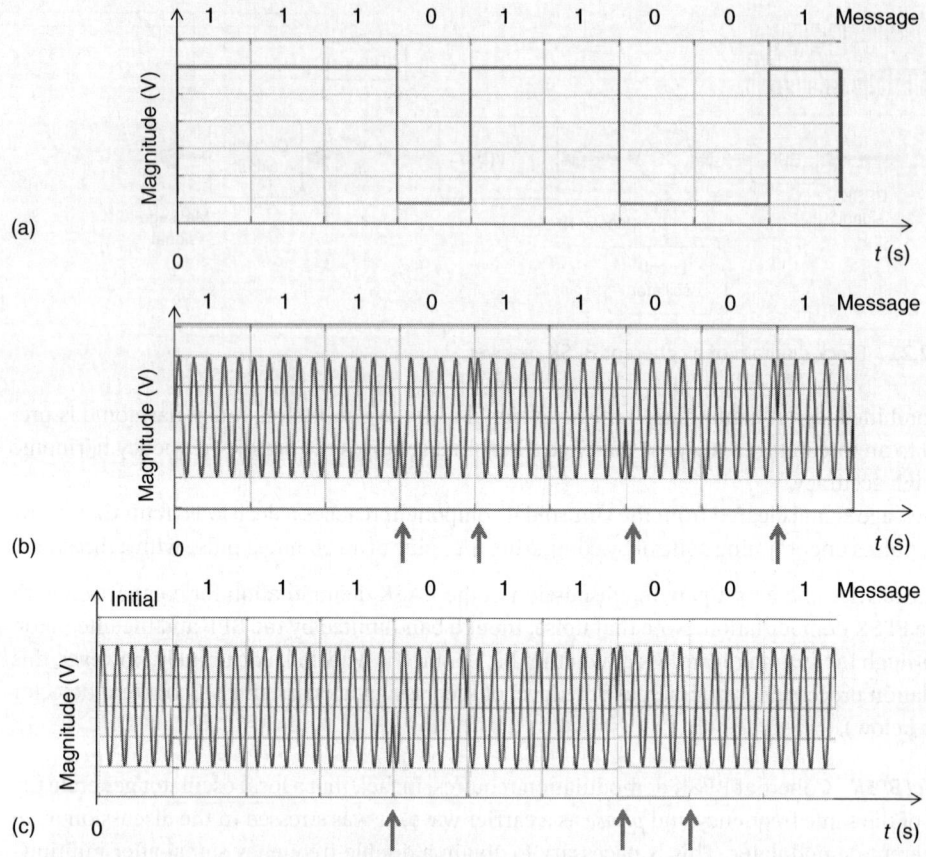

Figure 9.2.22 Modulation of BPSK and DPSK signals: (a) Pulse train delivering a message signal; (b) BPSK signal; and (c) DPSK signal (arrows indicate the phase changes).

A block diagram of a DPSK receiver is shown in Figure 9.2.23. A binary DPSK signal passes through a bandpass filter, BPF, to reject out-of-band noise. At the BPF output, a segment of the DPSK signal representing one bit is sent through the circuit providing a T_b delay. Two sinusoid segments – one representing the current bit and the other representing the preceding bit delayed by T_b – are multiplied and presented to an integrator. When there is no phase change, the integration yields

$$\int_0^{T_b} (A_c \cos(\omega_c t))(A_c \cos(\omega_c t))dt = \frac{A_C^2 \cdot T_b}{2} \quad (9.2.62a)$$

When there is a phase change, the integration yields

$$\int_0^{T_b} (A_c \cos(\omega_c t))(A_c \cos(\omega_c t + 180°))dt$$

$$\int_0^{T_b} (A_c \cos(\omega_c t))(-A_c \cos(\omega_c t))dt = -\frac{A_C^2 \cdot T_b}{2} \quad (9.2.62b)$$

Thus, an integrator's output is a negative pulse when the carrier phase changes, that is, when bit 0 is received. The decision circuit readily identifies the received bit 0 when the integrator's output is negative and bit 1 when the integrator's output is positive.

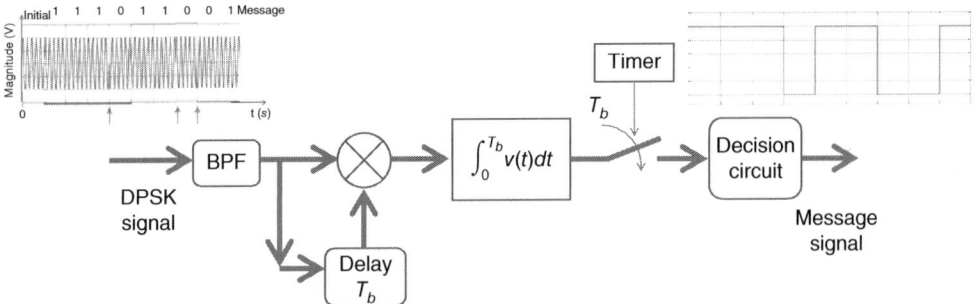

Figure 9.2.23 Demodulation of a binary DPSK signal.

The advantage of DPSK – no need for phase synchronization between the sent and received signals – comes, of course, with a price: DPSK requires greater SNR, or E_b/N_0, to sustain the same BER level, as we will now explain.

9.2.4.4 Bit Error Ratio, BER, of a BPSK Transmission

The general formula for BER is given by

$$\text{BER} = Q\left(\frac{A_c T_b}{\sigma_N}\right) \tag{9.2.63}$$

To calculate BPSK BER, we need to know the following BPSK parameters:

- The average amplitude of a received signal, $A_c = \sqrt{\frac{2E_b}{T_b}}$, which has twice the bit energy of an ASK transmission because both PSK bits 1 and 0 carry the same E_b energy, whereas ASK has zero energy for bit 0 and E_b for bit 1. This explains why (9.2.60) shows that the amplitude of each harmonic in PSK modulation is twice that of an ASK harmonic.
- The bit interval (duration), T_b, which is the same for all types of shift-keying modulation.
- The noise standard deviation, $\sigma_N = \sqrt{N_0 T_b}$, which is equal to σ_N of ASK.

Plugging these parameters into (9.2.63), we find that the BER of a binary PSK signal is given by

$$\text{BER}_{\text{BPSK}} = Q\left(\sqrt{\frac{2E_b}{N_0}}\right) \tag{9.2.64a}$$

Equation (9.2.64a) means that BPSK needs less bit energy, E_b (J), than ASK and FSK modulations to sustain the BER at the same level because $\text{BER}_{\text{ASK}} = \text{BER}_{\text{FSK}} = Q\left(\sqrt{\frac{E_b}{N_0}}\right)$, as (9.2.17) shows.

Therefore, a *PSK modulation is more advantageous than its ASK and FSK counterparts because its bandwidth is as small as that of ASK (and thus smaller than that of FSK by Δf (Hz)), and its BER is better than that of ASK and FSK)*.

At this point, we should pause and consider where this PSK advantage comes from. A comparison with ASK shows that PSK is the superior choice because both of its bits, 1 and 0, carry average energy, E_b, whereas ASK is On–Off modulation with bit 0 having zero energy. But a comparison with FSK is puzzling because both of the FSK bits also carry average energy, E_b; so, why is PSK still more advantageous? The answer lies not in bit energy but in *noise level*.

Review again coherent detection of an FSK signal in Figure 9.2.18. When CFSK bit 1 is presented to the detector, Branch 1 allows this signal to pass and Branch 2 cancels it by using the orthogonality of two carrier waves. But this cancellation is achieved only at the frequency ω_1, whereas noise

9 Digital Transmission with Binary Modulation

being spread over the wide range of frequencies (theoretically infinite, though realistically limited by an input bandpass filter) will penetrate through Branch 2. A similar process occurs when bit 0 is presented to the detector. In both branches, the integrators recover the baseband (message) signal but do not affect high-frequency components of noise. As a result, noise will pass through the detector at demodulation of both bit 1 and bit 0 and *be added at the detector's output*. This is why the standard noise deviation in FSK is defined as

$$\sigma_{\text{NFSK}} = \sqrt{2N_0 T_b} \tag{9.2.64b}$$

Here coefficient 2 appears because noise penetrates through a detector during both demodulation cycles. (Note that $\sigma_{\text{NASK}} = \sigma_{\text{NPSK}} = \sqrt{N_0 T_B}$.) Now, plugging $A_{\text{cFSK}} = \sqrt{\frac{2E_b}{T_b}}$ and $\sigma_{\text{NFSK}} = \sqrt{2N_0 T_b}$ into (9.2.63), we find

$$\text{BER}_{\text{FSK}} = Q\left(\frac{A_c T_b}{\sigma_N}\right) = Q\left(\sqrt{\frac{E_b}{N_0}}\right) \tag{9.2.17R}$$

This explains seeming contradiction between BER of FSK and that of PSK.

Finally, know that the BER of DPSK is given by

$$\text{BER}_{\text{DPSK}} = \frac{1}{2} e^{-\frac{E_b}{N_0}} \tag{9.2.65}$$

The derivation of this formula can be found in many textbooks. (See, for example, Stern and Mahmoud 2004, pp. 270–272.) Compare this formula with BER_{BPSK}, given in (9.2.64a).

We have mentioned that DPSK needs more bit energy than does BPSK to sustain the same BER level. Example 9.2.7 gives quantitative proof of that statement.

Example 9.2.7 Required Bit Energy for BPSK and DPSK Modulations

Problem

Required BER $= 2.28 \times 10^{-2}$ and $N_0 = 2 \times 10^{-7}$ W/Hz. Find the necessary E_b to sustain the BER for BPSK and DPSK modulations.

Solution

We have to use the BER formulas for each modulation to find the needed E_b. For BPSK, apply (9.2.64a) as

$$\text{BER}_{\text{BPSK}} = Q\left(\sqrt{\frac{2E_b}{N_0}}\right)$$

From Table 9.1.2, the value of the argument of the Q-function can be found:

For $\text{BER}_{\text{BPSK}} = Q\left(\sqrt{\frac{2E_b}{N_0}}\right) = 10^{-2}$, we have a $= \sqrt{\frac{2E_b}{N_0}} \approx 2$

Simple calculations produce

$$E_{\text{bBPSK}} = \frac{(a)^2 \cdot N_0}{2} = \frac{2^2 \times 2 \times 10^{-7}}{2} = 4 \times 10^{-7} (\text{J}) \tag{9.2.66}$$

To compute E_{bDPSK}, we need to use

$$\text{BER}_{\text{DPSK}} = \frac{1}{2} e^{-\frac{E_b}{N_0}} \tag{9.2.65R}$$

and express E_{bDPS}, as follows

$$E_{bDPSK} (J) = -N_0 \ln(2BER). \quad (9.2.67)$$

Computations yield

$$E_{bDPSK} = -N_0 \ln(2BER) = 2 \times 10^{-7} \times 3.09 = 6.18 \times 10^{-7} \text{ (J)}$$

Consider the ratio

$$\frac{E_{bDPSK}(J)}{E_{bBPSK}(J)} = \frac{6.18 \times 10^{-7}(J)}{4 \times 10^{-7}(J)} = 1.545$$

Therefore, DPSK needs approximately 1.5 times more bit energy than does BPSK to support the same BER.

The problem is solved.

Discussion

- Compare the *digital SNR* E_b/N_0 for E_{bBPSK} and E_{bDPSK}. It is easy to obtain from (9.2.66) and (9.2.67) the following equations for this comparison:

$$\frac{E_{bBPSK}(J)}{N_0(W/Hz)} = \frac{(a)^2}{2} \quad (9.2.68)$$

and

$$\frac{E_{bDPSK}(J)}{N_0(W/Hz)} = -\ln(2BER) \quad (9.2.69)$$

- Interestingly, the difference between the required bit energy for BPSK and DPSK tends to diminish with the increase in the requirement to BER. Table 9.2.3 demonstrates this tendency.

At BER = 10^{-1}, the E_{bDPSK}/N_0 is more than twice as great as that of BPSK, but – with the increase in BER – E_{bDPSK}/N_0 is getting closer and closer to that of E_{bBPSK}/N_0. Of course, we imply the same level of noise spectral density for each entry.

(*Question*: Table 9.2.3 shows that the better the required BER, the greater the energy, Eb (J), each bit must carry. Does this trend affect the transmission distance, L (km), that a given communication system can support? In other words, is there a trade-off between BER and L (km) because of the requirement to the value of Eb (J)?)

Table 9.2.3 Comparison of E_b/N_0 ratio for BPSK and DPSK modulations.

BER	$\sqrt{2} E_b/N_0$	$(E_b/N_0)_{BPSK}$	$(E_b/N_0)_{DPSK}$	E_{bDPSK}/E_{bBPSK}
1.59E−01	1	0.5	1.15	2.29
2.28E−02	2	2	3.09	1.54
1.35E−03	3	4.5	5.91	1.31
3.17E−05	4	8	9.67	1.21
2.87E−07	5	12.5	14.37	1.15
9.87E−10	6	18	20.04	1.11
1.28E−12	7	24.5	26.69	1.09

Table 9.2.4 Comparison of ASK, FSK, and PSK.

Binary transmission		Pulse-train bandwidth, BW (Hz)	Bit error rate, BER
Amplitude shift-keying, BASK		BW_{ASK} (Hz) = 2BR (b/s) *at first null, or @90% power.* (9.2.14a)	$BER_{ASK} = Q\left\{\sqrt{\frac{E_b}{N_0}}\right\}$ (9.2.17)
Frequency shift-keying, BFSK		BW_{FSK} (Hz) = $2\Delta f$ + 2BR (b/s) *at first null, or @90% power.* (9.2.31a) or BW_{FSK} (Hz) = BW_{ASK} (Hz) + $2\Delta f$ (9.2.32)	$BER_{FSK} = BER_{ASK} = Q\left(\sqrt{\frac{E_b}{N_0}}\right)$ (9.2.17)
Phase shift-keying	BPSK	BWB_{PSK} (Hz) = BW_{ASK} (Hz) = 2 BR (b/s) *at first null, or @90% power*)	$BER_{BPSK} = Q\left(\sqrt{\frac{2E_b}{N_0}}\right)$ (9.2.64)
	DPSK	BW_{DPSK} (Hz) = BW_{BPSK} (Hz)	$BER_{DPSK} = \frac{1}{2}e^{-\frac{E_b}{N_0}}$ (9.2.65)

9.2.4.5 BPSK Advantages and Applications

To reiterate, BPSK is a more advantageous modulation than either BASK or BFSK because its bandwidth is two times smaller than FSK's and is equal to ASK's, and its BER requires less bit energy than that of BASK and BFSK to sustain the same level. Due to its better immunity to noise, BPSK's main applications today lie in wireless transmission. Specifically, BPSK is used in radio-frequency identification (RFID) devices, wireless local area networks, some versions of Bluetooth, credit cards, etc. It should be emphasized that BPSK is typically used in low-bit-rate transmissions; the more sophisticated versions of PSK are used when transmission speed is high. What's more, those variants of PSK made it one of the main modulation formats not only in high-speed wireless communications but also in another major communication field – optical. *PSK is thus the basis for the most modern modulation techniques.*

9.2.4.6 Comparison of Binary ASK, FSK, and PSK

This section compares the main characteristics of three binary shift-keying modulations (see Table 9.2.4).

A close study of Table 9.2.4 will keep you apprised of the main points of this section and put all the material in perspective. (See also Stern and Mahmoud 2004, p. 235.)

Questions and Problems for Section 9.2

- Questions marked with an asterisk require a systematic approach to finding the solution.
- Many questions and problems, including those marked with an asterisk, imply that you, in addition to reading the textbook, will do your research to find the answers. Consider such questions as mini-projects.

Digital Signal over a Sinusoidal Carrier – Binary Shift Keying Modulation

1. What is shift keying modulation? Why do we need it?

2. Discussing binary shift keying, BSK, modulation:
   a) Sketch a pulse train of bits 10110001 and the waveforms of ASK, FSK, and PSK signaling.
   b) Why do we need to have three types of BSK modulation?
   c) Why are these types of modulation called "binary?" Can you imagine any other type, e.g. quaternary?

Binary Amplitude-Shift Keying (ASK)

3 Consider binary amplitude-shift keying, BASK, modulation:
   a) Why is it called so?
   b) Sketch a figure to show how we can produce BASK signal carrying a binary word 10110001.
   c) Figure 9.2.2b shows the block diagram of a BASK transmission system. We know, however, that a realistic block diagram of any communications system must include noise. How will the received information signal in Figure 9.2.2b change if the modulated signal is affected by noise?

4 Write the formula for an ASK signal carrying pulse train of 11010010 if $A_{C1} = 5$ V, $A_{C0} = 0$ V, and $\omega_C = 628$ rad/s.

5 The text says that "Most of the general properties of amplitude modulation discussed in Sections 8.1 and 8.2 are applicable to ASK modulation."
   a) List all these properties.
   b) Which properties are applicable to ASK modulation and which are not? Why?

6 *Determine the spectrum of an ASK signal carrying a periodic square wave message signal if the parameters of the signals are given as $A_C = 5$ V, $A_M = 2$ V, $\omega_C = 628$ rad/s, and $\omega_0 = 6.28$ rad/s:
   a) Write the formula for this $v_{ASK}(t)$.
   b) On the one set of axes, sketch the amplitude and phase spectra of the message signal and the ASK signal. Depict only first five harmonics for $v_M(t)$ and first 10 harmonics for $v_{ASK}(t)$ signals.

7 *The pulse train carrying 110010 bits modulates a carrier, $v_C(t) = 5 \sin(2\pi \times 400 \times 10^3 t)$. Sketch the waveforms of a message signal, $v_M(t)$, a carrier, $v_C(t)$, and the ASK signal against their corresponding amplitude spectra:
   a) Explain how you build the spectra of each signal.
   b) Show the first 5 zero crossings of the pulse train spectrum.
   c) How many cycles of the carrier wave are there within each message bit? Why?
   d) How many side lobes can exist between adjacent pulses in the ASK signal spectrum? Why?
   e) Do you want more or less carrier cycles per bit? Explain.
   (*Hint*: See Figure 9.2.4 and its accompanying discussion.)

8 Review the concept of signal's bandwidth: the definition given in (9.2.7) says BW (Hz) = $f_{max} - f_{min}$. We know, however, that the vast majority of signals in practice have infinite maximum and minimum frequencies. How, then, can we determine the bandwidth?

9 Derive Equation (9.2.12).

10 *Determine the bandwidth of the ASK signal with a square wave message if the signals' parameters are given as $A_C = 5$ V, $A_M = 2$ V, $\omega_C = 2\pi \times 10^2$ rad/s, and $\omega_0 = 2\pi \times 0.1$ rad/s. The communication system must deliver 95% of the ASK signal total power:
   a) Compute the power of individual harmonics of the ASK signal.
   b) How many harmonics do you need to meet the 95% criterion?
   c) Sketch a figure showing power spectrum and bandwidth of the ASK signal in question.

11 A BASK signal delvers bits and therefore its transmission speed (bit rate, BR) is measured in bit-per-second, (b/s), units. Why do we still need to know the BASK bandwidth?

12 A BASK signal operates at BR = 60 Mb/s. What is its bandwidth if the signal carries 90% of the total power? If 95% of the total power? If 98%?

13 What is the advantage of sending curved, smoothed pulses instead of rectangular pulses?

14 BER is an acronym that stands for two definitions: bit error rate and bit error ratio. Which definition is correct? Why does the communication industry use this confusing approach?

15 A receiver of a communication system detects 4 erroneous bits for 600 seconds of transmission at 200 gigabits per second. What is its BER?

16 *A communication system delivers a message signal at BR = 10 Gb/s by using BASK modulation. The required BER_{ASK} is 9.9×10^{-10}. The AWGN noise affecting the transmitted signal has $N_0 = 3.8 \times 10^{-12}$ W/Hz. What is the required receiver sensitivity, P_{rec} (dBm)?

17 Consider the preceding example: Will P_{rec} increase or decrease if BR becomes 5 Gb/s? If N_0 becomes 3.8×10^{-10} W/Hz? Explain.

18 The main drawback of ASK modulation is its susceptibility to noise. Why, however, had the optical-communication industry used BASK as its main modulation format until recently?

19 Compare the number of carrier cycles contained in one bit of BASK signals used in a wireless Bluetooth system and in an optical communication system, given that the Bluetooth carrier has a 2.4-GHz frequency and operates at 3 Mb/s bit rate, whereas optical communication system uses 200-THz carrier and operates at 2 Tb/s.

20 *Discussing the role of a carrier frequency in BASK modulation:
   a) Do you want the carrier frequency to be high or low? Explain
   b) When we refer to high carrier frequency or low carrier frequency, how to classify the frequency level? In other words, how do we know what frequency is high and what is low?
   c) What would happen with a BASK transmission if f_C (Hz) were equal to $\frac{1}{T_b(s)}$?

21 *Figure 9.2.9 shows a typical configuration of a coherent BASK receiver. Explanation of this operation contains a reference to the difference between coherent and incoherent BASK demodulations. Based on this explanation, sketch a block diagram of an incoherent BASK receiver.

Binary Frequency-Shift Keying (FSK)

22 Consider the concept of binary frequency shift keying, BFSK:
   a) What is the main difference between BASK and BFSK? Sketch the signal waveforms to support your explanations.
   b) How, in principle, can we generate a BFSK signal?

c) Figure 9.2.10 shows the block diagram of a BFSK transmission system: Explain how a demodulator can recover a binary message from the received BFSK signal.

23 Write the formula for a BFSK signal delivering the digital word 10010111.

24 *Take a frequency-domain view on a BFSK signal whose $A_c = 5$ V, $f_C = 100$ kHz, $\Delta f = 20$ kHz, $A_M = 2$ V, and $T = 20$ ms, where A_M and T are the parameters of a periodic square wave message signal:
a) Build the amplitude spectrum of this signal.
b) Build the power spectrum of this signal and find its bandwidths at 90%, 95%, and 98% of the total power.

25 A BFSK signal carries a rectangular pulse train whose BR = 2 Mb/s. The BFSK carrier wave has $A_c = 5$ V, $f_C = 100$ kHz, and $\Delta f = 20$ kHz.
a) Build the amplitude spectrum of this signal.
b) Build the power spectrum of this signal and find its bandwidths at 90%, 95%, and 98% of the total power.
c) Comment your results.

26 The BER of a BFSK signaling is given in (9.2.17) as $\text{BER}_{\text{FSK}} = Q\left(\sqrt{\frac{E_b}{N_0}}\right)$. What parameter of a BFSK transmission system can we control to sustain the required BER? How can we do this? Explain.

27 Find the allowed noise spectral density, N_0 (W/Hz), to provide BER = 10^{-3} at BR = 3 Mb/s in Bluetooth transmission if the required *receiver sensitivity* is $P_{\text{rec}} = -70$ dBm.

28 *BER is the ratio of the number of erroneous bits to the number of total bits transmitted. Errors can be caused by a number of factors, including technology, environment, design, human, and many others. Nevertheless, can you point out two decisive factors that determine the BER value?

29 *Compare two BFSK transmission systems: one operates at BR = 100 Gb/s and other does at BR = 400 Gb/s.
a) Provided that all other parameters of these systems are equal, find out which system requires the better BER and why.
b) Can you determine by how much one BER must be better than another?

30 *The text says that ASK is a linear modulation but FSK is nonlinear modulation:
a) Explain what that means. Why do these modulation types differ in this regard?
b) If FSK is nonlinear modulation, how then can we consider FSK as a superposition of two ASK signals, knowing that the principle of superposition is applied only to linear systems?

31 *Comparing ASK and FSK signaling, we expect that FSK should be a better modulation because it is less susceptible to noise. However, the BER is calculated for both modulations by the same formula, 9.2.17. Why?

32 A discontinuous-phase FSK, DPFSK, modulation is simple and effective. Why do we need another type of FSK signaling, such as continuous-phase FSK, CPFSK?

33 Contrast DPFSK and PFSK modulations: Why a VCO can control the phase of a carrier wave in transition from f_{C1} (Hz) to f_{C2} (Hz) and back but a switch cannot?

34 Which FSK modulation – DPFSK or CPFSK – is mostly used in modern communications? Why?

35 Prove that $5\cos 20t$ and $5\sin 20t$ are mathematically orthogonal signals over their period.

36 *To improve the FSK transmission, we set two requirements: orthogonality of two constituent FSK signals $v_{M1}(t)$ and $v_{M2}(t)$ and generation of a continuous-phase FSK signal. Do these requirements relate to one another? If yes, how? If no, why do we need to set both of them?

37 To generate a CPFSK signal, we need to meet two conditions simultaneously:
   a) What are those conditions?
   b) Explain their meaning in reference to FSK signaling.
   c) Derive these conditions mathematically.
   d) Does an FSK signal whose $f_C = 16$ GHz, BR = 2 Gb/s, and $\Delta f = 2.4$ GHz meet these conditions? Prove your answer.
   e) To meet these conditions, we need to choose constants n and k. What are the criteria and recommendations for these constants?

38 *Compare coherent and incoherent detection of a BFSK signal:
   a) What is the main difference between these two methods?
   b) If you were assigned to choose a demodulation method for BFSK transmission, which method would you choose? Why?

39 *Discuss incoherent detection of a CPFSK signal based on Figure 9.2.17:
   a) Show how the binary word 11101001 will be detected.
   b) Sketch the figures to show how BPF1 allows for the passing of frequency $f_{C1} = 120$ kHz and rejects $f_{C2} = 80$ kHz, whereas BPF2 passes frequency $f_{C2} = 80$ kHz and rejects $f_{C1} = 120$ kHz, based on the filters' amplitude (attenuation) characteristics.
   c) Show input and output signals of an envelope detector and explain how this signal processing occurs.
   d) Explain the operation of a comparator.
   e) How does an incoherent BFSK detector determine which bit – 1 or 0 – is received?

40 *Explain why an incoherent BFSK detector is an inherently inaccurate device and how this drawback is overcome by a coherent BFSK detector.

41 *Consider the block diagram of a coherent BFSK detector shown in Figure 9.2.18:
   a) What are the roles of local oscillators 1 and 2?
   b) Why do we need integrators in both branches of the detector?
   c) What operation is performed by the summing module?
   d) How does the coherent BFSK detector decide which bit – 1 or 0 – is received?

e) Why the integrators and the comparator are controlled by a timer, but other modules are not?
f) How does the coherent BFSK detector determine the duration of a received bit?

42 Provide qualitative comparative analysis of the operation of incoherent and coherent BFSK detectors in the presence of additive noise and show which detector will work better in a noisy environment.

43 Figure 9.2.19 shows application of FSK modulation in the Internet of Things, IoT: Comment on each specific application in all the categories.

Binary Phase-Shift Keying (PSK)

44 ASK modulation enables us to deliver a digital signal (bits 1 and 0) by using a sinusoidal carrier but suffers from susceptibility to noise. FSK performs the same task while being much more resilient to noise. Why, then, do we still need another shift-keying modulation, namely a phase-shift keying (PSK)?

45 *Consider the concept and details of PSK modulation are shown in Figure 9.2.20:
a) Figure 9.2.20a demonstrates that bit 1 is delivered by a carrier wave 1 whose $\Theta_1 = 0$ rad and bit 0 is assigned to wave 2 whose $\Theta_2 = \pi$ (rad). Can we change the assignments and make wave 1 carry bit 0? Can we use $\Theta_1 = \pi/2$ rad and $\Theta_2 = 3\pi/2$ rad? Can we take $\Theta_1 = 0$ rad and $\Theta_2 = \pi/2$ rad or any other combinations?
b) Figure 9.2.20b–d shows that the carrier-wave phase changes abruptly when bit 1 changes to bit 0. We know, however, from studying discontinuous-phase FSK (DPFSK) signaling, that the abrupt change in phase results in the increase of the required transmission bandwidth, which is a drawback. Can we, analogously to continuous-phase FSK, (CPFSK), make the phase change in PSK modulation continuous or, at least, smooth? Explain.
c) Figure 9.2.20a shows that a PSK modulator is simply a switch that assigns bit 1 and bit 0 to two carrier sinusoids whose amplitudes and frequencies are equal but whose phases differ by 180°. Can we use another circuitry – for example, VCO – to make the phase assignment more flexible? Explain.

46 Depict a BPSK signal delivering **01100111** digital word.

47 Determine the spectrum of a PSK signal carrying a periodic square wave message signal if the parameters of the signals are given as $A_C = 5$ V, $A_M = 2$ V, $\omega_C = 628$ rad/s, and $\omega_0 = 6.28$ rad/s:
a) Write the formula for this $v_{PSK}(t)$.
b) On one set of axes, sketch the amplitude and phase spectra of the message signal and the PSK signal. Depict only first five harmonics for $v_M(t)$ and first 10 harmonics for $v_{ASK}(t)$ signals.
c) Compare this spectrum with the spectrum of an ASK signal found in Problem 6.

48 Determine the bandwidth of the PSK signal with a square wave message if the signals' parameters are given as $A_C = 5$ V, $A_M = 2$ V, $\omega_C = 2\pi \times 10^2$ rad/s, and $\omega_0 = 2\pi \cdot 0.1$ rad/s. The communication system must deliver 95% of the PSK signal total power:

a) Compute the power of individual harmonics of the PSK signal.
b) How many harmonics do you need to meet the 95% criterion?
c) Sketch a figure showing power spectrum and the bandwidth of the PSK signal in question.
d) Compare this bandwidth with the ASK bandwidth found in Problem 10.

49 Compare block diagrams of coherent ASK and PSK receivers shown in Figures 9.2.9 and 9.2.21, respectively: What are the similarities and differences between these block diagrams?

50 *As our study shows, BASK and BPSK modulations have many common features and properties. Why?

51 Coherent BPSK demodulation requires that a local oscillator generates the sinusoid of the same frequency and *phase* as a carrier wave:
a) Why?
b) How can this issue be resolved in practice?

52 *The following pulse train, 110100 01, is sent by BPSK and DPSK transmission systems:
a) Depict both modulated signals.
b) Indicate the transitions from bit 1 to bit 0 and vice versa on both signals.
c) Comment on the differences between these two modulation types.

53 Is there any difference between demodulations of BPSK and DPSK signals? Base your answer on comparative analysis of Figures 9.2.21 and 9.2.23.

54 *Compare BER of ASK and FSK given in (9.2.17) as $\text{BER}_{\text{ASK}} = \text{BER}_{\text{FSK}} = Q\left(\sqrt{\frac{E_b}{N_0}}\right)$ with BER of PSK given in (9.2.64a) as $\text{BER}_{\text{BPSK}} = Q\left(\sqrt{\frac{2E_b}{N_0}}\right)$:
a) Which BER is better? What criterion do you choose for your answer?
b) Where does this BER advantage come from?

55 *There are two types of PSK modulation: BPSK and DPSK.
a) If $\frac{E_b}{N_0} = 16$, what will be the BER of each type?
b) Why does one modulation type have better BER than the other for the same systems' parameters?
c) Table 9.2.3 shows that the required E_b (J) for both BPSK and DPSK get closer to one another as BER improves. Why?

56 *Based on Table 9.2.3, build the graphs E_b/N_0 vs. BER for BPSK and DPSK modulations to see that the required digital SNR, E_b/N_0, increases with the decrease (improvement) of BER. On the other hand, it is intuitively clear that the required E_b/N_0 must increase with an increase in transmission distance. (Qualitatively sketch the graph E_b/N_0 vs. L (km).) How, then, does BER depend on transmission distance, L (km)?

57 Binary PSK would seem to be an obscure version of binary shift-keying modulation family. However, it appears that BPSK has several significant advantages over BASK and BFSK:
a) List those advantages.
b) List and discuss applications of BPSK available thanks to these advantages.

58 *Binary shift-keying modulations are essential for modern wireless and optical communications:
a) There are many metrics by which these modulations can be assessed and compared. What metrics do you consider the most important?
b) Compare bandwidth, BW (Hz), and BER of the *best versions* of ASK, FSK, and PSK modulations and add your results to Table 9.2.4.

59 This section discusses binary shift-keying modulations. Why has it been necessary to stress the word *binary*?

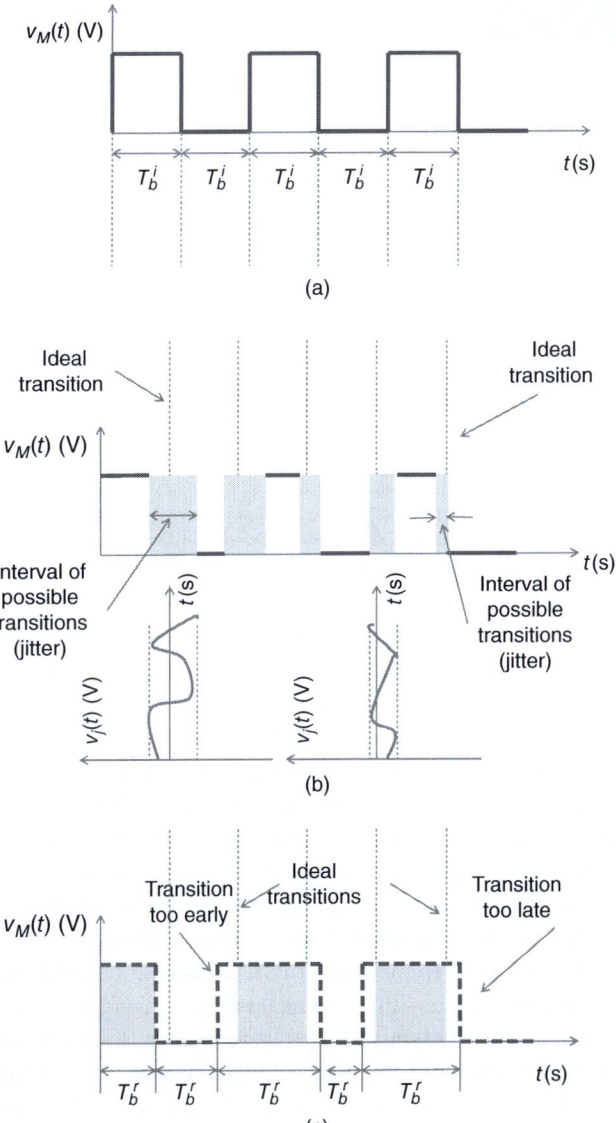

Figure 9.2.A.1 The concept of jitter: (a) A pulse train with ideal pulse intervals, T_b^i; (b) the pulse train with jitter; (c) the pulse train with various transition intervals, T_b^r, caused by jitter; and (d) measurable jitter's parameters in eye diagram.

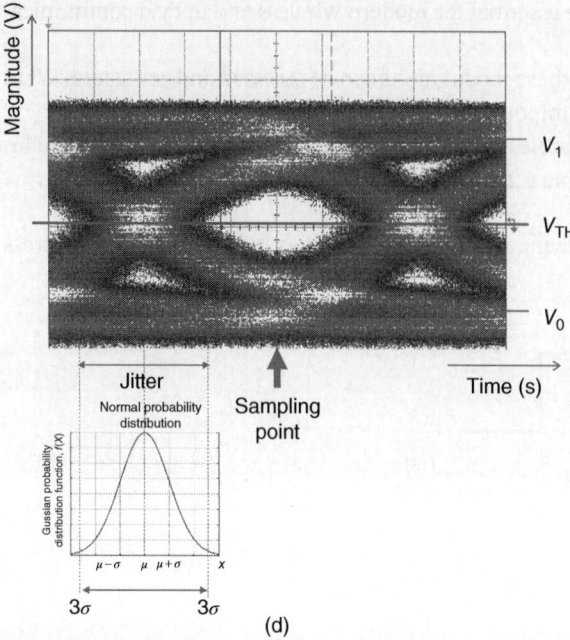

Figure 9.2.A.1 (*Continued*)

9.2.A Jitter

Why cannot an incoherent detector determine the bit interval (duration)? Because it is very difficult (realistically, it is impossible) to determine the borders of this interval. During transmission, the actual starting and ending points of each bit fluctuate in time. The perfect vertical lines that hypothetically separate adjacent bits in our figures exist only in our imagination; in reality, these lines are always blurred. As a result, a pulse width deviates from its theoretical value; this phenomenon is called *jitter*, and it is shown in Figure 9.2.A.1. The communication industry also uses term "jitter" to denote the measure of these deviations. An ideal rectangular pulse train with perfect transition lines is shown in Figure 9.2.A.1a, from which we can readily see that the ideal bit time, T_b^i, is the same for all transmitted bits. The concept of jitter is demonstrated in Figure 9.2.A.1b, where the transition lines separating adjacent bits are blurred. The possible waveforms of the actual transition lines are depicted beneath two of these lines. The train with real pulse widths resulting from the jitter is shown in Figure 9.2.A.1c by dotted lines. We can observe that the real bit times, T_b^r, caused by jitter are substantially different for each transmitted bit. Consequently, the actual areas under each bit presented to the processing unit vary. In incoherent FSK demodulation, both the envelope detectors and the processing unit are supplied with the same bit time, T_B^i, and therefore, these units always determine the same area for each arriving bit. Therefore, jitter causes a discrepancy between the actual and the computed areas under the received bits. Since the area under each bit represents the bit's energy, such discrepancy causes mistakes in the decision of the logic (1 or 0) of each bit. See Section 9.1.2 to brush up on decision procedure in a real stochastic digital transmission.

Bear in mind that jitter is a combination of random and deterministic processes; therefore, it does not reproduce itself at every cycle and cannot be conclusively predicted.

Be cognizant that modern transmission occurs at a very high bit rate, which imposes strict requirements on all parts of a transmission system, particularly on receivers. If we transmit 1 Gb/s, the duration of individual pulse is 1 ns. If jitter causes an 0.5-ns deviation of a bit duration, it will cause total failure of this transmission because adjacent pulses will overlap, and a receiver would not be able to decipher the message signal. (Remember the intersymbol interference phenomenon, ISI?) Now you should fully comprehend why incoherent detection is not widely used in today's communication technology.

Finally, we need to understand that jitter leads to closure of an eye diagram and eventually increases (which means, worsen) BER. This is because the intervals of possible transitions are formed from millions of random time-varying rising and falling edges of transmitted pulses. Figure 9.2.A.1d shows the eye diagram with jitter. The random jitter is described by Gaussian PDF; its measurable parameters, of course, are mean μ and standard deviation σ. In this regard, jitter is similar to eye level one and eye level zero shown in Figure 9.1.17 and discussed in Section 9.1. The difference is that the jitter is measured along time axis, whereas eye levels are measured along the magnitude axis. Figure 9.2.A.1d demonstrates that the jitter's mean value is measured at a crossing point; its rms value is σ, and the value of its width is $\pm 3\sigma$.

10

Digital Transmission with Multilevel Modulation

Objectives and Outcomes of Chapter 10

Objectives

- To understand the concept of multilevel (M-ary) modulation (signaling).
- To learn the basics of quadrature modulation techniques with a concentration on quadrature phase-shift keying, QPSK.
- To acquire knowledge of the multilevel (M-ary) phase-shift keying, M-PSK, and multilevel quadrature amplitude modulation, M-QAM.

Outcomes

- Understand that, besides the binary modulation, there are M-ary (multilevel) modulations, in which many bits can be mapped (packed) into one symbol for transmission.
- Learn that a multilevel modulation (signaling) can save the channel bandwidth (BW) without sacrificing the level of bit rate (BR) or enables to increase the bit rate without demanding more transmission bandwidth. Realize that either way the multilevel signaling improves the spectral efficiency (SE) – the integral measure of the efficiency of a communication system in achieving the maximum bit rate while utilizing the minimum transmission bandwidth.
- Get to know the basic types of M-ary signaling, such as QPSK, QAM, and their multilevel counterparts, M-PSK and M-QAM. Learn their modulation and demodulation techniques, characteristics, advantages, shortcomings, the bit error rates (BERs), and the areas of their applications.

10.1 Quadrature Modulation Systems

Objectives and Outcomes of Section 10.1

Objectives

- To understand the concept of multilevel (M-ary) modulation (signaling).
- To learn the basics of quadrature modulation techniques with a concentration on QPSK.
- To become familiar with modulation, demodulation, assessment, and application of QPSK signaling.

Outcomes

- Recall the concept of digital modulation in general and introduction to multilevel modulation in particular by reviewing Section 3.2.
- Brush up on the idea of packing several (preferably, many) bits into one symbol (pulse) and transmitting these symbols instead of individual bits through a communication system.
- Contrast a two-level (On–Off) modulation to multilevel level modulation (signaling).
- Become familiar with four-level pulse-amplitude modulation, PAM4, whose symbol (dibit) contains two bits. Know that in PAM4 signaling, there are four symbols with different amplitudes (levels) presenting all possible combinations of two bits 1 and 0.
- Be aware that PAM4 exists in two versions: PAM4-1 whose symbol duration, T_{S1}, is equal to the duration of two bits, i.e. T_{S1} (s) $= 2T_b$ and PAM4-2 whose T_{S2} (s) $= T_b$.
- Recall that the pulse's bandwidth is determined by the first null of its spectrum, and this null can be found as $1/T_X$. Understand that $\text{BW}_{\text{PAM4-1}}$ (Hz) $= 1/T_{S1}$, $\text{BW}_{\text{PAM4-2}}$ (Hz) $= 1/T_{S2}$, $\text{BW}_{\text{PAM-2}}$ (Hz) $= 1/T_b$, and therefore, $\text{BW}_{\text{PAM4-1}}$ (Hz) $= 0.5\text{BW}_{\text{PAM-2}}$.
- Remember that bit rate is the number of bits per second. Observe from Figure 10.1.1 that $\text{BR}_{\text{PAM2}} = 16 \, \text{b/s}$, $\text{BR}_{\text{PAM4-1}} = 16 \, \text{b/s}$, and $\text{BR}_{\text{PAM4-2}} = 32 \, \text{b/s}$.
- Compare bandwidths and bit rates of PAM4-1 and PAM4-2 and conclude that a *multilevel modulation provides saving the channel bandwidth without sacrificing the bit rate or enables to increase the bit rate without demanding more transmission bandwidth*. Refer to the Hartley capacity law, BR (b/s) $= 2 \cdot \text{BW (Hz)} \cdot \log_2 M$, as to a mathematical description of this statement.
- Discover that PAM4-1 modulation finds its application in communication areas where saving the bandwidth is the priority (e.g. wireless communications) and PAM4-2 signaling better serves in communication systems where achieving the highest bit rate is the main task (e.g. optical communications).
- Revisit spectral efficiency, SE (b/s/Hz) = BR (b/s)/BW (Hz) – the integral measure of efficiency of a communication system in achieving the maximum bit rate at the minimum transmission bandwidth – and find that $\text{SE}_{\text{PAM2}} = 2 \, \text{b/s/Hz}$ and $\text{SE}_{\text{PAM4}} = 4 \, \text{b/s/Hz}$ for both types of PAM4. Realize that this advantage of SE_{PAM4} is achieved only by using four-level modulation.
- Know that despite several disadvantages, *multilevel modulation formats are in the mainstream of today's digital transmission*. Since modern transmission is almost exclusively done digitally, the importance of multilevel signaling cannot be overstated.
- Remember that in *M*-ary signaling, we need to distinguish between *bit rate*, BR (b/s), which is the number of bits per second, and *symbol or baud rate*, BS (Bd), which is the number of symbols per second.
- Denote the number of bits per symbol as N_b and the number of symbols as M and derive that $M = 2^{N_b}$ or $\log_2 M = N_b$.
- Learn that one of the core multilevel modulation formats is a QPSK. Understand that this is a four-level variant of a phase-shift keying (PSK) modulation, in which one symbol, called *dibit*, represents two bits. Know that each symbol (dibit) is a segment of a carrier wave of the same amplitude and frequency, but with a 90°-phase shift with respect to one another.
- Grasp the fact that using the phasor presentation of a sinusoidal signal, we can depict QPSK signals in a *signal space*, where each of the four phasors is located in one of the quadrants. Learn that this space is divided by a horizontal *I* line termed *in phase* and a vertical *Q* axis called *quadrature*. Know that *Q* axis is shifted 90° counterclockwise with respect to the *I* axis and that the first symbol (dibit), $v_1(t)$, is shifted 45° counterclockwise from the *I* axis. Follow the fact that the next symbol, $v_2(t)$, is shifted 90° counterclockwise with respect to $v_1(t)$, that is, by 135° from

the I axis. The third and the fourth symbols are shifted further by 225° and 315° from the I axis, respectively.
- Get to know that a signal space figure leads to a constellation diagram, where the four constellation points (dots) represent four symbols and carry information about their amplitudes and phases.
- Understand that each pair of bits is assigned to one of the four QPSK symbols and that this assignment shown in the constellation diagram is called bit-to-symbol mapping.
- Observe that a constellation diagram can demonstrate the relationship among constellation points (symbols or dibits, that is), carrier phase changes, and I and Q projections, which is shown in Figures 10.1.2, 10.1.4, and 10.1.5. Learn that Figures 10.1.3 and 10.1.5 demonstrate the formation of a dibit's waveform from I and Q waveforms and show how a QPSK signal's waveform is composed.
- Realize that a QPSK signal is the sum of two independent components: the I component modulating the sine carrier and the Q component modulating the cosine carrier. Conclude that *QPSK can be viewed as two BPSK binary data streams.*
- Master the fact that QPSK modulators and demodulators are built based on this two-stream approach, as demonstrated in Figures 10.1.6–10.1.9.
- Learn that assessing the quality of QPSK transmission can be done by using BER_{QPSK}, which appears to be either equal or double to BER_{BPSK}, and SER_{QPSK}, which naturally surfaces to be either equal or half of BER_{QPSK}. (SER stands for symbol error rate.)
- Come to know that other versions of QPSK, such as offset quadrature phase-shift keying (OQPSK), differential quadrature phase-shift keying (DQPSK), and a minimum shift keying (MSK) have been developed to improve the performance of the QPSK basic model.

10.1.1 Multilevel (*M*-ary) Modulation Formats – What and Why

10.1.1.1 The Concept of Multilevel Modulation

As explained in Chapters 3 and 9, multilevel modulation formats are powerful tools for increasing the efficiency of digital transmission. Many modulation formats are in use in today's digital transmission, and they can easily be a subject of a specialized textbook. We, however, concentrate in this chapter on only two main types – QPSK and QAM. A solid understanding of these quadrature modulation formats gives you the key to mastering any other multilevel modulation you might encounter in your professional career. Refresh your memory of our discussion on digital modulation and introduction to multilevel modulation by reviewing Section 3.2.

The main idea of multilevel modulation formats is packing several (preferably, many) bits into one symbol (pulse) and transmitting these symbols instead of individual bits through a communication system. Figure 10.1.1 shows the concept of multilevel modulation. When a communication system transmits individual bits, there are only two levels of modulation: The pulse with amplitude A_1 carries bit 0 and the pulse with amplitude A_2 carries bit 1. A *two-level (On–Off) modulation* is shown in Figure 10.1.1a. We can, however, pack two bits into one symbol, as shown in Figure 10.1.1b,c, and make the pulse with amplitude A_1 (a symbol, that is) that transmits a combination of bits 0 and 0. Likewise, the pulse with amplitude A_2 (another symbol) can carry a pack of bits 0 and 1, the pulse with amplitude A_3 can deliver bits 1 and 0, and the pulse with amplitude A_4 can contain the combination of bits 1 and 1. Thus, instead of two levels (amplitudes) needed to deliver bit 0 and bit 1, we use in this example four levels (symbols) to transmit four combinations of bits, such as 0 and 0, 0 and 1, 1 and 0, and 1 and 1. This type of modulation is called *four-level*.

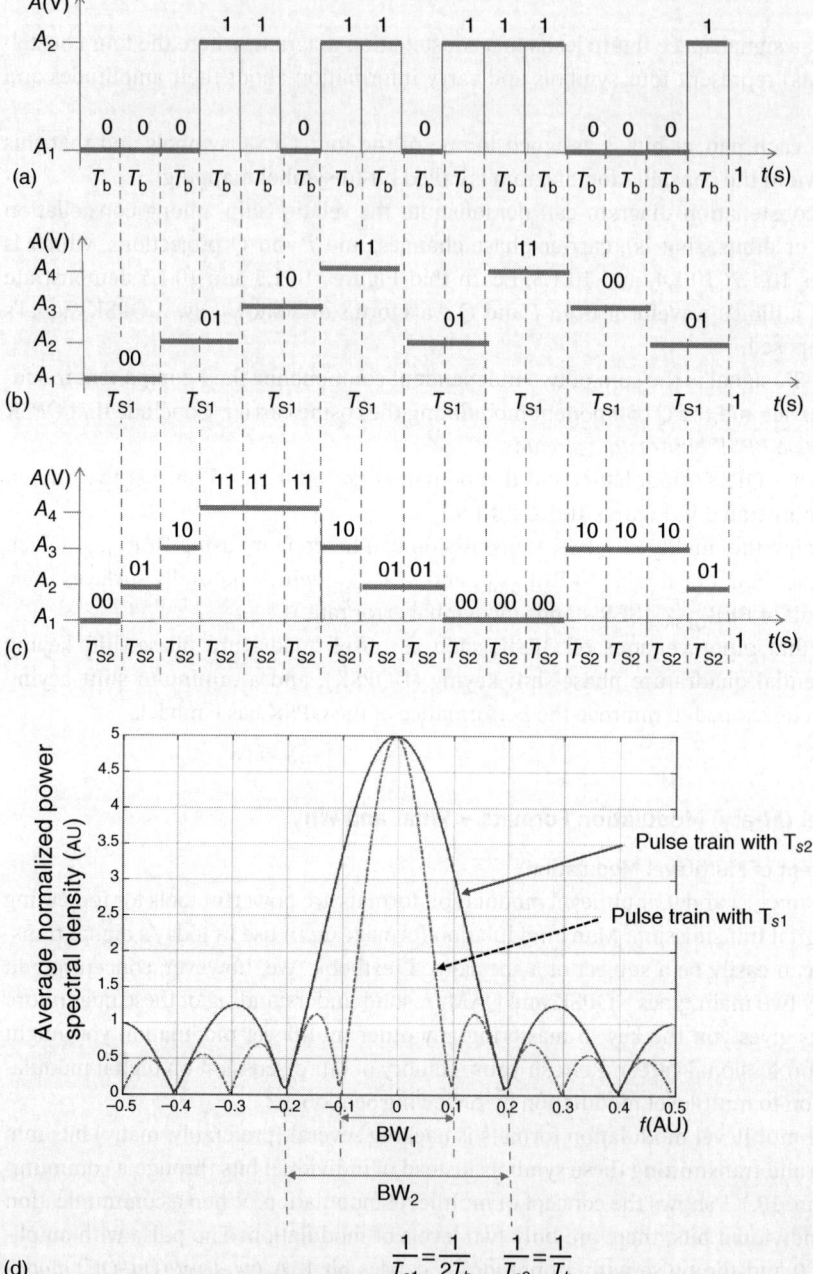

Figure 10.1.1 The concept of multilevel modulation: (a) two-level modulation with $T_{S1} = 2T_b$; (c) four-level modulation with $T_{S2} = T_b$; and (d) bandwidths of four-level modulations with T_{S1} and T_{S2} (AU – arbitrary units. Data for illustrative purpose only).

In general, we can map many (e.g. 256) bits into a single level (symbol). In such an arrangement, a pulse train contains pulses (symbols) with many levels; this is why this type of modulation is called *multilevel*.

Why do we need multilevel modulation? Compare Figures 10.1.1a and 10.1.1b: In this example, both two-level and four-level systems delivers 16 bits for one second, which means that their bit rate is the same. But notice a very important point: *A binary signaling needs to transmit 16 pulses to deliver 16 bits, whereas four-level signaling carries only eight pulses (symbols) to transmit the same amount of information*. This feature implies that four-level modulation requires twice less bandwidth than the two-level does. Why? First, pay attention to the duration of a bit, T_b (s), and a symbol, T_{S1} (s): Four-level modulation in Figure 10.1.1b has T_{S1} (s) = $2T_b$ (s). Secondly, refer to the spectrum of a rectangular pulse and its bandwidth shown in Figure 9.2.7, which is partly reproduced in Figure 10.1.1d. We recall from these figures that the pulse's bandwidth is determined by the first null of its spectrum. This null can be found as $1/T_X$, where T_X is the duration of a pulse. In our example, T_X is either T_b or T_{S1} and $BW_2 = 1/T_b$ and $BW_1 = 1/T_{S1}$. Thus, BW_1 (Hz) = $0.5 BW_2$ (Hz), as Figure 10.1.1d shows. Therefore, *a four-level modulation saves the bandwidth required for transmission of its signals*.

(Don't forget that a *first-null bandwidth* measured from $f = 0$ Hz to $1/T_b$ (s) is a *one-sided bandwidth*; Figure 10.1.1d shows, in fact, *null-to-null bandwidth*, which is *two-sided*.)

Wait a minute, you might correctly note, Figure 10.1.1c shows another version of four-level signaling that can deliver 32 b/s vs. 16 b/s of the two-level signaling. Thus, four-level modulation can generate greater bit rate than the two-level can. Yes, this is true, and the increasing of a bit rate is also an advantage of four-level signaling. However, the increase in BR (b/s) shown in Figure 10.2.1c is achieved by employing a smaller symbol duration, T_{S2}, namely, $T_{S2} = T_b = 0.5 T_{S1}$. This means that this type of four-level signaling requires greater transmission bandwidth because $BW_2 = 1/T_{S2} = 2BW_1$, as Figure 10.1.1d demonstrates.

A conclusion drawn from comparing two-level modulation and both types of four-level modulations can be generalized for any multilevel signaling as follows:

> *A multilevel modulation either provides saving the channel bandwidth without sacrificing the bit rate or enables to increase the bit rate without demanding more transmission bandwidth.*

This statement is, in fact, described by the *Hartley capacity law*,

$$BR \text{ (b/s)} = 2 \cdot BW(Hz) \cdot \log_2 M \qquad (1.3.4R)$$

where M is the level of modulation. (Be aware that in literature you can often meet another popular designation for M – *alphabet size*.) Thus, the Hartley law shows that at given M there is a trade-off between the bit rate of a signal being transmitted and the channel bandwidth required for its transmission. (Can you explain how? What if we change M?)

Can we increase the bit rate of multilevel signaling when a transmission bandwidth is fixed? Yes, we can. Hartley law shows explicitly that if BW (Hz) is constant, BR (b/s) can be increased only by increasing M. Can we increase BR (b/s) and decrease the required BW (Hz) simultaneously by increasing M? It will be a good exercise for you to find the answer to this question; so, let us leave it open for now. We will expand on this question later in this chapter.

We remember, of course, that BR (b/s)/BW (Hz) = SE (b/s/Hz), where SE is spectral efficiency (refer to Section 3.2 and specifically to (3.2.8)). In our example, spectral efficiency for the two-level signaling is given by $SE_{2\text{-level}} = BR_{2\text{-level}}/BW_{2\text{-level}} = 2$ b/s/Hz, whereas spectral efficiency of the four-level modulation is equal to $SE_{4\text{-level}} = BR_{4\text{-level}}/BW_{4\text{-level}} = 4$ b/s/Hz. In general, for multilevel modulation SE (b/s/Hz) = BR (b/s)/BW (Hz) = $\log_2 M$. Thus, the greater the M, the greater the SE. So, the spectral efficiency – the integral measure of efficiency in achieving the maximum bit rate at the minimum transmission bandwidth – conclusively demonstrates the advantage of multilevel modulation. We will elaborate on this topic in Section 10.2.

(**Question**: *We consider two types of four-level modulation, but the text states that* $SE_{4\text{-level}} = BR_{4\text{-level}}/BW_{4\text{-level}} = 4$ *b/s/Hz regardless of the type. Why?*)

So, we have two types of four-level modulation: One that saves the bandwidth, BW_1 (Hz), but does not increase the bit rate, BR_1 (b/s) = $1/T_{S1}$ (s), compared with two-level signaling and the other that increases the bit rate, BR_2 (b/s) = $1/T_{S2}$ (s) but requires the two-level transmission bandwidth, BW_2 (Hz). Which type of four-level signaling to choose? The answer is, it depends on applications of a communication system. In wireless communications, bandwidth is scarcity, and therefore, the BW_1-type is preferred. In optical communications, the bit rate is the main task, whereas bandwidth is still abundant; hence, BW_2-type is more suitable for the job in this field. These recommendations are, of course, too general; it is up to a communication system designer to choose the best type of modulation for a specific network. (See Section 1.2 to refresh your memory on the requirements for various types of modern communications.)

Be aware that all signaling shown in Figure 10.1.1 is called *pulse amplitude modulation*, *PAM*, because modulation is achieved by varying the pulse amplitudes. A two-level modulation (binary or On–Off) is called *PAM2*. This technique, also known as non-return-to-zero *encoding*, *NRZ*, is discussed in Section 3.2. Four-level signaling is called – you guessed this – PAM4. Review closely industrial documentation to understand to what types of four-level modulation – BW_1-type or BW_2-type – a specific document is referring. Again, in wireless communications, it will most likely be BW_1-type; so, PAM4 is widely considered as a synonym for the bandwidth-saving type of four-level signaling.

The communication industry often explains saving the bandwidth in multilevel signaling in terms of *Nyquist rate*. This rate is introduced in (3.2.4) as BR (b/s) $\leq$ 2BW (Hz). In such a PAM4, symbol rate (SR) measured in baud, BS_4 (Bd), is equal to ½BR (b/s) and therefore BS_4 (Bd) $\leq$ BW (Hz). This is why industrial documentations often stresses that four-level signaling uses only half of the Nyquist rate.

Get to know that, of course, there are other benefits, as well as shortcomings, of using multilevel signaling, which we will learn in this section.

We need to realize that

> *multilevel modulation formats are in the mainstream of today's digital transmission.*

Since modern transmission is almost exclusively done digitally, the importance of multilevel signaling cannot be overemphasized.

10.1.1.2 Symbols and Bits

Let us focus on four-level modulation, which means placing two bits into one symbol. What are those two bits? They can be one of the following combinations:

Symbol $v_1(t)$ represents 00

Symbol $v_2(t)$ represents 01

Symbol $v_3(t)$ represents 10

Symbol $v_4(t)$ represents 11 (10.1.1)

Therefore, *we must have four different symbols, each representing various combinations of two bits.* Why do we need four, not three, not six symbols? This is because we have four different combinations of bits 1 and 0, as shown in Figure 10.1.1 and Eq. (10.1.1).

The concept of multilevel modulation can be mathematically described as follows: The number of the required symbols, M, is equal to 2 to the power of N_b, where the number 2 shows that we use the binary system and N_b is the number of bits per symbol. In (10.1.1), N_b is two and therefore $2^{2\,\text{bits}} = 4$ (symbols). How many symbols do we need to squeeze three bits into one symbol? Count the number of possible combinations of three bits, such as 000, 001, 010, etc. You immediately found that there are eight possible combinations and therefore you need eight levels (amplitudes) to transmit all combinations of three bits. Mathematically, this result can be described as $2^{3\,\text{bits}} = 8$ (symbols). Using the properties of a logarithm, we can express this relationship as $\log_2 8$ (symbols) = 3 bits. In general, the rule governing the relationship between the number of symbols and the number of bits per symbol can be written in one of two forms:

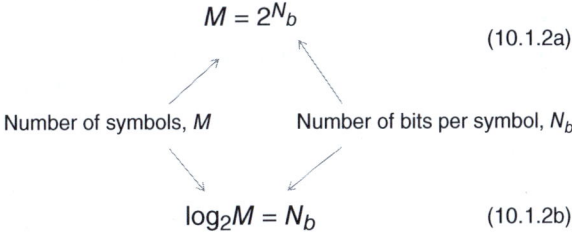

$$M = 2^{N_b} \quad (10.1.2a)$$

Number of symbols, M ⇄ Number of bits per symbol, N_b

$$\log_2 M = N_b \quad (10.1.2b)$$

This is why *multilevel modulation is often called M-ary modulation*. When $M = 4$ and $N_b = 2$, the modulation is called *quadrature*. Each pair of two bits is called a **dibit**; thus, there are four dibits in quadrature modulation, namely 00, 01, 10, 11, as shown in (10.1.1).

Why multilevel modulation is the main type of signaling employed today? Remember, *a multilevel modulation can either save bandwidth at a given bit rate or increase the bit rate at a given bandwidth*. How important are these features? This is one example: In the middle of 1990s, the exponential increase in the Internet usage required providing high-speed digital communications, especially in the access lines. At that time, the only universal connections from customer premises to the first Internet hubs were copper telephone lines. These lines were designed to deliver analog voice signal within 4-kHz bandwidth. How could engineers provide digital transmission of a megabit-per-second rate under such severe bandwidth restriction? (Refer to the Shannon limit given in (1.3.6) to understand this question.) Multilevel modulation was the solution to this seemed-to-be unresolvable problem. Packing more bits into one symbol, the transmission technology called digital subscriber line (DSL) provided acceptable transmission rate within enormously restricted bandwidth of the transmission line.

Example 10.1.1 Employing the Advantage of a Multilevel Modulation

Problem

The bandwidth of a telephone line is BW = 4 kHz. This line must support the digital transmission at 64 kb/s to provide access from customer premises to the nearest Internet hub. Questions and an assignment:

1. Can a binary signal at the required bit rate be directly transmitted through this telephone line?

2. If the answer to the previous question is no, what is the solution to this problem?
3. Give the numerical parameters of a proposed transmission.

Solution

1. We need to refer to Hartley's capacity law

$$\text{BR (b/s)} = 2 \cdot \text{BW (Hz)} \log_2 M \qquad (1.3.4\text{R})$$

where M is the number of modulation levels. Since $M = 2$ in a binary modulation, we find

$$\text{BR (b/s)} = 2 \cdot \text{BW (Hz)} = 8 \text{ kb/s}$$

(Remember that Hartley's law gives the maximum transmission capacity of a noiseless channel.) Thus, the digital signal in a binary format cannot be transmitted through this telephone line.

2. The solution to this problem is given by (1.3.4R), BR (b/s) = 2·BW (Hz)·$\log_2 M$: 64 kb/s would be equal to 2×4 kHz·$\log_2 M$ only if M would be properly increased. In other words, we must use a multilevel modulation, which is a general solution to transmitting a high-bit-rate signal over a bandlimited channel. This is because using a minimal bandwidth while providing a maximum bit rate is the main advantage of an M-ary modulation. (See Section 10.2.3.1 for limitations in the applicability of Hartley's law to finding M.)

3. Therefore, the problem boils down to the questions of how many symbols and how many bits per symbol we need to arrange. The number of bits per symbol can be found from (10.1.2b) and (1.3.4R) as

$$\log_2 M = N_b = \frac{\text{BR (b/s)}}{2\text{BW } (^1/_s)}$$

Plugging the required BR = 64 kb/s and given BW = 4 kHz into this formula yields

$$N_b = 8 \text{ (bit/symbol)}$$

To map 8 bits to one symbol, we need $M = 2^8 = 256$ levels.

The problem is solved.

Discussion

- Since we found that $N_b = 8$ and $M = 256$, the proposed solution is the use of 256-ary signaling transmitted at SR = 8 kBd. The symbol rate is found by the following manipulations: A symbol interval relates to a bit interval as T_S (s) = N_b (b/symbol)·T_b (s); therefore, a SR (Bd), pertains to a bit rate, BR (b/s) as SR (Bd) = BR (b/s)/N_b (b/symbol). Plugging in the required BR and needed N_b, we compute SR (Bd) = BR$_{\text{req}}$ (b/s)/N_b (b = 64 (kb/s)/8 (b/ksymbol) = 8 (ksymbol/s) = 8 (kBd)).
- The required bit rate of 64 kb/s stems from the digitizing the analog voice signal: The maximum voice frequency is $f_{\max} = 4$ kHz. Hence, the Nyquist frequency is 2×4 kHz = 8 (ksample/s). The telephone industry (the Bell Labs, in fact) puts 8 bits per sample; thus, BR = 8 (ksample/s)·8 (b/sample) = 64 kb/s. (Revisit Section 3.3 to brush up on analog-to-digital conversion, ADC.)

10.1.2 Quadrature Phase-Shift Keying, QPSK

10.1.2.1 Introduction to Quadrature Phase-Shift Keying, QPSK

How can multilevel modulation be achieved in practice? By varying not only the amplitude of a sinusoidal carrier but also its frequency and phase to transmit different combinations of bits 1 and 0. In this section, we will study a specific type of four-level signaling called *QPSK*. This is, obviously, a multilevel variant of a binary PSK format, the variant in which one symbol represents two bits, as shown in (10.1.1). We remember that PSK represents bit 1 by a carrier sinusoid with a zero-phase shift and bit 0 with an 180°-phase shift; that is

$\quad A \sin(\omega t)$ represents bit 1

and

$\quad A \sin(\omega t + 180°) = -A \sin(\omega t)$ represents bit 0

(We switch to a sine carrier wave instead of a cosine for the purpose that becomes clear later.)

Now, we need to deliver four symbols; how can we do this? Well, if for two dibits we need two phases, 0° and 180°, then for four dibits we will need four phases. Indeed, by assigning four carriers with four different phases to each dibit, we will achieve our goal. Obviously, we want to change our phases by equal increments. To do so, we implement the concept given in (10.1.1) as

$\quad v_1(t) = A \sin(\omega t + 45°)$ represents dibit 11

$\quad v_2(t) = A \sin(\omega t + 135°)$ represents dibit 01

$\quad v_3(t) = A \sin(\omega t + 225°)$ represents dibit 00

$\quad v_4(t) = A \sin(\omega t + 315°)$ represents dibit 10 $\hfill (10.1.3)$

As we can see, each symbol is a segment of a carrier wave of the same amplitude and frequency, but with a 90°-phase shift with respect to one another. (Be aware that mapping the dibits to the signals (phasors) in the way shown in (10.1.3) is not the rule but the example.)

Using the phasor presentation of a sinusoidal signal, we can depict (10.1.3) as shown in Figure 10.1.2. *Signal space*, where the layout of all four phasors is shown, is demonstrated in Figure 10.1.2a. The horizontal line is termed *in phase*, or *real*, or *I* axis; this axis, by convention, is considered the phase-reference (zero phase shift) axis. The vertical axis is called *quadrature* (also, the *Q* axis) because it is shifted 90° counterclockwise with respect to the *I* axis. The first symbol (phasor), $v_1(t)$, representing dibit 11, is shifted 45° counterclockwise from the *I* axis. The symbol $v_2(t)$, representing dibit 01, is shifted 90° counterclockwise with respect to $v_1(t)$, that is, by 135° from the *I* axis. The positions of $v_3(t)$ and $v_4(t)$ are clear from the figure.

Figure 10.1.2 demonstrates bit-to-symbol assignments. For example, the dibit 11 is assigned to the symbol located in the first quadrant of a signal space, or constellation diagram. In Figure 10.1.2, each pair of bits is mapped to one of the four QPSK symbols; hence, this figure shows *bit-to-symbol mapping*. The importance of such a mapping becomes apparent when we realize that *there are several options in assigning bits to symbols in QPSK*.

Typically, most of the information shown in Figure 10.1.2a is implied by default, and the graphical presentation of QPSK is reduced to the diagram demonstrated in Figure 10.1.2b. The four *constellation points* (dots in this figure) represent four symbols and carry information about their amplitudes and phases. All amplitudes are equal because the dots are equidistant from the zero point (the

952 | 10 Digital Transmission with Multilevel Modulation

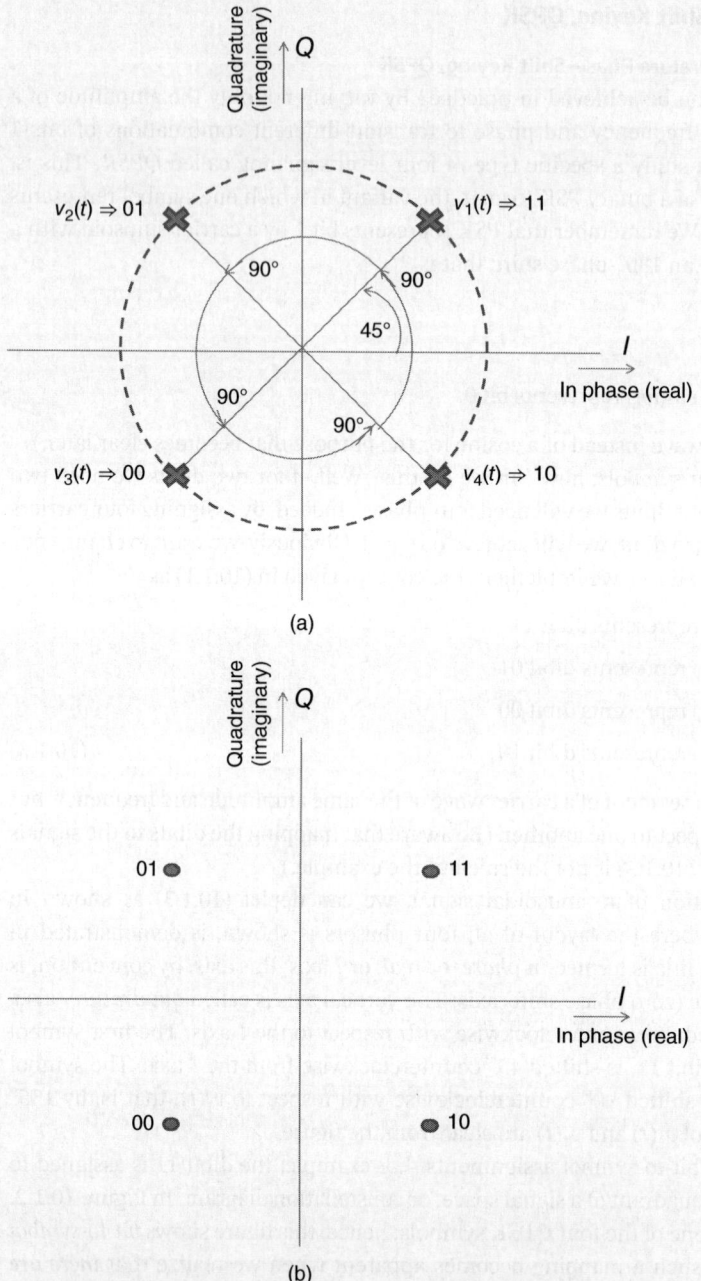

Figure 10.1.2 Graphical presentation of quadrature phase-shift keying with bit-to-symbol mapping: (a) signal space and (b) constellation diagram.

intersection of the I and Q axes), and every consecutive phase shift is 90°, which is clear from the locations of the dots. Such a presentation is called a constellation diagram, or simply a *constellation*, due to its resemblance to a map of stars visualized by astronomers.

The QPSK modulation scheme is widely used in today's communications. It is sufficed to say that QPSK is the main type of modulation used for communications in such a popular system as the Internet of Things (IoT).

10.1.2.2 QPSK Signal: Waveform and Constellation Diagram

How a quadrature-modulated signal is seen on an oscilloscope screen? This signal is the sum of two sinusoids, one is $v_I(t) = A_I \sin(\omega t)$ and the other is $v_Q(t) = A_Q \sin(\omega t + 90°) = A_Q \cos(\omega t)$. Thus, $v_{QPSK}(t) = v_I(t) + v_Q(t) = A_I \sin(\omega t) + A_Q \cos(\omega t)$. We remember, of course, that such a sum can be written as $v_{QPSK}(t) = A \sin(\omega t + \Theta)$, where $A = \sqrt{(A_I^2 + A_Q^2)}$ and $\Theta = \tan^{-1}(A_Q/A_I)$. Let $A_I = A_Q = A/\sqrt{2}$, then $v_{QPSK}(t) = A \sin(\omega t + \pi/4)$.

Assume using NRZ bipolar encoding, in which bits 1 and 0 are presented by positive and negative pulses, respectively. Then, the waveform of dibit **11** in QPSK modulation is the result of multiplying each of the signal constituents by bit 1; that is, $v^{11}_{QPSK}(t) = (1) \cdot A_I \sin(\omega t) + (1) \cdot A_Q \cos(\omega t)$. Therefore, $v^{11}_{QPSK} = A \sin(\omega t + 45°)$. The graph of this signal with the phase shift of 45° is shown in Figure 10.1.3a. Correspondingly, dibit **01** is presented by a QPSK signal as $v^{01}_{QPSK}(t) = (-1) \cdot A_I \sin(\omega t) + (1) \cdot A_Q \cos(\omega t) = A \sin(\omega t - 45°) = A \sin(\omega t + 135°)$; its graph is shown in Figure 10.1.3b. Consequently, Figure 10.1.3c,d shows the waveforms of signals $v^{00}_{QPSK}(t) = (-1) \cdot A \sin(\omega t) + (-1) \cdot A \cos(\omega t) = A \sin(\omega t + 225°)$ and $v^{10}_{QPSK}(t) = (1) \cdot A \sin(\omega t) + (-1) \cdot A \cos(\omega t) = A \sin(\omega t + 315°)$, respectively. Closely examine the phase shift of each QPSK signal. Compare these phase shifts with those given in (10.1.3).

Now, let us turn to QPSK constellation diagram. Recall PSK modulation: We have two bits and any transition from bit 1 to bit 0 and vice versa is signified by an 180°-phase change of a sinusoidal carrier. How can we modulate a sinusoidal carrier to obtain a QPSK signal carrying four dibits? Obviously, we need to signify each transition from one dibit to another by a 90°-phase change because now we have four possible transitions.

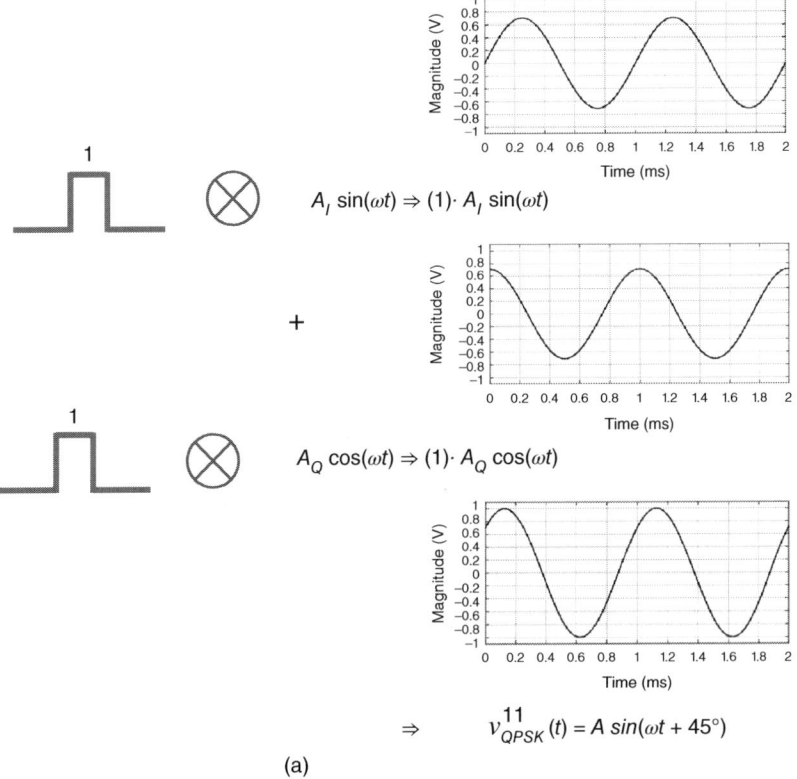

(a)

Figure 10.1.3 Formation of the waveforms of QPSK signals modulated by (a) **11** dibit, (b) **01** dibit, (c) **00** dibit, and (d) **10** dibit.

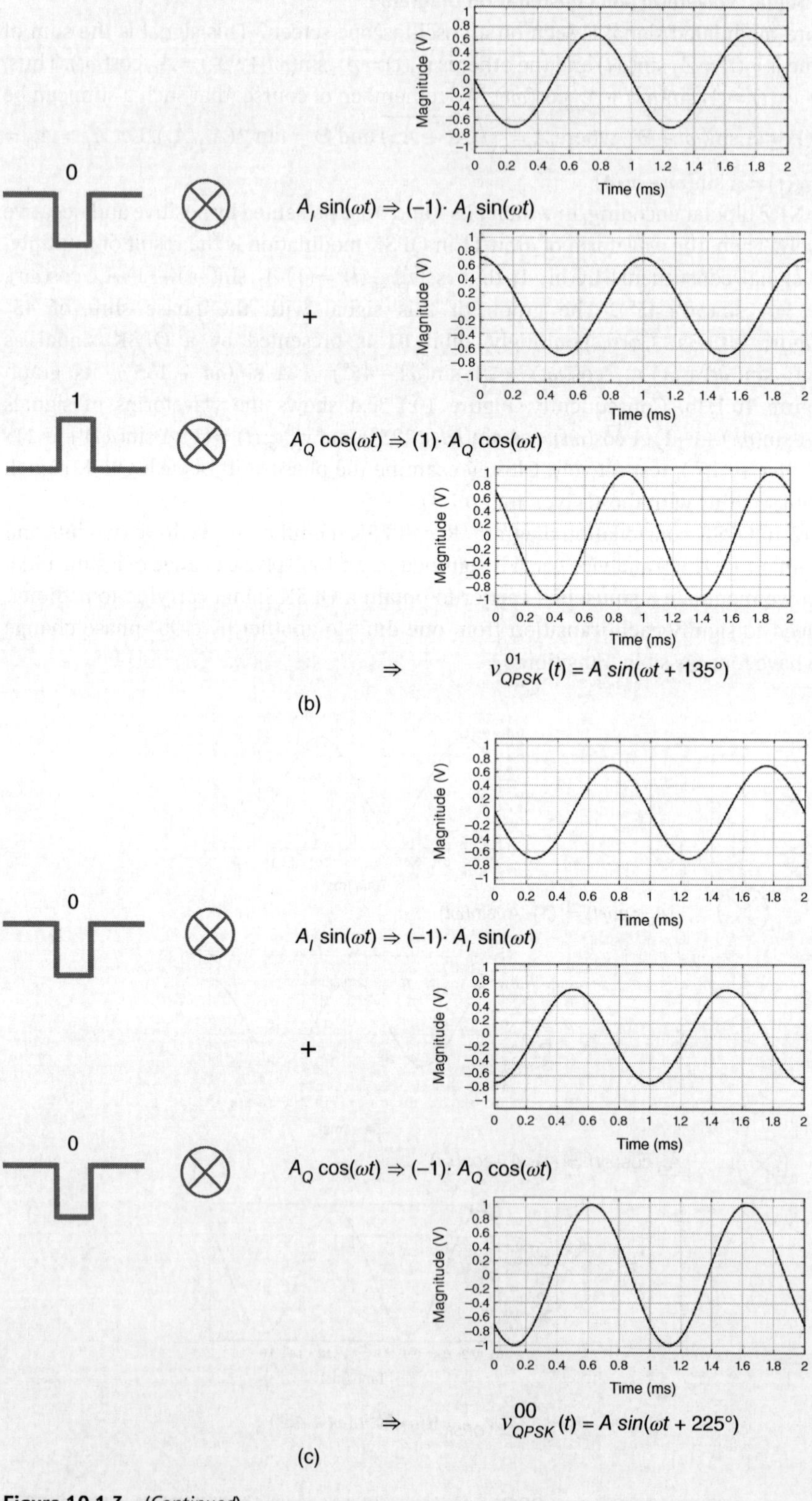

10 Digital Transmission with Multilevel Modulation

$A_I \sin(\omega t) \Rightarrow (-1) \cdot A_I \sin(\omega t)$

$A_Q \cos(\omega t) \Rightarrow (1) \cdot A_Q \cos(\omega t)$

$\Rightarrow \quad v_{QPSK}^{01}(t) = A \sin(\omega t + 135°)$

(b)

$A_I \sin(\omega t) \Rightarrow (-1) \cdot A_I \sin(\omega t)$

$A_Q \cos(\omega t) \Rightarrow (-1) \cdot A_Q \cos(\omega t)$

$\Rightarrow \quad v_{QPSK}^{00}(t) = A \sin(\omega t + 225°)$

(c)

Figure 10.1.3 (Continued)

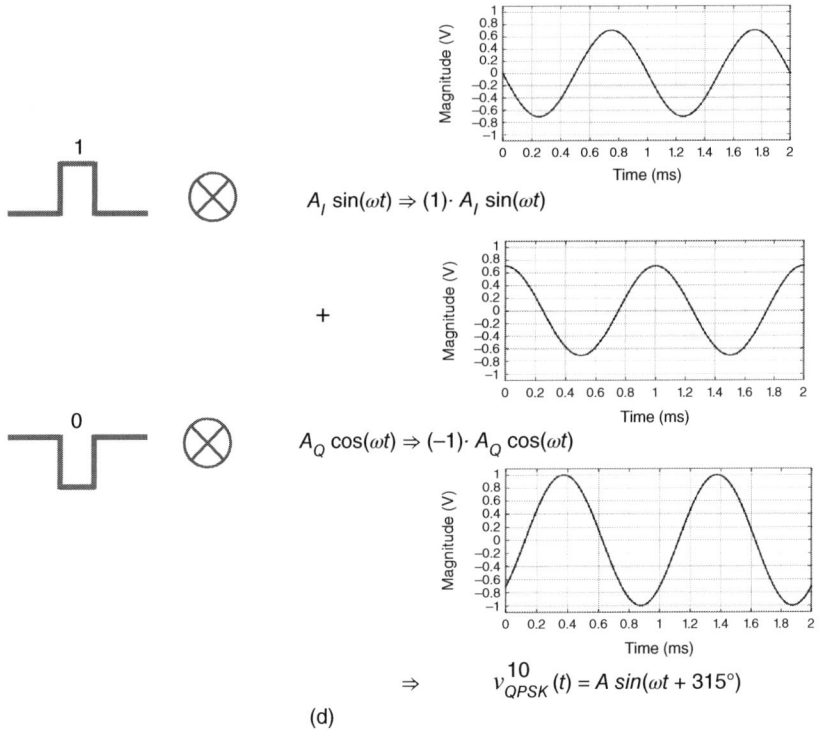

$$v_{QPSK}^{10}(t) = A\sin(\omega t + 315°)$$

(d)

Figure 10.1.3 (Continued)

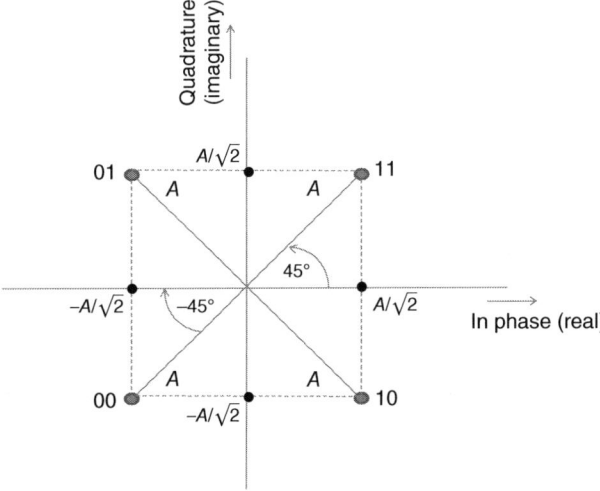

Figure 10.1.4 The *I* and *Q* components of dibit phasors in a QPSK constellation diagram.

To implement this idea, let us consider the QPSK signal space shown in Figure 10.1.2a. For all intents and purposes, *we have only one reference point – the I (in phase) axis.* (Here we use sine as a reference sinusoidal signal; hence, a cosine can be built by shifting the sine to 90°.) As soon as this axis is built, we can build a *Q* axis by rotating the *I* line counterclockwise by 90°. As soon as the *I* and *Q* axes are known, we can build the symbols for the dibits. Figure 10.1.4 shows all the dibits corresponding to their symbols.

We also need to find the I and Q components of all four symbols. For this, we apply the trigonometric identity

$$\sin(\alpha + \beta) = \sin \alpha \cdot \cos \beta + \cos \alpha \cdot \sin \beta \tag{10.1.4}$$

and calculate

$$A \sin(\omega t + 45°) = A \sin(\omega t) \cos(45°) + A \cos(\omega t) \sin(45°)$$
$$= \frac{A}{\sqrt{2}} \sin(\omega t) + \frac{A}{\sqrt{2}} \cos(\omega t) \text{ for dibit 11}$$

$$A \sin(\omega t + 135°) = A \sin(\omega t) \cos(135°) + A \cos(\omega t) \sin(135°)$$
$$= -\frac{A}{\sqrt{2}} \sin(\omega t) + \frac{A}{\sqrt{2}} \cos(\omega t) \text{ for dibit 01}$$

$$A \sin(\omega t + 225°) = A \sin(\omega t) \cos(225°) + A \cos(\omega t) \sin(225°)$$
$$= -\frac{A}{\sqrt{2}} \sin(\omega t) - \frac{A}{\sqrt{2}} \cos(\omega t) \text{ for dibit 00}$$

$$A \sin(\omega t + 315°) = A \sin(\omega t) \cos(315°) + A \cos(\omega t) \sin(315°)$$
$$= \frac{A}{\sqrt{2}} \sin(\omega t) - \frac{A}{\sqrt{2}} \cos(\omega t) \text{ for dibit 10} \tag{10.1.5}$$

Equation (10.1.5) shows that each dibit is transmitted by two carriers: A sine, which corresponds to the I component, and a cosine, corresponding to the Q component. See Figure 10.1.3. We can summarize this finding by showing the amplitudes of the I and Q projections of the given dibit and its symbol in Table 10.1.1.

Remember that in a signal-space diagram, we count the angle in the counterclockwise direction as a positive one and the angle in the clockwise direction as negative. See samples of this diagram in Figure 10.1.4. Also, observe that the values in the phasor diagram might be positive or negative, depending on the quarters, as shown in Figure 10.1.4. Now, you can verify that the results of the calculations in (10.1.5) comply with these conventions.

Equation (10.1.5) and Table 10.1.1 show the purpose of using sine as an in-phase carrier: With this arrangement, trigonometric calculations and a constellation diagram make perfect sense because they show that bit 1 is represented by a positive amplitude and bit 0 by a negative amplitude. What's more, this assignment complies with a traditional phasor presentation by quadrants of the phasor space. For example, for dibit 01, we calculate

Table 10.1.1 I and Q projections of dibit symbols.

Dibit	Symbol (phasor)	I amplitude	Q amplitude
11	$A \sin(\omega t + 45°)$	$\frac{A}{\sqrt{2}}$	$\frac{A}{\sqrt{2}}$
01	$A \sin(\omega t + 135°)$	$-\frac{A}{\sqrt{2}}$	$\frac{A}{\sqrt{2}}$
00	$A \sin(\omega t + 225°)$	$-\frac{A}{\sqrt{2}}$	$-\frac{A}{\sqrt{2}}$
10	$A \sin(\omega t + 315°)$	$\frac{A}{\sqrt{2}}$	$-\frac{A}{\sqrt{2}}$

$\Theta_{01} = \tan^{-1}\left(-\frac{A}{\sqrt{2}}/\frac{A}{\sqrt{2}}\right) = -\tan^{-1}(1) = -45° = 180° - 45° = 135°$ because this phasor is located in the second quadrant and we want to use the positive phase angle. Another example: The phase angle for dibit 00 is equal to $\Theta_{00} = \tan^{-1}\left(\frac{A}{\sqrt{2}}/-\frac{A}{\sqrt{2}}\right) = -\tan^{-1}(1) = -45° = 360° - 45° = 315°$ because this phasor is located in the fourth quadrant. (Refer to our discussion of complex numbers in Section 2.2.)

Having built a waveform and constellation diagram of a QPSK signal, we can combine both viewpoints in one picture. Figure 10.1.5 does just this. A constellation diagram of four QPSK dibits, their waveforms, and the corresponding formulas are presented in Figure 10.1.5a. Consider, for instance, dibit 01 located in the second quarter of the constellation diagram. This dibit is formed by I-signal representing bit 1 and Q-signal representing bit 0. How the dibit waveform is constructed? Consult Figure 10.1.3, where this operation is demonstrated in detail. Since we agreed to use an NRZ bipolar encoding shown in Figure 10.1.3, the dibit waveform is formed by the summation of $I(t) = (1) A_I \sin(\omega t)$ and $Q(t) = (-1) A_Q \cos(\omega t)$. Figure 10.1.5a shows that $v_{QPSK}^{01}(t) = (-1) \cdot A_I \sin(\omega t) + (1) \cdot A_Q \cos(\omega t) = A \sin(\omega t + 135°)$. Examine correspondence of every dibit and its waveform; compare the dibit formulas with the graphical presentation of their waveform and pay particular attention to the phase shifts.

A visual summary of our discussion of the relationship between signal-space (constellation diagram) and time-domain (waveforms) presentations of to-be-transmitted QPSK signal is shown in Figure 10.1.5b. The columns demonstrate the formation of the waveform of each dibit; the bottom row displays formation of the waveform of the QPSK signal prepared for transmission. The rows show that the modulated signal at each (I or Q) channel in Figure 10.1.5b is simply a BPSK signal, in which every transition from bit 1 to bit 0 and back is signified by an 180° phase shift.

(**Questions**: *What about the phase shifts for transitions from bit 1 to 1 and bit 0 to 0? What are they and why are they assume such values?

*Examine the QPSK signal's waveform to see how the phase changes at the transition from one dibit to the next: Do these changes make sense to you? If so, what rule would you derive from this observation?)

Thus, this figure shows how a mathematical model (the constellation diagram) relates to the real process (the plot of a QPSK signal in time domain, that is, the waveform). This view should help you to complete the picture of the formation of a QPSK signal.

With all these presentations in mind, we are ready to consider how to generate (modulate) a QPSK signal in practice.

10.1.2.3 Generating (Modulating) a QPSK Signal

Again, analyze Eqs. (10.1.3)–(10.1.5) and Figures 10.1.3–10.1.5 and note that each dibit is the sum of two independent components: The I component with the sine carrier and the Q component with the cosine carrier. This means that *QPSK can be considered as a sum of two BPSK binary data streams*, one modulating the sine carrier and the other modulating the cosine carrier. To create these two BPSK streams, an input binary data stream should be split into two streams. Then, each of these bit streams modulates its carrier (sine or cosine, respectively) and, at the output of a QPSK transmitter, they sum up to form the signal to be transmitted. Therefore, a QPSK modulator can be built as shown in Figure 10.1.6.

In Figure 10.1.6a, the input binary data stream, **11010010**, is encoded into an NRZ pulse train. This digital signal enters a serial-to-parallel converter, which creates two trains. The process of separating the input data into odd and even pulse trains is shown in Figure 10.1.6b. In the input binary data stream, **11010010**, the first bit 1 is an odd bit. The odd bits (physically, the pulses) are

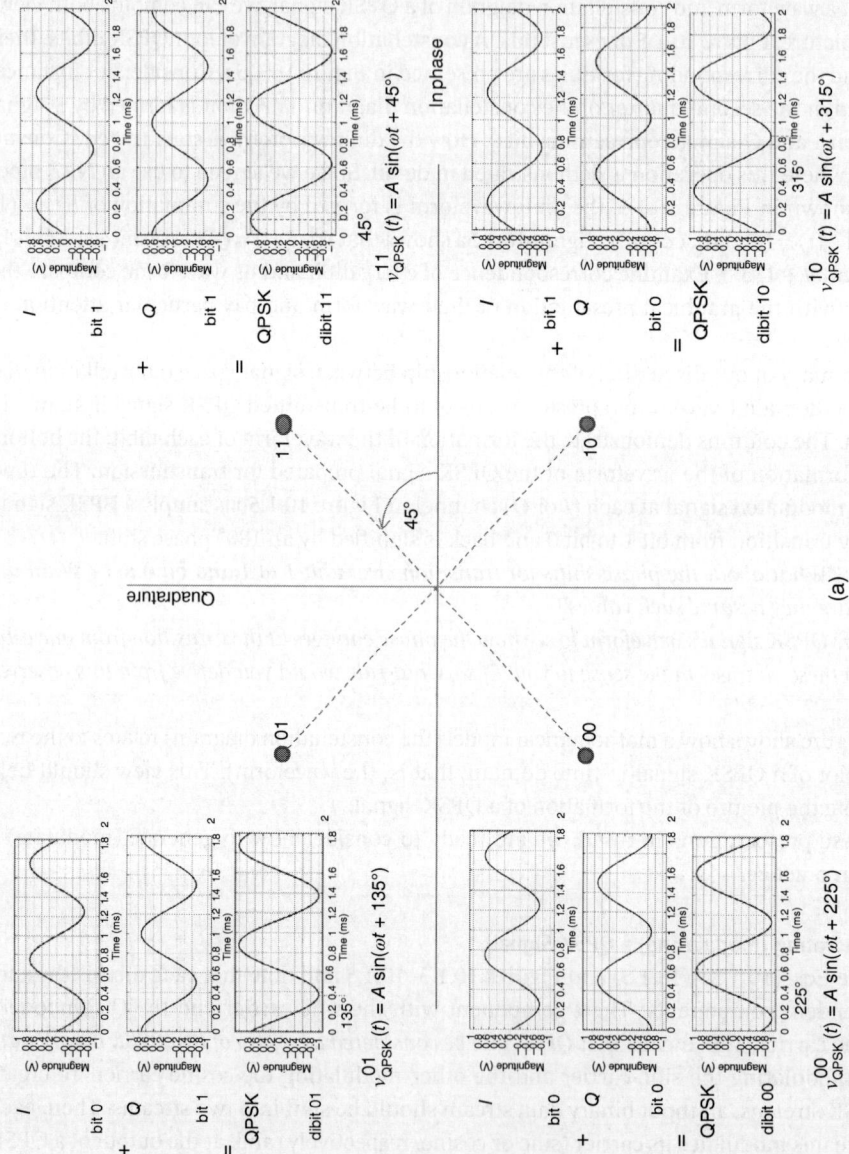

Figure 10.1.5 Constellation diagram and waveforms of I, Q, and QPSK signals: (a) four dibits and their waveforms on a constellation diagram and (b) formation of to-be-transmitted QPSK signal in time domain.

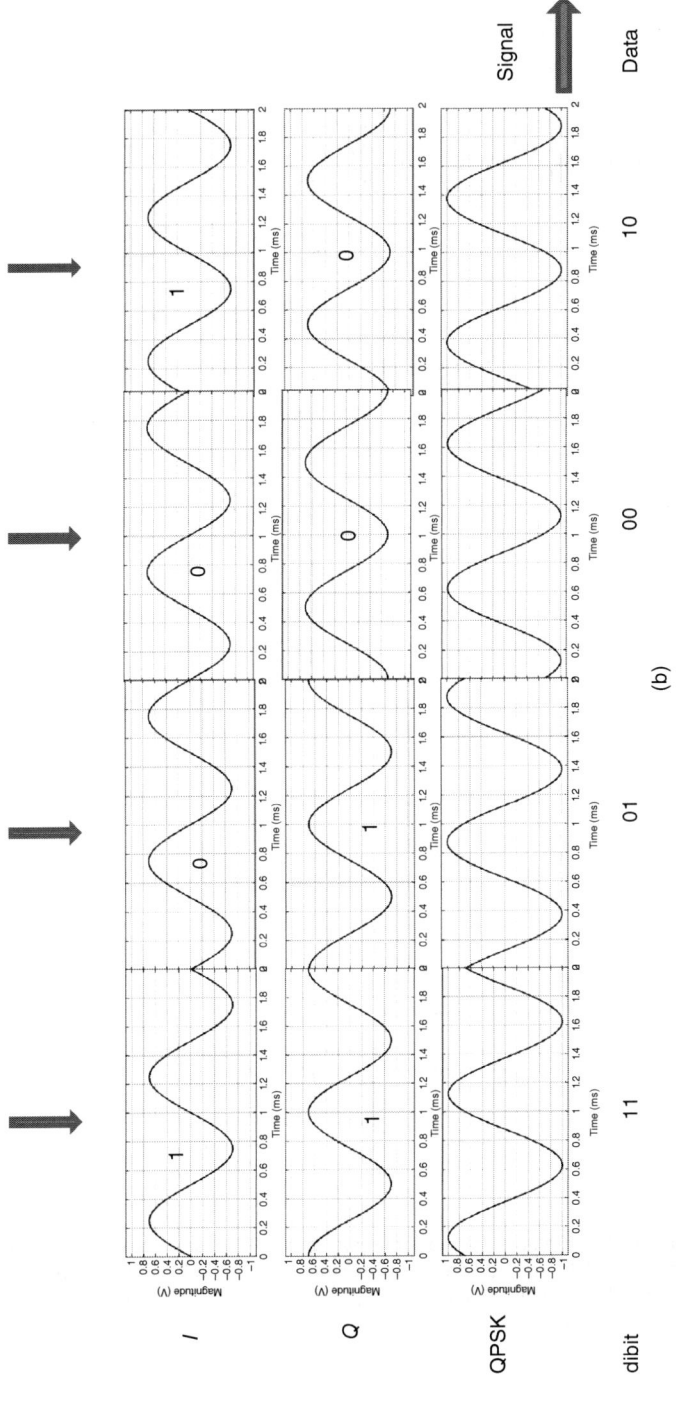

Figure 10.1.5 (*Continued*)

10 Digital Transmission with Multilevel Modulation

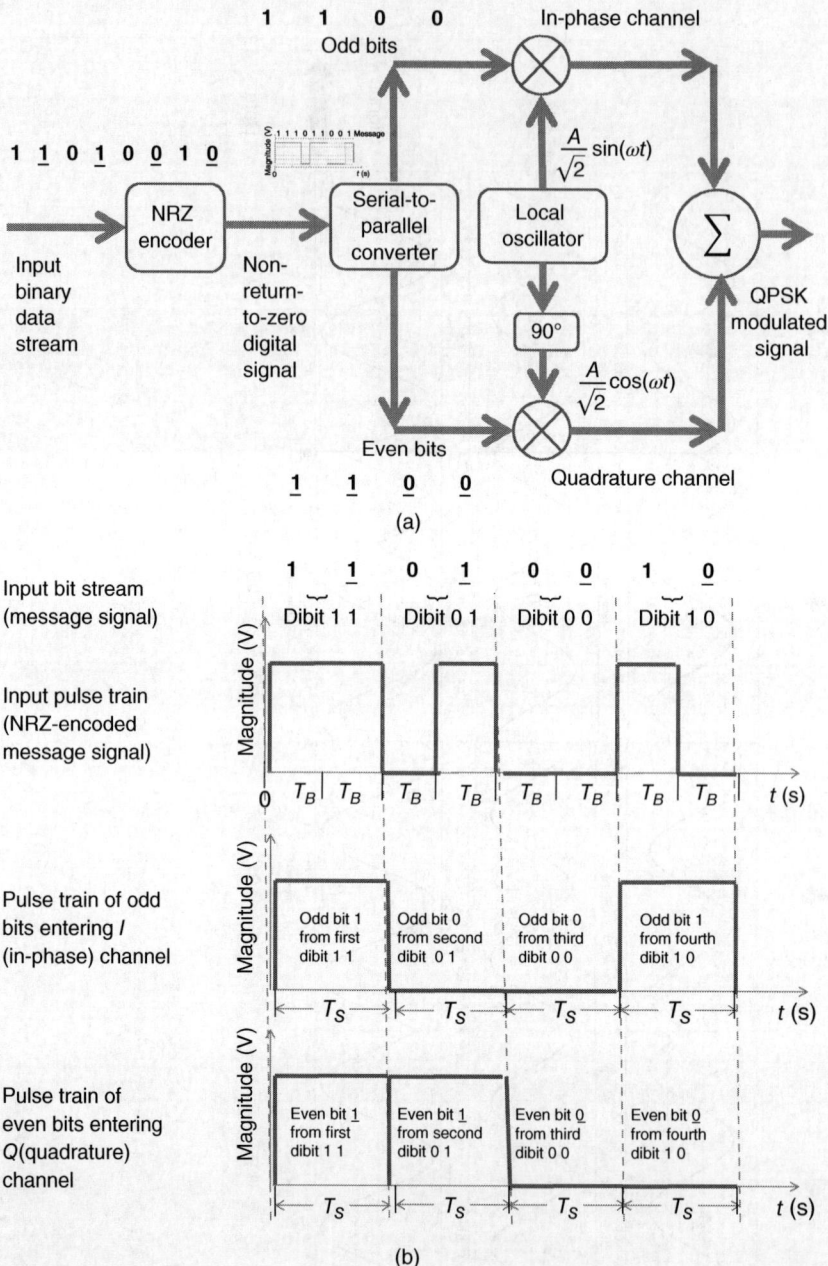

Figure 10.1.6 Generating a QPSK signal: (a) block diagram of a QPSK transmitter (modulator); (b) decomposition of the input data into odd and even streams; (c) input, odd, and even bit streams and their corresponding waveforms to be transmitted (by MATLAB/Simulink simulation); and (d) block diagram of the simulation setup.

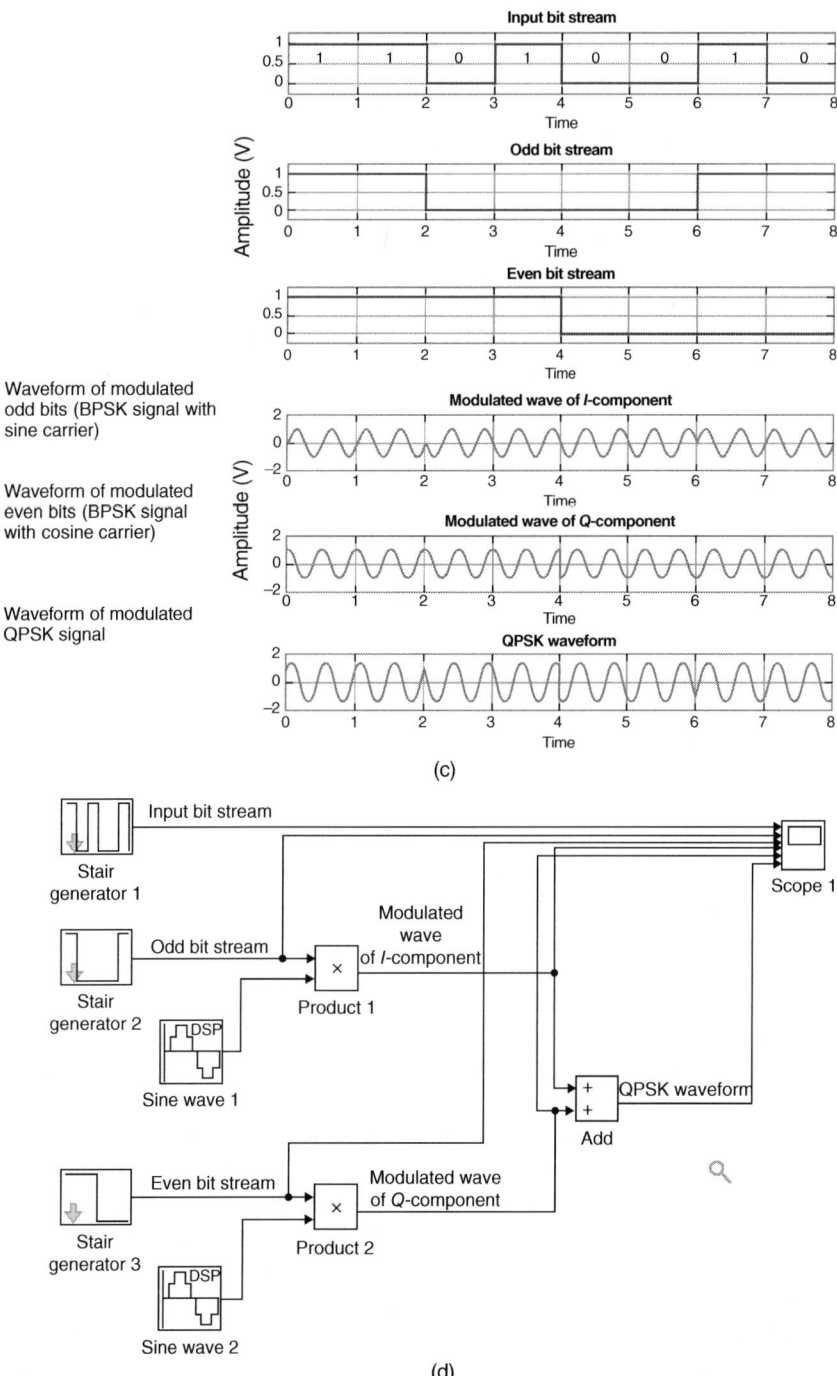

Figure 10.1.6 (*Continued*)

presented to the in-phase channel, where they modulate a sine carrier. The even bits, underscored in Figure 10.1.6a for clarity, enter the quadrature channel, where they modulate a cosine carrier. Pay particular attention to the timing: At both the I and Q channels, each bit occupies a $T_S = 2T_b$ interval, whereas in the input pulse train, each bit takes a T_b interval. This change of timing is done by a timer. (We opted not to show a timer in Figure 10.1.6a to avoid overloading the figure with details.)

The process of forming a QPSK output waveform is shown in Figure 10.1.6c. The input, even and odd bit streams are shown again for clarity. The sine signal modulated by the odd-numbered train is formed at the in-phase channel. The quadrature channel forms a cosine signal modulated by the even-numbered train. At the output, the waveforms from two channels are summing, and the resulting waveform is transmitted over the QPSK system. This is where timing again becomes a factor because any violation of synchronization between the I and Q channels will result in disrupting the output signal. Specifically, symbol intervals, T_S, must be synchronized between the I and Q channels to provide the correct formation of the symbols. These waveforms were obtained by using MATLAB/Simulink; the block diagram of the setup for this experiment is shown in Figure 10.1.6d.

We have to realize that Figure 10.1.6 presents an idealized process of generating a QPSK signal. In reality, the pulses are not rectangular, their durations are not constant, and so on. As we learned in Chapter 9, there is a powerful industrial tool for assessing the quality of real signals called *eye diagram*. This tool can be applied to multilevel signaling too. To understand the concept of using an eye diagram in *M*-ary modulation, consider the example of a *PAM4* and the corresponding eye diagram presented in Figure 10.1.7. (This example can be directly applied to the eye diagram of a QPSK signal.) Figure 10.1.7a shows the two-level and four-level encodings and their eye diagrams. It can be seen that a multilevel signal has an eye diagram containing several "eyes." Figure 10.1.7b demonstrates that an industrial eye diagram of a PAM4 signal contains four levels whose correspondence between dibits and levels are as follows: Level 1 → dibit 00, level 2 → dibit 01, level 3 → dibit 10, and level 4 → dibit 11. This figure also shows that the bottom eye is formed by the transitions between levels 0 and 1, the middle eye is formed by the transitions between levels 1 and 2, and the top eye is formed by the transitions between levels 2 and 3. These transitions are schematically shown in detail in Figure 10.1.7c. The last figure also compares the transitions among dibits of a four-level signal and bits of a two-level signal. Review Figure 10.1.7 again: It helps you to understand the structure and meaning of the eye diagram of a multilevel signal. It is important to realize that all the characteristics of the eye diagram of a two-level signal discussed in regard to Figure 9.1.31 are valid for a multilevel eye diagram.

Comparing PAM2 and PAM4, we can emphasize another advantage of four-level signaling: In reality, as discussed in Section 3.2, the waveform of a digital pulse is far from being rectangular. In particular, a digital pulse's waveform exhibits rise time. In PAM2, all hurdles associated with a rise time appear at the transmission of every bit, whereas in PAM4, the rise-time problems appear only at the transmission of every symbol. Figure 10.1.8 visualizes this point: In PAM2, the rise time occupies a significant portion of a bit duration, whereas in PAM4, the percentage of rise time with respect to the symbol duration is twice smaller. In this example, we assume that t_r is the same for both signaling and $T_S = 2T_b$. We need to remember that each pulse and symbol has also a fall time, which doubles the importance of this phenomenon.

(**Question**: *The text states that PAM4 signaling requires twice smaller bandwidth than the PAM2 provided that $T_S = 2T_b$. Does the rise-time issue relate to bandwidths of PAM4 and PAM2 consideration?*)

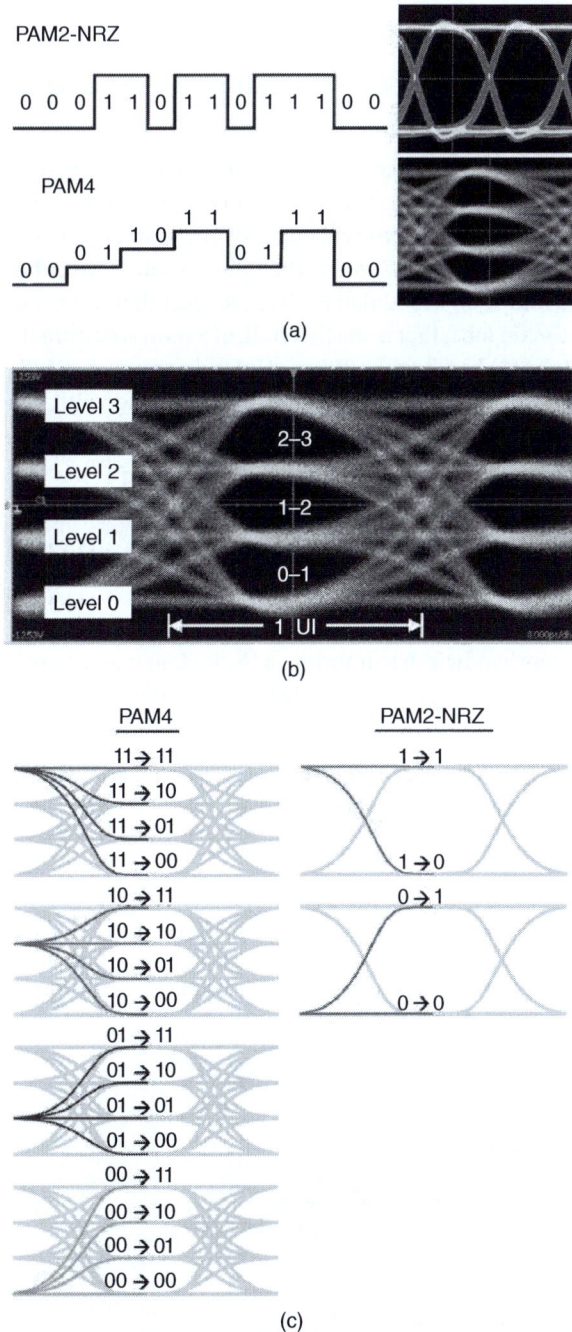

Figure 10.1.7 The eye diagrams of multilevel signals: (a) two-level (PAM2) and four-level (PAM4) encoding and the corresponding eye diagrams; (b) the practical eye diagram structure of a four-level signal; and (c) the concept of eye diagrams of a four-level (left) and two-level (right) signals with transitions among dibits (left) and bits (right) (UI stands for *a unit interval*, which is the normalized duration of one bit). Source: Reprinted with permission of Tektronix (2016).

10.1.3 Working with QPSK Signaling

10.1.3.1 Properties of a QPSK Signal

Analyzing Figures 10.1.3–10.1.8 and reviewing the discussion accompanying them, we can explicate the following properties of a QPSK signal:

- *Timing*: For the input pulse train, which is the encoded original data stream, the minimum time interval between the phase change is bit duration, T_b (s). For the I and Q signals and for the output QPSK signal, the minimum time interval for the phase change is symbol duration, T_S (s).
- *Timing, bandwidth and bit rate*: Remember that in a four-level modulation, we can have either $T_{S1} = 2T_b$ or $T_{S2} = T_b$. Comparing PAM2 and PAM4 modulations, we recollect that the bandwidth of a pulse train is determined by the main lobe, that is the first null, of a train spectrum. In one-sided spectrum, the *first-null a PSK* (PAM2) bandwidth, BW_{PSK} (Hz) is determined as $1/T_b$ (s) = BR (b/s), where T_b is the duration of a PSK bit. Application of the first-null-bandwidth rule to a PAM4 train yields: If $T_{S1} = 2T_b$, then BW_{PAM4-1} (Hz) = $1/T_{S1}$ (s) = ½ BW_{PSK} (Hz); if $T_{S2} = T_b$, then BW_{PAM4-2} (Hz) = $1/T_{S2}$ (s) = BW_{PSK} (Hz). Consequently, BR_{PAM4-1} (b/s) = $1/T_{S1}$ = BR_{PSK} and BR_{PAM4-2} (b/s) = $1/T_{S2}$ = $2BR_{PSK}$. In short,

$$BW_{PAM4-1} \text{ (Hz)} = \tfrac{1}{2}BW_{PSK} \text{ (Hz) and } BW_{PAM4-2} \text{ (Hz)} = BW_{PSK} \text{ (Hz)} \quad (10.1.6a)$$

$$BR_{PAM4-1} \text{ (b/s)} = BR_{PSK} \text{ (b/s) and } BR_{PAM4-2} \text{ (b/s)} = 2BR_{PSK} \text{ (b/s)} \quad (10.1.6b)$$

Thus, a PAM4-1 signaling needs two times less bandwidth to transmit at the PSK bit rate, whereas PAM4-2 modulation's bit rate is double of PSK's bit rate, but the bandwidths of both signaling are equal. (Examine Figure 10.1.1d again.)

- *Formula for a transmitted QPSK signal*: Analyzing (10.1.5) and Figures 10.1.3–10.1.6, we can derive the following general formula for a QPSK signal being transmitted:

$$v_{QPSK}(t) = A\,(\sin(\omega_c t) + (2k-1)\cdot 45°) \quad (10.1.7)$$

where $k = 1, 2, 3, 4$. Using (10.1.5), we can present (10.1.7) in the following form:

$$v_{QPSK}(t) = \frac{A}{\sqrt{2}}\cos((2k-1)\cdot 45°)(\sin(\omega_c t)) + \frac{A}{\sqrt{2}}\sin((2k-1)\cdot 45°)\cos(\omega_c t) \quad (10.1.8)$$

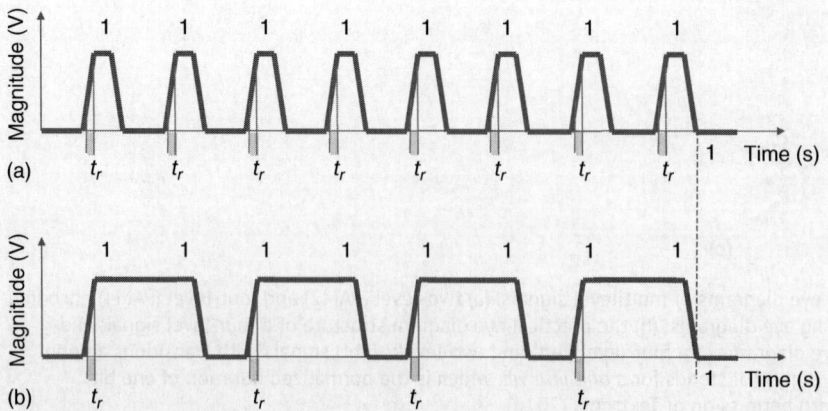

Figure 10.1.8 Rise time in (a) PAM2 and (b) PAM4 pulses.

where $k = 1, 2, 3, 4$. Equation (10.1.8) is the compact form of (10.1.5); the signs of their amplitudes, $A/\sqrt{2}$, are shown in Table 10.1.1.

- *The value of the* QPSK signal's amplitude: From (10.1.8), the average energy of the symbol is given by

$$E_s = \int_0^{T_s} v_{\text{QPSK}}^2(t)dt = \left(\frac{A}{\sqrt{2}}\right)^2 (T/2 + T/2) = \frac{A^2 T_s}{2} \tag{10.1.9}$$

where A is the amplitude of a carrier wave, as shown in (10.1.7). From (10.1.9), we can express a QPSK signal's amplitude through its energy as

$$A_{\text{QPSK}} = \sqrt{\frac{2E_s}{T_s}} \tag{10.1.10}$$

which is similar to that obtained for a PSK signal. Obviously, bit energy in a QPSK system is one-half of the symbol energy, that is, $E_b = E_s/2$.

These basic properties of a QPSK signal reflect the nature of QPSK modulation and stress the need for strong timing control in QPSK operation.

10.1.3.2 QPSK Demodulation

Analysis of a transmitted QPSK signal shown in Figures 10.1.5 and 10.1.6 leads to the concept of a QPSK demodulator (receiver). Since the QPSK signal is the sum of two PSK signals, the QPSK receiver should be the sum of two PSK receivers with a module combining two detected binary data streams in one output. With this approach in mind, we can build a QPSK coherent receiver shown in Figure 10.1.9. (We consider a *coherent* receiver because, as discussed, this is the most accurate type of such a unit.)

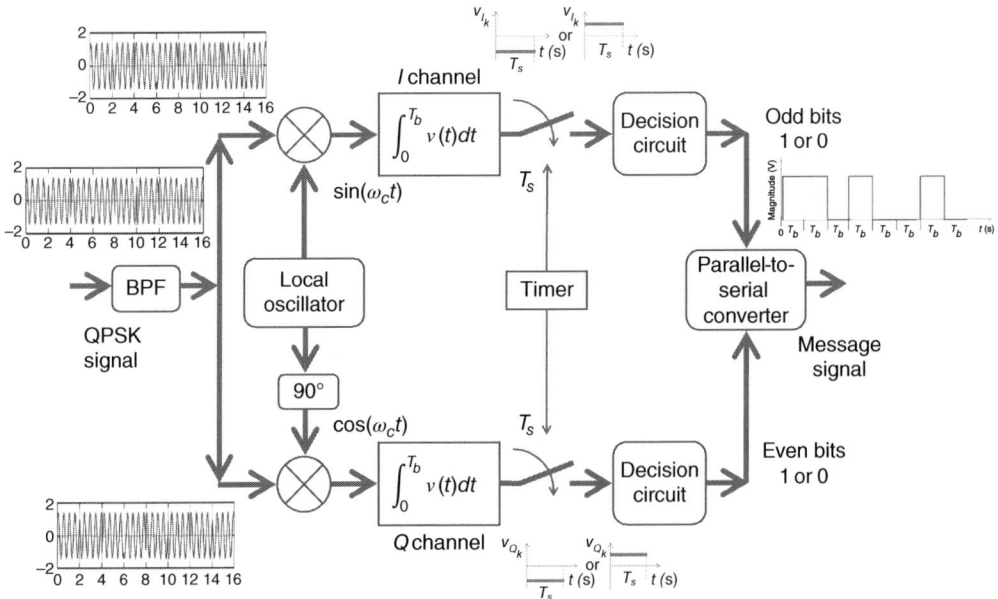

Figure 10.1.9 A coherent QPSK receiver.

Obviously, the receiver must first separate the streams of odd and even bits. To perform this task, it is necessary to make *the number of the carrier wave's cycles, N_c, per symbol's time, T_S, to be an integer*. We discussed the reasoning behind this condition in considering FSK and PSK signals in Section 9.2. To remind, the requirement is that both sine and cosine signals have the same number of cycles per the single time interval (T_S in this case). This is needed to obtain the orthogonality of the *I* and *Q* signals to be used for the separation of the odd and even bit streams. As was mathematically shown in Section 9.2.3.7, two signals are orthogonal if

$$\int_0^{T_s} (\cos(\omega_c t))(\sin(\omega_c t))dt = 0 \qquad (9.2.34R)$$

Under this condition, the QPSK receiver should operate as follows:

- The QPSK signal described in (10.1.8) enters the receiver and goes into two channels simultaneously.
- In the *I* channel, the signal is multiplied by $\sin(\omega_c t)$ and is presented to an *integrator* (a lowpass filter, remember). The output of the upper integrator is given by

$$\int_0^{T_s} (v_{QPSK}(t) \cdot \sin(\omega_c t))dt$$

$$= \frac{A}{\sqrt{2}} \int_0^{T_s} (I_k \sin(\omega_c t) \cdot \sin(\omega_c t))dt + \frac{A}{\sqrt{2}} \int_0^{T_s} (Q_k \sin(\omega_c t) \cdot \sin(\omega_c t))dt \qquad (10.1.11a)$$

$$= \frac{A}{\sqrt{2}} I_k \int_0^{T_s} (\sin^2(\omega_c t))dt + 0 = \frac{AT_S}{2\sqrt{2}} I_k = V_{I_k}$$

where $I_k = \cos((2k-1)\cdot 45°)$ with $k = 1, 2, 3, 4$.

- The signs of I_k and Q_k are determined by the integer k, as shown in Table 10.1.2. (You probably have noticed that Table 10.1.2 is, in essence, the other form of Table 10.1.1.)
- The upper integrator output (the stream of odd pulses) is the signals of constant values, $\frac{AT_S}{2\sqrt{2}} I_k$. These signals enter a decision circuit, where the value of each pulse received during symbol interval T_S is compared with a threshold. If the constant value, $V_{I_k} = \frac{AT_S}{2\sqrt{2}} I_k$, is positive (that is, above the threshold $V_{TH} = 0$), then the decision circuit considers the input as odd bit 1; if the constant value, $V_{I_k} = \frac{AT_S}{2\sqrt{2}} I_k$, is negative (that is, below the threshold $V_{TH} = 0$), then the decision circuit shows odd bit 0.
- Similar operations are performed in the *Q* channel, whose output is the stream of even bits; that is,

$$\int_0^{T_s} (v_{QPSK}(t) \cdot \cos(w_c t))dt$$

$$= \frac{A}{\sqrt{2}} \int_0^{T_s} (I_k \sin(w_c t) \cdot \cos(w_c t))dt + \frac{A}{\sqrt{2}} \int_0^{T_s} (Q_k \cos(w_c t) \cdot \cos(w_c t))dt \qquad (10.1.11b)$$

$$= 0 + \frac{A}{\sqrt{2}} Q_k \int_0^{T_s} (\cos^2(w_c))dt = \frac{AT_S}{2\sqrt{2}} Q_k = V_{Q_k}$$

where $Q_k = \sin((2k-1)\cdot 45°)$, $k = 1, 2, 3, 4$. For values of Q_k, see Table 10.1.2.

- The bottom integrator in Figure 10.1.9 performs the same operations as the upper integrator but with constant values, $V_{Q_k} = \frac{AT_S}{2\sqrt{2}} Q_k$. Hence, for $V_{Q_k} = \frac{AT_S}{2\sqrt{2}} Q_k > 0$, the decision circuit shows even bit 1; for $V_{Q_k} = \frac{AT_S}{2\sqrt{2}} Q_k < 0$, the decision circuit shows even bit 0.

Table 10.1.2 Carrier's phase angles and signs of I_k and Q_k components.

k	Angle	$I_k = \cos\left((2k-1)\frac{\pi}{4}\right)$	$Q_k = \sin\left((2k-1)\frac{\pi}{4}\right)$
1	$\pi/4 = 45°$	+1	+1
2	$3\pi/4 = 135°$	−1	1
3	$5\pi/4 = 225°$	−1	−1
4	$7\pi/4 = 315°$	+1	−1

- Both streams enter the parallel-to-serial converter, whose output is the demodulated message signal sent by a QPSK transmitter. Clearly, this converter also provides re-timing by allotting to each bit an interval, T_b, thus returning to the normal timing of individual bits.

This is how a coherent QPSK receiver recovers an original message signal. We have to always remember, however, that the described process is an idealization of reality. In practice, there are many fundamental and technological hurdles that cause the errors and restrict accuracy of the recovered message.

10.1.3.3 Assessing the Quality of QPSK Transmission

We recall that the digital transmission quality is assessed by its BER, which is the probability, P_e, of erroneously taking bit 1 for bit 0 or vice versa. On the other hand, we define BER as the ratio of the number of bits in error to the total number of transmitted bits. Hence,

$$\text{BER} = P_e/P_{\text{tot}} = \frac{\text{Average number of erroneous bits detected}}{\text{Total number of bits received}} \quad (9.1.33R)$$

We also recall that the BER of a BPSK signal is given as

$$\text{BER}_{\text{BPSK}} = Q\left(\sqrt{\frac{2E_b}{N_0}}\right) \quad (9.2.64R)$$

where E_b (J) is the average bit energy and N_0 (W/Hz) is the noise spectral density.

In our discussion, we consider the QPSK signaling as two PSK modulations. For each of these systems, the BER is determined by (9.2.64R). Since these two PSK systems operate independently (see the I and Q channels in Figures 10.1.5 and 10.1.6), we can conclude the BER, which refers to *errors in individual bits*, for a QPSK system is the same as for a BPSK. Thus,

$$\text{BER}_{\text{QPSK}} = \text{BER}_{\text{BPSK}} = Q\left(\sqrt{\frac{2E_b}{N_0}}\right) \quad (10.1.12)$$

(Formal proof of this statement can be found, for example, in Stern and Mahmoud 2004, pp. 281–283.)

However, strictly speaking, in regard to a QPSK signal, we need to consider a SER, which can be defined as

$$\text{SER} = \frac{\text{Average number of erroneously identified symbols, } N_{\text{SE}}}{\text{Total number of symbols received, } N_{\text{TM}}} \quad (10.1.13)$$

Rewriting BER given in (9.1.33R) in the similar notations yields

$$\text{BER} = \frac{\text{Average number of erroneous bits detected, } N_{\text{be}}}{\text{Total number of bits received, } N_{T_b}} \quad (9.1.33R)$$

Thus, to derive the SER formula for the QPSK, we can exercise the approach taken in calculating the probability of error in Section 9.1. Using symbols instead of bits, we obtain after similar manipulations

$$\text{SER}_{\text{QPSK}} \approx \text{erfc}\left(\sqrt{\frac{E_S}{2N_0}}\right) = 2Q\left(\sqrt{\frac{E_S}{N_0}}\right) = 2Q\sqrt{\frac{2E_b}{N_0}} \quad (10.1.14a)$$

Remember that $\text{erfc}(x) = 2Q(x\sqrt{2})$. (Derivation of (10.1.14a) can be found in many sources, for example, in https://www.embedded.com/print/4017668 or in http://www.dsplog.com/2007/11/06/symbol-error-rate-for-4-qam or in https://www.coursehero.com/file/7116016/m5l22.)

Equations (10.1.12) and (10.1.14a) show that QPSK has no advantage over BPSK in error performance. This is because the *underlying assumption in calculating SER is that a symbol's error is caused by the change of only one bit in the adjacent dibits*. For example, if dibit 00 is mistaken with dibit 10, this is because bit 1 of the second dibit is mistaken with bit 0 of the first dibit.

Remember that the errors are detected at the demodulation stage. A QPSK demodulator treats the I and Q received steams independently and therefore the exact error probability in the combined signal is given by

$$\text{SER}_{\text{QPSK}} = 2Q\left(\sqrt{\frac{2E_b}{N_0}}\right) - \left[Q\left(\sqrt{\frac{2E_b}{N_0}}\right)\right]^2 \quad (10.1.14b)$$

(Formal basis for (10.1.14b) is given in general theory of probability: For independent events A and B, we have $P(A \cup B) = P(A) + P(B) - P(A)P(B) = 2P - P^2$, provided that $P(A) = P(B)$. See, for example, https://dsp.stackexchange.com/questions/1186/matlab-plot-of-qpsk-system-doesnt-agree-perfectly-with-theoretical-ber-curves.

Neglecting $[Q(\sqrt{2E_b/N_0})]^2$ in (10.1.14b), we obtain (10.1.14a). This explains the use of approximation sign instead of an equality sign in (10.1.14a).

Notes:

1. Let us stress again that BER refers to errors in bits, whereas errors in QPSK transmission, by definition, are given in symbols. Hence, using the notation BER_{QPSK}, we want to stress that this is an error in bits for QPSK transmission.
2. Typically, (10.1.14a) is derived under the assumption that the most likely event is to mistakenly identify only one bit in the adjacent dibits. Therefore, this formula should be used with precaution.

10.1.3.4 Offset QPSK, Differential QPSK, and Minimum SK

In practice, the envelope of QPSK waveform experiences big variations when the phase changes at $\pm 180°$ (see Figure 10.1.5b, where the waveform of a QPSK signal is shown.) These amplitude fluctuations cause signal distortion, which decreases the opening of an eye diagram and brings the errors in demodulation. To overcome this obstacle, a version of QPSK called *offset (staggered) QPSK* has been developed. In OQPSK, the bit stream in the Q channel in a coherent QPSK receiver (as that shown in Figure 10.1.10) is delayed by a one-bit interval, T_b, with respect to I bit stream. With this arrangement, only one component of both streams changes at every transition, which eliminates the $\pm 180°$ phase change. Hence, the amplitude variations are reduced at OQPSK because the phase can change only by $\pm 90°$. Figure 10.1.10 compares the phase changes in QPSK and in OQPSK. Examine this figure carefully to see that the goal – reducing the maximum phase shift to $\pm 90°$ – is indeed achieved in OQPSK signaling.

Figure 10.1.10 Comparison of phase shifts between (a) QPSK and (b) OQPSK signaling. (*Note:* http://www.ni.com/tutorial/5487/en/ © 2019 National Instruments Corporation. All Rights Reserved.)

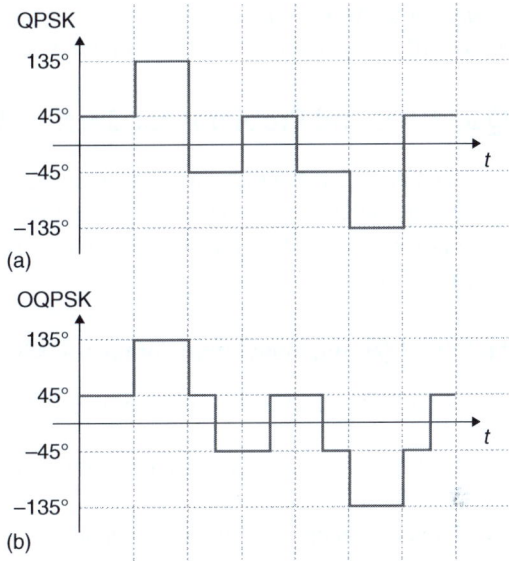

Table 10.1.3 Carrier's phase changes for DQPSK.

Symbol (dibit)	Carrier's phase for the current symbol relative to the carrier's phase for the previous symbol
11	No change
10	90°
00	180°
01	270°

The other issue in QPSK signaling is the choice between coherent and incoherent demodulation. We considered a coherent receiver in our discussion (see Figure 10.1.9) because it provides the better quality of demodulation. However, coherent detection requires more complex and more expensive equipment than a noncoherent device does. (Refer to Figure 9.2.23 to refresh your memory on the principles of incoherent demodulation.) To facilitate the use of an incoherent receiver when it is justified by the system design, engineers use the DQPSK. In this version of QPSK, the phase change is determined by the previous symbol rather than by the absolute value of a current symbol. Specifically, the phase shifts of all dibits can be organized as shown in Table 10.1.3.

DQPSK improves the performance of an incoherent QPSK, but it still cannot reach the accuracy level of a coherent QPSK. To achieve the coherent QPSK's BER level, DPSK signaling needs more than a 2 dB increase in signal-to-noise ratio (SNR) (E_b/N_0) (Stern and Mahmoud 2004, p. 286).

Can we further decrease the level of a phase transition to improve the performance of a QPSK system? Yes, it can be done with a system called *MSK*. Here, the rectangular pulses representing bits 1 and 0 are replaced by segments of sinusoids. Thus, a positive rectangular pulse showing bit 1 in QPSK is replaced by a positive half-wave in MSK, and a zero pulse in QPSK is replaced by a negative half-wave in MSK. (Similar approach is taken in Figure 10.1.3.) As a result, the phase change of an MSK signal in time becomes a ramp with zero change at the transition points. We leave further discussion of MSK transmission to specialized textbooks.

Questions and Problems for Section 10.1

- Questions marked with an asterisk require a systematic approach to finding the solution.
- Many questions and problems, including those marked with an asterisk, imply that you, in addition to reading the textbook, will do your research to find the answers. Consider such questions as mini-projects.

Multilevel (*M*-ary) Modulation Formats – What and Why

1. What is the difference between multilevel modulation and binary modulation?

2. Multilevel modulation is often called *M*-ary modulation. Why?

3. *Refer to Figure 10.1.1b showing modulation where every symbol contains two bits:
   a) Sketch a figure to show a multilevel modulation in which every symbol contains three bits.
   b) How many levels do you need for three-bit modulation?
   c) A symbol containing two bits is called dibit. Suggest the designation for the symbol containing three bits.
   d) What symbol durations, measured in T_b, can be arranged in this new modulation?
   e) What is the bandwidth of this new modulation compared with BW_2 (Hz)?

4. If we managed to pack three bits in one symbol, will the bit rate of this multilevel modulation increase three times as compared with PAM2? Prove your answer.

5. If a multilevel modulation, containing three bits per symbol, transmits eight symbols over one second, what are the bit rate and symbol rate of this transmission? In what units do you measure the symbol rate?

6. How does bandwidth of three-bit-per-symbol modulation change compare with BW_{PAM2} (Hz)?

7. The text says that "...*multilevel modulation formats are in the mainstream of today's digital transmission.*" Why is that so?

8. How many symbols do we need in order to pack four bits in one symbol? Show possible combinations of four bits and the symbol corresponding to each combination.

9. How many bits can we place in one symbol if we are given 16 symbols?

10. What is the difference between the terms "symbol" and "level"?

11. *The bandwidth of a coaxial cable is $BW = 2$ MHz and its signal-to-noise ratio is $SNR = 1000$:
    a) What is the maximum digital transmission rate that this channel can support?
    b) How much will a binary signal be distorted by transmission through this cable in practice? Assess it qualitatively and quantitatively.
    c) Propose a better solution for digital transmission over this bandlimited line.
    d) Give the numerical parameters of the proposed transmission.

Quadrature Phase-Shift Keying, QPSK

12. Example of PAM4 shows that we need four different amplitudes to pack two bits into one symbol. The text says, however, that we can place as many as 256 bits into a symbol. (In fact, modulation with 1024 bits per symbol exists.) How can such a multilevel modulation be achieved in practice?

13. The acronym QPSK stands for quadrature phase-shift keying. Explain the meaning of each word in this term.

14. Go online and find examples of applications of QPSK signaling in modern communications.

15. *Equation (10.1.3) shows every dibit assigned to a segment of a carrier wave:
    a) Develop similar assignments for the three-bit-per-symbol modulation. Write the formulas.
    b) Sketch the signal space for this new bit-to-symbol assignment.
    c) Sketch the constellation diagram to show this new bit-to-symbol mapping.

16. Show bit-to-symbol mapping in the constellation diagram for the following sequence of dibits: 00, 10, 11, and 01.

17. The sequence of binary numbers in which every successive number differs from the preceding by one bit only is called *Gray code* after Bell Lab's researcher Frank Gray. The dibit sequence shown in Figures 10.1.2–10.1.5 belongs to this type of code. Consider a new sequence of 00, 10, 11, and 01 dibits: Is it a Gray code? Sketch the constellation diagram for this group placing dibit 00 in the first quarter.

18. *Examine the formation of the waveforms of QPSK signals shown in Figure 10.1.3:
    a) Show the phase shift on the waveform of each dibit. (*Hint*: See Figures 2.1.9 and 2.2.3.)
    b) What are the amplitudes of I, Q, and QPSK waveforms?
    c) Prove that the sequence of the phase shifts of the waveforms of all dibits is shown correctly; i.e. the sequence corresponds to the dibit positions in the constellation diagram.

19. *Using Figure 10.1.5 as a prototype, develop similar figures for $M = 8$ modulation. Specifically,
    a) Sketch the constellation diagram, the bit-to-symbol mapping, and the corresponding segments of a carrier wave with their phases.
    b) Sketch the set of figures to show each symbol, its I and Q projections, and its segment of a carrier wave with the phase.
    c) In every figure, include the tables showing the transmitted symbol, carrier phase, and carrier amplitude.

20. Examine Figure 10.1.5a and verify that the waveforms and constellation points of every dibit are shown correctly.

21. Review Figure 10.1.5b:
    a) What is the phase shift of an in-phase signal when bit sequence changes from 1 to 0? From 0 to 1? From 1 to 1 and from 0 to 0?
    b) Repeat Question 21a for the quadrature signal.

c) What is the phase shift of a QPSK signal when dibit sequence changes from 11 to 01 to 00 and to 10? Why is this so?

22 The text says that QPSK can be viewed as two BPSK binary data streams. Why is that so? How can you distinguish between these two streams?

23 *Examine Figure 10.1.6a:
   a) Explain how the input binary data stream is converted in a QPSK modulated signal. Sketch the pertaining waveforms.
   b) Why are the amplitudes of a locally generated sine and cosine waves equal to $(A/\sqrt{2})$?
   c) Digital transmission always occurs under strong timing control. Should Figure 10.1.5a show a timer? If yes, where the timer should be placed and how it should be connected?

24 Consider Figure 10.1.6c:
   a) Explain how odd and even bit streams modulate their carrier waves.
   b) Combining two waveforms obtained in the above operation, show how the summation of two modulated I and Q BPSK signals produces QPSK signal.
   c) Compare QPSK signal obtained in Figure 10.1.6c by MATLAB/Simulink simulation with the QPSK signal obtained in Figure 10.1.5b manually. Comment on every discrepancy, if any.

25 A PSK signal changes its phase at every transition from bit 1 to bit 0 and vice versa. Does a QPSK signal change its phase at every transition from one dibit to another for all possible transitions? Prove your answer.

26 *How can we assess the quality of modulation by using Figure 10.1.7? (*Hint*: Consult Section 9.1.4.)

27 *Consider Figure 10.1.7:
   a) How can you assess the transmission quality by using Figure 10.1.7b? (*Hint*: See Figures 9.1.16 and 9.1.19.)
   b) Why does Figure 10.1.7b have only three openings, whereas in reality, it presents four-level modulation?
   c) By referring to Figure 10.1.7c, build the eye diagram for eight-level (three-bit-per-symbol) modulation.

28 Review Figure 10.1.8: If $t_r = t_f = 2.5$ ms, what percentage of the pulse duration does rise plus fall time occupy for PAM2 signal? for PAM4 signal? Explain why is this so.

Working with QPSK Signaling

29 *An input digital signal to a QPSK modulator is transmitted at $BR = 1$ Gb/s:
   a) What is the minimum time interval between the phase change in this signal?
   b) What are the minimum time intervals between the phase change in I and Q signals?
   c) What is the minimum time interval between the phase change in the QPSK signal? Show all your calculations. Sketch the figures.

30 A QPSK-modulated Wi-Fi signal has BW_{QPSK} (Hz) = 2.4 GHz:
   a) What is its bit rate, BR_{QPSK} (b/s)? Symbol rate, SR (Bd)?
   b) What is the transmission bandwidth for the input PSK signal? Its bit rate?

31 The general formula for a QPSK signal is given in (10.1.7) as $v_{QPSK}(t) = A(\sin(\omega_c t) + (2k-1) \cdot 45°)$, where $k = 1, 2, 3, 4$. Based on this formula, sketch the QPSK constellation diagram, show dibit-to-symbol mapping, and write the formula of every dot (symbol) given $A = 5$ mV and $f_C = 3$ MHz.

32 An input digital signal to a QPSK modulator is transmitted at BR = 1 Gb/s with amplitude 3 mV. What is the average energy of its symbol?

33 *Examine Figure 10.1.9 showing a coherent QPSK receiver:
   a) Why do we need to separate the input QPSK signal into two streams? What are these streams?
   b) How can we separate the input QPSK signal into two streams?
   c) What is the base timing interval?
   d) The receiver's output signal must be one stream of pulses. How can we combine two demodulated streams into one?

34 Consider the operation of a coherent QPSK detector: Based on Table 10.1.2, build the constellation diagram of the demodulated QPSK signal. Indicate all the parameters of each dot (symbol) shown. Assume $A = 5$ mV and $f_C = 3$ MHz.

35 *The objective of a QPSK demodulator is to recover a message signal, which must be a variable signal by its very nature:
   a) Why then does the text say, "The upper integrator output (the stream of odd pulses) is a signal of a constant value, $\frac{AT_s}{2\sqrt{2}} I_k$"?
   b) What will be this value if $v_{QPSK}(t)$ (mV) = $5\sin(2\pi \times 10^6 t)$, BR = 1 Gb/s, and $k = 3$?

36 *Equation (10.1.12), $BER_{QPSK} = BER_{BPSK} = Q(\sqrt{2E_b/N_0})$, states that bit error rates of PSK and QPSK signaling are equal. However, QPSK transmits symbols containing two bits, whereas PSK delivers one bit at a time. Therefore, it would be logical to assume that the probability of transmitting bits in error should be greater in QPSK than in PSK. How do you reconcile these seemingly contradictory statements?

37 BER of a BPSK signaling is 10^{-7}. What is the SER of the QPSK? Explain.

38 Using MATLAB, build the graph SER_{QPSK} vs. E_S/N_0 (dB). What is the SER_{QPSK} when E_S/N_0 =5 dB?

39 Consider error performance of a QPSK signaling:
   a) If this signaling transmits 100 Gb/s over one hour and its bit error rate is BER = 10^{-10}, how many erroneous bits does the system receive in average?
   b) Does the receiver know which received bits are erroneous?
   c) Does the receiver know how many erroneous bits it has received?

d) If the answers to Questions 39b and 39c are *no*, why do we still need to learn BER theory and how can we use its formula in practice?

40 *Discuss the versions of QPSK signaling:
a) Why do we need Offset QPSK (OQPSK)?
b) Figure 10.1.10 shows the phase shifts in QPSK and OQPSK signals. These shifts must be executed by a QPSK modulator shown in Figure 10.1.6a. Explain at what module of the modulator and how specifically these shifts are implemented.
c) Why do we need Differential QPSK (DQPSK)?
d) Sketch the waveform of a DQPSK signal based on Table 10.1.3.

10.2 Multilevel PSK and QAM Modulation

Objectives and Outcomes of Section 10.2

Objectives

- To understand the concept of multilevel (*M*-ary) modulation (signaling).
- To acquire knowledge of the *M*-PSK.
- To become familiar with the *M*-QAM.

Outcomes

- Understand the possibility of mapping more than two bits per symbol. Be aware that this is the only way to increase the actual bit rate of a given transmission without increasing the required channel's bandwidth or, alternatively, decrease the required bandwidth while keeping the needed bit rate. This and all other benefits of multilevel modulation will be enhanced with the increase in the bit-per-symbol number.
- Realize, however, that the greater the number of bits-per-symbol, the smaller the carrier phase change, and the smaller the distance between adjacent symbols, D_S, in a constellation diagram. Comprehend that this makes it difficult to implement such a signaling and, more importantly, result in two contradictory processes: On the one hand, increasing the number bits per symbol heightens the SE; on the other hand, increasing SE, worsen the BER because the distance D_S decreases.
- Analyze the error performance of *M*-PSK based on Eq. (10.2.7) and Figure 10.2.3 and understand that the greater the number of bits per symbol, N_b, (or the number of symbols, *M*), the greater SNR is needed to sustain the required BER. Deduce that the problem of sustaining the required BER boils down to keeping the proper distance, D_S, between two adjacent symbols when the number of bits per symbol (that is, the number of symbols) increases. Understand that an increase in the number of symbols causes a decrease in the distance D_S, which creates a problem in building a multilevel modulation system.
- Learn that the solution to the problem of *M*-ary PSK has been found in developing QAM. Be aware that QAM is a misnomer because this modulation changes both the amplitude and the phase of each symbol's carrier. Come to know that today QAM is one of the most widely used multilevel digital signaling methods.
- Compare the constellation diagrams of 16-QAM and 16-PSK shown in Figure 10.2.4b and observe that though the phase increments between adjacent symbols in both signaling are the same, the

10.2 Multilevel PSK and QAM Modulation

distance between the adjacent symbols is much greater in 16-QAM due to the changing of symbol amplitudes.
- Calculate minimum distances between the adjacent symbols in 16-QAM and 16-PSK and see that the minimum distance in 16-QAM is almost twice that in 16-PSK.
- Investigate BER of M-QAM by working with (10.2.9) and examining Figure 10.2.5 to understand that the higher the modulation level used in QAM, the greater the digital SNR, E_b/N_0, meaning that the more bit energy is required to sustain the same BER value.
- Become familiar with the example of industrial constellation diagrams shown in Figure 10.2.6 and understand that noise makes constellation dots look blurry.
- Compare error performance of M-PSK and M-QAM presented in Example 10.2.1 and Tables 10.2.2 and 10.2.3 and notice that for $M = 4$, both signaling performs evenly, but for $M = 64$, the QAM does about ten times better than the PSK.
- Investigate how spectral efficiency improves with an increase in M according to Eqs. (10.2.12)–(10.2.15) and Figure 10.2.8 and conclude that with an increase in M, the SE does not increase linearly but tends to saturate. Realize that this limitation stems from the bandwidth-power trade-off, which is one of the general telecommunications principles.
- Come to know that there is a minimum digital SNR supporting error-free communications, $\left(\frac{E_b}{N_0}\right) \min \approx -1.6$ dB, which is obtained from Shannon's law.
- Be aware that multilevel modulation is in the mainstream of modern communications and it is widely applied in wireless and wireline, starting from optical, communication systems.

10.2.1 Multilevel (M-ary) PSK

10.2.1.1 Introduction to M-ary PSK

As mentioned in Section 10.1, there are higher levels of PSK signaling other than QPSK. We can map three, four, or eight bits into one symbol. Referring to (10.1.2), we compute that for mapping three bits in one symbol, we need 8-ary PSK; with four bits per symbol, we have 16-ary PSK, and so on. The left three columns of Table 10.2.1 display the symbol mapping for M-ary PSK. The column on the right will be discussed shortly.

The advantage of using multibit-per-symbol mapping stems from the concept of multilevel signaling: This is the only way to increase the actual bit rate of a given transmission without significant increase of the required channel's bandwidth. This and all other benefits of multilevel modulation will be enhanced with the increase in the number of levels.

Table 10.2.1 Symbol mapping for M-ary PSK.

Number of bits per symbol, N_b	Number of symbols, $M = 2^{N_b}$	Designation of M-ary PSK	Carrier's phase increments, $\Delta\Theta = \dfrac{360°}{M}$
2	4	Quaternary PSK, QPSK	90°
3	8	8-ary PSK (8-PSK)	45°
4	16	16-ary PSK (16-PSK)	22.5°
5	32	32-ary PSK (32-PSK)	11.25°
6	64	64-ary PSK (64-PSK)	5.625°
7	128	128-ary PSK (128-PSK)	2.812 5°
8	256	256-ary PSK (256-PSK)	1.496 25°

976 | *10 Digital Transmission with Multilevel Modulation*

How can we generate an *M*-ary PSK? With our firm understanding of the modulation process of a QPSK signal presented in Figures 10.1.1–10.1.6 and the accompanying discussion, we can simply refer to constellation diagrams without the need to repeat the processes of changing phase shifts and composing waveforms. Extending the approach used in developing QPSK signaling to the greater number of bits-per-symbol, we can, for example, build 8-PSK signaling, as shown in Figure 10.2.1.

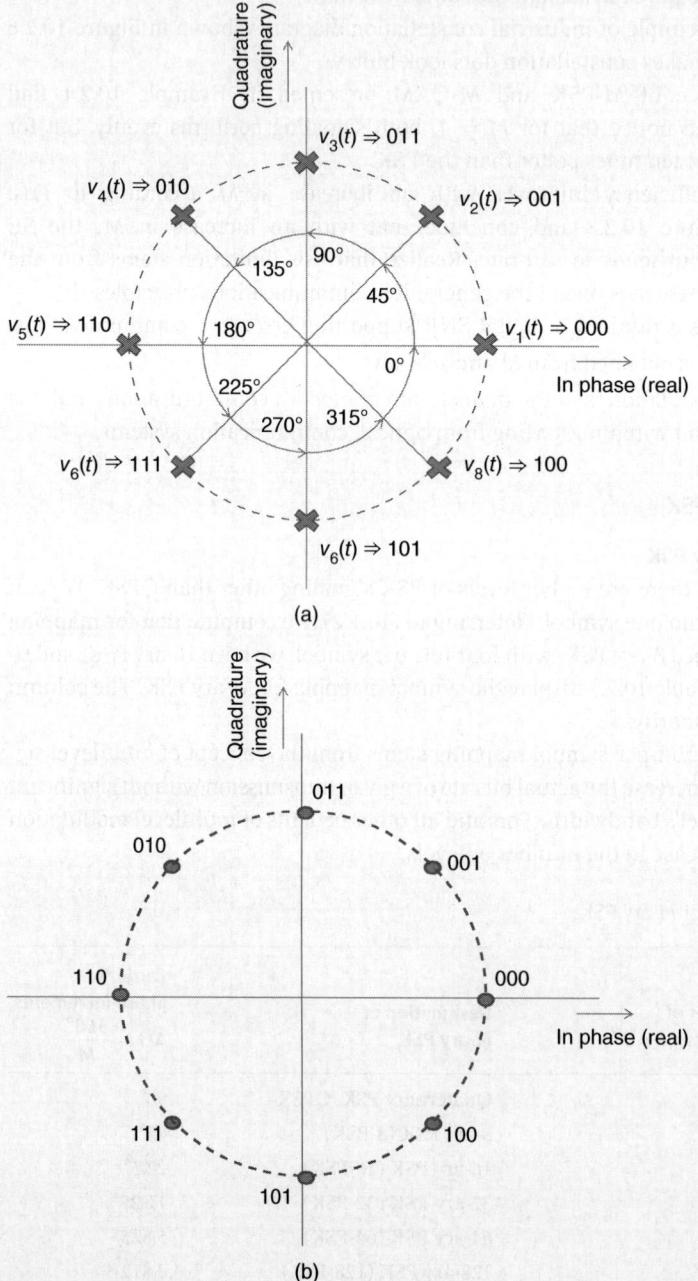

Figure 10.2.1 Graphical presentation of 8-PSK signaling: (a) signal space and (b) constellation diagram.

Mathematical description of the signal mapping shown in Figure 10.2.1 can be attained by generalizing (10.1.7) as follows:

$$v_{kMPSK}(t) = A \sin\left(w_c t + \frac{(k-1)\cdot 360°}{M}\right) \quad (10.2.1)$$

where $k = 1, 2, 3, 4, ..., M$. Thus, the formulas for each 8-ary PSK symbol obtained from (10.2.1) are

$$v_{1-8PSK}(t) = A \sin\left(\omega_c t + 0\frac{360°}{8}\right) = A \sin(\omega_c t + 0) \text{ for } k = 1, \text{ symbol } 000$$

$$v_{2-8PSK}(t) = A \sin\left(\omega_c t + 1\frac{360°}{8}\right) = A \sin(\omega_c t + 45°) \text{ for } k = 2, \text{ symbol } 001$$

$$v_{3-8PSK}(t) = A \sin\left(\omega_c t + 2\frac{360°}{8}\right) = A \sin(\omega_c t + 90°) \text{ for } k = 3, \text{ symbol } 011$$

$$v_{4-8PSK}(t) = A \sin\left(\omega_c t + 3\frac{360°}{8}\right) = A \sin(\omega_c t + 135°) \text{ for } k = 4, \text{ symbol } 010$$

$$v_{5-8PSK}(t) = A \sin\left(\omega_c t + 4\frac{360°}{8}\right) = A \sin(\omega_c t + 180°) \text{ for } k = 5, \text{ symbol } 110$$

$$v_{6-8PSK}(t) = A \sin\left(\omega_c t + 5\frac{360°}{8}\right) = A \sin(\omega_c t + 225°) \text{ for } k = 6, \text{ symbol } 111$$

$$v_{7-8PSK}(t) = A \sin\left(\omega_c t + 6\frac{360°}{8}\right) = A \sin(\omega_c t + 270°) \text{ for } k = 7, \text{ symbol } 101$$

$$v_{8-8PSK}(t) = A \sin\left(\omega_c t + 7\frac{360°}{8}\right) = A \sin(\omega_c t + 315°) \text{ for } k = 8, \text{ symbol } 100 \quad (10.2.2)$$

Using (10.2.2), we can write the formulas for each symbol of any M-ary PSK. It is important to note that in 8-ary PSK, the phase changes are reduced to 45° compared with the 90° phase changes of QPSK. Consequently, for 16-ary PSK, the adjacent phase increments will reduce to 22.5°, and so on. The carrier's phase increments for using M-ary PSK are shown in the right-hand column of Table 10.2.1.

10.2.1.2 BER of *M*-ary PSK

According to the *Hartley capacity law*, BR (b/s) = 2BW (Hz)·$\log_2 M$, with an increase in the number of bits per symbol and a *fixed transmission bandwidth,* the bit rate increases. There is, however, possibility to save the transmission bandwidth by sacrificing the increase in a bit rate, as shown in Section 10.1. Either way, the increase in M results in an increase in the *spectral efficiency*, SE (b/s/Hz) = BR (b/s)/BW (Hz), of a transmission system, which is our goal.

How many bits can we realistically map into one symbol? What is the price we pay for increasing SE by this approach? The answers to both questions follow from perusing the right-hand column of Table 10.2.1: *The greater the number of bits per symbol, the smaller the carrier phase change between adjacent symbols.*

We see from Table 10.2.1 that for 16-ary PSK, $\Delta\Theta_{16} = 22.5°$, but for 256-ary PSK, $\Delta\Theta_{256} \approx 1.5°$. Refer to Figures 10.1.3 and 10.1.4, where constellation diagrams and corresponding phase increments for QPSK signaling are shown. Now, imagine (or better yet, build) the constituents and the waveforms of 8-ary and 16-ary PSKs to see how small the phase increments become. Then imagine the scale of phase changes for a 256-ary PSK. Given that any real transmission takes place in a noisy environment, can we physically distinguish a 1.5° phase difference between two adjacent sinusoidal signals distorted by noise? In general, the answer is yes, given today's marvelous DSP circuitry, but the real question is, at what cost and, most importantly, at what *accuracy*? To answer the last question, we must turn back to our discussion of errors in digital transmission, a discussion presented in Sections 9.1 and 9.2.

Revisit (9.1.34), which can be shown as

$$\text{BER} = P_e = Q\left(\frac{V_1 - V_0}{2\sigma}\right) = \frac{1}{2}\text{erfc}\left(\frac{V_1 - V_0}{2\sqrt{2}\sigma}\right) \tag{9.1.34R}$$

We see that BER depends on $D_S = V_1 - V_0$, a distance between medians of bit 1 and bit 0. *In an M-ary PSK, D_S is the distance between medians of two adjacent symbols,*

$$D_S = V_{ks} - V_{(k+1)s} \tag{10.2.3}$$

where $k = 1, 2, 3, \ldots, M$. Obviously, D_S decreases as the number of bits per symbol – and therefore, the number of symbols, M – increases. In fact, we face two contradictory processes:

On the one hand, by increasing the number bits per symbol, we increase SE; on the other hand, by increasing SE, we worsen BER because the distance between the medians of two adjacent symbols, D_S, decreases.

This statement is illustrated in Figure 10.2.2.

Figure 10.2.2 shows the constellation diagrams and Gaussian PDFs for two adjacent symbols in the 8-ary and 16-ary PSKs. Median V_{s1} belongs to symbol $s1$ (logic 1) and median V_{s0} belongs to the symbol $s0$ (logic 0). With an increase in M, the distance between these two medians, $D_S = V_{s1} - V_{s0}$, decreases. As a result, the Gaussian PDFs of the adjacent symbols in 16-ary PSK overlap more than the PDFs in 8-ary PSK, which worsen BER of 16-ary PSK compared to that of the 8-ary. The ratio of distances $D_{16\text{PSK}}$ and $D_{8\text{PSK}}$ can be estimated as follows: In Figure 10.2.2a, $D_S = A_{s001} - A_{s000}$ is the length of the chord, which is given by

$$D_S = 2r \sin(\Theta/2) \tag{10.2.4}$$

where r is the circle's radius and Θ is the angle subtended by the chord at the center. In Figure 10.2.2a, the radius is the symbol's amplitude, A, and the angle is $2\pi/M$; therefore,

$$D_S = 2A \sin\left(\frac{\pi}{M}\right) = 2A \sin\left(\frac{180°}{M}\right) \tag{10.2.5}$$

Thus, the distances are determined for 8-ary PSK as

$$D_{8\text{PSK}} = 2A \sin(22.5°) \tag{10.2.6a}$$

and for 16-ary PSK as

$$D_{16\text{PSK}} = 2A \sin(11.25°) \tag{10.2.6b}$$

Calculating their ratio,

$$\frac{D_{16\text{PSK}}}{D_{8\text{PSK}}} = \frac{\sin(11.25°)}{\sin(22.5°)} = 0.51 \tag{10.2.6c}$$

we determine that the distance between adjacent constellation points for 16-ary PSK is less than a half that for 8-ary PSK.

The BER of M-ary PSK is given by

$$\begin{aligned}\text{BER}_{\text{MPSK}} &\approx \frac{1}{\log_2 M} 2Q\left(\sqrt{\frac{2E_b}{N_0} \cdot \log_2 M \cdot \sin^2\left(\frac{\pi}{M}\right)}\right) \\ &= \frac{1}{N_b} 2Q\left(\sqrt{\frac{2E_b}{N_0} \cdot N_b \cdot \sin^2\left(\frac{\pi}{M}\right)}\right)\end{aligned} \tag{10.2.7}$$

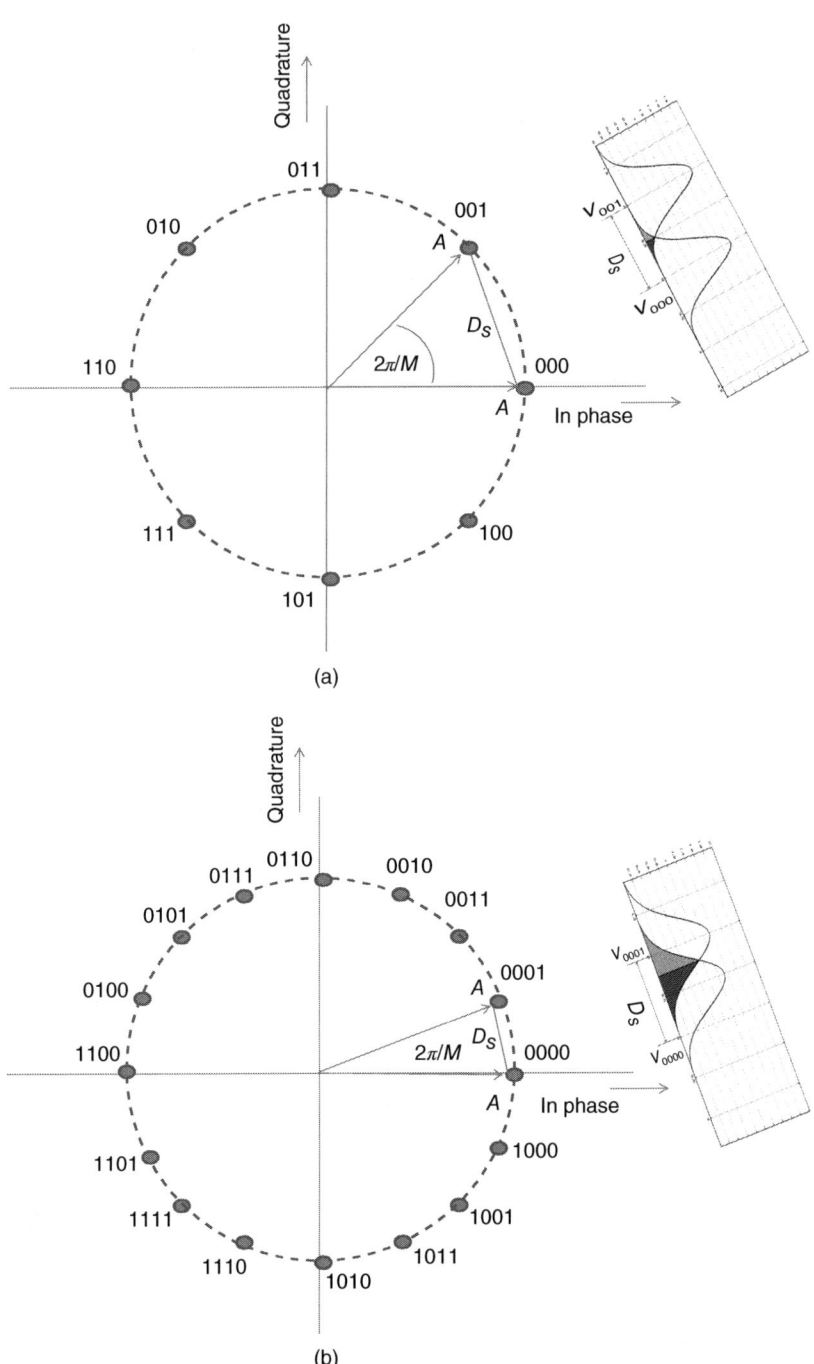

Figure 10.2.2 Constellation diagrams and BER performance of M-ary PSK: (a) 8-ary PSK and (b) 16-ary PSK.

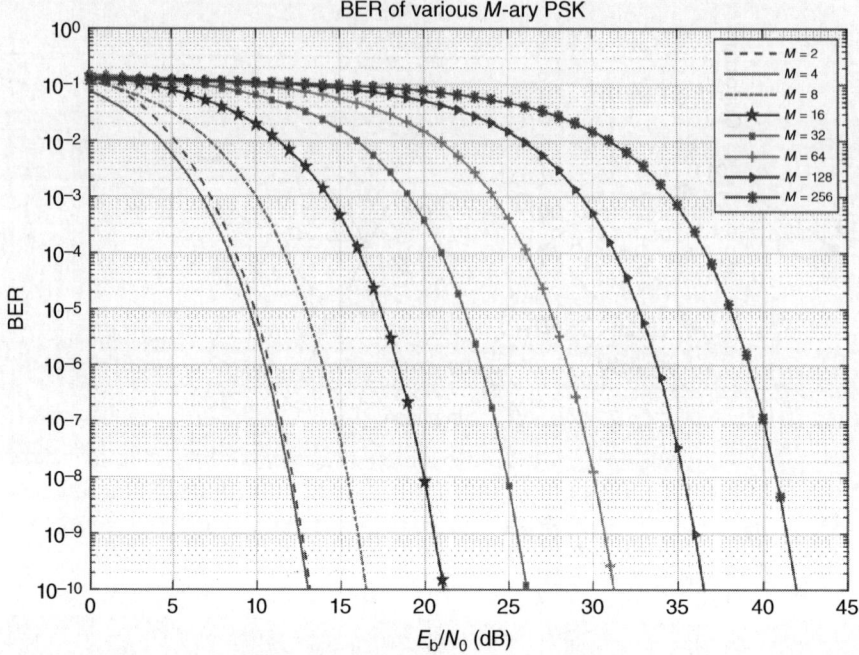

Figure 10.2.3 BER of various *M*-ary PSK.

Here, M is the number of symbols and N_b is the number of bits per symbol as before. (Derivation of (10.2.7) can be found, for example, in Mesiya 2013, p. 644–645 and Stern and Mahmoud 2004, p. 291.)

Equation (10.2.7) is visualized in Figure 10.2.3, which shows BER as a function of SNR (here E_b/N_0) for PSK of the various number of symbols. Examination of Figure 10.2.3 reveals that

> *the greater the number of bits per symbol, N_b, (or the number of symbols, M), the greater the SNR needed to sustain the required BER.*

Since the level of noise does not depend on M, the aforementioned requirement results in *the need to increase the bit energy, E_b, as N_b (or, equivalently, M) increases*. Compare (10.2.7) with (10.1.12),

$$\text{BER}_{\text{QPSK}} = \frac{1}{\log_2 M} 2Q\left(\sqrt{\frac{2E_b}{N_0}}\right) \qquad (10.1.12\text{R})$$

and see that each bit of MPSK signaling needs more energy than each bit of a QPSK in order to keep the same value of BER. Quantitatively, the argument of Q-function of BER_{MPSK} is greater than that of BER_{QPSK} by the factor of $\sqrt{(N_b \cdot \sin^2(\pi/M))}$. You are encouraged to perform specific calculations to compare, for example, $\text{BER}_{16\text{PSK}}$ and $\text{BER}_{64\text{PSK}}$. (Refer to Table 9.1.2 and accompanied MATLAB code for these calculations.) Figure 10.2.3 helps approximately but quickly clarify this point: For BER = 10^{-8}, for example, 16-PSK requires $E_b/N_0 = 20$ dB, which corresponds 100 in absolute numbers, and 64-PSK needs $E_b/N_0 = 30$ dB, which is 1000. Therefore, 64-PSK signaling demands 10 times greater bit energy than that of 16-PSK, provided that the noise level is fixed.

It would seem obvious that the amount of energy for symbol should be proportional to the number of bits per symbol and that the symbol energy should be the sum of the energy of all its bits. In reality, such a linear proportionality does not work. This is because, as (10.2.7) shows, *the energy needed for each bit packed into a symbol to sustain the same BER value increases with an increase in the modulating level*. Why? Physics behind this requirement is that the closer one symbol to another, the more energy from each bit of one symbol penetrates to the bits of an adjacent symbol. This means that SNR of the bit losing its energy becomes smaller; as for the adjacent bit, the one receiving energy from neighboring bits, this received energy is noise – and therefore its SNR becomes worse. Hence, both the losing and receiving bits need an increasing amount of energy to sustain the required SNR – and thus the BER. This reasoning is reflected in (10.2.7) and Figure 10.2.3.

Therefore, the problem of sustaining the required BER boils down to *keeping the proper* distance, D_S, between two adjacent symbols when the number of bits per symbol (and therefore, the number of symbols) increases. But remember, we *want to increase the number of symbols* to increase the SE. Such a problem – in which we need to satisfy contradictory requirements – is typical in engineering practice. And yes, astute engineers found the appropriate solution to this problem.

10.2.2 Multilevel Quadrature Amplitude Modulation, *M*-QAM

10.2.2.1 The Concept of Multilevel Quadrature Amplitude Modulation, *M*-QAM

To increase the number of bits per symbols and, at the same time, keep a considerable distance between the adjacent symbols, the modulation technique called *QAM* has been developed. Today, *QAM is one of the most widely used multilevel digital signaling methods*. In QAM, *the modulation is achieved by changing both the phase and the amplitude of each symbol's carrier*, as shown in Figure 10.2.4.

Figure 10.2.4a shows a circular constellation diagram of 16-QAM signaling with amplitude values, several bit assignments, and distances among symbols. Comparison of the 16-PSK and 16-QAM modulations shown in Figure 10.2.4b reveals that in many aspects, the QAM is a combination of several PSK modulations with different amplitudes. Hence, modulation and demodulation of *M*-QAM are similar to that of *M*-PSK; the detailed discussion of this topic we leave to specialized textbooks.

There are several types of constellations for 16-QAM signaling. Figure 10.2.4 gives two examples: A circular constellation (Figure 10.2.4a) and a rectangular constellation (Figure 10.2.4c). A rectangular constellation of 16-QAM is used mostly because of its simplicity.

Compare the circular constellation diagrams of 16-QAM and 16-PSK shown in Figure 10.2.4b: You can see that though the phase increments between adjacent symbols in both signaling are the same, the distances among the adjacent symbols are much greater in 16-QAM due to the changing of the symbol amplitudes. What's more, despite the fact that some symbols in the 16-QAM have the same phases, they can be readily distinguished thanks to their significantly different amplitudes. For example, even though symbols **1100** and **1111** in 16-QAM constellation diagram have the same phases of 45°, they can be easily differentiated because of $A_{1100} = 0.33 A_{QAM}$ and $A_{1111} = A_{QAM}$, where A_{QAM} is the 16-QAM symbol amplitude.

Examine the *I* and *Q* projections of the 16-QAM symbols shown in Figure 10.2.4c: Amplitudes of some symbols have equal *I* projections, whereas the amplitudes of the others have equal *Q* projections. Nevertheless, all symbols are uniquely identified thanks to the difference in their phases. Compare, for instance, symbols **1100** and **1101** in Figure 10.2.4c: Their *I* projections are equal to $0.23 A_{QAM}$, but their *Q* projections are different in values – that is, $0.23 A_{QAM}$ for **1100** and $0.72 A_{QAM}$, for **1101** – due to the different phases.

We can see from Figure 10.2.4a,c,d that the minimum distance between two adjacent symbols in 16-QAM signaling is either I or Q projection of their general distance. For example, the minimum distance between symbols **1100** and **1110** is computed as $d_{min}(16QAM) = 0.75A_{QAM} \cos 15° - 0.33A_{QAM} \cos 45° = 0.72A_{QAM} - 0.23A_{QAM} = 0.49A_{QAM}$. This calculation is illustrated in Figure 10.2.4d.

On the other hand, rewriting (10.2.5) with $A = A_{PSK}$

$$D_{MPSK} = 2(A_{PSK}) \sin\left(\frac{180°}{M}\right) \tag{10.2.5R}$$

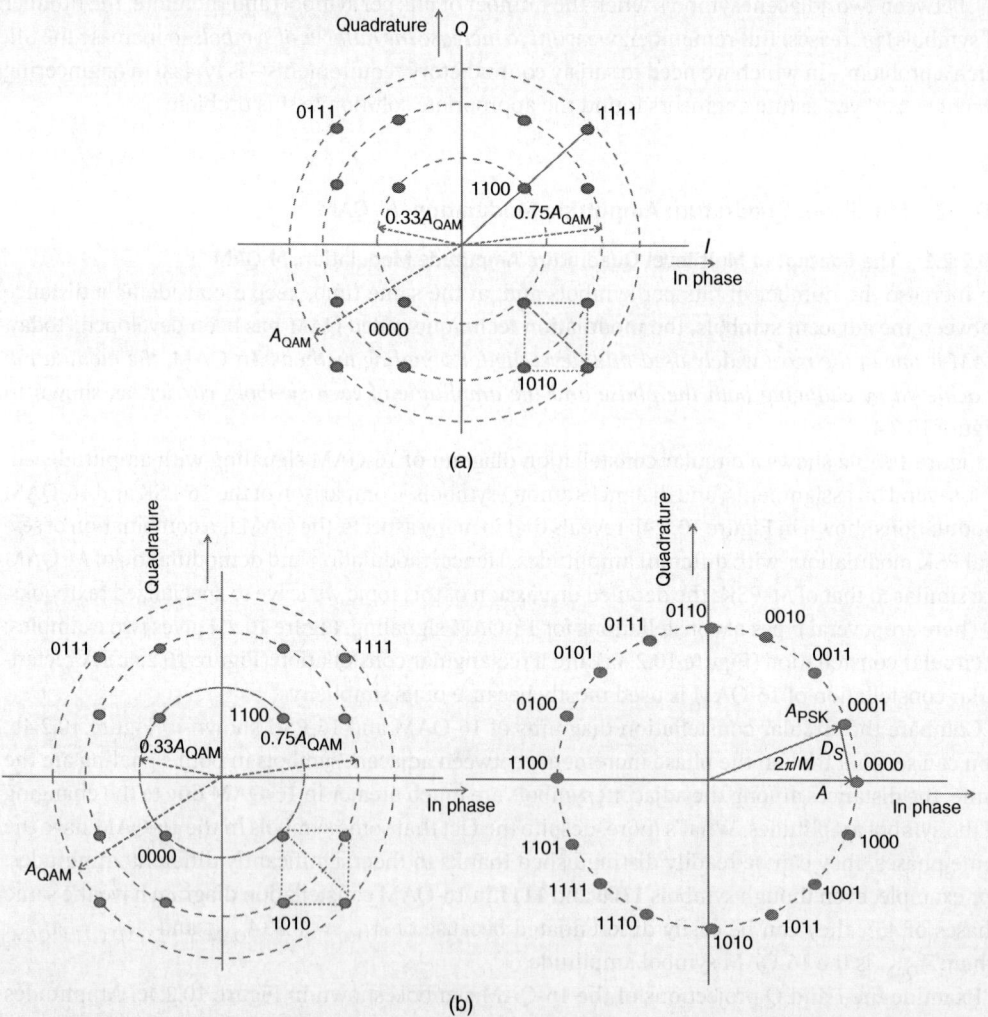

Figure 10.2.4 Constellation diagrams of multilevel quadrature amplitude modulation, M-QAM: (a) a circular constellation diagram of 16-QAM; (b) circular constellation diagrams of 16-QAM and 16-PSK; (c) a rectangular constellation diagram of 16-QAM with symbol mapping and I and Q amplitude projections; and (d) table of 16-QAM symbols, their carrier amplitudes, carrier phases, and example of the minimum distance (figures are not to scale).

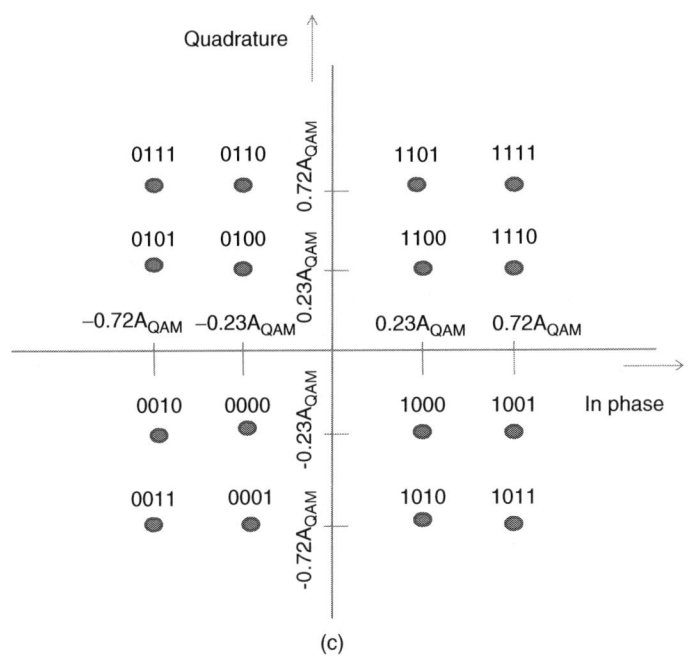

(c)

Symbol	Parameters of a carrier wave	
	A_{QAM}	Phase
0000	0.33	225°
0001	0.75	255°
0010	0.75	195°
0011	1.0	225°
0100	0.33	135°
0101	0.75	105°
0110	0.75	165°
0111	1.0	135°
1000	0.33	315°
1001	0.75	285°
1010	0.75	345°
1011	1.0	315°
1100	0.33	45°
1101	0.75	75°
1110	0.75	15°
1111	1.0	45°

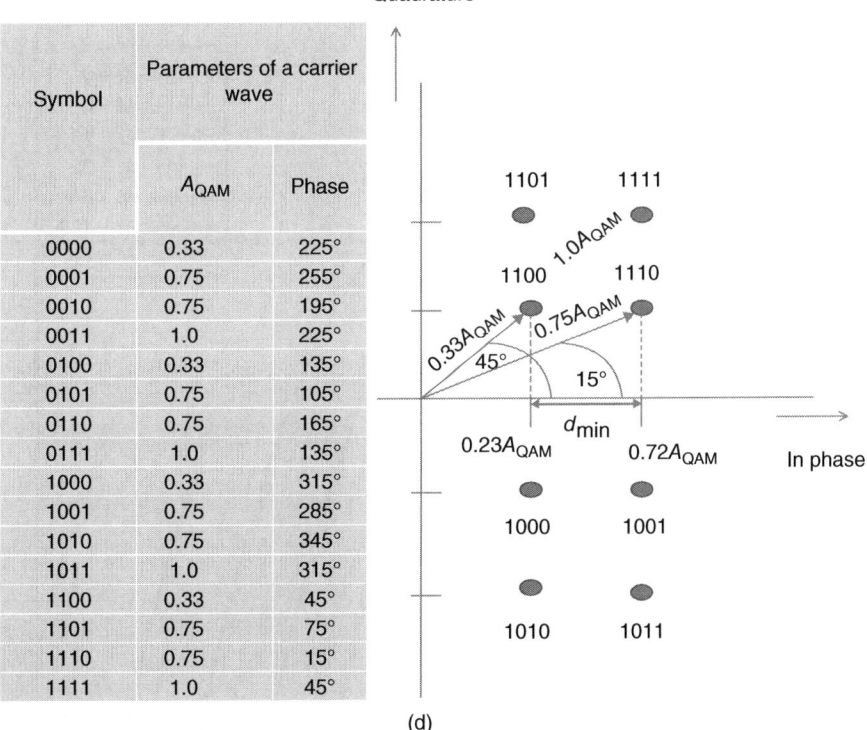

(d)

Figure 10.2.4 (*Continued*)

we find for $M = 16$

$$d_{\min}(16\text{PSK}) = 2(A_{\text{PSK}}) \sin\left(\frac{180°}{16}\right) = 0.39 A_{\text{PSK}}$$

We know that $A_{\text{PSK}} = 0.75 A_{\text{QAM}}$ (Stern and Mahmoud 2004, p. 299). Therefore,

$$\frac{d_{\min}(16\text{QAM})}{d_{\min}(16\text{PSK})} = \frac{0.49 A_{\text{QAM}}}{0.29 A_{\text{QAM}}} = 1.7$$

Thus, the minimum distance between the adjacent symbols in 16-QAM is almost twice that in 16-PSK. The implication of this result is that M-QAM needs a smaller SNR, E_b/N_0, than QPSK does to attain the same BER level because the BER depends on the distance between the adjacent symbols. (Remember that the smaller SNR means less bit energy, E_b at a given noise level, N_0)

10.2.2.2 BER of M-QAM

BER of M-QAM is determined by the following equation (Mesiya 2013, p. 656):

$$\text{BER}_{M-\text{QAM}} = \frac{1}{\log_2 M} \frac{4\sqrt{M}-1}{\sqrt{M}} Q\left(\sqrt{\left[\frac{3E_b(\log_2 M)}{N_0(M-1)}\right]}\right) = \frac{1}{N_b} \frac{4\sqrt{M}-1}{\sqrt{M}} Q\left(\sqrt{\left[\frac{3E_b(N_b)}{N_0(M-1)}\right]}\right) \tag{10.2.9}$$

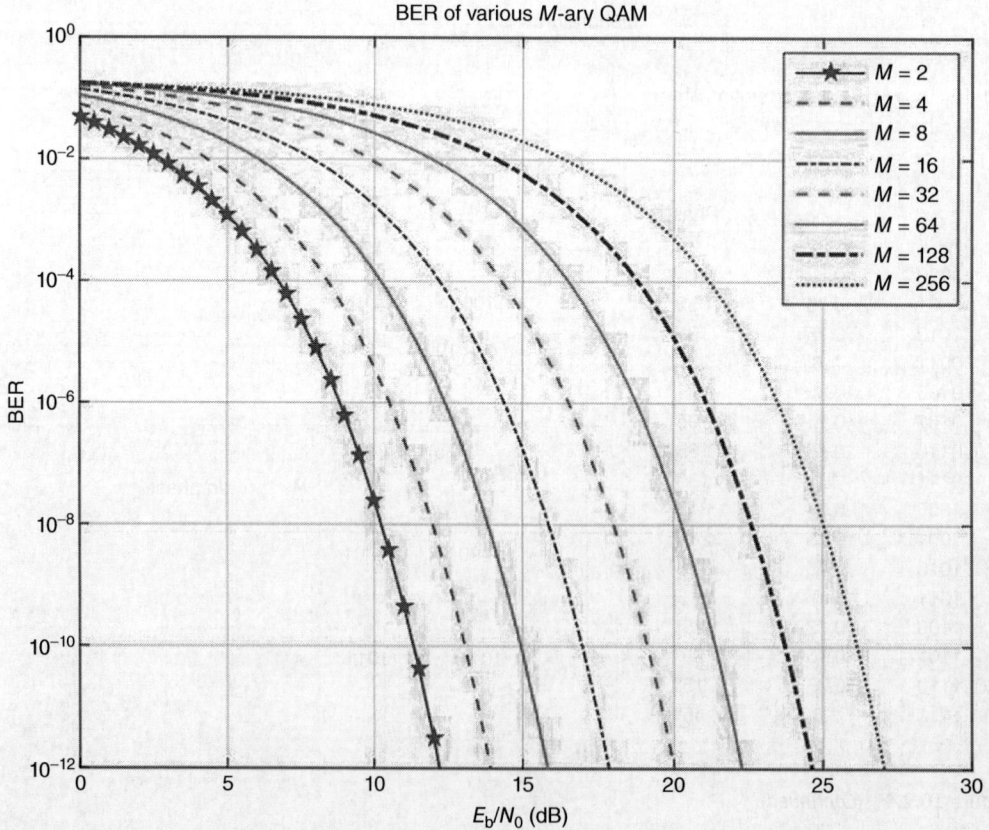

Figure 10.2.5 BER of various M-QAM signaling systems as a function of digital SNR.

where the symbol energy is assumed to be the sum of energy of its bits; that is,

$$E_s = (\log_2 M) \cdot E_b = N_b \cdot E_b \qquad (10.2.10)$$

(**Nota bene**: This is a very general assumption that works satisfactorily when the distance between adjacent signals is not very small.) Figure 10.2.5, built on (10.2.9), shows that the higher the modulation level used in QAM, the greater the needed digital SNR, E_b/N_0, meaning that the more bit energy is required to sustain the same BER value. (Sounds familiar? Yes, this is true for any M-ary signaling.)

The physics underlying this result is the same as that discussed previously in examining Figure 10.2.3: Energy from the adjacent symbols penetrating to a given symbol is considered noise by the receiving symbol. The more tightly the symbols are squeezed together in signaling, the noisier the environment for each symbol becomes and the greater the amount of its own energy is needed to keep the BER at the required level. We rightly consider this necessary increase in a symbol's energy to be a *penalty*.

We need to remember, of course, that, in addition to the *internal noise* produced by the adjacent symbols, each symbol is affected by the external noise produced by external man-made or natural sources. (To brush up on noise, review Section 9.1.1.) In the presence of internal noise – which is inevitable – symbol dots in a constellation diagram become blurry; the external noise makes them even blurrier.

Figure 10.2.6a gives us a glimpse of industrial transmission constellations with and without noise. This figure shows many tested parameters of a 32-QAM signal; constellation diagrams are shown at the left, eye diagrams of I and Q bit streams are shown at the center, and power and amplitude spectra along with phase error are shown at the right. Focus in particular on how blurry the dots in the constellation diagrams are. This is because of the external and internal noise cause errors in the values of symbol amplitude and its phase. Figure 10.2.6b exemplifies this point. It shows several measured dots of symbol **11** which are randomly off in their amplitudes and phases with respect to the ideal, targeted position of the dot. After transmission of millions of bits, these randomly distributed dots make kind of cloud, as shown in symbol **10**. It is this phenomenon that makes the dots in a constellation diagram look fuzzy and cloudy. (See our discussion of *error value magnitude*, EVM, in Section 9.1.)

Analyze Figure 10.2.6 carefully because it is an integrated visual presentation of many topics discussed in this section and in this book.

To further visualize the aforementioned discussion, review Figure 10.2.7, where the constellation and eye diagrams of 64-QAM signaling are shown. The constellation diagram in Figure 10.2.7a shows the symbol positions along with bit-to-symbol mapping. Analyze this diagram and examine the changes in the phases and amplitudes of each symbol. Based on your analysis, write down a formula for several symbols similar to (10.2.5) written for 8-PSK signaling. Examine the eye diagram shown in Figure 10.2.7b and ask yourself why you see overlapping "eyes" and why there are seven openings in the center of the figure.

Example 10.2.1 will help you understand the above discussion by providing a quantitative comparison of error performance of M-PSK and M-QAM signaling.

Example 10.2.1   Comparison of Bit Error Rates of M-PSK and M-QAM Signaling

Problem

Compare bit energy, E_b, and digital SNR, E_b/N_0, in (a) 4-PSK and 4-QAM and (b) 64-PSK and 64-QAM if the required BER = 10^{-12} and $N_0 = 10^{-8}$ W/Hz.

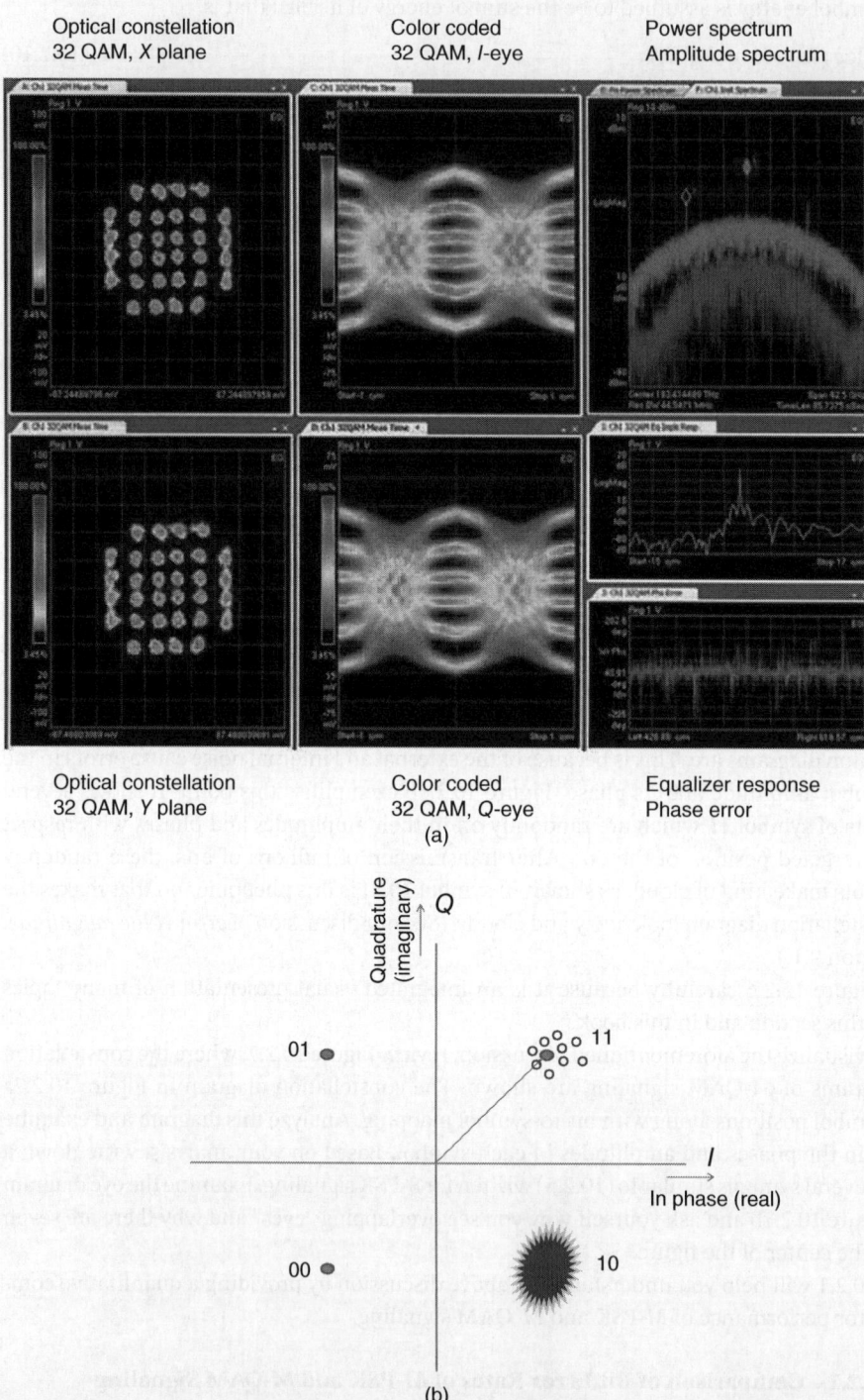

Figure 10.2.6 (a) Examples of industrial constellation diagrams, eye diagrams, spectrum, and phase error. Source: © Keysight Technologies, Inc. Courtesy of Keysight Technologies. (b) Explanation of noise effect on a constellation diagram.

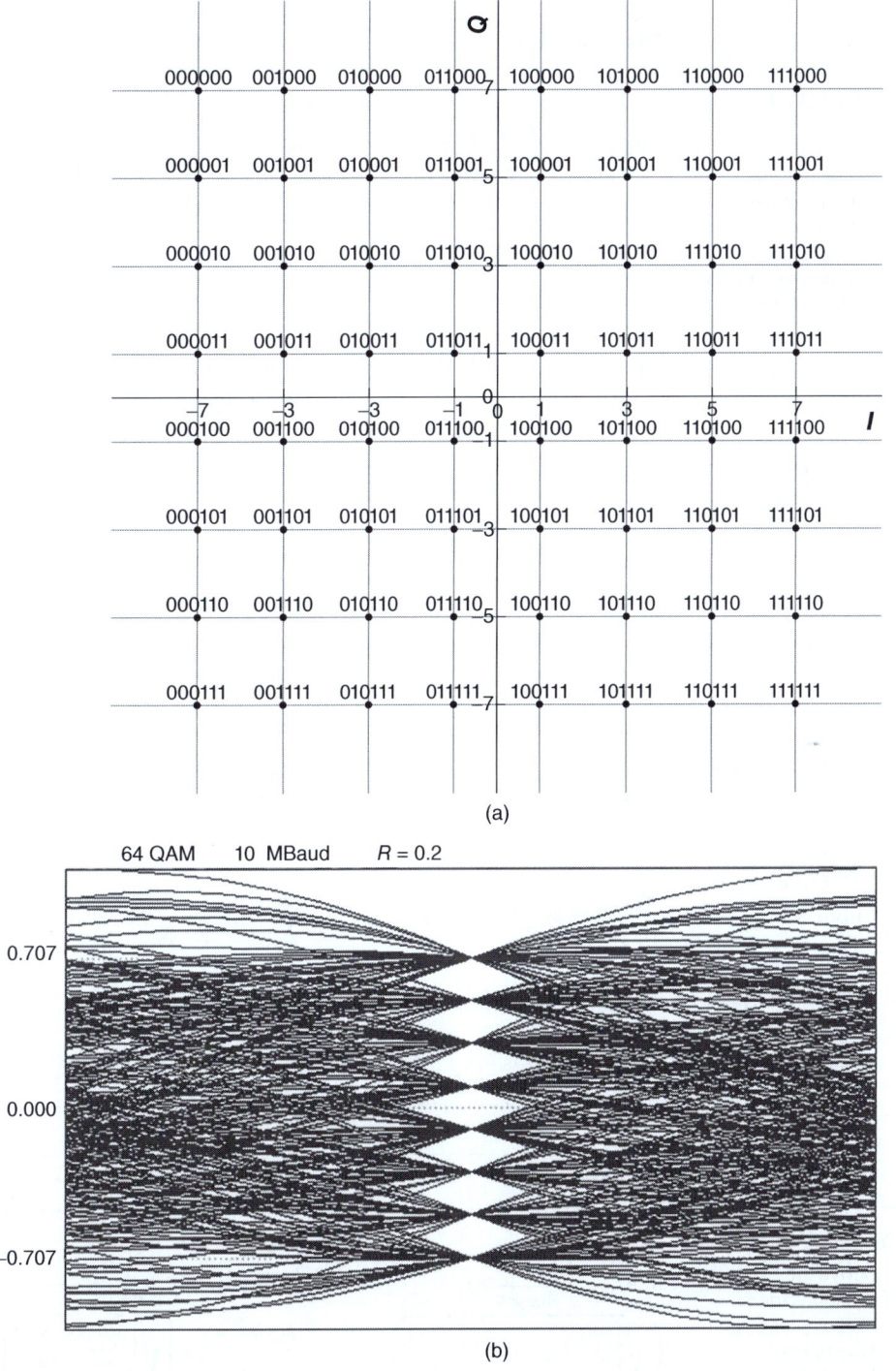

Figure 10.2.7 64-QAM signaling: (a) constellation diagram with symbol mapping and (b) eye diagram. (*Note*: www.radiosim.com/eye.htm.)

Table 10.2.2 Calculating BER parameters for M-PSK and M-QAM signaling.

	Steps	Results for 4-PSK	Results for 4-QAM
(a) 4-PSK and 4-QAM			
1.	Derive the equations for BER by plugging $M = 4$ into (10.2.7) for 4-PSK and into (10.2.9) for 4-QAM	$\text{BER}_{4\text{PSK}} = Q\left(\sqrt{\left[\dfrac{2E_b}{N_0}\right]}\right)$	$\text{BER}_{4\text{QAM}} = Q\left(\sqrt{\left[\dfrac{2E_b}{N_0}\right]}\right)$
2.	Find the values of the Q-function arguments by using the appendix to this section	Since $\text{BER}_{4\text{PSK}} = 10^{-12}$, then $\left(\sqrt{\left[\dfrac{2E_b}{N_0}\right]}\right) = 7$	Since $\text{BER}_{4\text{QAM}} = 10^{-12}$, then $\left(\sqrt{\left[\dfrac{2E_b}{N_0}\right]}\right) = 7$
3.	Obtain the equations for the digital SNR, $\dfrac{E_b}{N_0}$, find the values of $\dfrac{E_b}{N_0}$ in absolute numbers and in decibels, and compute the average bit energy, E_b	Since $\dfrac{2E_b}{N_0} = 49$, then $\dfrac{E_b}{N_0} = 24.5$, or $\dfrac{E_b}{N_0}$ (dB) $= 10 \cdot \log\left(\dfrac{E_b}{N_0}\right) = 13.89$ dB. We also compute $E_b = 24.5 \times N_0 = 24.5 \times 10^{-8}$ W/Hz $= 24.5 \times 10^{-8}$ J	Since $\dfrac{2E_b}{N_0} = 49$, then $\dfrac{E_b}{N_0} = 24.5$, or $\dfrac{E_b}{N_0}$ (dB) $= 10 \cdot \log\left(\dfrac{E_b}{N_0}\right) = 13.89$ dB. We also compute $E_b = 24.5 \times N_0 = 24.5 \times 10^{-8}$ W/Hz $= 24.5 \times 10^{-8}$ J
4.	*Answers*: The digital SNR, $\dfrac{E_b}{N_0}$, required to sustain the given BER is given by	$\dfrac{E_b}{N_0} = 24.5$ or $\dfrac{E_b}{N_0} = 13.9$ dB	$\dfrac{E_b}{N_0} = 24.5$ or $\dfrac{E_b}{N_0} = 13.9$ dB
	Bit average energy, E_b	$E_b = 24.5 \times 10^{-8}$ J	$E_b = 24.5 \times 10^{-8}$ J
	Steps	**Results for 64-PSK**	**Results for 64-QAM**
(b) 64-PSK and 64 QAM			
1.	Derive the equations for BER by plugging $M = 16$ into (10.2.7) for 16-PSK and (10.2.9) for 16-QAM	$\text{BER}_{64\text{PSK}} \approx \dfrac{1}{\log_2 64} 2Q\left(\sqrt{\left[\dfrac{2E_b}{N_0} \cdot \log_2 64 \cdot \sin^2\left(\dfrac{\pi}{64}\right)\right]}\right)$ $= \dfrac{Q}{3}\left(\sqrt{\left[\dfrac{E_b}{N_0} \cdot 0.029\right]}\right)$	$\text{BER}_{64\text{QAM}} = \dfrac{1}{\log_2 64} \dfrac{4(\sqrt{64}) - 1}{\sqrt{64}} Q\left(\sqrt{\left[\dfrac{3E_b(\log_2 64)}{(64-1)N_0}\right]}\right)$ $= \dfrac{7}{12} Q\left(\sqrt{\left[\dfrac{E_b}{N_0} \cdot 0.286\right]}\right)$

Table 10.2.2 (Continued)

	Steps	Results for 64-PSK	Results for 64-QAM
2.	Find the values of the Q-function arguments by using the appendix to this section	Since $3 \times 10^{-12} = Q\left(\sqrt{\left[\dfrac{E_b}{N_0} \cdot 0.029\right]}\right)$, then, from the appendix, $\sqrt{\left[\dfrac{E_b}{N_0} \cdot 0.029\right]} \approx 6.90$	Since $\dfrac{12}{7} \times 10^{-12} = Q\left(\sqrt{\left[\dfrac{E_b}{N_0} \cdot 0.286\right]}\right)$, then, from the appendix, $\sqrt{\left[\dfrac{E_b}{N_0} \cdot 0.286\right]} \approx 6.95$
3.	Obtain the equations for the digital SNR, $\dfrac{E_b}{N_0}$, find the values of $\dfrac{E_b}{N_0}$ in absolute numbers and in decibels, and compute the average bit energy, E_b	$\dfrac{E_b}{N_0} \cdot 0.029 = (6.90)^2 = 47.61$ and $\dfrac{E_b}{N_0} = 1641.72$ or $\dfrac{E_b}{N_0}$ (dB) = $10 \cdot \log(1641.72)$ = 32.15 dB We also compute $E_b = 1641.74 \times N_0$ (W/Hz) = 1641.74×10^{-8} J	$\dfrac{E_b}{N_0} \cdot 0.286 = (6.95)^2 = 48.30$ and $\dfrac{E_b}{N_0} = 168.88$ or $\dfrac{E_b}{N_0}$ (dB) = $10 \cdot \log(168.88)$ = 22.3 dB We also compute $E_b = 168.88 \times N_0$ (W/Hz) = 168.88×10^{-8} J
4.	*Answers*: The digital SNR, $\dfrac{E_b}{N_0}$, required to sustain the given BER for $M = 16$ are given by	$\dfrac{E_b}{N_0} = 1641.7$ or $\dfrac{E_b}{N_0}$ (dB) = 32.15 dB	$\dfrac{E_b}{N_0} = 168.9$ or $\dfrac{E_b}{N_0}$ (dB) = 22.3 dB
	Bit average energy, E_b, for $M = 16$ is equal to	$E_b = 1641.7 \times 10^{-8}$ J	$E_b = 168.9 \times 10^{-8}$ J

Solution

The solution is straightforward: Use (10.2.7),

$$\text{BER}_{\text{MPSK}} \approx \frac{1}{\log_2 M} 2Q\left(\sqrt{\left[\frac{2E_b}{N_0} \cdot \log_2 M \cdot \sin^2\left(\frac{\pi}{M}\right)\right]}\right) \quad (10.2.7\text{R})$$

and (10.2.9),

$$\text{BER}_{\text{MQAM}} = \frac{1}{\log_2 M} \frac{4(\sqrt{M} - 1)}{\sqrt{M}} Q\left(\sqrt{\left[\frac{3E_b(\log_2 M)}{(M-1)N_0}\right]}\right) \quad (10.2.9\text{R})$$

and the MATLAB code and Table 9.1.2 of Q-function given in Section 9.1 or MATLAB code for working with Q-function in Example 9.2.3.

Since for each case we need to perform the same type of calculations, it is reasonable to present them as a tabulated sequence of steps, which is shown in Table 10.2.2:

The problem is solved.

Discussion

- We need to analyze the results and compare M-PSK with M-QAM. To make this job easier, we summarize the results in Table 10.2.3.

  Let us recap: *Table 10.2.3 shows the values of digital SNR, E_b/N_0, and the average bit energy, E_b, needed to sustain the required BER = 10^{-12}.*

- The result of $M = 4$, $\text{BER}_{\text{QPSK}} = \text{BER}_{\text{4-QAM}}$, can be easily predicted based on the identical distance between adjacent symbols in both QPSK and 4-QAM signaling. Revisit Figure 10.2.2, build a constellation diagram of a 4-QAM, and recall that the BER depends on the distance between the adjacent symbols to understand why all the manipulations and the results of the calculations are identical for both modulations.

- We compare how the values of digital SNR, E_b/N_0, and the average bit energy, E_b, change with increasing numbers of bits per symbol for each type of modulation: The M-PSK column shows that the needed E_b/N_0 increases from 24.5 to 1641.7, when M grows from 4 to 64. Since the noise level is kept constant in this example, the aforementioned increase in E_b/N_0 values is provided by an increase in the average bit energy with a rise of M. For M-PSK, the results are $E_b = 24.5 \times 10^{-8}$ J for $M = 4$ and $E_b = 1641.7 \times 10^{-8}$ J for $M = 64$.

These results seem to be reasonable because we rely, based on our study, on the following logic:

✓ The more the symbols employed in signaling, the shorter the distance between the adjacent symbols.
✓ The shorter the distance, the greater the energy exchange between the adjacent symbols (a harmful phenomenon, remember), and the higher the SNR of each symbol is needed to sustain the required BER.
✓ Hence, this example is the numerical confirmation of our BER consideration in this section.

The same can be said about M-QAM, and we urge you to analyze the appropriate numbers in the M-QAM column of Table 10.2.3. (It helps to refer to the *power penalty*, which is the amount of power that must be added to a signal to keep the SNR at the level required to sustain the assigned BER. This power-penalty concept is applied here for energy.)

- To see how the different types of signaling respond to an increase in M, we need to compare the same rows in Table 10.2.3. For $M = 4$, the required E_b/N_0 and E_b are the same for both 4-PSK and 4-QAM modulations. For $M = 64$, the values of required E_b/N_0 and E_b for 64-PSK are ten times greater than that required for 64-QAM. (Make sure to write all the numbers and examine them closely.) This result is also predictable because the distances among the adjacent symbols in high-level QAM is much greater than the distances in PSK of the same level due to varying of the phase *and* the amplitude of the symbol vectors in QAM signaling. Refer to Figure 10.2.4 and the accompanying discussion.

- In conclusion, we strongly recommend reiterating the main points of this example: (i) The greater the number of bits per symbol (which means, the greater the M), the greater the digital SNR and the average bit energy that are required to sustain the assigned BER. (ii) In this regard, M-QAM performs better than M-PSK. Closely studying Figures 10.2.3–10.2.5 helps you to visually recognize these points.

Thus, the results of this example put on a solid numerical footing all the key points on BER of multilevel modulations covered in this section.

Table 10.2.3 Comparison of BER Parameters for M-PSK and M-QAM.

		M-PSK		M-QAM	
		$\dfrac{E_b}{N_0}$	E_b	$\dfrac{E_b}{N_0}$	E_b
$M = 4$	$\dfrac{E_b}{N_0}$	24.5 13.9		24.5 13.9	
	E_b		24.5×10^{-8} J		24.5×10^{-8} J
$M = 64$	$\dfrac{E_b}{N_0}$	1641.7 32.15 dB		168.9 22.3 dB	
	E_b		1641.74×10^{-8} J		168.9×10^{-8} J

10.2.3 Final Thoughts

10.2.3.1 Spectral Efficiency, Signal-to-Noise Ratio, and Multilevel Modulation

Spectral Efficiency and SNR Let us start with the statement that the maximum *spectral efficiency*, SE (bit/s/Hz), an integral measure of a transmission's efficiency, can be derived from Shannon's law as

$$\mathrm{SE}_{\max}(\text{bit/s/Hz}) \equiv \frac{C(\text{bit/s})}{\text{BW(Hz)}} = \log_2(1 + \mathrm{SNR}) \qquad (10.2.11)$$

see Eq. (1.3.24). It follows from (10.2.11) that when SNR $= P_{\text{signal}}/P_{\text{noise}}$ is 1, $\mathrm{SE}_{\max}$ equals one too; when SNR is zero, the $\mathrm{SE}_{\max}$ is, obviously, zero likewise. But what if $0 < \mathrm{SNR} < 1$? What if SNR > 1000? For deeper insight into relationship between SE and SNR, we need to consider a digital SNR, E_b/N_0. (Refer to Section 1.3, where we start discussing this topic.)

From the definition of SNR, we can derive the following formula:

$$\mathrm{SNR} = \frac{P_b(\text{W})}{P_N(\text{W})} = \frac{E_b(\text{J}) \cdot \text{BR(b/s)}}{N_0(\text{W/Hz}) \cdot \text{BW(Hz)}} = \frac{E_b(\text{J})}{N_0(\text{W/Hz})} SE(\text{b/s/Hz}) \qquad (10.2.12)$$

where, as usual, E_b (J) is energy of a bit and N_0 (W/Hz) is noise spectral density. For $\mathrm{SE}_{\max}$, we find from (10.2.11) and (10.2.12)

$$\mathrm{SE}_{\max}(\text{b/s/Hz}) = \log_2\left(1 + \frac{E_b(\text{J})}{N_0(\text{W/Hz})}\mathrm{SE}_{\max}(\text{b/s/Hz})\right) \qquad (10.2.13)$$

as in (1.3.14b). Obvious manipulations enable us to obtain

$$\frac{E_b(\text{J})}{N_0(\text{W/Hz})} = \frac{2^{\mathrm{SE}_{\max}} - 1}{\mathrm{SE}_{\max}} \qquad (10.2.14)$$

which is (1.3.14c) in another form. Remember that SE (b/s/Hz) and $\frac{E_b(\text{J})}{N_0(\text{W/Hz})}$ are, in fact, the dimensionless characteristic.

For BR (b/s) $< C$ (b/s), we introduce an actual spectral efficiency, SE_{act} (b/s/Hz) $=$ BR(b/s)/BW(Hz) and obtain

$$\frac{E_b(\text{J})}{N_0(\text{W/Hz})} > \frac{2^{\mathrm{SE}_{\text{act}}} - 1}{\mathrm{SE}_{\text{act}}} \qquad (10.2.15)$$

Equation (10.2.15) is considered as Shannon's law for digital transmission.

Using (10.2.14), we build Figure 10.2.8, which is a modified version of Figure 1.3.2.

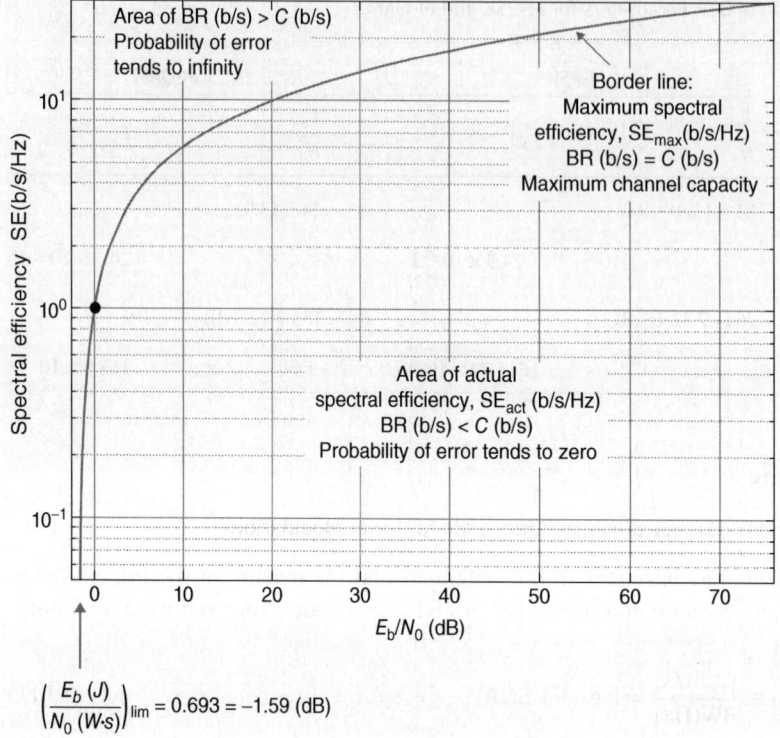

Figure 10.2.8 Spectral efficiency vs. digital SNR.

Now, we can investigate the behavior of SE when E_b/N_0 becomes less than 1. Letting BW $\to \infty$ in (10.2.15), it can be shown (Proakis and Salehi 2014, p. 366 and Sklar 2001, pp. 136–137) that

$$\lim_{\text{SE} \to 0} \left(\frac{E_b}{N_0} \right) = \ln 2 \approx 0.693 \to -1.6 \, \text{dB} \tag{10.2.16}$$

Since this value is calculated from the Shannon theorem, it does not depend on technology and can be considered as a *fundamental limitation for any digital communications system under the conditions which the Shannon theorem is to meet*. Condition (10.2.16) is often called the ultimate *Shannon limit* or *Shannon's power efficiency limit*.

Spectral Efficiency and Level of Modulation, M In this section, we continue to stress that the main advantage of M-ary signaling is increasing the spectral efficiency of communications systems. But how much the multilevel modulation can increase the spectral efficiency? This question is well discussed in the general theory of digital communications. (For detail, see Sklar 2001, pp. 219–236, Proakis and Salehi 2008, pp. 226–230, and Madhow 2008, pp. 272–277.) Yes, the SE (b/s/Hz) increases with the increase of M but tends to saturate with the growth of digital SNR, $\frac{E_b(J)}{N_0(\text{W/Hz})}$. It can be shown (Proakis and Salehi 2008, p. 228) that

$$\text{SE (b/s/Hz)} = \frac{\text{BR(b/s)}}{\text{BW(Hz)}} = \frac{2 \log_2 M}{N} \tag{10.2.17}$$

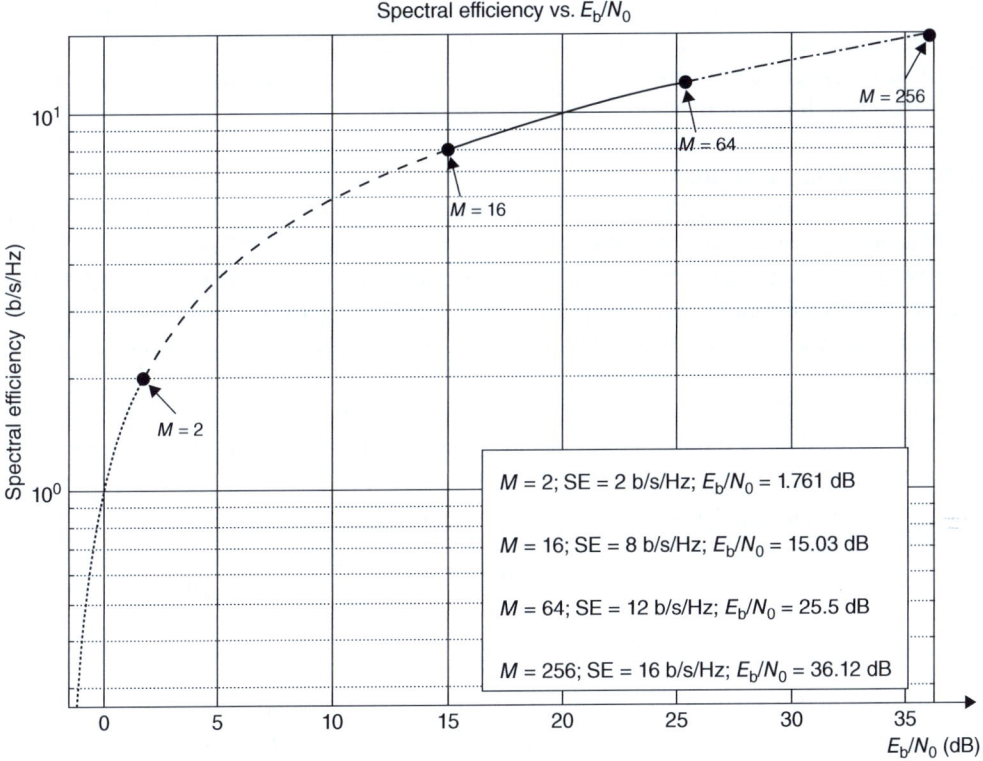

Figure 10.2.9 Spectral efficiency vs. digital signal-to-noise ratio for various modulation levels.

where M is the level of modulation and N is the dimensionality of the signal space. For QAM, for example, $N = 2$ and (10.2.17) turns to

$$\text{SE (b/s/Hz)} = \frac{\text{BR (b/s)}}{\text{BW (Hz)}} = 2 \log_2 M \tag{10.2.18}$$

as can be directly derived from the Hartley law (for Hartley law see Sections 1.3 and 10.1).

Combining (10.2.18) and (10.2.14), we can build the graph of spectral efficiency, SE (b/s/Hz), vs. digital SNR, $\frac{E_b(J)}{N_0(W/Hz)}$, for various modulation levels, M (Mynbaev 2016). This graph, shown in Figure 10.2.9, demonstrates that with an increase in M, the SE does not increase linearly but shows a tendency to saturation. This is because an increase in M causes the increase in the required $\frac{E_b(J)}{N_0(W/Hz)}$, as Figure 10.2.5 shows. In turn, the increase in digital SNR limits the growth of SE by restricting BR. Fundamentally, the limitation in an increase in SE with the increase in digital SNR stems from the *bandwidth-power trade-off*, which is one of the general telecommunications principles (see Section 1.3).

Equation (10.2.17) works for bandlimited communication systems, where bandwidth efficiency is a priority. For orthogonal signaling systems, such as orthogonal frequency division multiplexing, OFDM, discussed in Appendix 10.3, $N = M$ and (10.2.17) becomes

$$\text{SE (b/s/Hz)} = \frac{2 \log_2 M}{M} \tag{10.2.19}$$

Clearly, for these systems, the increase in M results in a steep decrease in SE; that is in their bandwidth efficiency. In these systems, however, the increase in M increases power efficiency, and therefore, the orthogonal signals should be employed in communication systems with limitation in power but with large bandwidth. Optical communication systems are the example of such systems.

The choice between bandwidth-efficient and power-efficient communication systems is, again, based on bandwidth-power trade-off, which we consider next.

10.2.3.2 Bandwidth-Power Trade-off

(We recommend reviewing Section 1.3.2, where this principle is discussed. Here, we take a deeper look at the principle based on information presented in the 10 chapters of this book.)

Since we are packing many bits into one symbol to increase the spectral efficiency, SE (b/s/Hz), this symbol needs more energy (or power) to be transmitted than does an individual bit. (For simplicity's sake, we assume that symbol energy is the sum of the energy of all its bits in the future considerations.)

Regarding the use of either power or energy, refer to Example 9.2.3, where we discussed the BER of the ASK modulation. To remind, energy in joules is equal to power in watts times the time of the operation; therefore, the SNR of a bit in energy terms takes the form

$$\text{SNR} = \frac{\text{bit power, } P_b(\text{W})}{\text{noise power, } P_N(\text{W})} = \frac{\text{bit energy, } E_b(\text{J}) = P_b(\text{W}) \cdot T_b(\text{s})}{\text{noise spectral density, } N_0(\text{W/Hz}) = \frac{\text{noise power, } P_N(\text{W})}{\text{bandwidth, } \text{BW(Hz)}}}$$

$$= \frac{E_b(\text{J})}{N_0(\text{W/Hz})} \qquad (10.2.20)$$

as was obtained in (9.2.17). Remember, we call the ratio $\frac{E_b(\text{J})}{N_0(\text{W/Hz})}$ a *digital SNR*.

Thinking about SNR, we need to recall from Section 1.3 its role in Shannon's law and *power-bandwidth trade-off* that follows from this law. Rewriting Shannon's law as given in (1.3.21),

$$C \text{ (b/s)} = \text{BW (Hz)} \log_2\left(1 + \frac{P_S(\text{W})}{P_N(\text{W})}\right) \qquad (1.3.21\text{R})$$

we can clearly see that the required bandwidth, BW (Hz), and needed signal power, P_S (W), must counterbalance each other in order to satisfy Eq. (1.3.21), where C (b/s) is a constant (limit) and P_N (W) is a given number.

In multilevel modulation, however, there is another aspect of the power-bandwidth trade-off caused by using pulses with various amplitudes. Consider, for example, transmission for one second two messages each consisting from 64 bits. The first message uses a *binary coding* with T_b (s) being its pulse duration. Assuming that 1s and 0s are equiprobable and letting the binary pulse amplitude to be 0.33 of a carrier amplitude A (V), we can estimate normalized signal power of this message as $P_{\text{NS}}^{\text{binary}}(\text{W}) = (0.33A)^2 \times 32 = 3.488$. The second message employs 16-QAM signaling shown in Figure 10.2.4 with T_S (s) = $4T_b$. From the table given in Figure 10.2.4d, we find that there are also 32 bits of logic 1. Calculating normalized power of individual bits as $(0.33A)^2$, $(0.75A)^2$, and $(1.0A)^2$ and summing these numbers up for the entire power, we compute $P_{\text{NS}}^{\text{QAM}}(\text{W}) = 16.186$. Thus, 16-QAM signaling requires much more power for signal transmission; specifically, $P_{\text{NS}}^{\text{QAM}}(\text{W})/P_{\text{NS}}^{\text{binary}}(\text{W}) = 4.64$.

On the other hand, since we pack four bits in one symbol, we transmit only 16 symbols whose duration, T_S (s), is four times greater than bit duration, that is T_S (s) = $4T_b$ (s). This means that the required bandwidth, $\text{BW}_{\text{QAM}} = 1/T_S$ (s), is four times less than that required for binary transmission. Therefore, feeding more power to 16-QAM transmission, we simultaneously save the bandwidth needed for this transmission. This is an example of power-bandwidth trade-off in M-ary modulation (see Lathi and Zhi 2009, pp. 8–10 for another example of this trade-off).

(**Question**: Consider the principle of power-bandwidth trade-off discussed above: What conditions (assumptions) are involved in this principle?)

This principle was introduced in Section 1.3 and brought up again in various forms throughout the course of this textbook. Review carefully the examples of this principle.

In general, an increase in spectral efficiency – and thus a saving in bandwidth – will always require additional technological resources, that is more complex, more advanced, and more costly equipment.

10.2.3.3 Applications of Multilevel Signaling

Today, practically all digital transmission systems, both wireless and wireline, rely on multilevel transmission. This is because M-ary modulation increases the spectral efficiency, SE (b/s/Hz), one of the most valuable transmission parameters. This advantage of M-ary signaling is extremely important for wireless transmission, where bandwidth is scarce. Since all types of mobile communications – Wi-Fi, Wi-Max, cellular telephony, wireless local area networks (WLANs), and all others – rely on wireless transmission, you can easily imagine the importance of multilevel signaling in modern communications.

What's more, since wireline transmissions today also face severe bandwidth limitations, this technology, too, has to turn to multilevel signaling. Example 10.1.1 demonstrates how multilevel modulation has helped in providing digital transmission over traditional telephone lines whose bandwidth is severely limited. Today, optical communication industry, which traditionally used only On–Off keying due to the vast amount of available bandwidth, has to employ M-ary modulation in its latest development called *coherent transmission*. Optical communications use today up to 1024-QAM signaling and add polarization multiplexing to further increase the spectral efficiency. Recently developed probabilistically shaped QAM technique varies the signal to match the characteristics of the fiber. This results in a dramatic increase in transmission rate and a decrease in power per bit consumption. In addition, this technique can bring optical fiber's transmitting capacity very close to the *Shannon limit*, discussed in Section 1.3.

Since almost every application uses a different version of multilevel modulation, it is impossible to discuss every specific transmission system. Nevertheless, the fundamentals of M-ary modulation presented here will enable you to quickly master every practical application of this theory.

Questions and Problems for Section 10.2

- Questions marked with an asterisk require a systematic approach to finding the solution.
- Many questions and problems, including those marked with an asterisk, imply that you, in addition to reading the textbook, will do your research to find the answers. Consider such questions as mini-projects.

Multilevel (*M*-ary) PSK

1   *Think about multilevel modulation in which nine bits are mapped in one symbol:
    a) How many symbols will we need?
    b) How would you designate this signaling?
    c) What will be the carrier's phase increment?
    d) Is it possible to implement such a phase increment in practice?

2 Sketch signal space and constellation diagram of 16-ary PSK. (Consider Figure 10.2.1 as a sample.)

3 Write the general formula for a symbol of 16-ary PSK, $v_{k16PSK}(t)$. Write formulas for specific symbols, $v_{4\text{-}16PSK}(t)$ and $v_{7\text{-}16PSK}(t)$, if $A = 5$ mV, and $f_C = 10$ MHz.

4 An 8-ary PSK is transmitting at BR = 1 Gb/s. What is its first-null bandwidth?

5 Compare the spectral efficiencies of a binary PSK and an 8-ary PSK if both transmit at BR = 1 Gb/s.

6 *Think about 128-ary PSK:
   a) Calculate the carrier phase change between its adjacent symbols.
   b) What is the distance between the medians of two adjacent symbols, D_{128PSK}, if each median is 3 mV?
   c) What is the ratio of D_{128PSK} to D_{64PSK}?

7 *Consider BER of 64-ary PSK:
   a) Derive the formula of BER of this signaling.
   b) Calculate the BER of this signaling.
   c) Compare BERs of QPSK and 128-PSK. Explain why these BERs are so different.
   d) How much additional energy do we need to supply to one bit of 128-ary PSK compared to that of a QPSK to sustain the same BER?

8 *Examine Figure 10.2.3:
   a) What must a bit energy, E_b (J), be for 64-ary signaling if its N_0 is 2×10^{-10} (W/Hz) and the required BER is 10^{-8}?
   b) For any BER other than 10^{-1}, the E_b/N_0 ratio required to support the given BER consistently increases with an increase in the number of symbols, M. Why?
   c) For BER = 10^{-8}, all the signaling from $M = 4$ to $M = 128$ are separated by approximately 5-dB intervals in the E_b/N_0 ratio. Is there any reason for this interval constancy?

9 In M-ary signaling we need, on the one hand, to decrease the distance between the adjacent symbols to accommodate more symbols in a signal space and, on the other hand, keep this distance as large as possible to support the required BER. How can we satisfy these two contradictory conditions?

Quadrature Amplitude Modulation, QAM

10 What do QPSK and QAM signaling have in common and what is the difference between them?

11 *Examine Figure 10.2.4 where 16-QAM parameters are shown:
   a) In 16-PSK signaling, as Figure 10.2.4b exemplifies, the distance between the adjacent symbols is the same for all symbols. In 16-QAM modulation, the distances between adjacent symbols vary. Why?
   b) Compare I and Q projections of the symbol amplitudes in 16-PSK and 16-QAM signaling. What conclusion can you make?

c) In Figure 10.2.4c, calculate the distances between adjacent symbols **1000** and **1011** and between **1001** and **1011**. Why are they different?

d) Calculate the distances between symbols **0000** and **0100** and between **0000** and **0010**. Your comments?

e) What is the minimum distance and what is the maximum distance among symbols in 16-QAM constellation diagram in Figure 10.2.4c?

f) Do you want a large or small minimum distance between adjacent symbols in any signaling? Explain.

12 Calculations show that d_{min}(16QAM) is 1.7 times greater than that of d_{min}(16PSK), which is considered an advantage of 16-QAM signaling. What is this advantage?

13 *Inspect Figure 10.2.5:

a) Approximately calculate the required digital SNR (E_b/N_0) for 2-QAM and for 128-QAM in absolute numbers. Why does 128-QAM need greater SNR? How can it be physically provided?

b) Compare this figure with Figure 10.2.3: To sustain BER $= 10^{-10}$, 16-PSK needs $E_b/N_0 \approx 32$ dB, whereas 16-QAM needs $E_b/N_0 \approx 18$ dB. Why is the difference? Which signaling is advantageous? Explain.

14 *Consider the effect of noise on the M-ary signaling:

a) What kind of noise – internal or external – do we need to take into account?

b) Figure 10.2.6a shows an experimental constellation diagram. Explain how the effect of noise can be seen in this diagram.

c) In a constellation diagram, noise transforms a focused, clear symbol dot into a fuzzy, cloudy spot, as shown in Figure 10.2.6b. What mechanism causes this transformation? Explain and sketch a figure to visualize your explanation.

d) Which signaling – M-PSK or M-QAM – performs better in noisy condition? Why?

15 *Inspect Figure 10.2.7:

a) Write the formulas for all 16 symbols in the first quarter of the constellation diagram shown in Figure 10.2.7a. (*Hint*: See Eq. (10.2.2).)

b) Demonstrate how the eye diagram shown in Figure 10.2.7b is formed by bit-to-bit transitions. (*Hint*: See Figure 10.1.6d.)

c) Why does the eye diagram of 64-QAM signaling presented in Figure 10.2.7b show seven openings? What is the height of an individual opening compared to 2-QAM signaling? (*Hint*: See Figure 10.1.6d. Do your search online.)

16 *Compare error performance of 32-PSK and 32-QAM signaling if the required bit error rate is BER $= 10^{-10}$ and the given noise spectral density is $N_0 = 10^{-10}$ W/Hz:

a) Find bit energy, E_b, and digital SNR, E_b/N_0, for each signaling.

b) Compare your results with the appropriate E_b/N_0 values given in Figures 10.2.3 and 10.2.5. If significant discrepancies are found, explain the possible sources of these discrepancies.

c) Based on the results, conclude which signaling performs better and explain why.

d) Refer to Example 10.2.2, insert your numbers in Table 10.2.3 and perform the comparative analysis of BER for the M-PSK and M-QAM signaling in three cases: $M = 4$, $M = 32$, $M = 64$.

17 Figures 10.2.3 and 10.2.5 demonstrate that the greater the number of bits assigned to each symbol, the higher the SNR needed to sustain the required BER. How would you quantify this statement?

Final Thoughts

18 *Given a communication system whose maximum transmission capacity is $C = 100$ Tb/s and available channel bandwidth is BW = 8.3 THz:
   a) What is the maximum spectral efficiency of this communication system?
   b) What signal-to-noise ratio, SNR, must this system maintain to keep this SE_{max}?
   c) How many modulation levels, M, are needed to support this SE_{max}? How many bits per symbol, N_b?
   d) If actual SE of the given system is $SE_{act} = 10$ b/s/Hz, what digital SNR is needed?

19 *Shannon's law states that C (b/s) = BW (Hz) $\log_2(1 + SNR)$. It would seem that if channel bandwidth tends to infinity, BW $\to \infty$, we can obtain infinite transmission capacity, C (b/s). In fact, we can't. Why not?

20 Equation (10.2.16) states that digital SNR, E_b/N_0, must be greater than -1.6 dB to support error-free communications:
   a) Since Shannon's law, C (b/s) = BW (Hz) $\log_2(1 + SNR)$, determines the condition for error-free communications, (10.2.16) can be derived from the Shannon formula. Do it. (*Hint*: Express SNR in Shannon's law as E_b/N_0 as shown in (10.2.12) and tend BW to infinity, see Sklar 2001, pp. 136–137.)
   b) Show how this condition is implemented in Figure 10.2.8.

21 The text states that P_{sig} (W) × BW (Hz) = constant, where P_{sig} (W) is the signal power and BW (Hz) is its bandwidth. How can you justify this statement?

22 Consider a 64-QAM transmitted at 100 Gb/s: If you use 128-QAM, you can increase the bit rate but use the same bandwidth. What price do you have to pay for this improvement?

23 Prove that normalized signal power of 64-QAM signaling is equal to 16.186 W. (See Figure 10.2.4. and Section "Final Thoughts.")

24 Multilevel modulation, as shown in this section, has a number of advantages over binary signaling, the main of which is increasing the system spectral efficiency. We know, however, that in a closed system an advantage in one aspect (or parameter) can be achieved only by additional contribution in another aspect of the system. This contribution can be power, time, system complexity, or anything else. What contribution into a multilevel modulation system has to be made to increase SE (b/s/Hz)?

25 *Today, both wireline and wireless communications rely on multilevel modulation:
   a) Why?
   b) Can you tell which specific M-ary modulation is the most popular and for what reason?

26 *Name and explain at least five areas of application of multilevel modulation in modern communications. (*Hint*: Go online.)

10.A Multiplexing

10.A.1 Multiplexing: Definition and Advantages

This appendix discusses the main multiplexing methods in a concise format.

Multiplexing is a technique enabling us to send many messages (data streams) across a single transmission channel. Put it more generally, multiplexing allows for sharing the resources of a communication system among many users. For example, even though an apartment building is typically connected to a central office by a single telephone cable, every apartment has its dedicated telephone service because all the telephone conversations are multiplexed and thus can be transmitted over the same transmission line. Another example is broadcasting radio: The programs from all stations are delivered to our radio receiver simultaneously; we choose one by tuning the radio set. The concept of multiplexing is depicted in Figure 10.A.1.1.

Every signal, as Figure 10.A.1.1 shows, is generated by its transmitter, Txi, and presented to a multiplexer, MUX; the latter unit combines all signals into one information stream. The stream propagates through a transmission link and reaches a demultiplexer, DEMUX, where every signal is separated and presented to its intended receiver, Rxi.

If all signals travel as a combined stream over one channel, how can they proceed without interference? There is a number of different multiplexing techniques, each based on using one of the signals' distinguishable properties. The main of these techniques employ *time-based* and *frequency-based* principles.

Figure 10.A.1.2 conceptually shows the idea of time-based (a) and frequency-based (b) principles of multiplexing. There are five messages in this example. For time-domain multiplexing, all messages occupy the same frequency band, f_2-f_1, but every signal is placed into its specific time slot. (In fact, they can hold different frequency bands; it would not change the concept in a discussion.) For frequency-domain multiplexing, messages are distinguished by their frequencies, whereas all of them may have been transmitted simultaneously, over a common time interval, t_1-t_2.

In addition to time-based and frequency-based principles, multiplexing can be done by assigning different parameters to multiplexed signals (*coding*) and can be based on *spatial (space)* distinction.

Why do we need multiplexing? In some cases, multiplexing is the only means to provide communication service to users. Consider an example with telephone service delivered to a building with the hundreds of apartments: How can a telephone connection be provided to every apartment without multiplexing? By connecting every apartment to a telephone building with a dedicated cable? Obviously, in this situation, multiplexing is the only solution to the problem. In other cases, multiplexing greatly improves the characteristics of existing communication systems. Consider, for instance, a submarine fiber-optic cable connecting American and European shores. Now imagine that the volume of information to be transmitted starts to exceed the cable capacity. There are two

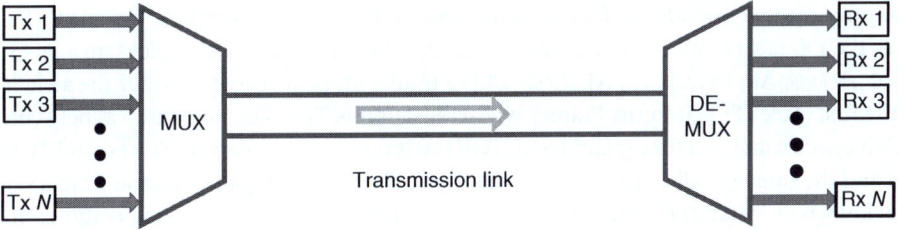

Figure 10.A.1.1 The concept of multiplexing (Tx, transmitter; Rx, receiver; MUX, multiplexer; and DEMUX, demultiplexer).

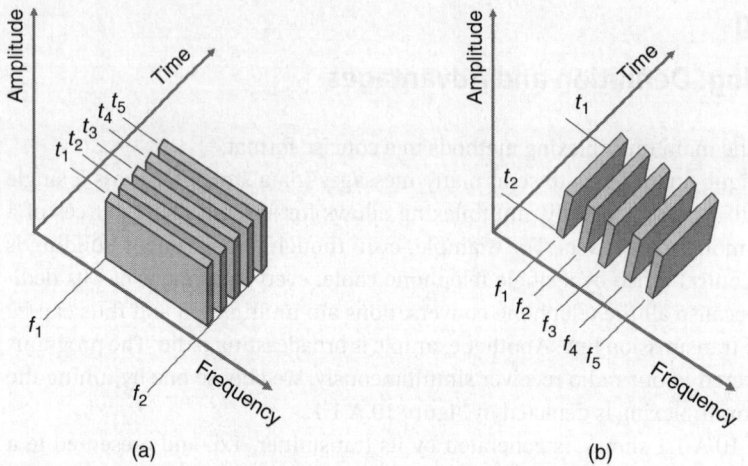

Figure 10.A.1.2 (a) Time-based and (b) frequency-based principles of multiplexing.[1]

possible solutions to this problem: Either to lay out a new submarine transatlantic cable, bearing all the cost and business troubles associated with such an undertaking, or to install several additional transmitters, receivers, multiplexers, and demultiplexers on both end sides of the existing cable. The choice is obvious. We will elaborate on the advantages of multiplexing in the sections that follow.

10.A.2 Time-Based Multiplexing Principles

The principle of signal multiplexing based on distinguishing them by time characteristics is called *time-division multiplexing, TDM*. We consider two main TDM versions, *synchronous TDM (sync-TDM)* and *statistical (asynchronous) TDM (stat-TDM)*.

(In literature, some authors use acronym *STDM* for synchronous TDM, and the others apply the same acronym to statistical TDM. To avoid confusion, we use *sync-TDM* for *synchronous* TDM and *stat-TDM* for *statistical (asynchronous)* TDM.)

Many terms and principles of synchronization that we discuss in this appendix were considered in Appendix 4.A.2 "Modes of digital transmission" and especially in Section 4.A.2.4 "The need for synchronization in digital transmission." We urge you to review this material.

10.A.2.1 Synchronous Time-Division Multiplexing, sync-TDM

Sync-TDM is a straightforward solution to the multiplexing problem: Allocate a fixed time slot to each signal for transmission and send the signals in a cogent sequence. Figure 10.A.2.1a presents the conceptual view of a sync-TDM system.

This conceptual system operates as follows: The independent transmitters TxA, TxB, TxC, …, TxN generate their data streams. A switch, rotating at a constant angular velocity, picks up a block of data called *packets,* Ak, Bk, Ck, …, Nk, from all the transmitters in sequence. After the switch completes the first cycle, TF1, it forms Frame$_1$ whose duration is TF1. The time interval between two consecutive cycles makes a time guard band, TGB (s). It is clear that intervals TFk (s) and TGB (s) are fixed and the same for all transmitters. (This approach to collecting information from a set of sources is known as *round robin* algorithm.) The structure of an individual frame is shown in Figure 10.A.2.1a (bottom).

---
1 After Stallings 2014, p. 239.

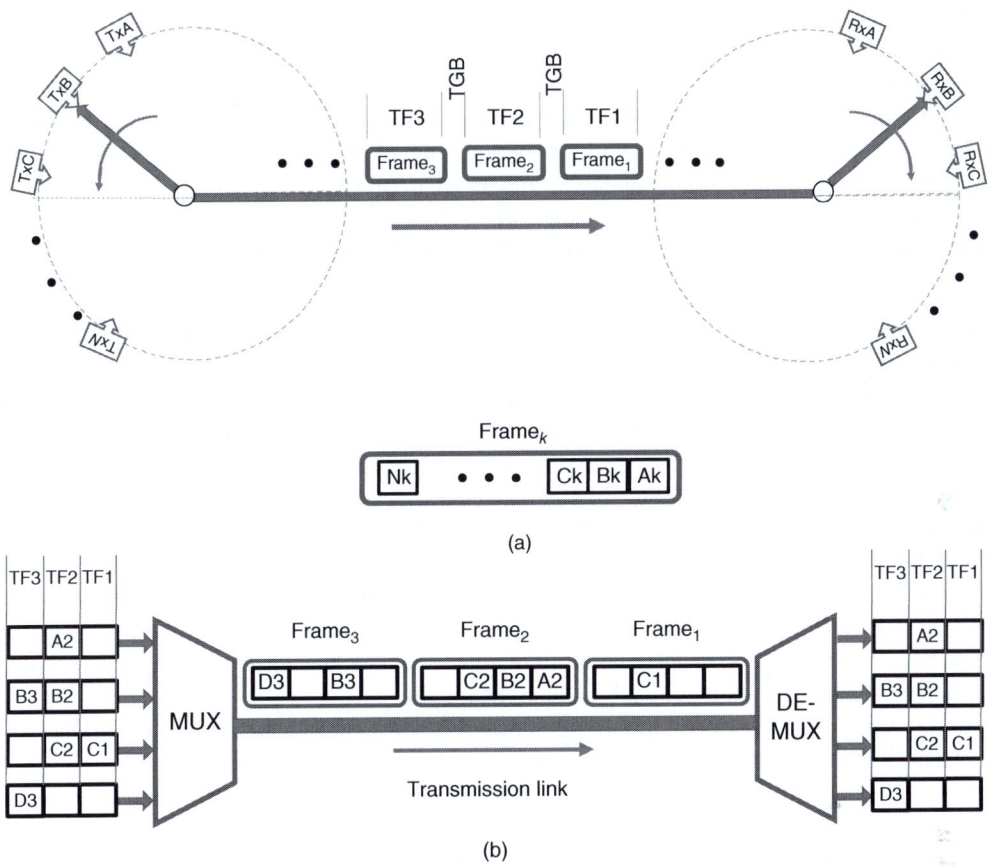

Figure 10.A.2.1 Synchronous time-domain multiplexing: (a) conceptual view and (b) transmission in sync-TDM (TFk, frame duration of k-th cycle, TGB, time guard band).

Thus, the data packets crammed into frames Frame$_1$, Frame$_2$, Frame$_3$, ..., Frame$_N$ travel across a single transmission link and do not interfere (despite being possibly in the same frequency range) because they occupy the different time slots. At the receiving end, the switch rotating in sync with the transmitter switch delivers the packet from a frame to the relevant Rx. As an example, Figure 10.A.2.1a shows that when packet Bk is sent from TxB, the receiver switch delivers this packet to receiver RxB, where k is the number of a cycle. A predetermined propagation delay is included, of course, in sync timing. This is how a sync-TDM system operates in principle.

Modern communication systems do not contain any rotating switches, of course, and digital electronics perform all operations. This modern approach is schematically shown in Figure 10.A.2.1b. This figure also demonstrates an example of how frames in sync-TDM are formed. For cycle TF1, a switch picks up in sequence two empty slots from TxA and TxB, one available packet, C1, from TxC, and another empty slot from TxD. These empty slots and the data packet form Frame$_1$. For cycle TF2, the switch picks up packets A2, B2, and C2 from their sources and makes Frame$_2$ with one empty slot for TxD. Finally, Frame$_3$ is composed of B3 and D3 and two empty slots from TxA and TxC.

At the receiving end, the operations are the reversal. At the first cycle TF1, the switch consequently delivers the empty slot to RxA, another empty slot to RxB, packet C1 to RxC, and another empty slot to RxD. You can easily trace delivery of all other data packets during cycles TF2 and TF3.

This is how sync-TDM operates.

(**Question**: *Why this type of TDM is called "synchronous?" How can we guarantee that receiver RxC gets packet C3?*)

Sync-TDM is a rather direct method of multiplexing; no wonder it has been used since the dawn of electrical communications, that is with the proliferation of telegraphy. However, it became the common multiplexing method only in the early 1960s when conventional telephony, then the main type of communications, turned from analog to digital format (see Sidebar 10.A.2.1.S).

There are, unhappily, some disadvantages of the sync-TDM method. The main of them are as follows:

- Sync-TDM is a serial transmission because, as Figure 10.A.2.1 shows, the data frames are transmitted one after another. The serial transmission is technologically restricted in its maximum bit rate by the minimum time interval because of BR (b/s) = $1/T_b$ (s). For today's bit rate readily reaching terabit-per-second range, sync-TDM would require a bit duration in the order of T_b (s) = 10^{-12} s. Unhappily, the direct generation of such short communication pulses is an arduous task. Compression and interleaving the frames enable system designers to reduce bit time intervals and thus alleviate the restriction on bit rate caused by the pulse duration, but the inherent limitation of a serial transmission remains.
- Sync-TDM requires high precision of timing to perform its task successfully. (Refer to Section 4.A.2.4 for the detailed discussion of timing requirements in digital communications.) If the duration of frame TFk (s) and TGB (s) unintendingly change, the adjacent frames can collide, which results in corruption of signals being transmitted.
- For a given sync-TDM system, the frame duration (size) is fixed. This one-size-fits-all feature creates difficulty in meeting the variety of demands of individual end users.
- The number of end users are fixed and must be known ahead of time; we cannot add or eliminate users for a given sync-TDM system, regardless of the business needs.
- Finally, sync-TDM utilizes its bandwidth inefficiently, as Figure 10.A.2.1b shows: Even if TxK does not have data to transmit, its idle slot remains taken; thus, a transmission link has a spare capacity but cannot use it even though other users might have data to transmit. This waste of bandwidth is considered to be the main sync-TDM problem.

Sync-TDM is intrinsically digital technology, and this is its main advantage. Thanks to this benefit, sync-TDM is still widely used by all branches of modern communications. Sidebar 10.A.2.S.1 exemplifies sync-TDM applications.

Sidebar 10.A.2.S Two sync-TDM Systems: T and Synchronous Optical Network (SONET)

This sidebar briefly discusses two synchronous TDM industrial systems – T and SONET.

10.A.2.S.1 T system

Introduction to T system

For many years following its invention in 1876, the wireline telephone was the principal means of telecommunications. Voice signal was transmitted in analog format, of course, and the Bell System used frequency-division multiplexing (FDM) for transmitting aggregate traffic over a long distance. As traffic volume grew, the need for new technology arose, and in 1962, the first digital transmission system, the so-called T system, was introduced. The T system samples analog voice signals at the 8000-Hz frequency. (Remember the Nyquist sampling frequency?

Then you know why 8 kHz is the voice sampling frequency.) Thus, each sample is taken during 1/8000 Hz = 125 μs. *This is where 125-μs time slot originated.* Bell System established that each sample in T system is mapped to eight bits. Consequently, the bit rate of an individual voice channel (vc) becomes equal to 8 (ksample/s) × 8 (bit/sample) = 64 kb/s. This signal is now called the *digital signal at level zero, DS-0,* and it is a building block of the entire T and SONET systems.

The first level of T-system hierarchy, T1, multiplexes 24 vc, picking eight bits (one sample) from each channel per cycle (frame). Thus, one frame carries 193 bits as the following computations show:

8 (b/vc) × 24 (vc/frame) + 1 (b/frame) = 193 (b/frame) where one bit is added for framing purposes. Since T1 transmits 8 kframes/s, its bit rate is given by

193 (bit/frame) × 8 (kframes/s) = 1.544 Mb/s

This signal is designated as *DS-1, which stands for the digital signal at level one.* These numbers characterize the T1 system. Again, it is important to remember that an individual frame is transmitted every 125 μs.

The higher levels of multiplexing hierarchy of the T system have been standardized too; the proper bit rates and signal designations are shown in Figure 10.A.2.S.1.

Observe that even DS-1 requires a higher bit rate than the simple multiplication (24 voice channels × 64 kb/s) shows because DS-1 adds one service bit to each frame at this level. The next stages of multiplexing add more service bits for framing, transport, and – mainly – synchronization purposes. This is why the transmission rates of the DS-1C, DS-2, DS-3, and DS-4 signals are not simple multiples of the bit rate of the DS-1 signal.

Important reminder: *The service bits are called the overhead, whereas the user data are called the payload* (data field, or information field). Since service bits and payload are carried over time, we count them in bits per second, b/s.

T System: First Problem

When the T system was introduced, it seemed that it would be able to transmit traffic at any desirable bit rate. And the T system did indeed work well – for a time. However, as the need to handle a higher level of traffic aggregation arose, the inefficiency of the T system became increasingly evident.

As Figure 10.A.2.S.1 shows, the amount of overhead for the DS-1, DS-3, and DS-4 signals increase faster than the transmission speed. In percentage to the total traffic, the overhead is equal to 0.52%, 3.86%, and 5.88%, respectively. These calculations mean that the useful payload of the DS-1, DS-3, and DS-4 signals drops from 99.48% to 97.34% to 96.14% to 94.12%, respectively. In other words, the higher the transmission speed, the less amount of useful information the T system can deliver in the total bit stream.

The increase in overhead percentage with an increase in transmission speed is the first of the major drawbacks of the T system. Indeed, if we were able to increase the transmission speed of this system to 10 Gbit/s, the overhead would occupy almost 30% of its transmission capacity.

(Continued)

Sidebar 10.A.2.S (Continued)

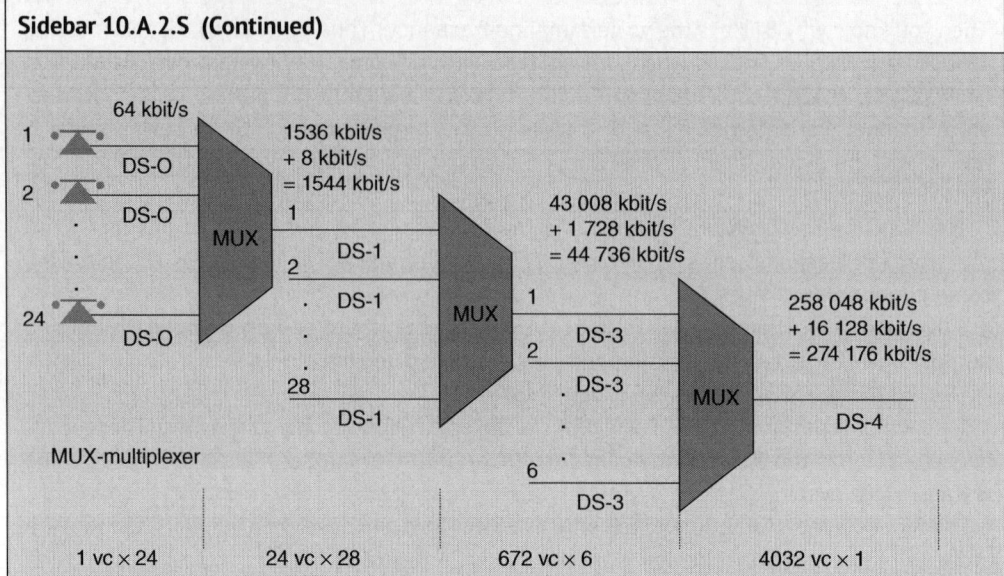

Figure 10.A.2.S.1 Multiplexing hierarchy of digital signals of T system.

(Question: Why does overhead percentage grow with the increase in data rate in the T system? Hint: See Section 10.A.2.S.2.)

T System: The Second Problem

Since a higher level of a T system results from multiplexing channels from a lower level, the access to the lower-level channel requires demultiplexing the entire transport unit. For example, to extract an individual voice channel at 64 kb/s from T4 signal at 274.176 Mb/s, we have to demultiplex T4 to the set of 64-kbit/s units. This is the second problem that T system meets at high-speed transmission: The more voice channels multiplexed, the more difficult it is to extract a specific channel.

Figure 10.A.2.S.2 shows how a specific voice channel can be extracted from and inserted into a T3 bit stream. This procedure is called an *add/drop operation*. As an example, the graphic shows how we can extract voice channel #536. We can add a new voice channel to fill the bit stream up to full capacity or, if there is no voice channel moving in the same direction, we can leave this channel idle.

Though conceptually simple, in practice the add/drop operation faces a problem, which stems from the method of synchronization used by the T system. Similar to all digital systems, this system operates under the control of a clock, which means that all operations are synchronized within the systems. However, in reality, there is no absolute synchronization, and we always have some time misalignment among the different clocks of a network. The T system allows relatively large margins of clock misalignment among the different nodes. This is why it is called *the plesiochronous digital hierarchy (PDH)* system, where the term *plesiochronous* means "pseudo-chronous" or "almost chronous."

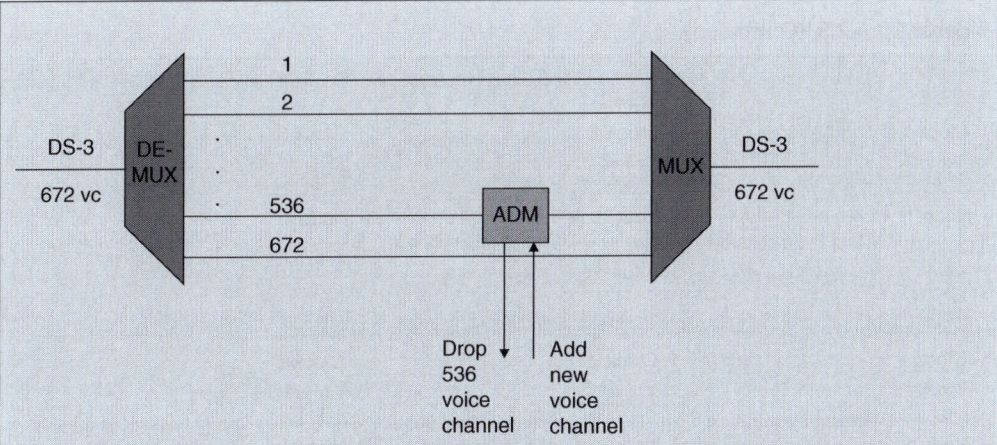

Figure 10.A.2.S.2 Add/drop multiplexing in the T system (MUX, multiplexer; DEMUX, demultiplexer; ADM, add/drop multiplexer).

10.A.2.S.2 Synchronization in Digital Telecommunications Networks

To reiterate, any digital circuitry works under the control of a clock because every pulse that represents a single bit has to be tied to the time scale. Therefore, timing is a necessary condition of operation of digital telecommunications networks. Since all digital circuitry that provides intelligent operations are located at network nodes, clocks must control nodes operation.

There are three types of timing control (synchronization) in data networks: Asynchronous, plesiochronous, and synchronous (see Figure 10.A.2.S.3). In asynchronous networks, a clock controls each node individually. All clocks operate independently, as shown in Figure 10.A.2.S.3a. Each clock is a very precise device; however, clocks have a finite accuracy, which means that over time they will accumulate difference in their readings. In other words, such a scheme inevitably results in independent timing operation of each node. Thus, even though each node works under the clock control, the entire network operates without general synchronization. This is why this type of network is called *asynchronous*.

Plesiochronous networks, as shown in Figure 10.A.2.S.3b, have the next level of synchronization. Timing information is transmitted along with data from one node to the other. Clocks extract this timing signal and use it to align their timing operation. Thus, a clock at one node is synchronized to the clock at the other node. However, this type of working provides synchronization only between two adjacent clocks. In the example shown in Figure 10.A.2.S.3b, Clock 1 is synchronized with Clock 2 and Clock 2 is synchronized with Clock 3, but Clock 1 is *not* synchronized with Clock 3. In other words, clocks of the entire network do not have the common reference, which implies that they are not entirely synchronized. This is why this type of network is called *plesiochronous*.

In the synchronous network, as shown in Figure 10.A.2.S.3c, all clocks operate under the control of one master clock, a clock with the highest achievable precision. Timing information is transmitted through a dedicated network, independent from a data transmission network. Thanks to this design, all clocks in an entire data network are governed by the same timing signal. In other words, all clocks have the same common reference. Ideally, such architecture

(Continued)

Sidebar 10.A.2.S (Continued)

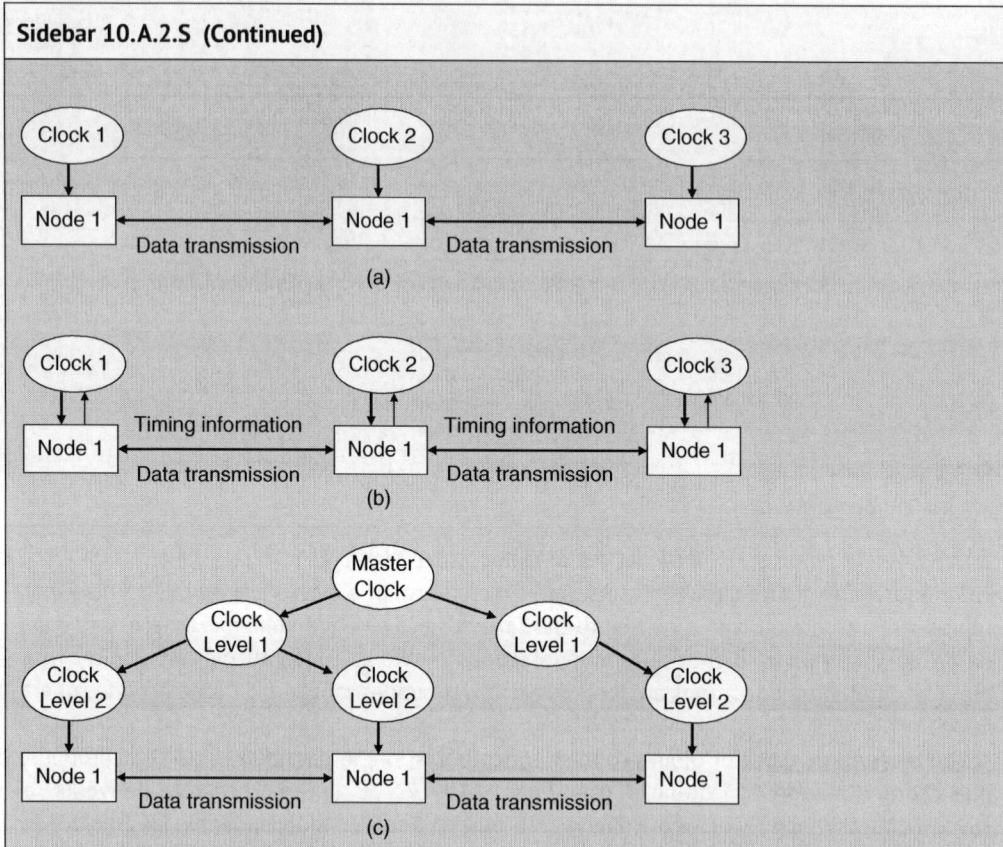

Figure 10.A.2.S.3 Types of synchronization in communication networks: (a) asynchronous; (b) plesiochronous; and (c) synchronous.

would result in complete synchronization of all network clocks. This is why this type of network is called *synchronous*. Practically, the finite precision of each component of the timing network and electromagnetic (EM) interferences result in some variations in readings of various clocks. However, this type of data networks provides the highest possible level of network synchronization.

10.A.2.S.3 SONET: Definition and Discussion
Two major needs spurred the development of new digital transmission technologies. The first involved described the aforementioned problems of using T system at high bit rates. The second was the advent of fiber-optic communications technology, which offered the unprecedented transmission capacity that cannot be used with the T system. Thus, the new system – SONET – was developed.

SONET stands for ***S**ynchronous **O**ptical **NET**work*. It is a standard that controls the way data are transmitted over a network. This feature implies that SONET controls, to some extent, the equipment that has to be used. The standard also specifies the architecture of the networks. Thus, we can say that SONET is a transmission system. Note that SONET's designation starts with the word "synchronous," which emphasizes the importance of this term.

SONET (first published in 1984) was developed by Bellcore (now known as Telcordia Technologies) in response, as mentioned previously, to the dire need for high-speed digital transmission stemming from the growth of traffic and stimulated by the advent of an optical fiber.

A hierarchy of digital signals in SONET is presented in Table 10.A.2.S.1:

Table 10.2.A.2.S.1 SONET digital hierarchy.

SONET		Bit rate (Mb/s)
Electrical signal	**Optical signal**	
STS-1	OC-1	51.84
STS-3	OC-3	155.52
STS-9	OC-9	466.56
STS-12	OC-12	633.08
STS-18	OC-18	933.12
STS-24	OC-24	1 244.16
STS-36	OC-36	1 866.24
STS-48	OC-48	2 488.32
STS-192	OC-192	9 953.28
	OC-768	39 813.12
	OC-3072	159 252.48

See, for example, Mynbaev and Scheiner (2001), p. 659.

STS stands for *synchronous transport signal*. The optical counterpart of STS-1 is called OC-1, where *OC* stands for *an optical carrier*. STS and OC have the same bit rate but differ in specific transmission formats. Note that STS-1 (or OC-1) is the basic building block of the SONET hierarchy, carrying 51.84 Mbit/s. It can accommodate one DS-3 signal. Compare the bit rates of SONET and T system to see the difference between these transmission systems (see Figure 10.A.2.S.1).

Multiplexing hierarchy in SONET is rather straightforward: The bit rate of each higher level is a multiple of the previous one and eventually can be tracked to STS-1. For example, STS-3 carries 155.52 Mbit/s, which is equal to 3×51.84 Mbit/s, that is $3 \times$ STS-1 bit rate. This hierarchy is in contrast to the T system hierarchy where the various number of overhead bits changes the transmission rate at every level of multiplexing. This is why a SONET is a *modular* system while the PDH system is called *nonmodular*.

In SONET, user data is called *synchronous payload envelope, SPE*, while service information is called *overhead*.

What if we need to transmit a signal at the bit rate lower than STS-1? SONET allows you to do that by using smaller transmission units called *virtual tributaries, VTs*.

SONET Framing
SONET frame consists of 9 rows and 90 columns. Each cell includes one octet (byte); therefore, each frame contains $9 \times 90 = 810$ octets (bytes) or 810 bytes $\times$ 8 bits = 6480 bits. The SONET frame's duration is 125 μs. (Remember where this number came from? See Section 10.A.2.S.1.)

(Continued)

Sidebar 10.A.2.S (Continued)

Each frame starts with two framing octets (bytes) called A1 and A2. These bytes serve for frame synchronization: When a receiver reads these bytes, it "understands" that this is the beginning of a frame. The framing bytes A1 and A2 must be the same for every SONET frame. They are **F628** in Hex or **1111 0110 0010 1000** in binary.

Bytes A1 and A2 must be provided in each STS-1 signal. Therefore, for the STS-3 signal, which is the triple multiplex of STS-1, there are three A1 and three A2 bytes.

Synchronous Multiplexing in SONET

SONET, as well as the T system, uses sync-TDM. However, the T system uses *bit-interleaving multiplexing*, which means that a single time slot is assigned to an individual bit. SONET uses *byte-interleaving multiplexing*, which means that a single time slot is allocated to an individual byte. It follows from the above explanation that SONET carries only signals at the predetermined bit rate.

In SONET, multiplexing is done synchronously that results in a fixed position of a tributary within a higher-level signal. True synchronous multiplexing is one of the key advantages of a SONET system since this type of multiplexing permits direct access to a lower-level signal without demultiplexing of the entire transport unit. This SONET's feature is in contrast to the T system, where, as we discussed previously, you need to demultiplex the entire transport unit to access a specific channel.

To achieve the required level of synchronization, the global timing system, whose block diagram is shown in Figure 10.A.2.S.3, has been built. The master clock of this timing system is the average time of national standard institutions of several developed countries – the absolute highest precision that can be achieved today. This signal is transmitted to the constellation of satellites. Each satellite carries an atomic clock whose precision is slightly lower than that of a master clock. (Still, we are talking about the accuracy of 10^{-16}, which means the probability of having one erroneous cycle out of 10^{16} total frequency cycles.) Satellites distribute their timing signals to the ground clocks usually located in telephone buildings (central offices). These clocks provide timing for SONET network equipment, thus supporting SONET synchronization described early.

This sidebar introduces the example of the industrial application of synchronous time-domain multiplexing technique. It is worth mentioning that, despite sharp declining in the use of landline telephone networks, both T and SONET systems are still widely employed. SONET, in particular, serves as a popular transport network thanks to all its advantages described previously. Besides, SONET can automatically recover after a failure of a segment of its transmission link or even its node.

10.A.2.2 Statistical (Asynchronous) Time-Division Multiplexing, stat-TDM

Statistical TDM is still a TDM system; it is intended to share a transmission link among many independent data streams over time. This type of TDM becomes invaluable when different sources generate to-be-multiplexed data streams at the various bit rate.

To clarify the difference between synchronous and statistical TDMs, let us compare their operations shown in Figure 10.A.2.2 (Figure 10.A.2.2a is a modified version of Figure 10.A.2.1b).

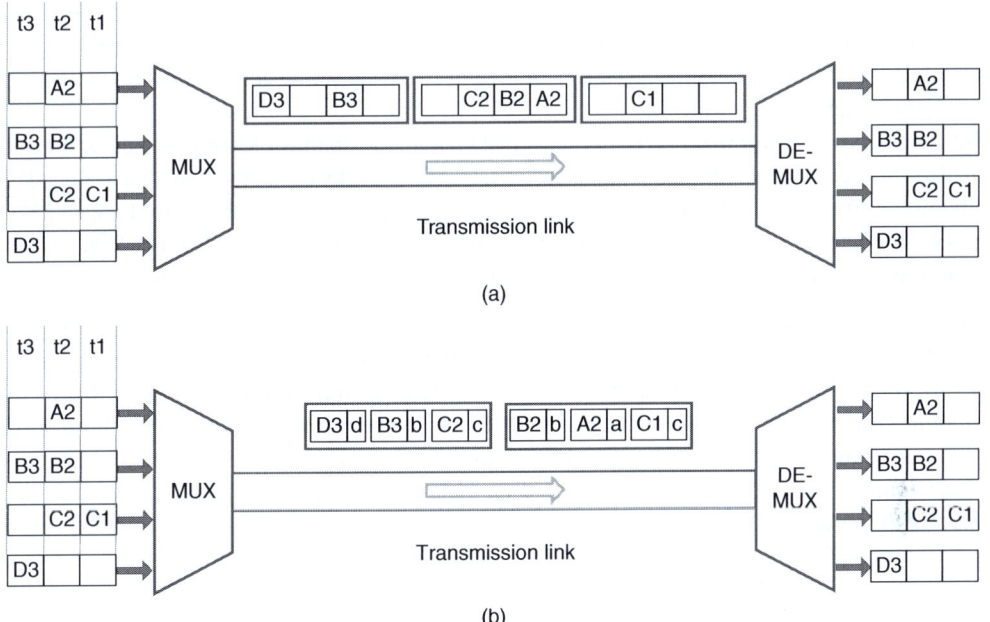

Figure 10.A.2.2 TDM transmission: (a) synchronous TDM and (b) statistical (asynchronous) TDM.

The glance at Figure 10.A.2.2 reveals that sync-TDM has to carry many empty slots, thus underutilizing the available channel bandwidth, whereas stat-TDM fully employs the capacity of a given channel by stuffing the frames without empty slots. How is this advantage achieved?

First and foremost, a stat-TDM is a sophisticated system whose multiplexer and demultiplexer are intelligent agents. As for the system's detailed description, its following distinguished features should clarify the system operation:

- The number of time slots on the frame of a stat-TDM is always fewer than the number of the input lines. In Figure 10.2.2b, there are three slots for the frame and four input lines.
- The system calculates the average data rate of the input streams to ensure that the channel is capable of supporting this rate.
- The multiplexer collects the input packets in the *round robin* fashion and assigns them to an available frame until the frame is filled. This way, the number of time slots allocated for every input line depends on the line's data rate. Figure 10.A.2.2b shows that the frames being transmitted contain four slots for lines B and C as each of them brings two packets for t1–t3 interval and only two slots for lines A and D because each of these lines presents only one packet over the same time interval. This arrangement means that a stat-TDM system allocates time slots on demand.
- If one line has a much higher data rate than the others, it can result in transmission the packets from this line only, while other lines must wait unspecified time. To avoid this situation, the multiplexer set an upper limit on the size of data flow (number of packets for a specified time interval) from each input line to make sure that every line can get its turn.
- If the multiplexer receives more data packets than it can accommodate for a given time, it buffers extra packets. Which packet from the buffer should be sent out first? The multiplexer decides

this on a packet-by-packet basis using a specific criterion. If the packets arrive faster than the multiplexer can process them, it will run out of the buffer space and will drop some packets.
- Thanks to all the above features, a stat-TDM system fills each frame completely, thus fully utilizing the channel bandwidth.
- Since there is no synchronization in the stat-TDM system, the demultiplexer has no other means to identify a received packet but the packet's address. This is why a stat-TDM must use a header containing the address of each packet. For example, Figure 10.A.2.2b shows that packets C1 and C2 each has a small attachment – a header designated by a lower-case letter c. This header guarantees that all input C packets after demultiplexing will be delivered to the output line C.

To sum up, a statistical TDM system sends input data not in a predetermined fixed order, as sync-TDM does, but on demand. This feature means that the frames are filled before being sent, which results in fully utilizing the capacity of a transmission link. Additionally, the principle of operation of the stat-TDM implies that data transmission is asynchronous because there is no need for slots and frames synchronization. This is why this TDM type is also called *asynchronous*.

(**Questions**: 1. Why is the number of time slots on the frame of a stat-TDM always fewer than the number of the input lines?
2. Why is this TDM type called "statistical?")

10.A.3 Frequency-Based Multiplexing Techniques

10.A.3.1 Frequency-Division Multiplexing, FDM

FDM is the method of transmitting many data streams (messages) through one link by dividing the whole bandwidth of a link into the set of frequency bands and sending each message over its frequency band. Refer to Figure 10.A.1.2.

In FDM, each message, $v_{Mi}(t)$, modulates a carrier wave with frequency f_{Ci}, so that a modulated carrier $v_{MC1}(t)$ transports message $v_{M1}(t)$, modulated signal $v_{MC2}(t)$ bears message $v_{M2}(t)$, and so on. Recall that a modulated signal occupies a specific band of frequencies; refer to the bandwidths of AM, FM, and PM signals discussed in Chapter 8. So, the modulated signal $v_{MC1}(t)$ occupies frequency band BW_{MC1} with center frequency f_{C1}, $v_{MC2}(t)$ takes next band, BW_{MC2}, with f_{C2}, and so on. Figure 10.A.3.1, where the FDM concept is demonstrated, shows that each signal band (channel) centers around its carrier frequency. A frequency interval Δf (Hz), separating one band from its adjacent bands, is constant for a given FDM system. How can we prevent interference between the neighboring channels? For this purpose, an FDM system introduces the guard bands shown in Figure 10.A.3.1.

In FDM, a carrier is a sinusoidal wave, and a message is an analog signal too. The message can be originated in digital form, but it must be converted into an analog signal for FDM transmission (see Section 4.2 for digital-to-analog conversion, DAC, techniques).

At a transmitter, an information (message) signal goes through a bandpass filter to restrict the message bandwidth. This filtering enables us to delimit the maximum frequency of the information signal. The filtered message modulates a carrier; the modulated signal passes through another bandpass filter that removes stray frequencies and bounds the signal bandwidth, BW_{MCi}. This two-step filtering ensures that the bandwidth of a transmitted signal is indeed restricted and that the guard bands truly separate the neighboring channels (see Chapter 5 to refresh your memory on filters and filtering process).

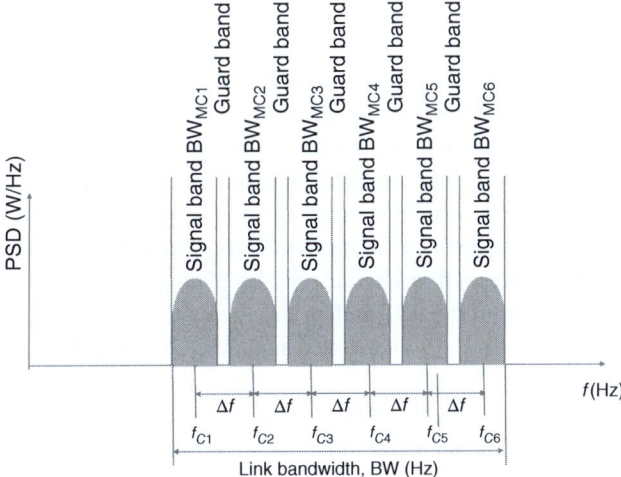

Figure 10.A.3.1 The principle of frequency division multiplexing (PSD, power spectral density).

How can a receiver identify an incoming message? Naturally, by its center frequency, f_{Ci}. Therefore, the receiver's front end must include the bank of bandpass filters whose center frequencies correspond to the proper carrier frequencies of incoming signals. The passbands of the receiving filters must be very narrow to allow only the intended messages to pass. This requirement translates into demand for the good steepness of the filter attenuation characteristic.

(**Exercise**: In the US, FCC allocated the frequency range from 88 to 108 MHz for 100 channels of FM broadcasting radio. What is a channel's bandwidth BW_{MCi}? What is the center frequency of Channel 1? Channel 99?)

The main advantage of an FDM is that many signals can be transmitted through one transmission link simultaneously. The main drawback of the FDM is obvious: The guard bands consume a significant portion of the precious link bandwidth, thus reducing the spectral efficiency and, eventually, the transmission throughput. It is necessary to add that implementation of the concept shown in Figure 10.A.2.1 requires the use of high-performance (read, expensive) bandpass filters and other specialized hardware.

(**Question**: Why FDM does need high-performance bandpass filters?)

The analog nature of the *FDM* technique limits its *area of application* as technology progresses from analog to digital formats. However, in the past, FDM systems played a vital role in telephony, broadcasting (and even cable) TV, and long-distance microwave transmissions that preceded fiber-optic communications. Today, FDM is still in use in broadcasting radio. Thanks to FDM, we have all stations available simultaneously at our radio set and choose one by simply tuning the receiving center frequency. (Refer to Figure 8.1.19 and its discussion to reprise the operation principle of a radio receiver.) Nevertheless, the FDM still forms the basis of such modern multiplexing techniques as OFDM and wavelength-division multiplexing (WDM), as the following sections show.

10.A.3.2 Orthogonal Frequency Division Multiplexing, OFDM

When we need to deliver a message, the natural option is to send the whole message by occupying the entire available bandwidth. The message, however, has to pass through a transmission link that has its characteristics, specifically, frequency-selective attenuation. In this case, the link operates as a filter that attenuates the signal at certain frequencies and, therefore, damages the integrity of the entire message. (Refer to Section 9.1 for discussion signal-link interaction.) As a result, the received signal will be distorted at certain frequencies, and some information might be lost.

A link, unhappily, distorts the signal not only by attenuating some spectral components. In wireless communications, due to the absence of a waveguide, one signal propagates in different directions – the situation called *multipath propagation* – and a receiver obtains several overlapping copies of the same signal. (Section 1.2 considers wireless communications.) Since these copies travel different distances, they arrive at the receiver with various delays. This phenomenon is called *multipath propagation delay*. The delay and other multipath hurdles (e.g. attenuation) produce the distorted received signal; recovering information from this corrupted signal will be difficult if not impossible.

In addition to the causes of signal distortion mentioned previously, there is a source of one more transmission problem: the dispersive property of any channel. Even the best transmission link, optical fiber, introduces the dispersion of a traveling pulse. Concerning transmission, *dispersion means the spread of pulse in time*. This spread leads to overlapping the adjacent pulses, the phenomenon known as *intersymbol interference, ISI*. This phenomenon is discussed in Chapters 3, 5, 6, 7, and 10.

To sum up, such features of a transmission link as the frequency attenuation, multipath propagation delay, and other multipath hurdles, and dispersion distort a transmitted message causing errors in the received information. (Understand that all the above problems related to pulse propagation and, therefore, to digital transmission.)

Can we build an ideal link with flat frequency response, single path wireless transmission, and zero-dispersion property? Obviously, not. Then, what is the solution to the abovementioned problems?

Consider the following approach: Let us (i) break the link's whole bandwidth into small bands (called *subchannels* or *sub-bands*) and evenly divide the entire message into small portions (the set of symbols for *M*-ary modulated signal) and (ii) assign each portion (symbol) to be transmitted by its carrier through a dedicated subchannel. Then, the variations in frequency response within each small subchannel can be considered negligible (or at least, small), the multipath delays become almost unnoticeable, and ISI gets relatively decreased. In other words, the features of each subchannel (small band) become close to that of an ideal channel. Figure 10.A.3.2 illustrates the concept of division of the entire bandwidth into small subchannels. Note that all the subchannels have the same bandwidth, Δf (Hz). Thus,

$$\Delta f \text{ (Hz)} = \frac{\text{BW (Hz)}}{N} \tag{10.2.A.3.1}$$

where BW (Hz) is the entire available bandwidth and N is the number of the subchannels.

The supposed solution – breaking the entire channel bandwidth into the set of the small subchannels – looks reasonable but how, then, can these subchannels be distinguished? Naturally, by frequencies. The center frequency of each subchannel is a carrier frequency; thus, the whole arrangement is FDM. However, we know that FDM utilizes a channel bandwidth very inefficiently because of a need for the guard bands. So, our solution does not seem effective until we manage to (i) place all individual subchannels very tightly to one another and (ii) at the same time, avoid interference between adjacent subchannels. Here is where *OFDM* comes into the play.

As the term suggests, the OFDM is a special type of an FDM, in which all the carrier waves are *orthogonal* to each other. We recall from Section 9.2 that two sinusoidal signals are orthogonal to one another if the integral of their product over a period is equal to zero. Mathematically, this

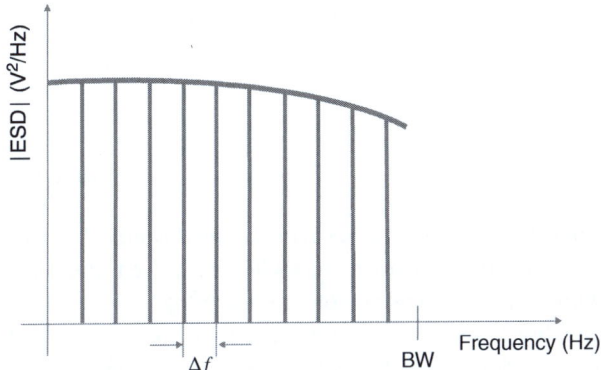

Figure 10.A.3.2 Division of the entire channel bandwidth into the set of small subchannels (ESD, energy spectral density).

condition is presented in (9.2.34); the equation is reproduced here in new notations as

$$\int_0^T v_{C1}(t) \cdot v_{C2}(t) dt = 0 \tag{9.2.34R}$$

To understand an OFDM operation, let us consider the transmission of a QPSK signal. (This signal is discussed in Section 10.1; see Figures 10.1.2 and 10.1.3.) First, suppose we transmit the 8-bit message consisting of four symbols (dibits), so that the first symbol is $S_1 = \mathbf{11}$, the second is $S_2 = \mathbf{10}$, $S_3 = \mathbf{00}$, and $S_4 = \mathbf{01}$. The whole message is transmitted for one second, which gives the symbol rate, SR = 4 (symbol/s) and a symbol duration $T_S = 1/\text{SR} = 1/4$ (s). The message occupies the entire bandwidth, BW_S. This is how a traditional QPSK system works.

Now, we want to transmit the same message by using an OFDM modulation. We divide the whole message into four symbols, transform them in a parallel stream by using a serial-to-parallel converter, and send these symbols in parallel through four subchannels. We want that the transmission still takes one second; thus, the total symbol rate remains the same, SR = 4 (symbol/s). An individual symbol rate, however, becomes four times slower, namely $\text{SR}_i = 1$ (symbol/s), where $i = 1, 2, 3, 4$. Consequently, a symbol duration increases four times and becomes $T_{Si} = 1/\text{SR}_i = 4\,T_S = 1$ s. Thus, using OFDM, we deliver the same amount of information for the same time but having four times slower individual symbol rate and employing four-time longer pulses at each subchannel.

(**Exercise**: Draw a sketch to visualize the above consideration.)

In OFDM, each subchannel is associated with a sinusoidal *subcarrier*, v_{Ci}, whose frequency is f_{Ci}; that is,

$$v_{Ci} = \cos(2\pi f_{Ci} t) \tag{10.A.3.2}$$

Thus, in our example, we have four subcarriers whose center frequencies are f_{C1}, f_{C2}, f_{C3}, and f_{C4}.

Remember, our goal is to place adjacent subchannels very closely to one another and even allow them to overlap but, at the same time, minimize or even eliminate their interference. These requirements can only be achieved if *all the subcarriers are mutually orthogonal over the symbol duration T_S*. Based on (9.2.34), this orthogonality can be written for an OFDM system as

$$\int_0^{T_S} v_{Ci}(t) \cdot v_{Cj}(t) dt = \int_0^{T_S} \cos(2\pi f_{Ci} t) \cdot \cos(2\pi f_{Cj} t) dt = 0 \tag{10.A.3.3}$$

Equations (10.A.3.3) hold only if the subcarrier frequencies satisfy the condition

$$f_{Ci} - f_{Cj} = \frac{n}{T_S} \text{ for } n = 1, 2, 3, \ldots, N \tag{10.A.3.4}$$

Clearly, for two adjacent subcarriers, (10.A.3.4) becomes

$$f_{Ci+1} - f_{Ci} = \Delta f \text{ (Hz)} = \frac{1}{T_S} \tag{10.A.3.5}$$

Equation (10.A.3.3) *along with* Conditions *(10.A.3.4) and (10.A.3.5) are key mathematical formulas for an OFDM system*; (10.A.3.3) is independent of the relative phases of the involved sinusoids. When (10.A.3.3)–(10.A.3.5) are met, the OFDM subchannels do not interfere. Why?

The physics behind sinusoid's orthogonality is that their waves cancel each other at every half-period interval, which eliminates the interference of two sinusoids. The concept of such cancelation, in general, is discussed in Sidebar 6.1.S.1. The similar geometrical presentation concerning the OFDM system can be found in literature.[2] We leave the proof of (10.A.3.3) under Condition (10.A.3.4) for you as an exercise.

Now, we can consider an OFDM operation in the frequency domain by examining Figure 10.A.3.3.

First, we need to recall that a communication system transmits rectangular pulses carrying digital, in general M-ary modulated signals. The amplitude spectra of these pulses obtained by the Fourier transform are the *sinc* functions in the frequency domain. This relationship is discussed in Section 7.1. See Figure 7.1.7 whose modified version is reproduced in Figure 10.A.3.3a. Recall that the duration of a pulse (symbol), T_S (s), determines the first-null bandwidth of a symbol's spectrum as $½ \text{ BW}_S \text{ (Hz)} = 1/T_S$ (s). This is how pulse presentations in the time domain and frequency domain are related. Figure 10.A.3.3b shows that in reality, the subchannels are not rectangular, as presented in Figure 10.A.3.2, but the graphs of sinc functions with all their lobes. From examining Figure 10.A.3.3, we conclude that the *first-null bandwidth of a modulated subcarrier* v_{Ci} is the bandwidth of its subchannel, BW_{SCi} (Hz) $= \Delta f$. (Obviously, null-to-null subchannel bandwidth is BW_{SCi} (Hz) $= 2\Delta f$.) Thus, Figure 10.A.3.3b demonstrates how OFDM signal looks in the frequency domain.

To sum up, even though OFDM supports the same symbol (bit) rate as its M-ary counterpart, it spread the load across several subcarriers. (In our example, BR = 4 (b/s) in cases OFDM and QPSK but the OFDM uses four subcarriers and QPSK uses just one.) Thanks to this arrangement, the OFDM eliminates the disadvantages of a traditional FDM modulation listed above. Also, OFDM provides high spectral efficiency of modern communication systems. There are many other OFDM advantages that we leave for your self-study.

To give you an idea of an industrial OFDM transmission system, consider the parameters of IEEE 802.11a/g standard for WLAN[3]:

- *Modulation technique*: OFDM
- *Bandwidth*: 16.25 MHz
- *Number of subcarriers*: 52
- *Subcarrier spacing*: 312.5 kHz
- *Subcarrier modulation types*: BPSK, QPSK, 16-QAM, or 64-QAM.

We encourage you to use these numbers to calculate all the OFDM parameters discussed in this section.

---

2 See, for example, http://complextoreal.com.
3 https://www.ieee.li/pdf/viewgraphs/introduction_to_orthogonal_frequency_division_multiplex.pdf.

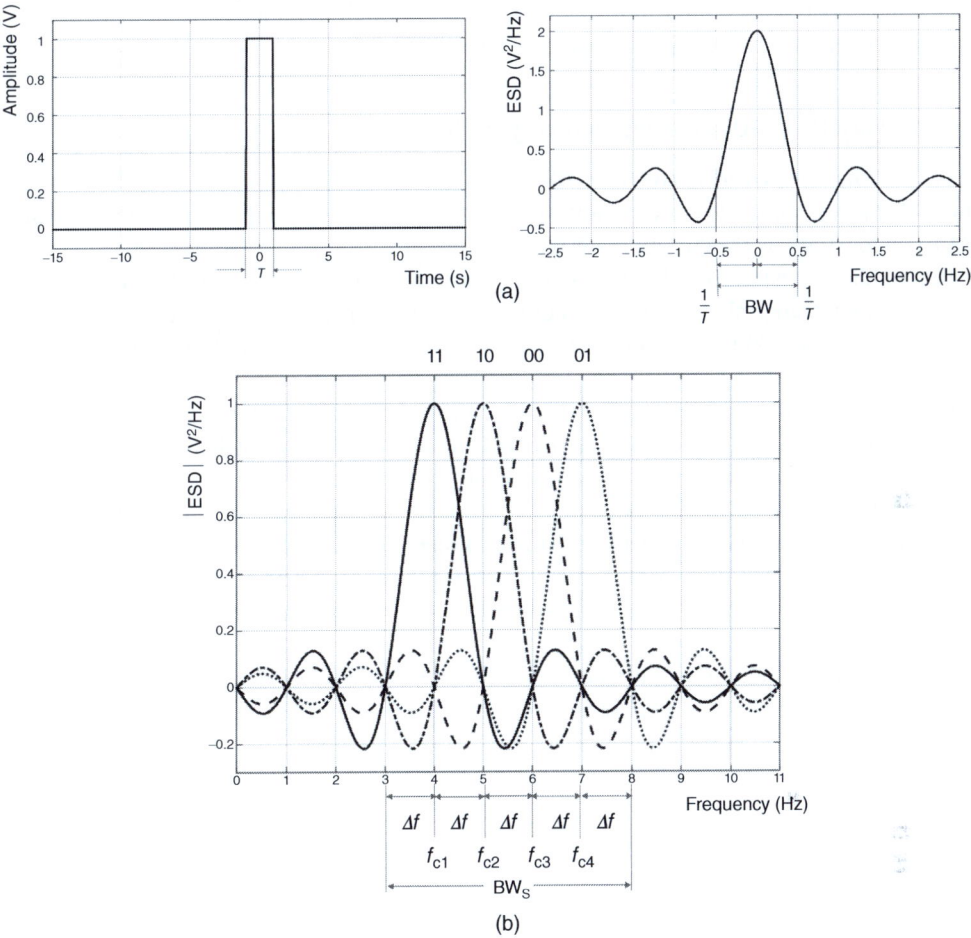

Figure 10.A.3.3 An example of an OFDM signal: (a) rectangular pulse and its energy spectrum and (b) an OFDM signal in the frequency domain (ESD, energy spectral density).

(*Question:* How is the bandwidth of 16.25 MHz composed from the subchannel bandwidths?)

Regarding OFDM applications, we cannot imagine all types of modern wireless communications without OFDM (see Section 1.2). Besides, OFDM plays a critical role in keeping in use ADSL, the technology providing high-speed Internet connections over a copper telephone line. OFDM is also in use in modern optical communications. (Sections 1.2 and 9.1 discuss these technologies.)

All the aforementioned discussion is just a first glance at the foundation of an OFDM transmission. We leave further exploration of this important communication technology for your independent study. Among numerous textbooks covering the topic, we would recommend Carlson and Crilly, pp. 697–703 and Proakis and Salehi, pp. 746–759 for a brief introduction to the subject. For a more in-depth study on OFDM application in optical communications, see Shieh and Djordjevic (2010). Consult, of course, online sources but be always critical to them and give preferences to that with extension "edu."

OFDM has been introduced in the article published by R.W. Chang in 1966 at Bell System Technical Journal. However, it becomes a practical technology only recently thanks to the tremendous progress in digital electronics and DSP technology. (See Section 7.3 for an introduction to DSP application in modern communications.) Today, OFDM is a well-developed area of modern communications supported by extensive research and development.

10.A.3.3 Wavelength-Division Multiplexing, WDM

10.A.3.3.1 Why We Need WDM and How WDM Works

Section 1.2.2 gives a brief overview of optical communications. Revisiting this section will better prepare you for studying the current material.

In the middle of 1990s, optical network operators started to experience a lack of optical fiber capacity. It was a strange phenomenon because theoretically, an optical fiber should be capable of supporting more than 100 Tb/s, but in practice, the network operators could barely achieve 10 Gb/s. Why?

We need to recall that for $BR = 1$ Tb/s the bit time is 1 Tb $= 1 \times 10^{-12}$ s $= 1$ ps. There are no modern photonic devices (laser diodes [LDs], optical modulators, optical switches, photoreceivers, etc.) that can operate at this bit rate. Today's transmitters and receivers mainly operate at 40 Gb/s; direct modulation at 100 Gb/s has been achieved, but the further increase is unlikely. Since an optical fiber can support more 100 Tb/s at a single wavelength – which means that an optical fiber can carry a thousand (!) of 100-Gb/s channels – and an LD can modulate its one wavelength at the rate no more than 100 Gb/s, the *capacity of optical fiber is tremendously underutilized*. Then WDM came to the rescue.

WDM is the method enabling simultaneous transmission of many wavelengths over a single fiber. A WDM is the version of FDM because a wavelength, λ (m) and a frequency, f (Hz), of the same wave are related as

$$\lambda \cdot f = c \tag{1.2.4R}$$

where $c = 3 \times 10^8$ m/s is the speed of light in vacuum. Since this technique is applied to optical communications, the industry prefers to use the term *WDM*. As it is true for any frequency-based multiplexing, the signals (wavelengths) travel through a link (optical fiber in this case) in parallel, that is, simultaneously. (Note that the optical communication industry prefers to use the term "channel" to denote the transmission of a single wavelength.)

Figure 10.A.3.4 shows an example of a WDM system: Each of the four LDs generates its single wavelength, and the MUX multiplexes them into one data stream. All wavelengths travel over an optical fiber simultaneously because, according to the law of quantum physics, many photons can occupy the same space if their wavelengths are different.

The idea to explore WDM was evident as early as the optical fiber became commercially available transmission medium. After all, the WDM conceptually is nothing more than an FDM. However,

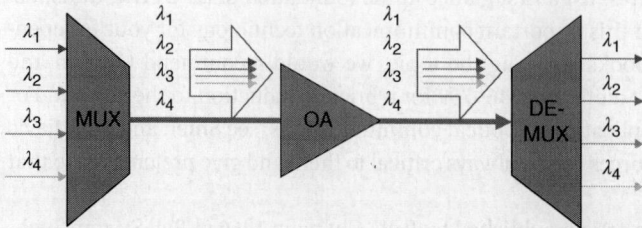

Figure 10.A.3.4 The concept of wavelength-division multiplexing, WDM, and the use of an optical amplifier, OA.

WDM has become a primary industrial technique for better utilizing the optical fiber capacity only with the advent of an optical amplifier (OA) capable to simultaneously amplify all multiplexed wavelengths. The use of an OA is shown schematically in Figure 10.A.3.4. Erbium-doped fiber amplifier, EDFA, was the first – and still is the most – a ubiquitous type of OA.

We remember that optical communications is a worldwide industry. How, then, a network in one corner of the globe can understand what wavelengths arrive from another corner? By the global convention based on ITU-T recommendation. This recommendation specifies the *frequency grid* anchored to 193.1 THz (1552.52 nm). This grid starts from 196.1 THz (1528.77 nm) and supports a variety of channel spacings ranging from 12.5 GHz (~0.1 nm) to 100 GHz (~0.8 nm) and wider. Figure 10.A.3.5 shows four wavelengths of the grid in the frequency domain. The first wavelength is centered at 193.4 THz; the channel spacing of this grid is 100 GHz. (Note that the WDM grid is given in frequencies but not in wavelengths because the frequency measurement is much more precise than that of wavelength.)

From the spectrum of a WDM signal shown in Figure 10.A.3.5, we can extract power-related information. Power of all these signals is about $P_S \approx -35$ dBm (~316.2 µW), and the power of noise (called *noise floor*) is approximately $P_N \approx -67.5$ dBm (~0.18 µW). Therefore, the optical signal-to-noise ratio (OSNR) of this WDM system is OSNR = P_S (dBm) $- P_N$ (dBm) = 32.5 dB (~1778.3). You are encouraged to verify this OSNR value directly by computing the ratio P_S (µW)/P_N (µW).

It is worth noting that the WDM not only gives a tremendous increase in achievable data rate but also provides new degrees of freedom in network design. This is because every wavelength is a huge transporting vehicle that theoretically can carry any information in any format, which enables the network's transporting flexibility. To mention only the main new features of an optical network, the WDM provides the following:

- Separation of the network physical and logical aspects;
- Creation of flexible logical topology over fixed physical topology, which allows for new types of switching, such as wavelength routing;
- The granularity at the wavelength and subwavelength levels, which enables a network to distribute a bandwidth on demand with the network fixed physical topology.

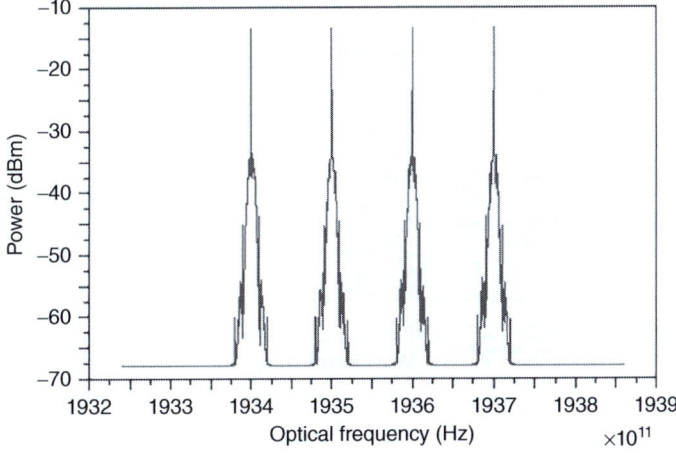

Figure 10.A.3.5 WDM signal: a spectral view. (*Note*: Computer simulation of a WDM system by using LinkSim of RSoft, Inc.)

All in all, the WDM has created a new type of networking in optical communications, which dramatically enhances the efficiency of optical networks.

(**Question**: *How can we identify the arriving wavelengths at the receiving end of a transmission link?*)

10.A.3.3.2 WDM Technology

Optical communication industry employs several multiplexing techniques. Within the traditional WDM discussed in this section so far, there are *dense WDM (DWDM)* used for long-distance transmission, *coarse WDM (CWDM)* employed in short-distance networks, and *short WDM (SWDM)* exploited for ultra-short-distance communications. Among all these WDM types, DWDM delivers most of the global telecommunication traffic; practically, there is no long-distance optical communication link without DWDM. This is why we concentrate here on this type of WDM.

DWDM Usable Bandwidth First, we need to consider the *wavelength bands* within which an optical fiber can transmit light. These bands are shown in Figure 10.A.3.6. Hypothetically, a silica optical fiber, billions of kilometers of which installed all over the globe, can transmit light of all wavelengths from visible to far-infrared. For communication application, however, this optical fiber can be used only within the wavelength bands where its *attenuation*, A (dB/km), is low enough to deliver optical signals over significant distance. For example, typical attenuation of a single-mode fiber, SMF, employed for long-distance communications is about 0.2 dB/km. This level of attenuation lasts only from ~1530 to ~1610 nm. Therefore, *the usable bandwidth*, BW_{DWDM}, *of the silica optical fiber for long-distance transmission is limited by* 1610 − 1530 nm = 80 nm, which corresponds to approximately 8 THz. Figure 10.A.3.6 illustrates this consideration.

Over what distance a transmission optical fiber can deliver an optical signal? If the loss of signal power is allowed to be 1000 (e.g. from 1 mW to 1 μW), then the transmission distance for this signal over the optical fiber with $A = 0.2$ dB/km is 150 km. (You can easily perform these computations

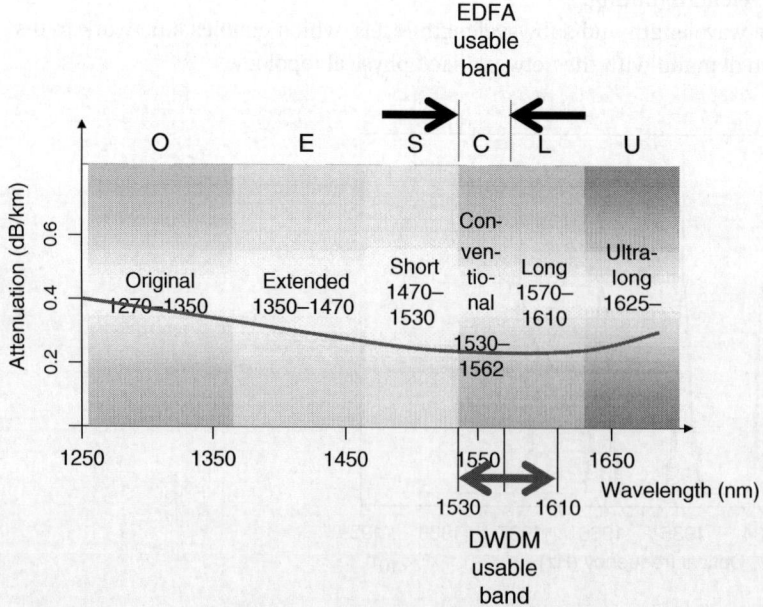

Figure 10.A.3.6 The spectrum of a transmission optical fiber.

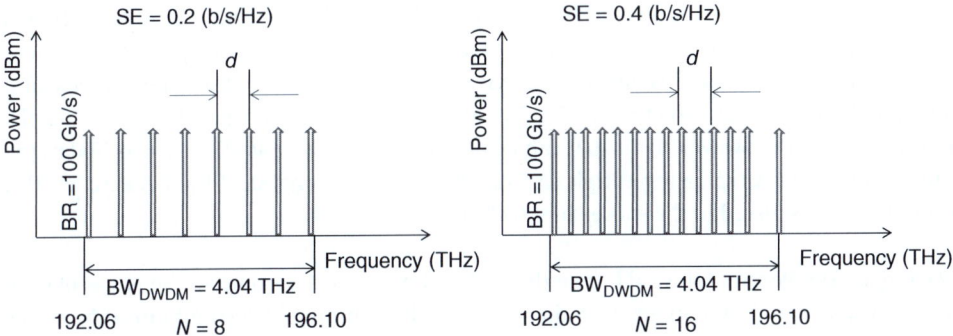

Figure 10.A.3.7 Spectral efficiency of DWDM systems.

by applying the basic formulas: A (dB/km) = −loss (dB)/distance (km) and loss (dB) = $10 \log(P_{out}$ (W)/P_{in} (W).) For longer distance, we have to use an OA, specifically EDFA. It is a gift of nature that an EDFA operates in the range (1530–1562 nm), where standard SMF exhibits minimum attenuation. This fortunate coincidence has triggered the wide use of DWDM technology. Unhappily, EDFA application comes with a price: First, EDFA's usable bandwidth is even smaller than that of DWDM; specifically, BW_{EDFA} = (1562−1530) = 32 nm. Secondly, as an amplifier, EDFA generates its noise assessed by *noise figure, NF*. Both drawbacks play a negative role in a DWDM operation, as we will see shortly.

DWDM and Spectral Efficiency How DWDM affects the system *SE* (b/s/Hz)? We need to recall that spectral efficiency, SE, is the number of bits per second transmitted over a usable bandwidth. In DWDM, the number of bits per second is called line rate, LR (b/s); it is defined as

$$\text{LR (b/s)} = \sum_{i=1}^{N} \text{BR}_i \text{ (b/s)} \qquad (10.A.3.6a)$$

where BR_i (b/s) is a bit rate of an individual channel (wavelength). If all BR_i (b/s) are equal, then (10.A.3.6a) simplifies as

$$\text{LR (b/s)} = \text{BR}_i \text{ (b/s)} \cdot N \qquad (10.A.3.6b)$$

where N is the number of channels. Equation (10.A.3.6a) reflects the *core value of multiplexing*: *The number of bits per second traveling through an optical fiber is the sum of bit rates of all multiplexed channels (wavelengths).*

Thus, the spectral efficiency of a DWDM system is

$$\text{SE}_{DWDM} \text{ (b/s/Hz)} = \frac{\text{LR(b/s)}}{\text{BW}_{DWDM}\text{(Hz)}} \qquad (10.A.3.7)$$

Consider an example: The DWDM usable bandwidth, BW_{DWDM}, is restricted by the bandwidth of EDFA and given by BW_{EDFA} (THz) = 195.9 − 191.9 = 4.0. Suppose the DWDM multiplexes eight channels with BR_i = 0.1 Tb/s each. The spectral efficiency of this system is SE_8 = LR (b/s)/BW_{DWDM} = 8 × 0.1 Tb/s/4.0 THz = 0.2 b/s/Hz. (Note that introducing a channel spacing as d (Hz/channel) = BW_{DWDM}/N, we can determine SE as SE (b/s/Hz) = BR_i/d.) If we multiplex 16 channels instead of eight, then SE increases two times and becomes SE = 0.4 b/s/Hz. Accordingly, the channel spacing reduces twice. Figure 10.A.3.7 illustrates the above discussion and show the meaning of all the above-calculated numbers.

How much can we improve the SE of a DWDM system? It depends on the number of wavelengths that can be multiplexed. This number, in turn, is based on the channel spacing d (Hz/channel). Consider the smallest channel spacing, $d = 12.5$ GHz, specified by ITU-T recommendations, and the entire usable optical fiber bandwidth, $BW_{DWDM} = 8000$ GHz, shown in Figure 10.A.3.6. Then the maximum number of the wavelengths that can be placed within this bandwidth is 640. Today, the communication industry typically multiplexes from 80 to 100 wavelengths. Why? Because DWDM has its limitations discussed in the succeeding section.

Limitations of DWDM Systems It would seem that DWDM plus EDFA resolve all the issues of optical communications, but, unhappily, they do not. Since the volume of telecommunication traffic continues to grow at an exponential rate, the capacity of optical fiber must be increased proportionally. The most direct approaches to meeting this demand would be (i) increasing the bit rate of each wavelength and (ii) adding to the number of multiplexed wavelengths in DWDM. However, are there limits on both measures?

We know that DWDM has increased the capacity of optical networks by 7 orders of magnitude (10^7 times) over the last 25 years; however, today WDM faces its limitations as shown below.

- Increasing the bit rate per wavelength (lambda) causes widening the bandwidth which this lambda occupies. The solution would seem obvious: Keep BR (b/s) per lambda restricted and increase the number of λs. Despite this limitation, the modern optical communication industry managed to transmit 24 Tb/s over 12 000 km per single wavelength. (Do you understand why we mentioned distance in conjunction with the bit rate in describing this achievement?)
- However, the greater is the number of wavelengths, the more EDFAs are needed because all lambdas compete for an EDFA's gain. The more EDFAs are employed, the faster the OSNR degrades. Increasing the power of the input signal to reduce the number of EDFAs is impossible because of nonlinear effects in an optical fiber. Keeping the number of wavelengths restricted contradicts to the solution of the first problem.
- Besides, the greater is the number of wavelengths, the better OSNR is required, and the higher quality of lasers diodes are needed.

The DWDM technique has been the most efficient method of increasing the transmission capacity of optical fiber. Today, however, DWDM, as it is shown above, cannot continue this rise at the required pace.

We need to realize that this capacity crunch is caused not by DWDM itself but the combined use of DWDM and EDFA. Mainly, DWDM is limited in both spectral and power aspects by the use of EDFA. The partial solution to this problem on the spectral side has been introducing a *flexible frequency grid*, which is now also supported by ITU-T. This flexible grid will enable us to utilize the existing bandwidth of EDFA better. The ultimate solution could be eliminating EDFA, which is impossible at this stage of existing technology. However, the recent advances in both technology and architecture of optical networks have enabled the providers to extend the transmission distances without amplification to thousands of kilometers.

Another solution will be the development and application of other types of multiplexing, which we will briefly discuss in the next section.

10.A.3.4 CWDM and Other Types of Multiplexing in Optical Communications

In this section, we briefly consider two classes of multiplexing in optical communications: versions of WDM and new principles of multiplexing.

Coarse WDM and Short WDM

As mentioned in the preceding section, there are two other versions of WDM: CWDM and SWDM.

CWDM is simply a version of WDM with greater channel spacing, as, for instance, shown in Figure 10.A.3.8. CWDM systems work for short-distance (≤20 km) applications, such as data centers and access optical networks. The distance limitation enables the network operators to exclude signal amplification and use inexpensive components. Thus, CWDM offers a cost-effective solution to the "fiber-exhaust" problem for short-distance transmissions.

The ITU-T recommendation specifies that the frequency grid for CWDM must range from 1271 to 1611 nm with 20-nm channel spacing. Such big channel-spacings allow for greater tolerances in determining the center frequencies and the pulse widths. This is why the channel-spacing specifications are given in nanometers in contrast to DWDM, where channel spacing is presented in frequency. Thanks to short transmission distances, CWDM enables network operators to use an optical fiber in the range of wavelengths where attenuation reaches 0.4 dB/km. As a result, CWDM employs the wide usable bandwidth of 340 nm, as seen in Figure 10.A.3.8.

Unhappily, the CWDM inexpensive components can support only relatively low-rate signals (typically, 10 Gb/s per wavelength). Given the CWDM huge bandwidth, we can expect that the spectral efficiency SE_{CWDM} will be low. Indeed, BW_{CWDM} is equal to 340 nm, or 43.18 THz. A CWDM link can accommodate 18 channels (wavelengths) transmitting at 10 Gb/s. Hence, the spectral efficiency of a typical CWDM system is SE_{CWDM} (b/s/Hz) = LR (b/s)/BW_{CWDM} (Hz) = 0.004 b/s/Hz. In contrast, the DWDM system with four wavelengths carrying 400 Gb/s each and occupying bandwidth from 193.4 to 193.7 THz shows SE_{DWDM} = 4.0 b/s/Hz.

Despite this drawback, CWDM finds extensive applications in its range of distances.

To consider *SWDM*, we need to explain that, apart from protection coating, an optical fiber consists of two concentric layers, core, and cladding. Light signals travel through a core, whereas cladding provides the optical condition for keeping the light signals within the core. If a core's diameter is small (≤10 μm), then there is only one path for a light beam to travel (this path is called *mode*). If the core diameter is large (50 μm), then there are many paths (many modes) within this optical

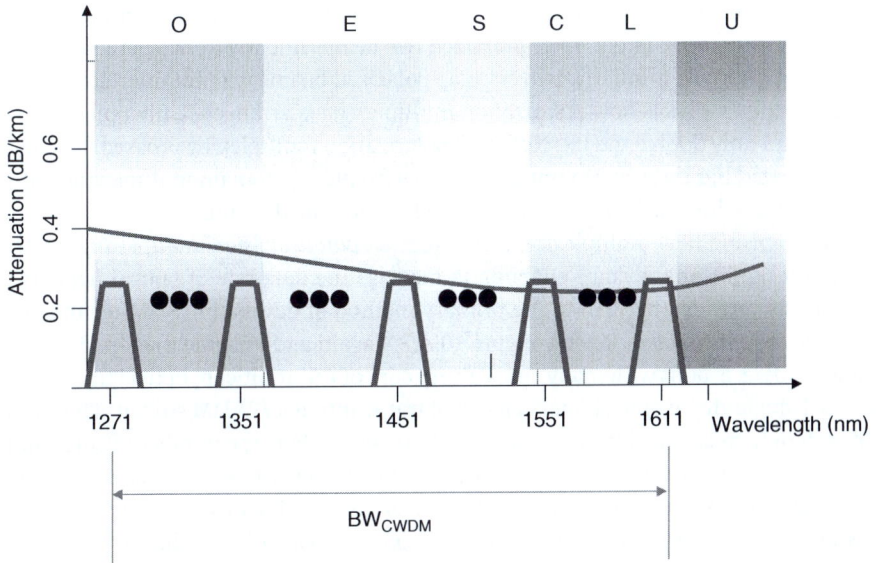

Figure 10.A.3.8 Bandwidth and channel spacing of coarse wavelength-division multiplexing.

fiber. Naturally, the former type of optical fiber is called *SMF* and the latter type is called *multimode fiber, MMF*. (This is an oversimplified description of the nature of modes; it, nevertheless, allows for proceeding with further explanations.) More than 95% of optical fiber employed in optical communications is an SMF because it possesses the best possible transmission characteristics. Nonetheless, MMF finds its applications in very-short distance communications even though it can only support a low bit rate.

To overcome the MMF drawback, the optical communication industry recently developed a new type of MMF called *wideband multimode fiber, WBMMF*. This MMF can support up to four wavelengths to provide SWDM. WBMMF's usable bandwidth is from 850 to 950 nm, where attenuation reaches 2.4 dB/km. This WBMMF can transmit four channels supporting 4×25 Gb/s $= 100$ Gb/s over 400 m.

We complete our discussion of WDM technology by the following summary:

- WDM is the method enabling simultaneous transmission of many wavelengths over a single fiber.
- WDM has increased transmission rate of optical networks by thousands of times.
- We use DWDM for long-distance, CWDM for short-distance, and SWDM for ultra-short-distance communications.
- WDM has enabled the 10-fold increase in spectral efficiency in recent years.
- The further significant increase of transmission capacity, as well as the spectral efficiency, by using WDM is limited by the nature of this technology.

Other Types of Multiplexing for Optical Communications

Since WDM technology approaches its limits, the optical communication industry is searching the alternative approaches to the multiplexing method. We briefly discussed the most promising of them (Mynbaev 2016).

Today, the industry is actively using *polarization multiplexing* in conjunction with *coherent transmission technology*. You will recall that light is an EM wave and oscillations of this wave occur perpendicularly to the direction of its propagation. Light becomes polarized if the wave oscillates either in a plane for linear-polarized light or circular for circular-polarized light. Mixing both polarizations produces elliptically polarized light. If we polarized two beams linearly in two perpendicular planes, they would not interfere. Similarly, two circular-polarized beams will not interfere if they rotate in opposite directions. This is how polarization multiplexing is arranged. Unhappily, polarization multiplexing can only double the LR. Note that polarization multiplexing is based on using two *orthogonally polarized* light waves. Do not confuse this frequency-domain multiplexing with orthogonal FDM, OFDM, where orthogonality is achieved in the time domain.

One of the most promising new multiplexing techniques is *spatial-division multiplexing, SDM*, which is considered a significant advance enabling to increase the data rate of optical networks equal to that previously provided by WDM. The primary method of developing SDM technology based on using *multicore optical fiber*. Review Figure 10.A.3.4 again and imagine that optical fiber linking MUX and DEMUX is built with many cores. Every core of this multicore optical fiber operates as an individual single-mode optical fiber capable of transmitting a DWDM stream. Thus, the LR of this SDM system increases as LR (b/s) $= M \cdot \text{LR}_k$ (b/s), where M is the number of cores and LR_k is the LR of k-th DWDM stream. By using a multicore optical fiber with single-mode cores, the LR at the level of petabit-per-second over thousand-kilometer distance is experimentally achieved.

Another SDM method *employs several EM modes in a single-core optical fiber*. Since we can consider a mode as a light path within an optical fiber, this method creates several virtual links inside an optical fiber.

We need to realize that the principle of spatial (space) division multiplexing can be used not only in optical but also in wireless and satellite communications.

Quite a different approach to improving the operation of the WDM system is based on making better *algorithms of DSP receiving systems*. These improvements enable the researchers to combat with nonlinear and other detrimental phenomena affecting the quality of recovering information from received noisy signals.[4] Unhappily, any improvement achievable with these methods is measured in percent, whereas we need to make better all the transmission metrics by the orders of magnitude.

Since the demand for the fast delivering of the ever-increasing volume of telecommunications traffic continues to grow, optical communications – the main transporting network for this traffic – needs to increase its transmission rate. Multiplexing remains one of the most potent technique enabling the network designers to meet this challenge. This is why wavelength-division and other methods of multiplexing in optical communications are actively developing, and we can expect new great achievements in this area in the near future.

10.A.4 Code-Division Multiplexing, CDM

10.A.4.1 CDM: The Principle of Operation

What if we need to transmit many signals across one transmission pass but simultaneously and within one frequency band, that is we cannot use either time-based or frequency-based multiplexing? Well, we need to think about another possible signal "dimensions" to resolve the problem. One of the approaches is changing the characteristics of a transmitted signal. We learned about modulation, but modulation does not help with multiplexing. Another possible method is to use *coding*. As it is pointed out in Section 3.1.2, code is nothing more than an agreement between communicating parties. (Recall the Morse code and character codes.) Thus, if we encode one signal to be transmitted in code A and the other signal in code B, these two signals will not interfere even when they travel across the same channel at the same time and in the same frequency range. This difference in the codes is the basis for *code-division multiplexing, CDM*, the method widely used in modern communications. Figure 10.A.4.1 shows the concept of CDM in application to wireless communications, where CDM finds its main exploitation.

At the transmitting end of a CDM system, a data flow and a unique code are presented to an encoder, which encodes the data stream (the pulse train carrying 1s and 0s). For example, data stream A and code A entered the top encoder, which produces encoded signal A. Clearly, every transmitter performs the similar operation. Then all encoded signals are sent through a single path, where they are naturally multiplexed. Thus, the first part of our problem – multiplexing – is solved. However, how can we demultiplex these signals? The input to every receiver is the stream of multiplexed signals, i.e. signals $A + B + \cdots + N$. To select its individual data stream, a decoder is fed with the same unique code that is used by the corresponding transmitter. For instance, the top decoder uses code A to select data A from the whole stream of multiplexed signals. How a decoder rejects all other signals? By making code A *orthogonal* to all other codes. Codes orthogonality implies that the attempts to extract data A by applying any other code except code A to decoder A must produce meaningless output. The same must hold for all other codes. In other words, all codes must be *mutually orthogonal*.

---

4 See, for example, Pan et al. (2016).

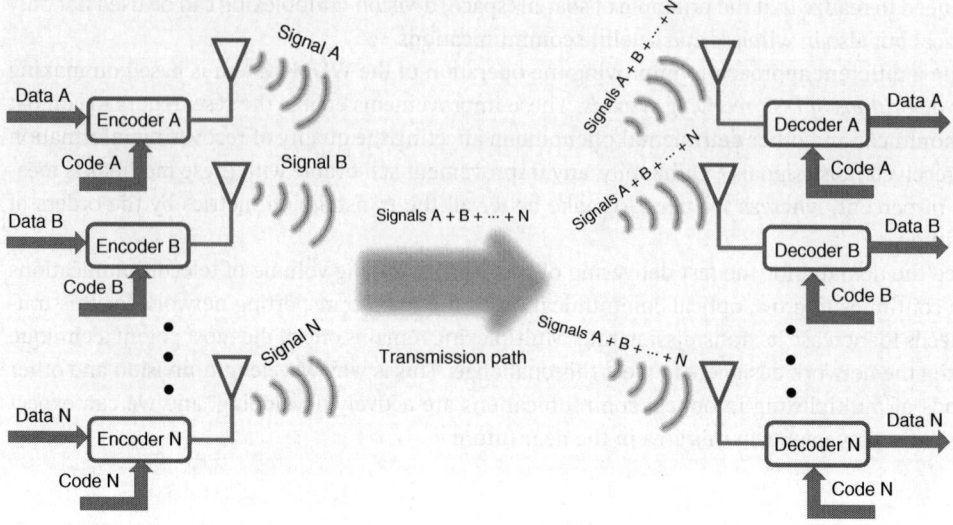

Figure 10.A.4.1 Concept of code-division multiplexing, CDM.

(*Question*: Where have we met the condition of orthogonality? What is the mathematical formula for this condition?)

10.A.4.2 Spread-Spectrum Technique

The general idea of CDM is now clear, but how can this idea be implemented? Modern CDM is based on the *spread spectrum*[5] technology.

Figure 10.A.4.2 shows how this technique works in principle. (In the explanations below, we follow Stern and Mahmoud 2004, pp. 370–381.) Generated by an information source, data stream A_D, which is high-energy, narrow-bandwidth, and low-bit-rate signal, enter a unit called *spreader* (see Figure 10.A.4.2a). The second input of the spreader is fed by the *spreading code*, A_S, locally generated at transmitter A. (A code generator is not shown). In contrast to the data stream, this is low-energy, wide-bandwidth, and high-bit-rate signal. The heart of the spreader is *exclusive-OR (XOR or EXOR) two-input digital logical gate* executing the following operations:

$$1 \oplus 1 = 0 \quad 1 \oplus 0 = 1 \quad 0 \oplus 1 = 1 \quad 0 \oplus 0 = 0 \quad (10.A.4.1)$$

The spreader performs XOR operation and thus generates new code $A_T = A_D \oplus A_S$. Now, A_T-coded low-energy, wide-bandwidth, and the high-bit-rate signal is prepared for transmission. This is a conventional digital signal, and it can be modulated by any digital modulation scheme, for instance, BPSK.

It is important to realize that *modulation* and *spreading* are independent operations; they can be performed separately. Modulation may be executed before the data signal enters the spreader or after that. The latter version is shown in Figure 10.A.4.2a.

---

5 Hollywood actress Hedy Lamarr invented the spread spectrum technique with the assistance of pianist and composer George Antheil as the means to keep military communications secret. They received US patent in 1942, and she donated this patent to US Navy. The invention, however, was forgotten until the 1960s when the military started to use it. After the technique was declassified in the mid-1980s, its commercial application throve.

The key point here is an operation performed by an XOR gate, an example of which is shown in Figure 10.A.4.2b. *While data bit **1** is remaining at the input of an XOR gate, four bits of a spreading code interact with it, and the XOR gate produces four bits of a transmitted signal.* The same holds for all other data bits entering a spreader. Timing disposition of the spreading operation can be seen in Figure 10.A.4.2c. During one period T of a data stream, four bits of the spreading code pass the XOR operations, which produces low-energy and high-bit-rate four bits of an output signal.

Why does the bandwidth of a coded transmitted signal become wide? Because the bit rate of a spreading signal is much higher than that of a data signal. Indeed, we know that BR (b/s) = 2BW (Hz); thus, the spreading signal must have greater bandwidth than the data signal. (The rule BR (b/s) = 2BW (Hz) is true for null-to-null bandwidth of a BPSK signal. See, for instance, Figure 10.A.3.3a. In general, refer to (1.3.3).) If we assume $BR_D = 1$ kb/s, then, for example shown in Figure 10.A.4.2, $BW_D = 2$ kHz and $BW_S = 8$ kHz. This is why Figure 10.A.4.2a shows the narrow bandwidth of the data signal and wide bandwidth of the transmitted signal.

Why does the spreader's output signal have low energy? Because this signal has the same physical characteristics as the spreading code whose bit rate and the level of energy are controlled by the CDM system's designer. For a spreading code, the designer wants to have a bandwidth substantially wider and energy significantly lower than that of the data signal.

Despreading and data recovery are operations reversal to data coding and spreading. You are encouraged to sketch the figures and timing table to see these operations explicitly. Compare Figure 10.A.4.2 and the new figure with Figure 10.A.4.1. Remember, the critical point

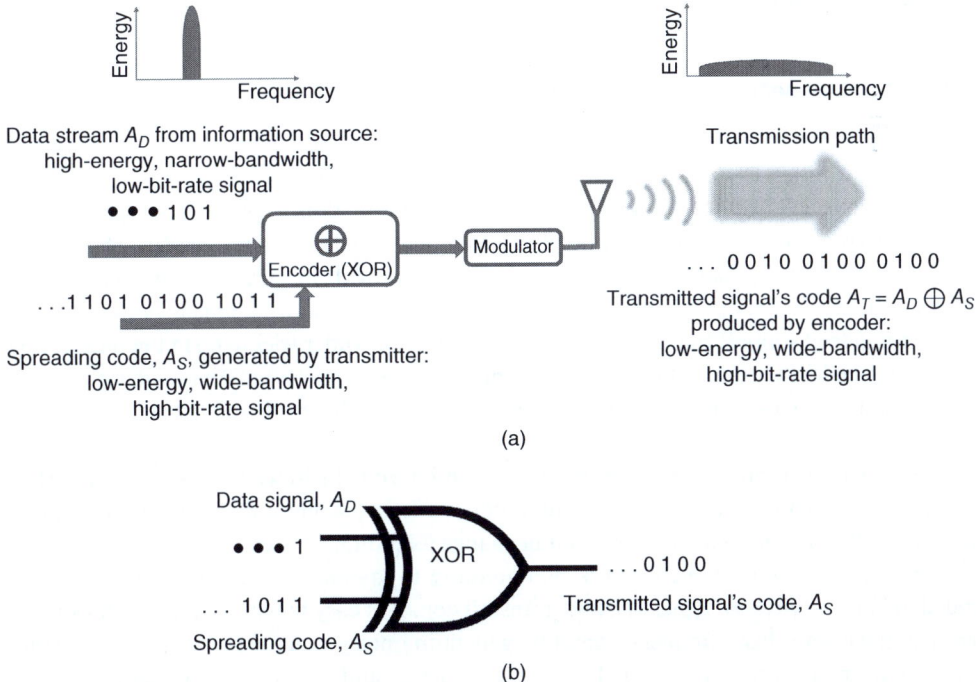

Figure 10.A.4.2 Spread spectrum operation: (a) transmitting part of a CDM system; (b) example of the operation of an XOR gate; and (c) example of a spread-spectrum coding.[6]

---

6 After Stallings 2014, p. 546.

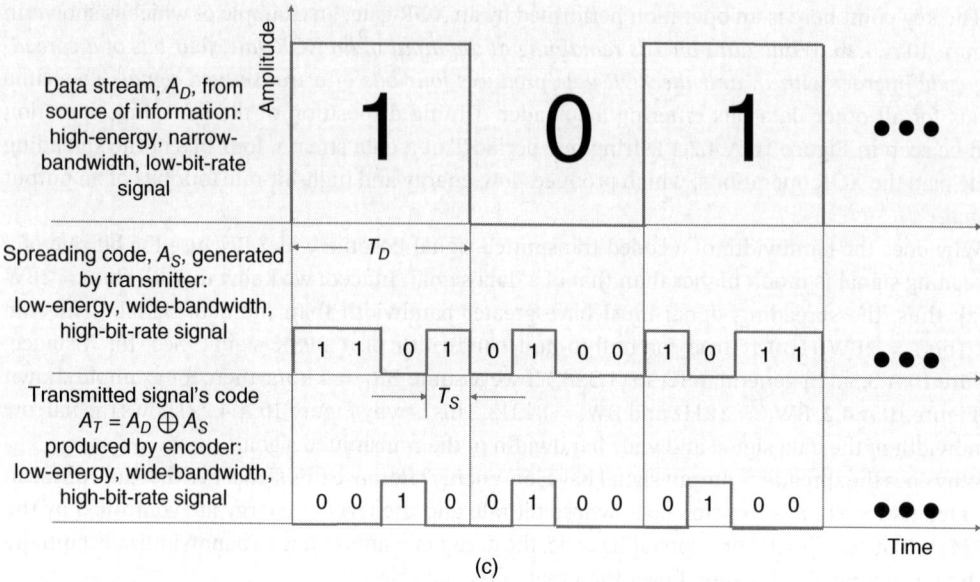

Figure 10.A.4.2 (*Continued*)

in a spread-spectrum operation is having the same code for coding and decoding for two communicating parties of a transmission system.

10.A.4.3 CDM: Benefits and Applications

What is the purpose of substantially widening the bandwidth and significantly decreasing the energy of a transmitted CDM signal? After all, throughout this book, we focus on the measures to minimize the usable bandwidth while increasing the bit rate, thus improving the spectral efficiency. We also want to raise energy (power) of a transmitted signal to increase an SNR. The answer is that the objective of a spread-spectrum transmission is to provide maximum security of communications at the expense of optimization of other system's characteristics. A spread-spectrum-based CDM system indeed support the high level of security: If an intruder tries to open the message by using a wrong code, the transmitted signal will appear just as noise. What's more, thanks to low energy of a transmitted signal, its signal-to-noise is below unity, $SNR < 1$, and the signal is buried in noise.

Another benefit of a spread-spectrum technique is immunity of a transmitted signal to interference from external sources. This is because interference affects the physical parameters of a signal, such as its amplitude and frequency, but cannot change its coding.

The other great benefit of the CDM is its immunity to *jamming*. Jamming is the operation intended to block a wireless, typically radio, transmission. It is easy to jam a regular radio station broadcasting within a fixed frequency band by generating noisy radio waves at the same band. In this arrangement, the listeners will hear mostly noise and cannot clearly distinguish the message meaning. (Incidentally, the authorities of the former Soviet Union practiced jamming of foreign radio stations such as Voice of America, BBC, and Radio Svoboda to deprive the country population of access to a free press.) Fortunately, it is difficult to jam a coded signal because it is practically immune to interference.

We need to realize that the spectral efficiency of a single channel in spread-spectrum transmission is very low because the channel's bit rate is low, and its bandwidth is wide. However, in total, a CDM transmission is rather efficient because low SE for one signal is compensated by sending many signals across the same path. So, a CDM system, as a whole, benefits from a spread-spectrum transmission. Also, even though the energy spectral density of a transmitted signal is low, the total energy of this signal is as high as that of a data signal. This is because the total original energy is simply spread over the wide bandwidth of a transmitted signal.

In wireless communications, where the spread-spectrum technique is mostly used, *multipath fading*, discussed previously, is one of the most detrimental effects. A spread-spectrum method eliminates it by rejecting all other copies of a transmitted signal after the first decoding is performed. What's more, the advanced versions of a CDM system capable of despreading all the reflected signals and adding them to the first one, thus increasing the reliability of decoding.

In this section, we considered the application of the spread-spectrum technique only for multiplexing. Bear in mind, however, that a CDM method can be used for developing multiple access systems (CDMA) and other applications.

It should be clear from our discussion that the vital point of a spread-spectrum technique is a code that must be known to both transmitter and receiver. Since a wirelessly transmitted signal can be easily intercepted and analyzed, there are chances that the code can be broken. Therefore, the measures must be taken to make the code as unbreakable as possible. These measures are to make the code as long as possible and as random as possible. Both of them, however, can be implemented to some extent. As the code becomes too long, the transmission also becomes prohibitively long. If a transmitter generates an entirely random code, the code cannot be reproduced by a receiver. The solution to these problems is to generate a reasonably long *pseudorandom number*, *PRN*, for coding, where pseudorandom means "nearly-random." This sequence of bits is indeed close to being random but remains reproducible by the receiver.

The CDM technique finds its applications in the satellite-based global positioning system, GPS, in mobile communications, in Bluetooth, in WLAN, and in many other sectors of wireless communications. We have mentioned that modern wireless communications cannot exist without CDM. Why? Because the main problem of wireless communications is the scarcity of bandwidth and the CDM systems capable of operating without demanding more bandwidth.

Finally, it should be apparent that CDM is a significant sector of the wireless communication industry. The brief introduction to this area presented here gives you an idea of its operational principle and importance. Being equipped with this knowledge, you will be able to delve into this technology as much as your professional duty will require.

Bibliography

This bibliography contains only sources concerning with general topics of modern communications. It also includes the entries cited in the text. Sources dedicated to specialized topics in communications – optical, wireless, and satellite – are collected in the next section. A reader should review both the bibliography sections to obtain the complete record of sources concerning with a specific area of communications.

Agarwal, A. and Lang, J.H. (2005). *Foundations of Analog and Digital Electronic Circuits*. San Francisco, CA: Morgan Kaufmann Publishers.

Agbo, S.O. and Sadiku, M.N. (2017). *Principles of Modern Communication Systems*. New York: Cambridge University Press.

Agrawal, G.P. (2012). *Fiber – Optic Communication Systems*, 4e. Hoboken, NJ: Wiley.

Alexander, C.K. and Sadiku, M.N. (2009). *Fundamentals of Electric Circuits*, 4e. New York: McGraw-Hill.

Anderson, J. (2005). *Digital Transmission Engineering*, 2e. Hoboken, NJ: Wiley.

Anderson, J.B. and Johannesson, R. (2005). *Understanding Information Transmission*. Piscataway, NJ: IEEE Press/Wiley Interscience.

Attaway, S. (2017). *MATLAB*, 4e. Oxford: BH/Elsevier.

Balck, U. (2002). *MPLS & Label Switching Networks*. Upper Saddle River, NJ: Prentice Hall PTR.

Balsamo, S., Marin, A., and Vicario, E. (eds.) (2018). *New Frontiers in Quantitative Methods in Informatics*. Cham, Switzerland: Springer.

Bayin, S.S. (2018). *Mathematical Methods in Science and Engineering*, 2e. Hoboken, NJ: Wiley.

Beard, C. and Stallings, W. (2016). *Wireless Communication Networks and Systems, Global Edition*. London: Pearson.

Beasley, J.S. and Miller, G.M. (2005). *Modern Electronic Communication*, 8e. Upper Saddle River, NJ: Prentice Hall.

Beasley, J.S., Hymer, J.D., and Miller, G.M. (2014). *Electronic Communications (A Systems Approach)*. Boston, MA: Pearson.

Bellamy, J.C. (2000). *Digital Telephony*, 3e. New York: Wiley-Interscience.

Benslama, M., Benslama, A., and Aris, S. (2017). *Quantum Communications in New Telecommunications Systems*. Hoboken, NJ: Wiley.

Beyda, W.J. (2005). *Data Communications*, 4e. Upper Saddle River, NJ: Prentice Hall.

Bhagyaveni, M.A., Kalidoss, R., and Vishvaksenan, K.S. (2016). *Introduction to Analog and Digital Communication*. Gistrup, Denmark: River Publishers.

Binh, L.N. (2015). *Advanced Digital Optical Communications*, 2e. Boca Raton, FL: CRC Press, Taylor & Francis Group.

Bowman, F. (2012). *Introduction to Bessel Function*. North Chelmsford, MA: Courier Corporation.
Boyle, R. (2011). *Applied Networking Labs*. Boston, MA: Prentice Hall.
Boylestad, R.L. (2016). *Introductory Circuit Analysis*, 13e. Upper Saddle River, NJ: Prentice Hall.
Brillouin, L. (2013). *Science and Information Theory*, 2e. Mineola, NY: Dover.
Buchwald, A. (2016). High-speed time interleaved ADCs. *IEEE Communications Magazine* 54 (4): 71–77.
Cai, Y., Cai, J.-X., Pilipetskii, A. et al. (2010). Spectral efficiency limits of pre-filtered modulation formats. *Optics Express* 18 (19): 20273–20281.
Carlson, A.B. and Crilly, P.B. (2010). *Communication Systems*, 5e. New York: McGraw-Hill.
Cavaiani, T.P. (2010). *IT Networking Lab*. Boston, MA: Prentice Hall.
Chang, F. (ed.) (2018). *Datacenter Connectivity Technologies: Principles and Practices*. Gistrup, Denmark: River Publishers.
Chaparro, L.F. (2015). *Signals and Systems Using MATLAB*. Amsterdam: Elsevier.
Chapell, L.A. and Tittel, E. (2007). *Guide to TCP/IP*, 3e. Boston, MA: Delmar.
Chapman, M.J., Goodball, D.P., and Steel, N.C. (1997). *Signal Processing in Electronic Communication*. Chichester: Harwood Publishing.
Chen, S., Qin, F., Hu, B. et al. (2016). User-centric ultra-dense networks for 5G: challenges, methodologies, and directions. *IEEE Communications Magazine* 54 (4): 79–85.
Cicconetti, C., Oliva de la, A., Chieng, D., and Zúñiga, J.C. (eds.) (2015). Extremely dense wireless networks. *IEEE Communications Magazine* 53 (1): 88–165.
Clark, C. (2002). *Network Cabling Handbook*. New York: McGraw-Hill/Osborne.
Comer, D.E. (2004). *Computer and Networks Internets with Internet Applications*, 4e. Upper Saddle River, NJ: Prentice Hall.
Cooley, J.W. and Tukey, J.W. (1965). An algorithm for the machine calculation of complex Fourier series. *Mathematics Computation* 19: 297–301.
Couch, L.W. (2012). *Digital and Analog Communication Systems*, 8e. Upper Saddle River, NJ: Pearson Prentice Hall.
Crystal, D. (1994). *Biographical Encyclopedia*. Cambridge: Cambridge University Press.
Cvijetic, M. and Djordjevic, I.B. (2013). *Advanced Optical Communication Systems and Networks*. Norwood, MA: Artech House.
Cisco Systems, Inc. (2001). *Dictionary of Internetworking Terms and Acronyms*. San Jose, CA: Cisco Press.
Das, A. (2012). *Signal Conditioning*. Berlin: Springer-Verlag.
Dineen, S. (2012). *Analysis – A Gateway to Understanding Mathematics*. Hackensack, NJ: World Scientific.
Dixit, S.S. (2003). *IP Over WDM*. Hoboken, NJ: Wiley.
Djordjevic, I.B. (2017). *Advanced Optical and Wireless Communications Systems*. Berlin: Springer-Verlag.
Dodd, A. (2019). *The Essential Guide to Telecommunications*, 6e. Upper Saddle River, NJ: Pearson Prentice Hall.
Dumas, M.B. and Schwartz, M. (2009). *Principles of Computer Networks and Communications*. Upper Saddle River, NJ: Pearson Prentice Hall.
Dunsmore, B. and Skander, T. (2003). *Telecommunications Technologies Reference*. Indianapolis, IN: Cisco Press.
Dutta, A.K., Dutta, N.K., and Fujiwara, M. (2004). *WDM Technologies: Optical Networks*. San Diego, CA: Academic Press.

Farrel, A. (2004). *The Internet and Its Protocols.* Amsterdam, The Netherlands: Morgan Kaumann/Elsevier.

Finney, R.L., Thomas, G.B., Demana, F., and Waits, B.K. (1994). *Calculus.* Reading, MA: Addison Wesley Publishing Company.

Floyd, T.L. (2005). *Electronic Devices,* 7e. Upper Saddle River, NJ: Prentice Hall.

Forouzan, B. and Fegan, S.C. (2007). *Data Communications and Networking.* Boston, MA: McGraw-Hill.

Fouli, K., Medard, M., and Shroff, K. (2018). Random liner networking coding: a technical feature overview. *IEEE ComSoc Technology News* (July 2018).

Franco, S. (1999). *Electric Circuits Fundamentals.* New York: Oxford University Press.

Frankel, J.L. (2016). *Binary Logic Levels.* Boston, MA: Harvard University sites.fas.harvard.edu/~cscie287/spring2016/slides/Binary%20Logic%20Levels.pdf.

Frenzel, L.E. Jr. (2014). *Contemporary Electronics.* New York: McGraw-Hill.

Frequency Devices, Inc (2009). *Analog Electronic Filter Design Guide.* Ottawa, IL: Frequency Devices, Inc.

Freude, W. et al. (2012). Quality metrics for optical signals: eye diagram, Q-factor, OSNR, EVM and BER. In: *14th International Conference on Transparent Optical Networks (ICTON).* Coventry, UK: IEEE https://doi.org/10.1109/ICTON.2012.6254380.

Gallager, R.G. (2008). *Principles of Digital Communications.* New York: Cambridge University Press.

Gaskin, S. and Lawson, R. (2008). *GO! With the Internet*, vol. 1. Upper Saddle River, NJ: Pearson Prentice Hall.

Gibson, J.D. (1997). *The Communications Handbook.* Boca Raton, FL: CRC Press.

Glover, I.A. and Grant, P.M. (2010). *Digital Communications*, 3e. Upper Saddle River, NJ: Pearson.

Goldman, J.E. and Rawles, P.T. (2005). *Applied Data Communications*, 4e. Hoboken, NJ: Wiley.

Grami, A. (2016). *Introduction to Digital Communications.* Cambridge, MA: Academic Press.

Granatstein, V.L. (2012). *Physical Principles of Wireless Communications*, 2e. Boca Raton, FL: CRC Press.

Greene, B. (2013). Mind over matter. *Smithsonian* 96: 25–28.

Grover, W.D. (2004). *Mesh-Based Survivable Networks.* Upper Saddle River, NJ: Prentice Hall PTR.

Hambley, A. (2009). *Electronics*, 4e. Pearson Education: Upper Saddle River, NJ.

Harnedy, S. (2002). *The MPLS Primer.* Upper Saddle River, NJ: Prentice Hall PTR.

Hayt, W.H. Jr., Kemmerly, J.E., Phillips, J.D., and Durbin, S.M. (2019). *Engineering Circuit Analysis*, 9e. New York: McGraw-Hill.

Haykin, S. and Moher, M. (2007). *Introduction to Analog and Digital Communications*, 2e. Hoboken, NJ: Wiley.

Held, G. (1995). *Dictionary of Communications Technology.* Macon, GE: Wiley.

Hioki, W. (2001). *Telecommunications*, 4e. Upper Saddle River, NJ: Prentice Hall.

Holmes, J.K. (2007). *Spread Spectrum System for GNSS and Wireless Communications.* Norwood, MA: Artech House.

Horowitz, P. and Hill, W. (1995). *The Art of Electronics*, 2e. Cambridge: Cambridge University Press.

Hsu, H.P. (2003). *Theory and Problems of Analog and Digital Communications*, Schaum's Outline Series, 2e. New York: McGraw-Hill.

Huitema, C. (2000). *Routing in the Internet*, 2e. Upper Saddle River, NJ: Prentice Hall PTR.

IEEE (2016). *IEEE Standard Dictionary of Electrical and Electronics Terms.* https://www.ieee.org/publications_standards/publications/subscriptions/prod/standards_dictionary.html (accessed 2 August 2016).

Irwin, J.D. and Helms, R.M. (2011). *Basic Engineering Circuit Analysis*, 10e. Hoboken, NJ: Wiley.

Israelohn, J. (2004). Noise 101 08/01/2004 and Noise 102 18/03/2004. www.edn.com (accessed 7 November 2017).

Ivaniga, T. and Ivaniga, P. (2014). Evaluation of the bit error rate and Q-factor in optical networks. *IOSR Journal of Electronics and Communication Engineering (IOSR-JECE)* 9 (6) https://doi.org/10.9790/2834-09610103.

James, G. (2015). *Modern Engineering Mathematics*, 5e. Harlow, England: Pearson (International).

James, G. and Dyke, P. (2018). *Advanced Modern Engineering Mathematics*, 5e. Harlow, England: Pearson (International).

Kartalopoulos, S.V. (2004). *Optical Bit Error Rate*. Hoboken, NJ: Wiley.

Kartalopoulos, S.V. (2009). *Security of Information and Communication Networks*. Hoboken, NJ: Wiley.

Kenyon, T. (2002). *High-Performance Data Network Design*. Boston, MA: Digital Press.

Kesisdis, G. (2007). *An Introduction to Communication Network Analysis*. Hoboken, NJ: Wiley-Interscience.

Kolimbiris, H. (2000). *Digital Communications Systems*. Upper Saddle River, NJ: Prentice Hall.

Kumar, S. (2015). *Wireless Communications Fundamental & Advanced Concepts: Design Planning and Applications*. Gistrup, Denmark: River Publishers.

Kumar, A., Manjunath, D., and Kuri, J. (2004). *Communication Networking – An Analytical Approach*. Amsterdam, The Netherlands: Morgan Kaufmann/Elsevier.

Kurose, J.F. and Ross, K.W. (2010). *Computer Networking*, 5e. New York: Addison-Wesley.

Landauer, R. (1966). Minimal energy requirements in communication. *Science* 272: 1914.

Langton, C. (2008). Tutorials on digital communications engineering. www.complextoreal.com/tutorials (accessed 11 June 2017).

Lathi, B.P. and Zhi, D. (2009). *Modern Digital and Analog Communication Systems*, 4e. New York: Oxford University Press.

Lawrence, D. (2008). Frequency modulation (FM) tutorial. http://www.silabs.com/Marcom%20Documents/Resources/FMTutorial.pdf (accessed 7 November 2015).

Lazarus, M. (2010). The great spectrum famine. *IEEE Spectrum* 47 (10): 26–31.

Lazarus, M. (2016). The troubled past and uncertain future of radio interference. *IEEE Spectrum* 53 (8): 40–56.

Leung, K.K. (2009). *Communications II – Lecture 5: Effects of Noise on FM*. London: Imperial College www.commsp.ee.ic.ac.uk/~kkleung/communications2_2009/Lecture 5.pdf.

Li, G., Bennis, M., and Yu, G. (eds.) (2015). Full duplex communications. *IEEE Communications Magazine* 53 (5): 90–152.

Liu, C.-P. and Seeds, A. (2010). Wireless-over-fiber technology – bringing the wireless world indoors. *Optics & Photonic News* 21 (11): 27–33.

Liu, S.-C., Kramer, I., Indiveri, G. et al. (2002). *Analog VLSI: Circuits and Principles*. Cambridge, MA: The MIT Press.

Lundberg, K.H. (1993). Noise sources in bulk CMOS. web.mit.edu/klund/www/papers/UNP_noise.pdf (accessed 21 March 2018).

Lutovac, M.D., Tosic, D.V., and Evans, B.L. (2001). *Filter Design for Signal Processing Using MATLAB and Mathematica*. Upper Saddle River, NJ: Prentice-Hall.

Lyons, R.G. and Fugal, D.L. (2014). *The Essential Guide to Digital Signal Processing*. Upper Saddle River, NJ: Pearson Prentice Hall.

Maas, S.A. (2005). *Noise in Linear and Nonlinear Circuits*. Boston, MA: Artech House.

Madhow, U. (2008). *Fundamentals of Digital Communications*. New York: Cambridge University Press.

Manolakis, D.G. and Ingle, V.K. (2011). *Applied Digital Signal Analysis*. New York: Cambridge University Press.
Manton, N. and Mee, N. (2017). *The Physical World – An Inspirational Tour of Fundamental Physics*. New York: Oxford University Press.
McClellan, J.H., Schafer, R.W., and Yoder, M.A. (2003). *Signal Processing First*. Upper Saddle River, NJ: Pearson Prentice Hall.
McNames, J. (n.d.). web.cecs.pdx.edu/~ece2xx/ECE222/Slides/ConvolutionIntegral.pdf (accessed 4 July 2015).
Mesiya, M.F. (2013). *Contemporary Communication Systems*. New York: McGraw-Hill.
Miao, G.J. (2007). *Signal Processing in Digital Communications*. Boston, MA: Artech House.
Minoli, D. and Cordovana, J. (2006). *Authoritative Computer & Network Security Dictionary*. Hoboken, NJ: Wiley.
Mitra, P.P. and Stark, J.B. (2001). Nonlinear limits to the information capacity of optical fibre communications. *Nature* (6841): 1027–1030.
Morrison, J.C. (2015). *Modern Physics for Scientists and Engineers*, 2e. Amsterdam, The Netherlands: Academic Press/Elsevier.
Muller, N.J. (2000). *IP Convergence: The Next Revolution in Telecommunications*. Norwood, MA: Artech House.
Murmann, B. (2016). The successive approximation register ADC: a versatile building block for ultra-low-power to ultra-high-speed applications. *IEEE Communications Magazine* 54 (4): 78–81.
Mynbaev, D.K. (2010). Modern photonics and optical communications. In: *Future Trends in Microelectronics: From Nanophotonics to Sensors to Energy* (eds. S. Luryi, J. Xi and A. Zaslavsky), 23–42. Hoboken, NJ: Wiley-IEEE Press.
Mynbaev, D.K. (2013). Will optical communications meet the challenges of the future? In: *Future Trends in Microelectronics* (eds. S. Luryi, J. Xu and A. Zaslavsky), 160–172. Hoboken, NJ: IEEE/Wiley.
Mynbaev, D.K. (2016). Fundamental and technological limitations of optical communications. *International Journal of High Speed Electronics and Systems* 25 (1 & 2): 1640010-1–1640010-21.
Mynbaev, D.K. and Scheiner, L.L. (2001). *Fiber-Optic Communications Technology*. Upper Saddle River, NJ: Prentice Hall.
Neamen, D. (2010). *Microelectronics: Circuit Analysis and Design*, 4e. New York: McGraw-Hill.
Newman, M.E.J. (2018). *Networks – An Introduction*. 2e. New York: Oxford University Press.
Nguyen, H.H. and Shwedyk, E. (2009). *A First Course in Digital Communications*. Cambridge: Cambridge University Press.
Nilsson, J.W. and Ridel, S.A. (2019). *Electric Circuits*, 11e. New York: Pearson.
Olenewa, J. and Ciampa, M. (2006). *Wireless Guide to Wireless Communications*. 2e. Boston, MA: Thomson.
Oppenheim, A.V. and Willsky, A.S. (2015). *Systems and Signals*, 2e. New York, NY: Pearson.
Palm, W.J. III (2011). *Introduction to MATLAB for Engineers*, 3e. New York: McGraw-Hill.
Pan, C., Frank, R., and Kschischang, F.R. (2016). Probabilistic 16-QAM shaping in WDM systems. *Journal of Lightwave Technology* https://doi.org/10.1109/JLT.2016.2594296.
Panwar, S.S., Mao, S., Ryco, J.-D., and Li, Y. (2004). *TCP/IP Essentials – A Lab-Based Approach*. New York: Cambridge University Press.
Paul, H.Y. (2004). *Electronic Communication Techniques*, 5e. Upper Saddle River, NJ: Prentice Hall.
Paulraj, A., Gore, D., and Nabar, R. (2003). *Introduction to Space-Time Wireless Communications*. Cambridge: Cambridge University Press.
Pearson, J. (1992). *Basic Communication Theory*. London: Prentice Hall International (UK).
Petersen, J.K. (2002). *The Telecommunications Illustrated Dictionary*, 2e. Boca Raton, FL: CRC Press.

Peterson, L.L. and Davie, B.S. (2003). *Computer Networks - A System Approach*, 3e. San Francisco, CA: Morgan Kaufmann Publishers/Elsevier Science.

Pratar, R. (2010). *Getting Started with MATLAB*. New York: Oxford University Press.

Proakis, J. and Salehi, M. (2002). *Communication System Engineering*, 2e. Upper Saddle River, NJ: Prentice Hall.

Proakis, J.B. and Salehi, M. (2008). *Digital Communications*, 5e. New York: McGraw-Hill Higher Education.

Proakis, J. and Salehi, M. (2014). *Fundamentals of Communication Systems*, 2e. Upper Saddle River, NJ: Pearson.

Proakis, J.G., Masoud, S., and Gerhard, B. (2013). *Contemporary Communication Systems Using MATLAB*, 3e. Boston, MA: Cengage.

Rappaport, T. (2002). *Wireless Communications: Principles and Practice*, 2e. Upper Saddle River, NJ: Prentice Hall.

Raymer, M.G. (2009). *The Silicon Web – Physics for the Internet Age*. Boston, MA: Taylor & Francis.

Razavi, B. (2008). *Fundamentals of Microelectronics*. Hoboken, NJ: Wiley.

Reiter, G. (2014). *Wireless Connectivity for the Internet of Things*. Texas Instruments White Paper.

Rice, M. (2009). *Digital Communications: Discrete-Time Approach*, 9e. Upper Saddle River, NJ: Prentice Hall.

Riley, K.F., Hobson, M.P., and Bence, S.J. (1998). *Mathematical Methods for Physics and Engineering*. Cambridge: Cambridge University Press.

Rizzoni, G. (2009). *Fundamentals of Electrical Engineering*. Boston, MA: McGraw-Hill Higher Education.

Roberts, M.J. (2008). *Fundamentals of Signals & Systems*. Boston, MA: McGraw-Hill Higher Education.

Roddy, D. and Coolen, J. (1995). *Electronic Communication*, 4e. Englewood Cliffs, NJ: Prentice Hall.

Roden, M.S. (2003). *Analog and Digital Communication Systems*, 5e. Los Angeles, CA: Discovery Press.

Rojo, A. and Bloch, A. (2018). *The Principle of Least Action*. New York: Cambridge University Press.

Rowe, S.H. and Schuh, M.L. (2005). *Computer Networking*. Upper Saddle River, NJ: Prentice Hall.

Sackinger, E. (2005). *Broadband Circuits for Optical Fiber Communication*. Hoboken, NJ: Wiley.

Sadiku, M.N.O. (2002). *Optical and Wireless Communications: Next Generation Networks*. Boca Raton, FL: CRC Press.

Schaumann, R., Xiao, H., and Van Valkenburg, M.E. (2010). *Design of Analog Filters*. New York: Oxford University Press.

Schecter, S. (n.d.). www4.ncsu.edu/~schecter/ma_341_sp06/varpar.pdf (accessed 12 April 2015).

Schiller, J. (2003). *Mobile Communications*, 2e. Harlow: Pearson.

Schmogrow, R., Nebendahl, B., and Winter, M. (2012). Error vector magnitude as a performance measure for advanced modulation formats. *IEEE Photonics Technology Letters* 24 (1): 61–63. https://doi.org/10.1109/LPT.2011.2172405.

Searle, S.J. (2002). A brief history of character codes in North America, Europe, and East Asia. tronweb.super-nova.co.jp/charactcodehist.html (accessed 23 July 2016).

Sedra, A.S. and Smith, K.C. (2004). *Microelectronic Circuits*, 5e. New York: Oxford University Press.

Sekar, V.C. (2010). *Analog Communication*. New York: Oxford University Press.

Shake, I., Takara, H., and Kawanishi, S. (2004). Simple measurement of eye diagram and BER using high-speed asynchronous sampling. *Journal of Lightwave Technology* 22 (5): 1296–1302.

Shenoi, K. (1995). *Digital Signal Processing in Telecommunications*. Upper Saddle River, NJ: Prentice Hall PTR.

Sherrick, J.D. (2001). *Concepts in Systems and Signals*. Upper Saddle River, NJ: Prentice Hall.

Shieh, W. and Djordjevic, I. (2010). *OFDM for Optical Communications*. San Diego, CA: Academic Press.

Siauw, T. and Alexandre, M.B. (2015). *An Introduction to MATLAB Programming and Numerical Methods for Engineers.* London: AP/Elsevier.

Sklar, B. (2001). *Digital Communications,* 2e. New Jersey, NJ: Prentice Hall PTR.

Smith, S.W. (2003). *Digital Signal Processing (A Practical Guide for Engineers and Scientists).* Newnes/Elsevier.

Smith, J.O. (2007). Introduction to digital filters with audio applications. http://ccrma.stanford.edu/~jos/filters/.

Smith, F.G. and King, T.A. (2007). *Optics and Photonics - An introduction,* 2e. New York: Wiley.

Stallings, W. (2004a). *Computer Networking with Internet Protocols and Technology.* Upper Saddle River, NJ: Prentice Hall.

Stallings, W. (2004b). *Cryptography and Network Security – Principles and Practice.* Upper Saddle River, NJ: Prentice Hall.

Stallings, W. (2008). *High-Speed Networks and Internets,* 2e. Upper Saddle River, NJ: Prentice Hall.

Stallings, W. (2014). *Data and Computer Communications,* 10e. Upper Saddle River, NJ: Pearson Prentice Hall.

Stanley, W.D. (2003). *Network Analysis with Applications,* 4e. Upper Saddle River, NJ: Prentice Hall.

Stanley, W.D. and Jeffords, J.M. (2005). *Electronic Communications: Principles and Systems.* Clifton Park, NY: Delmar.

Starr, T., Gioffi, J.M., and Silberman, P.J. (1999). *Understanding Digital Subscriber Line Technology.* Upper Saddle River, NJ: Prentice Hall PTR.

Stern, H.P.E. and Mahmoud, S.A. (2004). *Communication Systems: Analysis and Design.* Upper Saddle River, NJ: Prentice Hall.

Sundararajan, D. (2008). *A Practical Approach to Signals and Systems.* Singapore: Wiley (Asia) Pte Ltd.

Tanenbaum, A.S. and Wetherall, D.J. (2011). *Computer Networks,* 5e. Upper Saddle River, NJ: Prentice Hall.

Taub, H. and Schilling, D. (1986). *Principles of Communication Systems,* 2e. New York: McGraw-Hill.

Taylor, F.J. (1994). *Principles of Signals and Systems.* New York: McGraw-Hill.

Radcom Ltd. (2001). *Telecom Protocol Finder.* New York: McGraw-Hill.

Thierauf, S.C. (2011). *Understanding Signal Integrity.* Norwood, MA: Artech House.

Thomas, R.E., Rosa, A.J., and Toussaint, G.J., (2016). *The Analysis and Design of Linear Circuits.* 8e. Hoboken, NJ: Wiley.

Thompson, R.A., Tipper, D., Kabara, J. et al. (2006). *The Physical Layer of Communications Systems.* Norwood, MA: Artech House.

Tomar, G.S. and Bagwari, A. (2016). *Analog Communication.* PHI Learning Pvt. Ltd.

Tomasi, W. (2013). *Advanced Electronic Communications Systems,* 6e. Upper Saddle River, NJ: Pearson.

Trumper, D. (2007). *2.14 Analysis and Design of Feedback Control Systems.* Massachusetts Institute of Technology: MIT OpenCourseWare. http://ocw.mit.edu.License: Creative Commons BY-NC-SA.

U.S. Department of Transportation - Federal Highway Administration (2007). Weather Applications and Products Enabled through Vehicle Infrastructure Integration (VII). Internet. http://ops.fhwa.dot.gov/publications/viirpt/sec5.htm (21 November 2012).

Valdar, A. (2017). *Understanding Telecommunications Networks,* IET Telecommunications Series, 2e. Stevenage: The Institute of Engineering and Technology.

Vasseur, J.-P., Pickavet, M., and Demeester, P. (2004). *Network Recovery.* Amsterdam, The Netherlands: Morgan Kaufmann/Elsevier.

Weik, M.H. (1996). *Communication Standard Dictionary,* 3e. New York: Chapman & Hall.

Wheeler, T. (2006). *Electronic Communications for Technicians.* Upper Saddle River, NJ: Prentice Hall.

Williams, A. and Taylor, F. (2006). *Electronic Filter Design Handbook,* 4e. New York: McGraw-Hill.

Wilson, P. (2017). *The Circuit Designer's Companion,* 4e. Oxford: Newness/Elsevier.

Wing, S. (2002). *Analog and Digital Filter Design*. Amsterdam: Newnes/Elsevier.

Wozencraft, J.M. and Jacobs, I.M. (1990). *Principles of Communication Engineering*. Long Grove, IL: Waveland Press.

Xiong, F. (2000). *Digital Modulation Techniques*. Norwood, MA: Artech House.

Yevick, D. and Yevick, H. (2014). *Fundamental Math and Physics for Scientists and Engineers*. Hoboken, NJ: Wiley.

Young, P.H. (2004). *Electronic Communication Techniques*, 5e. Upper Saddle River, NJ: Prentice Hall.

Yu, J., Li, X., and Zhang, J. (2018). *Digital Signal Processing for High-Speed Optical Communication*. Singapore: World Scientific.

Yuen, C., Elkashian, M., Qian, Y. et al. (2015). Energy harvesting communications: Part III. *IEEE Communications Magazine* 53 (8): 90–91.

Zhou, X. and Xie, C. (eds.) (2016). *Enabling Technology for High Spectral-Efficiency Coherent Optical Communications Networks*. Hoboken, NJ: Wiley.

Ziemer, R.E. and Tranter, W.H. (2015). *Principles of Communications*, 7e. Hoboken, NJ: Wiley.

Ziemer, R.E., Tranfer, W.H., and Fannin, D.R. (1993). *Signals and Systems: Continuous and Discrete*. New York: Macmillan Publishing Company.

Zuidweg, J. (2002). *Next Generation Intelligent Networks*. Norwood, MA: Artech House.

Industrial Documents and the Internet References

A Filter Primer (2001). *Application Note 733*. Sunnyvale, CA: Maxim Integrated Products, Inc.

An Introduction to Spread-Spectrum Communications (2003). *Tutorial 1890*. Maxim Integrated, Inc.

Digital Modulation in Communications Systems — An Introduction (2001). *Application Note 1298*. Agilent Technologies (now Keysight Technologies).

Error Vector Magnitude (EVM). www.mathworks.com/help/comm/ug/error-vector-magnitude-evm.html (accessed 21 July 18).

Error Vector Magnitude . http://www.antenna-theory.com/definitions/evm.php (accessed 21 July 18).

Harvey Mudd College (n.d.). http://fourier.hmc.edu/e101/lectures/handout1/node6.html (accessed 12 July 2017).

Specifications Sheet (2002). *MAXIM 5th-order, Lowpass Switched-Capacitor Filters*. Sunnyvale, CA: Maxim Integrated Products, Inc.

Specifications Sheet (2007). *D68 & DP68 Series Fixed Frequency 8-Pole Low-Pass Filters*. Ottawa, IL: Frequency Device, Inc.

Spectrum Analysis Basics (2014). *Application Note 150*. Keysight Technologies.

Tektronix (2016). PAM4: Signaling in High Speed Serial Technology: Test, Analysis, and Debug. *Application Note*.

The Basics of Anti-Aliasing: Using Switched-Capacitor Filter (2005). *Application Note 928*. Sunnyvale, CA: Maxim Integrated Products, Inc.

Understanding Eye Pattern Measurements (2010). *Application Note No. 11410-00533*. Anritsu Company.

Universal Time (2013). The United States Naval Observatory. http://aa.usno.navy.mil/faq/docs/UT.html (accessed 27 September 2015).

Using Error Vector Magnitude Measurements to Analyze and Troubleshoot Vector-Modulated Signals (2000). *PN 89400-14*. Agilent Technologies (now Keysight Technologies).

XYZ of Signal Sources, Primer (2008). Signal Generator. Tektronix. www.tektronix.com/signal_generators (accessed 27 September 2016).

Specialized Bibliographies

Specialized bibliographies contain sources devoted to a specific area of communications – optical, wireless, or satellite. They are further references to the main bibliography; in other words, they don't duplicate, with the few exceptions, the entries from the primary reference list. Therefore, a reader should review both main and specialized bibliographies to obtain the complete record of sources concerning with a specific area of communications.

Optical Communications

Agrawal, G.P. (2012). *Nonlinear Fiber Optics*, 5e. Cambridge, MA: Academic Press.
Alwayn, V. (2004). *Optical Network Design and Implementation*. Indianapolis, IN: Cisco Press.
Born, M. and Wolf, E. (2000). *Principles of Optics*, 7e. Cambridge: Cambridge University Press.
Bouchet, O. (2013). *Wireless Optical Communications*. Hoboken, NJ: Wiley-ISTE.
Bouchet, O., Sizun, H., Boisrobert, C. et al. (2010). *Free-Space Optics: Propagation and Communication*. Hoboken, NJ: Wiley.
Buck, J.A. (2004). *Fundamentals of Optical Fibers*. Hoboken, NJ: Wiley.
Cai, Y., Cai, J.-X., Pilipetskii, A. et al. (2010). *Optics Express* 18 (19): 20281–20285.
Cvijetic, M. (1996). *Coherent and Nonlinear Lightwave Communications*. Norwood, MA: Artech House.
Cvijetic, M. (2004). *Optical Transmission*. Norwood, MA: Artech House.
Decusatis, C. (2002). *Handbook of Fiber Optic Data Communication*, 2e. Poughkeepsie, NY: Academic Press.
Desurvire, E. (2004a). *Broadband Access, Optical Components and Networks, and Cryptography*. Hoboken, NJ: Wiley-Interscience.
Desurvire, E. (2004b). *Signaling Principles, Network Protocols, and Wireless Systems*. Hoboken, NJ: Wiley-Interscience.
Dixit, S.S. (2003). *IP Over WDM*. Hoboken, NJ: Wiley.
Dutta, A.K., Dutta, N.K., and Fujiwara, M. (2002). *WDM Technologies. Active Optical Components*. San Diego, CA: Academic Press.
Dutta, A.K., Dutta, N.K., and Fujiwara, M. (2003). *WDM Technologies. Passive Optical Components*. San Diego, CA: Academic Press.
Forestieri, E. (ed.) (2005). *Optical Communication Theory and Techniques*. Berlin: Springer-Verlag.
Greenfield, D. (2002). *The Essential Guide to Optical Networks*. Upper Saddle River, NJ: Prentice Hall.
Gumaste, T. and Antony, T. (2003). *DWDM Network Designs and Engineering Solutions*. Indianapolis, IN: Cisco Press.
Hemmati, H. (ed.) (2006). *Deep Space Optical Communications*. Hoboken, NJ: Wiley.

Iannone, E., Matera, F., Mecozzi, A., and Settembre, M. (1998). *Nonlinear Optical Communication Networks*. New York: Wiley.

Kartalopoulos, S.V. (2000). *Introduction to DWDM Tecnology. Data in a Rainbow*. Murray Hill, NJ: Wiley.

Kartalopoulos, S.V. (2001). *Fault Detectability DWDM Toward Higher Signal Quality & System Reliability*. Murray Hill, NJ: Wiley.

Kartalopoulos, S.V. (2003). *DWDM Networks, Devices, and Technology*. Hoboken, NJ: Wiley.

Kasap, S., Ruda, H., and Boucher, Y. (2009). *Cambridge Illustrated Handbook of Optoelectronics and Photonics*. Cambridge: Cambridge University Press.

Keiser, G. (2000). *Optical Fiber Communications*, 3e. New York: McGraw-Hill.

Keiser, G. (2006). *FTTX Concepts and Applications*. Hoboken, NJ: Wiley.

Kenyon, I.R. (2011). *The Light Fantastic – A Modern Introduction to Classical and Quantum Optics*, 2e. New York: Oxford University Press.

Kolimbiris, H. (2004). *Fiber Optics Communications*. Upper Saddle River, NJ: Prentice Hall.

Kress, B.C. and Meyrueis, P. (2009). *Applied Digital Optics. From Micro-optics to Nanophotonics*. West Sussex: Wiley.

Kumar, S. and Deen, M.J. (2014). *Fiber Optic Communications*. West Sussex: Wiley.

Lachs, G. (1998). *Fiber Optic Communications*. New York: McGraw-Hill.

Laude, J. (2002). *DWDM Fundamentals, Components, and Applications*. Norwood, MA: Artech House.

Lecoy, P. (2013). *Fiber-Optic Communications*. Hoboken, NJ: Wiley-ISTE.

Lin, C. (2006). *Broadband Optical Access Networks and Fiber-to-the-Home*. West Sussex: Wiley.

Makazawa, M., Kikuchi, K., and Miyazaki, T. (eds.) (2010). *High Spectral Density Optical Communications Technologies*. Berlin: Springer-Verlag.

Mouftah, H.T. and Ho, P. (2003). *Optical Networks*. Norwell, MA: Kluwer Academic Publisher.

Murthy, C.S.R. and Gurusamy, M. (2002). *WDM Optical Networks*. Upper Saddle River, NJ: Prentice Hall.

Mynbaev, D.K. (2013). Will optical communications meet the challenges of the future? In: *Future Trends in Microelectronics* (eds. S. Luryi, J. Xu and A. Zaslavsky), 160–172. Hoboken, NJ: IEEE/Wiley.

Mynbaev, D.K. and Scheiner, L.L. (2001). *Fiber-Optic Communications Technology*. Upper Saddle River, NJ: Prentice Hall.

Mynbaev, D.K. and Sukharenko, V. (2013). Plasmonic-based devices for optical communications. In: *Frontiers in Electronics* (eds. S. Cristoloveanu and M.S. Shur), 201–221. Hackensack, NJ: World Scientific.

Palais, J.H. (2004). *Fiber Optic Communications*, 5e. Upper Saddle River, NJ: Pearson Prentice Hall.

Papadimitriou, G.I., Papazoglou, C., and Pomportsis, A.S. (2007). *Optical Switching*. Hoboken, NJ: Wiley.

Ramaswami, R. and Sivarajan, K. (2002). *Optical Network*. San Diego, CA: Academic Press.

Razavi, B. (2012). *Design of Integrated Circuits for Optical Communications*, 2e. Hoboken, NJ: Wiley.

Rogers, A. (2001). *Understanding Optical Fiber Communications*. Norwood, MA: Artech House.

Rogers, A. (2008). *Polarization in Optical Fibers*. Norwood, MA: Artech House.

Sackinger, E. (2005). *Broadband Circuits for Optical Fiber Communication*. Hoboken, NJ: Wiley.

Sadiku, M.N.O. (2002). *Optical and Wireless Communications: Next Generation Networks*. Boca Raton, FL: CRC Press.

Shieh, W. and Djordjevic, I. (2010). *OFDM for Optical Communications*. San Diego, CA: Academic Press.

Smith, F.G. and King, T.A. (2007). *Optics and Photonics An introduction*, 2e. New York: Wiley.

Somani, A.K. (2006). *Survivability and Traffic Grooming in WDM Optical Networks*. New York: Cambridge University Press.

Thyagarajan, K. and Ghatak, A. (2007). *Fiber Optic Essentials*. Hoboken, NJ: Wiley.

Tomsu, P. and Schmutzer, C. (2002). *Next Generation Optical Networks*. Upper Saddle River, NJ: Prentice Hall.

Vacca, J.R. (2007). *Optical Networking Best Practices Handbook*. Hoboken, NJ: Wiley.

Warier, S. (2017a). *Engineering Optical Networks*. Boston, MA: Artech House.

Warier, S. (2017b). *The ABCs of Fiber Optic Communication*. Boston, MA: Artech House.

Warren, D. and Hartmann, D. (2004). *Cisco Self-Study: Building Cisco Metro Optical Networks (Metro)*. Indianapolis, IN: Cisco Press.

Weik, M.H. (1997). *Fiber Optics Standard Dictionary*. New York: Chapman & Hall.

Weiner, J. and Nunes, F. (2013). *Light-Matter Interaction*. Oxford: Oxford University Press.

Yariv, A. and Yen, P. (2007). *Photonics Optical Electronics in Modern Communications*, 6e. New York: Oxford University Press.

Zang, H. (2003). *WDM Mesh Networks*. Norwell, MA: Kluwer Academic Publisher.

Zheng, J. and Mouftah, H.T. (2004). *Optical WDM Networks*. Hoboken, NJ: Wiley.

Wireless Communications

Biglieri, E., Calderbank, R., Constantinides, A. et al. (2007). *MIMO Wireless Communications*. Cambridge University Press.

Bing, B. (ed.) (2002). *Wireless Local Area Network*. New York: Wiely-Interscience.

Blake, R. (2001). *Wireless Communication Technology*. Albany, NY: Delmar.

Boccuzzi, J. (2008). *Signal Processing for Wireless Communications*. New York: McGraw-Hill.

Desurvire, E. (2004). *Signaling Principles, Network Protocols, and Wireless Systems*. Hoboken, NJ: Wiley-Interscience.

Djordjevic, I.B. (2017). *Advanced Optical and Wireless Communications Systems*. Berlin: Springer-Verlag.

Garg, V. (2007). *Wireless Communications & Networking*. Burlington, MA: Morgan Kaufmann.

Goldsmith, A. (2005). *Wireless Communications*. Cambridge: Cambridge University Press.

Guizani, M. (ed.) (2006). *Wireless Communications Systems and Networks*. Berlin: Springer-Verlag.

Haykin, S. and Moher, M. (2005). *Modern Wireless Communications*. Upper Saddle River, NJ: Prentice Hall.

Molisch, A.F. (2010). *Wireless Communications*. Hoboken, NJ: Wiley.

Oleneva, J. and Ciampa, M. (2006). *Wireless# Guide to Wireless Communications*. 2e. Boston, MA: Thomson.

Olenewa, J. (2017). *Guide to Wireless Communications*, 4e. Boston, MA: Cengage.

Paulraj, A., Gore, D., and Nabar, R. (2003). *Introduction to Space-Time Wireless Communications*. Cambridge: Cambridge University Press.

Pupolin, S. (ed.) (2007). *Wireless Communications 2007 CNIT Thyrrenian Symposium*. New York: Springer US.

Rappaport, T. (2002). *Wireless Communications: Principles and Practice*, 2e. Upper Saddle River, NJ: Prentice Hall.

Reiter, G. (2014). *Wireless Connectivity for the Internet of Things*. Texas Instruments White Paper.

Sadiku, M.N.O. (2002). *Optical and Wireless Communications: Next Generation Networks*. Boca Raton, FL: CRC Press.

Schiller, J. (2003). *Mobile Communications*, 2e. Harlow: Pearson.

Schwartz, M. (2004). *Mobile Wireless Communications*. Cambridge: Cambridge University Press.

Shankar, P.M. (2002). *Introduction to Wireless Systems*. New York: Wiley.
Sobot, R. (2013). *Wireless Communication Electronics by Example*. Berlin: Springer-Verlag.
Stallings, W. (2004). *Wireless Communications and Networking*, 2e. Upper Saddle River, NJ: Prentice Hall.
Tse, D. and Viswanath, P. (2010). *Fundamentals of Wireless Communication*. Cambridge: Cambridge University Press.
Wong, K.D. (2012). *Fundamentals of Wireless Communication Engineering Technologies*. Hoboken, NJ: Wiley.
Zheng, P., Peterson, L.L., Davie, B.S., and Farrel, A. (2009). *Wireless Networking Complete*. Burlington, MA: Morgan Kaufmann.

Satellite Communications

Armstrong-Smith, M. (2017). *Satellite Communications Technology: A guide to how satellite services work*. Nairn: Microcom Systems.
Banerjee, P. (2017). *Satellite Communication*. Delhi: PHI Learning Pvt. Ltd.
Braun, T.M. (2012). *Satellite Communications Payload and System*. Hoboken, NJ: Wiley/IEEE.
Cheruku, D.R. (2009). *Satellite Communication*. New Delhi: IK International Publishing House Pvt. Ltd.
Cocheti, R. (2014). *Mobile Satellite Communications Handbook*. Hoboken, NJ: Wiley.
Elbert, B. (2014). *The Satellite Communication Ground Segment and Earth Station Handbook*. Boston, MA: Artech House.
Ilčev, S.D. (2016). *Global Mobile Satellite Communications Theory: For Maritime, Land and Aeronautical Applications*. Berlin: Springer-Verlag.
Ippolito, L.J. Jr., (2017). *Satellite Communications Systems Engineering: Atmospheric Effects, Satellite Link Design and System Performance*, 2e. Hoboken, NJ: Wiley.
ITU (2002). *Handbook on Satellite Communications by International Telecommunications Union*. Hoboken, NJ: Wiley-Interscience.
Jo, K.Y. (2011). *Satellite Communications Networks Design and Analysis*. Boston, MA: Artech House.
Kapovits, A., Corici, M.-I., Gheorghe-Pop, I.-D. et al. (2018). Satellite communications integration with terrestrial networks. *China Communications* 15 (8): 22–38.
Kolawole, M.O. (2017). *Satellite Communication Engineering*. Boca Raton, FL: CRC Press.
Maral, G. and Bousquet, M. (2010). *Sattelute Communications Systems: Systems, Techniques and Techology*. Hobboken, NJ: Wiley.
Minoli, D. (2015). *Innovations in Satellite Communications and Satellite Technologies: The industry Implications of DVB-S2X, High Throughput Satellites, Ultra HD, M2M, and IP*. Hoboken, NJ: Wiley.
Park, H. and Cha, H. (2016). Electrical design of a solar array for LEO satellites. *International Journal of Aeronautical and Space Sciences* 17 (3): 401–408.
Pratt, T. and Bostian, C.W. (2016). *Satellite Communications*, 2e. Hoboken, NJ: Wiley.
Rao, R. (2013). *Satellite Communication: Concepts and Applications*. Delhi: PHI Learning Pvt. Ltd.
Razani, M. (2012). *Information, Communication, and Space Technology*. Boca Raton, FL: CRC Press.
Roddy, D. (2006). *Satellite Communications*, 4e. New York: McGraw-Hill.
Sharma, S.K., Chatzinotas, S., and Arapoglou, P.-D. (eds.) (2018). *Satellite Communications in the 5G Era (Telecommunications)*. UK: Institute of Engineering and Technology (IET).
Sheriff, R. and Hu, Y.F. (2001). *Mobile Satellite Communications Networks*. Hoboken, NJ: Wiley.
Sun, M. (2014). *Satellite Communications Systems: Systems, Techniques, and Technology*. Hoboken, NJ: Wiley.

Welti, C.R. (2012). *Satellite Basics for Everyone: An illustrated Guide to Satellites for Non-Technical and Technical People*. iUniverse.com (publisher).

Ya'acob, N., Johari, J., Zolkapli, M. et al. (2017). Link budget calculator system for satellite communication. In: *2017 International Conference on Electrical, Electronics and System Engineering (ICEESE)*. https://doi.org/10.1109/ICEESE.2017.8298397.

Index

5G – fifth generation of RF-based wireless communications   54

a
active
  component   410
ADC
  quality of   275
  the quality of vs. number of bits per sample   274
ADC - analog-to-digital conversion   950
ADC and DAC operations   262
ADC operation
  key point in determining accuracy of   276
  three major steps in   263
ADC system
  bit rate of a digital signal generated by   276
ADC unit
  analog-to-digital converter   263
additive white Gaussian noise, AWGN   80
Alferov, Zhores I.   47
aliasing
  aliased signal   267
AM - amplitude modulation   715
AM demodulation
  with superheterodyne detector   731
AM modulation
  example of   764
AM signal
  Fourier transform of   771
  spectrum of an arbitrary   768
  tone   766
AM transmission
  bandwidth efficiency of   774
  power efficiency of   774
  problems with full AM signal   774
AMI – alternate mark inversion   223
amplifier   840
  gain limitation   418
  in linear regime   566
  in nonlinear (saturation) regime   566
amplitude   105, 121
  and magnitude   105
  maximum and minimum of modulated signal   719
  peak-to-peak   122
amplitude and phase spectra   570
amplitude modulation, AM   168
amplitude spectrum   527
  of square-wave signal   514
analog
  natural and technological processes   261
  transmission   168
analog modulation   763
  important feature of   787
analog signal
  maximum frequency of   272
analog signals
  as the sources of information   262
analog-to-digital conversion, ADC   261
analog-to-digital converter, ADC   479
angle   115
angle (angular) modulation   750, 784
antenna
  size of   714
anti-aliasing filter   272
artificial intelligence   31
artificial intelligence, AI   6
ASCII   321

Essentials of Modern Communications, First Edition. Djafar K. Mynbaev and Lowell L. Scheiner.
© 2020 John Wiley & Sons, Inc. Published 2020 by John Wiley & Sons, Inc.

ASCII – **A**merican **S**tandard **C**bode for
Information **I**nterchange   208, 217
ASK - amplitude-shift keying   716, 882
ASK bandwidth   889
  relationship to bit rate   894
ASK demodulation
  coherent   900
ASK modulation
  advantage of   898
  drawbacks of   898
  formula of   883
  in wireless and optical communications   899
ASK spectrum
  example of   885
  formula of   885
  power   890
  with pulse train   888
ASK transmission
  BER of   895
ASK transmission system   882
attenuation   830, 1012
  of transmitted signal   715, 808
  spectral   832
automatic gain control   731
AWGN - additive white Gaussian noise   835

b

B8ZS – bipolar [binary] eight zero substitution   223
band-pass filter, BPF
  amplitude response of   373
  bandwidth of   374
  central frequency of   374
  concept and circuit of   371
  cutoff frequencies of   373
  designing   374
  quality factor of   374
band-stop filter, BSF
  bandwidth of   380
  central frequency of   380
  concept and circuit   378
  cutoff frequencies of   379
bandwidth   76, 156
  and bit rate   78
  as determined by transmission medium   87
  Carson's rule   795
  Carson's rule for FM signal   743

Carson's rule generalized   799
  first-null   1014
  in Wi-Fi Transmission   159
  of amplitude-modulated signal   726
  of frequency-modulated signal   743, 795
  of pulse   93
  of RC LPF   349
  of signal as sum of harmonics delivering the required power.   607
  of transmission link   727
bandwidth-length product
  principle of constancy   90
bandwidth-length trade-off.   247
bandwidth-power trade-off   993
baseband transmission   223
  drawbacks of   762
BASK - binary ASK   881
Bell Labs   9, 950
Bell System   1002
Bell Systems   9
Bell, Alexander   24, 710
BER
  bit error rate vs. bit error ratio   859
  of BPSK signal   967
  of QPSK signal   967
  parameters for M-PSK and M-QAM   988
BER - bit error rate (ratio)   855
Bessel functions   789, 801
  graphs of   790
  main properties of   792
  values of   790
Bessel, Friedrich Wilhelm   789
BFSK
  formula of   904
  modulation index of   916
BFSK - binary FSK   901
BFSK signal
  bandwidth of   906
BFSK transmission
  BER and receiver sensitivity for   908
  inaccuracy of incoherent detection   919
binary
  world   215
binary modulation   881
bit   75, 216
bit (transmission) rate   827
bit energy, E_b, and modulation level, M   980

bit error rate, BER
　and quality factor, QF   867
　as error probability   857, 858
　limitation of   859
bit rate   75, 77, 238, 314, 947
bit time
　for NRZ and RZ signals   247
　and pulse repetition interval   248
Bode plots   389
　accuracy of phase graph   392
　errors of   390
　for HPF   392
Boltzmann, Ludwig Eduard   837
BPF
　band-pass filter   329
BPSK - binary PSK   922
BPSK signal
　bandwidth of   926
　formula of   925
　spectrum of   926
BPSK transmission
　advantages of   932
　applications of   932
　BER of   929
　coherent demodulation of   927
BPSK vs. DPSK   930
BSF
　band-stop filter   329
Butterworth filter
　amplitude response of   451
　amplitude response of any type of   455
　phase response   457
　transfer function of   450
Butterworth polynomial   450
byte   216

C

capacitive reactance, Xc
　frequency response of   331
capacity
　of channel   78
　transmitting   7
carrier
　signal (wave)   713
carrier frequency   87
　and optical communications   89
　vs. transmission bandwidth   87

Carson's rule   743, 799
Carson, John Renshaw   743
CBR - Carson's bandwidth rule   743
CDM – code-division multiplexing   1023
　based on spread-spectrum technique   1024
　codes orthogonality in   1023
　decoder   1023
　efficiency of transmission   1027
　encoder   1023
channel capacity   77, 237
　vs. carrier frequency   88
channel coding   224
character code   217
circuit-switched networks   320
cloud   30
"Coordinated Universal Time" (UTC)   317
CMOS - Complementary Metal-Oxide
　Semiconductor   211
codeword   276
coherent detection
　and incoherent detection   921
communication network   31
　control plane   35
　data plane   35
　security   36
communication satellites   60
　orbits   62
communication system   6, 28, 710
　block diagram of   710, 827
　general block diagram of   6
　optical (fiber-optic)   43
　properties of   827
　satellite   61
　wireless   49
communications   5, 327, 710
　analog   169
　data and digital   214
　digital   169, 214
　digital – two major steps to implement   215
　electronic   5
　globalization of   22
　optical (fiber-optic)   42
　satellite   59
　security   13
　telecommunications   5
　wireless   49
comparator   900

computer
  binary machine   216
constellation
  diagram   952
  points   951
constellation diagram
  distance between symbols   978, 981, 982
  example of industrial   985
  of 64-QAM   985
  of M-QAM systems   982
Control field   323
converter
  parallel-to-serial   967
  serial-to-parallel   957, 1013
convolution integral   659
Cooley, J.W.   699
Corning Corporation   47
cosine signal
  spectrum of   654
CPFSK - continuous-phase FSK   911
CPFSK signal
  coherent detection of   919
  generating of   915
  incoherent detection of   916
cutoff frequency   342
  conditions for RC LPF   344
  criterion in logarithmic scale   345
  in filters   329
  key role in operation of RC LPF   344
  of RC LPF   343
  of real filter   342
  vs. passband and stopband   330
cutoff frequency of RC LPF
  the role of R and C   353
CWDM - coarse WDM   1018

d

data (information) field   324
data center   12
  computational operations   12
  energy consumption   12
  servers   12
data centers
  cloud   37
De Moivre, Abraham   843
decimal and binary number systems   299
decision circuit   966

delay
  propagation   835
delta (Dirac, or impulse) function   771
demodulation
  in AM transmission   730
  in FM transmission   748, 809
DEMUX - demultiplexer   999
detection of digital signal   855
detector
  coherent   919
  incoherent   918
DFT and DSP   682
DFT – discrete Fourier transform   672
dibit
  pair of two bits   949
differential equation
  for finding impulse response   665
digital modulation   763
digital signal processing, DSP
  basis functions of   684
  processing speed in   681
  two steps in   681
digital signals
  used to store, transmit, and process information   262
digital SNR   994
digital transmission   827
  asynchronous   320
  modes of   311
  synchronization in   316
  synchronization issues of   320
  synchronous   320
digital-to-analog conversion, DAC   261, 304
digital-to-analog converter, DAC   480
Dirac, Paul   97, 649
Dirichlet's conditions
  in Fourier transform   646
discrete Fourier transform, DFT   672
  as mathematical operation   684
  basis functions and frequency index in   688
  of train of Gaussian pulses   685
  operation of   683
dispersion
  of pulse   1012
DNS – Domain Name Server   41
DPFSK - discontinuous-phase FSK   910
DPSK - differential phase-shift keying   927

DPSK receiver 928
DPSK transmission
    BER of 930
DQPSK - differential QPSK 969
DSB AM - double-sideband suppressed carrier
    (DSB-SC) AM tarnsmission 774
DSL - digital subscriber line 949
DSP – digital signal processing 480, 677
DTFT and DSP 682
DTFT – discrete-time Fourier transform 672
duty cycle 250, 597, 617
Duty cycle, bit rate, and transmitting power 252
DWDM - dense WDM 1018

e

EBCDIC – **E**xtended **B**inary **C**oded **D**ecimal
    Interchange **C**ode 217
ECL - Emitter-Coupled Logic 211
EDFA - erbium-doped fiber amplifier 1017
Einstein, Albert 97
EM – electromagnetic 49
encoder and decoder 306
encoding 285
energy
    minimum amount for one bit 86
    of a single bit, Eb(J) 83
    of nonperiodic signal 641
energy spectral density 641
energy spectrum 642
envelope
    of a modulated signal 927
    of a signal 724
envelope detector 918
    in AM transmission 731
equivalent circuit 332
error free transmission
    according to Shannon's law 79
error function 851
    complementary 852
error probability
    formula of 856
    in digital transmission 853
    with Gaussian PDF in digital transmission
        854
error signal
    nature of 288
error vector magnitude, EVM 860

eye diagram 861, 962
    and error probability 864, 866
    and estimation of BER 868
    and quality factor, QF 865
    and rise and fall times 865
    and signal's parameters 864
    and transmission quality 862
    as statistical device 861
    sampling point in 865

f

fading 831
fall time 242
Faraday, Michael 27
fast Fourier transform, FFT
    as computational algorithm 700
    three main steps in 700
FCC – Federal Communications Commission
    53, 1011
FDM
    advantage of 1011
    drawback of 1011
    principle of 1010
FDM - frequency-division multiplexing 713,
    1010
FEC - forward error correction 859
FFT – fast Fourier transform 699
filter 327
    as frequency-dependent voltage divider 338
    band-pass 329, 1010
    band-stop 329
    difference equation for digital 484
    differential equation for analog 484
    high-pass 329
    ideal 328
    low-pass 329
    low-pass RL 400
    matched 900, 927
    order of 469
    order of digital 485
    prototypes 445
    transfer function of digital 489
filter design
    art and science 470
    automated with MATLAB 460
    automated with Multisim 462
    cascading 468

filter design (contd.)
    denormalization 467
    example of Butterworth filters 462
    normalization 465
    practical steps 470
    two main parts of 459
    using the poles of transfer function in 468
filter operation
    input-output view of 382
filter prototypes
    amplitude and phase responses of 447
    need for 446
    transfer functions of 449
filter specifications
    amplitude 355
    amplitude ripples 355
    cutoff frequency 357
    industrial 354
    normalized group delay 360
    passband frequency 357
    phase 359
    selectivity factor 357
    shape factor 357
    stopband amplitude 357
    stopband frequency 357
    time group delay 359
    time phase delay 359
    transition band 357
filtering 327, 328
    entire picture 560, 571
    in modern communications 566
    operation in frequency domain 555
    periodic signals 554, 571
    single harmonic 571
    square-wave signal 554
filters
    active 419
    active BPF 436
    active BSF 439
    adaptive 491
    advantages of active 424
    cascading 453
    choosing technology of 471
    comparison of analog and digital 492
    continuous-time and discrete-time 428
    difference between digital and analog 480
    digital 479
    digital – recursive and nonrecursive 486
    drawbacks of passive 410
    finite impulse response, FIR 487
    FIR means nonrecursive 487
    fixed, adaptive, and reconfigurable 444
    hardware of digital 481
    ideal and real 444
    IIR means recursive 487
    impulse response of digital 487
    infinite impulse response, IIR 487
    passive 410
    passive with load 410
    physical realization 445
    signal-processing devices 444
    specifications of digital 491
    summary of applications 492
    switched-capacitor - cutoff frequencies 425
    switched-capacitor - design of 425
    switched-capacitor - principle of operation of 424
    switched-capacitor – advantages and drawbacks 428
    switched-capacitor – universality of 429
    trade-off between amplitude and phase responses 445
    transfer functions of active 419
firewall 37
flag 323
FM - frequency modulation 715, 733
FM demodulation 748, 807
    with PLL device 748
    with slope detector 749
FM modulation 788
    example of 765
    nonlinear nature of 806
    with voltage-controlled oscillator, VCO 746
FM signal
    and amplitude and frequency of message signal 746
    bandwidth by the required power of 794
    bandwidth by the value of significant amplitudes 793
    bandwidth of 793
    bandwidth of a sinusoidal 743
    general formula for 787
    instantaneous frequency of 738
    instantaneous frequency of a single-tone 788

modulation index of   738
peak frequency deviation of   737
signal-to-noise ratio (SNR) of   807
spectral analysis of   740
spectrum and bandwidth of   796
spectrum of   792, 801
spectrum of a single-tone   790
waveform and bandwidth of   743
zero crossings of   738
FM transmission
    de-emphasis of FM signal   810
    figure of merit (FOM) of   807
    modulation index   788
    narrowband and wideband   793
    pre-emphasis of FM signal   809
    single-tone   788
    trade-off between FOM and BW   808
FOM - figure of merit   807
Fourier series   512, 527, 642
    amplitude-phase format of   537
    and signal spectrum   512
    as foundation of spectral analysis   537
    coefficients of   512, 535, 570
    convergence of   593, 608
    for odd and even signals   544
    for square-wave signal   514
    in exponential form   579
    in one-sided trigonometric form   608
    in traditional trigonometric form   607
    in two-sided exponential form   608
    of most common signals   547
    of various periodic signals - MATLAB script   545
    of various signals   570
    rate of convergence of   593
    three equivalent forms of   590
Fourier series
    Dirichlet conditions of convergence of   593
Fourier series (FS) and Fourier transform (FT)   676
Fourier theorem   511, 527, 534
Fourier transform   629, 642, 647
    and convolution integral   659
    and spectrum of non-periodic signal   632, 635
    and transfer function   659
    application of   671
    duality property of   655, 657
    example of   630
    existence of   646
    linearity property of   657
    main properties of   656
    modulation property of   657, 771
    of causal exponential decaying signal   635
    of cosine function   652
    of given signal   802
    of rectangular pulse   632, 638
    of single impulse   651
    table of   631, 655
Fourier transformations   672
    for DSP – choosing the best type of   681
    relationship among   698
Fourier, Joseph   677
frame   323
    for data packets   1001
frequencies
    low and high   329, 393
    of message and carrier signals in AM transmission   722
frequency   112
    fundamental   510
    linear (cyclic)   785
    radian   115
    radian (angular)   785
frequency deviation
    in BFSK modulation   904
frequency domain   151, 500
    cosine and sine signals in   151
    relationship with time domain   502
    vs. time domain   151
frequency modulation, FM   735
frequency of FM signal   734
frequency translation   713
    or frequency conversion   728
FSK - frequency-shift keying   882, 901
FSK signal
    continuous-phase version of   915
FSK transmission
    advantage of   921
    areas of application   921
    drawback of   921
FSK transmission system   901
FSL - free space loss   52
FSL – free space loss
    in satellite communications   64

full AM - double sideband transmitted carrier (DSB-TC) AM transmission   771
function   147, 646
  continuous vs. discrete   481
  delta (Dirac)   632, 648
  impulse   648
  unit-step (Heaviside)   630
functions
  odd and even   542
FWHM - full width half maximum   241

g

Gauss, Karl Frederick   843
Gaussian (bell) curve and Gaussian PDF   844
Gaussian PDF   843, 941
  normal distribution   843
GE – General Electric   98
Gibbs phenomenon   594, 608
Gibbs, Josiah   677
GSO – geostationary earth orbit   63

h

harmonic   520
  filtered   552
harmonic distortion   568, 571, 607
  amplitude   569
  example of   568
Hartley's capacity law   78
Hartley's information law   75
Hartley, Ralph   75
header   324
Henry, Joseph   24
Hertz, Heinrich   27
Higgs, Peter   98
high-pass filter, HPF
  active   422
  attenuation and phase shift of   369
  characterization of   369
  concept and circuit of   367
histogram   862
  and continuous probability distribution   862
HPF
  high-pass filter   329
HTML – Hypertext Markup Language   38
hypertext   38

i

IC
  voltage specifications of   211
IC – integrated circuit   210
IEEE - Institute of Electrical and Electronics Engineers   105
impulse
  frequencies of   652
impulse response   487, 661
  finding of   665
inductive reactance   400
information   5
  volume of   827
input-output relationship   339
integral
  definite   515
integration   515
integrator   900, 966
International Bureau of Weights and Measures   317
Internet   8, 28, 37
  operation   39
Internet of Things (IoT)   10, 921, 952
Internet transmission   324
intersignal interference, ISI   446
intersymbol interference, ISI   249, 308, 941
IP – Internet Protocol   40
IP address   40
IPv4 and IPv6 s IPv4 and IPv6 c 1   40
ISI – intersymbol interference   249, 1012
ISP – Internet Service Provider   41
ITU – International Telecommunications Union   53

j

jamming
  and spread-spectrum technique   1026
jitter   865, 919, 940
  and BER   941

k

Kao, Charles   47
Kroemer, Herbert   47

l

Lagrange, Joseph-Louis   677
Laplace transform   647

Laplace, Pierre Simon   843
Large Hadron Collider, LHC   94, 98
laser   841
   diodes   47
latency   834
LED - light emitting diode   47
Leibnitz, Gottfried   300
LEO – low earth orbit   64
Li-Fi – Light fidelity communications   56
limitations
   of principles, laws, and equations   94
LoS - Line-of-Sight   52
loss of signal power
   in an optical fiber   44
low-pass filter, LPF
   active   420
   and Fourier transform   661
   designing of active   421
   evaluation of its operation   352
   impulse response of   661
   phase shift and time delay of   382
   transfer function of   384, 387, 661
   typical RC   330
low-pass RC filter
   attenuation   336
   output phase shift   334
   output waveforms   334
   principle of operation   333
LPF
   low-pass filter   329
LPF, HPF, BPF, and BSF
   comparison of   380

m

M - level of modulation   947
   alphabet size   947
   number of symbols   949
M-ary PSK   975
   BER of various   980
   constellation diagrams and Gaussian PDFs for   978
M-ary QAM
   BER of various   984
M-QAM - multilevel quadrature amplitude modulation   981
magnitude
   and amplitude   105

Manchester code   223
Marconi, Guglielmo   26
MATLAB   460
MATLAB/Simulink   962
Maxwell, James Clark   27
mean of Gaussian PDF   843
MEO – medium earth orbit   64
message signals
   harmonically related and harmonically unrelated   802
model   96
   accuracy of   97
   optimization of   97
model and simulation
   accuracy of   98
models
   vs. laws and principles   99
modern communications   948
modulation   224, 712, 715, 882
   amplitude   715
   analog   715
   angle (angular)   715
   digital   945
   four-level   945
   in AM transmission   730
   linear   910
   multilevel   945, 947
   nonlinear   910
   quadrature   945
   tone (sinusoidal)   718
   two-level (On-Off)   945
modulation (deviation) constant
   of FM signal   734
modulation bandwidth   828
modulation index   720
modulation theorem   658
Morse telegraph
   shift keying   884
Morse, Samuel   23
MSK - minimum shift keying   916, 969
multilevel modulation   945
   increases spectral efficiency   995
   M-ary modulation   949
   saving bandwidth or increasing bit rate   949
multiplexing   238, 713, 999
   concept of   999

multiplexing (*contd.*)
  time-based and frequency-based principles of 1000
Multisim   462
MUX - multiplexer   999, 1016

n

National Institute of Standards and Technology (NIST)   317
natural binary code   276
natural processes
  examples of   261
network   29
  access   32
  circuit switching   33
  hierarchy   32
  long-distance   32
  metro   32
  packet switching (routing)   33
network topologies   31
noise   500
  external   836, 985
  flicker   838
  in modern communication systems   836
  internal   837, 985
  internal - filtering of   839
  internal - in transmission link (channel)   839
  internal - of amplifier   841
  internal - spectrum of   838
  of amplifier vs. transmission distance   842
  shot   837
  spectral density of   84
  thermal   837
noise figure, NF   842
  of amplifier   842
noise floor   1017
normalized group delay   360
NRZ – nonreturn-to-zero   222, 948
NRZ and RZ signals
  comparison of timing parameters of   249
NTIA – National Telecommunications and Information Administration   53
Nyquist (sampling) theorem   483
Nyquist rate   948
Nyquist theorem   266
  Nyquist – Shannon theorem   266
Nyquist, Harry   78

o

OA – optical amplifier   48, 1017
OC - optical carrier   1007
OFDM
  industrial transmission system   1014
OFDM - orthogonal frequency division multiplexing   1012
offset,
  or dc shift   124
Ohm's law   94
op-amp -operational amplifier   413
operational amplifier
  active device   414
  equivalent circuit of   414
  gain limitation   415
  gain-bandwidth trade-off   418
  open-loop and closed-loop   415
  working with source and load   415
optical communications
  coherent transmission technology   1022
optical fiber   710
  attenuation of   44
  bandwidth, BW(Hz)   44
OQPSK - offset QPSK   968
order of digital filter   485
overhead   1003
oversampling and undersampling   271

p

packet
  block of data   1000
PAM - pulse amplitude modulation   948, 962
parallel-to-serial converter   306
Parseval's theorem   595, 609, 618
passband   224, 330
passive
  component   410
  filters   410
payload   324
  data field, or information field   1003
PDF - probability distribution (or density) function   843
period   111
  of signal   499
periods
  of amplitude-modulated signal   719

phase
　initial, shift, or offset　785
　instantaneous　785
　of sinusoidal signal　132
phase distortion　569, 607
phase shift　117
　and time delay　137
　initial phase　117
　is angle　117
phase spectrum　527
　of square-wave signal　514
phase-shift keying
　binary　951
phasor
　form and polar form　180
phasor form　607
　of sinusoidal signal　579
phasors　178
　and complex numbers　191
　and phasor diagrams　178
　and waveforms　181
phone
　smart　710
pilot carrier
　in AM transmission　783
PLL - phase-locked loop　748
PM - phase modulation　715, 751, 785
PM modulation
　constant of　751
　example of　766
　index of　752
　signal-to-noise ratio of　786
PM signal
　instantaneous value of　754
　main feature of　751
point-to-point (PPT) link
　vs. network　28
polar form　607
　of sinusoidal signal　579
PON - passive optical network　99
power
　of frequency-modulated signal　734
power budget
　or link budget　830
power distribution
　in amplitude-modulated signal　728

power efficiency
　of amplitude-modulated transmission　729
power penalty　990
power spectrum　597
　of pulse train　597
power-bandwidth trade-off　91, 607
　in multilevel modulation　994
PRI - pulse repetition interval　248
principle of energy conservation　595
probability　843
　and Gaussian PDF　844
propagation delay　1012
　in satellite communications　65
PSK - phase-shift keying　882, 922
PSK modulation
　advantage of　929
PSTN – public switched telephone network
　36
pulse
　real – basic parameters of　242
pulse amplitude modulation, PAM　304
pulse code modulation, PCM　306
　transmission bandwidth of　308
　transmission system of　306
pulse train　617
　amplitude and phase spectra of　617
　bandwidth of　618
　ideal　618
　period and bandwidth of　626
　period and spectrum of　622
　period of　622
　power and bandwidth of　602
pulse width　241

q

Q-function
　and Gaussian (bell) curve　850
QAM - quadrature amplitude modulation　945
QPSK
　coherent receiver　965
　modulator　957
　receiver (demodulator)　965
QPSK - quadrature phase-shift keying　945, 951
QPSK signals
　amplitudes of　965
　constellation diagrams of　953, 977
　formation of　957

QPSK signals (*contd.*)
   general formula for   964
   waveforms of   953
QPSK symbols
   I and Q components of   956
quality factor, QF   865
quantization   272
   and resolution   282
   number of bits per sample   273
   operation (process)   273
   quantizer   273
   uniform and nonuniform   282
quantization errors   284, 286
   minimizing them   289
quantization levels   286
   and assigned values to samples   277
   assigning binary codewords to   279
   step sizes of   281
quantization noise   285, 286
   signal-to-noise ratio of   308
quantization operation
   choosing number of bits per sample in   276
quantum communications   37

r

radio   26
radio-frequency (RF) waves   49
Rayleigh energy theorem   641, 642
receiver sensitivity   829, 897
receiver, Rx   5, 7, 710
   characteristics of   829
   modulation bandwidth of   829
   spectral characteristic of   829
reciprocal spreading   93, 618
rectangular pulse   633
   energy of   642
resonance
   condition in RLC circuit   404
   frequency of   402
   quality factor at   405
   total impedance of   403
resonance circuit
   as band-pass filter   407
   as band-stop filter   408
   bandwidth of   405
   RLC   402

   tuning a   406
responsivity
   of a receiver   829
RF – radio frequency   49
RF waves
   propagation modes   50
RFID – radiofrequency identification   10
rise and fall times
   existence in communication systems   247
rise time   241, 829, 962
rise/fall time
   the use of   247
rise/fall time and bit rate   244
RLC circuit
   analysis of   186
root mean square (rms) value   596
round robin algorithm   1000
routing   31
RS-232 standard   207
RZ - return-to-zero   223

s

sample-and-hold (S&H) operation   263
sampler   900
sampling   481, 672
   frequency   266
   interval   266
   Nyquist theorem   266
Sampling (sample-and-hold, S&H) technique   265
satellite communications
   as related to optical and wireless communications   66
SDM - spatial-division multiplexing   1022
SDN – software-defined networking   36
semilogarithmic scale
   of graph axes   340
SER
   of QPSK signal   968
SER - symbol error rate   967
serial vs. parallel transmission   314
Shannon limit   992
Shannon's law   79
   for digital communications   83
   for digital transmission   991
   its three limits   95
   probability of errors   79

Shannon's limit   79
Shannon – Hartley theorem   80
Shannon, Claude   79, 827
signal
   amplitude-modulated - formula of   723
   amplitude of   785
   amplitude-modulated   716
   amplitude-modulated - envelope of   724
   amplitude-modulated - instantaneous value of   724
   analog   108, 214
   and function   146
   band-limited (bandpass)   767
   bandlimited   918
   cosine vs. sine   503
   delayed sawtooth   538
   digital   108, 205, 214
   digital – fundamental timing parameter   246
   digital – two types of variations it can withstand   207
   energy   639
   filtered   546
   frequency-modulated   733
   frequency-modulated - frequency of   734
   frequency-modulated - power of   734
   message, modulating, baseband, or intelligent   716
   non-periodic   622, 639
   normalized average power of   597
   overmodulated AM   721
   periodic   622, 690
   periodic non-sinusoidal   527
   power, bandwidth, and waveform of   601
   required power of   607
   sinusoidal   108, 110, 130
   sinusoidal - three parameters   130
   square-wave   512
   truncated (windowed)   672, 692
   waveform vs. spectrum of   693
signal bandwidth   249
signal multipath propagation
   in wireless communications   1012
signal space   860, 951, 956
signal's power vs. its bandwidth   600
signal-to-noise ratio, SNR
   in amplifier   841
   of transmission link (channel)   835

signals
   analog and digital   671
   continuous and discrete   671
   cosine and sine   138
   effect of filtering on   555
   energy and power   594
   four types of   671
   non-periodic   500, 526
   non-periodic - energy   595
   orthogonal   911
   periodic   499, 526
   periodic - power   595
   periodic and nonperiodic   671
   periodic non-sinusoidal   509
   power and energy   608
   random   500
   sinusoidal and non-sinusoidal   509
   symmetry of   542
signals, spectra, and Fourier transformations   675
sinusoidal signal
   and phasor   139
   phasor presentation of   951
slope efficiency
   of a transmitter   828
SNR - signal-to-noise ratio   79, 839
SNR – signal-to-noise ratio
   digital   84
SONET
   digital hierarchy of   1007
   framing in   1007
   synchronous multiplexing in   1008
SONET - synchronous optical network   1006
source coding   224
spectra
   two-sided and one-sided   588, 608
   two-sided and one-sided of pulse train   598
spectral analysis   497, 519, 527
spectral analysis and synthesis
   concept of   526
spectral density, $S(f)$ 768
spectral efficiency   947
   maximum   991
   of M-ary modulation system   977
Spectral efficiency and level of modulation   992
spectral efficiency, SE   84
   and quality of transmission system   92

spectral efficiency, SE (*contd.*)
　and Shannon's law   93
　concept of   92
spectral synthesis   497, 520, 527
spectrum   497, 710
　amplitude   509
　approximate vs. exact   693
　continuous   622, 642
　frequency   509
　linear (discrete)   622, 642
　of a signal   509
　of amplitude-modulated signal   725
　of delayed sawtooth signal   538
　of non-periodic signal   642
　of non-sinusoidal signal   510
　of pulse train   581
　of square-wave signal   512
　of truncated signal   692
　phase   509
spectrum
　power   609
spread-spectrum technique   1024
　applications of   1027
　benefits of   1026
　modulation and spreading   1024
　operation of   1025
　security of communications   1026
　spreader   1024
　XOR operation in   1024
SSB AM - single-sideband suppressed carrier (SSB-SC) AM transmission   779
SSB AM transmitter
　filtering method for building of   779
　phasing method for building of   779
standard deviation
　value of   846
standard deviation of Gaussian PDF   843
stopband   330
STS - synchronous transport signal   1007
subcarriers   1013
　mutually orthogonal   1013
subchannels   1012
SWDM - short WDM   1018
switching   31
synchronous optical network (SONET)   324

t
T system
　add/drop multiplexing in   1005
　digital signal at level zero, DS-O   1003
　drawbacks of   1003
　hierarchy of   1003
　plesiochronous   1005
TCP - Transmission Control Protocol   40
TCP/IP - Transmission Control Protocol/Internet Protocol   31
TDM
　statistical (asynchronous), stat-TDM   1000, 1008
　synchronous - disadvantages of   1002
　synchronous - serial transmission   1002
　synchronous - T system   1002
　synchronous - timing   1002
　synchronous - principle of operation   1001
　synchronous, sync-TDM   1000
TDM - time-division multiplexing   1000
technological processes
　examples of   262
telegraph   22
telephone   24
television   27
threshold   856
　in detection of digital signal   852, 855
time domain   500
　and signal's waveform   500
　relationship with frequency domain   500
time domain and frequency domain   526
　relationship between   503
Time Service Department of the US Naval Observatory   317
time slot   999, 1001
time-domain and frequency-domain
　presentation of filter operation in   347
timer   900, 962
transducer   262, 710
transfer function   659
　and differential equation   667
　and impulse response   667
　Bode plots   387
　characteristic equation of   457
　general form of   447

in factored form   457
of active filter   419
pole locations on complex plane   459
poles of   457
vs. difference equation   490
transform   647
transition band   467
transmission
   advantages of broadband   713
   analog   205
   baseband   711
   broadband   712
   channel   6
   code of RS-232 standard   207
   comparison of baseband and broadband   716
   digital   206
   digital – examples of   207
   digital – main advantage of   206
   link   5, 6
   quality of   607
   reliable (error free)   79
transmission channel
   bandlimited   607
transmission code   222
transmission link   710
   ideal   1012
transmission link (channel)
   attenuation of   831
   available bandwidth of   832
   bandwidth of   832
   input/output relationship   830
   latency of   834
   loss of   830
   modulation bandwidth of   831
   spectral attenuation of   832
transmission quality   827
   metrics for the assessment of   859
transmitter, Tx   5, 6, 710
   characteristics of   828
   input/output relationship of   828
   modulation bandwidth of   829
   slope efficiency of   828
TTL - Transistor-Transistor Logic   211
Tukey, J.W.   699
tuning
   in AM demodulation   731
two's complement code   280
two's complement binary code   286

u

uncertainty principle of Fourier analysis   693
unipolar NRZ transmission code   286

v

variable
   continuous random   843
VCO - voltage-controlled oscillator   736, 788
VLC – visual light communications   57
VSB - vestigial-sideband AM transmission   783

w

waveform   710
waveform of actual digital signal   239
waveform of ideal digital signal   233
waveforms   104, 181
   and phasors   181
   and signals' power   174
   continuous and discrete   170
   deterministic and random   170
   in transmission   168
   of analog and digital signals   104
   periodic and nonperiodic   173
   power and energy of   176
   practically realizable   178
   time-varying and spatial   169
WDM
   and network design   1017
   coarse and short   1021
   concept of   1016
   frequency grid in   1017
   limitations of   1020
   polarization multiplexing in   1022
   spectral efficiency of   1019
   spectrum of signal   1017
   usable bandwidth of   1018
WDM – wavelength-division multiplexing   48, 1016

wireless communications
   Bluetooth 55
   MIMO – Multiple-Input Multiple-Output 55
   mobile cellular communications 57
   networks 57
   optical 56
   problems 53
   scarcity of spectrum 53

Wi-Fi 54
ZigBee 56
WWW – World Wide Web 8, 38

z

z-transform 647
Zworykin, Vladimir 28

Printed and bound by CPI Group (UK) Ltd, Croydon, CR0 4YY